ISBN 978-1-5277-6302-9
PIBN 10888409

FOWNES'

MANUAL OF CHEMISTRY;

THEORETICAL AND PRACTICAL.

A NEW AMERICAN FROM THE TWELFTH ENGLISH EDITION,

EMBODYING

WATTS' "PHYSICAL AND INORGANIC CHEMISTRY."

WITH ONE HUNDRED AND SIXTY-EIGHT ILLUSTRATIONS.

PHILADELPHIA:
LEA BROTHERS & CO.
1885.

AMERICAN PUBLISHERS' PREFACE.

THE incessant changes which have characterized the science of Chemistry since the introduction of the new nomenclature have rendered successive editions of Professor Fowne's work more representative of the editor's than of the author's views. Under these circumstances, Mr. Watts very properly contemplated the issue of a work of his own, based upon that of Professor Fowne's, but his untimely death prevented the full accomplishment of the project. A large portion of the work, however, was completed by the issue of a volume on Physical and Inorganic Chemistry, and it has seemed to the publishers a duty to lay before chemists and students the latest thoughts of so renowned a scientist. The volume has, therefore, been reprinted with Professor Fowne's Organic Chemistry, in a form which experience with many previous editions has proved to be popular with those for whom the work is intended.

AMERICAN PUBLISHERS' PREFACE

TABLE OF CONTENTS.

PART I.

PHYSICS.

A*
(v)

PART II.

CHEMISTRY OF ELEMENTARY BODIES.

THE NON-METALLIC ELEMENTS.

B

PART III.

CHEMISTRY OF CARBON COMPOUNDS.

METHANE-DERIVATIVES, OR FATTY GROUP.

INDIGO GROUP.

DIPHENYL GROUP.

NAPHTHALENE GROUP.

APPENDIX.

LIST OF ILLUSTRATIONS.

MANUAL OF CHEMISTRY.

PHYSICAL AND INORGANIC.

INTRODUCTION.

THE Science of Chemistry has for its object the study of the nature and properties of the materials which enter into the composition of the earth, the sea, and the air, and of the various organized or living beings which inhabit them.

In ordinary scientific speech the term *chemical* is applied to changes which permanently affect the properties or characters of bodies, in opposition to effects termed p*hysical,* which are not attended by such consequences. Changes of decomposition or combination are thus easily distinguished from those temporarily brought about by heat, electricity, magnetism, and the attractive forces, whose laws and effects lie within the province of Physics.

Nearly all the objects presented by the visible world are of a compound nature, and capable of being resolved into simpler forms of matter. Thus, a piece of limestone or marble, by the application of a red heat, is decomposed into quicklime and a gaseous body, carbon dioxide. Both lime and carbon dioxide are in their turn susceptible of decomposition—though not by the action of heat alone—the former into calcium and oxygen, the latter into carbon and oxygen. Beyond this second step of decomposition the efforts of Chemistry have hitherto been found to fail : and the three bodies, calcium, carbon, and oxygen, having resisted all attempts to resolve them into simpler forms of matter, are admitted into the list of *elements ;* not from an absolute belief in their real oneness of nature, but from the absence of any evidence that they contain more than one description of matter.

The elementary bodies at present recognized are sixty-seven in number ; and about fifty of them belong to the class of *metals.* Several of these are of recent discovery, and are as yet very imperfectly known. This distinction between metals and non-metallic substances, although very convenient for purposes of description, is entirely arbitrary, the two classes not being separated by any exact line of demarkation.

The names of the elements are given in the following table. Opposite to them in the third column are placed certain numbers, which express the proportions in which they combine together, or simple multiples of those proportions ; these numbers, for reasons which will be afterwards explained,

2 (25)

are called Atomic or Indivisible Weights. In the second column are placed symbols by which these weights are denoted; these symbols are formed of the first letters of the Latin names of the elements, a second letter being added when the names of two or more elements begin with the same letter.

The names of the most important elements are distinguished by the largest and most conspicuous type; those next in importance, by small capitals: while the names of elements which are of rare occurrence, or of which our knowledge is still imperfect, are printed in ordinary type. The names with an asterisk are those of Non-metallic Elements, the others are names of Metals.

TABLE OF ELEMENTARY BODIES, WITH THEIR SYMBOLS AND ATOMIC
WEIGHTS.

Name.	Symbol.	Atomic Weight.	Name.	Symbol.	Atomic Weight.
ALUMINIUM	Al	27.3	Niobium	Nb	94
ANTIMONY (Stibium)	Sb	122	**NITROGEN***	N	14.01
ARSENIC	As	74.9	Osmium	Os	198.6
BARIUM	Ba	136.8	**OXYGEN***	O	15.96
Beryllium	Be	9	PALLADIUM	Pd	106.2
BISMUTH	Bi	210	**PHOSPHORUS***	P	30.96
BORON*	B	11	**PLATINUM**	Pt	196.7
BROMINE*	Br	79.75	**POTASSIUM**		
Cadmium	Cd	111.6	(Kalium)	K	39.04
Cæsium	Cs	133	Rhodium	Rh	104.1
CALCIUM	Ca	39.9	Rubidium	Rb	85.2
CARBON*	C	11.97	Ruthenium	Ru	103.5
Cerium	Ce	141.2	Scandium	Sc	44.9
CHLORINE*	Cl	35.37	Selenium*	Se	78
CHROMIUM	Cr	52.4	**SILICON**	Si	28
COBALT	Co	58.6	**SILVER** (Argen-		
COPPER (Cuprum)	Cu	63	tum)	Ag	107.66
Didymium	D	146.6	**SODIUM** (Natri-		
Erbium	E	166	um)	Na	23
FLUORINE*	F	19.1	STRONTIUM	Sr	87.2
Gallium	Ga	69.8	**SULPHUR***	S	31.98
GOLD (Aurum)	Au	196.2	Tantalum	Ta	182
HYDROGEN*	H	1	Tellurium*	Te	128
Indium	In	113.4	Terbium	Tb	148
IODINE*	I	126.53	Thallium	Tl	203.6
Iridium	Ir	196.7	Thorium	Th	231.5
IRON (Ferrum)	Fe	55.9	**TIN** (Stannum)	Sn	117.8
Lanthanum	La	139	Titanium	Ti	48
LEAD (Plumbum)	Pb	206.4	TUNGSTEN, or Wolf-		
Lithium	Li	7	ram	W	184
MAGNESIUM	Mg	23.94	URANIUM	U	240
MANGANESE	Mn	54.8	Vanadium	V	51.2
MERCURY (Hy-			Ytterbium	Yb	172.5?
drargyrum)	Hg	199.8	Yttrium	Y	89
Molybdenum	Mo	95.6	**ZINC**	Zn	64.9
NICKEL	Ni	58.6	Zirconium	Zr	90

It must be distinctly understood that the atomic or combining weights assigned to the elements are merely relative. The number 1 assigned to hydrogen may represent a grain, ounce, pound, gram, kilogram, etc., and the numbers assigned to the other elements will then represent so many grains, ounces, pounds, grams, kilograms, etc. Hydrogen is taken as the unit of the scale, because its combining weight is smaller than that of any other element; but this is merely a matter of convenience; in the older tables of atomic weights that of oxygen was assumed as 100, that of carbon being then 75, that of hydrogen 6.25, etc., etc.

By the combination of the elements in various proportions, and in groups of two, three, or larger numbers, all compound bodies are produced. And here it is important to state clearly the characters which distinguish true chemical combination from mechanical mixture, and from that kind of adhesion which gives rise to the solution of a solid in a liquid. Bodies may be mixed together in any proportion whatever, the mixture always exhibiting properties intermediate between those of its constituents, and in regular gradation, according to the quantity of each that may be present, as may be seen in the fusion together of metals to form alloys, in the mixture of water with alcohol, of alcohol with ether, and of different oils one with the other. A solid body may also be dissolved in a liquid—salt or sugar in water, for example—in any proportion up to a certain limit, the solution likewise exhibiting a regular gradation of physical properties, according to the quantity of the solid taken up. But a true chemical compound exhibits properties totally different from those of either of its constituent elements, and the proportion of these constituents which form that particular compound admits of no variation whatever. Water, for example, is composed of two elements, oxygen and hydrogen, which, when separated from one another, appear as colorless gases, differing widely in their properties one from the other, and from water in the state of vapor; moreover, water, whether obtained from natural sources or formed by direct combination of its elements, always contains in 100 parts by weight, 88.9 parts of oxygen and 11.1 of hydrogen. Common salt, to take another example, is a compound of chlorine and sodium, the former of which, in the separate state, is a yellowish-green gas, the latter a yellowish-white highly lustrous metal, capable of burning in the air, and decomposing water; moreover, from whatever part of the world the salt may be obtained, 100 parts of it invariably contain 39.6 parts of sodium and 60.4 parts of chlorine. Further, when two or more compounds are formed of the same elements, there is no gradual blending of one into the other, as in the case of mixtures, but each compound is sharply defined, and separated, as it were, from the others by an impassable gulf, exhibiting properties distinct from those of the others, and of the elements themselves in the separate state. Thus, there are two compounds of carbon and oxygen, one of which, containing 3 parts by weight of carbon with 4 of oxygen, is an inflammable gas, lighter than atmospheric air, and not absorbed by solution of potash; while the other, which contains 3 parts of carbon and 8 of oxygen, is non-inflammable, heavier than air, and easily absorbed by potash.

Oxygen is the most widely diffused of all the elementary bodies. It is under ordinary circumstances a gas, and, mixed with nitrogen, constitutes the air we breathe. In combination with hydrogen it forms water. In the pure state it is a colorless, inodorless gas, in which inflammable bodies, such, as wood, oil, sulphur, etc., burn much more rapidly than in common air.

Nitrogen, the other constituent of the air, is also a colorless gas, non-inflammable, and differing from oxygen in not being capable of maintaining the combustion of other bodies, so that a lighted taper immersed in it is immediately extinguished.

Hydrogen, the other constituent of water, is in the free state also a gas, and is the lightest of all known bodies. In presence of oxygen or common air, it is very inflammable, and burns with a light blue flame, producing water.

Carbon is a constituent of all vegetable and animal substances, existing therein in combination either with hydrogen alone, or with hydrogen and oxygen, or with hydrogen, oxygen, and nitrogen. When separated from these elements, it is for the most part a black solid body, well known under the names of charcoal, coke, lampblack, and plumbago, or black-lead. It occurs also naturally in the pure state, forming transparent crystals known as diamond.

Chlorine, Bromine, and Iodine exist in sea-water (hence called *Halogen-elements*, from ἁλς, the sea), in combination chiefly with sodium. Chlorine in the free state is a yellowish-green gas, bromine a red liquid, iodine a grayish-black crystalline solid. Of these elements, chlorine is by far the most abundant, its compound with sodium forming common sea-salt. Fluorine is an element closely allied in its chemical relations to the three last mentioned. It is not known in the free state, but in combination with calcium it constitutes the crystallized mineral called fluorspar.

Sulphur, a well-known substance, occurs in the free state in volcanic districts, and abundantly in various parts of the world in combination with iron, copper, lead, and other metals, forming compounds called *sulphides*. Selenium and Tellurium are rare elements, allied to sulphur, and likewise found in combination with metals.

Phosphorus, in the free state, is a highly inflammable solid body, which burns rapidly in contact with the air, being converted, by combination with oxygen, into *phosphoric acid*, a compound which, in combination with calcium, occurs very abundantly in nature, forming the minerals called phosphorite and apatite, and is the chief mineral constituent of the bones of animals. Arsenic, sometimes regarded as a metal, is an element closely allied in its chemical relations to phosphorus, and occurring in nature in combination with oxygen, sulphur, and various metals.

Silicon is a very abundant and important element, not occurring in nature in the free state, but forming, in combination with oxygen, the mineral called *silica*, which in the form of quartz, flint, and sandstone, and as a constituent of granite, gneiss, and other ancient rocks, forms a very large portion of the crust of the earth. Boron, an element analogous in many of its properties to silicon, occurs naturally in combination with oxygen, forming boric or boracic acid, which is a constituent of several minerals, and occurs somewhat abundantly in solution in certain natural waters.

Of the fifty-four metals, seven only were known in ancient times, viz., iron, copper, lead, tin, mercury, silver, and gold; of these, by far the most abundant and important is iron. Of the metals more recently discovered, the most abundant are sodium, potassium, calcium, and aluminium. Sodium, as already observed, forms, in combination with chlorine, the chief saline constituent of sea-water. Potassium, which resembles it in many respects, exists also in the sea, and is a constituent of all plants. Calcium is the metallic constituent of limestone, and aluminium of clay.

General Laws of Chemical Combination.—Chemical Nomenclature and Notation.—The compounds formed by the union of oxygen with other bodies bear the general name of oxides. They are conveniently divided into three principal groups or classes. The first division contains all those oxides which resemble in their chemical relations the oxides of potassium, sodium, silver, or lead; these are denominated alkaline or basic oxides. The oxides of the second group have properties opposed to those of the bodies mentioned; the oxides of sulphur and phosphorus may be taken as typical representatives of the class; they are called acid oxides, and are capable of uniting with the basic oxides and forming compounds called salts. Thus, when the oxide of sulphur, called sulphuric oxide, is passed in the state of vapor over heated barium oxide, combination takes place, attended with vivid incandescence, and a salt called barium sulphate is produced, containing all the elements of the two original bodies, namely, barium, sulphur, and oxygen.

Among salts there is a particular group, namely, the hydrogen salts, containing the elements of an acid oxide and water (hydrogen oxide), which are especially distinguished as acids, because many of them possess in an eminent degree the properties to which the term acid is generally applied, such as a sour taste, corrosive action, solubility in water, and the power of reddening certain blue vegetable colors. A characteristic property of these acids, or hydrogen salts, is their power of exchanging their hydrogen for a metal presented to them in the free state or in the form of oxide. Thus, sulphuric acid, which contains sulphur, oxygen, and hydrogen, readily dissolves metallic zinc, the metal taking the place of the hydrogen, which is evolved as gas, and forming a salt containing sulphur, oxygen, and zinc; in fact, a *zinc sulphate* produced from a *hydrogen sulphate* by substitution of zinc for hydrogen. The same substitution and formation of zinc sulphate take place when zinc oxide is brought into contact with sulphuric acid; but in this case the hydrogen, instead of being evolved as

3 *

gas, remains combined with the oxygen derived from the zinc oxide, forming water.

A series of oxides containing quantities of oxygen in the proportion of the numbers 1, 2, 3, united with a constant quantity of another element, are distinguished as *monoxide, dioxide,* and *trioxide* respectively, the Greek numerals indicating the several degrees of oxidation. A compound intermediate between a monoxide and a dioxide is called a *sesquioxide, e. g. :*—

	Chromium.	Oxygen.
Chromium monoxide	52.4 +	16
Chromium sesquioxide	52.4 +	24
Chromium dioxide	52.4 +	32
Chromium trioxide	52.4 +	48

When a metal forms two basic or salifiable oxides, they are distinguished by adjectival terms, ending in *ous* for the lower, and *ic* for the higher degree of oxidation, *e. g. :*—

	Iron.	Oxygen.
Iron monoxide, or Ferrous oxide . . .	56 +	16
Iron sesquioxide, or Ferric oxide . . .	56 +	24

The salts resulting from the action of acids on these oxides are also distinguished as ferrous and ferric salts respectively.

Acid oxides of the same element, sulphur for example, are also distinguished by the terminations *ous* and *ic,* applied as above; their acids, or hydrogen salts, receive corresponding names; and the salts formed from these acids are distinguished by names ending in *ite* and *ate* respectively. Thus, for the oxides and salts of sulphur:—

	Sulphur.	Oxygen.	Hydrogen.
Sulphuric oxide	32 +	32	
Hydrogen sulphite, or Sulphurous acid	32 +	48 +	2
Lead sulphite	32 +	48 +	206.4 (Lead.)
Sulphuric oxide	32 +	48	
Hydrogen sulphate, or Sulphuric acid	32 +	64 +	2
Lead sulphate	32 +	64 +	206.4 (Lead.)

The acids above spoken of are oxygen-acids; and formerly it was supposed that all acids contained oxygen—that element being, indeed, regarded as the acidifying principle. At present, however, we are acquainted with many bodies which possess all the characters above specified as belonging to an acid, and yet do not contain oxygen. For example, hydrochloric acid (formerly called muriatic acid, or spirit of salt)—which is a hydrogen chloride or compound of hydrogen and chlorine—is intensely sour and corrosive; reddens litmus strongly; dissolves zinc, which drives out the hydrogen and takes its place in combination with the chlorine, forming zinc chloride; and dissolves most metallic oxides, exchanging its hydrogen for the metal, and forming a metallic chloride and water.

Bromine, iodine, and fluorine also form, with hydrogen, acid compounds analogous in every respect to hydrochloric acid.

Compounds of chlorine, bromine, iodine, fluorine, sulphur, selenium, phosphorus, etc., with hydrogen and metals, are grouped, like the oxygen-compounds, by names ending in *ide*: thus we speak of zinc chloride, calcium fluoride, hydrogen sulphide, copper phosphide, etc. The numerical prefixes, *mono, di, tri,* etc., as also the terminations *ous* and *ic,* are applied to these compounds in the same manner as to the oxides, thus—

	Hydrogen.		Bromine.
Hydrogen bromide	1	+	80

	Potassium.		Sulphur.
Potassium monosulphide	78	+	32
Potassium disulphide	78	+	64
Potassium trisulphide	78	+	96
Potassium tetrasulphide	78	+	128
Potassium pentasulphide	78	+	160

	Iron.		Chlorine.
Ferrous chloride	56	+	71
Ferric chloride	56	+	106.5

	Tin.		Sulphur.
Stannous sulphide	117.8	+	64
Stannic sulphide	117.8	+	128

The Latin prefixes *uni, bi, ter, quadro,* etc., are often used instead of the corresponding Greek prefixes; there is no very exact rule respecting their use; but, generally speaking, it is best to employ a Greek or Latin prefix according as the word before which it is placed is of Greek or Latin origin. Thus, *di*oxide corresponds with *bi*sulphide; on the whole, however, the Greek prefixes are most generally employed.

The composition of these oxides and sulphides affords an illustration of a law which holds good in a large number of instances of chemical combination, viz., that *when two bodies,* A *and* B, *are capable of uniting in several proportions, the several quantities of* B *which combine with a given or constant quantity of* A *stand to one another in very simple ratios.* Thus, the several quantities of sulphur which unite with a given quantity (78 parts) of potassium are to one another as the numbers

$$1, \qquad 2, \qquad 3, \qquad 4, \qquad 5;$$

and the quantities of oxygen which unite with a given quantity of chromium are as the numbers

$$1, \qquad 1\tfrac{1}{2}, \qquad 2, \qquad 3,$$
$$\text{or } 2, \qquad 3, \qquad 4, \qquad 6.$$

It must be especially observed that no compounds are known intermediate in composition between those which are represented by these numbers. There is no oxide of chromium containing $1\tfrac{1}{4}$ or $1\tfrac{1}{3}$ or $2\tfrac{1}{7}$ times as much oxygen as the lowest; no sulphide of potassium the quantity of sulphur in which is expressed by any fractional multiple of the lowest. The quantities of the one element which can unite with a constant quantity of the other increase, not continuously, but by successive and well-defined steps or increments, standing to one another, for the most part, in simple numerical ratios.

This is called the "Law of Multiples." The observation of it has led to the idea that the elementary bodies are composed of ultimate or indivisible particles or atoms, each having a constant weight peculiar to itself (the atomic weights given in the table on page 26), and that combination between two elements takes place by the juxtaposition of these atoms. A collection of elementary atoms united together to form a compound constitutes a molecule, the weight of which is equal to the sum of the weights of its component atoms. Thus an atom of chlorine weighing 35.4 unites with an atom of hydrogen weighing 1, to form a molecule of hydrogen chloride weighing 36.4. An atom of oxygen weighing 16 unites with 2 atoms of hydrogen, each weighing 1, to form a molecule of water, weighing $16 + 2.1 = 18$. An atom of oxygen, weighing 16, unites with an atom of lead, weighing 206.4, to form a molecule of lead oxide, weighing 222.4. Two atoms of potassium, each weighing 39, unite with 1, 2, 3, 4, and 5 atoms of sulphur, each weighing 32, to form the several sulphides enumerated on page 31.

These combinations are represented symbolically by the juxtaposition of the symbols of the elementary atoms given in the table already referred to; thus the molecule of hydrogen chloride, composed of 1 atom of hydrogen and 1 atom of chlorine, is represented by the symbol or formula HCl; that of water (2 atoms of hydrogen and 1 atom of oxygen), by HHO, or more shortly H_2O. In like manner the different oxides, sulphides, acids, and salts above enumerated are represented symbolically as follows:—

Chromium monoxide	. CrO	
Chromium sesquioxide	. CrCrOOO	or Cr_2O_3
Chromium dioxide	. CrOO	or CrO_2
Chromium trioxide	. CrOOO	or CrO_3
Sulphurous oxide	. SOO	or SO_2
Hydrogen sulphite or Sulphurous acid .	. SOOOHH	or SO_3H_2
Lead sulphite	. SOOOPb	or SO_3Pb
Potassium monosulphide . . .	. KKS	or K_2S
Potassium disulphide	. KKSS	or K_2S_2
Potassium trisulphide	. KKSSS	or K_2S_3
Potassium tetrasulphide . . .	. KKSSSS	or K_2S_4
Potassium pentasulphide . . .	. KKSSSSS	or K_2S_5

A group of two or more atoms of the same element is denoted by placing a numeral either before the symbol, or, as in the preceding examples, a small numeral to the right of the symbol, and either above or below the line; thus OOO may be abbreviated into 3O, or O^3, or O_3.

The multiplication of a group of dissimilar atoms is denoted by placing a numeral to the left of the group of symbols, or by enclosing them in brackets, and placing a small numeral to the right: thus, $3HCl$ or $(HCl)_3$ denotes 3 molecules of hydrogen chloride; $2H_2SO_4$ denotes 2 molecules of hydrogen sulphate.

The combination of two groups or molecules is denoted by placing their symbols in juxtaposition, with a comma between them: thus ZnO,SO_3 denotes a compound of zinc oxide with sulphur trioxide; K_2O,H_2O, a compound of potassium oxide with hydrogen oxide or water. Sometimes the sign $+$ is used instead of the comma. To express the multiplication of such a group, the whole is enclosed in brackets, and a numeral placed on the left; *e. g.*, $2(ZnO,SO_3)$; $3(K_2O,H_2O)$, etc. If the brackets were omitted, the numeral would affect only the symbols to the left of the comma: thus $3K_2O,H_2O$ signifies 3 potassium oxide and 1 water, not 3 potassium oxide and 3 water.*

Equivalents.—It has been already stated that the elements can replace one another in combination; thus, when hydrogen chloride is placed in contact with zinc, the zinc dissolves and enters into combination with the chlorine, while a quantity of hydrogen is evolved as gas. Now this substitution of zinc for hydrogen always takes place in definite proportion by weight, 32.5 parts of zinc being dissolved for every 1 part of hydrogen expelled. In like manner when potassium is thrown into water, hydrogen is evolved and the potassium dissolves, 38 parts of the metal dissolving for every 1 part of hydrogen given off. Again, if silver be dissolved in nitric acid, and metallic mercury immersed in the solution, the mercury will be dissolved and will displace the silver, which will be separated in the metallic state; and for every 100 parts of mercury dissolved 108 parts of silver will be thrown down. In like manner copper will displace the mercury in the proportion of 31.5 parts of copper to 100 of mercury, and iron will displace the copper in the proportion of 28 parts of iron to 31.5 parts of copper.

These are particular cases of the general law, that, *when one element takes the place of another in combination, the substitution or replacement always takes place in fixed or definite proportions.* The relative quantities of different elements which thus replace one another are called c h e m i c a l e q u i v a-l e n t s or e q u i v a l e n t n u m b e r s; they are either identical with the atomic weights, or simple multiples, or submultiples of them. For example, in the substitution of potassium for hydrogen, and of copper for mercury, and of iron for copper, the equivalents are to one another in the same proportion as the atomic weights, as may be seen by comparing the numbers just given with those in the table on page 26. In the substitution of zinc for hydrogen, on the other hand, the quantity of zinc which takes the place of 1 part of hydrogen is only half the atomic weight; similarly in the substitution of mercury for silver.

* The neglect of this distinction often leads to considerable confusion in chemical notation.

2 *

All chemical reactions consist either in the direct addition or separation of elements, or in substitutions like those just noticed, the latter being by far the most frequent form of chemical change.

Chemical Equations.—Chemical reactions may be represented symbolically in the form of equations, the symbols of the reacting substances being placed on the left hand, and those of the new substances resulting from the change on the right : for example—

1. Resolution of mercuric oxide by heat into mercury and oxygen—

$$HgO = Hg + O.$$

2. Resolution of manganese dioxide by heat into manganoso-manganic oxide and oxygen—

$$3MnO_2 = Mn_3O_4 + O_2.$$

3. Action of zinc on hydrogen chloride, producing zinc chloride and free hydrogen—

$$2HCl + Zn = ZnCl_2 + H_2.$$

4. Action of zinc on hydrogen sulphate, producing zinc sulphate and hydrogen—

$$H_2SO_4 + Zn = ZnSO_4 + H_2.$$

5. Action of zinc oxide on hydrogen chloride or sulphate, producing zinc chloride or sulphate and water—

$$2HCl + ZnO = ZnCl_2 + H_2O,$$

and

$$H_2SO_4 + ZnO = ZnSO_4 + H_2O.$$

It need scarcely be observed that the test of corectness of such an equation is, that the number of atoms of each element on one side should be equal to the number of atoms of the same element on the other side.

Any such symbolical equation may be converted into a numerical equation by substituting for each of the chemical symbols its numerical value from the table of atomic weights.

The laws of chemical action and their expression by symbols and equations will receive abundant illustration in the special descriptions which follow ; their general consideration will also be more fully developed in a subsequent part of the work.

———————

Before proceeding with the detailed description of the several elements and their compounds, it is desirable to give a short sketch of certain branches of Physical Science, viz., the mechanical constitution of gases, and the chief phenomena of heat, light, electricity, and magnetism, the partial study of which greatly facilitates the understanding of chemical reactions.

PART I.—PHYSICS.

OF DENSITY AND SPECIFIC GRAVITY.

IT is of great importance to understand clearly what is meant by the terms *density* and *specific gravity*. By the *density of a body* is meant its *mass*, or *quantity of matter*, compared with the mass or quantity of matter of an *equal volume* of some standard body, arbitrarily chosen. *Specific gravity* denotes the *weight* of a body—or the force with which it tends to fall to the earth—as compared with the weight of an equal bulk, or volume, of the standard body, which is reckoned as unity.* In all cases of solids and liquids the standard of unity adopted in this country is pure water at the temperature of 15.5° C. (60° Fahr.). Anything else might have been chosen; there is nothing in water to render its adoption for the purpose mentioned indispensable; it is simply taken for the sake of convenience, being always at hand, and easily obtained in a state of perfect purity. An ordinary expression of specific weight, therefore, is a number expressing how many times the weight of an equal bulk of water is contained in the weight of the substance spoken of. If, for example, we say that concentrated oil of vitriol has a specific gravity equal to 1.85, or that perfectly pure alcohol has a density of 0.794 at 15.5° C., we mean that equal bulks of these two liquids and of distilled water possess weights in the proportion of the numbers 1.85, 0.794, and 1; or 1850, 794, and 1000. It is necessary to be particular about the temperature, for, as will be hereafter shown, liquids are extremely expansible by heat; otherwise a constant bulk of the same liquid will not retain a constant weight. It will be proper to begin with the description of the mode in which the specific gravity of liquids is determined; this is the simplest case, and the one which best illustrates the general principle.

To find the specific gravity of any particular liquid compared with that of water, it is only requisite to weigh equal bulks at the standard temperature, and divide the weight of the liquid by the weight of the water; the quotient will be greater or less than unity, as the liquid experimented on is heavier or lighter than water. Now, to weigh equal bulks of two liquids, the simplest and best method is, clearly, to weigh them in succession in the same vessel, taking care that it is equally full on both occasions.

The vessel commonly used for this purpose is a small glass bottle provided with a perforated conical glass stopper, very accurately fitted by grinding. By completely filling the bottle with liquid, and carefully removing the portion of liquid which is displaced when the stopper is inserted, an unalterable measure is obtained. The least possible quantity of grease applied to the stopper greatly promotes the exact fitting.

When the chemist has only a very small quantity of a liquid at his disposal, and wishes not to lose it, the little glass vessel (fig. 2) is particularly useful. It is formed by blowing a bulb on a glass tube. On that portion of the tube which is narrowed by drawing the tube out over a lamp, a fine scratch

---

* In other words, density means comparative *mass*, and specific gravity comparative *weight*. These expressions, although really relating to distinct things, are often used indifferently in chemical writings, and without practical inconvenience, since mass and weight are directly proportional to each other.

is made with a diamond.   The bulb is filled up to this mark with the liquid whilst it stands in water the temperature of which is exactly known.   A very fine funnel is used for filling the bulb, the stem of the funnel being drawn out so as to enter the tube, and the upper opening of the funnel being

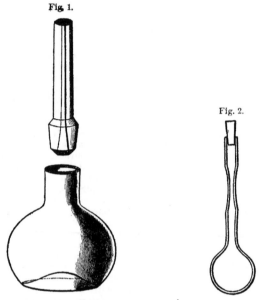

Fig. 1.

Fig. 2.

small enough to be closed by the finger.   The glass stopper is wanted only as a guard, and does not require to fit perfectly.

Another form of apparatus for determining the specific gravity of liquids, devised by Dr. Sprengel, consists of an elongated $\mathbf{U}$-tube (fig. 3), the ends

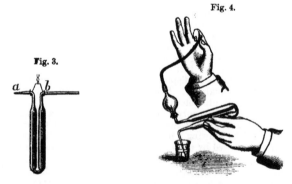

Fig. 4.

Fig. 3.

of which terminate in the two capillary tubes, a, b, bent at right angles in opposite directions.   The shorter one, a, is a good deal narrower (at least towards the end) than the longer one, the inner diameter of which is about half a millimeter.   The horizontal part of this wider tube is marked near the bend with a fine line, *b*. The U-tube is filled by suction—the little bulb-apparatus (fig. 4) having been previously attached to the narrow capillary tube by a piece of india-rubber tubing—then detached from the bulb, placed in water almost up to the bends of the capillary tubes, left there till it has assumed the temperature of the water, and, after careful adjustment of the volume of the liquid up to the mark *b* in the wider capillary tube, it is taken out, dried, and weighed.

The determination of the specific gravity of a solid body is also made according to the principles above explained, and may be performed with the specific-gravity bottle (fig. 1). The bottle is first weighed full of water; the solid is then placed in the same pan of the balance, and its weight is determined; finally, the solid is put into the bottle, displacing an equal bulk of water, the weight of which is determined by the loss on again weighing. Thus, the weights of the solid and that of an equal bulk of water, are obtained. The former divided by the latter gives the specific gravity.

For example, the weight of a small piece of silver wire
was found to be . . . . 98.18 grains.
Glass bottle filled with water . . . . . . 294.69 "

392.87 "

After an equal volume of water was displaced by the silver,
the weight was . . . . 383.54 "

Hence the displaced water weighed . . . . . 9.33 "
From this the specific gravity of the silver wire is $\dfrac{98.18}{9.33} = 10.523$ "

Another highly ingenious, but less exact method of determining the specific gravity of solids, is based on the well-known theorem of Archimedes, which may be thus expressed:

When a solid is immersed in a fluid, it loses a portion of its weight; and this portion is equal to the weight of the fluid which it displaces, that is, to the weight of its own bulk of that fluid.

Fig. 5.

This principle is applied as follows:—

Let it be required, for example, to know the specific gravity of a body of extremely irregular form, as a small group of rock-crystals: the first part of the operation consists in determining its absolute weight, or, more correctly speaking, its weight in air; it is next suspended from the balance-pan by a fine horsehair, immersed completely in pure water at 15.5°, and again weighed. It now weighs less, the difference being the weight of the water it displaces, that is, the weight of an equal bulk. This being known, nothing more is required than to find, by division, how many times the latter number is contained in the former; the quotient will be the density, water at the temperature of 15.5° being taken = 1. For example:—

The quartz-crystals weigh in air . . . . . 293.7 grains.
When immersed in water, they weigh . . . . 180.1 "

Difference, being the weight of an equal volume of water, 113.6 "
$\dfrac{293.7}{113.6} = 2.59$, the specific gravity required.

4

The rule is generally thus written : " Divide the weight in air by the loss of weight in water, and the quotient will be the specific gravity." In reality, it is not the weight in air which is required, but the weight the body would have in empty space : the error introdneed—namely, the weight of an equal bulk of air—is so trifling that it is usually neglected.

Sometimes the body to be examined is lighter than water, and floats. In this case, it is first weighed, and afterwards attached to a piece of metal heavy enough to sink it, and suspended from the balance. The whole is then exactly weighed, immersed in water, and again weighed. The difference between the two weighings gives the weight of a quantity of water equal in bulk to both together. The light substance is then detached, and the same operation of weighing in air, and again in water, repeated on the piece of metal. These data give the means of finding the specific gravity, as will be seen by the following example :—

Light substance (a piece of wax) weighs in air   .   . 133.7 grains.

Attached to a piece of brass, the whole now weighs   . 183.7   "
Immersed in water, the system weighs 38.8

Weight of water equal in bulk to brass and wax   .   . 144.9

Weight of brass in air   .   .   .   . 50.0   "
Weight of brass in water.   .   .   .   .   .   . 44.4   "

Weight of equal bulk of water   .   .   .   .   . 5.6   "

Weight of bulk of water equal to wax and brass   .   . 144.0   "
Weight of bulk of water equal to brass alone .   .   . 5.6   "

Weight of bulk of water equal to wax alone .   .   . 139.3   "

$$\frac{133.7}{139.3} = 0.9598.$$

In all such experiments it is necessary to pay attention to the temperature and purity of the water, and to remove with great care all adhering air-bubbles ;* otherwise a false result will be obtained.

Other cases require mention in which these operations must be modified to meet particular difficulties. One of these happens when the substance is dissolved or acted upon by water. This difficulty is easily overcome by substituting some other liquid of known density which experience shows is without action. Alcohol or oil of turpentine may generally be used when water is inadmissible. Suppose, for instance, the specific gravity of crystallized sugar is required, we proceed in the following way:—The specific gravity of the oil of turpentine is first determined ; let it be 0.87 ; the sugar is next weighed in the air, then suspended by a horsehair, and weighed in the oil ; the difference is the weight of an equal bulk of the latter ; a simple calculation gives the weight of a corresponding volume of water :—

Weight of sugar in air   .   .   .   . 400 grains.
Weight of sugar in oil of turpentine   .   .   .   . 182.5   "

Weight of equal bulk of oil of turpentine 217.5

$$87 : 100 = 217.5 : 250,$$

* A simple plan of avoiding altogether the adhesion of air-bubbles, which often are not easily perceived, consists in heating the water to ebullition, and introducing the body which has been weighed in the air into the still boiliug water, which is then allowed to cool to 15.5°, when the second weighing is performed.

the weight of an equal bulk of water: hence the specific gravity of the sugar,—

$$\frac{400}{250} = 1.6.$$

If the substance to be examined consists of small pieces, or of powder, the method first described, namely, that of the specific gravity bottle, can alone be used.

By this method the specific gravities of metals in powder, metallic oxides, and other compounds, and salts of all descriptions, may be readily determined. Oil of turpentine may be used with most soluble salts. The crystals should be crushed or roughly powdered to avoid errors arising from cavities in their substance.

The specific gravity of a solid can also be readily found by immersing it in a transparent liquid, the density of which has been so adjusted that the solid body remains indifferently at whatever depth it may be placed. The specific gravity of the liquid must now be determined, and it will, of course, be the same as that of the solid. It is necessary that the liquid chosen for this experiment do not dissolve or in any way act upon the solid. Solutions of mercuric nitrate, or corrosive sublimate, can be used for bodies heavier than water, whilst certain oils, and essences, and mixtures of alcohol and water, can be conveniently employed for such substances as have a lower specific gravity than water. This method is not only adapted to the exact determination of specific gravities, but also serves in many cases as a means of readily distinguishing substances much resembling one another. Suppose, for instance, a solution of mercuric nitrate to have a specific gravity 3; a red amethyst (2.67) will then float upon, and a topaz of the same color (3.55) will sink in, this liquid.

Hydrometers.—The theorem of Archimedes affords the key to the general doctrine of the equilibrium of floating bodies, of which an application is made in the common hydrometer,—an instrument for finding the specific gravities of liquids in a very easy and expeditious manner.

When a solid body is placed upon the surface of a liquid, specifically heavier than itself, it sinks down until it displaces a quantity of liquid equal to its own weight, at which point it floats.

Fig. 6.

Thus, in the case of a substance floating in water, whose specific weight is one-half that of the liquid, the position of equilibrium will involve the immersion of exactly one-half of the body, inasmuch as its whole weight is counterpoised by a quantity of water equal to half its volume. If the same body were put into a liquid of one-half the specific gravity of water, if such could be found, it would then sink beneath the surface, and remain indifferently in any part. A floating body of known specific gravity may thus be used as an indicator of the specific gravity of a liquid. In this manner little glass beads (fig. 6) of known specific gravities are sometimes employed in the arts to ascertain in a rude manner the specific gravity of liquids; the one that floats indifferently beneath the surface, without either sinking or rising, has of course the same specific gravity as the liquid itself; this is pointed out by the number marked upon the bead.

Fig. 7.

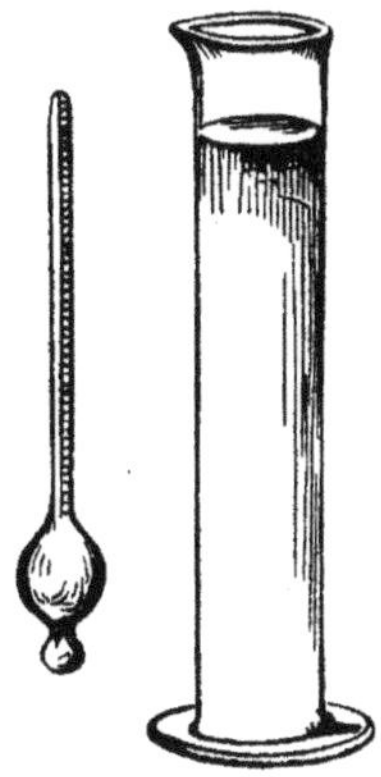

The hydrometer (fig. 7) in general use consists of a floating vessel of thin metal or glass, having a weight beneath to maintain it in an upright posi-tion, and a stem above bearing a divided scale. The liquid to be tried is put into a small narrow jar, and the instrument floated in it. It is obvious that the denser the liquid, the higher will the hydrometer float, because a smaller displacement of liquid will counterbalance its weight. For the same reason, in a liquid of less density, it sinks deeper. The hydrometer comes to rest almost immediately, and then the mark on the stem at the fluid-level may be read off.

Very extensive use is made of instruments of this kind in the arts: they sometimes bear different names, according to the kind of liquid for which they are intended ; but the principle is the same in all. The graduation is very commonly arbitrary, two or three different scales being unfortunately used. These may be sometimes reduced, however, to the true numbers expressing the specific gravity by the aid of tables of comparison drawn up for the purpose. (*See* APPENDIX.) Tables are likewise used to reduce the readings of the hydrometer at any temperature to those of the normal temperature.

It is much better, however, to use a hydrometer having the true scale of specific gravities marked upon its stem. To graduate such an instrument, a sufficient number of standard points may be determined by immersing it in liquids of known specific gravity, and the small interval between these points divided into equal parts.*

The determination of the specific gravity of gases and vapors of volatile liquids is a problem of very great practical importance to the chemist: the theory of the operation is as simple as when liquids themselves are con-cerned, but the processes are much more delicate, and involve certain cor-rections for differences of temperature and pressure, founded on principles yet to be discussed. It will be proper, therefore, to defer the consideration of these matters for the present.

* For an accurate method of dividing the hydrometer scale when only a few points are determined by actual observation, see the article " Hydrometer," by the late Pro-fessor Jevons, in Watts's *Dictionary of Chemistry*, vol. iii. p. 206.

THE PHYSICAL CONSTITUTION OF THE ATMOS-
PHERE AND OF GASES IN GENERAL.

It requires some little abstraction of mind to realize completely the condition in which all things at the surface of the earth exist. We live at the bottom of an immense ocean of gaseous matter, which envelops everything, and presses upon everything with a force which appears, at first sight, perfectly incredible, but whose actual amount admits of easy proof.

Gravity being, so far as is known, common to all matter, it is natural to expect that gases, being material substances, should be acted upon by the earth's attraction, as well as solids and liquids. This is really the case, and the result is the weight or pressure of the atmosphere, which is nothing more than the effect of the attraction of the earth on the particles of air.

Before describing the leading phenomena of the atmospheric pressure, it is necessary to notice one very remarkable feature in the physical constitution of gases, upon which depends the principle of the air-pump.

Gases are in the highest degree elastic; the volume or space which a gas occupies depends upon the pressure exerted upon it. Let the reader imagine a cylinder, a, closed at the bottom, in which moves a piston air-tight, so that no air can escape between the piston and the cylinder. Suppose now the piston be pressed downwards with a certain force; the air beneath it will be compressed into a smaller bulk, the amount of this compression depending on the force applied; if the power be sufficient, the bulk of the gas may be thus diminished to one-hundredth part or less. When the pressure is removed, the elasticity or *tension*, as it is called, of the included air or gas, will immediately force up the piston until it arrives at its first position.

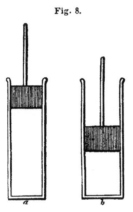

Fig. 8.

Again, take Fig. 8, b, and suppose the piston to stand about the middle of the cylinder, having air beneath in its usual state. If the piston be now drawn upwards, the air below will expand, so as to fill completely the increased space, and this to an apparently unlimited extent. A volume of air which, under ordinary circumstances, occupies the bulk of a cubic inch, might, by the removal of the pressure upon it, be made to expand to the capacity of a whole room, while a renewal of the former pressure would be attended by a shrinking down of the air to its former bulk. The smallest portion of gas introduced into a large exhausted vessel becomes at once diffused through the whole space, an equal quantity being present in every part; the vessel is *full*, although the gas is in a state of extreme tenuity. This power of expansion which air possesses may have, and probably has, in reality, a limit; but the limit is never reached in practice. We are quite safe in the assumption that, for all purposes of experiment, however refined, air is perfectly elastic.

4 *

The Air-Pump.—The ordinary air-pump, shown in section in fig. 9, con-sists essentially of a metallic cylinder, in which moves a tightly-fitting piston, by the aid of its rod. The bottom of the cylinder communicates with the

Fig. 9.

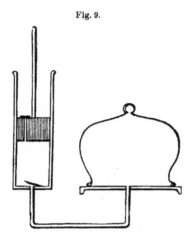

vessel to be exhausted, and is furnished with a valve opening upwards. A similar valve, also opening upwards, is fitted to the piston: three valves are made with slips of oiled silk. When the piston is raised from the bottom of the cylinder, the space left beneath it must be void of air, since the piston-

Fig. 10.

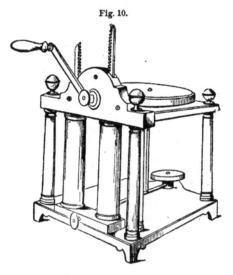

valve opens only in one direction; the air within the receiver having on that side nothing to oppose its elastic power but the weight of the little valve, lifts the latter, and escapes into the cylinder. So soon as the piston begins to descend, the lower valve closes, by its own weight, or by the pressure transmitted from above, and communication with the receiver is cut off. As the descent of the piston continues, the air enclosed in the cylinder becomes compressed, its elasticity is increased, and at length it forces open the upper valve, and escapes into the atmosphere. In this manner a cylinderful of air is removed from the receiver at every stroke of the pump. During the descent of the piston, the upper valve remains open, and the lower closed, and the reverse during the opposite movement.

In practice, it is very convenient to have two such barrels or cylinders, arranged side by side, the piston-rods of which are formed into racks, having a pinion or small-toothed wheel between them, moved by a winch. By this contrivance the operation of exhaustion is much facilitated, and the labor lessened. The arrangement is shown in fig. 10.

Fig. 11.

Atmospheric Pressure — The Barometer.

Air has weight; a light flask or globe of glass, furnished with a stopcock and exhausted by the air-pump, weighs considerably less than when full of air. If the capacity of the vessel be equal to 100 cubic inches, this difference may amount to nearly 30 grains.

After what has been said on the subject of fluid pressure, it will scarcely be necessary to observe that the law of equality of pressure in all directions also holds good in the case of the atmosphere. The perfect mobility of the particles of air permits the transmission of the force generated by their gravity. The sides and bottom of an exhausted vessel are pressed upon with as much force as the top.

If a glass tube of considerable length could be perfectly exhausted of air, and then held in an upright position, with one of its ends dipping into a vessel of liquid, the latter, on being allowed access to the tube, would rise in its interior until the weight of the column balanced the pressure of the air upon the surface of the liquid. Now, if the density of this liquid were known, and the height and area of the column measured, means would be furnished for exactly estimating the amount of pressure exerted by the atmosphere. Such an instrument is the barometer. To construct it, a straight glass tube, 36 inches long, is sealed by the blowpipe flame at one end; it is then filled with clean, dry mercury, care being taken to displace all air-bubbles, the open end is stopped with a finger, and the tube is inverted in a basin of mercury. On removing the finger, the mercury falls away from the top of the tube, until it stands at the height of about 30 inches above the level of that in the basin. Here it remains supported by, and balancing the atmospheric pressure, the space above the mercury in the tube being of necessity empty.

The pressure of the atmosphere is thus seen to be capable of sustaining a column of mercury 30 inches in height, or thereabouts. Now such a column, having an area of 1 inch, weighs between 14 and 15 lbs.: consequently such must be the amount of pressure exerted by the air upon every square inch of the surface of the earth, and of the objects situated thereon, at least near the level of the sea. This enormous force is borne without inconvenience by the animal frame, by reason of its perfect uniformity in every direction; and it may be doubled, or even tripled, without injury.

A barometer may be constructed with other liquids besides mercury; but, as the height of the column must always bear an inverse proportion to the density of the liquid, the length of tube required will be often considerable; in the case of water it will exceed 33 feet. It is seldom that any other liquid than mercury is employed in the construction of this instrument. The Royal Society of London possessed a water barometer at their apartments at Somerset House. Its construction was attended with great difficulties, and it was found impossible to keep it in repair.

Sprengel's Air-Pump.—If an aperture be made in the top of a barometer tube, the mercury will sink, and draw in air; and if the experiment be so arranged as to allow air to enter along with the mercury, and the supply of air is limited, while that of the mercury is unlimited, the air will be carried away and a vacuum produced. On this principle, Dr. Sprengel has contrived an apparatus by which a very complete exhaustion may be obtained. cd (fig. 12) is a glass tube longer than a barometer, open at both ends, and connected, by means of india-rubber tubing, with a funnel A filled with mercury, and supported on a stand. Mercury is allowed to fall through this tube at a rate regulated by a clamp at c; the lower end of the tube cd fits into the flask, B, which has a spout at its side, a little higher than the lower end of the tube cd; the upper part of this tube has a branch at x, to which a receiver R can be tightly fixed. When the clamp at c is opened, the first portion of the mercury that runs out closes the tube and prevents air from entering below. As the mercury is allowed to run down, the exhaustion begins, and the whole length of the tube from x to d is filled with cylinders of air and mercury having a downward motion. Air and mercury escape through the spout of the bulb B, below which is placed a basin H to receive the mercury; and this mercury is poured back from time to time into the funnel A, to be repassed through the tube till the exhaustion is complete. As this point is approached, the enclosed air between the mercury cylinders is seen to diminish, until the lower part of cd forms a continuous column of mercury about 30 inches high. The operation is complete when the column of mercury is quite free from air, and a drop of mercury falls on the top of it without enclosing the smallest air-bubble. The height of the column is then equal to that of the mercury in the barometer; in other words, the apparatus is a barometer whose vacuum is the receiver R. It may be advantageously combined with an exhausting

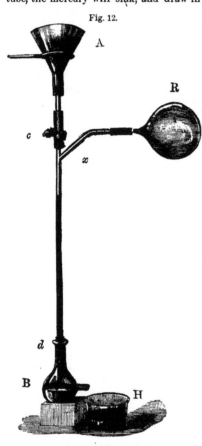

Fig. 12.

syringe, which removes the greater part of the air, the exhaustion being then completed as above.

Relations between Pressure, Elasticity, and Volume of Gases.

—It will now be necessary to consider a most important law which connects the volume occupied by a gas with the pressure made upon it, and is thus expressed :—

Fig. 13.

The volume of gas is *inversely* as the pressure; the density and elastic force are *directly* as the pressure, and *inversely* as the volume.

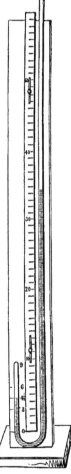

For instance, 100 cubic inches of gas under a pressure of 30 inches of mercury would expand to 200 cubic inches were the pressure reduced to one-half, and shrink, on the contrary, to 50 cubic inches if the original pressure were doubled. The change of density must necessarily be in inverse proportion to that of the volume, and the elastic force follows the same rule.

This, which is usually called the law of Mariotte, though really discovered by Boyle (1661), is easily demonstrable by direct experiment. A glass tube about 7 feet long is closed at one end, and bent into the form represented in fig. 13, the open limb of the syphon being the longer. It is next attached to a board furnished with a movable scale of inches, and enough mercury is introduced to fill the bend, the level being evenly adjusted, and marked upon the board. Mercury is now poured into the tube until the enclosed air is reduced to one-half of its former volume; and on applying the scale, it will be found that the level of the mercury in the open part of the tube stands very nearly 30 inches above that in the closed portion. The pressure of an additional "atmosphere" has consequently reduced the bulk of the contained air to one-half. If the experiment be still continued until the volume of air is reduced to a third, it will be found that the column measures 60 inches, and so in like proportion as far as the experiment is carried.

The above instrument is better adapted for the illustration of the principle than for furnishing rigorous proof of the law. This has, however, been done. MM. Arago and Dulong published, in the year 1830, an account of experiments made by them in Paris, in which the law in question had been verified to the extent of twenty-seven atmospheres. With rarefied air also, of whatever degree of rarefaction, the law has been found true.

All gases are alike subject to this law, and all vapors of volatile liquids, when remote from their points of liquefaction.* It is a matter of the greatest importance in practical chemistry, since it gives the means of making corrections for pressure, or determining by calculation the change of volume which a gas would suffer by any given change of external pressure.

Let it be required, for example, to solve the following problem : We have

---

* Near the liquefying point the law no longer holds; the volume diminishes *more rapidly* than the theory indicates, a smaller amount of pressure being then sufficient. (See further, p. 74.)

100 cubic inches of gas in a graduated jar, the barometer standing at 29 inches; how many cubic inches will it occupy when the column rises to 30 inches? Now the volume must be inversely as the pressure: consequently a change of pressure in the proportion of 29 to 30 must be accompanied by a change of volume in the proportion of 30 to 29, the 30 cubic inches of gas contracting to 29 cubic inches under the conditions imagined. Hence the answer—

$$30 : 29 = 100 : 96.67 \text{ cubic inches.}$$

The reverse of the operation will be obvious. The pupil will do well to familiarize himself with the simple calculations of correction for pressure.

Collection, Transference, and Preservation of Gases.—The management of gases is of great importance to the chemical student. We may illustrate it by describing the preparation of oxygen gas from the red oxide of mercury. This substance is placed in a short tube of hard glass, to which is fitted a perforated cork, furnished with a piece of narrow glass tube, bent as in fig. 14. The heat of a spirit-lamp or gas-lamp being applied to the substance, decomposition speedily commences, and globules of metallic mercury collect in the cool part of the wide tube, while gas issues in considerable quantity from the apparatus. This gas is collected and examined by the aid of the pneumatic trough, which consists of a vessel of water

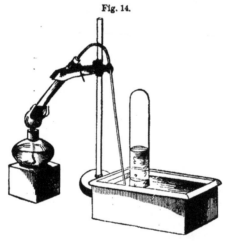

Fig. 14.

provided with a shelf, upon which stand the jars or bottles destined to receive the gas, filled with water and inverted. By keeping the level of the liquid above the mouth of the jar, the water is retained in the latter by atmospheric pressure, and entrance of air is prevented. When the jar is brought over the extremity of the gas-delivery tube, the bubbles of gas, rising through the water, collect in the upper part of the jar, and displace the liquid. As soon as one jar is filled, it may be removed, still keeping its mouth below the water-level, and another substituted.

The water-trough is one of the most indispensable articles of the laboratory, and by its aid all experiments on gases are carried on when the gases themselves are not sensibly acted on by water. The trough is best constructed of japanned copper, the form and dimensions being regulated by the magnitude of the jars. It should have a firm shelf, so arranged as to be always about an inch below the level of the water, and in the shelf a groove should be made about half an inch in width and depth, to admit the extremity of the gas-delivery tube, which stands firmly on the shelf.

For gases which are easily soluble in water, such as hydrogen chloride and ammonia, a small trough is used, made of hard wood, and filled with mercury.

Gases are transferred from jar to jar with the utmost facility, by first filling the vessel into which the gas is to be passed, with water, inverting it, carefully retaining its mouth below the water-level, and then bringing beneath it the aperture of the jar containing the gas. On gently inclining the latter, the gas passes by a kind of inverted decantation into the second vessel. When the latter is narrow, a funnel may be placed loosely in its neck, by which loss of gas will be prevented.

A jar wholly or partially filled with gas at the pneumatic trough may be removed by placing beneath it a shallow basin, or even a common plate, so as to carry away enough water to cover the edge of the jar; and many gases, especially oxygen, may be so preserved for many hours without material injury.

Gas-jars are often capped at the top, and fitted with a stop-cock for transferring gas to bladders or caoutchouc bags. When such a vessel is to be filled with water, it may be slowly sunk in an upright position in the well of the pneumatic trough, the stop-cock being open to allow the air to escape, until the water reaches the brass cap. The cock is then to be turned, and the jar lifted upon the shelf, and filled with gas in the usual way. If the trough be not deep enough for this method of proceeding, the mouth may be applied to the stop-cock, and the vessel filled by sucking out the air until the water rises to the cap. In all cases it is proper to avoid as much as possible wetting the stop-cocks and other brass apparatus.

Fig. 15.

The late Mr. Pepys contrived an admirable piece of apparatus for storing and retaining large quantities of gas. It consists of a drum or reservoir of sheet copper, surmounted by a shallow trough or cistern, the communication between the two being made by a couple of tubes furnished with stop-cocks, one of which passes nearly to the bottom of the drum, as shown in fig. 16. A short wide open tube is inserted obliquely near the bottom of the vessel, into which a plug may be tightly screwed. A stop-cock near the top serves to transfer gas to a bladder or tube apparatus. A glass water-gauge affixed to the side of the drum, and communicating with both top and bottom, indicates the level of the liquid within.

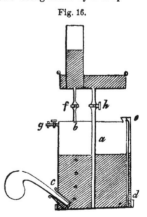

Fig. 16.

To use the gas-holder, the plug is first to be screwed into the lower opening, and the drum completely filled with water. All three stop-cocks are then to be closed and the plug removed. The pressure of the atmosphere retains the water in the gas-holder, and if no air-leakage occurs, the escape of water is inconsiderable. The extremity of the delivery-tube is now to be well pushed through the open aperture into the drum, so that the bubbles of gas may rise without hindrance to the upper part, displacing the water, which flows out in the same proportion into a vessel placed for its reception. When the drum is filled, or enough gas has been collected, the tube is withdrawn and the plug screwed into its place.

Fig. 17.

When a portion of the gas is to be transferred to a jar, the latter is to be filled with water at the pneumatic trough, carried by the help of a basin or plate to the cistern of the gas-holder, and placed over the shorter tube. On opening the cock of the neighboring tube, the pressure of the column of water will cause compression of the gas, and increase its elastic force, so that, on gently turning the cock beneath the jar, it will ascend into the latter in a rapid stream of bubbles. The jar, when filled, may again have the plate slipped beneath it, and may then be easily removed.

HEAT.

It will be convenient to consider the subject of heat under several sections, and in the following order :—

1. Expansion of bodies, or effects of variations of temperature in altering their dimensions.
2. Conduction, or transmission of heat.
3. Specific heat.
4. Change of state.
5. Sources of heat.
6. Dynamical theory of heat.

. Expansion.

If a bar of metal of such magnitude as to fit accurately to a gauge, when cold, be heated considerably, and again applied to the gauge, it will be found to have become enlarged in all its dimensions. When cold, it will once more enter the gauge.

Again, if a quantity of liquid contained in a glass bulb furnished with a narrow neck, be plunged into hot water, or exposed to any other source

Fig. 18.　　　　　　Fig. 19.　　　　　　Fig. 20.

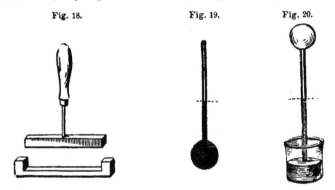

of heat, the liquid will mount in the stem, showing that its volume has been increased. The bulb, howeve. has likewise expanded by the heat, and its capacity has consequently been augmented. The rise of the liquid in the tube, therefore, denotes the difference between these two expansions.

Or, if a portion of air be confined in any vessel, the application of a slight degree of heat will suffice to make it occupy a space sensibly larger.

The most general of all the effects of heat furnishes in the outset a principle, by the aid of which an instrument can be constructed capable of taking cognizance of changes of temperature in a manner equally accurate and convenient ; such an instrument is the thermometer.

A capillary glass tube is chosen, of uniform diameter ; one extremity is closed and expanded into a bulb, by the aid of the blow-pipe flame, and the other somewhat drawn out, and left open. The bulb is now cautiously

3

heated by a spirit-lamp, and the open extremity plunged into a vessel of mercury, a portion of which rises into the bulb when the latter cools, replacing the air which had been expanded and driven out by the heat. By again applying the flame, and causing this mercury to boil, the remainder of the air is easily expelled, and the whole space filled with mercurial vapor. The open end of the tube must now be immediately plunged into the vessel filled with mercury. As the metallic vapors condense, the pressure of the external air forces the liquid metal into the instrument, until finally the tube is completely filled with mercury. The thermometer thus filled is now to be heated until so much mercury has been driven out by the

Fig. 21.

expansion of the remainder, that its level in the tube shall stand at common temperatures at the point required. This being satisfactorily adjusted, the heat is once more applied, until the column rises quite to the top; and then the extremity of the tube is hermetically sealed by the blow-pipe. The retraction of the mercury on cooling now leaves an empty space, which is essential to the perfection of the instrument.

The thermometer has yet to be graduated; and to make its indications comparable with those of other instruments, a scale, having at the least two fixed points, must be adapted to it.

It has been observed that the temperature of melting ice, that is to say, of a mixture of ice and water, is always constant; a thermometer, already graduated, plunged into such a mixture always marks the same degree of temperature, and a simple tube filled in the manner described and so treated, exhibits the same effect in the unchanged height of the little mercurial column, when tried from day to day. The freezing point of water, or melting point of ice, constitutes then one of the invariable temperatures demanded.

Another is to be found in the boiling point of water, or, more accurately, in the temperature of steam which rises from boiling water. In order to give this temperature, which remains perfectly constant whilst the barometric pressure is constant, to the mercury of the thermometer, distilled water is made to boil in a glass vessel with a long neck, when the pressure is at 30 inches (fig. 21). The thermometer is then so placed that all the mercury is surrounded with steam. It quickly rises to a fixed point, and there it remains as long as the water boils, and the height of the barometer is unchanged.

The tube having been carefully marked with a file at these two points, it remains to divide the interval into degrees; this division is entirely arbitrary. The scale now most generally employed is the Centigrade, in which the space is divided into 100 parts, the zero being placed at the freezing point of water. The scale is continued above and below these points, numbers below 0 being distinguished by the negative sign.

In England and North America the division of Fahrenheit is still in use; the above mentioned space is divided into 180 degrees; but the zero, instead of starting from the freezing point of water, is placed 32 degrees below it, so that the temperature of ebullition is expressed by 212°.

The plan of Reaumur is nearly confined to a few places in the north of Germany and to Russia; in this scale the freezing point of water is made 0°, and the boiling point 80°.

It is unfortunate that a uniform system has not been generally adopted ın graduating thermometers; this would render unnecessary the labor which now so frequently has to be performed of translating the language of one scale into that of another. To effect this, presents, however, no great difficulty. Let it be required, for example, to know the degree of Fahrenheit's scale which corresponds to 60° C.

$$100° \text{ C.} = 180° \text{ F.; or } 5° \text{ C.} = 9° \text{ F.}$$

Consequently,

$$5 : 9 = 60 : 108.$$

But then, as Fahrenheit's scale commences with 32° instead of 0°, that number must be added to the result, making 60°C. = 140° F.

The rule, then, is the following: To convert centigrade degrees into Fahrenheit degrees, multiply by 9, divide the product by 5, and add 32; to convert Fahrenheit degrees into centigrade degrees, subtract 32, multiply by 5, and divide by 9.

The reduction of negative degrees, or those below zero, of one scale into those of another scale, is effected in the same way. For example, to convert — 15° C. into degrees of Fahrenheit, we have—

$$-15 \times \frac{9}{5} + 32 = -27 + 32 = +5 \text{ F.}$$

In this work temperatures will always be given in centigrade degrees, unless the contrary is stated.

Mercury is usually chosen for making thermometers, on account of its regularity of expansion within certain limits, and because it is easy to have the scale of great extent, from the large interval between the freezing and boiling points of the metal. Other substances are sometimes used; alcohol is employed for estimating very low temperatures, because this liquid has not been frozen even at the lowest degree of cold which has been artificially produced.

Air-thermometers are also used for some few particular purposes; indeed, the first thermometer ever made was of this kind. There are two modifi-

Fig. 22.   Fig. 23.

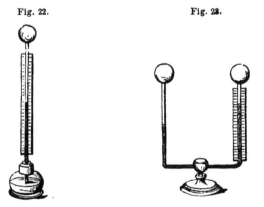

eations of this instrument; in the first, the liquid into which the tube dips is open to the air; and in the second, shown in Fig. 22, the atmosphere is completely excluded. The effects of expansion are in the one case complicated with those arising from changes of pressure, and in the other they cease to be visible at all when the *whole* instrument is subjected to alterations of temperature, because the air in the upper and lower reservoir, being equally affected by such changes, no alteration in the height of the fluid column can occur. Accordingly, such instruments are called *differential* thermometers, since they serve to measure differences of temperature between the two portions of air, while changes affecting both alike are not indicated. Fig. 23 shows another form of the same instrument.

The air-thermometer may be employed for measuring all temperatures from the lowest to the highest; M. Pouillet has described one by which the heat of an air-furnace could be measured. The reservoir of this instrument is of platinum, and it is connected with a piece of apparatus by which the increase of volume experienced by the included air is determined.

An excellent air-thermometer has been constructed and used by Rudberg, and more recently by Magnus and Regnault, for measuring the expansion of the air. Its construction depends on the law, that when air is heated and hindered from expanding, its tension increases in the same proportion in which it would have increased in volume if permitted to expand.

All bodies are enlarged by the application of heat, and reduced by its abstraction; or, in other words, contract on being artificially cooled; this effect takes place to a comparatively small extent with solids, to a larger amount in liquids, and most of all in the case of gases.

Each solid and liquid has a rate of expansion peculiar to itself; gases, on the contrary, expand nearly alike for the same increase of heat.

Expansion of Solids.—The actual amount of expansion which different solids undergo by the same increase of heat has been carefully investigated. The following are some of the results of the best investigations, more particularly those of Lavoisier and Laplace. The fraction indicates the amount of expansion in length suffered by the rods of the undermentioned bodies in passing from $0°$ to $100°$:—

Fir-wood*	$\frac{1}{1451}$	Tempered steel	$\frac{1}{807}$
English flint glass	$\frac{1}{1248}$	Soft iron	$\frac{1}{819}$
Platinum	$\frac{1}{1167}$	Gold	$\frac{1}{682}$
Common white glass	$\frac{1}{1160}$	Copper	$\frac{1}{584}$
Common white glass	$\frac{1}{1147}$	Brass	$\frac{1}{535}$
Glass without lead	$\frac{1}{1112}$	Silver	$\frac{1}{524}$
Another specimen	$\frac{1}{1090}$	Lead	$\frac{1}{351}$
Steel untempered	$\frac{1}{927}$	Zinc	$\frac{1}{333}$

From the *linear* expansion, the *cubic* expansion (or measure of volume), may be calculated. When the expansion of a body in different directions is equal, as, for example, in glass, hammered metals, and generally in most uncrystallized substances, it will be sufficient to triple the fraction expressing the increase in one dimension. This rule does not hold true for crystals belonging to irregular systems, for they expand unequally in the direction of the different axes.

Metals appear to expand pretty uniformly for equal increments of heat within the limits stated; but above the boiling point of water the rate of expansion becomes irregular and more rapid.

The force exerted in the act of expansion is very great. In laying down railways, building iron bridges, erecting long ranges of steam-pipes, and in executing all works of the kind in which metal is largely used, it is indispensable to make provision for these changes of dimensions.

* In the direction of the vessels.

In consequence of glass and platinum having nearly the same amount of expansion, a thin platinum wire may be fused into a glass tube, without any fear that the glass will break on cooling.

A very useful application of expansion by heat is that of the cutting of glass by a hot iron; this is constantly practised in the laboratory for a great variety of purposes. The glass to be cut is marked with ink in the required direction, and then a crack, commenced by any convenient method, at some distance from the desired line of fracture, may be led by the point of a heated iron rod along the latter with the greatest precision.

Expansion of Liquids.—The dilatation of a liquid may be determined by filling a thermometer with it, in which the relation between the capacity of the ball and that of the stem is exactly known, and observing the height of the column at different temperatures. It is necessary in this experiment to take into account the effects of the expansion of the glass itself, the observed result being evidently the *difference* of the two.

Liquids vary exceedingly in this particular. The following table is taken from Péclet's "Éléments de Physique:"—

Apparent Dilatation in Glass between 0° and 100°.

Water	$\frac{1}{22}$
Hydrochloric acid, sp. gr. 1.137	$\frac{1}{27}$
Nitric acid, sp. gr. 1.4	$\frac{1}{19}$
Sulphuric acid, sp. gr. 1.85	$\frac{1}{17}$
Ether	$\frac{1}{14}$
Olive oil	$\frac{1}{13}$
Alcohol	$\frac{1}{9}$
Mercury	$\frac{1}{64}$

Most of these numbers must be taken as representing mean results; for there are few liquids, which, like mercury, expand regularly between these temperatures. Even mercury above 100° shows an unequal and increasing expansion, if the temperature indicated by the air-thermometer be used for comparison. This is shown by the following abstract of a table given by Regnault:—

Reading of Air Thermometer.	Reading of Mercurial Thermometer.	Temperature deduced from the Absolute Expansion of Mercury.
0°	0°	0°
100	100	100
200	200	202.78
300	301	308.34
350	354	362.16

The absolute amount of expansion of mercury is, for many reasons, a point of great importance; it has been very carefully determined by a method independent of the expansion of the containing vessel. The apparatus employed for this purpose, first by Dulong and Petit, and later by Regnault, is shown in fig. 24, divested, however, of many of its subordinate parts. It consists of two upright glass tubes, connected at their bases by a horizontal tube of much smaller dimensions. Since a free communication exists between the two tubes, mercury poured into the one will rise to the same level in the other, provided its temperature is the same in both tubes; when this is not the case, the hotter column will be the taller, because the expansion of the metal diminishes its specific gravity, and the law of hydrostatic equilibrium requires that the height of such columns

5 *

should be inversely as their densities. By the aid of the outer cylinders, one of the tubes is maintained constantly at 0°, while the other is raised, by means of heated water or oil, to any required temperature. The perpendicular height of the columns may then be read off by a horizontal micrometer telescope, moving on a vertical divided scale.

These heights represent volumes of equal weight, because volumes of equal weight bear an inverse proportion to the densities of the liquids, so that the amount of expansion admits of being very easily calculated. Thus, let the column at 0° be 6 inches high, and that at 100°, 6.108 inches; the increase of height, 108 on 6000, or $\frac{1}{55.5}$ part of the whole, will represent the absolute cubical expansion.

Fig. 24.

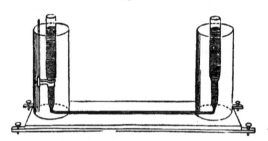

The indications of the mercurial thermometer are inaccurate when very high ranges of temperature are concerned, from the increased expansibility of the metal. The error thus caused is, however, nearly compensated, for temperatures under 204.5° C. (400° F.) by the expansion of the glass tube. For higher temperatures a small correction is necessary, as the above table shows.

To what extent the expansion of different liquids may vary between the same temperatures is obvious from a glance at fig. 25 which represents the

Fig. 25.

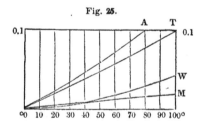

expansion of mercury (M), water (W), oil of turpentine (T), and alcohol (A). A column of these several liquids, equalling at 0° the tenfold height of the line 0-0.1 in the diagram, would exhibit, when heated to a temperature of 10°, 20°, 30°, etc., an expansion indicated by the distances at which the perpendicular lines drawn over the numbers 10, 20, 30, etc., are intersected by the curves belonging to each of these liquids. Thus it is seen that oil of turpentine, between 0° and 100°, expands very nearly $\frac{1}{10}$ of its volume, and that mercury between the same limits of temperature expands uniformly, while the rate of expansion of the other liquids increases with the rise of the temperature.

An exception to the regularity of expansion in liquids exists in the case of water ; it is so remarkable, and its consequences so important, that it is necessary to advert to it particularly.

Let a large thermometer-tube be filled with water at the common temperature of the air, and then artificially cooled. The liquid will be observed to contract, until the temperature falls to about 4° C. (39.2^O F., or 7.2^O above the freezing point). After this a further reduction of temperature causes expansion instead of contraction in the volume of the water, and this expansion continues until the liquid arrives at its point of congelation, when so sudden and violent an enlargement takes place that the vessel is almost invariably broken. At the temperature of 4° water is at its maximum density ; increase or diminution of heat produces upon it, for a short time, the same effect.

According to the latest researches of Kopp, the point of greatest density of water is 4.08^O C. (39.34^O F.). According to the determinations of this physicist, the volume of water $= 1$ at 0° C. changes, when heated, to the following volumes :—

2^O 0.99991	16° 1.00085	35° 1.00570	70° 1.02225
4° 0.99988	18° 1.00118	40° 1.00753	75° 1.02544
6° 0.99990	20° 1.00157	45° 1.00954	80° 1.02858
8° 0.99999	22° 1.00200	50° 1.01177	85° 1.03189
10° 1.00012	24° 1.00247	55° 1.01410	90° 1.03540
12° 1.00031	25° 1.00272	60° 1.01659	95° 1.03909
14° 1.00056	30° 1.00406	65° 1.01930	100° 1.04299

Sea-water has no point of maximum density above the freezing point. The more it is cooled the denser it becomes, until it solidifies at -2.6^O C.

The gradual expansion of pure water cooled below 4° C. must be carefully distinguished from the great and sudden increase of volume it exhibits in the act of freezing, in which respect it resembles many other bodies which expand on solidifying. The force thus exerted by freezing water is enormous. Thick iron shells quite filled with water, and exposed with their fuse-holes securely plugged, to the cold of a Canadian winter night, have been found split on the following morning. The freezing of water in the joints and crevices of rocks is a most potent agent in their disintegration.

Expansion of Gases.—The principal laws reiating to the expansion of gases are contained in the four following propositions :—

1. All gases expand nearly alike for equal increments of heat ; and all vapors, when remote from their condensing points, follow the same law.
2. The rate of expansion is not altered by a change in the state of compression, or elastic force, of the gas itself.
3. The rate of expansion is uniform for all degrees of heat.
4. The actual amount of expansion is equal to $\frac{1}{273}$ or $\frac{11}{3000}$ or 0.003666 of the volume of the gas at 0° centigrade, for each degree of the same scale.*

It will not be necessary to enter into any description of the methods of investigation by which these results have been obtained ; the advanced student will find in Pouillet's " Éléments de Physique," and in the papers of Magnus and Regnault,† all the information he may require.

* The fraction $\frac{11}{3000}$ is very convenient for calculation.
† Poggendorff's Annalen, iv. 1.—Ann. Chim. Phys. 3d series, iv. 5, and v. 52.
See also Watts's Dictionary of Chemistry, art. Heat, vol. iii p. 46 ; and Ganot's Éléments de Physique, translated by Dr. Atkinson, 4th edition, pp. 253-262.

In the practical manipulation of gases, it very often becomes necessary to make a correction for temperature, or to discover how much the volume of a gas would be increased or diminished by a particular change of temperature; this can be effected with great facility. Let it be required, for example, to find the volume which 100 cubic inches of any gas at 10° C. would become on the temperature rising to 20° C.

The rate of expansion is $\frac{1}{273}$ or $\frac{11}{3000}$ of the volume at 0° for each degree; or 3000 measures at 0° become 3011 at 1°, 3022 at 2°, 3110 at 10°, and 3220 at 20°. Hence

Meas. at 10°.		Meas. at 20°.		Meas. at 10°.		Meas. at 20°.
3110	:	3220	=	100	:	103.537.

If this calculation is required to be made on the Fahrenheit scale, it must be remembered that the zero of that scale is 32° below the melting-point of ice. Above this temperature the expansion for each degree of the Fahrenheit scale is $\frac{1}{460}$ of the original volume.

This, and the correction for pressure, are operations of very frequent occurrence in chemical investigations, and the student will do well to become familiar with them.

The following formula includes both these corrections: Let V and V′ be the volume of a gas at the temperatures t and t' centigrade, and under the pressure p and p', measured in millimeters of mercury: then

$$\frac{V}{V'} = \frac{1+0.003666t}{1+0.003666t'} \cdot \frac{p'}{p} \; .$$

The case which most frequently occurs is the reduction of a measured volume, V, of a gas at the temperature t and pressure p to the volume V_0, which it would occupy at 0° C., and under a pressure of 760 mm. In this case, we have $t' = 0$, and $p' = 760$, therefore

$$\frac{V}{V_0} = (1 + 0.003666t) \cdot \frac{760}{p} \; ,$$

and

$$V_0 = \frac{V}{1 + 0.003666t} \cdot \frac{p}{760} \; .$$

If the barometric pressure is measured in inches, the number 30 must be substituted for 760.

Note.—Of the four propositions stated in the text, the first and second have recently been shown to be true within certain limits only; and the third, although in the highest degree probable, would be very difficult to demonstrate rigidly; in fact, the equal rate of expansion of air is assumed in all experiments on other substances, and becomes the standard by which the results are measured.

The rate of expansion for the different gases is *not* absolutely the same, but the difference is so small that for most purposes it may safely be neglected. Neither is the state of elasticity altogether indifferent, the expansion being sensibly *greater* for an equal rise of temperature when the gas is in a compressed state.

It is important to notice that the greatest deviations from the rule are exhibited by those gases which, as will hereafter be seen, are most easily liquefied, such as carbon dioxide, cyanogen, and sulphur dioxide; and that the discrepancies become smaller and smaller as the elastic force is less-ened; so that, if means existed for comparing the different gases in states *equally distant* from their points of condensation, there is reason to believe that the law would be strictly fulfilled.

The experiments of Dalton and Gay-Lussac gave for the rate of expansion $\frac{1}{267}$ of the volume at 0°: this is no doubt too high. Those of Rudberg give $\frac{1}{274}$, those of Magnus and of Regnault $\frac{1}{273}$.

Conduction of Heat.

Different bodies possess very different conducting powers with respect to heat; if two similar rods, the one of iron, the other of glass, be held in the flame of a spirit-lamp, the iron will soon become too hot to be touched, while the glass may be grasped with impunity within an inch of the red-hot portion.

Experiments made by analogous but more accurate methods have established a numerical comparison of the conducting powers of many bodies. The following may be taken as a specimen :—

Silver	.	.	.	. 1000	Steel .	.	.	.	. 116
Copper	.	.	.	. 736	Lead .	.	.	.	. 85
Gold	.	.	.	. 532	Platinum .	.	.	. 84
Brass	.	.	.	. 236	German silver	.	.	. 63
Tin .	.	.	.	. 145	Bismuth .	.	.	. 18
Iron .	.	.	.	. 119					

As a class the metals are by very far the best conductors, although much difference exists between them; stones, dense woods, and charcoal follow next in order; then liquids in general, and lastly gases, whose conducting power is almost inappreciable.

Under favorable circumstances, nevertheless, both liquids and gases may become rapidly heated; heat applied to the bottom of the containing vessel is very speedily communicated to its contents; this, however, is not so much by conduction, as by convection, or carrying. A complete circulation is set up; the portions in contact with the bottom of the vessel get heated, become lighter, and rise to the surface, and in this way the heat becomes communicated to the whole. If these movements be prevented by dividing the vessel into a great number of compartments, the really low conducting power of the substance is made evident; and this is the reason why certain organic fabrics, as wool, silk, feathers, and porous bodies in general, the cavities of which are full of air, exhibit such feeble powers of conduction.

The circulation of heated water through pipes is now extensively applied to the warming of buildings and conservatories; and in chemical works a serpentine metal tube containing hot oil is often used for heating stills and evaporating pans; the two extremities of the tube are connected with the ends of another spiral built into a small furnace at a lower level, and an unintermitting circulation of the liquid takes place as long as heat is applied.

Specific Heat.

Equal weights of different substances having the same temperature require different amounts of heat to raise them to a given degree of temperature. If 1 lb. of water, at 100°, be mixed with 1 lb. at 40°, then, as is well known, a mean temperature of $\frac{100 + 40}{2} = 70^\circ$ is obtained. In the same way the mean temperature is found when warm and cold oil, or warm and cold mercury, etc., are mixed together. But if 1 lb. of water at 100° be shaken with 1 lb. of olive-oil at 40°, or with 1 lb. of mercury at 40°, then, instead of the mean temperature of 70°, the temperature actually

3*

obtained will be 80° in the first case, 98° in the second; 20 degrees of heat, which the water (by cooling from 100° to 80°) gave to the same weight of oil, were sufficient to raise the oil 40°, that is, from 40° to 80°; and 2°, which the water lost by cooling from 100° to 98°, sufficed to heat an equal quantity of mercury 58°, namely from 40° to 98°.

It is evident from these experiments that the quantities of heat which equal weights of water, olive-oil, and mercury, require to raise their temperature to the same height, are unequal, and that they are in the proportion of the numbers $1 : \frac{20}{40} : \frac{2}{58}$, or $1 : \frac{1}{2} : \frac{1}{29}$.

These quantities of heat, expressed relatively to the quantity of heat required to raise the temperature of an equal weight of water from 0° to 1° C., are called the specific heats of the various substances; thus the experiments just described show that the specific heat of olive-oil is $\frac{1}{2}$, that is to say, the quantity of heat which would raise the temperature of any given quantity of olive-oil from 0° to 1° would raise that of an equal weight of water only from 0° to $\frac{1}{2}$°, or of half that quantity of water from 0° to 1°.

The specific heats of bodies are sometimes said to measure their relative capacities for heat.

There are three distinct methods by which the specific heats of various substances may be estimated. The first of these is by observing the quantity of ice melted by a given weight of the substance heated to a particular temperature; the second is by noting the time which the heated body requires to cool down through a certain number of degrees; and the third is the method of mixture, on the principle illustrated; this latter method is preferred as the most accurate.

The determination of the specific heat of different substances has occupied the attention of many experimenters; among these Dulong and Petit, and recently Regnault and Kopp, deserve especial mention.

From the observations of these and other physicists, it follows that each body has its peculiar specific heat, and that the specific heat increases with increase of temperature. If, for example, the heat which the unit of water loses by cooling from 10° to 0° be marked at 10°, then the loss by cooling from 50° to 0° will be, not 50, corresponding to the difference of temperature, but 50.1. By cooling from 100° to 0° it is 100.5, and rises to 203.2 when the water is heated under great pressure to 200°, and afterwards cooled to 0°. Similar and even more striking differences have been found with other substances. It has also been proved that the specific heat of any substance is greater in the liquid than in the solid state. For example, the specific heat of ice is 0.504, that is, not more than half as great as that of liquid water.

It is remarkable that the specific heat of water is greater than that of all other solid and liquid substances, and is exceeded only by that of hydrogen. The specific heat of the solid parts of the crust of the globe is on an average $\frac{1}{5}$, and that of the atmosphere nearly $\frac{1}{4}$ that of water.

If the specific heat of any body within certain degrees of temperature be accurately known, then from the quantity of heat which this body gives out when quickly dipped into cold water, the temperature to which the body was heated may be determined. Pouillet has founded on this fact a method of measuring high temperatures, and for this purpose, with the help of the air-thermometer, he has determined the specific heat of platinum up to 1600° C.

The determination of the specific heat of gases is attended with peculiar difficulties, on account of the comparatively large volume of small weights of gases. For many gases, however, satisfactory results have been obtained by the method of mixing.

When a gas expands, heat becomes latent, that is to say, insensible to the thermometer; in fact, the molecular motion which constitutes heat is converted into another kind of motion, which overcomes the pressure to which the gas is subjected, and allows it to expand (see page 86). The amount of heat required, therefore, to raise a gas to any given temperature increases the more the gas in question is allowed to expand. The quantity of heat which the unit-weight of a gas requires in order to raise its temperature $1°$ without its volume undergoing any change (which can take place only by the pressure being simultaneously augmented) is called the specific heat of the gas *at constant volume*. The quantity of heat required by the unit-weight of a gas to raise its temperature $1°$, it being at the same time allowed to dilate to such an .extent that the pressure to which it is exposed remains unchanged, is called the specific heat of the gas *at constant ·pressure*. According to what has already been stated, the specific heat at constant pressure must be greater than that at constant volume. Dulong found, in the case of atmospheric air, of oxygen, of· hȳdrogen, and of nitrogen, that the two specific heats are in the proportion 1.421 : 1. For carbon monoxide, however, he obtained the proportion of 1.423, for carbon dioxide 1.337, for nitrogen dioxide 1.343, and for olefiant gas 1.24 to 1. The exact determination of these ratios is extremely difficult, and the results of different physicists by no means agree.

The first satisfactory comparison of the specific heat of air with that of water was made by Count Rumford ; later comparisons of the specific heat of various gases have been made by Delaroche and Bérard, Dulong and Regnault.

The first researches of Delaroche and Bérard furnished the results embodied in the following table :—

| | Equal volumes. | | Equal weights. | |
	The volumes constant.	The pressure constant.	Air=1.	Water=1.
Atmospheric air	1	1	1	0.2669
Oxygen	1	1	0.9045	0.2414
Hydrogen	1	1	14.4510	3.8569
Nitrogen	1	1	1.0295	0.2748
Carbon monoxide	1	1	1.0337	0.2759
Nitrogen monoxide	1.22	1.160	0.7607	0.2030
Carbon dioxide	1.24	1.175	0.7685	0.2051
Olefiant gas	1.75	1.531	1.5829	0.4225

The latest and most trustworthy determinations are those of Regnault, given in the subjoined table, in the second column of which the specific heats of the several gases and vapors are compared with that of an equal weight of water taken as unity, and in the third, with that of an equal volume of air referred to its own weight of water as unity. The latter series of numbers is obtained by multiplying the numbers in the second column by the specific gravities of the respective gases and vapors referred to air as unity :—

| Gases. | Specific gravity. Air = 1. | Specific heat at constant pressure. | |
		For equal weights. Water = 1.	For equal volumes.
Atmospheric air	1	0.2377	0.2377
Oxygen	1.1056	0.2175	0.2405
Nitrogen	0.9713	0.2438	0.2368
Hydrogen	0.0692	3.4090	0.2359
Chlorine	2.4502	0.1210	0.2965
Bromine vapor	5.4772	0.0555	0.3040

| | | *Specific heat at constant pressure.* | |
Gases.	*Specific gravity.* Air = 1.	For equal weights. Water = 1.	For equal volumes.
Carbon monoxide	0.9670	0.2450	0.2370
Carbon dioxide	1.5210	0.2169	0.3307
Nitrogen monoxide . . .	1.5241	0.2262	0.3447
Nitrogen dioxide	1.0384	0.2317	0.2406
Olefiant gas	0.9672	0.4040	0.4106
Marsh gas	0.5527	0.5929	·0.3277
Aqueous vapor	0.6220	0.4805	0.2989
Sulphuretted hydrogen . .	1.1746	0.2432	0.2857
Sulphur dioxide . . .	2.2112	0.1544	0.3414
Vapor of carbon bisulphide .	2.6258	0.1569	0.4122
Hydrochloric acid . .	1.2596	0.1852	0.2333
Ammonia	0.5894	0.5084	0.2996

The researches of Delaroche and Bérard led them to suppose that the specific heat of gases increased rapidly as the temperature was raised, and that for a given volume of gas it increased in proportion to the density or tension of the gas. Regnault found, however, the quantity of heat which a given volume of gas requires to raise it to a certain temperature, to be independent of its density; and that for each degree between —30° C. and 225° C. it is constant. Carbon dioxide, however, forms an exception to this rule, its specific heat increasing with the temperature. In the table mean values for temperatures between 10° C. and 200° C. have been given.

Several physicists have held that the specific heats of elementary gases, referred to equal volumes, are identical. The numbers which Regnault found for chlorine and bromine, however, show that the law does not hold good for all elementary gases.

It has been already stated that, when a gas expands, heat becomes latent. If a gas on expanding be not supplied with the requisite heat, its temperature falls on account of its own free heat becoming latent, that is to say, expended in overcoming pressure. On the other hand, if a gas be compressed, this latent heat becomes free, and causes an elevation of temperature, which, under favorable circumstances, may be raised to ignition; syringes by which tinder is kindled are constructed on this principle.

Change of State.

1. *Fusion and Solidification.*

Solid bodies when heated are expanded; many are liquefied, *i. e.*, they melt or fuse. The melting of solids is frequently preceded by a gradual softening, more especially when the temperature approaches the point of fusion. This phenomenon is observed in the case of wax or iron. In the case of other solids—of zinc and lead, for instance—and several other metals, this softening is not observed. Generally, bodies expand during the process of fusion; an exception to this rule is water, which expands during freezing (10 vol. of water produce nearly 11 vol. of ice), while ice when melting produces a proportionately smaller volume of water. The expansion of bodies during fusion, and at temperatures preceding fusion, or the contraction during solidification and further refrigeration, is very unequal. Wax expands considerably before fusing, and comparatively little during fusing itself. Wax, when poured into moulds, fills them perfectly during solidification, but afterwards contracts considerably.

Stearic acid, on the contrary, expands very little before fusion, but rather considerably during fusion, and consequently pure stearic acid when poured into moulds solidifies to a rough porous mass contracting little by further cooling. The addition of a little wax to stearic acid prevents the powerful contraction in the moment of solidification, and renders it more fit for being moulded.

The melting and solidifying point of each particular substance is constant under any given pressure, but those of different substances vary within very wide limits: thus sulphurous oxide (SO_2) melts at $-80°$, mercury at $-40°$, bromine at $-7.3°$, ice at $0°$, phosphorus at $+44°$, tin at $235°$, silver at $1000°$, platinum at $2000°$.

The ordinary variations of atmospheric pressure have no perceptible influence on the melting points of bodies, but greater variations produce very appreciable effects: thus at pressures of 8.1 and 16.8 atmospheres, the melting point of ice is lowered by 0.059° and 0.126° respectively. Some bodies, on the other hand, have their melting points raised by pressure, as shown in the following table:—

	Melting Points.			
Pressure in Atm.	Spermaceti.	Wax.	Sulphur.	Stearin.
1	51.1°	64.7°	107.2°	67.2
520	60.0	74.7	135.2	68.3
793	80.2	80.2	140.5	73.8

In general, those substances which expand on liquefying have their melting points raised by pressure; whereas those which, like water, contract on liquefying, have their melting points lowered by increased pressure.

Conversely, it has been shown by Dr. Carnelley* that ice, mercuric chloride, and certain other bodies, when subjected to pressures below a certain limit, called the "critical pressure," cannot be liquefied, however great may be the heat applied to them. Bodies under these conditions do not, however, appear to attain temperatures above their ordinary melting points, the whole of the heat applied to them being apparently absorbed as latent heat in the direct conversion of the solid body into vapor.

Determination of Melting Points.—The following is a convenient method for determining the melting points of bodies which fuse at moderate temperatures, such as commonly occur in the course of chemical investigations. Three or four glass tubes are drawn out until their sides become very thin and their bore nearly capillary, and into each is introduced a small quantity of the substance under investigation. They are then sealed at the bottom and placed in a glass vessel containing water (or paraffin or sulphuric acid if the substance melts above 100°), and standing in a small sand-bath, by which the temperature can be slowly raised. The liquid is heated till the substance melts, then allowed to cool slowly to the solidifying point, and again warmed, these observations being repeated, and the temperatures of liquefaction and solidification being each time noted, until several nearly concordant results have been obtained. The mean of all these is taken as the true melting point of the substance.

For determining the melting points of bodies which melt at higher temperatures, such as metallic salts, a method has lately been given by Dr. Carnelley, depending on the following principle: When a small quantity of

* Proceedings of the Royal Society, vol. **xxxi.** p. 284; Chem. Soc. Journal, xlii. 317.

a salt is placed in a weighed platinum crucible suspended in the flame of a Bunsen burner or a blowpipe, and the crucible, at the instant when the salt fuses, is dropped into a known weight of water of known temperature, the rise in temperature of the water being then noted, the temperature at which the fusion occurred may be found from the equation—

$$T = \theta + \frac{(W + w)\, r\, (\theta - t)}{Ms},$$

in which W denotes the weight of water in the calorimeter; w the thermal value of the calorimeter + mercury and glass of thermometer in grams of water; M the weight of the crucible; r the specific heat of water; s the specific heat of platinum (from $0°$ to T); t the initial and θ the final temperature of the water. Solidifying points may be determined in a similar manner, but with this difference, that when the salt has completely melted, the lamp is turned out, and the crucible dropped into the calorimeter at the instant when the salt begins to solidify.

The following table exhibits the melting and solidifying points of several salts determined in this manner:—

		M. P.	S. P.	Difference.
Potassium nitrate	353°	332°	21
" chlorate. . . .	372	351	21
" chloride. . . .	434	415	19
" iodide	639	622	17
" bromide. . ..	703	685	18
" carbonate . . .	838	832	6

Latent Heat of Fusion.—During fusion bodies absorb a certain quantity of heat which is not indicated by the thermometer.

If equal weights of water at $0°$ C. ($32°$ F.) and water at $79°$ C. ($174.2°$ F.) be mixed, the temperature of the mixture will be the mean of the two temperatures, viz., $39.5°$; but if the same experiment be made with snow or finely powdered ice at $0°$ and water at $79°$, the temperature of the mixture will be only $0°$, *but the ice will have been melted.*

$$\left.\begin{array}{l}\text{1 lb. of water at } 0° \\ \text{1 lb. of water at } 79°\end{array}\right\} = 2 \text{ lb. water at } 39.5°.$$

$$\left.\begin{array}{l}\text{1 lb. of ice at } 0° \\ \text{1 lb. of water at } 79°\end{array}\right\} = 2\text{lb. water at } 0°.$$

In the last experiment, therefore, as much heat has been apparently lost as would have raised a quantity of water equal to that of the ice through a range of $79°$ ($142.2°$ F.).

The heat, thus become insensible to the thermometer in effecting the liquefaction of the ice, is called l a t e n t h e a t, or better, h e a t o f f l u i d i t y.

Again, let a perfectly uniform source of heat be imagined, of such intensity that a pound of water placed over it would have its temperature raised $5°$ per minute. Starting with water at $0°$, in rather less than 16 minutes its temperature would have risen $79°$; but the same quantity of ice at $0°$, exposed for the same interval of time, would not have its temperature raised a single degree. But, then, it would have become water, the heat received being exclusively employed in effecting the change of state.

This heat is not lost, for when the water freezes it is again evolved. If a tall jar of water, covered to exclude dust, be placed in a situation where it shall be quite undisturbed, and at the same time exposed to great cold, the temperature of the water may be reduced $5°$ C. ($9°$ F.) or more below its freezing point without the formation of ice;* but then, if a little agitation be communicated to the jar, or a grain of sand dropped into the water, a portion instantly solidifies, and the temperature of the whole rises to 0°; the heat disengaged by the freezing of a small portion of the water is sufficient to raise the whole contents of the jar 5° C. (9° F.).

This curious condition of instable equilibrium shown by the very cold water in the preceding experiment, may be reproduced with a variety of solutions which tend to crystallize or solidify, but in which that change is for a while suspended. Thus, a solution of crystallized sodium sulphate in its own weight of warm water, left to cool in an open vessel, deposits a large quantity of the salt in crystals. If the warm solution, however, be filtered into a clean flask, which when full is securely corked and set aside to cool undisturbed, no crystals will be deposited, even after many days, until the cork is withdrawn and the contents of the flask are violently shaken. Crystallization then rapidly takes place in a very beautiful manner, and the whole becomes perceptibly warm. The law above illustrated in the case of water is perfectly general. Whenever a solid becomes a liquid, a certain fixed and definite amount of heat disappears, or becomes latent; and conversely, whenever a liquid becomes solid, heat to a corresponding extent is given out.

The following table exhibits the melting points of several substances, and their latent heats of fusion expressed in gram-degrees—that is to say, the numbers in the column headed "latent heat" denote the number of grams of water, the temperature of which would be raised 1° centigrade by the quantity of heat required to fuse one gram of the several solids:—

Substance.	Melting Point.	Latent Heat.	Substance.	Melting Point.	Latent Heat.
Mercury	—39°	2.82	Tin	235°	14.25
Phosphorus . . .	+44	5.0	Silver	1000	21.1
Lead	332	5.4	Zinc	433	28.1
Sulphur.	115	9.4	Calcium chloride ⎱ (CaCl₂ 6H₂O), ⎰	28.5	40.7
Iodine	107	11.7			
Bismuth	270	12.6	Potassium nitrate .	339	47.4
Cadmium	320	13.6	Sodium nitrate . .	310.5	63.0

Freezing Mixtures.—When a solid substance can be made to liquefy by a weak chemical attraction, cold results, from sensible heat becoming latent. This is the principle of frigorific mixtures. When snow or powdered ice is mixed with common salt, and a thermometer plunged into the mass, the mercury sinks to —17.7° C. (0° F.), while the whole after a short time becomes fluid by the attraction between the water and the salt; such a mixture is very often used in chemical experiments to cool receivers and condense the vapors of volatile liquids. Powdered crystallized calcium chloride and snow produce cold enough to freeze mercury. Even powdered potassium nitrate, or sal-ammoniac, or ammonium nitrate, dissolved in

* Fused bodies, when cooled down to or below their fusing point, frequently remain liquid, more especially when not in contact with solid bodies. Thus, water in a mixture of oil of almonds and chloroform, of specific gravity equal to its own, remains liquid to —10°; in a similar manner fused sulphur or phosphorus, floating in a solution of zinc chloride of appropriate concentration, retains the liquid condition at temperatures 40° below its fusing point. Liquid bodies, thus cooled below their fusing point, frequently solidify when touched with a solid substance, invariably when brought in contact with a fragment of the same body in the solid state. A body thus retained in the liquid state below its ordinary solidifying point, is said to be in a state of *surfusion* or *superfusion*.

water, occasions a very notable depression of temperature; in every case, in short, in which solution is unaccompanied by energetic chemical action, cold is produced.

No relation can be traced between the actual melting point of a substance and its latent heat when in the fused state.

2. *Vaporization and Condensation.*

A law of exactly the same kind as that above described affects universally the gaseous condition; change of state from solid or liquid to gas is accompanied by absorption of sensible heat, and the reverse by its disengagement. The latent heat of steam and other vapors may be ascertained by a mode of investigation similar to that employed in the case of water.

When water at 0° C. is mixed with an equal weight of water at 100° C., the whole is found to have the mean of the two temperatures, or 50°; on the other hand, 1 part by weight of *steam* at 100° C., when condensed in cold water, is found to be capable of raising 5.4 parts of the latter from the freezing to the boiling point, or through a range of 100° C. Now 100 × 5.4 = 540°; that is to say, steam at 100° C., in becoming water at 100° C., parts with enough heat to raise a weight of water equal to its own (if it were possible) 540° of the thermometer, or 540 times its own weight of water one degree. When water passes into steam the same quantity of sensible heat becomes latent.

The vapors of other liquids seem to have less latent heat than that of water. The following table, drawn up by Dr. Andrews of Belfast, serves well to illustrate this point. The latent heats are expressed, as in the last table, in gram-degrees:—

	Latent Heat.
Vapor of water .	535.90°
" alcohol	202.40
" ether .	90.45
" ethyl oxalate	72.72
" " acetate	92.68
" " iodide	46.87
" methyl alcohol .	263.70
" carbon bisulphide	86.67
" tin tetrachloride	30.35
" bromine .	45.66
" oil of turpentine	74.03

Boiling or Ebullition is occasioned by the formation of bubbles of vapor within the body of the evaporating liquid, which rise to the surface like bubbles of permanent gas. This occurs in different liquids at very different temperatures. Under the same circumstances, the boiling point is quite constant, and often becomes a physical character of great importance in distinguishing liquids which much resemble each other. A few cases may be cited in illustration:—

Substance.	Boiling Point.	
Aldehyde .	22° C.	(71.6° F.)
Ether .	35.6	(97)
Carbon bisulphide .	46.1	(116.4)
Alcohol .	78.4	(173.1)
Water .	100	(212)
Nitric acid (sp. gr. 1.414 at 15.5°) .	120.5	(248.9)
Oil of turpentine .	157	(314.6)
Sulphuric acid .	338	(604.4)
Mercury .	357	(674.6)

For ebullition to take place, it is necessary that the elasticity of the vapor should be able to overcome the cohesion of the liquid and the pressure upon its surface; hence the extent to which the boiling point may be modified.

Water, under the usual pressure of the atmosphere, boils at 100° C. (212° F.); in a partially exhausted receiver or on a mountain-top it boils at a much lower temperature; and in the best vacuum of an air-pump, over oil of vitriol, which absorbs the vapor, it will often enter into violent ebullition while ice is in the act of forming upon the surface.

On the other hand, water confined in a very strong metallic vessel may be restrained from boiling by the pressure of its own vapor to an almost unlimited extent; a temperature of 177° or 204° is very easily obtained, and, in fact, it is said that water may be made red-hot, and yet retain its liquidity.

The effect of diminished pressure in lowering the boiling point of a liquid may be illustrated by a very simple and beautiful experiment. A little water is made to boil for a few minutes in a flask or retort placed over a lamp, until the air has been chased out, and the steam issues freely from the neck. A tightly fitting cork is then inserted, and the lamp at the same moment withdrawn. When the boiling ceases, it may be renewed at pleasure for a considerable time by pouring cold water on the flask, so as to condense the vapor within, and occasion a partial vacuum.

The nature of the vessel, or, rather, the state of its surface, exercises an influence upon the boiling point, and this to a much greater extent than was formerly supposed. It has long been noticed that in a metallic vessel water boils, under the same circumstances of pressure, at a temperature one or two degrees below that at which ebullition takes place in glass; but by particular management a much greater difference can be observed. If two similar glass flasks be taken, the one coated in the inside with a film of shellac, and the other completely cleansed by hot sulphuric acid, water heated over a lamp in the first will boil at 99.4°, while in the second it will often rise to 105° or even higher; a momentary burst of vapor then ensues and the thermometer sinks a few degrees, after which it rises again. In this state, the introduction of a few metallic filings, or angular fragments of any kind, occasions a lively disengagement of vapor, while the temperature sinks to 100°, and there remains stationary.

When out of contact with solid bodies, liquids not only solidify with reluctance, but also assume the gaseous condition with greater difficulty. Drops of water or of aqueous saline solutions floating on the contact-surface of two liquids, of which one is heavier and the other lighter, may be heated from 10 to 20 degrees above the ordinary boiling point; explosive ebullition, however, is instantaneously induced by contact with a solid substance.

A cubic inch of water in becoming steam under the ordinary pressure of the atmosphere expands into 1696 cubic inches, or nearly a cubic foot.

Steam, *not in contact with water*, is affected by heat in the same manner as the permanent gases; its rate of expansion and increase of elastic force are practically the same. When water is present, the rise of temperature increases the quantity and density of the steam, and hence the elastic force increases in a far more rapid proportion.

This elastic force of steam in contact with water, at different temperatures, has been very carefully determined by Arago and Dulong, and lately by Magnus and Regnault. The force is expressed in atmospheres; the absolute pressure upon any given surface can be easily calculated, allowing 14.6 lb. per square inch to each atmosphere. The experiments were carried to twenty-five atmospheres—at which point the difficulties and danger became so great as to put a stop to the inquiry; the rest of the table is the result of calculations founded on the data so obtained.

6 *

Pressure of Steam in Atmospheres.	Corresponding Temperatures.		Pressure of Steam in Atmospheres.	Corresponding Temperatures.	
1	100° C.	212° F.	13	194° C.	381.2° F.
1.5	112	233.6	14	197	386.6
2	122	251.6	15	200.5	392.9
2.5	129	264.2	16	203	397.4
3	135	275	17	207	404.6
3.5	140.5	284	18	209	408.2
4	145.5	293.9	19	212	413.6
4.5	149	300.2	20	214	417.2
5	153	307.4	21	217	422.6
5.5	157	314.6	22	219	426.2
6	160	320	23	222	431.6
6.5	163	325.4	24	224	435.2
7	167	332.4	25	226	438.8
7.5	169	336.2	30	236	456.5
8	172	341.6	35	245	473
9	177	350.6	40	253	487.4
10	182	359.6	45	255	491
11	186	365.6	50	266	510.8
12	190	374			

It is very interesting to know the amount of heat requisite to convert water of any given temperature into steam of the same or another given temperature. The most exact experiments on this subject are due to Regnault. He arrived at this result, that when the unit-weight of steam at the temperature $t°$ is converted into water of the same temperature, and then cooled to $0°$, it gives out a quantity of heat T, represented by the formula:—

$$T = 606.5 + 0.305t.$$

This formula appears to hold good for temperatures both above and below the ordinary boiling point of water. The following table gives the values of T, corresponding to the respective temperatures in the first column:—

$t.°$	T.
0°	606.5°
50	621.7
100	637.0
150	652.2
200	667.5

T is called the total heat of steam, being the heat required to raise water from $0°$ to $t°$, together with that which becomes latent by the transformation of water at $t°$ into steam at $t°$. Regnault states, as a result of some very delicate experiments, that the heat necessary to raise a unit-weight of water from $0°$ to $t°$ is not exactly denoted by t; the discrepancy, however, is so small that it may be disregarded. Employing the approximate value, the latent heat of steam, L, at any temperature will be found by subtracting t from the total heat; or, according to the formula:—

$$L = 606.5 - 0.695t.$$

This equation shows us the remarkable fact that the latent heat of steam diminishes as the temperature rises. Before Regnault's experiments were made, two laws of great simplicity were generally admitted, one of which, however, contradicted the other. Watts concluded, from experiments of his own, as well as from theoretical speculations, that the total heat of steam would be the same at all temperatures. Were this true, equal weights of steam passed into cold water would always exhibit the same heating power, no matter what the temperature of the steam might be. Exactly the same *absolute* amount of heat, and consequently the same quantity of fuel, would be required to evaporate a given weight of water *in vacuo* at a temperature which the hand can bear, or under great pressure and at high temperature. Watts's law, though agreeing well with the rough practical results obtained by engineers, is only approximately true; and the same may be said of the deductions which have just been made from it. The second law in opposition to Watts's is that of Southern, stating the latent heat of steam to be the same at all temperatures. Regnault's researches have shown that neither Watts's law (T constant) nor Southern's law (L constant) is correct.

Distillation.—The process of distillation is very simple : its object is either to separate substances which rise in vapor at different temperatures, or to part a volatile liquid from a substance incapable of volatilization. The same process applied to bodies which pass directly from a solid to the gaseous

Fig. 26.

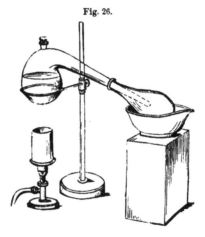

condition, and the reverse, is called *sublimation*. Every distillatory apparatus consists essentially of a boiler, in which the vapor is raised, and of a condenser, in which it returns to the liquid or solid condition. In the still employed for manufacturing purposes, the latter is usually a spiral metal tube immersed in a tub of water. The common retort and receiver constitute the simplest arrangement for distillation on the small scale; the retort is heated by a gas lamp, and the receiver is kept cool, if necessary, by a wet cloth, or it may be surrounded with ice (fig. 26).

Liebig's condenser (fig. 27) is a very valuable instrument in the laboratory; it consists of a glass tube tapering from end to end, fixed by perforated corks in the centre of a metal pipe, provided with tubes so arranged that a current of cold water may circulate through the apparatus. By putting ice into the little cistern, the water may be kept at 0°, and extremely volatile liquids condensed.

Fractional Distillation.—When a mixture of two liquids of different boiling points is distilled, the two always pass over together, the more volatile predominating in the earlier, and the less volatile in the later portions of the distillate. If the boiling points of the two liquids differ considerably, the earlier portions of the distillate will contain a large part of the more volatile liquid, but as the distillation proceeds, the boiling point of the remaining mixture gradually rises, and the less volatile passes over in larger and larger proportion; and it is only when the two liquids differ greatly in volatility, that anything approaching to complete separation can be effected by a single distillation; but on collecting separately the portions which distil between certain intervals of temperature, e. g., 5° to 10°, the first will

Fig. 27.

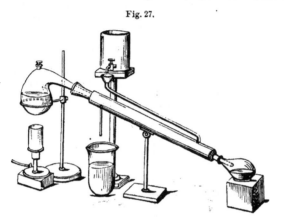

consist chiefly of the lower-boiling constituent, and the last of the less volatile, while the intermediate fractions will be very similar in composition to the original mixture, and must all be subjected to similar treatment in order to obtain more complete separation. This series of operations being generally long and tedious, various contrivances have been adopted for facilitating the separation, by causing the vapor of the less volatile constituent to condense in the upper part of the distilling apparatus, and run back again. One way of effecting this partial condensation is to surmount the flask or other distilling vessel with a vertical tube, having two or more bulbs blown on it, and a side-tube t (fig. 28) passing into the condenser; and still further separation may be effected by placing cups of platinum gauze c c in the vertical tube. The less volatile liquid then condenses in the meshes of these cups, and the mixed vapor in passing through them is washed, as it were, by this condensed liquid, a further portion of which is dissolved out of the vapor by the portion already condensed. A side-tube s may also be attached to one of the bulbs to facilitate this running back of the liquid into the flask.

Fig. 28.

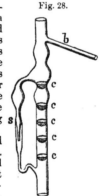

In some cases, however, the separation of the mixed liquids cannot be effected by fractional distillation. Wanklyn* has shown that when a mixture of equal parts by weight of two liquids of different boiling point is distilled, the quantity of each constituent in the distillate is proportional to the product of its vapor-density and vapor-tension at the boiling point of the fraction. Hence, if the vapor-tensions and vapor-densities of the two liquids

* Phil. Mag. [4], xlv. p. 129.

are proportional to one another, the mixture will distil unchanged. For example, a mixture of 91 parts carbon bisulphide (b. p. 46°) and 9 parts ethyl alcohol (78.4°), distils without alteration at 43° to 44°; and a mixture of equal volumes of carbon tetrachloride (b. p. 76.6°) and methyl alcohol (b. p. 65.2°) distils unaltered at 55.6° to 59°, that is to say, nearly 10° lower than the boiling point of the more volatile constituent.

Tension of Vapors.—Liquids evaporate at temperatures below their boiling points; in this case the evaporation takes place slowly from the surface. Water, or alcohol, exposed in an open vessel at the temperature of the air, gradually disappears; the more rapidly, the warmer and drier the air.

This fact was formerly explained by supposing that air and gases in general had the power of dissolving and holding in solution certain quantities of liquids, and that this power increased with the temperature; such an idea is incorrect.

If a barometer-tube be carefully filled with mercury and inverted in the usual manner, and then a few drops of water passed up the tube into the vacuum above, a very remarkable effect will be observed—the mercury will be depressed to a small extent, and this depression will increase with increase of temperature. Now, as the space above the mercury is void of air, and the weight of the few drops of water quite inadequate to account for this depression, it must of necessity be imputed to the vapor which instantaneously rises from the water into the vacuum; and that this effect is really due to the elasticity of the aqueous vapor, is easily proved by exposing the barometer to a heat of 100°, when the depression of the mercury will be complete, and it will stand at the same level within and without the tube; indicating that at that temperature the elasticity of the vapor is equal to that of the atmosphere—a fact which the phenomenon of ebullition has already shown.

Fig. 29.

By placing over the barometer a wide open tube dipping into the mercury below (fig. 29), and then filling this tube with water at different temperatures, the tension of the aqueous vapor for each degree of the thermometer may be accurately determined by its depressing effect upon the mercurial column; the same power which forces the latter *down* one inch against the pressure of the atmosphere, would of course *elevate* a column of mercury to the same height against a vacuum, and in this way the tension may be conveniently expressed. The following table was drawn up by Dalton, to whom we owe the method of investigation :—

Temperature. C.	F.	Tension in inches of mercury.	Temperature. C.	F.	Tension in inches of mercury.
0°	32°	0.200	54.4°	130°	4.34
4.4	40	0.263	60	140	5.74
10	50	0.375	65.5	150	7.42
15.5	60	0.524	71.1	160	9.46
21.1	70	0.721	76.6	170	12.13
26.6	80	1.000	82.2	180	15.15
32.2	90	1.360	87.7	190	19.00
37.7	100	1.860	93.3	200	23.64
43.3	110	2.530	100	212	30.00
48.9	120	3.330			

Another table representing the tension of the vapor of water, drawn up by Regnault, is given in the Appendix to this work.

Fig. 30.

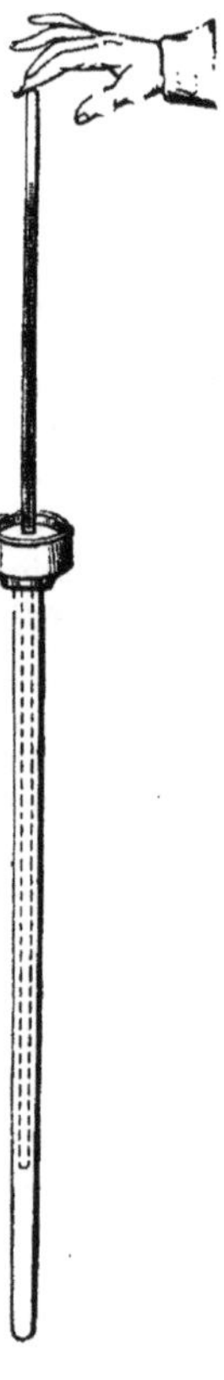

Other liquids tried in this manner are found to emit vapors of greater or less tension, for the same temperature, according to their different degrees of volatility; thus a little ether introduced into the tube depresses the mercury 10 inches or more at the ordinary temperature of the air; oil of vitriol, on the other hand, does not emit any sensible quantity of vapor until a much greater heat is applied; and that given off by mercury itself in warm summer weather, although it may be detected by very delicate means, is far too little to exercise any effect upon the barometer. In the case of water, the evaporation is quite distinct and perceptible at the lowest temperatures when frozen to solid ice in the barometer tube; snow on the ground, or on a housetop, may often be noticed to vanish, from the same cause, day by day in the depth of winter, when melting is impossible.

There exists for each vapor a state of density which it cannot pass without losing its gaseous condition, and becoming liquid; this is called the *condition of maximum density*. When a volatile liquid is introduced in sufficient quantity into a vacuum, this condition is always reached, and then evaporation ceases. Any attempt to increase the density of this vapor by compressing it into a smaller space will be attended by the liquefaction of a portion, the density of the remainder being unchanged. If a little ether be introduced into a barometer, and the latter slowly sunk into a very deep cistern of mercury (fig. 30), it will be found that the height of the column of mercury in the tube above that in the cistern remains unaltered until the upper extremity of the barometer approaches the surface of the metal in the column, and all the ether has become liquid. It will be observed also, that, as the tube sinks, the stratum of liquid ether increases in thickness, but no increase of elastic force occurs in the vapor above it, and consequently, no increase of density; for tension and density are always, under ordinary circumstances at least, directly proportionate to each other.

The maximum density of vapors is dependent upon the temperature; it increases rapidly as the temperature rises. This is well shown in the case of water. Thus, taking the spec. grav. of atmospheric air at $100° = 1000$, that of aqueous vapor in its greatest state of compression for the temperature will be as follows:—

| Temperature. | | Specific Gravity. | Weight of 100 Cubic |
C.	F.		Inches.
0°	32°	5.690	0.136 grains.
10	50	10.293	0.247 "
15.5	60	14.108	0.338 "
37.7	100	46.500	1.113
65.5	150	170.293	4.076
100	212	625.000	14.962

The last number was experimentally found by Gay-Lussac; the others are calculated from that by the aid of Dalton's table of tensions, on the assumption that steam, not in a state of saturation, that is below the point of greatest density, obeys Boyle's law (which is, however, only approximately true), and that when it is cooled it contracts like the permanent gases.

Thus, there are two distinct methods by which a vapor may be reduced to the liquid form—*pressure*, by causing increase of density until the point of maximum density for a given temperature is reached ; and *cold*, by which the point of maximum density is itself lowered. The most powerful effects are produced when both are conjoined.

For example, if 100 cubic inches of vapor of water at 37.7° C. (100° F.), in the state above described, had its temperature reduced to 10° C. (50° F.), not less than 0.89* grain of liquid water would necessarily separate, or very nearly eight-tenths of the whole.

Evaporation into a space filled with air or gas follows the same law as evaporation into a vacuum; as much vapor rises, and the condition of maximum density is assumed in the same manner, as if the space were perfectly empty ; the sole difference lies in the length of time required. When a liquid evaporates into a vacuum, the point of greatest density is attained at once, while in the other case some time elapses before this happens ; the particles of air appear to oppose a sort of mechanical resistance to the rise of the vapor. The ultimate effect is, however, precisely the same.

When to a quantity of perfectly dry gas confined in a vessel closed by mercury, a little water is added, the latter immediately begins to evaporate, and after some time as much vapor will be found to have risen from it as if no gas had been present, the quantity depending entirely on the temperature to which the whole is subjected. The tension of this vapor will add itself to that of the gas, and produce an expansion of volume, which will be indicated by an alteration of level in the mercury.

Vapor of water exists in the atmosphere at all times, and in all situations, and there plays a most important part in the economy of nature. The proportion of aqueous vapor present in the air is subject to great variation, and it often becomes important to determine its quantity. This is easily done by the aid of the foregoing principles.

Dew-point.—If the aqueous vapor be in its condition of greatest possible density for the temperature, or, as it is frequently but most incorrectly expressed, the air be saturated with vapor of water, the slightest reduction of temperature will cause the deposition of a portion in the liquid form. If, on the contrary, as is almost always in reality the case, the vapor of water be *below* its state of maximum density, that is, in an expanded condition, it is clear that a considerable fall of temperature may occur before liquefaction commences. The degree at which this takes place is called the dew-point, and it is determined with great facility by a very simple method. A little cup of thin tin-plate or silver, well polished, is filled with water of the temperature of the air, and a delicate thermometer is inserted. The water is then cooled by dropping in fragments of ice, or dissolving in it powdered sal-ammoniac, until moisture begins to make its appearance on the outside, dimming the bright metallic surface. The temperature of the dew-point is then read off upon the thermometer, and compared with that of the air.

Suppose, by way of example, that the latter were 21.1° C. (70° F.), and the dew-point 10° C. (50° F.), the elasticity of the watery vapor present would correspond to a maximum density proper to 10° C. (50° F.), and would support a column of mercury 0.375 inch high. If the barometer on the spot stood at 30 inches, therefore, 29.625 inches would be supported by the pressure of the dry air, and the remaining 0.375 inch by the vapor. Now, a cubic foot of such a mixture must be looked upon as made up of a cubic foot of dry air, and a cubic foot of watery vapor, occupying the same

---

* 100 cub. inch. aqueous vapor at 100° F., weighing 1.113 grain, would at 50° F. become reduced to 91.07 cub. inch., weighing 0.225 grain.

space, and having tensions indicated by the numbers just mentioned. A cubic foot, or 1728 cubic inches, of vapor at 70° F. would become reduced by contraction, according to the usual law, to 1662.8 cubic inches at 50° F.; this vapor would be at its maximum density, having the specific gravity pointed out in the table: hence, 1662 8 cubic inches would weigh 4.11 grains. The weight of the aqueous vapor contained in a cubic foot of air will thus be ascertained. In this country the difference between the temperature of the air and the dew-point seldom reaches 16.6° C. (30° F.); but in the Deccan, with a temperature of 32.2° C. (90° F.), the dew-point sinks as low as —1.7° C. (19° F.), making the degrees of dryness 33.9° C. (61° F.).*

Liquefaction and Solidification of Gases.—The perfect resemblance in every respect which vapors bear to permanent gases, led, very naturally, to the idea that the latter might, by the application of suitable means, be made to assume the liquid state, and this surmise has been verified. The first experiments in this direction were made by Faraday, who succeeded in liquefying eight gases previously regarded as permanent. The subjoined table shows the results of his first investigations, with the pressure in atmospheres, and the temperatures at which the condensation takes place.

	Atmosphere.	Temperatures. C.	F.
Sulphur dioxide . . .	2	7.2°	45°
Hydrogen sulphide . .	17	10	50
Carbon dioxide . . .	36	0	32
Chlorine	4	15.5	60
Nitrogen monoxide . .	50	7.2	45
Cyanogen	3.6	7.2	45
Ammonia	6.5	10	50
Hydrochloric acid . .	40	10	50

The method of proceeding was very simple: the materials were sealed up in a strong narrow tube, together with a little pressure-gauge, consist-ing of a slender tube closed at one

Fig. 31.

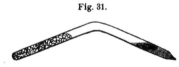

end, and having within it, near the open extremity, a globule of mer-cury. The gas being disengaged by heat, accumulated in the tube, and by its own pressure brought about condensation. The force required for this purpose was judged of by the diminution of volume of the air in the gauge.

By the use of narrow green glass tubes of great strength, powerful con-densing syringes, and an extremely low temperature, produced by means to be presently described, olefiant gas, hydriodic and hydrobromic acids, phosphoretted hydrogen, and the gaseous fluorides of silicon and boron were successfully liquefied. Oxygen, hydrogen, nitrogen, nitrogen dioxide, carbon monoxide, and marsh gas, refused to liquefy even at —110° C. (—166° F.) while subjected to pressures varying from 27 to 58 atmospheres.

Sir Isambard Brunel, and more recently M. Thilorier, of Paris, suc-ceeded in obtaining liquid carbon dioxide, (commonly called carbonic acid) in great abundance. Thilorier's apparatus (fig. 32) consists of a pair of extremely strong metallic vessels, one of which is destined to serve the purpose of a retort, and the other that of a receiver. They are made either of thick cast-iron or gun-metal, or, still better, of the best and heaviest boiler-plate, and are furnished with stop-cocks of a peculiar kind, the workmanship of which must be excellent. The generating vessel or

* D a n i e l l, Introduction to Chemical Philosophy, p. 154.

retort has a pair of trunnions upon which it swings in an iron frame. The joints are secured by collars of lead, and every precaution is taken to prevent leakage under the enormous pressure the vessel has to bear. The receiver resembles the retort in every respect; it has a similar stop-cock, and is connected with the retort by a strong copper tube and a pair of union screw-joints; a tube passes from the stop-cock downwards, and terminates near the bottom of the vessel.

Fig. 32.

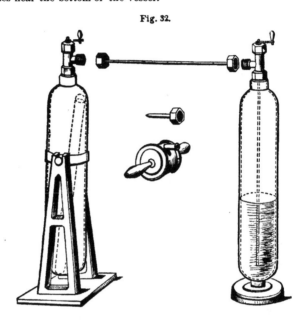

The operation is thus conducted: $2\frac{3}{4}$ lb. of acid sodium carbonate, and $6\frac{1}{4}$ lb. of water at 38° C. (100.4 F.), are introduced into the generator; oil of vitriol to the amount of $1\frac{1}{2}$ lb. is poured into a copper cylindrical vessel, which is lowered down into the mixture, and set upright; the stop-cock is then screwed into its place, and forced home by a spanner and mallet. The machine is next tilted up on its trunnions, that the acid may run out of the cylinder and mix with the other contents of the generator; and this mixture is favored by swinging the whole backwards and forwards for a few minutes, after which it may be suffered to remain a little time at rest.

The receiver, surrounded with ice, is next connected with the generator, and both cocks are opened; the liquefied carbon dioxide distils over into the colder vessel, and there again in part condenses. The cocks are now closed, the vessel is disconnected, the cock of the generator opened to allow the contained gas to escape; and, lastly, when the issue of gas *has quite ceased*, the stopcock itself is unscrewed, and the sodium sulphate turned out. This operation must be repeated five or six times before any considerable quantity of liquefied carbon dioxide will have accumulated in the receiver. When the receiver thus charged has its stopcock opened, a stream of the liquid is forcibly driven up the tube by the elasticity of the gas contained in the upper part of the vessel.

The experimenter incurs great personal danger in using this apparatus,

4

unless the utmost care be taken in its management. A dreadful accident occurred in Paris by the bursting of one of the iron vessels.

Liquid carbon dioxide is also very frequently prepared by means of an apparatus constructed by Natterer, of Vienna, which enables the experimentalist to work with less risk. The gas, disengaged by means of sulphuric acid from acid potassium carbonate, is pumped by means of a force-pump into a wrought-iron vessel, exactly as the air is pumped into the receiver of an air-gun. When a certain pressure has been reached, the gas is liquefied, and if the pressure be continued considerable quantities of the liquid carbon dioxide may be thus obtained. By this apparatus, nitrous oxide has been condensed to a liquid without the use of frigorific mixtures.

There still, however, remained six gases, viz., hydrogen, oxygen, nitrogen, nitrogen dioxide, carbon monoxide, and methane or marsh-gas, which down to the year 1877 had resisted all attempts to reduce them to the solid or liquid state; but towards the end of that year they were all liquefied by Cailletet* and Pictet,† who employed methods exactly similar in principle to those used by Faraday, consisting in the application of cold and pressure, the gases being condensed in thick-walled tubes of fine bore by means of forcing-pumps capable of producing a pressure of 300 to 1000 atmospheres, and at the same time subjected to the intense cold produced by a mixture of liquid sulphur dioxide and solid carbon dioxide.

Complete Vaporization of Liquids under Great Pressures.—When the temperature of a liquid is raised sufficiently high, vaporization occurs under the highest pressure to which the substance can be subjected. Alcohol, ether, or rock oil, inclosed in a tube of strong glass or iron, is completely converted into vapor, only when the space not occupied by the liquid is somewhat greater than the volume of the liquid itself. With rock oil the empty space may be somewhat smaller than with alcohol, and with ether still less. Alcohol when thus heated acquires increased mobility, expands to twice its original volume, and is then suddenly converted into vapor. This change takes place at 207° C. (404.6° F.), when the alcohol occupies just half the volume of the tube ; if the tube is more than half filled with alcohol, it bursts when heated. A glass tube one-third filled with water becomes opaque when heated, and bursts after a few seconds. If this chemical action of the water on the glass be diminished by the addition of a little carbonate of soda, the transparency of the glass will be much less impaired ; and if the space occupied by the water be $\frac{1}{4}$ of the whole tube, the liquid will be converted into vapor at about the temperature of melting zinc. These observations were made by Cagniard de Latour in 1822.

In like manner Dr. Andrews has observed that, when liquid carbon dioxide is gradually heated in a sealed tube to 31°, the surface of demarcation between the liquid and gas becomes fainter, loses its curvature, and at last disappears. The space is then occupied by a homogeneous fluid, which exhibits, when the pressure is suddenly diminished, or the temperature slightly lowered, a peculiar appearance of moving or flickering striæ throughout its entire mass. At temperatures above 31° no apparent liquefaction of carbon dioxide or separation into two distinct forms of matter can be effected, even under a pressure of 300 or 400 atmospheres. Similar results are obtained with nitrous oxide.

It appears indeed that there exists for every liquid a temperature, called by Andrews the "critical point," above which no amount of pressure is sufficient to retain it in the liquid form ; it is therefore not surprising that mere pressure, however intense, should fail to liquefy many bodies which usually exist in the form of gas.

Under the enormous pressures to which gases can be thus subjected,

* Comptes rendus, lxxxv. pp. 1016, 1213, 1217, 1270.
† Ibid., lxxxv. pp. 1214, 1276; lxxxvi. pp. 37, 106.

without liquefaction, they are found to deviate greatly from the laws of Boyle and Gay-Lussac (pp. 45, 56). Andrews has recently found that carbon dioxide, at 60.7°, under a pressure of 223 atmospheres, is reduced to $\frac{1}{317}$ of its original volume, or to less than one-half the volume it should occupy if it contracted according to Boyle's law. The coefficient of expansion of the same gas by heat increases rapidly with the pressure; between 6° and 64° it is $1\frac{1}{2}$ times as great under 22 atmospheres, and more than $2\frac{1}{2}$ times as great at 40 atmospheres, as at the pressure of 1 atmosphere.*

Cold produced by Evaporation.—This effect has been already adverted to: it arises from the conversion of sensible heat into latent by the rising vapor, and may be illustrated in a variety of ways. Ether dropped on the hand produces the sensation of great cold; and water contained in a thin glass tube, surrounded by a bit of rag, is speedily frozen when the rag is kept wetted with ether.

Ice-making machines and refrigerators are constructed on this principle. Harrison's apparatus for freezing water consists of a multitubular boiler containing about 10 gallons of ether and immersed in a trough of salt water. The boiler is connected with an exhausting pump, by the working of which the ether is rapidly volatilized, thereby cooling the boiler and the salt water surrounding it to about 24° F. (— 4.45° C.). This cold water is made to flow through a channel in which are placed a number of vessels containing the water to be frozen, and when its temperature has been thus raised to about 28° F., it is pumped back again into the trough containing the boiler, and then again cooled by the evaporation of the ether. In this manner a constant supply of cold salt water is kept up. The ether which is evaporated is condensed in a worm surrounded by cold water, and returned with very little loss to the boiler.†

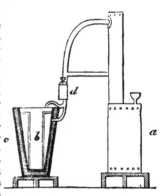

Fig. 33.

A simpler freezing apparatus is that of Carré, in which cold is produced by the rapid evaporation of liquefied ammonia gas. It consists essentially of a cylindrical boiler a, fig. 33, holding about two gallons, filled to about three-fourths of its capacity with a strong aqueous solution of ammonia, and connected by pipes with a wrought-iron annular condenser or freezer c. The boiler is first placed in a furnace, and the freezer in water cooled to 12° C. (53.6° F.). The boiler is heated to 130° C. (266° F.), whereupon ammonia gas is given off, and condenses in the freezer, together with about one-tenth of its own weight of water. This operation being completed, the boiler is removed from the fire and immersed in cold water; the freezer, wrapped in very dry flannel, is placed outside, and the vessel containing the water to be frozen is placed in the cylindrical space b. As the boiler cools, the ammonia gas with which it is filled is redissolved, and the pressure being thus diminished, the ammonia which has been liquefied in c is again volatilized, and passes over toward a, to redissolve in the water which has remained in the boiler. This rapid evaporation of the ammonia causes a great absorption of heat,

* Journal of the Chemical Society, 1876, vol. ii. p. 162.
† A figure of this apparatus is given in the Pharmaceutical Journal, vol. xvi. p. 477.

whereby the vessel c is reduced to a very low temperature, and the water contained in it is frozen. To obtain better contact between the sides of the vessel b and the freezer, alcohol is poured between them. This apparatus gives about 4 lb. of ice in an hour at the price of about a farthing a pound; but large continuously working apparatus have been constructed which produce as much as 800 lb. of ice in an hour.*

Another method of producing very low temperatures is by the rapid evaporation of methyl chloride, CH_2Cl, which is now prepared on the large scale from beetroot molasses. This compound is gaseous at ordinary temperatures and pressure, and when liquefied boils at $-23°$; and by blowing air through this liquid, or better by placing it in connection with a good air-pump, its temperature may in a very short time be reduced to $-55°$. Powerful refrigerating machines are now constructed on this principle.

Water may also be frozen by the cold resulting from its own evaporation. When a little water is put into a watch-glass, supported by a triangle of wire over a shallow glass dish of sulphuric acid placed on the plate of an air-pump, the whole covered with a low receiver, and the air withdrawn as perfectly as possible, the water is in a few minutes converted into a solid mass of ice. The absence of the air, and the rapid absorption of watery vapor by the oil of vitriol, induce such quick evaporation that the water has its temperature almost immediately reduced to the freezing point.

Fig. 34.

The same apparatus is constantly used in the laboratory for drying substances which cannot bear heating without decomposition. Frequently also the air-pump is dispensed with, and the substance to be dried is simply placed over a vessel containing strong sulphuric acid, quicklime, or some other substance which absorbs moisture very rapidly, and covered over with a bell-jar. Such an apparatus, with or without the air-pump, is called an *Exsiccator*. On the same principle a very powerful ice-making and refrigerating machine has lately been constructed, by which large quantities of ice can be rapidly and economically produced without the use of ether, or any other volatile liquid.†

Still lower temperatures are produced by the evaporation of liquefied carbon dioxide, nitrogen monoxide, and other liquefied gases. When a jet of carbon dioxide is allowed to issue into the air from a narrow aperture, so intense a degree of cold is produced by the vaporization of a part, that the remainder freezes to a solid, and falls in a shower of snow. By suffering this jet of liquid to flow into the metal box shown in fig. 32, a large quantity of the solid oxide may be obtained; it closely resembles snow in appearance, and when held in the hand occasions a painful sensation of cold, while it gradually disappears. When it is mixed with a little ether, and poured upon a mass of mercury, the latter is almost instantly frozen, and in this way pounds of the solidified metal may be obtained. The addition of the ether facilitates the contact of the carbon dioxide with the mercury.

The temperature of a mixture of solid carbon dioxide and ether in the air, measured by a spirit-thermometer, was found to be -76.7 C. ($-106°$ F.); when the same mixture was placed beneath the receiver of an air-pump, and exhaustion rapidly made, the temperature sank to 110° C. ($-166°$ F.) This was the method of obtaining extreme cold employed by Faraday in his last experiments on the liquefaction of gases, and recently extended, as

* See Richardson and Watts's Chemical Technology, part v. p. 296.

† See Chemical News, Oct. 27th, 1882, vol. xlii. p. 192.

already observed, to the liquefaction and solidification of other gases previously regarded as permanent.

Determination of the Specific Gravity of Gases and Vapors.

To determine the specific gravity of a gas, a large glass globe is filled with the gas to be examined, in a perfectly pure and dry state, having a known temperature, and an elastic force equal to that of the atmosphere at the time of the experiment. The globe so filled is weighed. It is then exhausted at the air-pump as far as possible, and again weighed. Lastly, it is filled with dry air, the temperature and pressure of which are known, and its weight once more determined. On the supposition that the temperature and elasticity are the same in both cases, the specific gravity is at once obtained by dividing the weight of the gas by that of the air.

The globe or flask must be made very thin, and fitted with a brass cap, surmounted by a small but well-fitting stop-cock. A delicate thermometer should be placed in the inside of the globe, secured to the cap. The gas must be generated at the moment, and conducted at once into the previously exhausted vessel, through a long tube filled with fragments of pumice moistened with oil of vitriol, or some other extremely hygroscopic substance, by which it is freed from all moisture. As the gas is necessarily generated under some pressure, the elasticity of that contained in the filled globe will slightly exceed the pressure of the atmosphere; and this is an advantage, since, by opening the stop-cock for a single instant, when the globe has attained an equilibrium of temperature, the tension becomes exactly that of the air, so that all barometrical correction is avoided, unless the pressure of the atmosphere should vary sensibly during the time occupied by the experiment. It is hardly necessary to observe that the greatest care must also be taken to purify and dry the air used as the standard of comparison, and to bring both gas and air as nearly as possible to the same temperature, so as to obviate the necessity of a correction, or at least to reduce almost to nothing the errors involved by such a process.

VAPORS.—1. *Dumas' Method.* This method consists in determining the weight of a given volume of the vapor at a known pressure and temperature. A glass globe about three inches in diameter, having its neck softened and drawn out in the blowpipe flame, is accurately weighed. About 6 or 7 grams of the volatile liquid are then introduced, by gently warming the globe and dipping the point into the liquid, which is then forced upward by the pressure of the air as the vessel cools. The globe is heated in a bath of water, oil, or melted paraffin to a temperature from 30° to 50° above the boiling point of the liquid, in order to bring the vapor as nearly as possible into the state in which it obeys the laws of gaseous expansion and contraction by alteration of pressure and temperature. The liquid is then rapidly converted into vapor, which escapes by the narrow orifice, chasing before it the air of the globe. When the issue of vapor has wholly ceased, and the temperature of the bath appears nearly uniform, the open extremity of the point is sealed by a small blowpipe flame. The globe is removed from the bath, suffered to cool, cleansed if necessary, and weighed, after which the neck is broken off beneath the surface of water which has been boiled and cooled out of contact of air, or (better) under mercury. The liquid enters the globe, and, if the expulsion of the air by

Fig. 35.

7 *

the vapor has been complete, fills it; if otherwise, an air-bubble is left, whose volume can be easily ascertained by pouring the liquid from the globe into a graduated jar, and then refilling the globe, and repeating the same observation, whereby also the capacity of the vessel is determined.

From these data the vapor-density (D) may be calculated by means of the formula:

$$D = \frac{P + Vn_t}{(V - v)\, n'_t}.$$

P = difference of weight (in grams) between the globe filled with air and when filled with vapor.

V = capacity of globe in cubic centimetres.

n_t = weight of one cubic centimetre of air at the temperature at which the globe filled with air was weighed.

n'_t = weight of one cubic centimetre of air at the temperature of sealing the globe.

The values of n_t and n'_t (in grams) for every five degrees Centigrade from 0° to 300° are given in the Appendix, Table VII.*

In very exact experiments account must be taken of the change of capacity of the glass globe by the high temperature of the bath. When this correction is neglected, the density of the vapor will come out a little too high. The error of the mercurial thermometer at high temperatures is, however, in the opposite direction.

The preceding method is applicable to the determination of the vapor-densities of all substances whose boiling points are within the range of the mercurial thermometer, that is to say, not exceeding 300° C., and therefore

Fig. 36.

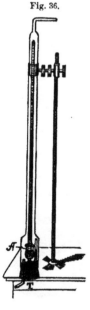

to nearly all volatile organic compounds; indeed, there are but few such compounds which can bear higher temperatures without decomposition. But for mineral substances, such as sulphur, iodine, volatile metallic chlorides, etc., it is often necessary to employ much higher temperatures; and for such cases a modification of the process has been devised by Deville and Troost. It consists in using a globe of porcelain instead of glass, heating it in the vapor of a substance whose boiling point is known and constant, and sealing the globe by the flame of the oxy-hydrogen blowpipe. The vapors employed for this purpose are those of mercury, which boils at 350° C.; of sulphur, which boils at 440°; of cadmium, boiling at 860°; and of zinc, boiling at 1040°. The use of these liquids of constant boiling point obviates the necessity of determining the temperature in each experiment, which at such degrees of heat would be very difficult.

2. *Gay-Lussac's Method.*—This method consists in ascertaining the volume occupied by a given weight of a substance when converted into vapor at a known temperature and pressure. It is most readily performed with the apparatus contrived by Dr. A. W. Hofmann (fig. 36). A graduated glass tube about a metre long and 15 to 20 mm. wide is filled with mercury and inverted in the little cup A, whereby a barometric vacuum 20 to 30 mm. high is formed at the top. The long tube is enclosed in another tube 30 to 40 mm. wide and 80 to

* The table here referred to is an abstract of one drawn up by C. Greville Williams for every degree of temperature between the above limits. (See Watts's Dictionary of Chemistry, vol. v. p. 370.)

90 mm. long, drawn out at the top to a conducting tube of moderate width, which is bent at right angles and connected with a glass or copper vessel in which water, aniline, or other liquid can be boiled. The lower part of the long tube is widened and rests upon a large cork, through which passes an escape tube T. By this arrangement, a stream of vapor of water, aniline, or other volatile liquid can be made to pass through the space between the two tubes, so as to keep the upper part of the barometer-tube at the temperature required for the determination. The substance whose vapor-density is to be determined, is introduced into the barometric vacuum in small glass tubes fitted with ground stoppers, which are forced out by the tension of the vapor. The great advantage of this method is that, under the very small pressure to which the enclosed vapor is subjected—which may be reduced to 20 or even 10 millimetres of mercury—the determinations may be made at comparatively low temperatures. Thus, in the case of liquids boiling under the ordinary pressure at 120° or even 150°, the vapor-density may be accurately determined at the temperature of boiling water.

By this mode of proceeding, we ascertain the volume which a known weight W' of substance occupies at a given temperature and pressure; and dividing this by W, the weight of an equal volume V of air at the same pressure and temperature, which is given by the formula—

$$W = 0.001293 \ V \cdot \frac{1}{1 + 0.00367 \ T} \cdot \frac{P}{760},$$

0.001293 being the weight in grams of a cubic centimetre of air at 0° C. and 760 mm. pressure, and 0.00367 the coefficient of the thermal expansion of gases,[*] we obtain, for the density of the vapor, the expression—

$$D = \frac{W'}{W} = W' \frac{1 + 0.00367 \ T}{0.001293} \cdot \frac{760}{P}.$$

3. *V. and C. Meyer's Method.*[†]—This is a very simple method of determining vapor-densities, founded on the same principle as the last, and applicable to substances either of low or high boiling point. Its simplicity consists in this, that it does not require the determination either of the temperature to which the vapor is heated, or of the volume of the vapor at that temperature—both these quantities being eliminated in the expression for the vapor-density—but only of the volume of air displaced by the vapor evolved, this volume being measured at the atmospheric temperature. The apparatus (fig. 37) consists of a cylindrical glass bulb, *b*, having a capacity of about 100 c.c., to which is attached by fusion a tube 600 mm. long, terminating in a thimble-shaped enlargement, and having attached to it a side exit-tube *a*, by which the displaced air is transferred to a graduated tube standing in a small water-trough. For substances of boiling point not exceeding 310°, the bulb is heated in a glass tube, which terminates below in a bulb *c* containing various liquids, according to the volatility of the substance under experiment, viz., aniline (b. p. 181.5°), toluidine (202°), ethyl benzoate (212°), amyl benzoate (261°), diphenylamine (310°). The mode of working is as follows:—The bulb,

Fig. 37.

---

[*] The values of the expression $\dfrac{1}{1 + 0.00367 \ T}$ have been calculated by C. Greville Williams for all temperatures from 1° to 150° C. (See Table VIII. in the Appendix.)

[†] Berichte der deutschen chemischen Gesellschaft, 1877, p. 2253.

into which a little ignited asbestos has been introduced, is fixed in position in the bath, the end of the exit-tube a dipping into the water-trough below the mouth of the graduated tube, which is filled with water and inverted. A cork is now inserted into the top of the vertical tube at d, and the extremity of the exit-tube a is watched, to see that the temperature in the bulb is uniform. The cork is now taken out, the small tube containing the substance is dropped in, and the cork is quickly replaced. The first few bubbles of air are suffered to escape, but immediately afterwards the inverted tube is placed quickly over the point of the delivery-tube a. The substance is soon volatilized and displaces air, which issues in a rapid stream into the graduated tube; and as soon as air ceases to come over, the cork is removed, and the air collected is cooled and measured.

The density of the vapor is calculated by the formula

$$\frac{S \cdot 760(1 + 0.00365t)}{(B - w)V \cdot 0.001293} \quad \text{or} \quad \frac{S(1 + 0.00365t) \cdot 587780}{(B - w) \, V},$$

in which S denotes the weight of substance, t the temperature of the water, B the barometric pressure reduced to $0°$, w the tension of aqueous vapor, and V the volume of air displaced.

For temperatures above $310°$, a bath of molten lead is employed; and for determining the vapor-densities of inorganic compounds which volatilize only at a red heat or at still higher temperatures, the glass vessel is replaced by one of porcelain or platinum heated in a gas-furnace.

For other recently proposed methods of determining vapor-densities, see Roscoe and Schorlemmer's Treatise on Chemistry, vol. iii. pp. 94–100; and Watts's Dictionary of Chemistry, vol. viii. p. 2096.

Sources of Heat.

The first and greatest source of heat, compared with which all others are totally insignificant, is the sun. The luminous rays are accompanied by heat-rays, which, striking against the surface of the earth, raise its temperature; this heat is communicated to the air by convection, as already described, air and gases in general not being sensibly heated by the passage of the rays.

A second source of heat is supposed to exist in the interior of the earth. It has been observed that, in sinking mine-shafts, boring for water, etc., the temperature rises, in descending, at the rate, it is said, of about $\frac{5}{9}°$ C. ($1°$ F.) for every 45 feet, or $65°$ C. ($117°$ F.) per mile. On the supposition that the rise continues at the same rate, the earth, at the depth of less than two miles, would have the temperature of boiling water; at nine miles, it would be red-hot; and at thirty or forty miles depth, all known substances would be in a state of fusion.[*]

According to this idea, the earth must be looked upon as an intensely heated fluid spheroid, covered with a crust of solid badly conducting matter, cooled by radiation into space, and bearing somewhat the same proportion in thickness to the ignited liquid within, that the shell of an egg bears to its fluid contents. Without venturing to offer any opinion on this theory, it may be sufficient to observe that it is not positively at variance with any known fact; that the figure of the earth is really such as would be assumed by a fluid mass; and, lastly, that it offers the best explanation we have of the phenomena of hot springs and volcanic eruptions, and agrees with the chemical nature of their products.

[*] The Artesian well at Grenelle, near Paris, has a depth of 1774.5 English feet; it is bored through the chalk basin to the sand beneath. The temperature of the water, which is exceedingly abundant, is $82°$ F.; the mean temperature of Paris is $51°$ F.; the difference is $31°$ F., which gives a rate of about $1°$ for 58 feet.

Among the other sources of heat are chemical combination and mechanical work.

The disengagement of heat in the act of combination is a phenomenon of the utmost generality. The quantity of heat given out in each particular case is fixed and definite; its intensity is dependent upon the time over which the action is extended. Many admirable researches on this subject have been published; but their results will be more advantageously considered at a later part of this work, in connection with the laws of chemical combination.

Heat produced by Mechanical Work.—Heat and motion are convertible one into the other. The powerful mechanical effects produced by the elasticity of the vapor evolved from heated liquids afford abundant illustration of the conversion of heat into motion; and the production of heat by friction, by the hammering of metals, and in the condensation of gases (p. 59), shows with equal clearness that motion may be converted into heat.

In some cases the rise of temperature thus produced appears to be due to a diminution of heat-capacity in the body operated upon, as in the case of a compressed gas just alluded to. Malleable metals, also, as iron and copper, which become heated by hammering or powerful pressure, are found thereby to have their density sensibly increased, and their capacity for heat diminished. A soft iron nail may be made red hot by a few dexterous blows on an anvil; but the experiment cannot be repeated until the metal has been *annealed*, and in that manner restored to its former physical state.

But the amount of heat which can be developed by mechanical force is, in most cases, out of all proportion to what can be accounted for in this way. Sir H. Davy melted two pieces of ice by rubbing them together in a vacuum at the temperature of 0°; and Count Rumford found that the heat developed by the boring of a brass cannon was sufficient to bring to the boiling point two and a half gallons of water, while the dust or shavings of metal cut by the borer weighed only a few ounces. In these and all similar cases the heat appears as a direct result of the force expended; the motion is converted into heat.

The connection between heat and mechanical force appears still more intimate when it is shown that they are related by an exact numerical law, a given quantity of the one being always convertible into a definite amount of the other. The first approximate determination of this most important numerical relation was made by Count Rumford in the manner just alluded to. A brass cylinder inclosed in a box containing a known weight of water at 60° F. was bored by a steel borer made to revolve by horse power, and the time was noted which elapsed before the water was raised to the boiling point by the heat resulting from the friction. In this manner it was found that the heat required to raise the temperature of a pound of water by 1° F. is equivalent to 1034 times the force expended in raising a pound weight one foot high, or to 1034 "foot-pounds," as it is technically expressed. This estimate is now known to be too high, no account having been taken of the heat communicated to the containing vessel, or of that which was lost by dispersion during the experiment.

For the most exact determinations of the mechanical equivalent of heat we are indebted to the careful and elaborate researches of Dr. J. P. Joule. From experiments made in the years 1840–43 on the relations between the heat and mechanical power generated by the electric current, Dr. Joule was led to conclude that the heat required to raise the temperature of a pound of water 1° F. is equivalent to 838 foot-pounds. This he afterwards reduced to 772; and a nearly equal result was afterwards obtained by

4 *

experiments on the condensation and rarefaction of gases; but this estimate has since been found to be likewise too great.

The most trustworthy results are obtained by measuring the quantity
of heat generated by the friction between solids and liquids. It was for a
long time believed that no heat was evolved by the friction of liquids and
gases; but in 1842 Meyer showed that the temperature of water may be
raised 22° or 23° F. by agitating it. The warmth of the sea after a few
days of stormy weather is also probably an effect of fluid friction.

The apparatus employed by Dr. Joule for the determination of this
important constant, by means of the friction of water, consisted of a brass

Fig. 38.   Fig. 39.

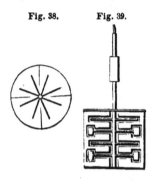

paddle-wheel furnished with eight sets of
revolving vanes, working between four sets
of stationary vanes. This revolving apparatus, of which fig. 38 shows a horizontal,
and fig. 39 a vertical section, was firmly
fitted into a copper vessel (see fig. 40) containing water, in the lid of which were two
necks, one for the axis to revolve in without touching, the other for the insertion of
a thermometer. A similar apparatus, but
made of iron, and of smaller size, having
six rotatory and eight sets of stationary
vanes, was used for the experiments on the
friction of mercury. The apparatus for the
friction of cast-iron consisted of a vertical
axis carrying a bevelled cast-iron wheel,
against which a bevelled wheel was pressed
by a lever. The wheels were inclosed in a cast-iron vessel filled with
mercury, the axis passing through the lid. In each apparatus motion
was given to the axis by the descent of leaden weights W (fig. 40) suspended by strings from the axis of two wooden pulleys, one of which is
shown at p, their axes being supported on friction wheels $d\,d$, and the

Fig. 40.

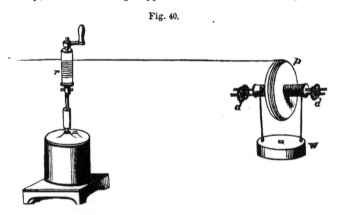

pulleys being connected by fine twine with a wooden roller r, which, by
means of a pin, could be easily attached to or removed from the friction
apparatus.

The mode of experimenting was as follows :—The temperature of the fric-
tional apparatus having been ascertained, and the weights wound up, the

roller was fixed to the axis, and the precise height of the weights ascertained; the roller was then set at liberty, and allowed to revolve till the weights touched the floor. The roller was then detached, the weights wound up again, and the friction renewed. This having been repeated twenty times, the experiment was concluded with another observation of the temperature of the apparatus. The mean temperature of the apartment was ascertained by observations made at the beginning, middle, and end of each experiment. Corrections were made for the effects of radiation and conduction; and, in the experiments with water, for the quantities of heat absorbed by the copper vessel and the paddle-wheel. In the experiments with mercury and cast-iron, the heat-capacity of the entire apparatus was ascertained by observing the heating effect which it produced on a known quantity of water in which it was immersed. In all the experiments, corrections were also made for the velocity with which the weights came to the ground, and for the friction and rigidity of the strings. The thermometers used were capable of indicating a variation of temperature as small as $\frac{1}{200}$ of a degree Fahrenheit.

The following table contains a summary of the results obtained by this method. The second column gives the results as they were obtained in air; in the third column the same results corrected for a vacuum :—

Material employed.	Equivalent in air.	Equivalent in vacuum.	Mean.
Water	773.640	772.692	772.692
Mercury . . .	{ 773.762 { 776.303	772.814 } 775.352 }	774.083
Cast-iron . . .	{ 776.997 { 774.880	776.045 } 774.930 }	774.987

In the experiments with cast-iron, the friction of the wheels produced a considerable vibration of the framework of the apparatus, and a loud sound; it was therefore necessary to make allowance for the quantity of force expended in producing these effects. The number 772.692, obtained by the friction of water, is regarded as the most trustworthy; but even this may be a little too high; because even in the friction of fluids it is impossible entirely to avoid vibration and sound. The conclusions deduced from these experiments are—

1. *That the quantity of heat produced by the friction of bodies, whether solid or liquid, is always proportional to the force expended.*

2. *That the quantity of heat capable of increasing the temperature of 1 lb. of water (weighed in vacuo, and between 55° and 60°) by 1° F., requires for its evolution the expenditure of a mechanical force represented by the fall of 772 lbs. through the space of 1 foot.*

Or, the heat capable of increasing the temperature of 1 gram of water by 1° C., is equivalent to a force represented by the fall of 423.65 grams through the space of 1 metre. This is consequently the effect of "a unit of heat."

Experiments made by other philosophers on the work done by a steam-engine, on the heat evolved by an electro-magnetic engine at rest and in motion, and on the heat evolved in the circuit of a voltaic battery and in a metallic wire through which an electric current is passing, have given values for the mechanical equivalent of heat very nearly equal to the above.

Dynamical Theory of Heat.

For a very long time two rival theories have been held regarding the nature of heat: on the one hand, heat has been viewed as having a material existence, though differing from ordinary matter in being without weight, and in other respects; on the other hand, it has been regarded as a state or condition of ordinary matter, and generally as a condition of motion. From the latter part of the last century, until the modern researches upon the mechanical equivalent, the former view had by far the greater number of adherents. Its popularity may be chiefly traced to the teaching of Black and Lavoisier. By the former of these philosophers, the various capacities for heat, or specific heats, of different bodies, seem to have been regarded as analogous to the various proportions of the same acid required to neutralize equal quantities of different bases, while the solid, liquid, and gaseous states were explained by Black as representing so many distinct proportions in which heat was capable of combining with ordinary matter. Very similar views were advocated by Lavoisier; he regarded all gases as compounds of a base characteristic of each, with *caloric*, and supposed that when, as the result of chemical action, they assumed the liquid or solid state, this caloric was set free, and appeared as sensible heat.

Heat was compared by these philosophers to a material substance, in order to explain its then known quantitative relations; and from this point of view the conception introduced by them had the great advantage of being more easily grasped than any which the advocates of the immaterial nature of heat had to offer in its place. It was much easier to conceive of definite quantities of an exceedingly subtile substance or fluid, than of definite quantities of motion, which was itself undefined as to its nature. It was a direct consequence of the material view, that heat should be considered as indestructible and as incapable of being produced, and therefore that the total quantity of heat in the universe should be regarded as at all times the same.

But, on the other hand, this hypothesis did not afford a satisfactory explanation of the production of heat by mechanical means. Here it was not easy to deny the actual generation of heat, or to explain the effects as depending merely on its altered distribution. Nevertheless, the evolution of heat by friction and percussion was generally considered, by the advocates of the material view, as in some way resulting from a diminution in the capacities for heat of the bodies operated upon; and this explanation derived considerable support from the remark, made by Black, that a piece of soft iron, which has been once made red hot by hammering (see p. 81), cannot be so heated a second time until it has been heated to redness in a fire and allowed to cool. In this case, certainly, it seemed as though the hammering forced out heat from the mass of iron, like water from a sponge, and that a fresh supply was taken up when the iron was put in the fire. This explanation, however, did not satisfy Rumford, who, in the investigation described above, made direct experiments upon the specific heat of the chips of metal detached by the friction, and found it to be identical with that of brass under ordinary circumstances. Still more decisive proof that the heat generated by friction cannot be ascribed to a diminution of specific heat in the substances operated on was afforded by Davy's experiment on the liquefaction of ice by friction; for in this case the ice was converted into a liquid having twice the specific heat of the ice itself. Hence Davy[*] drew the conclusion that "the immediate cause of the phenomena of heat is motion, and the laws of its communication are precisely the same as the laws of the communication of motion."

The mechanical, or dynamical theory, which regarded heat as consisting in a state of molecular motion, cannot, however, be said to have been definitely established, until it also was made quantitative,—until it was shown that exact numerical laws regulate the production of heat by work or of work by heat, equally with its production during solidification and disappearance during fusion.

[*] Elements of Chemical Philosophy, 1812, pp. 94, 95.

To illustrate the general nature of the dynamical theory of heat, we give an outline of the view of the constitution of gases, first put forward, in its present form, by Joule;[*] and subsequently developed by Krönig,[†] and Clausius,[‡] and of the explanation of the relation existing between solids, liquids, and gases, which has been deduced from it by the last-named philosopher.

First, then, it is assumed that the particles of all bodies are in constant motion, and that this motion constitutes heat, the kind and quantity of motion varying according to the state of the body, whether solid, liquid, or gaseous.

In gases, the molecules—each molecule being an aggregate of atoms—are supposed to be constantly moving forward in straight lines, and with a constant velocity, till they impinge against each other, or against an impenetrable wall. This constant impact of the molecules produces the expansive tendency or elasticity which is the peculiar characteristic of the gaseous state. The rectilinear movement is not, however, the only one with which the particles are affected. For the impact of two molecules, unless it takes place exactly in the line joining their centres of gravity, must give rise to a rotatory motion; and, moreover, the ultimate atoms of which the molecules are composed may be supposed to vibrate within certain limits, being, in fact, thrown into vibration by the impact of the molecules. This vibratory motion is called by Clausius, *the motion of the constituent atoms.* The total quantity of heat in the gas is made up of the progressive motion of the molecules, together with the vibratory and other motions of the constituent atoms; but the progressive motion alone, which is the cause of the expansive tendency, determines the *temperature.* Now, the outward pressure exerted by the gas against the containing envelope arises, according to the hypothesis under consideration, from the impact of a great number of gaseous molecules against the sides of the vessel. But at any given temperature, that is, with any given velocity, the number of such impacts taking place in a given time must vary inversely as the volume of the given quantity of gas: hence *the pressure varies inversely as the volume, or directly as the density,* which is Boyle's law.

When the volume of the gas is constant, the pressure resulting from the impact of the molecules is proportional to the sum of the masses of all the molecules multiplied into the squares of their velocities; in other words, to the so-called *vis viva* or *working force* of the progressive motion. If, for example, the velocity be doubled, each molecule will strike the sides of the vessel with a twofold force, and its number of impacts in a given time will also be doubled: hence the total pressure will be quadrupled.

Now, we know that when a given quantity of any perfect gas is maintained at a constant volume, it tends to expand by $\frac{1}{273}$ of its bulk at zero for each degree centigrade. Hence the pressure or elastic force increases proportionally to the temperature reckoned from —273° C.; that is to say, to the absolute temperature. Consequently, *the absolute temperature is proportional to the working force of the progressive motion.*

Moreover, as the motions of the constituent particles of a gas depend on the manner in which its atoms are united, it follows that in any given gas the different motions must be to one another in a constant ratio; and, therefore, the *vis vivo* or *working force* of the progressive motion must be an aliquot part of the entire working force of the gas; hence also the absolute temperature is proportional to the total working force arising from all the motions of the particles of the gas.

From this it follows that the quantity of heat which must be added to a gas of constant volume in order to raise its temperature by a given amount, is constant and independent of the temperature. In other words, the

* Ann. Ch. Phys. [3] 1. 381.   † Pogg. Ann. xcix. 315.   ‡ Ibid. 353.

specific heat of a gas referred to a given volume is constant, a result which agrees with the experiments of Regnault, mentioned at p. 59. This result may be otherwise expressed, as follows : *The total* or working *force of the gas is to the working force of the progressive motion of the molecules, which is the measure of the temperature, in a constant ratio.* This ratio is different for different gases, and is greater as the gas is more complex in its constitution ; in other words, as its molecules are made up of a greater number of atoms. The specific heat referred to a constant pressure is known to differ from the true specific heat only by a constant quantity.

The relations just considered between the pressure, volume, and temperature of gases, presuppose, however, certain conditions of molecular constitution, which are, perhaps, never rigidly fulfilled : and, accordingly, the experiments of Magnus and Regnault show (pp. 56–61) that gases do exhibit slight deviations from Gay-Lussac and Boyle's laws. What the conditions are which strict adherence to these laws would require, will be better understood by considering the differences of molecular constitution which must exist in the solid, liquid, and gaseous states.

A movement of molecules must be supposed to exist in all three states. In the *solid state*, the motion is such that the molecules oscillate about certain positions of equilibrium, which they do not quit, unless they are acted upon by external forces. This vibratory motion may, however, be of a very complicated character. The constituent atoms of a molecule may vibrate separately, the entire molecules may also vibrate as such about their centres of gravity, and the vibrations may be either rectilinear or rotatory. Moreover, when extraneous forces act upon the body as in shocks, the molecules may permanently alter their relative positions.

In the *liquid state*, the molecules have no determinate positions of equilibrium. They may rotate completely about their centres of gravity, and may also move forward into other positions. But the repulsive action arising from the motion is not strong enough to overcome the mutual attraction of the molecules, and separate them completely from each other. A molecule is not permanently associated with its neighbors, as in the solid state; it does not leave them spontaneously, but only under the influence of forces exerted upon it by other molecules, with which it then comes into the same relation as with the former. There exists, therefore, in the liquid state, a vibratory, rotatory, and progressive movement of the molecules, but so regulated that they are not thereby forced asunder, but remain within a certain volume without exerting any outward pressure.

In the *gaseous* state, on the other hand, the molecules are removed quite beyond the sphere of their mutual attractions, and travel onward in straight lines according to the ordinary laws of motion. When two such molecules meet, they fly apart from each other, for the most part with a velocity equal to that with which they came together. The perfection of the gaseous state, however, implies : 1. That the space actually occupied by the molecules of the gas be infinitely small in comparison with the entire volume of the gas ; 2. That the time occupied in the impact of a molecule, either against another molecule or against the sides of the vessel, be infinitely small in comparison with the interval between any two impacts ; 3. That the influence of the molecular forces be infinitely small. When these conditions are not completely fulfilled, the gas partakes more or less of the nature of a liquid, and exhibits certain deviations from Gay-Lussac and Boyle's laws. Such is, indeed, the case with all known gases: to a very slight extent with those which have not yet been reduced to the liquid state; but to a greater extent with vapors and condensable gases, especially near the points of condensation.

Let us now return to the consideration of the liquid state. It has been said that the molecule of a liquid, when it leaves those with which it is associated, ultimately takes up a similar position in regard to other molecules. This, however, does not preclude the existence of considerable irregularities in the actual movements. Now, at the surface of the liquid, it may happen that a particle, by a peculiar combination of the rectilinear, rotatory, and vibratory movements, may be projected from the neighboring molecules with such force as to throw it completely out of their sphere of action, before its projectile velocity can be annihilated by the attractive force which they exert upon it. The molecule will then be driven forward into the space above the liquid, as if it belonged to a gas, and that space, if originally empty, will, in consequence of the action just described, become more and more filled with these projected molecules, which will comport themselves within it exactly like a gas, impinging and exerting pressure upon the sides of the envelope. One of these sides, however, is formed by the surface of the liquid, and when a molecule impinges upon this surface, it will, in general, not be driven back, but retained by the attractive forces of the other molecules. A state of equilibrium, not static, but dynamic, will therefore be attained when the number of molecules projected in a given time into the space above is equal to the number which in the same time impinge upon and are retained by the surface of the liquid. This is the process of vaporization. The density of the vapor required to insure the compensation just mentioned, depends upon the rate at which the particles are projected from the surface of the liquid, and this again upon the rapidity of their movement within the liquid, that is to say, upon the temperature. It is clear, therefore, that the density of a saturated vapor must increase with the temperature.

If the space above the liquid is previously filled with a gas, the molecules of this gas will impinge upon the surface of the liquid, and thereby exert pressure upon it; but as these gas-molecules occupy but an extremely small proportion of the space above the liquid, the particles of the liquid will be projected into that space almost as if it were empty. In the middle of the liquid, however, the external pressure of the gas acts in a different manner. There, also, it may happen that the molecules may be separated with such force as to produce a small vacuum in the midst of the liquid. But this space is surrounded on all sides by masses which afford no passage to the disturbed molecules; and in order that they may increase to a permanent vapor-bubble, the number of molecules projected from the inner surface of the vessel must be such as to produce a pressure outwards equal to the external pressure tending to compress the vapor-bubble. The boiling of the liquid will, therefore, be higher as the external pressure is greater.

According to this view of the process of vaporization, it is possible that vapor may rise from a solid as well as from a liquid; but it by no means necessarily follows that vapor must be formed from all bodies at all temperatures. The force which holds together the molecules of a body may be too great to be overcome by any combination of molecular movements, so long as the temperature does not exceed a certain limit.

The *production and consumption of heat* which accompany changes in the state of aggregation, or of the volume of bodies, are easily explained, according to the preceding principles, by taking account of the *work* done by the acting forces. This work is partly *external* to the body, partly *internal*. To consider first the *internal* work:

When the molecules of a body change their relative positions, the change may take place either in accordance with or in opposition to the action of the molecular forces existing within the body. In the former case, the molecules, during the passage from one state to the other, have a certain velocity imparted to them, which is immediately converted into heat; in the latter case, the velocity of their movement, and consequently the temperature of the body, is diminished. In the passage from the solid to the liquid state, the molecules, although not removed from the spheres of their mutual attractions, nevertheless change their relative positions in oppositiou to the molecular forces, which forces have, therefore, to be overcome. In evaporation, a certain number of the molecules are completely separated from the remainder, which again implies the overcoming of opposing forces. In both cases, therefore, work is done, and a certain portion of the working force of the molecules, that is, of the heat of the body, is lost. But when once the perfect gaseous state is attained, the molecular forces are completely overcome, and any further expansion may take place without internal work, and, therefore, without loss of heat, provided there is no external resistance.

But in nearly all cases of change of state or volume, there is a certain amount of external resistance to be overcome, and a corresponding loss of heat. When the pressure of a gas, that is to say, the impact of its atoms, is exerted against a movable obstacle, such as a piston, the molecules lose just as much of their moving power as they have imparted to the piston, and, consequently, their velocity is diminished and the temperature lowered. On the contrary, when a gas is compressed by the motion of a piston, its molecules are driven back with greater velocity than that with which they impinged on the piston, and, consequently, the temperature uf the gas is raised.

When a liquid is converted into vapor, the molecules have to overcome the atmospheric pressure or other external resistance, and, in consequence of this, together with the internal work already spoken of, a large quantity of heat disappears, or is rendered *latent*, the quantity thus consumed being, to a considerable extent, affected by the external pressure. The liquefaction of a solid, not being attended with much increase of volume, involves but little external work ; nevertheless the atmospheric pressure does influence, to a slight amount, both the latent heat of fusion and the melting point.

LIGHT.

Two views have been entertained respecting the nature of light. New-ton imagined that luminous bodies emit, or shoot out, infinitely small par-ticles in straight lines, which, by penetrating the transparent parts of the eye and falling upon the nervous tissue, produce vision. Other philoso-phers drew a parallel between the properties of light and those of sound, and considered that, as sound is certainly the effect of undulations, or little waves, propagated through elastic bodies in all directions, so light might be nothing more than the consequence of similar undulations trans-mitted with inconceivable velocity through a highly elastic medium, of excessive tenuity, filling all space, and occupying the intervals between the particles of material substances. To this medium they gave the name of *ether*. The wave hypothesis of light is at present generally adopted. It is in harmony with all the known phenomena discovered since the time of Newton, not a few of which were first deduced from the undulatory theory, and afterwards verified by experiment. Several well-known facts are in direct opposition to the theory of emission.

A ray of light emitted from a luminous body proceeds in a straight line, and with extreme velocity. Certain astronomical observations afford the means of approximating to a knowledge of this velocity. The satellites of Jupiter revolve about the planet in the same manner as the moon about the earth, and the time of revolution of each satellite is exactly known from its periodical entry into or exit from the shadow of the planet. The time required by one is only 42 hours. Römer, the astronomer of Copen-hagen, found that this period appeared to be longer when the earth, in its passage round the sun, moved from the planet Jupiter; and, on the con-trary, he observed that the periodic time appeared to be shorter when the earth moved in the direction towards Jupiter. The difference, though very small for a single revolution of the satellite, increases, by the addi-tion of many revolutions, during the passage of the earth from its nearest to its greatest distance from Jupiter, that is, in about half a year, till it amounts to 16 minutes and 16 seconds. Römer concluded from this, that the light of the sun, reflected from the satellite, required that time to pass through a distance equal to the diameter of the orbit of the earth; and since this space is little short of 200 millions of miles, the velocity of light cannot be less than 200,000 miles in a second of time. It will be seen hereafter that this rapidity of transmission is rivalled by that of electricity. Another astronomical phenomenon, observed and correctly explained by Bradley, the aberration of the fixed stars, leads to the same result. Physi-cists have, moreover, succeeded in measuring the velocity of light for ter-restrial, and indeed comparatively small distances; the results of these experiments essentially correspond with those given by astronomical ob-servations.

REFLECTION.—When a ray of light falls upon a boundary between two media, a part of it, and, in exceptional cases, the whole, is reflected into the first medium, whilst the other part penetrates into the second medium.

The law of regular reflection is extremely simple. If a line be drawn perpendicular to the surface upon which the ray falls, and the angle con-tained between the ray and the perpendicular be measured, it will be found that the ray, after reflection, takes such a course as to make with the per-

8 *

pendicular an equal angle on the opposite side of the latter. A ray of
light, R, falling at the point P, will be reflected in the direction PR′, making

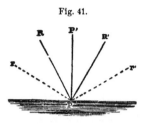

Fig. 41.

the angle R′PP′ equal to the angle RPP′;
and a ray from the point r falling upon
the same spot will be reflected to r′ in
virtue of the same law. Further, it is to
be observed that the incident and reflected rays are always contained in the
same normal plane.

The same rule holds good if the mirror
be curved, as a portion of a sphere, the
curve being considered as made up of a
multitude of little planes. Parallel rays
cease to be so when reflected from curved
surfaces, becoming divergent or converg-
ent according as the reflecting surface is convex or concave.

Bodies with rough and uneven surfaces, the smallest parts of which are
inclined toward each other without order, reflect the light diffused. The
perception of bodies depends upon the diffused reflected light.

REFRACTION.—It has just been stated that light passes in straight
lines; but this is true only so long as the medium through which it travels

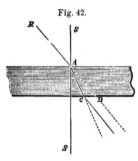

Fig. 42.

preserves the same density and the same
chemical nature: when this ceases to be the
case, the ray of light is bent from its course
into a new one, or is said to be *refracted*.

Let R (fig. 42) be a ray of light falling
upon a plate of some transparent substance
with parallel sides, such as a piece of thick
plate glass—in short, any transparent homogeneous material which is either non-crystalline, or crystallizes in the regular system—and
let A be its point of contact with the upper
surface. The ray, instead of holding a straight
course and passing into the glass in the direction A B, will be bent downward to C; and, on
leaving the glass, and issuing into the air on
the other side, it will again be bent, but in the opposite direction, so as to
make it parallel to the continuation of its former track, provided there be
one and the same medium on the upper and lower side of the plate. The
general law is thus expressed: When the ray passes from a rare to a denser
medium, it is usually refracted *toward* a line perpendicular to the surface
of the latter; and conversely, when it leaves a dense medium for a rarer
one, it is refracted *from* a line perpendicular to the surface of the denser
substance; in the former case the angle of incidence is greater than that
of refraction; in the latter, it is less. In both cases the direction of the
refracted ray is in the plane R A s, which is formed by the falling ray and
the perpendicular s A drawn from the spot where the ray is refracted; the
angle R A s = B A s′, is called the angle of incidence. The angle C A s′ is
called the angle of refraction. The difference of these two angles, that is,
the angle C A B, is the refraction.

The amount of refraction, for the same medium, varies with the obliquity
with which the ray strikes the surface. When perpendicular to the latter,
the ray passes without change of direction at all; and in other positions
the refraction increases with the obliquity.

Let R (fig. 43) represent a ray of light falling upon the surface of a mass
of plate glass at the point A. From this point let a perpendicular fall and

be continued into the new medium, and around the same point, as a centre, let a circle be drawn. According to the law just stated, the refraction must be toward the perpendicular; in the direction A R', for example. Let the lines a—a, a'—a', at right angles to the perpendicular, be drawn, and their length compared by means of a scale of equal parts, and noted: their length will, in the case supposed, be in the proportion of 3 to 2. These lines are termed the *sines* of the angles of incidence and refraction respectively.

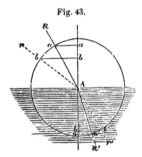

Fig. 43.

Now let another ray be taken, such as r; it is refracted in the same manner to r', the bending being greater from the increased obliquity of the ray; but what is very remarkable, if the sines of the two new angles of incidence and refraction be again compared, they will still be found to bear to each other the proportion of 3 to 2. The fact is expressed by saying, that so long as the light passes from one to the other of the same two media, the *ratio of the sines of the angles of incidence and refraction is constant.* This ratio is called the *index of refraction.*

Different bodies possess different refractive powers; generally speaking, the densest substances refract most. Combustible bodies have been noticed to possess greater refractive power than their density would indicate, and from this observation Newton predicted the combustible nature of the diamond long before anything was known respecting its chemical nature.

The method adopted for describing the comparative refractive power of different bodies, is to state the ratio borne by the sine of the angle of incidence in the first medium at the boundary of the second, to the sine of the angle of refraction in this second medium; this is called the *index of refraction* of the two substances; it is greater or less than unity, according as the second medium is denser or rarer than the first. In the case of air and plate glass the index of refraction is 1.5.

When the index of refraction of any particular substance is once known, the effect of the latter upon a ray of light entering it in any position can be calculated by the law of sines. The following table exhibits the indices of refraction of several substances, supposing the ray to pass into them from the air :—

Substances.	Index of refraction.		Substances.	Index of refraction.
Tabasheer*	. . 1.10		Garnet . . .	1.80
Ice . . .	. 1.30		Glass with much ox-	
Water .	. . 1.34		ide of lead . .	1.90
Fluor spar	. . 1.40		Zircon . . .	2.00
Plate glass	. . 1.50		Phosphorus . .	2.20
Rock-crystal	. . 1.60		Diamond . . .	2.50
Chrysolite	. . 1.69		Chromate of lead .	3.00
Bisulphide of carbon	1.70		Cinnabar . . .	3.20

When a luminous ray enters a mass of substance differing in refractive power from the air, and whose surfaces are not parallel, it becomes permanently deflected from its course and altered in its direction. It is upon this principle that the properties of prisms and lenses depend. To take an example: Figure 44 represents a triangular prism of glass, upon the

* A siliceous deposit in the joints of the bamboo.

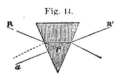

Fig. 44.

side of which the ray of light R may be supposed to fall. This ray will of course be refracted, on entering the glass, toward a line perpendicular to the first substance, and again, from a line perpendicular to the second surface, on emerging into the air. The result is the deflection *a c* R, which is equal to the sum of the two deflections which the ray undergoes in passing through the prism.

A convex lens is thus enabled to converge rays of light falling upon it, and a concave lens to separate them more widely ; each separate part of the surface of the lens producing its own independent effect.

DISPERSION.—The light of the sun and celestial bodies in general, as well as that of the electric spark and of all ordinary flames, is of a compound nature. If a ray of light from any of the sources mentioned be admitted into a dark room by a small hole in a shutter, or otherwise, and suffered to fall upon a glass prism, in the manner shown in fig. 45, it will

Fig. 45.

not only be refracted from its straight course, but will be decomposed into a number of colored rays, which may be received upon a white screen placed behind the prism. When solar light is employed, the colors are extremely brilliant, and spread into an oblong space of considerable length.

The prism being placed with its base upwards, as in fig. 45, the upper part of this image, or *spectrum*, will be violet and the lower red, the intermediate portion, commencing from the violet, being indigo, blue, green, yellow, and orange, all graduating imperceptibly into each other. This is the celebrated experiment of Sir Isaac Newton ; from it he drew the inference that white light is composed of seven primitive colors, the rays of which are differently refrangible by the same medium, and hence capable of being thus separated. The violet rays are most refrangible, and the red rays least.*

Bodies of the same refractive power do not always equally disperse or spread out the differently colored rays to the same extent ; because the principal yellow or red rays, for instance, are equally refracted by two prisms of different materials, it does not follow that the blue or the violet will be similarly affected. Hence, prisms of different varieties of glass, or

* The colors of natural objects are supposed to result from the power possessed by their surfaces of absorbing some of the colored rays, while they reflect or transmit, as the case may be, the remainder of the rays. Thus an object appears red because it absorbs or causes to disappear the yellow and blue rays composing the white light by which it is illuminated. Any color which remains after the deduction of another color from white light, is said to be *complementary* to the latter. Complementary colors, when acting simultaneously, reproduce white light. Thus, in the example already quoted, red and green are complementary colors. The fact of complementary colors giving rise to white light may be readily illustrated by mixing in appropriate quantities a rose-red solution of cobalt and green solution of nickel ; the resulting liquid is nearly colorless.

other transparent substances, give, under similar circumstances, very different spectra, both as respects the length of the image, and the relative extent of the colored bands.

The appearance of the spectrum may also vary with the nature of the source of light : the investigation of these differences, however, involves the use of a more delicate apparatus. Fig. 46 shows the principle of such

Fig. 46.

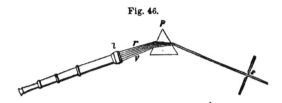

an apparatus, which is called a *spectroscope*. The light, passing through a fine slit, *s*, impinges upon a flint-glass prism, *p*, by which it is dispersed. The decomposed light emerges from the prism in several directions between r (red rays) and *v* (violet rays) ; and the spectrum thus produced is observed by the telescope, *t*, which receives only part of it at once ; but the several parts may be readily examined by turning slightly either the prism or the telescope.

If the solar spectrum be examined in this manner, numerous dark lines parallel with the edge of the prism are observed. They were discovered in 1802 by Dr. Wollaston, and subsequently more minutely investigated by Fraunhofer. They are generally known as Fraunhofer's lines. These dark lines, which exist in great numbers, and of varying strength, are irregularly distributed over the whole spectrum. Some of them, in consequence of their peculiar strength and their relative position, may always be easily recognized ; the more conspicuous are represented in fig. 46, and in the frontispiece. The same dark lines, though paler, and much more

Fig. 47.

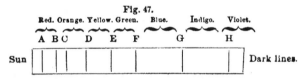

difficult to recognize, are observed in the spectrum of planets lighted by the sun ; for instance, in the light emanating from Venus. On the other hand, the dark lines observed in the spectra which are produced by the light emanating from fixed stars—from Sirius, for instance—differ in position from those previously mentioned.

Sources of light which contain no volatile constituents—incandescent platinum wire, for example—furnish continuous spectra, exhibiting no such lines. But if volatile substances be present in the source of light, bright lines are observed in the spectrum, which are frequently characteristic of the volatile substances.

Professor Plücker, of Bonn, has investigated the spectra which are produced by the electric light when developed in very rarefied gases. He found the bright lines and the dark stripes between the lines varying considerably with different gases. When the electric light was developed in a mixture of two gases, the spectrum thus obtained exhibited simultaneously the peculiar spectra belonging to the two gases of which the mixture consisted. When the experiment was made in gaseous compounds capable of being decomposed by the electric current, this decomposition was indicated by the spectra of the separated constituents becoming perceptible.

Many years ago, the spectra of colored flames were examined by Sir John Herschel, Fox Talbot, and W. A. Miller. Within the last few years results of the greatest importance have been obtained by Kirchhoff and Bunsen, who have investigated the spectra furnished by the incandescence of volatile substances ; these researches have enriched chemistry with a new method of analysis, the analysis by spectrum observations. In order to recognize one of the metals of the alkalies or of the alkaline earths, it is generally sufficient to introduce a minute quantity of a moderately volatile compound of the metal, on the loop of a platinum wire, into the edge of the very hot, but scarcely luminous flame, of a mixture of air and coalgas, and to examine the spectrum which is furnished by the flame containing the vapor of the metal or its compound. Fig. 48 exhibits the

Fig. 48.

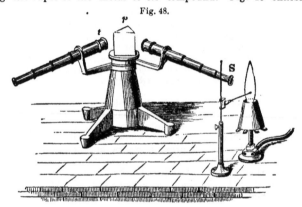

apparatus which is used in performing experiments of this description. The light of the flame in which the metallic compound is evaporated passes through the fine slit in the disk, s, into a tube, the opposite end of which is provided with a convex lens. This lens collects the rays diverging from the slit, and throws them parallel upon the prism, p. The light is decomposed by the prism, and the spectrum thus obtained is observed by means of the telescope, which may be turned round the axis of the stand carrying the prism. Foreign light is excluded by an appropriate covering.

The limits of this elementary treatise do not permit us to describe the ingenious arrangements which have been contrived for sending the light from different sources through the same prism at different heights, whereby their spectra, the solar spectrum, for instance, and that of a flame, may be placed in a parallel position, the one above the other, and thus be compared.* The spectra of flames in which different substances are volatilized frequently exhibit such characteristically distinct phenomena, that they may be used with the greatest advantage for the discrimination of these substances Thus the spectrum of a flame containing sodium (Na) exhibits a bright line on the yellow portion, the spectrum of potassium (K) a characteristic bright line at the extreme limit of the red, and another at the opposite violet limit of the spectrum. Lithium (Li) shows a bright brilliant line in the red, and a paler line in the yellow portion ; strontium

* See the article "Spectral Analysis," by Prof. Roscoe, in Watts's Dictionary of Chemistry, vol. v.

(Sr) a bright line in the blue, one in the orange, and six less distinct ones in the red portion of the spectrum. The frontispiece exhibits the most remarkable of the dark lines of the solar spectrum (Fraunhofer's lines), and the position of the bright lines in the spectra of flames containing the vapors of compounds of the metals of the alkalies and alkaline earths, also of the metals thallium and indium.

The delicacy of these spectral reactions is very considerable, but unequal in the case of different metals. The presence of $\dfrac{1}{200,000,000}$ grain of sodium in the flame is still easily recognizable by the bright yellow line in the spectrum. Lithium, when introduced in the form of a volatile compound, imparts to the flame a red color; but this coloration is no longer perceptible when a volatile sodium compound is simultaneously present, the yellow coloration of the flame predominating under such circumstances. But when a mixture of one part of lithium and 1000 parts of sodium is volatilized in a flame, the spectrum of the flame exhibits, together with the bright yellow sodium line, likewise the red line characteristic of lithium. The observation of bright lines not belonging to any of the previously known bodies has led to the discovery of new elements. Thus, Bunsen and Kirchhoff, when examining the spectrum of a flame in which a mixture of alkaline salt was evaporated, observed some bright lines, which could not be attributed to any of the known elements, and were thus led to the discovery of the two new metals, cæsium and rubidium. By the same method a new element, thallium, has been more recently discovered by Mr. Crookes; another, called indium, by Reich and Richter; and a third, called gallium, by Lecoq de Boisbaudran; and several new earth-metals by Cleve, Delafontaine, Marignac, and Nilson.

For the examination of the bright lines in the spectra of metals the electric spark may be conveniently employed as a source of light. Small quantities of the metal are invariably volatilized; and the spectrum developed by the electric light exhibits the bright lines characteristic of the metal employed. These lines were observed by Wheatstone as early as 1835. This method of investigation is especially applicable to the examination of the spectra of the heavy metals. The spark passes between two points of the metal or between two small cones of pure porous carbon impregnated with a solution of a compound of the metal.

By a series of theoretical considerations, Professor Kirchhoff has arrived at the conclusion that the spectrum of an incandescent gas is reversed—i. e., that the bright lines become dark lines—if there be behind the incandescent gas a very luminous source of light, which by itself furnishes a continuous spectrum. Kirchhoff and Bunsen have fully confirmed this conclusion by experiment. Thus a volatile lithium salt produces, as just pointed out, a very distinct bright line in the red portion of the spectrum; but if bright sunlight, or the light emitted by a solid body heated to the most powerful incandescence, be allowed to fall through the flame upon the prism, the spectrum exhibits, in the place of this bright line, a black line similar in every respect to Fraunhofer's lines in the solar spectrum. In like manner the bright strontium line is reversed into a dark line. Kirchhoff and Bunsen have expressed the opinion that all the Fraunhofer lines in the solar spectrum are bright lines thus reversed. In their conception, the sun is surrounded by a luminous atmosphere, containing a certain number of volatilized substances, which would give rise in the spectrum to certain bright lines, if the light of the solar atmosphere alone could reach the prism; but the intense light of the powerfully incandescent body of the sun, which passes through the solar atmosphere, causes these bright lines to be reversed and to appear as dark lines on the ordinary solar spectrum. Kirchhoff and Bunsen have thus been enabled to attempt the investigation of the chemical constituents of the solar atmosphere, by ascertaining the elements which, when in the state of incandescent vapor, develop bright spectral lines, coinciding with Fraunhofer's lines in the solar spectrum. Fraunhofer's line D (fig. 47) coincides most accurately with the bright spectral line of sodium, and may be artificially produced by reversing the latter ; sodium would thus appear to be a constituent of the solar atmosphere. Kirchhoff has proved, moreover, that sixty bright lines perceptible in the spectrum of iron correspond, both as to position and distinction, most exactly with the same number of dark lines in the solar spectrum, and, accordingly, he believes iron, in the state of vapor, to be present in the solar atmosphere. In a similar manner this physicist has endeavored to establish the presence of several other elements in the solar atmosphere.

Absorption Spectra.—The relative quantities of the several colored rays absorbed by a colored medium of given thickness may be observed by viewing a line of light through a prism and the colored medium ; the spectrum will then be seen to be diminished in brightness in some parts, and perhaps cut off altogether in others. This mode of observation is often of great use in chemical analysis, as many colored substances when thus examined afford very characteristic spectra, the peculiarities of which may often be distinguished, even though the solution of the substance under examination contains a sufficient amount of colored impurities to change its color very considerably. The following method of making the observation is given by Professor Stokes.*

A small prism is to be chosen of dense flint glass, ground to an angle of 60°, and just large enough to cover the eye comfortably. The top and bottom should be flat, for convenience of holding the prism between the thumb and forefinger, and laying it down on a table so as not to scratch or soil the faces. A fine line of light is obtained by making a vertical slit in a board six inches square, or a little longer in a horizontal direction, and adapting to the aperture two pieces of thin metal. One of the metal pieces is movable, to allow of altering the breadth of the slit. About the fiftieth of an inch is a suitable breadth for ordinary purposes. The board and metal pieces should be well blackened.

On holding the board at arm's length against the sky or a luminous flame, the slit, being, we will suppose, in a vertical direction, and viewing the line of light thus formed through the prism held close to the eye, with its edge vertical, a pure spectrum is obtained at a proper azimuth of the prism. Turning the prism round its axis alters the focus, and the proper focus is got by trial. The whole of the spectrum is not, indeed, in perfect focus at once, so that in scrutinizing one part after another, it is requisite to turn the prism a little. When daylight is used, the spectrum is known to be pure by its showing the principal fixed lines ; in other cases the focus is got by the condition of seeing distinctly the other objects, whatever they may be, which are presented in the spectrum. To observe the absorption spectrum of a liquid, an elastic band is put round the board near the top, and a test-tube containing the liquid is slipped under the band, which holds it in its place behind the slit. The spectrum is then observed just as before, the test-tube being turned from the eye.

To observe the whole progress of the absorption, different degrees of strength must be used in succession, beginning with a strength which does not render any part of the spectrum absolutely black, unless it be one or more very narrow bands, as otherwise the most distinctive features of the

---

* Chem. Soc. Journ., xvii. 306.

absorption might be missed. If the solution be contained in a wedge-shaped vessel instead of a test-tube, the progress of the absorption may be watched in a continuous manner by sliding the vessel before the eye. Some observers prefer using a wedge-shaped vessel in combination with the slit, the slit being perpendicular to the edge of the wedge. In this case each element of the slit forms an elementary spectrum corresponding with a thickness of the solution which increases in a continuous manner from the edge of the wedge, where it vanishes. This is the mode of observation adopted by Gladstone.*

Fig. 49 represents the effect produced in this way by a solution of chromic chloride, and fig. 50 that produced by a solution of potassium permanganate.

Fig. 49.   Fig. 50.

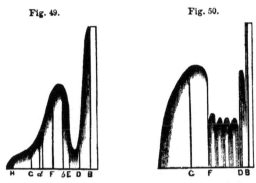

The right hand side of these figures corresponds with the red end of the spectrum; the letters refer to Fraunhofer's lines. The lower part of each figure shows the pure spectrum seen through the thinnest part of the wedge; and the progress of the absorption, as the thickness of the liquid increases, is seen by the gradual obliteration of the spectrum towards the upper part of the figures.

Fluorescence.—An examination into a peculiar mode of analysis of light, discovered by Sir John Herschel, in a solution of quinine sulphate, has within the last few years led to the discovery of a most remarkable fact. Professor Stokes has observed that light of certain refrangibility and color is capable of experiencing a peculiar influence in being dispersed by certain media, and of undergoing thereby an alteration of its refrangibility and color. This curious change, called fluorescence, can be produced by a great number of bodies, both liquid and solid, transparent and opaque. Frequently the change affects only the extreme limits; at other times larger portions; and in a few cases even the whole, or, at all events, the major part of the spectrum. A dilute solution of quinine sulphate, for instance, changes the violet and dark-blue light to sky-blue; by a decoction of madder in a solution of alum all rays of higher refrangibility than yellow are converted into yellow; by an alcoholic solution of the coloring matter of leaves, all the rays of the spectrum become red. In all cases in which this peculiar phenomenon presented itself in a greater or less degree, Mr. Stokes observed that it consisted in a diminution of the refrangibility. Thus, rays of so high a degree of refrangibility, that they extend far beyond the extreme limits of the spectrum visible under ordinary circumstances, may be rendered luminous, and converted into blue and even red light.

* Chem. Soc. Journ., x. 79.

DOUBLE REFRACTION AND POLARIZATION.—A ray of common light made to pass through certain crystals of a particular order is found to undergo a very remarkable change. It becomes split or divided into two rays, one of which follows the general law of refraction, while the other takes a new and extraordinary course, dependent on the position of the crystal. This effect, which is called d o u b l e r e f r a c t i o n, is beautifully illustrated in the case of Iceland spar, or crystallized calcium carbonate. On placing a rhomb of this substance on a piece of white paper on which a mark or line has been made, the object will be seen double.

Again, if a ray of light be suffered to fall on a plate of glass at an angle of 56° 45′, the portion of the ray which suffers reflection will be found to

Fig. 51.

have acquired properties which it did not before possess ; for on throwing it, at the same angle, upon a second glass plate, it will be observed that there are two particular positions of the latter, namely, those in which the planes of incidence are at right angles to one another, when the ray of light is no longer reflected, but entirely refracted. Light which has suffered this change is said to be *polarized*.

The light which passes through the first or polarizing plate is also, to a certain extent, in this peculiar condition, and by employing a series of similar plates held parallel to the first, this effect may be greatly increased ; a bundle of fifteen or twenty such plates may be used with great convenience for the experiment. It is to be remarked, also, that the light polarized by transmission in this manner is in an opposite state to that polarized by reflection : that is, when examined by a second or *analyzing* plate, held at the angle before mentioned, it will be seen to be reflected when the other is transmitted, and to be dispersed when the first is reflected.

It is not every substance that is capable of polarizing light in this manner ; glass, water, and certain other bodies bring about the change in question, each having a particular polarizing angle at which the effect is greatest. For each transparent substance the polarizing angle is that at which the reflected and refracted rays are perpendicular to each other. Metals can also polarize light, by reflection, but they do so very imperfectly.

The two rays into which a pencil of common light divides itself in passing through a doubly-refracting crystal are found, on examination, to be polarized in a very complete manner, and also transversely, the one being capable of reflection when the other vanishes or is transmitted. The two rays are said to be polarized in opposite directions. With a rhomb of transparent Iceland spar of tolerably large dimensions, the two oppositely polarized rays may be widely separated and examined apart.

Certain doubly-refracting crystals absorb one of these rays, but not the other. Through a plate of such a crystal one ray passes and becomes entirely polarized ; the other, which is likewise polarized, but in another plane, is removed by absorption. The best known of these media is tourmaline. When two plates of this mineral, cut parallel to the axis of the crystal, are held with their axes parallel, as in fig. 52, light traverses them both freely, but when one of them is turned round in the manner shown in fig. 53, so as to make the axes cross at right angles, the light is almost wholly stopped, if the tourmalines are good. A plate of the mineral thus becomes an excellent test for discriminating between polarized light and that which has not undergone the change.

Instead of the tourmaline plate, which is always colored, frequent use is made of two Nichol's. prisms, or conjoined prisms of calcium carbonate, which, in consequence of a peculiar cutting and combination, possess the property of allowing only one of the oppositely polarized rays to pass. A more advantageous method of cutting and combining prisms has been given by M. Foucault. His prisms are as serviceable as, and less expensive than, those of Nichol. If two Nichol's or Foucault's prisms be placed

Fig. 52.   Fig. 53.

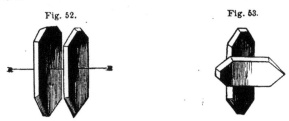

one behind the other, in precisely similar positions, the light polarized by the one goes through the other unaltered. But when one prism is slightly turned round in its setting, a cloudiness is produced ; and by continuing to turn the prism, this increases until perfect darkness ensues. This happens, as with the tourmaline plates, when the two prisms cross one another. The phenomenon is the same with colorless as with colored light.

Circular Polarization.—Supposing that polarized light, colored, for example, by going through a plate of red glass, has passed through the first Nichol's prism, and been altogether obstructed in consequence of the position of the second prism, then, if between the two prisms a plate of rock-crystal, formed by a section at right angles to the principal axis of the crystal, be interposed, the light polarized by the first prism will, by passing through the plate of quartz, be enabled partially to pass through the second Nichol's prism. Its passage through the second prism can then again be interrupted by turning the second prism round to a certain extent. The rotation required varies with the thickness of the plate of rock-crystal, and with the color of the light employed. It increases from red in the following order—yellow, green, blue, violet.

This property of rock-crystal was discovered by Arago. The kind of polarization has been called circular polarization. The direction of the rotation is with many plates towards the right hand ; in other plates it is towards the left. The one class is said to possess right-handed polarization, or to be *dextrorotatory* or *dextrogyrate;* the other class, to possess left-handed polarization, or to be *levorotatory* or *levogyrate.* For a long time quartz was the only solid body known to exhibit circular polarization. Others have since been found which possess this property in a far higher degree. Thus, a plate of cinnabar acts fifteen times more powerfully than a plate of quartz of equal thickness.

Biot observed that many solutions of organic substances exhibit the property of circular polarization, though to a far less extent than rock-crystal. Thus, solutions of cane-sugar, glucose, and tartaric acid, possess right-handed polarization ; whilst albumin, uncrystallizable sugar, and oil of turpentine, are left-handed. In all these solutions the amount of circular polarization increases with the concentration of the liquid, and the thickness of the column through which the light passes. Hence, circular polarization is an important auxiliary in chemical analysis. In order to determine the amount of polarization which any liquid exhibits, it is put into a glass tube not less than from ten to twelve inches long, which is closed with glass plates. This is then placed between the two Nichol's prisms, which have previously been so arranged with regard to each other that no light could pass through. An apparatus of this description, the

s a c c h a r i m e t e r, is used for determining the concentration of solutions
of cane-sugar.

The form of this instrument is shown in fig. 54.   The two Nichol's
prisms are inclosed in the corresponding fastenings *a* and *b*.   Between the
two there is a space to receive the tube, which is filled with the solution
of sugar.   If the prisms are crossed in the way above mentioned before

Fig. 54.

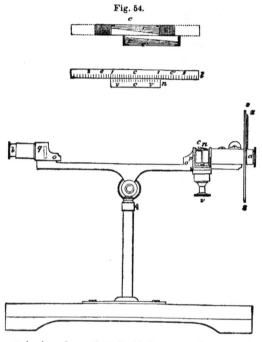

the tube is put in its place, that is, if they are placed so that no light
passes them, then, by the action of the solution of sugar, the light is
enabled to pass, and the Nichol's prism, *a*, must be turned through a
certain angle before the light is again perfectly stopped.   The magnitude
of this angle is observed on the circular disk *s s*, which is divided into
degrees, and upon which, by the turning of the prism, an index *z* is
moved along the division.   When the tube is exactly ten inches long, and
closed at both ends by flat glass plates, and when it is filled with solution
containing 10 per cent. by weight of cane-sugar, and free from any other
substance possessing an action on light, the angle of rotation for the
middle yellow ray is 19.6°.   Now, the magnitude of this angle is directly
proportional to the length of the column of liquid, and also to the quantity of sugar in solution.   If, therefore, a solution containing *z* per cent.
by weight of sugar in a tube *l* inches long, produce a rotation equal to *a*
degrees, the percentage of sugar will be given by the equation—

$$\frac{a}{19.6} = \frac{l}{10} \cdot \frac{z}{10},$$

whence,

$$z = \frac{100\,a}{19.6\,l}.$$

This process is not sufficient when the solution contains cane-sugar and uncrystallizable sugar ; for the latter rotates the ray to the left ; in that case only the difference of the two actions is obtained. But if the whole quantity of sugar be changed into uncrystallizable sugar, and the experiment be repeated, then from the results of the two observations the quantity of both kinds of sugar can easily be calculated.

It is difficult to find exactly that position of the Nichol's prisms in which the greatest darkness prevails. To make the measurements more exact and easy, Soleil has made some additions to the apparatus. At q, before the prism b, a plate of rock-crystal cut at right angles to the axis is placed. It is divided in the centre of the field of vision, half consisting of quartz rotating to the right hand, and half of the variety which rotates to the left ; it is 0.148 inch (3.75 millimetres) thick, this thickness being found by experiment to produce the greatest difference in the color of the two halves, when one prism is slightly rotated. The solution of sugar has precisely the same action on the rotation, since it increases the action of the half which has a right-handed rotation, and lessens the action of the half which rotates to the left. Hence the two halves will assume a different color when the smallest quantity of sugar is present in the liquid. By slightly turning the Nichol's prism a, this difference can be again removed. Soleil has introduced another more delicate means of effecting this, at the part l, which he calls the compensator. The most important parts of this are separately represented in fig. 54. It consists of two exactly equal right-angled prisms, of left-handed quartz, whose surfaces, c and c', are cut perpendicular to the optic axis. These prisms can, by means of the screw v and a rack and pinion, be made to slide on one another, so that, when taken together, they form a plate of varying thickness, bounded by parallel surfaces. One of the frames has a scale l, the other a vernier n. When this points to zero of the scale, the optical action of the two prisms is exactly compensated by a right-handed plate of rock-crystal, so that an effect is obtained as regards circular polarization, as if the whole system were not present. As soon, however, as the screw is moved, and thus the thickness of the plate formed by the two prisms is changed (we will suppose it increased), then a left-handed action ensues, which must be properly regulated, until it compensates the opposite action of a solution of sugar. Thus a convenient method is obtained of rendering the color of the double plate uniform, when it has ceased to be so by the action of the sugar.

Faraday made the remarkable discovery that, if a very strong electric current be passed round a substance which possesses the property of circular polarization, the amount of rotation is altered to a considerable degree.

HEATING AND CHEMICAL RAYS OF THE SOLAR SPECTRUM.—The luminous rays of the sun are accompanied, as already mentioned, by others which possess heating powers. If the temperature of the different colored spaces in the spectrum be tried with a delicate thermometer, it will be found to increase from the violet to the red extremity, and when the prism is of some particular kinds of glass, the greatest effect will be manifested a little beyond the visible red rays. The position of the greatest heating effect in the spectrum materially depends on the absorptive nature of the glass. Transparent though this medium is to the rays of light, it nevertheless absorbs a considerable quantity of the heat rays. Transparent rock-salt is almost without absorptive action on the thermal rays. In the spectrum obtained by passing the solar rays through prisms of rock-salt, the greatest thermal effect is found at a position far beyond the last visible

9 *

red rays. It is inferred from this that the chief mass of the heating rays of the sun are among the least refrangible components of the solar beam.

Again, it has long been known that chemical changes both of combination and of decomposition, but more particularly the latter, can be effected by the action of light. Chlorine and hydrogen combine at common temperatures only under the influence of light; and parallel cases occur in great numbers in organic chemistry. The blackening and decomposition of silver salts are familiar instances of the chemical powers of the same agent. Now, it is not always the luminous part of the ray which effects these changes; they are chiefly produced by certain invisible rays, which accompany the others, and are found most abundantly beyond the violet part of the spectrum. It is there that certain chemical effects are most marked, although the intensity of the light is exceedingly feeble. These chemically acting rays are sometimes called *actinic rays* (ἀκτὶς, a ray), and the chemical action of sunlight is called *actinism;* but these terms are not very well chosen. The chemical rays are thus directly opposed to the heating rays in the common spectrum in their degree of refrangibility, since they exceed all the others in this respect. The luminous rays, too, under peculiar conditions, exert decomposing powers upon silver salts. The result of the action of any ray depends, moreover, greatly on the physical state of the surface on which it falls, and on the chemical constitution of the body; indeed, for every kind of ray a substance may be found which under particular circumstances will be affected by it; and thus it appears that the chemical functions are by no means confined to any set of rays to the exclusion of the rest.

Upon the chemical changes produced by light is based the art of *photography*. In the year 1802 Mr. Thomas Wedgwood proposed a method of copying paintings on glass, by placing behind them white paper or leather moistened with a solution of silver nitrate, which became decomposed and blackened by the transmitted light in proportion to the intensity of the latter; and Davy, in repeating these experiments, found that he could thus obtain tolerably accurate representations of objects of a texture partly opaque and partly transparent, such as leaves and the wings of insects, and even copy with a certain degree of success the images of small objects obtained by the solar microscope. These pictures, however, required to be kept in the dark, and could only be examined by candle-light, otherwise they became obliterated by the blackening of the whole surface, from which the salt of silver could not be removed. These attempts at light-painting attracted but little notice till the year 1839, when Mr. Fox Talbot published his plan of "photogenic drawing." This consisted in exposing in the camera a paper soaked in a weak solution of common salt, and afterwards washed over with a strong solution of nitrate of silver; the image thus obtained was a *negative* one, the lights being dark and the shadows light, and the pictures were fixed by immersion in a solution of common salt.

Many improvements have been made in this process. In 1841 Fox Talbot patented the beautiful process known as the "Talbotype or Calotype process," in which the paper is coated with silver iodide by dipping it first in silver nitrate, then in potassium iodide.

Paper thus prepared is not sensitive *per se* to the action of light, but may be rendered so by washing it over with a mixture of silver nitrate and gallic or acetic acid. If it be exposed in the camera for two or three minutes, it does not receive a visible image (unless the light has been very strong); but still the compound has undergone a certain change by the influence of the light; for on subsequently washing it over with the mixture of silver nitrate and acetic or gallic acid, and gently warming it, a negative image comes out on it with great distinctness. This image is *fixed* by washing the paper with sodium hyposulphite, which removes the whole of the silver iodide not acted upon by the light, and thus protects the picture from further change by exposure to light. The negative picture thus obtained is rendered transparent by placing it between two sheets of blotting-paper saturated with white wax, and passing a moderately heated smoothing-iron over the whole. It may then be used for printing *positive pictures* by laying it on a sheet of paper prepared with chloride or iodide of silver and exposing it to the sun.

A most important step in the progress of photography is the substitution of a transparent film of iodized collodion or albumin spread upon glass, for the iodized paper used in Talbot's process, to receive the negative image in the camera. The process is thus rendered so much more certain and rapid, and the positive pictures obtained by transferring the negative to paper prepared with chloride or iodide of silver are found to be so much sharper in outline, than when the transference occurs through paper, as in the talbotype process, that this method is now universally employed. In this process, as in that of the Calotype, the image produced in the camera is a latent one, and requires development with substances such as pyrogallic acid, or ferrous sulphate, which, having a tendency to absorb oxygen, induce, in presence of silver nitrate, the reduction of the chloride or iodide to the metallic state. For a description of the best apparatus and latest processes used in the collodion method, the reader may consult Hardwich's "Manual of Photographic Chemistry."

Sir John Herschel has shown that a great number of other substances can be employed in these photographic processes by taking advantage of the deoxidizing effects of certain portions of the solar rays. Paper washed with a solution of ferric salt becomes capable of receiving impressions of this kind, which may afterwards be made evident by potassium ferricyanide, or gold chloride. Vegetable colors are also acted upon in a very curious and apparently definite manner by the different parts of the spectrum.

The *daguerreotype*, the announcement of which was first made in the summer of 1839, by M. Daguerre, who had been occupied with this subject from 1826, if not earlier, is another remarkable instance of the decomposing effects of the solar rays. A clean and highly polished plate of silvered copper is exposed for a certain time to the vapor of iodine, and then transported to the camera obscura. In the most improved state of the process, a very short time suffices for effecting the necessary change in the film of silver iodide. The picture, however, becomes visible only by exposing it to the vapor of mercury, which attaches itself, in the form of exceedingly minute globules, to those parts which have been most acted upon, that is to say, to the lights, the shadows being formed by the dark polish of the metallic plate. Lastly, the plate is washed with sodium hyposulphite, to remove the undecomposed silver iodide and render it permanent.

Since Daguerre's time this process has undergone considerable improvements; amongst these, we may mention the exposure of the plate to the vapor of bromine, by which the sensitiveness of the film is greatly increased, and the reduction of metallic gold upon the surface of the film during the process of fixing, by which the lights and shades of the picture are rendered more effective.

Etching and lithographic processes, by combined chemical and photographic agency, promise to be of considerable utility. The earliest is that of Nièpce: he applied a bituminous coating to a metal plate, upon which an engraving was superimposed. The light, being thus partially interrupted, acted unequally upon the varnish; a liquid hydro-carbon, *petroleum*, used as a solvent, removed the bitumen wherever the light had not acted; an engraving acid could now bite the unprotected metal, which could eventually be printed from in the usual way. Very successful results have also been obtained by M. Fizeau, who submits the daguerreotype to the action of a mixture of dilute nitric acid, common salt, and potassium nitrate, when the silver only is attacked, the mercurialized portion of the image resisting the acid; an etching is thus obtained following minutely the lights and shadows of the picture. To deepen this etching, the silver chloride formed is removed by ammonia, the plate is boiled in caustic potash and again treated with acid, and so on till the etching is of sufficient depth. Sometimes electro-gilding is resorted to, and an engraving acid is used to get still more powerful impressions.

Among recent results are those obtained by Mr. Talbot on steel plates: he uses a mixture of potassium bichromate and gelatin, which hardens by exposure to the light; the parts not affected are removed by washing. Platinum tetrachloride is used as an etching liquid; it has the advantage of biting with greater regularity than nitric acid.

The bitumen process of M. Nièpce has been applied to lithographic stone; and positives obtained from negative talbotypes have been printed off by a modification of the ordinary lithographic process. M. Nièpce finds that ether dissolves the altered bitumen, while naphtha, or benzol, attacks by preference the bitumen in its normal condition.

MAGNETISM.

A PARTICULAR species of iron ore has long been remarkable for its property of attracting small pieces of iron, and causing them to adhere to its surface ; it is called loadstone or magnetic iron ore.

If a piece of this loadstone be carefully examined, it will be found that the attractive force for particles of iron is greatest at certain particular points of its surface, while elsewhere it is much diminished, or even altogether absent. These attractive points are denominated poles, and the loadstone itself is said to be endowed with magnetic polarity.

If one of the pole-surfaces of a natural loadstone be rubbed in a particnlar manner over a bar of steel, its characteristic properties will be communicated to the bar, which will then be found to attract iron filings like the loadstone itself. Further, the attractive force will appear to be greatest at two points situated very near the extremities of the bar, and least of all towards the middle. The bar of steel so treated is said to be magnetized, or to constitute an artificial magnet.

When a magnetized bar or natural magnet is suspended at its centre in any convenient manner, so as to be free to move in a horizontal plane, it is always found to assume a particular direction with regard to the earth, one end pointing nearly north, and the other nearly south. This direction varies with the geographical position of the place, and is different also at the same place at different times. In London, at the present time, the needle points 19º 32′ west of the astronomical north. If the bar be moved from this position, it will tend to reassume it, and, after a few oscillations, settle at rest as before. The pole which points towards the astronomical north is usually distinguished as the north pole of the bar, and that which points southward as the south pole.

A magnet, either natural or artificial, of symmetrical form, suspended in the presence of a second magnet, serves to exhibit certain phenomena of attraction and repulsion which deserve particular attention. When a north pole is presented to a south pole, or a south pole to a north, attraction ensues between them ; the ends of the bars approach each other, and, if permitted, adhere with considerable force ; when, on the other hand, a north pole is brought near a second north pole, or a south pole near another south pole, mutual repulsion is observed, and the ends of the bars recede from each other as far as possible. *Poles of an opposite name attract, and poles of a similar name repel each other.* Thus, a small bar or needle of steel, properly magnetized and suspended, and having its poles marked, becomes an instrument fitted not only to discover the existence of magnetic power in other bodies, but to estimate the kind of polarity affected by their different parts.

A piece of soft iron brought into the neighborhood of a magnet acquires itself magnetic properties : the intensity of the power thus conferred depends upon that of the magnet, and upon the space which divides the two, becoming greater as that space decreases, and greatest of all in actual contact. The iron, under these circumstances, is said to be magnetized by *induction* or influence, and the effect, which reaches its maximum in an instant, is at once destroyed by removing the magnet.

. When steel is substituted for iron in this experiment, the inductive action is hardly perceptible at first, and becomes manifest only after the

5 *

lapse of a certain time: in this condition, when the steel bar is removed from the magnet, it retains a portion of the induced polarity. It becomes, indeed, a permanent magnet, similar to the first, and retains its peculiar properties for an indefinite time. This resistance, which steel always offers in a greater or less degree both to the development of magnetism and to its subsequent destruction, is called *specific coercive power*.

Fig. 55.

The rule which regulates the induction of magnetic polarity in all cases is exceedingly simple, and most important to be remembered. The pole produced is always of the opposite name to that which produced it, a north pole developing south polarity, and a south pole north polarity. The north pole of the magnet figured in the sketch induces south polarity in all the nearer extremities of the pieces of iron or steel which surround it, and a state similar to its own in all the more remote extremities. The iron thus magnetized is capable of exerting a similar inductive action on a second piece, and that upon a third, and so to a great number, the intensity of the force diminishing as the distance from the permanent magnet increases. It is in this way that a magnet is enabled to hold up a number of small pieces of iron; or a bunch of filings, each separate piece becoming for the time a magnet by induction.

Magnetic polarity, similar in degree to that which iron presents, has been found only in some of the compounds of iron, in nickel, and in cobalt.

Magnetic attractions and repulsions are not in the slightest degree interfered with by the interposition of substances destitute of magnetic properties. Thick plates of glass, shellac, metals, wood, or of any substances except those above mentioned, may be placed between a magnet and a suspended needle, or a piece of iron under its influence, the distance being preserved, without the least perceptible alteration in its attractive power, or force of induction.

One kind of polarity cannot be exhibited without the other. In other words, a magnetic pole cannot be isolated. If a magnetized bar of steel be broken at its neutral point, or in the middle, each of the broken ends acquires an opposite pole, so that both portions of the bar become perfect magnets ; and, if the division be carried still further, if the bar be broken into a hundred pieces, each fragment will be a complete magnet, having its own north and south poles.

This experiment serves to show very clearly that the apparent polarity of the bar is the consequence of the polarity of each individual particle, the poles of the bar being merely points through which the resultants of all these forces pass ; the largest magnet is made up of an immense number of little magnets regularly arranged side by side, all having their north poles looking one way, and their south poles the other. The middle portion of such a system cannot possibly exhibit attractive or repulsive effects on an external body, because each pole is in close juxtaposition with one of an opposite name and of equal power. Hence their forces will be exerted in opposite directions, and neutralize each other's influence.

Such will not be the case at the extremities of the bar; there uncompen. sated polarity will be found, capable of exerting its specific power.

Fig. 56.

This idea of regular polarization of particles of matter in virtue of a pair of opposite and equal forces, is not confined to magnetic phenomena; it is the leading principle in electrical science, and is constantly reproduced in some form or other in every discussion involving the consideration of molecular forces.

Artificial steel magnets are made in a great variety of forms; such as small light needles, mounted with an agate cap for suspension upon a fine point; straight bars of various kinds; bars curved into the shape of a horse-shoe, etc. All these have regular polarity communicated to them by certain processes of rubbing or touching with another magnet, which require care, but are not otherwise difficult of execution. When great power is wished for, a number of bars may be screwed together, with their similar ends in contact, and in this way it is easy to construct permanent steel magnets capable of sustaining great weights. To prevent the gradual destruction of magnetic force, which would otherwise occur, it is usual to arm each pole with a piece of soft iron or keeper, which, becoming magnetized by induction, serves to sustain the polarity of the bar, and in some cases even increases its energy.

Magnetism is not peculiar to these substances which have more especially been called magnetic, such as iron, nickel, cobalt, but it is the property of all metals, though to a much smaller degree. Very powerful magnets are required to show this remarkable fact. Large horse-shoe magnets, made by the action of the electric current, are best adapted for the purpose. The magnetic action on different substances which are capable of being easily moved, differs not only according to the size, but also according to the nature of the substance. In consequence of this, Faraday divides all bodies into two classes. He calls the one magnetic, or, better, *paramagnetic*, and the other *diamagnetic*.

The matter of which a paramagnetic (magnetic) body consists is attracted by both poles of the horse-shoe magnet; on the contrary, the matter of a diamagnetic body is repelled. When a small iron bar is hung by untwisted silk between the poles of the magnet, so that its long diameter can easily move in a horizontal plane, it arranges itself axially, that is, parallel to the straight line which joins the poles, or to the magnetic axis of the poles, assuming at the end which is nearest the north pole, a south pole, and at the end nearest the south pole, a north pole. Whenever the little bar is removed from this position, it returns, after a few oscillations, to its previous position. The whole class of paramagnetic bodies behave in a precisely similar way under similar circumstances, but in the intensity of the effects great differences occur.

Diamagnetic bodies, on the contrary, have their long diameters placed equatorially, that is, at right angles to the magnetic axis. They behave as if at the end opposite to each pole of the magnet the same kind of polarity existed.

In the first class of substances, besides iron, which is the best representative of the class, we have nickel, cobalt, manganese, chromium, cerium,

titanium, palladium, platinum, osmium, aluminium, oxygen, and alsc
most of the compounds of these bodies, most of them, even when in solution. According to Faraday, the following substances are also feebly paramagnetic (magnetic),—paper, sealing-wax, Indian-ink, porcelain, asbestos,
fluor-spar, minium, cinnabar, binoxide of lead, sulphate of zinc, tourmaline,
graphite, and charcoal.

In the second class are placed bismuth, antimony, zinc, tin, cadmium,
sodium, mercury, lead, silver, copper, gold, arsenic, uranium, rhodium,
iridium, tungsten, phosphorus, iodine, sulphur, chlorine, hydrogen, and
many of their compounds. Also, glass free from iron, water, alcohol,
ether, nitric acid, hydrochloric acid, resin, wax, olive oil, oil of turpentine,
caoutchouc, sugar, starch, gum, and wood. These are diamagnetic.

When diamagnetic and paramagnetic bodies are combined, their peculiar
properties are more or less neutralized. In most of these compounds,
occasionally in consequence of the presence of a very small quantity of
iron, the peculiar magnetic power remains more or less in excess. Thus
green bottle-glass, and many varieties of crown glass are magnetic in consequence of the iron they contain.

In order to examine the magnetic properties of liquids, they are placed
in very thin glass tubes, the ends of which are then closed by melting ;
they are then hung horizontally between the poles of the magnet. Under
the influence of poles sufficiently powerful, they begin to swing, and
according as the fluid contents are paramagnetic (magnetic) or diamagnetic,
they assume an axial or equatorial position.

Faraday has tried the magnetic condition of gases in different ways.
One method consisted in making soap-bubbles with the gas which he wished
to investigate, and bringing these near the poles. Soap and water alone
is feebly diamagnetic. A bubble filled with oxygen was strongly attracted
by the magnet. All other gases in the air are diamagnetic, that is, they
are repelled. But, as Faraday has shown, in a different way, this partly
arises from the paramagnetic (magnetic) property of the air. Thus he
found that nitrogen, when this differential action was eliminated, was perfectly indifferent, whether it was condensed or rarefied, whether cooled or
heated. When the temperature is raised, the diamagnetic property of
gases in the air is increased. Hence, the flame of a candle or of hydrogen
is strongly repelled by the magnet. Even warm air is diamagnetic in cold
air.

For some time it had been believed that crystallized bodies exhibited a
special and peculiar behavior when placed between the poles of a magnet.
It appeared as though the magnetic directing power of the crystal had
some peculiar relation to the position of its optic axis ; so that, independently of the magnetic property of the substance of the crystal, if the
crystal were positively optical, it possessed the power of placing its optic
axis parallel with the line which joined the poles of the magnet, while
optically negative crystals tried to arrange their axis at right angles to
this line. This supposition is disproved by the excellent investigation of
Tyndall and Knoblauch, who showed that exceptions to the above law are
furnished by all classes of crystals, and proved that the action, instead of
being independent of the magnetic nature of the mass, was completely
reversed where, in isomorphous crystals, a magnetic constituent was substituted for a diamagnetic one. Rejecting the various new forces assumed,
Tyndall and Knoblauch referred the observed phenomena to the modification of the magnetic force by structure, and they imitated the effects exactly
by means of substances whose structure had been modified by compression.
In a later investigation Tyndall demonstrated the fundamental principle
on which these phenomena depend, showing that the *entire mass* of a magnetic body is most strongly attracted when the attracting force acts parallel

to the line of compression ; and that a diamagnetic substance is most strongly repelled when the repulsion acts along the same line. Hence, when such a body is freely suspended in the magnetic field, the line of compression must set axially or equatorially, according as the mass is magnetic or diamagnetic. Faraday was the first to establish a differential action of this kind in the case of bismuth ; Tyndall extended it to several magnetic and diamagnetic crystals, and showed that it was not confined to them, but was a general property of matter. It was also proved that, for a fixed distance, the attraction of a magnetic sphere, and the repulsion of a diamagnetic sphere, followed precisely the same law, both being exactly proportioned to the square of the exciting current.

The phenomena of diamagnetism naturally suggest the inquiry, whether the repulsion exerted by a magnetic pole on diamagnetic bodies is a force distinct from that of magnetism as exerted upon iron and other bodies of the magnetic class ; or whether, on the other hand, the magnetic and diamagnetic conditions of matter are merely relative, so that all bodies are magnetic in different degrees, and the apparent repulsion of a diamagnetic body, such as bismuth, is merely the result of its being attracted by the magnet less than the particles of the surrounding medium, just as a balloon recedes from the earth because its weight is less than that of an equal bulk of the surrounding air. It is easy to show that the same body may appear magnetic or diamagnetic, according to the medium in which it is placed. Ferrous sulphate is a magnetic substance, and water is diamagnetic ; hence it is possible, by varying the strength of an aqueous solution of this salt, to make it either magnetic, indifferent, or diamagnetic, when suspended in air. Again, a tube containing a solution of ferrous sulphate suspended horizontally within a jar also filled with a solution of the same salt, and placed between the poles of two powerful electro-magnets, will place itself axially or equatorially, according as the solution contained in it is stronger or weaker than that in the jar. In the same manner, then, we may conceive that bismuth places itself equatorially between two magnetic poles, because it is less magnetic than the surrounding air. But the diamagnetism of bismuth and other bodies of the same class shows itself in a vacuum as well as in air ; hence, if diamagnetism is not to be regarded as a distinct force, we must suppose that the *ether* is also magnetic, and occupies in the magnetic scale the place intermediate between magnetic and diamagnetic bodies.

That a body suspended in a medium of greater magnetic susceptibility than itself will recede from a magnetic pole in its neighborhood, in consequence of the greater force with which the particles of the medium are impelled towards the magnet, is so obvious a consequence of mechanical laws, that we can scarcely avoid attributing the movements of diamagnetic bodies to the cause just mentioned ; at least, when the body is suspended in air or other magnetic gas. There is, however, some difficulty in reconciling the above-described phenomena of compressed and crystallized bodies with this view ; and, moreover, Tyndall has shown, by a method which we cannot here describe,* that diamagnetic bodies possess opposite poles, analogous to those of magnetic bodies, each of these poles being attracted by one pole of a magnet, and repelled by the other. This polarity shows decidedly that the properties of diamagnetic bodies cannot be wholly due to the differential action above mentioned ; for if they were, every part of a diamagnetic body would be repelled by either pole of a magnet. Diamagnetism must therefore, for the present at least, be regarded as a force distinct from magnetism.

* Phil. Trans., 1855 and 1856 ; see also Watts's Dictionary of Chemistry, vol. iii. p. 776.

ELECTRICITY.

WHEN glass, amber, or sealing-wax is rubbed with a dry cloth, it acquires the power of attracting light bodies, as feathers, dust, or bits of paper; this is the result of a new and peculiar condition of the body rubbed, called electrical excitation.

If a light downy feather be suspended by a thread of white silk, and a dry glass tube, excited by rubbing, be presented to it, the feather will be strongly attracted to the tube, adhere to its surface for a few seconds, and then fall off. If the tube be now excited anew, and presented to the feather, the latter will be strongly repelled.

The same experiment may be repeated with shellac or resin; the feather in its ordinary state will be drawn towards the excited body, and, after touching, again driven from it with a certain degree of force.

Now, let the feather be brought into contact with the excited glass, so as to be repelled by that substance, and let a piece of excited sealing-wax be presented to it; a degree of attraction will be observed far exceeding that exhibited when the feather is in its ordinary state. Or, again, let the feather be made repulsive for sealing-wax, and then the excited glass be presented: strong attraction will ensue.

The reader will at once see the perfect parallelism between the effects described and some of the phenomena of magnetism, the electrical excitement having a twofold nature, like the opposite polarities of the magnet. A body to which one kind of excitement has been communicated is attracted by another body in the opposite state, and repelled by one in the same state; the excited glass and resin being to each other as the north and south poles of a pair of magnetized bars.

To distinguish these two different forms of excitement, terms are employed which, although originating in some measure in theoretical views of the nature of the electrical disturbance, may be understood by the student as purely arbitrary and distinctive: it is customary to call the electricity manifested by glass rubbed with silk *positive* or *vitreous*, and that developed in the case of shellac, and bodies of the same class, rubbed with flannel, *negative* or *resinous*. The kind of electricity depends in some measure upon the nature of the surface and the quality of the rubber; smooth and perfectly clean glass, rubbed with silk, becomes positive, but when ground or roughened by sand or emery, it acquires under the same circumstances, a negative charge. Glass dried over a gas flame and rubbed with wool is generally also negative; when dried over a fire of wood-charcoal it remains positive.

The repulsion shown by bodies in the same electrical state is taken advantage of to construct instruments for indicating electrical excitement and pointing out its kind. Two balls of elder pith, hung by threads or very fine metal wires, serve this purpose in many cases: they open out when excited, in virtue of their mutual repulsion, and show by the degrees of divergence the extent to which the excitement has been carried. A pair of gold leaves suspended to a metal rod having a brass plate on its upper end constitute a much more delicate arrangement, and one of great value in all electrical investigations. The rod should be covered with a thick coating of shellac, and it must be fastened by means of a cork, air-tight, into a glass flask. The flask must have been perfectly dried previously by warming it. These instruments are called electroscopes or electrometers : when excited by the communication of a known kind of electricity, they show, by an increased or diminished divergence, the state of an electrified body brought into their neighborhood (fig. 58).

Fig. 57.   Fig. 58.

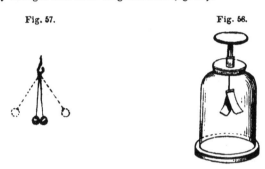

One kind of electricity can no more be developed without the other than one kind of magnetism : the rubber and the body rubbed always assume opposite states, and the positive condition on the surface of a mass of matter is invariably accompanied by a negative state in all surrounding bodies.

The induction of magnetism in soft iron has its exact counterpart in electricity ; a body already electrified disturbs or polarizes the particles of all surrounding substances in the same manner and according to the same law, inducing a state opposite to its own in the nearer portions, and a similar state in the more remote parts. A series of globes suspended by silk threads, in the manner represented in fig. 59, will each become electric by

Fig. 59.

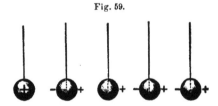

induction when a charged body is brought near the end of the series, like so many pieces of iron in the vicinity of a magnet, the positive half of each globe looking in one and the same direction, and the negative half in the opposite one. The positive and negative signs are intended to represent the opposite states.

The intensity of the induced electrical disturbance diminishes with the distance from the charged body ; if this be removed or discharged, all the effects cease at once.

So far, the greatest resemblance may be traced between these two sets of phenomena ; but here it seems in great measure to cease. The magnetic polarity of a piece of steel can awaken polarity in a second piece in contact with it by the act of induction, and in so doing loses nothing whatever of its power ; this is an effect completely different from the apparent transfer or discharge of electricity constantly witnessed, which in the air and in liquids often gives rise to the appearance of a bright spark of fire. Indeed, ordinary magnetic effects comprise two groups of phenomena only, those, namely, of attraction and repulsion, and those of induction. But in electricity, in addition to phenomena very closely resembling these, we have the effects of *discharge*, to which there is nothing analogous in magnetism, and which takes place in an instant when any electrified body is put in communication with the earth by any one of the class of substances called conductors of electricity, all signs of electrical disturbance then ceasing.

These conductors of electricity, which thus permit discharge to take place through their mass, are contrasted with another class of substances called non-conductors or insulators. The difference, however, is only one of degree, not of kind; the very best conductors offer a certain resistance to the electrical discharge, and the most perfect insulators permit it to a small extent. The metals are by far the best conductors; glass, silk, shellac, and dry gas, or vapor of any sort, the very worst; and between these there are bodies of all degrees of conducting power. Water is a moderately good conductor, and consequently the deposition of a new film of moisture on the glass pillars and handles of electric apparatus greatly impairs their insulating power. The best way of preventing this inconvenience is to varnish such supports with an alcoholic solution of shellac. Supports made of this substance, or of baked wood, do not readily condense the moisture of the air.

In good conductors of sufficient size electrical discharges take place silently and without disturbance. But if the charge be very intense, and the conductor very small, or imperfect from its nature, it is often destroyed with violence.

When a break is made in a conductor employed in effecting the discharge of a highly excited body, disruptive or spark-discharge takes place across the intervening air, provided the ends of the conductor be not too distant. The electrical spark itself presents many points of interest in the modifications to which it is liable.

The time of transit of the electrical wave through a chain of good conducting bodies of great length is so minute as to be altogether inappreciable to ordinary means of observation. Professor Wheatstone's very ingenious experiments on the subject give, in the instance of motion through a copper wire, a velocity surpassing that of light.

Electrical excitation is manifested only upon the surfaces of conductors, or those portions directed towards other objects capable of assuming the opposite state. An insulated ball charged with positive electricity, and placed in the centre of the room, is maintained in that state by the inductive action of the walls of the apartment, which immediately become negatively electrified; in the interior of the ball there is absolutely no electricity to be found, although it may be constructed of open metal gauze, with meshes half an inch wide. Even on the surface the distribution of electrical force is not always the same; it depends upon the figure of the body itself, and its position with regard to surrounding objects. The polarity is always highest in the projecting extremities of the same conducting mass, and greatest of all when these are attenuated to points; in which case the inequality becomes so great that discharge takes place to the air, and the excited condition cannot be maintained.

By the aid of these principles, the construction and use of the common electrical machine, and other pieces of apparatus of great utility, will become intelligible.

The electrophorus (fig. 60) is a simple and ingenious instrument, enabling

us to obtain by inductive action an unlimited
number of charges from one single charge.
It consists of a round tray or dish about twelve
inches in diameter, half an inch deep, and
filled with melted shellac, the surface of which
is rendered as even as possible. A brass
disc, with rounded edge, of about nine inches
diameter, is also provided, and fitted with an
insulating handle. The resinous plate is excited by striking it with a dry, warm piece
of fur or flannel, whereby it becomes charged

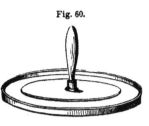

Fig. 60.

with negative electricity. If the cover be then placed upon it, the positiv
electricity is drawn to the under surface of the metal nearest to the negatively charged resinous cake, while the negative electricity is repelled to
the upper surface of the cover; on touching the cover with the finger, the
negative electricity passes away to the earth, while an additional quantity
of positive electricity is drawn into the plate; and if the finger be removed
and the cover then lifted by its insulating handle, it will be found so strongly
charged by induction with positive electricity as to give a bright spark; and
as the resin is not discharged by the cover, which merely touches it at a few
points, sparks may be drawn as often as may be wished.

For obtaining electricity in larger quantity, Electrical Machines are
used, consisting of a glass cylinder or plate, which is made to revolve and
press against a leather cushion.

In the *Cylinder Machine* (fig. 61) the cushion is made to press by a spring
against one side of the cylinder, while a large metallic conductor armed

Fig. 61.

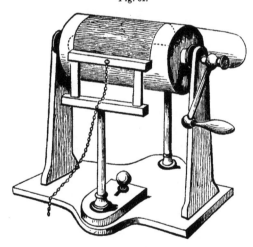

with a number of points next the glass, occupies the other: both cushion
and conductor are insulated by glass supports, and to the upper edge of the
former a piece of silk is attached long enough to reach half around the
cylinder. Upon the cushion is spread a quantity of soft amalgam of tin,
zinc, and mercury,* mixed up with a little grease: this substance is found

* 1 part tin, 1 zinc, and 6 mercury. An amalgam of permanent softness and great efficacy is obtained by mixing 65 parts mercury, 24 tin, and 11 zinc. It is better applied to
silk than to leather.

10 *

by experience to excite glass most powerfully. The cylinder, as it turns, becomes charged by friction against the rubber, and as quickly discharged by the row of points attached to the great conductor; and as the latter is also completely insulated, its surface speedily acquires a charge of positive electricity, which may be communicated by contact to other insulated bodies. The maximum effect is produced when the rubber is connected by a chain or wire with the earth. If negative electricity be wanted, the rubber must be insulated and the conductor discharged.

Another form of the electrical machine consists of a circular plate of glass (fig. 62) moving upon an axis, and provided with two pairs of cushions

Fig. 62.

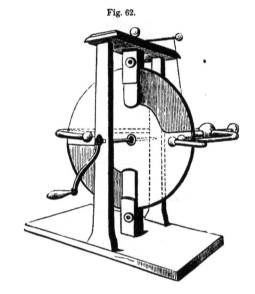

or rubbers, attached to the upper and lower parts of the wooden frame, and covered with amalgam, the plate moving between them with considerable friction. An insulated conductor, armed as before with points, discharges the plate as it turns, the rubber being at the same time connected with the ground by the woodwork of the machine or by a strip of metal. This form of machine is preferred in all cases where considerable power is required; but for demonstrating the principles of the science, as in lecture-experiments, the cylinder machine is by far the more convenient form, as it affords the means of obtaining a positive or negative charge at pleasure.

A machine of much greater power than either of the preceding is that of Holtz, in which the development of the charge takes place much in the same manner as in the electrophorus. This machine, which may indeed be regarded as a revolving electrophorus, consists of two glass plates (fig. 63), one, AA, fixed by means of four wooden rollers resting on glass axes and glass feet, while the other, BB, two inches less in diameter, turns on a horizontal glass axis passing through the centre of the fixed plate, and is set in motion by a winch M and a series of pulleys which give it a rotation of 12 to 15 turns in a second. The fixed plate A is perforated at the extremities of a diameter by two large apertures or windows, F, F'.

Two pieces of varnished paper, p, p', are fastened to the back of the fixed plate below the window on the left, and above that on the right. These pieces of paper, or *armatures*, are furnished with narrow tongues which project forward through the windows towards the movable plate, and nearly touch it with their blunt points. The plate must rotate in the direction opposite to that in which the tongues point, as shown by the arrow. In front of the movable plate B, and at the height of the armatures, are two

Fig. 63.

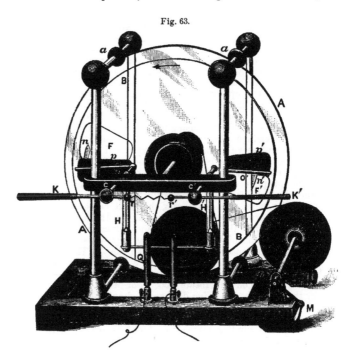

brass combs, *o, o'*, supported by two conductors of the same metal C, C'. At the front ends of these conductors are two brass knobs, through which pass two brass rods, terminated by smaller knobs, r, r', and provided with wooden handles, K, K'. These rods, besides moving with gentle friction in the knobs, can also be turned so as to be more or less approached and inclined to each other.

To work the machine, the knobs r, r' are brought into contact, and a small initial charge is given by an electrophorus or a rubbed glass rod to one of the armatures. The plate B is then quickly rotated, and it is found that, after a few turns, the exertion required to keep up the rotation increases greatly; at the same time pale blue brushes of light are seen to issue from the points of the combs, and if one of the rods K, K' be then drawn back, a torrent of brilliant sparks will dart between the brass knobs. The mode of action is as follows.

Suppose a small positive charge to be given to the left armature *p*; this charge acts inductively through the glass plates upon the brass comb, repelling positive electricity and attracting negative electricity, which is discharged by the points upon the front side of the movable plate, while the repelled positive electricity passes through the brass rods and balls, and is discharged through the right comb upon the front side of the plate. Here it acts inductively on the armature p', attracting negative electricity into the part opposite to itself and repelling positive electricity into its farthest part, viz., into the tongue, which being bluntly pointed, slowly discharges a positive charge upon the back of the movable plate. If now the plate be turned in the direction indicated by the arrow, the positive charge on its back comes over from its right to its left side, and when it gets opposite the combs it increases the inductive effect of the positive charge already existing on the armature p, and therefore repels more positive electricity through the brass rods and knobs into the right comb. Meanwhile the negative charge induced in the right armature has in its turn attracted positive electricity into the right comb, which has been discharged by the points upon the front of the moving plate, and has repelled negative electricity through the brass rods and knobs in the opposite direction, discharging it through the left comb upon the front of the same plate, there to neutralize the positive charge which is being conveyed over from the right on the front of the plate. These actions result in causing the upper half of the moving plate to be positively electrified on both sides, and the lower half negatively. The charges on the front serve, as they are carried round, to neutralize the electricities discharged from the points of the combs, while the charges on the back, induced in the neighborhood of each of the armatures, serve, when the motion of the plate carries them round, to increase the inductive influence of the charge of the other armature. Hence a very small initial charge is speedily raised to a maximum, the limit being reached when the electrization of the armatures becomes so great that the loss of electricity at their surface equals the gain by convection and induction.

The power of the machine may be increased by suspending to the conductors C, C', two condensers or small Leyden jars H, H' (see page 117), having their external coatings connected by a conductor G. One of these jars H becomes charged with + electricity on the inside and — on the outside, the other H' — on the inside and + on the outside. Becoming charged by the machine, and being discharged simultaneously with it by the knobs r, r', they strengthen the spark and increase its length.

The current of the machine is utilized by placing on the frame two brass uprights Q, Q', with binding screws in which are copper wires, then by means of the handles K, K' inclining the rods which support the knobs r, r', so as to bring them in contact with the uprights. The current being then directed by the wires, a battery can be quickly charged, water decomposed, a galvanometer deflected, and various other effects produced, as with the voltaic battery.

Very powerful electric excitation is produced by a jet of steam issuing at high pressure from a pipe provided with a nozzle of wood or metal, the effect being due, not to the pure steam itself, but to the friction of particles of condensed water against the inner surface of the exit-pipe. The steam is usually positive if the nozzle be constructed of wood or clean metal, but the slightest trace of oily matter produces a change of sign. The intensity of the charge increases, *cæteris paribus*, with the elastic force of the steam. A steam-boiler mounted on glass legs and provided with an exit-pipe, as above described, forms an electrical machine of very great power.

Condensers and Accumulators.—When the conductor of an electric machine is charged with electricity, it acts indirectly on all the surrounding conductors, in such a manner as to accumulate at their surfaces the electricity contrary to its own, producing the greatest effect on the conductor which is nearest to it and is in the best connection with the ground, whereby the electricity of the same kind as that of the machine may pass away. As the inducing electricity attracts the induced electricity of an opposite kind, so, on the other hand, is the former attracted by the latter. Hence, the electricity which the conductor receives from the machine must especially accumulate at that spot to which another good conductor of electricity is opposed. If a metal disc is in connection with the conductor of a machine, and if another similar disc, in good connection with the earth, is placed opposite to it, we have an arrangement by which tolerably large and good conducting surfaces can be brought close to one another, the positive condition of the first disc, as well as the negative condition of the other, being thereby increased in a very considerable degree. In this case, however, the limit is very soon reached, because the intervening air easily permits spark-discharge to take place through its substance; but with a solid insulating body, as glass or lac, this discharge takes place much less readily, even when the plate of insulating matter is very thin. It is on this principle that instruments for the *accumulation* of electricity depend, among which the Leyden jar is the most important.

A thin glass jar is coated on both sides with tinfoil, care being taken to leave several inches of the upper part uncovered (fig. 64); a wire, terminating in a metallic knob, communicates with the internal coating. When the outside of the jar is connected with the earth and the knob put in contact with the conductor of the machine, the inner and outer surfaces of the glass become respectively positive and negative, until a very great degree of intensity has been attained. On completing the connection between the two coatings by a metallic wire or rod, discharge occurs in the form of a very bright spark, accompanied by a loud snap; and if the human body be interposed in the circuit, the peculiar and disagreeable sensation of the electric shock is felt at the moment of its completion.

Fig. 64.

By enlarging the dimensions of the jar, or by connecting together a number of such jars in such a manner that all may be charged and discharged simultaneously, the power of the apparatus may be greatly augmented. By the discharge of such a combination, called an e l e c t r i c b a t t e r y, thin wires of metal may be fused and dissipated; pieces of wood may be shattered; many combustible substances set on fire; and all the well-known effects of lightning exhibited upon a small scale. The circumstances of a thunderstorm indeed exactly resemble those of the charge and discharge of a coated plate or jar, the cloud and the earth representing the two coatings, and the intervening air the conducting body, or *dielectric*. The polarities of the opposed surfaces and of the insulating medium between them become raised by mutual induction, until violent disruptive discharge takes place through the air itself, or through any other bodies which may happen to be in the interval. When these are capable of conducting freely, the discharge is silent and harmless; but in other cases it often proves highly destructive. These dangerous effects may in a great measure be obviated by the use of lightning-rods attached to buildings. The masts of ships may be gnarded in like manner by metal conductors. Sir W. Snow Harris devised a very ingenious plan for the purpose, which has been adopted, with complete success, in the Navy.

Condensing Electroscopes.—The delicacy of the gold-leaf electroscope may be greatly increased by laying on its cover a circular brass plate well polished; and on this another brass plate provided with an insulating handle and covered on its lower surface with copal varnish to prevent metallic contact between the two. Suppose now the lower plate *a* (fig. 65) to be connected with a source of electricity (say + E) not strong enough to produce a sensible divergence in the gold leaves, and the upper plate to be touched with the finger. The + E in a will produce a charge of — E on the lower surface of b, and drive the + E of that plate into the ground. The — E will then draw more + E into the lower plate a, and this again will draw more + E into the lower sur-

Fig. 65.

face of b, and so the action will go on till a quantity of + E will be accumulated on a much larger than that which it would otherwise have acquired; and if the plate a be then disconnected with the source of electricity, and b be lifted up, the + E accumulated on a will diffuse itself equally through the electroscope, causing the gold leaves to diverge. In this way decided indications of electric action may often be obtained from sources too feeble to produce a deflection of the electroscope by direct communication. The apparatus is therefore called a *condenser*.

In all these forms of condensing and accumulating apparatus the attainable degree of electric charge increases as the metallic surfaces are brought nearer together, or, in other words, as the intervening thickness of air, glass, or other insulator becomes less. It must not however be supposed that this intervening substance acts merely as an insulator; on the contrary, this medium itself becomes charged to a degree which for a given thickness is different for each particular substance; and this relative power or capability of receiving an electric charge, is called the *Specific Inductive Capacity*, or better *Specific Inductivity:* the insulating medium itself is called a *Dielectric*.

The specific inductivity of different dielectrics was first observed in 1775 by Cavendish; afterwards, in 1837, by Faraday, who obtained the following values, the inductivity of air being taken for unity: sulphur, 2.26; shellac, 2.0; glass, 1.76 or more. Later observations by Gordon have given the following numbers:—

Air	.	.	.	1	Gutta-percha	. . . 2.462
Paraffin (solid)	.	.	. 1.944	Sulphur	. . . . 2.58
Caoutchouc	.	.	2.220 to 2.497	Shellac .	. . . . 2.74
Ebonite	.	.	. . 2.284	Glass	. . . 3.013 to 3.258

A comparison of these numbers with those of Faraday shows that the determination of specific inductivity is attended with considerable uncertainty. For liquids Gordon finds: turpentine, 2.16; petroleum, 2.03 to 2.07; carbon disulphide, 1.81. All gases appear to have the same or nearly the same specific inductivity as atmospheric air.

The consideration of these differences of inductivity in various media, together with that of other phenomena, has led to the idea that electric charge and discharge consist, not, as was formerly supposed, in the actual transference of a substance—the so-called "electric fluid"—from one conductor to another across the air or other intervening insulator, but rather that electric *charge* consists in a state of strain or tension of the particles of an elastic medium pervading all space and interposed between the particles of bodies—in fact of the "ether," the vibrations of which give rise to the phenomena of light (p. 89)*—and that *discharge* is effected by the relief or

* See "Elementary Lessons in Electricity and Magnetism," by Silvanus Thompson, London, 1882, p. 61.

removal of that strain. The particles of bodies being surrounded by the ether, this strain is communicated to them, and its removal in the discharge occasions a more or less violent disturbance of those particles, sometimes resulting, in the case of solid dielectrics, in a rupture of the substance. Thus, when a Leyden jar is very highly charged, discharge sometimes takes place between the coatings through the substance of the glass, which is thereby cracked or perforated. The loud snap which accompanies the spark in air, and the rupture of wood and fusion of metallic wires by a powerful electric discharge, likewise indicate a violent disturbance of the particles of that medium.

Various Sources of Electricity.—As electric excitation is capable of producing disturbance in the particles of ponderable matter, so conversely is any disturbance in these particles, produced by mechanical or other causes, attended with development of electric power. The production of electricity by friction has been already considered. Other sources of electric excitation are:—

1. *Pressure.*—Solid bodies become oppositely electrified when simply pressed together (without friction) and afterwards separated. A slice of cork becomes positively electrified when pressed against caoutchouc, orange-peel, coal, amber, zinc, copper, silver, and heated Iceland spar; negative with dry animal substances, heavy spar, gypsum, fluorspar, and non-heated Iceland spar. Two pieces of the same substance do not become electrical by pressure unless one of them is heated, in which case the hotter becomes negative, the colder positive.

2. *Cleavage and Separation of Surfaces.*—When two laminæ of a crystal of mica are torn asunder, they appear oppositely electrified, and the separation is attended with a flash of light. Similar separation of electricities is exhibited in the cleavage of calcspar, fluorspar, and some other crystals, also when a playing card is torn into its two sheets, the separation being attended with sparks visible in the dark. Two sheets of paper stick fast together when rubbed with india-rubber, and on being pulled asunder appear strongly charged with opposite electricities. When melted sulphur is poured into a short conical glass vessel and a glass rod is inserted into it before it solidifies, no electricity is apparent so long as the sulphur remains in the glass, but on lifting it out by the glass handle, the sulphur appears positively, the inner surface of the glass negatively electrified. Similar effects are exhibited by chocolate and glacial phosphoric acid when left to solidify in glass vessels.

3. *Vibration.*—The vibrations set up in a metal rod coated at one end with resin or sulphur, and made to slide through an insulated metallic ring, give rise to a separation of the two electricities at the surface of contact of the metal and the non-conducting body.

4. *Heat.*—(a) In Crystals.—Pyro-electricity. Many hemihedral crystals,* while being heated or cooled, exhibit contrary electricities at their opposite ends, those extremities or poles which are positive while the temperature of the crystal is rising becoming negative as it falls. The effect depends entirely on change of temperature, no crystal exhibiting any electric polarity while its temperature remains constant. This effect was first observed in tourmaline, which has long been known to possess, when heated, the power of attracting light bodies. It has since been observed in boracite, zinc silicate or electric calamine, cane-sugar, and Brazilian topaz, and in a lower degree by many other hemihedral crystals.

(b.) In Metals.—Thermo-electricity. When two pieces of different metals are joined together at each end by soldering or otherwise, and

* See the chapter on Crystallography.

one of the joints is more heated than the other, an electric current is set up, the strength of which appears to be in direct proportion to the difference of temperature between the two joints, while its direction depends upon the nature of the two metals. The metals may be arranged in a thermo-electric series, such that each metal when connected with another will transmit positive electricity across the heated junction to those on its right hand, and negative to those on its left,—and *vice versâ* when the junction is cooled. The following is the thermo-electric series, according to Becquerel:—*Bismuth, platinum, lead, tin, gold, silver, copper, zinc, iron, antimony.* For a given difference of temperature, the current will be stronger the farther the two metals are separated in the series, the strongest combination being formed of bismuth and antimony; but the current is always very feeble.

Greater power may, however, be obtained by combining a number of thermo-electric pairs into a *battery, chain,* or *pile,* a number of bars of bismuth and antimony or of platinum and iron being alternately soldered together (figs. 66, 67) and heat applied to the first, third, fifth, etc. points of junction, while the alternate ones are kept cold.

Fig. 66.   Fig. 67.

With this arrangement, the thermo-electric current produces, not only deflection of the magnetic needle, but also chemical decomposition and heating effects, so that even when one of the conducting wires of the battery is cooled by immersion in ice, the point of junction of that wire situated without the ice becomes sensibly warmed. Melloni's *Thermo-multiplier* or *Thermoscope* is a pile formed in this manner of bars of bismuth and antimony, and connected with a galvanometer. It is capable of indicating very slight changes of temperature, and has rendered excellent service in researches on radiant heat.

5. *Magnetism.*—When a magnet is moved, or when magnetism is either developed or destroyed, in the neighborhood of a closed conducting circuit, a current of electricity is produced in that circuit, its direction depending upon the position of the magnetic poles relatively to the circuit (see page 126).

6. *Contact of Dissimilar Metals.*—If a bar made of two metals—zinc and copper for example—be held in the hand, and one end of it—say the zinc end—be made to touch the *lower* plate of a condensing electroscope (p. 117), the upper plate being at the same time touched with the hand, and if the hand be then removed, and the upper plate lifted up by its glass handle, the leaves of the electroscope will diverge with positive electricity, as may be shown by holding over the instrument a glass rod or tube excited by silk, which will increase the divergence. If a similar experiment be made with the copper end of the bar touching the lower plate of the condenser, the leaves will diverge with negative electricity. Here then it appears that a separation of the electricities has taken place, in consequence of mere contact of the zinc and copper, without any heating or cooling at the point of junction.

By experiments thus conducted it is found that the metals may be arranged in the following series, each metal becoming positive by contact with those which follow, negative with those which precede it.

+ Sodium.	Copper.
Magnesium.	Silver.
Zinc.	Gold.
Lead.	Platinum.
Tin.	— Carbon (graphite).
Iron.	

These results were first obtained by Volta, who attributed them solely to the contact of dissimilar metals. By other experimenters, however, this view has been disputed, especially by Faraday and De la Rive, who attributed the results obtained to the chemical action of water and acid vapors in the air on the zinc or other more oxidizable metal in the couple; and indeed their experiments led them to the conclusion that the more these causes of chemical action were got rid of, the smaller was the electrical effect produced, and that with sufficient care it might be reduced almost to nothing. Hence they inferred that electricity is not developed by mere contact of dissimilar metals. Later experiments, however, made by Sir W. Thomson and by Messrs. Ayrton and Perry, with great care and extremely delicate apparatus, tend to confirm Volta's view, and to show that contact of metals does give rise to electric charge even when all extraneous causes of disturbance are removed. Moreover, we may perhaps go so far as to say that, for the establishment of this point, no special experiments were required, the development of electric charge by the contact of dissimilar metals being sufficiently established by the phenomena of the thermo-electric circuit (p. 120), which in fact show that the development of electricity by such contact takes place *at all temperatures*, and in various degrees of intensity according to the temperature.

7. *Chemical Action.*—Voltaic Electricity. When two solid conducting bodies, generally metals, are immersed in a liquid which acts upon them unequally, the electric equilibrium is disturbed, the one acquiring the positive condition, and the other the negative. Thus, pieces of zinc and platinum put into dilute sulphuric acid, constitute an arrangement capable of generating electrical force: the zinc, which is the metal attacked, becomes negative; the platinum, which remains unaltered, assumes the positive condition: and on making a metallic communication in any way between the two plates, discharge ensues, evidenced by various electric phenomena, just as when the two surfaces of a coated and charged jar are put into connection. No sooner, however, has this occurred, than the disturbance is repeated; and as these successive charges and discharges take place through the fluid and metals with inconceivable rapidity, the result is an apparently continuous action, to which the term *electrical current* is given.

It is necessary, however, to guard against the idea, which this term naturally suggests, of an actual bodily transfer of something through the substance of the conductors, like water through a pipe: the real nature of these phenomena is not precisely known, but it probably consists in a movement or disturbance communicated from particle to particle throughout the whole chain. The word *current* is nevertheless convenient, and consecrated by long use; and with this caution the very dangerous error of applying figurative language to describe an effect, and then seeking the nature of the effect from the common meaning of words, may be avoided.

The original cause of the electric disturbance under the circumstances here considered is still a matter of dispute. Some authorities, following Volta, consider that the electricity is developed by the contact of the dissimilar metals, which, as already observed, is a real cause of electric excite-

6

ment; others again ascribe it solely to the chemical action between one of the metals and the liquid. In favor of the contact theory it may be urged that when pure zinc or zinc amalgamated with mercury is used, no action whatever is observable so long as the two metals remain unconnected, but as soon as they are made to touch or are connected by a wire, hydrogen begins to escape at the surface of the copper or platinum plate, and the entire circuit exhibits signs of electric action. On the other side, it is alleged that so slight a force as that which has been shown to be developed by the mere contact of metals can scarcely give rise to the powerful effects which are produced by the arrangements now under consideration; moreover that the electric activity of the circuit appears to be always in proportion to the intensity of the chemical action between the metal and the liquid, and that the direction of the current produced depends altogether upon the manner in which that action takes place. Thus when a plate of iron and a plate of copper connected together are immersed in dilute sulphuric acid, the current of positive electricity is directed from the iron through the liquid to the copper, and back again through the connecting wire to the iron; but if the plates be taken out, washed with water, and then immersed in a solution of an alkaline sulphide, the current will take the contrary direction, viz., from the copper through the liquid to the iron,—that is to say, in both cases its direction is from the metal chemically acted upon, through the liquid, to the unattacked metal. This question cannot with advantage be further discussed at present, but its consideration will be resumed in a subsequent part of the work, after the subject of chemical action in general has been more fully treated.

The intensity of the electrical excitement developed by a single pair of metals and a liquid is too feeble to be detected excepting by a delicate condensing electroscope; but, by arranging a number of such alternations in a connected series, in such a manner that the direction of the current shall be the same in each, the intensity may be very greatly exalted. The two instruments invented by Volta, called the p i l e and c r o w n of c u p s, depend upon this principle.

To construct a Voltaic pile, we lay upon a plate of zinc a piece of cloth, rather smaller than itself, steeped in dilute acid, or any liquid capable of

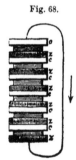

Fig. 68.

exerting chemical action upon the zinc; upon this is placed a plate of copper, silver, or platinum; then a second piece of zinc, another cloth, and a plate of inactive metal, until a pile of about twenty alternations has been built up. If the two terminal plates be now touched with wet hands, the sensation of the electrical shock will be experienced; but, unlike the momentary effect produced by the discharge of a jar, the sensation can be repeated at will by repeating the contact, and with a pile of one hundred such pairs, excited by dilute acid, it will be nearly insupportable. When such a pile is insulated, the two extremities exhibit strong positive and negative states; and when connection is made between them by wires armed with points of hard charcoal or plumbago, the discharge takes place in the form of a bright enduring spark or stream of fire.

The second form of apparatus, or "crown of cups," is precisely the same in principle, although different in appearance. A number of cups or glasses are arranged in a row or circle, each containing a piece of active and a piece of inactive metal, and a portion of exciting liquid—zinc, copper, and dilute sulphuric acid, for example. The copper of the first cup is connected with the zinc of the second, the copper of the second with the zinc of the third, and so to the end of the series. On establishing a communication between the first and last plates by means of a wire or otherwise, discharge takes place as before.

Fig. 69.

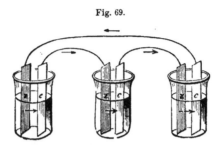

When any such electrical arrangement consists merely of a single pair of conductors and an interposed liquid, it is called a "simple circuit; when two or more alternations are concerned, the term "compound circuit" is applied: they are called also, indifferently, Voltaic batteries.

In every form of such apparatus, however complex it may appear, the direction of the current may be easily understood and remembered. When both ends of the series are insulated, the zinc end exhibits negative, the copper or platinum end positive electricity; consequently, when the two extremities or poles are joined by a conducting wire and a complete circuit formed, the current of positive electricity proceeds *without* the battery from the platinum or copper to the zinc, and *within* the battery from the zinc to the copper or platinum, as indicated by the arrows—just as in the common electrical machine, when the positive conductor and the rubber are joined by a wire, the positive current proceeds from the conductor through the wire to the rubber, and thence along the surface of the glass cylinder or plate to the conductor again.

In the modification of Volta's original pile, made by Cruikshank, the zinc and copper plates are soldered together, and cemented water-tight into a mahogany trough, which thus becomes divided into a series of cells or compartments capable of receiving the exciting liquid. This apparatus (fig. 70) is well fitted to exhibit effects of *tension*, to act upon the electro-

Fig. 70.

scope, and give shocks: hence its advantageous employment in the application of electricity to medicine.

A form of battery more convenient for many purposes is that contrived by Wollaston (fig. 71). In this the copper is made completely to encircle the zinc plate, except at the edges, the two metals being kept apart by pieces of cork or wood. Each zinc is soldered to the preceding copper, and the whole screwed to a bar of dry mahogany, so that the plates can be lifted into or out of the acid, which is contained in an earthenware trough divided into separate cells. The liquid consists of a mixture of 100 parts of water, $2\frac{1}{4}$ parts oil of vitriol, and 2 parts of commercial nitric acid, all by measure.

A number of such batteries are easily connected together by strips of sheet copper, and admit of being put into action with great ease.

Fig. 71.

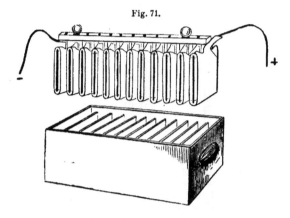

In this and other older forms of the Voltaic battery, however, the power rapidly decreases, so that, after a short time, scarcely the tenth part of the original action remains. This loss of power depends partly on the gradual change of the sulphuric acid into zinc sulphate, but still more on other causes, which, together with the more modern forms of the battery contrived to obviate them, will be more easily understood at a subsequent part of the work, when we come to consider the nature and effects of electro-chemical decomposition.

The term "galvanism," sometimes applied to this branch of electrical science, is used in honor of Galvani of Bologna, who, in 1790, observed that convulsions could be produced in the limbs of a dead frog when certain metals were made to touch the nerve and muscle at the same moment. It was Volta, however, who pointed out the electrical origin of these motions; and his name is very properly associated with the invaluable instrument his genius gave to science.

ELECTRO-MAGNETISM.

Although the fact that electricity is capable, under certain circumstances, both of inducing and of destroying magnetism has long been known from the effects of lightning on the compass-needle and upon small steel articles, as knives and forks, to which polarity has suddenly been given by the stroke, it was not until 1819 that the laws of these phenomena were discovered by Oersted of Copenhagen, and shortly afterwards fully developed by Ampère.

If a wire conveying an electrical current be brought near a magnetic needle, the latter will immediately alter its position, and assume a new one as nearly perpendicular to the wire as the mode of suspension and the magnetism of the earth will permit. When the wire, for example, is placed directly over the needle, and parallel to its length, while the current it carries travels from north to south, the needle is deflected from its ordinary direction, and the north pole driven to the eastward. When the current is reversed, the same pole deviates to an equal amount towards the west.

Placing the wire below the needle instead of above, produces the same effect as reversing the current.

The direction which the needle will assume when placed in any particular positiou relatively to the conducting wire may be determined by the following rule:— *Let the current be supposed to pass through a watch from the face to the back: the motion of the north pole will be in the direction of the hands.* Or, *let the observer imagine himself swimming in the direction of the current with his face towards the needle: the north pole of the needle will then be deflected towards his left hand.*

Fig. 72.　　　　　　Fig. 73.

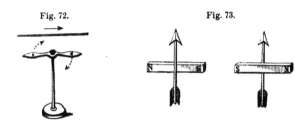

If reference is often required, a little piece of apparatus (fig. 73) may be used, consisting of a piece of pasteboard, or other suitable material, cut into the form of an arrow for indicating the current, crossed by a magnet having its poles marked, and arranged in the true position with respect to the current. The direction of the latter in the wire of the galvanoscope can at once be known by placing the representative magnet in the direction assumed by the needle itself.

When the needle is subjected to the action of two currents in opposite directions, the one above and the other below, they will obviously concur in their effects. The same thing happens when the wire carrying the current is bent upon itself, and the needle placed between the two portions, as in fig. 74; and since every time the bending is repeated a fresh portion of the current is made to act in the same manner upon the needle, it is easy to see how a current, too feeble to produce any effect when a simple straight wire is employed, may be made by this contrivance to exhibit a powerful action on the magnet. It is on this principle that instruments called *galvanometers, galvanoscopes,* or *multipliers,* are constructed; they serve not only to indicate the existence of electrical currents, but to show, by the effects upon the needle, the direction in which they are moving.

Fig. 74.

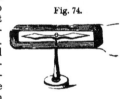

The delicacy of the instrument may be immensely increased by the use of a very long coil of wire and by the addition of a second needle. The two needles are of equal size, and magnetized as nearly as possible to the same extent; they are then immovably fixed together parallel, and with their poles opposed, and hung by a long fibre of untwisted silk, with their lower needle in the coil, and the upper one above it. The advantage thus gained is twofold: the system is *astatic,* unaffected, or nearly so, by the magnetism of the earth; and the needles, being both acted upon in the same manner by the current, are urged with much greater force than one alone would be, all the actions of every part of the coil being strictly concurrent. A divided circle is placed below the upper needle, by which the angular motion can be measured, and the whole is enclosed in a glass case, to shield

11 *

the needles from the agitation of the air. The arrangement is shown in fig. 75.

Fig. 75.

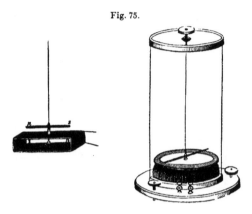

The direction of the current which deflects the galvanometer-needle in a particular way is easily determined by the rules given on page 124, when we know the direction in which the wire is coiled round the frame. For this purpose it is necessary to distinguish between *right-handed* and *left-handed* coils or helices. Suppose the wire to be coiled round a cylinder beginning at the left hand; then if the turns in front of the cylinder proceed from below upwards, as in fig. 76, the coil is left-handed; if, on the contrary, they proceed in front from above downwards, as in fig. 77, the coil is right-handed.

Fig. 76.   Fig. 77.

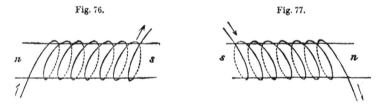

A magnetic needle placed with its centre in the axis of such a coil turns its north or south pole towards the end of the coil at which the current enters according as the coil is left- or right-handed.

The direction given to the needle is the same whether the coil is elongated, as in the above figures, or compressed, as in the galvanometer. As, however, in the galvanometer, when complete, it is not easy to see whether the coil is left- or right-handed, it is best to determine by experiment, once for all, the direction taken by the needle when the current enters at one particular end of the coil.

Action of the Magnet on the Electric Current.—The action between the current and the magnet is mutual, so that if the conductor conveying the current is free to move, it is deflected in the direction opposite to that which the magnet takes under its influence; in short, if the magnet and conducting wire are both free to move, they place themselves at right angles to each other, the magnet moving in the manner indicated at page 124, and the wire in the opposite direction.

The action of the magnet on the current may be shown by means of Ampère's apparatus (fig. 78). On holding a bar-magnet below the rectangular wire, and parallel to its lower horizontal arm, the wire turns round and places itself at right angles to the magnet, the position of equilibrium being determined by the rule just alluded to.

A simple apparatus for this purpose is De la Rive's floating battery, which consists of a pair of zinc and copper plates, contained in a wide glass tube attached to a cork float, and connected together by a rectangular wire, or a flat coil, or elongated helix of covered wire (fig. 79).

Fig. 78.   Fig. 79.

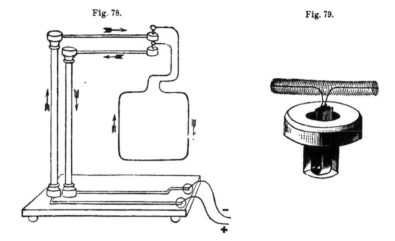

A movable electric current is deflected by the earth's magnetism in the same way as by an ordinary magnet. Thus the rectangular wire of Ampère's apparatus or of a floating battery, when left to itself, will take up a position at right angles to the magnetic meridian; and remembering that the north magnetic pole of the earth is analogous to the south pole of an ordinary magnet, it is easy to see that, in the position of stable equilibrium, the direction of the current will be from east to west in the lower horizontal branch, and from below upwards on the western vertical side. If the wire has the form of a long helix, it will, in like manner, place itself with the turns of the helix at right angles to the magnetic meridian, and therefore with its axis parallel to that meridian, the ends pointing north and south, just like those of an ordinary magnetic needle. If the helix is left-handed, the end connected with the copper plate of the battery will point to the north.

Mutual Action of Electric Currents.—If a conducting wire connecting the poles of a voltaic battery be brought near the movable wire of Ampère's apparatus or the wire of a floating battery, the movable wire will be attracted or repelled according to the relative direction of the two currents, the general law of the action being that *electric currents moving in parallel lines attract one another if they move in the same direction, and repel one another if they move in opposite directions.* From this it is easy to see that if a helix connecting the two poles of a battery be brought near the helix of a floating battery, and if the two helices are similar—that is, both right- or both left-handed—their similar ends, i. e., those by which the current enters or leaves the helix, will repel each other, and their dissimilar ends will attract each other, and consequently the movable helix will place itself parallel to the fixed helix with its poles or ends in the contrary direction to those of the fixed helix. In short, the two helices will act on one another exactly like two bar magnets; and if an ordinary bar magnet be substituted for the fixed helix, the effect will still be the same, each end of the movable helix being attracted by one pole of the magnet and repelled by the other.

This striking resemblance between the mutual action of electric currents and that of magnets has led to the idea, suggested and developed by Ampère, that magnetism is actually produced by electric currents circulating round the molecules of a magnet all in the same direction. These currents may be supposed to pre-exist in all magnetic bodies, even before the development of magnetic polarity, but to be disposed without regularity, so that they neutralize each other. Magnetization is the process by which these molecular currents are made to move in one direction, those situated at the surface yielding, as their resultant, a finite current circulating round the magnet, while the currents in the interior are neutralized by those in the next external layer, the contiguous portions of which move in a direction opposite to their own. The resultant action of all these molecular currents is equivalent to that of a number of currents circulating round the magnet in planes perpendicular to its axis (fig. 80); and from what has been said about the mutual action of magnets and helices traversed by electric currents, it is easy to see that, on looking along the axis of a magnet with its *south* pole towards the observer, the current moves in the direction of the hands of a watch—that is, upwards on the left side, and downwards on the right.

Fig. 80.

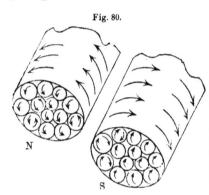

N

S

ELECTRO-DYNAMIC INDUCTION.

1. *Magnetization by the Current.*—When an electro-current is passed through a wire placed at right angles to a bar of iron or steel, the bar acquires magnetic polarity, temporary in the case of soft iron, permanent in the case of hard iron or steel, the position of the poles being determined by the direction of the current, according to the laws already explained.

This effect is prodigiously increased by coiling the conducting wire in a helix round the bar. A piece of soft iron worked into the form of a horseshoe (fig. 81), and surrounded by a coil of wire covered with silk or cotton for the purpose of insulation, furnishes an excellent illustration of the inductive energy of the current in this respect: when the ends of the wire are put into communication with a small Voltaic battery of a single pair of plates, the iron instantly becomes so highly magnetic as to be capable of sustaining a very heavy weight.

2. *Induction of Electric Currents by the action of Magnets, and of other Electric Currents.*—If the two extremities of the coil of the electro-magnet above described be connected with a galvanoscope, and the iron magnetized by the application of a permanent steel horse-shoe magnet to the ends of the bar, a momentary current will be developed in the wire, and pointed out by the movement of the needle. It lasts but a single instant, the needle

after a few oscillations, returning to a state of rest. On removing the magnet, whereby the polarity of the iron is at once destroyed, a second current or wave will become apparent, but in the opposite direction to that of the first. By employing a very powerful steel magnet, surrounding its iron keeper or armature with a very long coil of wire, and then making the armature itself rotate in front of the faces of the magnet, so that its induced polarity shall be rapidly reversed, magneto-electric currents may be produced of such intensity as to give bright sparks and most powerful shocks, and exhibit all the phenomena of Voltaic electricity. Fig. 82 represents a powerful arrangement of this kind.

Fig. 81.

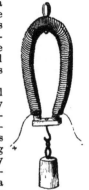

By using electro-magnets instead of permanent steel magnets, and rotating an armature between them by steam-power, very powerful machines, called "Dynamo-electric machines," are constructed, which are now extensively used for electric lighting and other applications of electricity. The light is produced by passing a strong current between two cylinders of hard carbon, whereby an arch of light of almost unendurable brilliancy is obtained—this is called the "Arc Light"—or by sending a current through a thin filament of carbon (prepared by heating cotton-thread or bamboo-fibre in a close vessel) placed in a glass vessel from which the air is exhausted by a Sprengel pump (p. 44), so that the carbon is made to glow without burning. This is called the "Incandescent Light." These, however, are matters the further discussion of which

Fig. 82.

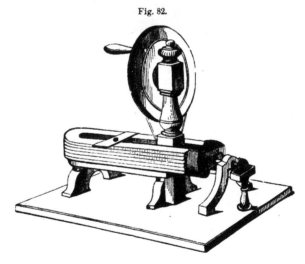

would be foreign to the object of this work. For information respecting them see Silvanus Thompson's "Elementary Lessons on Electricity and Magnetism," pp. 362–378.

Induction-coils.—When two covered wires are twisted together or laid side by side for some distance, and a current is transmitted through the one, a momentary electrical wave will be induced in the other in the reverse direction; and on breaking connection with the battery, a second single wave

6 *

will become evident by the aid of the galvanoscope, in the same direction as that of the primary current. The same effects are produced if the wires are twisted, one above the other, round a hollow wooden cylinder, and the intensity of the induced currents may be enormously increased by inserting within the cylinder a bar of soft iron, or, better, a bundle of iron wires, whereby magnetic and electric induction are made to co-operate. The more frequently contact is alternately made and broken, the greater will be the number of induced currents that follow each other, and the more powerful, within certain limits, will be the action. Instruments thus constructed are called induction-coils. The contact-breaker or commutator is usually made self-acting, an ordinary form consisting of a piece of thin steel which makes contact with a platinum point, and is drawn back by the attraction of the iron core on the passing of the current, and so makes and breaks circuit by vibrating backwards and forwards.

The most powerful form of induction-coil is that of Ruhmkorff (fig. 83),

Fig. 83.

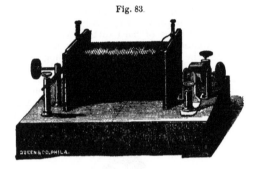

the cylinder of which is a foot or 18 inches long, and is surrounded by a coil of thick wire to convey the inducing current, and a much longer coil of thin wire for the development of the induced currents. The wires are insulated with resin. The direction of the battery-current can be changed at pleasure by a commutator, by which the battery poles are connected through the ends of the axis of a small ivory or ebonite cylinder to two cheeks of brass, which can be turned so as to place them either way in contact with two vertical springs, which are joined to the ends of the primary coil. This apparatus forms a very powerful electric machine, capable of yielding electricity of very high intensity, giving long sparks and shocks of unendurable force. It is capable also of firing gunpowder and other inflammable materials, and has been used for exploding charges of gunpowder and other inflammable substances for mining, engineering, and military purposes. It is also frequently used in the chemical laboratory for firing gaseous mixtures.

PART II.

CHEMISTRY OF ELEMENTARY BODIES.

1. The Non-metallic Elements.

We shall commence the study of these bodies with hydrogen, and arrange the rest according to the number of hydrogen-atoms with which they can unite, or to which they are equivalent (p. 33), designating them, according to this combining capacity, as—

Univalent Elements or Monads . . . H, Cl, Br, I, F.
Bivalent Elements or Dyads . . . . O, S, Se, Te.
Trivalent Elements or Triads . . . N, P, As, B.*
Quadrivalent Elements or Tetrads . . Si, C.

HYDROGEN.

Symbol, H. Atomic weight, 1. Density, 1.

Hydrogen does not occur naturally in the free state, but may be obtained by deoxidation of water, of which it is a component.†

If a tube of iron or porcelain, containing a quantity of filings or turnings of iron, be fixed across a furnace, and its middle portion be made red hot, and then the vapor of water passed over the heated metal, a large quantity of permanent gas will be disengaged from the tube, and the iron will be converted into oxide, and acquire an increase in weight. The gas is hydrogen; it may be collected over water and examined.

Hydrogen is, however, more easily obtained by decomposing hydrochloric or dilute sulphuric acid with zinc, the metal then displacing the hydrogen in the manner already explained (p. 33).

The simplest method of preparing the gas is the following :—A wide-necked bottle is fitted with a sound cork, perforated by two holes for the reception of a small tube-funnel reaching nearly to the bottom of the bottle, and a piece of bent glass ‡ tube to convey away the disengaged gas. Graun-

---

* Boron is not known to combine with hydrogen, but it is placed in this group because it unites with three atoms of other univalent elements and radicles, forming, for example, the compounds BCl_3, $B(CH_3)_3$, $B(C_2H_5)_3$.

† Hence the name, from ὕδωρ, water, and γεν, a root signifying production.

‡ A little practice will soon enable the pupil to construct and arrange a variety of useful forms of apparatus, in which bottles and other articles always at hand are made to supersede more costly instruments. Glass tube, purchased by weight of the maker, may be cut by scratching with a file, and then applying a little force with both hands. Glass may be softened and bent, when of small dimensions, by the flame of a spirit-lamp or a candle, or, better, by a gas-jet. Corks may be perforated by a heated wire, and the hole rendered smooth and cylindrical by a round file; or the ingenious cork-borer of Dr. Mohr, now to be had of all instrument-makers, may be used instead. Lastly, in the event of bad fitting or unsoundness in the cork itself, a little yellow wax melted over the surface, or even a little grease applied with the finger, renders it sound and air-tight when not exposed to heat.

lated zinc or scraps of the malleable metal are put into the bottle, together with a little water, and sulphuric acid slowly added by the funnel, the point of which should dip into the liquid. The evolution of gas is easily regulated by the supply of acid; and when enough has been discharged to expel

Fig. 84.   Fig. 85.

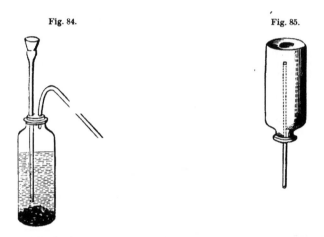

the air of the vessel, it may be collected over water in a jar, or passed into a gas-holder. In the absence of zinc, filings of iron or small nails may be used, but with less advantage.

Hydrogen is colorless, tasteless, and inodorous when quite pure. To obtain it in this condition, it must be prepared from the purest zinc that can be obtained, and passed in succession through solutions of potash and silver nitrate. When prepared from commercial zinc, it has a slight smell, due to impurity, and when iron has been used the odor is very strong and disagreeable. It is inflammable, and burns, when kindled, with a pale yellowish flame, evolving much heat, but very little light. The product of the combustion is water. Hydrogen is very slightly soluble in water. Although destitute of poisonous properties, it is incapable of sustaining life. By strong pressure and exposure to a very low temperature it may be liquefied and even solidified.

Hydrogen is the lightest substance known; Dumas and Boussingault place its density between 0.0691 and 0.0695, referred to that of air as unity. The weight of a litre of hydrogen at 0° C., and under a barometric pressure of 0.760 metre, is 0.089578 gram; consequently a gram of hydrogen occupies a space of 11.16345 litres.* At 15.5 C. (60° F.), and 30 inches barometric pressure, 100 cubic inches weigh 2.14 grains.

When a gas is much lighter or much heavier than atmospheric air, it may often be collected and examined without the aid of the pneumatic trough. A bottle or narrow jar may be filled with hydrogen without much admixture of air, by inverting it over the extremity of an upright tube delivering the gas. In a short time, if the supply be copious, the air will be wholly displaced, and the vessel filled. It may now be removed, the vertical position being carefully retained, and closed by a stopper or glass plate. If the mouth of the jar be wide, it must be partially closed by a piece of cardboard

---

* As a near approximation, it may be remembered that a litre of hydrogen weighs 0.09 gram, or 9 centigrams, and a gram of hydrogen occupies 11.2 litres.

during the operation. This method of collecting gases by displacement is often extremely useful.

Soap-bubbles and small collodion balloons ascend when filled with hydrogen. Caoutchouc balloons are filled and expanded by forcing hydrogen into them with a syringe. Hydrogen was formerly used also for filling air-balloons, being made for the purpose on the spot from zinc or iron and dilute sulphuric acid; but its use is now superseded by that of coal-gas, which may be made very light by employing a high temperature in the manufacture. Although far inferior to pure hydrogen in buoyant power, it is found in practice to possess advantages over that substance, while its greater density is easily compensated by increasing the size of the balloon.

Diffusion of Gases.—There is a very remarkable property possessed by gases and vapors in general, which is seen in a high degree of intensity in the case of hydrogen; that is what is called d i f f u s i v e p o w e r. If two bottles containing gases which do not act chemically upon each other at common temperatures be connected by a narrow tube and left for some time, the gases after a certain time, depending much upon the narrowness of the tube and its length, will be found uniformly mixed, even though they differ greatly in density, and the system has been arranged in a vertical position with the heavier gas downwards. Oxygen and hydrogen can thus be made to mix, in a few hours, against the action of gravity, through a tube a yard long and not more than one quarter of an inch in diameter: and the same is true of all other gases which do not act directly upon each other.

If a vessel be divided into two portions by a diaphragm or partition of porous earthenware or dry plaster of Paris, and each half of it filled with a different gas, diffusion will immediately commence through the pores of the dividing substance, and will continue until perfect mixture has taken place. All gases, however, do not permeate the same porous body, or, in other words, do not pass through narrow orifices, with the same degree of facility. Graham, to whom we are indebted for a very valuable investigation of this interesting subject, established the existence of a very simple relation between the rapidity of diffusion and the density of the gas, which is expressed by saying that the diffusive power varies inversely as the square root of the density of the gas itself. Thus, in the experiment supposed, if one-half of the vessel be filled with hydrogen and the other half with oxygen, the two gases will penetrate the diaphragm at very different rates, four cubic inches of hydrogen passing into the oxygen side, while one cubic inch of oxygen travels in the opposite direction. The densities of the two gases are to each other in the proportion of 1 to 16; their relative rates of diffusion are inversely as the square roots of these numbers, *i. e.*, as 4 to 1.

In order, however, that this law may be accurately observed, it is necessary that the porous plate be very thin; with plates of stucco an inch thick or more, which really consist of a congeries of long capillary tubes, a different law of diffusion is observed.[*] An excellent material for diffusion experiments is the artificially compressed graphite of Mr. Brockedon, of the quality used for making writing pencils. It may be reduced by cutting and grinding to the thickness of a wafer, but still retains considerable tenacity. The pores of this substance appear to be so small as entirely to prevent the transmission of gases in mass, so that, to use the language of Mr. Graham, it acts like a molecular sieve, allowing only molecules to pass through.

The simplest and most striking method of exhibiting the phenomenon of diffusion is by the use of Graham's diffusion-tube (fig. 86). This is merely a piece of wide glass tube ten or twelve inches long, having one of its extremities closed by a plate of plaster of Paris about half an inch thick, and well dried. When the tube is filled by displacement with hydrogen, and

[*] See Bunsen's Gasometry, p. 203; Graham's Elements of Chemistry, 2d ed. ii. 624; Watts's Dictionary of Chemistry, ii. 815.

then set upright in a glass of water, the level of the liquid rises in the tube so rapidly that its movement is apparent to the eye, and speedily attains a height of several inches above the water in the glass. The gas is actually rarefied by its superior diffusive power over that of the external air.

Fig. 86.

It is impossible to over-estimate the importance in the economy of Nature of this law affecting the constitution of gaseous bodies; it is the principal means by which the atmosphere is preserved in a uniform state, and the accumulation of poisonous gases and exhalations in towns and other confined localities prevented.

A partial separation of gases and vapors of unequal diffusibility may be effected by allowing the mixture to permeate through a plate of graphite or porous earthenware into a vacuum. This effect, called a t m o l y s i s, is best exhibited by means of an instrument called the *tube atmolyzer*. This is simply a narrow tube of unglazed earthenware, such as a tobacco-pipe stem, two feet long, which is placed within a shorter tube of glass, and secured in its position by corks. The glass tube is connected with an air-pump, and the annular space between the two tubes is made as nearly vacuous as possible. Air or other mixed gas is then allowed to flow along the clay tube in a slow stream, and collected as it issues. The gas or air atmolyzed is, of course, reduced in volume, much gas penetrating through the pores of the clay tube into the air-pump vacuum, and the lighter gas diffusing the more rapidly, so that the proportion of the denser constituent is increased in the gas collected. In one experiment the proportion of oxygen in the air, after traversing the atmolyzer, was increased from 20.8 per cent., which is the normal proportion, to 24.5 per cent. With a mixture of oxygen and hydrogen the separation is, of course, still more considerable.

A distinction must be carefully drawn between real diffusion through small apertures, and the apparently similar passage of gases through membranous diaphragms, such as caoutchouc, bladder, gold-beater's skin, etc. In this mode of passage, which is called o s m o s e, the rate of interchange depends partly on the relative diffusibilities of the gases; partly on the different degrees of adhesion exerted by the membrane on the different gases, by virtue of which the gas which adheres most powerfully penetrates the diaphragm most easily, and, attaining the opposite surface, mixes with the other; and partly on an actual liquefaction of the gases, which then penetrate the membrane (as ether and petroleum do), and may again evaporate on the surface and appear as gases. A sheet of caoutchouc tied over the mouth of a wide-mouthed bottle filled with hydrogen, is soon pressed inwards, even to bursting. If the bottle be filled with air, and placed in an atmosphere of hydrogen, the swelling and bursting take place outwards. If the membrane is moist, the result is likewise affected by the different solubilities of the gases in the water or other liquid which wets it. For example, the diffusive power of carbonic acid into atmospheric air is very small, but it passes into the latter through a wet bladder with the utmost ease, in virtue of its solubility in the water with which the membrane is moistened. It is by such a process that the function of respiration is performed: the aëration of the blood in the lungs, and the disengagement of the carbonic acid are effected through wet membranes; the blood is never brought into actual contact with the air, but receives its supply of oxygen, and disembarrasses itself of carbonic acid, by this kind of spurious diffusion. The high diffusive power of hydrogen against air renders it impossible to retain that gas for any length of time in a bladder or caoutchouc bag: it is even unsafe to keep it long in a gas-holder, lest it should become mixed with air by slight accidental leakage, and rendered explosive.

The unequal absorption of gases in the manner above mentioned often effects a much more complete separation of the components of a gaseous mixture than can be attained by the atmolytic method above described. Thus, Graham has shown that oxygen is absorbed and condensed by caoutchouc two and a half times more abundantly than nitrogen, and that when one side of a caoutchouc film is freely exposed to the air, while a vacuum is produced on the other side, the film allows 41.6 per cent. of oxygen to pass through, instead of the 21 per cent. usually present in the air, so that the air which passes through is capable of rekindling wood glowing without flame.

Even metals appear to possess this power of absorbing and liquefying gases. Deville and Troost have observed the remarkable fact that hydrogen gas is capable of penetrating platinum and iron tubes at a red heat, and Graham suggested that this effect may be connected with a power resident in these and certain other metals to absorb and liquefy hydrogen, possibly in its character as a metallic vapor. Platinum in the form of wire or plate, at a low red heat, can take up 3.8 volumes of hydrogen measured cold, and palladium foil condenses as much as 643 times its volume of hydrogen at a temperature below 100° C. In the form of sponge, platinum was found to absorb 1.48 times its volume of hydrogen, and palladium 90 volumes. This absorption of gases by metals is called occlusion.[*]

The meteoric iron of Lenarto contains a considerable quantity of occluded hydrogen. When placed in a good vacuum, it yields 2.85 times its volume of gas, of which 85.68 per cent. consists of hydrogen, with 4.46 carbon monoxide and 9.86 nitrogen. Now, hydrogen has been recognized by spectrum analysis in the light of the fixed stars, and constitutes, according to the observations of Father Secchi, the principal element in the atmosphere of a numerous class of stars. "The iron of Lenarto," says Mr. Graham, "has, no doubt, come from such an atmosphere, in which hydrogen greatly prevailed. This meteorite may therefore be looked upon as holding imprisoned within it, and bringing to us, the hydrogen of the stars."[†]

The rates of effusion of gases, that is to say, their rates of passage through a minute aperture in a thin plate of metal or other substance into a vacuum, follow the same law as their rates of diffusion, that is to say, they are inversely as the square roots of the densities of the gases. Nevertheless, the phenomena of diffusion and effusion are essentially different in their nature, the effusive movement affecting masses of a gas, whereas the diffusive movement affects only molecules; and a gas is usually carried by the former kind of impulse with a velocity many thousand times greater than by the latter. Mixed gases are effused at the same rates as one gas of the actual density of the mixture; and no separation of the gases occurs, as in *diffusion* into a vacuum.

The law of effusion just stated is true only under the condition that the gas shall pass through a minute aperture in a very thin plate. If the plate be thicker, so that the aperture becomes a tube, very different rates of efflux are observed; and when the capillary tube becomes considerably elongated, so that its length exceeds its diameter at least 400 times, the rates of flow of different gases into a vacuum assume a constant ratio to each other, following, however, a law totally distinct from that of effusion. The chief general results observed with relation to this phenomenon of "Capillary Transpiration" are as follows:—

1. The rate of transpiration of the same gas increases, *cæteris paribus*,

* G r a h a m, Phil. Trans., 1866; Journal of the Chemical Society [2], v. 235.
† Proceedings of the Royal Society, xv. 502.

directly as the pressure: in other words, equal volumes of gas at different densities require times inversely proportional to their densities. 2. With tubes of equal diameter, the volume transpired in equal times is inversely as the length of the tube. 3. As the temperature rises, the transpiration of equal volumes becomes slower. 4. The rates of transpiration of different gases bear a constant relation to each other, totally independent of their densities, or, indeed, of any known property of the gases. Equal *weights* of oxygen, nitrogen, and carbon monoxide are transpired in equal times; so likewise are equal weights of nitrogen, nitrogen dioxide, and carbon monoxide, and of hydrogen chloride, carbon dioxide, and nitrogen monoxide.*

CHLORINE.

Symbol, Cl.   Atomic weight, 35.37.   Density, 35.37.

THIS substance is a member of a very important natural group, containing also iodine, bromine, and fluorine.   So great a degree of resemblance exists between these bodies in all their chemical relations, especially between chlorine, bromine, and iodine, that the history of one will almost serve, with a few alterations, for that of the rest.   On account of the occurrence of chlorine, bromine, and iodine in sea-water, the elements of this group are called h a l o g e n   e l e m e n t s, and their metallic compounds, h a l o ï d   c o m p o u n d s.†

Fig. 87.

Chlorine is a very abundant substance: in common salt it exists in combination with sodium.   It is most easily prepared by pouring strong hydrochloric acid upon finely powdered black oxide of manganese (MnO_2) contained in a retort or flask (fig. 87), and applying a gentle heat; the chlorine then passes over in the form of a heavy yellow gas, which may be collected over warm water or by displacement: the mercurial trough cannot be employed, as chlorine rapidly acts upon the metal, and becomes absorbed.

The reaction consists in an interchange between the two atoms of oxygen of the manganese dioxide and 4 atoms of chlorine from the hydrochloric acid, the oxygen uniting with the hydrogen to form water, while, of the chlorine, one-half unites with the manganese, forming a chloride, $MnCl_2$, and the other half is given off as gas,

$$MnO_2 + 4HCl = 2H_2O + MnCl_2 + Cl_2.$$

The same process is used for the preparation of chlorine on the manufacturing scale, the hydrochloric acid which is evolved in large quantities by heating common salt with sulphuric acid, in Leblanc's soda-process, being utilized for the purpose.

The waste-liquor obtained in the chlorine manufacture consists of an im-

* G r a h a m, Phil. Trans., 1846, p. 591, and 1849, p. 349; also, Elements of Chemistry. 2d ed., i. 82.
† From ἁλς, the sea.

pure solution of manganous chloride, $MnCl_2$, from which the manganese may be separated by an alkali in the form of manganous oxide, MnO. But to render the manganese thus precipitated again available for the production of chlorine, it must first be brought to the state of dioxide, and this may be effected by mixing the manganese liquor with an excess of hot milk of lime or magnesia, and blowing hot air through the mixture. By this means, a compound of manganese dioxide with lime or magnesia, e. g., CaO,MnO_2, or $CaMnO_3$, called calcium or magnesium manganite, is formed, which, when heated with hydrochloric acid, gives off chlorine in the manner above described. This is Weldon's process for the regeneration of manganese, which is now largely used both in England and on the Continent.

A process for the separation of chlorine from hydrochloric acid, without the use of any manganese compound, was introduced by the late Mr. H. Deacon. It consists in passing a mixture of hydrochloric acid gas and oxygen, or air, over cupric sulphate or other cupric salt heated to 370-400° C. (698-752° F.), the hydrochloric acid being then decomposed, its hydrogen combining with the oxygen, and the chlorine being set free. The best way of conducting the process is to pass the mixed gases over pieces of brick soaked in solution of cupric sulphate and dried. The action of the copper salt is not well understood, but appears to belong to that class of phenomena called catalytic or contact actions. Other metallic salts act in a similar way, but less completely.

Chlorine is a yellow gaseous body, of intolerably suffocating properties, producing very violent cough and irritation when inhaled, even in exceedingly small quantity. It is soluble to a considerable extent in water, that liquid absorbing at 15.5° about twice its volume, and acquiring the color and odor of the gas. When this solution is exposed to light, it is slowly changed, by decomposition of water, into hydrochloric acid, the oxygen being at the same time liberated. When moist chlorine gas is exposed to a cold of 0°, yellow crystals are formed, which consist of a definite compound of chlorine and water, containing 35.5 parts of the former to 90 of the latter.

Chlorine has a specific gravity of 2.45; a litre of it weighs 3.167 grams; exposed to a pressure of about four atmospheres, it condenses to a yellow limpid liquid.

Chlorine has but little attraction for oxygen, its energies being principally exerted towards hydrogen and the metals. A lighted taper plunged into the gas continues to burn with a dull-red light, and emits a large quantity of smoke, the hydrogen of the wax being alone consumed, and the carbon separated. If a piece of paper be wetted with oil of turpentine, and thrust into a bottle filled with chlorine, the chemical action of the latter upon the hydrogen is so violent as to cause inflammation accompanied by a copious deposit of soot. Chlorine may, by indirect means, be made to combine with carbon; but this combination never occurs under the circumstances just described.

Phosphorus takes fire spontaneously in chlorine, burning with a pale and feebly luminous flame. Several of the metals, as copper-leaf, powdered antimony, and arsenic, undergo combustion in the same manner. A mixture of equal measures of chlorine and hydrogen explodes with violence on the passage of an electric spark or on the application of a lighted taper, hydrochloric acid gas being formed. Such a mixture may be kept in the dark for any length of time without change, but on exposing it to diffused daylight, the two gases slowly unite, while the direct rays of the sun induce instantaneous explosion.

The most characteristic property of chlorine is its bleaching power; the most stable organic coloring principles are instantly decomposed and destroyed by it; indigo, for example, which resists the action of strong oil of

12 *

vitriol, is converted by chlorine into a brownish substance, to which the blue color cannot be restored. The presence of water is essential to these changes, for the gas in a state of perfect dryness is incapable even of affecting litmus.

Chlorine is largely used in the arts for bleaching linen and cotton goods, rags for the manufacture of paper, etc. For these purposes it is employed, sometimes in the state of gas, sometimes in that of aqueous solution, but more frequently in combination with lime, forming the substance called bleaching-powder. It is also one of the best and most potent substances that can be used for the purpose of disinfection, but its employment requires care. Bleaching-powder mixed with water, and exposed in shallow vessels, is slowly decomposed by the carbonic acid of the air, and the chlorine is evolved: if a more rapid disengagement be wished, a little acid may be added. In the absence of bleaching powder, either of the methods for the production of the gas described may be had recourse to, always taking care to avoid an excess of acid.

Hydrogen Chloride, or Hydrochloric Acid, HCl; also called *Chlorhydric* and *Muriatic Acid.*—This substance, in the state of solution in water, has long been known. The gas is easily prepared by heating, in a flask fitted with a cork and bent tube, a mixture of common salt and oil of vitriol diluted with a small quantity of water; it must be collected by displacement or over mercury. It is a colorless gas, which fumes strongly in the air from condensing the atmospheric moisture; it has an acid, suffocating odor, but is much less offensive than chlorine. Exposed to a pressure of 40 atmospheres, it liquefies.

Hydrochloric acid gas has, according to experiment, a density of 1.278 compared with air, or 18.44 compared with hydrogen as unity, the calculated number being $\dfrac{35.37 + 1}{2} = 18.185$. It is exceedingly soluble in water, that liquid taking up, at the temperature of the air, about 418 times its bulk. The gas and solution are powerfully acid.

The action of sulphuric acid on common salt, or any analogous substance, is explained by the equation,

$$2NaCl + H_2SO_4 = Na_2SO_4 + 2HCl.$$

The composition of hydrochloric acid may be determined by synthesis: when a measure of chlorine and a measure of hydrogen are fired by the electric spark, two measures of hydrochloric acid gas result, the combination being unattended by change of volume. By weight the gas contains 35.37 parts of chlorine and 1 part of hydrogen.

Solution of hydrochloric acid, the liquid acid of commerce, is a very important preparation, and of extensive use in chemical pursuits: it is best prepared by the following arrangement:—

A large glass flask, containing a quantity of common salt, is fitted with a cork and bent tube, in the manner represented in fig. 88: this tube passes through and below a second tube into a wide-necked bottle containing a little water, into which the open tube dips. A bent tube is adapted to another hole in the cork of the wash-bottle, so as to convey the purified gas into a quantity of distilled water, by which it is instantly absorbed: the joints are made air-tight by melting a little yellow wax over the corks. A quantity of sulphuric acid, about equal in weight to the salt, is then slowly introduced by the funnel; the disengaged gas is at first wholly absorbed by the water in the wash-bottle, but when this becomes saturated it passes into the second vessel, and there dissolves. When all the acid has been added, heat may be applied to the flask by a charcoal chauffer, until its contents appear nearly dry, and the evolution of gas almost ceases, when the process may be stopped.   As much heat is given out during the condensation of the gas, it is necessary to surround the condensing vessel with cold water.

The simple wash-bottle, shown in fig. 88, will be found an exceedingly

Fig. 88.

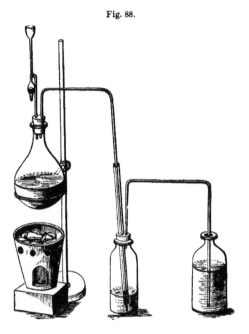

useful contrivance in a great number of chemical operations.   It serves in the present and in many similar cases to retain any liquid or solid matter mechanically carried over with the gas, and may be always employed when a gas of any kind is to be passed through an alkaline or other solution.   The open tube dipping into the liquid prevents the possibility of absorption, by which a partial vacuum would be occasioned, and the liquid of the second vessel lost by being driven into the first.   At present, however, bottles with two necks are more generally employed for this and similar purposes.

Fig. 89.

The arrangement by which the acid is introduced also deserves notice (fig. 89).   The tube is bent twice upon itself, and a bulb blown in one portion : the liquid poured into the funnel rises upon the opposite side of the first bend until it reaches the second; it then flows over and runs into the flask.   Any quantity can then be got into the latter without the introduction of air, and without the escape of gas from the interior.   The funnel acts also as a kind of safety-valve, and in both directions; for if by any chance the delivery-tube should be stopped, and the issue of gas prevented, its increased elastic force soon drives the little column of liquid out of the tube, the gas escapes, and the vessel is saved.   On the other hand, any absorption within is quickly compensated by the entrance of air through the liquid in the bulb.

Pure solution of hydrochloric acid is transparent and colorless :

,when strong it fumes in the air by evolving a little gas. It leaves no residue on evaporation, and gives no precipitate or opacity with diluted solution of barium chloride. When saturated with the gas, it has a specific gravity of 1.21, and contains about 42 per cent. of real acid. The commercial acid, which is obtained in immense quantity as a secondary product in the manufacture of sodium sulphate by the action of sulphuric acid upon common salt, has usually a yellow color, and is very impure, containing salts, sulphuric acid, chloride of iron, and organic matter. It may be rendered sufficiently pure for most purposes by diluting it to the density of 1.1, which happens when the strong acid is mixed with its own bulk or rather less of water, and then distilling it in a retort furnished with a Liebig's condenser.

On distilling an aqueous solution of hydrochloric acid, an acid is produced boiling at 110° C. (230° F.) which contains 20.22 per cent. of anhydrous hydrochloric acid: a more concentrated solution when heated gives off hydrochloric acid gas; a weaker solution loses water. Roscoe and Ditmar have proved that the composition of the distillate varies with the atmospheric pressure; it cannot, therefore, be viewed as a chemical compound.

A crystalline hydrate of hydrochloric acid, having the composition $HCl,2H_2O$, is formed by passing a stream of nearly dry hydrochloric acid gas through the concentrated aqueous acid cooled by a freezing mixture to — 22°. The crystals decompose rapidly in the air, emitting white fumes; they dissolve very quickly in water at ordinary temperatures; very slowly at — 18°.

Mixtures of snow and hydrochloric acid form very powerful and economical refrigerants. With two parts of snow and one part of the acid a lowering of temperature to — 32° is readily obtained.*

The presence of hydrochloric acid, or any other soluble chloride, is easily detected by solution of silver nitrate. A white curdy precipitate is formed, insoluble in nitric acid, freely soluble in ammonia, and blackening when exposed to light.

BROMINE.

Symbol, Br. Atomic weight, 79.75. Vapor-density, 79.75.

BROMINE† was discovered by Balard in 1826. It is found in sea-water, and is a frequent constituent of saline springs, chiefly as magnesium bromide: a celebrated spring of the kind exists near Kreuznach, in Prussia. Bromine may be obtained pure by the following process, which depends upon the fact that ether, agitated with an aqueous solution of bromine, removes the greater part of that substance.

The mother-liquor, from which the less soluble salts have separated by crystallization, is exposed to a stream of chlorine, and then shaken up with ether; the chlorine decomposes the magnesium bromide, and the ether dissolves the bromine thus set free. On standing, the ethereal solution, having a fine red color, separates, and may be removed by a funnel or pipette. Caustic potash is then added in excess, and heat applied, whereby bromide and bromate of potassium are formed. The solution is evaporated to dryness, and the saline matter, after ignition to redness to decompose the bro-

* Pierre and Puchot, Comptes rendus, lxxxii. 45.
† From βρῶμος, a noisome smell.

mate, is heated in a small retort with manganese dioxide and sulphuric acid diluted with a little water, the neck of the retort being plunged into cold water. The bromine volatilizes in the form of a deep red vapor, which condenses into drops beneath the liquid.

Bromine is at common temperatures a thin deep red very volatile liquid, which freezes at about $-24.5°$ C. ($-12°$ F.) and boils at $63°$ C. ($147.4°$ F.). The density of the liquid is 2.976, and that of the vapor 5.53 compared with air, and 79.75 compared with hydrogen. The odor of bromine is very suffocating and offensive, much resembling that of iodine, but more disagreeable. It is slightly soluble in water, more freely in alcohol, and most abundantly in ether. The aqueous solution bleaches.

Hydrogen Bromide, or Hydrobromic Acid, HBr.—This compound is obtained as a gas, together with phosphoric acid, by the action of water on bromine and phosphorus: $P + 5Br + 4H_2O = 5HBr + H_3PO_4$. It is colorless, pungent, has an acid taste and reaction, liquefies at $-73°$ C. ($-99°$ F.), and solidifies at $-87°$ C. ($-125°$ F.) to a colorless icy mass. The aqueous acid may be prepared by passing the gas into water, or directly by passing hydrogen sulphide through bromine-water: $SH_2 + Br_2 = 2BrH + S$. It is colorless, has a pungent odor, an acid taste and reaction, and when saturated at $0°$, a specific gravity of 1.78. By distillation the weak aqueous acid becomes stronger and the strong acid weaker, till an acid containing from 47.38 to 49.86 per cent. HBr passes over, under a pressure of 752 to 762 millimetres of mercury.

Chlorine and Bromine.—When chlorine gas is passed into liquid bromine, the two elements combine, forming a reddish-yellow, mobile, volatile liquid, which dissolves with yellow color in water, the solution when cooled below $0°$, depositing a crystalline hydrate which melts at $+7°$ C. ($45°$ F.).

IODINE.

Symbol, I.   Atomic weight, 126.53.   Vapor-density, 126.53.

THIS element was first noticed in 1812 by M. Courtois, of Paris. Minute traces are found in combination with sodium or potassium in sea-water, and occasionally a much larger proportion in that of certain mineral springs. It seems to be in some way beneficial to many marine plants, as these latter have the power of abstracting it from the surrounding water, and accumulating it in their tissues. It is from this source that all the iodine of commerce is derived. It has lately been found in minute quantity in some aluminous slates of Sweden, and in several varieties of coal and turf.

Kelp, or the half-vitrefied ashes of sea-weeds, prepared by the inhabitants of the Western Islands and the northern shores of Scotland and Ireland, is treated with water, and the solution filtered. The liquid is then concentrated by evaporation until it is reduced to a very small volume, the sodium chloride, sodium carbonate, potassium chloride, and other salts being removed as they successively crystallize. The dark-brown mother-liquor left contains very nearly the whole of the iodine, as iodide of sodium, magnesium, etc.: this is mixed with sulphuric acid and manganese dioxide, and gently heated in a leaden retort, when the iodine distils over and condenses in the receiver. The theory of the operation is exactly analogous to that of the preparation of chlorine; in practice, however, it requires careful management, otherwise the impurities present in the solution interfere with the general result:

$$MnO_2 + 2KI + 2H_2SO_4 = 2H_2O + K_2SO_4 + MnSO_4 + I_2.$$

The manganese is not absolutely necessary: potassium or sodium iodide, heated with an excess of sulphuric acid, evolves iodine. This effect is due to a secondary action between the hydriodic acid first produced and the excess of the sulphuric acid, in which both suffer decomposition, yielding iodine, water, and sulphurous acid.

Iodine crystallizes in plates or scales of a bluish-black color and imperfect metallic lustre, resembling that of plumbago; the crystals are sometimes very large and brilliant. Its density is 4.948. It melts between 113° and 115° C. (215° and 218° F.), solidifies at 113.6° C. (236.5° F.), and boils above 200° C. (360° F.), the vapor having an exceedingly beautiful violet color.* It is slowly volatile also at common temperatures, and exhales an odor much resembling that of chlorine. The density of the vapor is, by experiment, 8.716 compared with air, or 125.8 compared with hydrogen, the theoretical density being 126.53. Iodine requires for solution about 7000 parts of water, to which nevertheless it imparts a brown color; in alcohol it is much more freely soluble. Solutions of hydriodic acid and the iodides of the alkali-metals also dissolve it in large quantity: these solutions are not decomposed by water, which is the case with the alcoholic tincture.

Iodine stains the skin, but not permanently; it has a very energetic action upon the animal system, and is much used in medicine.

One of the most characteristic properties of iodine is the production of a splendid blue color by contact with starch. The iodine for this purpose must be free or uncombined. It is easy, however, to make the test available for the purpose of recognizing the presence of the element in question when a soluble iodide is suspected; it is only necessary to add a very small quantity of chlorine-water, when the iodine, being displaced from combination, becomes capable of acting upon the starch.

Hydrogen Iodide, or Hydriodic Acid, HI.—The simplest process for preparing hydriodic acid gas is to introduce into a test-tube a little iodine, then a small quantity of roughly-powdered glass moistened with water, upon this a few fragments of phosphorus, and lastly more glass, this order of iodine, glass, phosphorus, glass, being repeated until the tube is half or two-thirds filled. A cork and narrow bent tube are then fitted, and gentle heat is applied. The gas is best collected by displacement of air. The process depends on the formation of an iodide of phosphorus and its subsequent decomposition by water, whereby hydrogen phosphite, or phosphorous acid, H_3PO_3, and hydrogen iodide are produced: $P_2 + I_6 + 6H_2O = 6HI + 2H_3PO_3$. The glass merely serves to moderate the violence of the action, and may be dispensed with if amorphous phosphorus be used.

Fig. 90.

Hydriodic acid gas greatly resembles the corresponding chlorine compound; it is colorless and highly acid, fumes in the air, and is very soluble

* Whence the name, from ἰώδης, violet-colored.

in water. Its density is 4.3737 compared with air, or 63.13 compared with hydrogen; theoretic density 63.765. By weight, it is composed of 126.53 parts iodine and 1 part hydrogen; and by measure of equal volumes of iodine vapor and hydrogen united without condensation.

Solution of hydriodic acid may be prepared by a process much less troublesome than the above. Iodine in fine powder is suspended in water, and a stream of washed hydrogen sulphide is passed through the mixture; sulphur is then deposited, and the iodine is converted into hydriodic acid. When the liquid has become colorless, it is heated, to expel the excess of hydrogen sulphide, and filtered. The solution cannot be kept long, especially if it be concentrated; the oxygen of the air gradually decomposes the hydriodic acid, and iodine is set free, which, dissolving in the remainder, colors it brown.

Iodine and Chlorine.—These bodies unite directly, forming a monochloride and a trichloride. The *monochloride*, ICl, is obtained by passing dry chlorine gas over iodine till the whole is liquefied, but no longer; also by distilling iodine with potassium chlorate, oxygen being then evolved, the monochloride distilling over, and a mixture of chloride, iodate, and perchlorate of potassium remaining behind.

$$I_2 + 3KClO_3 = O_2 + ICl + KCl + KIO_3 + KClO_4.$$

Iodine monochloride is a reddish-brown oily liquid, having a suffocating odor and astringent taste; soluble in alcohol and ether; decomposed by water, with formation of hydrochloric and iodic acids, and separation of iodine. It sometimes solidifies in yellow needles. Sulphurous acid and hydrogen sulphide decompose it, with separation of iodine; with aqueous solutions of alkalies it yields a chloride and an iodate, together with free iodine; thus, with potash—

$$5ICl + 6KHO = 5KCl + KIO_3 + 3H_2O + I_4.$$

The *trichloride*, ICl_3, is produced by treating iodine at a gentle heat with chlorine in excess. It crystallizes in orange-yellow needles; melts at 20° to 25° C. (68° to 77° F.), giving off chlorine, which it reabsorbs on cooling; acts on most other substances like the monochloride.

In contact with a small quantity of water it is partly resolved into an insoluble yellowish body (probably a mixture of the trichloride with iodic oxide), and a solution containing hydrochloric acid and the monochloride—

$$4ICl_3 + 5H_2O = 10HCl + I_2O_5 + 2ICl.$$

A large quantity of water dissolves it, probably without decomposition, or perhaps as a mixture of hydrochloric and iodic acids containing free iodine—

$$5ICl_3 + 9H_2O = 15HCl + 3HIO_3 + I_2.$$

The *tetrachloride*, ICl_4, produced by spontaneous decomposition of the monochloride, $4ICl = ICl_4 + I_3$, crystallizes in red octahedrons.

Iodine and Bromine.—Iodine unites with bromine, forming a volatile crystalline compound, probably IBr, and a dark liquid, possibly IBr_3.

FLUORINE.

Symbol, F.   Atomic weight, 19.

THIS element exists in considerable quantity, in combination with calcium, forming the mineral called fluorspar, which crystallizes in fine
cubes of various colors.   Fluorine occurs also in small quantity as a constituent of bones and other animal substances.   Its intense affinities for
metals and for silicon, which is a constituent of glass, have hitherto baffled
all attempts to isolate it in a state fit for examination, but Dr. Brauner,* by
heating cerium tetrafluoride, has obtained a lower fluoride of cerium, together
with a gas smelling like chlorine or hypochlorous oxide.   The reaction is
probably $CeF_4 = CeF_3 + F$.   Anhydrous tetrafluoride of lead is also decomposed by heat, yielding the difluoride, and a gas smelling like chlorine.

Hydrogen Fluoride, or Hydrofluoric Acid, HF.—When powdered
calcium fluoride is heated with concentrated sulphuric acid in a retort of
platinum or lead connected with a carefully cooled receiver of the same
metal, a very volatile colorless liquid is obtained, which emits copious
white and highly suffocating fumes in the air.   This is the anhydrous acid,
not however quite pure.   It may be obtained in a state of perfect purity by
distilling hydrogen-potassium fluoride, HF,KF, to redness in a platinum
vessel.   As thus prepared, it is at ordinary temperatures a colorless, transparent, mobile liquid, having a specific gravity of 0.9879 at 12.8° C. (55° F.),
extremely volatile, boiling at 19.4° C. (67° F.), fuming densely at ordinary
temperatures, and absorbing water greedily from the air.
Gore, by measuring the volume of hydrogen required to combine with
the fluorine contained in a known weight of fluorspar, found that 1 volume
of hydrogen yields 2 volumes of gaseous hydrogen fluoride containing 19
parts by weight of fluorine to 1 part of hydrogen.   The density of the gaseous acid is 10 times that of hydrogen.
When hydrofluoric acid is put into water, it unites with the latter with
great violence; the dilute solution attacks glass with great facility.   The
concentrated acid, dropped upon the skin, occasions deep malignant ulcers,
so that great care is requisite in its management.
In a diluted state, this acid is occasionally used in the analysis of siliceous
minerals when alkali is to be estimated ; it is employed also for etching on
glass, for which purpose the acid may be prepared in vessels of lead, that
metal being but slowly attacked under these circumsances.   The vapor of
the acid is also very advantageously applied to the same purpose in the
following manner :—The glass to be engraved is coated with etching-ground
or wax, and the design traced in the usual way with a pointed instrument.
A shallow basin, made by beating up a piece of sheet lead, is then prepared,
a little powdered fluorspar placed in it, and enough sulphuric acid added to
form with the latter a thin paste.   The glass is placed upon the basin, with
the waxed side downwards, and gentle heat applied beneath, which speedily
disengages the vapor of hydrofluoric acid.   In a very few minutes the operation is complete; the glass is then removed, and cleaned by a little warm
oil of turpentine.   When the experiment is successful, the lines are very
clean and smooth.

* Journal of the Chemical Society, xlii. 8.

OXYGEN.

Symbol O.   Atomic weight, 15.96.   Density, 15.96.

OXYGEN was discovered in the year 1774, by Scheele, in Sweden, and Priestley, in England, independently of each other, and described under the terms *empyreal air* and *dephlogisticated air*.   The name oxygen* was given to it by Lavoisier some time afterwards.   Oxygen exists in a free and uncombined state in the atmosphere, mixed with another gaseous body, nitrogen.   No very good direct means exist, however, for separating it from the latter; and, accordingly, it is always obtained for purposes of experiment by decomposing certain metallic oxides.

The red oxide of mercury, or *red precipitate* of the old writers, may be employed for this purpose.   In this substance the attraction which holds together the mercury and the oxygen is so feeble that simple exposure to heat suffices to bring about decomposition.

The mode of conducting the decomposition and collecting the evolved gas has already been noticed (p. 46).   The experiment there described is, however, instructive rather as an instance of the resolution by simple means of a compound body into its constituents, than as a means of obtaining oxygen gas.   A better and more economical method is to expose to heat in a retort, or flask furnished with a bent tube, a portion of the salt called potassium chlorate.   A common Florence flask serves perfectly well, the heat of the spirit-lamp being sufficient.   The salt melts and decomposes with ebullition, yielding a very large quantity of oxygen gas, which may be collected over water.   The first portion of the gas often contains a little chlorine.   The white saline residue in the flask is potassium chloride.   This plan, which is very easy of execution, is always adopted when very pure gas is required for analytical purposes.

A third method, very good when perfect purity is not demanded, is to heat to redness, in an iron retort or gun-barrel, the black manganese oxide of commerce, which under these circumstances suffers decomposition, giving off part of its oxygen and leaving a lower oxide.

If a little of the black manganese oxide be finely powdered and mixed with potassium chlorate, and this mixture heated in a flask or retort by a lamp, oxygen will be disengaged with the utmost facility, and at a far lower temperature than when the chlorate alone is used.   All the oxygen comes from the chlorate, the manganese remaining quite unaltered.   The materials should be well dried in a capsule before their introduction into the flask. This experiment affords an instance of an effect by no means rare, in which a body seems to act by its mere presence, without taking any obvious part in the change brought about.

Methods for the preparation of oxygen on a large scale will be noticed under the heads of sulphuric acid and barium dioxide.

Oxygen is, bulk for bulk, a little heavier than atmospheric air, its specific gravity being 1.10563, referred to that of air as unity, and 15.96, referred to that of hydrogen as unity.   A litre of oxygen at the standard temperature and pressure, that is to say, at 0° C., and 760 millimetres barometric pressure, weighs 1.43028 gram.   At 15.5° C. (60° F.), and under a pressure of 30 inches, 100 cubic inches of the gas weigh 34.29 grains.

Combustion.—Oxygen is the sustaining principle of animal life and of all the ordinary phenomena of combustion.   Bodies which burn in the air burn with greatly increased splendor in oxygen gas.   If a taper be blown out,

---

* From ὀξύ, acid, and γεν, a root signifying production.

and then introduced while the wick remains red hot, it is instantly rekin
died : a slip of wood or a match is relighted in the same manner. This
effect is highly characteristic of oxygen, there being but one other gas which
possesses the same property ; and this is easily distinguished by other means.
The experiment with the match is also constantly used as a rude test of the
purity of the gas when it is about to be collected from the retort, or when it
has stood for some time in contact with water exposed to air.

When a bit of charcoal is affixed to a wire, and plunged with a single
point red hot into a jar of oxygen, it burns with great brilliancy, throwing
off beautiful scintillations, until, if the oxygen be in excess, it is completely
consumed. An iron wire, or, still better, a steel watch-spring, armed at its
extremity with a bit of lighted amadou, and introduced into a vessel of
oxygen gas, burns with great brilliancy. If the experiment be made in
a jar standing on a plate, the fused globules of black iron oxide fix themselves in the glaze of the latter, after falling through a stratum of water
half an inch in depth. Kindled sulphur burns with great beauty in
oxygen ; and phosphorus, under similar circumstances, exhibits a splendor
which the eye is scarcely able to support. In each case the burning body
enters into combination with the oxygen, forming a compound called an
oxide.

When a compound burns in atmospheric air, the same ultimate effect is
produced as in pure oxygen ; the action is, however, less energetic, in consequence of the presence of nitrogen, which dilutes the oxygen and weakens
its chemical powers. The process of respiration in animals is an effect of
the same nature as ordinary combustion. The blood contains substances
which slowly burn by the aid of the oxygen thus introduced into the system.
When this action ceases life becomes extinct.

It is necessary to bear in mind that combustion is not the same thing as
ignition. When any solid substance capable of bearing the fire is heated to
a certain point, it emits light, the character of which depends upon the
temperature. Thus a bar of platinum or a piece of porcelain, raised to a
particular temperature, becomes what is called red hot, or emissive of red
light : at a higher degree of heat, this light becomes whiter and more
intense, and when urged to the utmost, as in the case of a piece of lime
placed in the flame of the oxyhydrogen blowpipe, the light becomes exceedingly powerful, and acquires a tint of violet. Bodies in these states are said
to be *incandescent* or *ignited.*

If now the same experiment be made upon a piece of charcoal, similar
effects will be observed ; but something in addition ; for whereas the platinum and porcelain, when removed from the fire, or the lime from the blowpipe flame, immediately begin to cool, and emit less and less light until
they become quite dark, the charcoal maintains to a great extent its high
temperature. Unlike the other bodies, too, which suffer no change whatever, either of weight or substance, the charcoal gradually wastes away until
it disappears. This is what is called *combustion*, in contradistinction to mere
ignition ; the charcoal burns, and its temperature is kept up by the heat
evolved in the act of union with the oxygen of the air.

In the most general sense, a body in a state of combustion is one in the
act of undergoing intense chemical action : any chemical action whatsoever,
if its energy rise sufficiently high, may produce the phenomenon of combustion, by *heating the body to such an extent that it becomes luminous.*

In all ordinary cases of combustion the action lies between the burning
body and the oxygen of the air; and since the materials employed for the
economical production of heat and light consist of carbon chiefly, or that
substance conjoined with a certain proportion of hydrogen and oxygen, all
common effects of this nature are cases of the rapid and violent oxidation
of carbon and hydrogen by the aid of the free oxygen of the air. The

heat is due to the act of chemical union, and the light to the elevated temperature.

By this principle, it is easy to understand the means which must be adopted to increase the heat of ordinary fires to the point necessary to melt refractory metals, and to bring about certain desired effects of chemical decomposition. If the rate of consumption of the fuel can be increased by a more rapid introduction of air into the burning mass, the intensity of the heat will of necessity rise in the same ratio, the quantity of heat evolved being fixed and definite for the same constant quantity of chemical action. This increased supply of air may be effected by two distinct methods : it may be forced into the fire by bellows or blowing-machines, as in the common forge and in the blast and cupola furnaces of the iron-worker; or it may be drawn through the burning materials by the help of a tall chimney, the fireplace being closed on all sides, and no entrance of air allowed save between the bars of the grate. Such is the kind of furnace generally employed by the scientific chemist in assaying and in the reduction of metallic oxides by charcoal: the principle will be at once understood by the aid of the sectional drawing (fig. 91), in which a crucible is represented arranged in the fire for an operation of the kind mentioned.

Fig. 91.   Fig. 92.

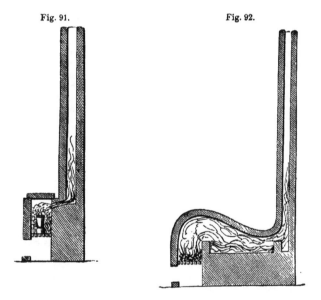

The "reverberatory" furnace (fig. 92) is one very much used in the arts when substances are to be exposed to heat without contact with the fuel. The fire-chamber is separated from the bed or hearth of the furnace by a low wall or *bridge* of brickwork, and the flame and heated air are reflected downwards by the arched form of the roof. Any degree of heat can be obtained in a furnace of this kind—from the temperature of dull redness to that required to melt very large quantities of cast iron. The fire is urged by a chimney provided with a sliding-plate, or damper, to regulate the draught.

Solids and liquids, as melted metal, possess, when sufficiently heated, the faculty of emitting light; and the same power is exhibited by gaseous bodies, but the temperature required to render a gas luminous is incomparably higher than in the cases already described. Gas or vapor in this condition constitutes *flame*, the actual temperature of which generally exceeds that of the white heat of solid bodies.

The light emitted from pure flame is often very faint, but the illuminating power may be immensely increased by the presence of solid matter. The flame of hydrogen, or of a mixture of hydrogen and oxygen, is scarcely visible in full daylight; in a dusty atmosphere, however, it becomes much more luminous by igniting to intense whiteness the floating particles with which it comes in contact. A piece of lime in the blowpipe flame (p. 146) cannot have a higher temperature than that of the flame itself, yet the light it throws off is infinitely greater.

It is possible, however, as pointed out by Dr. Frankland, to produce very bright flames in which no solid particles are present. Metallic arsenic burnt in a stream of oxygen produces an intense white flame, although both the metal itself and the product of its combustion (arsenious oxide) are gaseous at the temperature of the flame. The combustion of a mixture of nitrogen dioxide and carbon bisulphide also produces a dazzling white flame, without any separation of solid matter.

The conditions most essential to luminosity in a flame are a high temperature and the presence of gases or vapors of considerable density. The effect of high temperature is seen in the greater brightness of the flame of sulphur, phosphorus, and indeed all substances, when burnt in pure oxygen, as compared with that which results from their combustion in common air; in the former case the whole of the substances present take part in the combustion, and generate heat, whereas in the latter the temperature is lowered by the presence of a large quantity of nitrogen, which contributes nothing to the effect. The relation between the luminosity of a flame and the vapor-densities of its constituents may be seen from the following table, in which the vapor-densities are referred to that of hydrogen as unity.

Relative Densities of Gases and Vapors.

Hydrogen 1	Arsenious chloride . . . 90¾
Water 9	Phosphoric oxide . . 71, or 142
Hydrochloric acid . . . 18¼	Metallic arsenic . . . 150
Carbon dioxide . . . 22	Arsenious oxide . . . 198
Sulphur dioxide . . . 32	

A comparison of these numbers shows that the brightest flames are those which contain the densest vapors. Hydrogen burning in chlorine produces a vapor more than twice as heavy as that resulting from its combustion in oxygen, and accordingly the light produced in the former case is stronger than in the latter; carbon and sulphur burning in oxygen produce vapors of still greater density, namely, carbon dioxide and sulphur dioxide, and their combustion gives a still brighter light; lastly, phosphorus, which has a very dense vapor, and likewise yields a product of great vapor-density, burns in oxygen with a brilliancy which the eye can scarcely endure. Moreover the luminosity of a flame is increased by condensing the surrounding gaseous atmosphere, and diminished by rarefying it. The flame of arsenic burning in oxygen may be rendered quite feeble by rarefying the oxygen; and, on the contrary, the faint flame of an ordinary spirit-lamp becomes very bright when placed under the receiver of a condensing-pump. Frankland has also found that candles give much less light when burning on the top of Mont Blanc than in the valley below, although the rate of combustion in the two cases is nearly the same. The effect of condensation in increasing the brightness of a flame is also strikingly seen in the combustion of a mixture of oxygen and hydrogen, which gives but a feeble light when burnt under the ordinary atmospheric pressure, as in the oxyhydrogen blowpipe, but a very bright flash when exploded in the Cavendish eudiometer (p. 156), in which the water-vapor produced by the combustion is prevented from expanding.

Flameless Combustion.—Mr. Fletcher, of Warrington, has lately shown that it is possible to obtain extremely high temperatures by the combustion of hydrocarbons without the production of any visible flame. On directing a gas flame upon a large mass of iron wire till the metal becomes red hot, and then gradually turning on a blast of air till the flame totally disappeared, the temperature was observed to increase as the flame became smaller, until, on its disappearance, the iron fused and ran into drops, the whole mass of metal at the same time becoming intensely heated, and remaining so, although not a trace of flame could be detected in a perfectly dark room. Hence it appears that visible flame is not necessary for the production of high temperatures, but rather that its presence may be regarded as a proof of imperfect combustion.*

Lamps, Candles.—The Blowpipe.—Flames burning in the air, and not supplied with oxygen from another source, are hollow, the chemical action being necessarily confined to the spot where the two bodies unite. That of a lamp or candle, when carefully examined, is seen to con- Fig. 93. sist of three separate portions. The dark central part, easily rendered evident by depressing upon the flame a piece of fine wire gauze, consists of combustible matter drawn up by the capillarity of the wick, and volatilized by the heat. This is surrounded by a highly luminous cone or envelope, which, in contact with a cold body, deposits soot. On the outside, a second cone may be traced, feeble in its light-giving power, but having an exceedingly high temperature. The most probable explanation of these appearances is as follows: Carbon and hydrogen are very unequal in their attraction for oxygen, the latter greatly exceeding the former in this respect: consequently, when both are present, and the supply of oxygen is limited, the hydrogen takes up the greater portion of the oxygen, to the exclusion of a great part of the carbon. Now, this happens, in the case under consideration, at some little distance within the outer surface of the flame—namely, in the luminous portion, the little oxygen which has penetrated thus far inwards being mostly consumed by the hydrogen, and hydrocarbons being separated, rich in carbon and of great density in the state of vapor (naphthalene, chrysene, pyrene, etc.). These hydrocarbons, which would form smoke if they were cooler, and are deposited on a cold body held in the flame in the form of soot,† become intensely ignited by the burning hydrogen, and evolve a light whose whiteness marks a very elevated temperature. In the exterior and scarcely visible cone, these hydrocarbons undergo combustion.

A jet of coal-gas exhibits the same phenomena; but, if the gas be previously mingled with air, or if air be forcibly mixed with, or driven into, the flame, no such separation of carbon occurs; the hydrogen and carbon burn together, forming vapors of much lower density, and the illuminating power almost disappears.

The common mouth blowpipe is an instrument of great utility. It is made of various shapes, the simplest and most generally used consisting of a brass tube bent as in fig. 94, and terminating in a narrow aperture.

The flame so produced is very peculiar. Instead of the double envelope just described, two long pointed cones are observed (fig. 95), which, when the blowpipe is good, and the aperture smooth and round, are very well defined, the outer cone being yellowish, and the inner blue. A double com-

---

* Chemical News, vol. xlvi., pp. 22, 37.
† Soot is not pure carbon, but a mixture of heavy hydrocarbons.

13 *

bustion is, in fact, going on, by the blast in the inside, and by the external
air.   The space between the inner and outer cones is filled with exceedingly
hot combustible matter, possessing strong reducing or deoxidizing
Fig. 94. powers, while the highly heated air just beyond the point of the
exterior cone oxidizes with great facility.

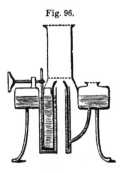

A small portion of matter, supported on a piece of charcoal, or
fixed in a ring at the end of a fine platinum wire, can thus in an
instant be exposed to a very high degree of heat under these contrasted circumstances, and observations of great value made in a very
short time.   The use of the instrument requires an even and unin-
terrupted blast of some duration, by a method easily acquired with
a little patience: it consists in employing
for the purpose the muscles of the cheeks
alone, respiration being conducted through
the nostrils, and the mouth from time to
time replenished with air, without inter-
mission of the blast.

Fig. 95.

The Argand lamp, adapted to burn either
oil or spirit, but especially the latter, is a
very useful piece of chemical apparatus.
In this lamp the wick is cylindrical, the
flame being supplied with air both inside
and outside: the combustion is greatly aided
by the chimney, which is made of copper when the lamp is used as a source
of heat.

Fig. 96 exhibits, in section, an excellent lamp of this kind for burning
alcohol or wood-spirit.   It is constructed of thin copper, and furnished with

Fig. 96.

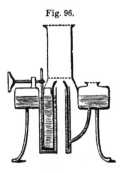

Fig. 97.

ground caps to the wick-holder and aperture * by which the spirit is introduced, in order to prevent loss when the lamp is not in use.   Glass spirit-
lamps (fig. 97), fitted with caps to prevent evaporation, are very convenient
for occasional use, being always ready and in order.

In London, and other large towns where coal-gas is to be had, it is constantly used with the greatest economy and advantage in every respect as a
source of heat.   Retorts, flasks, capsules, and other vessels, can be thus
exposed to an easily regulated and invariable temperature for many successive hours.   Small platinum crucibles may be ignited to redness by placing
them over the flame on a little wire triangle.   The arrangement shown in

* When in use, this aperture must always be open, otherwise an accident is sure to
happen; the heat expands the air in the lamp, and the spirit is forced out in a state of
inflammation.

fig. 98, consisting of a common Argand gas-burner fixed on a heavy and low foot, and connected with a flexible tube of caoutchouc or other material, is very convenient.

Fig. 98.

Fig. 99.

A higher temperature, and perfectly smokeless flame, is, however, obtained by burning the gas previously mixed with air. Such a flame is easily produced by placing a cap of wire gauze on the chimney of the Argand burner just described, and setting fire to the gas above the wire gauze. The flame does not penetrate below, but the gas, in passing up the chimney, becomes mixed with air, and this mixture burns above the cap with a blue, smokeless flame.

Another kind of burner for producing a smokeless flame has been contrived by Professor Bunsen, and is now used in all chemical laboratories. In this burner (fig. 99) the gas, supplied by a flexible tube t, passes through a set of small holes into the box at a, in which it mixes with atmospheric air, entering freely by a number of holes near the top of the box. The gaseous mixture passes up the tube b, and is inflamed at the top, where it burns with a tall, blue, smokeless flame, giving very little light, but much heat. By arranging two or more such tubes, together with an air-box containing a sufficient number of holes, a very powerful burner may be constructed.

Considerable improvements in this form of burner have been made by Mr. Griffin, who has also constructed, on the same principle, powerful gas-furnaces, affording heat sufficient for the decomposition of silicates and the fusion of considerable quantities of copper or iron.* (Mr. Fletcher, of Warrington, has also introduced several new and very powerful forms of gas-burners and furnaces, for use, not only in the laboratory, but also for manufacture and domestic purposes.) The principle of burning a mixture of gas and air is also applied in Hofmann's gas-furnace for organic analysis, which will be described under Organic Chemistry.

The *kindling-point*, or temperature at which combustion commences, is very different with different substances : phosphorus will sometimes take fire in the hand ; sulphur requires a temperature exceeding that of boiling water ; charcoal must be heated to redness. Among gaseous bodies similar differences are observed ; hydrogen is inflamed by a red-hot wire ; light carburetted hydrogen requires a white heat to kindle it. When flame is cooled by any means below the temperature at which the rapid oxidation of the combustible gas occurs, it is at once extinguished. Upon this depends the principle of Sir H. Davy's invaluable *Safety-lamp.*

* See the article on Gas-burners and Furnaces in Watts's Dictionary of Chemistry, ii 782.

Light carburetted hydrogen, or marsh-gas, is frequently disengaged in large quantities in coal-mines. This gas, mixed with seven or eight times its volume of atmospheric air, becomes highly explosive, taking fire at a light and burning with a pale-blue flame; and many fearful accidents have occurred from the ignition of large quantities of mixed gas and air occupying the extensive galleries and workings of a mine. Davy undertook an investigation with a view to discover some remedy for this constantly occurring calamity; and his labors resulted in some exceedingly important discoveries respecting flame, which led to the construction of the lamp which bears his name.

When two vessels filled with a gaseous explosive mixture are connected by a narrow tube, and the contents of one are fired by the electric spark or otherwise, the flame is not communicated to the other, provided the diameter of the tube, its length, and the conducting power for heat of its material, bear a certain proportion to each other; the flame is extinguished by cooling, and its transmission rendered impossible.

In this experiment high conducting power and diminished diameter compensate for diminution in length; and to such an extent can this be carried, that metallic gauze, which may be looked upon as a series of very short square tubes arranged side by side, when of sufficient degree of fineness, arrests in the most complete manner the passage of flame in explosive mixtures. Now the fire-damp mixture has an exceedingly high kindling point; a red heat does not cause inflammation; consequently, the gauze will be safe for this substance, when flame will pass in almost any other case.

The miner's safety lamp is merely an ordinary oil-lamp, the flame of

Fig. 100.

which is enclosed in a cage of wire gauze, made double at the upper part, and containing about 400 apertures to the square inch. The tube for supplying oil to the reservoir reaches nearly to the bottom of the latter, while the wick admits of being trimmed by a bent wire passing with friction through a small tube in the body of the lamp; the flame can thus be kept burning for any length of time, without the necessity of unscrewing the cage. When this lamp is taken into an explosive atmosphere, although the fire-damp may burn within the cage with such energy as sometimes to heat the metallic tissue to dull redness, the flame is not communicated to the mixture on the outside.

These effects may be conveniently studied by suspending the lamp in a large glass jar, and gradually admitting coal-gas below. The oil-flame is at first elongated, and then, as the proportion of gas increases, extinguished, while the interior of the gauze cylinder becomes filled with the burning mixture of gas and air. As the atmosphere becomes pure, the wick is once more relighted. These appearances are so remarkable that the lamp becomes an admirable indicator of the state of the air in different parts of the mine.*

Oxy-hydrogen Flame.—It has already been stated that though the light emitted by hydrogen burning in the air or in oxygen gas is very feeble, nevertheless the temperature of the flame is very high. A still hotter

* This is the true use of the lamp, namely, to permit the viewer or superintendent, without risk to himself, to examine the state of the air in every part of the mine; not to enable workmen to continue their labors in an atmosphere habitually explosive. which must be unfit for human respiration, although the evil effects may be slow to appear. Owners of coal-mines should be compelled either to adopt efficient means of ventilation, or to close workings of this dangerous character altogether.

flame may be produced by mixing the hydrogen with exactly the quantity of oxygen required for complete combination, that is, as will presently be seen, with half its own volume. Such a mixture burns like gunpowder, independently of the external air; and if a vessel or a bladder be filled with it, and a lighted match or a red hot wire applied to the mouth, the two gases instantly unite throughout their entire mass, producing a loud report, and an explosive force sufficient to shatter any glass vessel of ordinary strength. On the other hand, by bringing a jet of oxygen within a hydrogen flame burning from a platinum nozzle, a very hot flame may be produced without the slightest danger of explosion. This is effected by Daniell's oxy-hydrogen blowpipe (fig. 101), in which the two gases are supplied from separate vessels, large gas-holders for example, or air-tight bags of macintosh cloth pressed by weights, and suffered to mix only in the jet itself. The eye soon becomes accustomed to the peculiar appearance of the true oxy-hydrogen flame, so as to permit the supply of each gas to be exactly regulated by suitable stopcocks attached to the jet. The flame thus produced is *solid*, the combustion taking place simultaneously throughout its mass, whereas all ordinary flames, as already observed (p. 149), whether burning in air or in pure oxygen, are hollow, the combustion in them taking place only at the surface.

Fig. 101.

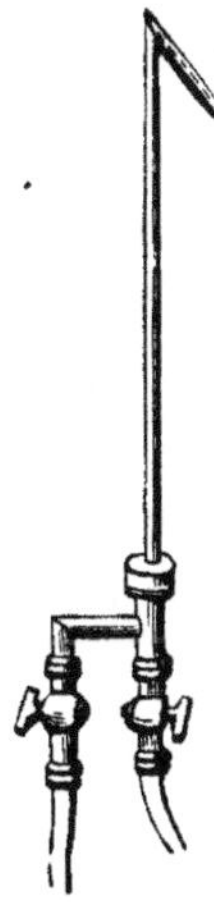

A piece of thick platinum wire introduced into the flame of the oxy-hydrogen blowpipe melts with the greatest ease; a watch-spring or a small steel file burns with the utmost brilliancy, throwing off showers of beautiful sparks; an incombustible oxidized body, as magnesia or lime, becomes so intensely ignited as to glow with a light insupportable to the eye, and to be susceptible of employment as a most powerful illuminator, as a substitute for the sun's rays in the solar microscope, and for night signals in trigonometrical surveys.

Slow Combustion.—Oxygen and hydrogen may be kept mixed at common temperatures for any length of time without combination taking place, but, under particular circumstances, they unite quietly and without explosion. Many years ago Döbereiner made the curious observation, that finely divided platinum possesses the power of determining the union of the gases; and, subsequently, Faraday showed that the state of minute division is by no means indispensable, since rolled plates of the metal have the same property, provided their surfaces are absolutely clean. Neither is the effect strictly confined to platinum; other metals, as palladium and gold, and even stones and glass, exhibit the same property, although in a far lower degree, since they often require to be aided by a little heat. When a piece of platinum foil, which has been cleaned by hot oil of vitriol and thorough washing with distilled water, is thrust into a jar containing a mixture of oxygen and hydrogen standing over water, combination of the two gases immediately begins, and the level of the water rapidly rises, whilst the platinum becomes so hot that drops of water accidentally falling upon it enter into ebullition. If the metal be very thin and quite clean, and the gases pure, its temperature rises after a time to actual redness, and the residue of the mixture explodes. But this is an effect altogether accidental, and dependent upon the high temperature of the platinum, which high temperature has been produced by the preceding quiet combination of the two bodies. When the platinum is reduced to a state of minute division, and its surface thereby much extended, it becomes immediately red hot in a mixture of hydrogen and oxygen, or hydrogen and air; a jet of hydrogen thrown upon a little of

7 *

the spongy metal, contained in a glass or capsule, is at once kindled, and on this principle machines for the production of instantaneous light have been constructed.   These, however, act well only when constantly used; the spongy platinum is apt to become damp by absorption of moisture from the air, and its power is then for the time lost.

The best explanation that can be given of these curious effects is to suppose that solid bodies in general have, to a greater or less extent, the property of condensing gases upon their surfaces, or even liquefying them (as shown p. 139), and that this faculty is exhibited pre-eminently by certain of the non-oxidizable metals, as platinum and gold.   Oxygen and hydrogen may thus, under these circumstances, be brought, as it were, within the sphere of their mutual attractions by a temporary increase of density, whereupon combination ensues.

Coal gas' and ether or alcohol vapor may also be made to exhibit the phenomenon of quiet oxidation under the influence of this remarkable surface action.   A close spiral of slender platinum wire, a roll of thin foil, or even a common platinum crucible, heated to dull redness, and then held in a jet of coal gas, becomes strongly ignited, and remains in that state as long as the supply of mixed gas and air is kept up, the temperature being maintained by the heat disengaged in the act of union.   Sometimes the metal becomes white hot, and then the gas takes fire.

If such a coil of wire be attached to a card, and suspended in a glass containing a few drops of ether, having previously been made red hot in the flame of a spirit-lamp, it will continue to glow until the oxygen of the air is exhausted, giving rise to the production of an irritating vapor which attacks the eyes. The combustion of the ether is in this case but partial; a portion of its hydrogen is alone removed, and the whole of the carbon left untouched.

Fig. 102.

A coil of thin platinum wire may be placed over the wick of a spirit-lamp, or a ball of spongy platinum sustained just above the cotton; on lighting the lamp, and then blowing it out as soon as the metal appears red hot, slow combustion of the spirit drawn up by the capillarity of the wick will take place, accompanied by the pungent vapors just mentioned, which may be modified, and even rendered agreeable, by dissolving in the liquid some sweet-smelling essential oil or resin.

Ozone or Active Oxygen.—It has long been known that dry oxygen or atmospheric air, when exposed to the action of a series of electric sparks, emits a peculiar and somewhat metallic odor, which may also be imparted to moist oxygen by allowing phosphorus to remain in it for some time, and by several other processes.   This odorous principle also possesses several properties not exhibited by oxygen in its ordinary state, one of the most characteristic being the liberation of iodine from potassium iodide.   It has been the subject of many researches, especially by Schonbein of Basle, who proposed for it the name of O z o n e.

An easy way of preparing ozone is to subject ordinary oxygen to the action of the silent electric discharge, or electric effluvium.   For this purpose a stream of the gas is passed through a tube into which is sealed a pair of very finely pointed platinum wires with their points at a little distance apart, one being connected with an electrical machine, and the other with the ground.   No sparks must be allowed to pass, as in that case a considerable portion of the ozone would be reconverted into ordinary oxygen.   Siemens prepares ozone by induction; he forms a sort of Leyden-jar by coating the inner surface of a long glass tube with tin-foil, and enclosing this within a wider tube coated with tin-foil on its outer surface. Between the two tubes a stream of pure dry oxygen is passed, which, when the inner and outer coatings are connected with terminal wires of an induction coil, becomes electrified by induction. By this means from 10 to 15 per cent. of the oxygen may be converted into ozone.

Ozone is also produced in small quantity by suspending a stick of phosphorus in a bottle filled with moist air; in the electrolytic decomposition of water, and in the action of strong sulphuric acid on potassium permanganate. There has been considerable discussion about its nature and composition, but the most trustworthy experiments seem to show that it is merely a modified form of oxygen.

Ozone is insoluble in water and in solutions of acids or alkalies, but is absorbed by a solution of potassium iodide. Air charged with it exerts an irritating action on the lungs. It is decomposed by heat gradually at 100° C., instantly at 290° C. (554° F.). It is an extremely powerful oxidizing agent, possesses strong bleaching and disinfecting powers; corrodes cork, caoutchouc, and other organic substances: oxidizes iron, copper, and even silver when moist, as well as dry mercury and iodine. Paper moistened with a mixture of starch and potassium iodide is instantly turned blue when exposed to its action. Now when paper thus prepared is exposed to the open air for five to ten minutes, it often acquires a blue tint, varying in intensity at different times. Hence it is supposed that ozone is present in the air in variable quantity.

Oxides of Hydrogen.—There are two oxides of hydrogen—namely, the m o n o x i d e, which is water, and the d i o x i d e, discovered in the year 1818 by Thénard.

It appears that the composition of water was first demonstrated in the year 1781 by Cavendish;[*] but the discovery of the exact proportions in which oxygen and hydrogen unite in generating that most important compound has, from time to time to the present day, occupied the attention of some of the most distinguished cultivators of chemical science. There are two distinct methods of research in chemistry—the *analytical*, or that in which the compound is resolved into its elements, and the *synthetical*, in which the elements are made to unite and produce the compound. The first method is of much more general application than the second; but in this particular instance both may be employed, although the results of the synthesis are the more valuable.

Fig. 103.

The decomposition of water may be effected by voltaic electricity. When water is acidulated so as to render it a conductor,[†] and a portion interposed between a pair of platinum plates connected with the extremities of a voltaic apparatus of moderate power, decomposition of the liquid takes place in a very interesting manner; oxygen, in a state of perfect purity, is evolved from the water in contact with the plate belonging to the copper end of the battery, and hydrogen, equally pure, is disengaged at the plate connected with the zinc extremity, the middle portions of liquid remaining apparently unaltered. By placing small graduated jars over the platinum plates, the gases can be collected, and their quantities determined. The whole arrangement is shown

* A claim to the discovery of the composition of water, on behalf of James Watt, has been very strongly urged, and supported by such evidence that the reader of the controversy may be led to the conclusion that the discovery was made by both parties, nearly simultaneously, and unknown to each other. See the article "Gas," by Dr. Paul, in Watts's Dictionary of Chemistry, ii. 780.
† See the section on "Electro-Chemical Decomposition."

in fig. 103; the conducting wires pass through the bottom of the glass cup, and away to the battery.

When this experiment has been continued for a sufficient time, it will be found that the volume of the hydrogen is a *very* little above twice that of the oxygen. Were it not for the accidental circumstance of oxygen being sensibly more soluble in water than hydrogen, the proportion of two to one by measure would come out exactly.

Fig. 104.

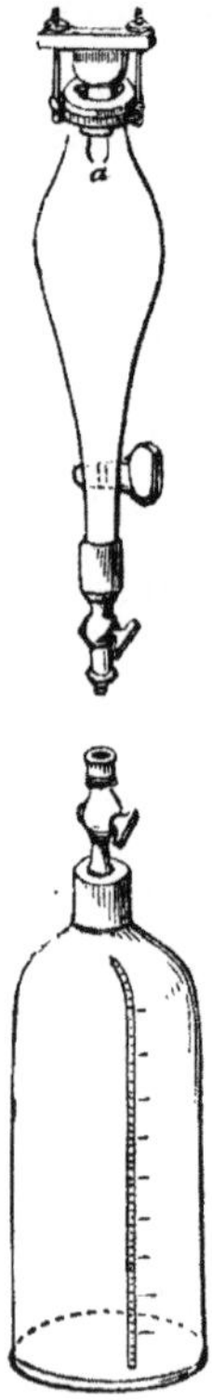

Water, as Sir W. Grove has shown, is likewise decomposed into its constituents by heat. This effect is produced by introducing platinum balls, intensely ignited by electricity or other means, into water or steam. The two gases are obtained in very small quantities at a time.

When oxygen and hydrogen, both as pure as possible, are mixed in the proportions mentioned, passed into a strong glass tube standing over mercury, and exploded by the electric spark, all the mixture disappears, and the mercury is forced up into the tube, filling it completely. The same experiment may be made with the explosion vessel or eudiometer of Cavendish (fig. 104). The instrument is exhausted at the air-pump, and then filled from a capped jar with the mixed gases; on passing an electric spark by the wires shown at *a*, explosion ensues, and the glass becomes bedewed with moisture; and if the stop-cock be then opened under water, the latter will rush in and fill the vessel, leaving merely a bubble of air, the result of imperfect exhaustion.

But the process upon which most reliance is placed for demonstrating the composition of water, is that in which pure copper oxide is reduced at a red heat by hydrogen, and the water so formed is collected and weighed. This oxide suffers no change by heat alone, but the momentary contact of hydrogen, or any common combustible matter, at a high temperature, suffices to reduce a corresponding portion to the metallic state. Fig. 105 will serve to convey some idea of the arrangement adopted in researches of this kind.

A copious supply of hydrogen is procured by the action of dilute sulphuric acid upon the purest zinc that can be obtained; and the gas is made to pass in succession through solutions of silver nitrate and strong caustic potash, by which its purification is completed. After this it is conducted through a tube three or four inches long, filled with fragments of pumice stone steeped in concentrated oil of vitriol, or with phosphoric anhydride. These substances have so great an attraction for aqueous vapor, that they dry

Fig. 105.

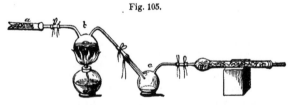

the gas completely during its transit. The extremity of this tube is shown at *a*. The dry hydrogen thus arrives at the part of the apparatus containing the copper oxide, represented at *b*; this consists of a two-necked flask of very hard white glass, maintained at a red heat by a lamp placed beneath. As the decomposition proceeds, the water produced by the reduction of the oxide begins to condense in the second neck of the flask, whence it drops into the receiver *c*. A second desiccating tube prevents the loss of aqueous vapor by the current of gas which passes in excess.

Before the experiment can be commenced, the copper oxide, the purity of which is well ascertained, must be heated to redness for some time in a current of dry air; it is then suffered to cool, and very carefully weighed with the flask. The empty receiver and second drying-tube are also weighed, the disengagement of gas set up, and when the air has been displaced, heat is slowly applied to the oxide. The action is at first very energetic; the oxide often exhibits the appearance of ignition; but as the decomposition proceeds it becomes more sluggish, and requires the application of a considerable heat to effect its completion.

When the process is at an end, and the apparatus perfectly cool, the stream of gas is discontinued, dry air is drawn through the whole arrangement, and, lastly, the parts are disconnected and reweighed. The loss of the copper oxide gives the oxygen; the gain of the receiver and its drying-tube indicates the water; and the difference between the two, the hydrogen.

A set of experiments, made in Paris in the year 1820, by Dulong and Berzelius, gave as a mean result, for the composition of water by weight, 8.009 parts oxygen to 1 part hydrogen; numbers so nearly in the proportion of 8 to 1, that the latter have usually been assumed to be true.

More recently the subject has been reinvestigated by Dumas, with the most scrupulous precision, and the above supposition fully confirmed. The composition of water may therefore be considered as established; it contains by weight 8 parts (or more exactly 7.99) oxygen to 1 part hydrogen, and by measure, 1 volume oxygen to 2 volumes hydrogen. The densities of the gases, as already mentioned, correspond very closely with these results.

Properties of Water.—Pure water is colorless and transparent, destitute of taste and odor, and an exceedingly bad conductor of electricity of low tension. It attains its greatest density towards 4.5° C. (40° F.), freezes at 0° C. (32° F.),* and boils under the ordinary atmospheric pressure at or near 100° C. (212° F.). It evaporates at all temperatures.

The weight of a cubic centimetre of water at the maximum density is chosen as the unit of weight of the metrical system; and called a g r a m; consequently a litre or cubic decimetre $= 1000$ cubic centimetres of water, at the same temperature, weighs 1000 grams, or 1 kilogram.

A cubic inch of water at 62° F. weighs 252.45 grains; a cubic foot weighs nearly 1000 ounces avoirdupois; and an imperial gallon weighs 70,000 grains, or 10 lbs. avoirdupois.

Water is 825 times heavier than air. To all ordinary observation it is incompressible; very accurate experiments have nevertheless shown that it does yield to a small extent when the power employed is very great, the diminution of volume for each atmosphere of pressure being about 51-millionths of the whole.

Clear water, although colorless in small bulk, is blue like the atmosphere when viewed in mass. This is seen in the deep ultramarine tint of the ocean, and perhaps in a still more beautiful manner in the lakes of Switzerland and other Alpine countries, and in the rivers which issue from them, the slightest admixture of mud or suspended impurity destroying the effect. The same magnificent color is visible in the fissures and caverns found in the ice of the glaciers, which is usually extremely pure and transparent within, although foul upon the surface.

The specific gravity of steam or vapor of water is found by experiment

---

* According to Dufour, the specific gravity of ice is 0.9175; water, therefore, on freezing, expands by $\frac{1}{11}$th of its volume.

14

to be 0.625, compared with air at the same temperature and pressure, or 9 as compared with hydrogen. Now, it has been already shown that water is composed of two volumes of hydrogen and one volume of oxygen; and if the weight of one volume of hydrogen be taken as unity, that of two volumes hydrogen ($= 2$) and one volume oxygen ($= 16$) will together make 18, which is the weight of two volumes of water-vapor. Consequently, *water in the state of vapor consists of two volumes of hydrogen and one volume of oxygen condensed into two volumes.* A method of demonstrating this important fact by direct experiment has been devised by Dr. Hofmann. It consists in exploding a mixture of two volumes hydrogen and one volume oxygen by the electric spark, in a eudiometer tube enclosed in an atmosphere of the vapor of a liquid (amylic alcohol) which boils at a temperature considerably above that of boiling water, so that the water produced by the combination of the gases remains in the state of vapor instead of at once condensing to the liquid form. It is then seen that the three volumes of mixed gas are reduced after the explosion to two volumes.*

Water seldom or never occurs in nature in a state of perfect purity : even the rain which falls in the open country contains a trace of ammoniacal salt, while rivers and springs are invariably contaminated to a greater or less extent with soluble matters, saline and organic. Simple filtration through a porous stone or a bed of sand will separate suspended impurities, but distillation alone will free the liquid from those which are dissolved. In the preparation of distilled water, which is an article of large consumption in the scientific laboratory, it is proper to reject the first portions which pass over, and to avoid carrying the distillation to dryness. The process may be conducted in a metal still furnished with a worm or condenser of silver or tin ; lead must not be used.

The ocean is the great recipient of the saline matter carried down by the rivers which drain the land: hence the vast accumulation of salts. The following table will serve to convey an idea of the ordinary composition of sea-water; the analysis is by Dr. Schweitzer, of Brighton,† the water being that of the British Channel :—

1000 grains contained—

Water.	964.745
Sodium chloride.	27.059
Potassium chloride	0.766
Magnesium chloride	3.666
Magnesium bromide	0.029
Magnesium sulphate	2.296
Calcium sulphate	1.406
Calcium carbonate	0.033
Traces of Iodine and Ammoniacal salts	....
	1000.000

Its specific gravity was found to be 1.0274 at 15.5° C. (60° F.). Sea-water is liable to variations of density and composition by the influence of local causes, such as the proximity of large rivers, or masses of melting ice, and other circumstances.

Natural springs are often impregnated to a great extent with soluble substances derived from the rocks they traverse ; such are the various mineral waters scattered over the whole earth, and to which medicinal virtues are attributed. Some of these hold ferrous oxide in solution, and are effervescent from carbonic acid gas ; others are alkaline, probably from traversing

---

* For a description of the apparatus, see Hofmann's Modern Chemistry (1865), p. 51.
† Philosophical Magazine, July, 1839.

rocks of volcanic origin; some contain a very notable quantity of iodine or bromine. Their temperatures also are as variable as their chemical nature.

Water acts on many oxides, both acid and basic, with great energy and considerable evolution of heat, producing compounds called h y d r o x i d e s, which contain hydrogen and oxygen in the proportion to form water, but not actually existing as water, the elements of the two bodies in combining having undergone a change of arrangement, thus :—

$$K_2O + H_2O = 2KHO \quad \text{Potassium hydroxide (potash).}$$
$$CaO + H_2O = CaH_2O_2 \quad \text{Calcium hydroxide (slaked lime).}$$
$$SO_3 + H_2O = SH_2O_4 \quad \text{Sulphur hydroxide (sulphuric acid).}$$
$$P_2O_5 + H_2O = 2PHO_3 \quad \text{Phosphorus hydroxide (metaphosphoric acid).}$$

In many of these compounds the elements of water are retained with great force, and require a high temperature to expel them: calcium hydroxide, for example, requires a red heat to convert it into anhydrous calcium oxide (quicklime), and the hydroxides of potassium, barium, sulphur, and phosphorus cannot be dehydrated by heat alone.

In other cases water appears to combine with other bodies—salts, for example—as such, or, in other words, without alteration of atomic arrangement. Such compounds are called h y d r a t e s, and the water contained in them—the presence of which has great influence on the crystalline form of the compound—is called w a t e r o f c r y s t a l l i z a t i o n. Water thus combined is easily expelled by heat, mostly at 100—120° C. (212—248° F.).

Many salts combine with different quantities of water, according to the temperature at which they separate from solution, the quantity thus taken up being for the most part greater as the temperature of solidification is lower: thus sodium carbonate crystallizes from solution at ordinary temperatures in oblique rhombic prisms containing 10 molecules of water ($CO_3Na_2 + 10H_2O$), whereas at higher temperatures it crystallizes as $CO_3Na_2 + 8H_2O$ or $5H_2O$, and from a boiling solution in rectangular plates containing $CO_3Na_2 + H_2O$.

There are also hydrates called c r y o h y d r a t e s,* which exist only at temperatures below the freezing point of water; thus sodium chloride (common salt), which at ordinary temperatures crystallizes in anhydrous cubes, solidifies at — 23° C. (— 9° F.) with $10\frac{1}{2}$ molecules of water, forming the hydrate $NaCl + 10\frac{1}{2}H_2O$, or $2NaCl + 21H_2O$, and ammonium chloride (sal-ammoniac), also anhydrous at ordinary temperatures, solidifies at — 15° C. (5° F.) to a hydrate containing $NH_4Cl + 12H_2O$.

In some cases water of crystallization is so feebly combined that it gradually separates when the substance containing it is exposed at ordinary temperatures to dry air, the salt at the same time losing its crystalline character and falling to powder. This change, called e f f l o r e s c e n c e, is strikingly exhibited by crystallized sodium carbonate and common alum. On the other hand, many substances which are very soluble in water attract water from moist air in such quantity as to form a solution; this change, which is exhibited by calcium chloride and potassium hydroxide (caustic potash), is called d e l i q u e s c e n c e.

Lastly, the solvent properties of water far exceed those of any other liquid known. Among salts a very large proportion are soluble to a greater or less extent, the solubility usually increasing with the temperature, so that a hot saturated solution deposits crystals on cooling. There are a few exceptions to this law, one of the most remarkable of which is common salt, the solubility of which is nearly the same at all temperatures: the hydroxide and certain organic salts of calcium dissolve more freely in cold than in·hot water.

* Guthrie, Phil. Mag. (Ser. 4) xlix. 1, 206; 1, 266;—(Ser. 5) i. 49; ii. 211.

Fig. 106 exhibits the unequal solubility of different salts in water of different temperatures. The *lines of solubility* cut the verticals raised from

Solubility of Salts in 100 parts of Water.

Fig. 106.

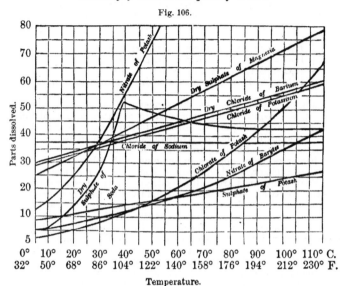

Temperature.

points indicating the temperatures upon the lower horizontal line, at heights proportioned to the quantities of salt dissolved by 100 parts of water. The diagram shows, for example, that 100 parts of water dissolve, of potassium sulphate 8 parts at 0° C. (32° F.), 17 parts at 50° C. (90° F.), and 25 parts at 100° C. (212° F.). There are salts which, like sodium chloride, possess, as already mentioned, very nearly the same degree of solubility in water at all temperatures; in others, like potassium sulphate or potassium chloride, the solubility increases directly with the increment of temperature; in others, again, like potassium nitrate or potassium chlorate, the solubility augments much more rapidly than the temperature. The differences in the deportment of these different salts are shown very conspicuously, by a straight horizontal line, by a straight inclined line, and lastly by curves, the convexity of which is turned towards the lower horizontal line.

The solubility of a salt is usually represented by the quantity of anhydrous salt dissolved by 100 parts of water. It is obvious, however, that salts containing water of hydration or water of crystallization cannot, within certain limits of temperature, dissolve in water in the anhydrous state, but must be dissolved as hydrates. The solubility of a hydrated salt frequently differs very considerably from that of the same salt in the anhydrous state. Again, many salts, as already observed, form more than one hydrate; and these several hydrates may also differ in their solubility. Sodium sulphate forms a hydrate, $SO_4Na_2 + 7H_2O$, consisting, in 100 parts, of 53 parts of anhydrous salt and 47 parts of water, which is obtained in crystals when a solution of sodium sulphate saturated at 100° C. is cooled out of contact with the air: this hydrate is much more soluble than the ordinary hydrate, $SO_4Na_2 + 10H_2O$ (Glauber's salt), which differs from the former in its crystalline form, and consists, in 100 parts, of 44.2 parts of anhydrous salt and 55 8 parts of water. When a solution of sodium sulphate is saturated at the boiling point of water, and cooled to the common temperature without depositing any crystals, the salt exists in the form of the more soluble hydrate. This salt, when coming in contact with the dust of the air, or with a small crystal of common Glauber's salt, is suddenly transformed into the less soluble hydrate, part of which separates from the solution in the form of Glauber's salt. From 0° to 33° C. (32° to 91° F.) sodium sulphate dissolves as Glauber's salt, the solubility of which increases with the temperature: hence the rapid rise of the curve representing the solubility of the salt. Above 33° C. (91° F.) the hydrate of sodium sulphate is decomposed, even in solution, being more and more thoroughly converted into the anhydrous salt as the temperature increases. Sodium sulphate appears, however, far less soluble in the anhydrous state, and hence the diminution of solubility of the salt when its solution is heated above 33° C. (91° F.).

Liquid Diffusion. Dialysis.—When a solution having a density greater than that of water is introduced into a cylindrical glass vessel, and then water very cautiously poured upon it, in such a manner that the two layers of liquid remain unmoved, the substance dissolved in the lower liquid will gradually pass into the supernatant water, though the vessel may have been left undisturbed, and the temperature remain unchanged. This gradual passage of a dissolved substance from its original solution into pure water, taking place notwithstanding the higher specific gravity of the substance which opposes this passage, is called the *diffusion of liquids*. The phenomena of this diffusion were elaborately investigated by Graham, who arrived at very important results. Different substances, when in solution of the same concentration, and under other similar circumstances, diffuse with very unequal velocity. Hydrochloric acid, for instance, diffuses with greater rapidity than potassium chloride, potassium chloride more rapidly than sodium chloride, and the latter, again, more quickly than magnesium sulphate; gelatin, albumin, and caramel diffuse very slowly. Diffusion is generally found to take place more rapidly at high than at low temperatures. It is more particularly rapid with crystallized substances, though not exclusively, for hydrochloric acid and alcohol are among the highly diffusive bodies; slow with non-crystalline bodies, which, like gelatin, are capable of forming a jelly, though even here exceptions are met with. Graham calls the substances of great diffusibility *crystalloïds*, the substances of low diffusibility *colloïds*. The unequal power of diffusion with which different substances are endowed frequently furnishes the means of separating them. When water is poured with caution, so as to prevent mixing, upon a solution containing equal quantities of potassium chloride and sodium chloride, the more diffusible potassium chloride travels more rapidly upwards than the less diffusible sodium chloride, and very considerable portions of potassium chloride will have reached the upper layers of the water before the sodium chloride has arrived there in appreciable quantity. The separation of rapidly diffusible crystalloïds and slowly diffusible colloïds succeeds still better.

A more complete separation of crystalloïds and colloïds may be accomplished in the following manner:—Graham made the important observation, that certain membranes, and also parchment paper, when in contact, on the one surface, with a solution containing a mixture of crystalloïdal and colloïdal substances, and on the other surface with pure water, will permit the passage to the water of the crystalloïds, but not of the colloïds. To carry out this important mode of separation, which is designated by the term *dialysis*, the lower mouth of a glass vessel, open on both sides (fig. 107), is tied over with parchment paper placed upon an appropriate support (fig.

14 *

108), and transferred, together with the latter, into a larger vessel filled with water (fig. 109); or the vessel may be suspended as shown in fig. 110.   The liquid containing the different substances in solution is then poured into the inner vessel, so as to form a layer of about half an inch in height upon

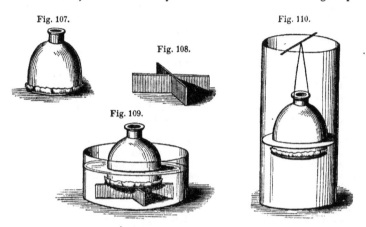

Fig. 107.

Fig. 108.

Fig. 109.

Fig. 110.

the parchment paper.   The crystalloïdal substances gradually pass through the parchment paper into the outer water, which may be renewed from time to time; the colloïdal substances are almost entirely retained by the liquid in the inner vessel.   In this manner Graham prepared several colloïds free from crystalloïds; he showed, moreover, that poisonous crystalloïds, such as arsenious acid or strychnine, even when mixed with very large proportions of colloïdal substances, pass over into the water of the dialyzer in such a state of purity that their presence may be established by re-agents with the utmost facility.

Osmose.—When two different liquids are separated by a porous diaphragm, as, for instance, by a membrane, and the liquids mix through this diaphragm, it is found that in most cases the quantities travelling in opposite directions are unequal.   Suppose three cylinders, the lower mouths of which are tied over with bladders, to be filled respectively with concentrated solutions of copper sulphate, sodium chloride, and alcohol, and let them be immersed in vessels containing water, to such a depth that the liquids inside and outside are level (fig. 111).   After some time the liquid within the tube is found to have risen appreciably above the level of the water (fig. 112).   On the other hand, if the cylinder filled with pure water be immersed in a solution of copper sulphate, or of sodium chloride, or in alcohol, the liquid in the cylinder is seen to diminish after some time (fig. 113).   A quantity of water passes through the bladder into the solution of copper sulphate or sodium chloride, or into alcohol larger than the amount of either of these three liquids which passes through the bladder into the water.   The mixing of dissimilar substances through a porous diaphragm is called *osmose.*   The passage in larger proportions of one liquid into another is designated by the term *exosmose.*

These phenomena are due to the attraction which the two liquids have for each other, and to the difference of the attraction exercised by the diaphragm upon these liquids.   Bladder takes up a much larger quantity of water than of a solution of salt or of alcohol.   Very rarely only one of the liquids traverses the diaphragm; generally two currents of unequal strength move in opposite directions. When water is separated by an animal membrane from a solution of salt or from alcohol, not only is a transition of

Fig. 111.   Fig. 112.   Fig. 113.

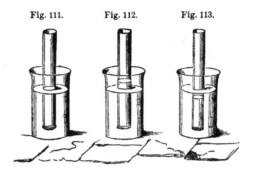

water to these liquids observed, but a small quantity of hydrochloric acid and of alcohol also passes over into the water. In some cases, however, when colloïdal substances in concentrated solutions are on one side of the diaphragm and water on the other, the latter alone traverses the diaphragm, not a trace of the former passing through to the water.

Water likewise dissolves gases. Solution of gases in water (or in other liquids) is called *absorption*, unless this solution gives rise to the formation of chemical compounds in definite proportions. The phenomena of absorption have been more particularly studied by Bunsen, to whom we are indebted for the most accurate examination of this subject.

Water dissolves very unequal quantities of the different gases, and very unequal quantities of the same gas at different temperatures. 1 vol. of water absorbs, at the temperatures stated in the table, and under the pressure of 30 inches of mercury, the following volumes of different gases, measured at 0° C. and 30 inches pressure :—

	Oxygen.	Nitrogen.	Hydrogen.	Nitrogen Monoxide.	Carbon Dioxide.
0° C. . . .	0.041	0.020	0.019	1.31	1.80
10°	0.033	0.016	0.019	0.92	1.18
20°	0.028	0.014	0.019	0.67	0.90

	Chlorine.	Hydrogen Sulphide.	Sulphurous Oxide.	Hydrochloric Acid.	Ammonia.
0° C. . . .	—	4.37	53.9	505	1180
10°	2.59	3.59	36.4	472	896
20°	2.16	2.91	27.3	441	680
30°	1.75	2.33	20.4	412	536
40°	1.37	1.86	15.6	387	444

When the pressure increases, a larger quantity of the gases is absorbed. Gases moderately soluble in water follow in their solubility the law of Henry and Dalton, according to which the quantity of gas dissolved is proportional to the pressure. At 10° C. (50° F.) 1 vol. of water absorbs, under a pressure of 1 atmosphere, 1.18 vol. of carbon dioxide, measured at 0° C. (32° F.) and under a pressure of 30 inches mercury. The quantity of carbon dioxide dissolved under a pressure of 2 atmospheres, and measured under conditions precisely similar to those of the previous experiments, equals 2.36 vols. Again, 1 vol. of water dissolves, under a pressure of ½ atmosphere, 0.59 vol. of carbon dioxide also measured at 0° C. (32° F.) and under 30 inches of mercury. Gases which are exceedingly soluble in water do not obey this law, except at higher temperatures, when the solubility has been already considerably diminished.

It deserves, however, to be noticed, that the pressure which determines the rate of absorption of a gas is by no means the general pressure to which the absorbing liquid is exposed, but that pressure which the gas under consideration would exert if it were alone present in the space with which the absorbing liquid is in contact. Thus, supposing water to be in contact with a mixture of 1 vol. of carbon dioxide and 3 vols. of nitrogen, under a pressure of 4 atmospheres, the amount of carbon dioxide dissolved by the water will be by no means equal to that which the water would have absorbed if it had been, at the same pressure of 4 atmospheres, in contact with pure carbon dioxide. In a mixture of carbon dioxide and nitrogen in the stated proportions, the carbon dioxide exercises only $\frac{1}{4}$, the nitrogen only $\frac{3}{4}$, of the total pressure of the gaseous mixture (4 atmospheres); the partial pressure due to the carbon dioxide is in this case 1 atmosphere, that due to the nitrogen 3 atmospheres; and water, though exposed to a pressure of 4 atmospheres, cannot, under these circumstances, absorb more carbon dioxide than it would if it were in contact with pure carbon dioxide under a pressure of 1 atmosphere.

It is necessary to bear this in mind in order to understand why the air which is absorbed by water out of the atmosphere differs in composition from atmospheric air. The latter consists very nearly of 21 vols. of oxygen and 79 vols. of nitrogen. In atmospheric air which acts under a pressure of 1 atmosphere, the oxygen exerts a partial pressure of $\frac{21}{100}$, the nitrogen a partial pressure of $\frac{79}{100}$ atmosphere. At 10° C. (50° F.) 1 vol. of water (see the above table) absorbs 0.033 vol. of oxygen and 0.016 vol. of nitrogen, supposing these gases to act in the pure state under a pressure of 1 atmosphere. But under the partial pressures just indicated, water of 10° C. (50° F.) cannot absorb more than $\frac{21}{100} \times 0.033 = 0.007$ of oxygen, and $\frac{79}{100} \times 0.016 = 0.13$ vol. of nitrogen. In $0.007 + 0.013 = 0.020$ vol. of the gaseous mixture absorbed by water there are consequently 0.007 vol. of oxygen, and 0.013 vol. of nitrogen, or in 20 vols. of this mixture, 7 vols. of oxygen, and 13 vols. of nitrogen, or in 100 vols. of the gaseous mixture, 35 vols. of oxygen and 65 vols. of nitrogen. The air contained at the common temperature in water is thus seen to be very much richer in oxygen than ordinary atmospheric air.

This property of water to absorb oxygen from the air more readily than nitrogen has been applied to the preparation of oxygen for industrial use.[*] Air is pressed into water by means of a forcing-pump, and the gases which escape on diminishing the pressure are subjected to the same treatment eight times in succession, by which time nearly pure oxygen is obtained. The following table shows the composition of the gaseous mixture at each successive stage:—

Atmospheric Air.		Composition after successive Pressures.							
		1	2	3	4	5	6	7	8
N . .	79	66.67	52.5	37.5	25.0	15.0	9.0	5.0	2.7
O . .	21	33.33	47.5	62.5	75.0	85.0	91.0	95.0	97.3

* Mallet, Dingler's Polyt. Journal, cxciv. 112.

Water containing a gas in solution, when exposed in a vacuum or in a space filled with another gas, allows the gas absorbed to escape until the quantity retained corresponds with the share of the pressure belonging to the gas evolved. If the latter be constantly removed by a powerful absorbent or by a good air-pump, it is in most cases easy to separate every trace of gas from the water. The same result is obtained when water containing a gas in solution is exposed in a space of comparatively infinite size filled with another gas. Water in which nitrogen monoxide is dissolved loses the latter entirely by mere exposure to the atmosphere, and the gas evolved cannot, at any moment, exert more than an infinitely small share of the pressure. If water be freed from gases by ebullition, the separation depends partly upon the diminution of the solubility by the increase of temperature, partly also upon the formation above the surface of the liquid of a constantly renewed atmosphere into which the gas still retained by the liquid may escape.

Some gases which are absorbed in large quantities and very quickly by water—hydrochloric acid, for instance—cannot be perfectly expelled either by the protracted action of another gas (exposure to the atmosphere) or by ebullition; in such cases the liquid, still charged with gas, evaporates as a whole when it has assumed a certain composition. This composition varies, however, with the temperature if the liquid be submitted to a current of air, and with the pressure if it be boiled.

Liquids also lose the gas they contain in solution by freezing: hence the air-bubbles in ice, which consist of the air which had been absorbed from the water. Gas is retained by liquids at the freezing temperature only when it forms a chemical combination in definite proportion with the liquid. Water containing chlorine or sulphurous acid in solution freezes without evolution of gas, with formation of a solid hydrate of chlorine or sulphurous acid.

Pure water generally dissolves gases more copiously than water containing solid bodies in solution (salt water, for instance). If in some few cases exceptions are observed to take place, they appear to depend upon the formation of feeble but true chemical compounds in definite proportion; the fact that carbon dioxide is more copiously absorbed by water containing sodium phosphate in solution than by pure water, may perhaps be explained in this manner.

When water is heated in a strong vessel to a temperature above that of the ordinary boiling point, its solvent powers are still further increased. Dr. Turner enclosed in the upper part of a high-pressure steam-boiler, worked at 149° C. (300° F.), pieces of plate and crown glass. At the expiration of four months the glass was found completely corroded by the action of the water; what remained was a white mass of silica, destitute of alkali, while stalactites of siliceous matter, above an inch in length, depended from the little wire cage which enclosed the glass. This experiment tends to illustrate the changes which may be produced by the action of water at a high temperature in the interior of the earth upon felspathic and other rocks. The phenomenon is manifest in the Geyser springs of Iceland, which deposit siliceous sinter.

Hydrogen Dioxide, H_2O_2.—To prepare this compound, barium dioxide (obtained by igniting pure baryta in a stream of oxygen) is decomposed by dilute sulphuric acid (1 part of the strong acid to at least 5 parts of water), the reaction being represented by the equation:

$$BaO_2 + H_2SO_4 = BaSO_4 + H_2O_2.$$

The resulting solution of hydrogen dioxide is concentrated by evaporation in a vacuum over sulphuric acid (p. 76); if it begins to give off oxygen during the evaporation, the decomposition may be checked by adding a drop or two of sulphuric acid. It may thus be concentrated until it acquires the specific gravity of 1.45. In this state it is a colorless, transparent, inodorous liquid, very prone to decomposition, the least rise of temperature causing effervescence, due to the escape of oxygen gas; near 100° C. (212° F.) it is decomposed with explosive violence. The dilute aqueous solution (*oxygenated water*) is however much more stable, and is prepared as an article of commerce. It possesses powerful oxidizing and bleaching properties, and is used as a disinfectant, and for cleansing old engravings and oil-paintings.

Hydrogen dioxide exhibits a remarkable reaction with certain easily reducible metallic oxides, the metal being reduced, and free oxygen evolved; *e.g.*, with silver oxide:

$$Ag_2O + H_2O_2 = Ag_2 + H_2O + O_2.$$

This reaction will be further discussed hereafter.

OXIDES AND OXY-ACIDS OF CHLORINE.

There are four oxy-acids of chlorine, which may be regarded as oxides of hydrochloric acid; thus—

	Formula.	Composition by weight.				
		Hydrogen.		Chlorine.		Oxygen.
Hydrochloric acid	HCl	1	+	35.4		
Hypochlorous acid	HClO	1	+	35.4	+	16
Chlorous acid	$HClO_2$	1	+	35.4	+	32
Chloric acid	$HClO_3$	1	+	35.4	+	48
Perchloric acid	$HClO_4$	1	+	35.4	+	64

The anhydrous chlorine oxides corresponding with hypochlorous and chlorous acids are also known, namely—

	Chlorine.		Chlorine.		Oxygen.
Chlorine monoxide, or Hypochlorous oxide, Cl_2O	35.4	+	35.4	+	16
Chlorine trioxide, or Chlorous oxide, Cl_2O_3	35.4	+	35.5	+	48

Also an oxide to which there is no corresponding acid, namely—

	Chlorine.		Oxygen.
Chlorine tetroxide, Cl_2O_4	2 × 35.4	+	64

The oxides corresponding with chloric and perchloric acid have not been obtained.

Hypochlorous and chloric acids are produced by the action of chlorine on certain metallic oxides in presence of water; hypochlorous and chlorous acids also by direct oxidation of hydrochloric acid. Perchloric acid and chlorine tetroxide result from the decomposition of chloric acid.

Hypochlorous Oxide, Acid, and Salts.—The oxide is best prepared by the action of chlorine gas upon dry mercuric oxide. This oxide, prepared by precipitation, and dried by exposure to a strong heat, is introduced into a glass tube kept cool, and well-washed dry chlorine gas is slowly passed over it. Mercuric chloride ($HgCl_2$) and hypochlorous oxide are thereby formed; the latter is collected by displacement. The reaction by which it is produced is represented by the equation,

$$HgO + Cl_4 = Cl_2O + HgCl_2.$$

The mercuric chloride, however, does not remain as such; it combines with another portion of the oxide when the latter is in excess, forming a peculiar brown compound, an oxychloride of mercury, $HgCl_2,HgO$. It is remarkable that the *crystalline* mercuric oxide prepared by calcining the nitrate, or by the direct oxidation of the metal, is scarcely acted upon by chlorine under the circumstances described.

Hypochlorous oxide is a pale-yellow gaseous body, containing, in every two measures, two measures of chlorine and one of oxygen, and is therefore analogous in constitution to water. It explodes, although with no great violence, by slight elevation of temperature. Its odor is peculiar, and quite different from that of chlorine. When the flask or bottle in which the gas is received is exposed to artificial cold by the aid of a mixture of ice and salt, the hypochlorous oxide condenses to a deep-red liquid, slowly soluble in water, and very subject to explosion.

Hypochlorous acid is produced by the solution of hypochlorous oxide in water; also by passing air saturated with hydrochloric acid gas through a solution of potassium permanganate acidulated with hydrochloric acid and heated in a water-bath: the distillate is a solution of hypochlorous acid, formed by oxidation of the hydrochloric acid; thirdly, by decomposing a metallic hypochlorite with sulphuric acid or other oxy-acid; fourthly, by passing chlorine gas into water holding in suspension a solution containing metallic oxides, hydroxides, carbonates, sulphates, phosphates, etc., the most advantageous for the purpose being mercuric oxide, or calcium carbonate (chalk), $CaCO_3$, the products in this case being carbon dioxide, calcium chloride, and hypochlorous acid:

$$CaCO_3 + H_2O + Cl_4 = CO_2 + CaCl_2 + 2HClO.$$

The aqueous solution of hypochlorous acid has a yellowish color, an acid taste, and a characteristic sweetish smell. The strong acid decomposes rapidly even when kept in ice. The dilute acid is more stable, but is decomposed by long boiling into chloric acid, water, chlorine, and oxygen. Hydrochloric acid decomposes it, with formation of chlorine:

$$HClO + HCl = H_2O + Cl_2.$$

It is a very powerful bleaching and oxidizing agent, converting many of the elements—iodine, selenium, and arsenic, for example—into their highest oxides, and at the same time liberating chlorine.

Metallic hypochlorites may be obtained in the pure state by neutralizing hypochlorous acid with metallic hydroxides, such as those of sodium, calcium, copper, etc.; but they are usually prepared by passing chlorine gas into solutions of alkalies or alkaline carbonates, or over the dry hydroxides of the earth-metals. In this process a metallic chloride is formed at the same time. With dry slaked lime, for example, which is a hydroxide of calcium, CaH_2O_2, the products are calcium hypochlorite, $CaCl_2O_2$, calcium chloride, and water:

$$2CaH_2O_2 + Cl_4 = CaCl_2O_2 + CaCl_2 + 2H_2O.$$

The salts thus obtained constitute the bleaching and disinfecting salts of commerce.   They will be more fully described under the head of Calcium Salts.

Chlorous Oxide, Acid, and Salts.—The oxide is prepared by heating in a flask filled to the neck, a mixture of four parts of potassium chlorate and three parts of arsenious oxide with 12 parts of nitric acid previously diluted with four parts of water.   During the operation, which must be performed in a water-bath, a greenish-yellow gas is evolved, which is permanent in a freezing mixture of ice and salt, but liquefiable by extreme cold.   It dissolves freely in water and in alkaline solutions, forming chlorous acid and metallic chlorites.   The reaction by which chlorous oxide is formed is somewhat complicated.   The arsenious oxide deprives the nitric acid of part of its oxygen, reducing it to nitrous acid, which is then reoxidized at the expense of the chloric acid, reducing it to chlorous oxide.

$$2HClO_3 + 2HNO_2 = 2HNO_3 + H_2O + Cl_2O_3.$$
Chloric　　Nitrous　　　Nitric　　　　　　Chlorous
acid.　　　acid.　　　　acid.　　　　　　oxide.

Chlorous Acid may be prepared by condensing chlorous oxide in water, or by decomposing a metallic chlorite with dilute sulphuric or phosphoric acid. Its concentrated solution is a greenish-yellow liquid having strong bleaching and oxidizing properties.   It does not decompose carbonates, but acts strongly with caustic alkalies and earths to form chlorites.

Chlorine Tetroxide, Cl_2O_4.—When potassium chlorate is made into a paste with concentrated sulphuric acid, and cooled, and this paste is very cautiously heated by warm water in a small glass retort, a deep yellow gas is evolved, which is the body in question ; it can be collected only by displacement, since mercury decomposes and water absorbs it.

Chlorine tetroxide has a powerful odor, quite different from that of the preceding compounds, and of chlorine itself.   It is exceedingly explosive, being resolved with violence into its elements by a temperature short of the boiling point of water.   Its preparation is, therefore, always attended with danger, and should be performed only on a small scale.   It is composed by measure of one volume of chlorine and two volumes of oxygen, condensed into two volumes.   It may be liquefied by cold.   The solution of the gas in water bleaches.

The *euchlorine* of Davy, prepared by gently heating potassium chlorate with dilute hydrochloric acid, is probably a mixture of chlorine tetroxide and free chlorine.

The production of chlorine tetroxide from potassium chlorate and sulphuric acid depends upon the spontaneous splitting of the chloric acid into chlorine tetroxide and perchloric acid, which latter remains as a potassium salt.

$$6KClO_3 + 3H_2SO_4 = 2Cl_2O_4 + 2HClO_4 + 3K_2SO_4 + 2H_2O.$$
Potassium　　Hydrogen　　　　　　Hydrogen　　Potassium
chlorate.　　sulphate.　　　　　　perchlorate.　sulphate.

When a mixture of potassium chlorate and sugar is touched with a drop of oil of vitriol, it is instantly set on fire, the chlorine tetroxide disengaged being decomposed by the combustible substance with such violence as to cause inflammation.   If crystals of potassium chlorate be thrown into a glass of water, a few small fragments of phosphorus added, and then oil of vitriol poured down a narrow funnel reaching to the bottom of the glass, the phosphorus will burn beneath the surface of the water, by the assistance of the oxygen of the chlorine tetroxide disengaged. The liquid at the same time becomes yellow, and acquires the odor of that gas.

Chloric Acid, $HClO_3$.—This is the most important compound of the series. When chlorine is passed to saturation into a moderately strong hot solution of potassium hydroxide or carbonate, and the liquid concentrated by evaporation, it yields, on cooling, flat tabular crystals of a colorless salt, consisting of potassium chlorate. The mother-liquor contains potassium chloride:

$$3K_2O \ + \ Cl_6 \ = \ 5KCl \ + \ KClO_3.$$
Potassium   Potassium   Potassium
oxide.   chloride.   chlorate.

From potassium chlorate, chloric acid may be obtained by boiling the salt with a solution of hydrofluosilicic acid, which forms an almost insoluble potassium salt, decanting the clear liquid, and digesting it with a little silica, which removes the excess of the hydrofluosilicic acid. Filtration through paper must be avoided.

By cautious evaporation, the acid may be so far concentrated as to assume a syrupy consistence; it is then very easily decomposed. It sometimes sets fire to paper, or other dry organic matter, in consequence of the facility with which it is deoxidized by combustible bodies.

The chlorates are easily recognized; they give no precipitate when in solution with silver nitrate; they evolve pure oxygen when heated, passing thereby into chlorides; and they afford, when treated with sulphuric acid, the characteristic explosive yellow gas already described. The dilute solution of the acid has no bleaching power.

Perchloric Acid, $HClO_4$.—When powdered potassium chlorate is thrown by small portions at a time into hot nitric acid, a change takes place of the same description as that which happens when sulphuric acid is used, but with this important difference, that the chlorine and oxygen, instead of being evolved in a dangerous state of combination, are emitted in a state of *mixture*. The result of the action is a mixture of potassium-nitrate and perchlorate, which may be readily separated by their difference of solubility.

Perchloric acid is obtained by distilling potassium perchlorate with sulphuric acid. Pure perchloric acid is a colorless liquid, of sp. gr. 1.782 at 15.5° C. (70° F.) not solidifying at — 35° C. (— 31° F.); it soon becomes colored, even if kept in the dark, and after a few weeks decomposes with explosion. The vapor of perchloric acid is transparent and colorless: in contact with moist air it produces dense white fumes. The acid, when cautiously mixed with a small quantity of water, solidifies to a crystalline mass, which is a compound of perchloric acid with one molecule of water, $HClO_4 + H_2O$. When brought in contact with carbon, ether, or other organic substances, perchloric acid explodes with nearly as much violence as chloride of nitrogen.

OXY-ACIDS OF BROMINE.

Hypobromous Acid, HBrO, is formed by agitating bromine-water with mercuric oxide, according to the equation:

$$HgO + 2Br_2 + H_2O = HgBr_2 + 2HOBr.$$

The greater part of the hypobromous acid contained in the resulting solution is resolved on distillation into bromine and oxygen, but the acid may be distilled in a vacuum at 40° C. (72° F.) without decomposition.

8

Aqueous hypobromous acid is a light straw-yellow liquid, which resembles hypochlorous acid in its properties, is a powerful oxidizing agent, and bleaches vegetable colors.

Bromic Acid, $HBrO_3$.—Caustic alkalies react with bromine, in the same manner as with chlorine, yielding a bromide and bromate, which may be separated by the inferior solubility of the latter. Bromic acid, obtained from barium bromate, closely resembles chloric acid, and is very unstable. The bromates when heated give off oxygen, and are converted into bromides.

Perbromic Acid, $HBrO_4$, is said by Kämmerer to be formed by the action of bromine on dilute perchloric acid, but its existence is doubtful.

No anhydrous oxide of bromine has yet been obtained.

COMPOUNDS OF IODINE AND OXYGEN.

Two of these compounds are known, viz., iodic oxide, I_2O_5, and periodic oxide, I_2O_7. Both these are acid oxides, uniting with water and metallic oxides.

Hydrogen Iodate, or Iodic Acid, H_2O,I_2O_5, or HIO_3, may be prepared by the direct oxidation of iodine with nitric acid of specific gravity 1.5. Five parts of dry iodine with 200 parts of nitric acid are kept at a boiling temperature for several hours, or until the iodine has disappeared. The solution is then cautiously distilled to dryness, and the residue dissolved in water and made to crystallize.

Iodic acid is a very soluble substance, crystallizing in colorless six-sided tables. At 107° C. (225° F.) it is resolved into water and iodic oxide, which forms tabular rhombic crystals, and when heated to the temperature of boiling olive oil, is completely resolved into iodine and oxygen. The solution of iodic acid is readily deoxidized by sulphurous acid. The iodates much resemble the chlorates: that of the potassium is decomposed by heat into potassium iodide and oxygen gas.

Hydrogen Periodate, or Periodic Acid, H_2O,I_2O_7, or HIO_4.—When solution of sodium iodate is mixed with caustic soda, and a current of chlorine is passed through the liquid, two salts are formed—namely, sodium chloride and a sparingly soluble compound of sodium periodate with sodium hydroxide and water, $NaIO_4,NaHO,H_2O$, or $Na_2H_3IO_6$, the reaction taking place as represented by the equation—

$$NaIO_3 + 3NaHO + Cl_2 = 2NaCl + Na_2H_3IO_6.$$

This sodium salt is separated, converted into a silver salt, and dissolved in nitric acid: the solution yields, on evaporation, crystals of yellow silver periodate, from which the acid may be separated by the action of water, which resolves the salt into free acid and an insoluble basic periodate.

Periodic acid crystallizes from its aqueous solution in deliquescent oblique rhombic prisms, which melt at 133° C. (261° F.), and are resolved at 140° C. (284° F.) into oxygen, water, and iodic oxide.

The solution of periodic acid is reduced by many organic substances, and instantly by hydrochloric acid, sulphurous acid, and hydrogen sulphide. With hydrochloric acid it forms water, iodine chloride, and free chlorine. The metallic periodates are resolved by heat into oxygen and metallic iodides.

SULPHUR.

Symbol, S.   Atomic weight, 31.98.   Vapor-density, 31.98.

SULPHUR occurs abundantly in nature both free and in combination. In the free state it is found in the neighborhood of volcanoes, both active and extinct, either in transparent yellow crystals or in opaque crystalline masses, and is separated from the rock or earth in which it is imbedded, by simple fusion. Large quantities of sulphur are thus obtained from Italy and Sicily.

Sulphur occurs, however, in still greater abundance in combination with metals, forming the *metallic sulphides*, *e.g.*, iron pyrites, FeS_2, copper pyrites, $CuFeS_2$, galena, PbS, cinnabar, HgS; or with metals and oxygen, forming *sulphates*, *e.g.*, gypsum, $CaSO_4, 2H_2O$, heavy spar, $BaSO_4$, bitter spar, $MgSO_4$, $7H_2O$, etc.

Large quantities of sulphur are obtained by the distillation of iron pyrites, which is thereby resolved into a lower sulphide of iron and free sulphur, thus: $3FeS_2 = Fe_3S_4 + S_2$. Smaller quantities are obtained as a by-product in the manufacture of coal-gas. The gas, as it issues from the retorts, contains hydrogen sulphide, and by passing it over oxide of iron, a sulphide of iron is formed, which, on exposure to the air, is oxidized, with separation of sulphur: $2FeS + O_3 = Fe_2O_3 + S_2$.

Lastly, sulphur may be obtained from the residue or waste of the soda-manufacture (see SODIUM) which consists of calcium sulphide mixed with chalk, lime, and alkaline sulphides. On exposing this mixture to the air, the calcium sulphide is partly oxidized to thiosulphate, CaS_2O_3, and on treating the resulting mass with hydrochloric acid, calcium chloride is formed and free sulphur is deposited:

$$2CaS + CaS_2O_3 + 6HCl = 4S + 3CaCl_2 + 3H_2O.$$

Properties.—Pure sulphur is a pale-yellow brittle solid, of well-known appearance. It melts when heated, and distils over unaltered, if air be excluded. The crystals of sulphur exhibit two distinct and incompatible forms—namely, first, an octahedron belonging to the rhombic system (fig. 114), which is the figure of native sulphur, and that assumed when sulphur separates from solution at common temperatures, as when a solution of sulphur in carbon bisulphide is exposed to slow evaporation in the air; and, secondly, a lengthened monoclinic prism; the form assumed when a mass of sulphur is melted, and, after partial cooling, the crust on the surface is broken, and the fluid portion poured out. Fig. 115 shows the result of such an experiment.

Fig. 114.

Fig. 115.

The specific gravity of sulphur varies according to the form in which it is crystallized. The octahedral variety has the specific gravity 2.045 at 4.5° C. (40° F.); the prismatic variety has the specific gravity 1.96.

Sulphur melts at 114.5° C. (218° F.), forming a thin amber-colored liquid of sp. gr. 1.803; when further heated, it begins to thicken, and acquire a deeper color; and between 221° and 249° C. (430° and 480° F.) it is so tenacious that the vessel in which it is contained may be inverted for a moment without the loss of its contents. If in this state it be poured into water, it retains for many hours a remarkably soft and flexible condition, which may be looked upon as the amorphous state of sulphur. After a while it again becomes brittle and crystalline. From the temperature last mentioned to the boiling-point—about 400° C. (752° F.)—sulphur again becomes thin and liquid. In the preparation of commercial flowers of sulphur, the vapor is conducted into a large cold chamber, where it condenses in minute crystals. The specific gravity of sulphur vapor is 2.22 referred to that of air as unity, or 31.98 compared with that of hydrogen (Deville).

Sulphur is insoluble in water and alcohol ; oil of turpentine and the fat oils dissolve it, but the best substance for the purpose is carbon bisulphide. In its chemical relations sulphur bears great resemblance to oxygen: to very many oxides there are corresponding sulphides, and the sulphides often unite among themselves, forming crystallizable compounds analogous to oxy-salts.

Sulphur is remarkable for the great number of modifications which it is capable of assuming. Of these, however, there are two principal well characterized varieties, one soluble, and the other insoluble in carbon bisulphide, and many minor modifications. The soluble variety is distinguished by Berthelot by the name of *electro-negative sulphur*, because it is the form which appears at the positive pole of the voltaic battery during the decomposition of an aqueous solution of hydrogen sulphide, and is separated from the combinations of sulphur with the electro-positive metals. The insoluble variety is distinguished as *electro-positive sulphur*, because it is the form which appears at the negative pole during the electro-decomposition of sulphurous acid, and separates from compounds of sulphur with the electro-negative elements, chlorine, bromine, oxygen, etc.

The principal modifications of soluble sulphur are the octahedral and prismatic varieties already mentioned, and an amorphous variety which is precipitated as a greenish-white emulsion, known as *milk of sulphur*, on adding an acid to a dilute solution of an alkaline polysulphide, such, for example, as is obtained by boiling sulphur with milk of lime. This amorphous sulphur changes by keeping into a mass of minute octahedral crystals. Sublimed sulphur appears also to be allied to this modification, but it always contains a small portion of one of the insoluble modifications.

The chief modifications of insoluble sulphur are:—1. The amorphous insoluble variety obtained as a soft magma by decomposing chlorine bisulphide with water, or by adding dilute hydrochloric acid to the solution of a thiosulphate (p. 181). 2. The plastic sulphur already mentioned as obtained by pouring viscid melted sulphur into water. A very similar variety is produced by boiling metallic sulphides with nitric or nitro-muriatic acid.

When solutions of hydrogen sulphide and ferric chloride are mixed together, a blue precipitate is sometimes formed, which is said to be a peculiar modification of sulphur.

Compounds of Sulphur and Hydrogen.

Hydrogen Monosulphide.—*Sulphydric Acid ; Hydrosulphuric Acid ; Sulphuretted Hydrogen*, H_2S.—There are two methods by which this important compound can be readily prepared; namely, by the action of dilute sulphuric acid upon iron monosulphide, and by the decomposition of antimony trisulphide with hydrochloric acid. The first method yields it most easily, the second in the purest state.

Iron monosulphide is put into the apparatus for hydrogen (p. 132), to-gether with water, and sulphuric acid is added by the funnel. Hydrogen sulphide is then evolved, and may be collected over tepid water, while ferrous sulphate remains in solution:

$$FeS + H_2SO_4 = H_2S + FeSO_4.$$

By the other plan, finely-powdered antimony trisulphide is put into a flask to which a cork and bent tube can be adapted, and strong liquid hydrochloric acid is poured upon it. On the application of heat, a double interchange occurs, hydrogen sulphide and antimony trichloride being formed. The action lasts only while the heat is maintained:

$$Sb_2S_3 + 6HCl = 3SH_2 + 2SbCl_3.$$

Hydrogen sulphide is a colorless gas, having the odor of putrid eggs; it is most offensive when in small quantity, when a mere trace is present in the air. It is not irritant, but, on the contrary, powerfully narcotic. When set on fire, it burns with a blue flame, producing sulphurous acid when the supply of air is abundant, and depositing sulphur when the oxygen is deficient. Mixed with chlorine, it is instantly decomposed, with separation of the whole of the sulphur.

This gas has a specific gravity of 1.171 referred to air, or 17 referred to hydrogen as unity; a litre weighs 1.51991 grams. A pressure of 17 atmospheres at 10° C. (50° F.) reduces it to the liquid form.

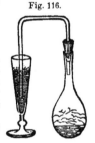

Fig. 116.

Cold water dissolves its own volume of hydrogen sulphide, and the solution is used as a test; it is, however, somewhat prone to decomposition by the oxygen of the air, and should therefore be kept in a tightly closed bottle. Another mode of testing with hydrogen sulphide is to keep a little apparatus for generating the gas always at hand. A small bottle or flask, to which a bit of bent tube is fitted by a cork, is supplied with a little iron sulphide and water; when it is required for use, a few drops of sulphuric acid are added, and the gas is at once evolved. The experiment completed, the liquid is poured from the bottle, replaced by a little clean water, and the apparatus is again ready for use.

Potassium heated in hydrogen sulphide burns with great energy, becoming converted into sulphide, while pure hydrogen remains, equal in volume to the original gas. Taking this fact into account, and comparing the density of the gas with those of hydrogen and sulphur-vapor, it appears that every volume of hydrogen sulphide contains one volume of hydrogen and half of a volume of sulphur-vapor, the whole condensed into one volume, a constitution precisely analogous to that of water-vapor. This corresponds with its composition by weight, determined by other means—namely, 16 parts sulphur and 1 part hydrogen.

When a mixture of 100 measures of hydrogen sulphide and 150 measures of pure oxygen is exploded by the electric spark, complete combustion ensues, and 100 measures of sulphurous oxide gas are produced.

Hydrogen sulphide is a frequent product of the putrefaction of organic matter, both animal and vegetable; it occurs also in certain mineral springs, as at Harrogate and elsewhere. When accidentally present in the atmosphere of an apartment it may be instantaneously destroyed by a small quantity of chlorine gas.

There are few re-agents of greater value to the practical chemist than this substance: when brought in contact with many metallic solutions, it

15 *

gives rise to precipitates, which are often exceedingly characteristic in appearance, and it frequently affords the means of separating metals from each other with the greatest precision and certainty. These precipitates are insoluble sulphides, formed by the mutual decomposition of the metallic oxides or chlorides and hydrogen sulphide, water or hydrochloric acid being produced at the same time. All the metals are, in fact, precipitated, whose sulphides are insoluble in water and in dilute acids.

Arsenic and cadmium solutions thus treated give bright yellow precipitates, the former soluble, the latter insoluble, in ammonium sulphide; tin-salts give a brown or a yellow precipitate, according as the metal is in the form of a stannous or a stannic salt, both soluble in ammonium sulphide. Antimony solutions give an orange-red precipitate, soluble in ammonium sulphide. Copper, lead, bismuth, mercury, and silver salts give dark-brown or black precipitates, insoluble in ammonium sulphide; gold and platinum salts, black precipitates, soluble in ammonium sulphide.

Hydrogen sulphide possesses the properties of an acid: its solution in water reddens litmus-paper.

The best test for the presence of this compound is paper wetted with solution of lead acetate, which is blackened by the smallest trace of the gas.

Hydrogen Persulphide.—This very unstable substance is prepared by the following means:—

Equal weights of slaked lime and flowers of sulphur are boiled with 5 or 6 parts of water for half an hour, whereby a deep orange-colored solution is produced, containing, among other things, calcium disulphide. This is filtered, and slowly added to an excess of dilute sulphuric acid, with constant agitation. A white precipitate of separated sulphur and calcium sulphate then makes its appearance, together with a quantity of yellow oily-looking matter, which collects at the bottom of the vessel: this is hydrogen persulphide.

This compound is generally regarded as a disulphide of hydrogen, H_2S_2, analogous to the dioxide, but its great instability prevents the determination of its composition by direct analysis. Hofmann, however, by treating strychnine in alcoholic solution with ammonium sulphide containing free sulphur, has obtained an orange-red crystalline compound, $C_{21}H_{22}N_2O_2,H_2S_3$, which is resolved by sulphuric acid into soluble strychnine sulphate and a yellow oily liquid resembling the persulphide of hydrogen obtained as above. Hence it might be inferred that this persulphide is really a sesqui-sulphide, H_2S_3; but it begins to decompose as soon as separated.[*] On the other hand, E. Schmidt, by treating an alcoholic solution of strychnine with sulphuretted hydrogen, has obtained a compound containing $2C_{21}H_{22}N_2O_2$,-$3H_2S_3$. The composition of hydrogen persulphide must therefore be regarded as still undecided

Hydrogen persulphide dissolves phosphorus and iodine, forming a phosphorus sulphide and hydrogen iodide respectively, with evolution of sulphuretted hydrogen. With chlorine it forms hydrochloric acid and sulphur chloride, S_2Cl_2. Bromine acts in a similar manner. Ammonia, either gaseous or in aqueous solution, decomposes it instantly, leaving sulphur in a peculiarly brittle, blistered state. It is remarkable that sulphurous acid, which rapidly decomposes hydrogen monosulphide, with separation of sulphur, has scarcely any action on the persulphide.[†]

Sulphur and Chlorine.—Three chlorides of sulphur are known, represented by the formulæ S_2Cl_2, SCl_2, and SCl_4.

The *monochloride*, SCl or S_2Cl_2, which may also be regarded as a disul-

---

[*] Berichte der Deutsch. Chem. Gesellschaft, 1868, p. 81.
[†] Liebig's Annalen, clxxx. 287.

phide of chlorine, analogous in composition to hydrogen dioxide, is prepared by passing dry chlorine gas into a retort in which sulphur is sublimed at a gentle heat. It then distils over, and may be collected in a receiver surrounded by cold water, and freed from excess of sulphur by rectification. It is also produced by distilling a mixture of 1 part sulphur with 9 parts stannic chloride, or 8.5 parts mercuric chloride.

Disulphide of chlorine is a mobile reddish-yellow liquid, having a peculiar, penetrating, disagreeable odor, and fuming strongly in the air. Specific gravity = 1.687. It boils at 136°. It dissolves in carbon disulphide, alcohol, and ether, not however without decomposition in the two latter. It dissolves sulphur in large quantities, especially when heated. When saturated with sulphur at ordinary temperatures, it forms a clear yellow liquid of specific gravity 1.7, and containing altogether 66.7 per cent. sulphur. The solution of chlorine disulphide with the excess of sulphur in crude benzol, is used for vulcanizing or sulphurizing caoutchouc. It is instantly decomposed by water, with formation of hydrochloric and thiosulphuric acids, and separation of sulphur, the thiosulphuric acid in its turn decomposing into sulphur and sulphurous acid: $2S_2Cl_2 + 3H_2O = 4HCl + S_2 + H_2S_2O_3$ (or $H_2SO_3 + S$).

The *dichloride*, SCl_2, is produced by passing chlorine to saturation into the preceding compound cooled by a mixture of ice and salt, and expelling the excess of chlorine by a stream of carbon dioxide. The product is a deep red liquid boiling at 164°, and containing 30.5 per cent. of sulphur, 69.5 of chlorine, agreeing nearly with the formula SCl_2, which requires 31.07 per cent. sulphur and 68.93 chlorine.

The *tetrachloride*, SCl_4, is prepared by saturating chlorine disulphide with chlorine at − 20° C. (− 4° F.). The product contains 81.59 per cent. chlorine and 18.41 sulphur, the numbers calculated from the formula SCl_4 being 81.61 and 18.39. Sulphur tetrachloride is acted upon by sulphuric oxide, producing sulphurous chloride together with chlorine and sulphurous oxide: thus,

$$SCl_4 + SO_3 = SOCl_2 + Cl_2 + SO_2$$

Sulphur and Bromine.—Bromine dissolves sulphur, forming a brown-red liquid probably containing a sulphur bromide analogous to sulphur monochloride; but it has not been obtained pure.

Sulphur and Iodine.—These elements combine when heated together, even under water. The resulting compound, S_2I_2, is a blackish-gray radio-crystalline mass, resembling native antimony sulphide. It decomposes at higher temperatures, gives off iodine on exposure to the air, and is insoluble in water. A *hexiodide*, SI_6, is deposited in crystals having the same crystalline form as iodine when a solution of iodine and sulphur in carbon disulphide is left to evaporate. By heating 254 parts of iodine with 32 parts of sulphur, a compound is obtained which smells like iodine, and is said to be a powerful remedy in skin diseases. A cinnabar-red sulphur iodide is obtained, according to Grosourdi, by precipitating iodine trichloride with hydrogen sulphide.

Sulphur and Oxygen.—There are two compounds of sulphur and oxygen, the names and composition of which are as follows:—

	Composition by weight.		
	Sulphur.		Oxygen.
Sulphur dioxide or sulphurous oxide, SO_2 .	32	+	32
Sulphur trioxide or sulphurous oxide, SO_3 .	32	+	48

Both these oxides unite with water and metallic oxides, or the elements thereof, producing salts; those derived from sulphurous oxide are called sulphites, and those derived from sulphuric acid, sulphates. The composition of the hydrogen salts or acids is as follows:

$$\text{Sulphurous acid, } H_2SO_3 = H_2O,SO_2$$
$$\text{Sulphuric acid, } H_2SO_4 = H_2O,SO_3$$

The replacement of half or the whole of the hydrogen by metals gives rise to metallic sulphites and sulphates.

By the combination of sulphuric oxide with sulphuric acid in the proportion of SO_3 to H_2SO_4 (or 80 parts by weight of the oxide to 98 of the acid) an acid is formed called disulphuric or pyrosulphuric acid, having the composition $H_2S_2O_7$.

There are also several acids of sulphur, with their corresponding metallic salts, to which there are no corresponding anhydrous oxides, viz.:—

1. *Hyposulphurous Acid*, H_2SO_2, having the composition of sulphurous acid *minus* one atom of oxygen.

2. *Thiosulphuric Acid*, $H_2S_2O_3$, having the composition of sulphuric acid in which one-fourth of the oxygen is replaced by sulphur.

Closely allied to this acid is—

3. *Seleniosulphuric Acid*, H_2SSeO_3, having the composition of sulphuric acid in which one-fourth of the oxygen is replaced by selenium.

4. A series of acids called *Polythionic Acids*,* in which the same quantities of oxygen and hydrogen are united with quantities of sulphur in the proportion of the numbers 2, 3, 4, 5, viz.:—

Dithionic or Hyposulphuric acid	$H_2S_2O_6$
Trithionic acid	$H_2S_3O_6$
Tetrathionic acid	$H_2S_4O_6$
Pentathionic acid	$H_2S_5O_6$

Sulphur Dioxide or Sulphurous Oxide, SO_2.—This is the only product of the combustion of sulphur in dry air or oxygen gas. It is most conveniently prepared by heating sulphuric acid with metallic mercury or copper clippings, whereby cupric sulphate, water, and sulphurous oxide are formed, the last being given off as gas.

Another very simple method of preparing sulphurous oxide consists in heating concentrated sulphuric acid with sulphur; a very regular evolution of sulphurous oxide is thus obtained.

Sulphurous oxide is a colorless gas, having the peculiar suffocating odor of burning brimstone; it instantly extinguishes flame, and is quite irrespirable. Its density is 2.21; a litre weighs 2.8605 grams; 100 cubic inches weigh 68.69 grains. At 0° F. (— 17.8° C.), under the ordinary pressure of the atmosphere, this gas condenses to a colorless, limpid liquid, very expansible by heat. Cold water dissolves more than thirty times its volume of sulphurous oxide. The solution, which contains hydrogen sulphite or sulphurous acid, H_2SO_3, may be kept unchanged so long as air is excluded, but access of oxygen gradually converts the sulphurous into sulphuric acid, although dry sulphurous oxide and oxygen gases may remain in contact for any length of time without change. When sulphurous oxide and aqueous vapor are passed into a vessel cooled below 17° or 21° F. (— 8.3° or — 6° C.), a crystalline body forms, which contains about 24.2 sulphurous oxide to 75.8 water.

One volume of sulphurous oxide gas contains one volume of oxygen and half a volume of sulphur-vapor, condensed into one volume.

* From πολύ, many, and θεῖον, sulphur.

Sulphurous acid has bleaching properties; it is used in the arts for bleaching woollen goods and straw-plait. A piece of blue litmus-paper plunged into the moist gas is first reddened and then slowly bleached.

The sulphites of the alkalies are soluble and crystallizable; they are easily formed by direct combination. The sulphites of barium, strontium, and calcium are insoluble in water, but soluble in hydrochloric acid. The stronger acids decompose them; nitric acid converts them into sulphates. The soluble sulphites act as powerful reducing agents, and are much used in that capacity in chemical analysis.

Sulphurous oxide unites, under the influence of sunlight, with chlorine, and with iodine, forming the compounds SO_2Cl_2 and SO_2I_2, which are decomposed by water. It also combines with dry ammoniacal gas; and with nitric oxide in presence of an alkali.

Sulphur Trioxide, or **Sulphuric Oxide**, SO_3 (also called *Anhydrous Sulphuric acid*, or *Sulphuric anhydride*).—This compound may be formed directly by passing a dry mixture of sulphurous oxide and oxygen gases over heated spongy platinum; or it may be obtained by distilling the most concentrated sulphuric acid with phosphoric oxide, which then abstracts the water and sets the sulphuric oxide free. It is usually prepared, however, from the fuming oil of vitriol of Nordhausen, which may be regarded as a solution of sulphuric oxide in sulphuric acid. On gently heating this liquid in a retort connected with a receiver cooled by a freezing mixture, the sulphuric oxide distils over in great abundance, and condenses into beautiful white silky crystals, resembling those of asbestos.

Sulphuric oxide is also produced in considerable quantity by the following process. Vapor of ordinary strong sulphuric acid is passed through a white-hot platinum tube, whereby it is almost completely resolved into water, oxygen, and sulphurous oxide:

$$H_2SO_4 = H_2O + O + SO_2.$$

These mixed gases, after passing through a leaden worm to condense the greater portion of the water, are dehydrated by passing them through a leaden tower filled with coke, over which a stream of concentrated sulphuric acid is allowed to trickle; and finally, the dry mixture of oxygen and sulphurous oxide is passed through platinum tubes heated to low redness and containing fragments of platinized pumice, whereupon they recombine to form sulphuric oxide, which is condensed in a series of Woulffe's bottles.[*]

Sulphuric oxide, when thrown into water, hisses like a red-hot iron, from the violence with which combination occurs: the product is sulphuric acid. When exposed to the air, even for a few moments, it liquefies by absorption of moisture. It unites with ammoniacal gas, forming a salt called *ammonium sulphamate*, the nature of which will be explained farther on.

Sulphuric oxide is susceptible of two modifications (a and β), differing in their properties, though identical in composition. Bodies thus related are said to be **isomeric** (from ισος, equal, and μερος, part). There are but few examples of this relation amongst inorganic compounds, but it is of frequent occurrence in organic chemistry. The a-modification, formed by cooling the liquid oxide, solidifies at $+16°$ in long colorless prisms which melt at the same temperature; it boils at 46°; and its vapor at 20° has a tension represented by 200 millimetres of mercury. *β-Sulphuric oxide* is produced from the a-modification at temperatures below 25°; above 27° the transformation does not go on. It forms extremely fine white needles; at temperatures above 50° it gradually liquefies and passes into the first modification.

[*] Messel and Squire, Chemical News, 1876, i. 177.

8 *

Liquid sulphuric oxide undergoes very great expansion by heat, its mean coefficient of expansion between 25° and 45° being 0.0027 for 1° C., that is to say, more than two-thirds as great as that of gases. The two modifications differ greatly in their relations to solvents. Liquid sulphuric oxide mixes in all proportions with hydrogen sulphate, H_2SO_4; the β-oxide dissolves in sulphuric acid with extreme slowness, and gradually separates from a mixture of 10 parts SO_3 and 1-2 parts H_2SO_4. With a larger proportion of the acid, no solid oxide separates, even on cooling to a low temperature; if the proportion of acid, H_2SO_4, amounts to 5 parts for 10 parts of oxide, the mixture may deposit crystals of pyrosulphuric acid, $H_2S_2O_7$ or H_2SO_4,SO_3.

Sulphuric Acid, H_2SO_4.—The preparation of this important acid depends upon the fact that, when sulphurous oxide, nitrogen tetroxide, and water are present together in certain proportions, the sulphurous oxide becomes oxidized at the expense of the nitrogen tetroxide, which, by the loss of one-half of its oxygen, sinks to the condition of nitrogen dioxide. The operation is thus conducted:

A large and very long chamber is built of sheet-lead supported by timber framing; on the outside, at one extremity, a small furnace or oven is constructed, having a wide tube leading into the chamber. In this, sulphur is kept burning, the flame of which heats a crucible containing a mixture of nitre and oil of vitriol. A shallow stratum of water occupies the floor of. the chamber, and a jet of steam is also introduced. Lastly, an exit is provided at the remote end of the chamber for the spent and useless gases. The effect of these arrangements is to cause a constant supply of sulphurous oxide, atmospheric air, nitric acid vapor, and water in the state of steam, to be thrown into the chamber, there to mix and react upon each other. The nitric acid immediately gives up a part of its oxygen to the sulphurous oxide, and is itself reduced to nitrogen tetroxide, N_2O_4 or NO_2; it does not remain in this state, however, but suffers further deoxidation until it becomes reduced to nitrogen dioxide, N_2O_2 or NO. That substance, in contact with free oxygen, absorbs a portion of the latter, and once more becomes tetroxide, which is again destined to undergo deoxidation by a fresh quantity of sulphurous oxide. A very small portion of nitrogen tetroxide mixed with atmospheric air and sulphurous oxide, may thus in time convert an indefinite amount of the latter into sulphuric acid, by acting as a kind of carrier between the oxygen of the air and the sulphurous oxide. The presence of water is essential to this reaction, which may be represented by the equation,

$$NO_2 + SO_2 + H_2O = NO + H_2SO_4.$$

Such is the simplest view that can be taken of the production of sulphuric acid in the leaden chamber; but it is too much to affirm that it is strictly true; the reaction may be more complex. When a little water is placed at the bottom of a large glass globe, so as to maintain a certain degree of humidity in the air within, and sulphurous oxide and nitrogen tetroxide are introduced by separate tubes, symptoms of chemical action become immediately evident, and after a little time a white crystalline matter is observed to condense on the sides of the vessel. This substance appears to be a compound of sulphuric acid, nitrous acid, and a little water.*

* Gaultier de Claubry assigned to this substance the composition expressed by the formula $2(N_2O_3,2H_2O),5SO_3$, and this view has generally been received by recent chemical writers. De la Provostaye has since shown that a compound possessing all the essential properties of the body in question may be formed by bringing together, in a sealed glass tube, liquid sulphurous oxide and liquid nitrogen tetroxide, both free from water. The white crystalline solid soon begins to form, and at the expiration of twenty-six hours the reaction appears complete. The new product is accompanied by an exceedingly volatile greenish liquid, having the characters of nitrous acid. The white substance, on analysis, was found to contain the elements of two molecules of sulphuric oxide, and one ·

When thrown into water, it is resolved into sulphuric acid, nitrogen dioxide, and nitric acid. This body is certainly very often produced in large quantity in the leaden chambers; but that its production is indispensable to the success of the process, and constant when the operation goes on well and the nitrogen tetroxide is not in excess, may perhaps admit of doubt.

The water at the bottom of the chamber thus becomes loaded with sulphuric acid. When a certain degree of strength has been reached, the acid is drawn off and concentrated by evaporation, first in leaden pans, and afterwards in stills of platinum, until it attains a density (when cold) of 1.84, or thereabouts; it is then transferred to carboys, or large glass bottles fitted in baskets, for sale. This manufacture is one of great importance, and is carried on to a vast extent, the annual production in Great Britain amounting to no less than 850,000 tons.*

Sulphuric acid is now more frequently made by burning iron pyrites, or poor copper ore, or zinc-blende, instead of Sicilian sulphur: as thus prepared it very frequently contains arsenic, from which it may be freed, however, by heating it with a small quantity of sodium chloride, or by passing through the heated acid a current of hydrochloric acid gas, whereby the arsenic is volatilized as trichloride.

The most concentrated sulphuric acid, or *oil of vitriol*, as it is often called, is a definite combination of 40 parts sulphuric oxide and 9 parts of water, and is represented by the formula H_2O,SO_3 or H_2SO_4. It is a colorless oily liquid, having a specific gravity of about 1.85, of intensely acid taste and reaction. Organic matter is rapidly charred and destroyed by it. At the temperature of $-26°$ C. ($-15°$ F.) it freezes; at $327°$ C. ($620°$ F.) it boils, and may be distilled without decomposition. Oil of vitriol has a most energetic attraction for water; it withdraws aqueous vapor from the air, and when it is diluted with water, great heat is evolved, so that the mixture always requires to be made with caution. Oil of vitriol is not the only hydrate of sulphuric oxide; three others are known to exist. When the fuming oil of vitriol of Nordhausen is exposed to a low temperature, a white crystalline substance separates, which is a hydrate containing half as much water as the common liquid acid. Further, a mixture of 98 parts of strong liquid acid and 18 parts of water, $2H_2O,SO_3$ or H_2SO_4,H_2O, congeals or crystallizes at a temperature above $0°$ C., and remains solid even at $7.2°$ C. ($45°$ F.). Lastly, when a very dilute acid is concentrated by evaporation in a vacuum over a surface of oil of vitriol, the evaporation stops when the sulphuric oxide and water bear to each other the proportion of 80 to 54, answering to the formula $3H_2O,SO_3$ or $H_2SO_4,2H_2O$.

When the vapor of sulphuric acid is passed over red-hot platinum, it is decomposed into oxygen and sulphurous acid. St. Claire Deville and Debray have recommended this process for the preparation of oxygen on the large scale, the sulphurous acid being easily separated by its solubility in water or alkaline solutions.

Sulphuric acid acts readily on metallic oxides; converting them into sulphates. It also decomposes carbonates with the greatest ease, expelling carbon dioxide with effervescence. With the aid of heat it likewise decomposes all other salts containing acids more volatile than itself. The sulphates are a very important class of salts, many of them being extensively used in the arts. Most sulphates are soluble in water, but they are all insoluble in alcohol. The barium, calcium, strontium, and lead salts are insoluble, or very

of nitrous oxide, or $N_2O_3,2SO_3$. De la Provostaye explains the anomalies in the different analyses of the leaden-chamber product, by showing that the pure substance forms crystallizable combinations with different proportions of sulphuric acid. (Ann. Chim. Phys. lxxiii. 362.) See also Weber (Jahresbericht für Chemie, 1863, p. 738; 1865, p. 93; Bull. Soc. Chim. de Paris, 1867, i. 15).

* For a fuller description of the process and figures of the apparatus, see Roscoe and Schorlemmer's Treatise on Chemistry, vol. i. pp. 319–338.

slightly soluble, in water, and are formed by precipitating a soluble salt of either of those metals with sulphuric acid or a soluble metallic sulphate. Barium sulphate is quite insoluble in water: consequently sulphuric acid, or its soluble salts, may be detected readily by solution of barium nitrate or chloride: a white precipitate is thereby formed, which does not dissolve in nitric or hydrochloric acid.

Disulphuric or **Pyrosulphuric Acid**, $H_2S_2O_7$ (also called *Fuming Sulphuric Acid* and *Nordhausen Sulphuric Acid*).—This acid contains the elements of one molecule of sulphuric oxide and one molecule of sulphuric acid, or of two molecules of sulphuric acid *minus* one molecule of water:

$$H_2S_2O_7 = SO_3,H_2SO_4 = 2H_2SO_4 - H_2O.$$

It may be obtained of definite composition and in the crystalline form by adding liquid sulphuric oxide to strong sulphuric acid, in the proportion above indicated. The resulting crystals melt at 35°. It is also prepared from the impure ferric sulphate obtained by exposing ordinary ferrous sulphate (green vitriol) to a moderate heat in contact with the air. This ferric sulphate is distilled in earthen retorts arranged in a reverberatory furnace, and the distillate, consisting chiefly of sulphuric oxide, is received in a small quantity of water, or more frequently in ordinary strong sulphuric acid. A brown fuming liquid is thus obtained, which agrees nearly in composition with the formula $H_2S_2O_7$, has a specific gravity of 1.9, solidifies at 0° in colorless crystals, and is resolved at a gentle heat into SO_3 which distils over, and H_2SO_4 which remains behind.

The manufacture of fuming sulphuric acid in the manner just described was first practised at Nordhausen, in Saxony, and appears to have been known since the fifteenth century; but it is now carried on almost exclusively in Bohemia. An easier and more productive method of obtaining the sulphuric oxide required for its formation is that of Messel and Squire, already described (p. 177). Fuming sulphuric acid was until lately employed only for dissolving indigo, but it is now used in very large quantities for dissolving anthraquinone for the manufacture of artificial alizarin. (See Organic Chemistry.)

The pyrosulphates, that of potassium, for example, which has the composition $K_2S_2O_7$, or K_2SO_4,SO_3, are prepared by the action of sulphuric oxide on the corresponding sulphates. When strongly heated, they give off sulphuric oxide, and are converted into sulphates.

Hyposulphurous Acid, H_2SO_2 (also called *Hydrosulphurous Acid*).— This acid is formed by the action of zinc on an aqueous solution of sulphurous acid. The zinc dissolves without evolution of hydrogen, merely removing an atom of oxygen. A yellow solution is thereby formed, which possesses much greater decolorizing power than sulphurous acid itself, and quickly reduces the metals from salts of silver and mercury. This solution is, however, very unstable, and quickly loses its bleaching power. A more definite product is obtained by immersing clippings of zinc in a concentrated solution of acid sodium sulphite, $NaHSO_3$, contained in a closed vessel, whereby sodium hyposulphite, Na_2SO_2, and zinc-sodium sulphite, $Na_2Zn(SO_3)_2$, are produced, the latter crystallizing out. To isolate the hyposulphite, the liquid is decanted, after about half an hour, into a flask three-fourths filled with strong alcohol, and the flask is sealed. A crystalline precipitate immediately forms, consisting for the most part of zinc-sodium sulphite, while nearly all the hyposulphite remains dissolved in the alcohol. The solution, decanted into a flask quite filled with it, well closed, and left in a cool place, solidifies in a few hours to a mass of slender colorless needles, consisting of sodium hyposulphite, which must be quickly pressed between folds of linen and dried in a vacuum, as it becomes very hot if exposed to the air in the moist state; when dry, however, it is not affected by oxygen. This salt is very soluble in water, soluble also in dilute alcohol, the solutions exhibiting all the bleaching and reducing properties above described. The crystals when exposed to the air are completely converted into acid sodium sulphite, $NaHSO_3$. By heating them with oxalic acid, hyposulphurous acid is obtained as a deep orange-colored strongly bleaching liquid, which quickly decomposes, becoming colorless, and depositing sulphur.[*]

Thiosulphuric Acid, $H_2S_2O_3$ (formerly called *Hyposulphurous Acid*).— By digesting sulphur with a solution of potassium or sodium sulphite, a portion of that substance is dissolved, and the liquid, by slow evaporation, yields crystals of thiosulphate: $Na_2SO_3 + S = Na_2S_2O_3$. The acid itself is scarcely known, for it cannot be isolated; when hydrochloric acid is added to a solution of a thiosulphate, the acid of the latter is almost instantly resolved into sulphur, which precipitates, and sulphurous acid, easily recognized by its odor. In a very dilute solution, however, it appears to remain undecomposed for some time. The alkaline thiosulphates readily dissolve certain salts of silver, as the chloride, which are insoluble in water—a property which has conferred upon them a considerable share of importance in relation to the art of photography. They are also much used as anti-chlores for removing the last traces of chlorine from bleached goods.

Seleniosulphuric Acid, H_2SeSO_3.—This acid, having the composition of sulphuric acid in which 1 atom of oxygen is replaced by selenium, is formed by direct addition of selenium to sulphurous acid. When selenium is digested with a solution of neutral potassium sulphite, and the easily decomposable liquid, after being filtered from the selenium which separates on cooling and dilution with water, is left to evaporate at ordinary temperatures, there crystallizes out, first a sparingly soluble seleniferous salt in small shining prisms, afterward a much more soluble salt, which is the chief product of the reaction, while the excess of sulphurite remains in the mother-liquor.

The most soluble seleniferous salt is potassium seleniosulphate, K_2SeSO_3. It is likewise formed, together with thiosulphate, when a solution of potassium selenide is mixed with sulphurous acid. It crystallizes readily, even from small quantities of solution, in large, very thin, six-sided tables belonging to the rhombic system, which deliquesce in moist air, and effloresce with partial loss of water over oil of vitriol. When heated they turn brown, and yield a polysulphide of potassium. Water separates selenium from them, and the filtered solution yields by evaporation crystals which again react in the same way with water, so that, by repeated crystallization, the whole of the seleniosulphate may be decomposed; the liquid then contains seleniotrithionate. Acids, even sulphurous acid, throw down the whole of the selenium from the aqueous solution; barium chloride and baryta-water precipitate barium sulphite and selenium; calcium and manganese salts give rise to a similar decomposition. With ammoniacal silver solution the seleniosulphate forms a precipitate of silver selenide, together with potassium sulphate:

$$K_2SeSO_3 + Ag_2O = Ag_2Se + K_2SO_4.$$

The sulphites of sodium, ammonium, and magnesium react with selenium in the same manner as the potassium salt, the magnesium salt, however,

[*] Schützenberger, Zeitschrift für Chemie, 1869, p. 545.

very slowly. The seleniosulphates of sodium and ammonium are very unstable.

Dithionic or Hyposulphuric Acid, $H_2S_2O_6$.—This acid is prepared by suspending finely divided manganese dioxide in water artificially cooled, and passing a stream of sulphurous acid gas through the liquid. The dioxide then becomes monoxide, half its oxygen converting the sulphurous into dithionic acid: $MnO_2 + 2SO_2 = MnS_2O_6$; the manganese dithionate thus prepared is decomposed by a solution of pure barium hydrate; and the barium salt, in turn, by enough sulphuric acid to precipitate the base. The solution of dithionic acid may be concentrated by evaporation in a vacuum, until it acquires a density of 1.347; on further concentration, it decomposes into sulphuric and sulphurous acids. It has no odor, is very sour, and forms soluble salts with baryta, lime, and lead oxide.

Trithionic Acid, $H_2S_3O_6$.—A substance accidentally formed by Langlois, in the preparation of potassium thiosulphate, by gently heating with sulphur a solution of potassium carbonate saturated with sulphurous acid. It is also produced by the action of sulphurous oxide on potassium thiosulphate: $2K_2S_2O_3 + 3SO_2 = 2K_2S_3O_6 + S$. Its salts bear a great resemblance to those of thiosulphuric acid, but differ completely in composition, while the acid itself is not quite so prone to change. It is obtained by decomposing the potassium salt with hydrofluosilicic acid: it may be concentrated under the receiver of the air-pump, but is gradually decomposed into sulphurous and sulphuric acids, and free sulphur.

Tetrathionic Acid, $H_2S_4O_6$.—This acid was discovered by Fordos and Gelis. When iodine is added to a solution of barium thiosulphate, a large quantity of that substance is dissolved, and a clear colorless solution obtained, which, besides barium iodide, contains barium tetrathionate: $2BaS_2O_3 + I_2 = BaI_2 + BaS_4O_6$. By suitable means, the acid can be eliminated, and obtained in a state of solution. It very closely resembles dithionic acid. The same acid is produced by the action of sulphurous acid on chlorine disulphide.

Pentathionic Acid, $H_2S_5O_6$.—This acid was discovered by Wackenroder, who formed it by the action of hydrogen sulphide on sulphurous acid: $5H_2SO_3 + 5H_2S = H_2S_5O_6 + 9H_2O + S_5$. It is colorless and inodorous, of acid and bitter taste, and capable of being concentrated to a considerable extent by cautious evaporation.

Under the influence of heat, it is decomposed into sulphur, sulphurous and sulphuric acids, and hydrogen sulphide. The salts of pentathionic acid are nearly all soluble. The barium salt crystallizes from alcohol in square prisms. The acid is also formed when lead dithionate is decomposed by hydrogen sulphide, and when chlorine monosulphide is heated with sulphurous acid.

Oxychlorides.—1. *Sulphurous Chloride*, $SOCl_2$.—This compound, also called *Chloride of Thionyl*, is derived from sulphurous acid, SO_3H_2 or $SO(HO)_2$, by the substitution of 2Cl for 2HO. It is formed by the action of water, alcohols, acids, etc., on the sulphides of chlorine; but is more easily prepared by the action of phosphorus pentachloride on sulphurous oxide, or by that of phosphorus oxychloride on calcium sulphite:

$$SO_2 + PCl_5 = POCl_3 + SOCl_2.$$
$$3CaSO_3 + 2POCl_3 = Ca_3P_2O_8 + 3SOCl_2.$$

It is separated by distillation from the fixed calcium phosphate produced simultaneously in the second, and by fractional distillation from the phosphorus oxychloride produced in the first reaction.

Sulphurous chloride is a colorless, strongly refracting liquid, which boils at 82° C. (179.6° F.) It is decomposed by water, yielding hydrochloric and sulphurous acids; and by alcohols with formation of alcoholic chlorides and sulphurous acid, thus:

$$SOCl_2 + 2H_2O = 2HCl + H_2SO_3$$
$$SOCl_2 + 2(C_2H_5C)HO = 2C_2H_5Cl + H_2SO_3.$$
Ethyl alcohol.   Ethyl chloride.

Sulphuric chloride, or *Sulphuryl chloride*, SO_2Cl_2 (also called *Chlorosulphuric acid*), is formed by prolonged exposure of a mixture of chlorine and sulphurous oxide gases to strong sunshine; also, together with phosphorus oxychloride, by the action of phosphorus pentachloride on sulphuric oxide:

$$SO_3 + PCl_5 = POCl_3 + SO_2Cl_2;$$

but it is best prepared by distilling strong sulphuric acid with the pentachloride, or lead sulphate with the oxychloride of phosphorus:

$$H_2SO_4 + 2PCl_5 = SO_2Cl_2 + 2POCl_3 + 2HCl$$
$$3PbSO_4 + 2POCl_3 = 3SO_2Cl_2 + Pb_3P_2O_8$$
Lead phosphate.

Sulphuric chloride is a colorless fuming liquid, of specific gravity 1.66. It boils at 77° C. (170.6° F.), and may be distilled unchanged over caustic lime or baryta. When poured into water, it sinks in the form of oily drops, which gradually disappear, being converted into hydrochloric and sulphuric acids:

$$SO_2Cl_2 + 2H_2O = 2HCl + H_2SO_4.$$

With alcohol it behaves in a similar manner, thus:

$$SO_2Cl_2 + 2(C_2H_5)HO = 2C_2H_5Cl + H_2SO_4.$$

In the actual reaction, however, the sulphuric acid is converted into ethyl-sulphuric acid by the intervention of another atom of alcohol:

$$H_2SO_4 + (C_2H_5)HO = H_2O + (C_2H_5)HSO_4.$$

Sulphuric Hydroxychloride or **Chlorhydrate**, $HClSO_3$ or SO_2 (HO)Cl (also called *Chlorhydrosulphurous Acid* and *Sulphuric Chlorhydrin*). —This compound, discovered by Williamson, is intermediate in composition between sulphuric acid and sulphuric chloride, and is derived from sulphuric acid, SO_4H_2, or $SO_2(HO)_2$, by the substitution of 1Cl for 1HO. It is the first product of the action of phosphorus pentachloride on strong sulphuric acid:

$$SO_2(HO)(HO) + PCl_5 = POCl_3 + HCl + SO_2(HO)Cl.$$

As thus prepared it is mixed with sulphuric chloride; but it may be obtained pure by treating sulphuric acid with phosphorus oxychloride, hydrochloric and metaphosphoric acid being produced at the same time:

$$2SO_2(HO)(HO) + POCl_3 = 2SO_2(HO)Cl + HCl + HPO_3.$$

It is also formed by the action of water on sulphuric chloride:

$$SO_2Cl_2 + H_2O = HCl + HClSO_3.$$

Sulphuric hydroxychloride is a colorless liquid, which boils at about 150°, being at the same time partially resolved into sulphuric acid and sulphuric chloride: $2HClSO_3 = H_2SO_4 + Cl_2SO_2$. When poured into water, it sinks to the bottom and gradually dissolves, with formation of hydrochloric and sulphuric acids. It has decided acid properties, and forms definite salts in which its hydrogen is replaced by metals. Thus it dissolves sodium chloride at a gentle heat, with evolution of hydrochloric acid, and formation of the salt $NaClSO_3$.

Pyrosulphuric Chloride, $S_2O_5Cl_2$ or $(SO_2Cl)_2O$.—This compound is formed on heating together phosphorus pentachloride and sulphuric oxide:

$$PCl_5 + 2SO_3 = S_2O_5Cl_2 + POCl_3,$$

and by the action of phosphorus pentachloride on sulphuric hydroxychloride:

$$PCl_5 + 2SO_2ClOH = (SO_2Cl)_2O + POCl_3 + 2HCl.$$

It is a colorless oily liquid of sp. gr. 1.819 at 18°, boiling at 146°. In contact with water it decomposes slowly and noiselessly, and is thus distinguished from sulphuric hydroxychloride, which is rapidly decomposed, with almost explosive violence, when thrown into water. When heated above its boiling point, it is resolved into chlorine, sulphurous oxide, and sulphuric oxide:

$$S_2O_5Cl_2 = SO_3 + SO_2 + Cl_2.$$

When submitted to the action of phosphorus pentachloride, it yields chlorine, sulphurous oxide, and phosphorus oxychloride:

$$S_2O_5Cl_2 + PCl_5 = POCl_3 + 2SO_2 + 2Cl_2.$$

SELENIUM.

Symbol, Se.   Atomic weight, 79.   Vapor-density, 79.

THIS substance, much resembling sulphur in its chemical relations, occurs frequently, associated in small quantities with that element, or replacing it in certain metallic combinations, as in the lead selenide of Clausthal in the Hartz. To separate it, the pulverized ore is treated with hydrochloric acid to dissolve earthy carbonates, and the washed and dried residue is ignited for some time with an equal quantity of black flux (a mixture of potassium carbonate and charcoal). The selenium is thereby converted into potassium selenide, which by treatment with boiling water is dissolved away from the oxides formed at the same time. This solution when exposed to the air absorbs oxygen, and yields the selenium as a gray deposit, which may be purified by washing, drying, and distillation.*

A new source of selenium has lately been discovered in the combustion

* For further details, and for other methods, see Gmelin's Hand-book of Chemistry English edition, vol. ii. p. 232.

of seleniferous pyrites for the manufacture of sulphuric acid, selenic oxide, SeO_3, being thereby formed, and reduced by sulphurous acid, in the course of the process, to free selenium, which partly dissolves in the sulphuric acid, and partly remains suspended. Now when this seleniferous sulphuric acid is used for the production of salt-cake in the soda-manufacture (see SODIUM CARBONATE), the selenium volatilizes with the escaping hydrochloric acid, and is deposited in the first condensers, the deposit thus formed containing from 41 to 45 per cent. of selenium. To extract this selenium, the deposit suspended in water is treated with chlorine, whereby selenium tetrachloride is formed, which is then decomposed by the water, yielding selenious acid, partly oxidized to selenic acid. The dark-colored liquid is then filtered through cloth, and boiled with excess of hydrochloric acid to reduce the selenic acid to selenious acid, which is then further reduced to selenium by the action of acid sodium sulphite. By this process large quantities of nearly pure selenium are easily and rapidly obtained.*

Selenium, like sulphur, exists in two allotropic modifications, one soluble, the other insoluble in carbon sulphide. _a._ The _soluble modification_ is obtained as a finely divided brick-red powder on passing a current of sulphur dioxide into a cold solution of selenious acid,—and as a black crystalline powder when the same gas is passed through a hot solution of the acid. It crystallizes from solution in carbon sulphide in small dark-red monoclinic crystals, isomorphous with monoclinic sulphur, and having a specific gravity of 4.5. It has no definite melting point, but softens gradually when heated.—β. _Insoluble_ or _Metallic Selenium_ is obtained by quickly cooling melted selenium to 210° C. (410° F.), and keeping the melted mass at this temperature for some time, whereupon it ultimately solidifies to a granular crystalline mass, the temperature then suddenly rising to 217° C. (422.6° F.). The solid selenium thus obtained has a density of 4.5, and is insoluble in carbon sulphide. This metallic selenium melts constantly at 217° C., and is converted by rapid cooling into the amorphous soluble variety. Both modifications dissolve in selenium chloride, the selenium separating therefrom in the metallic form. Selenium boils somewhat below 700°, forming a dark-red vapor, which condenses either in the form of scarlet "flowers of selenium," or in dark shining drops.

The vapor-density of selenium, like that of sulphur, diminishes very rapidly as its temperature rises, being 110.7 (hydrogen being taken as unity) at 860° C. (1580° F.), and 81.5 at 1420° C. (2588° F.), which does not differ greatly from the theoretical density.

Metallic selenium conducts electricity, and its conducting power is increased by exposure to light, and diminished by heating. Exposure to diffused daylight immediately diminishes the electrical resistance of selenium to one-half of what it was before, but on cutting off the light, the resistance slowly increases, and soon reaches its original amount.

Selenium when heated in the air burns with a pale-blue flame, emitting an odor of decayed horse-radish, due to the formation of an oxide.

Hydrogen Selenide, H_2Se.—This compound, also called _Selenhydric acid_ and _Selenietted hydrogen_, is formed by the action of dilute sulphuric acid on selenide of potassium or iron. It is very much like hydrogen sulphide, being a colorless gas, freely soluble in water, and decomposing metallic solutions, with formation of insoluble selenides. It acts very powerfully on the mucous membrane of the nose, exciting catarrhal symptoms and temporarily destroying the sense of smell.

Selenium Chlorides.—The _monochloride_, Se_2Cl_2, formed by passing

---

* Bulletin de la Société chimique de Paris, 2me série, **xxxvii.** 443; Journal of the Chemical Society, xliv. 16.

16 *

chlorine gas over selenium, is a brown oily liquid, which readily dissolves selenium, and deposits it in the metallic form on cooling. It is slowly decomposed by water, yielding selenious acid, hydrochloric acid, and free selenium: $2Se_2Cl_2 + 3H_2O = H_2SeO_3 + 3Se + 4HCl$.

The *tetrachloride*, $SeCl_4$, is obtained by the further action of chlorine on the monochloride, or by heating the dioxide with phosphorus pentachloride: $3SeO_2 + 3PCl_5 = 3SeCl_4 + P_2O_5 + POCl_3$. It is a white solid body, which when heated volatilizes without previous fusion, and sublimes in small crystals. It dissolves in water, forming hydrochloric and selenious acids: $SeCl_4 + 3H_2O = 4HCl + H_2SeO_3$.

Selenium Bromides.—The *monobromide*, Se_2Br_2, formed by heating together equal weights of bromine and selenium, is a black semi-opaque liquid, having a density of 3.6 at 15°. It smells like sulphur monochloride, and stains the skin brown-red. It is decomposed by heat and by the action of water, yielding in the latter case hydrobromic and selenious acids. The *tetrabromide*, $SeBr_4$, formed by the further action of bromine on the monobromide, is an orange-red crystalline powder, which volatilizes without decomposition between 75° and 80° C. (167° and 176° F.), and sublimes in black hexagonal scales. It smells like sulphur chloride, decomposes in moist air into bromine and monobromide, and dissolves in excess of water, with formation of hydrobromic and selenious acids.

Selenium Iodides.—The *monoiodide*, Se_2I_2, formed by direct combination, is a black shining crystalline body melting at 68–70° C. (154–158° F.), with slight evolution of iodine. It is resolved into its elements at a higher temperature, and is decomposed by water like the corresponding chloride and bromide.—The *tetraiodide*, SeI_4, is a dark-colored granular crystalline mass, melting at 75–80° C. (167–176° F.) to a brownish-black liquid, translucent in thin films. At a higher temperature it is resolved into its elements.

Fluoride.—This compound, formed by passing selenium vapor over melted lead fluoride, sublimes in crystals, which dissolve in hydrofluoric acid and are decomposed by water.

Oxides and Oxy-Acids of Selenium.—Two oxides of selenium are known. The one containing the smaller proportion of oxygen is formed by the imperfect combustion of selenium in air or oxygen gas. It is a colorless gas, which is the source of the peculiar horse-radish odor above mentioned. Its composition is not known.

The higher oxide, SeO_2, called **selenious oxide**, is produced by burning selenium in a stream of oxygen gas. It is a white solid substance, which absorbs water rapidly, forming a hydroxide.

Selenious acid, H_2SeO_3 or H_2O,SeO_2.—This acid, analogous in composition and properties to sulphurous acid, is likewise produced by dissolving selenium in nitric or nitromuriatic acid. It is deposited from its hot aqueous solution by slow cooling in prismatic crystals like those of saltpetre; but when the solution is evaporated to dryness, the selenious acid is resolved into water and selenious oxide, which sublimes at a higher temperature.

Selenious acid is a very powerful acid, approximating to sulphuric acid in the energy of its reactions. It reddens litmus, decomposes carbonates with effervescence, and decomposes nitrates and chlorides with aid of heat. Its solution precipitates lead and silver salts, and is decomposed by hydrogen sulphide, yielding a precipitate of selenium sulphide: $H_2SeO_3 + 2H_2S = 3H_2O + SeS_2$.

The metallic selenites resemble the sulphites. When heated with sodium carbonate in the inner blowpipe flame, they emit the characteristic odor of selenium. They are not decomposed by boiling with hydrochloric acid.

Selenic acid, H_2SeO_4, is analogous in composition to sulphuric acid. The corresponding anhydrous oxide is not known. Selenic acid is prepared by fusing potassium or sodium nitrate with selenium, precipitating the selenate so produced with a lead salt, and decomposing the resulting compound with hydrogen sulphide. The acid strongly resembles oil of vitriol; but when very much concentrated, it is decomposed by heat into selenious acid and oxygen. The selenates bear the closest analogy to the sulphates in almost every particular. They are decomposed by boiling with hydrochloric acid, with evolution of chlorine and formation of a metallic selenite.

TELLURIUM.

Symbol, Te. Atomic weight, 128. Vapor-density, 128.

THIS element possesses many of the characters of a metal, but it bears so close a resemblance to selenium, both in its physical properties and its chemical relations, that it is most appropriately placed in the same group with that body. Tellurium is found in a few scarce minerals, in association with gold, silver, lead, and bismuth, apparently replacing sulphur, and is most easily extracted from the bismuth sulpho-telluride of Chemnitz in Hungary. The finely powdered ore is mixed with an equal weight of dry sodium carbonate, and the mixture, made into a paste with oil, is heated to whiteness in a closely covered crucible. Sodium telluride and sulphide are thereby produced, and metallic bismuth is set free. The fused mass is dissolved in water, and the solution freely exposed to the air, when the sodium and sulphur oxidize to sodium hydroxide and thiosulphate, while the tellurium separates in the metallic state.

Tellurium has the color and lustre of silver; by fusion and slow cooling, it may be made to exhibit the form of rhombohedral crystals, similar to those of antimony and arsenic. It is brittle, and a comparatively bad conductor of heat and electricity; it has a density of 6.26, melts at a little below a red heat, and volatilizes at a higher temperature. Tellurium burns when heated in the air, and is oxidized by nitric acid.

Hydrogen Telluride, H_2Te.—*Tellurhydric Acid, Hydrotelluric Acid,* or *Telluretted Hydrogen.*—This compound is a gas, resembling sulphuretted and selenietted hydrogen. It is prepared by the action of hydrochloric acid on zinc telluride. It dissolves in water, forming a colorless liquid, which precipitates most metals from their solutions, and deposits tellurium on exposure to the air.

Tellurium Chlorides.—Tellurium forms a dichloride, $TeCl_2$, and a tetrachloride, $TeCl_4$, both volatile and decomposable by excess of water, the latter being completely resolved into tellurous and hydrochloric acids: $TeCl_4 + 3H_2O = 4HCl + H_2TeO_3$.

The tetrachloride unites with the chlorides of the alkali-metals, to form crystallizable double salts.

The *bromides* and *iodides of tellurium* correspond with the chlorides in properties and composition.

Oxides and Oxy-acids.—Tellurium forms two oxides, analogous in composition to the oxides of sulphur, and likewise forming acids by combination with water.

Tellurous Oxide, TeO_2, may be prepared by heating the precipitated acid to low redness. It also separates in semi-crystalline grains from the aqueous solution of the acid when gently heated; more abundantly and in well-defined octahedrons from the solution of tellurous acid in nitric acid. It is fusible and volatile, slightly soluble in water, but does not redden litmus. When fused with alkaline hydroxides or carbonates, it forms tellurites.

Tellurous Acid is best obtained by decomposing tellurium tetrachloride with water. It may also be prepared by dissolving tellurium in nitric acid of sp. gr. 1.25, and pouring the solution, after a few minutes, into a large quantity of water. By either process it is obtained as a somewhat bulky precipitate, which, when dried over oil of vitriol, appears as a light, white, earthy mass, having a bitter metallic taste. It is slightly soluble in water, more easily soluble in alkalies and acids, the nitric acid solution alone being unstable. Sulphurous acid, zinc, phosphorus, and other reducing agents, precipitate metallic tellurium from the acidified solution of tellurous acid. Like selenious acid, it is decomposed by hydrogen sulphide and alkaline hydrosulphides, with formation of a dark-brown tellurium sulphide, which dissolves readily in excess of alkaline hydrosulphide, forming a sulpho-tellurite.

Tellurous acid is a hydroxide in which the acid and basic tendencies are nearly balanced; in other words, the tellurium of the compound can replace the hydrogen of an acid to form tellurous salts, and the hydrogen can be replaced by the basylous metals to form metallic tellurites.

TELLURIUM SALTS.		TELLURITES.	
$Te(SO_4)_2$	Sulphate.	H_2TeO_3	Hydrogen tellurite.
$Te(NO_3)_4$	Nitrate.	K_2TeO_3	Potassium tellurite.
$Te(C_2O_4)_2$	Oxalate.	$HKTeO_3$	Hydrogen and potassium tellurite.
$TeCl_4$	Chloride.	$H_3K(TeO_3)_2$	Trihydropotassic tellurite.

The tellurites of potassium, sodium, barium, strontium, and calcium, are formed by fusing tellurous oxide or acid with the carbonates of the several metals in the required proportions. These tellurites are all more or less soluble in water. The tellurites of the other metals, which are insoluble, are obtained by precipitation.

Compounds of tellurous oxide with the halogen acids are also known. When this oxide is exposed to the action of gaseous *hydrogen bromide* in a vessel cooled to $-14°$ C. ($68°$ F.), the compound $TeO_2,3HBr$ is formed in groups of small nearly black scales resembling iodine. At $40°$ this compound gives off HBr, and is reduced to $TeO_2,2HBR$, which when heated to $300°$ C. ($572°$ F.) is resolved into water and a yellow o x y b r o m i d e: $TeO_2,2HBr = H_2O + TeOBr_2$; and at a still higher temperature this oxybromide is decomposed into the t e t r a b r o m i d e and tellurous oxide: $2TeOBr_2 = TeBr_4 + TeO_2$. The tetrabromide passes off in black vapors and crystallizes on cooling in dark-brown needles.

Tellurous oxide is decomposed by *hydriodic acid* at ordinary temperatures, but absorbs it at $-15°$ C. ($5°$ F.), forming a compound which decomposes as the temperature rises. Tellurous oxide likewise absorbs anhydrous *hydrofluoric acid.*

Telluric Oxide and Acid.—To prepare telluric acid, equal parts of tellurous oxide and sodium carbonate are fused, and the product is dissolved in water; a little sodium hydroxide is added, and a stream of chlorine passed through the solution. The liquid is next saturated with ammonia, and mixed with solution of barium chloride, by which a white insoluble precipitate of barium tellurate is thrown down. This is washed and digested with a quarter of its weight of sulphuric acid, and diluted with water. The filtered solution gives, on evaporation in the air, large crystals of telluric acid, which have the composition, $H_2TeO_4,2H_2O$.

Crystallized telluric acid is freely, although slowly, soluble in water: it has a metallic taste, and reddens litmus-paper. The crystals give off their water of crystallization at 100° C., and the remaining acid, H_2TeO_4, when strongly heated, gives off more water, and yields the anhydrous oxide, TeO_3, which is then insoluble in water, and even in a boiling alkaline liquid. At the temperature of ignition, telluric oxide loses oxygen, and passes into tellurous oxide.

The tellurates of the alkali-metals are soluble in water, and are prepared by dissolving the required quantities of telluric acid and an alkaline carbonate in hot water. The other tellurates are insoluble, and are obtained by precipitation.

The composition of the alkaline tellurates is exhibited by the following formulæ:—

Neutral or Bipotassic tellurate	K_2TeO_4	or K_2O,TeO_3
Acid or Hydro-potassic tellurate	$HKTeO_4$	or $H_2O,K_2O,2TeO_3$
Quadracid or Trihydropotassic tellurate . .	$HKTeO_4,H_2TeO_4$	or $3H_2O,K_2O,4TeO_3$
Anhydrous Quadritellurate	$K_2TeO_4,3TeO_3$	or $K_2O,4TeO_3$

Tellurium Sulphides.—Tellurium forms two sulphides, TeS_2 and TeS_3, analogous in composition to the oxides; they are obtained by the action of hydrogen sulphide on solutions of tellurous acid and telluric acid respectively. They are brown or black substances, which unite with metallic sulphides, forming salts called thiotellurites and thiotellurates.

NITROGEN.

Symbol, N. Atomic weight, 14.01. Density, 14.01.

NITROGEN * constitutes about four-fifths of the atmosphere, and enters into a great variety of combinations. It may be prepared by several methods. One of the simplest of these is to burn out the oxygen from a confined portion of air by phosphorus, or by a jet of hydrogen.

A small porcelain capsule is floated on the water of the pneumatic trough, and a piece of phosphorus is placed in it and set on fire. A bell-jar is then inverted over the whole, and suffered to rest on the shelf of the trough, so as to project a little over its edge. At first the heat causes expansion of the air of the jar, and a few bubbles are expelled, after which the level of the

* *i.e.*, Generator of nitre; also called *Azote*, from *a*, privative, and ξωή, life.

water rises considerably. When the phosphorus becomes extinguished by exhaustion of the oxygen, and time has been given for the subsidence of the cloud of finely divided snow-like phosphoric oxide which floats in the residual gas, the nitrogen may be transferred into another vessel, and its properties examined.

Fig. 117.

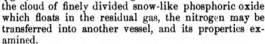

Prepared by the foregoing process, nitrogen is contaminated with a little vapor of phosphorus, which communicates its peculiar odor. A preferable method is to fill a porcelain tube with turnings of copper, or, still better, with the spongy metal obtained by reducing the oxide with hydrogen; to heat this tube to redness; and then pass through it a slow stream of atmospheric air, the oxygen of which is entirely removed, during its progress, by the heated copper.

When chlorine gas is passed into solution of ammonia, the latter substance, which is a compound of nitrogen with hydrogen, is decomposed; the chlorine combines with the hydrogen, and the nitrogen is set free, with effervescence. In this manner very pure nitrogen can be obtained. In making this experiment, it is necessary to stop short of saturating or decomposing the whole of the ammonia; otherwise there will be great risk of accident from the formation of an exceedingly dangerous explosive compound, produced by the contact of chlorine with an ammoniacal salt.

Another very easy and perfectly safe method of obtaining pure nitrogen is to decompose a solution of potassium nitrite with ammonium chloride (sal-ammoniac). The potassium nitrite is prepared by passing the red vapors of nitrous acid, obtained by heating dilute nitric acid with starch, into a solution of caustic potash. On boiling the resulting solution with sal-ammoniac, nitrogen gas is evolved, while potassium chloride remains in solution. The reaction is represented by the equation,

$$KNO_2 + NH_4Cl = KCl + 2H_2O + N_2.$$

Nitrogen is destitute of color, taste, and odor; it is a little lighter than air, its density being 0.972. A litre of the gas at 0° C. and 760 mm. barometric pressure weighs 1.25658 gram. 100 cubic inches, at 60° F. and 30 inches barometer, weigh 30.14 grains. Nitrogen is incapable of sustaining combustion or animal life, although, like hydrogen, it has no positive poisonous properties; neither is it soluble to any notable extent in water or in caustic alkali; it is, in fact, best characterized by negative properties.

Compounds of Nitrogen and Hydrogen.

Ammonia, NH_3.—When powdered sal-ammoniac is mixed with moist calcium hydrate (slaked lime), and gently heated in a glass flask, a large quantity of gaseous matter is disengaged, which must be collected over mercury, or by displacement, advantage being taken of its low specific gravity.

Ammonia gas thus obtained is colorless; it has a strong pungent odor, and possesses in an eminent degree those properties to which the term *alkaline* is applied; that is to say, it turns the yellow color of turmeric to brown, that of reddened litmus to blue, and combines readily with acids, neutralizing them completely. Under a pressure of 6.5 atmospheres at 15.5° C. (60° F.), ammonia condenses to the liquid form. Water dissolves about 700 times its volume of this gas, forming a solution which in a more dilute state has long been known under the name of *liquor ammoniæ;* by heat a great part is again expelled. The solution is decomposed by chlorine, sal-ammoniac being formed, and nitrogen set free.

Ammonia has a density of 0.589; a litre weighs 0.76271 gram. It cannot be formed by the direct union of its elements, although it is sometimes produced under rather remarkable circumstances by the deoxidation of nitric acid.* The great sources of ammonia are the feebly compounded azotized principles of the animal and vegetable kingdoms, which, when left to putrefactive change, or subjected to destructive distillation, almost invariably give rise to an abundant production of this substance.

The analysis of ammonia gas is easily effected. When a portion is confined in a graduated tube over mercury, and electric sparks are passed through it for a considerable time, the volume of the gas gradually increases until it becomes doubled. On examination, the tube is found to contain a mixture of 3 measures of hydrogen gas and 1 measure of nitrogen. Every two volumes of the ammonia, therefore, contained three volumes of hydrogen and one of nitrogen, the whole being condensed to one half. The weight of the two constituents is in the proportion of 3 parts hydrogen to 14 parts nitrogen.

Ammonia may also be decomposed into its elements by transmission through a red-hot tube.

The aqueous solution of ammonia is a very valuable reagent, and is employed in a great number of chemical operations, for some of which it is necessary to have it perfectly pure. The best mode of preparation is the following :—

Equal weights of sal-ammoniac (NH_4Cl), and quicklime (CaO), are taken; the lime is slaked in a covered basin, and the salt reduced to powder. These are mixed and introduced into a large flask connected with a wash-bottle and a receiver containing water. A little water is added to the mixture, just enough to damp it and cause it to aggregate into lumps. On cautiously applying heat to the flask, ammonia is disengaged very regularly and uniformly, and condenses in the water of the receiver. Calcium chloride ($CaCl_2$), with excess of calcium hydroxide (slaked lime) remains in the flask.

The decomposition of the salt is represented by the equation—

$$2NH_4Cl + CaO = 2NH_3 + CaCl_2 + H_2O.$$

Solution of ammonia should be perfectly colorless, leave no residue on evaporation, and when supersaturated by nitric acid, give no cloud or muddiness with silver nitrate. Its density diminishes as its strength increases, that of the most concentrated being about 0.875. The value in alkali of any sample of *liquor ammoniæ* is however most safely inferred, not from a knowledge of its density, but from the quantity of acid a given amount will saturate.

When solution of ammonia is mixed with acids, salts are generated exactly analogous to the corresponding potassium and sodium compounds: they will be discussed in connection with the latter. The ammonia salts may be regarded either as direct compounds of ammonia, NH_3, with acids (HCl, for example), or as resulting from the replacement of the hydrogen of an acid by the group NH_4, called *ammonium*, which in this sense is a compound metal, chemically equivalent to potassium, sodium, silver, etc. Thus :—

* A mode of converting the nitrogen of the atmosphere into ammonia, by a succession of chemical operations, will be noticed in connection with Cyanogen under Organic Chemistry.

Ammonia hydrochloride $NH_3,HCl = NH_4Cl$ Ammonium chloride
"   nitrate $NH_3,HNO_3 = NH_4NO_3$ "   nitrate
"   sulphate $(NH_3)_2,H_2SO_4 = (NH_4)_2SO_4$ "   sulphate

The formulæ in the second column are exactly analogous to those of the potassium salts, KCl, KNO_3, K_2SO_4.

The aqueous solution of ammonia may be supposed to contain *ammonium hydroxide*, $NH_4.HO$; but this compound is not known in the solid state.

Any ammoniacal salt can at once be recognized by the evolution of ammonia which takes place when it is heated with slaked lime, or solution of potash or soda.

Hydroxylamine, NH_3O.—This compound, intermediate in composition between ammonia, NH_3, and ammonium hydroxide, $NH_4.HO$, is formed by the direct union of hydrogen with nitrogen dioxide: $NO + H_3 = NH_3O$, and may be prepared by passing nitrogen dioxide through a series of flasks in which hydrogen is evolved by heating hydrochloric acid with tin. The resulting liquid is freed from tin by hydrogen sulphide; the filtered liquid evaporated to dryness; the residue washed with cold alcohol, and digested with boiling alcohol; the alcoholic solution mixed with platinic chloride to precipitate sal-ammoniac; and the filtered alcoholic liquid mixed with ether, which throws down pure hydrochloride of hydroxylamine.

Hydroxylamine is also formed by the action of hydrogen (evolved as above) on nitric acid or ethyl nitrate:

$$NO_3H + 6H =: 2H_2O + NH_3O.$$

Hydroxylamine is a very volatile and easily decomposable base, and can be obtained only in solution. Its salts are decomposed by potash, with evolution of nitrogen and formation of ammonia, quickly in concentrated, gradually in dilute solutions. Solutions of hydroxylamine may, however, be obtained by decomposing the salts in other ways, an alcoholic solution, for example, by decomposing the nitrate dissolved in alcohol with alcoholic potash. Alkaline carbonates also separate hydroxylamine, with evolution of carbon dioxide. The solutions have an alkaline reaction, and precipitate many metallic salts; with the salts of lead, iron, nickel, and zinc, and with chrome-alum and common alum, they form precipitates insoluble in excess of hydroxylamine. With aqueous cupric sulphate, hydroxylamine forms a grass-green precipitate, which, when boiled with water, is reduced, with evolution of gas, to cuprous oxide; an ammoniacal cupric solution is decolorized by it. Mercuric chloride is reduced to mercurous chloride, and if the hydroxylamine is in excess, to metallic mercury. Silver solutions yield a black precipitate, which is quickly reduced, with evolution of gas, to metallic silver. Hydroxylamine also reduces acid potassium chromate. In many of these reactions the hydroxylamine appears to be completely decomposed, with formation of nitrogen or its monoxide.

The salts of hydroxylamine decompose when heated, with copious and sudden evolution of gas; most of them easily form supersaturated solutions; none of those yet examined contain water of crystallization. The *hydrochloride*, NH_3O,HCl, crystallizes from alcohol in long spicular crystals resembling urea; from water in large irregular six-sided tables; it melts at $100°$, and then decomposes, with violent evolution of gas, into nitrogen, hydrochloric acid, water, and sal-ammoniac. The *nitrate*, NH_3O,HNO_3, solidifies slowly by spontaneous evaporation to a radio-crystalline, very deliquescent mass, easily soluble in absolute alcohol, decomposing at $100°$.

COMPOUNDS OF NITROGEN WITH THE HALOGEN ELEMENTS.

Nitrogen and Chlorine.—When sal-ammoniac or ammonia nitrate is dissolved in water, and a jar of chlorine is inverted, in the solution, the gas is absorbed, and a deep yellow oily liquid is observed to collect upon the surface of the solution, ultimately sinking in globules to the bottom. This is nitrogen chloride, one of the most dangerously explosive substances known. The following is the safest method of conducting the experiment:—

A somewhat dilute and tepid solution of pure sal-ammoniac in distilled water is poured into a clean basin, and a bottle of chlorine, the neck of which is quite free from grease, inverted in it. A shallow and heavy leaden cup is placed beneath the mouth of the bottle to collect the product. When enough has been obtained, the leaden vessel may be withdrawn with its dangerous contents, the chloride remaining covered with a stratum of water. The operator should protect his face with a strong wire-gauze mask when experimenting upon this substance.

The change may be explained by the equation—

$$NH_4Cl + 6Cl = NCl_3 + 4HCl.$$

Nitrogen chloride is very volatile, and its vapor is exceedingly irritating to the eyes. It has a specific gravity of 1.653. It may be distilled at $71°$ C. ($160°$ F.), although the experiment is attended with great danger. Between $93°$ and $105°$ C. ($200°$ and $221°$ F.) it explodes with fearful violence. Contact with almost any combustible matter, as oil or fat of any kind, determines the explosion at common temperatures, a vessel of porcelain, glass, or even of cast-iron, being broken by it to pieces, and the leaden cup receiving a deep indentation. This body has usually been supposed to have the composition NCl_3, but recent experiments upon the corresponding iodine compound induce a belief that it contains hydrogen, and may perhaps be represented by the formula $NHCl_2$ or NH_2Cl.

Nitrogen and Iodine.—When finely powdered iodine is added to caustic ammonia, it is in part dissolved, giving a deep-brown solution, and the residue is converted into a black powder, called n i t r o g e n   i o d i d e. The brown liquid consists of hydriodic acid holding iodine in solution, and is easily separated from the solid product by a filter. The latter, while still wet, is distributed in small quantities upon separate pieces of bibulous paper and left to dry in the air.

Nitrogen iodide is a black insoluble powder, which, when dry, explodes with the slightest touch—even that of a feather—and sometimes without any obvious cause. The explosion is, however, not nearly so violent as that of nitrogen chloride, and is attended with the production of violet fumes of iodine. According to Dr. Gladstone, this substance contains hydrogen, and may be viewed as NHI_2, that is, as ammonia in which two-thirds of the hydrogen are replaced by iodine. According to the researches of Bunset, it must be viewed as a combination of nitrogen tri-iodide with ammonia, NI_3,NH_3. It appears, however, that the substance called nitrogen iodide varies in composition. Gladstone, by changing the mode of preparation, obtained several compounds of nitrogen tri-iodide with ammonia.

COMPOUNDS OF NITROGEN AND OXYGEN.

There are five distinct compounds of nitrogen and oxygen, thus named and constituted:—

9

| | | Composition. | | | |
| | | By weight. | | By volume. | |
	Formula.	Nitrogen.	Oxygen.	Nitrogen.	Oxygen.
Monoxide . .	N_2O	28	16	2	1
Dioxide . .	N_2O_2 or NO	28	32	2	2
Trioxide, or Nitrous } oxide. . . }	N_2O_3	28	48	2	3
Tetroxide . .	N_2O_4 or NO_2	28	64	2	4
Pentoxide, or Nitric } oxide. . . }	N_2O_5	28	80	2	5

A comparison of these numbers will show that the quantities of oxygen which unite with a given quantity of nitrogen are to one another in the ratio of the numbers 1, 2, 3, 4, 5.

The first, third, and fifth of the compounds in the table are capable of taking up the elements of water and of metallic oxides to form salts (p. 29), called respectively h y p o n i t r i t e s, n i t r i t e s, and n i t r a t e s, the hydrogen salts being also called h y p o n i t r o u s, n i t r o u s, and n i t r i c a c i d.

The composition of these acids and of their potassium salts is represented by the following formulæ:—

Hydrogen Hyponitrite, or Hyponitrous acid,	.	H_2O,N_2O or HNO
Potassium Hyponitrite	.	K_2O,N_2O or KNO
Hydrogen Nitrite, or Nitrous acid . .	.	H_2O,N_2O_3 or HNO_2
Potassium Nitrite	.	K_2O,N_2O_3 or KNO_2
Hydrogen Nitrate, or Nitric acid . .	.	H_2O,N_2O_5 or HNO_3
Potassium Nitrate	.	K_2O,N_2O_5 or KNO_3

The dioxide and tetroxide of nitrogen do not form salts.

It will be convenient to commence the description of these compounds with the last on the list, viz., the pentoxide, as its salts, the nitrates, are the sources from which all the other compounds in the series are obtained.

Nitrogen Pentoxide, or Nitric Oxide, N_2O_5. (Also called *Anhydrous Nitric Acid*, or *Nitric Anhydride.*)—This compound was discovered in 1849 by Deville, who obtained it by exposing silver nitrate to the action of chlorine gas. Chlorine and silver then combine, forming silver chloride, which remains in the apparatus, while oxygen and nitrogen pentoxide separate:

$$Ag_2O,N_2O_5 + Cl_2 = 2AgCl + N_2O_5 + O.$$

It may also be prepared by slowly distilling pure and highly concentrated nitric acid at about 37° with phosphoric oxide, a substance which has a very powerful attraction for water. The distillate consists of two layers of liquid, the upper of which is nitrogen pentoxide mixed with nitrous and nitric acids; and on separating this upper layer, and cooling it with ice or a freezing mixture, the pentoxide separates in crystals.

Nitrogen pentoxide is a colorless substance, crystallizing in six-sided prisms, which melt at 30° C. (86° F.), and boil between 45° and 50° C. (113°-122° F.), when they begin to decompose. Its specific gravity in the solid state is above 1.64; in the liquid state below 1.636. Nitrogen pentoxide sometimes explodes spontaneously. It dissolves in water with great rise of temperature, forming hydrogen nitrate or nitric acid. It also unites with a smaller proportion of water, forming the *hemihydrate* $2N_2O_5,H_2O$, which constitutes the chief part of the lower layer of the distillate obtained in the manner just described. It is liquid at ordinary temperatures, but crystallizes in a freezing mixture.

NITRIC ACID, $HNO_3 = HO.NO_2$.—In certain parts of India, and in other hot dry climates, the surface of the soil is occasionally covered by a saline efflorescence, like that sometimes seen on newly plastered walls; this substance collected, dissolved in hot water, and crystallized from the filtered solution, furnishes the highly important salt known in commerce as nitre or saltpetre, and consisting of potassium nitrate. To obtain nitric acid, equal weights of powdered nitre and strong sulphuric acid are introduced into a glass retort, and heat is applied by means of a gas-lamp or charcoal fire. A flask, cooled by a wet cloth, is adapted to the retort to serve for a receiver. No luting of any kind must be used.

As the distillation advances, the red fumes which first arise disappear, but toward the end of the process they again become manifest. When this happens, and very little liquid passes over, while the greater part of the saline matter of the retort is in a state of tranquil fusion, the operation may be stopped; and when the retort is quite cold, water may be introduced to dissolve out the saline residue. The reaction consists in an interchange between the potassium of the nitre and half the hydrogen of the sulphuric acid (hydrogen sulphate), whereby there are formed hydrogen nitrate which distils over, and hydrogen and potassium sulphate which remains in the retort:

$$KNO_3 + H_2SO_4 = HNO_3 + HKSO_4.$$

In the manufacture of nitric acid on the large scale, the glass retort is replaced by a cast-iron cylinder, and the receiver by a series of earthen condensing vessels connected by tubes. Sodium nitrate, found native in Peru, is now generally substituted for potassium nitrate.

Nitric acid thus obtained has a specific gravity of from 1.5 to 1.52; it has a golden-yellow color, due to nitrogen trioxide, or tetroxide, which is held in solution, and, when the acid is diluted with water, gives rise by its decomposition to a disengagement of nitrogen dioxide. Nitric acid is exceedingly corrosive, staining the skin deep yellow, and causing total disorganization. Poured upon red-hot powdered charcoal, it causes brilliant combustion; and when added to warm oil of turpentine, acts upon that substance so energetically as to set it on fire.

Pure nitric acid, in its most concentrated form, is obtained by mixing the above with about an equal quantity of strong sulphuric acid, redistilling, collecting apart the first portion which comes over, and exposing it, in a vessel slightly warmed and sheltered from the light, to a current of dry air made to bubble through it, which completely removes the nitrous acid. In this state the product is as colorless as water; it has the sp. gr. 1.517 at 15.5°, boils at 84.5°, and consists of 54 parts nitrogen pentoxide and 9 parts water. Although nitric acid in a more dilute form acts very violently upon many metals, and upon organic substances generally, this is not the case with the most concentrated acid : even at a boiling heat, it refuses to attack iron or tin; and its mode of action on lignin, starch, and similar substances, is quite peculiar, and very much less energetic than that of an acid containing more water.

On boiling nitric acid of different degrees of concentration, at the ordinary atmospheric pressure, a residue is left, boiling at 120.5° C. (249° F.) and 29 inches barometer, and having the sp. gr. 1.414 at 15.5° C. (60. F.). This acid was formerly supposed to be a definite compound of nitric acid with water; but Roscoe has shown that this assumption is incorrect, the composition of the acid varying according to the pressure under which the liquid boils.

The nitrates form a very extensive and important group of salts, which are remarkable for being all soluble in water. Hydrogen nitrate is of great use in the laboratory, and in many branches of industry.

The acid prepared in the way described is apt to contain traces of chlorine from common salt in the nitre, and sometimes of sulphate from accidental splashing of the pasty mass in the retort. To discover these impurities, a portion is diluted with four or five times its bulk of distilled water, and divided between two glasses. Solution of silver nitrate is dropped into the one, and solution of barium nitrate into the other; if no change ensue in either case, the acid is free from the impurities mentioned.

Nitric acid has been formed in small quantity, by passing a series of electric sparks through a portion of air in contact with water or an alkaline solution. The amount of acid so formed after many hours is very minute; still it is not impossible that powerful discharges of atmospheric electricity may sometimes occasion a trifling p d   of nitric acid in the air. A very minute quantity of nitric acid is produced by the combustion of hydrogen and other substances in the atmosphere; it is also formed by the oxidation of ammonia.

Nitric acid is not so easily detected in solution in small quantities as many other acids. Owing to the solubility of all its compounds, no *precipitant* can be found for this acid. A good mode of testing it is based upon its power of bleaching a solution of indigo in sulphuric acid when boiled with that liquid. The absence of chlorine must be insured in this experiment; otherwise the result will be equivocal.

The best method for the detection of nitric acid is the following:—The substance to be examined is boiled with a small quantity of water, and the solution cautiously mixed with an equal volume of concentrated sulphuric acid; the liquid is then left to cool, and a strong solution of ferrous sulphate carefully poured upon it, so as to form a separate layer. If large quantities of nitric acid are present, the surface of contact first, and then the whole of the liquid, becomes black. If but small quantities of nitric acid are present, the liquid becomes reddish-brown or purple. The ferrous sulphate reduces the nitric acid to nitrogen dioxide, which, dissolving in the solution of ferrous sulphate, imparts to it a dark color.

Aqua regia.—This name is given to a mixture of nitric and hydrochloric acids, which has long been known for its property of dissolving gold and platinum. When heated it gives off nitrosyl chloride (p. 199) and free chlorine, the latter of which attacks any metal that may be present:

$$NO_3H + 3HCl = 2H_2O + NOCl + Cl_2.$$

Nitryl Chloride, Nitroxyl Chloride, or *Nitric Chloride,* NO_2Cl.—This compound, which may be regarded as derived from nitric acid, $NO_2.OH$, by substitution of Cl for OH, is produced: (1.) When a mixture of nitrogen tetroxide and chlorine is passed through a heated glass tube. (2.) By the action of chlorine on silver nitrate. (3.) Together with sulphuric acid, by the action of nitric acid on sulphuric hydroxychloride or chlorosulphonic acid, $Cl.SO_2.OH$ (p. 183):

$$Cl.SO_2.OH + NO_2.OH = NO_2Cl + HO.SO_2.OH.$$

(4.) Together with silver phosphate, by the action of phosphorus oxychloride on silver nitrate:

$$POCl_3 + 3AgNO_3 = Ag_3PO_4 + 3NO_2Cl.$$

This last reaction affords the best mode of preparing the compound.

Nitryl chloride is a heavy yellow liquid which boils at $5°$, and is decomposed by water into nitric and hydrochloric acids.

Nitrogen Monoxide, N_2O (sometimes called *Nitrous Oxide;* also *Laughing Gas*).—When solid ammonium nitrate is heated in a retort or flask (fig. 118), furnished with a perforated cork and bent tube, it is resolved into water and nitrogen monoxide: $NH_4NO_3 = 2H_2O + N_2O$ No particular precaution is required in the operation, save due regulation of the heat, and the avoidance of tumultuous disengagement of the gas.

Fig. 118.

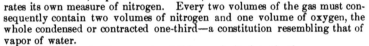

Nitrogen monoxide is a colorless, transparent, and almost inodorous gas, of distinctly sweet taste. Its specific gravity is 1.525; a litre of it weighs 1.97172 grams; 100 cubic inches weigh 47.29 grains. It supports the combustion of a taper or a piece of phosphorus with almost as much energy as pure oxygen: it is easily distinguished, however, from that gas by its solubility in cold water, which dissolves nearly its own volume: hence it is necessary to use tepid water in the pneumatic trough or gas-holder, otherwise great loss of gas will ensue.

Gaseous nitrogen monoxide mixed with an equal volume of hydrogen, and fired by the electric spark in the eudiometer, explodes with violence, and liberates its own measure of nitrogen. Every two volumes of the gas must consequently contain two volumes of nitrogen and one volume of oxygen, the whole condensed or contracted one-third—a constitution resembling that of vapor of water.

The most remarkable property of this gas is its intoxicating power upon the animal system. If quite pure, or merely mixed with atmospheric air, it may be respired for a short time without danger or inconvenience. The effect is very transient, and is not followed by depression. The gas is now much used as an anæsthetic in dental surgery.

Nitrogen monoxide has been liquefied, but with difficulty; it requires, at $45°$ F. ($7.2°$ C.), a pressure of 50 atmospheres: the liquid monoxide has a specific gravity of 0.9004, and, under a pressure of 763 mm., boils at $-87.9°$ C. ($-126°$ F.); it is not miscible with water. Faraday solidified it by exposing it in a sealed tube to the cold produced by a mixture of solid carbonic acid and ether, but he supposed that it could not be solidified by the cold produced by its own evaporation. This, however, may be effected if the evaporation be accelerated by a strong current of air. A very fine steel tube is directed into the axis of a thin brass cone, having a small opening, about the eighth of an inch, at its apex. On causing a stream of the liquid to issue from the jet, it is retained in the cone for a moment, and then forcibly blown out at the apex, together with a strong stream of air. The solid is in this way formed in some quantity, and may be collected in a dish lined with filter-paper, or other suitable vessel. Solid nitrogen monoxide is more compact in appearance than solid carbonic acid, and, unlike the latter, it melts and boils, if gently warmed, before passing into the gaseous state: hence if placed in contact with the skin, it produces a painful blister, like a burn.*

* W i l l s, Chem. Soc. Journ., 1874, p. 21.

17 *

HYPONITROUS ACID, N_2O,H_2O, or HNO.—When a solution of sodium nitrate, $NaNO_3$, or ammonium nitrate, $NH_4.NO_3$, is treated with sodium amalgam (a compound of sodium and mercury), the nitrate gives up 2 atoms of oxygen to the sodium, and is reduced to hyponitrite, NaNO. On neutralizing the excess of alkali in the liquid, by adding acetic acid till the solution no longer gives a brown or black precipitate (of silver oxide) with silver nitrate, a solution of sodium hyponitrite is obtained, which is alkaline to test-paper, and gives with silver nitrate a yellow precipitate of silver hyponitrite, AgNO. When the original alkaline liquid is acidified with acetic acid, and heated, the hyponitrous acid is resolved into water and nitrogen monoxide, which escapes as gas: $2HNO = H_2O + N_2O$.*

Nitrogen Dioxide, N_2O_2 or NO (sometimes called *Nitric Oxide*).— Clippings or turnings of copper are put into the apparatus employed for preparing hydrogen (p. 132), together with a little water, and nitric acid is added by the funnel until brisk effervescence is excited. The gas may be collected over cold water, as it is not sensibly soluble.

The reaction is a simple deoxidation of some of the nitric acid by the copper: the metal is oxidized, and the oxide so formed is dissolved by another portion of the acid, forming copper nitrate. Nitric acid is very prone to act thus upon certain metals:—

$$\underset{\text{Nitric acid.}}{8HNO_3} + Cu_3 = N_2O_2 + \underset{\text{Copper nitrate.}}{3Cu(NO_3)_2} + 4H_2O.$$

The gas obtained in this manner is colorless and transparent: in contact with air or oxygen gas it produces deep red fumes, which are readily absorbed by water: this character is sufficient to distinguish it from all other gaseous bodies. A lighted taper plunged into the gas is extinguished; lighted phosphorus, however, burns in it with great brilliancy.

The specific gravity of nitrogen dioxide is 1.039; a litre weighs 1.34343 grams. It contains equal measures of oxygen and nitrogen gases united without condensation. When this gas is passed into the solution of a ferrous salt, it is absorbed in large quantity, and a deep brown, or nearly black liquid produced, which seems to be a definite compound of the two substances (p. 196). The compound is decomposed by boiling.

Nitrogen Trioxide, or **Nitrous Oxide**, N_2O_3.—When four measures of nitrogen dioxide are mixed with one measure of oxygen, and the gases, perfectly dry, are exposed to a temperature of $-18°$, they condense to a thin mobile blue liquid, which emits orange-red vapors.

Nitrogen trioxide, sufficiently pure for most purposes, is obtained by pouring concentrated nitric acid on lumps of arsenious acid, and gently warming the mixture, in order to start the reaction. The trioxide is then evolved as an orange-red gas, arsenic acid remaining behind.

NITROUS ACID, $HNO_2 = HO.NO$, is obtained as a beautiful blue liquid on dissolving the trioxide in ice-cold water: $N_2O_3 + H_2O = 2HNO_2$. It is very unstable, even at ordinary temperatures, decomposing into nitric acid, nitrogen dioxide, and water, $3HNO_2 = HNO_3 + N_2O_2 + H_2O$. In consequence of this instability of the acid, the metallic nitrites cannot be obtained by treating the aqueous solution of the acid with metallic oxides, but they may be formed in other ways: potassium nitrite, for example, may be obtained by fusing the corresponding nitrate, whereby a third part of the oxygen is driven off; and a solution of potassium or sodium nitrite may be prepared by passing the vapor of nitrogen trioxide—obtained as above by

---

* D i v e r s, Proceedings of the Royal Society, xix. 425; Chem. Soc. Journ., 1871, p. 484.

heating nitric acid with arsenious oxide (or with starch)—into a solution of caustic potash or soda.

Nitrosyl Chloride or *Nitrous Chloride*, NOCl, the chlorine-analogue of nitrous acid, is formed by direct union of chlorine and nitrogen dioxide, and by the action of phosphorus pentachloride on potassium nitrite: $PCl_5 + NO.OK = NO.Cl + KCl + POCl_3$; also, together with free chlorine, when a mixture of nitric and hydrochloric acids (*aqua regia*) is slowly heated:

$$HNO_3 + 3HCl = NOCl + Cl_2 + 2H_2O.$$

Nitrosyl chloride in the pure state is an orange-yellow gas, which at the temperature of a freezing mixture, condenses to a deep orange-colored limpid liquid, boiling at about $-8°$. It unites with many metallic chlorides, and is decomposed by basic oxides, yielding a nitrite and a chloride: *e.g.*,

$$NOCl + 2KOH = KNO_2 + KCl + H_2O.$$

Nitrosyl Bromide, NOBr, formed by passing nitrogen dioxide into bromine at $-7°$ to $-15°$ C. ($19.4°$ to $5°$ F.), is a blackish-brown liquid, which begins to decompose at $-2°$ C. ($28.4°$ F.), giving off nitrogen dioxide. If the temperature is allowed to rise to $+20°$ C. ($68°$ F.), a dark brownish-red liquid remains, consisting of *nitrosyl tribromide*, NOBr$_3$, which is also formed when bromine is saturated with nitrogen dioxide at ordinary temperatures. It volatilizes undecomposed when quickly heated, but is resolved by slow heating into bromine and nitrogen dioxide.

Nitrogen Tetroxide, N_2O_4 or NO_2, also called *Nitric Peroxide.*—This is the principal constituent of the deep red fumes always produced when nitrogen dioxide escapes into the air.

It may be obtained in the pure state:—1. By exposing a mixture of 2 vols. nitrogen dioxide and 1 vol. oxygen incorporated by passing through a tube filled with broken porcelain, and thoroughly dried by transmission over pumice soaked in oil of vitriol, and then over recently fused stick potash, to the action of a freezing mixture of salt and ice; the tetroxide then condenses in transparent crystals, or if the slightest trace of moisture is present, into an almost colorless liquid. 2. By the direct combination of oxygen with the trioxide, as when a stream of oxygen is passed into the mixture of the trioxide and other oxides of nitrogen evolved by the action of fuming nitric acid on arsenious acid The liquid tetroxide thus obtained is pure enough for most purposes after one distillation. 3. By heating thoroughly dried lead nitrate in a retort, whereby a mixture of the tetroxide and oxygen is evolved, the former of which may be condensed as above, while the latter passes on:

$$(NO_3)_2Pb = PBO + O + N_2O_4.$$

The first portions of nitrogen tetroxide thus obtained do not solidify, doubtless owing to the presence of a trace of moisture; but if the receiver be changed in the midst of the operation, and if every care has been taken to avoid moisture, the later portions may be obtained in the crystalline form.

Nitrogen tetroxide at very low temperatures forms transparent, colorless, prismatic crystals which melt at $-9°$ C. ($15.8°$ F.), but when once melted do not resolidify till cooled down to $-30°$ C. ($22°$ F.). Above $-9°$ C. it forms a mobile liquid of specific gravity 1.451, the appearance of which varies greatly according to the temperature. When still liquid below $-9°$

C., it is almost colorless; at − 9° C. it has a perceptible greenish-yellow tint; at 0° C. the color is somewhat more marked; at + 10° C. (50° F.) it is decidedly yellow; and at 15° C. (59° F.) and upwards, orange-yellow, the depth of color increasing progressively with the temperature up to 22° C. (71.6° F.), the boiling point of the liquid. The vapor has a brown-red color, the depth of which also increases with the temperature, until at 40° C. (104° F.) it is so dark as to be almost opaque. This remarkable change of color is accompanied by a great diminution of density as the temperature rises, both phenomena pointing to a molecular change produced in the vapor by heat. Playfair and Wanklyn have determined the density of the vapor by Dumas' method, using nitrogen as a diluent, and find that the densities at different temperatures are as follows:—

Temperature.		Vapor-density.
97.5° C.	207.5° F.	1.783
24.5	76.1	2.520
11.3	52.3	2.655
4.2	39.5	2.588

The vapor is absorbed by strong nitric acid, which thereby acquires a yellow or red tint, passing into green, then into blue, and afterwards disappearing altogether on the addition of successive portions of water. The deep red fuming acid of commerce, called *nitrous acid*, is simply nitric acid impregnated with nitrogen tetroxide.

Nitrogen tetroxide is decomposed by water at very low temperatures in such a manner as to yield nitric and nitrous acids: $N_2O_4 + H_2O = HNO_3 + HNO_2$; but when added to excess of water at ordinary temperatures, it yields nitric acid and the products of decomposition of nitrous acid, namely, nitric acid and nitrogen dioxide. In like manner, when passed into alkaline solutions, it forms a nitrate and a nitrite of the alkali-metal; but it has been also supposed to unite directly, under certain circumstances, with metallic oxides—lead oxide, for example—forming definite crystalline salts, and has hence been called *hyponitric acid;* but it is most probable that these salts are compounds of nitrates and nitrites: *e. g.*,

$$2(PbO,N_2O_4) = Pb(NO_3)_2 + Pb(NO_2)_2.$$
Hyponitrate.   Nitrate.   Nitrite.

Atmospheric Air.—The exact composition of the atmosphere has repeatedly been made the subject of experimental research. Besides nitrogen and oxygen, the air contains a little carbon dioxide (carbonic acid), a very variable proportion of aqueous vapor, a trace of ammonia, and, perhaps, a little carburetted hydrogen. The oxygen and nitrogen are in a state of mixture, not of combination, yet their ratio is always uniform. Air has been brought from lofty Alpine heights, and compared with that from the plains of Egypt; it has been brought from an elevation of 21,000 feet by the aid of a balloon; it has been collected and examined in London and Paris, and many other places: still the proportion of oxygen and nitrogen remains unaltered, the diffusive energy of the gases being adequate to maintain this perfect uniformity of mixture. The carbon dioxide, on the contrary, being much influenced by local causes, varies considerably. In the following table the proportions of oxygen and nitrogen are given on the authority of Dumas, and the carbon dioxide on that of De Saussure: the ammonia, the discovery of which in atmospheric air is due to Liebig, is too small in quantity for direct estimation.

Composition of the Atmosphere.

						By weight.	By measure.
Nitrogen	.	.	.	.	.	. 77 parts	79.19
Oxygen .	.	.	.	.	.	. 23 "	20.81
						100	100.00

Carbon dioxide, from 3.7 measures to 6.2 measures in 10,000 measures of air.

Aqueous vapor variable, depending much upon the temperature.

Ammonia, a trace.

Dr. Frankland has analyzed samples of air taken by himself in the valley of Chamouni, on the summit of Mont Blanc, and at the Grands Mulets. The following are the results of his analyses:—

				Carbon Dioxide.	Oxygen.
Chamouni (3000 feet)	.	.	.	0.063	20.894
Grands Mulets (11,000 feet)	.	.	.	0.111	20.802
Mont Blanc (15,732 feet) .	.	.	.	0.061	20.963

A litre of pure and dry air at 0° C. and 760 mm. pressure weighs 1.29366 grams. 100 cubic inches at 60° F. and 30 inches barom. weigh 30.935 grains: hence a cubic foot weighs 536.96 grains, which is $\frac{1}{818}$ of the weight of a cubic foot of water at the same temperature.

The analysis of air is very well effected by passing it over finely divided copper contained in a tube of hard glass, carefully weighed and then heated to redness: the nitrogen is suffered to flow into an exhausted glass globe, also previously weighed. The increase in weight of the copper after the experiment gives the amount of oxygen, and the increase in weight of the exhausted globe gives the nitrogen.

An easier, but less accurate method, consists in introducing into a graduated tube, standing over water, a known quantity of the air to be examined, and then passing into the latter a stick of phosphorus affixed to the end of a wire. The whole is left for about twenty-four hours, during which the oxygen is slowly but completely absorbed, after which the phosphorus is withdrawn, and the residual gas read off.

Liebig proposed the use of an alkaline solution of pyrogallic acid (a substance which will be described in the department of Organic Chemistry) for the absorption of oxygen. The absorptive power of such a solution, which turns deep black on coming in contact with the oxygen, is very considerable. Liebig's method combines great accuracy with unusual rapidity and facility of execution.

Another plan is to mix the air with hydrogen and pass an electric spark through the mixture: after explosion the volume of gas is read off and compared with that of the air employed. Since the analysis of gaseous bodies by explosion is an operation of great importance, it may be worth while to describe the process in detail, as it is applicable, with certain obvious variations, to a number of analogous cases.

Instruments for this purpose are called eudiometers. The simplest, and, on the whole, the most convenient, consists of a straight graduated glass tube (fig. 120) closed at the top, and having platinum wires inserted near the closed end, to give passage to an electric spark. This tube is filled with mercury, and inverted in a mercurial pneumatic trough.

For the analysis of air, a quantity sufficient to fill about one-sixth of the tube is introduced, and its volume accurately ascertained by reading off with

9 *

a telescope the number of divisions on the tube to which the mercury reaches, whilst the height of the column of mercury in the tube above the

Fig. 119.   Fig. 120.

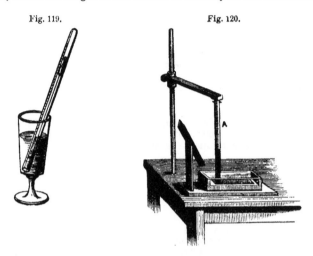

trough, together with that of the barometer, and the temperature of the air, are also read off. A quantity of pure hydrogen gas is now added, more than sufficient to unite with all the oxygen present (about half the volume of the air), and the volume of the gas and the pressure exerted upon it are determined as before. An electric spark is now passed through the mixture, care being taken to prevent any escape, by pressing the open end of the eudiometer against a piece of sheet caoutchouc under the mercury in the trough. After the explosion, the volume is again determined, and is found to be less than that before the explosion. The volume of gas read off must in each case be reduced to standard pressure and temperature by the method already given (p. 55).

Now, since the hydrogen is in excess, and 2 volumes of that gas unite with 1 volume of oxygen to form water, one-third of the diminution must be the volume of the oxygen contained in the air introduced. An example will render this clear:—

Air introduced   .   .   .   .   .   . 100 measures.
Air and hydrogen 160
Volume after explosion 97
Diminution   63

$$\frac{63}{3} = 21 = \text{oxygen in the 100 measures.}$$

NITROGEN WITH SULPHUR AND SELENIUM.

Nitrogen Sulphide, N_2S_2, is obtained, together with other compounds, by the action of dry ammonia on sulphur monochloride or on thionyl chloride (pp. 174, 182), as a yellow powder which crystallizes from carbon sulphide in yellowish-red rhombic prisms. It becomes dark-colored at 120° C. (248° F.) and emits pungent vapors; sublimes at 135° C. (275° F.) in yellowish-red crystals; begins to melt and give off gas at 158° C. (316° F.); decomposes rapidly at 160° C. (320° F.), and detonates violently when struck.

On adding sulphur monochloride, S_2Cl_2, to a solution of nitrogen sulphide in carbon sulphide, a yellow crystalline substance is formed, consisting of the compound $N_2S_2.S_2Cl_2$, which sublimes in needles when heated. On further addition of nitrogen sulphide, this yellow body is changed into a red compound $(N_2S_2)_2S_2Cl_2$, and this again, on still further addition, into $(N_2S_2)_3S_2Cl_2$, which forms a yellow powder permanent in the air.

Nitrogen Selenide, N_2Se_2, formed by the action of ammonia on selenium tetrachloride, is an orange-yellow mass which detonates strongly when heated to 200° C. (392° F.) or subjected to slight pressure.

Dinitrososulphuric Acid, $H_2SO_3(NO)_2$.—The alkali-salts of this acid (sulphuric acid, H_2SO_4, having 1 atom of oxygen replaced by 2 atoms of univalent nitrosyl, NO) are formed on passing nitrogen dioxide into the solution of an alkaline sulphide. They are colorless and crystalline, and are decomposed by all acids, even by carbonic acid, into sulphates and nitrogen monoxide, $K_2SO_3(NO)_2 = K_2SO_4 + N_2O$. The free acid (hydrogen salt) is not known. The potassium salt is resolved by heat into potassium sulphite, K_2SO_3, and nitrogen dioxide, N_2O_2.

Nitrosulphonic Acid, $NO_2.SO_2.OH$.*—This is the composition of the lead-chamber crystals, formed during the manufacture of sulphuric acid, when the supply of steam is insufficient to form that acid at once (p. 178). The same compound is formed by the action of strong nitric acid on sulphurous oxide: $SO_2 + NO_2.OH = NO_2.SO_2.OH$; by passing vapor of nitrosyl chloride into sulphuric acid: $NOCl + HO.SO_2.OH = HCl + NO_2.SO_2.OH$; and together with nitric acid by the action of nitrogen tetroxide on sulphuric acid: $N_2O_4 + HO.SO_2.OH = NO_2.OH + NO_2.SO_2.OH$. It crystallizes in rhombic p s , or in tabular or nodular masses, which begin to melt and give off vapor at 30°. It dissolves in small quantities of cold water, forming a blue liquid containing sulphuric and nitrous acids: $NO_2.SO_2.OH + HOH = HO.SO_2.OH + NO_2H$. It dissolves also in strong sulphuric acid, forming a solution which can be distilled without decomposition.

Nitrosulphonic chloride, $NO_2.SO_2.Cl$, formed by direct combination of SO_2 and NO_2Cl, is a white crystalline mass which melts when heated, with separation of nitrosyl chloride, and is decomposed by water, yielding sulphuric, hydrochloric, and nitrous acids.

Nitrosulphonic oxide or anhydride, $N_2S_2O_9 = (NO_2.SO_2)_2O$, is formed by the action of heat on nitrosulphonic acid, but is best prepared by passing dry nitrogen dioxide into sulphuric oxide, and heating the resulting solution nearly to the boiling point: $3SO_2O + N_2O_4 = (NO_2.SO_2)_2O + SO_2$. It is also formed by the action of sulphurous oxide on nitrogen tetroxide, and by heating the following compound. It crystallizes in hard colorless square prisms, melting at 217° C. (423° F.), to a yellow liquid which becomes darker at a higher temperature, and distils unchanged at 360° C. (680° F.). It dissolves readily in strong sulphuric acid, forming nitrosulphonic acid.

Oxynitrosulphonic anhydride, $\left\{ \begin{matrix} NO_2.SO_2 \\ O.NO_2.SO_2 \end{matrix} \right\} O$, is formed when sulphuric oxide and nitrogen tetroxide are brought together at a low temperature, and

---

* The name "sulphonic acid" is applied to derivatives of unsymmetrical sulphurous acid, $H.SO_2.OH$, formed by replacing the first hydrogen-atom by a univalent radicle: *e. g.,* methylsulphonic acid, $CH_3.SO_2.OH$; phenylsulphonic acid, $C_6H_5.SO_2.OH$.—(See ORGANIC CHEMISTRY.)

separates as a white crystalline mass, which gives off oxygen when heated, and is reduced to the preceding compound.

Nitroxypyrosulphuric acid, $SO_3H.O.SO_3(NO_2)$, separates out at a certain degree of concentration from the thick oily liquid formed on mixing sulphuric oxide with nitrogen tetroxide. It dissolves in warm dilute nitric acid, and separates on cooling in crystals containing 1 molecule of water.

Ammon-sulphonic Acids.—The acids of this series, known only as potassium salts, are formed on the type NH_5 (the nitrogen being quinquivalent as in the nitrates). *Potassium ammon-tetrasulphonate*, $NH(SO_3K)_4$, is obtained as a crystalline precipitate on adding potassium sulphite in excess to a solution of the nitrite, potassium hydroxide remaining dissolved:

$$4K_2SO_3 + KNO_2 + 3H_2O = NH(SO_3K)_4 + 5KOH.$$

By prolonged contact with water, or by boiling with dilute potash-ley, this salt is resolved into acid potassium sulphate and *potassium ammontrisulphonate*, according to the equation,

$$NH(SO_3K)_4 + H_2O = KHSO_4 + NH_2(SO_3K)_3.$$

The trisulphonate separates in needle-shaped crystals, insoluble in cold water, but soluble in hot dilute potash, from which it may be recrystallized. By boiling with acidulated water it is converted into *ammon-disulphonate*; thus:

$$NH_2(SO_3K)_3 + H_2O = HKSO_4 + NH_3(SO_3K)_2.$$

This last salt forms six-sided prisms, and is decomposed by prolonged boiling with a dilute acid, or by heating in the dry state at 200°, yielding ammonia sulphurous oxide and dipotassic sulphate: $NH_3(SO_3K)_2 = NH_3 + SO_2 + K_2SO_4$.

Hydroxylamine-disulphonic acid, $N(OH)(SO_3H)_2$ (also called *Sulphazotic acid*), is formed, as potassium salt, by mixing the solutions of potassium nitrite and sulphite in the proportion of one molecule of the first to less than four molecules of the second. This salt crystallizes in long bright transparent needles, and is converted by boiling with water into *potassium hydroxylaminemonosulphonate*:

$$N(OH)(SO_3K)_2 + H_2O = HKSO_4 + NH(OH)(SO_3K),$$

which may be crystallized from solution in hot water, but is decomposed by boiling with potash into ammonia, potassium sulphate, and nitrogen monoxide; thus:

$$4NH(OH)(SO_3K) + 4KOH = N_2O + 2NH_3 + 4K_2SO_4 + 3H_2O.$$

PHOSPHORUS.

Symbol, P.   Atomic weight, 30.96.   Vapor-density, 61.92.

PHOSPHORUS in the state of phosphoric acid is contained in the ancient unstratified rocks and in lavas of modern origin. As these disintegrate and crumble down into fertile soil, the phosphates pass into the organism of plants, and ultimately into the bodies of the animals to which the plants serve for food. The earthy phosphates play a very important part in the structure of the animal frame, by communicating stiffness and inflexibility to the bony skeleton.

Phosphorus was discovered in 1669 by Brandt, of Hamburg, who prepared it from urine. The following is an outline of the method of preparation now adopted :—Thoroughly calcined bones are reduced to powder, and mixed with two-thirds of their weight of sulphuric acid diluted with a considerable quantity of water : this mixture, after standing for some hours, is filtered, and the nearly insoluble calcium sulphate is washed. The liquid is then evaporated to a syrupy consistence and mixed with charcoal powder, and the desiccation is completed in an iron vessel exposed to a high temperature. The residue, when quite dry, is transferred to a stoneware retort to which a wide bent tube is luted, dipping a little way into the water contained in the receiver (fig. 121). A narrow tube serves to give issue to the gases, which are conveyed to a chimney. This manufacture is now conducted on a very large scale, the consumption of phosphorus, for the apparently trifling article of instantaneous-light matches, being something prodigious.

Fig. 121.

Phosphorus, when pure, very much resembles in appearance imperfectly bleached wax, and is soft and flexible at common temperatures. Its density is 1.77, and that of its vapor 4.35, air being unity, or 61.92 referred to hydrogen as unity. It melts at 44° C. (111° F.), and boils at 280° C. (536° F.). On slowly cooling melted phosphorus, well-formed dodecahedrons are sometimes obtained. Phosphorus is insoluble in water, and is usually kept immersed in that liquid, but dissolves in oil, in native naphtha, and especially in carbon bisulphide. When set on fire in the air, it burns with a bright flame, generating phosphorus oxide. It is exceedingly inflammable, sometimes taking fire at the heat of the hand, and demands great care in its management ; a blow or hard rub will very often kindle it. A stick of phosphorus held in the air always appears to emit a whitish smoke, luminous in the dark. This effect is chiefly due to a slow combustion which the phosphorus undergoes by the oxygen of the air, and upon it depends one of the methods employed for the analysis of air, as already described. It is singular that the slow oxidation of phosphorus may be entirely prevented by the presence of a small quantity of olefiant gas, or the vapor of ether, or some essential oil ; phosphorus may even be distilled in an atmosphere containing vapor of oil of turpentine in considerable quantity. Neither does the action go on in pure oxygen—at least, at the temperature of 15.5° C. (60° F.), which is very remarkable ; but if the gas be rarefied, or diluted with nitrogen, hydrogen, or carbonic acid, oxidation is set up.

A very remarkable modification of this element is known by the name of a m o r p h o u s p h o s p h o r u s. It was discovered by Schrötter, and may be made by exposing common phosphorus for fifty hours at a temperature of 240° to 250° C. (464° to 482° F.), in an atmosphere which is unable to act chemically upon it. At this temperature it becomes red and opaque, and insoluble in carbon bisulphide, whereby it may be separated from ordinary phosphorus. It may be obtained in compact masses when common phosphorus is kept for a week at a constant high temperature. It is a coherent, reddish-brown, infusible substance, of specific gravity between

18

2.089 and 2.106. It does not become luminous in the dark until the temperature is raised to about 200° C. (392° F.), nor has it any tendency to combine with the oxygen of the air. When heated to 260° C. (500° F.), it is reconverted into ordinary phosphorus.

When phosphorus is melted beneath the surface of hot water, and a stream of oxygen gas is forced upon it from a bladder, combustion ensues, and the phosphorus is converted in great part into a brick-red powder, which was formerly believed to be a peculiar oxide of phosphorus; but Schrötter has shown that it is a mixture consisting chiefly of amorphous phosphorus.

PHOSPHORUS AND HYDROGEN.

Phosphorus Trihydride.—*Phosphine.*—*Phosphoretted Hydrogen,* PH_3. —This body is analogous in some of its chemical relations to ammonia gas; its alkaline properties are, however, much weaker.

It may be obtained in a state of purity by heating phosphorous acid in a small retort, the acid being then resolved into phosphine and phosphoric acid:

$$4H_3PO_3 = PH_3 + 3H_3PO_4.$$

Thus obtained, the gas has a density of 1.20 (referred to air), or 16.98 (referred to hydrogen), which number is the half of its molecular weight. It contains three volumes of hydrogen and half a volume of phosphorus vapor, condensed into two volumes. It has a highly disagreeable odor of garlic, is slightly soluble in water, and burns with a brilliant white flame, forming water and phosphoric acid.

Phosphine may also be produced by boiling together, in a retort of small dimensions, caustic potash or slaked lime, water, and phosphorus; the vessel should be filled to the neck, and the extremity of the latter made to dip into the water of the pneumatic trough. In the reaction which ensues, the water is decomposed, and both its elements combine with the phosphorus.

$$P_8 + 3CaH_2O_2 + 6H_2O = 2PH_3 + 3CaH_4P_2O_4.$$

| Calcium | | Calcium |
| hydroxide. | | hypophosphite. |

The phosphine prepared by the latter process has the singular property of spontaneous inflammability when admitted into the air or into oxygen gas; with the latter, the experiment is very beautiful, but requires caution: the bubbles should be admitted singly. When kept over water for some time, the gas loses this property, without otherwise suffering any appreciable change; but if dried by calcium chloride, it may be kept unaltered for a much longer time. Paul Thénard has shown that the spontaneous combustibility of the gas arises from the presence of the vapor of a liquid hydrogen phosphide, PH_2, which may be obtained in small quantity, by conveying the gas produced by the action of water on calcium phosphide through a tube cooled by a freezing mixture. This substance forms a colorless liquid of high refractive power and very great volatility. It does not solidify at 0° F. (− 17.8° C.). In contact with air it inflames instantly, and its vapor in very small quantity communicates spontaneous inflammability to pure phosphine, and to all other combustible gases. It is decomposed by light into gaseous phosphine, and a solid phosphide, P_2H, which is often seen on the inside of jars containing gas which, by exposure to light, has lost the property of spontaneous inflammation. Strong acids occasion its instantaneous decomposition. It is as unstable as hydrogen dioxide. It is to be observed that pure phosphine gas itself becomes spontaneously inflammable if heated to the temperature of boiling water.

Phosphine decomposes several metallic solutions, giving rise to precipitates of insoluble phosphides. With hydriodic acid it forms a crystalline compounds, PH_4I, somewhat resembling sal-ammoniac.

PHOSPHORUS AND CHLORINE.

Phosphorus forms two chlorides, analogous in composition to the oxides, the quantities of chlorine combined with the same quantity of phosphorus being in the proportion of 3 to 5.

Phosphorus Trichloride, or Phosphorous Chloride, PCl_3, is prepared in the same manner as sulphur dichloride, by gently heating phosphorus in dry chlorine gas, the phosphorus being in excess; or by passing the vapor of phosphorus over fragments of calomel (mercurous chloride) contained in a glass tube, and strongly heated. It is a thin, colorless liquid, which fumes in the air, and has a powerful and offensive odor. Its specific gravity is 1.45. Thrown into water, it sinks to the bottom, and is slowly decomposed, yielding phosphorous acid and hydrochloric acid: $PCl_3 + 3H_2O = 3HCl + H_3PO_3$.

Phosphorus Pentachloride, or Phosphoric Chloride, PCl_5, is formed when phosphorus is burned in excess of chlorine. Pieces of phosphorus are introduced into a large tubulated retort, which is then filled with dry chlorine gas. The phosphorus takes fire, and burns with a pale flame, forming a white volatile crystalline sublimate, which is the pentachloride. It may be obtained in larger quantity by passing a stream of dry chlorine gas into the liquid trichloride, which becomes gradually converted into a solid crystalline mass. Phosphorus pentachloride is decomposed by water, yielding phosphoric and hydrochloric acids:

$$PCl_5 + 4H_2O = 5HCl + H_3PO_4.$$

Phosphorus Oxychloride, $POCl_3$, is produced, together with hydrochloric acid, when phosphorus pentachloride is heated with a quantity of water insufficient to convert it into phosphoric acid. It may also be prepared by distilling the pentachloride with dehydrated oxalic acid, or by distilling a mixture of phosphorus pentachloride and phosphoric oxide. It is a colorless liquid, of sp. gr. 1.7, having a very pungent odor, boiling at 110° C. (230° F.), readily decomposed by water into hydrochloric and phosphoric acids.

A *sulphochloride* of analogous composition is produced by the action of hydrogen sulphide on the pentachloride. It is a colorless oily liquid, decomposed by water.

Two *bromides of phosphorus*, an *oxybromide*, and a *sulphobromide*, are known, corresponding in composition and properties with the chlorine compounds, and obtained by similar processes.

Phosphorus forms also two *iodides*, PI_2 and PI_3. Both are obtained by dissolving phosphorus and iodine together in carbon bisulphide, and cooling the liquid till crystals are deposited. Whatever proportions of iodine and phosphorus may be used, these two compounds always crystallize out, mixed with excess either of iodine or of phosphorus.

The *di-iodide* melts at 110° C. (230° F.), forming a red liquid which condenses to a light red solid. The *tri-iodide* melts at 55° C. (131° F.), and crystallizes on cooling in well-defined prisms. Both are decomposed by water, yielding hydriodic and phosphorous acids, the di-iodide also depositing yellow flakes of phosphorus.

PHOSPHORUS AND OXYGEN.

There are two definite oxides of phosphorus, in which the quantities of oxygen united with the same quantity of phosphorus are to one another as 3 to 5, viz. :—

	Composition by weight.		
	Formula.	Phosphorus.	Oxygen.
Phosphorus Trioxide, or } Phosphorous oxide . }	P_2O_3	61.92	+ 48
Phosphorus Pentoxide, or } Phosphoric oxide . }	P_2O_5	61.92	+ 80

Both these are acid oxides, uniting with water and metallic oxides to form salts, called phosphites and phosphates respectively, the hydrogen salts being also called phosphorous and phosphoric acid. There are also two other oxygen-acids of phosphorus, viz., hypophosphorous acid containing a smaller proportion of oxygen than phosphorous acid, and hypophosphoric acid intermediate in composition between phosphorous and phosphoric acids.

Hypophosphorous Acid, H_3PO_2.—When phosphorus is boiled with a solution of lime or baryta, water is decomposed, giving rise to phosphine and hypophosphorous acid: the first escapes as gas, and the hypophosphorous acid remains as a barium salt:

$$P_8 + 3BaH_2O_2 + 6H_2O = 3BaH_4(PO_2)_2 + 2H_3P.$$

The soluble hypophosphite may be crystallized out by slow evaporation. On adding to the liquid the quantity of sulphuric acid necessary to precipitate the base, the hypophosphorous acid is obtained in solution, and may be reduced to a syrupy consistence by evaporation. The acid is very prone to absorb more oxygen, and is therefore a powerful deoxidizing agent. All its salts are soluble in water.

Phosphorous Oxide, P_2O_3, is formed by the slow combustion of phosphorus in the air, or by burning that substance by means of a very limited supply of dry air, in which case it is anhydrous, and presents the aspect of a white powder. *Phosphorous acid,* H_3PO_3 or $3H_2O,P_2O_3$, is most conveniently prepared by adding water to the trichloride of phosphorus, when mutual decomposition takes place, the oxygen of the water being transferred to the phosphorus, generating phosphorous acid, and its hydrogen to the chlorine, giving rise to hydrochloric acid :

$$PCl_3 + 3H_2O = 3HCl + H_3PO_3.$$

By evaporating the solution to the consistence of syrup, the hydrochloric acid is expelled, and the residue crystallizes on cooling.

Phosphorous acid is very deliquescent, and very prone to attract oxygen and pass into phosphoric acid. When heated in a close vessel, it is resolved into phosphoric acid and pure phosphine gas. It is composed of 110 parts of phosphorous oxide and 54 parts of water, or 31 phosphorus, 48 oxygen, and 3 hydrogen.

Phosphoric Oxide, P_2O_5 (also called *Anhydrous Phosphoric Acid,* or *Phosphoric Anhydride*).—When phosphorus is burned under a bell-jar by the aid of a copious supply of dry air, snow-like phosphoric oxide is produced in great quantity. This substance exhibits attraction for water stronger even than that of sulphuric oxide; in fact, sulphuric oxide may be prepared by heating strong sulphuric acid in contact with phosphoric oxide. Exposed to the air for a few moments, phosphoric oxide deliquesces to a liquid, and when thrown into water, combines with the latter with explosive violence, and is converted into phosphoric acid. The water thus taken up cannot again be separated.

When nitric acid of moderate strength is heated in a retort with which a receiver is connected, and fragments of phosphorus are added singly, taking care to suffer the violence of the action to subside after each addition, the phosphorus is oxidized to its maximum and converted into phosphoric acid. By distilling off the greater part of the nitric acid, transferring the residue in the retort to a platinum vessel, and then cautiously raising the heat to redness, the phosphoric acid may be obtained pure. This is the *glacial phosphoric acid* of the Pharmacopœia.

A third method of preparing phosphoric acid consists in taking the acid calcium phosphate produced by the action of sulphuric acid on bone-earth, precipitating it with a slight excess of ammonia carbonate, separating by a filter the insoluble calcium salt, and then evaporating and igniting in a platinum vessel the mixed phosphate and sulphate of ammonia. Phosphoric acid alone remains behind. The acid thus obtained is somewhat impure.

One of the most advantageous methods of preparing pure phosphoric acid on the large scale is to burn phosphorus in a two-necked glass globe through which a current of dry air is passed: in this way the process may be carried on continuously. The phosphoric oxide obtained may be preserved in that state, or converted into hydrate or glacial acid, by addition of water and subsequent fusion in a platinum vessel.

Glacial phosphoric acid, or metaphosphoric acid, is exceedingly deliquescent, and requires to be kept in a closely stopped bottle. It contains 142 parts of phosphoric oxide and 18 parts of water, or 31 phosphorus, 48 oxygen, and 1 hydrogen, and is represented by the formula H_2O,P_2O_5 or HPO_3. Phosphoric oxide likewise unites with 2 and 3 molecules of water, forming the compounds $2H_2O,P_2O_5$ or $H_4P_2O_7$, and $3H_2O,P_2O_5$ or H_3PO_4, called respectively pyrophosphoric acid and orthophosphoric acid. The last is formed by oxidizing phosphorus with nitric acid, and by the action of water on phosphorous pentachloride:

$$PCl_5 + 4H_2O = 5HCl + H_3PO_4.$$

The aqueous solution evaporated to a thin syrup, and left over oil of vitriol, deposits orthophosphoric acid in prismatic crystals. The same solution may be heated to 160° C. (320° F.) without change in the composition of the acid; but at 213° C. (415° F.) it gives off a molecule of water, and is converted into pyrophosphoric acid; and at a red heat it gives off another molecule of water, and leaves metaphosphoric acid. Each of these acids forms a distinct class of salts, exhibiting reactions peculiar to itself. They will be described in connection with the general theory of saline compounds.

Phosphoric oxide is readily volatilized, and may be sublimed by the heat of an ordinary spirit-lamp. The acid may be fused in a platinum crucible at a red heat; at this temperature it evolves considerable quantities of vapor, but is still far from its boiling point. Phosphoric acid is a very powerful acid: being less volatile than sulphuric acid, it expels the latter at higher temperatures, although it is displaced by sulphuric acid at common temperatures. Its solution has an intensely sour taste, and reddens litmus-paper; it is not poisonous.

18 *

Hypophosphoric Acid, $H_4P_2O_6 = (HO)_2OP . PO(OH)_2$.—This acid is one of the constituents of the acid liquid (Pelletier's *acide phosphatique*) formed when phosphorus partially covered with water is expo ed to the air, and may be separated as a sparingly soluble sodium salt bystreating that liquid with carbonate or acetate of sodium. The pure acid is best prepared by decomposing its lead salt suspended in water with hydrogen sulphide. Its aqueous solution is strongly acid, and may be boiled without decomposition, but when evaporated to a syrup it is resolved by further heating into phosphorous and phosphoric acids. It gives a white precipitate with silver salts, and is oxidized by potassium permanganate to phosphoric acid. It is quadribasic, yielding, for example, the sodium salts $Na_4(PO_3)_2$, $Na_3H(PO_3)_2$, $Na_2H_2(PO_3)_2$.

PHOSPHORUS WITH SULPHUR AND SELENIUM.

Sulphides.—When ordinary phosphorus and sulphur are heated together in the dry state, or melted together under water, combination takes place, attended with vivid combustion and often with violent explosion. When amorphous phosphorus is used, the reaction is not explosive, though still very violent.

Six compounds of sulphur and phosphorus have been prepared, containing the following proportions of their elements:—

		Composition by weight.	
		Phosphorus.	Sulphur.
Hemisulphide, P_4S .	. . . .	30.96 +	8
Monosulphide, P_2S .	. . . .	30.96 +	16
Sesquisulphide, P_4S_3	. . . .	30.96 +	24
Trisulphide, P_2S_3 .	. . . .	30.96 +	48
Pentasulphide, P_2S_5	. . . .	30.96 +	80
Dodecasulphide, P_2S_{12}	. . . .	30.96 +	192

The fourth and fifth are analogous to phosphorous and phosphoric oxides respectively; the others have no known analogues in the oxygen series. They may all be formed by heating the two bodies together in the required proportions; but the trisulphide and pentasulphide are more easily prepared by warming the monosulphide with additional proportions of sulphur. Moreover, the two lower sulphides exhibit isomeric modifications, each being capable of existing as a colorless liquid and as a red solid.

The mono-, tri-, and penta-sulphides of phosphorus unite with metallic sulphides, forming sulphur salts. The copper salts have the following composition:—

Hypothiophosphite, $CuS,P_2S = CuP_2S_2$.
Thiophosphite, $CuS,P_2S_3 = CuP_2S_4$.
Thiophosphate, $CuS,P_2S_5 = CuP_2S_6$.

Selenides of Phosphorus, analogous in composition to the first, second, fourth, and fifth of the sulphides above mentioned, are produced by heating ordinary phosphorus and selenium together in the required proportions in a stream of hydrogen gas. The hemi-selenide is a dark yellow, oily, fetid liquid, solidifying at 12° C. (53.6° F.); the other compounds are dark red solids. The mono-, tri-, and penta-selenides unite with metallic selenides, forming selenium salts analogous to the sulphur salts above mentioned.

Thiophosphoric Acid, H_3PSO_3 or $PS(OH)_3$ (phosphoric acid having 1 at. O replaced by S), is not known in the free state. Its sodium salt,

Na_3PSO_3, obtained by heating the corresponding chloride (*infra*) with caustic soda, forms crystals which have an alkaline reaction and are decomposed by the weakest acids, the hydrogen thiophosphate thereby produced being at once resolved into hydrogen sulphide and phosphoric acid: $PS(OH)_3 + H_2O = H_2S + PO(OH)_3$.

Thiophosphoryl Chloride, $PSCl_3$, formed by the action of phosphorus pentachloride on the pentasulphide, $(3PCl_5 + P_2S_5 = 5PSCl_3)$, is a colorless, mobile, strongly refractive, pungent-smelling liquid boiling at 125° C. (257° F.) and having a density of 1.16816 at 0°. It is decomposed by water into hydrochloric and thiophosphoric acids, the latter being then further decomposed as above.

Thiophosphoryl Bromide, $PSBr_3$, is prepared by distilling phosphorus tribromide with flowers of sulphur, or better by gradually adding bromine (8 pts.) to a well-cooled solution of 1 pt. sulphur and 1 pt. phosphorus in carbon sulphide. On distilling the liquid, the carbon sulphide passes over first and then the thiobromide, which may be purified by agitation with water, whereby a hydrate, $PSBr_3,H_2O$, is formed, which gives off its water at 35° C. (95° F.). The anhydrous compound is obtained, by evaporating its solution in carbon sulphide, as a yellow liquid, which when touched by a solid body immediately solidifies to a crystalline mass. It is slowly decomposed by heating with water, and is partly resolved by distillation into sulphur and the compound $PSBr_3,PBr_3$, which boils at 205° C. (401° F.).

Pyrothiophosphoryl Bromide, $P_2S_3Br_4$, is obtained by dropping a solution of bromine in carbon sulphide on phosphorus trisulphide moistened with the same liquid. It is an oily, yellow, fuming liquid, which has an aromatic pungent odor, and is resolved by heat into the ortho-compound and phosphoric sulphide: $3P_2S_3Br_4 = 4PSBr_3 + P_2S_5$.

The formation of this compound is accompanied by that of a thick yellow oily liquid—probably *metathiophosphoryl bromide*, PS_2Br, easily separated from the preceding by its insolubility in ether. Its formation is perhaps due to the reaction of the P_2S_5, formed as above, with the trisulphide and bromine, as shown by the equation, $P_2S_5 + P_2S_3 + 4Br = 4PS_2Br$.

PHOSPHORUS AND NITROGEN.

Phospham, PN_2H, formed by passing dry ammonia over PCl_5 and heating the product out of contact with the air as long as sal-ammoniac sublimes, is a light white powder insoluble in water and infusible at a red heat. When heated in the air, it slowly oxidizes, giving off white fumes; and when moistened and heated it yields ammonia and metaphosphoric acid: $PN_2H + 3H_2O = 2NH_3 + PO_3H$. By fusion with caustic alkali, it is decomposed, with evolution of light and heat, yielding ammonia and an ortho-phosphate: $PN_2H + 3KOH + H_2O = 2NH_3 + PO(OK)_3$.

Phosphamide, $PO(NH)(NH_2)$, is obtained, together with sal-ammoniac, by the action of ammonia on phosphorus pentachloride. It is insoluble in water, and is slowly converted by boiling with water into acid ammonium phosphate: $PO(NH)(NH_2) + 3H_2O = PO(OH)(ONH_4)_2$.

Phosphoryl Triamide, $PO(NH_2)_3$, the amide of orthophosphoric acid, is obtained by the action of water on phosphorus oxychloride: $POCl_3 + 6NH_3 = 3NH_4Cl + PO(NH_2)_3$. On dissolving out the sal-ammoniac by water, the triamide remains as a white amorphous powder, not acted upon by boiling with water or aqueous potash-solution, but decomposed by sulphuric acid into phosphoric acid and ammonia: $2PO(NH_2)_3 + 3H_2SO_4 + 6H_2O = 2PO(OH)_3 + 3(NH_4)_2SO_4$. By fusion with potash it yields ammonia and potassium phosphate.

Thiophosphoryl Triamide, $PS(NH_2)_3$, is formed together with sal-ammoniac by the action of ammonia on phosphorus thiochloride. It is a white amorphous mass, decomposing when heated, and converted by hot water into ammonium thiophosphate, $PS(ONH_4)_3$.

Phosphoryl Nitride, PON, obtained by heating phosphamide or phosphoryl triamide out of contact with the air, is a white amorphous powder, which melts at a red heat and resolidifies to a vitreous mass. It is not attacked by nitric acid, but is converted by fusion with potash into ammonia and potassium orthophosphate: $PON + 3KOH = NH_3 + PO(OK)_3$.

Pyrophosphotriamic Acid, $P_2O_3(OH)(NH_2)_3$. This compound, derived from pyrophosphoric acid, $P_2O_3(OH)_4$, by substitution of $3NH_2$ for 3HO, is obtained by passing ammonia into phosphorus oxychloride without cooling, and boiling the product with water: $2POCl_3 + 9NH_3 + 2H_2O = 6NH_4Cl + P_2O_3(OH)(NH_2)_3$. It is an amorphous tasteless powder, decomposed by boiling with water into ammonia and *pyrophosphodiamic acid*, $P_2O_3(OH)_2(NH_2)_2$, which is an amorphous mass soluble in water and decomposed by heat into ammonia and phosphoric oxide: $P_2O_3(OH)_2(NH_2)_2 = 2NH_3 + P_2O_5$. By continued boiling with water it is converted into pyrophosphamic acid, $P_2O_3(OH)_3NH_2$, which when similarly treated, yields ammonia phosphate and pyrophosphoric acid:

$$P_2O_3(OH)_3NH_2 + HOH = NH_3 + P_2O_3(OH)_4.$$

Nitrogen Chlorophosphide, $P_3N_3Cl_6$, is formed by distilling the product of the action of ammonia on phosphorus pentachloride, and by distilling 1 part of the pentachloride with 2 parts sal-ammoniac. It sublimes in thin transparent six-sided plates melting at 110° C. (230° F.) and boiling at 240° C. (284° F.). It is insoluble in water, but is slowly decomposed thereby, with formation of pyrophosphodiamic acid:

$$2P_3N_3Cl_6 + 15H_2O = 12HCl + 3P_2O_3(OH)_2(NH_2)_2.$$

ARSENIC.

Symbol, As. Atomic weight, 74.9. Vapor-density, 149.8.

ARSENIC is sometimes found native: it occurs in considerable quantity as a constituent of many minerals, combined with metals, sulphur, and oxygen. In the oxidized state it has been found in very minute quantity in a great many mineral waters. The largest proportion is derived from the roasting of natural arsenides of iron, nickel, and cobalt. The operation is conducted in a reverberatory furnace, and the volatile products are condensed in a long and nearly horizontal chimney, or in a kind of tower of brick-work, divided into numerous chambers. The crude arsenious oxide thus produced is purified by sublimation, and then heated with charcoal in a retort; the arsenic is reduced, and readily sublimes.

Arsenic has a steel-gray color and strong lustre: it is crystalline and very brittle; it tarnishes in the air, but may be preserved unchanged in pure water. Its density, in the solid state, is 5.7 to 5.9. When heated, it volatilizes without fusion, and if air be present, oxidizes to arsenious oxide. Its vapor-density, compared with that of hydrogen, is 149.8, which is twice its atomic weight, so that its molecule in the gaseous state, like that of phosphorus, occupies only half the volume of a molecule of hydrogen (p. 205). The vapor has the odor of garlic.

Arsenic combines with metals in the same manner as sulphur and phosphorus, which latter it resembles in many respects.

Arsenic, like nitrogen, behaves in most respects as a triad element, not being capable of uniting with more than three atoms of any one monad element. Thus it forms the compounds AsH_3, $AsCl_3$, $AsBr_3$, etc., but no compound analogous to the pentachloride of phosphorus or of antimony. But just as ammonia, NH_3, can take up the elements of hydrochloric acid to form sal-ammoniac, NH_4Cl, in which nitrogen appears quinquivalent, so likewise can arsenetted hydrogen or arsine, AsH_3, unite with the chlorides, bromides, etc., of the radicles, methyl, ethyl, etc., to form salts in which the arsenic appears to be quinquivalent, e. g.:

Arsenethylium bromide   .   . $AsH_3(C_2H_5)Br$,
Arsenmethylium chloride   . $AsH_3(CH_3)Cl$.

In like manner, arsentrimethyl, $As(CH_3)_3$, unites with the chlorides of methyl and ethyl, forming the compounds $As(CH_3)_4Cl$ and $As(CH_3)_3$-$(C_2H_5)Cl$.

Arsenic likewise forms two oxides, viz., arsenious oxide, As_2O_3, and arsenic oxide, As_2O_5, with corresponding acids and salts, analogous to the phosphorous and phosphoric compounds: the arsenates, in particular, are isomorphous with the orthophosphates, and resemble them closely in many other respects.

Hydrides.—Arsenic forms two hydrides, containing 2 and 3 atoms of hydrogen combined with 1 atom of arsenic.

The *trihydride, Arsenious hydride, Arsenetted hydrogen* or *Arsine*, AsH_3, analogous in composition to ammonia and phosphine, is obtained pure by the action of strong hydrochloric acid on an alloy of equal parts of zinc and arsenic, and is produced in greater or lesser proportion whenever hydrogen is set free in contact with arsenious acid. Arsenetted hydrogen is a colorless gas, of specific gravity 2.695 (air = 1) or 38.9 (hydrogen = 1), slightly soluble in water, and having the smell of garlic. It burns, when kindled, with a blue flame, generating arsenious acid. It is also decomposed by transmission through a red-hot tube. Many metallic solutions are precipitated by this substance. When inhaled, it is exceedingly poisonous, even in very minute quantity.

The *dihydride*, AsH_2, or rather As_2H_4, is produced by passing an electric current through water, the negative pole being formed of metallic arsenic; also when potassium or sodium arsenide is dissolved in water. It is a brown powder, which gives off hydrogen when heated in a close vessel, and burns when heated in the air. It is analogous in composition to arsendimethyl or cacodyl, $As_2(CH_3)_4$.

Arsenious Chloride, $AsCl_3$.—This, the only known chloride of arsenic, is produced, with emission of heat and light, when powdered arsenic is thrown into chlorine gas. It is prepared by distilling a mixture of 1 part of metallic arsenic and 6 parts of corrosive sublimate, and by distilling arsenious oxide with strong hydrochloric acid, or with a mixture of common salt and sulphuric acid. It is a colorless, volatile, highly poisonous liquid, decomposed by water into arsenious and hydrochloric acids.

Arsenious Bromide, $AsBr_3$, prepared by adding powdered arsenic to a solution of bromine in carbon sulphide, forms colorless transparent crystals having a strong arsenical odor, a density of 3.66, melting at 20° C. (68° F.), and boiling at 220° C. (418° F.); decomposed by water like the chloride.

Arsenious Iodide, AsI_3, prepared in like manner, or by passing gaseous hydrogen iodide into arsenious chloride, crystallizes in bright red hexagonal plates having a density of 4.39.

Arsenious Fluoride, AsF_3, prepared by distilling a mixture of 4 parts arsenious oxide and 5 parts fluorspar with 10 parts strong sulphuric acid, is a limpid liquid boiling at 63° C. (145° F.), and having a density of 2.73. It has a strong pungent odor, fumes in the air, and produces serious wounds on the skin. It attacks glass, and is decomposed by water into arsenious and hydrofluoric acids; dissolves in aqueous ammonia, and forms a crystalline compound with the gas. The *pentafluoride* is not known in the free state, but the compound AsF_5, KF is obtained in colorless crystals by dissolving potassium arsenate in hydrofluoric acid.

Arsenious Oxide, Acid, and Salts.—*Arsenious oxide*, As_2O_3, also called *white oxide of arsenic*, is produced in the manner already mentioned. It is commonly met with in the form of a heavy, white, glassy-looking substance, with smooth conchoïdal fracture, having evidently undergone fusion. When freshly prepared it is often transparent, but becomes opaque by keeping at the same time slightly diminishing in density, and acquiring a greater degree of solubility in water. 100 parts of water at 100° dissolve about 11.5 parts of the opaque variety: the larger portion separates, however, on cooling, leaving about 3 parts dissolved: the solution, which contains *arsenious acid*, feebly reddens litmus. Cold water, agitated with powdered arsenious oxide, takes up a still smaller quantity. It is much more soluble in hydrochloric acid, and separates from the solution in large crystals, the deposition of which in the dark is attended with a bright and continuous luminosity. It also occurs naturally crystallized, sometimes in octahedrons, sometimes in tetrahedrons.

Alkalies dissolve arsenious oxide freely, forming arsenites; compounds with baryta, strontia, lime, magnesia, and manganous oxide have also been formed: the silver salt is a lemon-yellow precipitate. The arsenites are, however, very unstable. Those which have the composition M_3AsO_3, or $3M_2O, As_2O_3$, are regarded as normal salts; there are also arsenites containing $M_4As_2O_5$, or $2M_2O, As_2O_3$, and $MAsO_2$, or M_2O, As_2O_3, besides acid salts. The oxide or acid has a slightly sweetish and astringent taste, and is a most fearful poison.

Arsenic Oxide, Acid, and Salts.—When powdered arsenious oxide is dissolved in hot hydrochloric acid, and oxidized by the addition of nitric acid, the latter being added as long as red vapors are produced, the whole then cautiously evaporated to complete dryness, and the residue heated to low redness, arsenic oxide, As_2O_5, remains in the form of a white anhydrous mass which has no action upon litmus. When strongly heated, it is resolved into arsenious oxide and free oxygen. In water it dissolves slowly but completely, giving a highly acid solution, which, on being evaporated to a syrupy consistence, deposits, after a time, hydrated crystals of arsenic acid, containing $2H_3AsO_4, H_2O$, or $3H_2O, As_2O_5 + H_2O$. These crystals, when heated to

100°, give off their water of crystallization and leave *trihydric arsenate*, H_3AsO_4, or $3H_2O,As_2O_5$; at 140°-160° C. (284°-320° F.), *dihydric arsenate*, $H_4As_2O_7$, or $2H_2O,As_2O_5$ is left; and at 260° C. (500° F.), *monohydric arsenate*, $HAsO_3$, or H_2O,As_2O_5. The aqueous solutions of the three hydrates and of the anhydrous oxide exhibit exactly the same characters, and all contain trihydric arsenate, the other hydrates being immediately converted into that compound when dissolved in water; in this respect the hydrates of arsenic oxide differ essentially from those of phosphoric oxide (p. 209).

Arsenic acid is a very strong acid, forming salts isomorphous with the corresponding phosphates; it is also tribasic. A *sodium arsenate*, Na_2HAsO_4,-$12H_2O$, undistinguishable in appearance from common sodium phosphate, may be prepared by adding the carbonate to a solution of arsenic acid, until an alkaline reaction is apparent, and then evaporating. This salt also crystallizes with 7 molecules of water. Another arsenate, $Na_3AsO_4,12H_2O$, is produced when sodium carbonate in excess is fused with arsenic acid, or when the preceding salt is mixed with caustic soda. A third, NaH_2AsO_4,-H_2O, is made by substituting an excess of arsenic acid for the solution of alkali. The alkaline arsenates which contain basic water lose the latter at a red heat, but, unlike the phosphates, recover it when again dissolved. The arsenates of the alkalies are soluble in water: those of the earths and other metallic oxides are insoluble, but are dissolved by acids. The reddish-brown precipitate with silver nitrate is highly characteristic of arsenic acid.

Sulphides.—Two sulphides of arsenic are known. The *disulphide*, As_2S_2, occurs native as *Realgar*. It is formed artificially by heating arsenic acid with the proper proportion of sulphur. It is an orange-red, fusible, and volatile substance, employed in painting and by the pyrotechnist in making *white fire*. The *trisulphide* or *arsenious sulphide*, As_2S_3, also occurs native as *Orpiment*, and is prepared artificially by fusing arsenic with the appropriate quantity of sulphur, or by precipitating a solution of arsenious acid with hydrogen sulphide. It is a golden-yellow, crystalline substance, fusible, and volatile by heat. A cold solution of arsenic acid is not immediately precipitated by hydrogen sulphide, but after some hours the solution, saturated with hydrogen sulphide, yields a light yellow deposit of sulphur, the arsenic acid being reduced to arsenious acid, which is then gradually converted into lemon-yellow arsenious sulphide. In boiling solutions the precipitation takes place immediately. The mixture of sulphur and trisulphide thus produced was formerly regarded as a pentasulphide analogous to arsenic oxide.

The disulphide and trisulphide of arsenic are sulphur acids, uniting with other metallic sulphides to form sulphur salts. Those of the disulphide are called hypothioarsenites; they are but little known. The salts of arsenious sulphide are called thioarsenites. Their composition may be represented by that of the potassium salts, viz., $KAsS_2$, or K_2S,As_2S_3: $K_4As_2S_5$, or $2K_2S,As_2S_3$; and K_3AsS_3, or $3K_2S,As_2S_3$. Of these the bibasic salts are the most common. The thioarsenites of the alkali metals and alkaline earth metals are soluble in water, and may be prepared by digesting arsenious sulphide in the solutions of the corresponding hydrates or sulphydrates; the rest are insoluble and are obtained by precipitation. Sulphur salts, called thioarsenates, analogous in composition to the arsenates, are produced, in like manner, by digesting the mixture of sulphur and arsenious sulphide precipitated, as above mentioned, from arsenic acid, in solutions of alkaline hydrates or sulphydrates; also by passing gaseous hydrogen sulphide through solutions of arsenates. There are three thioarsenates of potassium, containing $KAsS_3$, or K_2S,As_2S_5; $K_4As_2S_7$, or $2K_4S,As_2S_5$; and K_3AsS_4, or $3K_2S,As_2S_5$. The thioarsenates of the alkali

metals and alkaline earth metals are soluble in water; the rest are insoluble
and are obtained by precipitation.

Detection of Arsenic in Cases of Poisoning.

Arsenious acid is distinguished by characters which cannot be mistaken:

Silver nitrate, mixed with a solution of arsenious acid in water, occasions
no precipitate, or merely a faint cloud : but if a little fixed alkali or a drop
of ammonia be added, a yellow precipitate of silver arsenite immediately
falls. This precipitate is exceedingly soluble in excess of ammonia, which
must therefore be added with great caution ; it is likewise very soluble in
nitric acid.

Cupric sulphate gives no precipitate with solution of arsenious acid, until a
little alkali has been added, when a brilliant yellow-green precipitate
(Scheele's green) falls, which also is very soluble in excess of ammonia.

Hydrogen sulphide, passed into a solution of arsenious acid, to which a few
drops of hydrochloric or sulphuric acid have been added, throws down a
copious bright yellow precipitate of orpiment, which is easily dissolved by
ammonia, and reprecipitated by acids.

Solid arsenious oxide, heated by the blowpipe in a narrow glass tube with
small fragments of dry charcoal, affords a sublimate of metallic arsenic in
the shape of a brilliant steel-gray metallic ring. A portion of this, detached
by the point of a knife, and heated in a second glass tube, with access of air,
yields, in its turn, a sublimate of colorless, transparent, octahedral crystals
of arsenious oxide.

All these experiments, which *jointly* give demonstrative proof of the
presence of the substance in question, may be performed with perfect precision and certainty upon exceedingly small quantities of material.

The detection of arsenious acid in complex mixtures containing organic
matter and common salt, as beer, gruel, soup, etc., or the fluid contents of
the stomach in cases of poisoning, is a far more difficult problem, but one
which is, unfortunately, often required to be solved. These organic matters
interfere completely with the liquid tests, and render their indications worthless. Sometimes the difficulty may be eluded by a diligent search in the
suspected liquid, and in the vessel containing it, for fragments or powder of
solid arsenious oxide, which, from its small degree of solubility, often
escapes solution, and from the high density of the substance may be found
at the bottom of the vessels in which the fluids are contained.

Fig. 122.

If anything of the kind be found, it may be washed by decantation with a little cold water, dried, and then reduced with charcoal. For the latter purpose, a small glass tube is taken, having
the figure represented in the margin ; white German glass, free
from lead, is to be preferred. The arsenious oxide, or what is
suspected to be such, is dropped to the bottom, and covered with
splinters or little fragments of charcoal, the tube being filled to
the shoulder. The whole is gently heated, to expel any moisture
that may be present in the charcoal, and the deposited water is
wiped from the interior of the tube with bibulous paper. The
narrow part of the tube containing the charcoal from *a* to *b*, is
now heated by the blowpipe flame ; and the tube, when red hot,
is inclined so that the bottom also may become heated. The
arsenious oxide, if present, is vaporized, and reduced by the
charcoal, and a ring of metallic arsenic deposited on the cool
part of the tube. To complete the experiment, the tube may be
melted at *a* by the point of the flame, drawn off, and closed, and
the arsenic oxidized to arsenious oxide, by chasing it up and
down by the heat of a small spirit-lamp. A little water may

afterward be introduced, and boiled in the tube, by which arsenious oxide will be dissolved, and to this solution the tests of silver nitrate and ammonia, copper sulphate and ammonia, and hydrogen sulphide, may be applied.

When the search for solid arsenious oxide fails, the liquid itself must be examined; a tolerably limpid solution must be obtained, from which the arsenic may be precipitated by hydrogen sulphide, and the orpiment collected and reduced to the metallic state. It is in the first part of this operation that the chief difficulty is found: such organic mixtures refuse to filter, or filter so slowly as to render some method of acceleration indispensable.* Boiling with a little caustic potash or acetic acid will sometimes effect this object. The following is an outline of a plan which has been found successful in a variety of cases in which a very small quantity of arsenious acid had been purposely added to an organic mixture: Oil of vitriol, itself perfectly free from arsenic, is mixed with the suspected liquid in the proportion of about a measured ounce to a pint, having been previously diluted with a little water, and the whole is boiled in a flask for half an hour, or until a complete separation of solid and liquid matter becomes manifest. The acid converts any starch that may be present into dextrin and sugar: it completely coagulates albuminous substances, and casein, in the case of milk, and brings the whole in a very short time into a state in which filtration is both easy and rapid. Through the filtered solution, when cold, a current of hydrogen sulphide is passed, and the liquid is warmed, to facilitate the deposition of the arsenious sulphide, which falls in combination with a large quantity of organic matter, which often communicates to it a dirty color. This is collected upon a small filter, and washed. It is next transferred to a capsule, and heated with a mixture of nitric and hydrochloric acids, by which the organic impurities are in great measure destroyed, and the arsenic oxidized to arsenic acid. The solution is evaporated to dryness, the soluble part taken up by dilute hydrochloric acid, and then the solution saturated with sulphurous acid, whereby the arsenic acid is reduced to a state of arsenious acid, the sulphurous acid being oxidized to sulphuric acid. The solution of arsenious acid may now be precipitated by hydrogen sulphide without any difficulty. The liquid is warmed, and the precipitate washed by decantation, and dried. It is then mixed with *black flux*, and heated in a small glass tube, similar to that already described, with similar precautions; a ring of reduced arsenic is obtained, which may be oxidized to arsenious oxide, and further examined. The black flux is a mixture of potassium carbonate and charcoal, obtained by calcining cream of tartar in a close crucible; the alkali transforms the sulphide into arsenious acid, the charcoal subsequently effecting the deoxidation. A mixture of anhydrous sodium carbonate and charcoal may be substituted with advantage for the common black flux, as it is less hygroscopic.

Other methods of proceeding, different in principle from the foregoing, are also employed, as that of the late Mr. Marsh, which is exceedingly delicate. The suspected liquid is acidulated with sulphuric acid, and placed in contact with metallic zinc; the hydrogen reduces the arsenious acid and combines with the arsenic, if any be present. The gas is burned at a jet, and a piece of glass or porcelain held in the flame, when any admixture of arsenetted hydrogen is at once known by the production of a brilliant black metallic spot of reduced arsenic on the porcelain; or the gas is passed through a glass tube heated at one or two places to redness, whereby the arsenetted hydrogen is decomposed, a ring of metallic arsenic appearing behind the heated portion of the tube.

Antimonetted hydrogen, however, gives a similar result. In order to distinguish the two substances, the gas may be passed into a solution of

* Respecting the separation of the arsenious acid by dialysis, see page 161.

10

silver nitrate. Both gases give rise to a black precipitate, which, in the case of antimonetted hydrogen, consists of silver antimonide, Ag_3Sb, whilst in the case of arsenetted hydrogen it is pure silver, the arsenic being then converted into arsenious acid, which combines with a portion of silver oxide. The silver arsenite remains dissolved in the nitric acid, which is liberated by the precipitation of the silver, and may be thrown down with its characteristic yellow color by adding ammonia to the liquid filtered off from the black precipitate. The black silver antimonide, when carefully washed, and subsequently boiled with a solution of tartaric acid, yields a solution containing antimony only, from which hydrogen sulphide separates the characteristic orange-yellow precipitate of antimonious sulphide.

Fig. 123.

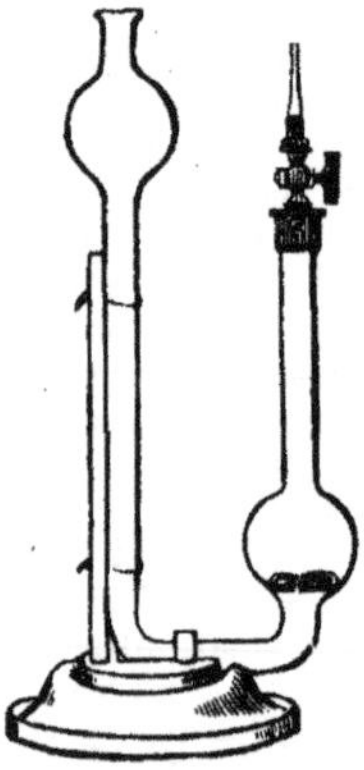

A convenient form of Marsh's instrument is that shown in fig. 123: it consists of a bent tube, having two bulbs blown upon it, fitted with a stop-cock and narrow jet. Slips of zinc are put into the lower bulb, which is afterward filled with the liquid to be examined. On replacing the stop-cock, closed, the gas collects and forces the liquid into the upper bulb, which then acts by its hydrostatic pressure, and expels the gas through the jet so soon as the stop-cock is opened. It must be borne in mind that both common zinc and sulphuric acid often contain traces of arsenic. Professor Bloxam[*] has proposed an important modification of Marsh's process for the detection of arsenic and antimony in organic substances, which is based on the behavior of their solutions under the influence of the electric current. Antimony is deposited in the metallic state, without any disengagement of antimonetted hydrogen, while arsenic is evolved as arsenetted hydrogen, which may be recognized by the characters already indicated.

A slip of copper-foil boiled in the poisoned liquid, previously acidulated with hydrochloric acid, withdraws the arsenic, and becomes covered with a white alloy. By heating the metal in a glass tube, the arsenic is expelled, and oxidized to arsenious acid. This is called Reinsch's test.

BORON.

Symbol, B.   Atomic weight, 11.

THIS element is not found in nature in the free state, but occurs somewhat abundantly in the form of boric acid, $B(OH)_3$, and some of its salts.

Pure boron is prepared by heating boric oxide, B_2O_3, or potassium borofluoride, with metallic potassium in an iron tube, and washing out the soluble salts with water. As thus obtained, it is a dark brown amorphous powder destitute of taste and smell.[†] It burns in the air when heated, producing

[*] Journal of the Chemical Society, xiii. 338.

[†] Wöhler and Deville, in 1858, by strongly igniting boric oxide with aluminium, in a crucible surrounded with powdered charcoal to prevent access of oxygen, obtained very hard transparent octahedral crystals and a graphite-like substance, which they regarded as modifications of boron analogous to the diamond and graphite varieties of carbon; but subsequent observations have shown that the octahedral crystals contained both carbon and aluminium in the proportions indicated by the formula $B_{48}C_2Al_3$, and that the graphite-like substance was a boride of aluminium having the composition AlB_2.

the trioxide, and is readily attacked by chlorine, nitric acid, fused alkalies, and other reagents.

Boron differs from all the other non-metallic elements in not combining with hydrogen; but the composition of its chloride, bromide, fluoride, and oxide shows that it is trivalent; and it further resembles the other elements of the nitrogen-group in being likewise capable of functioning as a quinquivalent element, as in fluoboric acid, HBF_4, and in certain organic compounds which will be described hereafter.

HALOGEN-COMPOUNDS OF BORON.

Boron Chloride, BCl_3, is most readily obtained by passing dry chlorine over a strongly heated mixture of boric oxide and charcoal, whereupon it passes over in vapor, together with gaseous carbon monoxide, and may be condensed to a liquid by passing it through a U-shaped tube cooled by a freezing mixture. It may be freed from excess of chlorine by agitation with mercury and subsequent distillation. It may also be prepared by heating finely-powdered boric chloride with twice its weight of phosphoric chloride for three days in a sealed tube at 150° C. (302° F.), and subsequently distilling the product at the heat of a water-bath. It is a liquid of sp. gr. 1.35 at 17° C. (62.6° F.), and boiling at 18.23° C. (64.81° F.). It is decomposed by water into hydrochloric and boric acids, and fumes strongly in the air. Heated with sulphuric anhydride in a sealed tube at 120°, it yields boric oxide and sulphuryl chloride: $3SO_3 + 2BCl_3 = B_2O_3 + 3SO_2Cl_2$. It unites with ammonia, forming a white crystalline body, having the composition, $2BCl_3,3NH_3$.

Boron Bromide, BBr_3, obtained by direct combination of its elements, or by passing bromine-vapor over a heated mixture of charcoal and boric oxide, is a colorless strongly fuming liquid, having a sp. gr. of 2.69 and boiling at 90.5° C. (194.9° F.). With water and with ammonia it reacts like the chloride.

Boron Fluoride, BF_3, is prepared by heating a mixture of 1 pt. boric oxide and 2 pts. fluorspar with 12 pts. strong sulphuric acid:

$$B_2O_3 + 3CaF_2 + 3H_2SO_4 = 2BF_3 + 3CaSO_4 + 3H_2O;$$

also by heating a mixture of potassium borofluoride and boric oxide with sulphuric acid:

$$6KBF_4 + B_2O_3 + 6H_2SO_4 = 8BF_3 + 6KHSO_4 + 3H_2O.$$

It is a colorless gas, having an extremely pungent odor and a density of 2.37 referred to air, or 34.25 referred to hydrogen; does not attack glass. It is decomposed by water, and must therefore be collected over mercury or by displacement.

Potassium and sodium burn brilliantly when heated in it. From certain organic bodies it withdraws the elements of water, carbonizing them at the same time like strong sulphuric acid.

With an equal volume of ammonia it forms the compound BF_3NH_3, which is a white opaque solid, sublimable without alteration. With 2 or 3 vol. ammonia to 1 vol. BF_3, colorless liquids are formed having respectively the formulæ $BF_3(NH_3)_2$ and $BF_3(NH_3)_3$. These liquids when heated give off ammonia and are converted into the solid compound.

One volume of water absorbs 700 volumes of boron fluoride, with great rise of temperature and formation of a fuming oily liquid, which has a density of 1.77, and cauterizes like strong sulphuric acid. When heated it gives off about a fifth of the absorbed gas, leaving a liquid which has nearly the composition $BF_3.H_2O$, and boils at 165°-200° C. (329°-392° F.), with partial decomposition and formation of boric acid.

Fluoboric acid, HBF_4, is formed, together with boric acid, when boron fluoride is brought in contact with water: $8BF_3 + 6H_2O = 6HBF_4 + 2B(OH)_3$, also when aqueous hydrofluoric acid is saturated with boric acid. It is a monobasic acid, forming with bases a series of salts called *fluoborides* or *borofluorides*, which are also produced, together with an alkaline hydroxide, or free alkali, on mixing an acid fluoride of potassium or sodium with boric acid: $B(OH)_3 + 2NaHF_2 = NaBF_4 + NaOH + 2H_2O$. This process exhibits the peculiar result of the formation of a liquid having an alkaline reaction by the mixing of two neutral liquids. The fluoborides are mostly soluble in water and crystalline. When heated they are resolved into gaseous boron fluoride and a metallic fluoride.

Boron and Oxygen.—There is but one oxide of boron, viz., the *trioxide*, B_2O_3, which unites with water and metallic oxides, forming boric acid and the metallic borates.

Boric or **Boracic Acid**, or **Hydrogen Borate**, $B(OH)_3$ or $3H_2O,B_2O_3$, is found in solution in the hot lagoons of Tuscany, which yield a large supply of it. It is easily prepared by decomposing with sulphuric acid a hot solution of borax, which is an acid borate of sodium, occurring abundantly in the salt lakes of India, Thibet, and other parts of Asia.

Boric acid crystallizes in transparent six-sided plates, soluble in about 25 parts of cold water, much more soluble in boiling water. It has but little taste, and differs considerably from the stronger acids in its action on vegetable colors, imparting to blue litmus-paper only a wine-red color, like that produced by carbonic acid, and to turmeric-paper a brown color, like that produced by alkalies, but easily distinguished therefrom by the fact of its not disappearing on the addition of an acid.

The crystallized acid, when heated, gives up its water and melts, with great tumefaction, to a glassy transparent mass of boric oxide, which readily dissolves many metallic oxides, with very characteristic colors, *e. g.*, deep blue with oxide of cobalt, amethyst with manganese, bright green with chromium, etc. The crystallized acid dissolves in alcohol, and the solution burns with a green flame.

Glassy boric oxide in a state of fusion requires for its dissipation in vapor a very intense and long-continued heat; the aqueous solution cannot, however, be evaporated without very appreciable loss by volatilization: hence it is probable that the acid is far more volatile than the anhydrous oxide.

Boric acid is susceptible of modifications analogous to those of phosphoric acid (p. 209), and forming corresponding salts. The acid $B(OH)_3$ or $B_2O_3,-3H_2O$, above described, which is analogous in constitution to the chloride, fluoride, etc., of boron, is distinguished as **orthoboric acid**. This acid, heated at 100° C., gives off $2H_2O$, and is reduced to **metaboric acid**, B_2O_3,H_2O or $BO(OH)$, which is a white powder, volatilizing slowly, but completely, at the same temperature. Lastly, when orthoboric acid is heated for a long time at 140° C. (284° F.), it gives off $5H_2O$, leaving **pyroboric acid**, $2B_2O_3,H_2O$ or $B_4O_5(OH)_2$, in the form of a brittle vitreous mass.

Boron Trisulphide, B_2S_3, is formed by heating boron in sulphur-vapor, but is more readily prepared by heating an intimate mixture of lamp-black and boric oxide in vapor of carbon bisulphide: $3CS_2 + 3C + 2B_2O_3 = 2B_2S_3 + 6CO$. This process yields it for the most part as a white glassy fusible mass, but sometimes in silky needles. It has a pungent odor and attacks the eyes; melts when heated, and may be distilled in a current of hydrogen sulphide. Water decomposes it immediately, forming boric acid and hydrogen sulphide: $B_2S_3 + 6H_2O = 3H_2S + 2B(OH)_3$.

Boron Nitride, BN, is produced by heating boric oxide with metallic cyanides, or by heating to bright redness a mixture of sal-ammoniac and pure anhydrous borax, or sodium biborate, $Na_2O,2B_2O_3$:

$$Na_2O,2B_2O_3 + 2NH_4Cl = 2BN + B_2O_3 + 2NaCl + 4H_2O.$$

It is a white amorphous powder, insoluble in water, infusible and non volatile. When heated in a current of steam, it yields ammonia and boric oxide: $2BN + 3H_2O = 2NH_3 + B_2O_3$, and likewise gives off a large quantity of ammonia when fused with potash. It dissolves slowly in hydrofluoric acid, forming ammonium fluoborate: $BN + 4HF = (NH_4)BF_4$.

SILICON or SILICIUM.

Symbol, Si. Atomic weight, 28.

THIS element in union with oxygen forms silica, or the earth of flints, which is one of the most abundant of natural minerals. To obtain silicon in the free state, potassium silicofluoride is heated in a glass tube with nearly its own weight of metallic potassium, whereupon violent reaction ensues, resulting in the formation of potassium fluoride and free silicon: $KF,SiF_4 + 4K = 5KF + Si$. On transferring the contents of the tube to cold water, the potassium fluoride dissolves, and silicon remains behind.

Silicon thus prepared is a dark-brown powder, destitute of lustre. Heated in the air, it burns, and becomes superficially converted into silica. It is also acted upon by sulphur and by chlorine. When silicon is strongly heated in a covered crucible, its properties are greatly changed; it becomes darker in color, denser, and incombustible, refusing to burn even when heated by the flame of the oxy-hydrogen blowpipe.

Silicon, like carbon (p. 226), is capable of existing in three different modifications. The modification above mentioned corresponds with the amorphous variety of carbon (lamp-black). The researches of Wöhler and Deville have established the existence of modifications corresponding with the diamond, and with the graphite variety of carbon. The diamond modification of silicon is most readily obtained by introducing into a red-hot crucible a mixture of 3 parts of potassium silicofluoride, 1 part of sodium in small fragments, and 1 part of granulated zinc, and heating to perfect fusion. On slowly cooling, there is formed a button of zinc, covered and interspersed with needle-shaped crystals, consisting of octahedrons joined in the direction of the axis. This crystallized silicon, which may be readily freed from zinc by treatment with acids, resembles crystallized hæmatite in color and appearance; it scratches glass, and fuses at a temperature approaching the melting point of cast iron. The graphite modification of silicon is prepared by fusing, in a Hessian crucible, 5 parts of soluble glass (potassium silicate), 10 parts of cryolite (sodium and aluminium fluoride), and 1 part of aluminium. On treating the resulting button of aluminium with hydrochloric acid, the silicon remains in the form of scaly crystals,

19 *

resembling graphite, but of somewhat brighter color, scratching glass, like the previous modification. It is infusible. Its specific gravity is 2.49.

Silicon Hydride, or *Silicomethane*, was discovered by Buff and Wöhler, who obtained it by passing an electric current through a solution of sodium chloride, the positive pole employed consisting of aluminium containing silicon. More recently, Wöhler and Martius produced this gas by treating magnesium containing silicon with hydrochloric acid. Both methods yield silicic hydride mixed with free hydrogen. Friedel and Ladenburg, however, by decomposing silicic triethyl-formate (see SILICIC ETHERS) in contact with sodium, have obtained it pure, and shown that it consists of 28 parts by weight of silicon and 4 parts of hydrogen, answering to the formula SiH_4. The reaction by which it is produced is represented by the following equation :—

$$4SiH(OC_2H_5)_3 = SiH_4 + 3Si(OC_2H_5)_4.$$
$$\underset{\text{formate.}}{\text{Silicic triethyl}} \quad \underset{\text{hydride.}}{\text{Silicic}} \quad \underset{\text{silicate.}}{\text{Tetrethylic}}$$

Silicon hydride is a colorless gas, having a density of 16 ($H = 1$). In the impure state, as obtained by the first two processes above given, it takes fire spontaneously on coming in contact with the air, and burns with a white flame, evolving clouds of silica. Pure silicic hydride, however, does not ignite spontaneously under the ordinary atmospheric pressure; but on passing a bubble of air into the rarefied gas standing over mercury, it takes fire, and yields a deposit of amorphous silicon mixed with silica. On passing silicic hydride through a red-hot tube, it is decomposed, silicon being deposited. When passed into chlorine gas, it takes fire, forming silicon tetrachloride and hydrochloric acid.

HALOGEN-COMPOUNDS OF SILICON.

Silicon Tetrachloride, or *Silicic Chloride*, $SiCl_4$, is formed by direct combination of its elements, but is most readily prepared by mixing finely-divided silica with charcoal-powder and oil, strongly heating the mixture in a covered crucible, and then exposing the mass so obtained, in a porcelain tube heated to full redness, to the action of perfectly dry chlorine gas. A good condensing arrangement, supplied with ice-cold water, must be connected with the porcelain tube. The product is a colorless and very volatile liquid, of sp. gr. 1.524 at 0°, boiling at 50° C. (122° F.), and having a pungent, suffocating odor. In contact with water, it yields hydrochloric acid and gelatinous silica.—*Silicon Trichloride*, Si_2Cl_6, is formed when the tetrachloride is passed over silicon, heated to whiteness in a porcelain tube. It is a colorless liquid, having a density of 1.58 at 0°, solidifying at − 1° C. (30.2° F.), and boiling at 146° C. (294.8° F.); vapor-density, 9.7. It fumes strongly in the air, and takes fire when heated. It begins to decompose at 350° C. (662° F.), and the decomposition goes on at an increasing rate till the temperature rises to 800° C. (1472° F.), at which point the compound is completely resolved into the tetrachloride and free silicon : $2Si_2Cl_6 = 3SiCl_4 + Si$. If, however, the temperature of the vapor be quickly raised beyond 1000° C. (1832° F.), no such decomposition takes place.

Silicon Hydrotrichloride, or *Silicic Chloroform*, $SiHCl_3$ (chloroform, $CHCl_3$, having its carbon replaced by silicon), is obtained by passing dry hydrogen chloride over crystallized silicon, heated to a temperature below redness in a porcelain tube. It is a very volatile strong-smelling liquid, which fumes in the air and boils between 34° and 37° C. (93° and 98° F.). It is very inflammable, and burns with a green-edged flame, evolving white clouds of silica. It is readily decomposed by water, with precipitation of a white powder, $Si_2H_2O_3$, called SILICOFORMIC ANHYDRIDE, formed as shown by the equation $2SiHCl_3 + 3H_2O = 6HCl + Si_2H_2O_3$. This body is very unstable, and is decomposed by dilute ammonia, yielding silicic acid and hydrogen.

With *bromine* and *iodine*, silicon forms compounds analogous in composition to the chlorine-compounds, and prepared by similar methods:—

$SiBr_4$ is a colorless heavy liquid, boiling at 154° C. (309° F.), and solidifying to a crystalline mass at 13° C. (55.4° F.). Si_2Br_6 forms large crystalline tablets, melts when heated, and distils without decomposition at 240° C. (464° F.). $SiCl_3Br$, prepared by heating silicochloroform with bromine at 100° C., is a colorless liquid, boiling at 80° C. (176° F.).

SiI_4, obtained by direct combination, crystallizes from carbon sulphide in regular octahedrons, melts at 120.5° C. (249.9° F.), and boils at 290° C. (554° F.). It takes fire when heated in the air, burning with a reddish flame, and is decomposed by water, with formation of hydriodic and silicic acids. Si_2I_6, obtained by heating the tetra-iodide with finely-divided silver at 280° C. (536° F.), crystallizes from carbon sulphide in splendid colorless hexagonal prisms or rhombohedrons, which, on exposure to moist air, are converted into a white mass, with formation of silicic and hydriodic acids. When heated, it melts and is reduced to a lower iodide, probably Si_2I_4. Ice-cold water decomposes it, with formation of a white substance, $Si_2O_4H_2$, called *silico-oxalic acid*, from its analogy to oxalic acid, $C_2O_4H_2$. This body is decomposed, even by weak bases, into hydrogen and silica. $SiHI_3$, formed on passing a mixture of hydrogen and hydriodic acid over silicon heated to low redness, is a colorless strongly refracting liquid, having a density of 3.362 at 0°, and boiling at 220° C. (428° F.). It is decomposed by water, like silicochloroform.

Silicon Tetrafluoride, SiF_4, is a gaseous compound, formed whenever dry hydrofluoric acid comes in contact with silica, either free or combined, and is easily prepared by heating white sand or pounded glass, with fluorspar and strong sulphuric acid:—

$$2CaF_2 + 2H_2SO_4 + SiO_2 = SiF_4 + 2CaSO_4 + 2H_2O.$$

The gas may be collected over mercury. It is colorless, and has a density of 3.57 (air = 1), or 52.2 (H = 1). It fumes strongly in the air, has a highly pungent odor, like that of hydrochloric acid, and condenses to a colorless liquid under a pressure of 30 atmospheres, or when cooled to − 107° C. (− 160.6° F.). According to Nattera, it solidifies at − 140° C. (− 220° F.). Fused sodium takes fire in the gas, and burns with a red flame. With dry ammonia-gas, it forms a white crystalline compound, $SiF_4,2NH_3$, which is decomposed by water.

Fig. 124.

Silicofluoric or **Hydrofluosilicic Acid**, H_2SiF_6 or $SiF_4,2HF$, is formed, together with silicic hydroxide or silicic acid, when silicon tetrafluoride comes in contact with water: $3SiF_4 + 4H_2O = 2H_2SiF_6 + Si(OH)_4$. To prepare it, silicon tetrafluoride obtained as above is passed into water through a tube, the lower end of which dips under mercury, as shown in the annexed figure; if the gas were passed immediately into water in the ordinary way, the tube would soon be stopped up by the separated silica.

To obtain the acid, the thick jelly is pressed through a linen filter, and the filtrate is concentrated by evaporation at a low temperature. The saturated solution is a strongly acid fuming liquid, which decomposes on boiling into silicon fluoride and hydrofluoric acid, so that it may be evaporated in platinum vessels, without leaving any residue.

Silicofluoric acid forms salts called *silicofluorides*, having the composition M_2SiF_6 or $M''SiF_6$ (M denoting a univalent and M'' a bivalent metal). Most of these are easily soluble in water, but the potassium, sodium, lithium, barium, calcium, and yttrium salts are more or less sparingly soluble. By prolonged ignition, they are resolved into silicon fluoride and a metallic fluoride: *e. g.*, $K_2SiF_6 = SiF_4 + 2KF$.

SILICON AND OXYGEN.

Silicon Dioxide or **Silica**, SiO_2, the only oxide of silicon, constitutes, either alone or in combination, a very large proportion of the crust of the earth. It occurs in various forms. Q u a r t z forms crystals, sometimes very large, belonging to the hexagonal system and usually having the form of a hexagonal prism terminated by pyramidal summits. Specific gravity 2.6. The transparent colorless variety is called rock-crystal. *Amethyst quartz* has a violet tint due to traces of manganese oxides. Other varieties are known as *milk-quartz, rose-quartz,* and *smoky quartz.* Sand and sandstone consist of an impure variety.—T r i d y m i t e is another hexagonal variety of silica, usually forming six-sided tables having their horizontal edges truncated by pyramidal faces; moreover the crystals usually occur in twins: hence the name. Sp. gr. 2.3.—A m o r p h o u s s i l i c a occurs as *Opal,* either colorless or variously colored, and having a density somewhat above 2. It contains water varying in amount from 0.1 to 12.9 per cent., but this water is not regarded as essential to its constitution. *Chalcedony, agate,* and *flint* are intimate mixtures of amorphous silica with quartz or tridymite.

Hydrated Amorphous Silica is formed as already mentioned (p. 223) by passing silicon tetrafluoride into water. The gelatinous mass washed with water, dried, and ignited, yields pure silica in the form of a white mobile powder.

Amorphous silica may also be prepared by decomposing a silicate of alkali-metal with an acid. Powdered quartz or fine white sand is mixed with about three times its weight of dry sodium carbonate, the mixture is fused in a platinum crucible, and the glassy mass when cold is boiled with water, by which it is softened and almost entirely dissolved. An excess of hydrochloric acid is then added to the filtered liquid, and the whole is evaporated to complete dryness. By this treatment the gelatinous silica thrown down by the acid becomes completely insoluble, and remains behind when the dry saline mass is treated with acidulated water, by which the alkaline salts, alumina, ferric oxide, lime, and many other bodies which may happen to be present, are removed. The silica is washed, dried, and heated to redness.

Silica thus prepared is a very fine, white, tasteless powder, having a density of about 2.66, fusible only by the oxy-hydrogen blowpipe. When once dried, it is not sensibly soluble in water or dilute acids (with the exception of hydrofluoric acid). But on adding hydrochloric acid to a very dilute solution of potassium silicate, the liberated silica remains in solution. From this mixed solution of silica and potassium chloride, the latter may be separated by diffusion (comp. p. 161), whereby a moderately concentrated solution of silica in water is obtained. This solution has a distinctly acid reaction; it presents, however, but little stability. When kept for some time, it gelatinizes, the silica separating in the insoluble modification. The same effect is produced by the addition of a few drops of sulphuric or nitric acid, or of a solution of salt.

Silica is essentially an acid oxide, forming salts with basic metallic oxides,* and decomposing all salts of volatile acids when heated with them. In strong alkaline liquids it is freely soluble. When heated with bases, especially those which are capable of undergoing fusion, it unites ₔwith them and forms salts, which are sometimes soluble in water, as in the case of the potassium and sodium silicates, when the proportion of base is considerable. Common glass is a mixture of several silicates, in which the reverse of this happens, the silica being in excess. Even glass, however, is slowly acted upon by water. Finely divided silica is highly useful in the manufacture of porcelain.

Silicon Oxychloride, Si_2Cl_6O, formed together with potassium chloride by passing the vapor of silicon tetrachloride over felspar, $Al_2K_2Si_6O_{16}$, heated to whiteness in a porcelain tube, is a colorless fuming liquid, easily decomposed by water into silicic and hydrochloric acids. By the further action of this compound on heated felspar, or better by repeatedly passing a mixture of its vapor and oxygen through a red-hot tube filled with fragments of porcelain, the following oxychlorides are produced: $Si_4Cl_4O_4$, boiling at 198–202° C. (388–396° F.); $Si_8Cl_{12}O_{10}$, at about 400° C. (752° F.); $(Si_2Cl_2O_3)_n$, above 300° C. (572° F.); and $(Si_4Cl_2O_7)_n$, above 440° C. (824° F.). The values of n in these last two formulæ—in other words, the moleenlar weights of the respective compounds—are not known: those of the first two have been determined by their vapor-densities.

SILICON AND SULPHUR.

Silicon Disulphide, SiS_2, obtained by passing the vapor of carbon disulphide over a strongly heated mixture of silica and carbon, such as is used in the preparation of silicon tetrachloride (p. 222), forms long silky needles which are decomposed by water, yielding hydrogen sulphide and silica. The *monosulphide*, SiS, and an *oxysulphide*, probably SiSO, are formed, together with other products, when sulphur-vapor is passed over white-hot silicon. The former is yellow, the latter yellowish, and both are volatile and condense in the cooler, but still hot, part of the tube.

Silicon Chlorhydrosulphide, $SiCl_3SH$, is formed by passing a mixture of hydrogen sulphide and vapor of silicon tetrachloride through a red-hot tube: $SiCl_4 + H_2S = HCl + SiCl_3SH$. It is a colorless fuming liquid boiling at 96° C. (210° F.). Vapor-density = 5.78 (air = 1).

SILICON AND NITROGEN.

A *silicon nitride* of unknown composition is formed by the action of ammonia on the tetrachloride, also when crystallized silicon is strongly heated in nitrogen gas. It is a white amorphous powder insoluble in all acids except hydrofluoric acid, by which it is converted into ammonium silicofluoride. By heating with potash it is converted into ammonia and potassium silicate.

---

* The composition of these salts will be discussed in the chapter on the general properties of the metals and their compounds.

10 *

CARBON.

Symbol, C. Atomic weight, 12.

THIS element occurs in a state of purity, and crystallized, in two distinct and very dissimilar forms—namely, as diamond, and as graphite or plumbago. It constitutes a large proportion of all organic structures, animal and vegetable: when these latter are exposed to destructive distillation in close vessels, a great part of their carbon remains, obstinately retaining some of the hydrogen and oxygen, and associated with the earthy and alkaline matter of the tissue, giving rise to the many varieties of charcoal, coke, etc. This residue, when perfectly separated from foreign matter, constitutes a third variety of carbon.

The diamond is one of the most remarkable substances known: long prized on account of its brilliancy as an ornamental gem, the discovery of its curious chemical nature confers upon it a high degree of scientific interest. Several localities in India, the Island of Borneo, South Africa, and Brazil, furnish this beautiful substance. It is always distinctly crystallized, often quite transparent and colorless, but now and then having a shade of yellow, pink, or blue. The origin and true geological position of the diamond are unknown; it is always found imbedded in gravel and transported materials whose history cannot be traced. The crystalline form of the diamond is that of the regular octahedron or cube, or some figure geometrically connected with these. Many of the octahedral crystals exhibit a very peculiar appearance, arising from the faces being curved or rounded, which gives to the crystal an almost spherical figure.

Fig. 125.

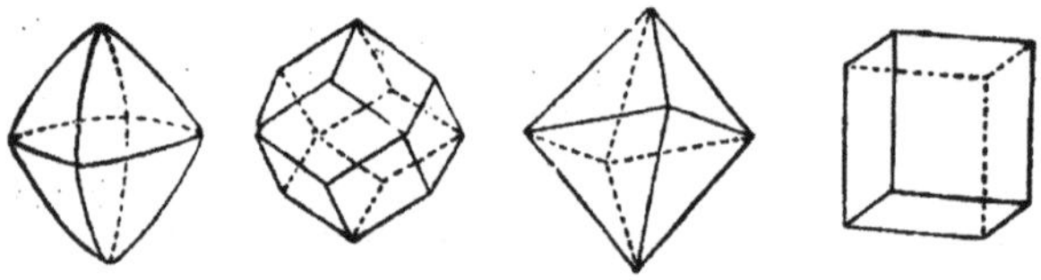

The diamond is the hardest substance known: it admits of being split or cloven without difficulty in particular directions, but can only be cut or abraded by a second portion of the same material; the powder rubbed off in this process serves for polishing the new faces, and is also highly useful to the lapidary and seal-engraver. One very curious and useful application of the diamond is made by the glazier; a *fragment* of this mineral, like a bit of flint, or any other hard substance, scratches the surface of the glass; a *crystal* of diamond, having the rounded octahedral figure spoken of, held in one particular position on the glass—namely, with an edge formed by the meeting of two adjacent faces—presented to the surface and then drawn along with gentle pressure, causes a split or cut, which penetrates to a considerable depth into the glass, and determines its fracture with perfect certainty.

The diamond is infusible and unalterable even by a very intense heat, provided air be excluded; but when heated, thus protected, between the poles of a strong galvanic battery, it is converted into coke or graphite; heated to whiteness in a vessel of oxygen, it burns with facility, yielding carbonic acid gas.

Graphite or plumbago appears to consist essentially of pure carbon, although most specimens contain iron, the quantity of which varies from a mere trace up to 5 per cent. Graphite is a somewhat rare mineral; the finest and most valuable for pencils was formerly obtained from Borrowdale, in Cumberland, where a kind of irregular vein is found traversing the ancient slate beds of that district, but the mine is now nearly exhausted. Large quantities of graphite are imported from Germany, the East Indies, and the United States.* Crystals are not common; when they occur, they have the figure of a short six-sided prism—a form bearing no geometric relation to that of the diamond.

Graphite is often formed artificially in certain metallurgic operations: the brilliant scales which sometimes separate from melted cast iron on cooling, called by the workmen "kish," consist of graphite.

Lamp-black, the soot produced by the imperfect combustion of oil or resin, is the best example that can be given of carbon in its uncrystallized or *amorphous* state. To the same class belong the different kinds of charcoal. That prepared from wood, either by distillation in a large iron retort, or by the smothered combustion of a pile of fagots partially covered with earth, is the most valuable as fuel. Coke, the charcoal of pit-coal, is much more impure; it contains a large quantity of earthy matter, and very often sulphur, the quality depending very much upon the mode of preparation. Charcoal from bones and animal matters in general is a very valuable substance, on account of the extraordinary power it possesses of removing coloring matters from organic solutions; it is used for this purpose by the sugar-refiner to a very great extent, and also by the manufacturing and scientific chemist. The property in question is possessed also in a small degree by all kinds of charcoal.

Charcoal made from box, or other dense wood, has the property of condensing gases and vapors into its pores; of ammoniacal gas it is said to absorb not less than ninety times its volume, while of hydrogen it takes up less than twice its own bulk, the quantity being apparently connected with the property in the gas of suffering liquefaction. This property of absorbing gases, as well as the decolorizing power, no doubt depends in some way upon the same peculiar action of surface so remarkable in the case of platinum in a mixture of oxygen and hydrogen. The absorbing power is, indeed, considerably increased by saturating charcoal with solution of platinum, and subsequently igniting it, so as to coat the charcoal with a thin film of platinum. Dr. Stenhouse, who suggested this plan, found that the gases thus absorbed undergo a kind of oxidation within the pores of the charcoal.

CARBON AND HYDROGEN.

The compounds of these elements are exceedingly numerous, and their complete study, together with that of their compounds and substitution-derivatives, constitutes a special branch of chemical science, called "Organic Chemistry," to which the latter part of this volume is devoted. They are obtained in nearly all cases by the decomposition of more complex bodies of organic origin, only one of them, viz., *acetylene*, C_2H_2, being pro-

---

* The graphite which can be directly cut for pencils, occurring only in limited quantity, powdered graphite, obtained from the inferior varieties of the mineral, is now frequently consolidated for this purpose. The mechanical division of graphite presents considerable difficulties, which, however, may be entirely obviated by adopting a chemical process suggested by the late Sir Benjamin Brodie, applicable, however, only to certain varieties, such as Ceylon and Siberian graphite. This process consists in introducing the coarsely powdered graphite, previously mixed with $\frac{1}{12}$ of its weight of potassium chlorate, into 2 parts of concentrated sulphuric acid, which is heated in a water-bath until the evolution of acid fumes ceases. The acid is then removed by water, and the graphite dried. Thus prepared, this substance, when heated to a temperature approaching a red heat, swells up to a bulky mass of finely divided graphite.

ducible by direct combination of its elements, namely, by passing a strea n of hydrogen gas through a tube in which the electric current is passing between two carbon poles. It is a colorless gas which burns with a bright smoky flame, producing water and carbon dioxide.

Two other hydrocarbons of simple constitution may here also be mentioned, viz., *methane*, CH_4, and *ethylene*, C_2H_4, both of which are constituents of coal-gas. Methane or marsh-gas is produced in the decomposition of vegetable matter under water, and is given off when the mud at the bottom of stagnant pools is stirred: it also constitutes the "fire-damp" of coal mines. It is not formed by direct combination of carbon and hydrogen, but may be prepared by strongly heating a mixture of sodium acetate ($NaC_2H_3O_2$) and sodium hydroxide (caustic soda) with quicklime: $NaC_2H_3O_2 + NaOH = CH_4 + Na_2CO_3$. It is a colorless gas which burns with a pale-blue flame, producing water and carbon dioxide. Ethylene, C_2H_4, is formed in the decomposition of numerous organic bodies, and is usually prepared by heating alcohol (C_2H_6O) with strong sulphuric acid, which abstracts the elements of water: $C_2H_6O - H_2O = C_2H_4$. It is a colorless gas which burns with a bright white flame, and unites with chlorine, forming an oily liquid, $C_2H_4Cl_2$: hence it was formerly called "Olefiant Gas."

Coal and Oil Gases.—The manufacture of coal-gas is a branch of industry of great interest and importance in several points of view. The process is one of great simplicity of principle, but requires, in practice, some delicacy in management to ensure a good result.

When pit-coal is subjected to destructive distillation, a variety of products show themselves— permanent gases, steam and volatile oils, besides a not inconsiderable quantity of ammonia from the nitrogen always present in the coal. These substances vary very much in their proportions with the temperature at which the process is conducted, the permanent gases becoming more abundant with increased heat, but, at the same time, losing much of their value for the purposes of illumination.

The coal is distilled in cast-iron retorts, maintained at a bright-red heat, and the volatilized product is conducted into a long horizontal pipe of large dimensions, always half filled with liquid, into which the extremity of each separate tube dips: this is called the *hydraulic main*. The gas and its accompanying vapors are next made to traverse a refrigerator—usually a series of iron pipes, cooled on the outside by a stream of water; here the condensation of the tar and the ammoniacal liquid becomes complete, and the gas proceeds onward to another part of the apparatus, in which it is deprived of the sulphuretted hydrogen and carbonic acid always present in the crude product. This separation was formerly effected by slaked lime, which steadily absorbs the compounds in question. The use of lime, however, has been almost superseded by that of a mixture of sawdust and iron oxide. This mixture, after having been used, is exposed for some time to the atmosphere, and is then fit for use a second time. The purifiers are large iron vessels filled either with slaked lime or with the iron oxide mixture. The gas is admitted at the bottom of the vessel, and made to pass over a large surface of the purifying agent. The last part of the operation, which, indeed, is often omitted, consists in passing the gas through dilute sulphuric acid, in order to remove ammonia. The quantity thus separated is very small, relatively, to the bulk of the gas, but in an extensive work becomes an object of importance.

Coal-gas thus manufactured and purified is preserved for use in immense cylindrical receivers, closed at the top, suspended in tanks of water by chains to which counterpoises are attached, so that the gas-holders rise and sink in the liquid as they become filled from the purifiers or emptied by the mains. These latter are made of large diameter, to diminish as much as possible the resistance experienced by the gas in passing through such a length of pipe.   The joints of these mains are still made in so imperfect a manner that immense loss is experienced by leakage when the pressure upon the gas exceeds that exerted by a column of water an inch in height.*

Coal-gas varies very much in composition, judging from its variable density and illuminating powers, and from the analyses which have been made. The difficulties of such investigations are very great, and unless particular precaution be taken, the results are merely approximate.   The purified gas is believed to contain the following substances, of which the first is the most abundant, and the second the most valuable :—

> Methane, or Marsh-gas.
> Ethylene, or Olefiant gas.
> Acetylene.
> Hydrogen.
> Carbon Monoxide.
> Nitrogen.
> Vapors of volatile liquid Hydrocarbons.†
> Vapor of Carbon Bisulphide.

> *Separated by Condensation and by the Purifiers.*

> Tar and Volatile Oils.
> Ammonium Sulphate, Chloride, and Sulphide.
> Hydrogen Sulphide.
> Carbon Dioxide.
> Hydrocyanic Acid, or Ammonium Cyanide.
> Thiocyanic Acid, or Ammonium Thiocyanate.

A far better illuminating gas may be prepared from oil, by dropping it into a red-hot iron retort filled with coke; the liquid is in great part decomposed and converted into permanent gas, which requires no purification, as it is quite free from the ammoniacal and sulphur compounds which vitiate gas from coal.   Many years ago this gas was prepared in London; it was compressed, for the use of the consumer, into strong iron vessels, to the extent of 30 atmospheres; but the manufacture proved unprofitable, and was therefore given up.

COMPOUNDS OF CARBON WITH THE HALOGEN ELEMENTS.—Several compounds of carbon and chlorine are known, viz., C_2Cl_2, C_2Cl_4, C_2Cl_6, and CCl_4, and similar compounds with bromine and iodine.   All these bodies are obtained indirectly by the action of chlorine, etc., upon organic compounds, and will be described under ORGANIC CHEMISTRY.

* It may give some idea of the extent of this species of manufacture to mention that in the year 1838, for lighting London and the suburbs alone, there were eighteen public gas-works, and £2,800,000 invested in pipes and apparatus. The yearly revenue amounted to £450,000, and the consumption of coal in the same period to 180,000 tons, 1460 *millions* of cubic feet of gas being made in the year.   There were 134,300 private lights, and 30,400 street lamps.   890 tons of coals were used in the retorts in the space of twenty-four hours at mid-winter, and 7,120,000 cubic feet of gas consumed in the longest night.—*Ure, Dictionary of Arts and Manufactures.*

Since that time, the production of gas has been enormously increased.   The amount of coal used in London for gas-making in the year ending June, 1852, is estimated at 408,000 tons, which on an average would yield 4000 *millions* of cubic feet of gas.   In the year 1857 the mains in the London streets had reached the extraordinary length of 2000 miles.

† These bodies increase the illuminating power, and confer on the gas its peculiar odor.

20

COMPOUNDS OF CARBON AND OXYGEN.

There are two direct compounds of carbon and oxygen, CO and CO_2.

Carbon Monoxide (commonly called *Carbonic Oxide*).—When carbon dioxide is passed over red-hot charcoal or metallic iron, one-half of its oxygen is removed, and it becomes converted into carbon monoxide. A very good method of preparing this gas is to introduce into a flask fitted with a bent tube some crystallized oxalic acid ($H_2C_2O_4$), and pour upon it five or six times as much strong oil of vitriol. On heating the mixture, the oxalic acid is resolved into water, carbon dioxide, and carbon monoxide: $H_2C_2O_4 = CO + CO_2 + H_2O$; and by passing the gases through a strong solution of caustic potash, the first is withdrawn by absorption, while the second remains unchanged. Another and perhaps preferable method is to heat finely-powdered yellow potassium ferrocyanide with 8 or 10 times its weight of concentrated sulphuric acid. The salt is entirely decomposed, yielding a copious supply of perfectly pure carbon monoxide, which may be collected over water in the usual manner. The reaction is represented by the equation—

$$K_4FeC_6N_6 + 6H_2O + 6H_2SO_4 = 6CO + 2K_2SO_4 +$$

Potassium ferro-   Sulphuric   Potassium
cyanide.   acid.   sulphate.

$$3(NH_4)_2SO_4 + FeSO_4.$$

Ammonium   Ferrous
sulphate.   sulphate.

Carbon monoxide is a combustible gas, which burns with a beautiful pale-blue flame, generating carbon dioxide. It has never been liquefied. It is colorless, has very little odor, and is extremely poisonous—much more so than carbon dioxide. Mixed with oxygen, it explodes by the electric spark, but with some difficulty. Its specific gravity is 14 (H = 1) or 0.973 (air = 1); a litre weighs 1.2515 grams; 100 cubic inches weigh 30.21 grains.

The relation by volume of the oxides of carbon is as follows:—Carbon dioxide contains its own volume of oxygen, that gas suffering no change of bulk by its conversion. One measure of carbon monoxide, mixed with half a measure of oxygen, and exploded, yields one measure of carbon dioxide: hence carbon monoxide contains half its volume of oxygen.

Carbon monoxide unites with chlorine under the influence of light, forming a pungent, suffocating compound, possessing acid properties, called p h o s g e n e gas or c a r b o n y l c h l o r i d e, $COCl_2$. It is made by mixing equal volumes of carbon monoxide and chlorine, both perfectly dry, and exposing the mixture to sunshine: the gases unite quietly, the color disappears, and the volume becomes reduced to one-half. A more convenient method of preparing this gas consists in passing carbon monoxide through antimony pentachloride. It must be received over mercury, as it is decomposed by water. When pure, it condenses to a liquid at 0°, or more quickly at the temperature of a mixture of ice and salt.

Carbon Dioxide or **Carbonic Anhydride,** CO_2 (commonly called *Carbonic Acid*).—This compound is always produced when charcoal or any other form of carbon burns in air or oxygen gas. It is, however, more conveniently prepared by decomposing a carbonate with one of the stronger acids. For this purpose the apparatus for generating hydrogen (p. 132), may again be employed: fragments of marble are put into the bottle with enough water to cover the extremity of the funnel-tube, and hydrochloric or nitric acid is added by the latter, until the gas is freely disengaged. Chalk-powder and dilute sulphuric acid may be used instead. The gas may be collected over water, although with some loss; or very conveniently by displacement, if it be required dry, as shown in fig. 126. The long dry-

Fig. 126.

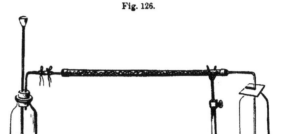

ing tube is filled with fragments of calcium chloride, and the heavy gas is conducted to the bottom of the vessel in which it is to be received, the mouth of the latter being lightly closed.

Carbon dioxide is a colorless gas having an agreeable pungent taste and odor. Its density is 22 (H = 1) or 1.524 (air = 1); a litre weighs 1.96664 grams, and 100 cubic inches weigh 47.26 grains. It is very hurtful to animal life, even when largely diluted with air, acting as a narcotic poison: hence the danger arising from imperfect ventilation, the use of fire-places and stoves of all kinds unprovided with proper chimneys, and the crowding together of many individuals in houses and ships without efficient means for renewing the air; for carbon dioxide is constantly disengaged during the process of respiration, which, as already mentioned (p. 146), is nothing but a process of slow combustion. This gas is sometimes emitted in large quantity from the earth in volcanic districts, and it is constantly generated where organic matter is in the act of undergoing fermentative decomposition. The fatal "after-damp" of the coal-mines contains a large proportion of carbon dioxide.

A lighted taper plunged into carbon dioxide is instantly extinguished even to the red-hot snuff. The gas, when diluted with three times its volume of air, still retains the power of extinguishing a light. It is easily distinguished from nitrogen, which is also incapable of supporting combustion, by its rapid absorption by caustic alkali, or by lime-water; the turbidity communicated to the latter from the production of insoluble calcium carbonate is very characteristic.

Cold water dissolves about its own volume of carbon dioxide, whatever be the density of the gas with which it is in contact (comp. p. 163); the solution temporarily reddens litmus-paper. Common soda-water and effervescent wines afford examples of the solubility of the gas. Even boiling water absorbs a perceptible quantity.

Some of the interesting phenomena attending the liquefaction of carbon dioxide have been already described (p. 73): it requires for the purpose a pressure of 38.5 atmospheres at 0° C. The liquefied oxide is colorless and limpid, lighter than water, and four times more expansible than air; it mixes in all proportions with ether, alcohol, naphtha, oil of turpentine, and carbon disulphide, and is insoluble in water and fat oils. In this condition it does not exhibit any of the properties of an acid.

Carbon dioxide exists, as already mentioned, in the air: relatively its quantity is but small; but absolutely, taking into account the vast extent of the atmosphere, it is very great, and fully adequate to the purpose of supplying plants with their carbon, these latter having the power, by the aid of their green leaves, of decomposing carbon dioxide, retaining the carbon, and expelling the oxygen. The presence of light is essential to this effect, but the manner of its production is not yet clearly understood.

The carbonates form a very large and important group of salts, some of which, as the carbonates of calcium and magnesium, occur very abundantly in nature.   They contain the elements of carbon dioxide and a metallic oxide: calcium carbonate, for example, being represented by the formula CaO,CO_2 or $CaCO_3$; but they are never formed by the direct union of dry carbon dioxide with a dry metallic oxide, the intervention of water being always required to bring about the combination.   Potassium carbonate (pearlash) is the chief constituent of wood-ashes: sodium carbonate is contained in the ashes of marine plants, and is manufactured on a very large scale by heating sodium sulphate with lime and coal.   These carbonates are soluble in water.   The other metallic carbonates, which are insoluble, may be formed by mixing a solution of potassium or sodium carbonate with a soluble metallic salt; thus, when solutions of lead nitrate and sodium carbonate are mixed together, the lead and sodium change places, forming sodium nitrate, which remains dissolved, and lead carbonate, which, being insoluble in water, is precipitated as a white powder: $Pb(NO_3)_2 + Na_2CO_3 = 2NaNO_3 + PbCO_3$.

The solution of carbon dioxide in water may be supposed to contain hydrogen carbonate or carbonic acid, H_2CO_3 or H_2O,CO_2; but this compound is not known in the separate state, only in aqueous solution.   According to Wroblewski,[*] however, carbon dioxide unites with water at $0°$ C. and under a pressure of about 16 atmospheres, forming the hydrate $CO_2,8H_2O$.

The amount of carbon dioxide contained in a carbonate may be estimated by the loss of weight which the salt undergoes when decomposed by an acid.

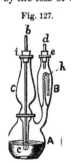

Fig. 127.

This may be determined by means of Geissler's apparatus (fig. 127).   It consists of three glass vessels, A, B, C, the last of which is ground air-tight into the neck of A; $b\,c$ is a glass tube open at both ends, ground water-tight into C at the lower end, and kept in position by the cork i, which can slide up and down on the tube $b\,c$.   The cork e must close air-tight in B, and the tube d in e.   The weighed substance to be decomposed is introduced into A, and water is added.   The vessel C is then filled with the aid of a pipette nearly to the top with dilute nitric or hydrochloric acid, the cork i having been previously moved upward without raising the tube b.   The cork i is then replaced; the vessel C again inserted into A; the vessel B is rather more than half filled with strong sulphuric acid; and the tube b is closed at the top by placing over it a small piece of caoutchouc tubing, with a glass rod fitted into the other end.   The apparatus having been weighed, the decomposition of the carbonate is effected by opening the tube b a little, so as to cause acid to pass from C into A.   The carbon dioxide then passes through the bent tube h into the sulphuric acid, by which it is dried, and finally leaves the apparatus through d.   When the decomposition is complete, the vessel A is gently heated, the stopper is removed from b, and the carbon dioxide still remaining in the apparatus is sucked out at d. The apparatus, after cooling, is again weighed, and the difference between the two weighings gives the weight of the carbon dioxide expelled.

* Comptes rendus, xciv. 954.

Carbon and Sulphur.

Two compounds of these elements are known, viz., the disulphide produced by the direct combination of its elements at a high temperature, and the monosulphide formed by reduction of the disulphide.

Carbon Disulphide or Bisulphide, CS_2.—To prepare this compound, a wide porcelain tube filled with pieces of charcoal which have been previously heated to redness in a covered crucible, is fixed across a furnace in a slightly inclined position. Into the lower extremity a tolerably wide tube is secured by a cork; this tube bends downward, and passes nearly to the bottom of a bottle filled with fragments of ice and a little water. The porcelain tube being heated to bright redness, fragments of sulphur are thrown into the open end, which is immediately closed by a cork. The sulphur melts, and is converted into vapor, which at that high temperature combines with the carbon, forming an exceedingly volatile compound, which is condensed by the ice, and collects at the bottom of the vessel. This is collected and redistilled at a very gentle heat in a retort connected with a good condenser.

For preparation on the large scale, a tubulated earthen retort is filled with charcoal, and the sulphur is dropped in through a porcelain tube passing through the tubulus, and reaching nearly to the bottom; or the charcoal is contained in a large iron cylinder, and the sulphur introduced through a pipe fitted into the lower part.

Carbon disulphide is a transparent colorless liquid of great refractive and dispersive power. Its density is 1.272, and that of its vapor is 2.67 (air $= 1$). It boils at 43° C. (109° F.), and emits vapor of considerable elasticity at common temperatures. In its ordinary state, it has a very repulsive odor, due perhaps to the presence of small quantities of other volatile sulphur-compounds; but when these are removed by agitating the liquid with mercury till it ceases to blacken the bright surface of the metal, it is said to have a pure ethereal odor. When set on fire in the air, it burns with a blue flame, forming carbon dioxide and sulphur dioxide; and when its vapor is mixed with oxygen it becomes explosive. Carbon disulphide, when heated with water in a sealed tube to about 153° C. (307° F.), is converted into carbon dioxide and hydrogen sulphide. In contact with nascent hydrogen (when heated with zinc and dilute sulphuric acid), it is converted into a white crystalline substance, having the composition CH_2S, crystallizing in square prisms, insoluble in water, alcohol, and ether, but soluble in carbon disulphide, subliming at 150° C. (302° F.), and decomposing at 200° C. (392° F.). Carbon disulphide freely dissolves sulphur, and by spontaneous evaporation deposits the latter in beautiful crystals; it also dissolves phosphorus, iodine, camphor, and caoutchouc, and mixes easily with oils. It is extensively used in the vulcanization of caoutchouc and in the manufacture of gutta percha; also for extracting bitumen from mineral substances, and oils from seeds.

Carbon Monosulphide, CS, is said by Sidot,* to be obtained by exposing the disulphide in sealed tubes for a considerable time to direct sunshine. It is then precipitated as a brown powder which may be purified by distilling off the undecomposed disulphide, and washing the residual mixture of monosulphide and free sulphur with pure disulphide till all the free sulphur is removed. It is a maroon-colored powder, without taste or smell, and having a specific gravity of 1.66, insoluble in water, alcohol, turpentine-oil, and benzine, slightly soluble at the boiling heat in carbon disulphide

* Comptes rendus, lxxxi. 32.

20 *

and in ether. It dissolves also in caustic potash and in boiling nitric acid; the strongest nitric acid ignites it. At about 200° C. (392° F.) it is resolved into its elements, a small quantity of the disulphide being formed at the same time.

According to S. Kern,[*] carbon monosulphide is also formed by the prolonged action of iron wire on the disulphide in sealed tubes.

Thiocarbonic Acid, $H_2CS_3 = CS(SH)_2$, also called *Sulphocarbonic Acid.*—The salts of this acid are formed, similarly to the carbonates, on adding carbon disulphide to the solution of an alkaline sulphide, e. g., $CS_2 + Na_2S = Na_2CS_3$. From the resulting solution alcohol precipitates the thiocarbonate as a heavy slightly brown liquid, which is decomposed by hydrochloric acid, yielding thiocarbonic acid in the form of a yellow oil having an offensive and pungent odor, and resolved by heat into CS_2 and H_2S. The thiocarbonates of the alkali metals and alkaline earth metals are soluble in water, and the solutions give a brown precipitate with copper salts, yellow with dilute silver nitrite, red with lead salts. These precipitates readily turn black, from separation of the corresponding sulphides.

Thiocarbonyl Chloride, $CSCl_2$, is obtained by the prolonged action of dry chlorine gas on carbon disulphide, and more readily by heating the disulphide in sealed tubes at 100° C. with phosphorus pentachloride: $CS_2 + PCl_5 = PSCl_3 + CSCl_2$. It is a colorless strong-smelling liquid, insoluble in water, boiling at 70° C. (158° F.).

Carbon Oxysulphide, or **Carbonyl Sulphide,** COS.—This compound, discovered by Than, is produced by direct combination when carbon monoxide mixed with sulphur-vapor is passed through a red-hot porcelain tube. As thus prepared it is mixed with free carbon monoxide; but on passing the gas through alcoholic potash, the oxysulphide is alone absorbed, and may be liberated in the pure state by treating the solution with hydrochloric acid.

Carbon oxysulphide is also produced by gently heating the disulphide with an equivalent quantity of sulphur trioxide: $CS_2 + SO_3 = CSO + SO_2 + S$; and by decomposing potassium thiocyanate with moderately dilute acids: thiocyanic acid, HCNS, is then liberated and decomposed by the water present in the manner represented by the equation, $HCSN + H_2O = NH_3 + CSO$.

Carbon oxysulphide is a gas of sp. gr. 2.1046, and may easily be poured from one vessel to another. It has an aromatic odor like that of some resins, slightly also that of hydrogen sulphide, and a feebler acid reaction than carbon dioxide. At a low red heat it is partly resolved into carbon monoxide and sulphur-vapor; by a fine platinum wire ignited by the electric current it is slowly but completely decomposed, yielding an equal volume of carbon monoxide. It burns in the air with a faint blue flame, producing carbon dioxide and sulphur dioxide; with $1\frac{1}{2}$ vol. oxygen it forms an explosive mixture burning with a shining bluish-white flame. It is not acted upon by chlorine or fuming nitric acid at ordinary temperatures, and does not form an explosive mixture with nitrogen dioxide.

Water absorbs about its own volume of carbon oxysulphide, acquiring a sweetish and afterward a pungent taste, and decomposing it after some time. It appears to exist in some sulphur springs and in the sulphurous gases of volcanoes. Potash-solution absorbs the gas as completely as carbon dioxide, though less quickly; the solution exhibits the reactions of metallic sulphides, and when treated with acids gives off hydrogen sulphide and carbon dioxide. Baryta-water and lime-water act in a similar manner. Neutral or

* Chemical News, xxxiii. 253.

acid solutions of lead, copper, cadmium, and silver salts are not precipitated by the gas, but when mixed with excess of ammonia they yield with it characteristic precipitates of metallic sulphides.*

Compounds of Carbon and Nitrogen. Cyanogen, CN.

When a stream of air is passed over a mixture of charcoal and potassium carbonate kept at a bright red heat, the nitrogen of the air unites with the carbon and the potassium, forming a compound called potassium cyanide, containing 39 parts of potassium, 12 of carbon, and 14 of nitrogen, and represented by the formula KCN. It is a crystalline salt, which dissolves easily in water, and decomposes mercuric oxide, forming potassium hydrate and mercuric cyanide:

$$2KCN + H_2O + HgO = 2KHO + HgC_2N_2.$$

Now, when dry mercuric cyanide, which is a white crystalline substance, is strongly heated in a glass tube, fitted up like that used for the evolution of oxygen from mercuric oxide (p. 46), it splits up, like the oxide, into metallic mercury, and a gaseous body called cyanogen, containing 12 parts by weight of carbon, and 14 of nitrogen, and represented by the formula CH. It must be collected over mercury, as it is rapidly absorbed by water.

Cyanogen is a colorless gas, having a pungent and very peculiar odor, remotely resembling that of peach-kernels. Exposed while at the temperature of 45° F. (7.2° C.) to a pressure of 3.6 atmospheres, it condenses to a thin, colorless, transparent liquid. It is inflammable, and burns with a beautiful purple or peachblossom-colored flame, generating carbon dioxide and liberating nitrogen. Its specific gravity is 1.801 referred to air, or 26 referred to hydrogen as unity. One volume of it exploded with 2 vols. oxygen, yields 1 vol. nitrogen and 2 vols. carbon dioxide. Now, the weights of equal volumes of cyanogen, nitrogen, and carbon dioxide are as 26 : 14 : 22. Consequently, 26 parts by weight of cyanogen yield by combustion 14 parts of nitrogen and 44 parts of carbon dioxide, containing 12 parts of carbon; or 26 cyanogen = 12 carbon + 14 nitrogen.

Water dissolves 4 or 5 times its volume of cyanogen gas, and alcohol a much larger quantity: the solution rapidly decomposes, yielding ammonium oxalate, a brown insoluble matter, and other products.

Cyanogen unites (though not directly) with hydrogen, forming the very poisonous compound called hydrocyanic or prussic acid; and with metals forming compounds called cyanides, analogous in composition and character to the chlorides, iodides, bromides, etc. In short, this group of elements, represented by the formula CN, combines with elementary bodies, and is capable of passing from one state of combination to another, just as if it were itself an elementary body. Such a group of elements is called a compound radicle. We have already had occasion to notice another such group, viz., ammonium, NH_4. Cyanogen, however, is analogous in its chemical relations to the non-metallic elements, chlorine, bromine, oxygen, etc., whereas ammonium is a quasi-metal analogous to potassium, etc.

The compounds of cyanogen will be further considered under ORGANIC CHEMISTRY.

Compounds containing Carbon and Silicon.

When crystallized silicon is heated nearly to whiteness in a porcelain tube through which a stream of carbon dioxide is passing, a greenish-white mass

* See further, Watts's Dictionary of Chemistry, First Supplement, 406 ; Second Supplement, p. 262.

is formed, which when freed from silica by hydrofluoric acid, leaves a greenish powder having the composition SiCO, and formed according to the equation $3Si + 2CO_2 = SiO_2 + 2SiCO$. It is oxidized, with incandescence, when heated with litharge or a mixture of litharge and lead chromate.

When crystallized silicon is heated to whiteness in a crucible lined with charcoal, the mass, after washing with boiling potash and with hydrofluoric acid, yields a bluish-green pulverulent residue, Si_2C_2N, which is insoluble in acids and in alkalies, and burns brilliantly, with formation of Cl_2 and nitrogen oxides, when heated with lead oxide or a mixture of lead oxide and chromate.*

When hydrogen gas saturated with benzene vapor at 50–60° is passed over silicon contained in two porcelain dishes heated to bright redness in a porcelain tube, the first dish, after a few hours, is found to contain a light black powder, and the second a gray substance, both of which may be purified by treatment with potash and hydrofluoric acid. The black powder then consists of uncombined carbon mixed with the dark green compound SiC_2, which does not burn in a stream of oxygen. The gray substance is somewhat variable in composition, and contains oxygen (derived from the porcelain dish) often in quantity larger than that required by the formula $SiCO_2$.

When powdered silicon is heated to whiteness in a gas-carbon crucible lined with lamp-black, a mass is obtained, which, when freed from adhering charcoal, powdered, and purified by treatment with potash and hydrofluoric acid, yields a bottle-green powder, having the composition $Si_2C_3O_2$. If the silicon is replaced by a mixture of rather thick iron wire with lamp-black and silica, a crystalline compound Fe_6Si_2C is obtained, having a density of 6.6. It is formed, however, only after prolonged heating at a very high temperature.

When vapor of carbon sulphide is passed over white-hot silicon, there are formed, besides the volatile sulphide and oxysulphide of silicon already mentioned (p. 225), a non-volatile substance, which when freed by boiling potash-solution from excess of silicon and its sulphur-compounds, and then digested for some time with warm hydrofluoric acid, yields a greenish powder, SiC_4S; and this, when heated in a stream of oxygen, is converted, without alteration of weight, into SiC_4O_2.†

Other compounds containing carbon and silicon will be described under ORGANIC CHEMISTRY.

* Schützenberger and Colson, Comptes rendus, xcii. 1508.
† A. Colson, Comptes rendus, xciv. 1316, 1526.

GENERAL LAWS OF CHEMICAL COMBINATION—ATOMIC THEORY.

Before proceeding further with the study of individual compounds, it is advisable to enter more fully into the consideration of certain general laws of chemical combination, and certain theoretical notions founded thereon, a sketch of which has already been given in the Introduction.

The laws in question are: (1.) The Law of Equivalents, according to which the replacement of elements one by another always takes place in definite proportion; (2.) The Law of Multiples, according to which the several quantities of an element A which can unite with a fixed quantity of another element B, stand to one another, for the most part, in simple numerical proportions. The observation of these laws has led to the idea that the elementary bodies are made up of indivisible particles called atoms, each having a constant weight peculiar to itself; and that chemical combination takes place by the juxtaposition of these atoms, 1 to 1, 1 to 2, 1 to 3, 2 to 3, etc., a group of atoms thus united being called a molecule. This is the atomic hypothesis of Dalton.

Equivalents.—The equivalent weight of an elementary body compared with that of hydrogen, taken as unity, may in many cases be determined by direct substitution. Thus, when a metal dissolves in hydrochloric or sulphuric acid, the quantity of the metal which takes the place of 1 part by weight of hydrogen is its equivalent weight. In this manner it is found that the equivalent of sodium is 23, of zinc 32.5, of magnesium 12, etc. Again, many organic compounds—acetic acid, for example—are acted upon by chlorine and bromine in such a manner that 1 part of the hydrogen is removed and its place supplied by chlorine or bromine, every 1 part by weight (gram, ounce, etc.) of hydrogen thus removed being replaced by 35.4 parts of chlorine or by 80 parts of bromine: these numbers are therefore the equivalent weights of chlorine and bromine.

When one element A unites with each of a number of others, B, C, D, etc., in one proportion only, the quantities of these latter which combine with or saturate a given quantity of A are clearly proportional to their equivalent weights. Thus 35.4 parts of chlorine are known to unite with 1 part of hydrogen, 23 of sodium, 39.1 of potassium, 32.5 of zinc: consequently the numbers 23, 39.1, and 32.5 are the equivalent weights of sodium, potassium, and zinc referred to hydrogen as unity. In this manner, the equivalent weights of elements may be determined without recourse to direct substitution, which is not always practicable.

The left-hand column of the following table contains a list of those metallic or basylous elements which unite, in one proportion only, with the four non-metallic or chlorous elements in the right-hand column, the numbers opposite to each element showing the proportions in which the combination takes place (e. g., 12 magnesium with 35.4 chlorine, 39.1 potassium with 80 bromine, etc.), or, in other words, the equivalent weights.

	Equiv.			Equiv.
Hydrogen . .	1	Fluorine . . .	19
Beryllium . .	4.7	Chlorine . . .	35.4
Aluminium . .	9.1	Bromine . . .	80
Lithium . . .	7	Iodine . . .	126.5
Magnesium . .	12		
Calcium . . .	20		
Sodium . . .	23		
Zinc . . .	32.5		
Indium . . .	37.7		
Potassium . .	39.1		
Strontium . .	43.6		
Barium . . .	68.4		
Rubidium . .	85.2		
Cæsium . . .	133		

These numbers, as will be explained farther on, are also the relative quantities of the several elements which would be separated from their compounds by an electric current of given strength: thus, if the same current were passed through solutions of sodium bromide, potassium iodide, and zinc chloride, the quantities of sodium, potassium, zinc, bromine, iodine, and chlorine simultaneously separated would be to one another in the proportion of the numbers in the table.

In most cases, however, combination between two elements takes place in more than one proportion, and in such cases the notion of equivalent value becomes less definite; in fact, such elements may be said to have as many equivalent weights as there are ways in which they can combine with others. Thus, tin forms two series of compounds, the stannous compounds, in which 59 parts of the metal unite with 35.4 of chlorine, 80 of bromine, 126.5 of iodine, etc., and the stannic compounds, in which half that quantity of tin, viz., 29.5 parts, discharges the same function; tin has therefore two equivalents, viz., 59 in the stannous and 29.5 in the stannic compounds. In like manner, the equivalent of iron is 28 in the ferrous and $18\frac{2}{3}$ in the ferric compounds.[*]

Atomic Weights.—Let us now compare the hydrogen compounds of chlorine, oxygen, nitrogen, and carbon with regard to the manner in which the hydrogen contained in them may be replaced by other elements. Compare first hydrochloric acid and water. When hydrochloric acid is acted upon by sodium, *the whole* of the hydrogen is expelled, and the chlorine enters into combination with an equivalent quantity of the metal; thus 36.4 parts hydrochloric acid (= 1 part hydrogen + 35.4 chlorine) and 23 sodium yield 1 part of free hydrogen and 23 + 35.4 (= 58.4) sodium chloride; there is no such thing as the expulsion of part of the hydrogen, or the formation of a compound containing both hydrogen and metal in combination with the chlorine.

With water, however, the case is different. When sodium is thrown upon water, 18 parts of that compound (= 2 hydrogen + 16 oxygen) are decomposed, in such a manner that half of the hydrogen is expelled by an equivalent quantity of sodium, 23, and sodium hydroxide is formed, containing—

---

[*] In such cases it is sometimes supposed that the two classes of compounds contain different metallic radicles, combined with the same quantity of a non-metallic or chlorous element, the stannous compounds, for example, being supposed to contain a radicle called *stannosum* (eq. 59), and the stannic compounds another radicle called *stannicum* (eq. 29.5). This, however, is a mere mode of expression, since, to take the two chlorides of tin for example, these two compounds might just as well be supposed to contain different chlorous radicles combined with the same quantity of tin.

Sodium.   Hydrogen.   Oxygen.

$$23 \quad + \quad 1 \quad + \quad 16$$

This compound remains in the solid state when the liquid is evaporated to dryness; and if it be further heated in a tube with sodium, the remaining half of the hydrogen is driven off, and anhydrous sodium oxide remains, composed of 46 parts sodium + 16 oxygen.

Water differs, therefore, from hydrochloric acid in this respect, that its hydrogen may be replaced by sodium in two equal portions, yielding successively a hydroxide and an anhydrous oxide, the relations of which to the original compound may be thus represented:—

Water.		Sodium hydroxide.		Sodium oxide.	
Hydrogen.	Ox.	Hyd. Sod.	Ox.	Sodium.	Ox.
$(1 + 1)$ +	16	$(1 + 23)$ +	16	$(23 + 23)$ +	16

Regarding these results in connection with the atomic hypothesis of the constitution of bodies, we may suppose: (1.) That each molecule of hydrochloric acid is composed of one atom of hydrogen and one atom of chlorine, and that when this compound is acted upon by sodium, each molecule is decomposed, its hydrogen-atom being driven out and replaced by an atom of sodium: thus—

$$\begin{array}{cccc} (H & + & Cl) \text{ and Na produce } (Na & + & Cl) \text{ and H} \\ 1 & & 35.4 \quad\quad 23 & & 23 \quad\quad 35.4 \quad\quad 1 \end{array}$$

The weights of the three atoms concerned in this reaction are to one another in the same proportion as the equivalent weights, or, taking the hydrogen as the unit in each case, we may say that the atomic weights of sodium and chlorine are identical with their equivalent weights.

(2.) Each molecule of water must be supposed to contain two atoms of hydrogen: for if it contained only one atom, then since the first action of the sodium is to expel only half the hydrogen, it would follow that each atom of hydrogen would be split into two, and that each molecule of sodium hydroxide would contain only half an atom of hydrogen; this, however, is at variance with the fundamental notion of atoms, namely, that they are indivisible. These two atoms of hydrogen are combined with a quantity of oxygen weighing 16, which is therefore the smallest quantity of oxygen capable of entering into the reaction under consideration:

$$\begin{array}{ccccccccc} H & H & O & + & Na & = & H & Na & O & + & H \\ 1 & 1 & 16 & & 23 & & 1 & 23 & 16 & & 1 \end{array}$$

and we shall hereafter find that the same is true with regard to all other well-defined reactions in which oxygen takes part. Hence this quantity of oxygen, 16 parts by weight (hydrogen being the unit), is regarded as the weight of the atom of oxygen.

This atomic weight of oxygen is not equal to the equivalent weight, as in the case of chlorine, but *twice as great*, 8 parts of oxygen being the quantity which is capable of replacing 1 part of hydrogen in combination, and may in many cases be directly substituted for it, as when alcohol, a compound of 12 parts carbon, 3 hydrogen, and 8 oxygen, is oxidized to acetic acid containing 12 carbon, 2 hydrogen, and 16 oxygen.

Let us now consider the hydrogen-compound of nitrogen, that is to say, *ammonia*. This is composed of 1 part of hydrogen united with $4\frac{2}{3}$ or $\frac{14}{3}$ of nitrogen, or 3 parts hydrogen with 14 parts nitrogen. Now in this compound the hydrogen is replaceable by thirds. When potassium is heated in ammonia gas, a compound called potassamine is formed, in which one-third of the hydrogen is replaced by potassium. Another compound, called tripotassamine, is also known, consisting of ammonia in which the whole of the hydrogen is replaced by an equivalent quantity of potassium.

There is also a large class of compounds derived from ammonia in like manner by the replacement of $\frac{1}{3}$, $\frac{2}{3}$, or the whole of the hydrogen by equivalent quantities of certain groups of elements called *compound radicles* (see page 254). Hence, by reasoning similar to that which was above applied to water, it is inferred that the molecule of ammonia contains 3 atoms of hydrogen, and that the atomic weight of nitrogen is 14. Moreover, in certain organic compounds this quantity of nitrogen may be substituted for 3 parts of hydrogen, or $\frac{1}{3}$ nitrogen for 1 hydrogen; consequently the atomic weight of nitrogen is *three* times its equivalent weight.

Next take the case of *marsh gas* or *methane*, a compound of 1 part hydrogen with 3 parts carbon, or 4 of hydrogen with 12 of carbon. When this gas is mixed with chlorine, and exposed to diffuse daylight, a new compound is formed, in which one-fourth of the hydrogen belonging to the marsh gas is replaced by an equivalent quantity of chlorine; and if the chlorine is in excess, and the mixture exposed to sunshine, three other compounds are formed, in which one-half, three-fourths, and all the hydrogen are thus replaced. The results may be thus expressed:—

Methane.

Carbon.	Hydrogen.			
12	+ 1	+ 1	+ 1	+ 1

Chloromethane.

Carbon.	Hydrogen.			Chlorine.
12	+ 1	+ 1	+ 1	+ 35.4

Dichloromethane.

Carbon.	Hydrogen.		Chlorine.	
12	+ 1	+ 1	+ 35.4	+ 35.4

Trichloromethane or Chloroform.

Carb.	Hyd.	Chlorine.		
12	+ 1	+ 35.4	+ 35.4	+ 35.4

Tetrachloromethane.

Carbon.	Chlorine.			
12	+ 35.4	+ 35.4	+ 35.4	+ 35.4

Hence, by reasoning similar to the above, it is inferred that the molecule of methane contains 4 atoms of hydrogen, and that the atomic weight of carbon is 12. Moreover, as this quantity of carbon can unite with 4×35.4 parts, or 4 atoms of chlorine, it follows that the atomic weight of carbon is equal to four times its equivalent weight.

We have thus shown in three cases how the atomic weight of an element may be determined by the proportion in which equivalent substitution takes place in its compounds with hydrogen. Sulphur, selenium, and tellurium form hydrogen-compounds exactly analogous in this respect to water, the hydrogen being replaceable by halves: their atomic weights are therefore double of their equivalent weights. Silicon forms with chlorine a compound containing 7 parts silicon with 35.4 parts chlorine; and in this one-fourth of the chlorine is replaceable by hydrogen or by bromine: hence the atomic weight of silicon is, like that of carbon, equal to four times the equivalent weight, its numerical value being 28. There are also some elements in which the atomic weight is equal to five times, others in which it is equal to six times, and others in which it is perhaps equal to seven times or eight times, the equivalent weight: higher ratios have not been observed.

It must not be supposed that the atomic weights of elementary bodies are always actually determined in the manner above described. There are several other methods of determining their numerical values, as will be presently explained, and the values obtained by different methods do not always exhibit exact agreement; but the atomic weights of all the more important elements may be regarded as definitely fixed within very small numerical errors. The equivalent value of an element, or the ratio of the equivalent to the atomic weight, is also subject to some variation, as will be presently explained, according to the view which may be taken of the constitution of particular compounds.

The values of the atomic weights on which chemists are now, for the most part, agreed, are given in the table on page 26.

Physical and Chemical Relations of Atomic Weights.

The atomic weights of the elementary bodies exhibit some remarkable relations to their physical properties, and to the proportions in which they unite by volume.

1. **To the Specific Heats of the Elementary Bodies.**—Dulong and Petit, in the course of their investigations on specific heat, observed that if the specific heats of bodies be computed upon equal weights, numbers are obtained all different, and exhibiting no simple relations amongst themselves; but if, instead of equal weights, quantities be taken in the proportion of the atomic weights, the resulting specific heats come out very nearly equal, at least in the case of solid and liquid elements, showing that some very intimate connection must exist between the relation of bodies to heat, and their chemical nature.

In the first table on p. 242 the solid and liquid elementary bodies are arranged in the order of their specific heats, as determined by Regnault, beginning with those whose specific heat is the greatest; and this order, it will be observed, is the inverse of that of the atomic weights in the third column.

A comparison of the numbers in the fourth column of this table shows that, for a considerable number of elementary bodies in the solid state, the specific heats are very nearly proportional to the atomic weights, so that the products of the specific heats of the elements into their atomic weights give nearly a constant quantity, the mean value being 6.4. This quantity may be taken to represent the *atomic heat* of the several elements in the solid state, or the quantity of heat which must be imparted to or removed from atomic proportions of the several elements, in order to produce equal variations of temperature.

11

Specific Heats of Elementary Bodies.

Elements.	Specific heat (that of water = 1).	Atomic weights.	Product of sp. heat × at. weight.
Lithium	0.9408	7	6.59
Sodium	0.2934	23	6.75
Aluminium	0.2143	27.3	5.85
Phosphorus { liquid	0.2120	} 31 {	6.57
{ solid	0.1887		5.85
Sulphur	0.2026	32	6.48
Potassium	0.1660	39	6.48
Iron	0.1138	55.9	6.36
Nickel	0.1080	58.6	6.33
Cobalt	0.1070	58.6	6.27
Copper	0.0952	63.0	6.00
Zinc	0.0956	64.9	6.20
Arsenic	0.0822	74.9	6.16
Selenium	0.0762	79	6.02
Bromine (solid)	0.0843	79.75	6.72
Palladium	0.0593	106.2	6.30
Silver	0.0570	107.7	6.13
Cadmium	0.0567	111.6	6.33
Indium	0.0570	113.4	6.46
Tin	0.0548	117.8	6.46
Antimony	0.0523	122	6.38
Iodine	0.0541	126.5	6.84
Tellurium	0.0475	128	6.08
Gold	0.0324	196.2	6.36
Platinum	0.0324 .	196.7	6.37
Mercury (solid)	0.0319	199.8	6.37
Thallium	0.0335	203.6	6.82
Lead	0.0314	206.4	6.50
Bismuth	0.0308	210	6.48

Carbon, boron, and silicon were formerly regarded as exceptions to this law, their atomic heats, calculated from the specific heats determined at ordinary temperatures, being considerably below the mean value of those of the other elements, as shown by the following table:—

Elements.	Specific heat.	Atomic weights.	Product of sp. heat × at. weight.
Boron	0.2500	11	2.75
Carbon { wood charcoal	0.2415	} 12 {	2.90
{ graphite	0.2008		2.41
{ diamond	0.1469		1.76
Silicon { crystallized	0.1774	} 28 {	4.97
{ fused	0.1750		4.70

F. Weber has, however, lately shown * that the specific heats of these

* Ann. Chim. Phys. [5], viii. 132.

three bodies increase rapidly at higher temperatures, and that at particular temperatures (about 600° C. for carbon) they become constant, giving for the atomic heats a mean value of about 6, which is nearly the same as that of other elements of small atomic weight, like aluminium and phosphorus, thus:—

	Sp. Heat.	At. Weight.	At. Heat.
Silicon	0.203	28	5.7
Carbon	0.467	12	5.6
Boron	0.5	11	5.5

The specific heats and molecular weights of similarly constituted compounds exhibit, for the most part, the same relation as that which is observed between the specific heats and atomic weights of the elements.

2. **To the Crystalline Forms of Compounds.**—It is found that in many cases two or more compounds which, from chemical considerations, are supposed to contain equal numbers of atoms of their respective elements, crystallize in the same or in very similar forms. Such compounds are said to be *isomorphous.** Thus the sulphates constituted like magnesium sulphate, $MgSO_4 + 7H_2O$, are isomorphous with the corresponding selenates, *e. g.*, $MgSeO_4 + 7H_2O$.

Accordingly, these isomorphous relations are often appealed to for the purpose of fixing the constitution of compounds, and thence deducing the atomic weights of their elements, in cases which would otherwise be doubtful. Thus aluminium forms only one oxide, viz., alumina, which is composed of 18.3 parts by weight of aluminium and 16 parts of oxygen. What, then, is the atomic weight of aluminium? The answer to this question will depend upon the constitution assigned to alumina, whether it is a monoxide, sesquioxide, dioxide, etc. Thus:—

			O.		Al.
Monoxide	AlO	=	16	+	18.3
Sesquioxide	Al_2O_3	=	48	+	$\begin{cases} 27.4 \\ 27.4 \end{cases}$
Dioxide	AlO_2	=	32	+	36.6
Trioxide	AlO_3	=	48	+	54.8

The numbers in the last column of this table are the weights which must be assigned to the atom of aluminium, according to the several modes of constitution indicated in the first column; but there is nothing in the constitution of the oxide itself that can enable us to decide between them. Now, iron forms two oxides, in which the quantities of oxygen united with the same quantity of iron are to one another as $1 : 1\frac{1}{2}$, or as $2 : 3$. These are therefore regarded as monoxide, FeO, and sesquioxide, Fe_2O_3, and this last oxide is known to be isomorphous with alumina. Consequently alumina is also regarded as a sesquioxide, Al_2O_3, and the atomic weight of aluminium is inferred to be 27.4.

3. **To the Volume-Relations of Elements and Compounds.**— The atomic weights of those elements which are known to exist in the state of gas or vapor are, with one or two exceptions, proportional to their specific gravities in the same state. Taking the specific gravity of hydrogen as unity, those of the following gases and vapors are expressed by numbers identical with their atomic weights:—

*Ἴσος, equal; μορφή, form.

Hydrogen	. .	1	Oxygen	. . .	16
Chlorine .	. .	35.4	Sulphur	. . .	32 .
Bromine .	. .	80	Selenium	. . .	79
Iodine	. .	127	Tellurium .	. .	128

The exceptions to this rule are exhibited by *phosphorus* and *arsenic*, whose vapor-densities are twice as great as their atomic weights, that of phosphorus being 62, and that of arsenic 150; and by *mercury* and *cadmium*, whose vapor-densities are the halves of their atomic weights, that of mercury being 100, and that of cadmium 56.

From these relations, considered in connection with above explained laws of combination by weight, it follows that the volumes of any two elementary gases which make up a compound molecule are to one another in the same ratio as the numbers of atoms of the same elements which enter into the compound, excepting in the case of phosphorus and arsenic, for which the number of volumes thus determined has to be halved, and of mercury and cadmium, for which it must be doubled: thus—

The molecule	HCl	contains	1 vol.	H	and	1 vol.	Cl			
"	H_2O	"	2	" H	"	1	" O			
"	H_3N	"	3	" H	"	1	" N			
"	H_3P	"$\{$	3 or 6	" H " H	" "	$\frac{1}{2}$ 1	" P " P			
"	Cl_3As	"$\{$	3 or 6	" Cl " Cl	" "	$\frac{1}{2}$ 1	" As " As			
"	Cl_2Hg	"	2	" Cl	"	2	" Hg			

If the smallest volume of a gaseous element that can enter into combination be called the combining volume of that element, the law of combination may be expressed as follows:—*The combining volumes of all elementary gases are equal, excepting those of phosphorus and arsenic, which are only half those of the other elements in the gaseous state, and those of mercury and cadmium, which are double those of the other elements.*

It appears, then, that in all cases the volumes in which gaseous elements combine together may be expressed by very simple numbers. This is the "Law of Volumes," first observed by Humboldt and Gay-Lussac in 1805, with regard to the combination of oxygen and hydrogen, and afterward established in other cases by Gay-Lussac, whose observations, published in his "Theory of Volumes," afforded new and independent evidence of the combination of bodies in definite and multiple proportions, in corroboration of that derived from the previously observed proportions of combination by weight.

Gay-Lussac likewise observed that the product of the union of two gases, when itself a gas, sometimes retains the original volume of its constituents, no contraction or change of volume resulting from the combination, but that when contraction takes place, which is the most common case, the volume of the compound gas always bears a simple ratio to the volumes of its elements; and subsequent observation, extended over a very large number of compounds, organic as well as inorganic, has shown that, with a few exceptions, probably only apparent, *the molecules of compound bodies in the gaseous state occupy twice the volume of an atom of hydrogen gas.* No matter what may be the number of atoms or volumes that enter into the compound, they all become condensed into two volumes; thus—

1 vol. H and 1 vol. Cl form 2 vol. HCl, hydrochloric acid.
1 " N " 1 " O   " 2 " NO, nitrogen dioxide.
2 " H " 1 " O   " 2 " H_2O, water.
3 " H " 1 " N   " 2 " H_3N, ammonia.
2 " H " $\frac{1}{2}$ " P   " 2 " H_3P, hydrogen phosphide.

Similarly in the union of compound gases, *e. g.*,

1 vol. ethyl, C_2H_5, and 1 vol. Cl form 2 vol. C_2H_5Cl, ethyl chloride.
2 " ethyl, C_2H_5, " 1 " O   " 2 " $(C_2H_5)_2O$, ethyl oxide.
2 " ethylene, C_2H_4, " 2 " Cl " 2 " $C_2H_4Cl_2$, ethylene chloride.
2 " ethylene, C_2H_4, " 1 " O   " 2 " C_2H_4O,   ethylene oxide.

It will presently be shown, as at least highly probable, that the molecule of an elementary gas in the free state is made up of two atoms, HH for example. The law just enunciated may therefore be generalized as follows. *The molecules of all gases, simple or compound, occupy equal volumes ; or, equal volumes of all gases contain equal numbers of molecules.*

This is called the "Law of Avogadro," having been first enunciated (in 1811) by an Italian physicist of that name. It is quite in accordance with the observed fact that all perfect gases, simple and compound, are equally affected by equal variations of pressure and temperature; and indeed it may be shown, by mathematical reasoning, to follow as a necessary consequence from the physical constitution of gases as explained in connection with the dynamical theory of heat (pp. 83-88) ; but the demonstration is not of a nature adapted for an elementary book. The law may, however, be considered as completely established by the relations between the combining proportions of the elements by weight and by volume as already explained, and it is now regarded as affording the surest method of fixing the molecular constitution of all compounds that can be obtained in the gaseous state, and the atomic weights of the elements contained in them.

Suppose, for example, it were required to determine the atomic weight of tin. This metal forms a volatile chloride (stannic chloride), in which 29.45 parts of tin are combined with 35.4 of chlorine. 29.45 is therefore the equivalent of tin in this compound. Now, the vapor-density of this chloride referred to hydrogen as unity, is 129.7 : consequently the weight of two volumes of its vapor is 259.4, and this contains 4×35.4, or 141.6 parts of chlorine and 117.8 parts of tin; and as this appears to be the chloride containing the largest proportion of chlorine or the smallest proportion of tin, it is regarded as a compound of 4 atoms of chlorine and 1 atom of tin, and the atomic weight of tin is thus found to be 117.8.

When an element does not form any volatile compounds whose vapor-densities can be exactly ascertained, its atomic weight may be determined by its specific heat, according to the law of Dulong and Petit (p. 241), or by its isomorphous relations with other elements, as already explained in the case of aluminium. To give an instance of the determination of the atomic weight of an element according to its specific heat, we may take the case of indium. This metal forms a chloride containing 35.4 parts of chlorine and 37.8 parts of indium, which latter number is therefore the equivalent weight of the metal. Now when indium was first discovered (in 1863), this chloride was, for reasons which need not here be specified, regarded as a dichloride, $InCl_2$, and consequently the atomic weight of indium was supposed to be 75.6. Subsequently, however, Bunsen showed that the specific heat of indium, referred to the unit of weight, is 0.057, and this number multiplied by 75.6 gives for the atomic heat the number 4.5, which does not agree with the law of Dulong and Petit; but if the chloride be regarded as a trichloride

21 *

MCl_3, making the atomic weight of the metal equal to three times its equivalent weight, or 113.4, the atomic heat becomes 6.15, which agrees very nearly with the general law. This number, 113.4, is therefore now adopted as the atomic weight of indium.

Specific or Atomic Volume.

These terms denote the quotient obtained by dividing the molecular weight of a body by its specific gravity. Now from the law of condensation in the combination of gases above detailed (p. 245) it follows that *the specific gravity of any compound gas or vapor referred to hydrogen as unity is equal to half its atomic or molecular weight;* hence also, *the specific volumes of compound gases or vapors referred to that of hydrogen as unity are, with a few exceptions, equal to* 2. It will presently be shown that the same law applies to the specific volumes of the elementary gases themselves.

Some compounds, however, exhibit a departure from this rule, their observed specific gravities being equal to only one-fourth their molecular weights, or their molecules occupying four times the volume of an atom of hydrogen. Such is the case with sal-ammoniac, NH_4Cl, phosphorus pentachloride, PCl_5, sulphuric acid, H_2SO_4, ammonium hydrosulphide, $(NH_4)SH$, and a few others. This anomaly is probably due, in some cases at least, to a decomposition or "dissociation" of the compound at the high temperature to which it is subjected for the determination of its vapor-density; NH_4Cl, for example, splitting up into NH_3 and HCl, each of which occupies two volumes, and the whole therefore four volumes; and in like manner H_2SO_4 may be supposed to separate into H_2O and SO_3; PCl_5 into PCl_3 and Cl_2; $(NH_4)SH$ into NH_3 and H_2S, etc.

On the other hand, some substances, both simple and compound, exhibit, at temperatures not far above their boiling points, vapor-densities considerably greater than they should have according to the general law, whereas when raised to higher temperatures they exhibit normal vapor-densities. Thus sulphur, which boils at 440° C. (826° F.), exhibits at 1000° C. (1832° F.), like elementary gases in general, a vapor-density equal to its atomic weight, viz., 32; but at 500° C. (932° F.) its vapor-density is nearly three times as great. Again, acetic acid, $C_2H_4O_2$, whose molecular weight is $24 + 4 + 16 = 60$, has, at temperatures considerably above its boiling point, a vapor-density nearly equal to 30; but at 125° C. (257° F.) (8 degrees above its boiling point), its vapor-density is rather more than 45, or $1\frac{1}{2}$ times as great. This anomalous increase of vapor-density appears to take place when the substance approaches its liquefying point, at which also it exhibits irregularities in its rate of expansion and contraction by variations of pressure and temperature—at which, in short, it begins to behave itself like a liquid; but at higher temperatures it exhibits the physical characters of a perfect gas, and then also its specific gravity becomes normal.

Specific Volumes of Liquids and Solids.—The following table exhibits the specific volumes of those solid and liquid elements whose specific gravities have been determined with sufficient accuracy. The elements are arranged in the order of their specific volumes, beginning with the smallest :—

Specific Volumes of Solid and Liquid Elements.

	Atomic weight.	Specific gravity.	Specific volume.		Atomic weight.	Specific gravity.	Specific volume.
Carbon, as *dia-*				Mercury, *liq.*	199.8	14.8	13.5
mond . . .	12	3.52	3.4	Sulphur, *tri-*			
Beryllium . .	9.0	2.1	4.4	metric . .	32	2.07	15.2
Carbon, as *gra-*				Indium . .	113.4	7.4	15.3
phite . . .	12	2.3	5.2	Phosphorus,			
Nickel . . .	58.6	8.6	6.8	red . . .	31	1.94	15.8
Manganese .	55	8.03	6.85	Sulphur, *mon-*			
Cobalt . . .	58.6	8.50	7.0	oclinic . .	32	1.98	16.2
Iron	56	7.8	7.2	Tin	117.8	7.3	16
Copper . . .	63.4	8.95	7.2	Selenium,			
Chromium .	52.4	7.01	7.4	granular .	79.0	4.80	16.4
Iridium . . .	196.7	21.8	9.1	Phosphorus,			
Platinum . .	196.7	21.5	9.2	yellow . .	31	1.84	16.8
Zinc	64.9	7.1	9.1	Antimony .	122	6.7	18.2
Palladium .	106.2	11.8	9.2	Lead	206.4	11.33	18
Rhodium . .	104.1	11.0	9.4	Selenium,			
Silver . . .	107.7	10.5	10.2	amorphous	79.4	4.28	18.1
Gold	196.2	19.34	10.1	Tellurium .	128	6.2	20.6
Aluminium .	27.3	2.67	10.26	Bismuth . .	210	9.8	21.2
Molybdenum	95.6	8.6	11.1	Sodium . .	23	0.97	23.7
Silicon, *graphi-*				Calcium . .	40	1.58	25
toïdal . . .	28	2.5	11,2	Iodine . . .	126.5	4.95	25.5
Lithium . .	7	0.59	11.9	Bromine, *liq.*	79.75	3.19	25.0
Cadmium . .	111.6	8.7	12.9	Chlorine, *liq-*			
Uranium . .	240	18.4	13.2	uid . . .	35.4	1.33	26.6
Arsenic . .	74.9	5.63	13.3	Strontium .	87.2	2.54	34.4
Magnesium .	23.9	1.74	13.8	Potassium .	39	0.86	45.6

The numbers in the third column of this table do not exhibit the simplicity of relation which exists between the specific volumes of gaseous bodies. There are, indeed, several causes which interfere with the existence, or at least with the observation, of such simple relations between the specific volumes of solid and liquid elements. In the first place, the densities of three of them, viz., mercury, bromine, and chlorine, are such as belong to them in the liquid state, whereas the densities assigned to all the others have been determined in the solid state. In solids, moreover, the density is greatly affected by the state of aggregation, whether crystalline or amorphons, and in dimorphous bodies each form has a density peculiar to itself. Further, as solids and liquids are variously affected by heat, each having a peculiar rate of expansion, and that rate being different at different temperatures, it is not to be expected that their specific volumes should exhibit simple relations, unless they are compared at temperatures at which they are similarly affected by heat. Even gases are found to exhibit abnormal specific volumes if compared at temperatures too near the points at which they pass into the liquid state. In liquids, the simplest relations of specific volume are found at those temperatures for which the tensions of the vapors are equal (Kopp); and in solids the melting points are most probably the comparable temperatures. Now, the specific gravities of most of the solid elements in the preceding table have been determined at mean temperatures, as at 15.5° C. (59.9° F.), which, in the case of potassium, sodium, phosphorus, and a few others, do not differ greatly from the melting points, but in other cases, as with gold, platinum, iron, etc., are removed from the melting points by very long intervals. In spite, however, of these causes of divergence, the specific volumes of certain analogous elements are very nearly equal to each other: viz., those of selenium and sulphur; of chromium, iron, cobalt, copper, manganese, and nickel; of molybdenum and tungsten; of iridium, platinum, palladium, and rhodium; and of gold and silver.

Specific Volumes of Solid and Liquid Compounds.—The most general relation that has been observed between the specific volumes of solid compounds is that *isomorphous compounds have equal specific volumes;* in other words, that their densities are proportional to their molecular weights; such is the case, for example, with the native carbonates of strontium (strontianite) and of lead (cerusite):

Formula.	Molecular weight.	Specific gravity.	Specific volume.
$SrCO_3$	147.2	3.60	40.4
$PbCO_3$	266.4	6.47	41.2

If the crystalline forms are only approximately similar, the specific volumes also are only approximately equal, the difference being less as the angles of the two crystalline forms are more nearly equal and their axes more nearly in the same ratio. In dimorphous compounds each modification has a density, and therefore a specific volume, peculiar to itself.

The hydrated sulphates of magnesium, zinc, nickel, cobalt, and iron, which have the general formula, $M''SO_4 + 7H_2O$ (M'' denoting a bivalent metal: see next page), and crystallize in similar forms, have specific volumes very nearly equal to 146; the double sulphates isomorphous with potassio-cupric sulphate, $K_2Cu(SO_4)_2 + 6H_2O$, have specific volumes ranging between 198 and 216; and the alums, *e. g.*, $KAl(SO_4)_2 + 12H_2O$, have specific volumes ranging between 276 and 281.

The specific volumes of liquid compounds have been studied chiefly with relation to organic compounds. The most general relation observed is that: *Differences of specific volume are in numerous instances proportional to the differences between the corresponding chemical formulæ.* Thus liquids whose formulæ differ by nCH_2 differ in specific volume by n times 22; for example, methyl formate, $CH_3.CHO_2$, and ethyl butyrate, $C_2H_5.C_4H_7O_2$, which differ by $4CH_2$, have specific volumes differing by nearly 4×22.

Atomicity, Quantivalence.

We have seen that the atomic weight of an element is in some cases equal to its equivalent weight, in others, twice, three times, four times, etc., as great as the equivalent weight; in other words, an atom of certain elements can replace or be substituted for only one atom of hydrogen, whereas the atoms of other elements can replace 1, 2, 3, 4, etc. atoms of hydrogen. Thus, when sodium dissolves in hydrochloric acid, each atom of sodium replaces 1 atom of hydrogen; but when zinc dissolves in the same acid, each atom of zinc takes the place of 2 atoms of hydrogen: thus

$$Na + HCl = NaCl + H$$
$$Zn + 2HCl = ZnCl_2 + H_2$$

Here it is seen that an atom of zinc is equal in combining, or saturating power to 2 atoms of hydrogen. In like manner, antimony and bismuth form trichlorides, $AlCl_3$ and $BiCl_3$, in which the atom of the metal performs the same chemical function as 3 atoms of hydrogen; that is to say, it saturates 3 atoms of chlorine; so also tin in the tetrachloride is equivalent to 4H, and phosphorus in the pentachloride, PCl_5, to 5H.

This difference of equivalent, combining or saturating power, is called quantivalence or atomicity, and is sometimes denoted by placing dashes or Roman numerals to the right of the symbol of an element, and at the top, as O'', B''', C^{iv}, etc.; and the several elements are designated as—

Univalent elements, or	Monads,	as	H
Bivalent "	Dyads,	"	O''
Trivalent "	Triads,	"	B'''
Quadrivalent "	Tetrads,	"	C^{iv}
Quinquivalent "	Pentads,	"	P^v
Sexvalent "	Hexads,	"	W^{vi}.

Elements of even equivalency, viz., the dyads, tetrads, and hexads, are also included under the general term a r t i a d s,* and those of uneven equivalency, viz., the monads, triads, and pentads, are designated generally as perissads.†

Another method of indicating the equivalent values of the elementary atoms, and the manner in which they are satisfied by combination, is to arrange the symbols in diagrams in which each element is connected with others by a number of lines or connecting bonds corresponding with its degree of equivalence; a monad being connected with other elements by only one such bond, a triad by three, a hexad by six, etc., as in the following examples:—

Water, H_2O . . . . . $\quad$ H — O — H

Carbon dioxide, CO_2 . . . $\quad$ O = C = O

Ammonium chloride, NH_4Cl . .

$$\begin{array}{c} \text{Cl} \\ \text{H} \mid \text{H} \\ >\text{N}< \\ \text{H} \quad \text{H} \end{array}$$

Sulphuric oxide, SO_3 . . .

$$\begin{array}{c} \text{O} \\ \| \\ \text{S} = \text{O} \\ \| \\ \text{O} \end{array}$$

Sulphuric acid, H_2SO_4 . . .

$$\begin{array}{c} \text{O} \\ \| \\ \text{H} - \text{O} - \text{S} - \text{O} - \text{H} \\ \| \\ \text{O} \end{array}$$

Nitric acid, HNO_3 . . . .

$$\begin{array}{c} \text{O} \\ \| \\ \text{N} - \text{O} - \text{H} \\ \| \\ \text{O} \end{array}$$

Zinc nitrate, ZnN_2O_6 . . .

$$\begin{array}{c} \text{O} \qquad\qquad \text{O} \\ \| \qquad\qquad \| \\ \text{N} - \text{O} - \text{Zn} - \text{O} - \text{N} \\ \| \qquad\qquad \| \\ \text{O} \qquad\qquad \text{O} \end{array}$$

* Ἄρτιος, even. $\qquad$ † Περισσός, uneven.

11 *

In most cases, however, these formulæ may be abridged by the use of dots instead of dashes, thus

$$O.H$$

H . O . H   O : C : O   O : N : O
Water.   Carbon dioxide.   Nitric acid.

It must be distinctly understood that these formulæ—which are called graphic, structural, or constitutional formulæ—are not intended to represent the actual arrangement of the atoms in a compound; indeed, even if we had a distinct notion of the manner in which the atoms of any compound are arranged, it could not be adequately represented on a plane surface. The lines connecting the different atoms indicate nothing more than the number of units of equivalency belonging to the several atoms, and the manner in which they are disposed of by combination with those of other atoms. Thus the formula for nitric acid indicates that two of the three constituent oxygen-atoms are combined with the nitrogen alone, and are consequently attached to that element by both their units of equivalency, whereas the third oxygen atom is combined both with nitrogen and with hydrogen.

By inspection of the preceding diagrams it will be observed that every atom of a compound has each of its units of equivalency satisfied by combination with a unit belonging to some other atom. Such, indeed, is the case in every saturated or normal compound. Accordingly, it is found that in all such compounds the sum of the perissad elements is always an even number. Thus a compound may contain two, four, six, etc. monad atoms, as ClH, OH_2, CH_4, C_2H_6, C_3H_8, SiH_3Cl; or one triad atom and three monads, as BCl_3; or one pentad and five monads, as NH_4Cl; but never an uneven number of perissad atoms. This is the "law of even numbers," announced some years ago by Gerhardt and Laurent as a result of observation. It was long received with doubt, but has now been confirmed by the analysis of so many well-defined compounds that a departure from it is looked upon as a sure indication of incorrect analysis.

For a similar reason, the atoms of elementary bodies rarely exist in the free state, but when separated from any compound, tend to combine with other atoms, either of the same or of some other element. Perissad elements, like hydrogen, chlorine, nitrogen, etc., separate from their compounds in pairs; their molecule contains two atoms, e. g., H — H. Artiad elements may unite in groups of two, three, or more; thus the molecule of oxygen, in its ordinary state, probably contains two atoms, that of ozone, three atoms; thus—

Oxygen . . . . $O = O$

Ozone . . . . $O—O$

$$\underset{O}{\bigvee}$$

The tendency of elementary atoms to separate in groups is shown in various ways. Thus when copper hydride, Cu_2H_2 (to be hereafter described), is decomposed by hydrochloric acid, a quantity of hydrogen is given off equal to twice that which is contained in the hydride itself; thus—

$$Cu_2H_2 + 2HCl = Cu_2Cl_2 + 2HH.$$

This action is precisely analogous to that of hydrochloric acid on cuprous oxide:

$$Cu_2O + 2HCl = Cu_2Cl_2 + H_2O.$$

In the latter case the hydrogen separated from the hydrochloric acid unites with oxygen, in the former with hydrogen. Again, when solutions of sulphurous acid and sulphydric acid are mixed, the whole of the sulphur is precipitated:

$$H_2SO_2 + 2H_2S = 3H_2O + SS_2,$$

the action being similar to that of sulphurous acid on selenhydric acid:

$$H_2SO_3 + 2H_2Se = 3H_2O + SSe_2.$$

In the one case, a sulphide of selenium is precipitated; in the other, a sulphide of sulphur. The precipitation of iodine, which takes place on mixing hydriodic acid with iodic acid, affords a similar instance of the combination of homogeneous atoms:

$$5HI + HIO_3 = 3H_2O + 3II.$$

Another striking illustration of this mode of action is afforded by the reduction of certain metallic oxides by hydrogen dioxide. When silver oxide is thrown into this liquid, water is formed, the silver is reduced to the metallic state, and a quantity of oxygen is evolved equal to twice that which is contained in the silver oxide:

$$Ag_2O + H_2O_2 = H_2O + Ag_2 + OO.$$

Further, elementary bodies frequently act upon others as if their atoms were associated in binary groups. Thus chlorine acting upon potassium oxide forms two compounds, the chloride and hypochlorite of potassium (p. 167).

$$ClCl + KKO = KCl + KClO.$$

Again, in the action of chlorine upon many organic compounds, one atom of chlorine removes one atom of hydrogen as hydrochloric acid, while another atom of chlorine takes the place of the hydrogen thus removed. For example, in the formation of chloracetic acid by the action of chlorine on acetic acid:

$$C_2H_4O_2 + ClCl = HCl + C_2H_3ClO_2.$$

Similarly, when metallic sulphides oxidize in the air, both the metal and the sulphur combine with oxygen; and sulphur acting upon potash forms both a sulphide and a thiosulphate. In all these cases the atoms of the elementary bodies act in pairs.

On the supposition that the molecules of elementary bodies in the gaseous state are made up of two atoms, the specific volumes of these gases will come under the same law as that which applies to compounds (p. 246); and it may then be stated generally, that, with the few exceptions already noticed, *the specific gravities of all bodies, simple and compound, in the gaseous state, are equal to half their molecular weights;* or the *specific volumes* (the quotients of the molecular weights by the specific gravities) *are equal to 2.*

There are, however, two elements, namely, phosphorus and arsenic, which at all temperatures hitherto attained exhibit a vapor-density twice as great as that which they should have according to the general law, that of phosphorus being always 61.9, and that of arsenic 149.8. This has been explained by supposing that the molecule of each of these two elements in the free state contains four atoms instead of two, as is the case with most elementary bodies; thus the molecule of phosphorus is supposed to be represented by the formula,

Variation of Equivalency—Multivalent elements often exhibit varying degrees of equivalency. Thus carbon, which is quadrivalent in marsh gas, CH_4, and in carbon dioxide, CO_2, is only bivalent in carbon monoxide, CO; nitrogen, which is quinquivalent in sal-ammoniac, NH_4Cl, and the other ammonium salts, and in nitrogen pentoxide, N_2O_5, is trivalent in ammonia, NH_3, and in nitrogen trioxide, N_2O_3, and univalent in nitrogen monoxide, N_2O; sulphur, also, which is sexvalent in sulphur trioxide, SO_3, is quadrivalent in sulphur dioxide, SO_2, and bivalent in hydrogen sulphide, H_2S, and in many metallic sulphides. In these cases, and in others of varying equivalency, the variation mostly takes place by two units of equivalency. It is not very easy to account for these variations; but it is observed in all cases that the compounds in which the equivalency of a polygenic element is most completely satisfied are more stable than the others, and that the latter tend to pass into the former by taking up the required number of univalent or bivalent atoms; thus, carbon monoxide, CO, easily takes up another atom of oxygen to form the dioxide, CO_2; nitrogen trioxide, N_2O_3, is readily converted into the pentoxide, N_2O_5; ammonia, NH_3, unites readily with hydrochloric acid to form sal-ammoniac, NH_4Cl, etc. Similar phenomena are exhibited by many organo-metallic bodies, as will be explained farther on.

From this it seems most probable that the true quantivalence of a polygenic element is that which corresponds with the maximum number of monad atoms with which it can combine, but that one or two pairs of its units of equivalency may, under certain circumstances, remain unsaturated. Whether a saturated or an unsaturated compound is formed, will depend on a variety of conditions, often in great measure on the relative quantities of the acting substances. Thus phosphorus, which is a pentad element, forms with chlorine either a trichloride, PCl_3, or a pentachloride, PCl_5, according as the phosphorus or the chlorine is in excess (p. 207).

In compounds containing two or more atoms of the same multivalent element, one or more units of equivalence belonging to each of these atoms may be neutralized by combination with those of another atom of the same kind, so that the element in question will appear to enter into the compound with less than its normal degree of equivalence. Thus in ethane, or dimethyl, C_2H_6, which is a perfectly stable compound, having no tendency to take up an additional number of atoms of hydrogen or any other element, the carbon appears to be trivalent instead of quadrivalent; similarly in propane, C_3H_8, its equivalence appears to be reduced to $\frac{8}{3}$; and in butane, or diethyl, C_4H_{10}, to $\frac{5}{2}$. In all these cases, however, the diminution of equivalent value in the carbon-atoms is only apparent, as may be seen from the following formulæ:

Ethane.   Propane.   Butane.

$$
\begin{array}{ccc}
\text{H} & \text{H} & \text{H} \\
| & | & | \\
\text{H}-\text{C}-\text{H} & \text{H}-\text{C}-\text{H} & \text{H}-\text{C}-\text{H} \\
| & | & | \\
\text{H}-\text{C}-\text{H} & \text{H}-\text{C}-\text{H} & \text{H}-\text{C}-\text{H} \\
| & | & | \\
\text{H} & \text{H}-\text{C}-\text{H} & \text{H}-\text{C}-\text{H} \\
 & | & | \\
 & \text{H} & \text{H}-\text{C}-\text{H} \\
 & & | \\
 & & \text{H}
\end{array}
$$

or, more shortly, omitting the equivalent marks of the monad atoms:

$$H_3C . CH_3 \qquad H_2C . CH_2 . CH_3 \qquad H_3C . CH_2 . CH_2 . CH_3$$
Ethane.   Propane.   Butane.

In each of these compounds, every carbon-atom, except the two outside ones, has two of its units of equivalence satisfied by combination with those of the neighboring carbon-atoms, while each of the two exterior ones has only one unit thus satisfied. Hence, in any similarly constituted compound containing n carbon-atoms, the number of units of equivalence remaining to be satisfied by the hydrogen-atoms is $4n - 2(n-2) - 2 = 2n + 2$. The general formula of this series of hydrocarbons is, therefore, C_nH_{2n+2}, and the equivalent value of the carbon is $\dfrac{2n+2}{n}$.

In other cases, multivalent atoms may be united by two or more of their units of equivalence, so that their combining power may appear to be still further reduced as in the hydrocarbon, C_2H_4, in which the carbon may be apparently bivalent, and in C_2H_2, in which it may appear to be univalent; thus—

$$
\begin{array}{cc}
\text{H}-\text{C}-\text{H} & \text{C}-\text{H} \\
\parallel & ||| \\
\text{H}-\text{C}-\text{H} & \text{C}-\text{H}
\end{array}
$$

Sometimes also the apparent alteration of combining capacity may be due to the fact that the simplest formula by which a compound may be represented is not its true molecular formula. For example, the two chlorides of tin are usually represented by the formulæ $SnCl_2$ and $SnCl_4$. The second of these is undoubtedly correct, that is to say, the molecule of stannic chloride, as shown by its vapor-density, contains 1 atom of tin and 4 atoms of chlorine (p. 245), and consequently tin is a tetrad or quadrivalent element. But the lower chloride, in which the tin appears to be only bivalent, should in all probability be represented by a higher formula; for this compound is much less volatile than stannic chloride, requiring a bright red heat to vaporize it, and at the same time undergoing partial decomposition, whereas stannic chloride volatilizes without decomposition at 120° C. (248° F.). Hence it may be inferred that the molecule of stannous chloride is heavier and more complex than that of stannic chloride, and should therefore be represented by the double formula Sn_2Cl_4 or $Cl_2Sn{=}SnCl_2$, in which the tin still figures as a tetrad, though relatively to the chlorine it is only bivalent.

In most cases, the quantivalence or atomicity of an element is most safely determined by the number of monad atoms with which it can combine. Of

22

dyad atoms, indeed, any element or compound may take up an indefinite
number, without alteration of its quantivalence or combining power: for
each dyad atom, possessing two units of equivalency, neutralizes one unit
in the compound which it enters, and introduces another, leaving, therefore,
the combining power of the compound just what it was before. Thus potassium forms only one chloride, KCl, and is therefore univalent or monadic;
but in addition to the oxide, K_2O, corresponding with this chloride, it likewise forms two others, viz., K_2O_2 and K_2O_4, in the former of which it might
be regarded as dyadic, and in the latter as tetradic; but the manner in which
dyad oxygen enters these compounds is easily seen by inspection of the following diagrams:—

K.O.K   K.O.O.K   K.O.O.O.OK

Monoxide.   Dioxide.   Tetroxide.

It is evident that any number of oxygen-atoms might, in like manner, be
inserted without disturbing the balance of equivalency. If, indeed, we turn
to the sulphides of potassium, in which the sulphur is dyadic, like oxygen,
we find the series, K_2S, K_2S_2, K_2S_3, K_2S_4, K_2S_5, the constitution of which
may be represented in a precisely similar manner. Hence the quantiv-
alence of any element is, for the most part, best determined by the composition of its chlorides, bromides, iodides, or fluorides, rather than by that
of its oxides or sulphides. In some cases, however, as will be seen farther
on, the combinations of an element with oxygen afford the best means of
determining its quantivalence or combining capacity.

Compound Radicles.— Suppose one or more of the component atoms
of a fully saturated molecule to be removed: it is clear that the remaining
atom or group of atoms will no longer be saturated, but will have a combining power corresponding with the number of units of equivalency removed.   Such unsaturated groups are called residues or radicles.
Methane, CH_4, is a fully saturated compound; but if one of its hydrogen-
atoms be removed, the residue CH_3 (called methyl), will be ready to combine with one atom of a univalent element, such as chlorine, bromine, etc.,
forming the compounds CH_3Cl, CH_3Br, etc.; two atoms of it unite in like
manner, with one atom of oxygen, sulphur, and other bivalent elements,
forming the compounds $O''(CH_3)_2$, $S''(CH_3)_2$, etc.; three atoms with nitrogen,
yielding $N'''(CH_3)_3$, etc.
The removal of two hydrogen-atoms from CH_4 leaves the bivalent radicle
CH_2, called methylene, which yields the compounds CH_2Cl_2, CH_2O, CH_2S,
etc. The removal of three hydrogen-atoms from CH_4 leaves the trivalent
radicle CH, which, in combination with three chlorine-atoms, constitutes
chloroform, $CHCl_3$. And, finally, the removal of all four hydrogen-atoms
from CH_4 leaves the quadrivalent radicle carbon, capable of forming the
compounds CCl_4, CS_2, etc.
In like manner, ammonia, NH_3, in which the nitrogen is trivalent, yields,
by removal of one hydrogen-atom, the univalent radicle amidogen, NH_2,
which with one atom of potassium forms potassamine, NH_2K, and when
combined with one atom of the univalent radicle methyl, CH_3, forms methylamine, $NH_2(CH_3)$, etc. The abstraction of two hydrogen-atoms from
the molecule NH_3 leaves the bivalent radicle imidogen, NH, which with
two methyl-atoms forms dimethylamine, $NH(CH_3)_2$, etc.; and the removal
of all three hydrogen-atoms from NH_3 leaves nitrogen itself, which frequently acts as a trivalent element or radicle, forming tripotassamine, NK_3,
trimethylamine, $N(CH_3)_3$, etc.
Finally, the molecule of water, OH_2, by losing an atom of hydrogen, is

converted into the univalent radicle h y d r o x y l, OH, which, in its relations to other bodies, is analogous to chlorine, bromine, and iodine,
and may be substituted in combination for one atom of hydrogen or other
monads.   Thus, water itself may be regarded as H.OH, analogous to hydrochloric acid, HCl; potassium hydroxide as K.OH, analogous to potassium
chloride; barium hydroxide, as Ba(OH)$_2$, analogous to barium chloride,
BaCl$_2$.

In a similar manner, the univalent radicle, p o t a s s o x y l, KO, may be derived from potassium hydroxide; the bivalent radicle, z i n c o x y l, ZnO$_2$, by
abstraction of H$_2$ from zinc hydroxide, Zn''H$_2$O$_2$.   The essential character
of these oxygenated radicles is that each of the oxygen-atoms contained in
them is united to the other atoms by one unit of equivalency only, so that
the radicle has necessarily one or two units unconnected; thus—

$$\text{Hydroxyl} \quad . \quad . \quad . \quad . \quad . \quad \text{H} - \text{O} -$$
$$\text{Potassoxyl} \quad . \quad . \quad . \quad . \quad . \quad \text{K} - \text{O} -$$
$$\text{Zincoxyl} \quad . \quad . \quad . \quad . \quad - \text{O} - \text{Zn} - \text{O} -$$

From the preceding explanations of the mode of derivation of compound
radicles, it is clear that there is no limit to the number of them which may
be supposed to exist; in fact, it is only necessary to suppose a number of
units of equivalency abstracted from any saturated molecule, in order to
obtain a radicle of corresponding combining power or equivalent value.
But unless a radicle can be supposed to enter into a considerable number
of compounds, thus forming them into a group, like the salts of the same
metal, there is nothing gained in point of simplicity or comprehensiveness
by assuming its existence.

It must also be distinctly understood that these compound radicles do not
necessarily exist in the separate state, and that those of uneven equivalency,
like methyl, cannot exist in that state, their molecules, if liberated from combination with others, always doubling themselves, as we have seen to be the
case with most of the elementary bodies.   Thus hydroxyl — O — H is not
known in the free state, the actually existing compound containing the same
proportions of hydrogen and oxygen being O$_2$H$_2$ or H — O — O — H.   In
like manner methyl, CH$_3$, has no separate existence, but dimethyl, C$_2$H$_6$, is
a known compound:—

Methyl.
CH$_3$
|

Dimethyl.
CH$_3$
|
CH$_3$

Relations between Atomic Weight and Quantivalence.—A
very remarkable relation has been shown to exist between the quantivalence
of the elements and the numerical order of their atomic weights.   Arranging
the elements in horizontal lines according to this order, as in the following
table (in which the atomic weights are, for the most part, represented by
their nearest whole numbers), we find that, with the exception of certain
metals belonging to the iron and platinum groups, they all arrange themselves in such a manner that the first vertical column is occupied by the
monad elements, the second by the dyads, the third by the triads, etc., as
indicated by the headings of the columns, in which the symbol R stands
for a metal.   Hydrogen stands alone, there being no known element intermediate between it and the monad metal lithium.   This relation of the elementary bodies, which is called the "periodic law," was first pointed out by

Newlands in 1864, and afterward developed by Odling, Lothar Meyer, and Mendelejeff.

Series.	Group I. R_2O.	II. RO.	III. R_2O_3.	IV. $RH_4 \cdot$ $RO^2 \cdot$	V. $RH_3 \cdot$ $RO_5 \cdot$	VI. $RH_2 \cdot$ $RO_3 \cdot$	VII. $RH_7 \cdot$ $R_2O_7 \cdot$	VIII. $RO_4 \cdot$
1	H. 1							
2	Li 7	Be — 9	B — 11	C 12	N — 14	O — 16	F 19	
3	Na 23	Mg 24	Al — 27.3	Si 28	P 31	S — 32	Cl 35.4	
4	K 39	— Ca 40	— Sc 44	Ti 48	— V 51	— Cr 52.4	— Mn 55	Fe 56　Ni 59　Co 59
5	— Cu 63	Zn — 65	Ga — 69.8	—	As — 75	Se — 79	Br — 80	
6	Rb — 85 —	— Sr 87	— Y 89	Zr 90	— Nb 94	— Mo 95.6	— —	Ru 103.5　Rh 104　Pd 106
7	— Ag 108	Cd — 111.6	In — 113.4	Sn 118	Sb 122	Te — 128?	I 126	
8	Cs 133 —	— Ba 137	— La 139	Ce 141.2	Di Ta 147 182	— W 184	— —	Os 199　Ir 197　Pt 197
9	— Au 196	Hg — 200	Tl — 204	Pb 206	Bi 210	—	—	
10	— —	— —	— —	Th 231.5		U 240		

The elements in each of the first seven groups of this table (except the fourth) range themselves naturally in two columns, each composed of elements closely related to one another in their chemical properties ; thus in Group I. the left-hand column contains the alkali-metals, lithium, sodium, etc., while the right-hand column includes the metals copper, silver, and gold, which exhibit considerable analogy in their chemical relations, each forming two chlorides, oxides, etc. Again, in Group VI. the first column is formed by the closely related non-metallic elements, oxygen, sulphur, selenium, and tellurium, while the second contains the metals chromium, molybdenum, tungsten and uranium, which also greatly resemble one another in their chemical relations.

Of the elements in the first group, the alkali-metals, Li, Na, K, etc., are shown to be univalent by their combination with chlorine and oxygen, e. g., NaCl, and $\frac{Na}{Na} > O$. Copper may be regarded as univalent in one series of its compounds, viz., the cuprous compounds, as in the chloride CuCl or Cu_2Cl_2, and the oxide Cu_2O, though in its most stable compounds it is bivalent, and appears to be more nearly related to the metals of the iron group. The place of gold in the series is somewhat exceptional, since, though univalent in the aurous compounds, as AuCl, it is trivalent in the more stable auric compounds, as $AuCl_3$.

The elements in Group II. are all dyads.

Of the elements in Group III., column 1, boron, gallium, indium, and thallium are undoubtedly triads. Aluminium forms a trichloride, $AlCl_3$, and a corresponding oxide, Al_2O_3 ; also a volatile methyl-compound, $Al(CH_3)_3$, the vapor-density of which indicates that the molecule, as represented by this formula, has the normal 2-volume condensation (p. 246). It is true that the chloride, which is also volatile, exhibits a vapor-density agreeing rather with the doubled formula, Al_2Cl_6, which would indicate that aluminium is a tetrad, the chloride having the constitution $Cl_3Al.AlCl_3$; but this chloride boils at a very high temperature, and it is therefore probable that the temperature at which its vapor-density was actually determined was not sufficiently raised above the boiling point to bring the compound into the state of a perfect gas (p. 246).

The rare earth-metals in Group III., column 2, have been lately shown to be triads, forming trichlorides and sesquioxides. All the elements in Group IV. have long been classed as tetrads, with the exception of cerium, the quantivalence of which was till lately regarded as doubtful; but the recent experiments of Dr. Brauner have decidedly shown that it is a tetrad.

Of the elements in Group V. phosphorus, antimony, niobium, and tantalum form pentachlorides. Nitrogen is quinquivalent in the ammonium-compounds, as in the chloride NH_4Cl. Vanadium, arsenic, and bismuth do not combine with more than 3 atoms of chlorine, bromine, or iodine; but bismuth forms an oxychloride, $BiOCl_3$, in which it is quinquivalent, and vanadium forms the analogous compound, $VOCl_3$. Arsenic does not form a similar oxychloride; but its highest oxide, As_2O_5, is the exact analogue of phosphoric oxide, P_2O_5, and vanadic oxide, V_2O_5, and forms a series of salts, the arsenates, which are isomorphous with the phosphates and vanadates. For these reasons arsenic is likewise regarded as a pentad.

Among the elements in Group VI. chromium forms a hexfluoride and tungsten a hexchloride; uranium forms an oxychloride, UO_2Cl_2, and a trioxide UO_3. Sulphur, selenium, and tellurium, so far as regards their hydrogen-compounds, H_2S, etc., are dyads; but with regard to their combinations with chlorine, they are tetrads, and sulphur is known to form certain organic compounds in which it is tetradic, and others in which it is hexadic.* Moreover, the chemical relations of the sulphates are much more clearly represented by formulæ in which sulphur is supposed to be hexadic, like that given for sulphuric acid on p. 249, than by formulæ into which it enters as a dyad, such as $H—O—O—S—O—O—H$, inasmuch as compounds in which dyadic elements are linked together in one row are for the most part very unstable, like the higher oxides and sulphides of potassium (p. 253). These three elements are therefore best regarded as hexads, though they sometimes enter into combination as tetrads, and vary frequently as dyads.

Oxygen, in its combinations with hydrogen and with most of the metals, is undoubtedly bivalent; but it appears also to be capable of higher degrees of combination; with silver, for example, it forms the two oxides Ag_2O and Ag_4O, in the latter of which it is quadrivalent; and from its close analogies to sulphur, and the place of its atomic weight in the series, it may be classed with the hexads.

Of the elements in Group VII. manganese appears to form a heptachloride, $MnCl_7$, though the composition of this chloride has not perhaps been very distinctly made out; but in the permanganates the metal appears to be decidedly septivalent; the potassium salt, $KMnO_4$, for example, may be represented by the structural formula:

---

* Sulphur triethiodide, $Sr \begin{cases} (C_2H_5)_3 \\ I \end{cases}$

Sulphur-diethylene dibromide, $Sri \begin{cases} (C_2H_4)'' \\ (C_2H_4)'' \\ Br_2 \end{cases}$

$$O = \overset{\overset{\displaystyle O}{\|}}{\underset{\underset{\displaystyle O}{\|}}{Mn}} - O - K$$

The perchlorates, *e. g.*, $KClO_4$, are similar to the permanganates in composition and in crystalline form, and may therefore be supposed to have a similar constitution, the chlorine in them being septivalent; in fact, the four oxy-acids of chlorine form a regular series in which the quantivalence of the chlorine varies by two units from 1 to 7; thus—

$$Cl-O-H \qquad O=Cl-O-H \qquad O=\overset{\overset{\displaystyle O}{\|}}{Cl}-O-H \qquad O=\overset{\overset{\displaystyle O}{\|}}{\underset{\underset{\displaystyle O}{\|}}{Cl}}-O-H$$

Hypochlorous.　　　Chlorous.　　　　Chloric.　　　　Perchloric.

Iodic acid, IO_3H, and periodic acid, IO_4H, are exactly similar in constitution to chloric and perchloric acids, and the corresponding oxides or anhydrides I_2O_5, I_2O_7 are likewise known (p. 170); hence iodine also may be regarded as a heptad. Bromic acid, BrO_3H, is similar to chloric acid, but the existence of perbromic acid is doubtful, and of fluorine no oxygen-compound is known; but from the close analogy in the reactions of these four elements, Cl, Br, I, and F, and the manner in which they replace one another in combination, there can be no doubt that they belong to the same group. In their combinations with hydrogen, and in the reactions in which they replace hydrogen and one another in combination, they invariably act as monads, the substitution taking place atom for atom.

Lastly, with regard to the elements (all metallic) which cannot be included in either of the seven groups above considered. The atomic weights of three of these metals, viz., iron, cobalt, and nickel, have values between those of manganese and copper; and of the other six, called p l a t i n u m m e t a l s, three, viz., ruthenium, rhodium, and palladium, have atomic weights intermediate between those of molybdenum (96) and silver (108), while those of the other three, viz., osmium, iridium, and platinum, are greater than that of tungsten, and nearly equal to that of gold.

These intermediate elements, Fe, Co, Ni, Ru, Rh, Pd, Os, Ir, constitute a group of themselves (the eighth), some of the members of which, viz., Ru and Os, form tetroxides (analogous to octachlorides) and may therefore be regarded as octads. None of these metals indeed form chlorides containing more than four atoms of chlorine, but four of them, viz., Pt, Ir, Ru, and Os, form with the alkali-metals double chlorides in which the platinum, etc., is united with eight univalent atoms, *e. g.*, potassium platinochloride,

$$Cl_4Pt(NaCl)_2.$$

The blank spaces in the preceding table indicate the places of elements which probably exist, but have not yet been discovered. Several anticipated discoveries of this kind have, however, been actually realized. When the table was drawn up, a blank in the space now occupied by gallium indicated the probable existence of a trivalent element intermediate in atomic weight between zinc and arsenic. This element was provisionally designated *Ekaluminium* by Mendelejeff, who predicted from its position in the series what its chief properties ought to be. The discovery of gallium with the atomic weight 69.8 has verified this prediction. Another example is afforded by scandium, which agrees closely in properties and atomic weight (44) with the element whose existence was predicted by Mendelejeff under the name of *Ekabor*.

CRYSTALLIZATION ; CRYSTALLINE FORM.

Almost every substance, simple or compound, capable of existing in the solid state, assumes, under favorable circumstances, a distinct geometrical figure, usually bounded by plane surfaces, and having angles of constant value. The faculty of crystallization seems to be denied only to a few bodies, chiefly highly complex organic principles, which stand, as it were, upon the very verge of organization, and which, when in the solid state, are frequently characterized by a kind of beady or globular appearance, well known to microscopical observers.

The most beautiful examples of crystallization are to be found among natural minerals, the results of exceedingly slow changes constantly occurring within the earth. It is invariably found that artificial crystals of salts, and other soluble substances which have been slowly and quietly deposited, surpass in size and regularity those of more rapid formation.

Solution in water or some other liquid is a very frequent method of effecting crystallization. If the substance be more soluble at a high than at a low temperature, then a hot and saturated solution left to cool slowly will generally be found to furnish crystals : this is a very common case with salts and various organic principles. If it be equally soluble, or nearly so, at all temperatures, then slow spontaneous evaporation in the air, or over a surface of oil of vitriol, often proves very effective.

Fusion and slow cooling may be employed in many cases : that of sulphur is a good example : the metals, when thus treated, usually afford traces of crystalline figure, which sometimes become very beautiful and distinct, as with bismuth. A third condition under which crystals very often form is in passing from the gaseous to the solid state, of which iodine affords a good instance. When by any of these means time is allowed for the symmetrical arrangement of the particles of matter at the moment of solidification, crystals are produced.

That crystals owe their figure to a certain regularity of internal structure is shown both by their mode of formation and also by the peculiarities attending their fracture. A crystal placed in a slowly evaporating saturated solution of the same substance, grows or increases by a continued deposition of fresh matter upon its sides, in such a manner that the angles formed by the meeting of the latter remain unaltered.

The tendency of most crystals to split in particular directions, called by mineralogists *cleavage*, is a certain indication of regular structure, while the optical properties of many among them, and their mode of expansion by heat, point to the same conclusion.

It may be laid down as a general rule that every substance has its own crystalline form, by which it may very frequently be recognized at once— not that each substance has a different figure, although very great diversity in this respect is to be found. Some forms are much more common than others, as the cube and six-sided prism, which are very frequently assumed by a number of bodies not in any way related.

The same substance may assume, under different sets of circumstances, as at high and low temperatures, two different crystalline forms, in which case it is said to be *dimorphous*. Sulphur and carbon furnish, as already noticed, examples of this curious fact ; another case is presented by calcium carbonate in the two modifications of calc spar and arragonite, both chemically the same, but physically different. A fourth example might be given in mercuric iodide, which also has two distinct forms, and even two distinct colors, offering as great a contrast as those of diamond and graphite.

Crystallographic Systems.—When a crystal of simple form is attentively considered, it becomes evident that certain directions can be pointed out in which straight lines may be imagined to be drawn, passing through the central point of the crystal from side to side, from end to end, or from one angle to that opposed to it, etc., about which lines the particles of matter composing the crystal may be conceived to be symmetrically built up. Such lines, or *axes*, are not always purely imaginary, however, as may be inferred from the remarkable optical properties of many crystals : upon their number, relative lengths, position, and inclination to each other, depends the outward figure of the crystal itself.

All crystalline forms may upon this plan be arranged in six classes or s y s t e m s ; these are the following.

1. **The monometric, regular, or cubic system** (fig. 128).—The crystals of this division have three equal axes, all placed at right angles to each other. The most important forms are the *cube* (1), *the regular octohedron* (2), and the *rhombic dodecahedron* (3).

The letters a—a, b—b, c—c (fig. 128) show the termination of the three axes, placed as stated.

Fig. 128.

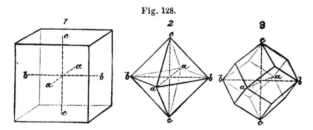

Very many substances, both simple and compound, assume these forms, as most of the metals, carbon in the state of diamond, common salt, potassium iodide, the alums, fluor-spar, iron bisulphide, garnet, spinelle, etc.

Fig. 129.

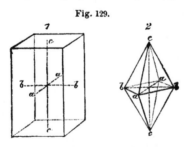

2. **The dimetric, quadratic, square prismatic, or pyramidal system.**—The crystals of this system (fig. 129) are also symmetrical about three axes at right angles to each other. Of these, however, two only are

of equal length, the third, c—c, being longer or shorter. The most important forms are, the *right square prism* (1), and the *right square-based octohedron* (2).

Examples of these forms are to be found in zircon, native stannic oxide, apophyllite, yellow potassium ferrocyanide, etc.

3. The rhombohedral system (fig. 130).—This is very important and extensive; it may be characterized by *four* axes,* three of which are equal, in the same plane, and inclined to each other at angles of 60°, while the fourth or principal axis is perpendicular to them all. The principal forms are,—the *regular six-sided prism* (1), the *regular double six-sided pyramid* (2), the *rhombohedron* (4), and the *scalenohedron* (5), a figure bounded by twelve scalene triangles.

Examples are found in ice, calcspar, sodium nitrate, beryl, quartz or rock-crystal, and the semi-metals, arsenic, antimony, and tellurium.

A combination of the regular six-sided prism and double six-sided pyramid (3) is a common form of quartz.

Fig. 130.

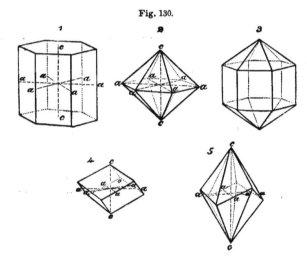

4. The trimetric, rhombic, or right prismatic system.—This is characterized by three axes of unequal lengths, placed at right angles to each other, as in the *right rectangular prism*, the *right rhombic prism*, the *right rectangular-based octohedron*, and the *right rhombic-based octohedron*.

Fig. 131.

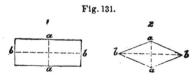

The bases of these forms are represented in fig. 131 (1 and 2). Let the reader imagine a straight line passing through the centre of each of

---

* This reference to *four* axes is a mere matter of convenience; *three* axes are sufficient for the determination of any solid figure whatever.

these figures, perpendicular to the plane of the paper; this will represent
the vertical axis. The octohedron will be formed by joining the ends of
this vertical line with the angles of the bases, and the prisms by vertical
planes passing through the sides of the base, and terminated by horizontal
planes passing through the extremities of the vertical axis. The perspective forms of these trimetric prisms and octohedrons are similar to those of
the dimetric system (fig. 129).

The system is exemplified in sulphur crystallized at a low temperature,
arsenical iron pyrites, potassium nitrate and sulphate, barium sulphate, etc.

5. **The monoclinic or oblique prismatic system.**—Crystals belonging to this group have also three axes, which may be all unequal; two of
these (the secondary) are placed at right angles, the third being so inclined
as to be oblique to one and perpendicular to the other. To this system may
be referred the four following forms: The *oblique rectangular prism*, the *oblique
rhombic prism*, the *oblique rectangular-based octohedron*, the *oblique rhombic-based
octohedron*.

The bases of these monoclinic forms are identical in form with those of
the trimetric system, fig. 131 (1) and (2). The principal axis may be represented by a line passing through the plane of the paper at the middle
point, perpendicular to *a a*, and oblique to *b b*. The perspective forms are
shown in fig. 132.

Fig. 132.

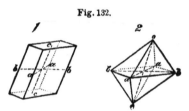

Such forms are taken by sulphur crystallized by fusion and cooling, by
realgar, sulphate, carbonate and phosphate of sodium, borax, green vitriol,
and many other salts.

6. **The triclinic, anorthic, or doubly oblique prismatic system.**
—The crystalline forms comprehended in this division are, from their great
apparent irregularity, exceedingly difficult to study and understand. In
them are traced three axes, which may be all unequal in length, and are
all oblique to each other, as in the *doubly-oblique prism*, and in the *doubly-
oblique octohedron*. The perspective forms are similar to those of the monoclinic system.

Copper sulphate, bismuth nitrate, and potassium quadroxalate afford
illustrations of these forms.

Primary and Secondary Forms.—If a crystal increase in magnitude by equal
additions on every part, it is quite clear that its figure must remain unaltered; but if, from some cause, this increase should be partial, the
newly deposited matter being distributed unequally, but still in obedience
to certain definite laws, then alterations of form are produced, giving rise
to figures which have a direct geometrical connection with that from which
they are derived. If, for example, in the cube, a regular omission of successive rows of particles of matter in a certain order be made at each solid
angle, while the crystal continues to increase elsewhere, the result will be

the production of small triangular planes, which, as the process advances, gradually usurp the whole of the surface of the crystal, and convert the cube into an octohedron. The new planes are called *secondary*, and their production is said to take place by regular *decrements* upon the solid angles. The same thing may happen on the edges of the cube; a new figure, the

Fig. 133.

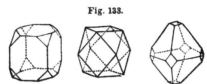

Passage of cube to octohedron.

rhombic dodecahedron, is then generated. The modifications which can thus be produced of the original or *primary* figure (all of which are subject to exact geometrical laws) are very numerous. Several distinct modifications may be present at the same time, and thus render the form exceedingly complex.

Crystals often cleave parallel to all the planes of the primary figure, as in calcspar, which offers a good illustration of this perfect cleavage. Sometimes one or two of these planes have a kind of preference over the rest in this respect, the crystal splitting readily in these directions only.

A very curious modification of the figure sometimes occurs by the excessive growth of each alternate plane of the crystal; the rest become at

Fig. 134.

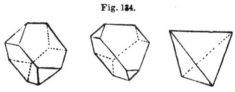

Passage of octohedron to tetrahedron.

length obliterated, and the crystal assumes the character called *hemihedral* or *half-sided*. This is well seen in the production of the tetrahedron from the regular octohedron (fig. 134), and of the rhombohedric form by a similar change from the double six-sided pyramid (fig. 130, 2).

Forms belonging to the same crystallographic system are related to each other by several natural affinities.

1. *It is only the simple forms of the same system that can combine into a complex form.*—For in all fully developed (holohedral) natural crystals, it is found that all the similar parts, if modified at all, are modified in an exactly similar manner (in hemihedral forms, half the similar edges and angles alternately situated are similarly modified). Now this can be the case only when the dominant form and the modifying form are developed according to the same law of symmetry. Thus, if a cube and a regular octohedron are developed round the same system of axes, each summit of the cube is cut off to the same extent by a face of the octohedron, or *vice versâ*. But a cube could never combine in this manner with a rhombic octohedron, because it would be impossible to place the two forms in such a manner that similar parts of the one should throughout replace similar parts of the other.

The crystals of each system are thus subject to a peculiar and distinct set of modifications, the observation of which very frequently constitutes an excellent guide to the discovery of the primary form itself.

2. *Crystals belonging to the same system are intimately related in their optical properties.* — Crystals belonging to the regular system (as the diamond, alum, rock-salt, etc.) refract light in the same manner as uncrystallized bodies ; that is to say, they have but one refractive index, and a ray of light passing through them in any direction is refracted singly. But all other crystals refract doubly, that is to say, a ray of light passing through them (except in certain directions) is split into two rays, the one called the *ordinary* ray, being refracted as it would be by an amorphous body, the other, called the *extraordinary* ray, being refracted according to peculiar and more complex laws (see Light). Now, the crystals of the d i m e t r i c and h e x a g o n a l systems resemble each other in this respect, that in all of them there is one direction, called the optic axis, or axis of double refraction (coinciding with the principal crystallographic axis), along which a ray of light is refracted singly, while in all other directions it is refracted doubly ; whereas, in crystals belonging to the other systems, viz., the trimetric and the two oblique systems, there are always *two* directions or axes, along which a ray is singly refracted.

3. *Crystals belonging to the same system resemble each other in their mode of conducting heat.* — Amorphous bodies and crystals of the regular system conduct heat equally in all directions, so that, supposing a centre of heat to exist within such a body, the isothermal surfaces will be spheres. But crystals of the dimetric and hexagonal systems conduct equally only in directions perpendicular to the principal axis, so that in such crystals the isothermal surfaces are ellipsoïds of revolution round that axis ; and crystals belonging to either of the three other systems conduct unequally in all directions, so that in them the isothermal surfaces are ellipsoïds with three unequal axes.

Relations of Form and Constitution; Isomorphism.

Certain substances, to which a similar chemical constitution is ascribed, possess the remarkable property of exactly replacing each other in crystallized compounds, without alteration of the characteristic geometrical figure. Such bodies are said to be *isomorphous.*[*]

For example, magnesia, zinc oxide, cupric oxide, ferrous oxide, and nickel oxide, are allied by isomorphic relations of the most intimate nature. The salts formed by these substances with the same acid and similar proportions of water of crystallization, are identical in their form, and, when of the same color, cannot be distinguished by the eye : the sulphates of magnesium and zinc may be thus confounded. These sulphates, too, all combine with potassium sulphate and ammonia sulphate, giving rise to double salts, whose figure is the same, but quite different from that of the simple sulphates. Indeed this connection between identity of form and parallelism of constitution runs through all their combinations.

In the same manner alumina and iron sesquioxide replace each other continually without change of crystalline figure : the same remark may be made of the oxides of potassium, sodium, ammonium. The alumina in common alum may be replaced by iron sesquioxide, the potash by ammonia or by soda, and still the figure of the crystal remains unchanged. These replacements may be partial only : we may have an alum containing both potash and ammonia, or alumina and chromium sesquioxide. By artificial management — namely, by transferring the crystal successively to

---

* From ἴσος, equal, and μόρφη, shape or form.

different solutions—we may have these isomorphous and mutually replacing compounds distributed in different layers upon the same crystal.

For these reasons mixtures of isomorphous salts can never be separated by crystallization, unless their difference of solubility is very great. A mixed solution of ferrous sulphate and nickel sulphate, isomorphous salts, yields on evaporation crystals containing both iron and nickel. But if before evaporation the ferrous salt be converted into ferric salt, by chlorine or other means, then the crystals obtained are free from iron, except that of the mother-liquor which wets them. The ferric salt is no longer isomorphous with the nickel salt, and easily separates from the latter.

Absolute identity of value in the angles of crystals is not always exhibited by isomorphous substances. In other words, small variations often occur in the magnitude of the angles of crystals of compounds which in all other respects show the closest isomorphic relations. This should occasion no surprise, as there are reasons why such variations might be expected, the chief perhaps being the unequal effects of expansion by heat, by which the angles of the same crystal are changed by alteration of temperature. A good example is found in the case of the carbonates of calcium, magnesium, manganese, iron, and zinc, which are found native crystallized in the form of obtuse rhombohedrons (fig. 130, 4), not distinguishable from each other by the eye, but exhibiting small differences in their angles when accurately measured. These compounds are isomorphous, and the measurements of the obtuse angles of their rhombohedrons are as follows :—

Calcium carbonate	.	.	.	.	.	. 105° 5'
Magnesium "	.	.	.	.	.	. 107° 25'
Manganous "	.	.	.	.	.	. 107° 20'
Ferrous "	.	.	.	.	.	. 107°
Zinc	.	.	.	.	.	. 107° 40'

Anomalies in the composition of various earthy minerals, which formerly threw much obscurity upon their chemical nature, have been in great measure explained by these discoveries. Specimens of the same mineral from different localities were found to afford very discordant results on analysis. But the proof once given of the extent to which substitution of isomorphous bodies may go, without destruction of what may be called the primitive type of the compound, these difficulties vanish.

Decision of a doubtful point respecting the constitution of a compound may sometimes be very satisfactorily made by reference to its isomorphous relations, as in the case of alumina, already mentioned, which is isomorphous with the sesquioxide of iron (p. 243).

The direct determination of the crystalline forms of the elementary bodies is often difficult, and the question of their isomorphism is complicated by the frequent dimorphism which they exhibit, but when compounds are found to correspond in chemical constitution and crystalline form, it may sometimes be inferred that the elements composing them are likewise isomorphous. Thus, the metals magnesium, zinc, iron, and copper are presumed to be isomorphous. Arsenic and phosphorus have not the same crystalline form ; nevertheless they are said to be isomorphous, because arsenic and phosphoric acids give rise to combinations which agree most completely in figure and constitution. The chlorides, iodides, bromides, and fluorides agree, whenever they can be observed, in the most perfect manner : hence the elements themselves are believed to be isomorphous.

The subjoined table, taken with slight modification from Graham's

12

"Elements of Chemistry,"* will serve to convey some idea of the most important families of isomorphous elements :—

Isomorphous Groups.

(1)	(3)	(6)
Sulphur	Barium	Sodium
Selenium	Strontium	Silver
Tellurium.	Lead.	Thallium
		Gold
(2)	(4)	Potassium
Magnesium	Platinum	*Ammonium.*
Calcium	Iridium	
Manganese	Osmium.	(7)
Iron		Chlorine
Cobalt	(5)	Iodine
Nickel	Tin	Bromine
Zinc	Titanium	Fluorine
Cadmium	Zirconium	*Cyanogen.*
Copper	Tungsten	
Chromium	Molybdenum	(8)
Aluminium	Tantalum	Phosphorus
Glucinum.	Niobium.	Arsenic
		Antimony
		Bismuth
		Vanadium.

A comparison of this table with that on page 256 will show that in many instances, isomorphous elements exhibit equal quantivalence or combining capacity, and more generally that the isomorphous groups consist wholly of perissad or wholly of artiad elements. The only apparent exception to this rule is afforded by tantalum and niobium, which, although pentads, are isomorphous with tin, tungsten, and other tetrad and hexad elements.

* Second edition, vol. i. p. 176.

CHEMICAL AFFINITY.

THE term Chemical Affinity, or Chemical Attraction, is used to describe that particular power or force, in virtue of which, union, often of a very intimate and permanent nature, takes place between two or more bodies, in such a way as to give rise to a *new* substance, having, for the most part, properties completely in discordance with those of its components.

The attraction thus exerted between different kinds of matter is to be distinguished from other modifications of attractive force which are exerted indiscriminately between all descriptions of substances, sometimes at enormous distances, sometimes at intervals quite inappreciable. Examples of the latter are to be seen in cases of what is called *cohesion*, when the particles of solid bodies are immovably bound together into a mass. Then there are other effects of, if possible, a still more obscure kind; such as the various actions of surface, the adhesion of certain liquids to glass, the repulsion of others, the ascent of water in narrow tubes, and a multitude of curious phenomena which are described in works on Physics, under the head of *molecular actions*. From all these, true chemical attraction may be at once distinguished by the deep and complete change of characters which follows its exertion: we might indeed define affinity to be a force by which new substances are generated.

Nevertheless, chemical combination graduates so imperceptibly into mere mechanical mixture, that it is often impossible to mark the limit. Solution is the result of a weak kind of affinity existing between the substance dissolved and the solvent—an affinity so feeble as completely to lose one of its most prominent features when in a more exalted condition—namely, power of causing elevation of temperature; for in the act of mere solution, the temperature falls, the heat of combination being lost and overpowered by the effects of change of state.

The force of chemical attraction thus varies greatly with the nature of the substances between which it is exerted; it is influenced, moreover, to a very large extent, by external or adventitious circumstances. An idea formerly prevailed that the relations of affinity were fixed and constant between the same substances, and great pains were taken in the preparation of tables exhibiting what was called the precedence of affinities. The order pointed out in these lists is now acknowledged to represent the order of precedence *for the circumstances* under which the experiments were made, but nothing more; so soon as these circumstances become changed, the order is disturbed. The ultimate effect, indeed, is not the result of the exercise of one single force, but rather the joint effect of a number, so complicated and so variable in intensity, that, in the greater number of cases, it is not possible to predict the consequences of a yet untried experiment.

It will be proper to examine shortly some of these extraneous causes to which allusion has been made, which modify to so great an extent the direct and original effects of the specific attractive force.

Alteration of temperature may be reckoned among these. When metallic mercury is heated nearly to its boiling point, and in that state exposed for a long time to the air, it absorbs oxygen, and becomes converted into a dark-red crystalline powder. This very same substance, when raised to a still higher temperature, separates spontaneously into metallic mercury and oxygen gas. It may be said, and probably with truth, that the latter change is greatly aided by the tendency of the metal to assume the vaporous state; but precisely the same fact is observed with another metal, palladium, which is not volatile, excepting at extremely high temperatures, but which oxidizes superficially at a red heat, and again becomes reduced when the temperature rises to whiteness.

Insolubility and the power of vaporization are, perhaps, beyond all other disturbing causes, the most potent; they interfere in almost every reaction which takes place, and very frequently turn the scale when the opposed forces do not greatly differ in energy. It is easy to give examples. When a solution of calcium chloride is mixed with a solution of ammonium carbonate, double interchange ensues, calcium carbonate and ammonium chloride being generated : $CaCl_2 + (NH_4)_2CO_3 = CaCO_3 + 2NH_4Cl$. Here the action can be shown to be in a great measure determined by the insolubility of the calcium carbonate. On the other hand, when dry calcium carbonate is powdered and mixed with ammonium chloride, and the whole heated in a retort, a sublimate of ammonium carbonate is formed, while calcium chloride remains behind. In this instance, it is no doubt the great volatility of the new ammoniacal salt which chiefly determines the kind of decomposition.

When iron filings are heated to redness in a porcelain tube, and vapor of water is passed over them, the water undergoes decomposition with the utmost facility, hydrogen being rapidly disengaged, and the iron converted into oxide. On the other hand, oxide of iron, heated in a tube through which a stream of dry hydrogen is passed, suffers almost instantaneous reduction to the metallic state, while the vapor of water, carried forward by the current of gas, escapes as a jet of steam from the extremity of the tube. In these experiments, the affinities between the iron and oxygen and the hydrogen and oxygen are so nearly balanced, that the difference of *atmosphere* is sufficient to settle the point. An atmosphere of steam offers little resistance to the escape of hydrogen; an atmosphere of hydrogen bears the same relation to steam; and this apparently trifling difference of circumstances is quite enough for the purpose.

What is called the *nascent state* is one very favorable to chemical combination. Thus, nitrogen refuses to combine with gaseous hydrogen; yet, when these substances are simultaneously liberated from some previous combination, they unite with great ease, as when organic matters are destroyed by heat, or by spontaneous putrefactive change.

There is a remarkable, and, at the same time, very extensive class of actions, grouped together under the general title of cases of *disposing affinity*. Metallic silver does not oxidize at any temperature : nay, more, its oxide is easily decomposed by simple heat; yet, if the finely-divided metal be mixed with siliceous matter and alkali, and ignited, the whole fuses to a yellow transparent glass of silver silicate. Platinum is attacked by fused potassium hydrate, hydrogen being probably disengaged while the metal is oxidized : this is an effect which never happens to silver under the same circumstances, although silver is a much more oxidable substance than platinum. The fact is, that potash forms with the oxide of the last-named metal a kind of saline compound, in which the platinum oxide acts as an acid ; and hence its formation under the *disposing* influence of the powerful base.

In the remarkable decompositions suffered by various organic bodies when heated in contact with caustic alkali or lime, we have other examples of the same fact. Products are generated which are never formed in the absence of the base ; the reaction is invariably less complicated, and its results fewer in number and more definite, than in the event of simple destruction by a graduated heat.

There is yet a still more obscure class of phenomena, called *catalytic*, in which effects are brought about by the mere *presence* of a substance which itself undergoes no perceptible change : the experiment mentioned in the chapter on oxygen, in which that gas is obtained, with the greatest facility, by heating a mixture of potassium chlorate and manganese dioxide, is a case in point. The salt is decomposed at a very far lower temperature than would otherwise be required, and yet the manganese oxide does not appear to undergo any alteration, being found after the experiment in the same state as before. It may, however, undergo a temporary alteration. We know, indeed, that this oxide when in contact with alkalies, is capable of taking up an additional proportion of oxygen and forming manganic acid ; and it is quite possible that in the reaction just considered it may actually take oxygen from the potassium chlorate, and pass to the state of a higher oxide, which, however, is immediately decomposed, the additional oxygen being evolved, and the dioxide returning to its original state. The same effect in facilitating the decomposition of the chlorate is produced by cupric oxide, ferric oxide, and lead oxide, all of which are known to be susceptible of higher oxidation. The oxides of zinc and magnesium, on the contrary, which do not form higher oxides, are not found to facilitate the decomposition of the chlorate ; neither is any such effect produced by mixing the salt with other pulverulent substances, such as pounded glass or pure silica.

The so-called catalytic actions are often mixed up with other effects which are much more intelligible, as the action of finely divided platinum on certain gaseous mixtures, in which the solid appears to condense the gas upon its greatly extended surface, and thereby to induce combination by bringing the particles within the sphere of their mutual attractions.

Influence of Pressure on Chemical Action.—When a body is decomposed by heat in a confined space, and one or more of the separated elements (ultimate or proximate) is gaseous, the decomposition goes on until the liberated gas or vapor has attained a certain tension, greater or less according to the temperature. So long as this temperature remains constant, no further decomposition takes place, neither does any portion of the separated elements recombine : but if the temperature be raised, decomposition recommences, and goes on till the liberated gas or vapor has attained a certain higher tension, also definite for that particular temperature ; if on the other hand the temperature be lowered, recomposition takes place, until the tension of the remaining gas is reduced to that which corresponds with the lower temperature. These phenomena, which are closely analogous to those exhibited in the vaporization of liquids, have been especially studied by Deville and Debray.[*] Deville designates decomposition under these conditions by the term " Dissociation."

When *calcium carbonate* is heated in an iron tube, from which the air has been exhausted by means of a mercury pump, no decomposition takes place at 300° C. (572° F.), and a scarcely perceptible decomposition at 440° C. (824° F.); but at 800° C. (1462° F.), in vapor of cadmium, it becomes very

[*] Watts's Dictionary of Chemistry, First Supplement, p. 425.

23 *

perceptible, and goes on till the tension of the evolved carbon dioxide becomes equivalent to 85 millimeters of mercury ; there it stops so long as the temperature remains constant; but on raising the temperature to 1040° C. (1904° F.), in vapor of zinc, more carbon dioxide is evolved until a tension equivalent to about 250 mm. is attained. If the tension be reduced by working the pump, it is soon restored to its former value by a fresh evolution of carbon dioxide. If, on the other hand, the apparatus be allowed to cool, the carbon dioxide is gradually reabsorbed by the quicklime, and a vacuum is re-established in the apparatus.

Similar phenomena are exhibited in the efflorescence of hydrated salts, and in the decomposition of the compounds of ammonia with metallic chlorides in closed spaces.

If the decomposed body, as well as one at least of its constituents, is gaseous, it is not possible to obtain an exact measurement of the maximum tension corresponding with the temperature; nevertheless, the decomposition is found to take place according to the same general law, ceasing as soon as the liberated gases have obtained a certain tension, which is greater as the temperature is higher.

It has long been known that chemical combination between any two bodies capable of uniting directly takes place only at and above a certain temperature, and that the combination is broken up at a higher temperature; but it is only in later years that we have become acquainted with the fact that bodies like water begin to decompose at temperatures considerably below that which they produce in the act of combining, and therefore that their combination at that temperature is never complete. Grove showed some years ago that water is resolved into its elements in contact with intensely ignited platinum. This reaction has been more closely studied by Deville, who finds that when vapor of water is passed through a heated platinum tube, decomposition commences at 960°–1000° C. (1760–1832° F.), (about the melting point of silver), but proceeds only to a limited extent; on raising the temperature to 1200° C. (2192° F.), further decomposition takes place, but again only to a limited amount, ceasing in fact as soon as the liberated oxygen and hydrogen have attained a certain higher tension. The quantity of these gases actually collected in this experiment is, however, very small, the greater portion of them recombining as they pass through the cooler part of the apparatus, till the tension of the remainder is reduced to that which corresponds with the lower temperature.

The recombination of the gases may be prevented to a certain extent by means of an apparatus consisting of a wide tube of glazed earthen-ware, through the axis of which passes a narrower tube of porous earthen-ware, the two being tightly fitted by perforated corks provided with gas-delivery tubes, and the whole strongly heated by a furnace. Vapor of water is passed through the inner tube, carbon dioxide through the annular space between the two, and the gases, after passing through the heated tubes, are received over caustic patash-solution. The vapor of water is then decomposed by the heat as before ; but the hydrogen, according to the laws of diffusion, passes through the porous earthen-ware into the surrounding atmosphere of carbon dioxide, being thus separated from the oxygen, which remains in the inner tube, and becomes mixed with carbon dioxide passing through the porous septum in the opposite direction to the hydrogen. As these gases pass through the alkaline water, the carbon dioxide is absorbed, and a mixture of hydrogen and oxygen collects in the receiver. A gram of water passed in the state of vapor through such an apparatus yields about a cubic centimetre of detonating gas.

The retarding influence of pressure is seen also in the action of acids upon zinc, or the electrolysis of water, in sealed tubes. In these cases the elimination of a gas is an essential condition of the change, and this being prevented, the action is retarded. On the other hand, there are numerous reactions which are greatly promoted by increased pressure—those, namely, which depend on the solution of gases in liquids, or on the prolonged contact of substances which under ordinary pressure would be volatilized by heat.

Relations of Heat to Chemical Affinity.—Whatever may be the real nature of chemical affinity, one most important fact is clearly established with regard to it; namely, that its manifestations are always accompanied by the production or annihilation of heat. Change of composition, or chemical action, and heat are mutually convertible: a given amount of chemical action will give rise to a certain definite amount of heat, which quantity of heat must be directly or indirectly expended, in order to reverse or undo the chemical action that has produced it. The production of heat by chemical action, and the definite quantitative relation between the amount of heat evolved and the quantity of chemical action which takes place, are roughly indicated by the facts of our most familiar experience; thus, for instance, the only practically important method of producing heat artificially consists in changing the elements of wood and coal, together with atmospheric oxygen, into carbon dioxide and water; and every one knows that the heat thus obtainable from a given quantity of coal is limited, and is, at least approximately, always the same. The accurate measurement of the quantity of heat produced by a given amount of chemical action is a problem of very great difficulty; chiefly because chemical changes very seldom take place alone, but are almost always accompanied by physical changes, involving further calorimetric effects, each of which requires to be accurately measured and allowed for, before the effect due to the chemical action can be rightly estimated. Thus the ultimate result has, in most cases, to be deduced from a great number of independent measurements, each liable to a certain amount of error. It is, therefore, not surprising that the results of various experiments should differ to a comparatively great extent, and that some uncertainty should still exist as to the exact quantity of heat corresponding with even the simplest cases of chemical action.

The experiments are made by inclosing the acting substances in a vessel called a calorimeter, surrounded by water or mercury, the rise of temperature in which indicates the quantity of heat evolved by the chemical action, after the necessary corrections have been made for the heat absorbed by the containing vessel and the other parts of the apparatus, and for the amount lost by radiation, etc. Combustions in oxygen and chlorine are made in a copper vessel surrounded by water; the heat evolved by the mutual action of liquids or dissolved substances is estimated by means of a smaller calorimeter containing mercury. The construction of these instruments and the methods of observation involve details which are beyond the limits of this work.*

The following table gives the quantities of heat, expressed in heat-units,† evolved in the combustion of various elements, and a few compounds, in oxygen, referred: (1) to 1 gram of each substance burned; (2) to 1 gram of oxygen consumed; (3) to one atom or molecule (expressed in grams), of the various substances :—

* See Miller's Chemical Physics, pp. 338, *et seq.*, and Watts's Dictionary of Chemistry, iii. 28, 103.
† The unit of heat here adopted is the quantity of heat required to raise 1 gram of water from 0° to 1° C.

Heat of Combustion of Elementary Substances in Oxygen.

Substance.	Product.	Units of heat evolved.			Observer.
		By 1 gram of substance.	By 1 gram of oxygen.	By 1 at. of sub-stance.	
Hydrogen　.　.	H_2O	{ 33881	4235	33881	Andrews.
		{ 34462	4308	34462	Favre & Silber-
Carbon—					[mann.
Wood-charcoal　.	CO_2	{ 7900	2962	94800	Andrews.
		{ 8080	3030	96960	Favre & Silber-
Gas-retort carbon .	"	8047	3018	96564	"[mann.
Native graphite　.	"	7797	2924	93564	"
Artificial graphite .	"	7762	2911	93144	"
Diamond　.　.	"	7770	2914	93940	"
Sulphur—					
Native .　.　.	SO_2	2220	2220	71040	"
Recently melted　.	"	2260	2260	72320	"
Flowers　.　.	"	2307	2307	73821	Andrews.
Phosphorus—					
(Yellow)　.　.	P_2O_5	5747	4454	178157	"
Zinc　.　.　.	ZnO	1330	5390	86450	"
Iron .　.　.　.	Fe_3O_4	1582	4153	88592	"
Tin .　.　.　.	SnO_2	1147	4230	135360	"
Copper　.　.　.	CuO	603	2394	38304	"

The following results have been obtained by the complete combustion of partially oxidized substances :—

Substance.	Product.	Units of heat evolved.		Observer.
		By 1 gram of substance.	In formation of 1 molecule of the ultimate product.	
Carbon monoxide, CO　.　.　. }	CO_2	{ 2403	67284	{ Favre & Sil-bermann.
		{ 2431	68064	Andrews.
Stannous oxide, SnO	SnO_2	519	69584	"
Cuprous oxide, Cu_2O	CuO	256	18304	"

The last three substances in this table contain exactly half as much oxygen as the completely oxidized products : and on comparing the amount of heat evolved in the formation of one molecule of stannic or cupric oxide from the corresponding lower oxide, with the quantity produced when a molecule of the same product is formed by the complete oxidation of the metal in one operation, we find that the combination of the second half of the oxygen contained in these bodies evolves sensibly half as much as the combination of the whole quantity. In the formation of carbon dioxide, however, the second half of the oxygen appears to develop more than two-thirds of the total amount of heat; but this result is probably due, in part at least, to the fact that when carbon is burned into carbon dioxide, a considerable but unknown quantity of heat is expended in converting the solid carbon into gas, and thus escapes measurement; while, in carbon monoxide, the carbon already exists in the gaseous form, and therefore no portion of the heat evolved in the combustion of this substance is similarly expended in producing a change of state.

It seems probable, also, that a similar explanation may be given of the inequalities in the quantities of heat produced by the combustion of different varieties of pure carbon and of sulphur—that is to say, that a portion of the heat generated by the combustion of diamond and graphite goes to assimilate their molecular condition to that of wood-charcoal, and that there is an analogous expenditure of heat in the combustion of native sulphur.

The quantities of heat evolved in the combination of chlorine, bromine, and iodine with other elements have been determined by Favre and Silbermann, Andrews, and others; but we must refer to larger works for the results.*

Reactions in Presence of Water.—The thermal effects which may result from the reaction of different substances on one another in presence of water, are more complicated than those resulting from direct combination. In addition to the different specific heats of the re-agents and products, and to the different quantities of heat absorbed by them in dissolving, or given out by them in combining with water, the conversion of soluble substances into insoluble ones, as a consequence of the chemical action, or the inverse change of insoluble into soluble bodies, are among the secondary causes to which part of the calorimetric effect may be due in these cases.

When a gas dissolves in water, the heat due to the chemical action is augmented by that due to the liquefaction of the gas; so also when a solid body is dissolved in water, the total thermal effect is due in part to the chemical action taking place between the water and the solid, and in part to the liquefaction of the substance dissolved. In the former case the chemical and physical parts of the phenomena both cause evolution of heat; in the latter case the physical change occasions disappearance of heat, and if this effect is greater than that due to the chemical action, the ultimate effect is the production of cold, and it is this which is generally observed.

Cold produced by Chemical Decomposition.—It is highly probable that the thermal effect of the reversal of a given chemical action is in all cases equal and opposite to the thermal effect of that action itself. A direct consequence of this proposition is that *the separation of any two bodies is attended with the absorption of a quantity of heat equal to that which is evolved in their combination.* The truth of this deduction has been experimentally established in various cases, by Wood,† Joule,‡ and Favre and Silbermann, by comparing the heat evolved in the electrolysis of dilute sulphuric acid, or solutions of metallic salts, with that which is developed in a thin metallic wire by a current of the same strength: also by comparison of the heat evolved in processes of combination accompanied by simultaneous decomposition, with that evolved when the same combination occurs between free elements.

By determining the heat evolved when different metals were dissolved

---

* Watts's Dictionary of Chemistry, iii. 109.
† Phil. Mag. [4] ii. 368; iv. 370.   ‡ Ibid. iii. 481.

in water or dilute acid, Wood found that it was less than that which would be produced by the direct oxidation of the same metals, by a quantity equal to that which would be obtained by burning the hydrogen set free, or which was expended in decomposing the water or acid ; and, therefore, that when this latter quantity was added to the results, they agreed with the numbers given by experiments of direct oxidation.

ELECTRO-CHEMICAL DECOMPOSITION ; CHEMISTRY OF THE VOLTAIC PILE.

When a voltaic current of considerable power is made to traverse various compound liquids, a separation of the elements of these liquids ensues : provided that the liquid be capable of conducting the current, its decomposition almost always follows.

The elements are disengaged solely at the limiting surfaces of the liquid, where, according to the common mode of speech, the current enters and leaves the latter, all the intermediate portions appearing perfectly quiescent. In addition, the elements are not separated indifferently and at random at these two surfaces ; but, on the contrary, make their appearance with perfect uniformity and constancy at one or the other, according to their chemical character, namely, oxygen, chlorine, iodine, acids, etc., at the surface connected with the *copper*, or *positive* end of the battery ; hydrogen, the metals, etc., at the surface in connection with the *zinc*, or *negative* extremity of the arrangement.

The terminations of the battery itself—usually, but by no means necessarily, of metal—are designated poles or *electrodes*,* as by their intervention the liquid to be experimented on is made a part of the circuit. The process of decomposition by the current is called *electrolysis*,† and the liquids which, when thus treated, yield up their elements, are denominated *electrolytes*.

When a pair of platinum plates are plunged into a glass of water to which a few drops of oil of vitriol have been added, and the plates connected by wires with the extremities of an active battery, oxygen is disengaged at the positive electrode, and hydrogen at the negative, in the proportion of one measure of the former to two of the latter nearly. This experiment has already been described.‡

A solution of hydrochloric acid mixed with a little Saxon blue (indigo), and treated in the same manner, yields hydrogen on the negative side and chlorine on the positive, the indigo there becoming bleached.

Potassium iodide dissolved in water is decomposed in a similar manner : the free iodine at the positive side can be recognized by its brown color, or by the addition of a little gelatinous starch.

All liquids are not electrolytes ; many refuse to conduct, and no decomposition can then occur ; alcohol, ether, numerous essential oils, and other products of organic chemistry, besides a few saline inorganic compounds, act in this manner, and completely arrest the current of a powerful battery.

One of the most important and indispensable conditions of electrolysis

* From ἤλεκτρον, and ὁδός, a way.
† From ἤλεκτρον, and λύειν, to loose.   ‡ Page 155.

is fluidity ; bodies which, when reduced to the liquid state, conduct freely, and as freely suffer decomposition, become insulators to the electricity of the battery when they become solid. Lead chloride offers a good illustration of this fact : when fused in a porcelain crucible, it gives up its elements with the utmost ease, and a galvanometer, interposed in the circuit, is strongly affected. But, when the source of heat is withdrawn, and the salt suffered to solidify, signs of decomposition cease, and at the same moment the magnetic needle reassumes its natural position. In the same manner, the thinnest film of ice arrests the current of a powerful voltaic apparatus ; but the instant the ice is liquefied, so that water communication is restored between the electrodes, the current again passes, and decomposition occurs. Fusion by heat, and solution in aqueous liquids, answer the purpose equally well.

Generally speaking, compound liquids cannot conduct the electric current without being decomposed ; but still there are a few exceptions to this statement, which perhaps are more apparent than real. Thus, Hittorf has shown that fused silver sulphide, which was formerly regarded as one of the exceptions, cannot be considered to be so, and Beetz has since proved the same to be the case as regards mercuric iodide and lead fluoride.

The quantity of any given compound liquid which can be decomposed by any given electric battery, depends on the resistance of the liquid : the more resistance, the less decomposition. Distilled water has only a small power of conduction, and is therefore only slightly decomposed by a battery of 30 to 40 pairs ; whilst diluted sulphuric acid is one of the best of fluid conductors, and undergoes rapid decomposition by a small battery.

When a liquid which can be decomposed, and a galvanometer, are included in the circuit of an electric current, if the needle of the galvanometer be deflected, it may be always assumed as certain that a portion of liquid, bearing a proportion to the strength of the current, is decomposed, although it may be impossible in many cases, without special contrivances, to detect the products of the decomposition, on account of their minuteness.

The metallic terminations of the battery, the poles or electrodes, have, in themselves, nothing in the shape of attractive or repulsive power for the elements separated at their surfaces. Finely-divided metal suspended in water, or chlorine held in solution in that liquid, shows not the least symptom of a tendency to accumulate around them ; a single element is altogether unaffected—directly, at least ; separation from previous combination is required, in order that this appearance should be exhibited.

It is necessary to examine the process of electrolysis a little more closely. When a portion of hydrochloric acid, for example, is subjected to decomposition in a glass vessel with parallel sides, chlorine is disengaged at the positive electrode, and hydrogen at the negative ; the gases are perfectly pure and unmixed. If, while the decomposition is rapidly proceeding, the intervening liquid be examined by a beam of light, or by other means, not the slightest disturbance or movement of any kind will be perceived ; nothing like currents in the liquid, or bodily transfer of gas from one part to another, can be detected ; and yet two portions of hydrochloric acid, separated perhaps by an interval of four or five inches, may be respectively evolving pure chlorine and pure hydrogen.

There is, it would seem, but one mode of explaining this and all similar cases of regular electrolytic decomposition : this is by assuming that *all* the particles of hydrochloric acid between the electrodes, and by which the current is conveyed, simultaneously suffer decomposition, the hydrogen travelling in one direction, and the chlorine in the other. The neighboring elements, thus brought into close proximity, unite and reproduce hydrochloric acid, again destined to be decomposed by a repetition of the same change. In this manner each particle of hydrogen may be made to travel in one direction, by becoming successively united to each particle of chlorine between itself and the negative electrode ; when it reaches the latter, finding no disengaged particle of chlorine for its reception, it is rejected, as it were, from the series, and thrown off in a separate state. The same thing happens to each particle of chlorine, which at the same time passes continually in the opposite direction, by combining successively with each particle of hydrogen, that moment separated, with which it meets, until at length it arrives at the positive plate or wire, and is disengaged. A succession of particles of hydrogen is thus continually thrown off from the decomposing mass at one extremity, and a corresponding succession of particles of chlorine at the other. The power of the current is exerted with equal energy in every part of the liquid conductor,

Fig. 135.

Hydrochloric acid in its usual state.

though its effects become manifest only at the very extremities. The action is one of a purely molecular or internal nature, and the metallic termina-

Fig. 136.

Hydrochloric acid undergoing electrolysis.

tions of the battery merely serve the purpose of completing the connection between the latter and the liquid to be decomposed. The figures 135 and 136 are intended to assist the imagination of the reader, who must at the same time avoid regarding them in any other light than that of a somewhat figurative mode of representing the phenomena described. The circles are intended to indicate the elements, and are distinguished by their respective symbols.

Like hydrochloric acid, all electrolytes, when acted on by electricity, are split into two constituents, which pass in opposite directions. Substances of the one class, like oxygen, chlorine, etc., are evolved at the positive electrode ; those of the other class, like hydrogen and the metals, at the negative electrode.

It is of importance to remark that oxygen-salts, such as sulphates and nitrates, when acted on by the current, do not divide into acid and basic oxide, but, as Daniell and Miller proved, into metal and a compound substance, or group of elements, which is transferred in such a state of association that, as regards its electrical behavior, it represents an element. Thus, cupric sulphate, $CuSO_4$, splits, not into SO_3 and CuO, but into metallic copper and *sulphione*, SO_4. Hydrogen sulphate, or sulphuric acid, H_2SO_4, divides into the same compound group and hydrogen. In a similar way, also, the part of the electrolyte which passes to the negative pole may consist of a group of elements. A solution of sal-ammoniac, NH_4Cl, furnishes a beautiful instance of this fact, since it is decomposed by the current in such a manner that the ammonium, NH_4, goes to the negative pole, where it is resolved into ammonia, NH_3, and free hydrogen, and the chlorine to the positive pole.

A distinction must be carefully drawn between true and regular electrolysis, and what is called secondary decomposition, brought about by the reaction of the bodies so eliminated upon the surrounding liquid, or upon the substance of the electrodes : hence the advantage of platinum for the latter purpose, when electrolytic actions are to be studied in their greatest simplicity, that metal being scarcely attacked by any ordinary agents. When, for example, a solution of lead nitrate or acetate is decomposed by the current between platinum plates, metallic lead is deposited at the negative side, and a brown powder, lead dioxide, at the positive : the latter substance is the result of a secondary action ; it proceeds, in fact, from the nascent oxygen, at the moment of its liberation, reacting upon the monoxide of lead present in the salt, and converting it into dioxide, which is insoluble in the dilute acid. When nitric acid is decomposed, no hydrogen appears at the negative electrode, because it is oxidized at the expense of the acid, which is reduced to nitrous acid. When potassium sulphate, K_2SO_4, is electrolized, hydrogen appears at the negative electrode, together with an equivalent quantity of potassium hydroxide, KHO, because the potassium which is evolved at the electrode immediately decomposes the water there present. At the same time, the sulphione, SO_4, which is transferred to the positive electrode, takes hydrogen from the water there present, forming sulphuric acid, H_2SO_4, and liberating oxygen. In like manner hydrogen sulphate, or sulphuric acid itself, is resolved by the current into hydrogen and sulphione, which latter decomposes the water at the positive electrode, reproducing hydrogen sulphate, and liberating oxygen, just as if the water itself were directly decomposed by the current into hydrogen and oxygen. A similar action takes place in the electrolytic decomposition of any other oxygen salt of an alkali-metal, or alkaline earth-metal, alkali and hydrogen gas making their appearance at the negative electrode, acid and oxygen gas at the positive electrode. This observation explains a circumstance which much perplexed the earlier experimenters upon the chemical action of the voltaic battery. In all experiments in which water was decomposed, both acid and alkali were liberated at the electrodes, even though distilled water was employed : and hence it was believed for some time that the voltaic current had some mysterious power of generating acid and alkaline matter. The true source of these compounds was, however, traced by Davy,[*] who showed that they proceeded from impurities either in the water itself, or in the vessels which contained it, or in the surrounding atmosphere. Having proved that ordinary distilled water always contains traces of saline matter, he redistilled it at a temperature below the boiling point, in order to avoid all risk of carrying over salts by splashing. He then found that when marble cups were used to contain the water used for decomposition, hydrochloric acid appeared at the positive electrode, soda at the negative, both being derived from sodium chloride present in the marble ; when agate cups were used, he obtained silica ; and when he used gold vessels, he obtained nitric acid and ammonia, which he traced to atmospheric air. By operating in a vacuum, indeed, the quantity of acid and alkali was reduced to a minimum, but the decomposition was almost arrested, although he operated with a battery of fifty pairs of 4-inch plates. Hence it is manifest that *water itself is not an electrolyte*, but that it is enabled to convey the current if it contains only traces of saline matter.[†]

Definite Chemical Action of the Electric Current.—If a number of different electrolytes, such as dilute sulphuric acid, cupric sulphate, potassium

* Philosophical Transactions, 1807.   † Miller's Chemical Physics, p. 454.

24

iodide, fused lead chloride, etc., be arranged in a series, and the same current be made to traverse the whole, all will suffer decomposition at the same time, but by no means to the same amount. If arrangements be made by which the quantities of the eliminated elements can be accurately ascertained, it will be found, when the decomposition has proceeded to some extent, that these latter have been disengaged exactly in the *ratio of their chemical equivalents.* The same current which decomposes 9 parts of water will separate into their elements 166 parts of potassium iodide, 139 parts of lead chloride, etc. Hence the very important conclusion : *The action of the current is perfectly definite in its nature, producing a fixed and constant amount of decomposition, expressed in each electrolyte by the value of its chemical equivalent.*

From a very extended series of experiments, based on this and other methods of research, Faraday was enabled to draw the general inference that effects of chemical decomposition are always proportionate to the quantity of circulating electricity, and may be taken as an accurate and trustworthy measure of the latter. Guided by this highly important principle, he constructed his *voltametre,* an instrument which has rendered the greatest service to electrical science. This is merely an arrangement

Fig. 137.

by which dilute sulphuric acid is decomposed by the current, the gas evolved being collected and measured. By placing such an instrument in any part of the circuit, the quantity of electric force necessary to produce any given effect can be at once estimated ; or, on the other hand, any required amount of the latter can be, as it were, measured out and adjusted to the object in view. The voltametre has received many different forms : one of the most extensively useful is that shown in fig. 137, in which the platinum plates are separated by a very small interval, and the gas is collected in a graduated jar standing on the shelf of the pneumatic trough, the tube of the instrument, which is filled to the neck with dilute sulphuric acid, being passed beneath the jar.

The decompositions produced by the voltaic battery can be effected by the electricity of the common machine, by that developed by magnetic action, and by that of animal origin, but to an extent incomparably more minute. This arises from the very small *quantity* of electricity set in motion by the machine, although its *tension*—that is, power of overcoming obstacles, and passing through imperfect conductors—is exceedingly great. A pair of small wires of zinc and platinum, dipping into a single drop of dilute acid, develops far more electricity, to judge from the chemical effects of such an arrangement, than very many turns of a large plate electrical machine in powerful action. Nevertheless, polar or electrolytic decomposition can be distinctly and satisfactorily effected by the latter, although on a minute scale.

Theory of the Voltaic Battery.—With a knowledge of the principles just laid down, the study of the voltaic battery may be resumed and completed. In the first place, two very different views have been held concerning the source of the electrical disturbance in that apparatus. Volta himself ascribed it to mere contact of dissimilar metals or other substances conducting electricity,—and, as formerly observed, electric force is actually developed in this way; but for reasons already given (p. 122) it is much more probable that the chief source of electric power in the battery is the chemical action of the liquid,—always an electrolyte,—on one of the plates, and that the direction of the positive electricity is in all cases from the metal acted upon, through the liquid to the other metal, as is strikingly shown by the experiment already described of plunging a copper and iron couple alternately into dilute sulphuric acid and solution of potassium sulphide (p. 122).

The experiments of Faraday and Daniell have moreover given very great support to the chemical theory, by showing that the contact of dissimilar metals is *not* necessary in order to call into being powerful electrical currents, and that the development of electrical Fig. 138.
force is not only in some way connected with the chemical action of the liquid of the battery, but is always in direct proportion to the latter. One very beautiful experiment, in which electrolytic decomposition of potassium iodide is performed by a current generated without any contact of dissimilar metals, can be thus made:—A plate of zinc is bent at a right angle, and cleaned by rubbing with sand-paper. A plate of platinum has a wire of the same metal attached to it by careful riveting, and the wire is bent into an arch. A piece of folded filter-paper is wetted with solution of potassium iodide, and placed upon the zinc; the platinum plate is arranged opposite to the latter, with the end of its wire resting upon the paper; and then the pair is plunged into a glass of dilute sulphuric, mixed with a few drops of nitric acid. A brown spot of iodine becomes in a moment evident beneath the extremity of the platinum wire—that is, at the positive side of the arrangement.

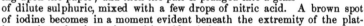

The metals employed in the construction of voltaic batteries are zinc for the active metal, and copper, silver, or, still better, platinum, for the inactive one: the greater the difference of oxidability, the better the arrangement. The liquid is either dilute sulphuric acid, sometimes mixed with a little nitric, or occasionally, where very slow and long-continued action is wanted, salt and water. To obtain the maximum effect of the apparatus with the least expenditure of zinc, that metal must be employed in the pure state, or its surface must be covered by an amalgam, which in its electrical relations closely resembles the pure metal. The zinc is easily brought into this condition by wetting it with dilute sulphuric acid, and then rubbing a little mercury over it, by means of a piece of rag tied to a stick.

The principle of the compound battery is, perhaps, best seen in the crown of cups: by each alternation of zinc, fluid, and copper, the current is urged forward with increased energy; its intensity is augmented, but the actual amount of electrical force thrown into the current form is not increased. The quantity, estimated by its decomposing power, is, in fact, determined by that of the smallest and least active pair of plates, the quantity of electricity in every part or section of the circuit being exactly equal. Hence large and small plates, batteries strongly and weakly charged, can never be connected without great loss of power.

When a battery, either simple or compound, constructed with pure or with amalgamated zinc, is charged with dilute sulphuric acid, a number of highly interesting phenomena may be observed. While the circuit remains broken, the zinc is perfectly inactive, no acid is decomposed, no hydrogen liberated; but the moment the connection is completed, torrents of hydrogen arise, not from the zinc, but from the copper or platinum surfaces alone, while the zinc undergoes tranquil and imperceptible oxidation and solution. Thus, exactly the same effects are seen to occur in every active cell of a closed circuit that are witnessed in a portion of sulphuric acid undergoing electrolysis: oxygen appears at the positive side, with respect to the current, and hydrogen at the negative; but with this difference, that the oxygen, instead of being set free, combines with the zinc. It is, in fact, a real case of electrolysis, and electrolytes alone are available as exciting liquids.

Common zinc is very readily attacked and dissolved by dilute sulphuric acid; and this is usually supposed to arise from the formation of a multitude of little voltaic circles, by the aid of particles of foreign metals or graphite, partially imbedded in the zinc. This gives rise in the battery to what is called local action, by which, in the common forms of apparatus, three-fourths or more of the metal is often consumed, without contributing in the least to the general effect, but, on the contrary, injuring it to some extent. This evil is got rid of by amalgamating the surface.

By careful experiments, in which local action was completely avoided, it has been distinctly proved that the quantity of electricity set in motion by the battery varies exactly with the zinc dissolved. Coupling this fact with that of the definite action of the current, it will be seen that when a perfect battery of this kind is employed to decompose hydrochloric acid, in order to evolve 1 gram of hydrogen from the latter, 32.5 grams of zinc must be dissolved as chloride, and its equivalent quantity of hydrogen disengaged in each active cell of the battery—that is to say, that the electrical force generated by the solution of an equivalent of zinc *in* the battery is capable of effecting the decomposition of an equivalent of hydrochloric acid or any other electrolyte *out* of it.

This is an exceedingly important discovery: it serves to show, in the most striking manner, the intimate nature of the connection between chemical and electrical forces, and their remarkable quantitative or equivalent relations. It almost seems, to use an expression of Faraday, as if a transfer of chemical force took place through the substance of solid metallic conductors; that chemical actions, called into play in one portion of the circuit, could be made at pleasure to exhibit their effects without loss or diminution in any other.

It is possible, however, to construct electric circuits containing no electrolytes, but composed entirely of elementary bodies, in which therefore the electric excitement appears to be due to direct combination between elements and to the heat thereby evolved. Messrs. Ayrton and Perry * have constructed such a circuit, capable of producing a current of considerable force, by dipping strips of platinum and magnesium into mercury, and connecting them as in an ordinary voltaic battery. In this case the current appears to be due to the amalgamation of the magnesium. In like manner Exner † has constructed circuits of considerable power consisting of carbon and a metal (magnesium, aluminium, zinc, etc.) immersed in bromine or in iodine either liquid or solid. In these circuits the positive current goes from the metal through the bromine or iodine to the carbon.

Electro-Chemical Theory.—There is an hypothesis, not of recent date, long countenanced and supported by the illustrious Berzelius, which refers all chemical phenomena to electrical forces—which supposes that bodies combine because they are in opposite electrical states; and that even the heat and light accompanying chemical union may be, to a certain extent, accounted for in this manner; that, in short, so far as our present knowledge goes, either electric or chemical action may be assumed as cause or effect: it may be that electricity is merely a form or modification of ordinary chemical affinity; or, on the other hand, that all chemical action is a manifestation of electrical force.

This electro-chemical theory is no longer received as a true explanation of chemical phenomena to the full extent intended by its author. Berzelius, indeed, supposed that the combining tendencies of elements, and their func-

* Proceedings of the Royal Society, xxvii. 222.
† Wiedermann's Annalen, xv. 412.

tions in compounds, depend altogether on their electric polarity; and accordingly he divided the elements into two classes, the electro-positive, which like hydrogen and the metals, move toward the negative pole of the battery, as if they were attracted by it, and the electro-negative, which, like oxygen, chlorine, and bromine, move toward the positive pole. We are, however, acquainted with a host of phenomena which show that the chemical functions of an element depend upon its position with regard to other elements in a compound quite as much as upon its individual character. Thus chlorine, the very type of an electro-negative element, can be substituted for hydrogen, one of the most positive of the elements, in a large number of compounds, yielding new products, which exhibit the closest analogy in composition and properties to the compounds from which they are derived. It is impossible, therefore, to admit that the chemical functions of bodies are determined exclusively by their electrical relations. Still, it is true in a general way that those elements which differ most strongly in their electrical characters, chlorine and potassium, for example, are likewise those which combine together with the greatest energy; and the division of bodies into electro-positive and electro-negative is therefore retained; the former are also called acid or chlorous, and the latter basylous or zincous.

Constant Batteries.—In all the older forms of the voltaic battery, such as those described on pages 123 and 124, the power rapidly decreases, so that after a short time, scarcely the tenth part of the original action remains. This loss of power depends partly on the gradual change of the sulphuric acid into zinc sulphate, but still more on the coating of hydrogen, and, at a later stage, on the precipitation of metallic zinc on the copper plates. It is self-evident that if the copper plate in the liquid became covered with zinc, it would act electrically like a zinc plate.

To prevent this alteration of the negative plate, and consequent loss of electric power, it is necessary either to remove the hydrogen as fast as it accumulates, or to prevent its evolution altogether, and for this purpose various forms of battery have been devised.

Fig. 139.

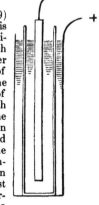

Daniell's Battery.—In this form of battery (fig. 139) each cell consists of a copper cylinder within which is placed a hollow cylinder of unglazed earthenware, animal membrane, or parchment-paper, in the axis of which is placed a rod of amalgamated zinc. This inner cylinder is filled with a mixture of 1 part by measure of oil of vitriol and 8 of water, and the exterior space with the same liquid saturated with copper sulphate. A sort of little colander is fitted to the top of the cell, in which crystals of the copper sulphate are placed, so that the strength of the solution may remain unimpaired. When a communication is made by a wire between the rod and the copper cylinder, a strong current is produced, the power of which may be increased to any extent by connecting a sufficient number of such cells into a series, on the principle of the crown of cups, the copper of the first being attached to the zinc of the second. Ten such alternations constitute a very powerful apparatus, which has the great advantage of retaining its energy undiminished for a long time.

By this arrangement of the voltaic battery, the accumulation of hydrogen and deposition of zinc on the surface of the copper plate is altogether avoided; the zinc in the porous cell, whilst it dissolves in the sulphuric

24 *

acid, decomposes it, but does not liberate any hydrogen; for by the progress of the decomposition (see p. 276), up to the boundary of the copper solution, the hydrogen takes the place of the copper, and thus ultimately the copper is precipitated on the copper plate. The copper plate therefore remains in its original state, so long as a sufficient quantity of copper sulphate is present in the solution.

Grove's Battery.—This form of battery is capable of producing a far greater degree of power than that last mentioned, and hence it has become one of the most important means of promoting electrical science in the present day. One of the cells is represented in fig. 140. The zinc plate is

Fig. 140.

bent round, so as to present a double surface, and is well amalgamated: within it stands a thin flat cell of porous earthenware, filled with strong nitric acid, and the whole is immersed in a mixture of 1 part by measure of oil of vitriol and 6 of water, contained either in one of the cells of Woollaston's trough, or in a separate cell of glazed porcelain, made for the purpose. The arrangement is completed by a strip of platinum foil, which dips into the nitric acid. In this arrangement there is no accumulation of hydrogen on the negative plate, and consequently no polarization: for the hydrogen, in passing through the nitric acid in its progress toward the platinum plate, decomposes the nitric acid, and is itself oxidized to water, while fumes of nitrogen peroxide are given off. This gas does not, however, form bubbles on the platinum plate, as it is very easily soluble in nitric acid; neither does it, like hydrogen, tend to set up a current in the contrary direction. With ten such pairs, experiments of decomposition, ignition of wires, the light between charcoal points, etc., can be exhibited with great brilliancy, while the battery itself is very compact and portable, and, to a great extent, constant in its action. The zinc, as in the case of Daniell's battery, is consumed only while the current passes, so that the apparatus may be arranged an hour or two before it is required for use, which is often a matter of great convenience; and local action from the precipitation of copper on the zinc is avoided.

Professor Bunsen has modified the Grove battery by substituting for the platinum dense charcoal or coke, which is an excellent conductor of electricity. By this alteration, at a very small expense, a battery may be made nearly as powerful and useful as that of Grove.

Fig. 141 shows the form of the round carbon cylinder which is used in these cells. It is hollowed so as to receive a porous earthenware cell, in which a round plate of zinc is placed. The upper edge of the cylinder of carbon is well saturated with wax, and is surrounded by a copper ring, by means of which it may be put in connection with the zinc of the adjoining pair.

Fig. 141.

Bunsen's carbon cylinder is likewise well adapted for the use of dilute sulphuric acid alone, without the addition of nitric acid. It is, however, better to saturate the dilute sulphuric acid with potassium bichromate. When this mixture contains at least double the amount of sulphuric acid which is necessary to decompose the chromate, a battery is formed which surpasses in power the nitric acid battery, but does not furnish currents of the same constancy.

Smee contrived an ingenious battery, in which silver, covered with a thin coating of finely divided metallic platinum, is employed in association with amalgamated zinc and dilute sulphuric acid.   The rough surface appears to permit the ready disengagement of the bubbles of hydrogen.

Zinc-carbon batteries with a single liquid are also employed.—α. In the *Bichromate battery*, an ordinary form of which is represented in fig. 142, a zinc plate passed between two gas-carbon plates is immersed in a mixture of dilute sulphuric acid and potassium bichromate, this salt acting as a depolarizer by oxidizing the hydrogen bubbles evolved at the surface of the zinc plate. This solution, however, acts upon the zinc, even when the poles are disconnected, and to prevent this, the zinc plate is fixed to a rod by which it can be drawn up out of the liquid when the battery is not in use.—β. In *Leclanche's battery* the exciting liquid is a solution of sal-ammoniac, NH_4Cl. The zinc dissolves in it, with formation of zinc-ammonium chloride, $Zn(NH_3Cl)_2$, and evolution of hydrogen; and to prevent polarization, the carbon plate is placed in a porous cell filled with fragments of carbon and powdered manganese dioxide, which slowly evolves oxygen, and thus destroys the hydrogen bubbles.—γ. In *Niaudet's battery* the zinc is immersed in a solution of common salt, NaCl, and the carbon is surrounded by bleaching powder, which readily yields chlorine and oxygen, both of which destroy the hydrogen bubbles.

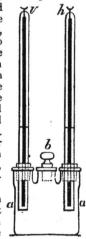

Fig. 142.

Silver Chloride Battery.—Dr. Warren de la Rue has constructed a perfectly constant battery, in which zinc and silver are the two metals, and the electrolyte is formed of solid silver chloride.   The zinc is immersed in a solution of sodium chloride or zinc chloride, and the silver, in the form of wire, is imbedded in a stick of fused silver chloride.   As the zinc dissolves, silver is continually deposited on the wire, just as copper is deposited in the Daniell cells.   The electromotive force of this battery is about equal to that of Daniell's.   A battery of 11,000 of these cells gives a spark not quite a quarter of an inch long.

Secondary Batteries.—*Storage of Electric Currents.*—If a voltameter which has been used for some time for decomposing aciduated water—so that the platinum electrodes have become coated with the electrolytic gases, the anode with oxygen and the cathode with hydrogen—be disconnected from the battery, and its wires then connected with a galvanometer, a current will be produced in the direction opposite to that of the original current,—that is to say, the positive electricity will proceed from the hydrogen plate through the liquid to the oxygen plate, the film of hydrogen, in fact, being oxidized by the acid liquid, and acting just like the zinc plate of an ordinary voltaic couple.   The same effect is produced by a pair of platinum plates immersed in hydrogen and oxygen gases respectively, and plunged into acidulated water, as in Grove's gas battery (fig. 143). The gases condense on the surfaces of the two plates, or are occluded by them, the plate immersed in the hydrogen then acting like the zinc of an ordinary voltaic battery, and producing a current the direction of which is indicated in the figure.

Fig. 143.

If the electrodes of a voltameter, after having been charged by the battery as above described, be left at rest for some time, and then connected through the medium of a galvanometer or otherwise, a reverse current will be produced, and in this way the power generated by a voltaic battery or other source of electricity may be stored up for use when required.

A more abundant storage of electric power may, however, be obtained by the electrolysis of a lead salt, whereby, as already observed (p. 277), metallic lead is deposited on the cathode and dioxide of lead, PbO_2, on the anode; and when the plates thus coated are immersed in dilute sulphuric acid, the metallic lead becomes oxidized and the dioxide reduced, and a current is thereby developed, the positive electricity proceeding from the former to the latter, i. e., in a direction opposite to that of the original battery current. If the plates are left unconnected, they may remain in the acid liquid for days without undergoing any change or showing any signs of electric charge; but on connecting them by a wire, a current is immediately produced in the direction just indicated. Such is the principle of the secondary battery, invented by Gustav Planté, the plates of which consist of two pieces of sheet lead rolled up together in cylindrical form, but not in metallic contact with each other. Its power may be greatly increased, as recommended by Faure, by covering the lead plates with a coating of minium or red lead, an oxide intermediate between PbO and PbO_2. On passing a current through the cell, the red lead is peroxidized at the positive, and reduced at the negative electrode, first to a lower oxide, then to metallic lead. In this way a greater thickness of the working substance is obtained, and the time of charging is much shortened.

H. Sutton* has devised a storage battery composed of amalgamated lead plates and plates of copper immersed in solution of copper sulphate. On connecting the lead plates with the positive pole of a battery, the lead becomes peroxidized, metallic copper is deposited on the copper plate, and sulphuric acid remains in solution. The battery is thus charged, and during its discharge the peroxide of lead is reduced, and the copper is oxidized and dissolved, reproducing copper sulphate. This form of battery is said to be very powerful and constant.

By means of these storage-batteries the power of an electric current generated in any way, by a dynamo-machine, for example (p. 131), may be stored up and transported to any place at which it is required for use, as for working a railway, driving machinery, electroplating, electric lighting, etc. Faure's batteries are made with flat plates, arranged in a rectangular trough.

Electrotype.—Within the last forty years several very beautiful and successful applications of voltaic electricity have been made, of which a brief notice must suffice. Spencer and Jacobi employed it in copying, or rather in multiplying, engraved plates and medals, by depositing upon their surfaces a thin coating of metallic copper, which when separated

Fig. 144.

from the original, exhibits, in reverse, a most faithful representation of the latter. By using this in its turn as a mould or matrix, an absolutely perfect *fac-simile* of the plate or medal is obtained. In the former case, the impressions taken on paper are quite undistinguishable from those directly derived from the work of the artist; and as there is no limit to the number of *electrotype* plates which can be thus produced, engravings of the most beautiful description may be multiplied indefinitely. The copper is very tough, and bears the action of the press perfectly well.

The apparatus used in this and many similar processes is of the simplest kind. A trough or cell of wood is divided by a porous diaphragm, made of a very thin piece of sycamore, into two parts; dilute sulphuric acid is put on one side, and a satu-

* Chemical News, xliv. 298.

rated solution of copper sulphate, sometimes mixed with a little acid, on the other. A plate of zinc is soldered to a wire or strip of copper, the other end of which is secured by similar means to the engraved copper plate. The latter is then immersed in the solution of sulphate, and the zinc in the acid. To prevent deposition of copper on the back of the copper plate, that portion is covered with varnish. For medals and small works, a porous earthenware cell, placed in a jelly-jar, may be used.

Other metals may be precipitated in the same manner, in a smooth and compact form, by the use of certain precautions which have been gathered by experience. Electro-gilding and plating are now carried on very largely and in great perfection by Messrs. Elkington and others. Even non-conducting bodies, as sealing-wax and plaster of Paris, may be coated with metal; it is only necessary, as shown by the late Mr. Robert Murray, to rub over them the thinnest possible film of plumbago. Seals may thus be copied in a very few hours with unerring truth.

The common but very pleasing experiment of the *lead tree* is greatly dependent on electro-chemical action. When a piece of zinc is suspended in a solution of lead acetate, the first effect is the decomposition of a portion of the latter, and the deposition of metallic lead upon the surface of the zinc; it is simply a displacement of a metal by a more oxidizable one. The change does not, however, stop here; metallic lead is still deposited in large and beautiful plates upon that first thrown down, until the solution becomes exhausted, or the zinc entirely disappears. The first portions of lead form with the zinc a voltaic arrangement of sufficient power to decompose the salt: under the peculiar circumstances in which the latter is placed, the metal is precipitated upon the negative portion—that is, the lead—while the oxygen and acid are taken up by the zinc.

Fig. 145.

Becquerel, several years ago, published an exceedingly interesting account of certain experiments in which crystallized metals, oxides, and other insoluble substances had been produced by the slow and continuous action of feeble electrical currents, kept up for months, or even years. These products exactly resemble natural minerals: and, indeed, the experiments throw great light on the formation of the latter within the earth.*

Heat Developed by the Electric Current.—All parts of the electric circuit, the plates, the liquid in the cells of the battery, the conducting wires, and any electrolytes undergoing decomposition, all become heated during the passage of the current. The rise of temperature in any part of the circuit depends partly on the strength of the current, partly on its resistance, those bodies which offer the greatest resistance, or are the worst conductors, being most strongly heated by a current of given strength. Thus, when a thick and a thin wire of the same metal are included in the same circuit, the latter becomes more strongly heated, and a platinum wire is much more strongly heated than a silver or copper wire of the same thickness.

By exact experiments it has been found that both in metallic wires and in liquids traversed by an electric current, the evolution of heat is directly proportional—1st, *to the resistance;* 2d, *to the strength of the current.* Joule has † also shown that the evolution of heat in each couple of the voltaic battery is subject to the same law, which, therefore, holds good in every part of the circuit, including the battery.

The strength of an electric current may be measured by the quantity of detonating gas (2 vols. H to 1 vol. O) which it can evolve from acidulated

* Traité de l'Electricité et du Magnétisme, iii. 239.
† Phil. Mag. [3] xix. 210.

water in a given time, the *unit of current strength being the current which elimi-nates one cubic centimetre of detonating gas at* 0° C. *and* 760 mm. *barometric pressure in a minute.** Now, Lenz has shown that when a current of this strength passes through a wire whose resistance is equal to that of a copper wire 1 metre long and 1 millimetre in diameter, it develops a quantity of heat sufficient to raise the temperature of 1 gram of water from 0° to 1° C. in 5¾ minutes; and assuming as the unit of heat the quantity required to raise the temperature of 1 gram of water from 0° to 1° C., the law may be thus expressed—

A current of the unit of strength passing through a conductor which exerts the unit of resistance, develops therein 1.057 *heat-units in an hour, or* 0.0176 *heat-unit in a minute.*

With a current of given strength, the sum of the quantities of heat evolved in the battery and in the metallic conductor joining its poles, is constant, the heat actually developed in the one part or the other varying according to the thickness of the metallic conductor. This was first shown by De la Rive, and has been confirmed by Favre.† De la Rive made use of a couple consisting of platinum and distilled zinc or cadmium, excited by pure and very strong nitric acid, the two metals being united by a platinum wire, more or less thick, which was plunged into the same quantity of strong nitric acid contained in a capsule, similar to that which held the voltaic couple. By observing the temperatures in the two vessels with delicate thermometers, the sum of these temperatures was found to be constant, the one or the other being greater according to the thickness of the connecting wire.

Favre,‡ by means of a calorimeter, similar to that which he used in his experiments on the development of heat by chemical action, has shown that, in a pair of zinc and platinum plates, excited by dilute sulphurie acid and connected by platinum wires of various length and thickness, for every 32.5 grams of zinc dissolved, a quantity of heat is developed in the entire circuit equal to 18,173 heat-units, but variously distributed between the battery-cell and the wire, according to the thickness of the latter. Now this quantity of heat is nearly the same as that which is evolved in the simple solution of 32.5 grams of zinc in dilute sulphuric acid, without the formation of a voltaic circuit, viz., 18,444 units. Hence Favre concludes that the heat developed by the resistance of a metallic or other conductor connecting the poles of the battery is simply borrowed from the total quantity of heat evolved by the chemical action taking place in the battery, and is rigorously complementary to that which remains in the cells of the battery, the heat evolved in the entire circuit being the exact equivalent of the chemical action which takes place. If any external work is performed by the current, such as electrolysis, or mechanical work, as by an electro-magnetic engine, the heat evolved in the circuit is diminished by the heat-equivalent of the decomposition or mechanical work done.

---

* Respecting the units of Current-strength and Resistance now adopted by practical electricians, see Thompson's "Elementary Lessons in Electricity and Magnetism" (pp. 378-380).
† Ann. Ch. Phys. [3] xl. 393.
‡ Comptes Rendus, xlv. 56.

CHEMISTRY OF THE METALS.

THE metals constitute the second and larger group of elementary bodies. A great number of them are of very rare occurrence, being found only in a few scarce minerals; others are more abundant, and some few are almost universally diffused throughout the globe. Some of these bodies are of most importance when in the metallic state; others, when in combination, chiefly as oxides, the metals themselves being almost unknown. Many are used in medicine and in the arts, and are essentially connected with the progress of civilization.

The number of metals at present known is fifty-three.

Physical Properties.—One of the most remarkable and striking characters possessed by the metals is their peculiar lustre: this is so characteristic, that the expression *metallic lustre* has passed into common speech. This property is no doubt connected with the great degree of opacity which the metals present in every instance, the thinnest leaves or plates of these bodies, and the edges of crystalline laminæ, completely arresting the passage of light. An exception to the rule is usually made in favor of gold-leaf, which, when held up to daylight, exhibits a greenish, and in some cases a purple color, as if it were really endued with a certain degree of translucency: the metallic film is, however, generally so imperfect that it is somewhat difficult to say whether the observed effect may not be in some measure due to multitudes of little holes, many of which are visible to the naked eye; but Faraday's experiments have established the translucency of gold beyond all doubt.

In point of *color*, the metals present a certain degree of uniformity: with two execptions—viz., copper, which is red, and gold, which is yellow—all these bodies are included between the pure white of silver and the bluishgray tint of lead: bismuth, it is true, has a pinkish color, and calcium and strontium a yellowish tint, but these tints are very feeble.

The differences of *density* are very wide, passing from lithium, potassium, and sodium, which are lighter than water, to platinum, which is more than twenty-one times heavier than an equal bulk of that liquid.

	Density.	Temp. C.		Density.	Temp. C.
Osmium	22.48		Cadmium	8.66	
Iridium	22.40		Molybdenum . .	8.63	
Platinum	21.50		Cobalt	8.51	
Gold	19.265	13°	Manganese . . .	8.02	
Uranium	18.40	0°	Iron	7.79	
Tungsten	18.30		Tin	7.29	13°
Mercury	13.60		Zinc	6.915	
Rhodium	12.1		Antimony . . .	6.72	14°
Thallium	11.8		Aluminium . . .	2.67	
Palladium . . .	11.4	22.5°	Magnesium . . .	1.74	5°
Lead	11.37	14°	Calcium	1.58	
Silver	10.47	13°	Rubidium . . .	1.52	
Bismuth	9.82	12°	Sodium	0.97	10°
Copper	8.95		Potassium . . .	0.865	
Nickel	8.82		Lithium	6.59	

The property of *malleability*, or power of extension under the hammer or between the rollers of the flatting-mill, is possessed by certain of the metals to a very great extent. Gold-leaf is a remarkable example of the tenuit to which a malleable metal may be brought by suitable means. The gilding on silver wire used in the manufacture of gold lace is even thinner, and yet presents an unbroken surface. Silver may be beaten out very thin—copper also, but to an inferior extent; tin and platinum are easily rolled out into foil; iron, palladium, lead, nickel, cadmium, and metals of the alkalies, and mercury when solidified, are also malleable. Zinc may be placed midway between the malleable and brittle division; then, perhaps, bismuth: and, lastly, such metals as antimony, which are altogether destitute of malleability.

The density of malleable metals is usually very sensibly increased by pressure or blows, and the metals themselves are rendered much harder, with a tendency to brittleness. This condition is destroyed and the former soft state restored by the operation of *annealing*, which consists in heating the metal to redness out of contact with air (if it will bear that temperature without fusion), and cooling it, quickly or slowly according to the circumstances of the case. After this operation, it is found to possess its original density.

Fig. 146.

Ductility is a property distinct from the last, inasmuch as it involves the principle of tenacity, or power of resisting tension. The art of wire-drawing is one of great antiquity; it consists in drawing rods of metal through a succession of trumpet-shaped holes in a steel plate, each being a little smaller than its predecessor, until the requisite degree of fineness is attained. The metal often becomes very hard and rigid in this process, and is then liable to break: this is remedied by annealing. The order of tenacity among the metals susceptible of being easily drawn into wire is the following: it is determined by observing the weights required to break asunder wires drawn through the same orifice of the plate:—

Iron.   Gold.
Copper.   Zinc.
Platinum.   Tin.
Silver.   Lead.

Metals differ as much in *fusibility* as in density. The following table will give an idea of their relations to heat:—

		Melting points.
	Mercury	$-39.44°$
	Gallium	$+30$
	Rubidium	$+38.5$
	Potassium	62.5
	Sodium	95.6
Fusible below a	Lithium	180
red heat.	Tin	235
	Bismuth	270
	Thallium	294
	Cadmium	315
	Lead	334
	Zinc	423
	Antimony	425

Melting points.

Silver 1000		
Copper 1090		
Gold 1102		
Cast iron 1530		

Infusible below a red heat. {

Pure iron,
Nickel,
Cobalt,   } Highest heat of forge.
Manganese,
Palladium,

Molybdenum,
Uranium,   } Agglomerate, but do not
Tungsten,   melt, in the forge.
Chromium,

Titanium,
Cerium,
Osmium,   Infusible in ordinary blast-
Iridium,   } furnaces; fusible by oxy-
Rhodium,   hydrogen blowpipe.
Platinum,
Tantalum,

Some metals acquire a pasty or adhesive state before becoming fluid : this is the case with iron and platinum and with the metals of the alkalies. It is this peculiarity which confers the very valuable property of welding, by which pieces of iron and steel are united with solder, and the finely-divided metallic sponge of platinum is converted into a solid and compact bar.

Some metals are *volatile*, and this character would perhaps be exhibited by all could sufficiently high temperatures be obtained. Mercury boils and distils below a red heat; potassium, sodium, zinc, magnesium, and calcium, rise in vapor when heated to bright redness.

CHEMICAL RELATIONS OF THE METALS.

METALLIC combinations are of two kinds—namely, those formed by the union of metals among themselves, which are called *alloys*, or, where mercury is concerned, *amalgams*, and those generated by combination with the non-metallic elements, as oxides, chlorides, sulphides, etc. In this latter case the metallic characters are almost invariably lost.

Alloys.—Most metals are probably, to some extent, capable of existing in a state of combination with each other in definite proportions; but it is difficult to obtain these compounds in the separate state, since they dissolve in all proportions in the melted metals, and do not generally differ so widely in their melting points from the metals they may be mixed with as to be separated by crystallization in a definite form. Exceptions to this rule are met with in the cooling of argentiferous lead, and in the crystallization of brass and of gun-metal.

The chemical force capable of being exerted between different metals is for the most part very feeble, and the consequent state of combination is therefore very easily disturbed by the influence of other forces. The sta-

13

bility of such metallic compounds is, however, greater in proportion to the general chemical dissimilarity of the metals they contain. But in all cases of combination between metals, the alteration of physical characters, which is the distinctive feature of chemical combination, does not take place to any great extent. The most unquestionable compounds of metals with metals are still metallic in their general physical characters, and there is no such transmutation of the individuality of their constituents as takes place in the combination of a metal with oxygen, sulphur, chlorine, etc. The alteration of characters in alloys is generally limited to the color, degree of hardness, tenacity, etc., and it is only when the constituent metals are capable of assuming opposite chemical relations that these compounds are distinguished by great brittleness. The formation of actual chemical compounds, in some cases, when two metals are melted together, is indicated by several phenomena, viz., the evolution of heat, as in the case of platinum and tin, copper and zinc, etc.

The density of alloys differs from that of mere mixtures of the metals. In the solidification of alloys, the temperature does not always fall uniformly, but often remains stationary at particular degrees, which may be regarded as the solidifying points of the compounds then crystallizing. Tin and lead melted together in any proportions always form a compound which solidifies at 187° C. (369° F.). The melting point of an alloy is often very different from the point of solidification, and it is generally lower than the mean melting point of the constituent metals.

But though metals may combine when melted together, it is doubtful whether they remain combined after the solidification of the mass, and the wide difference between the melting and solidifying points of certain alloys appears to indicate that the existence of these compounds is limited to a certain range of temperature. Matthiessen * regards it as probable that the condition of an alloy of two metals in the liquid state may be either that of—1. A solution of one metal in another; 2. Chemical combination; 3. Mechanical mixture; or, 4. A solution or mixture of two or all of the above.; and that similar differences may exist as to its condition in the solid state.

The chemical action of reagents upon alloys is something very different from their action upon metals in the separate state: thus, platinum alloyed with silver is readily dissolved by nitric acid, but is not affected by that acid when unalloyed. On the contrary, silver, which in the separate state is readily dissolved by nitric acid, is not dissolved by it when alloyed with gold in proportions much less than one-fourth of the alloy by weight.

Compounds of Metals with Metalloids—Classification of Metals.

A classification of the metals according to their quantivalence or atomicity, is given in the table on page 256. There are, however, several metals, especially among those of rare occurrence, whose position in the series is by no means definitely fixed; and partly on this account, but chiefly because metals of equal quantivalence or combining capacity do not, as a rule, resemble one another in their physical and chemical characters so closely as non-metallic elements of equal combining capacity (e. g., Cl, Br, I, F, and O, S, Se, Te, etc.), it is found advisable, in describing the metals, to classify them according to their general agreement in physical and chemical properties rather than according to their quantivalence.

* British Association Reports, 1863, p. 97.

The following arrangement based upon this principle does not, however, deviate greatly from the order of quantivalence:—

1. METALS OF THE ALKALIES:—
    Potassium—Sodium—Lithium—Rubidium—Cæsium.

2. METALS OF THE ALKALINE EARTHS:—
    Calcium—Strontium—Barium.

3. MAGNESIUM GROUP:—
    Beryllium—Magnesium—Zinc—Cadmium.

4. LEAD GROUP:—
    Lead—Thallium.

5. COPPER GROUP:—
    Copper—Silver—Mercury.

6. YTTRIUM GROUP:—
    Yttrium—Erbium—Terbium—Yterbium—Scandium—Decipium—Lanthanum—Cerium—Didymium.

7. ALUMINIUM GROUP:—
    Aluminium—Indium—Gallium.

8. IRON GROUP:—
    Manganese—Iron—Cobalt—Nickel.

9. CHROMIUM GROUP:—
    Chromium—Molybdenum—Tungsten—Uranium.

10. TIN GROUP:—
    Tin—Titanium—Zirconium—Thorium.

11. ANTIMONY GROUP:—
    Vanadium—Antimony—Bismuth—Tantalum—Niobium.

12. PLATINUM GROUP:—
    Gold—Platinum—Palladium—Rhodium—Iridium—Ruthenium—Osmium.

The degrees of quantivalence or combining capacity of the metals in these several groups are for the most part less easy to determine than those of the non-metallic elements. All the latter unite with hydrogen, forming volatile compounds, the vapor-densities of which are easily determined, and afford trustworthy data for fixing the molecular weight and constitution of these compounds (p. 131). Hydrogen silicide, for example, has a vapor-density of 16: consequently its molecular weight is 32 (p. 246), and, as analysis shows the compound to contain 7 pts. by weight of silicon to 1 pt. of hydrogen, its molecule must contain 28 pts. Si and 4H, and its molecular formula must be SiH_4, showing that silicon is a tetrad. But with the metals this mode of proceeding is not available: for only one of them, viz., antimony, forms a volatile compound with hydrogen; and if we endeavor to fix the quantivalence of a metal by its mode of combination with chlorine or other univalent element, we find that the results are not always accordant. With regard to some of the groups, indeed, this mode of determining the quantivalence of a metal appears to be satisfactory. Thus the metals of the first, second, third, sixth, tenth, and eleventh groups form chlorides related to one another as represented by the formulæ:—

1.	2.	3.	6.	10.	11.
RCl	RCl_2	RCl_2	RCl_3	RCl_4	RCl_5

but in the chromium group (9) anomalies present themselves; thus molybdenum and tungsten form pentachlorides, whence it might be inferred that they are pentads: but tungsten also forms a hexchloride, and both these metals, in their oxygen-compounds, exhibit so marked an analogy to sulphur that they must be regarded as hexads.

Moreover, we have no means of ensuring that the formulæ assigned to these several chlorides actually represent their molecules; for the only trustworthy method of determining the weight of a molecule is by means of its vapor-density, and in the case of the greater number of metallic chlorides (also bromides and iodides), this determination cannot be made, on account of the very high temperatures required to volatilize them.

In many cases, indeed, it seems probable that the molecular constitution of metallic chlorides, bromides, etc., should be represented, not by formulæ containing single atoms of the respective metals, like those above given, but by multiples thereof,—potassium chloride, for example, by the double formula K_2Cl_2 or $Cl — K — K — Cl$, into which the metal enters as a dyad, or even by some higher multiple, for example by

$$K_4Cl_4 = \begin{matrix} Cl — K — K — Cl \\ | \quad\; | \\ Cl — K — K — Cl \end{matrix} \quad \text{or} \quad \begin{matrix} Cl — K — K — Cl \\ \| \quad\; \| \\ Cl — K — K — Cl \end{matrix}, \text{ etc.,}$$

the potassium appearing as a triad or a tetrad, according to the manner in which we may imagine its atoms to be linked together. For, since both potassium and chlorine are easily volatile bodies, the chloride, if correctly represented by the simpler formula $K.Cl$, might be expected to volatilize at a comparatively low temperature, lower, for example, than those required to volatilize the more complex and heavier molecules $HgCl_2$ and $SbCl_3$, whereas it actually requires a strong red heat to convert it into vapor; and similar remarks might be made with regard to many other haloïd salts of the metals. In stating, therefore, that potassium chloride is represented by the formula KCl, and calcium chloride by $CaCl_2$, all that we mean to say is that for every atom of metal in these compounds, the first contains one atom of chlorine, and the second two atoms; and in this sense—whatever may be the number of atoms contained in the molecule—potassium is certainly univalent, or equivalent to one atom of hydrogen, and calcium bivalent, or equivalent to two atoms of hydrogen.

The metals of the alkalies and alkaline earths, on account of their inferior density, are often called light metals; the others, heavy metals.

Metallic Chlorides.—All metals combine with chlorine, and most of them in several proportions, as above indicated, forming compounds which may be regarded as derived from one or more molecules of hydrochloric acid, by substitution of a metal for an equivalent quantity of hydrogen; thus:—

From HCl are derived monochlorides like KCl
" H_2Cl_2 " dichlorides " $BaCl_2$
" H_3Cl_3 " trichlorides " $AuCl_3$
" H_4Cl_4 " tetrachlorides " $SnCl_4$, etc., etc.

Hydrochloric acid may, in fact, be regarded as the type of chlorides in general.

Several chlorides occur as natural products. Sodium chloride, or common salt, occurs in enormous quantities, both in the solid state as rock-salt, and dissolved in sea-water and in the water of rivers and springs. Potassium chloride occurs in the same forms, but in smaller quantity; the chlorides of lithium, cæsium, rubidium, and thallium also occur in small quantities in certain spring waters. Mercurous chloride, Hg_2Cl_2, and silver chloride, AgCl, occur as natural minerals.

1. Chlorides are generally prepared by one or other of the following processes: (1.) By acting upon the metal with chlorine gas. Antimony pentachloride and copper dichloride are examples of chlorides sometimes produced in this manner. The chlorides of gold and platinum are usually prepared by acting upon the metals with nascent chlorine, developed by a mixture of hydrochloric and nitric acids (p. 196). Sometimes, on the other hand, the metal is in the nascent state, as when titanic chloride is formed by passing a current of chlorine over a heated mixture of charcoal and titanic oxide. The chlorides of aluminium and chromium may be obtained by similar processes.

2. Chlorine gas, by its action upon metallic oxides, drives out the oxygen, and unites with the respective metals to form chlorides. This reaction sometimes takes place at ordinary temperatures, as is the case with silver oxide; sometimes only at a red heat, as is the case with the oxides of the alkalimetals and alkaline earth-metals. The hydroxides and carbonates of these last metals, when dissolved or suspended in hot water and treated with excess of chlorine, are converted, chiefly into chlorides, partly into chlorates.

3. Many metallic chlorides are prepared by acting upon the metals with hydrochloric acid. Zinc, cadmium, iron, nickel, cobalt, and tin dissolve readily in hydrochloric acid, with liberation of hydrogen; copper only in the strong boiling acid; silver, mercury, palladium, platinum, and gold not at all. Sometimes the metal is substituted, not for hydrogen, but for some other metal. Stannous chloride, for instance, is frequently made by distilling metallic tin with mercuric chloride; thus: $HgCl_2 + Sn = SnCl_2 + Hg$.

4. By dissolving a metallic oxide, hydroxide, or carbonate in hydrochloric acid.

All monochlorides and dichlorides are soluble in water, excepting silver chloride, AgCl, and mercurous chloride, Hg_2Cl_2; lead chloride, $PbCl_2$, is sparingly soluble; these three chlorides are easily formed by precipitation. Many metallic chlorides dissolve also in alcohol and in ether.

Most monochlorides, dichlorides, and trichlorides volatilize at high temperatures without decomposition; the higher chlorides when heated give off part of their chlorine. Some chlorides which resist the action of heat alone are decomposed by ignition in the air, yielding metallic oxides and free chlorine: this is the case with the dichlorides of iron and manganese; but most dichlorides remain undecomposed, even in this case. All metallic chlorides, excepting those of the alkali-metals and earth-metals, are decomposed at a red heat by hydrogen gas, with formation of hydrochloric acid: in this way metallic iron may be obtained in fine cubical crystals. Silver chloride, placed in contact with metallic zinc or iron under dilute sulphuric or hydrochloric acid, is reduced to the metallic state by the nascent hydrogen.

Sulphuric, phosphoric, boric, and arsenic acids decompose most metallic

25 *

chlorides, sometimes at ordinary, sometimes at higher temperatures. All metallic chlorides, heated with lead dioxide or manganese dioxide and sulphuric acid, give off chlorine, e. g.:

$$2NaCl + MnO_2 + 2H_2SO_4 = Na_2SO_4 + MuSO_4 + 2H_2O + Cl_2.$$

Chlorides distilled with sulphuric acid and potassium chromate, yield a dark bluish-red distillate of chromic oxychloride. Some metallic chlorides are decomposed by *water*, forming hydrochloric acid and an oxychloride, e. g.: $BiCl_3 + H_2O = 2HCl + BiClO$. The chlorides of antimony and stannous chloride are decomposed in a similar manner.

All soluble chlorides give with solution of *silver nitrate* a white precipitate of silver chloride, easily soluble in ammonia, insoluble in nitric acid. With *mercurous nitrate*, they yield a white curdy precipitate of mercurous chloride, blackened by ammonia; and with *lead-salts*, not too dilute, a white precipitate of lead chloride, soluble in excess of water.

Metallic chlorides unite with each other and with the chlorides of the non-metallic elements, forming such compounds as potassium chloromercurate, $2KCl,HgCl_2$, sodium chloroplatinate, $2NaCl,PtCl_4$, potassium chloriodate, KCl,ICl_3, etc. Metallic chlorides combine in definite proportions with ammonia and organic bases: the chlorides of platinum form with ammonia the compounds $2NH_3,PtCl_2$, $4NH_3,PtCl_2$, $2NH_3,PtCl_4$, and $4NH_3,PtCl_4$; mercuric chloride forms with aniline the compound $2C_6H,N,HgCl_2$, etc.

Chlorides also unite with oxides and sulphides, forming *oxychlorides* and *sulphochlorides*, which may be regarded as chlorides having part of their chlorine replaced by an equivalent quantity of oxygen or sulphur (Cl_2 by O or S). Bismuth, for example, forms an oxychloride having the composition $BiClO$ or $BiCl_3,Bi_2O_3$.

Bromides.—Bromine unites directly with most metals, forming compounds analogous in composition to the chlorides, and resembling them in most of their properties. The bromides of the alkali-metals occur in seawater and in many saline springs; silver bromide occurs as a natural mineral. Nearly all bromides are soluble in water, and may be formed by treating an oxide, hydroxide, or carbonate with hydrobromic acid, the solutions when evaporated giving off water for the most part, and leaving a solid metallic bromide; some of them, however—namely, the bromides of magnesium, aluminium, and the other earth-metals—are more or less decomposed by evaporation, giving off hydrobromic acid, and leaving a mixture of metallic bromide and oxide. Silver bromide and mercurous bromide are insoluble in water, and lead bromide is very sparingly soluble: these are obtained by precipitation.

Metallic bromides are solid at ordinary temperatures; most of them fuse at a moderate heat, and volatilize at higher temperatures. The bromides of gold and platinum are decomposed by mere exposure to heat; many others give up their bromine when heated in contact with the air. *Chlorine*, with the aid of heat, drives out the bromine and converts them into chlorides. *Hydrochloric acid* also decomposes them at a red heat, giving off hydrobromic acid. Strong *sulphuric* or *nitric acid* decomposes them, with evolution of hydrobromic acid, which, if the sulphuric or nitric acid is concentrated, and in excess, is partly decomposed, with separation of bromine and formation of sulphurous oxide or nitrogen dioxide. Bromides heated with *sulphuric acid* and *manganese dioxide* or *potassium chromate* give off free bromine.

Bromides in solution are easily decomposed by chlorine, either in the form of gas or dissolved in water, the liquid acquiring a red or reddish-yellow color, according to the quantity of bromine present; and on agitating the liquid with ether, that liquid dissolves the bromine, forming a red solution, which rises to the surface.

Soluble bromides give with *silver . nitrate* a white precipitate of silver bromide, greatly resembling the chloride, but much less soluble in ammonia, insoluble in hot nitric acid. *Mercurous nitrate* produces a yellowish-white precipitate; and *lead acetate* a white precipitate much less soluble in water than the chloride. *Palladium nitrate* produces in solutions of bromides not containing chlorine a black precipitate of bromide. Palladium chloride produces no precipitate; neither does the nitrate, if soluble chlorides are present.

Bromides unite with each other in the same manner as chlorides; also with oxides, sulphides, and ammonia.

Iodides.—These compounds are obtained by processes similar to those which yield the chlorides and bromides. Many metals unite directly with iodine. Potassium and sodium iodides exist in sea-water and in many salt-springs; silver iodide occurs as a natural mineral.

Metallic iodides are analogous to the bromides and chlorides in composition and properties. But few of them are decomposed by heat alone; the iodides of gold, silver, platinum, and palladium, however, give up their iodine when heated.

Most metallic iodides are perfectly soluble in water; but lead iodide is very slightly soluble, and the iodides of mercury and silver are quite insoluble.

Solutions of iodides evaporated out of contact of air, generally leave anhydrous metallic iodides, which partly separate in the crystalline form before the water is wholly driven off. The iodides of the earth-metals, however, are resolved, on evaporation, into the earthy oxides and hydriodic acid, which escapes. A very small quantity of *chlorine* colors the solution yellow or brown, by partial decomposition; and a somewhat larger quantity takes up the whole of the metal, forming a chloride, and separates the iodine, which then gives a blue color with starch; a still larger quantity of chlorine gives the liquid a paler color, and converts the separated iodine into trichloride of iodine, which does not give a blue color with starch, and frequently enters into combination with the metallic chloride produced. Strong *sulphuric acid* and somewhat concentrated *nitric acid* color the solution yellow or brown; and if the quantity of the iodide is large, and the solution much concentrated or heated, they liberate iodine, which partly escapes in violet vapors. *Starch* mixed with the solution, even if it be very dilute, is turned blue—permanently when the decomposition is effected by sulphuric acid; for a time only when it is effected by nitric acid, especially if that acid be added in large quantity.

The aqueous solution of an iodide gives a brown precipitate with salts of *bismuth;* orange-yellow with *lead* salts; dirty white with *cuprous* salts, also with *cupric* salts, especially on the addition of sulphurous acid; greenish-yellow with *mercurous* salts; scarlet with *mercuric* salts; yellowish-white with *silver* salts; lemon-yellow with *gold* salts; brown with *platinic* salts—first, however, turning the liquid dark brown-red; and black with salts of *palladium,* even when extremely dilute. All these precipitates consist of metallic iodides, many of them soluble in excess of the soluble iodide; the silver precipitate is insoluble in nitric acid and very little soluble in ammonia.

Metallic iodides unite with one another, forming double iodides, analogous to the double chlorides; they also absorb ammonia gas in definite proportions. Some of them, as those of antimony and tellurium, unite with the oxides of the corresponding metals, forming oxyiodides.

Fluorides.—These compounds are formed—1. By heating hydrofluoric acid with certain metals. 2. By the action of that acid on metallic oxides. 3.' By heating electro-negative metals—antimony, for example—with fluoride of lead or fluoride of mercury. 4. Volatile metallic fluorides may be prepared by heating fluor-spar with sulphuric acid and the oxide of the metal.

Fluorides have no metallic lustre; most of them are easily fusible, and for the most part resemble the chlorides. They are not decomposed by ignition, either alone or when mixed with charcoal. When ignited in contact with the air, in a flame which contains aqueous vapor, many of them are converted into oxides, while the fluoride is given off as hydrofluoric acid. All fluorides are decomposed by *chlorine*, and converted into chlorides. They are not decomposed by *phosphoric oxide* unless silica is present. They are decomposed at a gentle heat by strong *sulphuric acid*, with formation of metallic sulphate and evolution of hydrofluoric acid.

The fluorides of tin and silver are easily soluble in water; those of potassium, sodium, and iron are sparingly soluble; those of strontium and cadmium very slightly soluble, and the rest insoluble. The solutions of ammonium, potassium, and sodium fluoride have an alkaline reaction. The aqueous solutions of fluorides corrode glass vessels in which they are kept or evaporated. They form with soluble *calcium salts* a precipitate of calcium fluoride in the form of a transparent jelly, which is scarcely visible, because its refractive power is nearly the same as that of the liquid; the addition of ammonia makes it plainer. This precipitate, if it does not contain silica, dissolves with difficulty in hydrochloric or nitric acid, and is reprecipitated by ammonia. The aqueous fluorides give a pulverulent precipitate with lead acetate.

The fluorides of antimony, arsenic, chromium, mercury, niobium, osmium, tantalum, tin, titanium, tungsten, and zinc are volatile without decomposition.

Fluorine has a great tendency to form double salts, consisting of a fluoride of a basic or positive metal united with the fluoride of hydrogen, boron, silicon, tin, titanium, zirconium, etc., *e. g.*—

Potassium hydrofluoride	. . .	$KHF_2 = KF,HF$
Potassium borofluoride	. . .	$KBF_4 = KF,BF_3$
Potassium silicofluoride,	. . .	$K_2SiF_6 = 2KF,SiF_4$
Potassium titanofluoride	. . .	$K_2TiF_6 = 2KF,TiF_4$
Potassium stannofluoride	. . .	$K_2SnF_6 = 2KF,SnF_4$
Potassium zircofluoride	. . .	$K_2ZrF_6 = 2KF,ZrF_4$

The four classes of compounds just described, the chlorides, bromides, iodides, and fluorides, form a group often designated as **haloïd compounds** or **haloïd*** salts, from their analogy to sodium chloride or sea-salt, which may be regarded as a type of them all. The elements, chlorine, bromine, iodine, and fluorine, are called halogens.

Cyanides.—Closely related to these haloïd compounds are the cyanides, formed by the union of metals with the group CN, cyanogen, CN (p. 235).

Some metals—potassium among the number—are converted into cyanides by heating them in cyanogen gas or vapor of hydrocyanic acid. The cyanides of the alkali-metals are also formed (together with cyanates) by passing cyanogen gas over the heated hydroxides or carbonates of the same metals; potassium cyanide also, by passing nitrogen gas over a mixture of charcoal and hydroxide or carbonate of potassium at a bright red heat.

* From ἅλς, the sea.

Cyanides are formed abundantly when nitrogenous organic compounds are heated with fixed alkalies. Other modes of formation will be mentioned hereafter.

The cyanides of the alkali-metals, and of barium, strontium, calcium, magnesium, and mercury, are soluble in water, and may be produced by treating the corresponding oxides or hydroxides with hydrocyanic acid. Nearly all other metallic cyanides are insoluble, and are obtained by precipitation from the soluble cyanides.

The cyanides of the alkali-metals sustain a red heat without decomposition, provided air and moisture be excluded. The cyanides of many of the heavy metals, as lead, iron, cobalt, nickel, and copper, under these circumstances, give off all their nitrogen as gas, and leave a metallic carbonate; mercuric cyanide is resolved into mercury and cyanogen gas; silver cyanide gives off half its cyanogen as gas. Most cyanides, when heated with dilute acids, give off their cyanogen as hydrocyanic acid.

Cyanides have a strong tendency to unite with one another, forming double cyanides. The most important of these are the double cyanides of iron and potassium, namely, *potassio-ferrous cyanide*, $Fe''K_4(CN)_6$, commonly called yellow prussiate of potash; and *potassio-ferric cyanide*, $Fe''K_3(CN)_6$, commonly called red prussiate of potash. Both these are splendidly crystalline salts, which dissolve easily in water, and form highly characteristic precipitates with many metallic salts. These salts, with the other cyanides, will be more fully described under "Organic Chemistry;" but they are mentioned here, on account of their frequent use in the qualitative analysis of metallic solutions.

O x i d e s.—All metals combine with oxygen, and most of them in several proportions. In almost all cases oxides are formed corresponding in composition with the chlorides, one atom of oxygen taking the place of two atoms of chlorine. Many metals also form oxides to which no chlorine analogues are known; thus lead, which forms only one chloride, $PbCl_2$, forms, in addition to the monoxide, PbO, a dioxide, PbO_2, besides oxides of intermediate composition; osmium also, the highest chloride of which is $OsCl_4$, forms, in addition to the dioxide, a trioxide, and a tetroxide. This arises from the fact that any number of atoms of oxygen or other dyad element may enter into a compound without disturbing the balance of equivalency (p. 254).

Just as chlorides are derived by substitution from hydrochloric acid, HCl (p. 293), so likewise may oxides be derived from one or more molecules of water, H_2O; but as the molecule of water contains two hydrogen-atoms, the replacement of the hydrogen may, as already explained (p. 239), be either total or partial, the product in the first case being an anhydrous metallic oxide, and in the second a hydrated oxide or hydroxide, in which the oxygen is associated both with hydrogen and with metal; in this manner the following hydroxides and anhydrous oxides may be constituted:

Type.	Hydroxides.	Oxides.
H_2O	KHO	K_2O
		BaO
H_4O_2	BaH_2O_2	SnO_2
	$BiHO_2$	
H_6O_3	$AsHO_3$	Sb_2O_3
	SnH_2O_3	WO_3
H_8O_4	ZrH_4O_4	
$H_{10}O_5$		Sb_2O_5.

13*

It may be observed that the hydroxides of artiad metals contain the elements of a molecule of the corresponding anhydrous oxide, and of one or more molecules of water, and may, therefore, also be regarded as hydrates; thus—

Barium hydroxide or hydrate	.	.	$BaH_2O_2 = BaO,H_2O$
Stannic " "	.	.	$SnH_2O_3 = SnO_2,H_2O$
Zirconium " "	.	.	$ZrH_4O_4 = ZrO_2,2H_2O.$

But the hydroxide of a perissad metal contains in its molecule only half the number of atoms required to make up a molecule of oxide together with a molecule of water; thus—

| Potassium hydroxide | . | . | . | $KHO = \frac{1}{2}(K_2O,H_2O)$ |
| Bismuth " | . | . | . | $BiHO_2 = \frac{1}{2}(Bi'''_2O_3,H_2O).$ |

These perissad hydroxides cannot, therefore, be correctly regarded as hydrates, that is, as compounds of anhydrous oxide and water.

Many metallic oxides occur as natural minerals, and some, especially those of iron, tin, and copper, in large quantities, forming ores from which the metals are extracted.

All metals, except gold, platinum, iridium, rhodium, and ruthenium, are capable of uniting directly with oxygen. Some, as potassium, sodium, and barium, oxidize rapidly on exposure to the air at ordinary temperatures, and decompose water with energy. Most metals, however, when in the massive state, remain perfectly bright and unacted on in dry air or oxygen gas, but oxidize slowly when moisture is present; such is the case with iron, zinc, and lead. Some of the ordinarily permanent metals, when in a very finely-divided state, as lead when obtained by ignition of its tartrate, and iron reduced from its oxide by ignition in hydrogen gas, take fire and oxidize spontaneously as soon as they come in contact with the air. Lead, iron, copper, and the volatile metals, antimony, zinc, cadmium, and mercury, are converted into oxide when heated in air or oxygen. Many metals, especially at a red heat, are readily oxidized by water or steam. A very general method of preparing metallic oxides is to subject the corresponding hydroxides, carbonates, nitrates, sulphates, or any oxygen-salts containing volatile acids, to the action of heat.

Oxides are for the most part opaque earthy bodies, destitute of metallic lustre. The majority of them are fusible; those of lead and bismuth at a low red heat; those of copper and iron at a white heat; those of barium and aluminium before the oxyhydrogen blowpipe; while calcium oxide or lime does not fuse at any temperature to which it has yet been subjected. Oxides are, for the most part, much less fusible than the uncombined metals. Osmium tetroxide and antimony trioxide are readily volatile.

A greater or lesser degree of heat effects the decomposition of many metallic oxides. Those of gold, platinum, silver, and mercury are reduced to the metallic or reguline state by an incipient red heat. At a somewhat higher temperature, the higher oxides of barium, cobalt, nickel, and lead are reduced to the state of monoxides; while the trimetallic tetroxides of manganese and iron, Mn_3O_4 and Fe_3O_4, are produced by exposing manganese dioxide, MnO_2, and iron sesquioxide, Fe_2O_3, respectively to a still stronger heat. By gentle ignition, arsenic pentoxide is reduced to the state of trioxide, and chromium trioxide to sesquioxide.

The superior oxides of the metals are easily reduced to a lower state of oxidation by the action of a current of *hydrogen gas* at a more or less elevated temperature. At a higher degree of heat, hydrogen gas will transform to the reguline state all metallic oxides except the sesquioxides of aluminium and chromium, and the monoxides of manganese, magnesium, barium, strontium, calcium, lithium, sodium, and potassium. The temperature necessary to enable hydrogen to effect the decomposition of some oxides is comparatively low. Thus metallic iron may be reduced from its oxides by hydrogen gas at a heat considerably below redness. *Carbon*, at a red or white heat, is a still more powerful deoxidizing agent than hydrogen, and seems to be capable of completely reducing all metallic oxides whatsoever. The oxidizable metals in general act as reducing agents.

Chlorine decomposes all metallic oxides, except those of the earth-metals, converting them into chlorides, and expelling the oxygen. With silver oxide this reaction takes place at ordinary temperatures; with the alkalies and alkaline earths, at a full red heat. *Sulphur*, at high temperatures, can decompose most metallic oxides; with many oxides—those of silver, mercury, lead, and copper, for instance—metallic sulphides and sulphur dioxide are produced; with the highly basylous oxides the products are metallic sulphate and sulphide. There are some oxides upon which sulphur exerts no action. Of these the principal are magnesia, alumina, chromic, stannic, and titanic oxides. By boiling sulphur with soluble hydroxides, mixtures of polysulphide and thiosulphate are produced. With the exception of magnesia, alumina, and chromic oxide, most metallic oxides can absorb sulphuretted hydrogen to form metallic sulphide or hydrosulphide and water.

Oxygen-salts or **Oxysalts.**—It has been already explained in the Introduction that oxides may be divided into three classes, *acid, neutral,* and *basic,* the first and third being capable of uniting with one another in definite proportions, and forming compounds called s a l t s. The most characteristic of the acid oxides are those of certain metalloids, as nitrogen, sulphur, and phosphorus, which unite readily with water or the elements of water, forming compounds called o x y g e n - a c i d s, distinguished by sour taste, solubility in water, and the power of reddening certain vegetable blue colors. The most characteristic of the basic oxides, on the other hand, are those of the alkali-metals and alkaline earth-metals (p. 291), which likewise dissolve in water, but form alkaline solutions, possessing in an eminent degree the power of *neutralizing* acids and forming salts with them. The same power is exhibited more or less by the monoxides of most other metals, as zinc, iron, copper, manganese, etc., and by the sesquioxides of aluminium, iron, chromium, and others. The higher oxides of several of these metals —the trioxides of chromium, for example—exhibit acid characters, being capable of forming salts with the more basic oxides; and some metals, as antimony, niobium, and tantalum, form only acid oxides.

In some cases salts are formed by the direct combination of an acid and a basic oxide. Thus, when vapor of sulphuric oxide, SO_3, is passed over red-hot barium oxide, BaO, the two combine together and form barium sulphate, BaO,SO_3 or $BaSO_4$. Silicic oxide, SiO_2, phosphoric oxide, P_2O_5, boric oxide, B_2O_3, and other acid oxides capable of withstanding a high temperature without decomposing or volatilizing, likewise unite with basic oxides when heated with them, and form salts.

But in the majority of cases metallic salts are formed by substitution, or interchange of a metal for hydrogen, or of one metal for another. It is clear, indeed, that any metallic salt (zinc sulphate, ZnO,SO_3, for example) may be derived from the corresponding acid or hydrogen salt (H_2O,SO_3) by substitution of a metal for an equivalent quantity of hydrogen. Accordingly, metallic salts are frequently produced by the action of an acid on a metal or a metallic oxide or hydroxide, thus—

$$
\begin{aligned}
&(1.) \quad H_2SO_4 + Zn = ZnSO_4 + H_2. \\
&(2.) \quad 2HNO_3 + Ag_2O = 2AgNO_3 + H_2O. \\
&(3.) \quad HNO_3 + KHO = KNO_3 + H_2O.
\end{aligned}
$$

In the instances represented by these equations, the metallic salts formed are soluble in water. Insoluble salts are frequently prepared by interchange of the metals between two soluble salts, barium nitrate and sodium sulphate, for example, thus—

$$
(4.) \quad Ba(NO_3)_2 + Na_2SO_4 = BaSO_4 + 2NaNO_3.
$$

In this case the barium sulphate, being insoluble, is precipitated, while the sodium nitrate remains in solution.

In all these reactions, hydrochloric acid or a metallic chloride might be substituted for the oxygen-acid or oxygen-salt, without the slightest alteration in the mode of action, the product formed in each case being a chloride instead of a nitrate or sulphate; thus:—

$$
\begin{aligned}
&(1)' \quad 2HCl + Zn = ZnCl_2 + H_2 \\
&(2)' \quad 2HCl + Ag_2O = 2AgCl + H_2O \\
&(3)' \quad HCl + KHO = KCl + H_2O \\
&(4)' \quad AgNO_3 + NaCl = AgCl + NaNO_3.
\end{aligned}
$$

From all these considerations it appears that oxygen-salts may be regarded, either as compounds of acid oxides with basic oxides, or as analogous in composition to chlorides,—that is to say, as compounds of a metal with a radicle or group of elements, such as NO_3 (*nitrione*) in the nitrates, SO_4 (*sulphione*), in the sulphates, discharging functions similar to those of chlorine, and capable, like that element, of passing unchanged from one compound to another.

For many years, indeed, it was a subject of discussion among chemists, whether the former or the latter of these views should be regarded as representing the *actual* constitution of oxygen-salts. Berzelius divided salts into two classes: 1. H a l o ï d s a l t s, comprising, as already mentioned, the chlorides, bromides, iodides, and fluorides, which are compounds of a metal with a monad metallic element.—2. A m p h i d s a l t s, consisting of an acid or electro-negative oxide, sulphide, selenide, or telluride, with a basic or electro-positive compound of the same kind; such as potassium arsenate, $3K_2O,As_2O_5$; potassium sulpharsenate, $3K_2S,As_2S_5$; potassium seleniophosphate, $2K_2Se,P_2Se_5$, etc.

Davy, on the other hand, observing the close analogy between the reactions of chlorides, on the one hand, and of oxygen-salts, such as sulphates, nitrates, etc., on the other, suggested that the latter might be regarded, like the former, as compounds of metals with acid or electro-negative radicles, the only difference being, that in the former the acid radicle is an elementary body, Cl, Br, etc., whereas in the former it is a compound, as SO_4, NO_5, PO_4, etc. This was called the *binary theory of salts;* it was supported by many ingenious arguments by its proposer and several contemporary chemists; in later years also by Liebig, and by Daniell and Miller, who observed that the mode of decomposition of salts by the electric current is more easily represented by this theory than by the older one (p. 276).

At the present day, the relative merits of these two theories are not regarded as a point of very great importance. Chemists, in fact, no longer attempt to construct formulæ which shall represent the actual arrangement of atoms in a compound, the formulæ now in use being rather intended to exhibit, first, the balance of neutralization of the units of equivalency or combining capacity of the several elements contained in a compound (p. 249); and, secondly, the manner in which any compound or group of atoms splits up into subordinate groups under the influence of different reagents. According to the latter view, a compound containing three or more elementary atoms may be represented by different formulæ corresponding with the several ways in which it decomposes. Thus hydrogen sulphate or sulphuric acid, H_2SO_4, may be represented by either of the following formulæ:—

1. H_2SO_4, which represents the separation of hydrogen and formation of a metallic sulphate, by the action of zinc, etc.; this is the formula corresponding with the binary theory of salts.

2. SO_3,H_2O. This formula represents the formation of the acid by direct hydration of sulphuric oxide; the separation of water and formation of a metallic sulphate by the action of magnesia and other anhydrous oxides; and the separation of sulphuric oxide and formation of phosphoric acid by the action of phosphoric oxide:—

$$SO_3,H_2O + MgO = SO_3,MgO + H_2O$$
$$SO_3,H_2O + P_2O_5 = P_2O_5,H_2O + SO_3.$$

3. SO_2,O_2H_2, or $SO_2(OH)_2$. This formula represents such reactions as the elimination of hydrogen dioxide by the action of barium dioxide, BaO_2.

4. SH_2O_4. This formula represents the formation of sulphuric acid by direct oxidation of hydrogen sulphide, SH_2, and the elimination of the latter by the action of ferrous sulphide:—

$$SH_2O_4 + FeS = FeSO_4 + SH_2.$$

Formulæ of the third of these types, like $SO_2(OH)_2$, which represent oxygen-acids as compounds of hydroxyl with certain acid radicles, as SO_2'' (sulphuryl), CO'' (carbonyl), PO''' (phosphoryl), etc., correspond with a great variety of reactions, and are of very frequent use. They exhibit in particular the relation of the oxygen-acids (hydroxylates) to the corresponding chlorides, e. g. —

$(SO_2)(OH)_2$ $(SO_2)Cl_2$
Sulphuric acid.   Sulphuric chloride.

$(PO)(OH)_3$ $(PO)Cl_3$
Phosphoric acid.   Phosphoric chloride.

Basicity of Acids.—Normal, Acid and Double Salts.—Acids are monobasic, bibasic, tribasic, etc., according as they contain one or more atoms of hydrogen replaceable by metals; thus nitric acid, HNO_3, and hydrochloric acid, HCl, are monobasic; sulphuric acid, H_2SO_4, is bibasic; phosphoric acid, H_3PO_4, is tribasic.

Monobasic acids form but one class of salts by substitution, the metal taking the place of the hydrogen in one, two, or three molecules of the acid, according to its equivalent value: thus the action of hydrochloric acid on sodium, zinc, and aluminium is represented by the equations—

$$HCl + Na = NaCl + H$$
$$2HCl + Zn = ZnCl_2 + H_2$$
$$3HCl + Al = AlCl_3 + H_3,$$

and that of nitric acid on the hydroxides of the same metals by the equations—

26

$$HNO_3 + Na(HO) = NaNO_3 + H(HO)$$
$$2HNO_3 + Ba(HO)_2 = Ba(NO_3)_2 + 2H(HO)$$
$$3HNO_3 + Al(HO)_3 = Al(NO_2)_3 + 3H(HO).$$

Bibasic acids, on the other hand, form two classes of salts, viz., *primary* or *acid salts*, in which half the hydrogen is replaced by a metal; and *secondary salts*, in which the whole of the hydrogen is thus replaced, the salt being called *normal* or *neutral* if it contains one metal and *double* if it contains two metals, thus:—

From H_2SO_4 is derived		$KHSO_4$	{	hydro-potassic sulphate, primary, or acid potassium sulphate,
"	"	K_2SO_4	{	bipotassic sulphate, secondary, or normal sulphate,
" H_2SO_4	"	$BaSO_4$		barium sulphate,
" $2H_2SO_4$	"	$NaK_3(SO_4)_2$		sodio-tripotassic sulphate,
" "	"	$KAl(SO_4)_2$		potassio-aluminic sulphate,
" $3H_2SO_4$	"	$Al(SO_4)_3$		normal aluminium sulphate

Tribasic acids in like manner form two classes of acid salts, *primary* or *secondary*, according as one-third or two-thirds of the hydrogen is replaced by a *metal;* also *tertiary salts*, including *normal* and *double* or *triple salts*, in which the hydrogen is wholly replaced by one or more metals; in quadribasic acids the variety is of course still greater.

The use of the terminations *ous* and *ic*, as applied to salts, has already been explained. We have only further to observe in this place that when a metal forms but one class of salts, it is for the most part better to designate those salts by the name of the metal itself than by an adjective ending in *ic;* thus *potassium nitrate* and *lead sulphate* are mostly to be preferred to *potassic nitrate* and *plumbic sulphate*. But in naming double salts, and in many cases where a numeral prefix is required, the names ending in *ic* are more euphonious: thus *triplumbic phosphate* sounds better than *trilead phosphate*, and *hydrodisodic phosphate* is certainly better than *hydrogen* and *disodium phosphate;* but there is no occasion for a rigid adherence to either system.

All oxygen-salts may also be represented as compounds of an acid oxide with one or more molecules of the same or different basic oxides, including water, *e. g.:—*

Hydro-potassic sulphate,	$2HK(SO_4)$	$= H_2O,K_4O,2SO_3$
Sodio-tripotassic sulphate,	$2NaK_3(SO_4)_2$	$= Na_2O,3K_2O,4SO_3$
Potassio-aluminic sulphate,	$2KAl(SO_4)_2$	$= K_2O,Al_2O_3,4SO_3$
Hydrodisodic phosphate,	$2HNa_2(PO)_4$	$= H_2O,2Na_2O,P_2O_5.$

When a normal oxygen-salt is thus formulated, it is easy to see that the number of molecules of acid oxide contained in its molecule is equal to the number of oxygen-atoms in the base; thus:

Normal potassium sulphate,	K_2SO_4	$= K_2O,SO_3$
" barium sulphate,	$BaSO_4$	$= BaO,SO_3$
" stannic sulphate,	$Sn(SO_4)_2$	$= SnO_2,2SO_3$
" aluminium sulphate,	$Al_2(SO_4)_3$	$= Al_2O_3,3SO_3.$

When the proportion of acid oxide is less than this, the salt is called b a s i c; such salts may be regarded as compounds of a normal salt with one or more molecules of. basic oxide, or as derived from normal salts by substitution of oxygen for an equivalent quantity of the acid radicle; thus :—

Tribasic lead nitrate, $3PbO,N_2O_5 = Pb(NO_3)_2,2PbO$
$= Pb_3(NO_3)_2O_2$
Quadribasic aluminium sulphate, $4Al_2O_3,3SO_3, = 3Al_2O_3,Al_2(SO_4)_3$
$= Al_8(SO_4)_3O_9.$

The last mode of formulation exhibits the analogy of these basic oxy-salts to the oxychlorides, oxyiodides, etc.; thus the basic lead nitrate, $Pb_3(NO_3)_2O_2$, just mentioned, is analogous to the oxychloride of that metal, $Pb_3Cl_2O_2$, which occurs native as mendipite.

The terms basic and acid are sometimes applied to salts with reference to their action on vegetable colors. The normal salts formed by the union of the stronger acids with the alkalies and alkaline earths, such as potassium sulphate, K_2SO_4, barium nitrate, $Ba(NO_3)_2$, etc., are perfectly neutral to vegetable colors, but most other normal salts exhibit either an acid or an alkaline reaction; thus ferrous sulphate, cupric sulphate, silver nitrate, and many others, redden litmus, while the normal carbonates and phosphates of the alkali-metals exhibit a decided alkaline reaction. It is clear, then, that the action of a salt on vegetable colors bears no definite relation to its composition: hence the term *normal*, as applied to salts in which the basic hydrogen of the acid is wholly replaced, is preferable to *neutral*, and the terms *basic* and *acid*, as applied to salts, are best used in the manner above explained with reference to their composition.

When a normal salt containing a monoxide passes by oxidation to a salt containing a sesquioxide, dioxide, or trioxide, the quantity of acid present is no longer sufficient to saturate the base. Thus when a solution of ferrous sulphate, $FeSO_4$, or FeO,SO_3 (common green vitriol), is exposed to the air, it absorbs oxygen, and an insoluble ferric salt is produced containing an excess of base, while normal ferric sulphate remains in solution :—

$$4(FeO,SO_3) + O_2 = Fe_2O_3,3SO_3 + Fe_2O_3,SO_3.$$
Ferrous sulphate.   Normal ferric sulphate.   Basic ferric sulphate.

These basic salts are very often insoluble in water.

Salts containing a proportion of acid oxide larger than is sufficient to form a neutral compound are called a n h y d r o - s a l t s (sometimes, though improperly, acid salts); they may evidently be regarded as compounds of a normal salt with excess of acid oxide; *e. g.* :—

Sodium anhydrosulphate *
(bisulphate of soda), $\}$ $Na_2O(SO_3)_2 = Na_2(SO_4)SO_3$

Potassium anhydrochromate
(bichromate of potash), $\}$ $K_2O(CrO_3)_2 = K_2(CrO_4)CrO_3.$

The following is a list of the most important inorganic acids arranged according to their basicity :—

Monobasic Acids.

Hydrochloric	HCl	Boric		HBO_2
Hydrobromic	HBr	Antimonic		$HSbO_3$
Hydriodic	HI	Hypochlorous		HClO
Hydrofluoric	HF	Chlorous		$HClO_2$
Nitrous	HNO_2	Chloric		$HClO_3$
Nitric	HNO_3	Perchloric		$HClO_4$
Hyposulphurous	$H(SHO_2)$	Bromic		$HBrO_3$
Hypophosphorous	$H(PH_2O_2)$	Iodic		HIO_3
Metaphosphoric	HPO_3	Periodic		HIO_4

* The so-called "anhydrosulphates" are now regarded as salts of a distinct acid, pyro-sulphuric acid, $H_2S_2O_7$ (p. 180).

Bibasic Acids.

Hydric (water) . . . H_2O	Selenious . . . H_2SeO_3
Sulphydric . . . . H_2S	Selenic . . . . H_2SeO_4
Selenhydric . . . H_2Se	Tellurous . . . H_2TeO_3
Tellurhydric . . . H_2Te	Telluric . . . H_2TeO_4
Sulphurous . . . H_2SO_3	Manganic . . . H_2MnO_4
Sulphuric . . . . H_2SO_4	Permanganic . . . $H_2Mn_2O_8$
Pyrosulphuric . . . $H_2S_2O_7$	Chromic . . . H_2CrO_4
Thiosulphuric . . . $H_2S_2O_3$	Stannic . . . H_2SnO_3
Dithionic . . . . $H_2S_2O_6$	Metasilicic . . . H_2SiO_3
Trithionic . . . . $H_2S_3O_6$	Carbonic . . . H_2CO_3
Tetrathionic . . . $H_2S_4O_6$	Phosphorous . . . $H_2(PHO)_3$
Pentathionic . . . $H_2S_5O_6$	

Tribasic Acids.

Orthophosphoric . . . H_3PO_4 | Arsenic . . . . H_3AsO_4.

Quadribasic Acids.

Pyrophosphoric . . . $H_4P_2O_7$ | Orthosilicic . . . H_4SiO_4.

The general characters of most of the non-metallic acids and their salts have been already considered ; but the phosphates, borates, and silicates require further notice.

Phosphates.—There are three modifications of phosphoric acid : one being monobasic, the second tribasic, and the third quadribasic, as indicated in the preceding table.

Hydrogen phosphide, PH_3, burnt in air or oxygen gas, takes up four atoms of oxygen, and forms t r i h y d r i c p h o s p h a t e or t r i b a s i c p h o s-phoric acid, PH_3O_4. The same acid is produced by the oxidation of hypophosphorous or phosphorous acid ; by oxidizing phosphorus with nitric acid (p. 209) ; by the decomposition of native calcium phosphate (apatite) and other native phosphates ; and by the action of boiling water on phosphorus pentoxide, P_2O_5. This acid forms three distinct classes of metallic salts. With sodium, for example, it forms the three salts, NaH_2PO_4, Na_2HPO_4, and Na_3PO_4, the first two of which, still containing replaceable hydrogen, are acid salts, while the third is the normal or neutral salt.

If now the monosodic phosphate, NaH_2PO_4, be heated to redness, it gives off one molecule of water, and leaves an anhydrous monosodic phosphate, $NaPO_3$, the aqueous solution of which, when treated with lead nitrate, yields a lead-salt of corresponding composition ; thus :—

$$2NaPO_3 + Pb(NO_3)_2 = Pb(PO_3)_2 + 2NaNO_3 ;$$

and this lead-salt decomposed by sulphydric acid, yields a monohydric acid having the composition HPO_3, possessing properties quite distinct from those of the trihydric acid above mentioned :—

$$Pb(PO_3)_2 + H_2S = 2HPO_3 + PbS.$$

The trihydric acid which is produced by the oxidation of phosphorus, and by the decomposition of the ordinary native phosphates, is called o r t h o-p h o s p h o r i c a c i d or o r d i n a r y p h o s p h o r i c a c i d ; the monohydric acid is called m e t a p h o s p h o r i c a c i d. The former may be regarded as a trihydrate, the latter as a monohydrate of phosphoric oxide :—

$$2H_3PO_4 = P_2O_5,3H_2O, \text{ orthophosphoric acid,}$$
$$2HPO_3 = P_2O_5,H_2O, \text{ metaphosphoric acid.}$$

Both are soluble in water, and the former may be produced by the action of boiling water, the latter by that of cold water on phosphoric oxide. They are easily distinguished from one another by their reactions with albumin and with silver nitrate. Metaphosphoric acid coagulates albumin, and gives a white precipitate with silver nitrate; whereas orthophosphoric acid does not coagulate albumin, and gives no precipitate, or a very slight one, with silver nitrate, till it is neutralized with an alkali, in which case a yellow precipitate is formed.

Metaphosphoric acid and its salts differ from orthophosphoric acid and the orthophosphates by the want of one or two molecules of water or base; thus:—

Metaphosphates.		Orthophosphates.		
HPO_3	$=$	H_3PO_4	$-$	H_2O
$NaPO_3$	$=$	NaH_2PO_4	$-$	H_2O
$Ba(PO_3)_2$	$=$	$BaH_4(PO_4)_2$	$-$	$2H_2O$
$AgPO_3$	$=$	Ag_3PO_4	$-$	Ag_2O
$Pb(PO_3)_2$	$=$	$Pb_3(PO_4)_2$	$-$	$2PbO.$

Accordingly, we find that metaphosphates and orthophosphates are convertible one into the other by the loss or gain of one or two molecules of water or metallic base; thus:—

a. A solution of metaphosphoric acid is converted, slowly at ordinary temperatures, quickly at the boiling heat, into orthophosphoric acid, and the metaphosphates of sodium and barium are converted by boiling with water into the corresponding monometallic orthophosphates (see the first three equations above).—β. The metaphosphate of a heavy metal, silver or lead, for example, is converted by boiling with water into a trimetallic phosphate and orthophosphoric acid:—

$$3AgPO_3 + 3H_2O = Ag_3PO_4 + 2H_3PO_4.$$

γ. When any metaphosphate is fused with an oxide, hydroxide, or carbonate, it becomes a trimetallic orthophosphate, e. g. :—

$$NaPO_3 + Na_2CO_3 = Na_3PO_4 + CO_2.$$

On the other hand (*d*), when orthophosphoric acid is heated to redness, it loses water and becomes metaphosphoric acid; and when a monometallic orthophosphate is heated to redness, it also loses water and is transformed into a metaphosphate.

The metaphosphates are susceptible of five polymeric modifications.

a. Monometaphosphates, MPO_3 [M denoting a univalent metal]. These salts are produced by adding phosphoric acid in excess to solutions of sulphates or nitrates, and heating the evaporated residues to 316° or upward. The potassium salt is also formed by igniting potassium hydroxide with phosphoric acid in molecular proportion; the ammonium salt by heating ammonium dimetaphosphate to 250°. These metaphosphates are crystalline anhydrous powders, insoluble in water. There are no double salts of this modification.

β. *Dimetaphosphates,* $M_2(PO_3)_2$ and $M''(PO_3)_2$, are formed when aqueous phosphoric acid is heated to 350° with oxide of zinc, manganese, or copper; and the copper salt decomposed by sulphide of potassium or sodium yields potassium or sodium dimetaphosphate. These alkali-metal dimetaphosphates are soluble in water and crystallizable; the rest are insoluble or only slightly soluble.

γ. *Trimetaphosphates*, $M_3(PO_3)_3$.—The sodium salt of this modification is obtained, together with the monometaphosphate, by gently heating microcosmic salt $(NH_4)HNaPO_4$, till the fused mass becomes crystalline; and from this salt other trimetaphosphates may be formed by double decomposition. They are all soluble in water, including the silver salt, which may be obtained by slow deposition in large, transparent, monoclinic crystals.

δ. *Tetrametaphosphates*, $M_4(PO_3)_4$.—The lead salt, $Pb_2(PO_3)_4$, is formed by heating lead oxide at 300° with excess of phosphoric acid; and this salt decomposed by sodium sulphide yields the sodium salt, which is viscid and uncrystallizable. By fusing it with copper dimetaphosphate, and leaving the mass to cool gradually, a double salt is obtained having the composition $CuNa_2P_4O_{12}$.

ε. *Hexmetaphosphates*, $M_6(PO_3)_6$, discovered by Graham, and obtained by dehydration of dihydric orthophosphates (p. 305).

Pyrophosphates.—Intermediate between orthophosphates and metaphosphates, there are at least three distinct classes of salts, the most important of which are the pyrophosphates or paraphosphates, which may be derived from the tetrahydric or quadribasic acid, $H_4P_2O_7$, the normal sodium salt, for example, being $Na_4P_2O_7$, the normal lead salt, $Pb_2P_2O_7$, etc. These salts may be viewed as compounds of orthophosphate and metaphosphate, *e. g.* :—

$$Na_4P_2O_7 = Na_3PO_4 + NaPO_3.$$

Sodium pyrophosphate is produced by heating disodic orthophosphate to redness, a molecule of water being then given off :—

$$2Na_2HPO_4 = H_2O + Na_4P_2O_7.$$

The aqueous solution of this salt yields insoluble pyrophosphates with lead and silver salts; thus with lead nitrate :—

$$Na_4P_2O_7 + 2Pb(NO_3)_2 = 4NaNO_3 + Pb_2P_2O_7 ;$$

and lead pyrophosphate decomposed by hydrogen sulphide yields hydrogen pyrophosphate or pyrophosphoric acid :—

$$Pb_2P_2O_7 + 2H_2S = 2PbS + H_4P_2O_7.$$

Pyrophosphoric acid is distinguished from metaphosphoric acid by not coagulating albumin, and not precipitating neutral solutions of barium or silver salts, and from orthophosphoric acid by producing a white instead of a yellow precipitate with silver nitrate.

Pyrophosphates are easily converted into metaphosphates and orthophosphates, and *vice versâ*, by addition or abstraction of water or a metallic base.

α. The production of a pyrophosphate from an orthophosphate by loss of water has been already mentioned.—β. Conversely, when a pyrophosphate is heated with water or a base, it becomes an orthophosphate, *e. g.* :—

$$Na_4P_2O_7 + H_2O = 2Na_2HPO_4$$
$$Na_4P_2O_7 + 2NaHO = 2Na_3PO_4 + H_2O.$$

In like manner orthophosphoric acid heated to 215° is almost entirely converted into pyrophosphoric acid: $2H_3PO_4 - H_2O = H_4P_2O_7$; and conversely, when pyrophosphoric acid is boiled with water, it is transformed into orthophosphoric acid.

γ. Pyrophosphoric acid heated to dull redness is converted into metaphosphoric acid: $H_4P_2O_7 - H_2O = 2HPO_3$. The converse reaction is not easily effected, inasmuch as metaphosphoric acid by absorbing water generally passes directly to the state of orthophosphoric acid. Peligot, however, observed the formation of pyrophosphoric from metaphosphoric acid by very slow absorption of water —δ. When a metallic metaphosphate is treated with a proper proportion of a hydroxide, oxide, or carbonate, it is converted into a pyrophosphate; thus:—

$$2NaPO_3 \quad + \quad Na_2CO_3 \quad = \quad Na_4P_2O_7 \quad + \quad CO_2.$$
Metaphosphate.   Carbonate.   Pyrophosphate.   Carbon dioxide.

Fleitmann and Henneberg, by fusing together a molecule of sodium pyrophosphate, $Na_3PO_4,NaPO_3$, with two molecules of metaphosphate, $NaPO_3$, obtained a salt having the composition $Na_3PO_4,3NaPO_3 = Na_6P_4O_{13}$, which is soluble without decomposition in a small quantity of hot water, and crystallizes from its solution by evaporation over oil of vitriol. An excess of hot water decomposes it, but its cold aqueous solution is moderately permanent. Insoluble phosphates of similar composition may be obtained from the sodium salt by double decomposition. Fleitmann and Henneberg obtained another crystallizable but very insoluble salt, having the composition $Na_3PO_4,9NaPO_3 = Na_{12}P_{10}O_{31}$, by fusing together one molecule of sodium pyrophosphate with eight molecules of the metaphosphate; and insoluble phosphates of similar constitution were obtained from it by double decomposition.

The comparative composition of these different phosphates is best shown by representing them as compounds of phosphoric oxide with metallic oxide, and assigning to them all the quantity of base contained in the most complex member of the series; thus:—

Orthophosphate . . . . . . $6Na_2O,2P_2O_5 = 4Na_3PO_4$
Pyrophosphate . . . . . . $6Na_2O,3P_2O_5 = 3Na_4P_2O_7$
Fleitmann and Henneberg's phosphate (a) . $6Na_2O,4P_2O_5 = 2Na_6P_4O_{13}$
    "   "   "   (b) . $6Na_2O,5P_2O_5 = Na_{12}P_{10}O_{31}$
Metaphosphate . . . . . . $6Na_2O,6P_2O_5 = 12NaPO_3$.

B o r a t e s.—There are three series of these salts, analogous to the three principal modifications of the phosphates, and derived respectively from ortho-, meta-, and pyroboric acid (p. 220); thus [M denoting a univalent metal]:—

Orthoborates, $M_3BO_3 = B(OM)_3$
Metaborates, $MBO_2 = BO(OM)$
Pyroborates, $M_2B_4O_7 = B_4O_5(OM)_2$.

The metallic orthoborates are very unstable, the only well-defined orthoborate being the magnesium salt; but several volatile ethers of similar constitution are known, e. g., triethylic borate, $B(OC_2H_5)_3$, which will be described under ORGANIC CHEMISTRY.—The *metaborates* are much more stable, e. g., $NaBO_2$ and $Mg(BO_2)_2$.—The *pyroborates* are also stable salts, including sodium pyroborate or borax, $Na_2B_4O_7$, the calcium salt or borocalcite, CaB_4O_7, and boronatrocalcite, $Na_2B_4O_7,2CaB_4O_7$.

Borates of more complex constitution are also known, amongst which are the following minerals: Laderellite, $(NH_4)_2B_8O_{13}$, lagonite, $Fe_2B_6O_{12}$, and boracite, $2Mg_3B_8O_{15},MgCl_2$.

Silicates.—These salts occur very abundantly as natural minerals, and exhibit great diversity of composition; but the most important of them may be arranged in the following groups [*]:—

	Oxygen-ratio, $M_2O : SiO_2$.	Formula.	Examples.
Hemisilicates or Orthosilicates .	1 : 1	$\left\{ \begin{array}{l} 2M_2O,SiO_2 = \\ M_4SiO_4 \end{array} \right.$	$\left\{ \text{Olivine } \begin{array}{l} Mg'' \\ Fe'' \end{array} \right\} SiO_4.$
Monosilicates or Metasilicates .	1 : 2	$\left\{ \begin{array}{l} M_2O,SiO_2 = \\ M_2SiO \end{array} \right.$	Diopside $(Ca,Mg)''SiO_3.$
Sesquisilicates . .	1 : 3	$\begin{array}{l} 2M_2O,3SiO_2 =^3 \\ M_4Si_3O_8 \end{array}$	Orthoclase $(KAl''')Si_3O_8.$ Stilbite $(Ca''Al''')_2)Si_6O_{16} + 5H_2O.$
Di- or bi-silicates .	1 : 4	$\begin{array}{l} M_2O,2SiO_2 = \\ M_2Si_2O_5 \end{array}$	Okenite, $Ca''Si_2O_5 + 2H_2O.$

Silicates are sometimes also distinguished by names expressing directly the oxygen-ratio in the silica and metallic oxide, the ratio 1 : 1 giving *Singulosilicates;* 1 : 2, *Disilicates;* 1 : 3, *Trisilicates;* 1 : 4, *Quadrosilicates,* etc.

All silicates are insoluble in water, except those of the alkali-metals, which dissolve with greater facility in proportion as they contain a larger quantity of base. These salts, known as *soluble glass* or *water-glass,* are used for making artificial stone and preserving natural stone from decay; also for rendering muslin and other light fabrics uninflammable; and for a peculiar style of mural painting called *stereochromy.* [†]

Some silicates are entirely decomposed by heating them in the state of powder, with hydrochloric or nitric acid, the bases being dissolved and silica separated. Others, on the contrary, resist the action of all acids except hydrofluoric; but all without exception become soluble in dilute nitric or hydrochloric acid, after fusion with from 3 to 5 times their weight of hydroxide or carbonate of potassium or sodium, or with carbonate of barium, or calcium, or with lead oxide, the mineral being completely disintegrated, and the solution yielding on evaporation, first a jelly and then a dry residue, of which the part insoluble in hot hydrochloric acid exhibits the characters of silica.—Silicates heated in a platinum vessel with hydrofluoric acid, or with fluorspar and strong sulphuric acid, give off gaseous fluoride of silicon (p. 223).

Metallic Sulphides.—These compounds correspond, for the most part, in composition with the oxides: thus there are two sulphides of antimony, Sb_2S_3 and Sb_2S_5, corresponding with the oxides, Sb_2O_3 and Sb_2O_5; also two sulphides of mercury, Hg_2S and HgS, analogous to the oxides, Hg_2O and HgO. Occasionally, however, we meet with oxides to which there are no corresponding sulphides (manganese dioxide, for example), and more frequently with sulphides to which there are no corresponding oxides, the most remarkable of which are perhaps the alkaline polysulphides. Potassium, for example, forms the series of sulphides, K_2S, K_2S_2, K_2S_3,

[*] A more extended table is given in Watts's Dictionary of Chemistry, v. 243.
[†] See Richardson and Watts's Chemical Technology, vol. i. Part iv. pp. 69-104.

K_2S_4, and K_2S_5, the third and fifth of which have no analogues in the oxygen series.

There are also hydrosulphides analogous to the hydroxides, and containing the elements of a metallic sulphide and hydrogen sulphide, or sulphydric acid; e. g., potassium hydrosulphide, $K_2S,H_2S = 2KHS$; lead hydrosulphide, $PbS,H_2S = PbH_2S_2$. Hydrosulphides and sulphides may be derived from sulphydric acid by partial or total replacement of the hydrogen by metals, just as metallic hydroxides and oxides are derived from water.

Many metallic sulphides occur as natural minerals, especially the sulphides of lead, copper, and mercury, which afford valnable ores for the extraction of the metals, and iron bisulphide, or iron pyrites, FeS_2, which is largely used as a source of sulphur, and for the preparation of ferrous sulphate.

Sulphides are formed artificially by heating metals with sulphur; by the action of metals on gaseous hydrogen sulphide; by the reduction of sulphates with hydrogen or charcoal; by heating metallic oxides in contact with gaseous hydrogen sulphide or vapor of carbon bisulphide; and by precipitation of metallic solutions with hydrogen sulphide or a sulphide of alkali-metal. Some metals, as copper, lead, silver, bismuth, mercury, and cadmium, are precipitated from their acid solutions by hydrogen sulphide, passed into them as gas or added in aqueous solution, the sulphides of these metals being insoluble in dilute acids; others, as iron, cobalt, nickel, manganese, zinc, and uranium, form sulphides which are soluble in acids, and these are precipitated by hydrogen sulphide only from alkaline solutions, or by ammonium or potassium sulphide from neutral solutions. Many of these sulphides exhibit characteristic colors, which serve as indications of the presence of the respective metals in solution (p. 174).

Metallic sulphides are also formed by the reduction of sulphates with organic substances; many native sulphides have doubtless been formed in this way.

The physical characters of some metallic sulphides closely resemble those of the metals in certain particulars, such as the peculiar opacity, lustre, and density, especially when they are crystallized. They are generally crystallizable, brittle, and of a gray, pale yellow, or dark brown color. The sulphides of the alkali-metals are soluble in water; most of the others are insoluble. They are more frequently fusible than the corresponding oxides, and some are volatilizable, as mercury sulphide and arsenic sulphide.

Many sulphides, when heated out of contact with atmospheric air, do not undergo any decomposition; this is the case chiefly with those containing the smallest proportions of sulphur, such as the monosulphides of iron and zinc. Sulphides containing larger proportions of sulphur are partially decomposed by heat, losing part of their sulphur, and being converted into lower sulphides; as in the case of iron bisulphide. The sulphides of gold and platinum are completely reduced by heat.

Some sulphides may be decomposed by the simultaneous action of heat and of substances capable of combining with sulphur. Thus, for instance, silver, copper, bismuth, tin, and antimony sulphides are reduced by hydrogen; copper, lead, mercury, and antimony sulphides are reduced by heating with iron.

Sulphides which are not reduced by heat alone are always decomposed when heated in contact with oxygen or atmospheric air. Those of the alkali-metals and earth-metals are converted into sulphates by this means. Zinc, iron, manganese, copper, lead, and bismuth sulphides are converted into oxides, and sulphurous oxide is produced; but when the temperature is not above dull redness, a certain quantity of sulphate is also formed by direct oxidation. Mercury and silver sulphides are completely reduced to the metallic state. Some native sulphides gradually undergo alteration by mere exposure to the air; but the action is then generally limited to the production of sulphates, unless the oxidation takes place so rapidly that the heat generated is sufficient to decompose the sulphate first produced. In the production of some metals for use in the arts, the separation of sulphur from the native minerals is effected chiefly by means of this action in the operation of roasting.

Metallic sulphides are decomposed in like manner when heated with metallic oxides in suitable proportions, yielding sulphurons oxide and the metal of both the sulphide and oxide. Lead is reduced from the native sulphide in this manner.

Many metallic sulphides are decomposed by acids in presence of water, sulphuretted hydrogen being evolved, while the metal enters into combination with the chlorous radicle of the acid. Nitric acid when concentrated decomposes most sulphides, with formation of metallic oxide, sulphuric acid, sulphur, and a lower oxide of nitrogen. Nitromuriatic acid acts in a similar manner, but still more energetically.

Sulphur-Salts.—The sulphides of the more basylous metals unite with those of the more chlorous or electro-negative metals, and of the non-metallic elements, forming s u l p h u r - s a l t s, analogous in composition to the oxygen salts, *e. g.* :—

Carbonate,	K_2CO_3	Arsenate,	K_3AsO_4
Thiocarbonate,	K_2CS_3	Thioarsenate,	K_3AsS_4.
Antimonate,		$2K_3SbO_4 = 3K_2O,Sb_2O_5$	
Thioantimonate,		$2K_3SbS_4 = 3K_2S,Sb_2S_5$.	

Selenides.—These compounds are analogous in composition and in many of their properties to the sulphides, and unite one with the other, forming selenium-salts analogous to the oxygen and sulphur salts. They are prepared by fusing the metals with selenium; by precipitating a solution of a heavy metal with hydrogen selenide, or a dissolved selenide of alkali-metal; by heating selenium with a metallic oxide or carbonate; or by igniting a selenite or selenate with hydrogen or charcoal.

The selenides of the alkali-metals are soluble in water; their solutions are colorless when freshly prepared, but gradually turn red from separation of selenium. The selenides of barium, strontium, and calcium are flesh-colored, insoluble in pure water, but soluble in aqueous selenhydric acid. The selenides of aluminium, beryllium, magnesium, yttrium, and its allied metals (p. 291), and those of manganese and zinc, are also flesh-colored and insoluble in water. The other metallic selenides are mostly dark-colored, exhibit metallic lustre, and are mostly more fusible than the corresponding metals. When they are heated to redness in the air, the selenium burns slowly away with a reddish-blue flame and an odor of horseradish. Selenides are less soluble in nitric acid than the pure metals; mercury selenide almost insoluble. Chlorine with the aid of heat converts them into chloride of selenium and metallic chlorides. Heated in gaseous hydrogen chloride, they yield metallic chlorides and hydrogen selenide.

Tellurides are analogous in composition and properties to the sulphides and selenides, but they likewise exhibit some of the characters of metallic alloys. The tellurides of alkali-metal are prepared by heating powdered tellurium with alkaline carbonates mixed with finely-divided charcoal, or by passing gaseous hydrogen telluride through solutions of the alkalies. Their solutions have a port-wine color and deposit tellurium on exposure to the air. The other tellurides may be obtained by fusing the respective metals with tellurium, or by precipitating a metallic solution with hydrogen telluride or a telluride of alkali-metal. The tellurides of gold, silver, and bismuth are found native.

Metals also form definite compounds with nitrogen, phosphorus, silicon, boron, and carbon: but these compounds are comparatively unimportant, excepting the carbonides of iron, which form cast iron and steel.

METALS OF THE ALKALIES.

THIS group includes five metals, viz., potassium, sodium, rubidium, cæsium, and lithium. They are soft, easily fusible, volatile at higher temperatures; combine very energetically with oxygen; decompose water at all temperatures; and form strongly basic oxides, which are very soluble in water, yielding powerfully caustic and alkaline hydroxides, not decomposable by heat. Their carbonates are soluble in water, and each metal forms only one chloride. The hypothetical metal ammonium (p. 191) is usually added to the list of alkali-metals, on account of the general similarity of its compounds to those of potassium and sodium.

POTASSIUM.

Symbol K (Kalium).   Atomic weight, 39.04.

Potassium was discovered in 1807 by Sir H. Davy, who obtained it in very small quantity by exposing a piece of moistened potassium hydroxide to the action of a powerful voltaic battery, the alkali being placed between a pair of platinum plates connected with the apparatus. Processes have since been devised for obtaining this metal in almost any quantity that can be desired.

An intimate mixture of potassium carbonate and charcoal is prepared by calcining in a covered iron pot the crude tartar of commerce; when cold it is rubbed to powder, mixed with one-tenth part of charcoal in small lumps, and quickly transferred to a retort of stout hammered iron: the latter may be one of the iron bottles in which mercury is imported. The retort is introduced into a furnace a (fig. 147), and placed horizontally on supports of fire-brick, f, f. A wrought-iron tube d, four inches long, serves to convey the vapors of potassium into a receiver e, formed of two pieces of wrought iron $a\, b$ (fig. 148), which are fitted closely to each other so as to form a shallow box only a quarter of an inch deep, and are kept together by clamp-screws. The iron plate should be one-sixth of an inch thick, twelve inches long, and five inches wide. The receiver is open at both ends, the socket fitting upon the neck of the iron bottle. The object of giving the receiver this flattened form is to ensure the rapid cooling of the potassium, and thus to withdraw it from the action of the carbon monoxide, which is disengaged during the entire process, and has a strong tendency to unite with the potassium, forming a dangerously explosive compound. Before connecting the receiver with the tube d, the fire is slowly raised till the iron bottle attains a dull red heat. Powdered vitrefied borax is then sprinkled upon it, which melts and forms a coating, serving to protect the iron from oxidation. The heat is then to be urged until it is very intense, care being taken to raise it as equally as possible throughout every part of the furnace. When a full reddish-white heat is attained, vapors of potassium begin to appear and burn with a bright flame. The receiver is then adjusted to the end of the tube, which must not project more than a quarter of an inch through the iron plate forming the front wall of the furnace; otherwise the tube is liable to be obstructed by the accumulation of solid potassium or of the explosive compound above mentioned. Should any obstruction occur, it must be re-

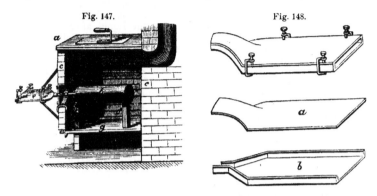

Fig. 147. Fig. 148.

moved by thrusting in an iron bar, and, if this fail, the fire must be immediately withdrawn by removing the bars from the furnace, with the exception of two which support the iron bottle. The receiver is kept cool by the application of a wet cloth to its outside. When the operation is complete, the receiver with the potassium is removed, and immediately plunged into a vessel of rectified petroleum provided with a cover, and kept cool by immersion in water. When the apparatus is sufficiently cooled, the potassium is detached and preserved under petroleum.

If the potassium be wanted absolutely pure, it must be afterward redistilled in an iron retort, into which some petroleum has been put, that its vapor may expel the air and prevent oxidation of the metal.

Potassium is a brilliant white metal, with a high degree of lustre; at the common temperature of the air it is soft, and may be easily cut with a knife, but at 0° C. it is brittle and crystalline. It melts completely at 62.5° C. (144.5° F.), and distils at a low red heat. It floats on water, its specific gravity being only 0.865.

Exposed to the air, potassium oxidizes instantly, a tarnish covering the surface of the metal, which quickly thickens to a crust of caustic potash. Thrown upon water, it takes fire spontaneously, and burns with a beautiful purple flame, yielding an alkaline solution. When it is brought into contact with a little water in a jar standing over mercury, the liquid is decomposed with great energy, and hydrogen is liberated. Potassium is always preserved under the surface of petroleum.

Potassium Chloride, KCl, occurs in sea-water and in many mineral springs, and is the chief constituent of the "potash-salt" of Stassfurt near Magdeburg, which forms a layer 80 to 100 feet thick, lying above the rock-salt, and consists mainly of carnallite, $KCl,MgCl_2 + 6H_2O$, and kieserite, $MgSO_4 + H_2O$, interspersed with veins of sylvine, KCl, and several calcium and magnesium salts. All these minerals are very deliquescent, and the entire deposit appears to have been formed by the drying up of an inland sea or salt-water lake. Similar beds occur at Kalusz in Gallicia.

The preparation of potassium chloride from the "potash-salt" depends upon the fact that carnallite forms only in solutions containing excess of magnesium chloride, so that when the salt is dissolved in hot water and the solution is left to cool, no double salt separates, but the more soluble chloride of magnesium remains in solution, while part of the potassium chloride crys-

14

tallizes out. The mother-liquors are then further treated for the recovery of the remaining quantity.

The "potash-salt" is first dissolved in cold water; the solution is heated by a steam pipe conveying steam at 120° C. (248° F.), and continually stirred by revolving arms to facilitate solution; and the liquid, after standing for ten hours, is decanted from the insoluble matter. The clear solution marking 32° B. is completely saturated with the chlorides of potassium and magnesium, and contains also small quantities of sodium chloride and magnesium sulphate. It is transferred to a series of crystallizing vessels in which crystals are deposited containing from 60 to 70 per cent. of potassium chloride, a charge of 20,000 kilograms of "potash-salt" yielding from 16,000 to 17,000 kilograms of this impure chloride of potassium. The crystals are washed with water to remove the adhering mother-liquor and the magnesium chloride, after which they contain 80 per cent. of potassium chloride.

Potassium chloride is obtained in like manner from the mother-liquor of sea-water and from certain mineral springs. In the salines on the west and south coasts of France, the mother-liquors remaining after the common salt has been deposited are preserved in reservoirs during the summer, when a mixture of magnesium sulphate and common salt separates out. The mother-liquor from this is evaporated in shallow pans, and thus converted into carnallite, which is worked up as above described.

Potassium chloride is also obtained in large quantities as a by-product in the manufacture of the chlorate, any residual portions of the latter being easily deoxidized by exposure to a red heat.

Potassium chloride closely resembles common salt in appearance, and crystallizes, like the latter, in cubes. The crystals dissolve in three parts of cold, and in a much smaller quantity of boiling water: they are anhydrous, have a simple saline taste, with slight bitterness, and fuse at a red heat. Potassium chloride is volatilized by a very high temperature.

Potassium chloride is used for the preparation of other potassium salts, as the nitrate, chlorate, carbonate, and chromate, and of potash-alum. The impure salt is employed in the composition of artificial manures.

Potassium Iodide, KI.—There are three different methods of preparing this important medicinal compound.

(1.) When iodine is added to a strong solution of caustic potash free from carbonate, it is dissolved in large quantity, forming a colorless solution containing potassium iodide and iodate; the reaction is the same as in the analogous case with chlorine. When the solution begins to be permanently colored by the iodine, it is evaporated to dryness, and cautiously heated to redness, by which the iodate is entirely converted into iodide. The mass is then dissolved in water, and, after filtration, made to crystallize.

(2.) Iodine, water, and iron-filings or scraps of zinc are placed in a warm situation until the combination is complete and the solution colorless. The resulting iodide of iron or zinc is then filtered, and exactly decomposed with solution of pure potassium carbonate, great care being taken to avoid excess of the latter. Potassium iodide and ferrous carbonate or zinc carbonate are thus obtained: the former is separated by filtration, and evaporated until the solution is sufficiently concentrated to crystallize on cooling, the washings of the filter being added to avoid loss.

$$FeI_2 + K_2CO_3 = 2KI + FeCO_3.$$

(3.) A very simple method for the preparation of potassium iodide was proposed by Liebig. One part of amorphous phosphorus is added to 40 parts of warm water; 20 parts of dry iodine are then gradually added and intimately mixed with the phosphorus by trituration. The dark-brown liquid thus obtained is now heated on the water-bath until it becomes color-less; it is then poured off from the undissolved phosphorus, and neutralized, first with barium carbonate and then with baryta-water, until it becomes slightly alkaline; and the insoluble barium phosphate is filtered off and washed. The filtrate now contains nothing but barium iodide, which, when treated with potassium sulphate, yields insoluble barium sulphate and potassium iodide in solution. Lime answers nearly as well as baryta.

Potassium iodide crystallizes in cubes, which are often, from some unexplained cause, milk-white and opaque. It is anhydrous, melts readily when heated, at 639° C. (1182° F.) according to Carnelly, and volatilizes at higher temperatures. The salt is very soluble in water, but when pure does not deliquesce in a moderately dry atmosphere: it is dissolved by alcohol.

Solution of potassium iodide, like those of all the soluble iodides, dissolves a large quantity of free iodine, forming a deep-brown liquid, not decomposed by water.

Potassium Bromide, KBr, may be obtained by processes exactly similar to those just described, substituting bromine for the iodine. It is a colorless and very soluble salt, quite undistinguishable in appearance and general characters from the iodide.

Potassium Oxides.—Potassium forms two oxides, K_2O and K_2O_4, also a hydroxide, KHO, corresponding with the monoxide.

The *monoxide*, K_2O, also called *anhydrous potash*, or *potassa*, is formed when potassium in thin slices is exposed at ordinary temperatures to dry air free from carbon dioxide; also when the hydroxide is heated with an equivalent quantity of metallic potassium.

$$2KHO + K_2 = 2K_2O + H_2.$$

It is white, very deliquescent and caustic, combines energetically with water, forming potassium hydroxide, and becomes incandescent when moistened with it; melts at a red heat, and volatilizes at very high temperatures.

The *tetroxide*, K_2O_4, or $\begin{array}{c} K-O-O \\ | \\ K-O-O \end{array}$, is produced when potassium is burnt in excess of dry air or oxygen gas. It is a chrome-yellow powder, which cakes together at about 280°. It absorbs moisture rapidly, and is decomposed by water, giving off oxygen, and forming a solution of the dioxide. When gently heated in a stream of carbon monoxide, it yields potassium carbonate and two atoms of oxygen:

$$K_2O_4 + CO = K_2CO_3 + O_2:$$

with carbon dioxide it acts in a similar manner, giving off 3 atoms of oxygen.—(Harcourt, *Chem. Soc. Jour.*, 1861, p. 267.)

POTASSIUM HYDROXIDE, KHO, commonly called *caustic potash*, or *potassa*, is a very important substance, and one of great practical utility. It is always prepared by decomposing the carbonate with calcium hydroxide (slaked lime). 10 parts of potassium carbonate are dissolved in 100 parts of water, and heated to ebullition in a clean untinned iron, or, still better, silver vessel: 8 parts of good quicklime are meanwhile slaked in a covered basin, and the resulting calcium hydroxide is added, little by little, to the boiling solution of carbonate, with frequent stirring. When all the lime has been introduced, the mixture is boiled for a few minutes, then removed from the fire, and covered up. In a very short time the solution will have become quite clear and fit for decantation, the calcium carbonate, with the excess of hydrate, settling down as a heavy, sandy precipitate. It should not effervesce with acids.

It is essential in this process that the solution of potassium carbonate be dilute, otherwise the decomposition becomes imperfect. The proportion of lime recommended is much greater than that required by theory, but it is always proper to have an excess.

The solution of potassium hydroxide may be concentrated by quick evaporation in the iron or silver vessel to any desired extent; when heated until vapor of water is no longer given off, and then left to cool, it yields the solid hydroxide, KHO or K_2O,H_2O. Pure potassium hydroxide is also easily obtained by heating to redness for half an hour in a covered copper vessel one part of pure powdered nitre with two or three parts of finely-divided copper foil. The mass, when cold, is treated with water.

Potassium hydroxide is a white solid substance, very deliquescent, and soluble in water; alcohol also dissolves it freely, which is the case with comparatively few potassium compounds: the solid hydroxide of commerce, which is very impure, may thus be purified. The solution of this substance possesses, in the very highest degree, the properties termed alkaline: it restores the blue color to litmus which has been reddened by an acid; neutralizes completely the most powerful acids; has a nauseous and peculiar taste; dissolves the skin and many other organic matters, and is frequently used by surgeons as a cautery. Its chief use, however, is for the manufacture of soft soap.

Potassium hydroxide, both in the solid state and in solution, rapidly absorbs carbonic acid from the air: hence it must be kept in closely-stopped bottles. When imperfectly prepared, or partially altered by exposure, it effervesces with an acid. It is not decomposed by heat, but volatilizes undecomposed at a very high temperature.

The following table of the densities and values in potassium hydroxide of different solutions of caustic potash has been calculated by Gerlach from the experiments of Tünnermann and Schiff:—

KOH per cent.				Density.	KOH per cent.				Density.
1	.	.	.	1.009	40	.	.	.	1.411
5	.	.	.	1.041	45	.	.	.	1.475
10	.	.	.	1.083	50	.	.	.	1.539
15	.	.	.	1.128	55	.	.	.	1.604
20	.	.	.	1.177	60	.	.	.	1.667
25	.	.	.	1.230	65	.	.	.	1.729
30	.	.	.	1.288	70	.	.	.	1.790
35	.	.	.	1.349					

The *liquor potassæ* of the Pharmacopœia contains about 5 per cent. KOH, and has a density of 1.058.

Potassium Chlorate, $KClO_3 = ClO_2(OK)$.—The theory of the production of chloric acid by the action of chlorine gas on a solution of caustic potash, has been already explained (p. 169). Chlorine gas is conducted by a wide tube into a strong and warm solution of potassium carbonate, until absorption of the gas ceases; and the liquid is, if necessary, evaporated, and then left to cool, in order that the slightly soluble chlorate may crystallize out. The mother-liquor affords a second crop of crystals, but they are much more contaminated with potassium chloride. It may be purified by one or two recrystallizations.

Potassium chlorate is soluble in about 20 parts of cold and 2 of boiling water: the crystals are anhydrous, flat, and tabular; in taste it somewhat resembles nitre. When heated it gives off the whole of its oxygen as gas and leaves potassium chloride. By arresting the decomposition when the evolution of gas begins to slacken, and redissolving the salt, potassium perchlorate and chloride may be obtained.

This salt deflagrates violently with combustible matter, explosion often occurring by friction or blows. When about 1 grain-weight of chlorate and an equal quantity of sulphur are rubbed in a mortar, the mixture explodes with a loud report: hence it cannot be used in the preparation of gunpowder instead of the nitrate. Potassium chlorate is now a large article of commerce, being employed, together with phosphorus, in making instantaneous-light matches.

Potassium Perchlorate, $KClO_4 = ClO_3(OK)$.—This salt has been already noticed under the head of perchloric acid. It is best prepared by projecting powdered potassium chlorate into warm nitric acid, when the chloric acid is resolved into perchloric acid, chlorine, and oxygen. The salt is separated from the nitrate by crystallization. Potassium perchlorate is a very slightly soluble salt: it requires 55 parts of cold water, but is more freely taken up at a boiling heat. The crystals are small, and have the figure of an octahedron with square base. It is decomposed by heat in the same manner as the chlorate.

Potassium Bromate, $KBrO_3$, is obtained by passing chlorine into a warm solution of potassium bromide and caustic potash: $KBr + 6KOH + 3Cl_2 = KBrO_3 + 6KCl + 3H_2O$. The solution on cooling deposits part of the bromate, and the rest may be separated by precipitation with alcohol. It crystallizes in six-sided tables or prisms, moderately soluble in hot, very slightly in cold water.

Potassium Iodate, KIO_3, is prepared by passing chlorine into cold water containing iodine in suspension till the whole is dissolved, then adding the calculated quantity of potassium chlorate and warming: $ICl + KClO_3 = Cl_2 + KIO_3$. It forms small cubic crystals, and requires a much stronger heat to decompose it than the chlorate. It unites with iodic acid, forming the salts KIO_3, HIO_3 and $KIO_3, 2HIO_3$. *Potassium periodate*, KIO_4, is formed when chlorine is passed through a mixture of the iodate and caustic potash, and separates in shining crystals isomorphous with the perchlorate.

Potassium Sulphates.—Potassium forms a normal or neutral sulphate two acid sulphates, and an anhydrosulphate.

Normal potassium sulphate, or *Dipotassic sulphate*, $K_2SO_4 = SO_2(OK)_2 = K_2O, SO_3$, is obtained by neutralizing the acid residue left in the retort when nitric acid is prepared with crude potassium carbonate. The solution yields, on cooling, hard transparent crystals of the normal sulphate, which may be redissolved in boiling water and recrystallized.

Potassium sulphate is soluble in about 10 parts of cold, and in a much smaller quantity of boiling water: it has a bitter taste, and is neutral to testpaper. The crystals are combinations of rhombic pyramids and prisms, much resembling those of quartz in figure and appearance: they are anhydrous, and decrepitate when suddenly heated, which is often the case with salts containing no water of crystallization. They are quite insoluble in alcohol.

Acid potassium sulphate, *Hydrogen and potassium sulphate*, or *Monopotassic sulphate*, $KHSO_4 = SO_2(OK)(OH)$, commonly called *bisulphate of potash*.—To obtain this salt, the normal sulphate in powder is mixed with half its

27 *

weight of oil of vitriol, and the whole evaporated to dryness in a platinum vessel placed under a chimney : the fused salt is dissolved in hot water and left to crystallize. The crystals have the figure of flattened rhombic prisms, and are much more soluble than the normal salt, requiring only twice their weight of water at 15.5° C. (59.9° F.), and less than half that quantity at 100° C. The solution has a sour taste and strongly acid reaction.

Potassium disulphate or *Pyrosulphate*, $K_2S_2O_7$, derived from Nordhausen sulphuric acid, $H_2S_2O_7$, and commonly called *anhydrous bisulphate of potash*, is obtained by dissolving equal weights of the normal sulphate and oil of vitriol in a small quantity of warm distilled water, and leaving the solution to cool.

The pyrosulphate crystallizes out in long delicate needles, which if left for several days in the mother-liquor disappear, and give place to crystals of the ordinary acid sulphate above described. This salt is decomposed by a large quantity of water, and is converted by strong fuming sulphuric acid into *hydropotassic pyrosulphate*, KHS_2O_7, which crystallizes in transparent prisms. The normal pyrosulphate in fine powder, heated with an alcoholic solution of potassium hydrosulphide, is converted into sulphate and thiosulphate, with evolution of hydrogen sulphide :—

$$K_2S_2O_7 + 2KHS = K_2SO_4 + K_2S_2O_3 + H_2S.$$

Potassium Nitrate ; *Nitre; Saltpetre,* $KNO_3 = NO_2(OK)$.—This important compound is a natural product, being disengaged by a kind of efflorescence from the surface of the soil in certain dry and hot countries. It may also be produced by artificial means, namely, by the oxidation of ammonia in presence of a powerful base.

In France, large quantities of artificial nitre are prepared by mixing animal refuse of all kinds with old mortar or slaked lime and earth, and placing the mixture in heaps, protected from the rain by a roof, but freely exposed to the air. From time to time the heaps are watered with putrid urine, and the mass is turned over to expose fresh surfaces to the air. When much salt has been formed, the mixture is lixiviated, and the solution, which contains calcium nitrate, is mixed with potassium carbonate; calcium carbonate is formed, and the nitric acid transferred to the alkali. The filtered solution is then made to crystallize, and the crystals are purified by resolution and crystallization, the liquid being stirred to prevent the formation of large crystals.

A large portion of the nitre used in this country comes from India: it is dissolved in water, a little potassium carbonate is added to precipitate lime, and then the salt is purified as above.

Considerable quantities of nitre are now manufactured by decomposing native sodium nitrate (Chili saltpetre) with carbonate or chloride of potassium. In Belgium the potassium carbonate obtained from the ashes of the beet-root sugar manufactories is largely used for this purpose; the potassium nitrate thus prepared is very pure, and is produced at a low price.

Potassium nitrate crystallizes in anhydrous six-sided prisms, with dihedral summits, belonging to the rhombic or trimetric system: it is soluble in 7 parts of water at 15.5° C. (59.9° F.), and in its own weight of boiling water. Its taste is saline and cooling, and it is without action on vegetable colors. It melts at a temperature below redness, and is completely decomposed by a strong heat.

When it is thrown on the surface of many metals in a state of fusion, or mixed with combustible matter and heated, rapid oxidation ensues at the expense of the oxygen of the nitric acid. Examples of such mixtures are found in common gunpowder, and in nearly all pyrotechnic compositions, which burn in this manner independently of the oxygen of the air, and even under water. Gunpowder is made by very intimately mixing together potassium nitrate, charcoal, and sulphur, in proportions which approach 2 molecules of nitre, 3 atoms of carbon, and 1 atom of sulphur.

These quantities give, reckoned to 100 parts, and compared with the proportions used in the manufacture of the English government powder, the following results:

	Theory.	Proportions in practice.
Potassium nitrate . . .	74.8	75
Charcoal . . .	13.3	15
Sulphur . . .	11.9	10
	100.0	100

The nitre is rendered very pure by the means already mentioned, freed from water by fusion, and ground to fine powder; the sulphur and charcoal —the latter being made from light wood, as. dogwood or alder—are also finely ground, after which the materials are weighed out, moistened with water, and thoroughly mixed by grinding under an edge-mill. The mass is then subjected to great pressure, and the mill-cake thus produced is broken in pieces, and placed in sieves made of perforated vellum moved by machinery, each containing, in addition, a round piece of heavy wood. The grains of powder broken off by attrition fall through the holes in the skin, and are easily separated from the dust by sifting. The powder is, lastly, dried by exposure to steam-heat, and sometimes glazed or polished by agitation in a kind of cask mounted on an axis.

It was formerly supposed that when gunpowder is fired, the whole of the oxygen of the potassium nitrate was transferred to the carbon, forming carbon dioxide, the sulphur combining with the potassium, and the nitrogen being set free. There is no doubt that this reaction does take place to a considerable extent, and that the large volume of gas thus produced, and still further expanded by the very high temperature, sufficiently accounts for the explosive effects. But recent investigations by Bunsen, Karolyi, and others, have shown that the actual products of the combustion of gunpowder are much more complicated than this theory would indicate, a very large number of products being formed, and a considerable portion of the oxygen being transferred to the potassium sulphide, converting it into sulphate, which in fact constitutes the chief portion of the solid residue and of the smoke formed by the explosion.*

Potassium Phosphates.—The *normal Orthophosphate*, K_3PO_4, formed by igniting phosphoric acid with excess of potassium carbonate, is easily soluble in water, and crystallizes in small needles. The *dipotassic salt*, K_2HPO_4, obtained by mixing aqueous phosphoric acid with a quantity of hydroxide or carbonate of potassium sufficient to produce a slight alkaline reaction, and evaporating, is easily soluble in water and uncrystallizable. The *monopotassic salt*, KH_2PO_4, obtained by using a slight excess of phosphoric acid, forms small needle-shaped crystals, belonging to the quadratic system, easily soluble in water, insoluble in alcohol.

Normal Potassium Pyrophosphate, $K_4P_2O_7$, formed by igniting dipotassic orthophosphate, is deliquescent, and separates from aqueous solution in fibrous crystals containing $3H_2O$. The *acid salt*, $K_2H_2P_2O_7$, separates as a deliquescent mass on adding alcohol to a solution of the normal pyrophosphate in acetic acid.

Potassium Monometaphosphate, KPO_3, obtained by igniting monopotassic orthophosphate, is almost insoluble in water. The *dimetaphosphate*, $K_2P_2O_6$-H_2O, prepared by decomposing the corresponding copper salt with potassium

* See Watts's Dictionary of Chemistry, vol. ii., p. 958.

sulphide, is soluble in water and crystallizable; converted by ignition into the monometaphosphate.

Potassium Arsenates.—The *normal salt*, K_3AsO_4, obtained by treating arsenic acid with excess of potash, is crystalline. The *dipotassic salt*, K_2HAsO_4, is difficult to crystallize. The *monopotassic salt*, KH_2AsO_4, forms large crystals.

Potassium Arsenites.—The salt, $KAsO_2,H_3AsO_3$, is obtained as a crystalline powder on adding alcohol to a solution of arsenious oxide in the minimum quantity of potash. This salt, heated with solution of potassium carbonate, is converted into the *meta-arsenite*, $KAsO_2$, and this latter heated with caustic potash yields the *diarsenite*, $K_4As_2O_5$, likewise precipitable by alcohol.

A solution of potassium arsenite, known in medicine as ."Fowler's solution," is prepared by boiling 1 pt. arsenious oxide and 1 pt. potassium carbonate in distilled water, and diluting to 90 parts of solution.

Borates.—The *metaborate*, $K.BO_2$, obtained by fusing a mixture of boric acid and potassium carbonate in the requisite proportions, is slightly soluble in water, and separates in small monoclinic crystals. It has an alkaline reaction, and absorbs carbonic acid from the air, being thereby converted into the *pyroborate* or *tetraborate*, $K_2B_4O_7 = K_2O,2B_2O_3$. This latter, which is also formed on mixing a solution of boric acid with a slight excess of potash, is easily soluble, and crystallizes in hexagonal prisms containing $5H_2O$. The *triborate*, $2KB_3O_5,5H_2O$, or $K_2O,3B_2O_3,5H_2O$, is formed on mixing the hot solutions of boric acid (1 mol.) and potassium carbonate (2 mols.), and separates in glittering rhombic crystals. The *pentaborate*, $KB_5O_8,4H_2O$, or $K_2O,5B_2O_3,4H_2O$, separates in rhombic octahedrons from a hot solution of caustic potash saturated with boric acid.

Silicates.—The *metasilicate*, K_2SiO_3 or K_2O,SiO_2, is formed when silica is fused with the calculated quantity of potassium carbonate, or with any larger quantity: for experiment shows that one molecule of silica cannot decompose more than one molecule of potassium carbonate, so that, for example, the reaction $2(K_2O,CO_2) + SiO_2 = 2K_2O,SiO_2 + 2CO_2$ cannot take place, and the orthosilicate, K_4SiO_4, does not exist. The *tetrasilicate*, $K_2Si_4O_9$ (Fuchs's soluble glass), is prepared by fusing 45 pts. quartz, 30 potash, and 3 powdered charcoal, boiling the resulting grayish-black glass with water, mixing the concentrated solution with strong alcohol, and washing the precipitate thereby formed with a little cold water. The potash water-glass thus obtained is soluble in water, and may be used for the same purposes as soda water-glass (p. 329), but is now almost superseded by the latter.

Potassium Fluosilicate or *Silicofluoride*, K_2SiF_6, is precipitated on adding hydrofluosilicic acid to a soluble potassium salt as a semi-transparent mass, which after washing and drying forms a fine white powder, sparingly soluble in cold, freely in hot water. By slow cooling it may be obtained in shining octahedrons.

Potassium Carbonates.—Potassium forms two well-defined carbonates, namely, a normal or neutral carbonate, K_2CO_3, and an acid salt containing $KHCO_3$.

Normal Potassium Carbonate, or *Dipotassic Carbonate*, $K_2CO_3 = CO(OK)_2 = K_2O,CO_2$.—Potassium salts of vegetable acids are of constant occurrence in plants, in the economy of which they perform important functions. The potassium is derived from the soil, which, when capable of supporting vegetable life, always contains that substance. When plants are burned, the organic acids are destroyed, and the potassium is left in the state of carbonate.

It is by these indirect means that the carbonate, and thence a large proportion of the salts of potassium, are obtained. The great natural depository of the alkali is the felspar of granitic and other unstratified rocks, where it is combined with silica and in an insoluble state. The extraction thence is attended with great difficulties, and many attempts at manufacturing it on a large scale from this source have failed; but experiments made by Mr. T. O. Ward appear to indicate that the object may be accomplished by fusing potassic rocks with a mixture of calcium carbonate and fluoride. There are, however, natural processes at work by which the potash is constantly being eliminated from these rocks. Under the influence of atmospheric agencies, the rocks disintegrate into soils, and as the alkali acquires solubility, it is gradually taken up by plants, and accumulates in their substance in a condition highly favorable to its subsequent applications.

Potassium-salts are always most abundant in the green and tender parts of plants, as may be expected, since from these evaporation of nearly pure water takes place to a large extent: the solid timber of forest trees contains comparatively little.

In preparing the salt on an extensive scale, the ashes are subjected to a process called *lixiviation:* they are put into a large cask or tun, having, near the bottom, an aperture stopped by a plug, and a quantity of water is added. After some hours the liquid is drawn off, and more water added, that the whole of the soluble matter may be removed. The weakest solutions are poured in place of water upon fresh quantities of ash. The solutions are then evaporated to dryness, and the residue is calcined, to remove a little brown organic matter: the product is the crude potash or pearlash of commerce, of which very large quantities are obtained from Russia and America. The salt is very impure, containing potassium silicate, sulphate, chloride, etc.

The purified potassium carbonate of pharmacy is prepared from the crude article by adding an equal weight of cold water, agitating and filtering, most of the foreign salts, from their inferior solubility, being left behind. The solution is then boiled down to a very small bulk, and left to cool, when the carbonate separates in small crystals containing 2 molecules of water, $K_2CO_3,2H_2O$; these are drained from the mother-liquor, and then dried in a stove.

A still purer salt may be obtained by exposing to a red heat purified cream of tartar (acid potassium tartrate), and separating the carbonate by solution in water and crystallization or evaporation to dryness.

Potassium carbonate is extremely deliquescent, and soluble in less than its own weight of water; the solution is highly alkaline to test-paper. It is insoluble in alcohol. By heat the water of crystallization is driven off, and by a temperature of full ignition the salt is fused, but not otherwise changed. This substance is largely used in the arts, and is a compound of great importance.

Acid Potassium Carbonate, Hydrogen and Potassium Carbonate, or *Monopotassic Carbonate,* $KHCO_3$, commonly called *Bicarbonate of Potash.*—When a stream of carbonic acid gas is passed through a cold solution of normal potassium carbonate, the gas is rapidly absorbed, and a white, crystalline, less soluble substance separated, which is the acid salt. It is collected, pressed, redissolved in warm water, and the solution is left to crystallize.

Acid potassium carbonate is much less soluble than the normal carbonate, requiring 4 parts of cold water to dissolve it. The solution is nearly neutral to test-paper, and has a much milder taste than the normal salt. When boiled it gives off carbon dioxide. The crystals, which are large and beautiful, derive their form from a monoclinic prism: they are decomposed by

14 *

heat, water and carbon dioxide being evolved, and normal carbonate left behind :—

$$2KHCO_3 = K_2CO_3 + H_2O + CO_2.$$

Potassium Sulphides.—Potassium heated in sulphur vapor burns with great brilliancy. It unites with sulphur in five different proportions, forming the compounds K_2S, K_2S_2, K_2S_3, K_2S_4, K_2S_5; also hydrosulphide or sulphydrate, KHS.

Monosulphide, K_2S.—It is doubtful whether this compound has been obtained in the pure state. It is commonly said to be produced by heating potassium sulphate in a current of dry hydrogen, or by igniting the same salt in a covered vessel with finely divided charcoal; but, according to Bauer, one of the higher sulphides is always formed at the same time, together with oxide of potassium. The product has a reddish-yellow color, is deliquescent, and acts as a caustic on the skin. When potassium sulphate is heated in a covered crucible with excess of lampblack, a mixture of potassium sulphide and finely divided carbon is obtained, which takes fire spontaneously on coming in contact with the air. The monosulphide might perhaps be obtained pure by heating 1 molecule of potassium sulphydrate, KHS, with 1 atom of the metal.

When sulphydric acid gas is passed to saturation into a solution of caustic potash, a solution of the sulphydrate is obtained, which is colorless at first, but if exposed to the air quickly absorbs oxygen and turns yellow, in consequence of the formation of bisulphide :—

$$2KHS + O = K_2S_2 + H_2O.$$

If a solution of potash be divided into two parts, one-half saturated with hydrogen sulphide, and then mixed with the other, a solution is formed which may contain potassium monosulphide :—

$$KHS + KHO = K_2S + H_2O.$$

But it is also possible that the hydroxide and hydrosulphide may mix without mutual decomposition. The solution when mixed with one of the stronger acids, gives off hydrogen sulphide without deposition of sulphur, a reaction which is consistent with either view of its constitution.

The *bisulphide*, K_2S_2, is formed, as already observed, on exposing a solution of the hydrosulphide to the air till it begins to show turbidity. By evaporation in a vacuum it is obtained as an orange-colored, easily fusible substance.

The *trisulphide*, K_2S_3, is obtained by passing the vapor of carbon bisulphide over ignited potassium carbonate, as long as gas continues to escape :—

$$2K_2CO_3 + 3CS_2 = 2K_2S_3 + 4CO + CO_2.$$

Also, together with potassium sulphate, forming one of the mixtures called *liver of sulphur*, by melting 552 parts (4 molecules) of potassium carbonate with 320 parts (10 atoms) of sulphur —:

$$4K_2CO_3 + S_{10} = K_2SO_4 + 3K_2S_3 + 4CO_2.$$

The *tetrasulphide*, K_2S_4, is formed by reducing potassium sulphate with the vapor of carbon bisulphide.

The *pentasulphide*, K_2S_5, is formed by boiling a solution of any of the preceding sulphides with excess of sulphur till it is saturated, or by fusing either of them in the dry state with sulphur. The excess of sulphur then separates and floats above the dark-brown pentasulphide.

Liver of sulphur, or *hepar sulphuris*, is a name given to a brownish substance, sometimes used in medicine, made by fusing together different proportions of potassium carbonate and sulphur. It is a variable mixture of the two higher sulphides with thiosulphate and sulphate of potassium.

When equal parts of sulphur and dry potassium carbonate are melted together at a temperature not exceeding 250° C. (482° F.), the decomposition of the salt is quite complete, and all the carbon dioxide is expelled. The fused mass dissolves in water, with the exception of a little mechanically mixed sulphur, with dark-brown color, and the solution is found to contain nothing besides pentasulphide and thiosulphate of potassium :—

$$3K_2O + S_{12} = 2K_2S_5 + K_2S_2O_3.$$

When the mixture has been exposed to a temperature approaching that of ignition, it is found, on the contrary, to contain potassium sulphate, arising from the decomposition of the thiosulphate which then occurs :—

$$4K_2S_2O_3 = K_2S_5 + 3K_2SO_4.$$

From both these mixtures the potassium pentasulphide may be extracted by alcohol, in which it dissolves.

When the carbonate is fused with half its weight of sulphur only, the trisulphide is produced, as above indicated, instead of the pentasulphide.

The effects described happen in the same manner when potassium hydroxide is substituted for the carbonate; also, when a solution of the hydroxide is boiled with sulphur, a mixture of sulphide and thiosulphate always results.

Potassium-salts are colorless when not associated with a colored metallic oxide or acid. They are all more or less soluble in water, and may be distinguished by the following characters :—

(1.) Solution of *tartaric acid*, added in excess to a moderately strong solution of a potassium-salt, gives, after some time, a white crystalline precipitate of cream of tartar; the effect is greatly promoted by strong agitation.

(2.) Solution of *platinic chloride*, with a little hydrochloric acid, if necessary, gives under similar conditions a crystalline yellow precipitate, which is a double salt of platinum, tetrachloride, and potassium chloride. Both this compound and cream of tartar are, however, soluble in about 60 parts of cold water. An addition of alcohol increases the delicacy of both tests.

(3.) *Perchloric acid* and *Silicofluoric acid*, added to a potassium-salt, give rise to slightly soluble white precipitates.

(4.) Potassium-salts usually color the outer blowpipe flame purple or violet; this reaction is clearly perceptible only when the potassium-salts are pure.

(5.) The spectral phenomena exhibited by potassium compounds are mentioned at page 96.

SODIUM.

Symbol, Na (Natrium).   Atomic weight, ~~13.~~ 23

SODIUM is a very abundant element, and very widely diffused. It occurs in large quantities as chloride, in rock-salt, sea-water, salt-springs, and many other mineral waters; more rarely as carbonate, borate, and sulphate in solution or in the solid state, and as silicate in many minerals.

Metallic sodium was obtained by Davy soon after the discovery of potassium, and by similar means. Gay-Lussac and Thénard afterward prepared it by decomposing sodium hydroxide with metallic iron at a white heat; and Brunner showed that it may be prepared with much greater facility by distilling a mixture of sodium carbonate and charcoal.

The preparation of sodium by this last-mentioned process is much easier than that of potassium, not being complicated, or only to a slight extent, by the formation of secondary products. Within the last few years it has been considerably improved by Deville and others, and carried out on the manufacturing scale, sodium being now employed in considerable quantity as a reducing agent, especially in the manufacture of aluminium and magnesium.

The sodium carbonate used for the preparation is prepared by calcining the crystallized neutral carbonate. It must be thoroughly dried, then pounded, and mixed with a slight excess of pounded charcoal or coal. An inactive substance, viz., pounded chalk, is also added to keep the mixture in a pasty condition during the operation, and prevent the fused sodium carbonate from separating from the charcoal. The following are the proportions recommended by Deville:—

For Laboratory Operations.		For Manufacturing Operations.	
Dry sodium carbonate	. 717 parts.	Dry sodium carbonate	. 30 kilogr.
Charcoal . . .	. 175 "	Coal	. 13 "
Chalk . . .	. 108 "	Chalk . . .	. 3 "

These materials must be very intimately mixed by pounding and sifting, and it is advantageous to calcine the mixture before introducing it into the distilling apparatus, provided the calcination can be effected by the waste heat of a furnace; the mixture is thereby rendered more compact, so that a much larger quantity can be introduced into a vessel of given size.

The distillation is performed, on the laboratory scale, in a mercury-bottle heated exactly in the manner described for the preparation of potassium. For manufacturing operations, the mixture is introduced into iron cylinders, which are heated in a reverberatory furnace, and so arranged that, at the end of the distillation, the exhausted charge may be withdrawn and a fresh charge introduced, without displacing the cylinders or putting out the fire. The receivers used in either case are the same in form and dimensions as those employed in the preparation of potassium (p. 312).

When the process goes on well, the sodium collected in the receivers is nearly pure; it may be completely purified by melting it under a thin layer of petroleum. This liquid is decanted as soon as the sodium becomes perfectly fluid, and the metal is run into moulds like those used for casting lead or zinc.

Sodium is a silver-white metal, greatly resembling potassium in every respect. Its specific gravity is 0.972. It is soft at common temperatures, melts at 97.6° C. (207.7° F.), and oxidizes very rapidly in the air. When placed on the surface of cold water, it decomposes that liquid with great violence, but seldom takes fire unless the motious of the fragment are restrained, and its rapid cooling is diminished by adding gum or starch to the water. With hot water it takes fire at once, burning with a bright yellow flame, and producing a solution of soda.

Sodium Chloride; Common Salt, NaCl.—This very important substance is found in many parts of the world in solid beds or irregular strata of immense thickness, as in Cheshire, Spain, Gallicia, at Stassfurt, and many other localities. An inexhaustible supply exists also in the waters of the ocean, and large quantities are obtained from saline springs.

Salt is obtained from sea-water by evaporation in salterns or brine-pans, with the aid of air and sunshine. This process is extensively practised on the coasts of France, Spain, and Portugal, also in this country at Hayling Island, near Portsmouth; at Lymington, in Dorsetshire; and in Scotland, at Saltcoats, on the Ayrshire coast. The salt thus separated is called b a y - s a l t ; the mother-liquor, called *bittern*, contains the sulphates, chlorides, and bromides of magnesium and potassium, and is utilized as a source of bromine.

Rock-salt is sometimes mined and brought to the surface in the solid state, but generally speaking it is too impure for use. It is therefore more usual to form an artificial brine-well by sinking a shaft into the rock-salt, and, if necessary, introducing water. This when saturated is pumped up, and evaporated more or less rapidly in large iron pans. As the salt separates, it is removed from the bottom of the vessel by means of a scoop, pressed while still moist into moulds, and then transferred to the drying-stove. When large crystals are required, as for the coarse-grained *salt* used in curing provisions, the evaporation is slowly conducted. Common salt is apt to be contaminated with sodium sulphate, calcium sulphate, and magnesium chloride, the last of which renders it liable to become damp in the air. These impurities may be removed by passing hydrochloric acid gas into a saturated solution of the salt, whereby pure sodium chloride is precipitated, the other salts remaining in solution. The precipitate is washed on a filter with strong hydrochloric acid, then dried and fused in a platinum basin.

Sodium chloride, when pure, is not deliquescent in moderately dry air. It crystallizes in anhydrous cubes, which are often grouped together into pyramids or steps. It requires about $2\frac{1}{2}$ parts of water at 60° F. for solution, and its solubility is not sensibly increased by heat; it dissolves to some extent in spirits of wine, but is nearly insoluble in absolute alcohol. It melts at a red heat, and volatilizes at a still higher temperature. The economical uses of common salt are well known.

The *iodide and bromide of sodium* much resemble the corresponding potassium-compounds; they crystallize in cubes, which are anhydrous, and very soluble in water.

Sodium Oxides.—Sodium forms a monoxide and a dioxide; also a hydroxide corresponding with the former.

Sodium Monoxide, or *Anhydrous Soda,* Na_2O, is produced, together with the dioxide, when sodium burns in the air, and may be obtained pure by exposing the dioxide to a very high temperature, or by heating sodium hydroxide with an equivalent quantity of sodium: $2NaHO + Na_2 = 2Na_2O + H_2$. It is a gray mass, which melts at a red heat, and volatilizes with difficulty.

Sodium Hydroxide, or *Caustic Soda,* NaHO, or Na_2O,H_2O.—This substance is prepared by decomposing a somewhat dilute solution of sodium carbonate with calcium hydroxide: the description of the process employed in the case of potassium hydroxide, and the precautions necessary, apply word for word to that of d   hydroxide.

2so ium

The solid hydroxide is a white fusible substance, very similar in properties to potassium hydroxide. It is deliquescent, but dries up again after a time in consequence of the absorption of carbonic acid. The solution is highly alkaline, and a powerful solvent for animal matter: it is used in large quantity for making soap.

The strength of a solution of caustic soda may be determined from a knowledge of its density, by the aid of the following table drawn up by Schiff* :—

NaOH per cent.	Density.	NaOH per cent.	Density.
1	. . . . 1.012	35	. . . . 1.384
5	. . . . 1.059	40	. . . . 1.437
10	. . . . 1.115	45	. . . . 1.488
15	. . . . 1.170	50	. . . . 1.540
20	. . . . 1.225	55	. . . . 1.591
25	. . . . 1.279	60	. . . . 1.643
30	. . . . 1.332		

Sodium Dioxide, Na_2O_2, formed by heating sodium to about 200° C. (392° F.) in a current of dry air, is white at ordinary temperatures, but becomes yellow when heated. It dissolves in water without decomposition. The solution may be evaporated under reduced pressure, depositing crystalline plates having the composition $Na_2O_2,8H_2O$. These crystals left to effloresce over oil of vitriol for nine days lose three-fourths of their water, and yield another hydrate containing $Na_2O_2,2H_2O$. The aqueous solution of sodium dioxide, when heated on the water-bath, is decomposed into oxygen and the monoxide.

Sodium Hypochlorite, $NaOCl$, is contained, together with the chloride, in the bleaching liquid (formerly known as chloride of soda, or *Eau de Labarraque*), obtained by passing chlorine into a solution of caustic soda.

Sodium Chlorate, $NaClO_3$, obtained by neutralizing sodium carbonate with chloric acid, or by boiling 9 parts of potassium chlorate with 7 parts sodium silicofluoride, crystallizes on cooling in regular tetrahedrons. It is much more soluble than potassium chlorate, 100 parts of water dissolving 81.9 parts of it at 0° C., and 232.6 parts at 100° C. Its great solubility renders it very useful to the calico-printer in the production of aniline-black.

Sodium Hyposulphite, $NaHSO_2$, is formed by the deoxidizing action of zinc on the sulphite. Its preparation has already been described (p. 180). It crystallizes in needles soluble in water and in weak spirit, the solution exhibiting strong bleaching and reducing properties: hence it is used as a reducing agent for indigo, and for the estimation of oxygen.

Sodium Sulphites.—The *normal salt*, Na_2SO_3, obtained by saturating a solution of sodium carbonate with sulphur dioxide, and then adding an equal quantity of the carbonate, crystallizes in transparent, monoclinic prisms, which contain seven molecules of water, and give up the whole of it at 150° C. (302° F.). The solution has a sharp taste and alkaline reaction. The *acid salt*, $NaHSO_3$, separates from a cold solution of sodium carbonate saturated with sulphur dioxide in turbid crystals, and is precipitated from its aqueous solution by alcohol as a white powder.

Sodium Sulphates.—The *normal salt*, $Na_2SO_4,10H_2O$, commonly called

* Liebig's Annalen, cvii. 300.

Glauber's salt, is a by-product in many chemical operations and an intermediate product in the manufacture of the carbonate (see p. 332): it may of course be prepared directly, if wanted pure, by adding dilute sulphuric acid to saturation to a solution of sodium carbonate. It crystallizes in forms derived from an oblique rhombic prism. The crystals contain 10 molecules of water, are efflorescent, and undergo watery fusion when heated, like those of the carbonate; they are soluble in twice their weight of cold water, and rapidly increase in solubility as the temperature of the liquid rises to 33° C. (91.4° F.), at which point a maximum is reached, 100 parts of water dissolving 117.9 parts of the salt, corresponding with 52 parts anhydrous sodium sulphate (see fig. 106, p. 160). When the salt is heated beyond this point, the solubility diminishes, and a portion of sulphate is deposited. A warm saturated solution evaporated at a high temperature, deposits opaque prismatic crystals, which are anhydrous. The salt has a slightly bitter taste, and is purgative. Mineral springs sometimes contain it, as that at Cheltenham.

Sodium and Hydrogen Sulphate, or *Acid Sodium Sulphate*, $2NaHSO_4,3H_2O$ or $Na_2SO_4,H_2SO_4,3H_2O$, commonly called *Bisulphate of Soda*, is prepared by adding to 10 parts of the anhydrous normal sulphate 7 of oil of vitriol, evaporating the whole to dryness, and gently igniting. The acid sulphate is very soluble in water, and has an acid reaction. It is not deliquescent. When very strongly heated the fused salt gives up sulphuric oxide, and is converted into normal sulphate; a change which necessarily supposes the previous formation of a pyrosulphate, $Na_2S_2O_7$ or Na_2SO_4,SO_3.

Sodium Thiosulphate, $Na_2S_2O_3$, formerly called *Hyposulphite*. This salt is formed from the sulphite, Na_2SO_3, by addition of sulphur. There are several modes of preparing it. One of the best is to form normal *sodium sulphite*, by passing a stream of well-washed sulphurous oxide gas into a strong solution of sodium carbonate, and then digesting the solution with sulphur at a gentle heat during several days. By careful evaporation at a moderate temperature, the salt is obtained in large regular crystals, which are very soluble in water. It is much used in photography as a solvent for the unaltered silver chloride, and in the paper-manufacture as an antichlore.

Sodium Nitrate, $NaNO_3$.—This salt, sometimes called *Cubic Nitre* or *Chili Saltpetre*, occurs native, and in enormous quantity, at Tarapaca in Southern Peru, where it forms a regular bed, of great extent, along with gypsum, common salt, and remains of recent shells. The pure salt commouly crystallizes in rhombohedrons, resembling those of calcareous spar. It is deliquescent, and very soluble in water. Sodium nitrate is employed for making nitric acid, but cannot be used for gunpowder, as the mixture burns too slowly and becomes damp in the air. It has been lately used with some success in agriculture as a superficial manure or top-dressing; also for preparing potassium nitrate (p. 318).

Sodium Phosphates.—The chemical relations of these salts have already been explained in speaking of the basicity of acids (p. 305).

(1.) ORTHOPHOSPHATES.—*Hydrodisodic Phosphate* or *Disodic Orthophosphate, Ordinary Phosphate of Soda*, $Na_2HPO_4,12H_2O$, is prepared by precipitating the acid calcium phosphate (obtained in decomposing bone-ash with sulphuric acid) with a slight excess of sodium carbonate, and evaporating the clear liquid. It crystallizes in large, transparent, monoclinic prisms, which are efflorescent, dissolve in 4 parts of cold water, and at a higher temperature give off water and are converted into metaphosphate. The salt is bitter and purgative; its solution is alkaline to test-paper.—*Tri-*

sodic orthophosphate, $Na_3PO_4,12H_2O$, is obtained by adding a solution of caustic soda to the preceding salt. The crystals are slender six-sided prisms, soluble in 5 parts of cold water. It is decomposed by acids, even carbonic, but suffers no change by heat, except the loss of its water of crystallization. Its solution is strongly alkaline. *Monosodic orthophosphate*, NaH_2PO_4,H_2O, often called superphosphate or biphospate, may be obtained by adding phosphoric acid to the ordinary phosphate, until it ceases to precipitate barium chloride, and exposing the concentrated solution to cold. The crystals are prismatic, very soluble, and have an acid reaction. When strongly heated, this salt gives off water and is converted into metaphosphate (*infra*).

(2.) *Pyrophosphates.*—The *normal pyrophosphate*, $Na_4P_2O_7,10H_2O$, is obtained as an anhydrous colorless vitreous mass by igniting disodic orthophosphate, and crystallizes from aqueous solution in monoclinic prisms. The solution is alkaline. The salt is converted into orthophosphate on adding an acid to its aqueous solution.—The *acid pyrophosphate*, $Na_2H_2P_2O_7$, is obtained by heating disodic orthophosphate to 150° C. (302° F.) with strong hydrochloric acid, or by heating monosodic orthophosphate to 190–204° C. (374–399° F.): $2NaH_2PO_4 = H_2O + Na_2H_2P_2O_7$; also by dissolving the normal pyrophosphate and precipitating with alcohol. This last process yields it as a white crystalline powder. It is easily soluble in water, and forms an acid solution.

Metaphosphates, $nNaPO_3$ (Mono-, Di-, Tri-, etc., p. 305).—$NaPO_3$ is formed by heating sodium hydroxide with a slight excess of phosphoric acid; by heating dihydrosodic orthophosphate to redness for a short time; and by fusing 2 parts sodium nitrate with 1 part of syrupy phosphoric acid. It is insoluble in water, but dissolves in acids, and is converted into orthophosphate by boiling with caustic soda. $Na_2P_2O_6 + 2H_2O$, obtained by boiling the corresponding copper-salt with sodium sulphide, crystallizes in slender needles, soluble in 7.2 parts of water. It has a strong tendency to form double salts, *e. g.*, $NaKP_2O_6,H_2O$, which crystallizes from a mixed solution of the two simple salts.—$Na_3P_3O_9,6H_2O$ may be prepared by exposing microcosmic salt, $Na(NH_4)HPO_4,4H_2O$, to a moderate heat, or by leaving the same salt in a state of fusion to cool slowly. It dissolves in 4.5 parts of cold water, and separates on spontaneous evaporation in large triclinic prisms.—$Na_4P_4O_8$, prepared by decomposing the corresponding lead-salt with sodium sulphide, forms a thick gummy solution, and remains on evaporation as a transparent colloïdal mass.—$Na_6P_6O_{12}$ is prepared by fusing either dihydric orthophosphate or microcosmic salt, and cooling the fused mass as quickly as possible, as otherwise the tetrametaphosphate may be formed. It dissolves easily in water and in alcohol, and remains on evaporation at 38° C. (100.4° F.) as a gummy mass. The aqueous solution is not altered by boiling with caustic soda.

Sodium Arsenates.—These salts closely resemble the corresponding phosphates. On adding sodium carbonate to a solution of arsenic acid till an alkaline reaction becomes apparent, and evaporating the salt, $Na_2HAsO_4,$ $12H_2O$, is produced, indistinguishable in appearance from common sodium phosphate. This salt also crystallizes with $7H_2O$. The *trisodic salt*, $Na_3AsO_4,12H_2O$, is formed when sodium carbonate in excess is fused with arsenic acid, or when the preceding salt is mixed with caustic soda. The *monosodic salt*, NaH_2AsO_4,H_2O, is made by substituting an excess of arsenic acid for the solution of alkali. The alkaline arsenates which contain basic water lose the latter at a red heat, but, unlike the phosphates, recover it when again dissolved. The arsenates of the alkalies are soluble in water: those of the earths and other metallic oxides are insoluble, but are dissolved by acids. The reddish-brown precipitate with silver nitrate is highly characteristic of arsenic acid.

An impure sodium arsenate, prepared by dissolving arsenious oxide in caustic soda, adding sodium nitrate, boiling the solution down, and heating the residual mass to dryness in a furnace, is largely used in calico-printing as a substitute for cowdung in clearing cloth after mordanting: it is known in the trade as "dung substitute."

Sodium Borates.—The *Orthoborate*, Na_3BO_3, is obtained by fusing boron trioxide with excess of caustic soda, according to the equation: $B_2O_3 + 6NaOH = 2Na_3BO_3 + 3H_2O$. It is very unstable, being resolved by solution in water into a hydrated metaborate and sodium hydroxide: $Na_3BO_3 + H_2O = NaBO_2 + 2NaOH$.

The *Pyroborate*, or *Borax*, $Na_2B_4O_7,10H_2O$ or $Na_2O,2B_2O_3,10H_2O$, occurs in the waters of certain lakes in Thibet and Persia, and is imported in a crude state from India under the name of *tincal*. When purified it constitutes the borax of commerce. Much borax is now, however, manufactured from the native boric acid of Tuscany, also from a native calcium borate called *hayesine*, which occurs in Southern Peru. Large quantities of borax are also obtained from the borax lake in California, the water of which contains a pound of crystallized borax in 13 gallons. Borax crystallizes in six-sided prisms, which effloresce in dry air, and require 20 parts of cold and six of boiling water for solution. On exposing it to heat, the 10 molecules of water of crystallization are expelled, and at a higher temperature the salt fuses, and assumes a glassy appearance on cooling: in this state it is much used for blowpipe experiments, the metallic oxides dissolving in it to transparent beads, many of which are distinguished by characteristic colors. By particular management, crystals of borax can be obtained with 5 molecules of water: they are very hard, and permanent in the air. Borax, though by constitution an acid salt, has an alkaline reaction to test-paper. It is used in the arts for soldering metals, its action consisting in rendering the surfaces to be joined metallic by dissolving the oxides; and it sometimes enters into the composition of the glaze with which stoneware is covered. *Octahedral borax*, $Na_2B_4O_7,5H_2O$, is deposited when a supersaturated solution of borax is allowed to evaporate in a warm place.

Sodium Metaborate, $NaBo_2$ or Na_2O,B_2O_3, is obtained by fusing common borax and sodium carbonate together in equivalent proportions, and dissolving the mass in water. It forms large monoclinic crystals having the composition $NaBO_2,3H_2O$.

Sodium Silicates.—The *Metasilicate*, Na_2SiO_3 or Na_2O,SiO_2, obtained by fusing silica and sodium carbonate together in equal numbers of molecules, is soluble in water and separates on evaporation in crystals containing $7H_2O$. With excess of sodium carbonate, the $\frac{4}{3}$-*silicate* $Na_8Si_3O_{10}$ or $4Na_2O,3SiO_2$ is obtained, as shown by the equation $3SiO_2 + 4(Na_2O,CO_2) = 4Na_2O,3SiO_2 + 4CO_2$.

The *quadrisilicate*, $Na_2Si_4O_9$ or $Na_2O,4SiO_2$, commonly known as silicate of soda or *soluble glass*, is prepared by heating white sand (180 parts), soda ash (100 parts), and charcoal (3 parts), in a reverberatory furnace, or by dissolving powdered flint under pressure in a hot concentrated solution of caustic soda. The first method yields it in the form of a transparent glassy mass, generally having a yellow, brown, or green color, and dissolving readily when pulverized, in boiling water, to a thick viscid liquid. It is used in the manufacture of artificial stone, which is made by mixing the soluble glass with sand and lime; also as a cement for joining broken surfaces of porcelain, stone, etc.; in fixing fresco colors by the process of stereochromy;* and for the manufacture of silicated soap, which is pre-

* See Richardson and Watts's Chemical Technology, vol. i. Part iv. pp. 69–104.

28 *

pared in large quantities by adding a solution of sodium silicate to the soap while setting.*

Manufacture of Glass.—Ordinary glass is a mixture of various insoluble silicates with excess of silica, altogether destitute of crystalline structure; the simple silicates, formed by fusing the bases with silicic acid in equivalent proportions, very often crystallize, which happens also with the greater number of the natural silicates included among the earthy minerals. Compounds identical with some of these are also occasionally formed in artificial processes, where large masses of melted glassy matter are suffered to cool slowly. The alkaline silicates, when in the fused state, have the power of dissolving a large quantity of silica.

Two principal varieties of glass are met with in commerce, namely, glass composed of silica, alkali, and lime; and glass containing a large proportion of lead silicate: *crown* and *plate glass* belong to the former division; *flint glass*, and the material of artificial gems, to the latter. The lead promotes fusibility, and confers also density and lustre. Common green bottle-glass contains no lead, but much silicate of iron, derived from the impure materials.

The principle of the glass manufacture is very simple. Silica, in the shape of sand, is heated with potassium or sodium carbonate, and slaked lime or lead oxide; at a high temperature fusion and combination occur, and the carbonic acid is expelled. Glauber's salt mixed with charcoal is sometimes substituted for soda. When the melted mass has become perfectly clear and free from air-bubbles, it is left to cool until it assumes the peculiar tenacious condition proper for working.

The operation of fusion is conducted in large crucibles of refractory fire-clay, which in the case of lead-glass are covered by a dome at the top, and have an opening at the side, by which the materials are introduced and the melted glass withdrawn. Great care is exercised in the choice of the sand, which must be quite white and free from iron oxide. Red lead, one of the higher oxides, is preferred to litharge, although immediately reduced to monoxide by the heat, the liberated oxygen serving to destroy any combustible matter that might accidentally find its way into the crucible, and stain the glass by reducing a portion of the lead. Potash gives a better glass than soda, although the latter is very generally employed from its lower price. A certain proportion of broken and waste glass of the same kind is always added to the other materials.

Articles of blown glass are thus made: The workman begins by collecting a proper quantity of soft pasty glass at the end of his *blowpipe*, an iron tube 5 or 6 feet in length, terminated by a mouth-piece of wood; he then begins blowing, by which the lump is expanded into a kind of flask, susceptible of having its form modified by the position in which it is held, and the velocity of rotation continually given to the iron tube. If an open-mouthed vessel is to be made, an iron rod, called a *pontil* or *puntil*, is dipped into the glass-pot and applied to the bottom of the flask, to which it thus serves as a handle, the blowpipe being removed by the application of a cold iron to the neck. The vessel is then reheated at a hole left for the purpose in the wall of the furnace, the aperture is enlarged, and the vessel otherwise altered in figure by the aid of a few simple tools until completed. It is then detached, and carried to the annealing oven, where it undergoes slow and gradual cooling during many hours, the object of which is to obviate the excessive brittleness always exhibited by glass which has been quickly cooled. The large circular *tables* of crown glass are made by a very curious process of this kind: the globular flask at first produced, transferred from the blowpipe to the pontil, is suddenly made to assume the form of a flat

* *Op. cit.,* vol. i. Part iii. pp. 710-716.

disc by the centrifugal force of the rapid rotatory movement given to the rod. *Plate glass* is cast upon a flat metal table, and after very careful annealing is ground true and polished by suitable machinery. Tubes are made by rapidly drawing out a hollow cylinder; and from these a great variety of useful small apparatus may be constructed with the help of a lamp and blowpipe, or, still better, the bellows-table of the barometer-maker. Small tubes may be bent in the flame of a spirit-lamp or gas-jet, and cut with a file, a scratch being made, and the two portions pulled or broken asunder in a way easily learned by a few trials.

Specimens of the two chief varieties of glass gave on analysis the following results:—

Bohemian plate glass (excellent).		English flint glass.	
Silica	60.0	Silica	51.93
Potassium oxide .	25.0	Potassium oxide .	13.77
Lime	12.5	Lead oxide . .	33.28
	97.5		98.98

The difficultly fusible white Bohemian tube, so valuable in organic analysis, has been found to contain in 100 parts—

Silica	72.80
Lime, with traces of alumina . . .	9.68
Magnesia	.40
Potassium oxide	16.80
Traces of manganese, etc., and loss . . .	.32

Different colors are often communicated to glass by metallic oxides. Thus, oxide of cobalt gives deep blue; oxide of manganese, amethyst; cuprous oxide, ruby-red; cupric oxide, green; the oxides of iron, dull green or brown, etc. These oxides are either added to the melted contents of the glass-pot in which they dissolve, or applied in a particular manner to the surface of the plate or other object, which is then reheated, until fusion of the coloring matter occurs: such is the practice of enamelling and glass-painting. An opaque white appearance is given by oxide of tin; the enamel of watch-faces is thus prepared.

Toughened Glass.—When ordinary glass is heated till it begins to soften, then plunged into melted paraffin, wax, or other substance melting at a comparatively low temperature, and left to cool gradually, it becomes very tough, so that it may be struck or thrown on the ground without breaking. It has also acquired greater power of resisting heat, and may be heated to redness, then dipped into cold water, and whilst wet again held in the flame without injury. Hence it is well adapted for lamp-chimneys and for culinary vessels. When it does break, however, it splits up into a multitude of minute angular fragments, indicating a crystalline structure, the existence of which is confirmed by the appearance of the toughened glass in polarized light.

Sodium Carbonates.—The *Normal* or *Disodic Carbonate*, $Na_2CO_3,$-$10H_2O$, was once exclusively obtained from the ashes of sea-weeds and of plants, such as the *Salsola soda*, which grow by the seaside, or, being cultivated in suitable localities for the purpose, are afterward subjected to incineration. The *barilla*, still employed to a small extent in soap-making, is thus produced in several places on the coast of Spain, as at Alicante, Carthagena, etc. That made in Brittany is called *varec*.

Sodium carbonate is now manufactured on a stupendous scale from common salt by a series of processes invented by M. Leblanc, which may be divided into two stages:—

(1.) Manufacture of sodium sulphate, or salt-cake, from sodium chloride (common salt); this is called the salt-cake process.

(2.) Manufacture of sodium carbonate, or soda-ash; called the soda-ash process.

(1.) *Salt-cake process.*—This process consists in the decomposition of common salt by sulphuric acid, and is effected in a furnace called the *Salt-cake furnace,* of which fig. 149 represents a section. It consists of a large covered

. Fig. 149.

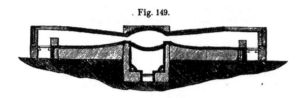

iron pan, placed in the centre, and heated by a fire underneath; and two roasters, or reverberatory furnaces, placed one at each end, and on the hearths of which the salt is completely decomposed. The charge of half a ton of salt is first placed in the iron pan, and then the requisite quantity of sulphuric acid is allowed to pass in upon it. Hydrochloric acid is evolved, and escapes through a flue, with the products of combustion, into towers or scrubbers, filled with coke and bricks moistened with a stream of water; the acid vapors are thus condensed, and the smoke and heated air pass up the chimney. After the mixture of salt and acid has been heated in the iron pan, it becomes converted into a solid mass of acid sodium sulphate and undecomposed sodium chloride:—

$$2NaCl + H_2SO_4 = NaCl + NaHSO_4 + HCl.$$

It is then raked on to the hearths of the furnaces at each side of the decomposing pan, where the flame and heated air of the fire complete the decomposition into neutral sodium sulphate and hydrochloric acid:—

$$NaCl + NaHSO_4 = Na_2SO_4 + HCl.$$

(2.) *Soda-ash process.*—The sulphate is next reduced to powder, and mixed with an equal weight of chalk or limestone, and half as much small coal, both ground or crushed. The mixture is thrown into a reverberatory furnace, and heated to fusion, with constant stirring. When the decomposition is judged complete, the melted matter is raked from the furnace into an iron trough, where it is allowed to cool. This crude product, called *black ash* or *ball-soda,* is broken up into little pieces when cold, and lixiviated with cold or tepid water. The solution is evaporated to dryness, and the salt calcined with a little sawdust in a suitable furnace. The product is the *Soda-ash* or *British alkali* of commerce, which, when of good quality, contains from 48 to 52 per cent. of anhydrous soda, Na_2O, partly in the state of carbonate and partly as hydroxide, the remainder being chiefly sodium sulphate and common salt, with occasional traces of sulphite or thiosulphate and cyanide of sodium. By dissolving soda-ash in hot water, filtering the solution, and then allowing it to cool slowly, the carbonate is deposited in large transparent crystals.

The reaction which takes place in the calcination of the sulphate with chalk and coal-dust seems to consist, first, in the conversion of the sodium sulphate into sulphide by the aid of the combustible matter, and, secondly, iɑ ᚢhe interchange of elements between that substance and the calcium carbonate:—

$$Na_2S + CaCO_3 = CaS + Na_2CO_3.$$

Ammonia-soda process.—Several other processes for the manufacture of soda have been proposed, but the only one which has attained commercial success is that above named, which was first suggested about forty years ago, and is now largely employed in England, on the continent of Europe, and in the United States. It consists in decomposing a solution of common salt with ammonium bicarbonate, whereby the greater part of the sodium is precipitated as bicarbonate, while the ammonia remains in solution as ammonium chloride. This latter salt is heated with lime to liberate ammonia, which is then reconverted into bicarbonate by the carbonic acid evolved in the conversion of the sodium bicarbonate into monocarbonate by heat; and the ammonium bicarbonate thus reproduced is employed to decompose fresh portions of sodium chloride, so that the process is made continuous. The chief advantages of this process are the direct conversion of the sodium chloride into carbonate, which is precipitated from the concentrated liquors uncontaminated with salts of other metals; the absolute freedom of the product from sulphur-compounds; and, lastly, simplicity of plant, saving of fuel, and freedom from noxious vapors and troublesome secondary products.

Its employment on the large scale was, however, for some time retarded by a loss of ammonia, which was found very difficult to obviate; but this loss has by recent improvements been to a great extent overcome, and the ammonia-process is now rapidly gaining ground on that of Leblanc. The following table, drawn up by Mr. Walter Weldon,[*] shows that the quantity of soda now made by the ammonia-process is nearly a fourth of the total quantity produced in the world:—

Present Soda Production of the World.

	Leblanc soda.	Ammonia soda.	Totals.	Ammonia soda per cent. of total soda.
	Tons.	Tons.	Tons.	
Great Britain	380,000	52,000	432,000	12.0
France	70,000	57,125	127,125	44.9
Germany	56,500	44,000	100,500	43.8
Austria	39,000	1,000	40,000	2.5
Belgium	—	8,000	8,000	100.0
United States . . .	—	1,100	1,100	100.0
Total	545,500	163,225	708,725	23.0

From various causes, indeed, which cannot here be discussed, the Leblanc process, so far as the production of soda itself is concerned, has ceased to yield any profit, and the hydrochloric acid which is evolved in the salt-cake process yields only a small profit; so that the process is now carried on mainly for the sake of certain secondary products which are obtained by it. The sulphuric acid used for decomposing the sodium chloride is now made exclusively from the sulphur of pyrites, and the pyrites used for the purpose in this country is either Spanish or Portuguese pyrites, both of which

[*] Chemical News, vol. xlvii. p. 67 (Feb. 9, 1883).

contain from 3 to 4 per cent. of copper, together with very small quantities of silver and gold ; and after the greater part of the sulphur has been burnt off, these metals, together with iron oxide, remain in the "burnt ore" or "pyrites cinders," as it is called, which is then treated by a wet process for the extraction of the copper (see COPPER), and the residue then left is an almost pure oxide of iron, which can be used for various purposes in the manufacture of iron and steel. It is to these secondary products that the profits still obtained by the Leblanc process are due. (See Mr. Weldon's Paper, above referred to.)

Properties of Normal Sodium Carbonate.—The ordinary crystals of sodium carbonate contain 10 molecules of water; but by particular management the same salt may be obtained with 15, 9, 7 molecules, or sometimes with only 1. The crystals are monoclinic; they effloresce in dry air, and crumble to a white powder. Heated, they fuse in their water of crystallization: when the latter has been expelled and the dry salt exposed to a full red heat, it melts without undergoing change. The common crystals dissolve in 2 parts of cold, and in less than their own weight of boiling water: the solution has a strong, disagreeable, alkaline taste, and a powerfully alkaline reaction.

Hydrogen and *Sodium Carbonate, Hydrosodic Carbonate, Monosodic Carbonate, Acid Sodium Carbonate,* $NaHCO_3$ or Na_2CO_3,H_2CO_3, commonly called *Bicarbonate of Soda.*—This salt is prepared by passing carbonic acid gas into a cold solution of the normal carbonate, or by placing the crystals in an atmosphere of the gas, which is rapidly absorbed, while the crystals lose the greater part of their water, and pass into the new compound.

Monosodic carbonate, prepared by either process, is a crystalline white powder, which cannot be redissolved in warm water without partial decomposition. It requires 10 parts of water at $15.5°$ for solution: the liquid is feebly alkaline to test-paper, and has a much milder taste than that of the normal carbonate. It does not precipitate a solution of magnesia. By exposure to heat, the salt is converted into normal carbonate.

Dihydro-tetrasodic Carbonate, $Na_4H_2(CO_3)_3,2H_2O$.—This salt, commonly called *Sesquicarbonate of soda,* may be regarded as a compound of the normal and acid salts ($Na_2CO_3,2NaHCO_3$). It occurs native on the banks of the soda lakes of Sokenna, near Fezzan, in Africa, where it is called *trona;* also as *urao,* at the bottom of a lake in Maracaibo, South America. It is produced artificially, though with some difficulty, by mixing the monosodic and disodic carbonates in the proportions above indicated, melting them together, drying, and exposing the dried mass in a cellar for some weeks; it then absorbs water, becomes crystalline, and contains spaces filled with the tetrasodic carbonate.

Sodium and *Potassium Carbonate,* $NaKCO_3,6H_2O$, separates in monoclinic crystals from a solution containing the two carbonates in equivalent proportions.

A mixture of these two carbonates in equivalent proportions melts at a much lower heat than either of the salts separately ; such a mixture is very useful in the fusion of silicates, etc.

Sodium Sulphide, Na_2S, is prepared like potassium sulphide, and separates from a concentrated solution in octahedral crystals, which are rapidly decomposed by contact with the air, yielding a mixture of hydroxide and thiosulphate of sodium. It forms soluble sulphur-salts with hydrogen sulphide, carbon bisulphide, and other electro-negative sulphides.

There is no good precipitant for sodium, all its salts being very soluble, excepting the metantimonate, which is precipitated on mixing a solution of a sodium salt with a solution of potassium metantimonate; the use of this reagent is, however, attended with some difficulties. The yellow color imparted by sodium salts to the outer blowpipe flame and to combustible matter is a character of considerable importance. The spectral reactions of sodium-compounds have been already noticed (p. 96).

LITHIUM.

Symbol, Li. Atomic weight, 7.

LITHIUM is found in petalite, spodumene, lepidolite, triphylline, and a few other minerals, and sometimes occurs in minute quantities in mineral springs. The most abundant source of it yet discovered is the mineral water of Wheal Clifford in Cornwall, in which it exists to the amount of 61 parts in a million.

The metal is obtained by fusing pure lithium chloride in a small, thick porcelain crucible, and decomposing the fused chloride by electrolysis. It is a white metal like sodium, and very oxidizable. Lithium melts at 180° C. (356° F.); its specific gravity is 0.59: it is, therefore, the lightest solid known.

A lithium salt may be obtained from petalite on the small scale by the following process: The mineral is reduced to an exceedingly fine powder, mixed with 5 or 6 times its weight of pure calcium carbonate, and the mixture is heated to whiteness, in a platinum crucible placed within a well-covered earthen one for 20 minutes or half an hour. The shrunken coherent mass is digested in dilute hydrochloric acid, the whole evaporated to dryness, acidulated water added, and the silica separated by a filter. The solution is then mixed with ammonium carbonate in excess, boiled, and filtered; the clear liquid is evaporated to dryness, and gently heated in a platinum crucible to expel the sal-ammoniac; and the residue is wetted with oil of vitriol, gently evaporated once more to dryness, and ignited: pure fused lithium sulphate then remains.

This process may serve to give a good idea of the general nature of the operation by which alkalies are extracted in mineral analysis, and their quantities determined.

Lithium hydroxide, or *Lithia*, LiHO, is much less soluble in water than the hydroxides of potassium and sodium; the *carbonate* and *phosphate* are also sparingly soluble salts. The *chloride* crystallizes in anhydrous cubes which are deliquescent. *Lithium sulphate* is a very beautiful salt, crystallizing in lengthened prisms which contain 1 molecule of water. It gives no double salt with aluminium sulphate.

The salts of lithium color the outer flame of the blowpipe carmine-red. The spectral phenomena exhibited by lithium compounds are mentioned on page 96.

CÆSIUM AND RUBIDIUM.

Cs = 133. Rb = 85.4.

THE two metals designated by these names were discovered by Bunsen and Kirchhoff by means of their spectrum apparatus mentioned on page 96.

the former in 1860 and the latter in 1861. These metals, it appears, are widely diffused in nature, but always occur in very small quantities: they have been detected in many mineral waters, as well as in some minerals, namely, lithia-mica or lepidolite, and petalite; lately also in felspar; they have also been found in the alkaline ashes of beetroot. The salt-spring of Dürkheim, which contains 0.17 part of cæsium chloride in a million parts of water, was till lately regarded as the richest source of cæsium; but from experiments by Colonel Yorke,* it appears that the hot spring of Wheal Clifford, already mentioned as a source of lithium, contains 1.71 parts of cæsium chloride in a million, or 0.12 grain in a gallon. According to Cossa,† a very rich source of these metals is found in the alum of "la Schicciola," in Vulcano, one of the Lipari Islands.

The best material for the preparation of rubidium is lepidolite, which has been found to contain as much as 0.2 per cent. of that metal. Both metals are closely allied to potassium in their deportment.

Rubidium and cæsium, like potassium, form double salts with tetrachloride of platinum, which are, however, much less soluble than the corresponding potassium salts: it is on this property that the separation of these metals from potassium is based. The mixture of platinochlorides is repeatedly extracted with boiling water, when a difficultly soluble residue, consisting chiefly of the platinochlorides of cæsium and rubidium, remains; and these two metals are finally separated by converting them into tartrates, rubidium tartrate requiring for solution eight times as much water as cæsium tartrate, and therefore crystallizing out first from the mixed solution. According to Cossa (loc. cit.), the separation is best effected by treating a solution of cæsium and rubidium alums with antimony trichloride, which throws down the whole of the cæsium, leaving the rubidium in solution.

The hydroxides of these metals are powerful bases, which attract carbonic acid from the air, passing first into normal and then into acid carbonates. Cæsium carbonate is soluble in absolute alcohol; rubidium carbonate is nearly insoluble in that liquid: this property is made use of for the separation of these two metals. Cæsium chloride crystallizes in cubes, and is somewhat more soluble in water than chloride of potassium.

Rubidium chloride, in the fused state, is easily decomposed by the electric current; the metal produced rises to the surface and burns with a reddish light. If this experiment be performed in an atmosphere of hydrogen to prevent oxidation, the separated metal is nevertheless lost, dissolving in the fused chloride, which is transformed into a subchloride having the blue color of smalt.

Rubidium, when separated under mercury by the electric current, forms a silvery crystalline amalgam, which oxidizes rapidly on exposure to the air, and decomposes water at ordinary temperatures.

Metallic cæsium cannot readily be obtained, either by the electrolysis of its chloride, or by heating its acid tartrate with sugar, charcoal, and chalk, but it may be prepared by electrolysis of its cyanide, which is formed on passing anhydrous hydrocyanic acid into a solution of cæsium hydroxide in absolute alcohol. To prepare the metal, a mixture of barium cyanide (1 pt.) and cæsium cyanide (4 pts.) is fused in a porcelain crucible, and an electric current is passed through the molten mass, aluminium poles being used. On warming the contents of the crucible.under petroleum, the metal separates in globules. Metallic cæsium is silver-white, like potassium and rubidium, and soft at ordinary temperatures; sp. gr. 1.88 at 15° C. (59° F.); melting point 26–27° C. (79–81° F.). It oxidizes quickly in the air, and takes fire when thrown on water (Setterberg).

* Journal of the Chemical Society, 1872, p. 273.
† Gazzetta chimica italiana, 1878, p. 225; Chem. Soc. J., xxxiv. 953. See also Setterberg, ibid., xliv. 464.

AMMONIUM.

THE ammonia salts are most conveniently studied in this place, on account of their close analogy to those of potassium and sodium. These salts are formed by the direct union of ammonia, NH_3, with acids, and, as already pointed out (p. 191), they may be regarded as compounds of acid radicles, Cl, NO_3, SO_4, etc., with a basylous radicle, NH_4, called a m m o n i u m, which plays in these salts the same part as potassium and sodium in their respective compounds; thus:

NH_3 Ammonia.	$+$	HCl Hydrochloric acid.	$=$	$NH_4.Cl$ Ammonium chloride.	
NH_3	$+$	HNO_3 Nitric acid.	$=$	$NH_4.NO_3$ Ammonium nitrate.	
NH_3	$+$	H_2SO_4 Sulphuric acid.	$=$	$NH_4.SO_4H$ Acid ammonium sulphate.	
$2NH_3$	$+$	H_2SO_4	$=$	$(NH_4)_2SO_4$ Normal ammonium sulphate.	

The radicle NH_4· is not capable of existing in the free state, inasmuch as it contains an uneven number of monad atoms: it is simply the residue which is left on removing the atom of chlorine from the saturated molecule, NH_4Cl. Whether the double molecule N_2H_8, or $\begin{matrix} NH_4 \\ | \\ NH_4 \end{matrix}$, has a separate existence, is a different question. Ammonium is said, indeed, to be capable of forming an amalgam with mercury; but even in this state it is quickly resolved into ammonia and free hydrogen.

When a globule of mercury is placed on a piece of moistened potassium hydroxide, and connected with the negative side of a voltaic battery of very moderate power, the circuit being completed through the platinum plate upon which the alkali rests, decomposition of the latter takes place, and an amalgam of potassium is rapidly formed. If this experiment be now repeated with a piece of sal-ammoniac instead of potassium hydroxide, a soft, solid, metalline mass is also produced, which has been called the *ammoniacal amalgam*, and considered to contain ammonium in combination with mercury. A simpler method of preparing this compound is the following:—A little mercury is put into a test-tube with a grain or two of potassium or sodium, and gentle heat is applied; combination ensues, attended by heat and light. When cold, the fluid amalgam is put into a capsule, and covered with a strong solution of sal-ammoniac. The production of an ammoniacal amalgam instantly commences, the mercury increasing prodigiously in volume, and becoming quite pasty. The increase of weight is, however, quite trifling, varying from $\frac{1}{1800}$ to $\frac{1}{1200}$ part. Left to itself, the amalgam quickly decomposes into fluid mercury, ammonia, and hydrogen. It is most probable, indeed, that the so-called amalgam may be nothing more than mercury which has absorbed a certain quantity of these gases; just as silver, when heated to a very high temperature, is capable of taking up about twenty times its volume of oxygen gas, which it gives up again on cooling.[*]

But whether ammonium has any separate existence or not, it is quite cer-

[*] See Watts's Dictionary of Chemistry, vol. vi., p. 718.

tain that many ammoniacal salts are isomorphous with those of potassium; and if from any two of the corresponding salts, as the nitrates KNO_3 and NH_4NO_3, we subtract the radicle NO_3 common to the two, there remain the metal K and the group NH_4, which are, therefore, supposed to be iso-morphous.

Ammonium Chloride, Sal-ammoniac, NH_4Cl or NH_3,HCl.—Sal-ammoniac was formerly obtained from Egypt, being extracted by subli-mation from the soot of camels' dung: it is now largely manufactured from the ammoniacal liquid of the gas-works, and from the condensed products of the distillation of bones and other animal refuse in the preparation of animal charcoal.

These impure and highly offensive solutions are treated with a slight excess of hydrochloric acid, by which the free alkali is neutralized, and the carbonate and sulphide are decomposed, with evolution of carbonic acid and sulphuretted hydrogen gases. The liquid is evaporated to dryness, and the salt carefully heated, to expel or decompose the tarry matter; it is then purified by sublimation in large iron vessels lined with clay, surmounted with domes of lead. Sublimed sal-ammoniac has a fibrous texture; and is tough and difficult to powder.

Sal-ammoniac separates from its aqueous solution in radiate groups of small but distinct cubes and octahedrons. It has a sharp saline taste, and dissolves in $2\frac{3}{4}$ parts of cold, and in a much smaller quantity of hot, water. By heat, it is sublimed without decomposition. The crystals are anhydrous. Ammonium chloride forms double salts with the chlorides of magnesium, nickel, cobalt, manganese, zinc, and copper.

Ammonium Sulphate, $(NH_4)_2SO_4$.—Prepared by neutralizing am-monium carbonate with sulphuric acid, or on the large scale, for use as a manure, by adding sulphuric acid in excess to the coal-gas liquor just men-tioned, and purifying the product by suitable means. It is soluble in 2 parts of cold water, and crystallizes in long, flattened, six-sided prisms. It is en-tirely decomposed and driven off by ignition, and, even to a certain extent, by long boiling with water, ammonia being expelled and the liquid ren-dered acid.

Ammonium Nitrate, $(NH_4)NO_3$, is easily prepared by adding am-monium carbonate to slightly diluted nitric acid until neutralization has been reached. By slow evaporation at a moderate temperature it crystal-lizes in six-sided prisms, like those of potassium nitrate; but as usually prepared for making nitrogen monoxide, by quick boiling until a portion solidifies completely on cooling, it forms a fibrous and indistinct crystalline mass.

Ammonium nitrate dissolves in 2 parts of cold water, producing consider-able depression of temperature; it is but feebly deliquescent, and deflagrates like nitre on contact with heated combustible matter. Its decomposition by heat has been already explained (p. 197).

Ammonium Phosphates.—The *Normal* or *Triammonic salt*, $(NH_4)_3$PO$_4$, is obtained as a crystalline mass when the di-ammonic salt is super-saturated with strong aqueous ammonia, and crystallizes from dilute am-monia in short prismatic needles containing 3 molecules of water. The salt decomposes when its aqueous solution is boiled, giving off two-thirds of its ammonia.—The *di-ammonic salt*, $(NH_4)_2HPO_4$, occurs in guano from Ichaboe, and is deposited in transparent monoclinic prisms, when an aque-ous solution of phosphoric acid containing excess of ammonia is left to evaporate.—The *mono-ammonic salt*, $(NH_4)_2HPO_4$, formed when aqueous phosphoric acid is added to aqueous ammonia till the solution becomes acid and is no longer precipitated by barium chloride, crystallizes in quadratic prisms isomorphous with the corresponding potassium salt.

Sodium, Ammonium, and Hydrogen Phosphate; Phosphorus Salt; Microcosmic Salt, $Na(NH_4)HPO_4,4H_2O$.—Six parts of common sodium phosphate are heated with 2 parts of water, until the whole is liquefied, and 1 part of powdered sal-ammoniac is added; common salt then separates, and may be removed by a filter; and from the solution, duly concentrated, the microcosmic salt is deposited in prismatic crystals, which may be purified by one or two recrystallizations. Microcosmic salt is very soluble. When gently heated, it parts with its 4 molecules of crystallization-water, and, at a higher temperature, the basic hydrogen is likewise expelled as water, together with ammonia, and a very fusible compound, sodium metaphosphate, remains, which is valuable as a flux in blowpipe experiments. Microcosmic salt occurs in decomposed urine.

Ammonium Carbonates.—There are three definite carbonates of ammonia, the composition of which is as follows:—

Normal or Di-ammonic carbonate,	$(NH_4)_2CO_3,H_2O$
Acid or Ammonio-hydric carbonate,	$(NH_4)HCO_3$
Half-acid, or Tetrammonio-dihydric carbonate,	$(NH_4)_4H_2(CO_3)_3,H_2O$.

(1.) The *normal carbonate* is prepared by addition of ammonia to one of the acid salts, or of water to the carbamate of ammonia, CON_2H_6 (p. 341), with certain special precautions, the description of which is too long for insertion in this work, to prevent the escape of a portion of the ammonia. It crystallizes in elongated plates or flattened prisms, having a caustic taste, a powerful ammoniacal odor, and easily giving off ammonia and water, whereby they are converted into the acid carbonate:—

$$(NH_4)_2CO_3,H_2O = NH_3 + H_2O + (NH_4)HCO_3.$$

(2.) *Ammonium and Hydrogen Carbonate,* or *Mono-ammonic Carbonate,* $(NH_4)HCO_3$, commonly called *Bicarbonate* or *Acid Carbonate of Ammonia.*—This salt is obtained by saturating an aqueous solution of ammonia, or of the half-acid carbonate, with carbonic acid gas; or by treating the finely-powdered half-acid carbonate with strong alcohol, which dissolves out normal or di-ammonic carbonate, leaving a residue of the mono-ammonic salt. Cold water may be used instead of alcohol for this purpose; but it dissolves a larger quantity of the mono-ammonic carbonate. All ammonium carbonates when left to themselves are gradually converted into mono-ammonic carbonate. This salt forms large crystals belonging to the trimetric system. According to Deville, it is dimorphous, but never isomorphous with mono-potassic carbonate; when exposed to the air it volatilizes slowly, and gives off a faint ammoniacal odor. It dissolves in 8 parts of cold water, the solution decomposing gradually at ordinary temperatures, quickly when heated above 30° C. (86° F.), with evolution of ammonia. It is insoluble in alcohol, but when exposed to the air under alcohol, it dissolves as normal carbonate, evolving carbon dioxide.

It has been found native in considerable quantity in the deposits of guano on the western coast of Patagonia, in white crystalline masses, having a strong ammoniacal odor.

(3.) *Tetrammonio-dihydric Carbonate,* $N_4H_{18}C_3O_9 = (NH_4)_4H_2(CO_3)_3$.—This salt, also called *Half-acid Carbonate* or *Sesquicarbonate of Ammonia,* contains the elements of 1 molecule of di-ammonic and 2 molecules of mono-ammonic carbonate, into which it is, in fact, resolved by treatment with water or alcohol :—

$$(NH_4)_4H_2(CO_3)_3 = (NH_4)_2CO_3 + 2(NK_4)HCO_3.$$

It is obtained by dissolving the commercial carbonate in strong aqueous ammonia, at about 30° C. (86° F.), and crystallizing the solution. It forms large transparent rectangular prisms, having their summits truncated by octahedral faces. These crystals decompose very rapidly in the air, giving off water and ammonia, and being converted into mono-ammonic carbonate.

Commercial carbonate of ammonia (*sal volatile, salt of hartshorn*) consists of half-acid carbonate more or less mixed with carbamate. It is prepared on the large scale by the dry distillation of bones, hartshorn, and other animal matter, and is purified from adhering empyreumatic oil by subliming it once or twice with animal charcoal in cast-iron vessels, over which glass receivers are inverted. Another method consists in heating to redness a mixture of 1 part ammonium chloride or sulphate, and 2 parts calcium carbonate (chalk) or potassium carbonate, in a retort to which a receiver is luted.

An elaborate description of the carbonates of ammonia has been published by Dr. Divers * to whom is due the discovery of the normal ammonium carbonate.

Ammonium Sulphides.—Several of these compounds exist, and may be formed by distilling with sal-ammoniac the corresponding sulphides of potassium or sodium.

Ammonium and Hydrogen Sulphide, or *Ammonium Hydrosulphide*, $(NH_4)HS$, is a compound of great practical utility ; it is obtained by saturating a solution of ammonia with well-washed sulphuretted hydrogen gas, until no more of the latter is absorbed. The solution is nearly colorless at first, but becomes yellow after a time, without, however, suffering material injury unless it has been exposed to the air. It gives precipitates with most metallic solutions, which are very often characteristic, and is of great service in analytical chemistry.

Ammoniacal salts are easily recognized ; they are all decomposed or volatilized at a high temperature ; and when heated with calcium hydroxide or solution of alkaline carbonate, they give off ammonia, which may be recognized by its odor and alkaline reaction. The salts are all more or less soluble, the acid tartrate and the platinochloride being, however, among the least soluble : hence ammonium salts cannot be distinguished from potassium salts by the tests of tartaric acid and platinum solution. When a solution containing an ammoniacal salt or free ammonia is mixed with potash, and a solution of *mercuric iodide* in *potassium iodide* is added, a brown precipitate or coloration is immediately produced, consisting of dimercurammonium iodide, NHg_2I :—

$$NH_3 + 2HgI_2 = NHg_2I + 3HI.$$

This is called Nessler's test ; it is by far the most delicate test for ammonia that is known.

Amic Acids and Amides.

Sulphamic Acid.—When dry ammonia gas is passed over a thin layer of sulphuric oxide, SO_3, the gas is absorbed, and a white crystalline powder

* Journal of the Chemical Society, 1870, pp. 171-279.

is formed, having the composition $N_2H_6SO_3$, that is, of ammonium sulphate minus one molecule of water:—

$$N_2H_6SO_3 = (NH_4)_2SO_4 - H_2O.$$

It is not, however, a salt of sulphuric acid: for its aqueous solution does not give any precipitate with baryta-water or soluble barium salts. It is, in fact, the ammonium salt of sulphamic acid, an acid derived from sulphuric acid, SO_4H_2 or $SO_2(HO)_2$, by substitution of the univalent radicle, NH_2 (p. 254), for one atom of hydroxyl, HO. The formula of this acid is $SO_2(NH_2)(OH)$, and that of its ammonium salt, $SO_2(NH_2)(ONH_4)$ or $SO_3N_2H_6$. Ammonium sulphamate is permanent in the air, and dissolves without decomposition in water. Its solution, evaporated in a vacuum over oil of vitriol, yields the salt in transparent colorless crystals.

The solution of the ammonium salt, mixed with baryta-water, gives off ammonia, and yields a solution of *barium sulphamate*, $[SO_2(NH_2)O]_2Ba$, which may be obtained by evaporation in well-defined crystals; and the solution of this salt, decomposed with potassium sulphate, yields *potassium sulphamate*, $SO_2(NH_2)(OK)$.

Carbamic Acid.—When dry ammonia gas is mixed with carbon dioxide, the mixture being kept cool, the gases combine in the proportion of 2 volumes of the former to 1 volume of the latter, forming a pungent, very volatile substance, which condenses in white flocks. This substance has the composition $CO_2N_2H_6$, that is, of normal ammonium carbonate, $CO_3(NH_4)_2$, *minus* one molecule of water. It exists, as already observed, in commercial carbonate of ammonia (p. 339). It was formerly called *anhydrous carbonate of ammonia;* but, like the preceding salt, it is not really a carbonate, but the ammonium salt of c a r b a m i c a c i d, $CO(NH_2)OH$, an acid derived from carbonic acid, CO_3H_2 or $CO(OH)_2$, by substitution of amidogen, NH_2, for one hydroxyl-group. Ammonium carbamate dissolves readily in water, and quickly takes up one molecule of that compound, whereby it is converted into normal ammonium carbonate. When treated with sulphuric oxide, it is converted into ammonium sulphamate.

Carbamide, CON_2H_4, or $CO(NH_2)_2$.—When - ammonia gas is mixed with carbon oxychloride or phosgene gas, $COCl_2$, a white crystalline powder is formed, having this composition:—

$$COCl_2 + 2NH_3 = 2HCl + CON_2H_4.$$

This compound, which is likewise formed in other reactions to be afterward considered, is derived from carbonic acid, $CO(OH)_2$, by substitution of 2 atoms of amidogen for 2 atoms of hydroxyl. It differs from carbamic acid in being a neutral substance, not containing any hydrogen easily replaceable by metals.

Other bibasic acids likewise yield an amic acid and a neutral amide by substitution of 1 or 2 atoms of amidogen for hydroxyl. Tribasic acids yield in like manner two amic acids and one neutral amide, and quadribasic acids may yield three amic acids and a neutral amide; thus, from pyrophosphoric acid, $P_2O_7H_4 = P_2O_3(OH)_4$, are obtained the three amic acids $P_2O_3(NH_2)(OH)_3$, $P_2O_3(NH_2)_2(OH)_2$ and $P_2O_3(NH_2)_3OH$.

Monobasic acids, which contain but one atom of hydroxyl, yield by this mode of substitution only neutral amides, no amic oxide: thus, from acetic acid, $C_2H_4O_2 = C_2H_3O(OH)$, is obtained acetamide, $C_2H_3O(NH_2)$.

By similar substitution of metals, or of basylous compound radicles, for the hydrogen of ammonia, basic compounds, called a m i n e s, are formed.

Thus, when potassium is gently heated in ammonia gas, *mono-potassamine*, NH_2K, is formed. It is an olive-green substance, which is decomposed by water into ammonia and potassium hydroxide:—

$$NH_2K + H_2O = NH_3 + KHO.$$

It melts at a little below 100° C., and when heated in a close vessel, is resolved into ammonia and *tri-potassamine*:—

$$3NH_2K = 2NH_3 + NK_3.$$

The latter effervesces violently with water, yielding ammonia and potassium hydroxide:—

$$NK_3 + 3H_2O = NH_3 + 3KHO.$$

The formation and properties of amides and amines will be further considered under Organic Chemistry.

Metallammoniums. — These are hypothetical radicles derived from ammonium, N_2H_8, by substitution of metals for hydrogen. Salts of such radicles are formed in several ways. Ammonia gas is absorbed by various metallic salts in different proportions, forming compounds, some of which may be formulated as salts of metallammoniums. Thus, platinous chloride, $PtCl_2$, absorbs two molecules of ammonia, forming *platosammonium chloride*, $N_2H_6Pt''Cl_2$, and platinum tetrachloride, $PtCl_4$, absorbs four molecules of ammonia, forming *platinammonium chloride*, $N_4H_{12}Pt^{iv}Cl_4$. In like manner, cupric chloride and sulphate form the *chloride* and *sulphate of cuprammonium*, $N_2H_6Cu''Cl_2$ and $N_2H_6Cu''SO_4$.

Similar compounds are formed in many cases by precipitating metallic salts with ammonia or ammoniacal salts: thus, ammonia added to a solution of mercuric chloride, $HgCl_2$, forms a white precipitate, consisting of *dimercurammonium chloride*, $N_2H_4Hg''_2Cl_2$; and by dropping solution of mercuric chloride into a boiling solution of sal-ammoniac mixed with free ammonia, crystals are obtained, consisting of *mercurammonium chloride*, $N_2H_6Hg''Cl_2$. Some of these compounds will be further considered in connection with the several metals.

METALS OF THE ALKALINE EARTHS.

THE three metals, calcium, strontium, and barium, constituting this group, form oxides less soluble in water than the true alkalies, but exhibiting similar taste, causticity, and action on vegetable colors. These metals form but one chloride; e. g., CaCl₂; their carbonates are insoluble in water; and barium sulphate is also insoluble, the sulphates of calcium and strontium slightly soluble.

CALCIUM.

Symbol, Ca. Atomic weight, 39.9.

CALCIUM is one of the most abundant and widely diffused of the metals, though it is never found in the free state. As carbonate, it occurs in a great variety of forms, constituting, as limestone, entire mountain ranges.

Calcium was obtained in an impure state by Davy, by means similar to those adopted for the preparation of barium (p. 351). Matthiessen prepared the pure metal by fusing a mixture of two molecules of calcium chloride and one of strontium chloride with some chloride of ammonium in a small porcelain crucible, in which an iron cylinder is placed as positive pole, and a pointed iron wire or a little rod of carbon connected with the zinc of the battery is made to touch the surface of the liquid. The reduced metal fuses and drops off from the point of the iron wire, and the bead is removed from the liquid by a small iron spatula. Liès-Bodart and Gobin prepare calcium by igniting the iodide with an equivalent quantity of sodium in an iron crucible having its lid screwed down.

Calcium is a light yellow metal of sp. gr. 1.5778. It is about as hard as gold, very ductile, and may be cut, filed, or hammered out into plates as thin as the finest paper. It tarnishes slowly in dry, more quickly in damp air; decomposes water quickly, and is still more rapidly acted upon by dilute acids. Heated on platinum foil over a spirit-lamp, it burns with a bright flash; with a brilliant light also when heated in oxygen or chlorine gas, or in vapor of bromine, iodine, or sulphur.

Calcium Chloride, CaCl₂, is usually prepared by dissolving marble in hydrochloric acid; it is also a by-product in several chemical manufactures. It separates from a strong solution in colorless, prismatic, and exceedingly deliquescent crystals, which contain six molecules of water. By heat this water is expelled, and by a temperature of strong ignition the salt is fused. The crystals reduced to powder are employed in the production of artificial cold by being mixed with snow or powdered ice; and the chloride, strongly dried or in the fused state, is of great use in desiccating gases, for which purpose the latter are slowly transmitted through tubes filled with fragments of the salt. Calcium chloride is also freely soluble in alcohol, which, when anhydrous, forms with it a definite crystallizable compound.

The *bromide*, CaBr$_2$, and *iodide*, CaI$_2$, closely resemble the chloride.

Calcium Fluoride, Fluor-Spar, CaF$_2$, is important as the most abundant natural source of hydrofluoric acid and the other fluorides. It occurs beautifully crystallized, of various colors, in lead veins, the crystals having commonly the cubic, but sometimes the octahedral form, parallel to the faces of which latter figure they always cleave. Some varieties, when heated, emit a greenish, and some a purple, phosphorescent light. The fluoride is quite insoluble in water, and is decomposed by oil of vitriol in the manner already mentioned (p. 144).

Calcium Oxides.—The *Monoxide* or *Lime*, CaO, may be obtained in a state of considerable purity by heating to full redness for some time fragments of the black bituminous marble of Derbyshire or Kilkenny. If required absolutely pure, it must be made by igniting to whiteness, in a platinum crucible, an artificial calcium carbonate, prepared by precipitating the nitrate with ammonia carbonate. Lime in an impure state is prepared for building and agricultural purposes by calcining, in a kiln, the ordinary limestones which abound in many districts; a red heat continued for some hours is sufficient to disengage the whole of the carbonic acid. In the best contrived lime-kilns the process is carried on continuously, broken limestone and fuel being constantly thrown in at the top, and the burned lime raked out at intervals from beneath. Sometimes, when the limestone contains silica, and the heat has been very high, the lime refuses to slake, and is said to be *overburned;* in this case a portion of silicate has been formed.

Pure lime is white, and often very hard: it is quite infusible and phosphoresces, or emits a pale light at a high temperature. When moistened with water, it slakes with great violence, evolving heat, and crumbling to a soft, white, bulky powder, which is a hydrate containing a single molecule of water, which can be again expelled by a red heat. This hydrate or hydroxide, CaH$_2$O$_2$ or CaO,H$_2$O, is soluble in water, but far less so than either the hydroxide of barium or of strontium, and, what is very remarkable, the *colder* the water the larger is the quantity of the compound that is taken up. A pint of water at 15.5° C. (59.9° F.) dissolves about 11 grains, while at 100° C. only 7 grains are retained in solution. The hydroxide has been obtained in thin delicate crystals by slow evaporation under the air-pump. Lime-water is always prepared for chemical and pharmaceutical purposes by agitating cold water with excess of calcium hydroxide in a closely-stopped vessel, and then, after subsidence, pouring off the clear liquid and adding a fresh quantity of water for another operation: there is not the least occasion for filtering the solution. Lime-water has a strong alkaline reaction, a nauseous taste, and when exposed to the air becomes almost instantly covered with a pellicle of carbonate, by absorption of carbonic acid. It is used, like baryta-water, as a test for carbonic acid, and in medicine. Lime-water prepared from some varieties of limestone may contain potash.

The hardening of mortars and cements is in a great measure due to the gradual absorption of carbonic acid; but even after a very great length of time, this conversion into carbonate is not complete. Mortar is known, under favorable circumstances, to acquire extreme hardness with age. Lime cements which resist the action of water contain iron oxides, silica, and alumina; they require to be carefully prepared, the stone not being overheated. When they are ground to powder and mixed with water, solidification speedily ensues, from causes not yet thoroughly understood, and the cement, once in this condition, is unaffected by wet. Parker's or Roman cement is made in this manner from the nodular masses of calcareo-argillaceous ironstone found in the London clay.

Lime is of great importance in agriculture: it is found more or less in every fertile soil, and is often very advantageously added by the cultivator. The decay of vegetable fibre in the soil is thereby promoted, and other important objects, as the destruction of certain hurtful compounds of iron in marsh and peat land, are often attained. The addition of lime probably serves likewise to liberate potassium from the insoluble silicate of that base contained in the soil.

Calcium Dioxide, CaO_2, is obtained in microscopic crystals, having the composition $CaO_2,8H_2O$, by precipitating lime-water with hydrogen dioxide.

Chloride of Lime, Bleaching Powder.—When calcium hydroxide, very slightly moist, is exposed to chlorine gas, the latter is absorbed, and a compound is produced—the bleaching powder of commerce—which is used on an immense scale for bleaching linen and cotton goods. At the commencement of the process, the chlorine must be supplied slowly, so as to avoid rise of temperature. The product when fresh and well prepared, is a soft white powder which attracts moisture slowly from the air, and exhales an odor sensibly different from that of chlorine. It is soluble in about 10 parts of water, the unaltered hydroxide being left behind: the solution is highly alkaline, and bleaches feebly. When calcium hydroxide is suspended in cold water, and chlorine gas transmitted through the mixture, the lime is gradually dissolved, and the same peculiar bleaching compound produced: the alkalies also, either caustic or carbonated, may by similar means be made to absorb a large quantity of chlorine, and give rise to corresponding compounds: such are the "disinfecting solutions" of Labarraque.

The composition of bleaching powder is represented by the formula $CaOCl_2$, and it was formerly supposed to be a direct compound of lime with chlorine. This view, however, is not consistent with its reactions, for when distilled with dilute nitric acid, it readily yields a distillate of aqueous hypochlorous acid, and when treated with water it is resolved into chloride and hypochlorite of calcium, the latter of which may be separated in crystals by exposing the filtered solution to a freezing mixture, or by evaporating it in a vacuum over oil of vitriol, and leaving the dense frozen mass to thaw upon a filter. A solution of calcium chloride mixed with hypochlorite then passes through, and feathery crystals remain on the filter, very unstable, but consisting, when recently prepared, of hydrated calcium hypochlorite, $Ca(OCl)_2,4H_2O$.[*] These results seem at first sight to show that the bleaching powder is a mixture of chloride and hypochlorite of calcium, formed according to the equation,

$$2CaO + Cl_4 = CaCi_2 + CaCl_2O_2$$

but if this were its true constitution, the powder when digested with alcohol, ought to yield a solution of calcium chloride containing half the chlorine of the original compound, which is not the case. Its constitution is therefore better represented by the formula $Cl.Ca.OCl$, suggested by Dr. Odling, this molecule being decomposed by water into chloride and hypochlorite in the manner just explained, and yielding, with dilute nitric or sulphuric acid, a distillate containing hydrochloric and hypochlorous acids:—

$$CaCl(OCl) + 2HNO_3 = Ca(NO_3)_2 + HCl + HClO.$$

[*] K i n g z e t t, Chem. Soc. Jour., 1875, p. 404.

15 *

When the temperature of the calcium hydroxide has risen during the absorption of the chlorine, or when the compound has been subsequently exposed to heat, its bleaching properties are impaired or altogether destroyed : it then contains chlorate and chloride of calcium; oxygen, in variable quantity, is usually set free. The same change seems to ensue by long keeping, even at the common temperature of the air. In an open vessel the compound is speedily decomposed by the carbonic acid of the air. Commercial bleaching powder thus constantly varies in value with its age, and with the care originally bestowed upon its preparation: the best may contain about 30 per cent. of available chlorine, easily liberated by an acid, which is, however, far short of the theoretical quantity.

The general method in which this substance is employed for bleaching is the following :—The goods are first immersed in a dilute solution of chloride of lime, and then transferred to a vat containing dilute sulphuric acid. Decomposition ensues; the calcium both of the hypochlorite and of the chloride is converted into sulphate, while the free hypochlorous and hydrochloric acids yield water and free chlorine :—

$$CaOCl_2 + H_2SO_4 = CaSO_4 + HClO + HCl, \text{ and}$$
$$HClO + HCl = H_2O + Cl_2.$$

The chlorine thus disengaged in contact with the cloth causes destruction of the coloring matter. The process is repeated several times, since it is unsafe to use strong solutions.

On the same principle, white patterns are imprinted upon colored cloth, the figures being stamped with tartaric acid thickened with gum-water, and then the stuff immersed in the chlorine bath, when the parts to which no acid has been applied remain unaltered, while the printed portions are bleached.

For purifying an offensive or infectious atmosphere, *as an aid to proper ventilation*, the bleaching powder is very convenient. The solution is exposed in shallow vessels, or cloths steeped in it are suspended in the apartment, when the carbonic acid of the air slowly decomposes it in the manner above described. Addition of a strong acid causes rapid disengagement of chlorine.

The value of any sample of bleaching powder may be easily determined by the following method, in which the feebly combined chlorine is estimated by its effect in oxidizing a ferrous salt to ferric salt, 2 molecules of ferrous oxide, FeO, requiring for the purpose 2 atoms of chlorine, according to the equation: $2FeSO_4 + H_2SO_4 + Cl_2 = 2HCl + Fe_2(SO_4)_3$. To make the estimation, 0.784 gram of pure crystallized ferrous sulphate is dissolved in water, a little sulphuric acid is added, and the solution of bleaching powder is dropped in from a burette till all the ferrous salt is converted into ferric salt. The completion of the reaction is ascertained by bringing a drop of the solution, together with a drop of potassium ferricyanide, on a white plate, until no further coloration is observed. A simple calculation gives the result, the above quantity of ferrous sulphate, $FeSO_4,7H_2O$, being equivalent to 1 gram of chlorine.

Calcium Sulphate, $CaSO_4$.—Crystalline native calcium sulphate, containing 2 molecules of water, is found in considerable abundance in some localities as *gypsum:* it is often associated with rock-salt. When regularly crystallized, it is termed *selenite.* Anhydrous calcium sulphate is also occasionally met with. The salt is formed by precipitation, when a moderately concentrated solution of calcium chloride is mixed with sulphuric acid. Calcium sulphate is soluble in about 500 parts of cold water, and its solubility is a little increased by heat. It is more soluble in water containing ammonium chloride or potassium nitrate. The solution is precipitated by alcohol. Gypsum, or native hydrated calcium sulphate, is largely employed for the purpose of making casts of statues and medals, also for moulds in the porcelain and earthenware manufactures, and for other applications. It is exposed to heat in an oven where the temperature does not exceed 127° C. (261° F.), by which the water of crystallization is expelled, and it is afterward reduced to a fine powder. When mixed with water, it solidifies after a short time, from the re-formation of the same hydrate; but this effect does not happen if the gypsum has been overheated. It is often called plaster of Paris. Artificial colored marbles, or *scagliola*, are frequently prepared by inserting pieces of natural stone in a soft stucco containing this substance, and polishing the surface when the cement has become hard. Calcium sulphate is one of the most common impurities of spring water.

The peculiar property water acquires by the presence of calcium salts is termed *hardness*. It manifests itself by the effect such waters have upon the palate, and particularly by its peculiar behavior with soap. Hard water yields a lather with soap only after the whole of the calcium salts have been thrown down from the water in the form of an insoluble lime-soap. The hardness produced by calcium sulphate is called *permanent hardness*, since it cannot be remedied.

Calcium and Potassium Sulphate, $CaSO_4, K_2SO_4 + H_2O$, is formed by the solutions of the two salts mixed together. An intimate mixture of equal weights of the anhydrous salts, stirred up with less than their weight of water, coagulates suddenly to a solid mass. With 4 to 5 parts of water, solidification takes place more slowly, and the mixture may be used for taking casts.—*Calcium and Sodium Sulphate*, $CaSO_4, Na_2SO_4$, occurs native as *Glauberite*. Needle-shaped crystals, having the composition $CaSO_4, Na_2SO_4, 2H_2O$ are obtained on mixing 1 part precipitated calcium sulphate and 50 parts Glauber's salt, and heating the mixture at 80° C. (176° F.) with 25 parts water. When further heated they are transformed into small crystals of glauberite.

Calcium Nitrate, $Ca(NO_3)_2$.—This salt, prepared by dissolving chalk in nitric acid, is in the anhydrous state a white porous deliquescent mass, very soluble in water and in alcohol. When heated and then exposed to sunshine, it appears luminous in the dark, and is hence called "Baldwin's phosphorus," from the name of the alchemist Baldewein or Balduinus, who first prepared it. It is often found as an efflorescence on the walls of stables and other places through which urine or other organic liquids percolate: hence it has been called *lime-saltpetre* or *wall-saltpetre*.

Calcium Phosphates.—The *tricalcic salt*, $Ca_3(PO_4)_2$ or $3CaO, P_2O_5$, occurs, combined with chloride and fluoride of calcium, in apatite, which crystallizes in the rhombohedral system; and its massive varieties, *phosphorite* and *estremadurite*, which occur in Estremadura in Spain, have the composition $3Ca_3(PO_4)_2 + CaF_2$, the fluorine being sometimes partly replaced by chlorine. Tricalcic phosphate occurs pure in osteolite, a mineral found near Hanau and at Amberg in the Erzgebirge; also pure in ornithite, a crystallized mineral found in the guano of Sombrero, a small island of the Antilles group; and mixed with an aluminium phosphate in sombrerite, occurring also in the same islands. Tricalcic phosphate is also the chief mineral constituent of bones, in the ash of which it occurs to the amount of about 80 per cent., together with magnesium phosphate, calcium carbonate, and calcium fluoride.

Tricalcic phosphate is obtained as a white precipitate by adding disodic orthophosphate (ordinary phosphate of soda) to an ammoniacal solution of calcium chloride. It is nearly insoluble in water, but is decomposed by prolonged boiling with water into an insoluble basic salt, $Ca_3(PO_4)_2 + Ca_2(PO_4)$·OH, and a soluble acid salt. It dissolves also in wåter containing ammoniacal salts, chloride and nitrate of sodium, and other salts; easily also in all acids, even in aqueous carbonic acid.

Monocalcic Orthophosphate, $CaHPO_4$, separates with 2 mols. H_2O, as a white crystalline precipitate, on treating a solution of calcium chloride with ordinary sodium phosphate. It occurs in urinary concretions, and is sometimes deposited from urine in rosettes or stellate groups of microscopic crystals.

Tetrahydro-calcic Phosphate, $H_4Ca(PO_4)_2$, separates in rhombic tablets when a solution of either of the preceding salts in the requisite quantity of phosphoric acid is left to evaporate. By treatment with boiling water, it is converted into the anhydrous monocalcic salt, and with cold water into the same with $2H_2O$.

Superphosphate of Lime.—This name is applied to a mixture of monocalcic orthophosphate and calcium sulphate prepared on the large scale by treating bone-ash, coprolites, phosphorites, etc. with two-thirds of their weight of sulphuric acid. It is used for the preparation of phosphorus, and very largely as a manure, especially for root-crops.

Calcium Hypophosphite, $Ca(PO_2H_2)_2$, a salt used in medicine, is prepared by boiling phosphorus with milk of lime, and separates from the clear solution on evaporation in bright flexible four-sided prisms, insoluble in alcohol.

Calcium Carbonate, *Chalk; Limestone; Marble;* $CaCO_3$.—Calcium carbonate, often more or less contaminated with iron oxide, clay, and organic matter, forms rocky beds, of immense extent and thickness, in almost every part of the world. These present the greatest diversities of texture and appearance, arising, in a great measure, from changes to which they have been subjected since their deposition. The most ancient and highly crystalline limestones are destitute of visible organic remains, while those of more recent origin are often entirely made up of the shelly exuviæ of once-living beings. Sometimes these latter are of such a nature as to show that the animals inhabited fresh water; marine species and corals are, however, most abundant. Cavities in limestone and other rocks are very often lined with magnificent crystals of calcium carbonate or calcareous spar, which have evidently been slowly deposited from a watery solution. Calcium carbonate is always precipitated when an alkaline carbonate is mixed with a solution of that base.

Although this substance is not sensibly soluble in pure water, it is freely taken up when carbonic acid happens at the same time to be present. If a little lime-water be poured into a vessel of that gas, the turbidity first produced disappears on agitation, and a transparent solution of calcium carbonate in excess of carbonic acid is obtained. This solution is decomposed completely by boiling, the carbonic acid being expelled and the carbonate precipitated. Since all natural waters contain dissolved carbonic acid, it is to be expected that calcium in this state should be of very common occurrence; and such is really found to be the fact, river, and more especially spring water, almost invariably containing calcium carbonate thus dissolved. In limestone districts this is often the case to a great extent. The *hardness* of water due to the presence of calcium carbonate is called *temporary*, since it is diminished to a very considerable extent by boiling, and may be nearly removed by mixing the hard water with lime-water, when both the dissolved carbonate and the dissolved lime, which thus becomes carbonated, are precipitated. Upon this principle Dr. Clark's process of softening water is based. This process is of considerable importance, since a supply of hard water to towns is in many respects a source of great inconvenience. As already mentioned. the use of such water for the purposes of washing is attended with a great loss of soap. Boilers, in which such water is heated, speedily become lined with a thick stony incrustation.* The beautiful stalactitic incrustations of limestone caverns, and the deposit of calc sinter or travertin upon various objects, and upon the ground, in many places, are explained by the solubility of calcium carbonate in water containing carbonic acid.

Crystallized calcium carbonate is dimorphous; calc-spar and arragonite, although possessing exactly the same chemical composition, have different crystalline forms, different densities, and different optical properties. Rose has observed that calcium carbonate appears in the form of calc-spar when deposited from its solution in water containing carbonic acid at the ordinary temperature. At 90° C. (194° F.), however, and at the boiling heat, it is chiefly deposited in the form of arragonite; at lower temperatures the formation of arragonite decreases, whilst that of calc-spar increases, the limit for the formation of the former variety being between 30° and 50° C. (86-90° F.).

Calc-spar occurs very abundantly in crystals derived from an obtuse rhombohedron, whose angles measure 105° 5′ and 74° 55′: its density varies from 2.5 to 2.8. The rarer variety, or arragonite, is found in crystals whose primary form is a right rhombic prism, a figure having no geometrical relation to the preceding: it is, besides, heavier and harder. Sp. gr. = 2.92 to 3.28.

Calcium Sulphides.—The monosulphide, CaS, is obtained by reducing the sulphate at a high temperature with charcoal or hydrogen: it is nearly colorless, and but little soluble in water. By boiling together calcium hydrate, water, and flowers of sulphur, a red solution is obtained, which, on cooling, deposits crystals of the bisulphide, CaS_2, containing water. When the sulphur is in excess, and the boiling long continued, a pentasulphide is generated: thiosulphuric acid is also formed during these reactions:—

$$3CaO + S_6 = 2CaS_2 + CaS_2O_3.$$

When the yellow solution obtained by boiling lime with excess of sulphur is poured into an excess of hydrochloric acid, sulphur is precipitated together with a yellow oily liquid, which is hydrogen persulphide (p. 174); but if the acid be poured into the solution of calcium sulphide, gaseous hydrogen sulphide is given off, and the whole precipitate formed consists wholly of finely-divided sulphur, the *sulphur precipitatum* of the Pharmacopœia. If dilute sulphuric acid is used, the precipitate also contains gypsum.

Calcium Phosphide.—When vapor of phosphorus is passed over fragments of lime heated to redness in a porcelain crucible, a chocolate-brown compound, the so-called *phosphuret of lime*, is produced. This substance is probably a mixture of calcium phosphide and phosphide. When thrown into water, it yields spontaneously inflammable hydrogen phosphide. According to Paul Thénard, the calcium phosphide in this compound has the composition P_2Ca_2. In contact with water it yields liquid hydrogen phosphide, P_2H_4 (p. 206),

$$P_2Ca_2 + 2H_2O = 2CaO + P_2H_4;$$

* Many proposals have been made to prevent the formation of boiler deposits. The most efficient appears to be the method of Dr. Ritterband, which consists in throwing into the boiler a small quantity of sal-ammoniac, whereby carbonate of ammonia is formed, which is volatilized with the steam, calcium chloride remaining in solution. It need scarcely be mentioned that this plan is inapplicable in the case of permanently hard waters.

and the greater portion of this liquid phosphide is immediately decomposed into solid and gaseous hydrogen phosphide: $5P_2H_4 = P_4H_2 + 6PH_3$.

Respecting the reactions of calcium salts see page 353.

STRONTIUM.

Symbol, Sr.   Atomic weight, 87.2.

· This element occurs as carbonate in the mineral called *Strontianite*, found at Strontian, in Argyleshire, also as sulphate or *Cœlestin;* in small quantity as chloride or sulphate in many brine-springs and mineral waters, in sea-water, and in the ash of *Fucus vesiculosus.*

Metallic strontium was discovered by Davy, who obtained it by elec-trolysis of the moistened peroxide or of the aqueous chloride, but it is more readily prepared by electrolysis of the fused anhydrous chloride. A small porcelain crucible having a porous cell is filled with strontium chloride mixed with a little sal-ammoniac; a fine iron wire placed within the cell constitutes the negative pole, and the positive pole is an iron cylinder placed in the crucible round the cell. The heat is regulated so that a crust forms in the cell, and the metal collects under this crust (Matthiessen).

Strontium is a yellow metal, somewhat harder than calcium, malleable, and having a density of 2.5. It melts at a red heat, and is more electro-negative than calcium or the alkali-metals. It oxidizes quickly in the air, burns brilliantly when heated, and decomposes water at common temperatures.

Strontium Chloride, $SrCl_2$, prepared by dissolving the carbonate in hydrochloric acid, crystallizes in colorless needles or prisms, $SrCl_2·6H_2O$, which are slightly deliquescent, dissolve in 2 parts of cold and a smaller quantity of boiling water; also in alcohol, forming a solution which burns with a crimson flame.

Oxides.—The *monoxide* or *Strontia*, prepared by ignition of the nitrate, is a grayish-white porous infusible mass. It unites readily with a small quantity of water, forming a white powder of strontium hydroxide, $Sr(OH)_2$, which dissolves readily in hot water, and separates on cooling as a hydrate, $Sr(OH)_2·8H_2O$, soluble in 50 parts of cold and 2.4 of boiling water. The solution is alkaline and caustic, but less so than those of the alkalies or of baryta.—The *dioxide*, SrO_2, is prepared by the action of hydrogen dioxide on strontia-water, and separates in pearly scales containing 2 molecules of water, which they give off when heated.

Strontium Sulphate, $SrSO_4$, occurs native in rhombic crystals and fibrous masses having a light-blue color: hence called *cœlestin*. The same compound is thrown down as a white precipitate, of sp. gr. 3.707, when sul-phuric acid or a soluble sulphate is added to a solution of strontium salt. It is intermediate in solubility between the sulphates of calcium and barium, and less soluble in boiling than in cold water.

· **Strontium Nitrate,** $Sr(NO_3)_2$, crystallizes in anhydrous octahedrons, soluble in 5 parts of cold, and about half their weight of boiling water. It is chiefly of value to the pyrotechnist, who employs it in the composition of the well-known "red fire." *

Strontium Carbonate, $SrCO_3$, occurs as strontianite in crystals isomorphous with those of arragonite, and is obtained by precipitation as a white impalpable powder, having a density of 3.62. When boiled with sal-ammoniac it is converted into the chloride.

All strontium compounds impart a deep crimson color to the flame of alcohol. Their spectral reactions have been already noticed (p. 96). For their reactions with liquid reagents, see page 353.

BARIUM.†

Symbol, Ba. Atomic weight, 136.8.

THIS metal occurs abundantly as sulphate and carbonate, forming the *veinstone* in many lead-mines. Davy obtained it in the metallic state by means similar to those described in the case of lithium. Bunsen subjects barium chloride mixed to a paste with water and a little hydrochloric acid, at a temperature of 100° C., to the action of the electric current, using an amalgamated platinum wire as the negative pole. In this manner the metal is obtained as a solid, highly crystalline amalgam, which, when heated in a stream of hydrogen, yields barium in the form of a tumefied mass, tarnished on the surface, but often exhibiting a silver-white lustre in the cavities. Barium may also be obtained, though impure, by passing vapor of potassium over the red-hot chloride or oxide of barium. It is malleable, melts below a red heat, decomposes water, and gradually oxidizes in the air.

Barium Chloride, $BaCl_2, 2H_2O$, is prepared by dissolving the native carbonate in hydrochloric acid, filtering the solution, and evaporating until a pellicle begins to form at the surface: the solution on cooling deposits crystals. When native carbonate cannot be procured, the native sulphate may be employed in the following manner:—It is reduced to fine powder, and intimately mixed with one-third of its weight of powdered coal; the mixture is pressed into an earthen crucible to which a cover is fitted, and exposed for an hour or more to a high red heat, by which the sulphate is converted into sulphide at the expense of the combustible matter of the coal; the black mass thus obtained is powdered and boiled in water, by which the sulphide is dissolved; and the solution, filtered hot, is mixed with a slight excess of hydrochloric acid. Barium chloride and hydrogen sulphide are then produced, the latter escaping with effervescence. Lastly,

---

* RED FIRE:	Grains.	GREEN FIRE:	Grains.
Dry strontium nitrate	800	Dry barium nitrate	450
Sulphur	225	Sulphur	150
Potassium chlorate	200	Potassium chlorate	100
Lampblack	50	Lampblack	25

The strontium or barium salt, the sulphur, and the lampblack must be finely powdered and intimately mixed, after which the potassium chlorate should be added in rather coarse powder, without rubbing, with the other ingredients. The red fire composition has been known to ignite spontaneously.

† From βαρύς, heavy, in allusion to the great specific gravity of the native carbonate and sulphate.

the solution is filtered to separate any little insoluble matter, and evaporated to the crystallizing point.

The crystals of barium chloride are flat four-sided tables, colorless and transparent. They contain two molecules of water, easily driven off by heat. 100 parts of water dissolve 43.5 parts at 15.5° C. (59.9° F.), and 78 parts at 104.5° C. (220.1° F.), which is the boiling-point of the saturated solution.

Barium Oxides.—The *Monoxide* or *Baryta*, BaO, is best prepared by decomposing the crystallized nitrate by heat in a capacious porcelain crucible until red vapors are no longer disengaged: the nitric acid is then resolved into nitrous acid and oxygen, and the baryta remains behind in the form of a grayish spongy mass, fusible at a high degree of heat. When moistened with water it combines into a hydrate, with great elevation of temperature. The *Hydroxide* or *Hydrate*, $BaH_2O_2 = BaO,H_2O$, is prepared on the large scale by decomposing a hot concentrated solution of barium chloride with a solution of caustic soda; on cooling, crystals of barium hydrate are deposited, which may be purified by recrystallization. In the laboratory barium hydrate is often prepared by boiling a strong solution of the sulphide with small successive portions of black oxide of copper until a drop of the liquid ceases to form a black precipitate with lead salts: the filtered liquid on cooling yields crystals of the hydrate. The crystals of barium hydrate contain $BaH_2O_2,8H_2O$; they fuse easily, and lose their water of crystallization when strongly heated, leaving the hydroxide, BaH_2O_2, in the form of a white, soft powder, having a great attraction for carbonic acid, and soluble in 20 parts of cold and 2 parts of boiling water. The solution is a valuable reagent: it is highly alkaline to test-paper, and instantly rendered turbid by the smallest trace of carbonic acid.

The *Dioxide*, BaO_2, may be formed, as already mentioned, by exposing baryta, heated to full redness in a porcelain tube, to a current of pure oxygen gas. It is gray, and forms with water a white hydrate, which is not decomposed by that liquid in the cold, but dissolves in small quantity. Barium hydroxide, when heated to redness in a current of dry atmospheric air, gives off water, and is converted by absorption of oxygen into barium dioxide, from which the second atom of oxygen may be expelled at a higher temperature. These reactions are utilized for the preparation of oxygen upon a large scale. The dioxide may also be made by heating pure baryta to redness in a platinum crucible, and then gradually adding an equal weight of potassium chlorate, whereby barium dioxide and potassium chloride are produced. The latter may be extracted by cold water, and the dioxide left in the state of hydrate. It is used for the preparation of hydrogen dioxide (p. 166). When dissolved in dilute acid, it is decomposed by potassium dichromate, and by the oxide, chloride, sulphate, and carbonate of silver.

Oxysalts of Barium.—The *Nitrate*, $Ba(NO_3)_2$, is prepared by methods exactly similar to those adopted for preparing the chloride, nitric acid being substituted for hydrochloric. It crystallizes in transparent colorless anhydrous octahedrons, requiring for solution 8 parts of cold and 3 parts of boiling water. This salt is much less soluble in dilute nitric acid than in pure water. Errors sometimes arise from such a precipitate of crystalline barium nitrate being mistaken for sulphate. It disappears on heating or by large affusion of water.

The *Sulphate*, $BaSO_4$, is found native as *heavy spar* or *barytes*, often beautifully crystallized: its specific gravity is as high as 4.4 to 4.8. This compound is always produced when sulphuric acid or a soluble sulphate is mixed with a solution of a barium salt. It is not sensibly soluble in water or in dilute acids: even in nitric acid it is almost insoluble; hot oil of vitriol dissolves a little, but the greater part separates again on cooling. Barium sulphate is now produced artificially on a large scale, and is used as a substitute for white lead in the manufacture of oil paints. The sulphate to be used for this purpose is precipitated from very dilute solutions: it is known in commerce as *blanc fixe*. Powdered native barium sulphate, being rather crystalline, has not sufficient body. For the production of sulphate, the chloride of barium is first prepared, which is dissolved in a large quantity of water, and then precipitated by dilute sulphuric acid.

The *Carbonate*, $BaCO_3$, is found native as *witherite*, and may be formed artificially by precipitating the chloride or nitrate with an alkaline carbonate, or carbonate of ammonia. It is a heavy white powder, very sparingly soluble in water, and chiefly useful in the preparation of other barium salts.

Barium Sulphides.—The *Monosulphide*, BaS, is obtained in the manner already described (P. 352); the higher sulphides may be formed by boiling it with sulphur. Barium monosulphide crystallizes from a hot solution in thin, nearly colorless plates, which contain water, and are not very soluble: they are rapidly altered by the air. A strong solution of this sulphide may be employed, as already described, in the preparation of barium hydroxide.

Reactions of the Alkaline Earth-metals in Solution.—Barium, strontium, and calcium are distinguished from all other substances, and from each other, by the following characters:

Caustic Potash, when free from carbonate, and caustic *ammonia* occasion no precipitates in *dilute* solutions of the alkaline earths, especially of the last two, the hydrates being soluble in water.

Alkaline Carbonates and *Carbonate of Ammonia* give white precipitates, insoluble in excess of the precipitant, with all three.

Sulphuric Acid, or a *Sulphate*, added to very dilute solutions of the salts of these metals, gives an immediate white precipitate with barium salts; and a similar precipitate after a short interval with strontium salts; no change with calcium salts. The precipitates with barium and strontium salts are insoluble in nitric acid.

Solution of *calcium sulphate* gives an instantaneous cloud with barium salts; also with strontium salts after a little time.

Strontium sulphate is itself sufficiently soluble to occasion turbidity when mixed with barium chloride.

Lastly, the soluble *Oxalates* give, in the most dilute solutions of calcium salts, a white precipitate, which is not dissolved by a drop or two of hydrochloric, or by an excess of acetic, acid. This is an exceedingly characteristic test.

The *Chlorides of Strontium* and *Calcium*, dissolved in alcohol, color the flame of the latter red or purple: *barium salts* communicate to the flame a pale green tint.

Silicofluoric acid gives a white precipitate with barium salts, none with salts of strontium or calcium.

30 *

METALS OF THE MAGNESIUM GROUP.

Beryllium.	Zinc.
Magnesium.	Cadmium.

THESE metals are all dyads. They are volatile, and burn brightly when heated in the air. They decompose water at high temperatures only, but dissolve readily at ordinary temperatures in hydrochloric and dilute sulphuric acid, with evolution of hydrogen. Each forms only one oxide and one sulphide.. The oxides are insoluble in water, except that of magnesium, which is slightly soluble. The sulphates are soluble in water; the normal carbonates and phosphates insoluble.

BERYLLIUM, or GLUCINUM.

Symbol, Be.   Atomic weight, 9.

THIS somewhat rare metal occurs as a silicate, either alone as in phenacite, or associated with other silicates, in beryl, emerald, euclase, leucophane, helvite, and several varieties of gadolinite; also as an aluminate in chrysoberyl or cymophane.

Metallic beryllium is obtained by passing the vapor of the chloride over melted sodium. It is a white metal of specific gravity 2.1; it may be forged and rolled into sheets like gold; its melting point is below that of silver. It does not decompose water at the boiling heat. Sulphuric and hydrochloric acids dissolve it, with evolution of hydrogen.

Beryllium forms but one class of compounds, and there is some doubt as to its atomic weight and equivalent value. On the one hand, it is regarded as a dyad, like magnesium, with the atomic weight 9.0, its chloride being $BeCl_2$, its oxide, BeO; on the other hand, as a triad, like aluminium, on which supposition its chloride would be $BeCl_3$, its oxide, Be_2O_3, and its atomic weight, 13.5; but the former view appears, on the whole, to be most in accordance with observed facts.

Beryllium Chloride, $BeCl_2$, is formed by heating the metal in chlorine or hydrochloric acid gas, or by the action of aqueous hydrochloric acid on the metal or its oxide.

The anhydrous chloride is prepared by passing chlorine over an ignited mixture of beryllia and charcoal. It is less volatile than aluminium chloride, very deliquescent, and easily soluble in water.

Beryllium Oxide.—Beryllia, BeO.—This earth may be prepared from beryl, or either of the other beryllium silicates, by fusing the finely-powdered mineral with potassium carbonate or quicklime; treating the fused mass with hydrochloric acid; evaporating to dryness; then moistening the residue with hydrochloric acid and treating it with water, whereby everything is dissolved except the silica. The filtered liquid is then mixed with excess of ammonia solution, which throws down a bulky precipitate containing both alumina and beryllia; this precipitate is well washed, and the beryllia is dissolved out from the alumina by digestion in a cold strong solution of ammonium carbonate. The liquid is again filtered, and on boiling it, beryllium carbonate is deposited as a white powder, which, when ignited, leaves pure beryllia.

Beryllia is very much like alumina in physical characters, and further resembles that substance in being readily dissolved by caustic potash or soda; but it is distinguished from alumina by its solubility, when recently precipitated in a cold solution of ammonium carbonate. Beryllium salts have a sweet taste, whence the former name of the metal, *glucinum* (from γλυκύς). They are colorless, and are distinguished from those of aluminium by not yielding an alum with potassium sulphate; or a blue color when heated before the blowpipe with cobalt nitrate; also by their reaction with ammonium carbonate.

MAGNESIUM.

Symbol, Mg. Atomic weight, 23.94.

THIS metal was formerly classed with the metals of the alkaline earths, but it is much more nearly related to zinc by its properties in the free state as well as by the volatility of its chloride, the solubility of its sulphate, and the isomorphism of several of its compounds with the analogously constituted compounds of zinc.

Magnesium occurs in the mineral kingdom as hydroxide, carbonate, borate, phosphate, sulphate, and nitrate, sometimes in the solid state, sometimes dissolved in mineral waters: magnesium limestone or dolomite, which forms entire mountain masses, is a carbonate of magnesium and calcium. Magnesium also occurs as silicate, combined with other silicates, in a variety of minerals, as steatite, hornblende, augite, talc, etc.; also as aluminate in spinelle and zeilanite. It likewise occurs in the bodies of plants and animals, chiefly as carbonate and phosphate, and in combination with organic acids.

Metallic magnesium is prepared —

1. By the electrolysis of fused magnesium chloride, or, better, of a mixture of 4 molecules of magnesium chloride and 3 molecules of potassium chloride with a small quantity of sal-ammoniac. A convenient way of effecting the reduction is to fuse the mixture in a common clay tobacco-pipe over an argand spirit-lamp or gas-burner, the negative pole being an iron wire passed up the pipe-stem, and the positive pole a piece of gas-coke just touching the surface of the fused chlorides. On passing the current of a battery of ten Bunsen's cells through the arrangement, the magnesium collects round the extremity of the iron wire.

2. Magnesium may be prepared in much larger quantity by reducing magnesium chloride, or the double chloride of magnesium and sodium or potassium, with metallic sodium. The double chloride is prepared by dissolving magnesium carbonate in hydrochloric acid, adding an equivalent quantity of sodium or potassium chloride, evaporating to dryness, and fusing the

residue. This product, heated with sodium in a wrought-iron crucible, yields metallic magnesium, containing certain impurities from which it may be freed by distillation. This process is now carried out on the manufacturing scale, and the magnesium is drawn out into wire or formed into ribbon for burning.*

Magnesium is a brilliant metal almost as white as silver, somewhat more brittle at common temperatures, but malleable at a heat a little below redness. Its specific gravity is 1.75. It melts at a red heat, and volatilizes at nearly the same temperature as zinc. It retains its lustre in dry air, but in moist air it becomes covered with a crust of magnesia.

Magnesium in the form of wire or ribbon takes fire at a red heat, burning with a dazzling bluish-white light. The flame of a candle or spirit-lamp is sufficient to inflame it, but to ensure continuous combustion, the metal must be kept in contact with the flame. For this purpose lamps are constructed, provided with a mechanism which continually pushes three or more magnesium wires into a small spirit-flame.

The magnesium flame produces a continuous spectrum, containing a very large proportion of the more refrangible rays: hence it is well adapted for photography, and has, indeed, been used for taking photographs in the absence of the sun or in places where sunlight cannot penetrate, as in caves or subterranean apartments.

Magnesium Chloride, $MgCl_2$.—When magnesia, or its carbonate, is dissolved in hydrochloric acid, magnesium chloride and water are produced; but when this solution is evaporated to dryness, the last portions of water are retained with such obstinacy that decomposition of the water is brought about by the concurring attractions of magnesium for oxygen and of chlorine for hydrogen, hydrochloric acid being expelled and magnesia remaining. If, however, sal-ammoniac, potassium chloride, or sodium chloride is present, a double salt is produced, which is easily rendered anhydrous. The best mode of preparing the chloride is to divide a quantity of hydrochloric acid into two equal portions, to neutralize one with magnesia, and the other with ammonia or carbonate of ammonia; to mix these solutions, evaporate them to dryness, and then expose the salt to a red heat in a loosely covered porcelain crucible. Sal-ammoniac sublimes, and fused magnesium chloride remains; the latter is poured out upon a clean stone, and when cold transferred to a well-stoppered bottle.

The chloride so obtained is white and crystalline. It is very deliquescent and highly soluble in water, from which it cannot again be recovered by evaporation for the reasons just mentioned. When long exposed to the air in the melted state, it is converted into magnesia. It is soluble in alcohol.

Magnesium chloride occurs in sea-water, in many brine-springs and salt-beds, and is prepared in large quantities at Stassfurt.

Magnesium Oxide, or **Magnesia**, MgO, is easily prepared by exposing the *magnesia alba* of pharmacy, which is a hydrocarbonate, to a full red heat in an earthen or platinum crucible. It forms a soft, white powder, which slowly attracts moisture and carbonic acid from the air, and unites quietly with water to a slightly soluble hydrate, requiring for solution about 5000 parts of water at 15.5° C. (59.9° F.) and 36,000 parts at 100° C. The alkalinity of magnesia can be observed only by placing a small portion in the moist state upon test-paper, but it neutralizes acids completely. It is infusible.

Magnesium sulphide is formed by passing vapor of carbon sulphide over magnesia, in capsules of coke, at a strong red heat.

* For details of the manufacturing process, see Richardson and Watts's Chemical Technology, vol. i., part v., pp. 336–339.

Oxysalts of Magnesium.—The *Sulphate*, $MgSO_4,7H_2O$, commonly called Epsom salt, occurs in sea-water and in many mineral springs, and is now manufactured in large quantities by acting on magnesian limestone with dilute sulphuric acid, and separating the magnesium sulphate from the greater part of the slightly soluble calcium sulphate by filtration. The crystals are derived from a right rhombic prism; they are soluble in an equal weight of water at 5.5° C. (9.9° F.), and in a still smaller quantity at 100° C. The salt has a nauseous bitter taste, and, like many other neutral salts, possesses purgative properties. When it is exposed to heat, 6 molecules of water readily pass off, the seventh being energetically retained. Magnesium sulphate forms with the sulphates of potassium and ammonium, beautiful double salts which contain 6 molecules of crystallization-water, their formulæ being $MgK_2(SO_4)_2,6H_2O$ and $Mg(NH_4)_2(SO_4)_2,6H_2O$. These salts are isomorphous, and form monoclinic crystals.

Magnesium Phosphate, $MgHPO_4,7H_2O$, separates in small colorless prismatic crystals when solutions of sodium phosphate and magnesium sulphate are mixed and left at rest for some time. According to Graham, it is soluble in about 1000 parts of cold water. Magnesium phosphate exists in the grain of the cereals, and can be detected in considerable quantity in beer.

Magnesium and Ammonium Phosphate, $Mg(NH_4)PO_4,6H_2O$.—When ammonia, or its carbonate, is mixed with a magnesium salt, and a soluble phosphate is added, a crystalline precipitate having the above composition subsides—immediately if the solutions are concentrated, and after some time if they are very dilute: in the latter case the precipitation is promoted by stirring. This salt is slightly soluble in pure water, but nearly insoluble in saline and ammoniacal liquids. When heated it gives off water and ammonia, and is converted into *magnesium pyrophosphate*, $Mg_2P_2O_7$:

$$2Mg(NH_4)PO_4 = Mg_2P_2O_7 + H_2O + 2NH_3.$$

At a strong red heat it fuses to a white enamel-like mass. Magnesium and ammonium phosphate sometimes forms a urinary calculus, and occurs also in guano.

In analysis, magnesium is often separated from solutions by bringing it into this state. The liquid, free from alumina, lime, etc., is mixed with sodium phosphate and excess of ammonia, and gently heated for a short time. The precipitate is collected upon a filter, and thoroughly washed with water containing a little ammonia, after which it is dried, ignited to redness, and weighed. The proportion of magnesia is then easily calculated.

Silicates.—The following natural componuds belong to this class: *Chrysolite*, $Mg_2SiO_4 = 2MgO,SiO_2$, a crystallized mineral, sometimes employed for ornamental purposes; a portion of the magnesia is commonly replaced by ferrous oxide, which communicates a green color. *Meerschaum*, $2MgSiO_3,SiO_2 = 2MgO,3SiO_2$, is a soft, sectile mineral, from which pipe-bowls are made. *Talc*, $4MgSiO_3,SiO_2,4H_2O$ (called *steatite* when massive), is a soft, white, sectile, transparent or translucent mineral, used as firestones for furnaces and stoves, and in thin plates for glazing lanterns, etc.; also in the state of powder for diminishing friction. *Soapstone*, also called steatite, is a silicate of magnesium and aluminium of somewhat variable composition. *Serpentine* is a combination of silicate and hydroxide of magnesium. *Jade*, an exceedingly hard stone brought from New Zealand, is a silicate of magnesium and aluminium: its green color is due to chromium. *Augite* and *hornblende* are essentially double silicates of magnesia and lime, in which the magnesia is more or less replaced by its isomorphous substitute; ferrous oxide.

Carbonates.—The *normal carbonate*, $MgCO_3$ or MgO,CO_2, occurs native in rhombohedral crystals, resembling those of calc-spar, imbedded in talc-slate; a soft earthy variety is sometimes met with.

When magnesia alba is dissolved in aqueous carbonic acid, and the solution left to evaporate spontaneously, small prismatic crystals are deposited, consisting of trihydrated magnesium carbonate, $MgCO_3,3H_2O$.

The magnesia alba itself, although often called carbonate of magnesium, is not so in reality; it is a compound of carbonate with hydroxide. It is prepared by mixing hot solutions of potassium or sodium carbonate and magnesium sulphate, the latter being kept in slight excess, boiling the whole for a few minutes, during which time much carbonic acid is disengaged, and well washing the precipitate so produced. If the solution is very dilute, the magnesia alba is exceedingly light and bulky; if otherwise, it is denser. The composition of this precipitate is not perfectly constant. In most cases it contains $4MgCO_3,MgH_2O_2,6H_2O$.

Magnesia alba is slightly soluble in water, especially when cold.

Magnesium salts are isomorphous with zinc salts, ferrous salts, cupric salts, cobalt salts, and nickel salts, etc.; they are usually colorless, and are easily recognized by the following characters:—A gelatinous white precipitate with *caustic alkalies*, including *ammonia*, insoluble in excess, but soluble in solution of sal-ammoniac. A white precipitate with *potassium* and *sodium carbonates*, but none with ammonium carbonate in the cold. A white crystalline precipitate with soluble *phosphates*, on the addition of a little ammonia.

ZINC.

Symbol, Zn.   Atomic weight, 64.9.

ZINC is a somewhat abundant metal: it is found in the state of carbonate, silicate, and sulphide, associated with lead ores in many districts, both in Britain and on the Continent; large supplies are obtained from Silesia and from the neighborhood of Aachen. The native carbonate, or *calamine*, is the most valuable of the zinc ores, and is preferred for the extraction of the metal: it is first roasted to expel water and carbonic acid, then mixed with fragments of coke or charcoal, and distilled at a full red heat in a large earthen retort; carbon monoxide then escapes, while the reduced metal volatilizes and is condensed by suitable means, generally with minute quantities of arsenic.

Zinc is a bluish-white metal, which slowly tarnishes in the air: it has a lamellar, crystalline structure, a density varying from 6.8 to 7.8, and is, under ordinary circumstances, brittle. Between 120° and 150° C. (248–302° F.) it is, on the contrary, malleable, and may be rolled or hammered without danger of fracture: and, what is very remarkable after such treatment, it retains its malleability when cold; the sheet-zinc of commerce is thus made. At 210° C. (410° F.) it is so brittle that it may be reduced to powder. It melts at 412° C. (764° F.); boils at 930° C. (1706° F.) (Violle); 932° C. (1719° F.) (Becquerel); 1040° C. (1904° F.) (Deville and Troost);

and, if air be admitted, burns with a splendid greenish light, generating the oxide. Dilute acids dissolve zinc very readily: it is constantly employed in this manner for preparing hydrogen gas.

Zinc chloride, $ZnCl_2$, may be prepared by heating metallic zinc in chlorine; by distilling a mixture of zinc-filings and corrosive sublimate; or, more easily, by dissolving zinc in hydrochloric acid. It is a nearly white, translucent, fusible substance, very soluble in water and alcohol, and very deliquescent. A strong solution of zinc chloride is sometimes used as a bath for obtaining a graduated heat above 100° C. Zinc chloride unites with sal-ammoniac and potassium chloride to form double salts: the former of these, made by dissolving zinc in hydrochloric acid, and then adding an equivalent quantity of sal-ammoniac, is very useful in tinning and soft soldering copper and iron.

The *oxide*, ZnO, is a strong base, forming salts isomorphous with the magnesium salts. It is prepared either by burning zinc in atmospheric air, or by heating the carbonate to redness. Zinc oxide is a white, tasteless powder, insoluble in water, but freely dissolved by acids. When heated it is yellow, but turns white again on cooling. It is sometimes used as a substitute for white lead. To prepare zinc-white on a large scale, metallic zinc is volatilized in large earthen muffles, whence the zinc vapor passes into a small receiver (*guérite*), where it comes in contact with a current of air and is oxidized. The zinc oxide thus formed passes immediately into a condensing chamber divided into several compartments by cloths suspended within it.

The *sulphate*, $ZnSO_4,7H_2O$, commonly called *White Vitriol*, is hardly to be distinguished by the eye from magnesium sulphate: it is prepared either by dissolving the metal in dilute sulphuric acid, or, more economically, by roasting the native sulphide, or *blende*, which, by absorption of oxygen, becomes in great part converted into sulphate. The altered mineral is thrown hot into water, and the salt is obtained by evaporating the clear solution. Zinc sulphate has an astringent metallic taste, and is used in medicine as an emetic. The crystals dissolve in $2\frac{1}{2}$ parts of cold, and in a much smaller quantity of hot water. Crystals containing 6 molecules of water have been observed. Zinc sulphate forms double salts with the sulphates of potassium and ammonium, namely, $ZnK_2(SO_4)_2,6H_2O$ and $Zn(NH_4)_2,6H_2O$, isomorphous with the corresponding magnesium salts.

The *carbonate*, $ZnCO_2$, is found native; the white precipitate obtained by mixing solutions of zinc and of alkaline carbonates, is a combination of carbonate and hydroxide. When heated to redness, it yields pure zinc oxide.

The *sulphide*, ZnS, occurs native, as *blende*, in regular tetrahedrons, dodecahedrons, and other monometric forms, and of various colors, from white or yellow to brown or black, according to its degree of purity: it is a valuable ore of zinc. A variety called *Black Jack* occurs somewhat abundantly in Derbyshire, Cumberland, and Cornwall. A *hydrated sulphide*, ZnS,H_2O, is obtained as a white precipitate on adding an alkaline sulphide to the solution of a zinc salt.

Zinc salts are distinguished by the following characters:—*Caustic potash, soda,* and *ammonia* give a white precipitate of hydroxide, freely soluble in excess of the alkali. *Potassium* and *sodium carbonates* give white precipitates, insoluble in excess. *Ammonium carbonate* gives also a white precipitate, which is redissolved by an excess. *Potassium ferrocyanide* gives a white precipitate. *Hydrogen sulphide* causes no change in zinc solutions containing free mineral acids; but in neutral solutions, or with zinc salts of organic acids, such as the acetate, a white precipitate is formed. *Ammonium sulphide* throws down white sulphide of zinc, insoluble in caustic alkalies.

The formation of this precipitate in a solution containing excess of caustic alkali serves to distinguish zinc from all other metals.

All zinc compounds, heated on charcoal with sodium carbonate in the inner blowpipe flame, give an incrustation of zinc oxide, which is yellow while hot, but becomes white on cooling. If this incrustation be moistened with a dilute solution of cobalt nitrate, and strongly heated in the outer flame, a fine green color is produced.

The applications of metallic zinc to the purposes of roofing, the construction of water-channels, etc., are well known; it is sufficiently durable, but inferior in this respect to copp r. It is much used also for protecting iron and copper from oxidation when immersed in saline solutions, such as sea-water, or exposed to damp air. This it does by forming an electric circuit, in which it acts as the positive or more oxidizable metal (p. 280). *Galvanized iron* consists of iron having its surface coated with zinc.

CADMIUM.

Symbol, Cd. Atomic weight, 112.3.

THIS metal was discovered in 1817 by Stromeyer and by Hermann: it accompanies the ores of zinc, especially those occurring in Silesia, and, being more volatile than that substance, rises first in vapor when the calamine is subjected to distillation with charcoal. Cadmium resembles tin in color, but is somewhat harder; it is very malleable, has a density of 8.7, melts below 260° C. (500° F.), and is nearly as volatile as mercury. It tarnishes but little in the air, but burns when strongly heated. Dilute sulphuric and hydrochloric acids act but little on cadmium in the cold; nitric acid is its best solvent.

The observed vapor-density of cadmium is 3.94 compared with air, or 56.3 compared with hydrogen, which latter number does not differ greatly from the half of 112.3, the atomic weight of the metal: hence it appears that the *atom* of cadmium in the state of vapor occupies twice the space of an atom of hydrogen (p. 245).

Cadmium Oxide, CdO, may be prepared by igniting either the carbonate or the nitrate: in the former case it has a pale-brown color, and in the latter a much darker tint, and forms octahedral microscopic crystals. Cadmium oxide is infusible: it dissolves in acids, producing a series of colorless salts: it attracts carbonic acid from the air, and turns white. The *sulphate*, $CdSO_4,4H_2O$, is easily obtained by dissolving the oxide or carbonate in dilute sulphuric acid: it is very soluble in water, and forms double salts with the sulphates of potassium and ammonium, which contain respectively $CdK_2(SO_4)_2,6H_2O$ and $Cd(NH_4)_2(SO_4)_2,6H_2O$. The *chloride*, $CdCl_2$, is a very soluble salt, crystallizing in small four-sided prisms. The *sulphide*, CdS, is a very characteristic compound, of a bright yellow color, forming microscopic crystals, fusible at a high temperature. It is obtained by passing sulphuretted hydrogen gas through a solution of the sulphate, nitrate, or chloride. This compound is used as a yellow pigment, of great beauty and permanence. It occurs native as *Greenockite*.

Fixed caustic *alkalies* give with cadmium salts a white precipitate of hydroxide, insoluble in excess. *Ammonia* gives a similar white precipitate, readily soluble in excess. The *fixed alkaline carbonates* and *ammonia carbonate* throw down white cadmium carbonate, insoluble in excess of either precipitant. *Hydrogen sulphide* and *ammonium sulphide* precipitate the yellow sulphide of cadmium.

16

METALS OF THE LEAD GROUP.

Lead—Thallium.

THESE metals are soft, heavy, and form basic oxides having an alkaline reaction. Thallium in many of its properties closely resembles the alkali-metals, and lead resembles the metals of the alkaline earths in forming an insoluble carbonate and sulphate; but lead and thallium are distinguished from both these groups by the black color and insolubility of their sulphides and the sparing solubility of their dichlorides.

LEAD.

Symbol, Pb (Plumbum). Atomic weight, 206.4.

THIS abundant and useful metal is wholly obtained from the native sulphide, or *galena*, no other lead-ore being found in large quantity. The reduction is effected in a reverberatory furnace, into which the crushed lead-ore is introduced and roasted for some time at a dull red heat, by which much of the sulphide is oxidized to sulphate. The contents of the furnace are then thoroughly mixed and the temperature raised, whereupon the sulphate and oxide react with the remaining sulphide in such a manner as to produce sulphur dioxide and metallic lead:—

$$2PbO + PbS = SO_2 + 3Pb$$
$$\text{and } PbSO_4 + PbS = 2SO_2 + 2Pb.$$

The lead thus produced sometimes contains antimony and other metals, which render it hard. They may, however, be removed, and the lead rendered soft by melting and partially oxidizing it in a reverberatory furnace with a cast-iron bottom. The lead so far purified still contains silver in sufficient quantity to render it worth extracting; in fact, a considerable proportion of the silver which now comes into the market is obtained from argentiferous galena. The method of separating the silver will be described under SILVER.

Lead is a soft bluish metal, possessing very little elasticity; its specific gravity is 11.45. It may be easily rolled out into plates or drawn out into coarse wires, but has very little tenacity. It melts at 315.5° C. (600° F.), or a little above, and boils and volatilizes at a white heat. By slow cooling, it may be obtained in octahedral crystals. In moist air it becomes coated with a film of gray matter, thought to be suboxide, and when exposed to the atmosphere in the melted state it rapidly absorbs oxygen. Dilute acids, with the exception of nitric acid, act but slowly upon lead.

·Lead is a tetrad, as shown by the composition of plumbic ethide, $Pb(C_2H_5)_4$, and of the perchloride, probably $PbCl_4$, and the peroxide, PbO_2. In the greater number of its compounds, however, it is bivalent, as in the dichloride, $PbCl_2$, the corresponding bromide and iodide, and the monoxide, PbO.

Chlorides.—The *dichloride*, $PbCl_2$, is prepared by precipitating a solution of lead nitrate or acetate with hydrochloric acid or common salt, and separates as a heavy white crystalline precipitate, which dissolves in about 33 parts of boiling water, and separates again, on cooling, in needle-shaped crystals.

The *perchloride*, probably $PbCl_4$, is obtained by dissolving the dioxide in strong well-cooled hydrochloric acid, whereby a yellow strongly oxidizing solution is obtained, from which water and alkalies throw down the dioxide.

Oxychlorides.—The compound $Pb_3Cl_2O_2$ or $PbCl_2,2PbO$, occurs crystallized in rhombic prisms on the Mendip Hills, and is thence called *Mendipite*. Another oxychloride, constituting Pattinson's white oxychloride, Pb_2Cl_2O or $PbCl_2,PbO$, is prepared for use as a pigment by grinding galena with strong hydrochloric acid, dissolving the resulting chloride in hot water, and precipitating with lime-water. A third oxychloride, $PbCl_2 7PbO$, called *Patent yellow* or *Turner's yellow*, is prepared by heating 1 part of sal-ammoniac with 13 parts of litharge.

Lead Iodide, PbI_2, is precipitated on mixing the nitrate or acetate with potassium iodide, as a bright yellow p   d   , which dissolves in boiling water, and crystallizes therefrom in beautiful yellow iridescent spangles.

Oxides.—*Diplumbic oxide*, or *Lead suboxide*, Pb_2O, is formed when the monoxide is heated to dull redness in a retort; a gray pulverulent substance is then left, which is resolved by acids into monoxide and metal. It absorbs oxygen with great rapidity when heated, and even when simply moistened with water and exposed to the air.

The *monoxide*, PbO, called *Litharge* or *Massicot*, is the product of the direct oxidation of the metal. It is most conveniently prepared by heating the carbonate to dull redness; common *litharge* is impure monoxide which has undergone fusion. Lead oxide has a delicate straw-yellow color, is very heavy, and slightly soluble in water, giving an alkaline liquid. It is soluble in potash, and crystallizes from the solution in rhombic prisms. At a red heat it melts, and tends to crystallize on cooling. In the melted state it attacks and dissolves silicious matter with astonishing facility, often penetrating an earthen crucible in a few minutes. It is easily reduced when heated with organic substances of any kind containing carbon or hydrogen. It forms a large class of salts, often called *plumbic salts*, which are colorless if the acid itself is not colored.

A *basic hydroxide*, $Pb_2O(OH)_2$ or HO.Pb.O.Pb.OH, is obtained as a white precipitate by the action of air and water free from carbonic acid on metallic lead, and is thrown down as a white precipitate on addition of ammonia or a fixed alkali to a lead-salt, the precipitate in the latter case being redissolved by excess of the reagent. The hydroxide gives off part of its water at 130° C. (266° F.), and the whole at 145° C. (293° F.). Both the hydroxide and the oxide are somewhat soluble in water, and turn reddened litmus-paper blue. They act as strong bases and also as acids, combining with certain metallic oxides, the lead oxide when fused with these bases forming a glass.

The *sesquioxide*, Pb_2O_3, is formed by the action of sodium hypochlorite on a solution of the monoxide in potash, and when a solution of red lead in acetic acid is precipitated by very dilute ammonia. The reddish-yellow powder thus formed contains water, part of which it retains even at 150° C. (302° F.). It is resolved by acids into PbO and PbO_2.

Triplumbic tetroxide, or *Red lead,* is not of very constant composition, but generally contains Pb_3O_4 or $2PbO,PbO_2$. It is prepared by exposing the monoxide, which has not been fused, for a long time to the air at a very faint red heat. It is a brilliant red and extremely heavy powder, decomposed, with evolution of oxygen, by a strong heat, and converted by acids into a mixture of monoxide and dioxide. It is used as a cheap substitute · for vermilion.

The *dioxide,* PbO_2, often called *Puce* or *Brown lead-oxide,* is easily obtained by digesting red lead in dilute nitric acid, whereby lead nitrate is dissolved out, and insoluble dioxide left behind in the form of a deep-brown powder. It is also formed by the action of chlorine on lead-salts in presence of alkalies, and, as already observed (p. 277), by the electrolysis of lead-salts being deposited on the anode. The dioxide is decomposed by a red heat, yielding up one-half of its oxygen. By hydrochloric acid it is converted into lead chloride, with disengagement of chlorine, and by hot sulphuric acid into sulphate, with liberation of oxygen. The dioxide is very useful in separating sulphurous oxide from certain gaseous mixtures, lead sulphate being then produced: $PbO_2 + SO_2 = PbSO_4$. The dioxide possesses acid properties. When fused in a silver crucible with excess of potassium hydroxide, it yields crystalline *potassium plumbate,* $K_2PbO_3,2H_2O$, the solution of which gives with most metallic salts precipitates of the corresponding plumbates.

Lead Sulphate, $PbSO_4$, occurs native as *Lead-vitriol* or *Anglesite* in transparent rhombic crystals isomorphous with cœlestin, $SrSO_4$, and heavy spar, $BaSO_4$, and is obtained as a white powder soluble in alkalies by precipitating a lead-salt with sulphuric acid or a soluble sulphate.

Lead Nitrate, $Pb(NO_3)_2$ or PbO,N_2O_5, may be obtained by dissolving lead carbonate in nitric acid, or by acting directly upon the metal with the same agent aided by heat: it is, as already noticed, a by-product in the preparation of dioxide. It crystallizes in anhydrous octahedrons, which are usually milk-white and opaque. It dissolves in $7\frac{1}{2}$ parts of cold water, and is decomposed by heat, yielding nitrogen tetroxide, oxygen and lead monoxide, which obstinately retains traces of nitrogen. When a solution of this salt is boiled with an additional quantity of lead oxide, a portion of the latter is dissolved, and a basic nitrate generated, which may be obtained in crystals. Carbonic acid separates this excess of oxide in the form of a white compound of carbonate and hydrate of lead.

By boiling a solution of lead nitrate for several hours with lead-turnings, N. v. Lorenz[*] has lately obtained a number of basic nitrates and nitrites of lead, the composition of which is shown in the following table, from which it will be seen that, as the action progresses, the amounts of nitrite and oxide or hydroxide in the salts produced increase continually in proportion to the nitrate. The symbol A stands for the basic nitrate, $NO_3.Pb.OH$, and B for the nitrite, NO_2PbOH :—

$NO_3.Pb.OH$	$4A + 6B + 5PbO + Pb(OH)_2$
$3(2A + B) + H_2O$	$2A + 4B + 3PbO + Pb(OH)_2$
$3(A + B) + 2H_2O$	$\left\{ \begin{array}{l} Pb(NO_2)_2 + 2PbO \text{ or} \\ 5Pb(NO_2)_2 + NO_3.Pb.OH \end{array} \right.$
$6A + 7B + 5H_2O$	
$3A + 5B + H_2O$	$\quad\quad\quad\quad\quad + 10PbO$
$3A + 6B + H_2O$	
$3A + 9B + H_2O$	$NO_2.Pb.OH + PbO$
$A + 4B$	
$A + 5B$	

* Monatshefte für Chemie, 1881, pp. 810–841, and Journal of the Chemical Society, 1882, vol. xlii. p. 364.

Respecting the mode of formation and the physical characters of these successive products, reference must be made to the paper above cited.

Lead Phosphates.—The *Normal Orthophosphate*, $Pb_3(PO_4)_2$, is obtained as a white precipitate on adding ordinary phosphate of soda to a solution of lead acetate. A boiling solution of lead nitrate mixed with phosphoric acid yields a glittering white crystalline precipitate of an acid phosphate, $HPbPO_4$. A phosphato-chloride of lead, $Pb_3PO_4,Pb_2(PO_4)Cl$, occurs native as *pyromorphite*, and the corresponding arsenato-chloride as *mimetesite*. In both these minerals a portion of the chlorine is usually replaced by fluorine.

Lead Silicates.—Silica fuses with lead oxide to a yellow glass. Lead silicate forms a constituent of flint glass.

Lead Carbonates.—The *normal salt*, $PbCO_3$, occurs native as *cerusite* in white rhombic crystals isomorphous with arragonite; also in pseudomorphs after galena and lead sulphate. The same compound is formed by precipitating a cold solution of lead acetate with carbonate of ammonia, and by passing carbon dioxide into a dilute solution of normal lead acetate.

Lead forms several basic carbonates, the most important of which, known as *white lead*, is manufactured on a very large scale for the use of the painter. Of the many methods put in practice, or proposed, for making white lead, the two following are the most important. The first consists in forming a basic nitrate or acetate of lead by boiling finely-powdered litharge with the normal salt. This solution is then brought into contact with carbonic acid gas, whereby all the excess of oxide previously taken up by the normal salt is at once precipitated as white lead, and the solution strained or pressed from the latter is again boiled with litharge and treated with carbonic acid, these processes being susceptible of indefinite repetition, whereby the small loss of normal salt left in the precipitates is compensated. The second, and by far the more ancient method, is rather more complex, and at first sight not very intelligible. A great number of earthen jars are prepared, into each of which is poured a few ounces of crude vinegar; a roll of sheet lead is then introduced in such a manner that it shall neither touch the vinegar nor project above the top of the jar. The vessels are next arranged in a large building, side by side, upon a layer of stable manure, or better, spent tan, and closely covered with boards. A second layer of tan is spread upon the top of the latter, and then a second series of pots; these are in turn covered with boards and decomposing bark, and in this manner a pile of many alternations is constructed. After the lapse of a considerable time the pile is taken down, and the sheets of lead are removed and carefully unrolled; they are then found to be in great part converted into carbonate, which merely requires washing and grinding to be fit for use. The nature of this curious process is generally explained by supposing that the vapor of vinegar raised by the high temperature due to the fermentation of the dung or tan gives rise to the formation of basic acetate of lead, which by contact with the carbon dioxide likewise evolved from the decomposing manure is converted into basic carbonate.

White lead is a white, heavy, amorphous powder, appearing under the microscope as an aggregate of round transparent globules, having a diameter of one to four hundred-thousandths of an inch. The product obtained by the method above described is generally preferred to those obtained by other processes, on account of its superior opacity or "body." Commercial white lead, however prepared, always contains a certain proportion of hydroxide.

When clean metallic lead is put into pure water and exposed to the air, a white crystalline scaly powder begins to show itself in a few hours, and

31 *

very rapidly increases in quantity. This substance may consist of lead hy-droxide, formed by the action of the oxygen dissolved in the water upon the lead. It is slightly soluble, and may be readily detected in the water. In most cases, however, the formation of this deposit is due to the action of the carbonic acid dissolved in the water: it consists of carbonate in com-bination with hydroxide, and is nearly insoluble in water. When common river- or spring-water is substituted for the pure liquid, this effect is less ob-servable, the little sulphate almost invariably present causing the deposition of a very thin but closely adherent film of lead sulphate upon the surface of the metal, which protects it from further action. It is on this account that leaden cisterns are used with impunity, at least in most cases, for hold-ing water: if the latter were quite pure, it would be speedily contaminated with lead, and the cistern would be soon destroyed. Natural water, highly charged with carbonic acid, cannot under any circumstances be kept in lead or passed through leaden pipes with safety, the carbonate, though insoluble in pure water, being slightly soluble in water containing carbonic acid.

Lead Sulphide, PbS, occurs native as *Galena* in cubes or other forms of the regular system, having a bluish-gray color and a density of 7.25 to 7.7. The same compound is obtained in crystalline form by passing sul-phur vapor over metallic lead, the combination being attended with vivid combustion; also by fusing lead oxide with excess of sulphur; and as a black amorphous precipitate by passing hydrogen sulphide into a solution of nitrate or acetate of lead. A *chlorosulphide*, $PbCl_2,3PS$, is obtained as a yellowish-red precipitate on passing a small quantity of hydrogen sulphide into the solution of a lead salt containing hydrochloric acid. An excess of hydrogen sulphide converts it into the sulphide.

The soluble salts of lead behave with reagents as follows :—
Caustic *potash* and *soda* precipitate a white hydroxide freely soluble in excess. *Ammonia* gives a similar white precipitate not soluble in excess. The *carbonates of potassium, sodium,* and *ammonium* precipitate lead car-bonate, insoluble in excess. *Sulphuric acid* or a *sulphate* causes a white pre-cipitate of lead sulphate, insoluble in nitric acid. *Hydrogen sulphide* and *ammonium sulphide* throw down black lead sulphide. Lead is readily de-tected before the blowpipe by fusing the compound under examination on charcoal with sodium carbonate, when a bead of metal is easily obtained, which may be recognized by its chemical as well as by its physical properties.

An alloy of 2 parts of lead and 1 of tin constitutes *plumbers' solder ;* these proportions reversed give a more fusible compound, called *fine solder.* The lead employed in the manufacture of shot is combined with a little arsenic.

THALLIUM.

Symbol, Tl. Atomic weight, 203.6.

THIS element was discovered by Crookes, in 1861, in the seleniferous deposit of a lead-chamber of a sulphuric-acid factory in the Hartz Moun-tains, where iron pyrites is used for the manufacture of sulphuric acid. The name is derived from θαλλός, "green," because the existence of this metal was first recognized by an intense green line appearing in the spectrum of a flame in which thallium was volatilized. It was at first suspected to be a metalloïd, but further examination proved it to be a true metal. It was first obtained in a distinct metallic form by Crookes toward the end of the year 1861, and soon afterward by Lamy, who prepared it from the deposit in the lead-chamber of M. Kuhlmann of Lille, where Belgian pyrites is employed for the manufacture of sulphuric acid.

Thallium appears to be very widely diffused as a constituent of iron and copper pyrites, though it never constitutes more than the 4000th part of the bulk of the ores. It has also been found in lepidolite from Moravia, in mica from Zinnwald in Bohemia, and in the mother-liquors of the salt-works at Nauheim.

Thallium is most economically prepared from the flue-dust of pyrites burners. This substance is stirred up in wooden tubs with boiling water, and the clear liquor, siphoned off from the deposit, is mixed with excess of strong hydrochloric acid, which precipitates impure thallium monochloride. To obtain a pure salt, this crude chloride is added by small portions at a time to half its weight of hot oil of vitriol in a porcelain or platinum dish, the mixture being constantly stirred, and the heat continued till the whole of the hydrochloric acid and the greater portion of the excess of sulphuric acid are driven off. The fused acid sulphate is now to be dissolved in an excess of water, and an abundant stream of hydrogen sulphide passed through the solution. The precipitate, which may contain arsenic, antimony, bismuth, lead, mercury, and silver, is separated by filtration, and the filtrate is boiled till all free hydrogen sulphide is removed. The liquid is now to be rendered alkaline with ammonia, and boiled ; the precipitate of iron oxide and alumina which generally appears in this place is filtered off, and the clear solution evaporated to a small bulk. Thallium sulphate then separates on cooling in long clear prismatic crystals.

Metallic thallium may be reduced from the solution of the sulphate either by electrolysis or by the action of zinc.

Thallium is a heavy metal, resembling lead in its physical properties. When freshly cut, it exhibits a brilliant metallic lustre and grayish color, somewhat between that of silver and that of lead, assuming a slight yellowish tint by friction with harder bodies. It is very soft, being readily cut with a knife and making a streak on paper like plumbago. It is very malleable, is not easily drawn into wire, but may be readily squeezed into that form by the process technically called "squirting." It has a highly crystalline structure, and crackles like tin when bent. It melts at 294° C. (561° F.).

In contact with the air, thallium tarnishes more rapidly than lead, becoming coated with a thin layer of oxide, which preserves the rest of the metal.

The most characteristic property of thallium is the bright-green color which the metal or any of its compounds impart to a colorless flame; and this color, when viewed by the spectroscope, is seen to be absolutely monochromatic, appearing as one intensely brilliant and sharp green line.

Thallium dissolves in hydrochloric, sulphuric, and nitric acids, the latter attacking it very energetically with copious evolution of red vapors.

Thallium forms two classes of compounds—namely, the t h a l l i o u s c o m - p o u n d s, in which it is univalent, and the t h a l l i c c o m p o u n d s, in which it is trivalent. Thus it forms two oxides, Tl_2O and Tl_2O_3, with corresponding chlorides, bromides, iodides, and oxygen-salts. In some of its chemical relations it resembles the alkali-metals, forming a readily soluble and highly alkaline monoxide, a soluble and alkaline carbonate, an insoluble platinochloride, a thallio-aluminic sulphate, similar in form and composition to common potash-alum, and several phosphates exactly analogous in composition to the phosphates of sodium. In most respects, how-ever, it is more nearly allied to the heavy metals, especially to lead, which it resembles closely in appearance, density, melting-point, specific heat, and electric conductivity; also in giving a black precipitate with hydrogen sulphide, and a white precipitate with soluble chlorides.

Thallium Chlorides.—Thallium forms four chlorides, represented by the formulæ $TlCl$, Tl_4Cl_6, Tl_2Cl_4, and $TlCl_3$, the second and third of which may be regarded as compounds of the monochloride and trichloride.

The *monochloride* or *Thallious chloride*, $TlCl$, is formed by direct combina-tion, the metal burning when heated in chlorine gas, or as a white curdy precipitate resembling silver chloride, by treating the solution of any thallious salt with a soluble chloride. When boiled with water it dissolves like lead chloride, and separates in white crystals on cooling. It forms double salts with trichloride of gold and tetrachloride of platinum. The *platinum salt*, $2TlCl,PtCl_4$, separates as a pale yellow, very slightly soluble, crystalline powder on adding platinic chloride to thallious chloride.

The *trichloride* or *Thallic chloride*, $TlCl_3$, is obtained by dissolving the trioxide in hydrochloric acid, or by acting upon thallium, or one of the lower chlorides, with a large excess of chlorine at a gentle heat. It is soluble in water, and separates by evaporation in a vacuum in hydrated crystals; melts easily, and decomposes at a high temperature. It forms crystalline double salts with the chlorides of the alkali-metals.

The *Sesquichloride*, $Tl_4Cl_6 = TlCl_3,3TlCl$, is produced by dissolving thal-lium or the monochloride in nitromuriatic acid, and separates on cooling in yellow crystalline scales. By aqueous ammonia, potash, or even by thallious oxide, it is instantly decomposed into sesquioxide and monochloride, accord-ing to the equation,

$$2Tl_4Cl_6 + 3KHO = Tl_2O_3 + 6TlCl + 3KCl + 3HCl.$$

The *Dichloride*, $Tl_2Cl_4 = TlCl_3,TlCl$, is formed by carefully heating thal-lium or the monochloride in a slow current of chlorine. It is a pale-yellow substance reduced to sesquichloride by further heating.

The BROMIDES of thallium resemble the chlorides.

Iodides.—*Thallious iodide*, TlI, is formed by direct combination of its elements, or by double decomposition. It forms a beautiful yellow powder, rather darker than sulphur, and melting below redness to a scarlet liquid, which, as the mass cools, remains scarlet for some time after solidification, then changes to bright yellow. The dried precipitate, when spread on paper with a little gum-water, undergoes a similar but opposite change to that experienced by mercuric iodide when heated, the yellow surface when held over a flame suddenly becoming scarlet, and frequently remaining so after cooling for several days; hard friction with a glass rod, however, changes the scarlet color back to yellow. It is very slightly soluble in water, requiring, according to Crookes, 4453 parts of water at $17.2°$ C. ($63°$ F.), and 842.4 parts at $100°$ C., to dissolve it.

Thallic Iodide, TlI_3, is formed by the action of thallium on iodine dis-solved in ether, as a brown solution which gradually deposits rhombic prisms. It forms crystalline compounds with the iodides of the alkali-metals.

Thallium Oxides.—Thallium forms a monoxide and a trioxide.

The *monoxide* or *Thallious oxide*, Tl_2O, constitutes the chief part of the crust which forms on the surface of the metal when exposed to the air. It may be prepared by allowing granulated thallium to oxidize in warm moist air, and then boiling with water. The filtered solution first deposits white needles of thallium carbonate, and, on further cooling, yellow needles of the hydroxide, TlHO or Tl_2O,H_2O, which, when left over oil of vitriol in a vacuum, yields the anhydrous monoxide as a reddish-black mass retaining the shape of the crystals. It is partially reduced to metal by hydrogen at a red heat. When fused with sulphur it yields thallious sulphide. It dissolves readily in water, forming a colorless strongly alkaline solution, which reacts with metallic salts very much like caustic potash. This solution, treated with zinc or subjected to electrolysis, yields metallic thallium.

The *trioxide*, or *Thallic oxide*, is the chief product obtained by burning thallium in oxygen gas. It is best prepared by adding potash to the solution of a thallic salt, and drying the precipitate at 260° C. (500° F.). It is also formed by electrolysis of thallious sulphate. It is a dark-red powder reduced to thallious oxide at a red heat; neutral, insoluble in water and in alkalies. Thallic hydrate, $Tl'''HO_2$, is obtained by drying the above-mentioned precipitate at 100° C.

Oxygen-salts.—Both the oxides of thallium dissolve readily in acids, forming crystalline salts, soluble in water; there are also a few insoluble thallium salts formed by double decomposition.

Sulphates.—Thallious sulphate, Tl_2SO_4, obtained by evaporating the chloride or nitrate with sulphuric acid, or by heating metallic thallium with that acid, crystallizes in anhydrous rhombic prisms, isomorphous with potassium sulphate. It forms, with aluminium sulphate, the salt, $AlTl(SO_4)_2,12H_2O$, isomorphous with common alum; and with the sulphates of magnesium, nickel, etc., double salts containing 6 molecules of water, and isomorphous with magnesium and potassium sulphate, etc. (p. 357).—*Thallic sulphate*, $Tl_2(SO_4)_3,7H_2O$, separates by evaporation from a solution of thallic oxide in dilute sulphuric acid in thin colorless laminæ, which are decomposed by water, even in the cold, with separation of brown thallic oxide.

Phosphates.—The thallious phosphates form a series nearly as complete as those of the alkali-metals, which they also resemble in their behavior when heated. There are three *orthophosphates*, containing respectively H_2TlPO_4, HTl_2PO_4, and Tl_3PO_4. The first two are soluble in water; the second is obtained by neutralizing dilute phosphoric acid at boiling heat with thallious carbonate, and the first by mixing the dithallious salt with excess of phosphoric acid. The *trithallious salt*, Tl_3PO_4, is very sparingly soluble, and is formed as a crystalline precipitate on mixing the saturated solutions of ordinary disodic phosphate and thallious sulphate; also, together with ammonio-thallious phosphate, by treating the monothallious or dithallious salt with excess of ammonia. There are two *thallious pyrophosphates*, $H_2Tl_2P_2O_7$ and $Tl_4P_2O_7$, both very soluble in water: the first produced by carefully heating monothallious orthophosphate, the second by strongly heating dithallious orthophosphate. Of *thallious metaphosphate*, $TlPO_3$, there are two modifications: the first remaining as a slightly soluble vitreous mass when monothallious orthophosphate is strongly ignited, the second obtained as an easily soluble glass by igniting ammonio-thallious ortho-phosphate.

Thallic orthophosphate, $Tl'''PO_4,2H_2O$, separates as an insoluble gelatinous precipitate on diluting a solution of thallic nitrate mixed with phosphoric acid.

Thallious carbonate, Tl_2CO_3, is deposited in crystals, apparently tri-

16 *

metric, when a solution of thallious oxide is exposed to the air. It is soluble in water, and the solution has a slightly caustic taste and alkaline reaction.

Thallium Sulphide, Tl_2S.—This compound is precipitated from all thallious salts by ammonium sulphide, and from the acetate, carbonate, or oxalate by hydrogen sulphide (incompletely also from the nitrate, sulphate, or chloride) in dense flocks of a grayish or brownish-black color. Thallic salts appear to be reduced to thallious salts by boiling with ammonium sulphide. Thallium sulphate projected into fused potassium cyanide is reduced to sulphide, which then forms a brittle metal-looking mass having the lustre of plumbago, and fusing more readily than metallic thallium.

Reactions of Thallium Salts.—The reactions of thallious salts with hydrogen sulphide and ammonium sulphide have just been mentioned. From their aqueous solutions thallium is rapidly precipitated in metallic crystals by *zinc*, slowly by *iron*. *Soluble chlorides* precipitate difficultly soluble white thallious chloride; soluble *bromides* throw down white, nearly insoluble bromide; soluble *iodides* precipitate insoluble yellow thallious iodide. Caustic *alkalies* and *alkaline carbonates* form no precipitate; *sodium phosphate* forms a white precipitate, insoluble in ammonia, easily soluble in acids. *Potassium chromate* gives a yellow precipitate of thallious chromate, insoluble in cold nitric or sulphuric acid, but turning orange-red on boiling in the acid solution. *Platinic chloride* precipitates a very pale-yellow insoluble double salt.

Thallic salts are easily distinguished from thallious salts by their behavior with alkalies and with soluble chlorides or bromides. Their solutions give with *ammonia*, and with *fixed alkalies* and their *carbonates*, a brown gelatinous precipitate of thallic oxide, containing the whole of the thallium. Soluble *chlorides* or *bromides* produce no precipitate in solutions of pure thallic salts; but if a thallious salt is likewise present, a precipitate of sesquichloride or sesquibromide is formed. *Oxalic acid* forms in solutions of thallic salts a white pulverulent precipitate; *phosphoric acid*, a white gelatinous precipitate; and *arsenic acid*, a yellow gelatinous precipitate. Thallic nitrate gives with *potassium ferrocyanide* a green, and with the *ferricyanide* a yellow, precipitate.

In examining a mixed metallic solution, thallium will be found in the precipitate thrown down by ammonium sulphide, together with iron, nickel, manganese, etc. From these metals it may be easily separated by precipitation with potassium iodide or platinic chloride, or by reduction to the metallic state with zinc.

Thallium salts are reduced before the blowpipe with charcoal and sodium carbonate or potassium cyanide. The green color imparted to flame by thallium, and the peculiar character of its spectrum, have already been mentioned.

METALS OF THE COPPER GROUP.

Copper—Mercury—Silver.

EACH of these metals forms several oxides and two series of salts, the constitution of which is best represented, as will be presently explained, by regarding the metals as dyads, although two of them, viz., copper and silver, figure in Mendelejeff's series as monads. These metals do not decompose water at a red heat. They are precipitated from their solutions in the metallic state by zinc, iron, and the more oxidizable metals, and their lower chlorides, like those of the preceding group, are insoluble in water.

COPPER.

Symbol, Cu (Cuprum). Atomic weight, 63.1.

COPPER is a metal of great value in the arts; it sometimes occurs in the metallic state, crystallized in octahedrons, or more frequently in dodecahedrons, but is more abundant in the form of red oxide and in that of sulphide combined with sulphide of iron, as *yellow copper ore* or *copper pyrites*. Large quantities of the latter substance are annually obtained from the Cornish mines and taken to South Wales for reduction, which is effected by a somewhat complex process. The principle of this may, however, be easily made intelligible. The ore is roasted in a reverberatory furnace, by which much of the iron sulphide is converted into oxide, while the copper sulphide remains unaltered. The product of this operation is then strongly heated with silicious sand: the latter combines with the iron oxide to a fusible *slag*, and separates from the heavier copper compound. When the iron has, by a repetition of these processes, been got rid of, the copper sulphide begins to decompose in the flame-furnace, losing its sulphur and absorbing oxygen; the temperature is then raised sufficiently to reduce the oxide thus produced by the aid of carbonaceous matter. The last part of the operation consists in thrusting into the melted metal a pole of birch-wood, the object of which is probably to reduce a little remaining oxide by the combustible gases thus generated. Large quantities of extremely valuable ore, chiefly carbonate and red oxide, have lately been obtained from South Australia and Chili.

Another process, known as the Mansfeld process, and especially adapted for working the cupriferous schists of Mansfeld in Prussia, which are mostly poor in copper, has long been practised in that locality and in other parts of Europe. The ore, called "Kupferschiefer," is first roasted in heaps, whereby the bituminous matter contained in it is burnt away, while water and arsenic are expelled, together with part of the sulphur. The roasted ore is then mixed with 5 to 8 parts of slag and fluor-spar, and heated in a blast- or cupola-furnace from 16 to 30 feet high, the coarse metal or "Rohstein" thereby reduced running into basins placed to receive it. This coarse metal, containing from 20 to 60 per cent. of copper according to the nature of the ore, is next roasted, and the product is melted for fine metal or "Spurstein."

Extraction in the Wet Way.—This mode of treatment is applied to copper ores which are too poor in copper to yield a profitable return by either of the methods above described, especially to the oxide of iron, technically called burnt pyrites or *blue billy*, enormous quantities of which are obtained as residue in the burning of iron pyrites for the manufacture of sulphuric acid. When this oxide, which contains a small proportion of copper, is mixed with coarsely crushed rock-salt and calcined, the copper is converted into soluble cupric chloride, $CuCl_2$, and on lixiviating the calcined mass with water a solution is obtained, from which the copper may be thrown down in the metallic state by scrap iron. According to another process, invented by Hunt and Douglas, the ores, if they consist wholly of oxides and carbonates, are simply heated, or if they contain sulphides they are roasted. In either case the product is treated with a solution of common salt or calcium chloride and ferrous sulphate, whereby the copper oxides are converted into chlorides. If cuprous oxide is present, metallic copper is at once precipitated, according to the equation $3Cu_2O + 2FeCl_2 = 2Cu_2Cl_2 + Fe_2O_3 + Cu_2$. Cupric oxide, on the other hand, yields cupric and cuprous chlorides, thus: $3CuO + 2FeCl_2 = CuCl_2 + Cu_2Cl_2 + Fe_2O_3$. The cuprous chloride, though insoluble in pure water, is held in solution in presence of the other chlorides, and is decomposed by iron at the same time as the cupric chloride.

Properties of Copper.—Copper has a well-known yellowish-red color, a specific gravity of 8.95, and is very malleable and ductile: it is an excellent conductor of heat and electricity; it melts at a bright red heat (about 1090° C.), and seems to be slightly volatile at a very high temperature.

Copper undergoes no change in dry air; exposed to a moist atmosphere, it becomes covered with a strong adherent green crust, consisting in a great measure of carbonate. Heated to redness in the air, it is quickly oxidized, becoming covered with a black scale. Dilute sulphuric and hydrochloric acids scarcely act upon copper; boiling oil of vitriol attacks it, with evolution of sulphurous oxide; nitric acid, even dilute, dissolves it readily, with evolution of nitrogen dioxide.

Copper in its most stable compounds, the c u p r i c compounds, is bivalent, these compounds containing 1 atom of the metal combined with 2 atoms of a univalent, or 1 atom of a bivalent negative radicle, *e. g.*, $CuCl_2$, CuO, $Cu(NO_3)_2$, $CuSO_4$, etc. It also forms another series, the c u p r o u s compounds, in which it may be regarded as univalent, *e. g.*, $CuCl$, Cu_2O, etc., like silver in the argentic compounds. On the other hand, the cuprous compounds may be supposed to be formed by addition of copper to the cupric compounds, the metal still remaining bivalent; thus cuprous chloride,

$$Cu_2Cl_2 = \begin{array}{l} CuCl \\ | \\ CuCl \end{array} \; ; \text{ cuprous oxide, } Cu_2O = \begin{array}{l} Cu \\ | \\ Cu \end{array}\!\!\Big\rangle O. \text{ These compounds are very}$$

unstable, being easily converted into cupric compounds by the action of oxidizing agents.

Cuprous Hydride, Cu_2H_2.—When a solution of cupric sulphate is heated to about 70° C. (158° F.), with hypophosphorous acid, this compound is deposited as a yellow precipitate, which soon turns red-brown. It gives off hydrogen when heated, takes fire in chlorine gas, and is converted by hydrochloric acid into cuprous chloride, with evolution of a double quantity of hydrogen, the acid giving up its hydrogen as well as the copper hydride :—

$$Cu_2H_2 + 2HCl = Cu_2Cl_2 + 2H_2.$$

This reaction affords a remarkable instance of the union of two atoms of the same element to form a molecule (see page 250).

Copper Chlorides.—*Cupric Chloride*, $CuCl_2$, is most easily prepared by dissolving cupric oxide in hydrochloric acid, and concentrating the green solution thence resulting. It forms green crystals, $CuCl_2,2H_2O$, very soluble in water and in alcohol: it colors the flame of the latter green. When gently heated, it parts with its water of crystallization and becomes yellowish-brown; at a high temperature it loses half its chlorine and becomes converted into *cuprous chloride*. The latter is a white fusible substance, but little soluble in water, and prone to oxidation: it is formed when copperfilings or copper-leaf are put into chlorine gas; also by precipitating a solution of cupric chloride or other cupric salt with stannous chloride:—

$$2CuCl_2 + SnCl_2 = Cu_2Cl_2 + SnCl_4.$$

A plate of copper immersed in hydrochloric acid in a vessel containing air, becomes covered with white tetrahedrons of cuprous chloride. This compound dissolves in hydrochloric acid, forming a colorless solution, which gradually turns blue on exposure to the air.

A *hydrated cupric oxychloride*, $CuCl_2,3CuH_2O_2$, occurs native as *atacamite*.

Both the chlorides of copper form double salts with the chlorides of the alkali-metals.

Copper Oxides.—Two oxides of copper are known, corresponding with the chlorides; also a *tetrantoxide* or *quadrantoxide*, Cu_4O; and a very unstable dioxide or peroxide, CuO_2, is said to be formed, as a yellowish-brown powder, by the action of hydrogen dioxide on cupric hydroxide.

Copper Monoxide, Cupric Oxide, or *Black Oxide of Copper*, CuO, is prepared by calcining metallic copper at a red heat, with full exposure to air, or, more conveniently, by heating the nitrate to redness, which then suffers complete decomposition. Cupric salts mixed with caustic alkali in excess, yield a bulky pale-blue precipitate of hydrated cupric oxide, or cupric hydroxide, CuH_2O_2, or CuO,H_2O, which, when the whole is raised to the boiling point, becomes converted into a heavy dark-brown powder: this also is anhydrous oxide of copper, the hydroxide suffering decomposition, even in contact with water. The oxide prepared at a high temperature is perfectly black and very dense. Cupric oxide is soluble in acids, and forms a series of very important salts, isomorphous with the salts of magnesium.

Cuprous Oxide, Cu_2O, also called *Red Oxide* and *Suboxide of Copper*.— This oxide may be obtained by heating in a covered crucible a mixture of 5 parts of black oxide and 4 parts of fine copper-filings; or by adding grape-sugar to a solution of cupric sulphate, and then putting in an excess of caustic potash; the blue solution, heated to boiling, is reduced by the sugar and deposits cuprous oxide. This oxide often occurs native in beautiful transparent ruby-red crystals, associated with other ores of copper, and can be obtained in the same state by artificial means. It communicates to glass a magnificent red tint, while that given by cupric oxide is green.

Cuprous oxide dissolves in excess of hydrochloric acid, forming a solution of cuprous chloride, from which that compound is precipitated on dilution with water. Most oxygen-acids, namely, sulphuric, phosphoric, acetic, oxalic, tartaric, and citric acids, decompose cuprous oxide, forming cupric salts, and separating metallic copper; nitric acid converts it into cupric

32

nitrate. Hence there are but few cuprous oxygen-salts, none indeed except-ing the sulphites, and certain double sulphites formed by mixing a cupric solution with the sulphite of an alkali-metal, *e. g.*, ammonio-cuprous sul-phite, $Cu(NH_4)SO_3$ or $Cu_2(NH_4)_2(SO_3)_2$.

Copper tetrantoxide, Cu_4O, is formed, according to H. Rose,[*] by the action of a dilute solution of stannous chloride and caustic potash on a solution of copper sulphate, the whole being well cooled. Cupric hydroxide is then first formed, and afterward reduced by the stannous chloride, with the forma-tion of potassium stannate:—

$$4Cu(OH)_2 + 12KOH + 3SnCl_2 = Cu_4O + 6KCl + 3K_2SnO_3 + 10H_2O.$$

This oxide is an olive-green powder which may be kept unchanged under water if air be excluded, but oxidizes quickly on exposure to air. It is decomposed by dilute hydrochloric acid, according to the equation $Cu_4O + 2HCl = Cu_2Cl_2 + 2Cu + H_2O$, and in like manner by sulphuric acid.

Cupric Oxysalts.—The *sulphate*, $CuSO_4,5H_2O$, commonly called *blue vitriol*, is prepared by dissolving cupric oxide in sulphuric acid, or, at less expense, by oxidizing the sulphide. It forms large blue triclinic prisms, soluble in 4 parts of cold and two parts of boiling water; when heated to 100° C. it readily loses four molecules of crystallization-water, but the fifth is retained with great pertinacity, and is expelled only at a low red heat. At a very high temperature cupric sulphide is entirely converted into cupric oxide, with evolution of sulphurous oxide and oxygen. Cupric sulphate combines with the sulphates of potassium and of ammonium, forming pale-blue salts, $CuK_2(SO_4)_2,6H_2O$ and $Cu(NH_4)_2(SO_4)_2,6H_2O$, isomorphous with the corresponding magnesium salts.

Cupric Nitrate, $Cu(NO_3)_2,3H_2O$, is easily made by dissolving the metal in nitric acid: it forms deep-blue crystals, very soluble and deliquescent. It is highly corrosive. There is also an insoluble basic nitrate having a green color.

Cupric Arsenite is a bright-green insoluble powder, prepared by mixing the solution of a cupric salt with an alkaline arsenite.

Cupric Carbonates.—When sodium carbonate is added in excess to a solu-tion of cupric sulphate, the precipitate is at first pale-blue and flocculent, but by warming it becomes sandy, and assumes a green tint; in this state it contains $CuCO_3,CuH_2O_2 + H_2O$. This substance is prepared as a pigment. The beautiful mineral *malachite* has a similar composition, but contains no water of crystallization, its composition being $CuCO_3,CuH_2O_2$. Another natural compound, called *azurite*, not yet artificially imitated, occurs in large transparent crystals of the deepest blue: it contains $2CuCO_3,CuH_2O_2$. *Ver-diter*, made by decomposing cupric nitrate with chalk, is said, however, to have a somewhat similar composition.

Copper Sulphides.—There are two well-defined copper sulphides, analogous in composition to cupric and cuprous oxides, and four others, containing larger proportions of sulphur, but of less defined constitution; these latter are precipitated from solutions of cupric salts by potassium pentasulphide.

Cupric Sulphide, CuS, occurs native as *Indigo-copper* or *Covellin*, in soft bluish-black hexagonal plates and spheroïdal masses, and is produced artifi-cially by precipitating cupric salts with hydrogen sulphide.

Cuprous Sulphide, Cu_2S, occurs native as *Copper-glance* or *Redruthite* in lead-gray hexagonal prisms, belonging to the rhombic system; it is produced artificially by the combustion of copper-foil in sulphur vapor, by igniting cupric oxide with sulphur, and by other methods. It is a powerful sulphur-base, uniting with the sulphides of antimony, arsenic, and bismuth, to form several natural minerals. The several varieties of *fahl-ore* or *tetrahydrite* consist of cuprous thioantimonite or thioarsenite, in which the copper is more or less replaced by equivalent quantities of iron, zinc, silver, and mercury. The important ore called *Copper pyrites* is a cuprosoferric sulphide, $Cu'Fe'''S_2$ or Cu_2S,Fe_2S_3, occurring in tetrahedral crystals of the quadratic system or in irregular masses. Another species of copper and iron sulphide, containing various proportions of the two metals, occurs native, as *Purple copper* or *Erubescite*, in cubes, octahedrons, and other mono-metric forms.

Ammoniacal Copper-compounds.—The chlorides, sulphate, nitrate, and other salts of copper unite with one or more molecules of ammonia, forming, for the most part, crystalline compounds of blue or green color, some of which may be regarded as salts of metallammoniums (p. 341). Thus cupric chloride forms with ammonia the compounds $2NH_3,CuCl_2$, $4NH_3,CuCl_2$, and $6NH_3,CuCl_2$, the first of which may be formulated as *cupro-diammonium chloride*, $(N_2H_6Cu'')Cl_2$. Cupric sulphate forms, in like manner, *cupro-diammonium sulphate*, $(N_2H_6Cu'')SO_4$, which is a deep-blue crystalline salt. Cuprous iodide forms with ammonia the compound $4NH_3,Cu_2I_2$.

Caustic *potash* gives with cupric salts a pale-blue precipitate of cupric hydroxide, changing to a blackish-brown anhydrous oxide on boiling.— *Ammonia* also throws down the hydroxide; but, when in excess, redissolves it, yielding a deep purplish-blue solution.— *Potassium* and *sodium carbonates* give pale-blue precipitates of cupric carbonate, insoluble in excess.—*Ammonium carbonate* the same, but soluble with deep-blue color.—*Potassium ferrocyanide* gives a fine red-brown precipitate of cupric ferrocyanide.— *Hydrogen sulphide* and *ammonium sulphide* afford black cupric sulphide, insoluble in ammonium sulphide. Metallic *iron* or *zinc* immersed in a solution of a copper salt quickly becomes coated with metallic copper.

Copper and its compounds impart a green color to flame. Any compound of copper fused with borax in the oxidizing blowpipe flame forms a transparent glass, which is green while hot, but assumes a fine blue color on cooling. In the reducing flame the glass becomes opaque, and covered on the surface with liver-colored streaks of cuprous oxide or metallic copper. This last reaction is facilitated by fusing in the bead a small piece of metallic tin. Copper-compounds, mixed with sodium carbonate or potassium cyanide and heated on charcoal before the blowpipe, yield metallic copper.

The alloys of copper are of great importance. *Brass* consists of copper alloyed with from 28 to 34 per cent. of zinc; the latter may be added directly to the melted copper, or granulated copper may be heated with calamine and charcoal-powder, as in the old process. *Gun-metal*, a most valuable alloy, consists of 90 parts copper and 10 tin. *Bell* and *speculum metals* contain a still larger proportion of tin; these are brittle, especially the last-named. A good bronze for statues is made of 91 parts copper, 2 parts tin, 6 parts zinc, and 1 part lead. The *brass* or *bronze* of the ancients is an alloy of copper with tin, often also containing lead, and sometimes zinc.

SILVER.

Symbol, Ag (Argentum).   Atomic weight, 107.66.

SILVER is found in the metallic state, as sulphide, in union with sulphide of antimony and sulphide of arsenic, also as chloride, iodide, and bromide. Among the principal silver-mines may be mentioned those of the Hartz Mountains in Germany, of Kongsberg in Norway, and, more particularly, of the Andes, in both North and South America.

The greater part of the silver of commerce is extracted from ores so poor as to render any process of *smelting* or fusion inapplicable, even where fuel could be obtained, and this is often difficult to be procured.   Recourse, therefore, is had to another method—that of *amalgamation*—founded on the easy solubility of silver and many other metals in metallic mercury.

In Mexico the ore containing small quantities of sulphide and chloride of silver is stamped or ground, together with water, in mills worked by horses or mules; and the fine mud thus produced is mixed on a floor with 3 to 5 per cent. of common salt, which is thoroughly incorporated with the mass by the treading of mules.   The mass is then left to itself for a day, after which mercury is added, together with an impure mixture of cupric and ferric salts called "magistral;" then more mercury; and the heap is again trodden by mules till thoroughly incorporated.   The amalgamation being completed, the slimy mass is washed in buddles worked by mules, whereby the lighter particles are washed away and the heavier amalgam is deposited.   The amalgam is next filtered through canvas bags and finally distilled, and the silver which is left behind is melted into bars.

The reactions wich take place in this amalgamation process are probably as follows.   The sulphates of copper and iron (in the magistral) are converted by the chloride of sodium into the corresponding chlorides, which then react with the silver sulphide, forming silver chloride:—

$$2CuCl_2 + Ag_2S = Cu_2Cl_2 + 2AgCl + S,$$

the cuprous chloride thus produced acting further on the silver sulphide and forming more silver chloride:—

$$Cu_2Cl_2 + Ag_2S = Cu_2S + 2AgCl.$$

The resulting silver chloride is held in solution by the sodium chloride, and on addition of metallic mercury is decomposed, yielding calomel and metallic silver.   All the mercury thus converted into calomel—amounting to about twice the weight of the silver obtained—is lost.

In Nevada, where the ores are of various degrees of richness—the richer ores are usually roasted with common salt, whereby the silver sulphide is converted into chloride; the roasted and pulverized mass is then introduced, together with water, scrap-iron, and mercury, into barrels rotating on a horizontal axis; and the metallic silver thereby set free is dissolved out by mercury.   This barrel-process was first worked at Schemnitz in Hungary, and afterward at Freiberg and Mansfeld, but is now no longer practised in Europe.

Considerable loss often occurs in the amalgamation process from the combination of a portion of the mercury with sulphur, oxygen, etc., whereby it is brought into a pulverulent condition known as "flouring," and is then liable to be washed away, together with the silver it has taken up. This inconvenience may be prevented, as suggested by Mr. Crookes, by amalgamating the mercury with 1 or 2 per cent. of sodium, which, by its superior affinity for sulphur and oxygen, prevents the mercury from becoming floured.

A considerable quantity of silver is obtained from argentiferous galena; in fact, almost every specimen of native lead sulphide is found to contain traces of this metal. When the proportion rises to a certain amount, it becomes worth extracting. The ore is reduced in the usual manner, the whole of the silver remaining with the lead; the latter is then remelted in a large vessel and allowed to cool slowly until solidification commences. The portion which first crystallizes is nearly pure lead, the alloy with silver being *more fusible than lead itself*: this is drained away by means of a perforated ladle, and is found to contain nearly the whole of the silver [Pattinson's process]. This rich mass is next exposed to a red heat on the shallow hearth of a furnace, while a stream of air is allowed to impinge upon its surface; oxidation then takes place with great rapidity, the fused oxide or litharge being constantly swept from the metal by the blast. When the greater part of the lead has been thus removed, the residue is transferred to a cupel or shallow dish made of bone-ashes, and again heated: the last portion of the lead is now oxidized, and the oxide sinks in a melted state into the porous vessel, while the silver, almost chemically pure and exhibiting a brilliant surface, remains behind.

Extraction of Silver in the Wet Way.—Silver may be extracted from argentiferous copper pyrites: (1.) By roasting the ore, whereby the sulphides of copper and iron are converted into insoluble oxides, while the silver is obtained as soluble sulphate, from the solution of which, obtained by lixiviating the roasted ore with hot water, the silver is easily precipitated in the metallic state. (2.) By roasting the ore with sodium chloride to convert the silver into chloride, dissolving out this compound with sodium thiosulphate; precipitating the silver from the resulting solution as sulphide by means of sodium sulphide; and strongly heating the precipitated sulphide in a muffle furnace, whereby the sulphur is burnt away and the silver is left. (3.) By Claudet's process, which is applied to the extraction of the small quantity of silver contained in the burnt pyrites of the sulphuric acid works (pp. 179, 333). The cupreous tank-liquors obtained in this process contain the silver as chloride, held in solution by sodium chloride; and from this solution the silver may be thrown down in the form of iodide, by adding a solution of kelp; a very small quantity of gold is precipitated at the same time.

Purification.—Pure silver is easily obtained. The impure metal is dissolved in nitric acid: if it contains copper, the solution will have a blue tint; gold will remain undissolved as a black powder. The solution is mixed with hydrochloric acid or with common salt, and the white, insoluble curdy precipitate of silver chloride is washed and dried. This is then mixed with about twice its weight of anhydrous sodium carbonate, and the mixture, placed in an earthen crucible, is gradually raised to a temperature approaching whiteness, during which the sodium carbonate and the silver chloride react upon each other; carbon dioxide and oxygen escape, while metallic silver and silver chloride result : the former melts into a button at the bottom of the crucible, and is easily detached. The following is perhaps the most simple method for the reduction of silver chloride. The silver-salt is covered with water, to which a few drops of sulphuric acid are added; a plate of zinc is then introduced; the silver chloride soon begins to decompose, and after a short time is entirely converted into metallic silver. The silver thus obtained is gray and spongy; it is ultimately purified by washing with slightly aciduated water.

Properties.—Pure silver has a perfect white color and a strong lustre: it is exceedingly malleable and ductile, and is probably the best conductor of both heat and electricity known. Its specific gravity is 10.5. In hard-

32 *

ness it lies between gold and copper. It melts at a bright red heat. Silver is unalterable by air and moisture: it refuses to oxidize at any temperature, but possesses the remarkable faculty already noticed of absorbing many times its volume of oxygen when strongly heated in an atmosphere of that gas or in common air. The oxygen is again disengaged at the moment of solidification, and gives rise to the peculiar arborescent appearance often remarked on the surface of masses or buttons of pure silver. The addition of 2 per cent. of copper is sufficient to prevent the absorption of oxygen. Silver oxidizes when heated with fusible silicious matter, as glass, which it stains yellow or orange from the formation of a silicate. It is but slightly attacked by hydrochloric acid; boiling oil of vitriol converts it into sulphate, with evolution of sulphurous oxide; nitric acid, even dilute and in the cold, dissolves it readily. The tarnishing of surfaces of silver exposed to the air is due to hydrogen sulphide, the metal having a strong attraction for sulphur.

Silver Chlorides.—Two of these compounds are known, containing respectively 1 and 2 atoms of silver to 1 atom of chlorine; the second, however, is a very unstable compound.*

The *Monochloride* or *Argentic Chloride*, AgCl, or rather Ag_2Cl_2 $= Cl.Ag.Ag.Cl$, is almost invariably produced when a soluble silver salt and a soluble chloride are mixed. It falls as a white curdy precipitate, quite insoluble in water and nitric acid; one part of silver chloride is soluble in 200 parts of hydrochloric acid when concentrated, and in about 600 parts when diluted with double its weight of water. When heated it melts, and on cooling becomes a grayish crystalline mass, which cuts like horn: it is found native in this condition, constituting the mineral called *horn silver*. Silver chloride is decomposed by light, both in the dry and in the wet state, very slowly if pure, and quickly if organic matter be present: it is reduced also when put into water with metallic zinc or iron. It dissolves very easily in ammonia and in a solution of potassium cyanide. In analysis the proportion of chlorine or hydrochloric acid in a compound is always estimated by precipitation with silver solution. The liquid is acidulated with nitric acid, and an excess of silver nitrate added; the chlorine is collected on a filter, or better by subsidence, then washed, dried, and fused; 100 parts correspond to 24.7 of chlorine, or 25.43 of hydrochloric acid.

Argentous Chloride, $Ag_4Cl_2 = Cl.Ag.Ag.Ag.Ag.Cl$, is obtained by treating the corresponding oxide with hydrochloric acid, or by precipitating an argentous salt—the citrate, for example—with common salt. It is easily resolved by heat or by ammonia into argentic chloride and metallic silver.

Silver Bromide, AgBr or Ag_2Br_2, occurs native in Chili and Mexico, as *bromargyrite*, usually in small yellow or greenish masses; also mixed with the chloride as *embolite*. It is obtained on adding hydrobromic acid or a bromide of alkali-metal to a solution of silver nitrate, as a white curdy precipitate insoluble in nitric acid, nearly insoluble in dilute ammonia—whereby it is distinguished from the chloride—but easily soluble in concentrated ammonia. When suspended in water it is easily decomposed by chlorine; hydrochloric acid gas also decomposes it at 700° C. (1292° F.), with evolution of hydrobromic acid. The fused bromide is scarcely acted

---

* If the argentic compounds stood alone, silver might be regarded as a monad, like the alkali-metals, the chloride AgCl, for example, being analogous to NaCl; but the constitution of the argentous compounds is utterly inconsistent with this view: argentous chloride, for example, cannot be represented by the formula Ag_2Cl, into which both the silver and the chlorine enter as monads; but by doubling the formula as above, and regarding the silver as a dyad, the constitution of the compound appears perfectly normal.

on by light, but the precipitated bromide when exposed to light quickly assumes a grayish-white color.

Silver Fluoride, AgF or Ag_2F_2, is produced by dissolving argentic oxide or carbonate in aqueous hydrofluoric acid, and separates on evaporation in transparent quadratic octahedrons, which contain AgF,H_2O or $Ag_2F_2,2H_2O$, and give off their water when fused. Their solution gives, with hydrochloric acid, a precipitate of argentic chloride.

Silver Iodide, AgI or Ag_2I_2, is a pale yellow insoluble precipitate, produced by adding silver nitrate to potassium iodide; it is insoluble, or nearly so, in ammonia, and in this respect forms an exception to silver-salts in general. Deville obtained a crystalline silver iodide by the action of concentrated hydriodic acid upon metallic silver, which it dissolves with disengagement of hydrogen. Hydriodic acid converts silver chloride into iodide.

Silver Oxides.—There are three oxides of silver, only one of which, however, can be regarded as a well-defined compound, namely:—

The *Monoxide*, or *Argentic Oxide*, $Ag_2O = \begin{matrix} Ag \\ Ag \end{matrix}\Big\rangle O$. This oxide is a strong base, yielding salts isomorphous with those of the alkali-metals. It is obtained as a pale-brown precipitate on adding caustic potash to a solution of silver nitrate:—

$$2AgNO_3 + KHO = Ag_2O + KNO_3 + HNO_3.$$

It is very soluble in ammonia, and is dissolved also to a small extent by pure water; the solution is alkaline. Recently precipitated silver chloride, boiled with a solution of caustic potash of specific gravity 1.25, is converted, according to Gregory, although with difficulty, into argentic oxide, which in this case is black and very dense. Argentic oxide neutralizes acids completely, and forms, for the most part, colorless salts. It is decomposed by a red heat, with evolution of oxygen, spongy metallic silver being left: the sun's rays also effect its decomposition to a small extent.

Argentous Oxide, $Ag_4O = \begin{matrix} Ag - Ag \\ Ag - Ag \end{matrix}\Big\rangle O$. When dry argentic citrate is heated to 100° in a stream of hydrogen gas, it loses oxygen and becomes dark-brown. The product, dissolved in water, gives a dark-colored solution containing free citric acid and argentous citrate, which when mixed with potash yields a precipitate of argentous oxide. This oxide is a black powder, very easily decomposed, and soluble in ammonia. The solution of argentous citrate is rendered colorless by heat, being resolved into argentic citrate and metallic silver.

Silver Dioxide, $Ag_2O_2 = \begin{matrix} Ag - O \\ Ag - O \end{matrix}$, is a black crystalline substance, which forms upon the positive electrode of a voltaic arrangement employed to decompose a solution of silver nitrate. It is reduced by heat; evolves chlorine when acted upon by hydrochloric acid; explodes when mixed with phosphorus and struck; and decomposes solution of ammonia, with great energy and rapid disengagement of nitrogen gas.

Oxysalts of Silver.—*Silver Sulphate*, Ag_2SO_4, may be prepared by boiling together oil of vitriol and metallic silver, or by precipitating a concentrated solution of silver nitrate with an alkaline sulphate. It dissolves in

88 parts of boiling water, and separates in great measure in the crystalline form on cooling, being but slightly soluble at a low temperature. It forms with ammonia a crystallizable compound, which is freely soluble in water, contains $2NH_3,Ag_2SO_4$, and may therefore be regarded as *argentammonium sulphate*, $(NH_3Ag)_2SO_4$.

Silver Dithionate, $Ag_2S_2O_6$, is a soluble crystallizable salt, permanent in the air. The *thiosulphate*, $Ag_2S_2O_3$, is insoluble, white, and very prone to decomposition: it combines with the alkaline thiosulphates, forming soluble compounds distinguished by an intensely sweet taste. The alkaline thiosulphates dissolve both oxide and chloride of silver and give rise to similar salts, an oxide or chloride of the alkali-metal being at the same time formed: hence the use of alkaline thiosulphates in fixing photographic pictures (p. 102).

Silver Nitrate, $AgNO_3$, is prepared by dissolving silver in nitric acid, and evaporating the solution to dryness, or until it is strong enough to crystallize on cooling. The crystals are colorless, transparent, anhydrous tables, soluble in an equal weight of cold and in half that quantity of boiling water; they also dissolve in alcohol. They fuse when heated like those of nitre, and at a high temperature suffer decomposition: the *lunar caustic* of the surgeon is silver nitrate, which has been melted and poured into a cylindrical mould. The salt blackens when exposed to light, more particularly if organic matters of any kind are present, and is frequently employed to communicate a dark stain to the hair; it enters into the composition of the "indelible" ink used for marking linen. The black stain has been thought to be metallic silver; it may possibly be argentous oxide. Pure silver nitrate may be prepared from the metal alloyed with copper: the alloy is dissolved in nitric acid, the solution evaporated to dryness, and the mixed nitrates cautiously heated to fusion. A small portion of the melted mass is removed from time to time for examination: it is dissolved in water, filtered, and ammonia added to it in excess. While any copper-salt remains undecomposed the liquid will be blue, but when that no longer happens the nitrate may be left to cool, dissolved in water, and filtered from the black oxide of copper.

Silver Carbonate is a white, insoluble substance, obtained by mixing solutions of silver nitrate and sodium carbonate. It is blackened and decomposed by boiling.

Silver Sulphide, Ag_2S, is a soft, gray, and somewhat malleable substance, found native in the crystallized state, and easily produced by melting together its constituents, or by precipitating a solution of silver with hydrogen sulphide. It is a strong sulphur-base, and combines with the sulphides of antimony and arsenic: examples of such compounds are found in the beautiful minerals, *dark-* and *light-red silver ore.*

Ammonia-Compound of Silver; *Berthollet's Fulminating Silver.*—This is a black, explosive compound, formed by digesting precipitated argentic oxide in ammonia. While moist, it explodes only when rubbed with a hard body; but when it is dry, the touch of a feather is sufficient. The ammonia retains some of this substance in solution, and deposits it in small crystals by spontaneous evaporation. A similar compound exists containing oxide of gold.

Argentic chloride slowly absorbs ammonia-gas, forming the compound $3NH_3,Ag_2Cl_2$, which readily gives off its ammonia when heated, and may be used for the preparation of pure ammonia-gas; or of liquid ammonia, if heated in a sealed tube. Argentic iodide likewise absorbs ammonia-gas, forming the compound NH_3,Ag_2I_2, which gives off its ammonia on exposure to the air.

Soluble silver-salts are perfectly characterized by the white curdy precipitate of silver chloride, darkening on exposure to light, and insoluble in hot nitric acid, which is produced by the addition of any soluble *chloride*. Lead and mercury are the only metals which can be confounded with silver in this respect; but lead chloride is soluble to a great extent in boiling water, and is deposited in brilliant acicular crystals when the solution cools; and mercurous chloride is instantly blackened by ammonia, whereas silver chloride is dissolved thereby.

Solutions of silver are reduced to the metallic state by *iron, copper, mercury,* and other metals. They give with *hydrogen sulphide* a black precipitate of argentic sulphide, insoluble in ammonium sulphide; with *caustic alkalies,* a brown precipitate of argentic oxide; and with *alkaline carbonates,* a white precipitate of argentic carbonate, both precipitates being easily soluble in ammonia. Ordinary *sodium phosphate* forms a yellow precipitate of argentic orthophosphate; *potassium chromate* or *bichromate,* a red-brown precipitate of argentic chromate.

The economical uses of silver are many; it is admirably adapted for culinary and other similar purposes, not being attacked in the slightest degree by any of the substances used for food. It is necessary, however, to diminish the softness of the metal by a small addition of copper. The standard silver of England contains 222 parts of silver and 18 parts of copper.

MERCURY.

Symbol, Hg (Hydrargyrum). Atomic weight, 199.8. Vapor-density, 99.9.

THIS very remarkable metal, sometimes called *quicksilver,* has been known from early times, and perhaps more than all others has excited the attention and curiosity of experimenters by reason of its peculiar physical properties. Mercury is of great importance in several of the arts, and enters into the composition of many valuable medicaments.

Metallic mercury is occasionally met with in globules disseminated through the native sulphide, which is the ordinary ore. This latter substance, sometimes called *cinnabar,* is found in considerable quantity in several localities, of which the most celebrated are Almaden in Spain and Idria in Austria. Recently it has been discovered in great abundance and of remarkable purity in California and Australia.

The metal is obtained by heating the sulphide in an iron retort with lime or scraps of iron, or by roasting it in a furnace and conducting the vapors into a large chamber, where the mercury is condensed, while the sulphurous acid is allowed to escape. Mercury is imported into this country in bottles of hammered iron containing seventy-five pounds each, and in a state of considerable purity. When purchased in smaller quantities it is sometimes found adulterated with tin and lead, which metals it dissolves to some extent without much loss of fluidity. Such admixture may be known by the foul surface which the mercury exhibits when shaken in a bottle containing air, and by the globules, when made to roll upon the table, leaving a train or tail.

Mercury has a nearly silver-white color, and a very high degree of lustre: it is liquid at all ordinary temperatures, and solidifies only when cooled to — 40° C. In this state it is soft and malleable. It boils at 357.25° C.

(643.3° F.) (Regnault), yielding a transparent, colorless vapor, of great density. The metal volatilizes, however, to a sensible extent at all temperatures above 19° or 21° C.; below this point its volatility is imperceptible. The volatility of mercury at the boiling heat is much retarded by the presence of minute quantities of lead or zinc. The specific gravity of mercury at 15.5° C. is 13.59; that of frozen mercury about 14, great contraction taking place in the act of solidification.

Pure mercury is quite unalterable in the air at common temperatures, but when heated to near its boiling point it slowly absorbs oxygen, and becomes converted into a crystalline dark-red powder, which is the highest oxide. At a dull red heat this oxide is again decomposed into its constituents. Hydrochloric acid has little or no action on mercury, and the same may be said of sulphuric acid in a diluted state: but when the acid is concentrated and boiling hot it oxidizes the metal, converting it into mercuric sulphate, with evolution of sulphurous oxide. Nitric acid, even dilute and in the cold, dissolves mercury freely, with evolution of nitrogen dioxide.

The observed vapor-density of mercury referred to air as unity is 6.7; this referred to hydrogen is nearly 100; * that is to say, half the atomic weight of the metal: consequently the atom of mercury, like that of cadmium, occupies in the gaseous state twice the volume of an atom of hydrogen (see p. 244).

Mercury forms two series of compounds, namely, the m e r c u r i c c o m-p o u n d s, in which it is bivalent, as $HgCl_2$, HgO, $HgSO_4$, etc., and the m e r-e u r o u s c o m p o u n d s, in which it is apparently univalent, as Hg_2Cl_2, Hg_2O, etc. These compounds are analogous in constitution to the cupric and cuprous compounds; and the mercurous compounds, like the latter, are easily converted into mercuric compounds by the action of oxidizing agents, which remove one atom of mercury; but they are, on the whole, much more stable than the cuprous compounds.

Mercury Chlorides.—*Mercuric chloride*, $HgCl_2$, commonly called *Corrosive Sublimate.*—This compound may be obtained by several different processes: (1.) When metallic mercury is heated in chlorine gas, it takes fire and burns, producing this substance. (2.) It may be made by dissolving mercuric oxide in hot hydrochloric acid, crystals of corrosive sublimate then separating on cooling. (3.) Or, more economically, by subliming a mixture of equal parts of mercuric sulphate and dry common salt; and this is the plan generally followed. The decomposition is represented by the equation,

$$HgSO_4 + 2NaCl = HgCl_2 + Na_2SO_4.$$

Sublimed mercuric chloride forms a white transparent crystalline mass, of specific gravity 5.43; it melts at 265° C. (509° F.), boils at 295° C. (563° F.), and volatilizes somewhat more easily than calomel, even at ordinary temperatures. Its observed vapor-density, referred to hydrogen as unity, is 140: and the density calculated from the formula $HgCl_2$, supposing that the molecule occupies the same space as a molecule or two atoms of hydrogen (p. 246), is $\dfrac{199.8 + 2 \cdot 35.37}{2} = 135.27$; the near agreement of this number with the observed result shows that the vapor is in the normal state of condensation.

Mercuric chloride dissolves in 16 parts of cold and 3 parts of boiling water, and crystallizes from a hot solution in long white prisms. Alcohol

* $\dfrac{6.7}{0.6926} = 98.3.$

and ether also dissolve it with facility; the latter even withdraws it from a watery solution.

Mercuric chloride combines with a great number of other metallic chlorides, forming a series of beautiful double salts, of which the ancient *sal alembroth* may be taken as a good example: it contains $HgCl_2,2NH_4Cl,H_2O$. *The chloride* absorbs ammoniacal gas with great avidity, generating the compound $HgCl_2,NH_3$.

Mercuric chloride forms several compounds with mercuric oxide. These are produced by several processes, as when an alkaline carbonate is added in varying proportions to a solution of mercuric chloride. They differ greatly in color and physical character, and are mostly decomposed by water.

Mercuric chloride forms insoluble compounds with many of the azotized organic principles, as albumin, etc. It is perhaps to this property that its strong antiseptic properties are due. Animal and vegetable substances are preserved by it from decay, as in Kyan's method of preserving timber and cordage. Albumin is on this account an excellent antidote to corrosive sublimate in cases of poisoning.

Mercurous Chloride, $HgCl$ or Hg_2Cl_2, commonly called *Calomel.*—This very important substance may be easily and well prepared by pouring a solution of mercurous nitrate into a large excess of dilute solution of common salt. It falls as a dense white precipitate, quite insoluble in water; it must be thoroughly washed with boiling distilled water, and dried. Calomel is, however, generally procured by another and more complex process. Dry mercuric sulphate is rubbed in a mortar with as much metallic mercury as it already contains, and a quantity of common salt, until the globules disappear and a uniform mixture has been produced. This is subjected to sublimation, and the vapor of the calomel being carried into an atmosphere of steam, or into a chamber containing air, is thus condensed in a minutely divided state, and the laborious process of pulverization of the sublimed mass is avoided. The reaction is thus explained:—

$$HgSO_4 + Hg + 2NaCl = Hg_2Cl_2 + Na_2SO_4.$$

Pure calomel is a heavy white insoluble tasteless powder: it rises in vapor at a temperature below redness, and is obtained by ordinary sublimation as a yellowish-white crystalline mass. It is as insoluble in cold dilute nitric acid as silver chloride; boiling-hot strong nitric acid oxidizes and dissolves it. Calomel is instantly decomposed by an alkali, or by lime-water, with production of mercurous oxide. It is sometimes apt to contain a little mercuric chloride, which would be a very dangerous contamination in calomel employed for medical purposes. This is easily discovered by boiling with water, filtering the liquid, and adding caustic potash. Any corrosive sublimate is indicated by a yellow precipitate.

The observed vapor-density of calomel, referred to hydrogen as unity, is 119.2. Now the formula Hg_2Cl_2, if it represents a molecule occupying in the gaseous state two volumes (*i. e.*, twice the volume of an atom of hydrogen, p. 246), would give a density nearly double of this: for $\dfrac{400 + 2 \cdot 35.5}{2} = 235.5$.

Hence it might be inferred that the composition of calomel should rather be represented by the simpler formula $HgCl$, which would give for the vapor-density the number 117.75. The frequent decomposition of mercurous salts into mercuric salts and free mercury tends, however, to favor the supposition that their molecules contain two atoms of mercury; and the anomaly in the vapor-volume of calomel may be explained by supposing that the vapor of this compound, like that of many others, undergoes at high temperatures the change known as *dissociation* (p. 246), the two volumes of mercurous chloride,

Hg_2Cl_2, being resolved into two volumes of mercuric chloride, $HgCl_2$, and two volumes of mercury, Hg. This supposition has been, to some extent, supported by the observation that calomel vapor amalgamates gold-leaf, and that corrosive sublimate may be detected in resublimed calomel. Recent experiments, however, have thrown doubts upon these statements, and tend to show that the true molecular formula of mercurous chloride is HgCl.*

Iodides.—*Mercuric iodide*, HgI_2, is formed when solution of potassium iodide is mixed with mercuric chloride, as a precipitate which is at first yellow, but in a few moments changes to a most brilliant scarlet, this color being retained on drying. This is the normal iodide: it may be made, although of rather duller tint, by triturating equivalent quantities of iodine and mercury with a little alcohol. In preparing it by precipitation, it is better to weigh out the proper proportions of the two salts, as the iodide is soluble in an excess of either, more especially in excess of potassium iodide. Mercuric iodide exhibits a very remarkable case of dimorphism, attended with difference of color, which is red or yellow according to the figure assumed. Thus, when the iodide is suddenly exposed to a high temperature, it becomes bright yellow throughout, and yields a copious sublimate of minute but brilliant yellow crystals. If in this state it be touched by a hard body, it instantly becomes red, and the same change happens spontaneously after a certain lapse of time. On the other hand, by a very slow and careful heating, a sublimate of red crystals, having a totally different form, may be obtained, which are permanent. The same kind of change happens with the freshly precipitated iodide, the yellow crystals first formed breaking up in a few seconds from the passage of the salt to the red modification.

Mercuric iodide forms double salts with the more basic or positive metallic iodides, as those of the alkali-metals and alkaline earth-metals; thus it dissolves in aqueous potassium iodide, and the hot solution deposits, on cooling, crystals of potassio-mercuric iodide, $2(KI,HgI_2),3H_2O$. They are decomposed by water, with separation of about half the mercuric iodide, the solution then containing the salt $2KI,HgI_2$, which remains as a saline mass on evaporation.

Mercurous Iodide, Hg_2I_2, is formed when a solution of potassium iodide is added to mercurous nitrate: it then separates as a dirty yellow insoluble precipitate, with a tinge of green. It may also be prepared by rubbing mercury and iodine together in a mortar in the proportion of 1 atom of the former to 1 atom of the latter, the mixture being moistened from time to time with alcohol.

Oxides.—*Monoxide*, or *Mercuric Oxide*, HgO, commonly called *Red Oxide of Mercury*, or *Red Precipitate*.—There are numerous methods by which this compound may be obtained. The following may be cited as the most important: (1.) By exposing mercury in a glass flask with a long narrow neck, for several weeks, to a temperature approaching 315°. The product has a dark-red color, and is highly crystalline; it is the *red precipitate* of the old writers. (2.) By cautiously heating any of the mercuric or mercurous nitrates to complete decomposition, whereby the acid is decomposed and expelled, oxidizing the metal to the maximum if it happens to be in the state of mercurous salt. The product thus obtained is also crystalline and, very dense, but has a much paler color than the preceding; while hot, it is nearly black. It is by this method that the oxide is generally prepared: it is apt to contain undecomposed nitrate, which may be discovered by strongly heating a portion in a test-tube: if red fumes are produced or the odor of nitrous acid exhaled, the oxide has been insufficiently heated in the process of manufacture. (3.) By adding caustic potash in excess to a solution of corrosive sublimate, by which a bright yellow precipitate of mercuric oxide

* Fileti (*Gazz. chim. ital.*, 1881, p. 431; *Chem. Soc. Journal*, xlii. 466).

is thrown down, which differs from the foregoing preparations merely in being destitute of crystalline texture and much more minutely divided. It must be well washed and dried.

Mercuric oxide is slightly soluble in water, communicating to the latter an alkaline reaction and metallic taste: it is highly poisonous. When strongly heated it is decomposed, as before observed, into metallic mercury and oxygen gas.

Mercurous Oxide, Hg_2O; *Suboxide*, or *Gray Oxide of Mercury*.—This oxide is easily prepared by adding caustic potash to mercurous nitrate, or by digesting calomel in solution of caustic alkali. It is a dark-gray, nearly black, heavy powder, insoluble in water, slowly decomposed by the action of light into metallic mercury and red oxide. The preparations known in pharmacy by the names *blue pill, gray ointment, mercury with chalk*, etc., often supposed to owe their efficacy to this substance, merely contain the finely divided metal.

Mercury Nitrates.—Nitric acid varies in its action upon mercury according to the temperature. When cold and somewhat diluted, it forms only mercurous salts, and these are normal or basic—*i. e.*, oxynitrates (p. 302)—as the acid or the metal happens to be in excess. When, on the contrary, the nitric acid is concentrated and hot, the mercury is raised to its highest state of oxidation, and a mercuric salt is produced. Both classes of salts are apt to be decomposed by a large quantity of water, giving rise to insoluble or sparingly soluble basic compounds.

Mercuric Nitrates.—By dissolving mercuric oxide in excess of nitric acid, and evaporating gently, a syrupy liquid is obtained, which, enclosed in a bell-jar over lime or sulphuric acid, deposits bulky crystals and crystalline crusts, both having the composition $2Hg(NO_3)_2,H_2O$. The same substance is deposited from the syrupy liquid as a crystalline powder by dropping it into concentrated nitric acid. The syrupy liquid itself appears to be a definite compound, containing $Hg(NO_3)_2,H_2O$. By saturating hot dilute nitric acid with mercuric oxide, a salt is obtained on cooling, which crystallizes in needles permanent in the air, containing $Hg(NO_3)_2,HgO,H_2O$. The preceding crystallized salts are decomposed by water, with production of compounds more and more basic as the washing is prolonged or the temperature of the water raised.

Mercurous Nitrate, $(Hg_2)(NO_3)_2,2H_2O$, forms large colorless crystals soluble in a small quantity of water without decomposition; it is made by dissolving mercury in an excess of cold dilute nitric acid.

When excess of mercury has been employed, a finely crystallized basic salt is deposited after some time, containing $(Hg_2)(NO_3)_2,2Hg_2O,3H_2O$, or $3Hg_2O,N_2O_5,3H_2O$; this is also decomposed by water. The two salts are easily distinguished when rubbed in a mortar with a little sodium chloride; the normal compound gives sodium nitrate and calomel; the basic salt, sodium nitrate and a black compound of calomel with mercurous oxide. A black substance, called *Hahnemann's soluble mercury*, is produced when ammonia in small quantity is dropped into a solution of mercurous nitrate: it contains $2NH_3,3Hg_2O,N_2O_5$, or, according to Kane, $2NH_3,2Hg_2O,N_2O_5$; the composition of this preparation evidently varies according to the temperature and the concentration of the solutions.

Mercury Sulphates.—*Mercuric Sulphate*, $HgSO_4$, is readily prepared by boiling together oil of vitriol and metallic mercury until the latter is wholly converted into a heavy white crystalline powder, which is the salt in question; the excess of acid is then removed by evaporation carried to perfect dryness. Equal weights of acid and metal may be conveniently employed. Water decomposes the sulphate, dissolving out an acid salt, and

17

leaving an insoluble, yellow, basic compound, formerly called *turpith* or *turbeth mineral,* containing, according to Kane's analysis, $HgSO_4,2HgO$, or $3HgO,SO_3$. Long-continued washing with hot water entirely removes the remaining acid and leaves pure mercuric oxide.

Mercurous Sulphate, Hg_2SO_4, falls as a white crystalline powder when sulphuric acid is added to a solution of mercurous nitrate: it is but slightly soluble in water.

Mercury Sulphides.—*Mercuric Sulphide,* HgS, occurs native as cinnabar, a dull-red mineral, which is the most important ore of mercury. Hydrogen sulphide passed in small quantity into a solution of mercuric nitrate or chloride, forms a white precipitate, which is a compound of mercuric sulphide with the salt itself. An excess of the gas converts the whole into sulphide, the color at the same time changing to black. When this black sulphide is sublimed, it becomes dark-red and crystalline, but undergoes no change of composition: it is then *cinnabar* or *vermilion.* Mercuric sulphide is most easily prepared by subliming an intimate mixture of 6 parts of mercury and 1 part of sulphur, and reducing the resulting cinnabar to very fine powder, the beauty of the tint depending much upon the extent to which division is carried. The red or crystalline sulphide may also be formed directly, without sublimation, by heating the black precipitated substance in a solution of potassium pentasulphide. Mercuric sulphide is soluble to a certain extent in the alkaline sulphides, and forms with them crystallizable compounds.

When vermilion is heated in the air, it yields metallic mercury and sulphurous oxide: it resists the action both of caustic alkalies in solution and of strong mineral acids, even nitric, and is attacked only by nitro-muriatic acid.

Mercurous Sulphide, Hg_2S, is obtained by passing hydrogen sulphide into a solution of mercurous nitrate as a black precipitate, which is resolved at a gentle heat into mercuric sulphide and metallic mercury.

Ammoniacal Mercury Compounds.—Mercurammonium Salts. —By the action of ammonia and its salts on mercury compounds, a variety of substances are formed, which may be regarded as salts of mercurammoniums—that is, of ammonium-molecules in which the hydrogen is more or less replaced by mercury, in the proportion of 100 or 200 parts of mercury to 1 part of hydrogen, according as the compound is formed from a mercurous or a mercuric salt. The following are the most important of these compounds :—

Mercuric Compounds.—*Mercuro-diammonium chloride,* $(N_2H_8Hg'')Cl_2$, known in pharmacy as *fusible white precipitate,* is produced by adding potash to a solution of ammonio-mercuric chloride, $(2NH_4Cl,HgCl_2)$, or by dropping solution of mercuric chloride into a boiling solution of sal-ammoniac containing free ammonia, as long as the resulting precipitate redissolves : it then separates on cooling in regular dodecahedrons. At a gentle heat it gives off ammonia, having a chloride of mercurammonium and hydrogen, $(NH_2Hg'')Cl,HCl$:—

$$N_2H_6Hg''Cl_2 = NH_3Hg''Cl_2 + NH_3.$$

Mercurammonium choride, $(NH_2Hg'')Cl$.—This salt, known in pharmacy as *infusible white precipitate,* is formed by adding ammonia to a solution of mercuric chloride. When first produced, it is bulky and white, but by contact with hot water, or by much washing with cold water, it is converted into hydrated dimercurammonium chloride, NHg''_2Cl,H_2O.

Trimercuro-diammonium nitrate, $(N_2H_2Hg''_3)(NO_3)_2,2H_2O$, is formed as a white precipitate on mixing a dilute and very acid solution of mercuric nitrate with very dilute ammonia.

Trimercuro-diamine, $N_2Hg''_3$, a compound derived from a double molecule of ammonia, N_2H_6, by substitution of 3 atoms of bivalent mercury for 6 atoms of hydrogen, is formed by passing dry ammonia gas over dry precipitated mercuric oxide:—

$$3HgO + 2NH_3 = N_2Hg_3 + 3H_2O.$$

The excess of oxide being removed by nitric acid, the trimercuro-diamine is obtained as a dark-brown powder, which explodes by heat, friction, percussion, or contact with oil of vitriol almost as violently as nitrogen chloride.

Dimercurammonium chloride, NHg''_2Cl,H_2O, is obtained, as already observed, by boiling mercuro-diammonium chloride (infusible white precipitate) with water. It is a heavy, granular, yellow powder, which turns white again when treated with sal-ammoniac.

Dimercurammonium iodide, NHg''_2I,H_2O.—This compound may be formed by digesting the corresponding chloride in a solution of potassium iodide, or by heating mercuric iodide with excess of aqueous ammonia:—

$$2HgI_2 + 4NH_3 + H_2O = NHg''_2I,H_2O + 3NH_4I;$$

also by passing ammonia gas over mercuric oxy-iodide:—

$$Hg_4I_2O_3 + 2NH_3 = 2(NHg''_2I,H_2O) + H_2O;$$

and, lastly, by adding ammonia to a solution of potassio-mercuric iodide mixed with caustic potash:—

$$2(2KI,HgI_2) + NH_3 + 3KHO = NHg''_2I,H_2O + 7KI + 2H_2O.$$

This last reaction affords an extremely delicate test for ammonia. A solution of potassio-mercuric iodide is prepared by adding potassium iodide to a solution of corrosive sublimate, till a portion only of the resulting red precipitate is redissolved, then filtering, and mixing the filtrate with caustic potash. The liquid thus obtained forms, with a very small quantity of ammonia, either free or in the form of an ammoniacal salt, a brown precipitate, soluble in excess of potassium iodide. This is called Nessler's test for ammonia.

Dimercurammonium Hydroxide, $NHg''_2(HO)$.—This compound is formed by treating precipitated mercuric oxide with aqueous ammonia, or by treating either of the dimercurammonium salts with a caustic alkali. It is a brown powder, which dissolves in acids, yielding salts of dimercurammonium.

Dimercurammonium Sulphate, $(NHg''_2)_2SO_4,2H_2O$, formerly called *ammoniacal turpethum,* is prepared by dissolving mercuric sulphate in ammonia, and precipitating the solution with water. It is a heavy white powder, yellowish when dry, resolved by heat into water, nitrogen, ammonia, and mercurous sulphate.

Mercurous Compounds.—*Mercurosammonium Chloride,* $NH_3Hg'Cl$, is the black precipitate formed when dry calomel is exposed to the action of ammonia gas. When exposed to the air, it gives off ammonia and leaves white mercurous chloride.—*Dimercurosammonium chloride,* $NH_4Hg'_2Cl$, is formed, together with sal-ammoniac, by digesting calomel in aqueous ammonia:—

$$Hg_2Cl_2 + 2NH_3 = NH_2Hg_2Cl + NH_4Cl.$$

It is gray when dry, and is not altered by boiling water.

Dimercurosammonium nitrate, $2(NH_2Hg_2)NO_3,H_2O$.—This, according to Kane, is the composition of the velvet-black precipitate known as Hahnemann's soluble mercury, which is produced on adding ammonia to a solution of mercurous nitrate. According to C. G. Mitscherlich, on the other hand, the precipitate thus formed has the composition $2NH_3,N_2O_5,3Hg_2O$, which is that of a hydrated *trimercurosammonium nitrate*, $2(NHHg_3)NO_3,2H_2O$.

Reactions of Mercury Salts.—All mercury compounds are volatilized or decomposed by a temperature of ignition: those which fail to yield the metal by simple heating may in all cases be made to do so by heating in a test-tube with a little dry sodium carbonate. The metal is precipitated from its soluble combinations by a plate of *copper*, also by a solution of *stannous chloride* used in excess.

Hydrogen sulphide and *Ammonium sulphide* produce in solutions, both of mercuric and of mercurous salts, black precipitates insoluble in ammonium sulphide. In mercuric salts, however, if the quantity of the reagent added is not sufficient for complete decomposition, a white precipitate is formed, consisting of a compound of mercuric sulphide with the original salt, and often colored yellow or brown by excess of mercuric sulphide. An excess of hydrogen sulphide, or ammonium sulphide, instantly turns the precipitate black. This reaction is quite characteristic of mercuric salts.

M e r c u r i c s a l t s are further distinguished by forming a yellow precipitate with caustic *potash* or *soda*: white with *ammonia* or *ammonium carbonate*, insoluble in excess; red-brown with *potassium* or *sodium carbonate*. With *potassium iodide* they yield a bright-scarlet precipitate, soluble in excess, either of the mercuric salt or of the alkaline iodide.

M e r c u r o u s s a l t s are especially characterized by forming, with *hydrochloric acid* or *soluble chlorides*, a white precipitate which is turned black by ammonia. They also yield black precipitates with *caustic alkalies;* white with *alkaline carbonates*, soon turning black; greenish-yellow with *potassium iodide.*

Alloys of mercury with other metals are termed *Amalgams:* mercury dissolves in this manner many of the metals, as gold, silver, tin, lead, etc. These combinations sometimes take place with considerable violence, as in the case of potassium, in which light and heat are produced; besides this, many of the amalgams after a while become solid and crystalline. The amalgam of tin used in silvering looking-glasses, and that of silver and of copper, sometimes employed for stopping hollow teeth, are examples. The solid amalgams appear to be, for the most part, definite compounds, while the liquid amalgams may be regarded, in many instances, as solutions of definite compounds in excess of mercury,· inasmuch as, when they are pressed between chamois leather, mercury, containing only a small quantity of the other metal, passes through, while a solid amalgam, frequently of definite atomic constitution, remains behind. A native compound of mercury and silver, called "amalgam" by mineralogists, and having the composition Ag_2Hg_2 or Ag_3Hg_3, is found crystallized in octahedrons, rhombic dodecahedrons, and other forms of the regular system.

METALS OF THE YTTRIUM GROUP.

This group includes a considerable number of metals occurring in gado-linite, cerite, samarskite, and other rare minerals. The extraction and sepa-ration of the oxides of these metals is complex and difficult, and the number of distinct oxides or earths thus obtainable is not exactly known, the dis-covery of some of them, announced by certain chemists, having been denied by others; but the existence of the ten following may perhaps be considered as established:—

	Symbol.	At. w.			Symbol.	At. w.
Scandium	. Sc	44	Samarium .	. .	. Sm	150
Yttrium	. Y	89	Terbium .	. .	. Tb	149
Lanthanum	. La	138	Decipium .	. .	. Dp	159
Cerium .	. Ce	141.6	Erbium	. .	. Er	169
Didymium	. Di	145.4	Ytterbium	. .	. Yb	173

YTTRIUM, ERBIUM, TERBIUM, YTTERBIUM, SCANDIUM, DECIPIUM.

These metals exist as silicates in the gadolinite or ytterbite of Ytterby in Sweden, and in a few other rare minerals; also as niobates and tantalates in the samarskite of North Carolina.

To obtain the earths yttria and erbia (including terbia and ytterbia), gado-linite is digested with hydrochloric acid, and the solution separated from the silica is treated with oxalic acid, which throws down the oxalates of erbium and yttrium together with those of calcium, cerium, lanthanum, and didym-ium. These oxalates are converted into nitrates; the solution is treated with excess of solid potassium sulphate, to separate the cerium metals; the erbia and yttria, which still remain in solution, are again precipitated by oxalic acid; and the same treatment is repeated till the solution of the mixed earths, when examined by the spectroscope, no longer exhibits the absorption-bands characteristic of didymium. To separate the erbia and yttria, they are again precipitated by oxalic acid. The oxalates are con-verted into nitrates, and the nitrates of erbium and yttrium are separated by a series of fractional crystallizations, the erbium salt being the less solu-ble of the two and crystallizing out first.

This process, for the details of which we must refer to larger works,* effects the separation of yttria from erbia, terbia, and ytterbia; but no method of completely separating the last three earths one from the other has yet been devised. The proportions of these earths in gadolinite, are however, variable, some specimens containing nearly pure erbia, whilst others are nearly free from that earth. Ytterbia has been obtained in very small quantity only.

Erbia, Eb_2O_3, obtained by ignition of erbium nitrate or oxalate, has a faint rose color. It does not melt at the strongest white heat, but aggregates to a spongy mass glowing with an intense green light, which, when examined

* Roscoe and Schorlemmer's Treatise on Chemistry, vol. II. pt. i. p. 420; and Watts's Dictionary of Chemistry, vol. V. p. 721.

by the spectroscope, exhibits a *continuous* spectrum intersected by a number of bright bands. Solutions of erbium salts, on the other hand, give an absorption-spectrum exhibiting dark bands, and *the points of maximum intensity of the light bands in the emission-spectrum of glowing erbia coincide exactly in position with the points of greatest darkness in the absorption-spectrum.* The position of these bands is totally different from those in the emission and absorption-spectra of didymium.

Erbium salts have a rose-red color, deeper in the hydrated than in the anhydrous state; they have an acid reaction and sweet astringent taste. The *sulphate*, $EbSO_4,8H_2O$, forms light rose-colored crystals, isomorphous with the sulphates of yttrium and didymium.

Yttria, Y_2O_3, is a soft, nearly white powder, which when ignited glows with a pure white light, and yields a spectrum not containing any bright bands like that of erbia. It does not unite directly with water, but is precipitated as a hydroxide by alkalies from solutions of yttrium salts. It dissolves slowly but completely in hydrochloric, nitric, and sulphuric acids, forming colorless solutions which do not exhibit any absorption-spectrum. *Yttrium sulphate*, $Y_2(SO_4)_3,8H_2O$, forms small colorless crystals.

Metallic erbium has not yet been isolated. Yttrium (containing erbium) was obtained by Berzelius as a blackish-gray powder by igniting yttrium chloride with potassium.

Terbia, Tb_2O_3, is most advantageously prepared from the samarskite of North Carolina in which it exists as niobate, mixed with only small quantities of erbia and yttria. The mixture of oxides obtained from this mineral is dissolved in nitric acid; the solution is treated with potassium sulphate to precipitate the cerium metals; and the solution of the remaining nitrates is mixed with a saturated solution of sodium sulphate, whereby a crystalline sulphate is obtained, which, when treated with ammonium oxalate, yields an insoluble oxalate, leaving on ignition a dark-yellow oxide. This oxide is dissolved in a large excess of strong nitric acid; a hot concentrated solution of oxalic acid is dropped in till a permanent precipitate is formed; the liquid, after 24 hours, is decanted, and again fractionally precipitated with oxalic acid; and this treatment is repeated three times more. The united precipitates are then calcined and redissolved in nitric acid; the resulting solution is likewise subjected to a series of fractional precipitations with oxalic acid, and the precipitated oxalates are calcined. When the earths thus purified are dissolved by small quantities in dilute formic acid, and the clear solution is heated and slightly concentrated, a white precipitate is obtained, yielding on ignition a deep orange-colored base; and on repeatedly dissolving this base in formic acid, and precipitating it therefrom by concentration till the percentage of base in the salt amounts to 60, a product is obtained consisting of pure terbia.[*]

Terbia, after ignition at a moderate heat, has a deep orange color, but becomes quite colorless after very strong ignition, or when heated in a stream of hydrogen. It dissolves slowly, but completely, in the most dilute acids; in hydrochloric acid, with evolution of chlorine. Its solutions are colorless, and show no absorption-spectrum. The *sulphate*, $Tb_2(SO_4)_3,8H_2O$, forms colorless crystals isomorphous with the sulphates of yttrium, erbium, and didymium. The *formate* separates from solution on the sides of the vessels as a non-crystalline crust. The *acetate* crystallizes in small colored transparent prisms less soluble than acetate of didymium.

Metallic terbium has not been isolated.

Ytterbia, Yb_2O_3, has been obtained, together with erbia, from gadolinite

[*] Delafontaine, Ann. Ch. Phys. [5], xiv. 238.

and from euxenite. It is a colorless, infusible earth, of sp. gr. 9.175, insoluble in water, easily soluble in dilute acids, and attacked with difficulty even by strong acids in the cold. Its solutions have a sweet, astringent taste, are colorless, and give no absorption-spectrum. The salts impart no color to the Bunsen flame, but with the electric light the chloride gives a brilliant spectrum. The *sulphate*, $Yb_2(SO_4)_3,8H_2O$, forms large brilliant prisms permanent in the air, but giving off their water at 100°. It dissolves slowly in boiling water, completely in a saturated solution of potassium sulphate.

Scandium, Sc. Atomic weight, 44.03 (Nilson); 44.91 (Cleve).—This metal, discovered by Nilson in 1879 and further examined by Cleve, exists only in gadolinite and yttrotitanite. The metal has not been isolated.

Scandia, Sc_2O_3, may be separated from ytterbia by the readier decomposability of its nitrate by heat, and by its property of forming an insoluble double salt with potassium sulphate. It is a light, infusible, white powder, resembling magnesia, easily soluble in hot nitric and hydrochloric acids, nearly insoluble in the same acids when cold. It is not volatile and gives no color to flame, but the chloride gives a very brilliant spectrum with the electric spark.

Scandium salts are colorless and have a sour, astringent taste, quite different from the sweet taste of the salts of yttrium and erbium. With potash and ammonia they give bulky white precipitates, insoluble in excess; tartaric acid prevents the precipitation in the cold, but a precipitate forms on heating. Sodium carbonate gives a precipitate soluble in excess. Ammonium sulphide throws down the hydroxide. Sodium orthophosphate gives a gelatinous precipitate; oxalic acid, a curdy precipitate quickly becoming crystalline.

Scandium nitrate crystallizes from a strong solution in small prisms. The *sulphate*, $Sc_2(SO_4)_3,8H_2O$, is permanent at ordinary temperatures, but gives off $4H_2O$ at 100° C., the rest on gentle ignition; decomposes at a high temperature. The *double sulphate*, $K_2Sc_2(SO_4)_4,xH_2O$, crystallizes in groups of small prisms, very slightly soluble in water, and quite insoluble in a saturated solution of potassium sulphate.

The spectrum of scandium obtained by passing a powerful induction spark between aluminium poles moistened with a solution of the chloride, is very complicated, containing more than a hundred lines.

Scandium agrees very closely, in its atomic weight and in the properties of its oxide and salts, with the element whose existence was predicted by Mendelejeff under the name of *ekabor*—*e. g.*, in forming a white sesquioxide, infusible, soluble with difficulty in acids after ignition, insoluble in alkalies; also in forming colorless salts, and a sulphate which unites with potassium sulphate, forming a double salt analogous in composition to common alum.

Decipium, Dp. Atomic weight, 159.—This metal, discovered (as oxide) by Delafontaine in the samarskite of North Carolina, is intermediate in character between the metals of the yttrium and cerium groups. Its oxide has not yet been sufficiently separated from that of didymium to exhibit its true color. Decipium salts are colorless; the *acetate* crystallizes easily, is less soluble than didymium acetate, more soluble than the terbium salt. *Decipio-potassic sulphate* is but slightly soluble in a saturated solution of potassium sulphate, but dissolves easily in pure water. *Decipium nitrate* gives in direct solar light an absorption-spectrum containing at least three bands in the blue and indigo.

Four other earths are also mentioned as yielding characteristic spectra, viz., *Yttria a* and *Yttria β*, obtained by Marignac from samarskite; *Thulia* and *Holmia*, obtained by Cleve in attempting to prepare pure erbia; but further experiments are required to establish their separate identity.*

CERIUM—LANTHANUM—DIDYMIUM—SAMARIUM.

Ce = 141.2.—La = 139.—Di = 145.4.—Sm = 150.

THE first three of these metals occur together as silicates in the Swedish mineral cerite, also in allanite, orthite, and a few others; and as phosphates in monazite, edwardsite, and cryptolite, a mineral occurring disseminated through apatite and through certain cobalt ores.

Cerium was discovered in 1803 by Klaproth, and by Hisinger and Berzelius, who obtained it in the form of oxide from cerite. This mineral is completely decomposed by boiling with strong hydrochloric acid, silica being separated, and the cerium, together with iron and other metals, dissolving as chloride. On treating the acid solution thus obtained, with oxalic acid, cerium oxalate is precipitated as a white crystalline powder, which, when ignited, leaves a brown oxide. The product thus obtained was for some time regarded as the oxide of a single metal, c e r i u m; but in 1839 and 1841, Mosander showed that it contained the oxides of two other metals, which he designated as l a n t h a n u m† and d i d y m i u m.‡

Cerium oxide may be separated from the oxides of lanthanum and didymium by treating the crude brown oxide above mentioned, first with dilute and then with strong nitric acid, which gradually removes the whole of the lanthanum and didymium oxides.

The separation of these two oxides one from the other is much more difficult, and can be effected only by successive crystallization of their sulphates. If the lanthanum salt is in excess, in which case the solution of the mixed sulphates has only a faint amethyst tinge, the liquid is evaporated to dryness, and the residue heated to a temperature just below redness, to render the sulphates anhydrous. The residue thus obtained is then to be added by small portions to ice-cold water, which dissolves it easily, and the resulting solution heated in a water-bath to about 40° C. (104° F.). Lanthanum sulphate then crystallizes out, containing only a small quantity of didymium, and may be further purified by repeating the whole process. If, on the other hand, the didymium salt is in excess, in which case the liquid has a decided rose-color, separation may be effected by leaving the acid solution in a warm place for a day or two. Didymium sulphate then separates in large rhombohedral crystals.

Metallic cerium, lanthanum, and didymium are obtained by reducing the chlorides with sodium, in the form of gray powders, which decompose water at ordinary temperatures, and dissolve rapidly in dilute acids with evolution of hydrogen.

Atomic Weights and *Quantivalence.*—These three metals were originally regarded as dyads; afterward from Hildebrand's determinations of their specific heats, and from the fact that they all three form salts in which the metal is trivalent, they were supposed to be triads; § but recent experiments by Dr. Brauner ¶ have shown that lanthanum only is a triad, forming a single series of salts, LaR_3 (R denoting an acid radicle); whereas cerium is a tetrad, inasmuch as, besides the chloride Ce_2Cl_6, the oxide Ce_2O_3, etc., it also forms a tetrafluoride, CeF_4 and a dioxide CeO_2; and lastly that didymium is a pentad, forming a pentoxide Di_2O_5, and some double fluorides,

---

* For further details respecting the newly-discovered earths above described, see Watts's Dictionary of Chemistry, vol. viii. pp. 2154-2157.
† From λανθανειν, to lie hid.   ‡ From διδυμοι, twins.
§ See 12th Edition of this Manual, vol i. p. 398.
¶ Journal of the Chemical Society, vol. xli. p. 68.

like DiF_3,KF or $Di(F_4K)$, in which the didymium may be regarded as quin-quivalent. The three metals thus fall regularly into their places in Mendelejeff's series (p. 256).

Lanthanum, as already observed, forms only one set of compounds, viz., $LaCl_3$, La_2O_3, $La_2(SO_4)_3$. There is, however, a higher oxide, the composition of which is not exactly known. Lanthanum salts are colorless; their solutions yield, with alkalies, a precipitate of *lanthanum hydroxide*, LaH_3O_3, which, when ignited, leaves the white anhydrous sesquioxide. Both the hydroxide and the anhydrous oxide dissolve easily in acids. *Lanthanum sulphate* forms small prismatic crystals, containing $La_2(SO_4)_3,9H_2O$. *Lanthanum and potassium sulphate*, $LaK_3(SO_4)_3$ is formed, on mixing the solution of a lanthanum salt with potassium sulphate, as a white crystalline precipitate, resembling the corresponding cerium salt.

Cerium forms two series of compounds, viz., the ce r o u s componuds, in which it is trivalent, as above stated, e. g., Ce_2Cl_6, Ce_2O_3, $Ce_2(SO_4)_3$;* and the c e r i c compounds, in which it is quadrivalent, e. g., ceric oxide, CeO_2, ceric sulphate, $Ce(SO_4)_2$, etc.

Cerous oxide, Ce_2O_3, is obtained by igniting the carbonate or oxalate in a current of hydrogen as a grayish-blue powder, quickly converted into ceric oxide on exposure to the air. Its salts are colorless. The *sulphate*, $Ce_2(SO_4)_3$, crystallizes with various quantitities of water, according to the temperature at which it is deposited. *Cerium and potassium sulphate*, $CeK_3(SO_4)_3$, separates as a white powder on immersing solid potassium sulphate in a solution of a cerous salt. It is slightly soluble in pure water, but insoluble in a saturated solution of potassium sulphate. The formation of this salt affords the means of separating cerium from most other metals.

Ceric oxide, CeO_2, is produced when cerous hydroxide, carbonate, or nitrate is ignited in an open vessel. It is yellowish-white, acquires a deep orange-red color when heated, but recovers its original tint on cooling. It is not converted into a higher oxide by ignition in oxygen. *Hydrated ceric oxide*, $2CeO_2,3H_2O$, obtained by passing chlorine into aqueous potash in which cerous hydroxide is suspended, is a bright-yellow precipitate, which dissolves readily in sulphuric and nitric acids, forming yellow solutions of ceric salts; and in hydrochloric acid, with evolution of chlorine, forming colorless cerous chloride.

The solution of the sulphate yields by spontaneous evaporation, first, brown-red crystals of a *ceroso-ceric salt*, $\left.\begin{array}{l}Ce_2(SO_4)_3 \\ 2Ce(SO_4)_2\end{array}\right\} + 4H_2O$, and afterward yellow indistinctly crystalline *ceric sulphate*, $Ce(SO_4)_2 + 4H_2O$.

Cerium Fluorides.—The mineral *fluocerite*, found at Finnbo in Sweden, is generally regarded as CeF_4; but the only published analysis of it, by Berzelius, agrees better with the formula CeF_6,H_2O. When hydrated ceric oxide is treated with aqueous hydrofluoric acid, a brownish powder is obtained, consisting of a h y d r a t e d t e t r a f l u o r i d e, CeF_4,H_2O. This compound, when moderately heated, gives off its water and part of its fluorine as hydrofluoric acid, and at a higher temperature it gives off a gas (fluorine) which decomposes potassium iodide, with liberation of free iodine, these decompositions apparently taking place according to the following equations:—

$$(a.)\ 2(CeF_4,H_2O) = Ce_2F_6 + 2HF + O + H_2O$$
$$(b.)\ 2(CeF_4,H_2O) = Ce_2F_6 + 2H_2O + F_2.$$

---

* These compounds, like those of aluminium (p. 396), may be represented by formulæ in which the metal figures as a tetrad, e. g., the chloride as $Cl_3Ce.CeCl_3$, the oxide as $O<^{Ce=O}_{Ce=O}$, etc.

17 *

On heating the hydrated tetrafluoride for some time in contact with moist air, pure ceric oxide is left: $CeF_4,H_2O + H_2O = CeO_2 + 4HF$. A *potassioceric fluoride*, $3KF,2CeF_4,H_2O$, is obtained, as a yellowish-white crystalline powder, on treating hydrated ceric oxide with a solution of potassium and hydrogen fluoride.

Didymium.—The salts of this metal are rose-colored, and their solutions give, with alkalies, a pale rose-colored precipitate of the *hydroxide*, DiH_3O_3, which, when ignited in a covered crucible, leaves the anhydrous trioxide, Di_2O_3, in white, hard lumps. When, however, the hydroxide, nitrate, carbonate, or oxalate of didymium is heated in contact with the air, and not very strongly, a dark-brown peroxide is left, which, according to Brauner, is a pentoxide, Di_2O_5. This oxide is decomposed at a higher temperature with evolution of oxygen, and when heated to dull redness in hydrogen gas it is reduced with formation of water. It dissolves in strong nitric or sulphuric acid, with evolution of oxygen and formation of a didymium salt, $Di(NO_3)_3$ or $Di_2(SO_4)_3$.

Didymium sulphate separates from an acid solution, by spontaneous evaporation, in well-defined rhombohedral crystals, exhibiting numerous secondary faces, and containing $Di_2(SO_4)_3,8H_2O$: they are isomorphous with the similarly constituted sulphates of yttrium and erbium. The sulphate is more soluble in cold than in hot water, and a solution saturated in the cold, deposits when heated to the boiling point, a crystalline powder containing $Di_2(SO_4)_3,6H_2O$. *Didymium and potassium·sulphate*, $DiK_3(SO_4)_3$, resembles the lanthanum salt.

Solutions of didymium salts exhibit a well-marked absorption spectrum,[*] containing two black lines enclosing a very bright space. One of these black lines is in the yellow, immediately following Fraunhofer's line D; the other is situated between E and b. These characters can be distinctly recognized in a solution half an inch deep, containing only 0.10 per cent. of didymium salt. Lanthanum salts do not exhibit an absorption spectrum.

From recent experiments by Brauner[†] and by Cleve, it appears probable that cerite contains at least two more metals closely resembling didymium, and hitherto not completely separated from it. By fractionally precipitating with ammonia a solution of didymium sulphate prepared as above described, Brauner has obtained an oxide which he designates as Diγ, of higher molecular weight than the trioxide of didymium; and the mother-liquor of the didymium preparation, after being freed from lanthanum and yttrium by repeated precipitation with potassium sulphate, yielded the oxide of another metal, Diβ, of lower atomic weight. The atomic weights of these allied metals are as follows:—

Didymium β.	Didymium.	Didymium γ.
144.13	145.4	146.6

The higher number 147 formerly attributed to didymium was due to admixture of Diγ. The existence of the two new elements Diβ and Diγ is corroborated by spectroscopic observation.

Samarium.—This metal, discovered by Lecoq de Boisbaudran in samarskite, by means of its characteristic spectrum, has quite recently been further examined by Cleve,[‡] who obtained its oxide, samaria, from the thorite of Arendal in Norway, in which it exists together with the oxides of thorium, cerium, lanthanum, and didymium.

---

[*] See LIGHT, p. 98.   [†] Monatshefte für Chemie, 1882, p. 486.
[‡] Journal of the Chemical Society, August, 1883, vol. xliv.

The spectrum of samarium is distinguished by several bands, amongst which four in the blue region are most characteristic. Its atomic weight is 150.02, supposing its oxide to have the composition Sm_2O_3, a view which is in accordance with the close analogy of the samarium salts to those of didymium. Samarium is in fact more nearly related to didymium than to any other element.

Samarium Chloride, $SmCl_3,6H_2O$, forms large tabular deliquescent crystals. The *oxide*, Sm_2O_3, is a white powder with a scarcely perceptible tinge of yellow, and easily soluble in acids, forming salts whose solutions have a deep yellow color. The crystallized salts are topaz-colored, but in the state of powder they are yellowish-white. The *hydroxide* is a gelatinous white or faintly yellowish precipitate, insoluble in alkalies. It is a stronger base than yttria, but not so strong as didymia.

The *acetate*, $Sm(C_2H_3O_2)_3 + 4H_2O$, forms short, well-defined prisms, moderately soluble in water. The *sulphate*, $Sm_2(SO_4)_3,8H_2O$, forms well-defined shining crystals, less soluble in water than in sulphate of didymium. The *double sulphate*, $2Sm_2(SO_4)_3,9K_2SO_4 + 3H_2O$?, was slowly deposited as a white powder on adding samarium acetate to excess of potassium sulphate: it is but very slightly soluble in excess of potassium sulphate. The *oxalate*, $Sm_2(C_2O_4)_3,10H_2O$, is a white or pale-yellowish crystalline precipitate.

METALS OF THE ALUMINIUM GROUP.

Aluminium—Indium—Gallium.

THESE metals are trivalent, forming compounds of the forms MCl_3, M_2O_3, etc. Their oxides are weak bases, and their sulphates unite with those of the alkali-metals, forming double salts, called alums, which crystallize in octahedrons and other forms of the regular system.

ALUMINIUM.

Symbol, Al. Atomic weight, 27.3.

THIS metal occurs very abundantly in nature in the state of silicate, as in felspar and its associated minerals; also in the various modifications of clay thence derived. It was first isolated by Wöhler, who obtained it as a gray powder by decomposing aluminium chloride with potassium; and H. Sainte-Claire Deville, by an improved process founded on the same principle, succeeded in obtaining it in the compact form and on the manufacturing scale. The process consists in decomposing the double chloride of aluminium and sodium, $AlCl_3, NaCl$, by heating it with metallic sodium, fluor-spar or cryolite being added as a flux. The reduction is effected in crucibles, or on the large scale on the hearth of a reverberatory furnace. Sodium is used as the reducing agent in preference to potassium: first, because it is more easily prepared; and, secondly, because it has a lower atomic weight, and, consequently, a smaller quantity of it suffices to do the same amount of chemical work.

Aluminium is also prepared directly from cryolite by reduction with sodium, but the metal thus obtained is said to be more contaminated with iron and silicon than that prepared by Deville's process.

Aluminium is remarkable for its low specific gravity, which is 2.67; it is nearly as white as silver, and is capable of assuming a high polish. It is employed in the manufacture of delicate apparatus and ornamental articles. Some of the alloys of aluminium are more generally applicable, more especially the alloy with copper, which is remarkable for being similar in appearance to gold: this alloy is found in commerce under the name of *aluminium bronze*.

Aluminium forms only one class of compounds, in which it is trivalent, one atom of the metal being equivalent to three atoms of hydrogen; thus the chloride is $AlCl_3$, the oxide, Al_2O_3, etc. These compounds might indeed be represented by formulæ into which the metal enters as a tetrad, *e. g.*, the

chloride $Cl_3Al - AlCl_3$, the oxide $O{<}{\begin{smallmatrix} Al = O \\ | \\ Al = O \end{smallmatrix}}$, etc., but the simpler formulæ are justified, as already observed (p. 256), by the vapor-density of aluminium methide, which is in accordance with the formula $Al(CH_3)_3$, and by the position of aluminium in Mendelejeff's series.

Aluminium Chloride, $AlCl_3$.—This compound is obtained in solution by dissolving alumina or aluminium hydrate in hydrochloric acid; but the solution, when evaporated, gives off hydrochloric acid and leaves alumina. The anhydrous chloride may be prepared by heating a mixture of alumina and finely divided carbon in chlorine gas.

Pure precipitated alumina is dried and mixed with oil and lampblack, and the mixture, after being strongly calcined in a covered crucible, is introduced into a porcelain tube or tubulated earthen retort, placed in a furnace and connected at one end with an apparatus for evolving chlorine, and at the other with a dry receiver. On raising the heat to bright redness, and passing chlorine through the apparatus, aluminium chloride distils over, together with carbon monoxide, and condenses as a solid mass in the receiver:—

$$Al_2O_3 + C_3 + Cl_6 = 2AlCl_3 + 3CO.$$

Aluminium chloride is a transparent waxy substance, having a crystalline structure, colorless when pure, but generally exhibiting a yellow color, due perhaps to the presence of iron. It boils at about 180°, fumes in the air, and smells of hydrochloric acid. It is very deliquescent, and dissolves readily in water; the solution when left to evaporate yields the hydrated chloride, $AlCl_3,6H_2O$, in six-sided prisms, which when heated are resolved into alumina and hydrochloric acid.

Aluminium and Sodium Chloride, $AlCl_3,NaCl$, is obtained by melting together the component chlorides in proper proportions, or by adding the requisite quantity of sodium chloride to the mixture of alumina and charcoal used for the preparation of aluminium chloride, igniting the mass in chlorine or hydrochloric acid, and condensing the vapor in a receiver. It is a crystalline mass, less deliquescent than aluminium chloride, and, therefore, more convenient for the preparation of aluminium.

Aluminium Fluoride, AlF_3, is produced by the action of gaseous silicon fluoride on aluminium, and forms cubic crystals, volatilizing at a red heat, insoluble in water, and resisting the action of all acids.

Aluminium and Sodium Fluoride, $AlF_3,3NaF$, occurs abundantly as cryolite, at Evigtok in Greenland, and is prepared artificially by pouring hydrofluoric acid in excess on a mixture of calcined alumina and sodium carbonate. Cryolite forms quadratic crystals, colorless, transparent, softer than felspar, and of specific gravity 2.96. It is used, as already mentioned, for the preparation of aluminium, and in Germany for the manufacture of soda for the use of soap-boilers.

Aluminium Oxide; Alumina, Al_2O_3.—This substance is inferred to be a sesquioxide from its isomorphism with ferric oxide. It is prepared by mixing a solution of alum with excess of ammonia, by which an extremely bulky, white, gelatinous precipitate of aluminium hydrate is thrown down. This is washed, dried, and ignited to whiteness. Thus obtained, alumina constitutes a white, tasteless, coherent mass, very little acted upon by acids. It is fusible before the oxyhydrogen blowpipe. The mineral called *corundum*, of which the ruby and sapphire are transparent varieties, consists of nearly pure alumina in a crystallized state, with a little coloring oxide: emery, used for polishing glass and metals, is a coarse variety of corundum. Alumina is a very weak base, and many of its salts have an acid reaction.

Aluminium Hydrates.—Aluminium forms three hydrates, namely:—

34

Monohydrate. . . . . Al_2O_3,H_2O or $AlO(OH)$
Dihydrate . . . . . $Al_2O_3,2H_2O$ or $Al_2O(OH)_4$
Tryhidrate . . . . . $Al_2O_3,3H_2O$ or $Al(OH)_3$.

The first is found native, as *diaspore*, in translucent masses which crumble to powder when heated, and give off the whole of their water at 360°.

The third is the ordinary gelatinous precipitate obtained by treating solutions of aluminium salts—alum, for example—with ammonia or alkaline carbonates. When dried at a moderate heat, it forms a soft, friable mass, which adheres to the tongue and forms a stiff paste with water, but does not dissolve in that liquid. At a strong red heat it gives off water and undergoes a very great contraction of volume. It dissolves with great facility in acids and in the fixed caustic alkalies. When a solution of alumina in caustic potash is exposed to the air, the potash absorbs carbonic acid, and the aluminium trihydrate is then deposited in white crystals, which are but sparingly soluble in acids.

Aluminium trihydrate has a very powerful attraction for organic matter, and when digested in solutions of vegetable coloring matter, combines with and carries down the coloring matter, which is thus removed entirely from the liquid if the alumina is in sufficient quantity. The pigments called *lakes* are compounds of this nature. The fibre of cotton impregnated with alumina acquires the same power of retaining coloring matters: hence the great use of aluminous salts as *mordants* to produce fast colors.

Aluminium trihydrate occurs native as *Gibbsite*, a stalactitic, translucent, fibrous mineral, easily dissolved by acids, occurring at Gumuch-dagh in Asia Minor, at Richmond in Massachusetts, and other localities in the United States; also as *hydrargyllite*, near Slatoust in the Ural, in small hexagonal prisms.

Dihydrate.—When a dilute solution of aluminium diacetate is exposed for several days to a temperature of 100° in a close vessel, the acetic acid appears to be set free, although no precipitation of alumina takes place. The liquid acquires the taste of acetic acid, and if afterward boiled in an open vessel gives off nearly the whole of its acetic acid, the alumina, nevertheless, remaining in solution. This solution is coagulated by mineral acids and by most vegetable acids, by alkalies, and by decoctions of dye-woods. The alumina contained in it is, however, no longer capable of acting as a mordant. Its coagulum with dye-woods has the color of the infusion, but is translucent and totally different from the dense opaque lakes which ordinary alumina forms with the same coloring matters. On evaporating the solution to dryness at 100° C. (212° F.), the alumina remains in the form of dihydrate retaining only a trace of acetic acid. In this state it is insoluble in the stronger acids, but soluble in acetic acid, provided it has not been previously coagulated in the manner just mentioned. Boiling potash converts it into the trihydrate.*

An aluminium dihydrate, called *Bauxite*, in which the aluminium is partly replaced by iron, viz., $(AlFe)_2O_3,2H_2O$ or $(AlFe)_2O(OH)_4$, occurs in roundish masses at Beaux, near Arles in France, and is largely employed for the preparation of alumina and metallic aluminium.

Aluminates.—The hydrogen in aluminium trihydrate may be replaced by an equivalent quantity of various metals; such compounds are called *aluminates.* According to Frémy, a solution of alumina in potash slowly evaporated out of contact with the air deposits granular crystals of potassium aluminate, $AlKO_2$ or $Al_2K_2O_4 = Al_2O_3,K_2O$. Similar compounds occur native; thus *Spinell* is an aluminate of magnesium, Al_2MgO_4; *Gahnite*, an aluminate of zinc, Al_2ZnO_4.

* Walter Crum, Chem. Soc. Journ., vi. 225.

Aluminium Sulphide, Al_2S_3.—When the vapor of carbon bisulphide is passed over alumina at a bright red heat a glassy melted mass remains, which is instantly decomposed by water, with evolution of sulphuretted hydrogen.

Aluminium Sulphate, $Al_2(SO_4)_3,18H_2O$ or $Al_2O_3,3SO_3,18H_2O$.—Prepared by saturating dilute sulphuric acid with aluminium hydrate, and evaporating, or, on the large scale, by heating clay with sulphuric acid. It crystallizes in thin pearly plates, soluble in 2 parts of water; it has a sweet and astringent taste and an acid reaction. Heated to redness it is decomposed, leaving pure alumina. Two other aluminium sulphates, with excess of base, are also described, one of which is insoluble in water.

Aluminium sulphate combines with the sulphates of potassium, sodium, and ammonium, and the other alkali-metals, forming double salts of great interest, the *alums*. Common alum, the source of all the preparations of alumina, contains $AlK(SO_4)_2,12H_2O$. It is manufactured on a very large scale from a kind of slaty clay loaded with iron bisulphide which abounds in certain localities. This is gently roasted, and then exposed to the air in a moistened state; oxygen is absorbed; the sulphur becomes acidified; ferrous sulphate and aluminium sulphate are produced, and afterward separated by lixiviation with water. The solution is next concentrated and mixed with a quantity of potassium chloride, which decomposes the iron-salt, forming ferrous chloride and potassium sulphate, and the latter combines with the aluminium sulphate to form alum. By crystallization the alum is separated from the highly soluble iron chloride, and afterward easily purified by a repetition of the process.

Potassium-alum crystallizes in colorless, transparent octahedrons, which often exhibit the faces of the cube. It has a sweetish and astringent taste, reddens litmus-paper, and dissolves in 18 parts of water at 15.5° C. (59.9° F.), and in its own weight of boiling water. Exposed to heat, it is easily rendered anhydrous, and by a very high temperature it is decomposed. The crystals have little tendency to change in the air. Alum is largely used in the arts in preparing skins, dyeing, etc.: it is occasionally contaminated with iron oxide, which interferes with some of its applications. The celebrated Roman alum made from *alum-stone*, a felspathic rock altered by sulphurous vapors, was once much prized on account of its freedom from this impurity. A mixture of dried alum and sugar carbonized in an open pan, and then heated to redness in a glass flask, contact with air being avoided, furnishes *Homberg's pyrophorus*, which ignites spontaneously on exposure to the air. The essential ingredient is, in all probability, finely divided potassium sulphide.

A process has been patented by Messrs. Duncan and Newlands for separating the potash from syrups of beet- and cane-sugar by means of aluminium sulphate, the potash being thereby precipitated in the form of alum. A solution of aluminium sulphate of a density corresponding with about 27° of Baumé's hydrometer is added to the cold syrup, having a density of about 36° B., in quantity sufficient to precipitate the whole of the potash present; the mixture is well stirred for fifteen minutes to an hour; and the whole left at rest for four or five hours till the deposit has completely subsided. This deposit consists of small crystals, technically known as "alum meal." Considerable quantities of alum are now made by this process.

Sodium-alum, in which sulphate of sodium replaces sulphate of potassium, has a form and constitution similar to that of common alum: it is, however, much more soluble and difficult to crystallize.

Ammonium-alum, containing NH_4 instead of K, very closely resembles common potassium-alum, having the same figure, appearance, and constitution, and nearly the same degree of solubility as that substance. As the value of potassium salts is continually increasing, ammonium-alum, which may be used in dyeing with the same advantage as the corresponding potassium salt, has almost entirely replaced the potassium-alum in that process When heated to redness, ammonium-alum yields pure alumina.

Cæsium-alum, $AlCs(SO_4)_2,12H_2O$, and *Rubidium-alum*, $AlRb(SO_4)_2,12H_2O$, resemble potassium-alum. A *silver-alum*, $AlAg(SO_4)_2,12H_2O$, is formed by heating equivalent quantities of argentic and aluminium sulphates till the former is dissolved. It crystallizes in regular octahedrons, and is resolved by water into its component salts. There is also a *thallium-alum*, $AlTl(SO_4)_2,12H_2O$, which crystallizes in regular octahedrons.

Lastly, there are alums isomorphous with those just described, in which the aluminium is replaced by iron, chromium, and manganese: for example, *potassio-ferric sulphate* or *potassium-iron alum*, $FeK(SO_4)_2,12H_2O$; *ammonio-chromic sulphate*, $Cr(NH_4)(SO_4)_2,12H_2O$. These will be described further on.

Phosphates.—The *normal salt*, $AlPO_4$ or $Al_2(PO_4)_2$, is obtained on adding a neutral solution of alum to a solution of ordinary sodium phosphate as a gelatinous precipitate, which dissolves in caustic potash or soda and in mineral acids, but not in ammonia or in acetic acid. On adding ammonia to a solution of this salt in hydrochloric acid a precipitate is formed, consisting of a basic salt, $3Al_2(PO_4)_2,Al_2(OH)_6 + 15H_2O$. The mineral *wavellite*, which occurs in trimetric crystals or in hemispherical or globular concretions having a radiate structure, is likewise a basic aluminium phosphate having the composition $2Al_2(PO_4)_2, Al_2(OH)_6 + 9H_2O$: it also contains small quantities of fluorine and sometimes of chlorine. *Turquois* is another basic aluminium phosphate containing $Al_2(PO_4)_2,Al_2(OH)_6 + 2H_2O$.

Silicates.—Silicates of aluminium enter into the composition of a number of crystallized minerals, among which felspar, by reason of its abundant occurrence, occupies a prominent place. Granite, porphyry, trachyte, and other ancient unstratified rocks consist in great part of this mineral, which, under particular circumstances by no means well understood, and particularly by the action of the carbonic acid of the air, suffers complete decomposition, being converted into a soft friable mass of earthy matter. This is the origin of clay: the change itself is seen in great perfection in certain districts of Devonshire and Cornwall, the felspar of the fine white granite of those localities being often disintegrated to a great depth and the rock altered to a substance resembling soft mortar. By washing, this finely-divided substance is separated from the quartz and mica ; and the milk-like liquid, being stored up in tanks, deposits the suspended clay, which is afterward dried, first in the air and afterward in a stove, and employed in the manufacture of porcelain. The composition assigned to unaltered felspar is $AlKSi_3O_8$, or $AlKSiO_4,2SiO_2$, or $Al_2O_3,K_2O,6SiO_2$. The nature of the change by which felspar passes into porcelain clay is not exactly understood, but it consists essentially in the abstraction of silica and alkali.

When the decomposing rock contains iron oxide, the clay produced is colored. The different varieties of shale and slate result from the alteration of ancient clay-beds, apparently in many instances by the infiltration of water holding silica in solution: the dark appearance of some of these deposits is due to bituminous matter. Clays containing an admixture of calcium carbonate are called *marls*, and may be recognized by their property of effervescing with acids. Pure clays are but little affected by dilute acids, but on boiling them with strong sulphuric acid alumina is dissolved out and finely divided silica left behind.

A basic aluminium silicate, Al_2O_3,SiO_2, is found crystallized, constituting the beautiful mineral called *cyanite*. The compounds formed by the union of the aluminium silicates with other silicates are almost innumerable. A sodium felspar, *albite*, containing that metal in place of potassium is known, and there are two somewhat similar lithium compounds, *spodumene* and *petalite*. The *zeolites* are hydrated silicates belonging to this class; *analcime, nepheline, mesotype*, etc., are double silicates of sodium and aluminium, with water of crystallization. *Stilbite, heulandite, laumontite, prehnite*, etc., consist of calcium silicate combined with silicate of aluminium. The *garnets, axinite, mica*, etc., have a similar composition, but are anhydrous. Iron sesquioxide is very often substituted for alumina in these minerals.

Salts of aluminium, when moistened with *cobalt nitrate* and heated before the blowpipe, assume a characteristic blue color.

Alumina when in solution is readily distinguished. Caustic *potash* and *soda* occasion a white gelatinous precipitate of aluminium hydrate, freely soluble in excess of the alkali. *Ammonia* produces a similar precipitate, insoluble in excess of the reagent. The *alkaline carbonates* and *carbonate of ammonia* precipitate the hydrate, with escape of carbonic acid. The precipitates are insoluble in excess.

Ammonium sulphide also produces a white precipitate of aluminium hydrate.

Porcelain and Earthenware.—The plasticity of natural clays, and their hardening when exposed to heat, are properties which suggested in very early times their application to the making of vessels for the various purposes of daily life: there are few branches of industry of higher antiquity than that exercised by the potter.

True porcelain is distinguished from earthenware by very obvious characters. In porcelain the body of the ware is very compact and translucent, and breaks with a conchoïdal fracture, indicative of incipient fusion. The glaze, too, applied for giving a perfectly smooth surface, is closely adherent, and, in fact, graduates by insensible degrees into the substance of the porcelain. In earthenware, on the contrary, the fracture is open and earthy, and the glaze may be detached with greater or less facility. The compact and partly glassy character of porcelain is the result of the admixture with the clay of a small portion of some substance, such as felspar or a calcic or alkaline silicate, which is fusible at the temperature to which the ware is exposed when baked or fired, and, being absorbed by the less fusible portion, binds the whole into a solid mass on cooling. The clay employed in porcelain-making is always directly derived from decomposed felspar, none of the clays of the secondary strata being pure enough for the purpose; it must be white and free from iron oxide. To diminish the contraction which this substance undergoes in the fire, a quantity of finely-divided silica, carefully prepared by crushing and grinding calcined flints or chert, is added, together with a proper proportion of felspar or other fusible material, also reduced to impalpable powder. The utmost pains are taken to effect perfeet uniformity of mixture and to avoid the introduction of particles of grit or other foreign bodies. The ware itself is fashioned either on the potter's wheel—a kind of vertical lathe—or in moulds of plaster of Paris, and dried first in the air, afterward by artificial heat, and at length completely hardened by exposure to the temperature of ignition. The porous *biscuit* is now fit to receive its glaze, which may be either ground felspar, or a mixture of gypsum, silica, and a little porcelain clay, diffused through water. The piece is dipped for a moment into this mixture and withdrawn; the water sinks into its substance, and the powder remains evenly spread upon its surface; it is once more dried, and, lastly, fired at an exceedingly high temperature.

The porcelain-furnace is a circular structure of masonry, having several

34 *

fireplaces, and surmounted by a lofty dome. Dry wood or coal is consumed as fuel, and its flame is directed into the interior, and made to circulate around and among the earthen cases, or *seggars*, in which the articles to be fired are packed. Many hours are required for this operation, which must be very carefully managed. After the lapse of several days, when the furnace has completely cooled, the contents are removed in a finished state, so far as regards the ware.

The ornamental part, consisting of gilding and painting in enamel, has yet to be executed; after which the pieces are again heated, in order to flux the colors. The operation has sometimes to be repeated more than once.

The manufacture of porcelain in Europe is of modern origin: the Chinese have possessed the art from the commencement of the seventh century, and their ware is, in some respects, altogether unequalled. The materials employed by them are known to be *kaolin* or decomposed felspar; *petuntze*, or quartz reduced to fine powder; and the ashes of fern, which contain potassium carbonate.

Stoneware.—This is a coarse kind of porcelain, made from clay containing oxide of iron and a little lime, to which it owes its partial fusibility. The glazing is performed by throwing common salt into the heated furnace; this is volatilized, and decomposed by the joint agency of the silica of the ware and of the vapor of water always present; hydrochloric acid and soda are produced, the latter forming a silicate, which fuses over the surface of the ware, and gives a thin, but excellent glaze.

Earthenware.—The finest kind of earthenware is made from a white secondary clay, mixed with a considerable quantity of silica. The articles are thoroughly dried and fired; after which they are dipped into a readily fusible glaze mixture, of which lead oxide is usually an important ingredient, and, when dry, reheated to the melting point of the latter. The whole process is much easier of execution than the making of porcelain, and demands less care. The ornamental designs in blue and other colors, so common upon plates and household articles, are printed upon paper in enamel pigment mixed with oil, and transferred while still wet to the unglazed ware. When the ink becomes dry, the paper is washed off, and the glazing completed.

The coarser kinds of earthenware are sometimes covered with a whitish opaque glaze, which contains the oxides of lead and tin; such glaze is very liable to be attacked by acids, and is dangerous for culinary vessels.

Crucibles, when of good quality, are very valuable to the chemist. They are made of clay free from lime, mixed with sand or ground ware of the same description. The Hessian and Cornish crucibles are among the best. Sometimes a mixture of plumbago and clay is employed for the same purpose; and powdered coke has been also used with the earth: such crucibles bear rapid changes of temperature with impunity.

INDIUM.

Symbol, In. Atomic weight, 114.4.

THIS metal was discovered in 1863 by Reich and Richter in the zinc-blende of Freiberg, and has since been found in a few other zinc-ores and in the flue-dust of zinc furnaces. The metallic sponge which remains when the crude zinc of the Freiberg works is dissolved in dilute sulphuric acid contains the whole of the indium (0.045 per cent.), together with lead, arsenic, cadmium, and iron. It is dissolved in nitric acid; the lead, arsenic, and cadmium are precipitated by hydrogen sulphide; and the filtered solution is heated with potassium chlorate to bring all the iron to the state of ferric salt, and then treated with ammonia, which precipitates the indium as a hydroxide, together with iron and zinc. The precipitate is dissolved in acetic acid; and the indium, precipitated as sulphide by hydrogen sulphide, is freed from traces of iron and zinc by dissolving the precipitate in dilute hydrochloric acid, precipitating the indium by agitation with barium carbonate, dissolving out the indium hydroxide by dilute sulphuric acid, and reprecipitating with ammonia.

Indium, reduced from its oxide by ignition with sodium, is a silver-white metal, soft and ductile, has a specific gravity of 7.421, melts at 176° C. (348.8° F.), and is less volatile than cadmium or zinc. When heated to redness in the air, it burns with a violet flame, and is converted into the yellow sesquioxide. Heated in chlorine, it burns with a yellow-green light, and forms a chloride which sublimes without fusion at an incipient red heat, in soft white laminæ.

Indium dissolves in the strong mineral acids, and is precipitated from the solutions by *zinc* and *cadmium*. *Hydrogen sulphide* precipitates it completely as a yellow sulphide from a solution of its acetate, and from neutral solutions of indium salts in general, but not from solutions containing excess of the strong mineral acids. *Ammonia* and *sodium carbonate* produce white precipitates insoluble in excess: *caustic potash* or *soda* throws down white indium hydroxide soluble in excess of the alkali. *Ammonium carbonate* forms a white precipitate soluble in excess, and reprecipitated on boiling. *Barium carbonate* precipitates indium completely.

The spectrum of indium is characterized by two indigo-colored lines, one very bright and more refrangible than the blue line of strontium, the other fainter and still more refrangible, and approaching the blue line of potassium. It was the observation of this peculiar spectrum that led to the discovery of the metal.

The formulæ of the principal normal indium compounds are as follows:—

Chloride	$InCl_3$
Indium and ammonium chloride,	$2NH_4Cl,InCl_3 + H_2O$
Yellow oxide	In_2O_3
Hydroxide	$In_2H_6O_3$
Nitrate	$In(NO_3)_3$
Sulphate	$In_2(SO_4)_3 + 9H_2O.$

The yellow oxide, heated in a stream of hydrogen, is successively reduced to a green, a gray, and a black oxide, and at a low red heat to the metallic state.

GALLIUM.

Symbol, Ga. Atomic weight, 69.8.

THIS metal was discovered in 1875 by Lecoq de Boisbaudran[*] in a zinc-blende from the mine of Pierrefitte, in the valley of Argeles, Pyrenees, and

[*] Comptes rendus, lxxxi. 493.

has likewise been found, though always in very small quantity, in blendes from other localities. It is separated by dissolving the blende in nitro-muriatic acid, immersing plates of zinc in the solution till the disengage-ment of hydrogen becomes slow—whereby copper, lead, cadmium, and other metals are precipitated—and then boiling the clear liquid for several hours with a large excess of zinc, which throws down alumina, basic zinc-salts, and oxide of gallium. This precipitate is redissolved in hydrochloric acid; the solution again boiled with zinc; the resulting precipitate, which contains the gallium in a more concentrated form, is redissolved in hydrochloric acid; the solution mixed with ammonium acetate is treated with hydrogen sulphide, which throws down the zinc and gallium as sulphides, leaving the aluminium in solution; and this treatment is repeated to ensure the com-plete separation of the alumina. The white sulphides of zinc and gallium are then dissolved in hydrochloric acid; the solution is fractionally precipi-tated with sodium carbonate, the gallium going down chiefly in the first portions; and, to complete the separation of the zinc, the gallium oxide is dissolved in sulphuric acid and reprecipitated by excess of ammonia; this dissolves all the zinc oxide and the greater part of the gallium oxide, which may be reprecipitated by boiling the liquid to expel the free am-monia.

Metallic gallium is obtained by electrolyzing a solution of the oxide in potash or ammonia with platinum electrodes, the gallium being deposited on the negative plate as a compact, closely adhering crust, which may be detached by bending the plate backward and forward under cold water.

Gallium is a hard metal somewhat whiter than platinum, and acquires a good polish by pressure; it is sectile and somewhat malleable; its specific gravity is 5.9, which is intermediate between that of aluminium (2.7) and that of indium (7.4). Its melting point is 30.1° C. (54.2° F.), so that it liquefies when pressed between the fingers; frequently also it remains liquid for a long time even when cooled nearly to 0° C. The melted metal adheres to glass, forming a mirror whiter than that produced by mercury. When heated to bright redness in contact with the air, it oxidizes merely on the surface, and does not volatilize.

Gallium forms a very bright electric spectrum, exhibiting a brilliant line and a fainter band in the violet. In a gas-flame only the line is exhibited, and even this is very faint and fugitive. It was by these spectroscopic cha-racters that the existence of gallium was first recognized.

Gallium chloride is very soluble and deliquescent. Its aqueous solution is clear when highly concentrated, but becomes turbid on addition of water; the precipitate (probably an oxychloride) dissolves but very slowly in hy-drochloric acid. A slightly acidulated solution of the chloride evaporated at a gentle heat deposits needles which act strongly on polarized light. The *sulphate* is not deliquescent.

Gallium forms an *ammonia-alum*, which crystallizes in octahedrons like common alum; it dissolves in cold water, but the solution becomes turbid on boiling, and is decomposed by heating with acetic acid. The existence of this alum shows that the oxide of gallium is represented by the formula Ga_2O_3, and its chloride by $GaCl_3$ or Ga_2Cl_6.

Solutions of gallium salts give with *ammonium sulphide* a white precipitate of gallium sulphide insoluble in excess of the reagent. The same precipitate is formed by *hydrogen sulphide* in presence of acetic, but not of hydrochloric acid. *Ammonia* and *carbonate of ammonia* give white precipitates soluble in excess. Slightly acid solutions of the chloride and sulphate are not precip-itated in the cold by *ammonium acetate*, but the neutral solutions are clouded thereby. Gallium oxide is easily precipitated by *barium carbonate, cupric* *hydroxide*, and *potassium ferrocyanide*, the last-mentioned reagent affording an extremely delicate test of its presence.*

In a mixed solution of gallium and aluminium, the latter is precipitated before the gallium, and in a mixed solution of gallium and indium, the gallium goes down first; in point of basicity, therefore, gallium is intermediate between aluminium and indium.

The existence of a metal having the atomic weight and properties of gallium was predicted, as already observed, by Mendelejeff (p. 259).

* On the separation of gallium from other metals, see Lecoq de Boisbaudran: Comptes rendus, xciv. 1154, 1439, 1625; xcv. 157, 410, 503, 703, 1192, 1332; xcvi. 152; and Journal of the Chemical Society, xlii. 897, 1323; xliv. 21, 153, 156, 293, and 715.

METALS OF THE IRON GROUP.

Manganese.   Cobalt.
Iron.   Nickel.

THESE metals form basic oxides, MO, in which they are bivalent, and corresponding sulphates, which unite with the sulphates of the alkali-metals, forming double sulphates isomorphous with those of the magnesium group (p. 356), e. g., $K_2Fe(SO_4)_2 + 6H_2O$. Manganese and iron also form sesquioxides, which, like alumina, act as weak bases, and their sulphates, like aluminium sulphate, unite with sulphates of alkali-metal, forming alums isomorphous with common alum; e. g., *potassio-ferric sulphate*, $KFe'''(SO_4)_2 + 12H_2O$ or $K_2(Fe_2)^{vi}(SO_4)_4 + 24H_2O$. The sesquioxides of nickel and cobalt on the other hand, act as peroxides, dissolving in acids with evolution of oxygen or chlorine. Manganese and iron likewise form salts which may be supposed to contain the trioxides MO_3, or in which the metal is sexvalent, analogous therefore in composition to the sulphates, e. g., *potassium ferrate*, K_2FeO_4 or K_2O,FeO_3. Lastly, manganese forms a class of salts, the *permanganates*, e. g., $KMnO_4$ or K_2O,Mn_2O_7, analogous in composition to the perchlorates, and isomorphous therewith; in these salts the manganese must be supposed to be septivalent, like chlorine in the perchlorates.

MANGANESE.

Symbol, Mn.   Atomic weight, 54.8.

MANGANESE is tolerably abundant in nature in the state of oxide, occurring chiefly as dioxide or pyrolusite, MnO_2; also as braunite, Mn_2O_3, and hausmannite, Mn_3O_4; and as carbonate in rhodochrosite or manganese spar, $MnCO_3$, which also occurs frequently as an isomorphous constituent in ferrous carbonate and other similar minerals. Manganese likewise enters, though only in small quantity, into the composition of many other minerals, and traces of it, derived from the soil, are often found in the ashes of plants.

Metallic manganese, or perhaps strictly, manganese carbonide, may be prepared by the following process:—The carbonate is calcined in an open vessel, by which it becomes converted into a dense brown powder: this is intimately mixed with a little charcoal and about one-tenth of its weight of anhydrous borax. A charcoal crucible is next prepared by filling a Hessian or Cornish crucible with moist charcoal powder, introduced a little at a time, and rammed as hard as possible. A smooth cavity is then scooped in the centre, into which the above-mentioned mixture is compressed, and covered with charcoal powder. The lid of the crucible is then fixed, and the whole arranged in a very powerful wind-furnace. The heat is slowly raised until the crucible becomes red hot, after which it is urged to its maximum for an hour or more. When cold, the crucible is broken up, and the metallic button of manganese extracted.

Deville has lately prepared pure manganese by reducing pure manganese oxide with an insufficient quantity of sugar-charcoal in a crucible made of caustic lime. Thus prepared, metallic manganese possesses a reddish lustre like bismuth : it is very hard and brittle, and, when powdered, decomposes water, even at the lowest temperature. Dilute sulphuric acid dissolves it with great energy, evolving hydrogen. Brunner produced metallic manganese from manganese and sodium fluoride by means of sodium. The metal obtained by this process scratches glass and hardened steel, and has a specific gravity of 7.13.

Manganese Chlorides.—The *dichloride*, or *Manganous chloride*, may be prepared in a state of purity from the dark-brown liquid residue of the preparation of chlorine from manganese dioxide and hydrochloric acid, which often accumulates in the laboratory to a considerable extent: from the pure chloride, the carbonate and all the other salts can be conveniently obtained. The liquid referred to consists chiefly of the mixed chlorides of manganese and iron : it is filtered, evaporated to perfect dryness, and the residue is slowly heated to dull redness in an earthen vessel, with constant stirring. The iron chloride is thus either volatilized or converted by the remaining water into insoluble sesquioxide, while the manganese salt is unaffected. On treating the grayish-looking powder thus obtained with water, the manganese chloride is dissolved out, and may be separated by filtration from the iron oxide. Should a trace of the latter yet remain, it may be got rid of by boiling the liquid for a few minutes with a little manganese carbonate. The solution of the chloride has usually a delicate pink color, which becomes very manifest when the salt is evaporated to dryness. A strong solution deposits rose-colored tabular crystals, which contain 4 molecules of water ; they are very soluble and deliquescent. The chloride is fusible at a red heat, is decomposed slightly at that temperature by contact with air, and is dissolved by alcohol, with which it forms a crystallizable compound.

The *trichloride*, or *Manganic chloride*, Mn_2Cl_6, is formed when precipitated manganic oxide is immersed in cold concentrated hydrochloric acid, the oxide then dissolving quietly without evolution of gas. Heat decomposes the trichloride into dichloride and free chlorine.

Heptachloride, $MnCl_7$ (?).—When potassium permanganate is dissolved in strong sulphuric acid, and fused sodium chloride is added by small portions at a time, a greenish-yellow gas is given off, which condenses at $0°$ C. to a greenish-brown liquid. This compound, when exposed to moist air, gives off fumes colored purple by permanganic acid, and is instantly decomposed by water into permanganic and hydrochloric acids. It is regarded by Dumas, who discovered it, as the heptachloride of manganese; but H. Rose regarded it as an oxychloride, $MnCl_2O_2$, analogous to chromic oxychloride—a view which is in accordance with its mode of formation.

Fluorides of manganese have been formed analogous to each of these chlorides.

Manganese Oxides and **Oxysalts.**—Manganese forms four well-defined oxides, constituted as follows:—

Monoxide, or Manganous oxide . . . . . . MnO
Trimangano-tetroxide, or Manganoso-manganic oxide . Mn_3O_4
Sesquioxide, or Manganic oxide. . . . . . Mn_2O_3
Dioxide, or Peroxide . . . . MnO_2

The first is a strong base, the third a weak base; the second and fourth are neutral; the second may be regarded as a compound of the first and third, MnO,Mn_2O_3. There are also several oxides intermediate between the monoxide and dioxide, occurring as natural minerals or ores of manganese. Manganese likewise forms two series of oxygen-salts, called *manganates* and *permanganates*, the composition of which has been already noticed (p. 406). The oxides, MnO_3 and Mn_2O_7, corresponding with these salts, are not known.

Monoxide, or *Manganous oxide*, MnO.—When manganese carbonate is heated in a stream of hydrogen gas, or vapor of water, carbon dioxide is disengaged, and a greenish powder left behind, which is the monoxide. Prepared at a dull red heat only, the monoxide is so prone to absorb oxygen from the air that it cannot be removed from the tube without change; but when prepared at a higher temperature it appears more stable. This oxide is a very strong base, isomorphous with magnesia and zinc oxide; it dissolves quietly in dilute acids, neutralizing them completely, and forming salts, which have often a beautiful pink color. When alkalies are added to solutions of these compounds, the white hydrated oxide first precipitated speedily becomes brown by passing into a higher state of oxidation.

Sesquioxide, or *Manganic oxide*, Mn_2O_3.—This compound occurs in nature as *braunite*, and in the state of hydrate as *manganite;* a very beautiful crystallized variety is found at Ilefeld, in the Hartz. It is produced artificially, by exposing the hydrated monoxide to the air, and forms the principal part of the residue left in the iron retort when oxygen gas is prepared by exposing the native dioxide to a moderate red heat. The color of the sesquioxide is brown or black, according to its origin or mode of preparation. It is a weak base, isomorphous with alumina: for when gently heated with diluted sulphuric acid, it dissolves to a red liquid, which, on the addition of potassium or ammonium sulphate, deposits octahedral crystals having a constitution similar to that of common alum: these are, however, decomposed by water. Strong nitric acid resolves this oxide into a mixture of monoxide and dioxide, the former dissolving, and the latter remaining unaltered, while hot oil of vitriol decomposes it by forming manganous sulphate and liberating oxygen gas. On heating it with hydrochloric acid, chlorine is evolved, as with the dioxide, but in smaller amount.

Dioxide, MnO_2; *Peroxide of Manganese;* *Pyrolusite.*—The most common ore of manganese; it is found both massive and crystallized. It may be obtained artificially in the anhydrous state by gently calcining the nitrate, or in combination with water, by adding solution of bleaching powder to a salt of the monoxide. Manganese dioxide has a black color and is insoluble in water. It is decomposed by hot hydrochloric acid and by oil of vitriol in the same manner as the sesquioxide. It unites with the stronger bases, potash, lime, etc., forming salts called m a n g a n i t e s, e. g., CaO,MnO_2 or $CaMnO_3$, which are produced by precipitating a solution of a manganous salt with the corresponding base in presence of an oxidizing agent, such as a stream of air or oxygen gas. Such are the manganites of calcium and magnesium formed in Weldon's process for the recovery of manganese dioxide from waste chlorine-liquors (p. 136). According to Frémy,[*] manganese dioxide likewise acts as a base, forming definite salts with acids, e. g., a sulphate, MnO_2,SO_3.

The proportion of real dioxide contained in a commercial sample of the black oxide may be estimated by determining the quantity of carbon dioxide evolved on gently heating a weighed quantity of oxalic acid, $C_2H_2O_4$, with strong hydrochloric acid in presence of a known weight of the manganese ore. The following reaction then takes place:—

[*] Comptes rendus, lxxxii. 1231; Chem. Soc. Journ., 1877, I. 52.

$$MnO_2 + C_2H_2O_4 + 2HCl = MuCl_2 + 2H_2O + 2CO_2.$$

This equation shows that every two molecules of carbon dioxide evolved correspond with one molecule of manganese dioxide decomposed. Now the molecular weight of this oxide, 87, is so nearly equal to twice that of carbon dioxide, 44, that the loss of weight suffered by the apparatus when the reaction has become complete and the residual gas has been driven off by momentary ebullition, may be taken to represent the quantity of real dioxide in the 50 grains of the sample. Geissler's apparatus (p. 232) may be advantageously used in this process.

Trimangano-tetroxide, or *Red manganese oxide*, Mn_3O_4, or probably MnO,-Mn_2O_3.—This oxide is also found native as *hausmannite*, and is produced artificially by heating the dioxide or sesquioxide to whiteness, or by exposing the monoxide or carbonate to a red heat in an open vessel. It is a reddish-brown substance incapable of forming salts, and acted upon by acids in the same manner as the two other oxides already described. Borax and glass in the fused state dissolve it, and acquire the color of amethyst.

Varvicite, Mn_4O_7,H_2O or $MnO,3MnO_2,H_2O$, is a natural mineral discovered by Phillips among certain specimens of manganese ore from Warwickshire: it has also been found at Ilefeld in the Hartz. It much resembles the dioxide, but is harder and more brilliant. By a strong heat varvicite is converted into red oxide, with disengagement of aqueous vapor and oxygen gas.

Several other oxides, intermediate in composition between the monoxide and dioxide, also occur native; they are probably mere mixtures, and in many cases the monoxide is more or less replaced by the corresponding oxides of iron, cobalt, and copper.

Manganous Sulphate, $MnSO_4,7H_2O$ or $MnO,SO_3,7H_2O$.—A beautiful rose-colored and very soluble salt, isomorphous with magnesium sulphate. It is prepared on the large scale for the use of the dyer by heating in a close vessel manganese dioxide and coal, and dissolving the impure monoxide thus obtained in sulphuric acid, with addition of a little hydrochloric acid toward the end of the process. The solution is evaporated to dryness and again exposed to a red heat, by which ferric sulphate is decomposed. Water then dissolves out the pure manganese sulphate, leaving ferric oxide behind. This salt is used to produce a permanent brown dye, the cloth steeped in the solution being afterward passed through a solution of bleaching powder, by which the monoxide is changed to insoluble hydrate of the dioxide. Manganous sulphate sometimes crystallizes with 5 molecules of water. It forms a double salt with potassium sulphate, containing $MnK_2(SO_4)_2,6H_2O$.

Manganous Carbonate, $MnCO_3$ or MnO,CO_2.—Prepared by precipitating the dichloride with an alkaline carbonate. It is an insoluble white powder, sometimes with a buff-colored tint. Exposed to heat, it loses carbon dioxide and absorbs oxygen.

Manganates.—When an oxide of manganese is fused with potash, oxygen is taken up from the air and a deep green saline mass results, which contains *potassium manganate*, K_2MnO_4 or K_2O,MnO_3. The addition of potassium nitrate or chlorate facilitates the reaction. Water dissolves this compound very readily, and the solution, concentrated by evaporation in a vacuum, yields green crystals. *Barium manganate*, $BaMnO_4$, is formed in a similar manner. In these salts manganese is sexvalent, like chromium in the chromates.

Permanganates.—When potassium manganate, free from any great excess of alkali, is put into a large quantity of water, it is resolved into hydrated manganese dioxide, which subsides, and *potassium permanganate*,

18

$K_2Mn_2O_8$ or K_2O,Mn_2O_7, which remains in solution, forming a deep purple liquid:—

$$3K_2MnO_4 + 2H_2O \rightleftharpoons MnO_2 + K_2Mn_2O_8 + 4KHO.$$

This effect is accelerated by heat. The changes of color accompanying this decomposition are very remarkable, and have procured for the manganate the name *mineral chameleon;* excess of alkali hinders the reaction in some measure by conferring greater stability on the manganate. Potassium permanganate is easily prepared on a considerable scale. Equal parts of very finely-powdered manganese dioxide and potassium chlorate are mixed with rather more than 1 part of potassium hydroxide dissolved in a little water, and the whole is exposed, after evaporation to dryness, to a temperature just short of ignition. The mass is treated with hot water, the insoluble oxide separated by decantation, and the deep purple liquid concentrated by heat until crystals form upon its surface: it is then left to cool. The crystals have a dark purple color, and are not very soluble in cold water. The manganates and permanganates are decomposed by contact with organic matter. The green and red disinfecting agents, known as Condy's fluids, are alkaline manganates and permanganates.

Hydrogen Permanganate, or *Permanganic acid,* $H_2Mn_2O_8$, is obtained by dissolving potassium permanganate in hydrogen sulphate, H_2SO_4, diluted with 1 molecule of water, and distilling the solution at 60°–70°. Permanganic acid then passes over in violet vapors and condenses to a g   -
black liquid, which has a metallic lustre, absorbs moisture greedily from the air, and acts as a most powerful oxidizing agent, instantly setting fire to paper and to alcohol.

Manganous salts are very easily distinguished by reagents. The *fixed caustic alkalies* and *ammonia* give white precipitates, insoluble in excess, quickly becoming brown. The *carbonates of the fixed alkalies* and *carbonate of ammonia* give white precipitates but little subject to change and insoluble in excess of carbonate of ammonia. *Hydrogen sulphide* gives no precipitate, but *ammonium sulphide* throws down insoluble flesh-colored sulphide of manganese, which is very characteristic. *Potassium ferrocyanide* gives a white precipitate.

Manganese is also easily detected by the blowpipe: it gives with borax an amethyst-colored bead in the outer or oxidizing flame, and a colorless bead in the inner flame. Heated upon platinum foil with sodium carbonate, it yields a green mass of sodium manganate.

IRON.

Symbol, Fe (Ferrum).   Atomic weight, 56.

THIS is the most important of all the metals: there are few substances to which it yields in interest when we consider how very intimately the knowledge of its properties and uses is connected with human civilization.

Metallic iron occurs but rarely on the earth, being found only at Canaan in Connecticut, where it forms a vein about two inches thick in mica-slate, and in a few other localities, but it enters into the composition of many *meteorites.* Isolated masses of soft malleable iron also, of large dimensions, lie loose upon the surface of the earth in South America, Greenland, and elsewhere, and are presumed to have had a similar origin: these latter, in common with the iron of the undoubted meteorites, contain nickel. In the state of oxide the presence of iron may be said to be universal: it constitutes a great part of the common coloring-matter of rocks and soils; it is contained in plants, and forms an essential component of the blood of the animal body. It is also very common in the state of bisulphide.

Pure iron may be prepared, according to Mitscherlich, by introducing into a Hessian crucible 4 parts of fine iron wire cut small and 1 part of black iron oxide. This is covered with a mixture of white sand, lime, and potassium carbonate, in the proportions used for glass-making, and a cover being closely applied the crucible is exposed to a very high degree of heat. A button of pure metal is thus obtained, the traces of carbon and silicon present in the wire having been removed by the oxygen of the oxide.

Pure iron has a white color and perfect lustre: it is extremely soft and tough, and has a specific gravity of 7.8. Its crystalline form is probably the cube, to judge from appearances occasionally exhibited. In good bar-iron or wire a distinct fibrous texture may always be observed when the metal has been attacked by rusting or by the application of an acid; and upon the perfection of this fibre much of its strength and value depends. Iron is the most tenacious of all the metals, a wire $\frac{1}{18}$ of an inch in diameter bearing a weight of 60 lbs. It is very difficult of fusion, and before becoming liquid passes through a soft or pasty condition. Pieces of iron, pressed or hammered together in this state, cohere into a single mass: this operation, termed *welding*, is usually performed by sprinkling a little sand over the heated metal, which combines with the superficial film of oxide, forming a fusible silicate, which is subsequently forced out from between the pieces of iron by the pressure applied: clean surfaces of metal are thus presented to each other, and union takes place without difficulty.

Iron does not oxidize in dry air at common temperatures: heated to redness it becomes covered with a scaly coating of black oxide, and at a high white heat burns brilliantly, producing the same substance. In oxygen gas the combustion occurs with still greater ease. The finely-divided spongy metal, prepared by reducing the red oxide with hydrogen gas, takes fire spontaneously in the air. Pure water, free from air and carbonic acid, does not tarnish a surface of polished iron, but the combined agency of free oxygen and moisture speedily leads to the production of rust, which is a hydrate of the sesquioxide. The rusting of iron is wonderfully promoted by the presence of a little acid vapor. At a red heat iron decomposes water, evolving hydrogen and passing into the black oxide. Dilute sulphuric and hydrochloric acids dissolve it freely, with separation of hydrogen. Iron is strongly magnetic up to a red heat, but at that temperature it loses all traces of magnetism.

Iron forms two classes of compounds: namely, the ferrous compounds, in which it is bivalent, *e. g.*, $FeCl_2$, FeO, $FeSO_4$, etc., and the ferric compounds, in which, like aluminium, it may be regarded either as a triad or as a tetrad: ferric chloride, for example, may be either $FeCl_3$ or Fe_2Cl_6 $= Cl_3Fe.FeCl_3$; the vapor density of this compound, as determined by Deville, is in favor of the latter formula.

Chlorides.—The *dichloride* or *Ferrous chloride*, $FeCl_2$, is formed by transmitting dry hydrochloric acid gas over red-hot metallic iron or by dissolving iron in hydrochloric acid. The latter solution yields, when duly concentrated, green crystals of the hydrated dichloride, $FeCl_2,4H_2O$; they are very soluble and deliquescent, and oxidize rapidly in the air.

The *trichloride* or *Ferric chloride*, $FeCl_6$, is usually prepared by dissolving ferric oxide in hydrochloric acid. The solution, evaporated to a syrupy consistence, deposits red hydrated crystals, which are very soluble in water and alcohol. It forms double salts with potassium chloride and sal-ammoniac. When evaporated to dryness and strongly heated, much of the chloride is decomposed, yielding sesquioxide and hydrochloric acid: the remainder sublimes, and afterward condenses in the form of small brilliant red crystals, which deliquesce rapidly. Anhydrous ferric chloride is also produced by the action of chlorine upon the heated metal. The solution of ferric chloride is capable of dissolving a large excess of recently precipitated ferric hydroxide, by which it acquires a much darker color.

Iodides.—*Ferrous iodide*, FeI_2, is an important medicinal preparation: it is easily made by digesting iodine with water and metallic iron. The solution is pale-green, and yields, on evaporation, crystals resembling those of the chloride, which rapidly oxidize on exposure to air. It is best preserved in solution in contact with excess of iron.—*Ferric iodide*, Fe_2I_6, is yellowish-red and soluble.

Iron Oxides and Oxysalts.—Three oxides of iron are known, namely, ferrous oxide, FeO, and ferric oxide, Fe_2O_3, analogous to the chlorides, and an intermediate oxide, usually called magnetic iron oxide, containing Fe_3O_4 or FeO,Fe_2O_3. A trioxide, FeO_3, may be supposed to exist in the ferrates, but it has not been isolated.

Monoxide, or *Ferrous oxide*, FeO.—This is a very powerful base, neutralizing acids, and isomorphous with magnesia, zinc, oxide, etc. It is almost unknown in the separate state from its extreme proneness to absorb oxygen and pass into the sesquioxide. When a ferrous salt is mixed with caustic alkali or ammonia, a bulky whitish precipitate of ferrous hydroxide falls, which becomes nearly black when boiled, the water being separated. This hydroxide changes very rapidly when exposed to the air, becoming green and ultimately red-brown. The soluble ferrous salts have commonly a delicate pale-green color and a nauseous metallic taste.

Sesquioxide, or *Ferric oxide*, Fe_2O_3.—A weak base, isomorphous with alumina. It occurs native, most beautifully crystallized, as specular iron ore, in the island of Elba and elsewhere; also as red and brown *hæmatite*, the latter being a hydrate. It is artificially prepared by precipitating a solution of ferric sulphate or chloride with excess of ammonia, and washing, drying, and igniting the yellowish-brown hydrate thus produced: fixed alkali must not be used in this operation, as a portion is retained by the oxide. In fine powder this oxide has a full red color, and is used as a pigment, being prepared for the purpose by calcination of ferrous sulphate; the tint varies somewhat with the temperature to which it has been exposed. The oxide is unaltered in the fire, although easily reduced at a high temperature by carbon or hydrogen. It dissolves in acids with difficulty after strong ignition, forming a series of reddish salts, which have an acid reaction and an astringent taste. Ferric oxide is not acted upon by the magnet.

Triferro-tetroxide, *Ferroso-ferric oxide*, $Fe_3O_4 = FeO,Fe_2O_3$, also called *black iron oxide*, *magnetic oxide*, and *loadstone*, occurs as a natural product, one of the most valuable of the iron-ores, and is often found in regular octahedral crystals, which are magnetic. It may be prepared by mixing due proportions of ferrous and ferric salts, precipitating them with excess of alkali, and then boiling the mixed hydroxides; the latter then unite to a black sandy substance consisting of minute crystals of the magnetic oxide. This oxide is the chief product of the oxidation of iron at a high temperature in the air and in aqueous vapor.

Ferrates.—When a mixture of one part of pure ferric oxide and four parts of dry nitre is heated to full redness for an hour in a covered crucible, and the resulting brown, porous, deliquescent mass is treated when cold with ice-cold water, a deep amethystine-red solution of potassium ferrate is obtained. The same salt may be more easily prepared by passing chlorine gas through a strong solution of potash in which recently precipitated ferric hydroxide is suspended; it is then deposited as a black powder, which may be drained upon a tile. It consists of K_2FeO_4 or K_2O,FeO_3, and is therefore analogous in composition to the sulphate and chromate of potassium. The solution of this salt gradually decomposes, even in the cold, and rapidly when heated, giving off oxygen and depositing sesquioxide. The solution of potassium ferrate gives no precipitate with salts of calcium, magnesium, or strontium, but when mixed with a barium salt it yields a deep crimson, insoluble *barium ferrate*, $BaFeO_4$ or BaO,FeO_3, which is very permanent. Neither the hydrogen-salt nor ferric acid, H_2FeO_4, nor the corresponding anhydrous oxide, FeO_3, is known in the separate state.

Ferrous Sulphate, $FeSO_4,7H_2O$ or $FeO,SO_3,7H_2O$.—This beautiful and important salt, commonly called *green vitriol*, *iron vitriol*, or *copperas*, may be obtained by dissolving iron in dilute sulphuric acid: it is generally prepared, however, and on a very large scale, by contact of air and moisture with common iron pyrites, which, by absorption of oxygen, readily furnishes the substance in question. Heaps of this material are exposed to the air until the decomposition is sufficiently advanced: the salt produced is then dissolved out by water, and the solution is left to crystallize. It forms large green crystals of the composition above stated, which slowly effloresce and oxidize in the air: it is soluble in about twice its weight of cold water. Crystals containing 4 and also 2 molecules of water have been obtained. Ferrous sulphate forms double salts with the sulphates of potassium and ammonium, containing $FeK_2(SO_4)_2,6H_2O$ and $Fe(NH_4)_2(SO_4)_2,6H_2O$, isomorphous with the corresponding magnesium salts.

Ferric Sulphate, $Fe_2(SO_4)_3$ or $Fe_2O_3,3SO_3$, is prepared by adding to a solution of the ferrous salt exactly one-half as much sulphuric acid as it already contains, raising the liquid to the boiling point, and then dropping in nitric acid until the solution ceases to blacken by such addition. The red liquid thus obtained furnishes on evaporation to dryness a buff-colored amorphous mass, which dissolves very slowly when put into water. With the sulphates of potassium and ammonium, this salt yields compounds having the form and constitution of alums; the potassium salt, for example, has the composition $Fe'''K(SO_4)_2,12H_2O$. The crystals are nearly colorless; they are decomposed by water, and sometimes by long keeping in the dry state. These iron-alums are best prepared by exposing to spontaneous evaporation a solution of ferric sulphate to which potassium or ammonium sulphate has been added.

Ferrous Nitrate, $Fe(NO_3)_2$.—When dilute cold nitric acid is made to act to saturation upon iron monosulphide, and the solution is evaporated in a vacuum, pale-green and very soluble crystals of ferrous nitrate are obtained, which are very subject to alteration. *Ferric nitrate* is readily formed by pouring nitric acid, slightly diluted, upon iron: it is a deep-red liquid, apt to deposit an insoluble basic salt, and is used in dyeing.

Ferrous Carbonate, $FeCO_3$ or FeO,CO_2, is obtained as a whitish precipitate on mixing solutions of ferrous salt and alkaline carbonate: it cannot be washed and dried without losing carbonic acid and absorbing oxygen. This compound occurs in nature as *spathose iron ore* or *iron spar*, associated with variable quantities of calcium and magnesium carbonates; also in the common *clay ironstone*, from which nearly all the British iron is made. It is often found in mineral waters, being soluble in excess of carbonic acid:

35 *

such waters are known by the rusty matter they deposit on exposure to the air. No ferric carbonate is known.

The *phosphates* of iron are all insoluble.

Iron Sulphides.—Several compounds of iron and sulphur are known: of these the two most important are the following: The *monosulphide* or *ferrous sulphide*, FeS, is a blackish brittle substance, attracted by the magnet, formed by heating together iron and sulphur. It is dissolved by dilute acids, with evolution of sulphuretted hydrogen gas, and is constantly employed for that purpose in the laboratory, being made by projecting into a red-hot crucible a mixture of $2\frac{1}{2}$ parts of sulphur and 4 parts of iron filings or borings of cast iron, and excluding the air as much as possible. The same substance is formed when a bar of white-hot iron is brought in contact with sulphur. The *bisulphide*, FeS$_2$, or iron pyrites, is a natural product, occurring in rocks of all ages, and evidently formed in many cases by the gradual deoxidation of ferrous sulphate by organic matter. It has a brass-yellow color, is very hard, not attracted by the magnet, and not acted upon by dilute acids. When it is exposed to heat, sulphur is expelled, and an intermediate sulphide, Fe$_3$S$_4$, analogous to the black oxide, is produced. This latter substance also occurs native, under the name of *magnetic pyrites*. Iron pyrites is the material now chiefly employed for the manufacture of sulphuric acid; for this purpose the mineral is roasted in a current of air, and the sulphurous acid formed is passed into the lead chambers; the residue consists of iron oxide, frequently containing a quantity of copper large enough to render the extraction of the metal remunerative.

Compounds of iron with *phosphorus, carbon*, and *silicon* exist, but little is known respecting them in a definite state. The carbonide is contained in cast iron and in steel, to which it communicates ready fusibility; the silicon-compound is also found in cast iron. Phosphorus is a very hurtful substance in bar iron, rendering it brittle or *cold-short*:

REACTIONS OF IRON SALTS.—**Ferrous salts** are thus distinguished:

Caustic alkalies and *ammonia* give nearly white precipitates, insoluble in excess of the reagent, rapidly becoming green, and ultimately brown, by exposure to the air. The *carbonates* of *potassium, sodium*, and *ammonium* throw down whitish ferrous carbonate, also very subject to change. *Hydrogen sulphide* gives no precipitate, but *ammonium sulphide* throws down black ferrous sulphide, soluble in dilute acids. *Potassium ferrocyanide* gives a nearly white precipitate, becoming deep-blue on exposure to air; the *ferricyanide* gives at once a deep-blue precipitate.

Ferric salts are thus characterized:

Caustic fixed alkalies and *ammonia* give foxy-red precipitates of ferric hydrate, insoluble in excess. The *carbonates* behave in a similar manner, the carbonic acid escaping. *Hydrogen sulphide* gives a nearly white precipitate of sulphur, and reduces the sesquioxide to monoxide. *Ammonium sulphide* gives a black precipitate, slightly soluble in excess. *Potassium ferrocyanide* yields Prussian blue. Tincture or infusion of *gall-nuts* strikes a deep bluish-black with the most dilute solutions of ferric salts.

Iron Manufacture.—This most important branch of industry consists, as now conducted, of two distinct parts—viz., the production from the ore of a fusible carbonide of iron, and the subsequent decomposition of the carbonide, and its conversion into pure or malleable iron.

Clay iron ore is found in association with coal, forming thin beds or nodules: it consists, as already mentioned, of ferrous carbonate mixed with clay; sometimes lime and magnesia are also present. It is broken in pieces, and exposed to heat in a furnace resembling a lime-kiln, by which the water and carbonic acid are expelled, and the ore is rendered dark-colored, denser, and magnetic: it is then ready for reduction. The furnace in which this operation is performed is usually of very large dimensions, 50 feet or more in height, and constructed of very solid brickwork, the interior being lined with excellent fire-bricks. A general idea of the shape may be gained from the section shown in Fig. 150. In modern blast-furnaces, however, the sides of the shaft are nearly vertical, and the lower part, called the *hearth*, is proportionally wider than in the figure. The furnace is closed at the bottom, the fire being maintained by a powerful artificial blast introduced by two or three *twyere-pipes*. The materials, consisting of due proportions of coke or coal, roasted ore, and limestone, are constantly supplied from the top, the operation proceeding continuously night and day, often for years, or until the furnace is judged to require repair. In the upper part or shaft of the furnace, where the temperature is still very high, and where combustible gases abound, the iron of the ore is probably reduced to the metallic state, being disseminated through the earthy matter of the ore. As the whole sinks down and attains a still higher degree of heat, the iron becomes converted into carbonide by *cementation*, while

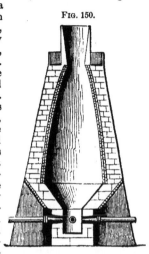

FIG. 150.

the silica and alumina unite with the lime, purposely added, forming a kind of glass or *slag*, nearly free from iron oxide. The carbonide and slag, both in the melted state, reach at last the bottom or hearth of the furnace, where they arrange themselves in the order of their densities. The slag flows out at certain apertures contrived for the purpose, and the iron is discharged from time to time, and suffered to run into rude moulds of sand by opening an orifice at the bottom of the recipient, previously stopped with clay. Such is the origin of *crude, cast*, or *pig iron*, of which there are several varieties, distinguished by differences of color, hardness, and composition, and known by the names of *gray, black*, and *white* iron. The first is for most purposes the best, as it admits of being filed and cut with perfect ease. The black and gray kinds probably contain a mechanical admixture of graphite, which separates during solidification.

A great improvement in the original mode of conducting the process was the substitution of raw coal for coke, and the blowing of hot air instead of cold into the furnace. This is effected by causing the air, on leaving the blowing-machine, to circulate through a system of red-hot iron pipes, until its temperature becomes high enough to melt lead. This alteration effects a prodigious saving in fuel, without injury to the quality of the product.

The conversion of cast into bar iron is effected chiefly by an operation called *puddling*, previous to which, however, it sometimes undergoes a process called *refining*, which consists in remelting it, in contact with the fuel, in small low furnaces called *refineries*, while air is blown over its surface by means of twyeres. The effect of this operation is to deprive the iron of a great part of the carbon and silicon associated with it. The metal thus purified is run out into a trench, and suddenly cooled, by which it becomes white, crystalline, and exceedingly hard: in this state it is called *fine metal*. The *puddling* process is conducted in a reverberatory furnace, into which the charge of crude or of fine metal is introduced by a side aperture. This is speedily melted by the flame, and its surface covered with a crust or oxide,

The workman then, by the aid of an iron tool, diligently stirs the melted mass, so as intimately to mix the oxide with the metal; he now and then also throws in a little water, with the view of promoting more rapid oxidation. Small jets of blue flame soon appear upon the surface of the iron, and the latter, after a time, begins to lose its fluidity, and acquires, in succession, a pasty and a granular condition. At this point the fire is strongly urged, the sandy particles once more cohere, and the contents of the furnace now admit of being formed into several large balls or masses, which are then with-drawn, and placed under an immense hammer, moved by machinery, by which each becomes quickly fashioned into a rude bar. This is reheated, and passed between grooved cast-iron rollers, and drawn out into a long bar or rod. To make the best iron, the bar is cut into a number of pieces, which are afterwards piled or bound together, again raised to a welding heat, and hammered or rolled into a single bar; and this process of *piling* or *fagoting* is sometimes twice or thrice repeated, the iron becoming greatly improved thereby.

The general nature of the change in the puddling furnace is not difficult to explain. Cast iron consists essentially of iron in combination with carbon and silicon, and these compounds, when strongly heated with iron oxide, undergo decomposition, the carbon and silicon becoming oxidized at the expense of the oxygen of the oxide. As this change takes place, the metal gradually loses its fusibility, but retains a certain degree of adhesiveness, so that when at last it comes under the tilt-hammer or between the rollers, the particles of iron become agglutinated into a solid mass, while the readily fusible silicate of the oxide is squeezed out and separated.

All these processes are, in Great Britain, performed with coal or coke; but the iron obtained is, in many respects, inferior to that made in Sweden and Russia from the magnetic oxide by the use of wood charcoal—a fuel too dear to be extensively employed in England. Plate iron is, however, sometimes made with charcoal.

A method of producing malleable iron directly from the ore has been invented by C. W. Siemens.* The furnace consists of a rotatory iron cylinder, which, by means of wheel-gearing, may be made to revolve either four or five times or from 60 to 80 times an hour. The ore to be smelted is broken into fragments not exceeding the size of peas or beans; and to it is added lime or other fluxing material, in such proportion that the gangue contained in the ore and flux combines with only a little ferrous oxide into basic and fluid slag. If the ore is hæmatite or contains silica, it is best to add alumina in the shape of aluminous iron ore; manganiferous iron ore may also be added with advantage. A charge of about 20 cwts. of ore is put into the furnace when fully heated, while it is revolving slowly. In about forty minutes this charge of ore and fluxing material will have been heated to bright redness, and at this time from 5 to 6 cwts. of small coal of uniform size (not larger than nuts) is added to the charge, whilst the rotative velocity is increased in order to accelerate the mixture of coal and ore. A quick reaction follows: the ferric oxide being reduced to magnetic oxide begins to fuse, and at the same time metallic iron is precipitated by each piece of carbon, while the fluxing materials form a fluid slag with the silicious gangue of the ore. The slow rotative action is then again resorted to, whereby the mass is turned over and over, presenting continually new surfaces to the heated lining and to the flame within the rotator.

When the reduction of the iron ore is nearly completed, the rotator is stopped in the proper position for tapping off the fluid cinder; after this the quick speed is imparted to the rotator, whereby the loose masses of iron contained in it are rapidly collected into two or three metallic balls. These are

* Chem. Soc. Journ., 1874, p. 671; Watts's Dictionary of Chemistry, 2d Suppl., p. 701.

taken out and shingled in the usual way of consolidating pudd.ed balls; the furnace is tapped again, and is ready to receive another charge of ore.

S t e e l is prepared by heating iron in contact with charcoal. Bars of Swedish iron are imbedded in charcoal powder, contained in a large rectangular crucible or chest of some substance capable of resisting the fire, and exposed for many hours to a full red heat. The iron takes up, under these circumstances, from 1.3 to 1.7 per cent. of carbon, becoming harder, and at the same time fusible, but with a certain diminution of malleability. The active agent in this cementation process is probably carbon monoxide: the oxygen of the air in the crucible unites with the carbon to form that compound, which is afterwards decomposed by the heated iron, one-half of its carbon being abstracted by the latter. The carbon dioxide thus formed takes up an additional dose of carbon from the charcoal, and again becomes monoxide, the oxygen, or rather the carbon dioxide, acting as a carrier between the charcoal and the metal. The product of this operation is called *blistered* steel, from the blistered and rough appearance of the bars: the texture is afterwards improved and equalized by welding a number of these bars together, and drawing the whole out under a light tilt-hammer.

Some chemists have recently asserted that nitrogen is necessary for the production of steel, and have, in fact, attributed to its presence the peculiar properties of this material; others, again, have disputed this assertion, and believe that the transformation of iron into steel depends upon the assimilation of carbon only; experimentally, the question remains undecided.

Excellent steel is obtained by fusing gray cast iron with tungstic oxide; the carbon of the iron reduces the tungstic oxide to tungsten, which forms with the iron an alloy possessing the properties of steel. The quantity of tungsten thus absorbed by the iron is very small, and some chemists attribute the properties of the so-called tungsten steel to the general treatment rather than to the presence of tungsten.

The most perfect kind of steel is that which has undergone fusion, having been cast into ingot-moulds, and afterwards hammered: of this all fine cutting instruments are made. It is difficult to forge, requiring great skill and care on the part of the operator.

Steel may also be made directly from some particular varieties of cast iron, as that from spathose iron ore containing a little manganese. The metal is retained, in a melted state, on the hearth of a furnace, while a stream of air plays upon it and causes partial oxidation: the oxide produced reacts, as before stated, on the carbon of the iron, and withdraws a portion of that element. When a proper degree of stiffness or pastiness is observed in the residual metal, it is withdrawn, and hammered or rolled into bars. The *wootz*, or native steel of India, is probably made in this manner. Annealed cast iron, sometimes called *run steel*, is now much employed as a substitute for the more costly products of the forge: the articles when cast, are imbedded in powdered iron ore, or some earthy material, and, after being exposed to a moderate red heat for some time, are allowed to cool slowly, by which a very great degree of softness and malleability is attained. It is possible that some little decarbonization may take place during this process.

Cast steel may also be made in Siemens's rotatory furnace above described, the balls being transferred from the rotator to the bath of a steel-melting furnace in their heated condition, and without subjecting them to previous consolidation under a hammer or shingling machine. It is possible, however, to push the operation within the rotator to the point of obtaining cast steel. For this purpose the relative amount of carbonaceous matter is somewhat increased in the first instance, so that the ball, if shingled, would be

18 *

of the nature of puddled steel, or even contain some carbon mechanically mixed.

Bessemer steel is produced by forcing atmospheric air into melted cast iron. The carbon, being oxidized more readily than the iron, is converted into carbon monoxide, which escapes in a sufficiently heated state to take fire on coming in contact with the air. Considerable heat is generated by the oxidation of the carbon and iron, so that the temperature is kept above the melting point of steel during the whole of the operation. When the decarburation has been carried far enough, the current of air is stopped, and a small quantity of white pig-iron, containing a large amount of manganese, is dropped into the liquid metal. This serves to facilitate the separation of any gas retained within the melted metal, which, after a few minutes' rest, is run into ingot moulds.

Another method of steel-making is that known as the Siemens-Martin process, which consists in dissolving scrap-iron in molten pig-iron heated in a reverberatory furnace.*

The most remarkable property of steel is that of becoming exceedingly hard when quickly cooled. When heated to redness, and suddenly quenched in cold water, steel, in fact, becomes capable of scratching glass with facility: if reheated to redness, and once more left to cool slowly, it again becomes nearly as soft as ordinary iron; and between these two conditions any required degree of hardness may be attained. The articles, forged into shape, are first hardened in the manner described; they are then *tempered*, or *let down*, by exposure to a proper degree of annealing heat, which is often judged of by the color of the thin film of oxide which appears on the polished surface. Thus, a temperature of about 221° C. (430° F.), indicated by a faint straw color, gives the proper temper for razors; that for scissors, penknives, etc. is comprised between 243° and 254° C. (470–490° F.), and is indicated by a full yellow or brown tint. Swords and watch-springs require to be softer and more elastic, and must be heated to 288° or 293° C. (550–560° F.), or until the surface becomes deep blue. Attention to these colors has now become of less importance, as metal baths are often substituted for the open fire in this operation.

COBALT.

Symbol, Co. Atomic weight, 58.8.

THIS metal occurs in combination with sulphur in linnæite, Co_3S_4; with arsenic in skutterudite, $CoAs_3$; with sulphur and arsenic in speiss-cobalt, $(Co,Ni,Fe)As_2$, and cobalt-glance, $(Co,Fe)(AsS)_2$; as arsenate in erythrin or cobalt-bloom, $Co_3(AsO_4)_2,8H_2O$; and associated with manganese in earthy cobalt or wad,

$$(Co,Mn)O,2MnO_2,4H_2O.$$

It is an almost invariable constituent of meteoric iron, and has been detected spectroscopically in the atmosphere of the sun.

Cobalt-compounds may be prepared from speiss or any arsenical cobalt-ore

* See Journal of the Chemical Society, 1868, pp. 278–310; Roscoe and Schorlemmer's Treatise on Chemistry, vol. 2, part ii. p. 77; Watts's Dictionary of Chemistry, vol. vi. p. 736.

by the following process:—The mineral is broken into small fragments, mixed with from one-fourth to half its weight of iron filings, and the whole is dissolved in nitro-muriatic acid. The solution is gently evaporated to dryness, the residue treated with boiling water, and the insoluble iron arsenate removed by a filter. The liquid is then acidulated with hydrochloric acid treated with hydrogen sulphide in excess, which precipitates the copper, and after filtration boiled with a little nitric acid to bring back the iron to the state of sesquioxide. To the cold and largely diluted liquid solution acid sodium carbonate is gradually added, by which the ferric oxide may be completely separated without loss of nickel-salt. Lastly, the filtered solution, boiled with sodium carbonate in excess, yields an abundant precipitate of cobalt carbonate more or less mixed with nickel carbonate.

The separation of cobalt from nickel is by no means easy, but is best effected by the following process devised by H. Rose. The mixed carbonates having been dissolved in excess of hydrochloric acid, the solution, largely diluted with water, is supersaturated with chlorine gas, whereby the cobalt monoxide is converted into sesquioxide, while the nickel monoxide remains unaltered. The liquid is next mixed with excess of recently precipitated barium carbonate left at rest for twelve to eighteen hours, and shaken up from time to time. The whole of the cobalt is thereby thrown down as sesquioxide, while the nickel remains in solution. The precipitate, consisting of cobalt sesquioxide mixed with barium carbonate, is then boiled with hydrochloric acid in order to reduce the cobalt sesquioxide to monoxide, and dissolve it as chloride together with the barium. The latter metal is precipitated by sulphuric acid, and from the filtered liquid the cobalt may be precipitated as hydroxide by potash. A solution of cobalt free from the nickel may also be obtained by precipitating the mixed solution with oxalic acid: the whole of the nickel is thereby precipitated together with a small portion of the cobalt, leaving pure cobalt in solution.[*]

Cobalt is a white, brittle, very tenacious metal, having a specific gravity of 8.5, and a very high melting point. It is unchanged in the air, and but feebly attacked by dilute hydrochloric and sulphuric acids. It is strongly magnetic.

Cobalt forms two classes of salts, analogous in composition to the ferrous and ferric salts; but the cobaltic salts, in which the metal is apparently trivalent, are very unstable.

Chlorides.—The *dichloride* or *Cobaltous chloride*, $CoCl_2$, is easily prepared by dissolving the oxide in hydrochloric acid; or it may be prepared directly from *cobalt-glance*, the native arsenide, by the process above described. It forms a deep rose-red solution, which, when sufficiently strong, deposits hydrated crystals of the same color; when the liquid is evaporated by heat to a very small bulk it deposits anhydrous crystals, which are blue: these latter by contact with water again dissolve to a red liquid. A dilute solution of cobalt chloride constitutes the well-known *blue sympathetic ink:* characters written on paper with this liquid are invisible from their paleness of color, until the salt has been rendered anhydrous by exposure to heat, when the letters appear blue. On laying it aside moisture is absorbed, and the writing once more disappears. Green sympathetic ink is a mixture of the chlorides of cobalt and nickel.

The *trichloride*, or *Cobaltic chloride*, Co_2Cl_6, is obtained in solution by dissolving the sesquioxide in hydrochloric acid, and in small quantity by saturating a solution of the dichloride with chlorine gas. The liquid has a dark-brown color, but easily decomposes, giving off chlorine and leaving the rose-colored dichloride.

[*] For other methods of separating cobalt and nickel, see Gmelin's Handbook, English edition, v. 355–360, and Watts's Dictionary of Chemistry, i. 1046.

Oxides and **Oxysalts.**—Cobalt forms a monoxide and a sesquioxide, also two or three oxides of intermediate composition, but not very well defined. The *monoxide*, or *Cobaltous oxide*, CoO, is a gray powder, very soluble in acids, and is a strong base, isomorphous with magnesia, affording salts of a fine red tint. It is prepared by precipitating cobaltous sulphate or chloride with sodium carbonate, and washing, drying, and igniting the precipitate. When the cobalt solution is mixed with caustic potash, a beautiful blue precipitate falls, which, when heated, becomes violet, and at length dirty red, from absorption of oxygen and a change in the state of hydration.

The *sesquioxide*, or *Cobaltic oxide*, Co_2O_3, is a black, insoluble, neutral powder, obtained by mixing solutions of cobalt and chloride of lime. It dissolves in acids, yielding the cobaltic salts.

Cobaltoso-cobaltic oxide, Co_3O_4, analogous to the magnetic oxide of iron, is formed when cobaltous nitrate or oxalate, or hydrated cobaltic oxide, is heated in contact with the air. According to Frémy, it is a salifiable base.

Another oxide, of acid character, is said to be obtained in the form of a potassium salt by fusing the monoxide or sesquioxide with potassium hydroxide. A crystalline salt is thus formed, consisting, according to Schwarzenberg, of $K_2O,3Co_3O_4,3H_2O$.

Cobaltous Salts.—The *sulphate*, $CoSO_4,7H_2O$, forms red crystals, requiring for solution 24 parts of cold water: they are identical in form with those of magnesium sulphate. It combines with the sulphates of potassium and ammonium, forming double salts containing 6 molecules of water.

A solution of oxalic acid added to cobaltous sulphate occasions, after some time, the separation of nearly the whole of the base in the state of oxalate.

The *carbonate* is thrown down by the alkaline carbonates from solutions of cobalt as a pale peach-blossom-colored precipitate of combined carbonate and hydroxide, containing $2CoCO_3,3CoH_2O_2 + H_2O$.

Cobaltic Salts.—Cobaltic oxide is a weak base dissolving in cold acids, and forming brown solutions which are easily decomposed, the oxysalts evolving oxygen and the chloride evolving chlorine. The most stable of these salts is the *acetate*, the solution of which gives brown precipitates with alkalies and sodium phosphate, and a black precipitate with ammonium sulphide.

Greater stability is exhibited by certain double cobaltic salts, the most important of which is—

Potassio-cobaltic Nitrite, $K_6Co_2(NO_2)_{12},3H_2O$, which is obtained as a yellow precipitate on adding potassium nitrite to the solution of a cobaltous salt, the chloride, for example, acidulated with acetic acid, the reaction being represented by the equation—

$$2CoCl_2 + 10KNO_2 + 4HNO_2 = K_6Co_2(NO_2)_{12} + 4KCl + N_2O_2 + 2H_2O.$$

This compound, called *Cobalt-yellow*, is a bright yellow powder composed of microscopic pyramids or stellate forms. It is usually anhydrous, but may be obtained, according to the strength of the solution, with 1 to 4 molecules of water, its color then varying from bright yellow to a dark greenish-yellow. It is decomposed by nitric and hydrochloric acids with aid of heat, slowly by caustic potash, but quickly by caustic soda or baryta at a gentle heat, yielding a precipitate of brown cobaltic hydroxide. When suspended in water it is slowly attacked by hydrogen sulphide, quickly by ammonium sulphide, with formation of black cobalt sulphide. Corresponding double salts are also known, containing sodium, ammonium, and thallium.

Ammoniacal Cobalt Compounds.—Cobaltous salts, treated with ammonia in a vessel protected from the air, unite with the ammonia, forming compounds which may be called ammonio-cobaltous salts. Most of them contain 6 molecules of ammonia to 1 molecule of the cobalt salt; thus the chloride contains $CoCl_2,6NH_3 + H_2O$; the nitrate, $Co(NO_3)_2,6NH_3 + 2H_2O$. They are generally crystallizable and of a rose color, soluble without decomposition in ammonia, but decomposed by water with formation of a basic salt. H. Rose, by treating dry cobalt chloride with ammonia gas, obtained the compound $CoCl_2,4NH_3$; and in like manner an ammonio-sulphate has been formed containing $CoSO_4,6NH_3$.

When an ammoniacal solution of cobalt is exposed to the air oxygen is absorbed, the liquid turns brown and new salts are formed, containing a higher oxide of cobalt (either Co_2O_3 or CoO_2), and therefore designated generally as peroxidized ammonio-cobalt salts. Several of them, containing different bases, are often formed at the same time. Most of the peroxidized ammonio-cobalt salts are composed of cobaltic salts united with two or more molecules of ammonia. The composition of the normal salts may be illustrated by the chlorides, as in the following table:—

Tetrammonio-cobaltic chloride	$Co_2Cl_6(NH_3)_4$
Hexammonio-cobaltic chloride	$Co_2Cl_6(NH_3)_1$
Octammonio-cobaltic (or praseo-cobaltic) chloride . .	$Co_2Cl_6(NH_3)_1$
Decammonio-cobaltic (roseo- and purpureo-cobaltic) chloride	$Co_2Cl_6(NH_3)_{12}$
Dodecammonio-cobaltic (or luteo-cobaltic) chloride . .	$Co_2Cl_6(NH_3)_{15}$

The formulæ of the corresponding normal nitrates are deduced from the preceding by substituting NO_3 for Cl; those of the sulphates, oxalates, and other bibasic salts by substituting SO_4, C_2O_4, etc., for Cl_2, e. g., decammonio-cobaltic sulphate = $Co_2(SO_4)_3(NH_3)_{10}$. There are also acid and basic salts of the same ammonia-molecules, such as the oxyoctammonio-cobaltic or fusco-cobaltic salts, which may be regarded as basic prasco-cobaltic salts, e. g., the hydroxynitrate, $Co_2(NO_3)_4(OH)_2(NH_3)_8$. Further, there are salts containing the radicles NO and NO_2 in addition to ammonia, e. g., decammonio-nitroso-cobaltic or xantho-cobaltic chloride, $Co_2Cl_4(NO_2)_2(NH_3)_2$, which may be regarded as roseo- or purpureo-cobaltic salts, in which one-third of the chlorine or other acid radicle is replaced by NO_2. Lastly, Frémy has obtained ammoniacal compounds (oxycobaltic salts) containing salts of cobalt corresponding with the dioxide.*

Cobaltous salts have the following characters:—
Solution of potash gives a blue precipitate, changing by heat to violet and red. Ammonia gives a blue precipitate, soluble with difficulty in excess with brownish-red color. Sodium carbonate forms a pink precipitate. Ammonium carbonate, a similar compound, soluble in excess. Potassium ferrocyanide gives a grayish-green precipitate. Potassium cyanide forms a yellowish-brown precipitate, which dissolves in an excess of the precipitant. The clear solution, after boiling, may be mixed with hydrochloric acid without giving a precipitate. Hydrogen sulphide produces no change, if the cobalt is combined with a strong acid. Ammonium sulphide throws down black sulphide of cobalt, insoluble in dilute hydrochloric acid.

Cobaltic salts, formed by dissolving cobaltic oxide in acids, give, with potash, a dark-brown precipitate of hydrated cobaltic oxide; with ammonia, a brownish-red solution; with the fixed alkaline carbonates, a green solution, which deposits a small quantity of cobaltic oxide; with ammonium sulphide (after saturation of the free acid by ammonia), a black precipitate.

* For the preparation and properties of all these salts, see Watts's Dictionary of Chemistry, vol. i. 1051, first Supplement, p. 479, and second Supplement, p. 363. Their rational formulæ are similar to those of the ammoniacal platinum salts (q. v.).

Oxide of cobalt is remarkable for the magnificent blue color it communicates to glass: indeed, this is a character by which its presence may be most easily detected, a very small portion of the substance to be examined being fused with borax on a loop of platinum wire before the blowpipe; the production of this color both in the inner and in the outer flame distinguishes cobalt from all other metals.

The substance called *smalt*, used as a pigment, consists of glass colored by cobalt: it is thus made:—The cobalt ore is roasted until nearly free from arsenic, and then fused with a mixture of potassium carbonate and quartz-sand free from oxide of iron. Any nickel that may happen to be contained in the ore then subsides to the bottom of the crucible as arsenide: this is the *speiss* already mentioned. The glass, when complete, is removed and poured into cold water; it is afterward ground to powder and elutriated. *Cobalt ultramarine* is a fine blue color, prepared by mixing 16 parts of freshly-precipitated alumina with 2 parts of cobalt phosphate or arsenate: this mixture is dried and slowly heated to redness. By daylight the color is pure blue, but by artificial light it is violet. A similar compound, of a fine green color, is formed by igniting zinc oxide with cobalt salts. *Zaffre* is the roasted cobalt ore mixed with silicious sand, and reduced to fine powder: it is used in enamel painting. A mixture in due proportions of the oxides of cobalt, manganese, and iron is used for giving a fine black color to glass.

NICKEL.

Symbol, Ni. Atomic weight, 58.8.

THIS metal, as already observed, is a constant constituent of meteoric iron. It is found in tolerable abundance in some of the metal-bearing veins of the Saxon mountains, in Westphalia, Hesse, Hungary, and Sweden, chiefly as arsenide, the *kupfernickel* of mineralogists, so called from its yellowish-red color. The word *nickel* is a term of detraction, having been applied by the old German miners to what was looked upon as a kind of false copper ore. The artificial, or perhaps merely fused product, called *speiss*, has nearly the same composition.

From either of these substances a pure salt of nickel may be obtained in the manner already described in connection with cobalt-salts (p. 419). The cobalt having been thrown down by Rose's process, as sesquioxide, the nickel remains in solution, and may be precipitated as hydroxide by potash, after the barium also contained in the solution has been thrown down by sulphuric acid.

Metallic nickel is easily prepared by exposing the oxalate to a high white heat in a crucible lined with charcoal, or by reducing one of the oxides by means of hydrogen at a high temperature. It is a white, malleable metal, having a density of 8.8, a high melting point, and a less degree of oxidability than iron, since it is but little attacked by dilute acids. Nickel is strongly magnetic, but loses this property when heated to 350° C. (662° F.).

Nickel Chloride, $NiCl_2$, is easily prepared by dissolving oxide or carbonate of nickel in hydrochloric acid. A green solution is obtained which furnishes crystals of the same color containing water. When rendered anhydrous by heat, the chloride is yellow, unless it contains cobalt, in which case it has a tint of green.

Nickel Oxides and Oxysalts.—Nickel forms two oxides analogous to the two principal oxides of iron.

The *monoxide*, NiO, is prepared by heating the nitrate to redness, or by precipitating a soluble nickel salt with caustic potash, and washing, drying, and igniting the apple-green hydroxide thrown down. It is an ashy-gray powder, freely soluble in acids, which it completely neutralizes, forming salts isomorphous with those of magnesium and the other members of the same group. Nickel salts, when hydrated, have usually a beautiful emerald-green color; in the anhydrous state they are yellow.

The *sesquioxide*, Ni_2O_3, is a black insoluble substance, prepared by passing chlorine through the hydroxide suspended in water; nickel chloride is then formed, and the oxygen of the oxide decomposed is transferred to a second portion. It is also produced when a salt of nickel is mixed with a solution of bleaching powder. The sesquioxide is decomposed by heat, and evolves chlorine when treated with hot hydrochloric acid.

Nickel Sulphate, $NiSO_4,7H_2O$.—This is the most important of the nickel salts. It forms green prismatic crystals, which require 3 parts of cold water for solution. Crystals with 6 molecules of water have also been obtained. It forms with the sulphates of potassium and ammonium beautiful double salts, $NiK_2(SO_4)_2,6H_2O$, and $Ni(NH_4)_2(SO_4)_2,6H_2O$, isomorphous with the corresponding magnesium salts.

When a strong solution of oxalic acid is mixed with sulphate of nickel, a pale bluish-green precipitate of oxalate falls after some time, very little nickel remaining in solution. The oxalate can thus be obtained for preparing the metal.

Nickel Carbonate, $NiCO_3$.—When solutions of nickel sulphate or chloride and of sodium carbonate are mixed, a pale-green precipitate falls, which is a combination of nickel carbonate and hydroxide. It is readily decomposed by heat.

Nickel-salts are well characterized by their behavior with reagents.

Caustic alkalies give a pale apple-green precipitate of hydroxide, insoluble in excess. *Ammonia* affords a similar precipitate, which is soluble in excess, with deep purplish-blue color. *Potassium* and *sodium carbonates* give pale-green precipitates. *Ammonium carbonate*, a similar precipitate, soluble in excess, with blue color. *Potassium ferrocyanide* gives a greenish-white precipitate. *Potassium cyanide* produces a green precipitate, which dissolves in an excess of the precipitant to an amber-colored liquid, and is reprecipitated by addition of hydrochloric acid. *Hydrogen sulphide* occasions no change if the nickel be in combination with a strong acid. *Ammonium sulphide* produces a black precipitate of nickel sulphide, which dissolves slightly in excess of the precipitant, with dark-brown color. Nickel sulphide when once precipitated, is insoluble in dilute hydrochloric acid; it is soluble in nitro-muriatic and in hot nitric acid.

Nickel is used for the preparation of a white alloy, sometimes called German silver, made by melting together 100 parts copper, 16 of zinc, and 40 of nickel. This alloy is very malleable, and takes a high polish.

Alloys of copper with nickel and zinc, or with nickel alone, are also used in Germany, Belgium, Switzerland, and the United States, for the manufacture of small coin. The advantages of nickel coining are that, nickel being dearer than copper, the coins can be made smaller for the same value; that the alloy is hard and therefore wears well; and further, that its manufacture requires experienced workmen and the use of powerful machinery.

Another application, of recent introduction, is the electrolytic deposition of nickel on iron, steel, copper, brass, and other metals, from a solution of nickel sulphate or the double sulphate or chloride of nickel and ammonium or nickel and potassium. The nickel is deposited in dense layers, capable of receiving a good polish.

METALS OF THE CHROMIUM GROUP.

Chromium.	Tungsten.
Molybdenum.	Uranium.

THESE metals are hexads, their highest oxides, which are of acid character, having the composition MO_3; they also form lower oxides, MO_2, M_2O_3, M_3O_4, and MO, more or less basic, and analogous in composition to the basic oxides of manganese, iron, cobalt, and nickel. Chromium, molybdenum, and tungsten form hexfluorides analogous to the highest oxides, and uranium forms an oxychloride, UO_2Cl_2, also of analogous composition.

CHROMIUM.

Symbol, Cr. Atomic weight, 52.4.

CHROMIUM is found in the state of oxide, in combination with iron oxide, somewhat abundantly in the Shetland Islands and elsewhere; as lead chromate it constitutes a very beautiful mineral, from which it was first obtained. The metal itself is prepared in a half-fused condition by mixing the oxide with half its weight of charcoal powder, enclosing the mixture in a crucible lined with charcoal, and then subjecting it to the very highest heat of a powerful furnace.

Deville prepared metallic chromium by subjecting a mixture of the sesquioxide and sugar to an intense heat in a lime crucible. Thus prepared, metallic chromium is less fusible than platinum and as hard as corundum. It is readily acted upon by dilute hydrochloric acid, less so by dilute sulphuric acid, and not at all by concentrated nitric acid. Frémy obtained a chromium in small cubic crystals, by the action of sodium vapor on chromium trichloride at a red heat. The crystalline chromium resists the action of concentrated acids, even of nitro-muriatic acid.

Chromium forms a hexfluoride, CrF_6, and a corresponding oxide, CrO_3, analogous to sulphuric oxide; also, an acid, H_2CrO_4, analogous to sulphuric acid, with corresponding salts, the chromates, which are isomorphous with the sulphates. In its other compounds chromium resembles iron, forming the chromic compounds Cr_2Cl_6, Cr_2O_3, etc., in which it is apparently trivalent, but more probably quadrivalent, and the chromous compounds $CrCl_2$, CrO, etc., in which it is bivalent.

Chlorides.—The *dichloride*, or *Chromous chloride*, $CrCl_2$, is prepared by heating the violet-colored trichloride, contained in a porcelain or glass tube, to redness in a current of perfectly dry and pure hydrogen gas; hydrochloric acid is then disengaged, and a white foliated mass is obtained, which dissolves in water with great elevation of temperature, yielding a blue

36 *

solution, which, on exposure to the air, absorbs oxygen with extraordinary energy, acquiring a deep-green color, and passing into the state of chromic oxychloride, Cr_2Cl_6,Cr_2O_3. Chromous chloride is one of the most powerful reducing or deoxidizing agents known, precipitating calomel from a solution of mercuric chloride, instantly converting tungstic acid into blue tungsten oxide, and precipitating gold from a solution of auric chloride. It forms, with ammonia, a sky-blue precipitate which turns green on exposure to the air; with ammonia and sal-ammoniac, a blue solution turning red on exposure to the air; and with ammonium sulphide, a black precipitate of chromous sulphide.

The *trichloride*, or *Chromic chloride*, $CrCl_3$, or more probably $Cr_2Cl_6 = Cl_3Cr$ — $CrCl_3$, is obtained in the anhydrous state by heating to redness in a porcelain tube a mixture of chromium sesquioxide and charcoal, and passing dry chlorine gas over it. The trichloride sublimes, and is deposited in the cool part of the tube in the form of beautiful crystalline plates of a pale-violet color. It is totally insoluble in water under ordinary circumstances, even at the boiling heat. It dissolves, however, and assumes the deep-green hydrated state in water containing an exceedingly minute quantity of the dichloride in solution. The hydration is marked by great rise of temperature.

The green hydrated chromic chloride is easily formed by dissolving chromic hydroxide in hydrochloric acid, or by boiling lead chromate, or silver chromate, or a solution of chromic acid, with hydrochloric acid and a reducing agent, such as alcohol or sulphurous acid, or even with a hydrochloric acid alone:—

$$2CrO_3 + 12HCl = Cr_2Cl_6 + 6H_2O + Cl_6.$$

The solution thus obtained exhibits the same characters as the chromic oxygen-salts. When evaporated it leaves a dark-green syrup, which, when heated to 100° C. in a stream of dry air, yields a green mass containing $Cr_2Cl_6,9H_2O$. The same solution evaporated in a vacuum yields green granular crystals containing Cr_2Cl_6,H_2O.

Fluorides.—The *trifluoride*, or *Chromic fluoride*, Cr_2F_6, is obtained by treating the dried sesquioxide with hydrofluoric acid, and strongly heating the dried mass as a dark-green substance which melts at a high temperature, and sublimes when still more strongly heated in shining regular octahedrons.

The *hexfluoride*, CrF_6, is formed by distilling lead chromate with fluorspar and fuming oil of vitriol in a leaden retort, and condensing the vapors in a cooled and dry leaden receiver. It then condenses to a blood-red fuming liquid, which volatilizes when its temperature rises a few degrees higher. The vapor is red, and, when inhaled, produces violent coughing and severe oppression of the lungs. The hexfluoride is decomposed by water, yielding hydrofluoric acid and chromium trioxide. A fluoride, intermediate in composition between the two just described, is obtained in solution by decomposing the brown dioxide in hydrofluoric acid. The solution is red, and yields by evaporation a rose-colored salt, which is redissolved without alteration by water, and precipitated brown by ammonia.

O x i d e s.—Chromium forms five oxides, containing CrO, Cr_3O_4, Cr_2O_3, CrO_2, and CrO_3, the first three being analogous in composition to the three oxides of iron.

The *monoxide*, or *Chromous oxide*, CrO, is formed on adding potash to a solution of chromous chloride as a brown precipitate, which speedily passes to deep foxy-red, with disengagement of hydrogen, being converted into a higher oxide. Chromous oxide is a powerful base, forming pale-blue salts, which absorb oxygen with extreme avidity. Potassio-chromous sulphate has the composition $CrK_2(SO_4)_2,6H_2O$, like the other members of the same group.

Trichromic tetroxide, $Cr_3O_4 = CrO,Cr_2O_3$, is the above-mentioned brownish-red precipitate produced by the action of water upon the monoxide. The decomposition is not complete without boiling. This oxide corresponds with the magnetic oxide of iron, and is not salifiable.

Sesquioxide, or *Chromic oxide*, Cr_2O_3.—When mercurous chromate, prepared by mixing solutions of mercurous nitrate and potassium chromate or dichromate, is exposed to a red heat, it is decomposed, pure chromium sesquioxide, having a fine green color, remaining. In this state the oxide, like alumina after ignition, is insoluble in acids. The anhydrous sesquioxide may be prepared in a beautifully crystalline form by heating potassium pyrochromate, $K_2O,2CrO_3$, to full redness in an earthen crucible. One-half of the chromium trioxide contained in that salt then suffers decomposition, oxygen being disengaged and sesquioxide left. The melted mass is treated with water, which dissolves out normal potassium chromate, and the oxide is washed and dried. Chromium sesquioxide communicates a fine green tint to glass, and is used in enamel painting. The crystalline sesquioxide is employed in the manufacture of razor-strops. From a solution of chromium sesquioxide in potash or soda, green gelatinous chromic hydroxide separates on standing. When finely powdered and dried over sulphuric acid, it consists of $Cr_2O_3,6H_2O$ or $Cr_2(HO)_6,3H_2O$. A hydroxide may also be prepared by boiling a somewhat dilute solution of potassium pyrochromate strongly acidulated with hydrochloric acid, with small successive portions of sugar or alcohol. In the former case carbon dioxide escapes: in the latter, aldehyde and acetic acid are formed, and the chromic acid of the salt becomes converted into chromium trichloride, the color of the liquid changing from red to deep green. The reduction may also be effected, as already observed, by hydrochloric acid alone. A slight excess of ammonia precipitates the hydroxide from its solution. It has a pale purplish-green color, which becomes full green on ignition; a great shrinking of volume and sudden incandescence are observed when the hydroxide is decomposed by heat.

Chromium sesquioxide is a weak base, resembling iron sesquioxide and alumina, with which it is isomorphous; its salts (chromic salts) have a green or purple color, and are said to be poisonous.

Chromic sulphate, $Cr_2(SO_4)_3$, is prepared by dissolving the hydroxide in dilute sulphuric acid. It unites with the sulphates of potassium and ammonium, giving rise to magnificent double salts, which crystallize in regular octahedrons of a deep claret-color, and are analogous in constitution to common alum, the aluminium being replaced by chromium. The ammonium-salt, for example, has the composition $Cr'''(NH_4)(SO_4)_2,12H_2O$. The finest crystals are obtained by spontaneous evaporation, the solution being apt to be decomposed by heat.

The *dioxide*, CrO_2, possibly a chromic chromate, CrO_3,Cr_2O_3, is a brown substance obtained by digesting chromic oxide with excess of chromic acid, or by partial reduction of chromic acid with alcohol, sulphurous acid, etc.

Chromium Trioxide, CrO_3; in combination with water, forming *Chromic acid*, $CrO_3,H_2O = H_2CrO_4 = (CrO_2)(OH)_2$.—Whenever chromium sesquioxide is strongly heated with an alkali in contact with air, oxygen is absorbed and the trioxide is generated. Chromium trioxide may be obtained nearly pure and in a state of great beauty by mixing 100 measures of a cold saturated solution of potassium pyrochromate with 150 measures of oil of vitriol, leaving the whole to cool, pouring off the mother-liquor, and leaving the crystals to drain upon a tile, closely covered by a glass or bell-jar. It is also formed by decomposing the hexfluoride with a small quantity of water. Chromium trioxide crystallizes in brilliant crimson-red prisms, very deliquescent and soluble in water: the solution is instantly reduced by contact with organic matter.

Chromic acid is bibasic and analogous in composition to sulphuric acid: its salts are isomorphous with the corresponding sulphates.

Potassium Chromate, K_2CrO_4 or $(CrO_2)(OK)_2$.—This salt is made directly from the native *chrome-iron ore*, which is a compound of chromium sesquioxide and ferrous oxide, analogous to magnetic iron ore, by calcination with nitre or with potassium carbonate, or with caustic potash, the ore being reduced to powder and heated for a long time with the alkali in a reverberatory furnace. The product, when treated with water, yields a yellow solution, which, on evaporation, deposits anhydrous crystals of the same color, isomorphous with potassium sulphate. Potassium chromate has a cool, bitter, and disagreeable taste, and dissolves in 2 parts of water at 15.5°.

Potassium Dichromate or *Pyrochromate*, K_2CrO_4,CrO_3, analogous in composition to the pyrosulphate (p. 318), is obtained by treating the preceding salt with a moderate quantity of sulphuric acid, whereby one-half of the base is removed and the chromate is converted into pyrochromate. This salt, of which immense quantities are manufactured for use in the arts, crystallizes by slow evaporation in beautiful red tabular crystals derived from a triclinic prism. It melts when heated, and is soluble in 10 parts of water; the solution has an acid reaction.

Potassium Trichromate, $K_2O,3CrO_3$ or $K_2CrO_4,2CrO_3$, may be obtained in crystals by dissolving the dichromate in an aqueous solution of chromic acid, and leaving it to evaporate over sulphuric acid.

Lead Chromate, $PbCrO_4$.—This salt, the *chrome-yellow* of the painter, is obtained as a brilliant yellow precipitate on mixing solutions of potassium chromate or pyrochromate with lead nitrate or acetate. On boiling it with lime-water, one-half of the acid is withdrawn and a basic lead chromate of an orange-red color left. The basic chromate is also formed by adding lead chromate to fused nitre, and afterward dissolving out the soluble salts by water: the product is crystalline, and rivals vermilion in beauty of tint. The yellow and orange chrome colors are fixed upon cloth by alternate application of the chromium and lead solutions, and in the latter case by passing the dyed stuff through a bath of boiling lime-water.

Silver Chromate, Ag_2CrO_4, is precipitated as a reddish-brown powder when solutions of potassium chromate and silver nitrate are mixed. It dissolves in hot dilute nitric acid, and separates on cooling in small ruby-red platy crystals. The chromates of *barium, zinc*, and *mercury* are insoluble; the first two are yellow, the last is brick-red.

Perchromic Acid is obtained, according to Barreswil, by mixing chromic acid with dilute hydrogen dioxide, or potassium pyrochromate with a dilute but very acid solution of barium dioxide in hydrochloric acid; a liquid is then formed of a blue color, which is removed from the aqueous solution by ether. This very unstable compound has perhaps the composition $H_2Cr_2O_8$ or Cr_2O_7,H_2O, analogous to that of permanganic acid.

Chromium Dioxydichloride, or *Chromyl dichloride*, CrO_2Cl_2, commonly called *Chlorochromic acid*.—When 3 parts of potassium pyrochromate and 3 parts of common salt are intimately mixed and introduced into a small glass retort, 9 parts of oil of vitriol then added, and heat applied as long as dense red vapors arise, this compound passes over as a heavy deep-red liquid resembling bromine: it is decomposed by water, with production of chromic and hydrochloric acids. It is analogous to the so-called chloro-molybdic, chlorotungstic, and chlorosulphuric acids in composition, and in the products which it yields when decomposed. It may be regarded as formed from the trioxide by substitution of Cl_2 for O, or from chromic acid, $(CrO_2)(OH)_2$, by substitution of Cl_2 for $(OH)_2$; also as a compound of chromium hexchloride (not known in the separate state) with chromium trioxide: $CrCl_6.2CrO_3 = 3CrO_2Cl_2$.

Trichromyl dichloride, $(CrO_2)_3Cl_2$ or $CrO_2Cl.CrO_2.CrO_2Cl$, is formed by heating the preceding compound to 180–190° in a sealed tube: $3CrO_2Cl_2 = (CrO_2)_3Cl_2 + Cl_4$. It is a black non-crystalline powder, which deliquesces rapidly in the air to a dark reddish-brown syrupy liquid smelling of free chlorine. When gently heated in hydrogen gas it takes fire, and is resolved into chromium sesquioxide, hydrochloric acid, and water:—

$$2Cr_3O_6Cl_2 + 10H = 3Cr_2O_3 + 4HCl + 3H_2O.$$

Reactions of Chromium Compounds.—A solution of chromic chloride, or a chromic oxygen salt, is not precipitated or changed in any way by hydrogen sulphide. *Ammonium sulphide* throws down a grayish-green precipitate of chromic hydroxide. *Caustic fixed alkalies* also precipitate the hydroxide, and dissolve it easily when added in excess. *Ammonia* the same, but nearly insoluble. The *carbonates of potassium, sodium,* and *ammonium* also throw down a green precipitate of hydroxide, slightly soluble in a large excess.

Chromous salts are but rarely met with; for their reactions, see Chromium Dichloride, p. 426.

Chromic acid and its salts are easily recognized in solution by forming a pale-yellow precipitate with *barium salts*, bright yellow with *lead salts*, brick-red with *mercurous salts*, and crimson with *silver salts ;* also by their capability of yielding the green sesquioxide by reduction.

All chromium compounds, ignited with a mixture of nitre and an alkaline carbonate, yield an alkaline chromate, which may be dissolved out by water, and on being neutralized with acetic acid will give the reactions just mentioned.

The oxides of chromium and their salts, fused with borax in either blowpipe flame, yield an emerald-green glass. The same character is exhibited by those salts of chromic acid whose bases do not of themselves impart a decided color to the bead. The production of the green color in both flames distinguishes chromium from uranium and vanadium, which give green beads in the inner flame only.

MOLYBDENUM.

Symbol, Mo. Atomic weight, 96.

THIS metal occurs in small quantity as sulphide or *molybdenite*, and as lead molybdate or *wulfenite*. Metallic molybdenum is obtained by exposing molybdic oxide in a charcoal-lined crucible to the most intense heat that can be obtained. It is a white, brittle, and exceedingly infusible metal, having a density of 8.6, and oxidizing, when heated in the air, to molybdic oxide.

Chlorides.—Molybdenum forms four chlorides, containing $MoCl_2$, $MoCl_3$ or Mo_2Cl_6, $MoCl_4$, and $MoCl_5$ or Mo_2Cl_{10}.

The *pentachloride* is produced when metallic molybdenum (previously freed from oxide by ignition in hydrogen chloride) is heated for some time in a stream of dry chlorine gas.

The pentachloride, heated to about 250° in a stream of hydrogen, is reduced to the red, difficultly volatile *trichloride*, $MoCl_3$ or Mo_2Cl_6; and this compound, heated to redness in an atmosphere of carbon dioxide free from oxygen, is resolved according to the equation $Mo_2Cl_6 = MoCl_2 + MoCl_4$ into the yellow *dichloride* which remains in the tube, and the brown *tetrachloride* which sublimes or is carried forward by the stream of gas.

Of these four chlorides the pentachloride is the only one which crystallizes distinctly, and melts and volatilizes without decomposition. The pure pentachloride is black. Its vapor has a dark brown-red color. The sulphur-yellow dichloride and the red trichloride, which is deceptively like amorphous phosphorus, have been obtained only in the amorphous state; the tetrachloride is an indistinctly crystalline brown sublimate.   In an atmosphere of carbon dioxide the dichloride bears a bright red heat without melting or volatilizing; the trichloride under the same circumstances is resolved into di- and tetra-chloride, which when again heated splits up into pentachloride which sublimes, and trichloride which remains behind.

The di- and tri-chloride are quite permanent in the air at ordinary temperatures and insoluble in water; the tetra- and penta-chloride, on the other hand, are extremely susceptible of the action of oxygen and more particularly of moisture.

The dichloride is insoluble in *nitric acid*, which, however, dissolves all the other chlorides.   The dichloride dissolves easily in hot *hydrochloric acid* with aid of heat, and crystallizes therefrom on cooling in long, shining, yellow needles, $Mo_2Cl_4,3H_2O$, which give off $2H_2O$ at 100°.*

The *bromides* of molybdenum correspond in composition with the chlorides; there is also an oxybromide containing $MoBr_2O_2$.

Fluorides.—Molybdenum forms three fluorides, MoF_2, MoF_4, and MoF_6, which are obtained by dissolving the corresponding oxides in hydrofluoric acid.   The *hexfluoride* is not known in the free state, but only in combination with basic metallic fluorides and molybdates; thus there is a potassium salt described by Berzelius, containing $2KF,MoF_6 + K_2O,MoO_3$.

Oxides.—Molybdenum forms the three oxides, MoO, MoO_2, and MoO_3, besides several oxides intermediate between the last two, which may be regarded as molybdic molybdates.

The *monoxide*, or *Molybdous oxide*, MoO, is produced by bringing the dioxide or trioxide in presence of one of the stronger acids in contact with any of the metals which decompose water.   Thus when zinc is immersed in a concentrated solution of an alkaline molybdate mixed with a quantity of hydrochloric acid sufficient to redissolve the precipitate first thrown down, zinc chloride and molybdous chloride are formed.   The dark-colored solution thus obtained is mixed with a large quantity of caustic potash, which precipitates a black hydrated molybdous oxide and retains the zinc oxide in solution.   The freshly-precipitated hydroxide is soluble in acids and ammonium carbonate; when heated in the air it burns to dioxide, but when dried in a vacuum it leaves the black anhydrous monoxide.

The *dioxide*, or *Molybdic oxide*, MoO_2, is obtained in the anhydrous state by heating sodium molybdate with sal-ammoniac, the molybdic trioxide being reduced to dioxide by the hydrogen of the ammoniacal salt; or, in the hydrated state, by digesting metallic copper in a solution of molybdic acid in hydrochloric acid until the liquid assumes a red color, and then adding a large excess of ammonia.   The anhydrous dioxide is deep brown,

* Liechti and Kempi, Liebig's Annalen, clxix. 344.

and insoluble in acids; the hydroxide resembles ferric hydroxide and dissolves in acids, yielding red solutions. It is converted into molybdic acid by strong nitric acid.

Trioxide, MoO_3.—To obtain this oxide (commonly called *Molybdic acid*) native molybdenum sulphide is roasted, at a red heat, in an open vessel, and the impure molybdic trioxide thence resulting is dissolved in ammonia. The filtered solution is evaporated to dryness, and the salt is taken up by water and purified by crystallization. It is, lastly, decomposed by heat and the ammonia expelled. The trioxide may also be prepared by decomposing native lead molybdate with sulphuric acid. It is a white, crystalline powder, fusible at a red heat and slightly soluble in water. The solution contains *molybdic acid*, but this acid or hydroxide is not known in the solid state. The trioxide is easily dissolved by alkalies and forms two series of salts, viz., *normal* or *neutral molybdates*, R_2MoO_4 or R_2O,MoO_3, and *anhydromolybdates*, *bimolybdates*, or *pyromolybdates*, R_2MoO_4,MoO_3 or $R_2O,2MoO_3$, the symbol R denoting a univalent metal. The neutral molybdates of the alkali-metals are easily soluble in water, and their solutions yield, with the stronger acids, a precipitate, either of a less soluble bimolybdate or of the anhydrous trioxide. The other molybdates are insoluble and are obtained by precipitation. *Lead Molybdate*, $PbMoO_4$, occurs native in yellow quadratic plates and octahedrons.

Sulphides.—Molybdenum forms three sulphides, MoS_2, MoS_3, and MoS_4, the last two of which are acid sulphides forming sulphur salts. The *disulphide*, or *Molybdic sulphide*, MoS_2, occurs native as *molybdenite* in crystallo-laminar masses or tabular crystals, having a strong metallic lustre and lead-gray color and forming a gray streak on paper like plumbago. The same compound is produced artificially by heating either of the higher sulphides or by igniting the trioxide with sulphur. When roasted in contact with the air it is converted into trioxide.

The *trisulphide*, MoS_3, is obtained by passing hydrogen sulphide into a concentrated solution of an alkaline molybdate and precipitating with an acid. It is a black-brown powder which is dissolved slowly by alkalies, more easily by alkaline sulphides and sulphydrates, forming sulphur salts called *thiomolybdates*. Most of these salts have the composition R_2MoS_4 or R_2S,MoS_3, analogous to that of the molybdates. The thiomolybdates of the alkali-metals, alkaline earth-metals, and magnesium are soluble in water, forming solutions of a fine red color; the rest are insoluble.

Tetrasulphide, MoS_4.—This is also an acid sulphide, forming salts called *perthiomolybdates*, the general formula of which is R_2MoS_5 or R_2S,MoS_4. The *potassium salt* is obtained by boiling the thiomolybdate with molybdenum trisulphide.

Molybdous salts, obtained by dissolving molybdous oxide in acids, are opaque and almost black. They yield with *hydrogen sulphide* a brown-black precipitate soluble in ammonium sulphide; with *alkalies* and *alkaline carbonates*, a brownish-black precipitate of. molybdous hydroxide, easily soluble in acid potassium carbonate or in ammonium carbonate; with *potassium ferrocyanide*, a dark-brown precipitate; with *sodium phosphate*, a white precipitate.

Solutions of **molybdic salts** have a reddish-brown color. When heated in the air they have a tendency to become blue by oxidation. In contact with metallic *zinc*, they first blacken and then yield a black precipitate of molybdous hydroxide. Their reactions with *alkalies, hydrogen, sulphide*, etc, are similar to those of molybdous salts; but the precipitates are lighter in color.

Molybdates are colorless unless they contain a colored base. Solutions of the alkaline molybdates yield with *acids* a precipitate of molybdic trioxide, soluble in excess of the precipitant. They are colored yellow by *hydrogen sulphide* from formation of a thiomolybdate of the alkali-metal, and then yield with acids a brown precipitate of molybdenum trisulphide. This is an extremely delicate test for molybdic acid. They form white precipitates with the salts of the *earth-metals,* and precipitates of various color with salts of the *heavy metals.* When *orthophosphoric acid,* or a liquid containing it, is added to the solution of ammonium molybdate together with an excess of hydrochloric acid the liquid turns yellow, and after a while deposits a yellow precipitate of molybdic trioxide combined with small quantities of phosphoric acid and ammonia. This precipitate is soluble in ammonia and likewise in excess of the phosphate. The reaction is therefore especially adapted for the detection of small quantities of phosphoric acid. The pyrophosphates and metaphosphates do not produce the yellow precipitate. *Arsenic acid* gives a similar reaction.

All the oxides of molybdenum form with *borax* in the outer blowpipe flame a bead which is yellow while hot and colorless on cooling; in the inner flame a dark-brown bead, which is opaque if excess of molybdenum is present. By long-continued heating, the molybdic oxide may be separated in dark-brown flakes floating in the clear yellow glass. With *phosphorus salt* in the outer flame, all oxides of molybdenum give a bead which is greenish while hot and colorless on cooling; in the inner flame, a clear green bead from which molybdic oxide cannot be separated by continued heating.

TUNGSTEN, or WOLFRAM.

Symbol, W.   Atomic weight, 184.

TUNGSTEN is found as ferrous tungstate in the mineral *wolfram,* tolerably abundant in Cornwall; occasionally also as calcium tungstate (*scheelite* or *tungsten*), and as lead tungstate (*scheeletine*).

Metallic tungsten is obtained in the state of a dark-gray powder by strongly heating tungstic oxide in a stream of hydrogen, but requires for fusion an exceedingly high temperature. It is a white metal, very hard and brittle, and has a density of 17.4. Heated to redness in the air, it takes fire and reproduces tungstic oxide.

Tungsten forms two classes of compounds in which it is quadrivalent and sexvalent respectively, and a third class of intermediate composition in which it is apparently quinquivalent.

Chlorides.—These compounds are formed by heating metallic tungsten in chlorine gas. The *hexchloride,* or *tungstic chloride,* WCl_6, is also produced together with oxychlorides by the action of chlorine on an ignited mixture of tungstic oxide and charcoal. The oxychlorides, being more volatile than the hexchloride, may be separated from it by sublimation. The hexchloride forms dark-violet scales or fused crusts having a bluish-black metallic iridescence. By contact with water or moist air, it is converted into hydrochloric and tungstic acids. The chlorides, WCl_5, WCl_4, and WCl_2, are formed when the hexchloride is heated in hydrogen gas. The two former are crystalline: the dichloride is a loose, gray powder, destitute of crystalline structure.*

* R o s c o e, Journal of the Chemical Society, 1872, p. 287.

A *pentabromide* and *hexbromide* are formed by the action of bromine in excess of tungsten.—The *hexfluoride*, WF_6, is obtained by evaporating a solution of tungstic acid in hydrofluoric acid.

Oxides.—Tungsten forms three oxides, WO_2, WO_3, and W_2O_5, neither of which exhibits basic properties, so that there are no tungsten salts in which the metal replaces the hydrogen of an acid or takes the electro-positive part. The trioxide exhibits decided acid tendencies, uniting with basic metallic oxides, and forming crystallizable salts called *tungstates*. The pentoxide may be regarded as a compound of the other two.

The *dioxide*, or *Tungstous oxide*, WO_2, is most easily prepared by exposing tungstic oxide to hydrogen, at a temperature not exceeding dull redness. It is a brown powder, sometimes assuming a crystalline appearance and an imperfect metallic lustre. It takes fire when heated in the air, and burns, like the metal itself, to tungstic oxide. It forms a definite compound with soda.

The *trioxide*, or *Tungstic oxide*, WO_3, is most easily prepared from native calcium tungstate by digestion in nitric or hydrofluoric acid, the soluble calcium salt thereby produced being washed out with water, and the remaining tungstic acid ignited. From wolfram it may be prepared by repeatedly digesting the mineral in strong hydrochloric acid, ultimately with addition of a little nitric acid, to dissolve out the iron and manganese; dissolving the remaining tungstic acid in aqueous ammonia; evaporating to dryness; and heating the residual ammonium tungstate in contact with the air. Tungstic oxide is a yellow powder insoluble in water and in most acids, but soluble in alkalies. The hot solutions of the resulting alkaline tungstates, when neutralized with an acid, yield a yellow precipitate of *tungstic monohydrate* or *tungstic acid*, H_2WO_4 or H_2O,WO_3. Cold dilute solutions, on the other hand, yield with acids a white precipitate, consisting of *tungstic dihydrate* or *hydrated tungstic acid*, $2H_2O,WO_3$ or H_2WO_4,H_2O. Tungstic acid reddens litmus and dissolves easily in alkalies.

Tungstates.—Tungstic acid unites with bases in various and often in very unusual proportions. It is capable of existing also in two isomeric modifications, viz.: 1. *Ordinary tungstic acid*, which is insoluble in water, and forms insoluble salts with all metals, except the alkali-metals and magnesium; 2. *Metatungstic acid*, which is soluble in water, and forms double salts with nearly all metals. Ordinary tungstic acid forms normal salts containing M_2WO_4 or M_2O,WO_3, and acid salts containing $3M_2O,7WO_3$, which may perhaps be regarded as double salts composed of diacid and triacid tungstates, that is, as $2(M_2O,2WO_3) + M_2O,3WO_3$. The tungstates of potassium and sodium, especially the latter, are sometimes used as mordants in dyeing, in place of stannates, also for rendering muslin and other light fabrics uninflammable. Tungstous tungstate, WO_2,WO_3, which has the composition of *tungsten pentoxide*, W_2O_5, is a blue substance formed by reducing tungstic oxide or tungstic acid with zinc and hydrochloric acid; also by heating ammonium tungstate to redness in a retort.

Metatungstates.—These salts, which have the composition of quadracid tungstates, $M_2O,4WO_3$, are formed from ordinary tungstates by addition of tungstic acid, or by removing part of the base by means of an acid. They are for the most part soluble and crystallizable. By decomposing barium metatungstate with dilute sulphuric acid, and evaporating the filtrate in a vacuum, hydrated metatungstic acid is obtained in quadratic octahedrons, apparently containing $H_2W_4O_{13},31H_2O$; it is very soluble in water.

Silicotungstates.[*]—By boiling gelatinous silica with acid potassium tungstate, a crystalline salt is obtained, having the composition of a diacid

---

[*] **Marignac**, Ann. Chim. Phys. [4], iii. 5; Watts's Dictionary of Chemistry, v. 915.

potassium tungstate, $6(K_2O,2WO_3)$, or $K_{12}O_6,12WO_3$, in which one-third of the potassium is replaced by silicon, viz., $K_8SIO_6,12WO_3$, so that the silicon here enters as a *basylous* element. The resulting solution yields with mercurous nitrate a precipitate of *mercurous silicotungstate;* this, when decomposed by an equivalent quantity of hydrochloric acid, yields a solution of *hydrogen silicotungstate,* or *silicotungstic acid;* and the other silicotungstates, which are all soluble, are obtained by treating the acid with carbonates.

Silico-decitungstic acid, $H_8SIO_6,10WO_3$, is obtained as an ammonium salt by boiling gelatinous silica with solution of acid ammonium tungstate; and from this the acid and its other salts may be obtained in the same manner as the preceding. The silico-decitungstates are very unstable, and the acid is decomposed by mere evaporation, depositing silica, and being converted into *tungstosilicic acid,* which is isomeric with silicotungstic acid, and likewise decomposes carbonates. All three of these acids are capable of exchanging either one-half or the whole of their basic hydrogen for metals, thereby forming acid and neutral salts; silicotungstic acid also forms an acid sodium salt in which only one-fourth of the hydrogen is replaced by sodium.

Tungsten Sulphides.—The *disulphide,* or *Tungstous sulphide,* WS_2, is obtained in soft, black, needle-shaped crystals by igniting tungsten, or one of its oxides, with sulphur.

The *trisulphide,* or *Tungstic sulphide,* WS_3, is formed by dissolving tungstic acid in ammonium sulphide, and precipitating with an acid, or by adding hydrochloric acid to the solution of an alkaline tungstate saturated with hydrogen sulphide. It is a light-brown precipitate, turning black when dry. It unites easily with basic metallic sulphides, forming the *thiotungstates,* M_2WS_4, analogous to the normal tungstates.

Reactions of Tungsten Compounds.—Soluble tungstates, or metatungstates, supersaturated with sulphuric, hydrochloric, phosphoric, oxalic, or acetic acid, yield, on the introduction of a piece of *zinc,* a beautiful blue color, arising from the formation of blue tungsten oxide. A soluble tungstate, mixed with *ammonium sulphide,* and then with excess of acid, yields a light-brown precipitate of tungstic sulphide soluble in ammonium sulphide. *Hydrogen sulphide* does not precipitate the acidulated solution of a tungstate, but turns it blue, owing to the formation of a blue oxide. Ordinary tungstates give, with *potassium ferrocyanide,* after addition of hydrochloric acid, a brown flocculent precipitate, soluble in pure water free from acid; metatungstates give no precipitates. *Acids,* added to solutions of ordinary tungstates, throw down a white or yellow precipitate of tungstic acid; with metatungstates no precipitate is obtained.

All tungsten compounds form colorless beads with borax and phosphorus salt, in the outer blowpipe flame. With *borax,* in the inner flame, they form a yellow glass, if the quantity of tungsten is somewhat considerable, but colorless with a smaller quantity. With *phosphorus salt* in the inner flame they form a glass of a pure blue color, unless metallic oxides are present, which modify it; in presence of iron the glass is blood-red, but the addition of metallic tin renders it blue.

Steel, alloyed with a small quantity of tungsten, acquires extraordinary hardness. Wootz, or Indian steel, contains tungsten. Tungsten has also a remarkable effect on steel in increasing its power of retaining magnetism when hardened. A horseshoe magnet of ordinary steel, weighing two pounds, is considered of good quality when it bears seven times its own weight; but, according to Siemens, a similar magnet made with steel containing tungsten may be made to carry twenty times its weight suspended from the armature.*

URANIUM.

Symbol, U.   Atomic weight, 240.

THIS metal is found in a few minerals, as *pitchblende*, which is an oxide, and *uranite*, which is a phosphate; the former is its principal ore. The metal itself is isolated by decomposing the chloride with potassium or sodium, and is obtained as a black coherent powder, or in fused white malleable globules, according to the manner in which the process is conducted. It is permanent in the air at ordinary temperatures, and does not decompose water; but in the pulverulent state it takes fire at 207° C. (405° F.), burning with great splendor and forming a dark-green oxide. It unites also very violently with chlorine and with sulphur.

Uranium forms two classes of compounds: viz., the u r a n o u s c o m-p o u n d s, in which it is quadrivalent, e. g., UCl_4, UO_2, $U(SO_4)_2$, etc., and the u r a n i c c o m p o u n d s, in which it is sexvalent, e. g.,

$$UO_3, \ UO_2Cl_2, \ UO_2(NO_3)_2, \ UO_2(SO_4).$$

There are also two oxides intermediate between uranous and uranic oxide. There is no chloride, bromide, iodide, or fluoride corresponding with uranic oxide, such as UCl_6: neither are there any normal uranic oxysalts, such as $U(NO_3)_6$, $U(SO_4)_3$, etc.; but all the uranic salts contain the group UO_2, which may be regarded as a bivalent radicle (uranyl), uniting with acids in the usual proportions and forming normal salts; thus—

Uranic oxide or Uranyl oxide	$(UO_2)O$
Uranic oxychloride or Uranyl chloride . .	$(UO_2)Cl_2$
Uranic nitrate or Uranyl nitrate . . .	$(UO_2)(NO_3)_2$
Uranic sulphate or Uranyl sulphate . . .	$(UO_2)(SO_4)$.

This view of the composition of the uranic salts is not, however, essential, since they may also be formulated as basic salts in the manner above illustrated.

Chlorides.— *Uranous chloride*, UCl_4, is formed, with vivid incandescence, by burning metallic uranium in chlorine gas, also by igniting uranous oxide in hydrochloric acid gas, and, lastly, by heating uranoso-uranic oxide with charcoal in a current of chlorine. It crystallizes in dark-green regular octahedrons, and dissolves easily in water, forming an emerald-green solution, which is decomposed when dropped into boiling water, giving off hydrochloric acid, and yielding a brown precipitate of hydrated uranous oxide. It is a powerful deoxidizing agent, reducing gold and silver, converting ferric salts into ferrous salts, etc. Its vapor-density, determined in V. Meyer's apparatus (P. 79) was found to be 13.33 (hydrogen = 1), which agrees closely with the theoretical number 13.21 deduced from the atomic weight of uranium, 240.†

* Journal of the Chemical Society, 1868, vol. xxi. p. 284.
† Z i m m e r m a n n, Berichte der deutschen chemischen Gesellschaft, xiv. 1934.

Uranic oxychloride, or *Uranyl chloride*, UO_2Cl_2, is formed when dry chlorine gas is passed over red-hot uranous oxide as an orange-yellow vapor, which solidifies to a yellow crystalline fusible mass, easily soluble in water. It forms double salts with the chlorides of the alkali-metals, the potassium salt, for example, having the composition $UO_2Cl_2,2KCl,2H_2O$.

Uranous bromide, UBr_4, is prepared by heating a mixture of uranoso-uranic oxide and charcoal in a current of carbon dioxide laden with bromine-vapor, and collects on the cooler parts of the tube in black glistening leaflets, which are converted by heat into a brown vapor and may be sublimed unchanged. It is extremely hygroscopic. Vapor-density, by experiment 19.46, by calculation 19.36 (Zimmermann).

Fluorides.—Uranoso-uranic oxide, U_3O_8, treated with aqueous hydrofluoric acid yields a bulky, green powder consisting of u r a n o u s   f l u o r i d e, UF_4, and a yellow solution which on evaporation leaves u r a n i c   o x y-f l u o r i d e or u r a n y l   f l u o r i d e, UO_2F_2. When uranous fluoride is heated in a closed crucible, a white sublimate is formed; and if the crucible be then left to cool and this sublimate be removed, a fresh white sublimate will be formed, and so on till nothing is left in the crucible but uranous oxide, UO_2. The white sublimate, for the formation of which access of air is of course necessary, is isomeric with the yellow oxyfluoride above mentioned, and is distinguished as a-UO_2F_2.*

Oxides.—*Uranous oxide*, UO_2, formerly mistaken for metallic uranium, is obtained by heating the oxide, U_3O_8, or uranic oxalate in a current of hydrogen. It is a brown powder, sometimes highly crystalline. In the finely-divided state it is pyrophoric. It dissolves in acids forming green salts.

Uranoso-uranic oxide, $U_3O_8 = UO_2,2UO_3$.—This oxide forms the chief constituent of pitchblende. It is obtained artificially by igniting the metal or uranous oxide in contact with the air, or by gentle ignition of uranic oxide or uranic nitrate. It forms a dark-green velvety powder of specific gravity 7.1 to 7.3. When ignited in hydrogen, or with sodium, charcoal, or sulphur, it is reduced to uranous oxide. When ignited alone it yields a black oxide, U_2O_5. Uranoso-uranic oxide dissolves in strong sulphuric or hydrochloric acid, yielding a mixture of uranous and uranic salt; by nitric acid it is oxidized to uranic nitrate.

Uranic oxide, or *Uranyl oxide*, UO_3.—Uranium and its lower oxides dissolve in nitric acid, forming uranic nitrate; and when this salt is heated in a glass tube till it begins to decompose at 250°, pure uranic oxide remains in the form of a chamois-yellow powder. *Hydrated uranic oxide*, $UO_3,2H_2O$, cannot be prepared by precipitating a uranic salt with alkalies, inasmuch as the precipitate always carries down alkali with it; but it may be obtained by evaporating a solution of uranic nitrate in absolute alcohol at a moderate heat, till at a certain degree of concentration nitrous ether, aldehyde, and other vapors are given off, and a spongy yellow mass remains, which is the hydroxide. In a vacuum at ordinary temperatures or at 100° C. in the air it gives off half its water, leaving the *monohydrate*, UO_3,H_2O, or *Uranyl dihydroxide*, $UO_2(OH)_2$. This hydrate cannot be deprived of all its water without exposing it to a heat sufficient to drive off part of the oxygen and reduce it to uranoso-uranic oxide.

Uranic oxide and its hydrates dissolve in acids, forming the u r a n i c   s a l t s. The *nitrate*, $(UO_2)(NO_3)_2,6H_2O$, may be prepared from pitchblende by dissolving the pulverized mineral in nitric acid, evaporating to dryness, adding water, and filtering; the liquid yields by due evaporation crystals

---

* A. S m i t h e l l s, Chem. Soc. Journal, xliii. 125.

of uranic nitrate, which are purified by a repetition of the process, and, lastly, dissolved in ether. This latter solution yields the pure nitrate.

Uranates.—Uranic oxide unites with the more basic metallic oxides. The uranates of the alkali-metals are obtained by precipitating a uranic salt with a caustic alkali: those of the earth-metals and heavy metals by precipitating a mixture of a uranic salt and a salt of the other metal with ammonia, or by igniting a double carbonate or acetate of uranium and the other metal (calcio-uranic acetate, for example) in contact with the air. The uranates have, for the most part, the composition $M_2O,2UO_3$. They are yellow, insoluble in water, soluble in acids. Those which contain fixed bases are not decomposed at a red heat; but at a white heat the uranic oxide is reduced to uranoso-uranic oxide or by ignition in hydrogen to uranous oxide: the mass obtained by this last method easily takes fire in contact with the air. *Sodium uranate*, $Na_2O,2UO_3$, is much used for imparting a yellowish or greenish color to glass and as a yellow pigment in the glazing of porcelain. The "uranium yellow" for these purposes is prepared on the large scale by roasting pitchblende with lime in a reverberatory furnace; treating the resulting calcium uranate with dilute sulphuric acid; mixing the solution of uranic sulphate thus obtained with sodium carbonate, by which the uranium is first precipitated together with other metals, but then redissolved tolerably free from impurity by excess of the alkali; and treating the liquid with dilute sulphuric acid, which throws down hydrated sodium uranate, $Na_2O,-2UO_3,6H_2O$. *Ammonium uranate* is but slightly soluble in pure water, and quite insoluble in water containing sal-ammoniac; it may, therefore, be prepared by precipitating a solution of sodium uranate with that salt. It occurs in commerce as a fine deep-yellow pigment, also called "uranium yellow." This salt, when heated to redness, leaves pure uranoso-uranic oxide, and may, therefore, serve as the raw material for the preparation of other uranium compounds.

Uranous salts form green solutions, from which *caustic alkalies* throw down a red-brown gelatinous precipitate of uranous hydrate; *alkaline carbonates*, green precipitates, which dissolve in excess, especially of ammonium carbonate, forming green solutions. *Ammonium sulphide* forms a black precipitate of uranous sulphide; *hydrogen sulphide*, no precipitate.

Uranic salts are yellow, and yield with *caustic alkalies* a yellow precipitate of alkaline uranate, insoluble in excess of the reagent. *Alkaline carbonates* form a yellow precipitate, consisting of a carbonate of uranium and the alkali-metal, soluble in excess, especially of acid ammonium or potassium carbonate. *Ammonium sulphide* forms a black precipitate of uranic sulphide. *Hydrogen sulphide* forms no precipitate, but reduces the uranic to a green uranous salt. *Potassium ferrocyanide* forms a red-brown precipitate.

All uranium compounds, fused with *phosphorus salt* or *borax* in the outer blowpipe flame, produce a clear yellow glass, which becomes greenish on cooling. In the inner flame the glass assumes a green color, becoming still greener on cooling. The oxides of uranium are not reduced to the metallic state by fusion with sodium carbonate on charcoal.

Uranium compounds are used, as already observed, in enamel painting and for the staining of glass, uranous oxide giving a fine black color, and uranic oxide a delicate greenish-yellow, highly fluorescent glass. Uranium salts are also used in photography.

37 *

METALS OF THE TIN GROUP.

Tin—Titanium—Zirconium—Thorium.

THESE metals are tetrads like silicon, forming volatile tetrachlorides ; also tetrafluorides, which unite with other metallic fluorides, forming double salts analogous in composition to the silicofluorides and isomorphous therewith.

TIN.

Symbol, Sn (Stannum).   Atomic weight, 117.8°.

THIS valuable metal occurs in the state of oxide, and more rarely as sulphide; the principal tin-mines are those of Saxony and Bohemia, Malacca, and more especially Cornwall.   In Cornwall the tin stone is found as a constituent of metal-bearing veins, associated with copper ore in granite and slate rocks, and as an alluvial deposit mixed with rounded pebbles in the beds of several small rivers.   The first variety is called *mine* and the second *stream tin*.   Tin oxide is also found disseminated through the rock itself in small crystals.

To prepare the ore for reduction it is stamped to powder, washed to separate as much as possible of the earthy matter, and roasted to expel sulphur and arsenic: it is then strongly heated with coal, and the metal thus obtained is cast into large blocks.   Two varieties of commercial tin are known, called *grain* and *bar tin;* the first is the best, it is prepared from the stream ore.

Pure tin has a white color, approaching that of silver: it is soft and malleable, and when bent or twisted emits a peculiar crackling sound; it has a density of 7.3°.   At 200° C. (392° F.) it becomes so brittle that it may be powdered, and at 235° C. (455° F.) it melts.   Tin is but little acted upon by air and water, even conjointly; when heated above its melting-point it oxidizes rapidly, becoming converted into a whitish powder, used in the arts for polishing under the name of *putty powder*.   The metal is attacked and dissolved by hydrochloric acid, with evolution of hydrogen; nitric acid acts with great energy, converting it into a white hydrate of the dioxide.

Tin forms two well-defined classes of compounds, namely, the s t a n n o u s c o m p o u n d s, in which it is bivalent, as $SnCl_2$, SnI_2, SnO, etc., and the s t a n n i c c o m p o u n d s, in which it is quadrivalent, as $SnCl_4$, SnO_2, etc. ; also a few compounds called s t a n n o s o - s t a n n i c c o m p o u n d s of intermediate composition, *e. g.,* Sn_2Cl_6, Sn_2O_3, etc.

Chlorides.—The *dichloride,* or *Stannous chloride,* $SnCl_2$, or more probably $Sn_2Cl_4 = Cl_2Sn{=\!\!\!=}SnCl_2$ (p. 253), is obtained in the anhydrous state by distilling a mixture of calomel and powdered tin, prepared by agitating the melted metal in a wooden box until it solidifies.   It is a gray, resinous-looking substance, fusible below redness and volatile at a high temperature.

The *hydrated chloride*, commonly called *tin salt*, is easily prepared by dissolving metallic tin in hot hydrochloric acid. It crystallizes in needles containing $SnCl_2,2H_2O$, which are freely soluble in a small quantity of water, but are apt to be partly decomposed when put into a large mass unless hydrochloric acid be present in excess. Solution of stannous chloride is employed as a deoxidizing agent; it reduces the salts of mercury and other metals of the same class. It is also extensively employed as a mordant in dyeing and calico-printing, sometimes also as an antichlore.

Stannous chloride unites with the chlorides of the alkali-metals, forming crystallizable double salts, $SnCl_2,2KCl$, etc., called *Stannoso-chlorides* or *Chlorostannites*.

The *tetrachloride*, or *Stannic chloride*, $SnCl_4$, formerly called *fuming liquor of Libavius*, is made by exposing metallic tin to the action of chlorine, or, more conveniently, by distilling a mixture of 1 part of powdered tin with 5 parts of mercuric chloride. It is a thin, colorless, mobile liquid, boiling at 120° C. (248° F.), and yielding a colorless invisible vapor. It fumes in the air, and when mixed with a third part of water solidifies to a soft fusible mass, called *butter of tin*. The solution of stannic chloride is much employed by the dyer for the brightening and fixing of red colors, and is sometimes designated by the old names "composition, physic, or tin solution;" it is commonly prepared by dissolving metallic tin in a mixture of hydrochloric and nitric acids, care being taken to avoid too great elevation of temperature. The solution when evaporated yields a deliquescent crystalline hydrate, $SnCl_4,5H_2O$.

Stannic chloride forms, with the chlorides of the alkali-metals and alkaline earth-metals, crystalline double salts, called *Stannochlorides* or *Chlorostannates*, e. g., $SnCl_4,2NH_4Cl$, $SnCl_4,BaCl_2$, etc. It also forms crystalline compounds with the pentachloride and oxychloride of phosphorus, viz., $SnCl_4,PCl_5$ and $SnCl_4,POCl_3$, and a solid compound with phosphine, containing $3SnCl_4,2PH_3$.

The *trichloride*, or *Stannoso-stannic chloride*, known only in solution, is produced by dissolving the sesquioxide in hydrochloric acid. The solution acts like a mixture of the dichloride and tetrachloride.

Fluorides.—*Stannous Fluoride*, SnF_2, obtained by evaporating the solution of stannous oxide in hydrofluoric acid, crystallizes in small shining opaque prisms. *Stannic fluoride*, SnF_4, is not known in the free state, but unites with other metallic fluorides, forming crystalline compounds, called *stannofluorides* or *fluostannates*, isomorphous with the corresponding silicofluorides, titanofluorides, and zircofluorides. The potassium salt contains $SnF_4,2KF,H_2O$, the barium salt, SnF_4,BaF_2, etc.

Oxides.—The *monoxide*, or *Stannous oxide*, SnO, is produced by heating stannous oxalate out of contact with the air; also by igniting stannous hydrate. This *hydrate*, $2SnO,H_2O$ or $Sn_2H_2O_3 = OS{<}{Sn(OH) \atop Sn(OH)}$, is obtained as a white precipitate by decomposing stannous chloride with an alkaline carbonate, carbon dioxide being at the same time evolved. When carefully washed, dried, and heated in an atmosphere of carbon dioxide, it leaves anhydrous stannous oxide as a dense black powder, which is permanent in the air, but when touched with a red-hot body, takes fire and burns like tinder, producing the dioxide. The hydrate is freely soluble in caustic potash; the solution decomposes by keeping into metallic tin and dioxide. It dissolves also in sulphuric acid, forming *stannous sulphate*, $SnSO_4$, which crystallizes in needles.

The *sesquioxide*, Sn_2O_3, is produced by the action of hydrated ferric oxide upon stannous chloride: it is a grayish, slimy substance, soluble in hydrochloric acid and in ammonia.

The *dioxide*, or *Stannic oxide*, SnO_2, occurs native as tin-stone or cassiterite, the common ore of tin, and is easily formed by heating tin, stannous oxide, or stannous hydrate in contact with the air. As thus prepared it is a white, or yellowish amorphous powder; but by passing the vapor of stannic chloride mixed with aqueous vapor through a red-hot porcelain tube, it may be obtained in crystals. It is not attacked by acids, even in the concentrated state.

Stannic oxide forms two hydroxides, differing from one another in composition and properties; both, however, being acids, and capable of forming salts by exchanging their hydrogen for metals. These hydroxides or acids are s t a n n i c a c i d, SnO_2,H_2O, or H_2SnO_3 or $O=Sn(OH)_2$, and m e t a - s t a n n i c a c i d, $Sn_5O_{10},5H_2O$, or $H_{10}Sn_5O_{15}$ or $Sn_5O_5(OH)_{10}$, the former being capable of exchanging the whole of its hydrogen for metal, and forming the stannates containing M_2SnO_3; while the latter exchanges only one-fifth of its hydrogen, forming the m e t a s t a n n a t e s, $H_8M_2Sn_5O_{15}$.

Stannic acid is precipitated by acids from solutions of alkaline stannates, also from solution of stannic chloride, by calcium or barium carbonate not in excess; alkaline carbonates throw down an acid stannate. When dried in the air at ordinary temperatures it has, according to Weber, the composition $SnO_2,2H_2O$; in a vacuum half the water is given off, leaving SnO_2,H_2O.

Stannic hydroxide dissolves in the stronger acids, forming the stannic salts; thus with sulphuric acid it forms *stannic sulphate*, $Sn(SO_4)_2$ or $SnO,2SO_3$. *Hydrochloric acid* converts it into the tetrachloride. The stannic salts of oxygen-acids are very unstable.

Stannates.—Stannic hydroxide exhibits acid much more decidedly than basic properties. It forms easily soluble salts with the alkalies, and from these the insoluble stannates of the earth-metals and heavy metals may be obtained by precipitation. *Sodium stannate*, Na_2SnO_3, which is much used in calico-printing as a "preparing salt" or mordant, is produced on the large scale by fusing tin-stone with hydrate, nitrate, chloride, or sulphide of sodium; by boiling the tin ore with caustic soda-solution; by fusing metallic tin with a mixture of sodium nitrate and carbonate; or by heating it with soda-solution mixed with sodium and nitrate and chloride.*

Metastannic acid is produced by the action of nitric acid upon tin. When dried in the air at ordinary temperatures, it contains $5SnO_2,10H_2O$ or $H_{10}Sn_5O_{15},5H_2O$, but at 100° it gives off 5 molecules of water and is reduced to $H_{10}Sn_5O_{15}$. It is a white crystalline powder insoluble in water and in acids. It dissolves slowly in alkalies, forming the metastannates, but is gradually deposited in its original state as the solution absorbs carbonic acid from the air. The *potassium salt*, $K_2H_8Sn_5O_{15}$ or $\left.\begin{array}{c}K_2O\\4H_2O\end{array}\right\}(SnO_2)_5$, may be precipitated in the solid state by adding pieces of solid potash to a solution of metastannic acid in cold potash. It is gummy, uncrystallizable, and strongly alkaline. The *sodium salt*, $Na_2H_8Sn_5O_{15}$, prepared in like manner, is crystallo-granular, and dissolves slowly, but completely, in water. The metastannates exist only in the hydrated state, being decomposed when deprived of their basic water.

Sulphides.—The *monosulphide*, SnS, is prepared by fusing tin with excess of sulphur, and strongly heating the product. It is a lead-gray, brittle substance, fusible at a red heat, and soluble with evolution of sulphuretted hydrogen in hot hydrochloric acid. A *sesquisulphide* may be formed by gently heating the above compound with a third of its weight of

* Richardson and Watts's Chemical Technology, vol. i. Part iv. p. 35, and Part v. p. 342.

sulphur: it is yellowish-gray, and easily decomposed by heat. The *bisulphide*, SnS_2, or *Mosaic gold*, is prepared by exposing to a low red heat, in a glass flask, a mixture of 12 parts of tin, 6 of mercury, 6 of sal-ammoniac, and 7 of flowers of sulphur. Sal-ammoniac, cinnabar, and stannous chloride sublime, while the bisulphide remains at the bottom of the vessel in the form of brilliant gold-colored scales; it is used as a substitute for gold powder. The same compound is obtained as an amorphous light-yellow powder by passing hydrogen sulphide into a solution of stannic chloride.

Stannous salts give with:

Fixed caustic alkalies: white hydroxide, soluble in excess.
Ammonia: carbonates of potassium, sodium and ammonium . . } white hydroxide, nearly insoluble in excess.

Hydrogen sulphide . *Ammonium sulphide* . } black-brown precipitate of monosulphide, soluble in ammonium sulphide containing excess of sulphur, and reprecipitated by acids as yellow bisulphide.

Stannic salts give with

Fixed caustic alkalies: white hydroxide, soluble in excess.
Ammonia: white hydroxide, slightly soluble in excess.
Alkaline carbonates: white hydroxide, slightly soluble in excess.
Ammonium carbonate: white hydroxide, insoluble.
Hydrogen sulphide: yellow precipitate of bisulphide.
Ammonium sulphide: the same, soluble in excess.

Trichloride of gold, added to a dilute solution of stannous chloride, gives rise to a brownish-purple precipitate, called *purple of Cassius.* See GOLD.

The useful applications of tin are very numerous. *Tinned plate* consists of iron superficially alloyed with this metal; *pewter* of the best kind is chiefly tin, hardened by the admixture of a little antimony, etc. Cooking-vessels of copper are usually tinned in the interior. The use of tin solutions in dyeing and calico-printing has been already mentioned.

TITANIUM.

Symbol, Ti. Atomic weight, 48.

THIS is one of the rarer metals, and is never found in the metallic state. The most important titanium minerals are *rutile*, *brookite*, and *anatase*, which are different forms of titanic oxide, and the several varieties of *titaniferous iron*, consisting of ferrous titanate, sometimes alone, but more generally mixed with ferric or ferroso-ferric oxide. Occasionally in the slag adhering to the bottom of blast-furnaces in which iron ore is reduced, small brilliant copper-colored cubes are found hard enough to scratch glass, and in the highest degree infusible. This substance, of which a single smelting-furnace in the Hartz produced as much as 80 pounds, was formerly believed to be metallic titanium. Wöhler, however, has shown it to be a compound of titanium cyanide with titanium nitrite. When these crystals are powdered,

19 *

mixed with potassium hydroxide, and fused, ammonia is evolved and **potassium** titanate is formed.

Metallic titanium in a finely divided state may be obtained by heating titanium and potassium fluoride with potassium. This element is remarkable for its affinity for nitrogen: when heated in the air, it simultaneously absorbs oxygen and nitrogen.

Titanium is tetradic, like tin, and forms two classes of compounds: the titanic compounds in which it is quadrivalent, e. g., $TiCl_4$, TiO_2, and the titanous compounds, in which it is apparently trivalent, but really also quadrivalent, e. g., Ti_2Cl_6 or $Cl_3Ti.TiCl_3$.

Chlorides.—*Titanous chloride*, Ti_2Cl_6, is produced by passing the vapor of titanic chloride mixed with hydrogen through a red-hot tube; it forms dark-violet scales having a strong lustre. *Titanic chloride*, $TiCl_4$, is prepared by passing chlorine over an ignited mixture of titanic oxide and charcoal. It is a colorless, volatile, fuming liquid, having a specific gravity of 1.7609 at 0° C., vapor-density = 6.836, and boiling at 135° C. (275° F.). It unites very violently with water, and forms definite compounds with ammonia, ammonium chloride, hydrogen cyanide, cyanogen chloride, phosphine, and sulphur tetrachloride.

Fluorides.—*Titanous fluoride*, Ti_2F_6, is obtained as a violet powder by igniting potassio-titanic fluoride in hydrogen gas, and treating the resulting mass with hot water. *Titanic fluoride*, TiF_4, passes over as a fuming colorless liquid when titanic oxide is distilled with fluorspar and fuming sulphuric acid in a platinum apparatus. It unites with hydrofluoric acid and metallic fluorides, forming double salts called titanofluorides or fluotitanates, isomorphous with the silicofluorides, zircofluorides, etc., e. g., $TiF_4,2KF$, TiF_4,CaF_2.

Oxides,—The *sesquioxide*, or *Titanous oxide*, Ti_2O_3, is obtained by igniting the dioxide in hydrogen as a black powder, which, when heated in the air to a very high temperature, oxidizes to titanic oxide.

The *dioxide*, or *Titanic oxide*, occurs native in three different forms, viz., as rutile and anatase, which are dimetric; and brookite, which is trimetric; of these anatase is the purest, and rutile the most abundant. To obtain pure titanic oxide, rutile or titaniferous iron ore reduced to fine powder is fused with twice its weight of potassium carbonate, and the fused mass is dissolved in dilute hydrofluoric acid, whereupon titano-fluoride of potassium soon begins to separate. From the hot aqueous solution of this salt ammonia throws down snow-white ammonium titanate, which is easily soluble in hydrochloric acid, and when ignited gives reddish-brown lumps of titanic oxide. This oxide is insoluble in water and in all acids, except strong sulphuric acid. By fusing it with six times its weight of acid potassium sulphate a clear yellow mass is obtained, which dissolves perfectly in warm water.

Titanic oxide appears to form two hydroxides or acids analogous to stannic and metastannic acids. One of these, called titanic acid, is precipitated by ammonia from a solution of titanic chloride as a white powder, which dissolves easily in sulphuric, nitric, and hydrochloric acids, even when these acids are rather dilute; but these dilute solutions when boiled deposit metatitanic hydrate as a soft white powder, which, like the anhydrous oxide, is insoluble in all acids, except strong sulphuric acid.

The titanates have not been much studied; most of them may be represented by the formulæ $M_4TiO_4 = 2M_2O,TiO_2$, and $M_2TiO_3 = M_2O,TiO_2$ (the symbol M denoting a univalent metal). The titanates of calcium and iron occur as natural minerals. The titanates of the alkali-metals are formed by fusing titanic oxide with alkaline hydroxides, carbonates, or acid sulphates, some of them also in the wet way. When finely-pulverized and levigated, they dissolve in moderately warm concentrated hydrochloric acid; but the greater part of the dissolved titanic acid is precipitated on boiling the solution with dilute acid. The normal titanates of the alkali-metals, M_2TiO_3, are insoluble in water, but soluble in acids. The titanates of the earth-metals and heavy metals are insoluble, and may be obtained by precipitation.

In a solution of titanic acid in hydrochloric acid containing as little free acid as possible, *tincture of galls* produces an orange-colored precipitate; *potassium ferrocyanide*, a dark-brown precipitate.

Titanic oxide fused with borax, or better with microcosmic salt in the inner blowpipe flame, forms a glass which is yellow while hot, but becomes violet on cooling. The delicacy of this reaction is much increased by melting a little metallic zinc in the bead.

ZIRCONIUM.

Symbol, Zr. Atomic weight, 90.

THIS is a tetrad metal, intermediate in many of its properties between aluminium and silicon. Its oxide, zirconia, was first obtained by Klaproth in 1789, from zircon, which is a silicate of zirconium. It has since been found in fergusonite, eudialyte, and two or three other rare minerals.

Zirconium, like silicon, is capable of existing in three different states, amorphous, crystalline, and graphitoïdal.* The amorphous and crystalline varieties are obtained by processes similar to those described for preparing the corresponding modifications of silicon (p. 222); graphitoïdal zirconium was once obtained by Troost in attempting to decompose sodium zirconate with iron in light scales of a steel-gray color. Amorphous zirconium when heated in the air takes fire at a heat somewhat below redness, and burns with a bright light forming zirconia. Crystalline zirconium forms very hard brittle scales resembling antimony in color and lustre; it burns in the air only at the heat of the oxyhydrogen blowpipe, but takes fire at a red heat in chlorine gas. Zirconium is but little attacked by the ordinary acids; but hydrofluoric acid dissolves it readily, with evolution of hydrogen.

Zirconium Oxide, or **Zirconia**, ZrO_2, is prepared by strongly igniting zircon (zirconium silicate) with four times its weight of dry sodium carbonate and a small quantity of sodium hydrate. The silica is separated from the fused mass by hydrochloric acid as described in the case of beryllia; the resulting solution is treated with ammonia, which throws down zirconia, generally mixed with ferric oxide; the precipitate is redissolved in hydrochloric acid; and the solution is boiled with excess of sodium thiosulphate as long as sulphurous acid continues to escape, whereby pure zirconia is precipitated, the whole of the iron remaining in the solution. Zirconia thus obtained forms a white powder, or hard lumps, of specific gravity 4.35 to 4.9. By fusing it with borax in a pottery-furnace and dissolving out the soluble salts with hydrochloric acid, zirconia is obtained in small quadratic prisms isomorphous with the native oxides of tin and titanium.

Zirconium hydroxides are obtained by precipitating the solution of a zir-

* T r o o s t, Bull. Soc. Chim., 1866, i. 213.

conium salt with ammonia: the precipitate contains $ZrH_2O_3 = ZrO_2,H_2O$ or $ZrH_4O_4 = ZrO_2,2H_2O$, according to the temperature at which it is dried.

Zirconia acts both as a base and as an acid. After ignition it is insoluble in all acids except hydrofluoric and very strong sulphuric acid, but the hydroxide dissolves easily in acids, forming the zirconium salts; the normal sulphate has the composition $Zr(SO_4)_2$ or $ZrO_2,2SO_3$.

Compounds of zirconia with the stronger bases, called zirconates, are obtained by precipitating a zirconium salt with potash or soda, or by igniting zirconia with an alkaline hydrate. *Potassium zirconate* dissolves completely in water. Three *sodium zirconates* have been formed, containing $Na_2ZrO_3 = Na_2O,ZrO_2$; $Na_4ZrO_4 = 2Na_2O,ZrO_2$; and $Na_2Zr_8O_{17} = Na_2O,8ZrO_2$.

Zirconium Fluoride, ZrF_4, is obtained by dissolving zirconia or the hydroxide in hydrofluoric acid; or in the anhydrous state, by igniting zirconia with ammonium and hydrogen fluoride till all the ammonium fluoride is driven off. It unites with other metallic fluorides, forming double salts, called zircofluorides or fluozirconates, which are isomorphous with the corresponding silicofluorides, stannofluorides, and titanofluorides, and are mostly represented by the formulae—

$$4MF,ZrF_4; \ 3MF,ZrF_4; \ 2MF,ZrF_4; \ MF,ZrF_4,$$

in which M denotes a monad metal. The sodium salt, however, has the composition $5NaF,3ZrF_4$.

THORIUM.

Symbol, Th.   Atomic weight, 231.5.

THIS very rare metal was discovered in 1828 by Berzelius in thorite, a mineral from the Norwegian island Lovön, in which it exists as a silicate. It has since been found in euxenite, pyrochlore, and a few other minerals, all very scarce.

Metallic thorium is obtained by reducing the chloride with potassium or sodium as a gray powder, which acquires metallic lustre by pressure, and has a density of 7.66 to 7.795. It is not oxidized by water, dissolves easily in nitric, slowly in hydrochloric acid, and is not attacked by caustic alkalies.

Thorium forms but one class of compounds, in all of which it is quadrivalent.

Thorium Oxide, or **Thoria**, ThO_2, is prepared by decomposing thorite with hydrochloric acid, separating the silica in the usual way, treating the filtered solution with hydrogen sulphide to separate lead and tin, and precipitating the thoria by ammonia, together with small quantities of the oxides of iron, manganese, and uranium. To get rid of these the precipitate is redissolved in hydrochloric acid, and the hot saturated solution is boiled with a solution of normal potassium sulphate. The thorium is thereby precipitated as thorium and potassium sulphate; and from the solution of this salt in hot water the thorium is precipitated by alkalies as a hydroxide, which on ignition yields pure thoria.

Thoria is white and very heavy, its specific gravity being 9.402. After ignition it is insoluble in nitric and hydrochloric acids, and dissolves in strong sulphuric acid only after prolonged heating. The *hydroxide*, precipitated from thorium salts by alkalies, dissolves easily in acids.

Thorium Chloride, $ThCl_4$, prepared by igniting an intimate mixture of thoria and charcoal in chlorine gas, sublimes in white shining crystals. It forms double salts with the chlorides of the alkali-metals.

Thorium Sulphate, $Th(SO_4)_2$, crystallizes with various quantities of water, according to the temperature at which its solution is evaporated. *Thorium and potassium sulphate*, $ThK_4(SO_4)_4, 2H_2O$, separates as a crystalline powder when a crust of potassium sulphate is suspended in a solution of thorium sulphate. It is easily soluble in water, but insoluble in alcohol and in solution of potassium sulphate.

38

METALS OF THE ANTIMONY GROUP.

Vanadium.   Tantalum.
Antimony.   Niobium.
Bismuth.

VANADIUM.

Symbol, V.   Atomic weight, 51.2.

VANADIUM is found in small quantity in some iron ores, also as *vanadate of lead*. It has likewise been discovered in the iron slag of Staffordshire, and more recently by Roscoe,* in larger quantity in the copper-bearing beds at Alderley Edge and Mottram St. Andrews in Cheshire.

Metallic vanadium is obtained by prolonged ignition of the dichloride in pure dry hydrogen as a grayish-white powder, appearing under the microscope as a crystalline mass, with a strong silver-white lustre. It is non-volatile; decomposes water at $100°$; does not tarnish in the air; burns with brilliant scintillations when thrown into a flame; burns vividly when quickly heated in oxygen, forming the pentoxide; is insoluble in hydrochloric acid; dissolves slowly in hydrofluoric acid with evolution of hydrogen, rapidly in nitric acid, forming a blue solution. In a current of chlorine it takes fire, and is converted into the tetrachloride.

Vanadium was, till lately, regarded as a hexad metal, analogous to tungsten and molybdenum; but Roscoe has shown that it is a pentad, belonging to the phosphorus and arsenic group. This conclusion is based upon the composition of the oxides and oxychlorides, and on the isomorphism of the vanadates with the phosphates.

Vanadium Oxides.—Vanadium forms five oxides, represented by the formulæ V_2O, V_2O_2, V_2O_3, V_2O_4, V_2O_5, analogous, therefore, to the oxides of nitrogen.

The *monoxide*, V_2O, is formed by prolonged exposure of metallic vanadium to the air at ordinary temperature, more quickly at a dull red heat. It is a brown substance, which, when heated in the air, is gradually converted into the higher oxides.

The *dioxide*, V_2O_2, which was regarded by Berzelius as metallic vanadium, is obtained by reducing either of the higher oxides with potassium, or by passing the vapor of vanadium oxytrichloride ($VOCl_3$), mixed with excess of hydrogen, through a combustion tube containing red-hot charcoal. As obtained by the second process, it forms a light-gray glittering powder, or a metallically lustrous crystalline crust, having a specific gravity of 3.64, brittle, very difficult to fuse, and a conductor of electricity. When heated to redness in the air, it takes fire and burns to black oxide. It is insoluble in sulphuric, hydrochloric, and hydrofluoric acid, but dissolves easily in nitro-muriatic acid, forming a dark-blue liquid,

The dioxide may be prepared in solution by the action of nascent hydrogen (evolved by metallic zinc, cadmium, or sodium-amalgam), on a solution

* Proceedings of the Royal Society, xvi. 223.

of vanadic acid in sulphuric acid. After passing through all shades of blue and green, the liquid acquires a permanent lavender tint, and then contains the vanadium in solution as dioxide, or as hypovanadious salt. This compound absorbs oxygen more rapidly than any other known substance, and bleaches indigo and other vegetable colors as quickly as chlorine.

Vanadium dioxide may be regarded as entering into many vanadium compounds, as a bivalent radicle (just like uranyl in the uranic compounds), and may therefore be called *vanadyl*.

Vanadium trioxide, V_2O_3, or *Vanadyl monoxide*, $(V_2O_2)''O$, is obtained by igniting the pentoxide in hydrogen gas, or in a crucible lined with charcoal. It is a black powder, with a almost metallic lustre, and infusible; by pressure it may be united into a coherent mass which conducts electricity. When exposed warm to the air, it glows, absorbs oxygen, and is converted into pentoxide. At ordinary temperatures it slowly absorbs oxygen, and is converted into tetroxide. By ignition in chlorine gas it is converted into vanadyl trichloride and vanadium pentoxide. It is insoluble in acids, but may be obtained in solution by the reducing action of nascent hydrogen (evolved from metallic magnesium) on a solution of vanadic acid in sulphuric acid.

Vanadium tetroxide, Hypovanadic oxide, or *Vanadyl dioxide*, $V_2O_4 = (V_2O_2)O_2$. —This oxide is produced either by oxidation of the dioxide or trioxide, or by partial reduction of the pentoxide; also by heating hypovanadic chloride, $V_2O_4Cl_2$, to redness in an atmosphere of carbon dioxide.* By allowing the trioxide to absorb oxygen at ordinary temperatures, the tetroxide is obtained in blue shining crystals. It dissolves in acids, the more easily in proportion as it has been less strongly ignited, forming solutions of hypovanadic salts, which have a bright-blue color. The same solutions are produced by the action of moderate reducing agents, such as sulphurous, sulphydric, or oxalic acid, upon vanadic acid in solution; also by passing air through acid solutions of the dioxide till a permanent blue color is attained. With the *hydrates* and *normal carbonates of the fixed alkalies*, they form a grayish-white precipitate of hydrated oxide, V_2O_4,H_2O, which dissolves in a moderate excess of the reagent, but is reprecipitated by a large excess in the form of a vanadate of the alkali-metal.

Ammonia in excess produces a brown precipitate, soluble in pure water, but insoluble in water containing ammonia.—*Ammonium sulphide* forms a black-brown precipitate, soluble in excess.—*Tincture of galls* forms a finely divided black precipitate, which gives to the liquid the appearance of ink.

Hypovanadic trisulphate, $V_2O_4,3SO_3 + 6H_2O$, is obtained as a blue deliquescent crystalline powder by dissolving vanadic oxide in strong sulphuric acid and reducing the solution with sulphurous acid. A salt of similar character, containing $4H_2O$, is obtained by precipitating the concentrated solution with strong sulphuric acid. A *disulphate*, $V_2O_4,2SO_3 + 7H_2O$, is obtained as a light-blue crystalline powder, when the pure trisulphate, or the residue left on evaporating a solution of the tetroxide in sulphuric acid, is treated with absolute alcohol (Crow). Berzelius by similar means obtained a disulphate with $4H_2O$.

Vanadium tetroxide also unites with the more basic metallic oxides, forming salts called hypovanadates, all of which are insoluble, except those of the alkali-metals. The solutions of the alkaline hypovanadates are brown, and when treated with *hydrogen sulphide* they acquire a splendid red-purple color, arising from the formation of a sulphur salt. *Acids* color them blue, by forming a double hypovanadic salt; *tincture of galls* colors them blackish-blue. The insoluble hypovanadates, when moistened or covered with water, become green, and are converted into vanadates.

---

* J. K. Crow, Chem. Soc. Jour., 1876, ii. 453.

Crow has obtained the following hypovanadates by treating hypovanadic chloride with the corresponding bases :—

Potassium salt	.	.	.	.	$K_2O,2V_2O_4$	$+ 7H_2O$
Sodium salt .	.	.	.	$Na_2O,2V_2O_4$	$+ 7H_2O$
Ammonium salt	.	.	.	$(NH_4)_2O,2V_2O_4$	$+ 3H_2O$
Barium salt .	.	.	.	$BaO,2V_2O_4$	$+ 5H_2O.$

The *lead salt*, PbO,V_2O_4, is formed, together with potassium acetate and free acetic acid, by precipitating a solution of lead acetate with potassium hypovanadate :—

$$K_2O,2V_2O_4 + 2Pb(C_2H_3O_2)_2 + H_2O = 2(PbO,V_2O_4) + 2KC_2H_3O_2 + 2C_2H_4O_2.$$

The *silver salt*, Ag_2O,V_2O_4, is formed by a precisely similar reaction from potassium hypovanadate and silver nitrate.

Vanadium pentoxide, Vanadic oxide, or *Vanadyl trioxide,* $V_2O_5 = (V_2O_2)O_3$. —This is the highest oxide of vanadium. It may be prepared from native lead vanadate. This mineral is dissolved in nitric acid, and the lead and arsenic are precipitated by hydrogen sulphide, which at the same time reduces the vanadium pentoxide to tetroxide. The blue filtered solution is then evaporated to dryness and the residue digested in ammonia, which dissolves out the vanadic oxide reproduced during evaporation. Into this solution a lump of sal-ammoniac is put; as that salt dissolves, ammonium vanadate subsides as a white powder, being scarcely soluble in a saturated solution of ammonium chloride. By exposure to a temperature below redness in an open crucible, the ammonia is expelled and vanadic oxide left. By a similar process, Roscoe has prepared vanadic oxide from a lime precipitate containing 2 per cent. of vanadium, obtained in working up a poor cobalt ore from Mottram in Cheshire.

Vanadium pentoxide has a reddish-yellow color, and dissolves in 1000 parts of water, forming a light yellow solution. It dissolves also in the stronger acids, forming red or yellow solutions, some of which yield crystalline compounds (vanadic salts) by spontaneous evaporation. It unites, however, with bases more readily than with acids, forming salts called vanadates. When fused with alkaline carbonates, it eliminates 3 molecules of carbon dioxide, forming *orthovanadates* analogous to the orthophosphates; thus:—

$$3(Na_2O,CO_2) + V_2O_5 = 3Na_2O,V_2O_5 + 3CO_2.$$

It also forms *metavanadates* and *pyrovanadates* analogous to the meta- and pyro-phosphates, and two series of *acid vanadates* or *anhydrovanadates,* viz. :—

Lead orthovanadate	$Pb_3(VO_4)_2$	or	$3PbO,V_2O_5$
Barium pyrovanadate,	$Ba_2V_2O_7$	or	$2BaO,V_2O_5$
Strontium metavanadate,	$Sr(VO_3)_2$	or	SrO,V_2O_5
Strontium divanadate,	$Sr(VO_3)_2,V_2O_5$	or	$SrO,2V_2O_5$
Strontium trivanadate,	$Sr(VO_3)_2,2V_2O_5$	or	$SrO,3V_2O_5.$

Lead metavanadate occurs native as *dechenite;* the orthovanadate also, combined with lead chloride, as *vanadinite* or *vanadite*, $PbCl_2,3Pb_3(VO_4)_2$, the mineral in which vanadium was first discovered. *Descloizite* is a diplumbic vanadate, $Pb_2V_2O_7$, or $2PbO,V_2O_5$, analogous in composition to a pyrophosphate.

The metavanadates are mostly yellow; some of them, however, especially those of the alkaline earth-metals, and of zinc, cadmium, and lead, are converted by warming—either in the solid state, or under water, or in aqueous solution, especially in presence of a free alkali or alkaline carbonate—into isomeric colorless salts. The same transformation takes place also, though more slowly, at ordinary temperatures. The metavanadates of alkali-metal are colorless. The acid vanadates are yellow or yellowish-red, both in the solid state and in solution : hence the solution of a neutral vanadate becomes yellowish-red on addition of an acid. The metavanadates of ammonium, the alkali-metals, barium, and lead, are but sparingly soluble in water; the other metavanadates are more soluble. The alkaline vanadates are more soluble in pure water than in water containing free alkali or salt : hence they are precipitated from their solutions by addition of alkali in excess or of salts. The vanadates are insoluble in alcohol. The aqueous solutions of vanadates form yellow precipitates with *antimony, copper, lead*, and *mercury* salts ; with *tincture of galls* they form a deep-black liquid, which has been proposed for use as vanadium ink.

Hydrogen sulphide reduces them to hypovanadates, changing the color from red or yellow to blue, and forming a precipitate of sulphur.

Ammonium sulphide colors the solutions brown-red, and, on adding an acid, a light-brown precipitate is formed, consisting of vanadic sulphide mixed with sulphur, the liquid at the same time turning blue. *Hydrochloric acid* decomposes the vanadates, with evolution of chlorine and formation of vanadium tetroxide.

Vanadium Chlorides.—Three of these compounds have been obtained, viz. : VCl_2, VCl_3, and VCl_4.

The *tetrachloride*, VCl_4, is formed when metallic vanadium or the mononitride is heated in a current of chlorine, or when the vapor of the oxytrichloride, $VOCl_3$, mixed with chlorine, is passed several times over red-hot charcoal. It is a dark yellowish-brown liquid, having a specific gravity of 1.8384 at 0° C., boiling at 154° C. (309° F.), not solidifying at 18° C. (64° F.). Its vapor-density referred to hydrogen is 96.6, which is half the molecular weight ($= \dfrac{51.5 + 4 \times 35.4}{2}$), showing that the molecule VCl_4 exhibits the normal condensation to 2 volumes of vapor. The tetrachloride is quickly decomposed by water, forming a blue solution of vanadious acid. It does not take up bromine or an additional quantity of chlorine when heated therewith in sealed tubes : hence it appears that vanadium does not readily form pentad compounds with the monatomic chlorous elements.

The *trichloride*, VCl_3, is obtained by decomposition of the tetrachloride, slowly at ordinary temperatures, quickly at the boiling heat; also by warming the trisulphide in a current of chlorine. It crystallizes in peach-blossom-colored shining plates resembling chromic chloride. It is slowly decomposed by water, forming a green solution of hypovanadic acid.

The *dichloride*, VCl_2, obtained by passing the vapor of the tetrachloride mixed with hydrogen through a red-hot tube, crystallizes in green micaceous plates which are decomposed by water, forming a violet solution of hypovanadious acid.

Vanadium Oxychlorides, or **Vanadyl Chlorides.**—Four of these compounds are known, viz. : $VOCl_3$, $VOCl_2$, $VOCl$, and V_2O_2Cl.

The *oxytrichloride*, $VOCl_3$ (formerly regarded as vanadium trichloride), is prepared :

(1) By the action of chlorine on the trioxide :—

$$3V_2O_3 + Cl_{12} = V_2O_5 + 4VOCl_3.$$

(2) By burning the dioxide in chlorine gas, or by passing that gas over an ignited mixture of the trioxide, tetroxide, or pentoxide, and condensing the vapors in a cooled U-tube.

Vanadium oxytrichloride, or vanadyl trichloride, is a golden-yellow liquid, of specific gravity 1.841 at 14.5° C. (58.1° F.).   Boiling-point, 127° C. (260.6° F.).   Vapor-density by experiment, 6.108; by calculation, 6.119.   When exposed to the air it emits cinnabar-colored vapors, being resolved by the moisture of the air into hydrochloric and vanadic acids.   It oxidizes magnesium and sodium.   Its vapor, passed over perfectly pure carbon at a red heat, yields carbon dioxide; and when passed, together with hydrogen, through a red-hot tube it yields vanadium trioxide.   These reactions show that the compound contains oxygen.

The other oxychlorides of vanadium are solid bodies obtained by partial reduction of the oxytrichloride with zinc or hydrogen.

The second, $VOCl_2$ or $V_2O_2Cl_4$ (hypovanadic chloride), is also produced by dissolving the pentoxide with aid of heat in hydrochloric acid, and reducing the green solution with sulphurous acid.   The resulting blue liquid leaves, on evaporation over the water-bath, a brown deliquescent residue which yields a blue solution with water (Crow).

The *tribromide*, VBr_3, and the *oxybromides*, $VOBr_3$ and $VOBr_2$, have also been obtained.   The first is a grayish-black amorphous solid; the second a dark-red liquid; the third a yellowish-brown deliquescent solid.

Vanadium Sulphides.—Two of these compounds are known, analogous to the tetroxide and pentoxide: both are sulphur-acids.   The *tetrasulphide*, or *Vanadious sulphide*, V_2S_4, is a black substance formed by heating the tetroxide to redness in a stream of hydrogen sulphide; also as a hydrate by dissolving a vanadious salt in excess of an alkaline monosulphide, and precipitating with hydrochloric acid.   The *pentasulphide*, or *Vanadic sulphide*, V_2S_5, is formed in like manner by precipitation from an alkaline vanadate.

Vanadium Nitrides.—The *mononitride*, VN, is formed by heating the compound of vanadium oxytrichloride with ammonium chloride to whiteness in a current of ammonia gas.   It is a greenish-white powder, unalterable in the air.   The *dinitride*, VN_2 or V_2N_4, is obtained by exposing the same double salt in ammonia gas to a moderate heat.   It is a black powder strongly acted upon by nitric acid.

All vanadium compounds heated with borax or phosphorous salt in the outer blowpipe flame produce a clear bead, which is colorless if the quantity of vanadium is small, yellow when it is large; in the inner flame the bead acqnires a beautiful green color.

Vanadic and chromic acids are the only acids whose solutions are red; they are distinguished from one another by the vanadic acid becoming blue and the chromic acid green by deoxidation.

When a solution of vanadic acid, or an acidulated solution of an alkaline vanadate, is shaken up with ether containing hydrogen dioxide, the aqueous solution acquires a red color like that of ferric acetate, while the ether remains colorless.   This reaction will serve to detect the presence of 1 part of vanadic acid in 40,000 parts of liquid.   The other reactions of vanadium in solution have already been described.

ANTIMONY.

Symbol, Sb (Stibium). Atomic weight, 122.

THIS important metal is found chiefly in the state of sulphide. To obtain it in the free state the ore is freed by fusion from earthy impurities, and is afterward decomposed by heating with metallic iron or potassium carbonate, which retains the sulphur.

Antimony has a bluish-white color, a strong lustre, and a granular or coarsely-laminated crystalline fracture, according as it is quickly or slowly cooled. By melting it in a crucible, then leaving it to cool partially, and pouring out the still liquid portion, it may be obtained in rhombohedral crystals. It is extremely brittle and easily pulverized. Its specific gravity is 6.8. It melts just below redness, and boils and volatilizes at a white heat.

On electrolyzing a solution of 1 part of tartar-emetic in 4 parts of antimonious chloride with a small battery of two elements, antimony forming the positive and metallic copper the negative pole, crusts of antimony are obtained, which possess the remarkable property of exploding and catching fire when scratched with a metal point or touched with a red-hot wire.

Antimony is not oxidized by the air at common temperatures; but when strongly heated it burns with a white flame producing oxide, which is often deposited in fine crystals. It is dissolved by hot hydrochloric acid, with evolution of hydrogen and production of antimonious chloride. Nitric acid oxidizes it to antimonic acid, which is insoluble in that liquid.

Antimony forms two classes of compounds—the antimonious compounds, in which it is trivalent, as $SbCl_3$, Sb_2O_3, Sb_2S_3, etc., and the antimonic compounds, in which it is quinquivalent, as $SbCl_5$, Sb_2O_5, Sb_2S_5, etc.

Antimonious Hydride; Antimonetted Hydrogen; Stibine, SbH_3.—When zinc is put into a solution of antimonious oxide and sulphuric acid added, part of the hydrogen combines with the antimony, and the resulting gas, which is a mixture of stibine with free hydrogen, burns with a greenish flame, giving rise to white fumes of antimonious oxide. When the gas is conducted through a red-hot glass tube of narrow dimensions, or burned with a limited supply of air, as when a cold porcelain surface is pressed into the flame, metallic antimony is deposited. On passing a current of antimonetted hydrogen through a solution of silver nitrate, a black precipitate is obtained, containing $SbAg_3$: from the formation of this compound it is inferred that the gas has the composition SbH_3, analogous to ammonia, phosphine, and arsine. There are also several analogous compounds of antimony with alcohol-radicles, such as *trimethylstibine,* $Sb(CH_3)_3$, *triethylstibine,* $Sb(C_2H_5)_3$, etc.

Chlorides.—The *trichloride,* or *Antimonious chloride,* $SbCl_3$, formerly called *butter of antimony,* is produced when hydrogen sulphide is prepared by the action of strong hydrochloric acid on antimonious sulphide. The impure and highly acid solution thus obtained is put into a retort and distilled until each drop of the condensed product, on falling into the aqueous liquid of the receiver, produces a copious white precipitate. The receiver is then changed and the distillation continued. Pure antimonious chloride then passes over and solidifies on cooling to a white, highly crystalline mass, from which the air must be carefully excluded. The same compound is formed by distilling metallic antimony in powder with $2\frac{1}{2}$ times its weight of mercuric chloride. Antimonious chloride is very deliquescent: it dissolves in strong hydrochloric acid without decomposition, and the solution when poured into water gives rise to a white bulky precipitate, which, after a short time, becomes highly crystalline and assumes a pale-fawn color. This is the old *powder of Algaroth*. It has the composition SbOCl, and may be regarded as a compound of the trichloride and trioxide $SbCl_3,Sb_2O_3$. Alkaline solutions extract the chloride and leave the oxide. On heating this oxychloride, the trichloride is given off and another oxychloride, $Sb_4O_5Cl_2$, remains behind: $5SbOCl = SbCl_3 + Sb_4O_5Cl_2$. When antimonious oxide is dissolved in boiling antimonious chloride, a pearl-gray crystalline mass separates on cooling, having the composition $SbOCl,7SbCl_3$, and this when treated with absolute alcohol yields the compound $Sb_2O_3,2SbOCl$.

The *pentachloride*, $SbCl_5$, is formed with brilliant combustion when finely-powdered antimony is thrown into chlorine gas. It may be prepared by passing dry chlorine over pulverized antimony gently heated in a tubulated retort or over the trichloride. It is a colorless volatile liquid, which forms a crystalline compound with a small quantity of water, but is decomposed by a larger quantity, yielding antimonic and hydrochloric acids.

Tribromide and *Tri-iodide of Antimony* are also formed by direct combination, the reaction being attended with evolution of light and heat. The bromide sublimes in colorless, deliquescent needles, melts at 95° C. (203° F.), sublimes at 275° C. (527° F.), and is decomposed by water with formation of an oxybromide. The tri-iodide is a cinnabar-red powder, soluble in carbon sulphide, and crystallizing therefrom in six-sided tablets.

Antimony Trifluoride, SbF_3, is obtained as a dense snow-white powder by distilling antimony with mercuric fluoride, and in rhombic pyramids by evaporating a solution of antimonious oxide in excess of hydrofluoric acid. It is deliquescent, and is not decomposed by water. It forms crystalline double salts with fluorides of alkali-metal. The *pentafluoride*, SbF_5, obtained by dissolving antimonic acid in hydrofluoric acid is a gummy mass, which is decomposed by heat. It unites with the fluorides of the alkali-metals, forming difficultly crystallizable double salts.

Oxides.—Antimony forms two oxides, Sb_2O_3 and Sb_2O_5, analogous to the chlorides, the first being a basic and the second an acid oxide, also an intermediate neutral oxide, Sb_2O_4. The *trioxide*, or *Antimonious oxide*, Sb_2O_3, occurs native, though rarely, as *valentinite*, or *white antimony*, in shining white trimetric crystals; also as *senarmontite* in regular octahedrons: it is therefore dimorphous. It may be prepared by several methods: as by burning metallic antimony at the bottom of a large red-hot crucible, in which case it is obtained in brilliant crystals; or by pouring solution of antimonious chloride into water and digesting the resulting precipitate with a solution of sodium carbonate. The oxide thus produced is anhydrous; it is a pale buff-colored powder, fusible at a red heat, and volatile in a closed vessel; but in contact with air at a high temperature it absorbs oxygen and becomes changed into the tetroxide. When boiled with cream of tartar (acid potassium tartrate) it is dissolved, and the solution yields on evaporation crystals of *tartar emetic*, which is almost the only antimonious salt that can bear a mixture with water without decomposition. An impure oxide for this purpose is sometimes prepared by carefully roasting the powdered sulphide in a reverberatory furnace, and raising the heat at the end of the process, so as to fuse the product: it has long been known under the name *glass of antimony*, or *vitrum antimonii*.

Antimonious oxide likewise acts as a weak acid, forming salts called *Antimonites*, which, however, are very unstable.

The *tetroxide*, or *Antimonoso-antimonic oxide*, Sb_2O_4 or Sb_2O_3,Sb_2O_5, occurs native as *cervantite*, or *antimony ochre*, in acicular crystals or as a crust or powder. It is the ultimate product of the oxidation of the metal by heat and air: it is a grayish-white powder, infusible and non-volatile, insoluble in water and acids, except when recently precipitated.

On treating it with tartaric acid (acid potassium tartrate), antimonious oxide is dissolved, antimonic acid remaining behind; and when a solution of the tetroxide in hydrochloric acid is gradually dropped into a large quantity of water, antimonious oxide is precipitated, while antimonic acid remains dissolved. From these and similar reactions it has been inferred that the tetroxide is a compound of the trioxide and pentoxide. On the other hand, it is sometimes regarded as a distinct oxide, because it dissolves without decomposition in alkalies, forming salts (often called *antimonites*) which may be obtained in the solid state. Two potassium salts, for example, have been formed, containing K_2O,Sb_2O_4 and $K_2O,2Sb_2O_4$; and a calcium salt, $3CaO,2Sb_2O_4$, occurs as a natural mineral, called *romeine*. These salts may, however, be regarded as compounds of antimonates and antimonites (containing Sb_2O_3): thus, $2(K_2O,Sb_2O_4) = K_2O,Sb_2O_5 + K_2O,Sb_2O_3$.

The *pentoxide*, or *Antimonic oxide*, Sb_2O_5, is formed as an insoluble hydrate or hydroxide when strong nitric acid is made to act upon metallic antimony; and on exposing this hydrate to a heat short of redness, it yields the anhydrous pentoxide as a pale straw-colored powder, insoluble in water and acids. It is decomposed by a red heat, yielding the tetroxide.

Antimonic hydroxide is likewise obtained by decomposing antimony pentachloride with an excess of water, hydrochloric acid being formed at the same time. The hydroxides or antimonic acids produced by the two processes mentioned differ in many of their properties, and especially in their deportment with bases. The acid produced by nitric acid, called antimonic acid, is monobasic, producing normal salts of the form M_2O,Sb_2O_5 or $MSbO_3$, and acid salts containing $M_2O,2Sb_2O_5$ or $2MSbO_3,$-Sb_2O_5. The other, called metantimonic acid, is bibasic, forming normal salts containing $2M_2O,Sb_2O_5$ or $M_4Sb_2O_7$, and acid salts containing $2M_2O,2Sb_2O_5$ or M_2O,Sb_2O_5, so that the acid metantimonates are isomeric or polymeric with the normal antimonates. Among the metantimonates an acid potassium salt, $K_2O,Sb_2O_5,7H_2O$, is to be particularly noticed as yielding a precipitate with sodium salts: it is, indeed, the only reagent which precipitates sodium. It is obtained by fusing antimonic oxide with an excess of potash in a silver crucible, dissolving the fused mass in a small quantity of cold water, and allowing it to crystallize in a vacuum. The crystals consist of normal potassium metantimonate, $2K_2O,Sb_2O_5$, and, when dissolved in pure water, are decomposed into free potash and acid metantimonate.

Sulphides.—The *trisulphide*, or *Antimonious sulphide*, Sb_2S_3, occurs native as a lead-gray, brittle substance, having a radiated crystalline texture, and easily fusible. It may be prepared artificially by melting together antimony and sulphur. When a solution of tartar-emetic is precipitated by hydrogen sulphide, a brick-red precipitate falls, which is the same substance combined with a little water. If the precipitate be dried and gently heated, the water may be expelled without other change of color than a little darkening, but at a higher temperature it assumes the color and aspect of the native sulphide. This remarkable change probably indicates a passage from the amorphous to the crystalline state. When powdered antimonious sulphide is boiled in a solution of caustic potash, it is dissolved, antimonious oxide and potassium sulphide being produced; and the latter unites with an additional quantity of antimonious sulphide to form a soluble sulphur-salt, in which the potassium sulphide is the sulphur-base, and the antimonious sulphide is the sulphur-acid:—

$$3K_2O + 2Sb_2S_3 = Sb_2O_3 + 3K_2S,Sb_2S_3.$$

The antimonious oxide separates in small crystals from the boiling solution

when the latter is concentrated, and the sulphur-salt dissolves an extra portion of antimonious sulphide, which it again deposits on cooling as a red
amorphous powder, containing a small admixture of antimonious oxide and
potassium sulphide. This is the *kermes mineral* of the old chemists. The
filtered solution mixed with an acid gives a potassium salt, hydrogen sulphide, and precipitated antimonious sulphide. Kermes may also be made
by fusing a mixture of 5 parts of antimonious sulphide and 3 of dry sodium
carbonate, boiling the mass in 80 parts of water, and filtering while hot : the
compound separates on cooling. The compounds of antimonious sulphide
with basic sulphides are called *thio-antimonites;* many of them occur as
natural minerals. For example : zinkenite, PbS,Sb_2S_3 ; feather-ore, $2PbS,$-
Sb_2S_3 ; boulangerite, $3PbS,Sb_2S_3$; fahlore, or tetrahedrite, $4Cu_2S,Sb_2S_3$, the
antimony being more or less replaced by arsenic, and the copper by silver,
iron, zinc, and mercury.

The *pentasulphide*, or *Antimonic sulphide*, Sb_2S_5, formerly called *sulphur
auratum*, is also a sulphur-acid, forming salts called *thio-antimonates*, most
of which have the composition $3M_2S,Sb_2S_5$ or M_3SbS_4, analogous to the
normal orthophosphates and arsenates. When 18 parts of finely powdered
antimonious sulphide, 17 parts dry sodium carbonate, 13 parts slaked lime,
and 3¼ parts sulphur are boiled in water for some hours, calcium carbonate,
sodium antimonate, antimony pentasulphide, and sodium sulphide are produced. The first is insoluble, and the second partially so : the two last-
named bodies, on the contrary, unite to form soluble sodium thio-antimonate,
Na_3SbS_4, which may be obtained by evaporation in beautiful crystals. A
solution of this substance, mixed with dilute sulphuric acid, furnishes sodium
sulphate, hydrogen sulphide, and antimony pentasulphide, which falls as a
golden-yellow flocculent precipitate.

The thio-antimonates of the alkali-metals and alkaline earth-metals are
very soluble in water, and crystallize for the most part with several molecules of water. Those of the heavy metals are insoluble, and are obtained
by precipitation.

The few salts of antimony soluble in water are distinctly characterized
by the orange or brick-red precipitate with *hydrogen sulphide*, which is
soluble in a solution of ammonium sulphide, and again precipitated by an
acid.

Antimonious chloride, as already observed, is decomposed by *water*, yielding a precipitate of oxychloride. The precipitate dissolves in hydrochloric
acid, and the resulting solution gives, with *potash*, a white precipitate of
trioxide, soluble in a large excess of the reagent; with *ammonia* the same,
insoluble in excess ; with *potassium* or *sodium carbonate*, also a precipitate
of trioxide, which dissolves in excess, especially of the potassium salt, but
reappears after a while. If, however, the solution contains *tartaric acid*, the
precipitate formed by potash dissolves easily in excess of the alkali ; ammonia forms but a slight precipitate, and the precipitates formed by alkaline
carbonates are insoluble in excess. The last-mentioned characters are likewise exhibited by a solution of tartar-emetic (potassio-antimonious tartrate).
Zinc and *iron* precipitate antimony from its solutions as a black powder.
Copper precipitates it as a shining metallic film, which may be dissolved off
by potassium permanganate, yielding a solution which will give the characteristic red precipitate with hydrogen sulphide.

Solid antimony compounds, fused upon charcoal with sodium carbonate
or potassium cyanide, yield a brittle globule of antimony, a thick white
fume being at the same time given off, and the charcoal covered to some
distance around with a white deposit of oxide.

Besides its application to medicine, antimony is of great importance in the arts, inasmuch as, in combination with lead, it forms *type-metal*. This alloy expands at the moment of solidifying and takes an exceedingly sharp impression of the mould. It is remarkable that both its constituents shrink under similar circumstances, and make very bad castings.

Britannia metal is an alloy of 9 parts tin and 1 part antimony, frequently also containing small quantities of copper, zinc, or bismuth. An alloy of 12 parts tin, 1 part antimony, and a small quantity of copper forms a superior kind of pewter. Alloys of antimony with tin, or tin and lead, are now much used for machinery-bearings in place of gun-metal. Alloys of antimony with nickel and with silver occur as natural minerals.

Antimony trisulphide enters into the composition of the blue or Bengal lights used at sea, which contain dry nitre 6 parts, sulphur 2 parts, antimony trisulphide 1 part; all in fine powder and intimately mixed.

BISMUTH.

Symbol, Bi. Atomic weight, 210.

BISMUTH is found chiefly in the metallic state disseminated through various rocks, from which it may be separated by simple fusion. It occurs as the trioxide or *bismuth ochre*, Bi_2O_3, less frequently as *bismuthite*, Bi_2S_3, and still more sparingly in a few other minerals, as *tetradymite*, or *telluric bismuth*, Bi_2Te_3, *eulytin* or bismuth silicate, $Bi(SiO_4)_3$, etc.

Extraction.—The chief sources of bismuth are the Saxon smalt-works, where ores containing bismuth and cobalt are worked. Formerly the bismuth was extracted by simply heating these ores in sloping tubes; but in this way only the portion existing in the metallic state was obtained, and not the whole even of that. The residue was used in the preparation of smalt, and the bismuth again extracted from the cobalt-speiss. At present, however, all bismuth ores are roasted, and afterward smelted in the pots of the smalt-furnaces, with addition of iron, carbon, and slag. Two layers are thus obtained, the lower consisting of nearly pure bismuth, which, owing to its low melting point, can be drawn off in the liquid state after the upper layer of cobalt-speiss has solidified. The crude bismuth thus obtained contains only small traces of iron, cobalt, lead, and a few other metals, from which it may be purified by heating it on a slightly inclined iron plate placed over a fire, so that the metal melts and runs down.[*] A considerable quantity of bismuth from various parts of the world is now worked up in England at the magnesium works at Patricroft.

Bismuth is highly crystalline and very brittle; it has a grayish-white color, with distinct reddish tinge, a specific gravity of 9.823 at 12° C. (64.4° F.), and melts at 270° C. (518° F.), expanding in the act of solidification. It may be obtained in a mass of very fine crystals by melting a considerable quantity, leaving it to cool till it begins to solidify, then piercing the crust and pouring out the liquid residue. The crystals thus obtained are rhombohedrons with angles of nearly 90°, so that they present. the appearance of cubes. Bismuth volatilizes at a high temperature. It is remarkable as being the most diamagnetic of all known bodies. It is a little oxidized by

---

[*] For further details and description of other processes, see Winckler (Bericht über die Entwickelung der chemischen Industrie, i. 953); and Roscoe and Schorlemmer's Treatise on Chemistry, vol. ii. Part ii. p. 330.

the air, but burns with a bluish flame when strongly heated. Nitric acid somewhat diluted dissolves it freely.

Bismuth forms three classes of compounds, in which it is bi-, tri-, and quinqui-valent respectively. The tri-compounds are the most stable and the most numerous. The only known compounds in which bismuth is quinquivalent are indeed the pentoxide, Bi_2O_5, together with the corresponding acid and metallic salts. Nevertheless, bismuth is regarded as a pentad on account of the analogy of its compounds with those of antimony. Several bismuth compounds are known in which the metal is apparently bivalent, but really trivalent, as—

$$Bi_2Cl_4, \text{ or } \begin{matrix} BiCl_2 \\ | \\ BiCl_2 \end{matrix}; \qquad Bi_2O_2, \text{ or } \begin{matrix} Bi\!=\!\!\!=\!O \\ | \\ Bi\!=\!\!\!=\!O \end{matrix}, \text{ etc.}$$

Chlorides.—The *trichloride*, or *Bismuthous chloride*, is formed when bismuth is heated in a current of chlorine gas, and passes over as a white, easily fusible substance, which readily attracts moisture from the air and is converted into a crystallized hydrate. The same substance is produced when bismuth is dissolved in nitro-muriatic acid and the solution evaporated. Bismuthous chloride dissolves in water containing hydrochloric acid, but is decomposed by pure water, yielding a white precipitate of oxychloride:—

$$BiCl_3 + H_2O = BiClO + 2HCl.$$

The *dichloride*, Bi_2Cl_4, produced by heating the trichloride with metallic bismuth, is a brown, crystalline, easily fusible mass, decomposed by water. At a high temperature it is resolved into the trichloride and metallic bismuth.

Oxides.—The *trioxide*, or *Bismuthous oxide*, is a straw-yellow powder, obtained by gently igniting the neutral or basic nitrate. It is fusible at a high temperature, and in that state acts toward siliceous matter as a powerful flux.

The *hydroxide*, $BiHO_2$ or Bi_2O_3,H_2O, is obtained as a white precipitate when a solution of the nitrate is decomposed by an alkali. Both the hydroxide and the anhydrous oxide dissolve in the stronger acids, forming the bismuthous salts, which have the composition BiR_3, where R denotes an acid radicle, e. g., $BiCl_3$, $Bi(NO_3)_3$, $Bi_2(SO_4)_3$. Many of these salts crystallize well, but cannot exist in solution unless an excess of acid is present. On diluting the solutions with water, a basic salt is precipitated and an acid salt remains in solution.

The *normal nitrate*, $Bi(NO_3)_3,5H_2O$ or $Bi_2O_3,3N_2O_5,10H_2O$, forms large, transparent, colorless crystals, which are decomposed by water in the manner just mentioned, yielding an acid solution containing a little bismuth and a brilliant white, crystalline powder, which varies to a certain extent in composition according to the temperature and the quantity of water employed, but frequently consists of a basic nitrate, $Bi_2O_3,N_2O_5,2H_2O$ or $Bi(NO_3)_3,Bi_2O_3,3H_2O$. A solution of bismuth nitrate free from any great excess of acid, poured into a large quantity of cold water, yields an insoluble basic nitrate very similar in appearance to the above, but containing rather a larger proportion of bismuth oxide. This basic nitrate was once extensively employed as a cosmetic, but it is said to injure the skin, rendering it yellow and leather-like. It is used in medicine.

Bismuth pentoxide, or *Bismuthic oxide*, Bi_2O_5.—When bismuth trioxide is suspended in a strong solution of potash and chlorine passed through the liquid, decomposition of water ensues, hydrochloric acid being formed and the trioxide being converted into the pentoxide. To separate any trioxide that may have escaped oxidation, the powder is treated with dilute nitric acid when the bismuthic oxide is left as a reddish powder, which is insoluble in water. This substance combines with bases, but the compounds are not very well known. According to Arppe, there is an acid potassium bismuthate containing Bi_2KHO_6 or $2Bi_2O_5, \begin{cases} K_2O \\ H_2O \end{cases}$. The pentoxide when heated loses oxygen, an intermediate oxide, Bi_2O_4, being formed, which may be considered as *bismuthous bismuthate*, $2Bi_2O_4 = Bi_2O_3, Bi_2O_5$.

Bismuth is sufficiently characterized by the decomposition of its nitrate and chloride by water, and by the black precipitate of bismuth sulphide, insoluble in ammonium sulphide, which its solutions yield when exposed to the action of hydrogen sulphide.

A mixture of 8 parts of bismuth, 5 parts of lead, and 3 of tin is known under the name of *fusible metal*, and is employed in taking impressions from dies and for other purposes: it melts below 100°.

Bismuth is used in conjunction with antimony in the construction of thermo-electric piles, these two metals forming the opposite extremes of the thermo-electric series.

TANTALUM.

Symbol, Ta. Atomic weight, 182.

THIS metal was discovered in 1803 by Ekeberg in two Swedish minerals, tantalite and yttrotantalite. A very similar metal, *columbium*, had been discovered in the preceding year by Hatchett in columbite from Massachusetts; and Wollaston in 1807, on comparing the compounds of these metals, concluded that they were identical, an opinion which was for many years received as correct; but their separate identity has been completely established by the researches of H. Rose (commenced in 1846), who gave to the metal from the American and Bavarian columbites the name *Niobium*, by which it is now universally known. More recently, Marignac has shown that nearly all tantalites and columbites contain both tantalum and niobium (or columbium), some tantalates, from Kimito in Finland, being, however, free from niobium, and some of the Greenland columbites containing only the latter metal unmixed with tantalum. In all these minerals tantalum exists as a tantalate of iron and manganese; yttrotantalite is essentially a tantalate of yttrium, containing also uranium, calcium, iron, and other metals. Tantalum is also contained in some varieties of wolfram.

Metallic tantalum is obtained by heating the fluotantalate of potassium or sodium with metallic sodium in a well-covered iron crucible, and washing out the soluble salts with water. It is a black powder, which, when heated in the air, burns with a bright light, and is converted, though with difficulty, into tantalic oxide. It is not attacked by sulphuric, hydrochloric, nitric, or even nitro-muriatic acid. It dissolves slowly in warm, aqueous, hydrofluoric acid, with evolution of hydrogen, and very rapidly in a mixture of hydrofluoric and nitric acids.

Tantalum in its principal compounds is quinquivalent, the formula of tantalic chloride being $TaCl_5$, that of tantalic fluoride, TaF_5, and that of tantalic oxide (which in combination with bases forms the tantalates), Ta_2O_5. There is also a tantalous oxide said to have the composition TaO_2, and a corresponding sulphide, TaS_2.

20

Tantalic Chloride, $TaCl_5$, is obtained as a yellow sublimate, by igniting an intimate mixture of tantalic oxide and charcoal in a stream of chlorine gas. It begins to volatilize at 144° C. (291° F.), and melts to a yellow liquid at 221° C. (430° F.). The vapor-density between 350° and 400° C. (662–824° F.) has been found by Deville and Troost to be 12.42 referred to air, or 178.9 referred to hydrogen: by calculation for the normal condensation to two volumes it is 179.75. Tantalic chloride is decomposed by water, yielding hydrochloric and tantalic acids; but the decomposition is not complete even at the boiling heat.

Tantalic Fluoride, TaF_5, is obtained in solution by treating tantalic hydroxide with aqueous hydrofluoric acid. The solution, mixed with alkaline fluorides, forms soluble crystallizable salts, called t a n t a l o - f l u o r i d e s or f l u o t a n t a l a t e s. The potassium salt, TaK_2F_7 or $TaF_5,2KF$, crystallizes in monoclinic prisms, isomorphous with the corresponding fluoniobate.

Tantalic Oxide, Ta_2O_5, is produced when tantalum burns in the air, also by the action of water on tantalic chloride, and may be separated as a hydroxide from the tantalates by the action of acids. It may be prepared from tantalite, which is a tantalate of iron and manganese, by fusing the finely-pulverized mineral with twice its weight of potassium hydroxide, digesting the fused mass in hot water, and supersaturating the filtered solution with hydrochloric or nitric acid: tantalic hydroxide is then precipitated in white flocks, which may be purified by washing with water.*

Anhydrous tantalic oxide, obtained by igniting the hydroxide or sulphate, is a white powder, varying in density from 7.022 to 8.264, according to the temperature to which it has been exposed. Heated in ammonia gas, it yields tantalum nitride; heated with carbon bisulphide, it is converted into tantalum bisulphide. It is insoluble in all acids, and can be rendered soluble only by fusion with potassium hydroxide or carbonate.

Tantalum hydroxide, or *Tantalic acid*, obtained by precipitating an aqueous solution of potassium tantalate with hydrochloric acid, or by decomposing the chloride with water, is a snow-white, bulky powder, which dissolves in hydrochloric and hydrofluoric acids; when strongly heated, it glows and gives off water.

Tantalic oxide unites with basic metallic oxides, forming the tantalates, which are represented by the formulæ, M_2O,Ta_2O_5 and $4M_2O,3Ta_2O_5$, the first including the native tantalates, such as ferrous tantalate, and the second certain easily crystallizable tantalates of the alkali-metals. The tantalates of the alkali-metals are soluble in water, and are formed by fusing tantalic oxide with caustic alkalies; those of the earth-metals and heavy metals are insoluble, and are formed by precipitation.

Tantalum dioxide, or *Tantalous oxide*, TaO_2, may be represented by the

formula $\begin{matrix} O = Ta = O \\ | \\ O = Ta = O \end{matrix}$, in which the metal is still quinquivalent. It is produced by exposing tantalic oxide to an intense heat in a crucible lined with charcoal. It is a hard, dark-gray substance, which, when heated in the air, is converted into tantalic oxide.

Hydrochloric or *sulphuric acid*, added in excess to a solution of alkaline tantalate, forms a precipitate of tantalic acid, which redissolves in excess of the hydrochloric, but not of the sulphuric acid. *Potassium ferrocyanide*, added to a very slightly acidulated solution of an alkaline tantalate, forms a yellow precipitate; the *ferricyanide*, a white precipitate. *Infusion of galls* forms a light-yellow precipitate, soluble in alkalies. When tantalic chloride

* For more complete methods of preparation, see Watts's Dictionary of Chemistry, vol. v. p. 665.

is dissolved in strong sulphuric acid, and then water and metallic *zinc* are added, a fine blue color is produced, which does not turn brown, but soon disappears.

Tantalic oxide fused with microcosmic salt in either blowpipe flame forms a clear, colorless glass, which does not turn red on addition of a ferrous salt. With borax it also forms a transparent glass, which may be rendered opaque by interrupted blowing or *flaming*.

NIOBIUM, or COLUMBIUM.

Symbol, Nb.   Atomic weight, 94.

THIS metal, discovered in 1801 by Hatchett in American columbite, exists likewise, associated with tantalum, in columbites from other sources and in most tantalates; also associated with yttrium, uranium, iron, and small quantities of other metals in Siberian samarskite, uranotantalite, or yttroilmenite; also in pyrochlore, euxenite, and a variety of pitchblende from Satersdälen in Norway.

The metal, obtained in the same manner as tantalum, is a black powder, which oxidizes with incandescence when heated in the air. It dissolves in hot hydrofluoric acid, with evolution of hydrogen, and at ordinary temperatures in a mixture of hydrofluoric and nitric acid; slowly also when heated with strong sulphuric acid. It is oxidized by fusion with acid potassium sulphate, and gradually converted into potassium niobate by fusion with potassium hydrate or carbonate.

Niobium is quinquivalent, and forms only one class of compounds, namely, a chloride, $NbCl_5$; oxide, Nb_2O_5; oxychloride, $NbOCl_3$, etc.

Niobic Oxide, Nb_2O_5, is formed when the metal burns in the air. It is prepared from columbite, etc. by fusing the levigated mineral in a platinum crucible with 6 or 8 parts of acid potassium sulphate, removing soluble salts by boiling the fused mass with water, digesting the residue with ammonium sulphide to dissolve tin and tungsten, boiling with strong hydrochloric acid to remove iron, uranium, and other metals, and finally washing with water. As thus obtained it is generally mixed with tantalic oxide, from which it may be separated by means of hydrogen potassium fluoride, which converts the tantalum into sparingly soluble potassium tantalofluoride, $3KF,TaF_5$, and the niobium into easily soluble potassium nioboxyfluoride, $2KF,NbOF_3 + H_2O$.

Niobic oxide is a white, amorphous, infusible powder, having a density of 4.53 and becoming crystalline when strongly heated. By fusion with boric oxide or with borax, it may be obtained in prismatic crystals.

Niobium Hydroxide, or *Niobic acid*, $HNbO_3$, formed by decomposing the pentachloride or oxychloride with water, is a white powder resembling tantalic acid in its principal properties. It is slightly soluble in hot hydrochloric acid, and the solution on addition of zinc turns blue and yields a precipitate, probably of Nb_2O_4.

Niobates.—Some of these salts occur as natural minerals, columbite, for example, being a ferromanganous niobate. The *potassium niobates* crystallize readily and in well-defined forms. The *hexniobate*, $K_8Nb_6O_{19},16H_2O$ or $4K_2O,3Nb_2O_5,16H_2O$, obtained by fusing niobic oxide with twice its weight of potassium carbonate, dissolving the melt in water and evaporating under the air-pump, forms large glistening monoclinic crystals which effloresce on exposure to the air. On adding caustic potash to the solution of this salt and evaporating slowly, the salt, $K_6Nb_4O_{13},13H_2O$, is obtained in efflorescent rhombic pyramids. By fusing niobic oxide and potassium carbonate together in molecular proportions, a crystalline mass is obtained which when treated with water yields the salt,

$$2K_2Nb_4O_{11},11H_2O \text{ or } 2(K_2O,2Nb_2O_5) + 11H_2O.$$

With a larger proportion of potassium carbonate the salt

$$K_4Nb_6O_{17},11H_2O \text{ or } 2K_2O,3Nb_2O_5 + 11H_2O$$

is obtained.* The *sodium niobates* are crystalline powders, which decompose during washing. There is also a sodium and potassium niobate, containing $Na_2O,3K_2O,3Nb_2O_5,9H_2O$.

Niobic Chloride, $NbCl_5$, is obtained, together with the oxychloride, by heating an intimate mixture of niobic oxide and charcoal in a stream of chlorine gas. It is yellow, volatile, and easily fusible. Its observed vapor-density, according to Deville and Troost, is 9.6 referred to air, or 138.6 referred to hydrogen as unity; by calculation for a two-volume condensation, it is $\dfrac{94 + 5 \cdot 35.5}{2} = 135.75$. The *oxychloride*, $NbOCl_3$, is white, volatile, but not fusible: its density, referred to hydrogen, is by observation 114.06; by calculation, $\dfrac{94 + 16 + 3 \cdot 35.5}{2} = 108.25$. Both these compounds are converted by water into niobic oxide.

Niobic Oxyfluoride, $NbOF_3$, is formed by dissolving niobic oxide in hydrofluoric acid. It unites with the fluorides of the more basic metals, forming salts isomorphous with the titanofluorides, stannofluorides, and tungstofluorides, 1 atom of oxygen in these salts taking the place of 2 atoms of fluorine. Marignac has obtained five potassium nioboxyfluorides, all perfectly crystallized, namely—

$2KF,NbOF_3,H_2O$, crystallizing in	monoclinic plates,	
$3KF,NbOF_3$,	"	cuboïd forms (system undetermined),
$3KF,HF,NbOF_3$,	"	monoclinic needles,
$5KF,3NbOF_3,H_2O$,	"	hexagonal prisms,
$4KF,3NbOF_3,2H_2O$,	"	triclinic prisms.

Potassium niobofluoride, $3KF,NbF_5$, separates in shining monoclinic needles from a solution of the first of the above-mentioned nioboxyfluorides in hydrofluoric acid. Nioboxyfluorides of ammonium, sodium, zinc, and copper have also been obtained.

The isomorphism of these salts with the stannofluorides, titanofluorides, and tungstofluorides shows clearly that the existence of isomorphism between the corresponding compounds of any two elements must not be taken as a decided proof that those elements are of equal quantivalence: for in the case now under consideration we have isomorphous salts formed by tin and titanium, which are tetrads, niobium, which is a pentad, and tungsten, which is a hexad.

The compounds of niobium cannot easily be mistaken for those of any

* Santesson, Bull. Soc. Chim. [2], xxiv. 53.

other metal except tantalum. The most characteristic reactions of niobates and tantalates with liquid reagents are the following:—

	Niobates.	Tantalates.
Hydrochloric acid . .	White precipitate, insoluble in excess.	White precipitate, soluble in excess.
Ammonium chloride .	Precipitation slow and incomplete.	Complete precipitation as acid ammonium tantalate.
Potassium ferrocyanide	Red precipitate.	Yellow precipitate.
" ferricyanide	Bright yellow precipitate.	White precipitate.
Infusion of galls . . .	Orange-red precipitate.	Light-yellow precipitate.

Niobic oxide, heated with borax in the outer blowpipe flame, forms a colorless bead, which, if the oxide is in sufficient quantity, becomes opaque by interrupted blowing or flaming. In microcosmic salt it dissolves abundantly, forming a colorless bead in the outer flame and in the inner a violet-colored; or if the bead is saturated with the oxide, a beautiful blue bead, the color disappearing in the outer flame.

39 *

METALS OF THE PLATINUM GROUP.

Gold.	Palladium.	Ruthenium.
Platinum.	Rhodium.	Osmium.
	Iridium.	

THESE metals may be divided into three classes, according to their be-havior with oxygen: (1.) Gold and platinum do not combine directly with oxygen under any circumstances. (2.) Palladium, rhodium, and iridium oxidize when they are heated in air or oxygen, but their oxides decompose on strong ignition into oxygen and metal. (3.) Ruthenium and osmium unite with oxygen to form volatile oxides, which do not decompose even at the highest temperatures.

GOLD.

Symbol, Au (Aurum).   Atomic weight, 196.2.

GOLD, in small quantities, is a very widely-diffused metal; traces of it are constantly found in the iron pyrites of the more ancient rocks.   It is always met with in the metallic state, sometimes beautifully crystallized in the cubic form, associated with quartz, iron oxide, and other substances in regular min-eral veins.   The sands of various rivers have long furnished gold derived from this source, and separable by a simple process of washing; such is the *gold-dust* of commerce.[*]   When a veinstone is wrought for gold, it is stamped to powder and shaken in a suitable apparatus with water and mercury; an amalgam is thus formed, which is afterward separated from the mixture and decomposed by distillation.   Formerly, the chief supply of gold was obtained from the mines of Brazil, Hungary, and the Ural mountains; but California and Australia now yield by far the largest quantity.   The new gold-field of British Columbia is also very productive.

Native gold is almost always alloyed with silver.   The purest specimens have been obtained from Schabrowski, near Katharinenburg in the Ural. A specimen analyzed by Gustav Rose was found to contain 98.96 per cent. of gold.   The Californian gold averages from 87.5 to 88.5 per cent., and the Australian from 96 to 96.6 per cent.   In some specimens of native gold, as in that from Linarowski in the Altai mountains, the percentage of gold is as low as 60 per cent., the remainder being silver.

Pure gold is obtained from its alloys by solution in nitro-muriatic acid, and precipitated with a ferrous salt, which reduces the gold and is converted into a ferric salt, thus:—

[*] For a description of the methods of Gold-washing and Hydraulic Gold-mining prac-tised in California, see Roscoe and Schorlemmer's Treatise on Chemistry, vol. ii. part ii., pp. 365–369.

$$6FeSo_4 + 2AuCl_3 = 2Fe_2(SO_4)_3 + Fe_2Cl_6 + Au_2.$$

The gold falls as a brown powder, which acquires the metallic lustre by friction.

Gold is a soft metal, having a beautiful yellow color. It surpasses all other metals in malleability, the thinnest gold leaf not exceeding, it is said, $\frac{1}{200000}$ of an inch in thickness, while the gilding on the silver wire used in the manufacture of *gold-lace* is still thinner. It may also be drawn into very fine wire. Gold has a density of 19.5: it melts at a temperature a little above the fusing point of silver. Neither air nor water affects it in the least at any temperature; the ordinary acids fail to attack it singly. A mixture of nitric and hydrochloric acid dissolves gold, however, with ease, the active agent being the liberated chlorine.

Gold forms two series of compounds: the a u r o u s c o m p o u n d s, in which it is univalent, as $AuCl$, Au_2O, etc., and the a u r i c c o m p o u n d s, in which it is trivalent, as $AuCl_3$, Au_2O_3, etc.

Chlorides.—The *mono-chloride* or *Aurous chloride*, $AuCl$, is produced when the trichloride is evaporated to dryness, and exposed to a heat of 227° C. (440.6° F.) until chlorine ceases to be evolved. It forms a yellowish-white mass, insoluble in water. In contact with that liquid it is decomposed slowly in the cold, and rapidly by the aid of heat, into metallic gold and trichloride.

The *trichloride* or *Auric chloride*, $AuCl_3$, is the most important compound of gold: it is always produced when gold is dissolved in nitro-muriatic acid. The deep-yellow solution thus obtained yields, by evaporation, yellow crystals of the double chloride of gold and hydrogen; when this is cautiously heated, hydrochloric acid is expelled, and the residue, on cooling, solidifies to a red crystalline mass of auric chloride, very deliquescent, and soluble in water, alcohol, and ether. Auric chloride combines with a number of metallic chlorides, forming a series of double salts, called c h l o r o a u r a t e s, of which the general formula in the anhydrous state is $MAuCl_4$ or $MCl,-AuCl_3$, M representing an atom of a monad metal. These compounds are mostly yellow when in crystals, and red when deprived of water. The *ammonium salt*, $(NH_4)AuCl_4,H_2O$, crystallizes in transparent needles; the sodium salt, $NaAuCl_4,2H_2O$, in long four-sided prisms. Auric chloride also forms crystalline double salts with the hydrochlorides of many organic bases.

A mixture of auric chloride with excess of acid potassium or sodium carbonate is used for gilding small ornamental articles of copper: these are cleaned by dilute nitric acid, and then boiled in the mixture for some time, by which means they acquire a thin but perfect coating of reduced gold.

Oxides.—The *monoxide*, or *Aurous oxide*, is produced when caustic potash in solution is poured upon the monochloride. It is a green powder, partly soluble in the alkaline liquid; the solution rapidly decomposes into metallic gold, which subsides, and auric oxide, which remains dissolved.

Trioxide or *Auric oxide*, Au_2O_3.—When magnesia is added to auric chloride, and the sparingly soluble aurate of magnesium is well washed and digested with nitric acid, auric oxide is left as an insoluble reddish-yellow powder, which when dry becomes chestnut-brown. It is easily reduced by heat and by mere exposure to light; it is insoluble in oxygen-acids, with the exception of strong nitric acid, insoluble in hydrofluoric acid, easily dissolved by hydrochloric and hydrobromic acids. Alkalies dissolve it freely; indeed its acid properties are very strongly marked; it partially decomposes a solution of potassium chloride when boiled with that liquid, potassium hydroxide being produced. When digested with ammonia, it yields fulminating gold, consisting, according to Berzelius, of Au_2O_3, $4NH_3,H_2O$.

The compounds of auric oxide with alkalies are called au r a t e s. The *potassium salt*, $K_2O,Au_2O_3,6H_2O$ or $KAuO_2,3H_2O$, is a crystalline salt, the solution of which is sometimes used as a bath for electro-gilding. A compound of aurate and acid sulphite of potassium, or *potassium aurosulphite*, $2(KAuO_2,4KHSO_3),H_2O$, is deposited in yellow needles when potassium sulphite is added, drop by drop, to an alkaline solution of potassium aurate.

Gold shows but little tendency to form oxygen-salts. Auric oxide dissolves in strong nitric acid, but the solution is decomposed by evaporation or dilution. A *sodio-aurous thiosulphate*, $Na_3Au(S_2O_3)_2,2H_2O$, is prepared by mixing the concentrated solutions of auric chloride and sodium thiosulphate, and precipitating with alcohol. It is very soluble in water, and crystallizes in colorless needles. Its solution is used for fixing daguerreotype pictures. With barium chloride it yields a gelatinous precipitate of *bario-aurous thiosulphate*, $Ba_3Au_2(S_2O_3)_4$.

Sulphides.—*Aurous sulphide*, Au_2S, is formed as a dark-brown, almost black precipitate when hydrogen sulphide is passed into a boiling solution of auric chloride. It forms sulphur-salts with the monosulphides of potassium and sodium. *Auric sulphide*, Au_2S_3, is precipitated in yellow flocks when hydrogen sulphide is passed into a cold dilute solution of auric chloride. Both these sulphides dissolve in ammonium sulphide.

The presence of gold in solution may be detected by the brown precipitate with *ferrous sulphate*, fusible before the blowpipe to a bead of metallic gold; also by the brownish-purple precipitate, called "Purple of Cassius," formed when *stannous chloride* is added to dilute gold solutions. The composition of this precipitate is not exactly known, but after ignition it doubtless consists of a mixture of stannic oxide and metallic gold.[*] It is used in enamel painting.

Oxalic acid slowly reduces gold to the metallic state: to insure complete precipitation, the gold-solution must be digested with it for 24 hours. For the quantitative analysis of a solution containing gold and other metals, oxalic acid is in most cases a more convenient precipitant than ferrous sulphate; inasmuch as, if the quantities of the other metals are also to be determined, the presence of a large quantity of iron salt may complicate the analysis considerably.

Gold intended for coin and most other purposes is always alloyed with a certain proportion of silver or copper to increase its hardness and durability: the first-named metal confers a pale-greenish color. English standard gold contains $\frac{1}{12}$ of alloy, now always copper. Gold when alloyed with copper may be estimated by fusion in a cupel with lead, in the same way as silver (p. 377). If the alloy be free from silver, the weight of the globule of gold left in the cupel will, after repeated fusions, accurately represent the quantity of gold which is present in the alloy. But if the alloy contains silver, that metal remains with the gold after cupellation. In this case the original alloy, consisting of gold, silver, and copper, is fused in the muffle, together with lead and silver; the alloy of gold and silver remaining after cupellation is then boiled with nitric acid, which dissolves the silver, the gold being left behind. By this treatment, however, an accurate separation of the two metals is obtained only when they are present in certain proportions. If the alloy contains but little silver, that metal is protected from

* G r a h a m 's Elements of Chemistry, 2d edit., vol. ii. p. 353,

the action of the nitric acid by the gold; and if it contains too much silver, the gold is left as a powder when the silver is dissolved out. Experience has shown that the most favorable proportions are ¼ gold to ¾ silver; the gold is then left pure, retaining the original shape of the alloy, and can be easily dried and weighed. The quantity of silver which is added to the alloy must therefore vary with the amount of gold which it contains.

Gold-leaf is made by rolling out plates of pure gold as thin as possible, and then beating them between folds of membrane with a heavy hammer, until the requisite degree of tenuity has been reached. The leaf is made to adhere to wood, etc., by size or varnish.

Gilding on copper is sometimes performed by dipping the articles into a solution of mercury nitrate, and then shaking them with a small lump of a soft amalgam of gold, which thus becomes spread over their surfaces: the articles are subsequently heated to expel the mercury, and then burnished. Gilding on steel may be effected either by applying a solution of auric chloride in ether, or by roughening the surface of the metal, heating it, and applying gold-leaf with the burnisher. Gilding by electrolysis—an elegant and simple method, now rapidly superseding many of the others—has already been noticed (p. 285). The solution usually employed is obtained by dissolving oxide or cyanide of gold in a solution of potassium cyanide.

PLATINUM.

Symbol, Pt.   Atomic weight, 196.7.

PLATINUM, palladium, rhodium, iridium, ruthenium, and osmium form a group of metals, allied in some cases by properties in common, and still more closely by their natural association. *Crude platinum*, a native alloy of platinum, palladium, rhodium, iridium, and a little iron, occurs in grains and rolled masses, sometimes of tolerably large dimensions, mixed with gravel and transported materials, on the slope of the Ural Mountains in Russia, also in Brazil, Ceylon, and a few other places. It has never been seen in the rock, which, however, is judged from the accompanying materials to have been serpentine. It is stated to be always present in small quantities with native silver.

From this substance platinum is prepared by the following process:—The crude metal is acted upon as far as possible by *aqua regia* containing an excess of hydrochloric acid and slightly diluted with water, in order to dissolve as small a quantity of iridium as possible: to the deep yellowish-red and highly acid solution thus produced, sal-ammoniac is added, by which nearly the whole of the platinum is thrown down in the state of ammonium platinochloride. This substance, washed with a little cold water, dried, and heated to redness, leaves metallic platinum in the spongy state. This metal cannot be fused into a compact mass by ordinary furnace-heat, but the same object may be accomplished by taking advantage of its property of welding, like iron, at a high temperature. The spongy platinum is made into a thin uniform paste with water, introduced into a slightly conical mould of brass, and subjected to a graduated pressure by which the water is squeezed out, and the mass at length rendered sufficiently solid to bear handling. It is then dried, very carefully heated to whiteness, and hammered or subjected to powerful pressure. If this operation is properly

20 *

conducted, the platinum will then be in a state to bear forging into a bar,' which can afterward be rolled into plates or drawn into wire at pleasure.

A method of refining platinum has been devised by Deville and Debray, consisting in submitting the crude metal to the action of an intensely high temperature in a crucible of lime. The apparatus they employ is as follows:—The lower part of the furnace consists of a piece of lime hollowed out in the centre to the depth of about a quarter of an inch; a small notch is filed at one side of this basin through which the metal is introduced and poured out. A cover made of another piece of lime fits on the top of this basin; it is also hollowed to a small extent, and has a conical perforation at the top, into which is inserted the nozzle of an oxyhydrogen blowpipe. The whole arrangement is firmly bound with iron wire. To use the apparatus, the stop-cock supplying the hydrogen (or coal gas) is opened, and the gas lighted at the notch in the crucible; the oxygen is then gradually supplied; and when the furnace is sufficiently hot, the metal is introduced in small pieces through the orifice. By this arrangement as much as 50 pounds of platinum and more may be fused at once. All the impurities in the platinum, except the iridium and rhodium, are separated in this manner: the gold and palladium are volatilized; the sulphur, phosphorus, arsenic, and osmium oxidized and volatilized: and the iron and copper are oxidized and absorbed by the lime of the crucible.

Platinum is a little whiter than iron: it is exceedingly malleable and ductile both hot and cold, and is very difficult to fuse, melting only before the oxyhydrogen blowpipe or in the powerful blast-furnace just described. It is the heaviest substance known, except osmium, its specific gravity being 21.5. Neither air, moisture, nor the ordinary acids attack platinum in the slightest degree at any temperature: hence its great value in the construction of chemical vessels. It is dissolved by *aqua regia*, and superficially oxidized by fused potassium hydroxide, which enters into combination with the oxide.

The remarkable property of the spongy metal to determine the union of oxygen and hydrogen has been already noticed (p. 154). There is a still more curious state in which platinum can be obtained—that of *platinum-black*, in which the division is carried much farther. It is easily prepared by boiling a solution of platinic chloride to which an excess of sodium carbonate and a quantity of sugar have been added until the precipitate formed after a little time becomes perfectly black and the supernatant liquid colorless. The black powder is collected on a filter, washed, and dried by gentle heat. This substance appears to possess the property of condensing gases, more especially oxygen, into its pores to a very great extent: when placed in contact with a solution of formic acid, it converts the latter, with copious effervescence, into carbonic acid; alcohol, dropped upon the platinum-black, becomes changed by oxidation to acetic acid, the rise of temperature being often sufficiently great to cause inflammation. When exposed to a red heat the black substance shrinks in volume, assumes the appearance of common spongy platinum and loses these peculiarities, which are no doubt the result of its excessively comminuted state.

Platinum forms two series of compounds: the platinous compounds, in which it is bivalent, *e. g.*, $PtCl_2$, PtO, and the platinic compounds, in which it is quadrivalent, *e. g.*, $PtCl_4$, PtO_2, etc.

Chlorides.—The *dichloride*, or *Platinous chloride*, $PtCl_2$, is produced when platinic chloride, dried and powdered, is exposed for some time to a heat of about 200° C. (392° F.), whereby half the chlorine is expelled; also, when sulphurous acid gas is passed into a solution of the tetrachloride until the latter ceases to give a precipitate with sal-ammoniac. It is a greenish-gray powder, insoluble in water, but dissolved by hydrochloric acid. The latter solution, mixed with sal-ammoniac or potassium chloride, deposits a double salt in fine red prismatic crystals, containing, in the last case, K_2PtCl_4 or $2KCl,PtCl_2$. The corresponding sodium compound is very soluble and difficult to crystallize. These double salts are called *platinoso-chlorides* or *chloroplatinites*. Platinous chloride is decomposed by heat into chlorine and metallic platinum.

Platinous chloride unites with carbon monoxide, forming the three compounds—

$$Cl_2Pt\!=\!\!CO, \qquad Cl_2Pt\!\!<\!\!\begin{array}{l} CO \\ | \\ CO \end{array}, \qquad Cl_2Pt\!\!<\!\!\begin{array}{l} CO - PtCl_2 \\ | \\ CO - CO \end{array}$$

all of which are produced by heating platinous chloride in a stream of carbon monoxide. The first and third crystallize in yellow needles, the second in white needles.

Platinous chloride also unites with phosphorus trichloride, forming p h o sp h o - p l a t i n i c c h l o r i d e, $Cl_2Pt\!=\!\!PCl_3$, which is obtained by heating spongy platinum with phosphorus pentachloride to 250° C. (482° F.). It crystallizes in maroon-colored needles, melting at 170° C. (338° F.). When heated with excess of phosphorus trichloride, it is converted into d i p h o s p h o - p l a t i n i c

c h l o r i d e, $Cl_2Pt\!\!<\!\!\begin{array}{l} PCl_3 \\ | \\ PCl_3 \end{array}$, which forms canary-yellow crystals, melting at

160° C. (320° F.). These two chlorides are converted by water, the latter on exposure to moist air at a low winter temperature, into p h o s p ho- p l a t i n i c and d i p h o s p h o - p l a t i n i c a c i d s, $Cl_2Pt\!=\!\!P(OH)_3$ and

$Cl_2Pt\!\!<\!\!\begin{array}{l} P(OH)_3 \\ | \\ P(OH)_3 \end{array}$, the former of which is tribasic, the latter sexbasic.

Platinum tetrachloride, or *Platinic chloride*, $PtCl_4$, is always formed when platinum is dissolved in *aqua regia*. The acid solution yields, on evaporation to dryness, a red or brown residue, deliquescent, and very soluble both in water and in alcohol; the aqueous solution has a pure orange-yellow tint. Platinic chloride unites with a great variety of metallic chlorides, forming double salts called *platino-chlorides* or *chloro-platinates;* the most important of these compounds are those containing the metals of the alkalies and ammonium. *Potassium platinochloride*, $2KCl,PtCl_4$ or K_2PtCl_6, the platinum being octavalent, forms a bright yellow crystalline precipitate, being produced whenever solutions of the chlorides of platinum and of potassium are mixed or a potassium salt mixed with a little hydrochloric acid is added to platinum tetrachloride. It is feebly soluble in water, still less soluble in dilute alcohol, and is decomposed with some difficulty by heat. It is easily reduced by hydrogen at a high temperature, yielding a mixture of potassium chloride and platinum-black; the latter substance may thus, indeed, be very easily prepared. The *sodium salt*, $2NaCl,PtCl_4,6H_2O$ or $Na_2PtCl_6,6H_2O$, is very soluble, crystallizing in large, transparent, yellow-red prisms. The *ammonium salt*, $2NH_4Cl,PtCl_4$, is undistinguishable, in physical characters, from the potassium salt; it is thrown down as a precipitate of small, transparent, yellow, octahedral crystals when sal-ammoniac is mixed with platinic chloride; it is but feebly soluble in water, still less so in dilute alcohol, and is decomposed by heat, yielding spongy platinum, while sal-ammoniac, hydrochloric acid, and nitrogen are driven off. Platinic chloride also forms crystallizable double salts with the hydrochlorides of many organic bases; with ethylamine, for example, the compound—

$$2[NH_2(C_2H_5),HCl]_2\dot{P}tCl_4 \text{ or } [NH_3(C_2H_5)_2]_2PtCl_6.$$

The *bromides* and *iodides of platinum* are analogous in composition to the *chlorides*, and likewise form double salts with alkaline bromides and iodides.

Oxides.—The *monoxide*, or *Platinous oxide*, PtO, is obtained by digesting the dichloride with caustic potash as a black powder, soluble in excess of alkali. It dissolves also in acids with brown color, and the solutions are not precipitated by sal-ammoniac. When platinum dioxide is heated with solution of oxalic acid it is reduced to monoxide, which remains dissolved. The liquid has a dark-blue color, and deposits fine copper-red needles of platinous oxalate.

The *dioxide*, or *Platinic oxide*, PtO_2, is best prepared by adding barium nitrate to a solution of platinic sulphate; barium sulphate and platinic nitrate are then produced, and from the latter caustic soda precipitates one-half of the platinum as *platinic hydroxide*. The sulphate is itself obtained by acting with strong nitric acid upon platinum bisulphide, which falls as a black powder when a solution of the tetrachloride is dropped into potassium sulphide. Platinic hydroxide is a bulky brown powder, which, when gently heated, becomes black and anhydrous. It may also be formed by boiling platinic chloride with a great excess of caustic soda, and then adding acetic acid. It dissolves in acids and combines with bases: the salts have a yellow or red tint, and a great disposition to unite with salts of the alkalies and alkaline earths, giving rise to a series of double compounds, which are not precipitated by excess of alkali. A combination of platinic oxide with ammonia exists, which is explosive. Both oxides of platinum are reduced to the metallic state by ignition.

Sulphides.—The compounds, PtS and PtS_2, are produced by the action of hydrogen sulphide, or the hydrosulphide of an alkali-metal, on the dichloride and tetrachloride of platinum respectively; they are both black substances, insoluble in water. Platinic sulphide, heated in a close vessel, gives off half its sulphur, and is reduced to platinous sulphide. It dissolves in alkaline hydroxides, carbonates, and sulphides, forming salts called *thio-platinates*, which are decomposed by acids.

Ammoniacal Platinum Compounds.—The chlorides, oxides, sulphates, etc. of platinum are capable of taking up two or more molecules of ammonia, and forming compounds analogous in many respects to the ammoniacal mercury compounds already described.

The nitrogen in all these compounds is quinquivalent, and consequently the groups *ammonia*, NH_3, and *diammonia*, N_2H_6 or $NH_3 — NH_3$ or $NH_2(NH_4)$, are bivalent, having two free combining units.

The platinum in some of these compounds is bivalent (*plato* or *platoso*), and unites by two of its combining units with the bivalent groups, NH_3 or N_2H_6, each of which retains one combining unit free. In others the platinum is quadrivalent (*platino*), and unites by some of its combining units with ammonia or diammonia, the remaining units being satisfied by combination with electro-negative radicles. In others again, the platinum accumulates in such a manner as to form compounds containing $(Pt_2)''$, $(Pt^{iv} — Pt^{iv})^{vi}$, $(Pt^{iv} — Pt^{iv} — Pt^{iv} — Pt^{iv})^x$, etc. The bivalent groups, NH_3, N_2H_6, always go by pairs, excepting in the semi-diammoniums, in which half or a quarter of the combining units of the platinum is satisfied by once N_2H_6.

The names and constitution of the several groups are given in the following table, the symbol R denoting a univalent chlorous radicle, such as Cl, NO_2, etc.

1. Platosammonium compounds, $\quad Pt<{}^{NH_3R}_{NH_3R}$

2. Platosemidiammonium compounds, $\quad Pt<{}^{NH_2(NH_4)R}_{R}$

3. Platomonodiammonium compounds, $\quad Pt<{}^{NH_2(NH_4)R}_{NH_3R}$

4. Platosodiammonium compounds, $\quad Pt<{}^{NH_2(NH_4)R}_{NH_2(NH_4)R}$

5. Platinammonium compounds, $\quad R_2Pt<{}^{NH_3R}_{NH_3R}$

6. Platinosemidiammonium compounds, $\quad R_2Pt<{}^{NH_2(NH_4)R}_{R}$

7. Platinomonodiammonium compounds, $\quad R_2Pt<{}^{NH_2(NH_4)R}_{NH_3R}$

8. Platinodiammonium compounds, $\quad R_2Pt<{}^{NH_2(NH_4)R}_{NH_2(NH_4)R}$

9. Diplatinammonium compounds,

$$RPt<{}^{NH_3R}_{NH_3R}$$
$$\mid$$
$$RPt<{}^{NH_2R}_{NH_3R}$$

10. Diplatosodiammonium compounds,

$$Pt - NH_2(NH_4)R$$
$$\mid$$
$$Pt - NH_2(NH_4)R$$

11. Diplatinodiammonium compounds,

$$R_2Pt - NH_2(NH_4)R$$
$$\mid$$
$$R_2Pt - NH_2(NH_4)R$$

12. Diplatinotetradiammonium compounds,

$$RPt<{}^{NH_2(NH_4)R}_{NH_2(NH_4)R}$$
$$\mid$$
$$RPt<{}^{NH_2(NH_4)R}_{NH_2(NH_4)R}$$

We shall here describe the most characteristic compounds of each group, referring for more complete description to larger works.*

1. *Platosammonium Compounds.*—These compounds are formed by abstraction of the elements of ammonia, NH_3, from the corresponding platosodiammonium compounds. They are for the most part insoluble in water, but dissolve in ammonia, reproducing the platosodiammonium compounds. They detonate when heated.

The *chloride*, $Pt<{}^{NH_3Cl}_{NH_3Cl}$ or $N_2H_6PtCl_2$, is formed by heating platosodiammonium chloride to 220–270° C. (428–518° F.), or by heating the same salt with hydrochloric acid, or by boiling the green salt of Magnus (p. 471)

* See Watts's Dictionary of Chemistry, iv. 673, and 2d Suppl. 992.

with nitrate or sulphate of ammonium, and is deposited as a yellow crystalline powder or in rhombohedral scales. It dissolves in 4472 parts of water at 0° C. and in 130 parts of boiling water. At 270° C. (518° F.) it decomposes in the manner represented by the equation,

$$3N_2H_6PtCl_2 = 3Pt + 4NH_4Cl + 2HCl + N_2.$$

Silver nitrate added to its solution throws down all the chlorine. This salt is isomeric with the green salt of Magnus, with the yellow chloride of platosemidiammonium, and with the chloroplatinite of platosomonodiammonium.

The corresponding *iodide*, $N_2H_6PtI_2$, is a yellow powder, obtained by heating the aqueous solution of the compound $N_4H_{12}PtI_2$. It dissolves in ammonia, reproducing the latter compound. The *oxide*, N_2H_6PtO, obtained by heating platosodiammonium oxide (p. 471) to 110°, is a grayish mass, which, when heated to 100° in a close vessel, gives off water, ammonia, and nitrogen, and leaves metallic platinum. The *hydroxide*, $N_2H_6Pt(HO)_2$, obtained by decomposing the sulphate with baryta-water, is a strong base, soluble in water, having an alkaline reaction, absorbing carbonic acid from the air, and liberating ammonia from its salts (Odling). The *sulphate*, $N_2H_6PtSO_4,H_2O$, and the *nitrate*, $N_2H_6Pt(NO_3)_2$, are obtained by boiling the iodide with sulphate and nitrate of silver: they are crystalline, and have a strong acid reaction. The sulphate retains a molecule of crystallization-water, which cannot be removed without decomposing the salt.

2. *Platososemidiammonium Compounds.*—These compounds, isomeric with the preceding, are formed by direct addition of ammonia to platinous salts. The *chloride*, $Pt{<}{\substack{N_2H_6Cl \\ Cl}}$, is obtained by adding ammonia to a cold solution of platinous chloride in hydrochloric acid, filtering after 24 hours, and treating the yellowish-green residue with boiling water, which dissolves the platosemidiammonium salt, and leaves the green salt of Magnus formed at the same time. The solution on cooling deposits the platososemidiammonium chloride in small prisms, differing in form from the chloride above described, and much more soluble in water, requiring for solution 387 parts of cold and 26 parts of boiling water. The other salts of this base are obtained by decomposing the chloride with the corresponding silver salts. The *bromide* and *iodide* crystallize in yellow needles, the *nitrite* in silky needles, which detonate when heated; the *nitrate* and *sulphate* form yellowish crystalline crusts.

3. *Platosomonodiammonium Compounds*, $Pt{<}{\substack{N_3H_6R \\ NH_3R}}$.—The *chloroplatinite* of this series, $2N_3H_9PtCl_2,PtCl_2$, formed in small quantity on adding ammonia to a solution of platinous chloride, crystallizes in brown square laminæ, slightly soluble in cold, more soluble in boiling water. Treated with silver nitrate it is converted into platosomonodiammonium *nitrate*, and this, when heated with hydrochloric acid, yields the corresponding *chloride*, $N_3H_9PtCl_2$, which is very soluble, and crystallizes in colorless needles or nacreous scales.

4. *Platosodiammonium Compounds*, $Pt{<}{\substack{N_2H_6R \\ N_2H_6R}}$.—The *chloride*, $N_4H_{12}PtCl_2$, one of the earliest discovered of the ammoniacal platinum compounds, is obtained by the action of ammonia on the green salt of Magnus or on the chloride of platosammonium. When platinous chloride is boiled with excess of ammonia till the green precipitate formed in the

first instance is redissolved, a solution is obtained, which, when filtered and evaporated, yields the chloride of platosodiammonium in splendid yellow crystals containing one molecule of water, which they give off at 110° C. (230° F.). It is soluble in water, and its solution mixed with platinous chloride yields *platosodiammonium chloroplatinite*, $N_4H_{12}PtCl_2,PtCl$, isomeric with platosammonium chloride, and constituting the *green salt of Magnus*, the first discovered of the ammonia-platinum compounds. This last salt may also be prepared by passing sulphurous acid gas into a boiling solution of platinic chloride till it is completely converted into platinous chloride (and is therefore no longer precipitated by sal-ammoniac), and neutralizing the solution with ammonia. It forms dark-green needles, insoluble in water, alcohol, and hydrochloric acid.

The *bromide* and *iodide* of this series are obtained by treating the solution of the sulphate with bromide or iodide of barium: they crystallize in cubes. The *oxide*, $N_4H_{12}PtO$, is obtained as a crystalline mass by decomposing the solution of the sulphate with an equivalent quantity of baryta-water, and evaporating the filtrate in a vacuum. It is strongly alkaline and caustic, like potash, absorbs carbonic acid rapidly from the air, and precipitates silver oxide from the solution of the nitrate. It is a strong base, neutralizing acids completely, and expelling ammonia from its salts. It melts at 110° C. (230° F.), giving off water and ammonia, and leaving platosammonium oxide. Its aqueous solution does not give off ammonia, even when boiled. The oxide absorbs carbon dioxide rapidly from the air, forming first a normal carbonate, $N_4H_{12}PtCO_3,H_2O$, and afterward an acid salt, $N_4H_{12}PtCO_3,H_2CO_3$. The *sulphate*, $N_4H_{12}PtSO_4$, and the *nitrate*, $N_4H_{12}Pt(NO_3)_2$, are obtained by decomposing the chloride with silver sulphate or nitrate; they are neutral, and crystallize easily.

5. *Platinammonium Compounds.*—The chloride, $Cl_2Pt{<}{NH_3Cl \atop NH_3Cl}$, is obtained by the action of chlorine on platosammonium chloride suspended in boiling water. It is a lemon-yellow crystalline powder, made up of quadratic octahedrons with truncated summits. It is insoluble in cold water, very slightly soluble in boiling water or in water containing hydrochloric acid. It dissolves in ammonia at a boiling heat, and the solution on cooling deposits a yellow precipitate, consisting of platinodiammonium chloride. It dissolves in boiling potash without evolving ammonia.

Nitrates.—An *oxynitrate*, $N_2H_6Pt(NO_3)_2O$, is obtained by boiling the chloride, $N_2H_6PtCl_4$, for several hours with dilute silver nitrate. It is a yellow crystalline powder, sparingly soluble in cold, more soluble in boiling water. The *normal nitrate*, $N_2H_6Pt(NO_3)_4$, is obtained by dissolving the oxynitrate in nitric acid: it is yellowish, insoluble in cold water, soluble in hot nitric acid.

The *oxide*, $N_2H_6PtO_2$, is obtained by adding ammonia to a boiling solution of platinammonium nitrate; it is then precipitated in the form of a heavy, yellowish, crystalline powder, composed of small shining rhomboidal prisms; it is nearly insoluble in boiling water, and resists the action of boiling potash. Heated in a close vessel, it gives off water and ammonia, and leaves metallic platinum. It dissolves readily in dilute acids, even in acetic acid, and forms a large number of crystallizable salts, both neutral and acid, having a yellow color, and sparingly soluble in water. Another compound of platinic oxide with ammonia, called *fulminating platinum*, whose composition has not been exactly ascertained, is produced by decomposing ammonium platino-chloride with aqueous potash. It is a straw-colored powder, which detonates slightly when suddenly heated, but strongly when exposed to a gradually increasing heat.

6. *Platinosemidiammonium Compounds.*—Isomeric with the preceding. The *chloride*, $Cl_2Pt{<}{N_2H_6Cl \atop Cl}$, formed by the action of chlorine on platososemidiammonium chloride, crystallizes in yellow six-sided plates belonging to the rhombic system, turning green at 100° C., and dissolving in   without evolution of ammonia. A *basic nitrate*, $(OH)_2Pt{<}{N_2H_6.NO_3 \atop OH}$, is obtained as an amorphous yellow precipitate by treating the chloride with silver nitrate. A *chloronitrate*, $Cl_2Pt{<}{N_2H_6.NO_2 \atop NO_2}$, obtained by the action of chlorine on platososemidiammonium nitrate, crystallizes in small yellow needles.

7. *Platinomonodiammonium Compounds.*—The *chloride* $Cl_2Pt{<}{N_2H_6Cl \atop NH_3Cl}$, formed by the action of nitro-muriatic acid on platosomonodiammonium chloride, crystallizes in rhombic or hexagonal plates. A *bromonitrate*, $Br_2Pt{<}{N_2H_6(NO_3) \atop NH_3(NO_3)} + H_2O$, obtained by adding bromine to the nitrate of platosomonodiammmonium, forms yellow soluble crusts.

8. *Platinodiammonium Compounds.*—The chloride, $PtCl_2{<}{N_2H_6Cl \atop N_2H_6Cl}$, is obtained by passing chlorine gas into a solution of platosodiammonium chloride; by dissolving platinammonium chloride in ammonia, and expelling the excess of ammonia by evaporation; or by precipitating a solution of platinodiammonium oxynitrate or nitratochloride with hydrochloric acid. It is white, and dissolves in small quantity in boiling water, from which solution it is deposited in the form of transparent regular octahedrons, having a faint-yellow tint. When a solution of this salt is treated with silver nitrate, one-half of the chlorine is very easily precipitated, but to remove even a small portion of the remainder requires a long-continued action of the silver-salt. The *chlorobromide*, ${Br \atop Cl}{>}Pt{<}{N_2H_6Br \atop N_2H_6Cl}$, is obtained as a yellow precipitate by treating platinodiammonium chloride with bromine. A *basic nitrate*, ${HO \atop NO_3}{>}Pt{<}{N_2H_6.NO_3 \atop N_2H_6.NO_3}$, is obtained by the action of nitric acid on platosodiammonium nitrate, as a white crystalline powder, converted by ammonia into the salt $(HO)_2Pt(N_2H_6.NO_3)_2$. The *sulphatochloride*, $Cl_2Pt{<}{N_2H_6 \atop N_2H_6}{>}SO_4$, formed by the action of sulphuric acid on the chloride, crystallizes in slender transparent needles. An *oxalochloride*, $Cl_2Pt(N_2H_6)_2$-C_2O_4, obtained by treating the chloride with ammonium oxalate, is a very soluble crystalline powder.

9. *Diplatinammonium iodide*, $\begin{matrix} IPt(NH_3I)_2 \\ | \\ IPt(NH_3I)_2 \end{matrix}$, or $I_2(Pt_2)^{vi}(NH_3I)_4$, the only term of this series at present known, is obtained by treating platosammonium iodide with boiling potash, and the resulting yellow powder with hydriodic acid. It is a black amorphous substance, which when again treated with potash and hydriodic acids yields the compound $I_2(Pt_4)^x(NH_3I)_8$, and this by similar treatment may be converted into the still more condensed compound $I_2(Pt_8)^{xviii}(NH_3I)_{16}$.

10. *Diplatosodiammonium Compounds.*—The *hydroxide*,

$$\begin{array}{l} Pt\ -N_2H_6-OH \\ | \\ Pt\cdot-N_2H_6-OH \end{array}$$, formed by the action of caustic soda on the chloride of platososemidiammonium, is a grayish crystalline insoluble powder, which detonates violently when heated. Treated with hydrochloric acid, it yields a yellow powder, which is converted by boiling water into the *chloride*, $Pt_2(N_2H_6Cl)_2$.

11. *Diplatinodiammonium Chloride,* $\begin{array}{l} Cl_2Pt-NH_3-NH_2 \\ | \\ Cl_2Pt-NH_3-NH_2 \end{array}$, is a yellow amorphous powder formed by the action of nitro-muriatic acid on the hydroxide of the preceding series.

12. *Diplatino-tetradiammonium Compounds,* $\begin{array}{l} RPt(N_2H_6R)_2 \\ | \\ RPt(N_2H_6R)_2 \end{array}$ or

$N_8H_{24}Pt_2R_6$.—An *oxynitrate,* $O\!\!<\!\!\begin{array}{l} Pt(N_2H_6.NO_3)_2 \\ | \\ Pt(N_2H_6.NO_3)_2 \end{array}$ or $N_8H_{24}Pt_2(NO_3)_4O$, is produced by boiling platosodiammonium nitrate with nitric acid. It is a colorless, crystalline, detonating salt, slightly soluble in cold water, more soluble in boiling water, insoluble in nitric acid (Gerhardt). A *nitratoxychloride,* $N_8H_{24}Pt_2(NO_3)_4OCl_2$, discovered by Raewsky, is formed when Magnus's green salt is boiled with a large excess of nitric acid. Red fumes are then evolved, and the resulting solution deposits the nitratoxychloride in small brilliant needles, which deflagrate when heated, giving off water and sal-ammoniac and leaving metallic platinum. The nitric acid in this salt may be replaced by an equivalent quantity of carbonic or oxalic acid, yielding the compounds, $N_8H_{24}Pt_2(CO_3)_2OCl_2$ and $N_8H_{24}Pt_2(C_2O_4)_2OCl_2$, both of which are crystallizable and sparingly soluble. A *basic oxalonitrate,* $N_8H_{24}Pt_2(C_2O_4)_2(NO_3)_2O$, insoluble in water, is obtained by adding ammonium oxalate to the oxynitrate.

Reactions of Platinum Salts.—Platinic chloride or a platinic oxygen-salt may be recognized in solution by the yellow precipitate which it forms with *sal-ammoniac,* decomposable by heat, with production of spongy metal.

Hydrogen sulphide and *ammonium sulphide* generally form a brown precipitate of platinic sulphide, soluble in excess of ammonium sulphide. *Zinc* precipitates metallic platinum.

Platinic chloride and sodium platinochloride are employed in analytical investigations to detect the presence of potassium and separate it from sodium. For the latter purpose the alkaline salts are converted into chlorides, and in this state mixed with four times their weight of sodium platinochloride in crystals, the whole being dissolved in a little water. When the formation of the yellow salt appears complete, alcohol is added, and the precipitate collected on a weighed filter, washed with weak spirit, carefully dried, and weighed. The potassium chloride is then easily reckoned from the weight of the double salt; and this, subtracted from the weight of the mixed chlorides employed, gives that of the sodium chloride by difference; 100 parts of potassium platinochloride correspond with 30.55 parts of potassium chloride.

Capsules and crucibles of platinum are of great value to the chemist; the latter are constantly used in mineral analysis for fusing silicious matter with alkaline carbonates. They suffer no injury in this operation, although caustic alkali roughens and corrodes the metal. The experimenter must

40 *

be particularly careful to avoid introducing any oxide of an easily fusible metal, as that of lead or tin, into a platinum crucible. If reduction should by any means occur, these metals will at once alloy themselves with the platinum, and the vessel will be destroyed. A platinum crucible must never be put naked into a coke or charcoal fire, but always placed within a covered earthen crucible.

PALLADIUM.

Symbol, Pd.   Atomic weight, 106.2.

WHEN the solution of crude platinum, from which the greater part of that metal has been precipitated by sal-ammoniac, is neutralized by sodium carbonate, and mixed with a solution of mercuric cyanide, palladium cyanide separates as a whitish insoluble substance, which, on being washed, dried, and heated to redness, yields metallic palladium in a spongy state. The palladium may then be welded into a mass in the same manner as platinum.

Palladium closely resembles platinum in color and appearance ; it is also very malleable and ductile. Its density differs very much from that of platinum, being only 11.4. Palladium is more oxidable than platinum. When heated to redness in the air, especially in the state of sponge, it acquires a blue or purple superficial film of oxide, which is reduced at a white heat. Palladium is slowly attacked by nitric acid ; its best solvent is nitro-muriatic acid.

Palladium is remarkable for its power of occluding hydrogen. When heated to redness in a stream of that gas, or brought in contact with nascent hydrogen, as when made to form the negative electrode in an electrolytic cell containing acidulated water, it can absorb as much as 643 times its volume of the gas, whereby its specific gravity is reduced to 11.06.

The absorption of the gas is attended with evolution of heat, and from this fact, together with the supposed constancy of composition of the hydrogenized palladium, it has been inferred that this product is a definite chemical compound ; but this view can scarcely be regarded as satisfactorily demonstrated ; at all events, Graham assigned to the supposed compound the formula PdH_2, whereas, according to Troost and Hautefeuille, it is Pd_2H.[*]

Palladium, like platinum, forms two classes of compounds ; namely, the palladious compounds, in which it is bivalent, and the palladic compounds, in which it is quadrivalent.

Chlorides.—The *dichloride*, or *Palladious chloride*, $PdCl_2$, is obtained by dissolving the metal in nitro-muriatic acid and evaporating the solution to dryness. It is a dark-brown mass, which dissolves in water if the heat has not been too great, and forms double salts with many metallic chlorides. The palladic-chlorides of ammonium and potassium are much more soluble than the corresponding platinochlorides ; they have a brownish-yellow tint.

The *tetrachloride*, or *Palladic chloride*, $PdCl_4$, exists only in solution and in combination with the alkaline chlorides. It is formed when the dichloride is digested in nitro-muriatic acid. The solution has a deep-brown color

---

[*] Watts's Dictionary of Chemistry, vi. 634; vii. 658.

and is decomposed by evaporation. Mixed with potassium chloride or with sal-ammoniac, it gives rise to a red crystalline precipitate, which is but slightly soluble in water.

Palladious Iodide, PdI_2, is precipitated from the chloride or nitrate by soluble iodides, as a black mass, which gives off its iodine between 300° and 360° C. (570–678° F.). Palladium-salts are employed for the quantitative estimation of iodine, and its separation from chlorine and bromine, which are not precipitated by them.

Oxides.—The *monoxide*, or *Palladious oxide*, PdO, is obtained by evaporating to dryness, and cautiously heating the solution of palladium in nitric acid. It is black, and but little soluble in acids. The hydroxide falls as a dark-brown precipitate when sodium carbonate is added to the above solution. It is decomposed by a strong heat.

The *dioxide*, or *Palladic oxide*, PdO_2, is not known in the separate state. From a solution of palladic chloride, alkalies and alkaline carbonates throw down a brown precipitate, consisting of hydrated palladic oxide combined with the alkali. This compound gives off half its oxygen at a moderate heat, and the whole at a higher temperature. From hot solutions a black precipitate is obtained, containing the anhydrous dioxide. The hydroxide dissolves slowly in acids, forming yellow solutions. In strong hydrochloric acid it dissolves without decomposition, forming *potassio-palladic chloride*, arising from admixed potash; with dilute hydrochloric acid, on the contrary, it gives off chlorine.

Palladious Sulphide, PdS, is formed by fusing the metal with sulphur or by precipitating a solution of a palladious salt with hydrogen sulphide. It is insoluble in ammonium sulphide.

Ammoniacal Palladium Compounds.—A moderately concentrated solution of palladium dichloride, treated with a slight excess of ammonia, yields a beautiful flesh-colored or rose-colored precipitate, consisting of $N_2H_6PdCl_2$. This precipitate dissolves in a large excess of ammonia, and the ammoniacal solution, when treated with acids, yields a yellow precipitate having the same composition. This yellow modification is likewise obtained by heating the red compound in the moist state to 100° C., or in the dry state to 200° C. (392° F.). The yellow compound dissolves abundantly in aqueous potash, forming a yellow solution, but without giving off ammonia, even when the liquid is heated to the boiling point; the red compound behaves in a similar manner, but, before dissolving, is converted into the yellow modification. These compounds, discovered by Hugo Müller,[*] are analogous in their modes of formation, and probably therefore in constitution, to the two modifications of the platinum compound, $N_2H_6PtCl_2$ (pp. 469, 470); the red compound being *palladiosemidiammonium chloride*, $Cl.Pd.NH_2(NH_4)Cl$, and the yellow compound, *palladammonium chloride*, $Pd(NH_3Cl)_2$. The yellow compound, digested with water and silver oxide, yields *palladammonium oxide*, N_2H_6PdO, which is a strong base, soluble in water, having an alkaline taste and reaction, and absorbing carbonic acid from the air. *Palladammonium sulphite*, $N_2H_6PdSO_3$, is formed by the action of sulphurous acid on the oxide or chloride; it crystallizes in orange-yellow octahedrons. The *sulphate, chloride, iodide*, and *bromide* have likewise been formed.

The compound $(NH_3)_4PdCl_2$, or *Palladiodiammonium chloride*, $Pd[NH_2-(NH_4)Cl]_2$, separates from an ammoniacal solution of palladammonium chloride in oblique rhombic prisms.

[*] Ann. Ch. Pharm., lxxxvi. 341.

The *oxide*, $N_4H_{12}PdO$, obtained by decomposing the solution of this chloride with silver oxide, is also a strong base yielding crystallizable salts.

Palladious salts are well marked by the pale yellowish-white precipitate which they form with solution of mercuric cyanide. It consists of palladious cyanide, $PdCy_2$, and is converted by heat into the spongy metal.
Hydriodic acid and *potassium iodide* throw down a black precipitate of palladium iodide, visible even to the 500,000th degree of dilution.

Palladium, on account of its unalterability in the air, and its bright silver-white color, which is not affected by exposure to sulphuretted hydrogen, is used for preparing the graduated surfaces of astronomical instruments and for coating silver goods.
An amalgam of palladium is used by dentists for stopping teeth.
A native alloy of gold with palladium is found in Brazil.

RHODIUM.

Symbol, Rh. Atomic weight, 104.1.

THE solution from which platinum and palladium have been separated in the manner already described is mixed with hydrochloric acid and evaporated to dryness. The residue is treated with alcohol of specific gravity 0.837, which dissolves everything except the double chloride of rhodium and sodium. This is well washed with spirit, dried, heated to whiteness, and then boiled with water, whereby sodium chloride is dissolved out, and metallic rhodium remains. Thus obtained, rhodium is a white, coherent, spongy mass, still less fusible and less capable of being welded than platinum. Its specific gravity is 12.1.
Rhodium is very brittle: reduced to powder and heated in the air, it becomes oxidized, and the same alteration happens to a greater extent when it is fused with nitrate or bisulphate of potassium. None of the acids, singly or conjoined, dissolve this metal, unless it be in the state of alloy, as with platinum, in which state it is attacked by nitro-muriatic acid.
Rhodium forms but one chloride, containing $RhCl_3$, in which, like iron in ferric chloride, it may be regarded as either tri- or quadrivalent. This chloride is prepared by adding silicofluoric acid to the double chloride of rhodium and potassium, evaporating the filtered solution to dryness, and dissolving the residue in water. It forms a brownish-red deliquescent mass, soluble in water, with a fine red color. It is decomposed by heat into chlorine and metallic rhodium.
Rhodium and Potassium Chlorides.—The salt $K_6Rh_2Cl_{12},6H_2O$, formed by mixing a solution of rhodic oxide in hydrochloric acid with a strong solution of potassium chloride, crystallizes in sparingly soluble efflorescent prisms. Another double salt, containing $K_4Rh_2Cl_{10},2H_2O$, is prepared by heating in a stream of chlorine a mixture of equal parts of finely powdered metallic rhodium and potassium chloride. The salt has a fine red color, is soluble in water, and crystallizes in four-sided prisms. *Rhodium and sodium chloride*, $Na_6Rh_2Cl_{12},24H_2O$, is also a very beautiful red salt, prepared like the last. The *ammonium salt*, $(NH_4)_6Rh_2Cl_{12},3H_2O$, obtained by decomposing the sodium salt with sal-ammoniac, crystallizes in fine rhombohedral prisms.

Rhodium Oxides.—Rhodium forms three oxides, containing RhO, Rh_2O_3, RhO_2.

The *monoxide*, RhO, is formed with incandescence when the hydrated sesquioxide, $Rh_2O_3,3H_2O$, is heated in a platinum crucible. It is a dark-gray substance, perfectly indifferent to acids.

The *sesquioxide*, or *Rhodic oxide*, Rh_2O_3, obtained by heating the nitrate, is a gray porous mass, with metallic iridescence; insoluble in acids, easily reduced by hydrogen.

The *trihydroxide* or *Rhodic hydroxide*, $Rh_2(OH)_6$, is formed as a black gelatinous precipitate on heating a solution of sodio-rhodic chloride with caustic potash. When dry it forms a heavy dark-brown metallically lustrous mass having a conchoïdal fracture. It is scarcely attacked by acids. A solution of sodio-rhodic chloride treated with potash in the cold becomes opaque, and after some time deposits thin lemon-yellow crystals of the hydrate $Rh_2(OH)_6,2H_2O$, which dissolves readily in acids, and when moist in caustic potash.

The *dioxide*, RhO_2, obtained by fusing pulverized rhodium or the sesquioxide with nitre and potash, and digesting the fused mass with nitric acid to dissolve out the potash, is a dark-brown substance insoluble in acids. When chlorine is passed into a solution of rhodic pentahydrate, $Rh_2O_3,5H_2O$, a black-brown gelatinous precipitate of the trihydrate, $Rh_2O_3,3H_2O$, is formed at first; but this compound gradually loses its gelatinous consistence, becomes lighter in color, and is finally converted into a green hydrate of the dioxide, $RhO_2,2H_2O$. The alkaline solution at the same time acquires a deep violet-blue color.

The *tetrahydroxide*, $Rh(OH)_4$, separates, on passing chlorine for a long time into an alkaline solution of the trihydroxide, with occasional addition of potash, in the form of a green powder, the liquid at the same time assuming a blue or violet tint. The green powder dissolves in hydrochloric acid, yielding a blue solution gradually changing to dark-red, with evolution of chlorine. The violet-blue solution probably contains the potassium salt of a *rhodic acid*, which latter separates after some time as a blue powder, and is converted on drying into the tetrahydroxide.

Rhodic sulphate, $Rh_2(SO_4)_3 12H_2O$, formed by oxidizing the sulphide with nitric acid, is a yellowish-white crystalline mass. *Potassio-rhodic sulphate*, $RhK_3(SO_4)_3$, is a reddish-yellow crystalline powder, formed by adding sulphuric acid to a solution of rhodium and potassium chloride.

Ammoniacal Rhodium Compounds.—An *ammonio-chloride*, $(NH_3)_{10}$-$Cl_2Rh_2Cl_6$, or
$$\begin{array}{c} Cl_2Rh - (NH_3)_5 - Cl \\ | \\ Cl_2Rh - (NH_3)_5 - Cl \end{array}$$
, is obtained as a yellow crystalline powder on mixing a dilute solution of rhodium and ammonium chloride with excess of ammonia, and leaving the filtered solution to evaporate. The corresponding oxide, $(NH_3)_{10}Rh_2O_3$, obtained by heating the chloride with silver oxide, is a strong base, from which the sulphate and oxalate may be obtained in crystalline form.

Rhodic salts are, for the most part, rose-colored, and exhibit in solution the following reactions: with *hydrogen sulphide* and *ammonium sulphide*, a brown precipitate of rhodic sulphide, insoluble in excess of ammonium sulphide; with soluble *sulphites*, a pale-yellow precipitate, affording a characteristic reaction; with *potash*, a yellow precipitate of rhodic oxide, soluble in excess; with *ammonia* and with *alkaline carbonates*, a yellow precipitate after a while; no precipitate with alkaline chlorides or mercuric cyanide. *Zinc* precipitates metallic rhodium.

An alloy of steel and rhodium in equal parts has a density of 9.176, a very fine color and lustre for metallic mirrors, and does not tarnish in the air.

IRIDIUM.

Symbol, Ir. Atomic weight, 196.7.

WHEN crude platinum is dissolved in nitro-muriatic acid, a small quantity of a gray scaly metallic substance usually remains behind, having altogether resisted the action of the acid: this is a native alloy of iridium and osmium, called *osmiridium* or *iridosmine;* it is reduced to powder, mixed with an equal weight of dry sodium chloride, and heated to redness in a glass tube, through which a stream of moist chlorine gas is transmitted. The farther extremity of the tube is connected with a receiver containing solution of ammonia. The gas, under these circumstances, is rapidly absorbed, iridium chloride and osmium chloride being produced: the former remains in combination with the sodium chloride; the latter, being a volatile substance, is carried forward into the receiver, where it is decomposed by the water into osmic and hydrochloric acids, which combine with the alkali. The contents of the tube when cold are treated with water, by which the iridium and sodium chloride is dissolved out; this is mixed with an excess of sodium carbonate, and evaporated to dryness. The residue is ignited in a crucible, boiled with water, and dried; it then consists of a mixture of ferric oxide and a combination of iridium oxide with soda: it is reduced by hydrogen at a high temperature, and treated successively with water and strong hydrochloric acid, by which the alkali and the iron are removed, while metallic iridium is left in a finely divided state. By strong pressure and exposure to a white heat a certain degree of compactness may be communicated to the metal.*

Iridium is a white brittle metal, fusible with great difficulty before the oxyhydrogen blowpipe. Deville and Debray, by means of their powerful oxyhydrogen blast-furnace, have fused it completely into a pure white mass resembling polished steel, brittle in the cold, somewhat malleable at a red heat, and having a density equal to that of platinum, viz., 21.15. By moistening the pulverulent metal with a small quantity of water, pressing it tightly, first between filtering paper, then very forcibly in a press, and calcining it at a white heat in a forge fire, it may be obtained in the form of a compact, very hard mass, capable of taking a good polish, but still very porous, and of a density not exceeding 16.0. After strong ignition it is insoluble in all acids, but when reduced by hydrogen at low temperature, it oxidizes slowly at a red heat, and dissolves in nitro-muriatic acid. It is usually rendered soluble by fusing it with nitre and caustic potash, or by mixing it with common salt, or, better, with a mixture of the chlorides of potassium and sodium, and igniting it in a current of chlorine, as above described.

* Osmiridium, however, generally contains platinum, ruthenium, and other metals of the same group, which are not effectually separated by the method above described. The complete separation of the several metals of the platinum group has of late years formed the subject of several elaborate investigations, into which the limits of this work will not permit us to enter. (See Watts's Dictionary of Chemistry, iii. 35; iv. 241, 680; v. 101, 124; and Roscoe and Schorlemmer's Treatise on Chemistry, vol. ii. Part ii. pp. 390-396 and 438-441.)

Iridium forms three series of compounds, namely, the h y p o i r i d i o u s compounds, in which it is bivalent, as $IrCl_2$, IrO; the i r i d i o u s compounds, in which it is trivalent or quadrivalent, e. g., $IrCl_2$ or $Ir_2Cl_6 = Cl_3Ir.IrCl_3$; and the i r i d i c compounds, in which it is also quadrivalent, as in $IrCl_4$, IrO_2, etc.* It forms also a trioxide, IrO_3, in which it appears to be sexvalent.

Chlorides.—Iridium appears to form three chlorides, but only two of them—namely, the trichloride and tetrachloride—have been obtained in definite form.

The *dichloride*, $IrCl_2$, is not known in the separate state, but appears to exist in certain double salts, called *hypochloriridites*.

The *trichloride*, or *Iridious chloride*, $IrCl_3$ or Ir_2Cl_6, is prepared by strongly heating iridium with nitre, adding water and enough nitric acid to saturate the alkali, warming the mixture, and then dissolving the precipitated hydrate of the sesquioxide in hydrochloric acid; it forms a dark yellowish-brown solution. This substance combines with other metallic chlorides, forming compounds called *iridoso-chlorides* or *chloriridites*, which may be prepared by reducing the corresponding chloriridiates with sulphurous acid, hydrogen sulphide, or potassium ferrocyanide. Claus has obtained the compounds $IrCl_3,3NH_4Cl,3H_2O$, $IrCl_3,3KCl,3H_2O$, and $IrCl_3,3NaCl,12H_2O$. They are olive-green pulverulent salts, soluble in water.

The *tetrachloride*, or *Iridic chloride*, $IrCl_4$, is obtained in solution by dissolving very finely divided iridium, or one of its oxides, or the trichloride, in nitro-muriatic acid, and heating the liquid to the boiling point. On evaporating the solution, it remains in the form of a black, deliquescent, amorphous mass, translucent, with dark-red color at the edges; soluble, with reddish-yellow color, in water. It unites with alkaline chlorides, forming compounds called *iridiochlorides* or *chloriridiates*, analogous in composition to the chloroplatinates. The *ammonium salt*, $(NH_4)_2IrCl_6,H_2O$, and the *potassium salt*, K_2IrCl_6, or $2KCl,IrCl_4$, are formed, as dark-brown crystalline precipitates, on mixing the solutions of the component chlorides. The potassium salt may also be prepared by passing chlorine over a gently ignited and finely divided mixture of iridium with potassium chloride. It is soluble in boiling water, and crystallizes in black octahedrons, yielding a red powder. The *sodium salt*, $Na_2IrCl_6,6H_2O$, prepared like the potassium salt, forms easily soluble black tables and prisms, isomorphous with the corresponding platinum salt.

Iodides.—Iridium forms three iodides, IrI_2, IrI_3, and IrI_4, analogous to the chlorides, and yielding similar double salts with the iodides of the alkali-metals.†

Oxides.—Iridium forms three oxides, IrO, Ir_2O_3, and IrO_2. The *monoxide*, or *hypoiridious oxide*, IrO, is but little known. It is obtained by precipitating an alkaline hypochloriridite with caustic alkali in an atmosphere of carbon dioxide; but on exposure to the air it is quickly converted into a higher oxide.

The *sesquioxide*, or *Iridious oxide*, Ir_2O_3, was formerly regarded as the most easily formed and most stable of the oxides of iridium; but, according to Claus, it has a great tendency to take up oxygen and pass to the state of dioxide. It may be prepared by gently igniting a mixture of potassium

---

* A hexchloride, $IrCl_6$, was said by Berzelius to be obtained in combination with potassium chloride by fusing iridosmine with nitre, distilling the product with nitro-muriatic acid, and treating the residue with successive portions of water; but, according to Claus, the salt formed was really a ruthenium compound, having been prepared by Berzelius from iridosmine containing ruthenium.

† O f f l e r, Ueber die Iodverbindungen des Iridiums. Göttingen, 1857.

chloriridite ($K_6Ir_2Cl_3$) with sodium carbonate in an atmosphere of carbon dioxide; on treating the product with water, the sesquioxide remains in the form of a black powder insoluble in acids. The *trihydroxide* or *Iridious hydroxide*, $Ir_2(OH)_6$, is prepared like the corresponding rhodium compound, which it closely resembles.

Iridious oxide unites with bases, forming salts which may be called *iridites.* A solution of a chloriridite in excess of lime-water deposits, after standing for some time out of contact of air, a dirty yellow precipitate containing $3CaO,Ir_2O_3$.

The *dioxide,* or *Iridic oxide*, IrO_2, is a black powder obtained by heating the tetrahydroxide in a current of carbon dioxide. It is insoluble in acids.—The *tetrahydroxide* or *Iridic hydroxide*, $Ir(OH)_4$, formed by oxidation of iridious hydroxide in the air, or by precipitating the tetrachloride with an alkali, is a heavy indigo-blue powder, almost insoluble in dilute sulphuric and nitric acids, but dissolving slowly in hydrochloric acid. The indigo-blue solution when heated turns green and afterward brown.

Iridium, like the other platinum metals, shows but little tendency to form oxygen salts. The oxides dissolve in acids, but no definite salts are obtained in this way. The solution of iridic oxide in sulphuric acid has a dark-brown color, which is not modified by potash in the same manner as that of the dichloride; neither does it yield any blue precipitate on boiling.

The only definite oxygen-salts of iridium that have been obtained are an iridious sulphite and double salts containing sulphurous and dithionic acids.

Iridious Sulphite, $Ir_2(SO_3)_3,6H_2O$, is formed on passing sulphur dioxide into water, holding the trioxide in suspension, and separates from solution on evaporation as a yellow crystalline precipitate.

Hypo-iridoso-potassic Sulphite, $K_6Ir_2(SO_3)_4$, is obtained as a white crystalline powder, when the mother-liquor obtained in preparing potassium chloriridite by passing sulphurous oxide through a solution of the chloriridate, is evaporated to a small bulk.

Sulphides.—Three sulphides of iridium are known, analogous to the first three oxides above described. The *sesquisulphide* and *disulphide* are obtained as brown-black precipitates by treating the solutions of the trichloride and tetrachloride respectively with hydrogen sulphide. The *monosulphide* is a grayish-black substance obtained by decomposing either of the higher sulphides in a close vessel.

Ammoniacal Compounds of Iridium.—The *chlorides of iridosammonium* and *iridosodiammonium*, $Ir(NH_3Cl)_2$ and $Ir[NH_2(NH_4)Cl]_2$, together with the correspondiing sulphates, are prepared like the platinous compounds of analogous composition, which they also resemble in their properties. The *nitratochloride*, $Ir{<}{}^{NH_2(NH_4)NO_3}_{NH_2(NH_4)Cl}$, is formed by heating the chloride, $Ir(NH_3Cl)_2$, with strong nitric acid. *Iridiodiammonium chloride*, $Cl_2Ir[NH_2(NH_4)Cl]_2$, is obtained as a violet precipitate by treating the nitrate just mentioned with hydrochloric acid.[*]

The compound $10NH_3,Ir_2Cl_6$, analogous to the rhodium compound above described (p. 477), but having an analogue in the platinum series, is obtained as a flesh-colored crystalline powder by prolonged digestion of ammonium chloriridate with warm aqueous ammonia. The corresponding carbonate, nitrate, and sulphate have also been prepared.[†]

---

[*] Skoblikoff, Ann. Ch. Pharm., lxxxiv. 275.
[†] Claus, Beiträge zur Chemie der Platinmetalle. Dorpat, 1854.

Iridic solutions (containing the dioxide or tetrachloride) are of a dark brown-red color; iridious solutions (containing the sesquioxide or trichloride) have an olive-green color. The characters of an iridic solution are best observed with sodium chloriridate, all the other iridic compounds being but slightly soluble.

Iridic solutions give with *ammonium* or *potassium chloride* a crystalline precipitate of ammonium or potassium chloriridate, which is distinguished from the corresponding platinum precipitate by its dark brown-red color, and further by its reduction to soluble chloriridite when treated with solution of hydrogen sulphide. This reaction serves for the separation of iridium from platinum.

RUTHENIUM.

Symbol, Ru. Atomic weight, 103.5.

THIS metal, discovered by Claus in 1846, occurs in platinum ore, and chiefly in osmiridium, of which there are two varieties—one scaly, consisting almost wholly of osmium, iridium, and ruthenium; while the other, which is granular, contains but mere traces of osmium and ruthenium, but is very rich in iridium and rhodium. To obtain ruthenium, scaly osmiridium is heated to bright redness in a porcelain tube, through which a current of air (freed from carbonic acid by passing through potash, and from organic matter by passing through oil of vitriol) is drawn by means of an aspirator. The osmium and ruthenium are thereby oxidized, the former being carried forward as tetroxide and condensed in caustic potash solution, while the ruthenium oxide remains behind, together with iridium; and by fusing this residue with potassium hydroxide, treating the mass with water, and leaving the liquid in a corked bottle for about two hours to clarify, an orange-colored solution of potassium rutheniate is obtained, which, when neutralized with nitric acid, deposits velvet-black ruthenium sesquioxide, and this, when washed, dried, and ignited in hydrogen, yields the metal.

Ruthenium, thus prepared, forms porous lumps very much like iridium, and is moderately easy to pulverize. It is the most refractory of all metals except osmium. Deville and Debray have, however, fused it by placing it in the hottest part of the oxyhydrogen flame. After fusion it has a density of 11.4; that of the porous metal is 8.6.

Ruthenium is scarcely attacked by nitro-muriatic acid. It is, however, more easily oxidized than platinum, or even than silver. When pure it is easily oxidized by fusion with potassium hydroxide, still more easily on addition of a small quantity of nitrate or chlorate, producing potassium rutheniate, which dissolves in water with orange-yellow color.

Chlorides.—Ruthenium forms three chlorides, $RuCl_2$, $RuCl_3$, and $RuCl_4$.

The *dichloride*, $RuCl_2$, is produced, together with the trichloride, by igniting pulverized ruthenium in a stream of chlorine, the trichloride then volatilizing, while the dichloride remains in the form of a black crystalline powder, insoluble in water and in all acids, even nitro-muriatic acid, and only partially decomposed by alkalies. A soluble dichloride is formed by passing sulphydric acid gas into a solution of the trichloride, a brown sulphide being then precipitated, and the solution acquiring a fine blue color.

21

The *trichloride*, or *Ruthenious chloride*, $RuCl_3$, prepared by precipitating a solution of potassic rutheniate with an acid, dissolving the precipitated black oxide in hydrochloric acid, and evaporating, is a yellow-brown, crystalline, very deliquescent mass, becoming dark green and blue at certain points when strongly heated. It dissolves easily in water and in alcohol, leaving a small quantity of a yellow insoluble salt.

The concentrated solution of ruthenious chloride, mixed with concentrated solutions of the chlorides of potassium and ammonium, yields the double salts, $K_4Ru_2Cl_{10}$ and $(NH_4)_2Ru_2Cl_{10}$, in the form of crystalline precipitates, with violet iridescence, very slightly soluble in water, insoluble in alcohol.

The *tetrachloride*, or *Ruthenic chloride*, $RuCl_4$, is known only in its double salts. The *potassium salt*, or $K_2Ru^{iv}Cl_6$, is prepared by mixing a solution of ruthenic hydroxide in hydrochloric acid with potassium chloride, and evaporating to the crystallizing point. It is brown, with rose-colored iridescence, very soluble in water, but insoluble in alcohol. The *ammonium salt*, $(NH_4)_2RuCl_6$, is prepared like the potassium salt, which it closely resembles.

Oxides.—Ruthenium forms six oxides, viz., RuO, Ru_2O_3, RuO_2, RuO_3, Ru_2O_7, and RuO_4, the fourth and fifth, however, being known only in combination.

The *monoxide*, RuO, obtained by calcining the dichloride with sodium carbonate in a current of carbon dioxide, and washing the residue with water, has a dark-gray color and metallic lustre; is not acted upon by acids, but is reduced by hydrogen at ordinary temperatures.—The *sesquioxide*, or *Ruthenious oxide*, Ru_2O_3, is a bluish-black powder, formed by heating the metal in the air. The corresponding hydroxide, $Ru(OH)_3$, is obtained by precipitating ruthenious chloride with an alkaline carbonate, as a blackish-brown substance which dissolves with yellow color in acids. The *dioxide*, or *Ruthenic oxide*, RuO_2, is a black-blue powder, obtained by roasting the disulphide. *Ruthenic hydroxide*, $Ru(OH)_4$, is obtained as a gelatinous precipitate by decomposing potassium chloro-rutheniate with sodium carbonate. —The *trioxide*, RuO_3, commonly caled *ruthenic acid*, is known only as a potassium salt, which is obtained by igniting ruthenium with caustic potash and nitre: it forms an orange-yellow solution.—The *oxide*, Ru_2O_7, called *Ruthenium peroxide*, is not known in the free state, neither has the corresponding acid, "per-ruthenic acid," been obtained; but the potassium salt is formed by the action of chlorine on potassium ruthenate, and separates from the resulting dark-green solution in black lustrous rhombic pyramids, isomorphous with potassium permanganate. The blackish-green solution soon turns ‘yellow, being resolved into potassium ruthenate and the following oxide.—The *tetroxide*, RuO_4, is a volatile compound, analogous to osmic tetroxide, obtained by heating ruthenium with potash and nitre, in a silver crucible, dissolving the fused mass in water, and passing chlorine through the solution in a tubulated retort, connected by a condensing tube with a receiver containing potash. The tetroxide then passes over and condenses in the neck of the retort and in the tube as a golden-yellow crystalline crust, which melts between 50° and 60° C. (122–140° F.). It is heavier than oil of vitriol, dissolves slightly in water, readily in hydrochloric acid, forming a solution easily decomposed by alcohol, sulphurous acid, and other reducing agents.

Sulphides.—Hydrogen sulphide, passed into a solution of either of the chlorides of ruthenium, usually forms a precipitate consisting of ruthenium sulphide and oxysulphide mixed with free sulphur. The blue solution of the dichloride yields a dark-brown sesquisulphide, Ru_2S_3. When hydrogen sulphide is passed for a long time into a solution of the trichloride, ruthenium disulphide, RuS_2, is formed, as a brown-yellow precipitate, becoming dark-brown by calcination.

Ammoniacal Ruthenium Compounds. — *Tetrammonio-hyporuthenious chloride*, $(NH_3)_4RuCl_2,3H_2O$ or $Ru[NH_3(NH_4)Cl]_2,3H_2O$, is formed by boiling the solution of ammonium chloro-rutheniate $(RuCl_4,2NH_4Cl)$ with ammonia. It forms golden-yellow oblique rhombic crystals, very soluble in water, insoluble in alcohol. Treated with silver oxide, it yields the corresponding oxide, $(NH_3)_4RuO$, which, however, is decomposed by evaporation of its solution, giving off half its ammonia, and leaving the compound $(NH_3)_2RuO$. The carbonate, nitrate, and sulphate obtained by treating this last-mentioned oxide with the corresponding silver salts form yellow crystals.

The compounds of ruthenium may readily be distinguished from those of the other platinum metals by fusing a few milligrams of the substance in a platinum spoon, with a large excess of nitre, leaving it to cool when it ceases to froth, and dissolving the cooled mass in a little distilled water. An orange-yellow solution of potassium rutheniate is thus formed, which on addition of a drop or two of nitric acid yields a bulky, black precipitate; and on adding hydrochloric acid to the liquid, with the precipitate still in it, and heating it in a porcelain crucible, the oxide dissolves, forming a solution which has a fine orange-yellow color when concentrated, and when treated with *hydrogen sulphide* till it becomes nearly black yields a filtrate of a splendid sky-blue color. Characteristic reactions are also obtained with *potassium thiocyanate*, which colors the liquid deep red, changing to violet on heating, and with *lead acetate*, which forms a purple-red precipitate.

OSMIUM.

Symbol, Os. Atomic weight, 198.6.

THE separation of this metal from iridium, ruthenium, and the other metals with which it is associated in native osmiridium and in platinum residues, depends chiefly on its ready oxidation with nitric or nitro-muriatic acid, or by ignition in air or oxygen, and the volatility of the oxide thus produced.

To prepare metallic osmium, the solution obtained by condensing the vapor of osmium tetroxide in potash (P. 481) is mixed with excess of hydrochloric acid, and digested with mercury in a well-closed bottle at 40° C. (104° F.). The osmium is then reduced by the mercury, and an amalgam is formed, which, when distilled in a stream of hydrogen till all the mercury and calomel are expelled, leaves metallic osmium in the form of a black powder (Berzelius). The metal may also be obtained by igniting ammonium chloro-osmite with sal-ammoniac. Deville and Debray prepare it by passing the vapor of the pure tetroxide mixed with carbon monoxide and carbon dioxide through a red-hot porcelain tube.

The properties of osmium vary according to its mode of preparation. In the pulverulent state it is black, destitute of metallic lustre, which, however, it acquires by burnishing; in the compact state, as obtained by Berzelius's method above described, it exhibits metallic lustre. As obtained by Deville and Debray's method, it is an amorphous powder, which may, however, be rendered crystalline by fusing it with three or four parts of tin in a charcoal-lined crucible, treating the crystalline alloy with hydrochloric acid to remove the tin, and heating the residue in a current of hydrochloric acid gas. The crystals are cubes or very obtuse rhombohedrons, having a bluish-white color with violet lustre; they are harder than glass, and have a density of 22.477, which is higher than that of any other known substance. At a very high temperature, capable of melting ruthenium and iridium and volatilizing platinum, osmium likewise volatilizes, but still does not melt; in fact, it is the most refractory of all metals.

Osmium in the finely-divided state is highly combustible, continuing to burn when set on fire, till it is all volatilized as tetroxide. In this state also it is easily oxidized by nitric or nitro-muriatic acid, being converted into tetroxide. But after exposure to a red heat it becomes less combustible, and is not oxidized by nitric or nitro-muriatic acid. Osmium which has been heated to the melting point of rhodium does not give off any vapor of tetroxide when heated in the air to the melting point of zinc, but takes fire at higher temperatures.

Osmium Chlorides.—Osmium forms three chlorides, analogous to those of iridium and ruthenium. When it is heated in dry chlorine gas there is formed, first a blue-black sublimate of the dichloride, then a red sublimate of the tetrachloride. The *dichloride* or *hypo-osmious chloride*, dissolves in water with dark violet-blue color. It is likewise formed by the action of reducing agents on either of the higher chlorides, into which, on the other hand, it is easily converted by oxidation. The addition of potassium chloride renders it more stable by forming a double salt. The *trichloride* has not been isolated, but is contained in the solution obtained by treating the sesquioxide with hydrochloric acid. It forms double salts with alkaline chlorides. The *potassium salt*, $K_6Os_2Cl_{12},6H_2O$, is produced, together with potassium chlorosmate, when a mixture of pulverized osmium and potassium chloride is ignited in chlorine gas; it forms dark red-brown crystals.

The *tetrachloride*, or *Osmic chloride*, $OsCl_4$, is the red compound which constitutes the principal part of the product obtained by igniting osmium in chlorine gas. It dissolves with yellow color in water and alcohol, and is decomposed quickly in dilute solution, more slowly in presence of hydrochloric acid or metallic chlorides, yielding a black precipitate of osmic oxide and a solution of osmium tetroxide in hydrochloric acid.

Osmic chloride unites with the chlorides of the alkali-metals, forming salts sometimes called osmiochlorides, or chlorosmates. From the solutions of these salts *hydrogen sulphide* and *ammonium sulphide* slowly precipitate a yellow-brown sulphide, insoluble in alkaline sulphides; *silver nitrate* forms an olive-green, *stannous chloride* a brown precipitate. *Tannic acid*, on heating, produces a blue color, but no precipitate; *potassium ferrocyanide*, first a green, then a blue color; *potassium iodide*, a deep purple-red color. *Potash* gives a black, *ammonia* a brown precipitate, slowly in the cold, immediately on boiling. Metallic *zinc* and *sodium formate* throw down metallic osmium.

Sodium osmiochloride, Na_2OsCl_6, prepared by heating a mixture of osmium sulphide and sodium chloride in a current of chlorine, crystallizes in orange-colored rhombic prisms, an inch long, easily soluble in water and in alcohol. The *potassium* and *ammonium salts*, of analogous composition, are obtained as red-brown crystalline precipitates on adding sal-ammoniac or potassium chloride to the solution of the sodium salt.

Oxides.—Osmium forms five oxides analogous to those of ruthenium.

The *monoxide*, or *Hypo-osmious oxide*, OsO, is obtained by igniting hypo-osmious sulphite in a stream of carbonic acid gas; also as a blue-black hydrate, by heating the same salt with strong potash solution in a closed vessel. *Hypo-osmious sulphite*, $OsSO_3$ or OsO,SO_2, is a black-blue salt, produced by mixing the aqueous solution of osmium tetroxide with sulphurous acid.—The *sesquioxide* or *Osmious oxide*, Os_2O_3, is obtained by heating either of the double salts of the trichloride with sodium carbonate in a stream of carbonic acid gas. It is a black powder, insoluble in acids. The *hydroxide*, obtained by precipitation, has a dirty brown-red color; it is soluble in acids, but does not yield pure salts.

The *dioxide*, or *Osmious oxide*, OsO_2, is obtained as a black insoluble powder by heating potassium osmiochloride with sodium carbonate in a stream of carbonic acid gas, or in copper-red metallic shining lumps by heating the corresponding hydroxide. *Osmious hydroxide*, or *Osmious acid*, $Os(OH)_4$, is obtained by precipitating a solution of potassium osmiochloride with potash at the boiling heat, or in greater purity by mixing a solution of potassium osmate, K_2OsO_4, with dilute nitric acid according to the equation,

$$2K_2OsO_4 + 4HNO_3 = Os(OH)_4 + OsO_4 + 4KNO_3.$$

The *trioxide*, OsO_3, and the corresponding hydroxide or *Osmic acid*, H_2OsO_4, are not known, osmium in this respect resembling manganese; but the osmates of alkali-metals are formed by the action of reducing agents in presence of alkalies on the tetroxide. The *potassium salt*, $K_2OsO_4,2H_2O$, is a rose-colored crystalline powder.

The *tetroxide*, OsO_4, commonly called *Osmic acid*, is the volatile, strong-smelling compound formed when osmium or either of its lower oxides is heated in the air or treated with nitric or nitro-hydrochloric acid. It may be prepared by heating osmium in a current of oxygen gas, and condenses in the cool part of the apparatus in colorless transparent crystals. It melts below 100°, and boils at a temperature a little above its melting point. Its vapor has an intolerably pungent odor, attacks the eyes strongly and painfully, and is excessively poisonous. Osmium tetroxide is dissolved slowly, but in considerable quantity, by water, forming an acid solution. It is a powerful oxidizing agent, decolorizing indigo-solution, separating iodine from potassium iodide, converting alcohol into aldehyde and acetic acid, etc. It dissolves in alkalies, forming yellow-red solutions, which are inodorous when cold, but when heated give off the tetroxide and free oxygen, leaving a residue of alkaline osmite.

Sulphides.—Osmium burns in sulphur-vapor. Five sulphides of osmium are said to exist, analogous to the oxides, the first four being produced by decomposing the corresponding chlorides with hydrogen sulphide, and the tetrasulphide by passing that gas into a solution of the tetroxide. The last is a sulphur-acid, perfectly soluble in water, whereas the others are sulphur-bases, slightly soluble in water, and forming deep-yellow solutions.

Ammoniacal Osmium Compounds.—A cold solution of potassium osmite, mixed with sal-ammoniac, yields a yellow crystalline precipitate, consisting, according to Claus, of *hydrated osmammonium chloride*, $Os(NH_3Cl)_2$. An aqueous solution of the tetroxide treated with ammonia yields a brown-black powder, consisting of $N_2H_8OsO_3$, or $O{-}Os{<}^{NH_3}_{NH_3}{>}O + H_2O$.

OSMIAMIC ACID, $H_2Os_2N_2O_5$.—The potassium salt of this bibasic acid, $K_2Os_2N_2O_5$, is produced by the action of ammonia on a hot solution of osmium tetroxide in excess of potash:—

$$6OsO_4 + 8NH_3 + 6KHO = 3K_2Os_2N_2O_5 + 15H_2O + N_2.$$

It separates as a yellow crystalline powder, and its solution, treated with silver nitrate, yields a precipitate of silver osmiamate, $Ag_2Os_2N_2O_5$, from which the aqueous acid may be prepared by decomposition with hydrochloric acid. It is a strong acid, decomposing not only the carbonates, but also the chlorides, of potassium and sodium. The osmiamates of the alkali-metals and alkaline earth-metals are soluble in water; the lead, mercury, and silver salts are insoluble.

Osmyl-ditetramine Chloride, $OsO_2(NH_3)_4Cl$, is obtained as an orange-colored crystalline precipitate on mixing the concentrated solutions of ammonium chloride and potassium osmate. It is slightly soluble in cold water, but is quickly decomposed by hot water, with evolution of osmium tetroxide and precipitation of a black powder. It dissolves in hot water containing a little hydrochloric acid, forming a deep orange-yellow solution, from which it separates on cooling in brown-yellow crystals. It is completely decomposed by ignition, leaving pure metallic osmium as a gray porous mass. The solution of the chloride gives with potassium ferrocyanide a fine violet color, affording a very delicate test for osmium. The *platino-chloride*, $OsO_2(NH_3)_4Cl_2,PtCl_4$, forms orange-yellow crystals slightly soluble in cold water. The *sulphate*, $OsO_2(NH_3)_4SO_4$, forms small orange-yellow crystals, freely soluble in hot, slightly in cold, water. The *nitrate* forms orange-yellow crystals paler than the sulphate. The *oxalate*, $OsO_2(NH_3)_4C_2O_4$, is a very stable salt, forming yellow or orange-yellow crystals slightly soluble in cold water.

The constitution of the osmyl-ditetramine compounds—the chloride, for example—is most probably represented by the formula

$$\begin{array}{c} O \\ | \\ O \end{array} \!\!\! > Os < \!\!\! \begin{array}{c} NH_3.NH_3.Cl \\ \\ NH_3.NH_3.Cl \end{array}$$

All osmium compounds heated with excess of nitric acid give off the disagreeable odor of osmium tetroxide. By ignition in hydrogen they are reduced to metallic osmium, which, as well as the lower oxides, emits the same odor when heated in contact with the air. The reactions of osmium salts in solution have already been described.

PART III.

CHEMISTRY OF CARBON COMPOUNDS,

OR

ORGANIC CHEMISTRY.

PART III.

CHEMISTRY OF CARBON COMPOUNDS

OR

ORGANIC CHEMISTRY.

INTRODUCTION.

THE term "Organic Chemistry" originally denoted the chemistry of compounds formed in the bodies of plants and animals. The peculiar characters of the compounds thus formed, and the failure of the earlier attempts to produce them by artificial means, led to the erroneous idea that their formation was due to a mysterious power, called "vital force," supposed to reside in the living organism, and to govern all the changes and processes taking place within it. In accordance with this idea, the chemistry of organic compounds, including those which were formed by artificial processes from the products of vegetable and animal life, was erected into a special branch of chemical science.

Later researches have, however, shown that a large number of compounds, formerly regarded as producible only under the influence of the so-called vital force, may be formed either by direct combination of their elements, or by chemical transformation of inorganic compounds.

The first step in the formation of organic compounds from their elements was made by Wöhler, who showed, in 1828, that urea, the characteristic constituent of urine, can be produced by molecular transformation of ammonium cyanate. This experiment, viewed in connection with the fact established about twelve years afterwards, that cyanogen (CN) can be formed by direct combination of its elements, is conclusive of the possibility of forming a product of the living organism from inorganic materials. More recently it has been shown that ethine, or acetylene, C_2H_2, can be produced by the direct combination of carbon and hydrogen; that this compound can be made to take up two additional atoms of hydrogen to form ethene, C_2H_4; and that this latter compound can be converted into alcohol, C_2H_6O, a body formerly supposed to be producible only by the fermentation of sugar; and from this a large number of other compounds can be produced by the action of various reagents. The researches of Bertholet, Kolbe, Wurtz, and other distinguished chemists, have led to the discovery of a large number of other cases of the formation of organic compounds, often of great complexity, from substances of purely mineral origin, and ultimately from the elements themselves. The division of compounds into two distinct branches, inorganic and organic—formed according to distinct laws, the former being artificially producible by direct combination of their elements, the latter only under the influence of a supposed vital force—must therefore be abandoned. There is, indeed, but

(488)

one science of chemistry, of which the study of the compounds called organic forms a part.

Organic chemistry is in fact the chemistry of carbon-compounds, and, in a strictly systematic arrangement, these compounds should be described in connection with the element carbon itself. But the compounds into which carbon enters are so numerous, their constitution and the transformations which they undergo under the influence of heat and of chemical reagents, are, in many instances, so complicated, that it is found best, for the purposes of instruction, to defer their consideration till the other elements and their compounds have been studied.

It is important, in this place, to mark the distinction between o r g a n i c c o m p o u n d s and o r g a n i z e d b o d i e s. Organic bodies, such as marsh gas, ethene, benzene, alcohol, sugar, morphine, etc., are definite chemical compounds, many of which, as already observed, may be formed by artificial methods; those which are solid can, for the most part, be crystallized; those which are liquid exhibit constant boiling points. Organized bodies, on the contrary, always consist of mixtures of several definite compounds. They never crystallize, but exhibit a fibrous or cellular structure, and cannot be reduced to the liquid or gaseous state without complete decomposition. Lastly, they are organs, or parts of organs, which are essentially products of vitality, and there is not the slightest prospect of their ever being produced by artificial means.

The study of the composition and chemical relations of organized bodies belongs to a special department of the science called "Physiological Chemistry," which bears the same relation to Organic Chemistry that Chemical Geology bears to Mineralogy.

DECOMPOSITIONS AND TRANSFORMATIONS OF ORGANIC COMPOUNDS.

Organic bodies are, generally speaking, distinguished by the facility with which they decompose under the influence of heat or of chemical reagents; the more complex the body, the more easily does it undergo decomposition or transformation.

1. *Action of Heat.*—Organic bodies of simple constitution and of some permanence, but not capable of subliming unchanged, like many of the organic acids, yield, when exposed to a high but regulated temperature, in a retort, new compounds, perfectly definite and often crystallizable, which partake, to a certain extent, of the properties of the original substance. Carbon dioxide and water are often eliminated under these circumstances. If the heat be suddenly raised to redness, the regularity of the decomposition vanishes, while 'the products become more uncertain and more numerous; carbon dioxide and watery vapor are succeeded by inflammable gases, as carbon monoxide and hydrocarbons; oily matter and tar distil over, and increase in quantity until the close of the operation, when the retort is found to contain, in most cases, a residue of charcoal. Such is *dry* or *destructive distillation.*

If the organic substance contains nitrogen, and it is not of a kind capable of taking a new and permanent form at a moderate degree of heat, then that nitrogen is in most instances partly disengaged in the shape of ammonia, or substances analogous to it, partly left in combination with the carbonaceous matter in the distillatory vessel. The products of dry distillation thus become still more complicated.

21 *

A much greater degree of regularity is observed in the effects of heat on fixed organic matters, when these are previously mixed with an excess of strong alkaline base, as potash or lime. In such cases an acid, the nature of which is chiefly dependent upon the temperature applied, is produced, and remains in union with the base, the residual element or elements escaping in some volatile form. Thus benzoic acid distilled with calcium hydrate, at a dull red heat, yields calcium carbonate and benzene ; woody fibre and caustic potash, heated to a very moderate temperature, yield free hydrogen, and a brown, somewhat indefinite substance called *ulmic acid;* with a higher degree of heat, oxalic acid appears in the place of the ulmic ; and, at the temperature of ignition, carbon dioxide, hydrogen being the other product.

2. *Action of Oxygen.*—Oxygen, either free or in the nascent state, in which latter condition it is most active, may act on organic compounds in four different ways :—

a. By simple addition, as

$$\underset{\text{Aldehyde.}}{C_2H_4O} + O = \underset{\text{Acetic acid.}}{C_2H_4O_2}$$

β. By simply removing hydrogen :

$$\underset{\text{Alcohol.}}{C_2H_6O} + O = H_2O + \underset{\text{Aldehyde.}}{C_2H_4O}$$

γ. By removing hydrogen and taking its place, 2 atoms of hydrogen being replaced by one of oxygen ; *e. g. :*

$$\underset{\text{Alcohol.}}{C_2H_6O} + O_2 = H_2O + \underset{\text{Acetic acid.}}{C_2H_4O_2}$$

δ. By removing both carbon and hydrogen. In this manner complex organic bodies containing large numbers of carbon and hydrogen atoms are reduced to others of simpler constitution, and ultimately the carbon and hydrogen are wholly converted into carbon dioxide and water. Nitrogen, chlorine, bromine, and iodine, if present, are at the same time disengaged, for the most part in the free state, and sulphur is oxidized.

Moist organic substances, especially those containing nitrogen, undergo, when exposed to the air, a slow process of oxidation, by which the organic matter is gradually burned and destroyed without sensible elevation of temperature : this process is called *Decay,* or *Eremacausis.* Closely connected with this change are those called *Fermentation* and *Putrefaction,* consisting in a new arrangement of the elements of the compound (often with assimilation of the elements of water), and the consequent formation of new products. The change is called *putrefaction,* when it is accompanied by an offensive odor ; *fermentation,* when no such odor is evolved, and especially if the change results in the formation of useful products , thus the decomposition of a dead body, or of blood or urine, is putrefaction ; that of grape-juice or malt-wort, which yields alcohol, is fermentation. Putrefaction and fermentation are not processes of oxidation ; nevertheless, the presence of oxygen appears to be indispensable to their commencement ; but the change, when once begun, proceeds without the aid of any other substance external to the decomposing body, unless it be water or its elements. Every case of putrefaction thus begins with decay ; and if the decay, or its cause, namely, the absorption of oxygen, be prevented, no putrefaction occurs. The most putrescible substances, as milk, highly azotized vegetables, and animal flesh intended for food, may be preserved indefinitely, by inclosure in metallic cases from which the air has been *completely* removed and excluded.

Fermentation and putrefaction are always accompanied by the develop-- ment of certain living organisms of the fungous class ; but whether the growth of these is a cause or a consequence of the chemical change is a point not yet decided. We shall return to this subject in speaking of the fermentation of sugar.

3. *Action of Chlorine, Bromine, and Iodine.*—Chlorine and bromine exert precisely similar actions on organic bodies ; that of chlorine is the more energetic of the two. The reactions consist :

a. In simple addition of chlorine or bromine to the organic molecule ; *e. g.:*

$$C_4H_4O_4 \;+\; Br_2 \;=\; C_4H_4Br_2O_4$$

Fumaric acid.   Dibromosuccinic acid.

ß. In removal of hydrogen without substitution :

$$C_2H_6O \;+\; Cl_2 \;=\; 2HCl \;+\; C_2H_4O$$

Alcohol.   Aldehyde.

γ. In substitution of chlorine or bromine for hydrogen :

$$C_2H_4O_2 \;+\; Cl_2 \;=\; HCl \;+\; C_2H_3ClO_2$$

Acetic acid.   Chloracetic acid.

$$C_2H_4O_2 \;+\; 3Cl_2 \;=\; 3HCl \;+\; C_2HCl_3O_2$$

Acetic acid.   Trichloracetic acid.

The substitution-products thus formed undergo transformations closely analogous to those of the original compounds, under the influence of similar reagents ; but they are always more acid, or less basylous, in proportion to the quantity of chlorine or bromine substituted for hydrogen. Thus aniline, C_6H_7N, which is a strong base, may be converted, by processes to be hereafter described, into the chlorinated compounds, C_6H_6ClN, $C_6H_5Cl_2N$, and $C_6H_4Cl_3N$, the first and second of which are less basic than aniline itself, while the third does not show any tendency to form salts with acids.

δ. In presence of water they remove the hydrogen of that liquid, and set free the oxygen : hence, chlorine-water and bromine-water act as powerful oxidizing agents.

Iodine may also act in this manner as an oxidizing agent ; and it sometimes attaches itself directly to organic molecules ; but it never acts directly by substitution. Iodine substitution-products may, however, be obtained in some cases by treating organic bodies with chloride of iodine, the chlorine then removing hydrogen, and the iodine taking its place.

4. *Action of Nitric Acid.*—This acid acts very powerfully on organic substances. The action may be of three kinds:

a. Direct combination, as with organic bases ; *e.g.:*

$$C_2H_7N \;+\; NO_3H \;=\; C_2H_7N.NO_3H$$

Ethylamine.   Nitric acid.   Ethylamine nitrate.

ß. Oxidation. This mode of action is most frequently observed with the somewhat diluted acid.

γ. Substitution of nitryl (NO_2) for hydrogen ; *e.g.:*

$$C_6H_6 \;+\; NO_2(OH) \;=\; H_2O \;+\; C_6H_5(NO_2)$$

Benzene.   Nitric acid.   Nitrobenzene.

$$C_6H_{10}O_5 \;+\; 3NO_2(OH) \;=\; 3H_2O \;+\; C_6H_7(NO_2)_3O_5$$

Cellulose.   Nitric acid.   Trinitrocellulose (gun-cotton).

This action takes place most readily with the strongest nitric acid (pure hydrogen nitrate). The products (called n i t r o-c o m p o u n d s) are always easily combustible, and in many cases highly explosive.

5. *Action of Alkalies.*—The hydroxides of potassium and sodium act on organic bodies in a great variety of ways, the most important and general of which are the following :—

a. By direct combination :—

$$CO \quad + \quad HKO \quad = \quad CHKO_2$$

Carbon monoxide.   Potassium hydroxide.   Potassium formate.

$$C_{10}H_{16}O \quad + \quad HKO \quad = \quad C_{10}H_{17}KO_2$$

Camphor.   Potassium hydroxide.   Potassium campholate.

β. By double decomposition with acids, water being eliminated, and a salt produced :

$$C_2H_4O_2 + HKO = H_2O + C_2H_3KO_2$$

Acetic acid.   Potassium acetate.

γ. Oxidation, with elimination of hydrogen :

$$C_2H_6O + HKO = C_2H_3KO_2 + 2H_2$$

Alcohol.   Potassium acetate.

δ. From chlorinated compounds they remove a part or the whole of the chlorine :

$$C_2H_4Cl_2 + HKO = C_2H_3Cl + KCl + H_2O$$

Ethene chloride.   Chlor- ethene.

$$C_5H_{11}Cl + HKO = C_5H_{10} + KCl + H_2O$$

Amyl chloride.   Amylene.

ι. Amides (p. 340) are decomposed by them in such a manner that the whole of the nitrogen is given off as ammonia, and a potassium or sodium salt of the corresponding acid is produced :

$$NH_2.C_2H_3O + HKO = NH_3 + C_2H_3O.OK$$

Acetamide.   Potassium acetate.

Many other azotized organic compounds, when heated with alkaline hydroxides, likewise give up the whole of their hydrogen in the form of ammonia.

6. *Action of Reducing Agents.*—This name is given to bodies whose action is the inverse of that of oxygen, chlorine, bromine, and iodine ; such are nascent hydrogen, obtained by the action of sodium-amalgam on water, or by that of zinc on aqueous acids or alkalies ; also hydrogen sulphide, ammonium sulphide, sulphurous acid, and metals, especially potassium and sodium—all of which either give up hydrogen, or abstract oxygen, chlorine, etc.

Reducing agents may act in the following ways :—

a. By adding hydrogen to an organic body :—

$$C_2H_4O \quad + \quad HH \quad = \quad C_2H_6O$$

Ethene oxide.   Alcohol.

ß. By removing oxygen, chlorine, bromine, or iodine, without introducing anything in its place; thus :— ·

$$C_7H_6O_2 \ + \ HH \ = \ H_2O \ + \ C_7H_6O$$
Benzoic   Benzoic
acid.   aldehyde.

γ. By substituting hydrogen for oxygen, chlorine, etc. This process is called *inverse substitution*. It may take place either in equivalent quantities, *e. g.* :—

$$C_7H_5O.OH \ + \ 2HH \ = \ H_2O \ + \ C_7H_7.OH \ ;$$
Benzoic   Benzylic
acid.   alcohol.

or it may happen that the quantity of hydrogen introduced is only half that which is equivalent to the oxygen removed. This mode of substitution takes place with nitro-compounds, which are thereby reduced to others containing amidogen (NH_2) in place of nitryl (NO_2); thus :—

$$C_6H_5(NO_2) \ + \ 3H_2 \ = \ 2H_2O \ + \ C_6H_5(NH_2)$$
Nitrobenzene.   Amidobenzene
                                      (aniline).

A large number of organic bases are formed in this manner from nitro-compounds.

7. *Action of Dehydrating Agents.*—Strong sulphuric acid, sulphuric oxide, phosphoric oxide, and zinc chloride, remove oxygen and hydrogen from organic bodies in the form of water, the elements of which are derived, sometimes from a single molecule of the organic body, sometimes from two molecules :—

$$C_2H_6O \ - \ H_2O \ = \ C_2H_4$$
Alcohol.   Ethene.

$$2C_2H_6O \ - \ H_2O \ = \ C_4H_{10}O$$
Alcohol.   Ether.

Compounds which, like sugar, starch, and woody fibre, consist of carbon united with hydrogen and oxygen in the proportions to form water, are often reduced by these dehydrating agents to black substances consisting mainly of carbon.

Other reactions of less generality than those above described will be sufficiently illustrated by special cases in the sequel. ·

THE ELEMENTARY OR ULTIMATE ANALYSIS OF ORGANIC COMPOUNDS.

Organic compounds contain, for the most part, only a small number of elements. Many consist only of carbon and hydrogen. A very large number, including most of those which occur ready-formed in the bodies of plants and animals, consist of carbon, hydrogen and oxygen; others consist of carbon, hydrogen, and nitrogen. Others, again, including most of the proximate principles of the animal organism, consist of four elements, carbon, hydrogen, oxygen, and nitrogen. Some contain sulphur, phosphorus, chlorine, and metallic elements; in fact, artificially prepared carbon compounds may contain any elements whatever. Moreover, even those which contain only a small number of elements often exhibit great complexity of structure, in consequence of the accumulation of a large number of carbon atoms in the same molecule.

Determination of Carbon and Hydrogen.—The quantities of these elements are determined by heating a known weight of the body to be analyzed in contact with some easily reducible metallic oxide, black oxide of copper being the substance generally used. The organic body then undergoes complete combustion at the expense of the oxygen of the copper oxide, the carbon being converted into carbon dioxide, and the hydrogen into water. These products are collected and their weights determined, and from the data thus obtained the quantities of carbon and hydrogen present in the organic substance are calculated. When nothing but carbon and hydrogen, or those bodies together with oxygen, is present, one experiment suffices; the carbon and hydrogen are determined directly, and the oxygen by difference.

The substance to be analyzed, if solid, must be carefully freed from moisture. If it will bear the application of a moderate heat, this desiccation is very easily accomplished by a water- or steam-bath : in other cases, exposure at common temperatures to the absorbent powers of a large surface of oil of vitriol in the vacuum of an air-pump must be substituted.

The copper oxide is best made from the nitrate by complete ignition in an earthen crucible ; it is reduced to powder and reheated just before use, to expel hygroscopic moisture, which it absorbs with avidity, even while warm. The combustion is performed in a tube of hard white Bohemian glass, having a diameter of 0.4 or 0.5 inch, and varying in length from 14 to 18 inches : this kind of glass bears a moderate red heat without becoming soft enough to lose its shape. One end of the tube is drawn out to a point as shown in fig. 151, and closed ; the other is simply heated to fuse and

Fig. 151.

Copper oxide.   Mixture.   Copper oxide.

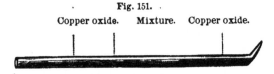

soften the sharp edges of the glass. To prevent absorption of moisture by the copper oxide during the filling of the tube, the oxide, while still hot, is poured into a long-necked flask closed by a cork ; it is then left to cool, and afterwards mixed with the organic substance in the combustion-tube itself. A small quantity of the cooled oxide is first introduced into the tube, then a portion of the organic substance is added, and upon this a column of oxide about 3 inches long is poured in the same manner as before. This portion of oxide is mixed with the organic substance by stirring it with a clean iron rod, the extremity of which is bent into two turns of a screw ; and when it is thoroughly incorporated, the rest of the substance is introduced, then another column of oxide of the same length as before, and the stirring is repeated. Lastly, the rest of the tube is filled with pure oxide.

The tube is then ready to be placed in the furnace or chauffer : this, when charcoal is the fuel employed, is constructed of thin sheet-iron, and is furnished with a series of supports of equal height, which serve to prevent flexure of the combustion-tube when softened by heat. The chauffer is placed upon flat bricks or a piece of stone, so that but little air can enter the grating, unless the whole be purposely raised. A slight inclination is also given towards the extremity occupied by the mouth of the combustion-tube, which passes through a hole provided for that purpose.

To collect the water produced in the experiment, a small light tube of the form represented in fig. 153, or a U-tube, as in fig. 156, filled with fragments of spongy calcium chloride, is attached by a perforated cork, thoroughly dried, to the open extremity of the combustion-tube. The carbon

Fig. 152.

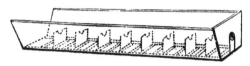

dioxide is absorbed by a solution of caustic potash, of specific gravity 1.27, which is contained in a small glass apparatus on the principle of a Woulfe's bottle, shown in fig. 154. The connection between the latter and the cal-

Fig. 154.

Fig. 153.

cium-chloride tube is completed by a little tube of caoutchouc, secured with silk cord. The whole is shown in fig. 155, as arranged for use.

Fig. 155.

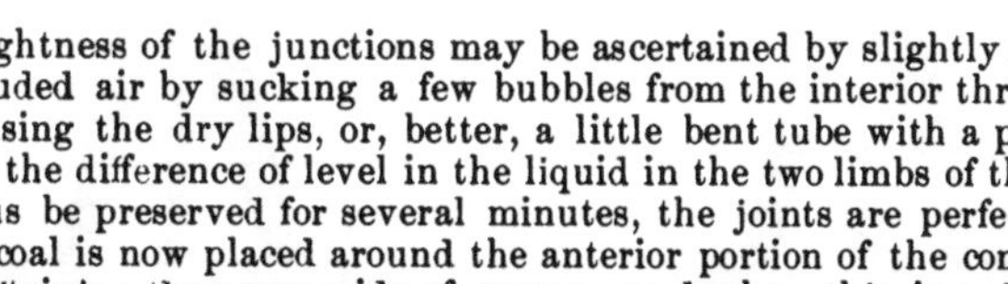

Drawing of the whole arrangement.

The tightness of the junctions may be ascertained by slightly rarefying the included air by sucking a few bubbles from the interior through the liquid, using the dry lips, or, better, a little bent tube with a perforated cork ; if the difference of level in the liquid in the two limbs of the potash apparatus be preserved for several minutes, the joints are perfect. Red-hot charcoal is now placed around the anterior portion of the combustion-tube, containing the pure oxide of copper ; and when this is red-hot, the fire is slowly extended towards the farther extremity by shifting the movable screen represented in the drawing. The experiment must be so conducted, that a uniform stream of carbon dioxide shall enter the potash-apparatus by bubbles which may be easily counted ; when no nitrogen is present, these bubbles are, towards the termination of the experiment, almost completely absorbed by the alkaline liquid, the little residue of air alone escaping. In the case of an azotized body, on the contrary, bubbles of nitrogen gas pass through the potash-solution during the whole process.

When the tube has been completely heated from end to end, and no more gas is disengaged, but, on the other hand, absorption begins to be evident, the coals are removed from the farthest extremity of the combustion-tube, and the point of the latter broken off. A little air is drawn through the whole apparatus, by which the remaining carbon dioxide and watery vapor are secured. The parts are, lastly, detached, and the calcium-chloride tube and potash-apparatus re-weighed.

The mode of heating the combustion-tube with red-hot charcoal is the original process, and is still employed where gas is not available. But since the use of coal gas has been universally adopted in laboratories, many contrivances have been suggested, by means of which this convenient fuel may be employed also in organic analysis. An apparatus of this kind* is the one represented in fig. 156, in which the combustion-tube

Fig. 156.

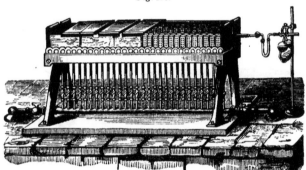

is heated by a series of perforated clay-burners. These clay-burners are fixed on pipes provided with stopcocks, so that the gas may be lighted according to the requirements of the case. The stopcocks being appropriately adjusted, the gas burns on the surface of the burners with a smoke-

Fig. 157.   Fig. 158.

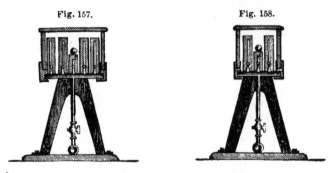

* H o f m a n n, Journal of Chemical Society, vol. xi. p. 30.

less blue flame, which renders them in a short time incandescent. The construction of this furnace is readily intelligible by a glance at figures 157 and 158, which exhibit the different parts of the apparatus in section, fig. 157 representing a large furnace with five rows, and fig. 158 a smaller furnace with three rows of clay-burners.

Gas-furnaces of simpler construction, in which the combustion-tube is heated by a row of Bunsen burners, are also much used. Such a furnace is that of von Babo, represented in fig. 160.

The following account of a real experiment will serve to illustrate the calculation of the result obtained in the combustion of crystallized sugar:

Quantity of sugar employed	4.750 grains.
Potash-apparatus weighed after experiment .	781.13
" " before experiment .	773.82
Carbon dioxide	7.31
Calcium-chloride tube after experiment . .	226.05
" before experiment .	223.30
Water	2.75

7.31 gr. carbon dioxide $=$ 1.994 gr. carbon : and 2.75 gr. water $=$ 0.3056 gr. hydrogen ; or in 100 parts of sugar,[*]

Carbon	. 41.98
Hydrogen	. 6.43
Oxygen, by difference	. 51.59
	100.000

When the organic substance cannot be mixed with the copper oxide in the manner described, the process must be modified. If, for example, a volatile liquid is to be examined, it is inclosed in a little glass bulb with a narrow stem, which is weighed before and after the introduction of the liquid, the point being hermetically sealed. A little copper oxide is put into the combustion-tube, then the bulb, with its stem broken at a, a file-scratch having been previously made ; and, lastly, the tube is filled with the cold and dry copper oxide. It is arranged in the furnace, the calcium-chloride tube and potash apparatus adjusted, and then some 6 or 8 inches of oxide having been heated to redness, the liquid in the bulb is, by the approximation of a hot coal, expelled, and slowly converted into vapor, which, in passing over the hot oxide, is completely burned. The experiment is then terminated in the usual manner. Fatty and waxy substances, and volatile solid bodies, as camphor, are placed in little boats of glass or platinum.

Fig. 159.

Copper oxide which has been used, may be easily restored by moistening with nitric acid, and igniting to redness ; it becomes, in fact, rather improved than otherwise, as, after frequent employment, its density is

---

[*] The theoretical composition of sugar, $C_{12}H_{22}O_{11}$, reckoned to 100 parts, gives—

Carbon	. 42.11
Hydrogen	. 6.43
Oxygen	. 51.46
	100.00

increased, and its troublesome hygroscopic powers diminished. For substances which are very difficult of combustion, from the large proportion
of carbon they contain, and for compounds into which chlorine enters as
a constituent, fused and powdered lead chromate may be substituted for
the copper oxide, as it freely gives up oxygen to combustible matters, and
even evolves, when strongly heated, a little of that gas, which thus insures
the perfect combustion of the organic body.

Lead chromate is, however, troublesome to prepare, and always destroys
the glass tubes. A more convenient method of insuring the complete combustion of substances which burn with difficulty is to finish the combustion
in a stream of oxygen. This may be effected either by placing a small
quantity of fused potassium chlorate at the closed end of the combustion-
tube, or by connecting the tube with a gas-holder containing oxygen gas.
The latter method is to be preferred, as it enables the operator to regulate
the stream of gas at his pleasure, whereas the ignition of potassium chlorate is apt to give rise to a rapid evolution of gas, which may force a portion of liquid out of the potash-bulbs and render the analysis worthless.

A form of apparatus for supplying a stream of oxygen is represented in
figure 160. The combustion-tube is open at both ends, and the end
farthest from the potash-bulbs and calcium-chloride tube is connected
with two gas-holders, one filled with air, the other with oxygen gas. The

Fig. 160.

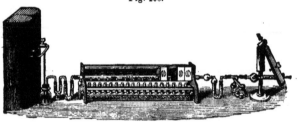

communication is made by means of a T-tube provided with a stopcock,
so that either gas-holder may be connected with the combustion-tube and
the other shut off at the same time. The air and gxygen, before entering
the combustion-tube, are made to pass through two U-tubes, one containing
lumps of pumice soaked in sulphuric acid to dry the gas, the other containing lumps of caustic potash to free it from carbonic acid.

The combustion-tube may be heated either with gas or with charcoal
(the figure represents a Babo's gas apparatus), and the mixture may be
disposed in the tube in the way above described—or better, in the case of
difficultly combustible bodies—the substance may be placed by itself in a
small boat of platinum or porcelain. In this case, the combustion-tube is
first filled to two-thirds of its length with copper oxide, which need not be
previously ignited, the remaining third, nearest to the gas-holder, being
left free to receive the boat. It is then laid in the furnace, and connected
in the manner just described with the gas-holder containing atmospheric
air; the copper oxide is heated to redness; and a stream of dry air is
passed through the tube so as to remove every trace of moisture. The
tube is then left to cool; the boat containing the substance is introduced,
a plug of recently ignited asbestos having been previously introduced to
prevent the copper oxide from coming in contact with it; the calcium-
chloride tube and potash-bulbs are attached in the usual manner; and the
tube is connected with the gas-holder containing oxygen. The copper

oxide is now once more heated to redness, and as soon as it is thoroughly ignited, heat is very cautiously applied to the part of the tube containing the boat, a slow stream of oxygen being passed through the apparatus, sufficient to prevent any backward passage of the gases, but not to cause any free oxygen to pass through the solution of potash. If the oxide of copper exhibits a red color, indicating reduction, the heating of the substance in the boat must be discontinued till the copper is reoxidized. When at length there is nothing left of the organic substance but black charcoal, the heat may be increased and the stream of oxygen accelerated. In this manner the combustion is soon completed, and when the bubbles of gas appear to pass through the potash without absorption, the process is continued in the same manner for a few minutes longer, and the potash bulbs and calcium-chloride tube are then detached, after air has been passed through the apparatus for a little time to displace the oxygen. Lastly, the stream of air is continued for a sufficient time to effect the compiete reoxidation of the copper, which is then ready, without further preparation, for another experiment.

After the tube has cooled, the boat is taken out and re-weighed. If any inorganic matter remains in it (as in the case of a salt) the quantity of this is at once ascertained, if the weight of the boat itself is previously known.

As the stream of hot gas is likely to carry vapor of water with it in passing through the potash-bulbs, whence loss of weight would ensue, a second calcium-chloride tube is attached to the potash-bulbs, as shown in the figure, to arrest any water-vapor thus carried over. This tube is weighed, together with the potash-bulbs, before and after the experiment.

The method just described is capable of giving very exact results ; it insures the complete combustion of the carbon, and obviates all danger of an excess of hydrogen arising from moisture in the copper oxide. It likewise saves the trouble of igniting this oxide before the experiment and afterwards treating it with nitric acid. But to insure a good result especial care must be taken not to heat the substance in the boat too suddenly ; otherwise combustible gases will be given off faster than they can be burnt, and the analysis will be worthless.

Analysis of Azotized Substances.—The presence of nitrogen in an organic compound is easily ascertained by heating a small portion with solid potassium hydrate in a test-tube : the nitrogen, if present, is converted into ammonia, which may be recognized by its odor and alkaline reaction.

In determining the carbon and hydrogen in such bodies, by combustion with copper oxide, as above described, a longer tube than usual must be employed, and 4 or 5 inches of its anterior portion filled with copper-turnings rendered perfectly metallic by ignition in hydrogen. This serves to decompose any nitrogen oxides formed in the process of combustion, which, if suffered to pass off undecomposed, would be absorbed by the potash and vitiate the determination of the carbon.

The nitrogen may be estimated either by converting it into ammonia, by igniting the substance with an alkaline hydrate, as above mentioned, or by evolving it in the free state and measuring its volume.

1. By *conversion into Ammonia:* Will and Varrentrapp's method.—An intimate mixture is made of 1 part caustic soda and 2 or 3 parts quicklime, by slaking lime of good quality with the proper proportion of strong caustic soda, drying the mixture in an iron vessel, and then heating it to redness in an earthen crucible. The ignited mass is rubbed to powder in a warm mortar, and carefully preserved from the air. The lime is useful in many ways : it diminishes the tendency of the alkali to deliquesce, facilitates mixture with the organic substance, and prevents fusion and

liquefaction. A proper quantity of the substance to be analyzed, namely, from 5 to 10 grains, is dried and accurately weighed out : this is mixed in a warm porcelain mortar with enough of the soda-lime to fill two-thirds of an ordinary combustion-tube, the mortar being rinsed with a little more of the alkaline mixture, and, lastly, with a small quantity of powdered glass, which completely removes everything adherent to its surface; the tube is then filled to within an inch of the open end with the lime-mixture, and arranged in the chauffer in the usual manner. The ammonia is collected in a little apparatus of three bulbs (fig. 161), containing mode-

Fig. 161.

rately strong hydrochloric acid, attached by a cork to the combustion-tube. Matters being thus adjusted, fire is applied to the tube, commencing with the interior extremity. When it is ignited throughout its whole length, and when no gas issues from the apparatus, the point of the tube is broken, and a little air drawn through the whole. The acid liquid is then emptied into a capsule, the bulbs rinsed into the same, first with a little alcohol, and then repeatedly with distilled water ; an excess of pure platinic chloride is added ; and the whole evaporated to dryness in a water-bath. The dry mass, when cold, is treated with a mixture of alcohol and ether, which dissolves out the superfluous platinum chloride, but leaves untouched the yellow crystalline ammonium platinochloride. The latter is collected upon a small weighed filter, washed with the same mixture of alcohol and ether, dried at 100⁰, and weighed ; 100 parts correspond to 6.272 parts of nitrogen. Or, the salt with its filter may be very carefully ignited, the filter burned in a platinum crucible, and the nitrogen reckoned from the weight of the spongy metal, 100 parts of that substance being equivalent to 14.18 parts of nitrogen. The former plan is to be preferred in most cases.

Bodies very rich in nitrogen, as urea, must be mixed with about an equal quantity of pure sugar, to furnish uncondensable gas, and thus diminish the violence of the absorption which otherwise occurs ; and the same precaution must be taken, for a different reason, with those which contain little or no hydrogen.

A modification of this process has been suggested by Peligot, which is very convenient if a large number of nitrogen-determinations is to be made. By this plan, the ammonia, instead of being received in hydrochloric acid, is conducted into a known volume ($\frac{1}{2}$ to 1 cubic inch) of a standard solution of sulphuric acid contained in the ordinary nitrogen-bulbs. After the combustion is finished, the acid containing the ammonia is poured out into a beaker, colored with a drop of tincture of litmus, and then neutralized with a standard solution of soda in water, or of lime in sugar-water, the point of neutralization becoming perceptible by the sudden appearance of a blue tint. The lime-solution is conveniently poured out from an alkalimeter. The volume of lime-solution necessary to neutralize the same amount of acid that is used for condensing the ammonia, having been ascertained by a preliminary experiment, it is evident that the difference of the quantities used in the two experiments gives the ammonia collected in the acid during the combustion. The amount of nitrogen may thus be calculated. If, for instance, an acid be prepared, containing 20 grains of pure hydrogen sulphate (H_2SO_4) in 1000 grain-measures, then 200 grain-measures of this acid—the quantity introduced into the bulbs —will correspond with 1.38 grains of ammonia, or 1.14 grains of nitrogen. The alkaline solution is so graduated that 1000 grain-measures will exactly neutralize the 200 grain-measures of the standard acid. If we now find that the acid, partly saturated with the ammonia disengaged during the combustion of a nitrogenous substance, requires only 700 grain-measures of the alkaline solution, it is evident that $\dfrac{200 \times 300}{1000} = 60$ grain-measures were saturated by the ammonia, and the quantity of nitrogen is obtained by the proportion—200 : 1.14 = 60 : x, wherefore $x = \dfrac{1.14 \times 60}{200} = 0.342$ grain of nitrogen.

2. *By measurement as free Nitrogen.*—When the nitrogen exists in the organic substance in the form of an oxide, as in nitrobenzene, $C_6H_5(NO_2)$, ethyl nitrite, $C_2H_5(NO)O$, etc., the preceding method cannot be employed, because these nitrogen oxides are not completely converted into ammonia by heating with alkaline hydrates: it fails also in the case of certain organic bases. In such cases the nitrogen must be evolved in the free state by heating the organic body with copper oxide; and its volume determined by collecting it over mercury in a graduated jar. There are several ways of effecting this : the one most frequently employed is that of Dumas, as simplified by Melsens.

A tube of Bohemian glass, 28 inches long, is securely sealed at one end ; into this enough dry acid sodium carbonate is put to occupy 6 inches. A little pure copper oxide is next introduced, and afterwards the mixture of oxide and organic substances, the weight of the latter, between 4.5 and 9 grains, in a dry state, having been correctly determined. The

Fig. 162.

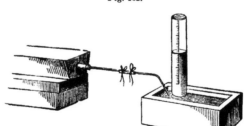

remainder of the tube, amounting to nearly one-half of its length, is then filled up with pure copper oxide and spongy metal, and a round cork, perforated by a piece of narrow tube, is securely adapted to its mouth. This tube is connected by means of a caoutchouc joint with a bent delivery-tube, a, and the combustion-tube is arranged in the furnace. Heat is now applied to the further end of the tube, so as to decompose a portion of the acid sodium carbonate, the remainder of the carbonate, as well as of the other part of the tube, being protected from the heat by a screen, n. The current of carbon dioxide thus produced is intended to expel all the air from the apparatus. In order to ascertain that this object, on which the success of the whole operation depends, is accomplished, the delivery-tube is depressed under the level of a mercurial trough, and the gas, which is evolved, collected in a test-tube filled with concentrated potash-solution. If the gas be perfectly absorbed, or, if after the introduction of a considerable quantity, only a minute bubble be left, the air may be considered as expelled. The next step is to fill a graduated glass jar two-thirds with mercury and one-third with a strong solution of potash, and to invert it over the delivery-tube, as represented in fig. 162.

This done, heat is applied to the tube, commencing at the front end, and gradually proceeding to the closed extremity, which still contains some undecomposed acid sodium carbonate. This, when the heat at length reaches it, yields up carbon dioxide, which chases forward the nitrogen lingering in the tube. The carbon dioxide generated during the combustion is wholly absorbed by the potash in the jar, and nothing is left but the nitrogen. When the operation is at an end, the jar, with its contents, is transferred to a vessel of water, and the volume of the nitrogen read off. This is properly corrected for temperature, pressure, and aqueous vapor, and its weight determined by calculation. When the operation has been very successful, and all precautions minutely observed, the result still leaves an error in excess, amounting to 0.3 or 0.5 per cent., due to the residual air of the apparatus, or that condensed in the pores of the copper oxide.

A modification of the process, by which this error is considerably diminished, has been devised by Dr. Maxwell Simpson.*

The method just described is applicable to the estimation of nitrogen in the oxides and oxygen-acids of nitrogen, in metallic nitrates and nitrites, and, in fact, to the analysis of all nitrogenous bodies whatever.

Analysis of Chlorinated Compounds.—In the case of a volatile liquid containing chlorine, the combustion with copper oxide must be very carefully conducted, and 2 or 3 inches of the anterior portion of the tube kept cool enough to prevent volatilization of the copper chloride into the calcium-chloride tube. Lead chromate is much better for the purpose.

The chlorine is determined by placing a small weighed bulb of liquid in a combustion-tube, which is afterwards filled with fragments of pure quicklime. The lime is brought to a red heat, and the vapor of the liquid driven over it, when the chlorine displaces oxygen from the lime, and gives rise to calcium chloride. When cold, the contents of the tube are dissolved in dilute nitric acid, the liquid is filtered, and the chlorine precipitated by silver nitrate.

Bromine and *iodine* are estimated in a similar manner.

Analysis of Organic Compounds containing Sulphur.—When a body of this nature is burned with copper oxide, a small tube containing lead oxide may be interposed between the calcium-chloride tube and the potash apparatus, to retain any sulphurous acid that may be formed. It is better, however, to use lead chromate in such cases. The proportion of sulphur is determined by oxidizing a known weight of the substance with strong nitric acid, or by fusion in a silver vessel with ten or twelve times its weight of pure potassium hydrate and half as much nitre. The sulphur is thus converted into sulphuric acid, the quantity of which can be determined by dissolving the fused mass in water, acidulating with nitric acid, and adding a barium salt. *Phosphorus* is, in like manner, oxidized to phosphoric acid, the quantity of which may be determined by precipitation as ammonio-magnesium phosphate.

An easier method of estimating sulphur, phosphorus, chlorine, etc., in organic compounds, consists in heating the substance with nitric acid, of specific gravity about 1.2, in a sealed tube. *Sulphur* is thereby, in nearly all cases, completely converted into sulphuric acid, and may be precipitated by chloride of barium ; *phosphorus* and *arsenic* are converted into phosphoric and arsenic acids, and may be precipitated as ammonio-magnesium salts ; *chlorine* is partly oxidized, partly separated in the free state, but may be completely converted into hydrochloric acid by means of a dilute solution of sulphurous acid or sulphite of sodium, and then precipitated by nitrate of silver ; *bromine* and *iodine* are completely separated in the free state, and may be estimated in like manner ; lastly, *metals* are converted into oxides or nitrates, and may be estimated by the ordinary methods of mineral analysis.

This method of oxidation by nitric acid in sealed tubes, is likewise applicable to many inorganic compounds, the sulphides of arsenic, for example.

EMPIRICAL AND MOLECULAR FORMULÆ.

A chemical formula is termed *empirical* when it merely gives the simplest possible expression of the composition of the substance to which it refers. A *molecular* formula, on the contrary, expresses the absolute number of atoms of each of its elements supposed to be contained in the molecule, as well as the mere numerical relations existing between them. The empirical formula is at once deduced from the analysis of the substance, reckoned to 100 parts.

The case of sugar, already cited, may be. taken as an example. This substance gives by analysis—

Carbon	41.98
Hydrogen	6.43
Oxygen	51.59
	100.00

If each of these quantities be divided by the atomic weight of the corresponding element, the quotients will express the relations existing between the numbers of atoms of the three elements : these are afterwards reduced to their simplest expression. This is the only part of the calculation attended with any difficulty. If the numbers were rigidly correct, it would only be necessary to divide each by the greatest divisor common to the whole ; but as they are only approximative, something is of necessity left to the judgment of the experimenter.

In the case of sugar, we have—

$$\frac{41.98}{12} = 3.50 ; \quad \frac{64.3}{1} = 6.43 ; \quad \frac{51.59}{16} = 3.42,$$

or 350 atoms carbon, 643 atoms of hydrogen, and 342 atoms oxygen. Now it is evident, in the first place, that the hydrogen and oxygen are present nearly in the proportions to form water, or twice as many atoms of the former as of the latter. Again, the atoms of carbon and hydrogen are nearly in the proportion of 12 : 22, so that the formula $C_{12}H_{22}O_{11}$ appears likely to be correct. It is now easy to see how far this is admissible, by reckoning it back to 100 parts, comparing the result with the numbers given by the actual analysis, and observing whether the difference falls fairly, in direction and amount, within the limits of error of what may be termed a good experiment, viz., two or three-tenths per cent. *deficiency* in the carbon, and not more than one-tenth or two-tenths per cent. *excess* in the hydrogen :—

$$
\begin{array}{lllll}
\text{Carbon} & . & . & . & . & 12 \times 12 = 144 \\
\text{Hydrogen} & . & . & . & . & 1 \times 22 = 22 \\
\text{Oxygen} & . & . & . & . & 16 \times 11 = 176 \\
\end{array}
$$

$$342$$

$$342 : 144 = 100 : 42.11$$
$$342 : 222 = 100 : 6.43$$
$$342 : 176 = 100 : 51.46$$

To determine the molecular formula, several considerations must be taken into account, namely, the combining or saturating power of the compound, if it is acid or basic; the number of atoms of any one of its elements (generally hydrogen) which may be replaced by other elements; the law of even numbers, which requires that the sum of the numbers of atoms of all the perissad elements (hydrogen, nitrogen, chlorine, etc.) contained in the compound shall be divisible by 2; and the vapor-density of the compound (if it be volatile without decomposition) which, in normally constituted compounds, is always half the molecular weight (p. 246).

The molecular formula may either coincide with the empirical formula, or it may be a multiple of the latter. Thus, the composition of *acetic acid* is expressed by the formula CH_2O, which exhibits the simplest relations of the three elements; but if we want to express the quantities of these, in atoms, required to make up a molecule of acetic acid, we have to adopt the formula $C_2H_4O_2$: for only one-fourth of the hydrogen in this acid is replaceable by metals to form salts, $C_2H_3KO_2$, for example: and its vapor-density, compared with hydrogen, is nearly 30, which is half the weight of the molecule, $C_2H_4O_2 = 2 \times 12 + 4 \times 1 + 2 \times 16$. Again, the empirical formula of benzene is CH; but this contains an uneven number of hydrogen-atoms; moreover, if it expressed the weight of the molecule of benzene, the vapor-density of that compound should be $\dfrac{12 + 1}{2} = 6.5$, whereas experiment shows that it is six times as great, or equal to 39: hence the molecular formula of benzene is C_6H_6.

Organic acids and salt-radicles have their molecular weights most frequently determined by an analysis of their lead and silver salts, by burning these latter, with suitable precautions, in a thin porcelain capsule, and noting the weight of the lead oxide or metallic silver left behind. If the lead oxide be mixed with globules of reduced metal, the quantity of the latter must be ascertained by dissolving away the oxide with acetic acid. Or the lead salt may be converted into sulphate, and the silver compound into chloride, and both metals thus estimated. An organic base, on the contrary, has its molecular weight fixed by observation of the quantity of a mineral acid or organic salt-radicle, required to form with it a compound having the characters of neutrality.

The rational and constitutional formulæ of organic compounds will be considered further on.

It is scarcely necessary to observe that the methods just described for determining the empirical and molecular formula of an organic compound from the results of its analysis, together with its physical properties and chemical reactions, are equally applicable to inorganic compounds.

CLASSIFICATION OF ORGANIC COMPOUNDS.—ORGANIC SERIES.

The classification of organic compounds is based upon the quantivalence or atomicity of carbon. This element is a tetrad, being capable of uniting with at most four atoms of hydrogen or other monatomic elements. Methane or marsh gas, CH_4, is therefore a saturated hydrocarbon, not capable of uniting directly with chlorine, bromine, or other monad elements, but only of exchanging a part or the whole of its hydrogen for an equivalent quantity of another monad element. It may, however, as already explained (p. 254), take up any number of dyad elements or radicles, because such a radicle introduced into any group of atoms whatever, neutralizes one unit of equivalency, and adds another, leaving therefore the combining capacity or equivalence of the group just the same as before. Accordingly, the hydrocarbon, CH_4, may take up any number of molecules of the bivalent radicle, CH_2, thereby giving rise to the series of saturated hydrocarbons,

$$CH_4, \quad C_2H_6, \quad C_3H_8, \quad C_4H_{10} \ldots C_nH_{2n+2}.$$

A series of compounds, the terms of which differ from one another by CH_2, is called a h o m o l o g o u s series. There are many such series besides that of the hydrocarbons just mentioned; thus methyl chloride, CH_3Cl, gives, by continued addition of CH_2, the series of chlorides,

$$CH_3Cl, \quad C_2H_5Cl, \quad C_3H_7Cl, \quad C_4H_9Cl \ldots C_nH_{2n+1}Cl;$$

and from methyl alcohol, CH_4O, is derived in like manner the series of homologous alcohols,

$$CH_4O, \quad C_2H_6O, \quad C_3H_8O, \quad C_4H_{10}O \ldots C_nH_{2n+2}O.$$

The terms of the same homologous series resemble one another in many respects, exhibiting similar transformations under the action of given reagents, and a regular gradation of properties from the lowest to the highest; thus, of the hydrocarbons, C_nH_{2n+2}, the lowest terms CH_4, C_2H_6, and C_3H_8, are gaseous at ordinary temperatures, the highest containing 20 or more carbon atoms, are solid, while the intermediate compounds are liquids, becoming more and more viscid and less volatile, as they contain a greater number of carbon atoms, and exhibiting a constant rise of about 20° C. (36° F.) in their boiling points for each addition of CH_2 to the molecule.

The saturated hydrocarbons C_nH_{2n+2}, may, under various circumstances, be deprived of two atoms, or one molecule, of hydrogen, thereby producing a new homologous series,

$$CH_2, \quad C_2H_4, \quad C_3H_6, \quad C_4H_8 \ldots C_nH_{2n}.$$

These are unsaturated molecules, having two units of equivalency uncombined, and therefore acting as bivalent radicles, capable of taking up 2 atoms of chlorine, bromine, or other univalent radicles, and 1 atom of oxygen or other bivalent radicle.

The first term of this last series cannot give up 2 atoms of hydrogen without being reduced to the atom of carbon; but the remaining terms may each give up 2 atoms of hydrogen, and thus give rise to the series,

$$C_2H_2, \quad C_3H_4, \quad C_4H_6 \ldots C_nH_{2n-2},$$

each term of which is a quadrivalent radicle.

22

And, in like manner, by successive abstraction of H_2, a number of homologous series may be formed, whose general terms are

$$C_nH_{2n}+_2, \quad C_nH_{2n}, \quad C_nH_{2n-2}, \quad C_nH_{2n-4} \ldots \text{etc.}$$

The individual series, as far as C_6, are given in the following table, together with the names proposed for them by Dr. Hofmann :*

CH_4 CH_2
Methane   Methene

C_2H_6 C_2H_4 C_2H_2
Ethane   Ethene   Ethine

C_3H_8 C_3H_6 C_3H_4 C_3H_2
Propane   Propene   Propine   Propone

C_4H_{10} C_4H_8 C_4H_6 C_4H_4 C_4H_2
Quartane   Quartene   Quartine   Quartone   Quartune

C_5H_{12} C_5H_{10} C_5H_8 C_5H_6 C_5H_4 C_5H_2
Quintane   Quintene   Quintine   Quintone   Quintune

C_6H_{14} C_6H_{12} C_6H_{10} C_6H_8 C_6H_6 C_6H_4 C_6H_2
Sextane   Sextene   Sextine   Sextone   Sextune

Each vertical column of this table forms a homologous series, in which the terms differ by CH_2, and each horizontal line an isologous series, in which the successive terms differ by H_2. The bodies of these last series are designated as the monocarbon, dicarbon group, etc.

The formulæ in the preceding table represent hydrocarbons, all of which are capable of existing in the separate state, and many of which have been actually obtained. They are all derived from saturated molecules, $C_nH_{2n}+_2$, by abstraction of one or more *pairs* of hydrogen-atoms.

But a saturated hydrocarbon, CH_4, for example, may give up 1, 2, 3, or any number of hydrogen-atoms in exchange for other elements; thus, marsh gas, CH_4, subjected to the action of chlorine under various circumstances, yields the substitution-products,

$$CH_3Cl, \quad CH_2Cl_2, \quad CHCl_3, \quad CCl_4,$$

which may be regarded as compounds of chlorine with the radicles,

$$(CH_3)', \quad (CH_2)'', \quad (CH_3)''', \quad C^{iv};$$

and in like manner each hydrocarbon of the series, $C_nH_{2n}+_2$, may yield a series of radicles of the forms,

$$(C_nH_{2n}+_1)', \quad (C_nH_{2n})'', \quad (C_nH_{2n-1})''', \quad (C_nH_{2n-2})^{iv}, \text{etc.,}$$

each of which has an equivalent value, or combining power, corresponding with the number of hydrogen-atoms abstracted from the original hydrocarbon. Those of even equivalence contain even numbers of hydrogen-atoms, and are identical in composition with those in the above table; but those of uneven equivalence contain odd numbers of hydrogen-atoms, and are incapable of existing in the separate state, except, perhaps, as double molecules (p. 254).

These hydrocarbon radicles of uneven equivalence are designated by names ending in *yl*, those of the univalent radicles being formed from methane, ethane, etc., by changing the termination *ane* into *yl*; those of the trivalent radicles by changing the final *e* in the names of the bivalent

* Proceedings of the Royal Society, xv. 57. Names with Greek prefixes are, however, more generally used; *e. g.*, pentane, hexane, and heptane, rather than quintane, sextane, and septane.

radicles, methene, etc., into *yl;* and similarly for the rest.   The names of the whole series will therefore be as follows :—

CH_4	$(CH_3)'$	$(CH_2)''$	$(CH)'''$
Methane	Methyl	Methene	Methenyl

C_2H_6	$(C_2H_5)'$	$(C_2H_4)''$	$(C_2H_3)'''$	$(C_2H_2)^{iv}$	$(CH_2)^{v}$
Ethane	Ethyl	Ethene	Ethenyl	Ethine	Ethinyl

C_3H_8	$(C_3H_7)'$	$(C_3H_6)''$	$(C_3H_5)'''$	$(C_3H_4)^{iv}$	$(C_3H_3)^{v}$	$(C_3H_2)^{vi}$	$(C_3H)^{vii}$
Propane	Propyl	Propene	Propenyl	Propine	Propinyl	Propone	Proponyl
	etc.	etc.		etc.			etc.

From these hydrocarbon-radicles, called alcohol-radicles, because they enter into the composition of alcohols, others of the same degree of equivalence may be derived by partial or total replacement of the hydrogen by other elements, or compound radicles.   Thus from propyl, C_3H_7, may be derived the following univalent radicles :

C_3H_6Cl	$C_3H_3Cl_4$	C_3H_5O	$C_3H_3Cl_2O$	$C_3H_6(CN)'$
Chloropropyl.	Tetrachloro- propyl.	Oxypropyl.	Trichlor- oxypropyl.	Cyanopropyl.

$C_3H_6(NO_2)$	$C_3H_4(NH_2)O$	$C_3H_6(CH_3)$	$C_3H_5(C_2H_5)_2$
Nitropropyl.	Amidoxypropyl.	Methyl-propyl.	Diethyl-propyl.

From the radicles above mentioned, all well-defined organic compounds may be supposed to be formed by combination and substitution, each radicle entering into combination just like an elementary body of the same degree of quantivalence.

Organic compounds may thus be arranged in the following classes :—

1. Hydrocarbons containing even numbers of hydrogen atoms.—These are the compounds tabulated on page 506; they are sometimes regarded as hydrides of radicles containing uneven numbers of hydrogen atoms, *e.g.,*

Methane, $CH_4 = CH_3.H$ Methyl hydride.

2. Haloïd Ethers.—Compounds of hydrocarbons with halogen elements, *e.g.,*

CH_3Cl	$C_2H_4Br_2$	$C_3H_5I_3$
Methyl chloride.	Ethene bromide.	Propenyl iodide.

3. Compounds of hydrocarbons with nitryl, NO_2 (the radicle of nitric acid, $NO_2.OH$), *e.g.,*

$CH_3.NO_2$	$C_2H_4.(NO_2)_2$
Nitro-methane.	Dinitro-methene.

4. Amines and Nitrils.—Compounds of alcohol-radicles with amidogen (NH_2), imidogen (NH)'', and nitrogen (N'''): `e.g.,`

$C_2H_5.NH_2$	$(C_2H_5)_2.NH$	$(C_2H_5)_3N$
Ethylamine.	Diethylamine.	Triethylamine.

$C_2H_4.(NH_2)_2$	$(C_2H_4)_2(NH)_2$	$(C_2H_4)_3.N_2$
Ethene-diamine.	Diethene-diamine.	Triethene-diamine.

These bodies are mostly of basic character, and capable of forming salts with acids, like ammonia, H_3N, from which they may, in fact, be derived by substitution of alcohol-radicles for part or the whole of the hydrogen. Those in which the hydrogen is wholly thus replaced are called nitrils; and among these special mention must be made of a group consisting of nitrogen combined with a trivalent hydrocarbon-radicle, such as—

(CH)N	$(C_2H_3)N$	$(C_3H_5)N$
Methenyl nitril.	**Ethenyl nitril.**	**Propenyl nitril.**

These nitrils have no basic properties, but are all neutral, except the first, which is a monobasic acid, capable of exchanging its hydrogen for metals, and this character may be regarded as a compound of hydrogen with the equivalent radicle cyanogen —$C\equiv N$; it is accordingly named h y d r o g e n   c y a n i d e, or h y d r o c y a n i c   a c i d, and the other nitrils homologous with it are the ethers of this acid; thus:

$$\text{Methenyl nitril, (CH)N} = \text{CN.H, Hydrogen cyanide,}$$
$$\text{Ethenyl nitril, (C}_2\text{H}_3\text{)N} = \text{CN.CH}_3\text{, Methyl cyanide,}$$
$$\text{Propenyl nitril, (C}_3\text{H}_5\text{)N} = \text{CN.C}_2\text{H}_5\text{, Ethyl cyanide.}$$

By the combination of amines with the chlorides, hydroxides, etc., of alcohol radicles, compounds are formed containing pentad nitrogen, and having the composition of ammonium salts, in which the hydrogen is more or less replaced by alcohol radicles; thus:

$$N(C_2H_5)H_3Cl, \text{ Ethylammonium chloride,}$$
$$N(C_2H_5)_2H_2Cl, \text{ Diethylammonium chloride,}$$
$$N(C_2H_5)_3HCl, \text{ Triethylammonium chloride,}$$
$$N(C_2H_5)_4Cl, \text{ Tetrethylammonium chloride,}$$
$$N(C_2H_5)_4OH, \text{ Tetrethylammonium hydroxide.}$$

This last compound and its analogues, containing methyl, amyl, etc., are powerful alkalies, obtainable, by evaporation of their aqueous solution, as white deliquescent crystalline masses resembling caustic potash.

Analogous to the amines are the p h o s p h i n e s, a r s i n e s, and s t i-
b i n e s,—*e. g.*,

$$(C_2H_5)_2.PH \qquad (C_2H_5)_3As \qquad (C_2H_5)Sb$$
$$\text{Diethyl-phosphine.} \quad \text{Triethyl-arsine.} \quad \text{Triethyl-stibine.}$$

5. A l c o h o l s.—Compounds of hydrocarbons with hydroxyl: *e. g.*,

$$C_2H_5.OH \qquad C_2H_4.(OH)_2 \qquad C_3H_5.(OH)_3$$
$$\text{Ethyl-alcohol.} \qquad \text{Ethene alcohol.} \qquad \text{Propenyl alcohol}$$
$$\text{(Glycol).} \qquad \text{(Glycerine).}$$

6. The replacement of the hydroxylic hydrogen in an alcohol by hydro-carbon-radicles (alcohol-radicles), or oxygenated hydrocarbon radicles (acid radicles), gives rise respectively to o x y g e n - e t h e r s and c o m -
p o u n d   e t h e r s, *e. g.*,

$$C_2H_5.O.C_2H_5 \qquad C_2H_5.O.CH_3 \qquad C_2H_4O$$
$$\text{Ethyl oxide.} \qquad \text{Ethyl-methyl} \qquad \text{Ethene oxide.}$$
$$\text{oxide.}$$

$$C_2H_5.O.C_2H_3O \qquad C_2H_4{<}^{OH}_{OC_2H_3O} \qquad C_2H_4{<}^{OC_2H_3O}_{OC_2H_3O}$$
$$\text{Ethyl acetate.} \qquad \text{Ethene mono-acetate.} \qquad \text{Ethene di-acetate.}$$

7. S u l p h u r   and   S e l e n i u m   A l c o h o l s   and   E t h e r s.—Compounds analogous to the oxygen alcohols and ethers—the oxygen being replaced by sulphur or selenium. The sulphur and selenium alcohols are called m e r c a p t a n s.

8. Compounds of hydrocarbon radicles with various elements, metallic and non-metallic: *e. g.*,

$$NaC_2H_5 \qquad Zn(C_2H_5)_2 \qquad Sn(CH_3)_4$$
$$\text{Sodium ethide.} \qquad \text{Zinc ethide.} \qquad \text{Stannic methide.}$$

$$B(C_2H_5)_3 \qquad Si(C_2H_5)_4$$
$$\text{Triethylic boride.} \qquad \text{Tetrethylic}$$
$$\text{silicide.}$$

Those containing metals are called O r g a n o - m e t a l l i c   **Compounds.**

9. A l d e h y d e s .—Compounds intermediate between alcohols and acids.
Thus :

$$C_2H_6O \qquad C_2H_4O \qquad C_2H_4O_2$$
Ethyl   Acetic   Acetic
alcohol.   aldehyde.   acid.

10. K e t o n e s .—Bodies derived from aldehydes by the replacement of 1
atom of hydrogen by an alcohol-radicle ; *e. g.*,

Acetone, $C_3H_6O = C_2H_3(CH_3)O$. .

O r g a n i c   A c i d s .—Compounds of oxygenated radicles with hydroxyl ;
e. g.,

$$C_2H_3O.HO \qquad (C_4H_4O_2).(HO)_2 \qquad (C_6H_5O_4).(HO)_3$$
Acetic acid.   Succinic acid.   Citric acid.

The hydrogen in the radicles of these acids may be more or less replaced
by chlorine, bromine, nitryl, amidogen, etc. : thus from benzoic acid,
$C_7H_5O.OH$, are derived :

$$C_7H_4ClO.OH \qquad C_7H_4(NO_2)O.OH \qquad C_7H_3(NH_2)_2O.OH$$
Chlorobenzoic   Nitrobenzoic acid.   Diamidobenzoic
acid.   acid.

11. A c i d   H a l i d e s , also called C h l o r - a n h y d r i d e s , B r o m a n -
h y d r i d e s , etc.—Compounds of oxygenated radicles (acid radicles) with
halogen-elements, and derived from the acids by substitution of those
elements for the hydroxyl ; *e. g.*,

$$C_2H_3O.Cl \qquad C_2H_4O_2.Cl_2 \qquad C_6H_5O_4.Cl_3$$
Acetyl chloride. Succinyl chloride.   Citryl chloride.

12. A c i d   O x i d e s , sometimes called *Anhydrous Acids* or *Anhydrides ;*
e. g.,

$$(C_2H_3O)_2O \qquad C_4H_4O_2.O \qquad \left.\begin{array}{c}C_2H_3O \\ C_7H_5O\end{array}\right\} O$$
Acetic oxide.   Succinic acid.   Aceti-benzoic
oxide.

13. A m i d e s.—Compounds analogous to the amines, but containing
acid-radicles instead of alcohol-radicles : those which contain bivalent
radicles combined with imidogen, NH, are called i m i d e s ; *e. g.*,

Acetamide, $C_2H_3O.NH_2$	Succinamide, $C_4H_4O_2(NH_2)_2$
Diacetamide, $(C_2H_3O)_2.NH$	Trisuccinamide, $(C_4H_4O_2)_3N_2$
Succinimide, $C_4H_4O_2.NH$	Citramide, $(C_6H_5O_4)N$

Each of the classes of carbon compounds above enumerated may be
divided into homologous and isologous series, though in most cases these
series are far from being complete.

Further, organic compounds may be divided into two great groups,
called the F a t t y and A r o m a t i c groups, each including hydrocarbons,
alcohols, acids, bases, etc., those of the first group being derived from
m e t h a n e , CH_4, and those of the second from b e n z e n e , C_6H_6. The
corresponding compounds in these two groups are distinguished from one
another by well-marked characters, supposed to depend upon the arrangement of the carbon-atoms in their molecules.

The preceding classes, most of which have their analogues amongst
inorganic compounds, include nearly all artificially prepared organic
bodies, and the majority of those produced in the living organism. There
are still, however, many compounds formed in the bodies of plants and
animals, the chemical relations of which are not yet sufficiently well made

out to enable us to classify them with certainty. Such is the case with many vegetable oils and resins, with most of the alkaloïds or basic nitrogenized compounds found in plants, such as morphine, quinine, strychnine, etc., and several definite compounds formed in the animal organism, as albumin, fibrin, casein, and gelatin.

Rational Formulæ of Organic Compounds.—It must be distinctly under. derstood that the formulæ above given are not the only ones by which the constitution of the several classes of organic compounds may be repre. sented. Rational formulæ are intended to represent the mode of formation and decomposition of compounds, and the relation which allied compounds bear to one another : hence, if a compound can, under varying circumstances, split up into different atomic groups or radicles, or if it can be formed in various ways by the combination of such radicles, different rational formulæ must be assigned to it. This point has been already noticed in connection with the constitution of metallic salts, and illustrated especially in the case of the sulphates (p. 301); but organic compounds, which, for the most part, contain larger numbers of atoms, and are therefore capable of division into a greater number of groups, afford much more abundant illustration of the same principle. Take, for example, acetic acid, the molecular formula of which is $C_2H_4O_2$. This may be resolved into the following rational formulæ :—

1. $C_2H_3O_2.H$.—This formula, analogous to that of hydrochloric acid, Cl.H, indicates that a molecule of acetic acid can give up one atom of hydrogen in exchange for a univalent metal or alcohol-radicle, forming, for example, sodium acetate, $C_2H_3O_2.Na$, ethyl acetate, $C_2H_3O_2.C_2H_5$, etc.; that two molecules of the acid may give up two hydrogen atoms in exchange for a bivalent metal or alcohol-radicle, forming barium acetate, $(C_2H_3O_2)_2Ba$, ethene acetate, $(C_2H_3O_2)_2.(C_2H_4)$, etc. ; in other words, that acetic acid is a monobasic acid (p. 301).

2. $C_2H_3O.HO$.—This formula, analogous to that of water, H.HO, indicates such reactions as the formation of acetic acid from acetic chloride by the action of water :—

$$C_2H_3O.Cl + H.HO = HCl + C_2H_3O.HO.$$

3. $C_2H_3O.H.O$.—This formula, also comparable with that of water, H.H.O, indicates the conversion of acetic acid into acetic chloride, by the action of phosphorus pentachloride :—

$$C_2H_3O.H.O + PCl_3.Cl_2 = C_2H_3O.Cl + HCl + PCl_3O ;$$

also, the formation of thiacetic acid, $C_2H_3O.H.S$, by the action of phos. phorus pentasulphide on acetic acid :—

$$5(C_2H_3O.H.O) + P_2S_5 = 5(C_2H_3O.H.S) + P_2O_5.$$

4. $(C_2H_3)'''.HO.O$.—This represents the formation of acetic acid from ethenyl nitril, $(C_2H_3)'''N$, by heating with caustic alkalies :—

$$(C_2H_3)N + \left. \begin{matrix} HH.O \\ H.HO \end{matrix} \right\} = NH_3 + (C_2H_3).O.HO.$$

Ethenyl   Water.
nitril.

5. $(CO.CH_3).HO$.—This formula, in which the radicle acetyl, C_2H_3O, is resolved into carbonyl, $(CO)''$, and methyl, represents :—

a. The decomposition of acetic acid by electrolysis, in which hydrogen is evolved at the positive pole, while carbon dioxide and ethane, C_2H_6, appear at the negative :—

$$2(CO.CH_3.HO) = H_2 + C_2H_6 + 2CO_2.$$

ß. The production of methane (marsh-gas) by heating potassium acetate with excess of potassium hydroxide (p. 228).

$$CO.CH_3.KO \; + \; HKO \; = \; CH_4 \; + \; (CO)''.(KO)_2.$$

Potassium acetate.   Potassium   Methane.   Potassium
hydroxide.   carbonate.

γ. The production of acetone and barium carbonate by the dry distillation of barium acetate :—

$$(CO.CH_3)_2.BaO_2 \; = \; (CO)(CH_3)_2 \; + \; (CO).BaO_2.$$

Barium acetate.   Acetone.   Barium
carbonate.

Now, on comparing these several rational formulæ, it will be seen that they are all included under the constitutional formula,

in which the molecule is resolved into its component atoms, and these atoms are grouped, as far as possible, according to their different equivalences, or combining capacities. These constitutional formulæ are the nearest approach to the representation of the true constitution of a compound that our knowledge of its reactions enables us to give; but the student cannot too carefully bear in mind that they are not intended to represent the actual arrangement of the atoms in space, but only, as it were, their relative mode of combination, showing which atoms are combined together directly, and which only indirectly, that is, through the medium of others. Thus, in the formula of acetic acid, it is seen that three of the hydrogen atoms are united directly with the carbon, while the fourth is united to it only through the medium of oxygen ; that one of the two oxygen atoms is combined with carbon alone, the other both with carbon and with hydrogen; and that one of the carbon atoms is combined with the other carbon atom and with hydrogen ; the second with carbon and with oxygen. Abundant illustration of these principles will be afforded by the special descriptions of organic compounds in the following pages.

Isomerism.—Two compounds are said to be isomeric, when they have the same empirical formula or percentage composition, but exhibit different properties. A few examples of isomerism are met with amongst inorganic compounds ; but they are much more numerous amongst organic or carbon compounds.

Isomeric bodies may be divided into two principal groups, namely :—

A.—Those which have the same molecular weight; and these are subdivided into:—

α. Isomeric bodies, strictly so called, namely, those which exhibit analogous decompositions and transformations when heated, or subjected to the action of the same reagents, and differ only in physical properties. Such is the case with the volatile oils of turpentine, lemons, juniper, etc., all of which have the composition $C_{10}H_{16}$, resemble each other closely in their chemical reactions, and are distinguished chiefly by their odor and their action on polarized light.

ß. Metameric bodies, which, with the same percentage composition and molecular weight, exhibit dissimilar transformations under similar circumstances. Thus the molecular formula, $C_3H_6O_2$, represents three different bodies, all exhibiting different modes of decomposition under the influence of caustic alkalies, viz. :—

(1) Propionic acid, $C_3H_5O.OH$, which is converted by caustic potash, at ordinary temperatures, into potassium propionate, $C_3H_5O.OK$.

(2) Methyl acetate, $C_2H_3O.OCH_3$, a neutral liquid not acted upon by potash at common temperatures, but yielding, when heated with it, potassium acetate and methyl alcohol :—

$$C_2H_3O.OCH_3 + KOH = C_2H_3O.OK + CH_3.OH.$$

(3) Ethyl formate, $CHO.OC_2H_5$, converted in like manner, by heating with potash, into potassium formate, $CHO.OK$, and ethyl alcohol, $C_2H_5.OH.$

These three compounds may be represented by the following constitutional formulæ, the dotted lines marking the division into radicles indicated by the rational formulæ above given :—

| Propionic acid. | Methyl acetate. | Ethyl formate. |

Another kind of metamerism is exhibited by the normal and iso-alcohols and their derivatives, the structure of which will be explained hereafter.

B.—Compounds which have the same percentage composition, but differ in molecular weight; such bodies are called **polymeric**. The most striking example of polymerism is exhibited by the hydrocarbons C_nH_{2n}, all of which are multiples of the lowest, namely, methene, CH_2. Another example is afforded by certain natural volatile oils, which are polymeric with oil of turpentine, $C_{10}H_{16}$, and have the formulæ, $C_{20}H_{32}$, $C_{30}H_{48}$, etc. All polymeric compounds exhibit regular gradations of boiling point, vapor density, and other physical characters, from the lowest to the highest. Some are chemically isomeric, exhibiting analogous transformations under similar circumstances, while others are metameric, exhibiting dissimilar reactions under given circumstances.

PHYSICAL PROPERTIES OF CARBON-COMPOUNDS.

1. Density and Specific Volume.—It has been already pointed out (p. 244) that—with a few apparent exceptions depending on decomposition or dissociation at high temperatures—the densities of all compounds in the gaseous state are proportional to their molecular weights ; further that, taking hydrogen as the unit, both of density and of atomic weight, the density of any compound gas or vapor is equal to half its molecular weight ; and consequently that the specific volume of any compound gas or vapor—that is to say, the quotient of its molecular weight by its density —is equal to 2.

In the liquid and solid states the relations between density and molecular weight are less simple ; nevertheless some very remarkable laws have been made out, depending, in the case of solid bodies, chiefly on isomorphism (p. 246).

In the case of carbon compounds, it is principally with reference to the liquid state that general relations between density, molecular weight, and atomic constitution have been discovered.

The specific volumes of liquids are comparable only at those temperatures for which their vapor-tensions are equal, as at the boiling points. If the molecular weights are compared with the densities at equal temperatures, no regular relations can be perceived; but when the same comparison is made at the boiling points of the respective liquids, several remarkable laws become apparent. The density of a liquid at its boiling point cannot be ascertained by direct experiment, but when the density at any one point, say at 15.5° C., has been determined, and the rate of expansion is also known, the density at the boiling point may be calculated. The most important of these relations are the following:—

1. In homologous series a difference of CH_2 in the composition answers to a difference of 22 in the specific volume; thus:

		Molecular Weight.	Specific Volume.	Difference.
Formic acid,	CH_2O_2	46	42	22
Acetic acid,	$C_2H_4O_2$	60	64	22
Propionic acid,	$C_3H_6O_2$	74	86	22
Butyric acid,	$C_4H_8O_2$	88	108	22
Valeric acid,	$C_5H_{10}O_2$	102	130	22

Hence it follows that the group CH_2 has the specific volume 22.

2. The substitution of 1 atom of carbon for 2 atoms of hydrogen makes no alteration in the specific volume of a liquid compound:

		Mol. Wt.	Specific Volume.
Octane,	C_8H_{18}	114	187
Cymene,	$C_{10}H_{14}$	134	187
Ethylic ether,	$C_4H_{10}O$	74	106.8
Phenol,	C_6H_6O	94	106.8

Now, since the specific volume of CH_2 is 22, and that of 1 carbon-atom is the same as that of 2 hydrogen-atoms, it follows that the specific volume (or atomic volume) of c a r b o n is 11 and that of h y d r o g e n 5.5.

Calculating in a similar manner the specific volume of o x y g e n in various carbon-compounds, it is found that this element has two specific or atomic volumes, according as it is united by both its units of affinity to one carbon-atom, as in the group $C=O$, in which case its specific volume is 12.2; or to two different atoms, in methyl alcohol, $O{<}^{CH_3}_{H}$, and methyl ether, $O{<}^{CH_3}_{CH_3}$, in which case the specific volume is 7.8. In acetic acid,

$$O=\overset{\overset{\displaystyle CH_3}{|}}{C}-O-H$$

, the O-atom which is connected with a carbon-atom alone has the specific volume 12.2, while that joined to both C and H has the specific volumes 7.8. If this compound be represented as acetyl-hydroxide, $\left.\begin{matrix}CH_3CO\\H\end{matrix}\right\}O$, or $\left.\begin{matrix}C_2H_3O\\H\end{matrix}\right\}O$, that is to say, as water in which 1 atom of hydrogen is replaced by the radicle acetyl, C_2H_3O, we may say that the specific volume of the intra-radical oxygen is 12.2, and that of the extraradical oxygen 7.8

The specific volume of dyad s u l p h u r varies in a similar manner, being 28.6 or 23 accordingly as the sulphur-atom is united to one atom of another element by both its combining units, or to two separate atoms, as in thiocarbonic acid, $\left.\begin{matrix}CS\\H_2\end{matrix}\right\}S_2$, or $S=C{<}^{S-H}_{S-H}$, where the sulphur

22 *

atom united with the C alone has the specific volume 28.6, and the two others, each of which is connected both with C and H, have the specific volume 23.

The specific volume of nitrogen in the amines, e. g., H_3C-NH_2 (methyl amine), is 2.3; in cyanogen-compounds, 17; and in nitroxyl, NO_2, 17.4. The specific volumes of the haloïd elements are the same in all their compounds, viz., Cl = 22.8; Br = 27.8; I = 37.5.

From these data the specific volumes of compounds may be calculated with very near approach to the values directly obtained by dividing the molecular weights by the specific gravities. A comparison of the observed and calculated values of a few compounds is given in the following table:—

Substance.	Formula.	Specific Volumes at the Boiling Point.		Quotient of Mol. Wt. by Sp. Gr.
		Calculated.		
Water . . .	H_2O	$2 \times 5.5 + 7.8$	= 18.8	18.8
Methyl alcohol	$CH_3.O.H$	$11 + 4 \times 5.5 + 7.8$	= 40.8	42
Ethyl alcohol .	$C_2H_5.O.H$	$2 \times 11 + 6 \times 5.5 + 7.8$	= 62.8	62.5
Acetic acid .	$C_2H_3O.O.H$	$2 \times 11 + 4 \times 5.5 + 7.8 + 12.2$	64	63.6
Acetone . .	$CO(CH_3)_2$	$3 \times 11 + 6 \times 5.5 + 12.2$	= 78.2	77.6
Butyric acid .	$C_4H_7O.OH$	$4 \times 11 + 8 \times 5.5 + 7.8 + 12.2$	= 108	107
Ethyl acetate .	$C_2H_3O.O.C_2H_5$	$4 \times 11 + 8 \times 5.5 + 7.8 + 12.2$	= 108	107.6
Mercaptan . .	$C_2H_5S.H$	$2 \times 11 + 6 \times 5.5 + 23$	= 78	76.1
Aniline . . .	$C_6H_5.NH_2$	$6 \times 11 + 7 \times 5.5 + 2.3$	= 106.8	106.8
Ethyl cyanide	$C_2H_5.CN$	$3 \times 11 + 5 \times 5.5 + 17$	= 77.5	77.2
Ethyl nitrate .	$C_2H_5.O.NO_2$	$2 \times 11 + 5 \times 5.5 + 17.4 + 3 \times 7.8$ =	90.3	90.1
Chloroform .	$CHCl_3$	$11 \quad 5 \times 5 + 3.2 \quad 2 \times 8$	= 84.9	84.8
Ethyl bromide	C_2H_5Br	$2 \times 11 + 5 \times 5.5 + 27.8$	= 77.3	78.4
Ethyl iodide .	C_2H_5I	$2 \times 11 + 5 \times 5.5 + 37.5$	= 87.0	86.4

Melting Point and Boiling Point.—Solid carbon-compounds are for the most part capable of melting without decomposition, and exhibit fixed melting points. Many of those which are liquid at ordinary temperatures boil and volatilize without decomposition under the ordinary atmospheric pressure, and their boiling temperatures are constant under any given pressure. Many others, on the contrary, when heated under the ordinary pressure, decompose before they boil, but some of these are found to boil without decomposition under pressures more or less reduced.

Generally speaking, the boiling point of a compound is higher as its constitution is more complex. This is seen (1) in polymeric compounds: for example, formic aldehyde, CH_2O, is a gas; but acetic acid, $C_2H_4O_2$, is a liquid boiling at 118° C. (244.4° F.); lactic acid, $C_3H_6O_3$, boils at 200°C. (392° F.), but at the same time decomposes to a great extent; and grape sugar, $C_6H_{12}O_6$, is not volatile at all, but undergoes complete decomposition when strongly heated.

2. In the successive terms of a homologous series, the boiling point rising successively for every addition of CH_2. In some cases this increase

is very regular : thus in the series of normal alcohols,* $C_nH_{2n}+_2O$, the successive members, up to the 8-carbon alcohol, differ in boiling point by nearly 19° C. (34.2^O F.), and in the normal fatty acids, $C_nH_{2n}O_2$, the difference from the second to the sixth term is very nearly 22°, but afterwards becomes less ; as shown by the following table :—

Alcohols, $C_nH_{2n}+_2O$.

Ethylic,	C_2H_6O	78.4^O C. (173.1^O F.)
Propylic,	C_3H_8O	97° C. (206.6^O F.)
Butylic,	$C_4H_{10}O$	116° C. (240.8 F.)
Pentylic,	$C_5H_{12}O$	137° C. (278.6^O F.)
Hexylic,	$C_6H_{14}O$	156 6° C. (313.9^O F.)
Heptylic,	$C_7H_{16}O$	177° C. (350.6^O F.)
Octylic,	$C_8H_{18}O$	$190–192^O$ C. ($374–337.6^O$ F.)
Nonylic,	$C_9H_{20}O$	200° C. (392° F.)

Acids, $C_nH_{2n}O_2$.

Acetic,	$C_2H_4O_2$	118° C. (244.4^O F.)
Propionic,	$C_3H_6O_2$	140.6^O C. (285° F.)
Butyric,	$C_4H_8O_2$	163.2^O C. (325.9^O F.)
Valeric,	$C_5H_{10}O_2$	184.5^O C. (364.1^O F.)
Caproic,	$C_6H_{12}O_2$	204.5^O C. (408.2^O F.)
Oenanthylic,	$C_7H_{14}O$	$223–224^O$ C. ($434.2–435.2^O$ F.)
Rutic,	$C_8H_{16}O_2$	$232–234^O$ C. ($449.6–453.2^O$ F.)
Pelargonic,	$C_9H_{18}O_2$	$253–254^O$ C ($487.4–489.2^O$ F.)

In other cases, the differences between the boiling points of the successive members of a series decrease by a regular amount up to a certain term, beyond which they become constant. Thus, in the normal paraffins, $C_nH_{2n}+_2$, the differences decrease regularly by about 4° C. (7.2^O F.) till a constant difference of 19° C. 34.2^O F.) is attained ; thus—

		Boiling point.	Difference.
Butane,	C_4H_{10}	1^O	
Pentane,	C_5H_{12}	38°	37°
Hexane,	C_6H_{14}	70°	32°
Heptane,	C_7H_{16}	99°	29°
Octane,	C_8H_{18}	124°	25°
Dodecane,	$C_{12}H_{26}$	202°	4×19°
Hexdecane,	$C_{16}H_{34}$	271°	4×19°

In the haloïd ethers and acetic ethers of the alcohol-radicles, $C_nH_{2n}+_1$, the differences of boiling point decrease by about 2^O, as shown in the following table :—

Radicles.	Chlorides, $C_nH_{2n}+_1Cl$.		Bromides, $C_nH_{2n}+_1$ Br.		Iodides, $C_nH_{2n}+_1$ I		Acetates, $C_nH_{2n}+_1.C_2H_3O_2$	
Methyl, CH_3	..	..	..	..	$40^OC.$	$104^OF.$	..	..
Ethyl, C_2H_5	$12.5^OC.$	$54.5^OF.$	$39^OC.$	$102.2^OF.$	72°	161.6^O	$74^OC.$	$165.2^OF.$
Propyl, C_3H_7	46.4^O	115.5^O	71°	159.8^O	102°	$215\ 6^O$	102°	215.6^O
Butyl, C_4H_9	77.6^O	171.7^O	$100\ 4^O$	212.7^O	129.6^O	265.3^O	$125\ 1^O$	257.2^O
Pentyl, C_5H_{11}	105.6^O	222°	$128\ 7^O$	263.7^O	153.4^O	308.1^O	143.4^O	299.1^O
Hexyl, C_6H_{13}	..	..	..	..	179.4^O	$354\ 9^O$	168.7^O	334.5^O
Octyl, C_8H_{17}	180°	356°	$199\ ^o$	390.1^O	221°	429.8^O	207°	404.6^O

* Normal carbon-compounds are those in which all the carbon-atoms are linked together in a single chain, e. g., normal butyl alcohol, $CH_3–CH_2–CH_2–CH_2OH$. (See HYDROCARBONS, ALCOHOLS, etc.)

Metameric compounds containing similarly constituted radicles boil for the most part at nearly equal temperatures, *e. g.*,

Formula, $C_6H_{12}O_2$.	Boiling Point.	Formula, $C_7H_{14}O_2$.	Boiling Point.
Butyl acetate,	124.5° C. 256.1° F.	Pentyl acetate,	148.4° C. 299.1° F.
Propyl proprion-		Butyl propion-	
ate,	122.4° 252.3°	ate,	146.0° 294.8°
Ethyl butyrate,	121.6° 250.6°	Propyl butyrate,	143.4° 290.1°

Those, on the other hand, which contain differently constituted radicles, boil at different temperatures, the boiling point being lower as the compound has a more complex constitution, or, in other words, as it contains ı greater number and variety of radicles: in the butyl alcohols, for example :—

			Boiling point.
Normal butyl alcohol,	$C_4H_9.OH$	116° C.	240.8° F.
Isobutyl alcohol,	$C_2H_3(CH_3)_2.OH$	108°	226.4°
Secondary butyl alcohol,	$(CH_3)(C_2H_5)CH.OH$	96°	203°
Tertiary butyl alcohol,	$C(CH_3)_3.OH$	92°	197.6°

Subtraction of hydrogen generally raises the boiling point, *e. g.*,

Heptane, C_7H_{16}	Heptene, C_7H_{14}	Heptine, C_7H_{12}
99° C. (210.2° F.)	100° C. (212° F.)	107° C. (224.6° F.)

Hydrocarbons always boil at lower temperatures than their substitution-derivatives, the boiling point rising in proportion as a greater amount of hydrogen is displaced.

Ethane, C_2H_6 gas.
Ethyl chloride, C_2H_5Cl	Ethene chloride, $C_2H_4Cl_2$
12° C. (53.6° F.)	82.5° C. (180.5° F.)
Ethyl alcohol, $C_2H_5.OH$	Ethene alcohol, $C_2H_4(OH)_2$
78.4° C. (173.1° F.)	197.5° C. (387.5° F.)
Acetic acid, $C_2H_3O.OH$	Chloracetic acid, $C_2H_2ClO.OH$
118° C. (244.4° F.)	186° C. (366.8° F.)

Benzene, C_6H_6 82°
Chlorobenzene,	Dichlorobenzene,	Trichlorobenzene,
C_6H_5Cl	$C_6H_4Cl_2$	$C_6H_3Cl_3$
135° C. (275° F.)	172° C. (341.6° F.)	210° C. (410° F.)
Amidobenzene, $C_6H_5(NH_2)$		Diamidobenzene, $C_6H_4(NH_2)_2$
182° C. (259.6° F.)		287° C. (548.6° F.)

OPTICAL PROPERTIES.

Refractive Power.—In the chapter on LIGHT (p. 91) it has been explained that the index of refraction of any transparent substance, that is to say, the ratio of the sine of the angle of incidence to the sine of the angle of refraction, is a constant quantity, so long as the density of the substance remains constant :—

$$\frac{\sin i}{\sin r} = n.$$

With variation of density (*d*), and therefore also with variation of temperature, the index of refraction likewise varies; but it has been found, by exact observation made on a large number of substances at different temperatures, that the quantity

$$\frac{n-1}{d},$$

called the s p e c i f i c r e f r a c t i v e e n e r g y, is constant for all tempera- tures; and this quantity muliplied by the molecular weight of the sub- stances, gives the m o l e c u l a r r e f r a c t i v e e n e r g y or r e f r a c t i o n- e q u i v a l e n t.

The relations between chemical composition and refractive energy have been studied chiefly in bodies of the fatty group (methane-derivatives), and the following general laws have been established:—

1. *Isomeric bodies have in many instances equal refraction-equivalents.*

2. *In compounds belonging to the same homologous series, the refraction-equivalent increases, in all cases, for a difference of* CH_2, *by a nearly equal amount; the mean increment in the fatty alcohols and acids, etc., being* 7.60.

3. *The refraction-equivalent of a mixture or compound is equal to the sum of the refraction-equivalents of its components.*

This last law gives the means of determining the refraction-equivalents of elements from those of their compounds, the method of calculation being similar to that adopted in the determination of the specific volumes of the elements (p. 513). In the homologous fatty alcohols and acids, the mean values of the refraction-equivalents of carbon, hydrogen, and oxygen are found to be—

C	H	O
5.00	1.30	3.00

By means of these values the refraction-equivalent of a compound of car- bon, hydrogen, and oxygen, may be calculated from the formula

$$R = mr + m'r' + m''r'',$$

where m, m', m'', denote the numbers of atoms of the three elements, and r, r', r'', their refraction-equivalents.

Thus, for e t h y l - a l c o h o l, C_2H_6O, the calculated value is

$$2 \times 5 + 6 \times 1.3 + 3 = 20.8.$$

On the other hand, direct observation gives

$$d = 0.7964; \quad n = 1.3606: \quad \text{therefore} \quad \frac{n-1}{d} = 0.4526;$$

and multiplying this number by 46, the molecular weight of alcohol, we obtain for the refraction-equivalent the number 20.8, agreeing exactly with the calculated value.

In like manner for a c e t i c a c i d, $C_2H_4O_2$, we have

$$2 \times 5 + 4 \times 1.3 + 2 \times 3 = 21.2,$$

and for the direct determination—

$$d = 1.053; \quad n = 1.372: \quad \text{mol. wt.} = 60.$$

Therefore— $\frac{n-1}{d} \times 60 = 0.3533 \times 60 = 21.2.$

The specific refractive energy of a mixture is the mean of that of its constituents, so that if we know what compounds are present in a mixture, the determination of the specific refractive energy of the mixture (those of the components being known) affords the means of estimating the quantity of each.   For example, a mixture of 81.3 per cent. amyl alcohol, $C_5H_{12}O$, and 18.7 ethyl alcohol, C_2H_6O, was found to have an index of refraction which gave $\dfrac{n-1}{d} = 0.4940$.   Now the mode of calculation above explained gives for ethyl alcohol $\dfrac{n-1}{d} = 0.4528$, and for amyl alcohol, 0.5033. Hence, putting x and y for the relative quantities of ethyl and amyl alcohols in the mixture, we have the two equations—

$$x + y = 100$$
$$0.4528\,x + 0.5033\,y = 49.4:$$

whence $x = 18.4$; $y = 81.5$, values agreeing very nearly with the quantities actually mixed.

This method of optical analysis affords a convenient method of determining the proportions in which any two substances are mixed, in cases when a separation of the two cannot easily be effected by chemical or physical means, as when they differ but slightly in boiling point, solubility, or other quality which might otherwise be available for their separation.

The composition of chemical compounds may, in many cases, be determined in a similar manner.

In certain series of carbon compounds, viz., the hydrocarbons of the aromatic group, including benzene, C_6H_6, and its homologues—in the series of organic bases homologous with pyridine, C_5H_5N, and chinoline, C_9H_7N—and in certain oxidized essential oils, the refraction-equivalents are found to be considerably higher than those calculated as above from the values of the elements.   The deviations from the normal values exhibit a certain regularity, depending on the composition, which, for the typical hydrocarbons of the several series, may be represented as follows :—

		Refraction equivalent.	
Paraffins,	C_nH_{2n+2}	Normal	
Olefines,	C_nH_{2n}	"	
Terpenes,	C_nH_{2n-4}	"	$+ 3$
Benzene and its homologues,	C_nH_{2n-6}		$+ 6$
Naphthalene,	$C_{10}H_8$		$+14$
Anthracene,	$C_{14}H_{10}$		$+17$

Similar relations are exhibited by a series of oxidized compounds, differing from one another only in their amount of hydrogen :—

Peppermint camphor,	$C_{10}H_{20}O$	Normal	
Dihydrate of cajeputene,	$C_{10}H_{18}O$	"	
Wormwood oil,	$C_{10}H_{16}O$	"	$+ 1$
Carvol,	$C_{10}H_{14}O$	"	$+ 6$
Anethol,	$C_{10}H_{12}O$	"	$+13$

Circular Polarization.—The power of turning the plane of polarization of a ray of light to the right or to the left, possessed by many carbon compounds, even in the liquid state or in solution, has already been mentioned in the chapter on LIGHT (p. 100), which also contains a description and figure of an instrument used for measuring the amount of this power. Substances possessing this power are said to be o p t i c a l l y active, or to possess o p t i c a l r o t a t o r y p o w e r, d e x t r o-, or l æ v o -; such are amyl alcohol, turpentine oil, camphor, various kinds of sugar, and several vegetable acids and alkaloids.

The angle of rotation (a) for any particular ray of the spectrum, the line D, for example—produced by any particular liquid, is proportional to the length of the column of liquid traversed by the ray, and to the quantity of active substance contained in it; and the quantity [a], given by the expression

$$[a] = \frac{a}{l \delta l},$$

in which a is the observed angle of rotation, l the weight of substance in 1 gram of the solution, δ the specific gravity of the solution, and l the length of the column (in decimeters), is called the s p e c i f i c r o t a t o r y p o w e r. It is constant for each substance at a given temperature.

For example, by dissolving 11.347 grams of grape-sugar in 88.653 grams of water, a solution is obtained, having a specific gravity of 1.048, and producing in a tube 2 decimeters long, a rotation of 13.7°. Hence, the molecular rotatory power of grape-sugar is given by the equation,

$$[a] = \frac{13.7}{0.11347 \times 2 \times 1.048} = 57.6.$$

The power of circular polarization appears to depend on a certain want of symmetry in the molecules of the active substance, and in crystallized bodies it is usually associated with the existence of hemihedral faces. This connection is strikingly exhibited by the isomeric bodies, tartaric and racemic acid, $C_4H_6O_6$. Racemic acid, the crystals of which are holohedral, has no action or polarized light; but, by certain processes to be explained hereafter (see TARTARIC ACID), it may be separated into two isomeric acids, called dextro- and lævo-tartaric acids, the former of which (ordinary tartaric acid) turns the plane of polarization to the right, the latter to the left; and the crystals of these acids are perfectly similar in form, excepting that they possess certain hemihedral faces, occupying opposite positions on the crystal, so that the two crystals are not superposible, but are related to one another like the two hands, or the two sides of the face, or an object and its reflected image. Moreover, the union of these two oppositely active tartaric acids preproduces racemic acid with its original properties, optically inactive, and forming holohedral crystals.

Artificially prepared carbon compounds are, for the most part, active or inactive, according to the character of the compounds from which they are derived. Thus malic acid, prepared by the action of hydriodic acid on tartaric acid, or by that of nitrous acid on asparagin or on optically active aspartic acid, is itself optically active, like the malic acid of fruits; but inactive aspartic acid heated with nitrous acid, yields a malic acid which has no action on polarized light. Recently, however, it has been found that inactive tartaric acid may be converted by a strong heat into dextro-tartaric acid.

Methane-Derivatives, or Fatty Group.

HYDROCARBONS.

FIRST SERIES, C_nH_{2n+2}.—PARAFFINS.*

THIS series, as already observed, consists of saturated hydrocarbons. The names and formulæ of the first six are given in the table on page 506; the following terms may be called, *heptane, octane, nonane, decane, endecane, dodecane*, etc.

Occurrence and Formation.—Many of the paraffins occur ready formed in American petroleum and other mineral oils of similar origin. They are formed artificially by the following processes :—

1. By the simultaneous action of zinc and water on the alcoholic iodides (p. 507), compounds derived from these same hydrocarbons by the substitution of one atom of iodine for hydrogen.

This reaction, which appears to be applicable to the formation of the whole series of paraffins, is represented by the general equation :—

$$2C_nH_{2n+1}I + Zn_2 + 2H_2O = ZnH_2O_2 + ZnI_2 + 2C_nH_{2n+2}$$
$$\text{Alcoholic} \qquad\qquad\qquad\qquad\qquad\qquad\qquad\qquad\qquad \text{Paraffin.}$$
$$\text{iodide.}$$

As an example, we may take the formation of ethane from ethyl iodide :—

$$2C_2H_5I = Zn_2 + 2H_2O = ZnH_2O_2 + ZnI_2 + 2C_2H_6.$$

2. All the paraffins may be produced by heating the alcoholic iodides with zinc alone. Generally speaking, however, two of these hydrocarbons are obtained together, the first product of the reaction being a paraffin containing twice as many carbon atoms as the alcoholic iodide employed; and this compound being then partly resolved into the paraffin containing half this number of carbon atoms, and the corresponding olefine, C_nH_{2n} ; thus :—

$$2C_2H_5I + Zn = ZnI_2 + C_4H_{10}$$
$$\text{Ethyl iodide.} \qquad\qquad\qquad\qquad \text{Quartane.}$$

and

$$C_4H_{10} = C_2H_4 + C_2H_6$$
$$\text{Quartane.} \qquad \text{Ethene.} \quad \text{Ethane.}$$

Generally :—

$$2C_nH_{2n+1}I + Zn = ZnI_2 + C_{2n}H_{4n+2}$$

and,

$$C_{2n}H_{4n+2} = C_nH_{2n} + C_nH_{2n+2}.$$

3. By the electrolysis of the fatty acids ($C_nH_{2n}O_2$). For example, a solution of potassium acetate, divided into two parts by a porous diaphragm yields pure hydrogen, together with potash, at the negative electrode, and at the positive electrode (if of platinum) a mixture of carbon dioxide and ethane gases :—

$$2C_2H_4O_2 = 2CO_2 + C_2H_6 + H_2.$$

---

* From *parum affinis*, indicating their chemical indifference. The name paraffin has long been applied to the solid compounds of the series, on account of this character; and many of the liquid compounds of the same series are known commercially as *paraffin oils*. It is convenient, therefore, to employ the term paraffin as a generic name for the whole series.

We may suppose that the two molecules of acetic acid are resolved by the current into H_2 and $C_4H_6O_4$, and that the latter then splits up into $2CO_2$ and C_2H_6. The general reaction is :—

$$2C_nH_{2n}O_2 \ + \ 2CO_2 \ + \ C_{2n-2}H_{4n-2} \ + \ H_2.$$

4. Some of the paraffins are obtained from the acids of the series $H_nH_{2n}O_2$ and $C_nH_{2n-2}O_4$, by the action of alkalies, which abstract carbon dioxide from those acids, the hydrocarbon thus eliminated containing one or two atoms of carbon less than the acid from which it is produced. In this manner methane (marsh-gas) is obtained by heating potassium acetate with potassium hydroxide (p. 164).

$$C_2H_3O_2K \ + \ HKO \ + \ CO_3K_2 \ + \ CH_4.$$

Also, hexane and octane, by similar treatment of the potassium salts of suberic acid, $C_8H_{14}O_4$, and sebacic acid, $C_{10}H_{18}O_4$:—

$$\underset{\text{Suberate.}}{C_8H_{12}O_4} \ + \ 2HKO \ = \ 2CO_3K_2 \ + \ \underset{\text{Hexane.}}{C_6H_{14}}$$

$$\underset{\text{Sebate.}}{C_{10}H_{18}O_4K_2} \ + \ 2HKO \ = \ 2CO_3K_2 \ + \ \underset{\text{Octane.}}{C_8H_{18}}$$

Generally speaking, however, a further decomposition takes place, resulting in the formation of hydrocarbons containing a smaller proportion of hydrogen than the paraffins.

5. The paraffins may also be produced from the olefines, C_nH_{2n}, by combining the latter with bromine, and heating the resulting compound, $C_nH_{2n}Br_2$, with a mixture of potassium iodide, water, and metallic copper. The bromine compound is then decomposed, and the hydrocarbon, C_nH_{2n}, is partly reproduced in the free state, partly converted, by addition of hydrogen, into a paraffin.

6. Several of the paraffins are produced by the dry or destructive distillation of butyrates and acetates.

7. They are also found amongst the products of the dry distillation of coal, especially Boghead and Cannel coal, and, as already observed, they constitute the principal portion of many mineral oils, called *petroleum*, *naphtha*, or *rock-oil*, formed by the gradual decay or decomposition of vegetable matter beneath the earth's surface. By far the largest quantities of these oils are obtained from Canada, Pennsylvania, and other parts of North America. Abundant petroleum springs exist, also, on the northwest of the Caspian Sea, near Baku, at Rangoon in Burmah, and in various parts of Italy. The American petroleum consists almost wholly of paraffins. Burmese tar contains, also, small quantities of hydrocarbons belonging to other series, especially homologues of benzene.

Properties and Reactions of the Paraffins.—Methane, ethane, propane, and butane are gaseous at ordinary temperatures ; most of the others are liquids, regularly increasing in specific gravity, viscosity, boiling point, and vapor-density, as their molecular weight becomes greater ; those containing 20 carbon-atoms or more are crystalline solids.

The paraffins are saturated hydrocarbons, incapable of uniting directly with monatomic elements or radicles, but they easily yield substitution derivatives. When subjected to the action of *chlorine* or *bromine*, they give up a part, or in some cases the whole, of their hydrogen in exchange for the halogen element. Thus equal volumes of chlorine and methane, CH_4, exposed to diffused daylight, yield the compound CH_3Cl, called c h l o r o - m e t h a n e or m e t h y l c h l o r i d e ; and by further subjecting this product to the action of an excess of chlorine in direct sunshine, it may be successively converted into the more highly chlorinated compounds CH_2Cl_2, $CHCl_3$, CCl_4. Ethane, C_2H_6, also yields, by a series of processes to be pre-sently described, the products C_2H_5Cl, $C_2H_4Cl_2$, $C_2H_3Cl_3$, $C_2H_2Cl_4$, C_2HCl_5, and C_2Cl_6; and similarly for the other compounds of the series. These bodies, which may be regarded as compounds of chlorine and other halogen elements with the radicles $(CH_3)'$, $(CH_2)''$, $(CH)'''$, etc., are called haloïd ethers. When treated with water or aqueous alkalies, they exchange the haloïd element for an equivalent quantity of hydroxyl, (HO), thereby producing alcohols (p. 508); and, on the other hand, they may be formed from the alcohols by the action of the chlorides, bromides, and iodides of hydrogen or phosphorus.

Nitric acid attacks the higher members of the paraffin series, forming nitro-compounds; octane, C_8H_{18}, thus treated, yields the compound, $C_8H_{17}(NO_2)$. The lower paraffins, on the other hand, are not affected by nitric acid; but by indirect means compounds may be formed, having the composition of paraffins in which the hydrogen is more or less replaced by nitril: for example, *nitro-methane*, $CH_3(NO_2)$; *trinitromethane* or *nitroform*, $CH(NO_2)_3$.

The first three hydrocarbons of the series, viz., CH_4, C_2H_6, C_3H_8, exhibit exactly the same physical and chemical properties in whatever way they may be prepared; and indeed the constitutional formulæ of these bodies, viz:

$$CH_4 \qquad \begin{array}{c} CH_3 \\ | \\ CH_3 \end{array} \qquad \begin{array}{c} CH_3 \\ | \\ CH_2 \\ | \\ CH_3 \end{array}$$

show that they are not susceptible of isomeric modifications, inasmuch as there is but one way in which the carbon-atom in either of them can be grouped; in ethane each carbon-atom is directly combined with three hy-drogen-atoms and the other carbon-atom; and whether we regard it as ethyl hydride, $H—CH_2CH_3$, or as dimethyl, $H_3C—CH_3$, this arrangement remains the same. In propane, C_3H_8, each carbon-atom is directly com-bined with at most two other carbon-atoms, and there is no other way in which the atoms can be arranged.

But if we look at the formula of the 4-carbon paraffin, C_4H_{10}, we see that it may be written in either of the following forms:—

$$\begin{array}{c} CH_3 \\ | \\ CH_2 \\ | \\ CH_2 \\ | \\ CH_3 \end{array} \qquad\qquad \begin{array}{c} H_3C \quad CH_3 \\ \vee \\ CH \\ | \\ CH_3 \end{array}$$

in the first of which, neither of the carbon-atoms is directly united with more than two others, whereas, in the second, one of the carbon-atoms is directly combined with three others. The first may be represented, either as *propyl-methane*, $C \begin{cases} CH_2CH_2CH_3 \\ H_3 \end{cases} = C \begin{cases} CH_2C_2H_5 \\ H_3 \end{cases} = C \begin{cases} C_3H_7 \\ H_3 \end{cases}$, or as *diethyl*, $H_5C_2.C_2H_5$, according to the manner in which we may suppose it to be di-vided; the second as *trimethyl-methane*, $C \begin{cases} (CH_3)_3 \\ H \end{cases}$, or as *isopropyl-methane*, $C \begin{cases} CH(CH_3)_2 \\ H_3 \end{cases}$, the radicle $CH(CH_3)_2$ being called isopropyl, to distin-guish it from normal propyl, $CH_2(C_2H_5)$.

A comparison of the modes of formation and decomposition of all the paraffins which have been obtained by definite reactions shows that they may be arranged in the four following groups :—*

1. Normal Paraffins, in which each carbon-atom is directly conuected with, at most, *two* other carbon-atoms, or which contain the group or residue CH_2 (methene), associated with two monatomic alcohol-radicles, $C_nH_{2n} + 1$, $e.\,g.$,

Dimehthyl-methane, $\quad C \begin{cases} CH_3 \\ CH_3 \\ H_2 \end{cases}$ or $H_2C \Big\langle \begin{matrix} CH_3 \\ CH_3 \end{matrix}$

Methyl-propyl methane, $\quad C \begin{cases} C_3H_7 \\ CH_3 \\ H_2 \end{cases}$ or $H_2C \Big\langle \begin{matrix} CH_2CH_2CH_3 \\ CH_3 \end{matrix}$

2. Isoparaffins, in which one carbon-atom is directly united with *three* other carbon-atoms, or in which the trivalent group or residue, CH ;(methenyl), is associated with three monatomic radicles, $C_nH_{2n} + 1$, $e.g.$,

Trimethyl-methane $\quad \begin{cases} (CH_3)_3 \\ H \end{cases}$ or $HC \Big\langle \begin{matrix} CH_3 \\ CH_3 \\ CH_3 \end{matrix}$

Ethyldimethyl-methane, $\quad C \begin{cases} C_2H_5 \\ (CH_3)_2 \\ H \end{cases}$ or $HC \Big\langle \begin{matrix} CH_2CH_3 \\ CH_3 \\ CH_3 \end{matrix}$

3. Neoparaffins, in which one carbon-atom is directly united with *four* other carbon-atoms, or, in other words, with four monatomic radicles, $C_nH_{2n}+1$, $e.g.$,

Tetramethyl-methane, $\quad C(CH_3)_4 \quad$ or $\quad H_3C—\overset{\displaystyle CH_3}{\underset{\displaystyle CH_3}{\overset{|}{\underset{|}{C}}}}—CH_3$

Dimethyl-diethyl-methane, $\quad C \begin{cases} (C_2H_5)_2 \\ (CH_3)_2 \end{cases}$ or $\quad H_3C—\overset{\displaystyle CH_2CH_3}{\underset{\displaystyle CH_2CH_3}{\overset{|}{\underset{|}{C}}}}—CH_3$

The paraffins of these three classes are constructed on the methane type.

4. Mesoparaffins, in which two methenyl groups, having their carbon-atoms linked together by one combining unit, are each associated with *two* monatomic radicles, $C_nH_{2n}+1$. These paraffins are, therefore, con-structed on the ethane type, $\begin{matrix} HCH_2 \\ | \\ HCH_2 \end{matrix}$, $e.g.$:

Tetramethylethane . . $\begin{matrix} HC(CH_3)_2 \\ | \\ HC(CH_3)_2 \end{matrix}$.

The paraffins of the first and second class have for some time been distinguished as *normal* and *iso-paraffins*, and Dr. Odling has lately proposed to distinguish the third class by the prefix *neo* (from νεος, new, as being the latest discovered), and the fourth by the prefix *meso*, intimating their in-

* Schorlemmer, Proceedings of the Royal Society, xvi. 34, 367.

termediate character, as associated with the iso-paraffins in containing the residue CH, and with the neo-paraffins in containing four alcohol radicles.[*] He also suggests an abbreviated notation, consisting in the use of the Greek letters σ, ν, and μ, as indices of the iso-, neo-, and meso-paraffins respectively, for example:

C_6H_{14}	$C_6H_{14}^{\sigma}$	$C_6H_{14}^{\nu}$	$C_6H_{14}^{\mu}$
Hexane.	Isohexane.	Neohexane.	Mesohexane.

$$H_2C \begin{cases} CH_2CH_2CH_3 \\ CH_3 \end{cases} \qquad HC \begin{cases} CH_2CH_2CH_3 \\ CH_3 \\ CH_3 \end{cases} \qquad C \begin{cases} CH_2CH_3 \\ CH_3 \\ CH_3 \\ CH_3 \end{cases} \qquad HC \begin{cases} CH_3 \\ CH_3 \end{cases} \\ HC \begin{cases} CH_3 \\ CH_3 \end{cases}$$

Isomerism in the Substitution-derivatives of the Paraffins.—It has already been stated that the paraffins can exchange one or more of their hydrogen-atoms for various elements and compound radicles, Cl, Br, O, OH, NO₂, etc., giving rise to alcohols and ethers. Confining our attention for the present to the monatomic derivatives, that is to say, those in which 1 atom of hydrogen is replaced by a univalent radicle X, it is obvious that the first two hydrocarbons of the series

C_nH_{2n+2}, viz., CH_4 and $\begin{matrix} CH_3 \\ | \\ CH_3 \end{matrix}$, can each give rise to only one derivative

containing any particular monatomic radicle, these derivatives being respectively represented by the formulæ—

$$CH_3X \qquad \text{and} \qquad \begin{matrix} CH_3 \\ | \\ CH_2X \end{matrix}$$

for supposing, as is most probable, that all the hydrogen-atoms have the same value, and are attached to their respective carbon-atoms in the same way, the result of the substitution must be the same, whichever of these hydrogen-atoms may be thus replaced. But with all the paraffins containing more than two atoms of carbon, the case is different. Thus, in propane, CH_3—CH_2—CH_3, the substitution may take place either in one of the exterior groups CH_3, or in the middle group CH_2, giving rise to two derivatives of different structure, distinguished by the terms primary and secondary, viz.,

Primary.	Secondary.
CH_2X	CH_3
$\|$	$\|$
CH_2	CHX
$\|$	$\|$
CH_3	CH_3

In the primary derivative, the carbon atom joined to the radicle X is connected immediately with only one other carbon atom; in the secondary derivative, it is linked to two other carbon atoms. These are the only possible modifications of a monatomic derivative of the 3-carbon paraffin, C_3H_8; they may be more shortly represented by the formulæ

$$H_2C {\Large<}^{CH_2X}_{CH_3} \qquad\qquad HXC{\Large<}^{CH_3}_{CH_3.}$$

* Philosophical Magazine [5], 1. 205.

The 4-carbon paraffin, butane, admits of a greater number of modes of substitution. In the first place, the hydrocarbon itself is susceptible of two modifications, viz :—

$$CH_3—CH_2—CH_2—CH_3 \qquad\qquad CH_3—CH\underset{CH_3}{\overset{CH_3}{<}}$$

Normal butane.   Isobutane or
           Trimethyl-methane.

From the first may be formed one primary and one secondary derivative, these terms having the meaning above explained ; while the second yields another primary derivative, and likewise a *tertiary* derivative, in which the carbon atom joined to the radicle X is joined also to three other atoms of carbon. These four derivatives are represented by the following formulæ :—

Primary,
{ Normal $CH_3—CH_2—CH_2—CH_2X$ or $H_2C\left\{{C_2H_4X \atop CH_3}\right.$.

   Iso- $\overset{H_3C}{\underset{H_3C}{>}}CH—CH_2X$ or $HC\left\{{CH_2X \atop (CH_3)_2}\right.$

Secondary, $CH_3—CHX—CH_2—CH_3$ or $HXC\left\{{C_2H_5 \atop CH_3}\right.$

Tert $\overset{H_3}{\underset{H_3}{>}}CX—CH_3$ or $XC\left\{{CH_3 \atop {CH_3 \atop CH_3}}\right.$

The two primary derivatives are distinguished by containing the radicles normal propyl, $CH_3—CH_2—CH_2—$ and isopropyl $—CH(CH_3)_2$ respectively.

Of the four monatomic butane derivatives, the normal primary and the secondary are derived from normal butane, the iso-primary and the tertiary from isobutane.

The higher paraffins yield a larger number of monatomic derivatives according to the nature of the radicles which enter into their constitution ; in other words, according as they are either normal, *i. e.*, have all their carbon atoms in a single chain, or contain one or more isopropyl groups, $HC\underset{CH_3}{\overset{CH_3}{<}}$; but these derivatives must all be either primary, secondary, or tertiary ; for the carbon atom joined to the radicle X, having one of its combining units thus disposed of, has only three remaining, and cannot therefore be joined to a number of other carbon atoms greater than three. In other words, the replacement of an H-atom by the radicle X must take place either in a methyl residue, CH_3, a methene residue, CH_2, or a methenyl residue, CH, producing respectively a primary, secondary, or tertiary derivative.

Dr. Odling denotes the secondary and tertiary derivatives by the prefixes *pseudo* and *kata*, distinguishing also the latter by the index x, and the former by the indices π and $\downarrow$, according as they contain a normal or an iso-radicle, thus—

    Pentyl (C_5H_{11}).   Isopentyl $(C_5H_{11}^{\sigma})$.   Neopentyl $(C_5H_{11}^{\gamma})$.

Primary, $H_2C\left\{{C_3H_6X \atop CH_3}\right.$ $HC\left\{{C_2H_4X \atop {CH_3 \atop CH_3}}\right.$ $C\left\{{CH_2X \atop {CH_3 \atop {CH_3 \atop CH_3}}}\right.$

    Pseudopentyl $(C_5H_{11}^{\pi})$.   Pseudisopentyl $(C_5H_{11}^{\downarrow})$.

Pseudo, $XHC\left\{{C_3H_7 \atop CH_3}\right.$ $HC\left\{{CHX.CH_3 \atop {CH_3 \atop CH_3}}\right.$

Katapentyl ($C_5H_{11}^a$).

Kata . . $XC\begin{cases} C_2H_5 \\ CH_3 \\ CH_3 \end{cases}$

We now proceed to describe the more important of the individual paraffins.

Methane, CH_4, also called *Methyl hydride* and *Marsh-gas*, is formed by passing a mixture of hydrogen sulphide and vapor of carbon disulphide over red-hot copper :—

$$2H_2S + CS_2 + 4Cu_2 = CH_4 + 4Cu_2S ;$$

also, by the action of water on zinc-methyl :—

$$Zn(CH_3)_2 + H_2O = 2CH_4 + ZnO.$$

The easiest method of preparing it, however, is to heat a mixture of dry sodium acetate and soda-lime (p. 228):—

$$\begin{matrix} CH_3 \\ | \\ CO_2Na \end{matrix} + \begin{matrix} H \\ Na \end{matrix} \Big\rangle O = CH_4 + CO_3Na_2 .$$

Methane is also produced in the slow decay of vegetable substances, and is found in coal mines as fire-damp, also in marshes and stagnant pools, from which it may be evolved as gas by stirring the mud. Lastly, it is formed by the dry distillation of various organic substances, and forms the chief constituent of coal-gas.

Methane is a colorless inodorous gas which burns with a yellow flame, producing water and carbon dioxide. Its density compared with hydrogen as unity is 8, showing that the molecule CH_4 has the normal 2-volume condensation, $\dfrac{12 + 4 \times 1}{2} = 8$. When a mixture of equal volumes of methane and chlorine is exposed to sunshine, an explosion takes place, hydrogen chloride being formed, and carbon separated ; but in diffused daylight the action goes on slowly, and substitution products are formed, viz., monochloromethane or methyl chloride, CH_3Cl, and with excess of chlorine, also the compounds, CH_2Cl_2 and $CHCl_3$.

Ethane, C_2H_6 or $H_3C{-}CH_3$.—This compound, which may also be regarded as *dimethyl*, or as *ethyl hydride*, $C_2H_5.H$ (p. 522), is formed by the general reactions already indicated (p. 520), viz., by the action of zinc and water on ethyl iodide; of zinc alone on the same compound, and on methyl iodide; and by the electrolysis of acetic acid, or rather of its potassium salt. It may be prepared in the pure state by decomposing zinc-ethyl with water, or more easily by heating acetic anhydride with barium dioxide :—

$$2(C_2H_3O)_2O + BaO_2 = C_2H_6 + (C_2H_3O_2)_2Ba + 2CO_2.$$

Ethane is a colorless and inodorous gas which has not been liquefied. It is nearly insoluble in water, soluble in about two-thirds of its volume of alcohol. Mixed with an equal volume of chlorine, and exposed to diffused daylight, it forms chlorethane or ethyl chloride, C_2H_5Cl; with excess of chlorine higher substitution products are formed.

Propane, $C_3H_6 = CH_3$—CH_2—CH_3, also called *Methyl-ethyl*, is one of the constituents of petroleum, and may be produced by the action of zinc and hydrochloric acid on propyl iodide or isopropryl iodide. It is a gas which liquefies at -20^0 C. (-4^0 F.), and dissolves in one-sixth of its volume of alcohol.

Butanes, C_4H_{10}.—Of these compounds, also called *tetranes* or *quartanes*, there are two modifications, viz. :—

$$CH_3—CH_2—CH_2—CH_3 \qquad CH_3—CH \Big\langle {}^{CH_3}_{CH_2}$$
Normal butane.   Isobutane or Trimethyl-methane.

1. *Normal butane, Diethyl*, or *Methyl-propyl*, occurs in natural petroleum, and in the distillation-products of Cannel and Boghead coal. It may be formed synthetically by heating ethyl iodide to 100^0 in sealed tubes with zinc:

$$2C_2H_5I + Zn = ZnI_2 + C_4H_{10}.$$

It is a colorless gas, which condenses below 0^0 to a liquid boiling at $+1$. Mixed with an equal volume of chlorine, and exposed to light, it yields butyl chloride and other substitution-products.

Isobutane, Trimethyl-methane, or *Methyl-isopropyl*, is formed from tertiary butyl iodide, $(CH_3)_3CI$ (p. 497), by the action of zinc and hydrochloric acid. It is a gas which liquefies at -17^0 C. (1.4^0 F.).

Pentanes, C_5H_{12}.—Of these hydrocarbons there are three modifications, viz. :—

1. Normal Pentane   .   . $CH_3—CH_2—CH_2—CH_2—CH_3$

2. Isopentane .   .   .   . $CH_3—CH_2—CH \Big\langle {}^{CH_3}_{CH_3}$

3. Neopentane   .   .   . ${}^{H_3C}_{H_3C} \Big\rangle C \Big\langle {}^{CH_3}_{CH_3}$

1. *Normal Pentane*, or *Ethyl-propyl*, C_2H_6—C_3H_7, occurs in petroleum, and in the light oils of Boghead and Cannel coal, but has not been prepared synthetically. It is a liquid of specific gravity 0.6, and boiling at $37–39^0$ C. (98.6^0–102.3^0 F.). On passing chlorine into its vapor, a primary and a secondary chloride are formed, from which, by a reaction to be hereafter described, the corresponding alcohols may be prepared.

2. *Isopentane* likewise occurs in petroleum, and may be prepared by dehydration of isopentyl alcohol (ordinary amyl alcohol), or by heating the corresponding iodide with water and zinc.

3. *Neopentane*, or *Tetramethyl-methane*, is formed by the action of zinc-methyl on tertiary butyl iodide :—

$$2C(CH_3)_3I + Zn(CH_3)_2 = ZnI_2 + 2C(CH_3)_4.$$

It is a colorless mobile liquid, which boils at 9.5^0 C. (48.6^0 F.), and solidifies at -20^0 C. (-4^0 F.) to crystals resembling sal ammoniac.

Hexanes, C_6H_{14}.—Of these paraffins, five are possible, and four are known :—

1. Normal Hexane $CH_3—CH_2—CH_2—CH_2—CH_2—CH_3$

2. Isohexane, $\begin{cases} (a) & CH_3—CH_2—CH_2—CH{<}_{CH_3}^{CH_3} \\ (\beta) & CH_3—CH{<}_{CH_2—CH_3}^{CH_2—CH_3} \end{cases}$

3. Neohexane . . $_{H_3C}^{H_3C}{>}C{<}_{CH_3}^{CH_2—CH_3}$

4. Mesohexane. $\begin{matrix} HC(CH_3)_2 \\ | \\ HC(CH_3)_2 \end{matrix}$

1. *Normal Hexane,* or *Dipropyl,* occurs in the light oils of Boghead and Cannel coal, and abundantly in Pennsylvanian petroleum. It is formed by the action of sodium on normal propyl iodide, or by that of zinc and hydrochloric acid on secondary hexyl iodide ; also by distilling suberic acid with barium oxide :—

$$C_8H_{14}O_4 + 2BaO = 2CO_3Ba + C_6H_{14}.$$

It has a density of 0.63 at 17O C. (62.6O F.), and boils at 70–71O C. (158O –159.8O F.).

2. *Isohexane, Propyl-dimethyl-methane,* or *Ethyl-isobutyl,* is formed by the action of sodium on a mixture of ethyl iodide and isobutyl iodide, $CH_2I—CH(CH_3)_2$. It has a density of 0.702 at 0O, and boils at 92O C. (143.6O F.).

3. *Neohexane,* or *Ethyl-trimethyl-methane,* produced by the action of zinc-methyl on tertiary butyl iodide, boils at 45O C.(113O F.).

4. *Mesohexane,* or *Di-isopropyl,* formed by the action of sodium on isopropyl iodide, $IHC(CH_3)_2$, has a density of 0.677 at 0O, and boils at 58O C. (136.4O F.).

Heptanes, C_7H_{16}. Of these hydrocarbons nine are possible and four are known, viz. :

1. *Normal heptane,* $CH_3—(CH_2)_5—CH_3$, is contained in Pennsylvanian petroleum, and in the light oils of Boghead and Cannel coal, and may be formed by distilling azelaic acid, $C_9H_{16}O_4$, with barium oxide. It has a density of 0.712 at 16O C. (60.8O F.), and boils at 98O C. (208.4O F.).

2. *Ethyl-isopentyl,* or *Ethyl-amyl,* $CH_3.CH_2.CH_2.CH_2.CH(CH_3)_2$, a variety of isoheptane, is formed by the action of sodium on a mixture of ethyl-iodide and isopentyl iodide :

$$C_2H_5I + CH_2I.CH_2.CH(CH_3)_2 + Na_2 = 2NaI + C_2H_5.CH_2.CH_2.CH(CH_3)_2.$$

It has a density of 0.683 at 18O C. (64.4O F.), and boils at 90O C. (194O F.).

3. *Triethyl-methane,* $CH_3—CH_2—CH{<}_{CH_2—CH_3}^{CH_2—CH_3}$, another variety of isoheptane, is formed by the action of zinc-ethyl on orthoformic ether (see Formic Ethers).

$$2CH(OC_2H_5)_3 + 3Zn(C_2H_5)_2 = 3Zn(OC_2H_5)_2 + 2CH(C_2H_5)_3.$$
Orthoformic ether.   Zinc-ethyl.   Zinc-ethylate.   Triethyl-methane.

4. *Dimethyl-diethyl-methane,* or *Neoheptane,*
$_{H_3C}^{H_3C}{>}C{<}_{CH_2—CH_3}^{CH_2—CH_3}$, is formed by the action of zinc-ethyl on acetone chloride, $(CH_3)_2CCl_2$. It has a density of 0.689 at 27O C. (80.6O F.), and boils at 96O C. (204.8O F.).

The higher paraffins have been but little examined. Normal octane, C_8H_{18}, boiling at 124° C. (255.2° F.), and having a density of 0.703, occurs in petroleum, and is formed—(1) by the action of zinc and hydrochloric acid on normal octyl iodide ; (2) by the action of sodium on normal butyl iodide ; hence it may be regarded as *dibutyl*, $C_4H_9.C_4H_9$.—*Di-isobutyl*, $(CH_3)_2CH—CH_2—CH_2—CH(CH_3)_2$, obtained by the action of sodium on isobutyl iodide, has a density of 0.705, and boils at 109° C. (228.2° F.).

Di-isopentyl, $C_{10}H_{22}$, obtained from the iodide of isopentyl (ordinary amyl), has a density of 0.770, and boils at 158° C. (316.4° F.).

The paraffins of the higher orders are found, together with those already mentioned, in natural petroleum, and in the light oils obtained by the action of heat on various kinds of coal, especially Boghead and Cannel coal.

In most of these products, however, they are mixed with hydrocarbons belonging to other series ; and to separate them from these, the crude petroleum or coal-tar oil is mixed with strong sulphuric acid, which removes the olefines and other non-saturated hydrocarbons, and afterwards with fuming nitric acid, or a mixture of strong nitric and sulphuric acids. The hydrocarbons of the benzene series, and all other compounds except the paraffins, are thereby oxidized or converted into nitro-compounds, which are either dissolved by the acid, or are much less volatile than the hydrocarbons. The oils are then washed with water, dried over caustic potash, and rectified over sodium. The distillate thus obtained consists wholly of paraffins, which are separated by fractional distillation.

When thus purified the petroleum and coal-oil paraffins, which boil between 0° and 130° C. (266° F.) are found to consist of two series ; those of the first series, which have the higher boiling points, being normal, while those of the second agree for the most part in boiling point with the corresponding synthetically prepared isoparaffins. The boiling points of the two series are as follows :—

		Normal.		Iso.	
Butane,	C_4H_{10}	0°	C. (32° F.)	....	
Pentane,	C_5H_{12}	38°	100.4°	30°	C. (86° F.)
Hexane,	C_6H_{14}	69°	156.2°	61°	141.8
Heptane,	C_7H_{16}	98°	208.4°	91°	195.8
Octane,	C_8H_{18}	124°	255.2°	118°	244.4

The boiling points and specific gravities of the higher paraffins of un known structure, obtained from the same sources, are as follows :—

		Boiling point.		Specific gravity.
Nonane,	C_9H_{20}	136—138° C. (276.3—280.2° F.)		0.741 at 15° C.
Decane,	$C_{10}H_{22}$	160—162°	320 —323.6°	0.757 " 15°
Endecane,	$C_{11}H_{24}$	180—184°	356 —363.2°	0.765 " 16°
Dodecane,	$C_{12}H_{26}$	196—200°	384.8—392°	0.776 " 20°
Tridecane,	$C_{13}H_{28}$	216—218°	420.8—421°	0.792 " 20°
Tetradecane,	$C_{14}H_{30}$	236—240°	456.8—464°	....
Pentadecane,	$C_{15}H_{32}$	256—260°	490 —500°	....

American petroleum likewise yields a quantity of liquid boiling above 300°, and doubtless containing paraffins of still higher order. Some specimens of the crude oil, as it issues from the ground, contain ethane, C_2H_6, and propane, C_3H_8, which are given off from it as gas at ordinary temperatures. In boring for oil also, large quantities of gas escape, exhibiting the characters of methane : hence, it is probable that, in the great geological changes which have given rise to the separation of the petroleum, the whole series of paraffins have been formed, from marsh-gas upwards.

23

Solid paraffin is a colorless crystalline fatty substance, probably con-sisting of a mixture of several of the higher members of the series $C_nH_{2n}+_2$. When heated for some time in a sealed tube, it is resolved, with little or no evolution of gas, into a mixture of olefines and paraffins of lower mole-cnlar weight, which remain liquid at ordinary temperatures. This trans-formation is easily understood: the hydrocarbon, $C_{20}H_{42}$, for example, might be resolved into $C_5H_{12} + C_{15}H_{30}$, or $C_6H_{14} + C_{14}H_{28}$, or $C_7H_{16} + C_{13}H_{26}$, etc., the general equation of the decomposition being,

$$C_nH_{2n}+_2 = C_{n-p}H_{2(n-p)}+_2 + C_pH_{2p}.$$
Paraffin.   Paraffin.   Olefine.

The product actually obtained is a mixture of several paraffins and several olefines.*

Paraffin is found native in the coal-measures, and other bituminous strata, constituting the minerals known as *fossil wax, ozocerite, hatchettin,* etc. It exists also in the state of solution in many kinds of petroleum, and may be separated by distilling off the more volatile portions, and ex-posing the remainder to a low temperature. In a similar manner also may solid paraffin be obtained from the tar of wood, coal, and bituminous shale. It was first prepared by Reichenbach from wood-tar. It is tasteless and inodorous, insoluble in water, slightly soluble in alcohol, freely in ether, and miscible in all proportions, when melted, with fixed or volatile oils. It burns with a very bright flame, and those varieties of it which melt at temperatures above 45° C. (113° F.) are very hard, and well adapted for making candles. Paraffin is largely used also as a substitute for sulphur for dipping matches; and Dr. Stenhouse has patented its application to woollen cloths, to increase their strength and make them waterproof. More extensive, however, are the uses of the liquid compounds of the paraffine series, known in commerce as *paraffin oil, photogene, solar oil, eupione,* etc. These oils are largely used for burning in lamps; and, when mixed with fatty oils, such as rape and cotton-seed oils, form excellent materials for lubricating machinery. For the former purpose they are exceedingly well adapted, as, with a proper supply of air, they give a much brighter light than that obtained from fatty oils containing oxygen, and are much cleaner in use.

It is necessary to observe, however, that natural petroleum and the oils obtained by the dry distillation of coal, etc., at low temperatures, are mixtures of a great number of paraffins differing greatly in volatility, and that to render them safe for burning in lamps of ordinary construction, they must be freed by distillation from the more volatile members of the series; otherwise they will take fire too easily, and when they become heated, will give off highly inflammable vapors, which, mixing with the air in the body of the lamp, may easily produce dangerously explosive mixtures; serious accidents have indeed arisen from this cause. It has been found by experience that it is not safe to use paraffin oil which will take fire on the application of a match and burn continuously at a tem-perature below 38 C. (100.4^O F.).

Second Series, C_nH_{2n}.—Olefines.

The hydrocarbons of this series are polymeric, as well as homologous with one another, inasmuch as their formulæ are all exact multiples of that of the lowest, CH_2. The lower members of the series are gaseous at

* **Thorpe** and **Young** (Berichte der deutschen chemischen gesellsc., 1872, p. 436).

ordinary temperatures, the higher members are solid, and the intermediate compounds liquid. The names and formulæ of the known members of the olefine series are given in the following table :—

Ethene	or	Ethylene	.	.	.	.	C_2H_4
Propene	or	Propylene	.	.	.	.	C_3H_6
Tetrene, Butene, or Butylene	.	.	.	.	C_4H_8
Pentene	or	Amylene	.	.	.	.	C_5H_{10}
Hexene	or	Hexylene	.	.	.	.	C_6H_{12}
Heptene	or	Heptylene	.	.	.	.	C_7H_{14}
Octene	or	Octylene	.	.	.	.	C_8H_{16}
Nonene	or	Nonylene	.	.	.	.	C_9H_{18}
Decene	or	Paramylene	.	.	.	.	$C_{10}H_{20}$
Hexdecene	or	Cetene	.	.	.	.	$C_{16}H_{32}$
Cerotene	.	.	.	.	.	.	$C_{27}H_{54}$
Melene	or	Melisseue	.	.	.	.	$C_{30}H_{60}$

Methene, CH_2, the lowest term of the series, does not appear to be capable of existing in the separate state.

Formation of the Olefines.—1. By abstraction of the elements of water from the alcohols of the series $C_nH_{2n+2}O$, homologous with common alcohol, under the influence of powerful dehydrating agents, such as oil of vitriol, phosphoric oxide, or zinc chloride; thus :—

$$\underset{\text{Ethyl alcohol.}}{C_2H_6O} \quad - \quad \underset{\text{Water.}}{H_2O} \quad = \quad \underset{\text{Ethene.}}{C_2H_4}$$

2. By heating the monatomic haloïd ethers, $C_nH_{2n+1}Br$, etc., with alcoholic solution of potash, or by passing their vapors over lime at a dull red heat :—

$$\underset{\text{Ethyl bromide.}}{C_2H_5Br} \quad + \quad KOH \quad = \quad \underset{\text{Ethene.}}{C_2H_4} \quad + \quad H_2O \quad + \quad KBr.$$

$$\underset{\text{Pentyl chloride.}}{2C_5H_{11}Cl} \quad + \quad CaO \quad = \quad \underset{\text{Pentene.}}{2C_5H_{10}} \quad + \quad H_2O \quad + \quad CaCl_2.$$

The secondary and tertiary alcohols, and their haloïd ethers are converted into olefines, by this and the preceding reactions, more readily than the primary alcohols. The higher alcohols of the series, $C_nH_{2n+2}O$, undergo this change when merely heated; cetyl alcohol, $C_{16}H_{34}O$, for example, is resolved by distillation into H_2O and cetene, $C_{16}H_{32}$.

3. By the decomposition of the paraffins, at the moment of their formation, by the action of zinc or sodium on the alcoholic iodides of the monad alcohol-radicles, C_nH_{2n+1} (see p. 520).

4. By the action of the same iodides on the sodium-compounds of the same radicles; for example:

$$\underset{\substack{\text{Ethyl} \\ \text{iodide.}}}{C_2H_5I} \quad + \quad \underset{\substack{\text{Sodium} \\ \text{ethyl.}}}{C_2H_5Na} \quad = \quad \underset{\substack{\text{Sodium} \\ \text{iodide.}}}{NaI} \quad + \quad \underset{\text{Ethene.}}{C_2H_4} \quad + \quad \underset{\text{Ethane.}}{C_2H_6}.$$

5. By decomposition of the hydroxides of ammonium bases containing four atoms of a monad alcohol-radicle (p. 508), these compounds when heated splitting up into a tertiary monamine (p. 515) and an olefine; thus :

$$\underset{\substack{\text{Tetrethylammon-} \\ \text{ium hydroxide.}}}{N(C_2H_5)_4(HO)} \quad = \quad \underset{\substack{\text{Triethyl-} \\ \text{amine.}}}{N(C_2H_5)_3} \quad + \quad \underset{\text{Water.}}{H_2O} \quad + \quad \underset{\text{Ethene.}}{C_2H_4}.$$

6. Olefines are formed by the decomposition of acetates and butyrates at a red heat, distilling over, together with several other products, from which

they are separated by combining them with bromine, and heating the resulting bromine-compounds, $C_nH_{2n}Br_2$, to 275° with copper, water, and
potassium iodide. In this manner Berthelot has obtained ethene, propene,
butene, and pentene.

7. By electrolysis of the alkali-salts of bibasic acids: e. g., :

$$\underset{\text{Succinic acid.}}{C_4H_6O_4} = C_2H_4 + 2CO_2 + H_2.$$

8. By heating the diatomic haloïd ethers, $C_nH_{2n}Cl_2$, etc., with sodium:

$$C_2H_4Cl_2 + Na_2 = 2NaCl + C_2H_4.$$

9. Several of the olefines may be produced by direct synthesis from other
hydrocarbons of simpler constitution, or their haloïd derivatives.

α. Ethene is formed by the action of nascent hydrogen upon ethine or
acetylene:

$$C_2H_2 + H_2 = C_2H_4$$

β. Propene, C_3H_6, is formed by passing a mixture of methane and carbon
monoxide (oxymethene) through a red-hot tube:

$$2CH_4 + CO = H_2O + C_3H_6.$$

Also by the action of trichloromethane (chloroform) on zinc ethide :-

$$2CHCl_3 + 3Zn(C_2H_5)_2 = 3ZnCl_2 + 4C_3H_6 + 2CH_4.$$

γ. Pentene, C_5H_{10}, or a compound isomeric or polymeric with it, is formed
by the action of zinc ethide on propenyl (allyl) iodide:

$$2C_3H_5I + Zn(C_2H_5)_2 = ZnI_2 + 2C_5H_{10}.$$

Constitution and Reactions of the Olefines.—These hydrocarbons in the free
state may be supposed to have one pair of their carbon-atoms linked together by two combining units, thus:

$$
\begin{array}{cc}
 & CH_3 \\
CH_2 & \| \\
\| & CH \\
CH_2 & \| \\
 & CH_2 \\
\text{Ethene.} & \text{Propene.}
\end{array}
$$

In all the higher members of the series this structure admits of isomeric
modifications; three for butene, five for pentene, etc., as will presently be
further considered.

With two carbon-atoms thus doubly linked, the olefines present the appearance of saturated molecules incapable of uniting with monatomic
elements or radicles. Under the influence of chlorine, bromine, and iodine,
however, they undergo a change of structure, the double linking of the
two carbon-atoms being partly broken up, so that these atoms remain
united by one combining unit only, and the hydrocarbon becomes a bivalent radicle, capable of taking up 2 atoms of chlorine, bromine, etc., thus:

$$
\begin{array}{ccc}
CH_2 & & CH_2Cl \\
\| & + Cl_2 = & | \\
CH_2 & & CH_2Cl
\end{array}
$$

The resulting diatomic ethers, treated with silver acetate or potassium
acetate, exchange their bromine for an equivalent quantity of the halogenic
residue of the acetate, $C_2H_3O_2$ (p. 509), giving rise to diatomic acetic ethers;
thus:

$$(C_2H_4)Br_2 \quad + \quad 2C_2H_3O_2K \quad = \quad 2KBr \quad + \quad (C_2H_4)(C_2H_3O_2)_2;$$
Ethene   Potassium   Potassium   Ethene
bromide.   acetate.   bromide.   diacetate.

and these ethers, distilled with a caustic alkali, yield diatomic alcohols or glycols; for example:

$$(C_2H_4)(C_2H_3O_2)_2 \quad + \quad 2KHO \quad = \quad 2C_2H_3O_2K \quad + \quad (C_2H_4)(OH)_2.$$
Ethene   Potassium   Ethene
diacetate.   acetate.   alcohol.

5. The bromides, $C_nH_{2n}Br_2$, heated to 275° C. (527° F.) with a mixture of potassium iodide, copper, and water, give up their bromine and reproduce the original olefine, together with other hydrocarbons (p. 521).

6. Some olefines, when briskly shaken up with strong *sulphuric acid*, unite with it, forming acid ethers of sulphuric acid, which contain the monatomic alcohol-radicles analogous to the olefines; thus:

$$C_2H_4 \quad + \quad H_2SO_4 \quad = \quad C_2H_5.HSO_4;$$
Ethene.   Sulphuric acid.   Ethyl-sulphuric acid.

and these acid ethers distilled with water reproduce sulphuric acid and the monatomic alcohol analogous to the olefine:

$$C_2H_5.HSO_4 \quad + \quad H(OH) \quad = \quad H_2SO_4 \quad + \quad C_2H_5(OH).$$
Ethyl-sulphuric acid.   Water.   Ethyl alcohol.

With fuming sulphuric acid (which contains sulphuric oxide in solution) the olefines yield sulpho-acids which are isomeric with the preceding, but are not decomposed by water with formation of an alcohol.

7. Olefines unite with *hydrochloric, hydrobromic, and hydriodic acids*; and the resulting compounds, treated with silver oxide in presence of water, give rise to two different reactions which go on simultaneously, one part of the compound exchanging its halogen element for hydroxyl, and thereby producing an alcohol, while another portion gives up hydrochloric, hydrobromic, or hydriodic acid, reproducing the original olefine:—

$$2(C_6H_{12}.HI) \quad + \quad Ag_2O \quad + \quad H_2O \quad = \quad 2AgI \quad + \quad 2C_6H_{14}O$$
Hexene   Hexyl
hydriodide.   alcohol.

$$2(C_6H_{12}.HI) \quad + \quad Ag_2O \quad = \quad 2AgI \quad + \quad H_2O \quad + \quad 2C_6H_{12}$$
Hexene hydriodide.   Hexene.

Olefines also unite with *hypochlorous acid*, forming compounds called glycolic chlorhydrins; *e. g.*:—

$$\begin{matrix} CH_2 \\ \| \\ CH_2 \end{matrix} \quad + \quad ClOH \quad = \quad \begin{matrix} CH_2Cl \\ | \\ CH_2OH \end{matrix}$$

By *nascent hydrogen* (evolved by zinc and hydrochloric acid) they are converted into paraffins; *e. g.*, $C_2H_4 + H_2 = C_2H_6$.

They also yield paraffins when heated with concentrated *hydriodic acid*, the moniodoparaffin formed in the first instance being reduced by a second molecule of hydriodic acid; *e. g.*:—

$$C_2H_4 + HI = C_2H_5I; \text{ and } C_2H_5I + HI = C_2H_6 + I_2.$$

Condensation of the Olefines.—These bodies, under the influence of sulphuric acid, zinc chloride, boron fluoride, and other reagents, even at ordinary temperatures, become polymerized by the linking together of two or more molecules. In this manner, isopentene or amylene, C_5O_{10}, may be converted into $C_{10}H_{20}$, $C_{15}H_{30}$, etc. Propene and butene may also be polymerized, but ethene is not polymerized either by sulphuric acid or by boron fluoride. The polymeric hydrocarbons thus formed are dyad radicles, capable of uniting directly with chlorine, etc.

The following diagram exhibits the conversion of isopentene or amylene into di-amylene, showing that the latter also has two of its carbon atoms doubly linked, and may therefore act as a dyad radicle :—

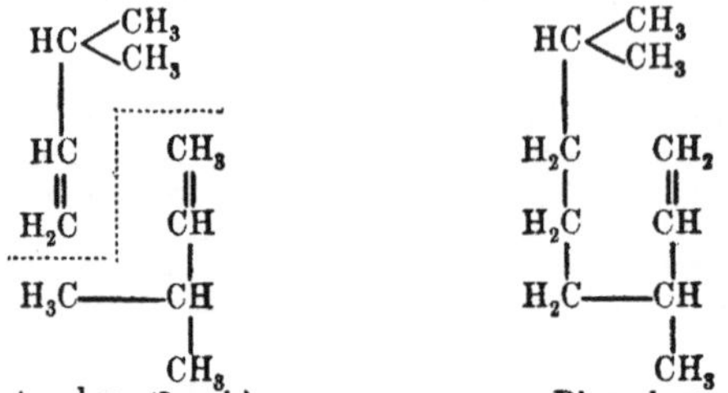

$$
\begin{array}{ll}
HC{<}^{CH_3}_{CH_3} & HC{<}^{CH_3}_{CH_3} \\
| & | \\
HC \quad CH_3 & H_2C \quad CH_2 \\
\| \quad \| & | \quad \| \\
H_2C \quad CH & H_2C \quad CH \\
| \quad | & | \quad | \\
H_3C{-}{-}CH & H_2C{-}{-}CH \\
| & | \\
CH_3 & CH_3 \\
\text{Amylene (2 mol.)} & \text{Diamylene.}
\end{array}
$$

Ethene or **Ethylene**, C_2H_4, also called *Olefiant gas*, is produced in the dry distillation of many organic bodies, and is found, to the amount of about 6 per cent., in coal gas. It is most easily prepared by heating strong alcohol with three or four times its weight of strong sulphuric acid (p. 228). It is a colorless inflammable gas, which burns with a bright white flame. When exposed to a strong pressure at a temperature of —110° C. (—166° F.), it condenses to a liquid which does not solidify. Its specific gravity, referred to hydrogen as unity, is 14, showing that the molecule C_2H_4 has the normal 2-volume condensation :—

$$\frac{2 \times 12 + 4 \times 1}{2} = 14.$$

Ethene unites readily with chlorine, bromine, and iodine, forming oily liquids. Fuming hydriodic acid absorbs it, with formation of ethyl iodide C_2H_5I. It dissolves also in strong sulphuric acid, after prolonged agitation at ordinary temperatures, easily and completely at 160–175° C. (320–347° F.). The product is ethylsulphuric acid, $C_2H_5.SO_4H$, which, when boiled with water, yields ethyl alcohol. Ethene is oxidized by chromic acid to aldehyde, C_2H_4O, and by potassium permanganate to oxalic and formic acid.

Propene, $C_3H_6 = CH_3{-}CH{=}CH_2$, is formed (1) By heating propyl iodide or isopropyl iodide with alcoholic potash solution :—

$$C_3H_7I + KOH = C_3H_6 + KI + H_2O ;$$

(2) By the action of nascent hydrogen (from zinc and hydrochloric acid), or of hydriodic acid, on allyl iodide :—

$$C_3H_5I + HI = C_3H_6 + I_2 ;$$

(3) Together with other products, by passing the vapor of ordinary amyl alcohol or butyl alcohol through a red-hot tube. It also occurs in coal gas.

Propene is a colorless gas, condensable to a liquid by strong pressure. It dissolves readily in strong hydriodic acid, forming secondary propyl iodide :

$$CH_3{-}CH{=}CH_2 + IH = CH_3{-}CHI{-}CH_3.$$

Butene, or **Butylene,** C_4H_8. Of this hydrocarbon there are three modifications, represented by the following formulæ :—

1. Butene,	CH_3—CH_2—$CH{=}CH_2$;
2. Pseudobutene,	CH_3—$CH{=}CH$—CH_3;
3. Isobutene,	$\begin{matrix}H_3C\\H_3C\end{matrix}{>}C{=}CH_2$.

1. *Normal butene* is produced by abstraction of HI (action of alcoholic potash) from normal primary butyl iodide, CH_3—CH_2—CH_2—CH_2I, and by the action of zinc-ethide on bromethene :

$$2(CH_3\text{—}CHBr) + Zn{<}{\begin{matrix}CH_2CH_3\\CH_2CH_3\end{matrix}} = ZnBr_2 + 2(CH_3\text{—}CH_2\text{—}CH{=}CH_2).$$

It is gaseous at ordinary temperatures, and condenses at —10° C. (14° F.) to a liquid which boils at —5° C. (23° F.). It unites with hydriodic acid, forming secondary butyl iodide, CH_3—CH_2—CHI—CH_3.

2. *Pseudobutene* is formed by heating secondary or pseudobutyl iodide, or the corresponding alcohol, with alcoholic potash, or with silver iodide. It boils at +3° C. (37.4° F.), and solidifies at low temperatures. It unites with HI, reproducing secondary butyl iodide.

3. *Isobutene* is produced by the action of alcoholic potash on isobutyl iodide, $\begin{matrix}H_3C\\H_3C\end{matrix}{>}CH$—$CH_2I$, or on tertiary butyl iodide, $\begin{matrix}H_3C\\H_3C\end{matrix}{>}CI$ — CH_3, or by that of sulphuric acid on the corresponding alcohols. It is also formed by passing the vapor of isopentyl alcohol through a red-hot tube, and by the electrolysis of isovaleric (ordinary valeric) acid.

Isobutene is gaseous at ordinary temperatures, but condenses at very low temperatures to a liquid, which boils at —6°. Strong hydriodic acid absorbs it, with formation of tertiary butyl iodide. With sulphuric acid it forms, together with polymeric butenes, tertiary butylsulphuric acid, $\begin{matrix}H_3C\\H_3C\end{matrix}{>}C{<}\begin{matrix}CH_3\\SO_4H\end{matrix}$,which, when boiled with water, yields tertiary butyl alcohol.

The dibromides of the three isomeric butenes, $C_4H_8Br_2$, boil at the following temperatures :—

Normal.	Iso.	Pseudo.
166° C. (330.8° F.).	159° C. (318.2° F.).	149° C. (300.2° F.).

Pentenes, C_5H_{10}. Of the four possible modifications of these hydrocarbons, viz.,

1.	2.	3.	4.
CH_3	$H_3C\quad CH_3$	CH_3	$H_3C\quad CH_3$
$\mid$	$\diagdown\diagup$	$\mid$	$\diagdown\diagup$
CH_2	CH	CH_2	C
$\mid$	$\mid$	$\mid$	$\parallel$
CH_2	CH	CH	CH
$\mid$	$\parallel$	$\parallel$	$\mid$
CH	CH_2	CH	CH_3
$\parallel$		$\mid$	
CH_2		CH_3	

1. *Normal Pentene,* or *Ethyl-allyl,* $C_2H_5.C_3H_5$, is formed by the action of sodium on a mixture of the iodides of ethyl and allyl, or of zinc-ethyl on allyl iodide. It boils at 37° C. (98.6° F.) ; unites with hydri-

odic acid to form the secondary iodide, $CH_3-CH_2-CH_2-CHI-CH_3$ or $IHC<^{CH_3}_{C_3H_7}$. Its bromide, $C_5H_{10}Br_2$, boils at 175° C. (347° F.).

2. *Isopentene*, or *Amylene*, is obtained, together with isopentane, by distilling ordinary amyl alcohol, $^{H_3C}_{H_3C}>CH-CH_2-CH_2OH$, with sulphuric acid or zinc chloride. It is a colorless mobile liquid having a fragrant odor, a density of 0.663 at 0°, and boiling at 35° C. (95° F.). Its bromide boils at 170–180° C (338–356° F.), with partial decomposition. Isopentene unites with hydriodic acid, forming another secondary pentyl iodide, $(CH_3)_2CH-CHI-CH_3$.

Isopentene shaken up with sulphuric acid (previously diluted with half its bulk of water, and cooled) is converted into several polymerides, viz., diamylene, $C_{10}H_{20}$, boiling at 156° C. (312.8° F.); triamylene, $C_{15}H_{30}$, at 240–250° C. (460–482° F.); tetramylene, $C_{20}H_{40}$, at 360° C. (644° F.). They are oily liquids, uniting directly with bromine.

3. The third modification is not known with certainty, but is perhaps identical with a pentene, boiling at 25° C. (77° F.), obtained by the action of concentrated alcoholic potash on isopentyl iodide.*

4. The fourth modification is obtained by the action of very strong alcoholic potash on tertiary pentyl iodide:

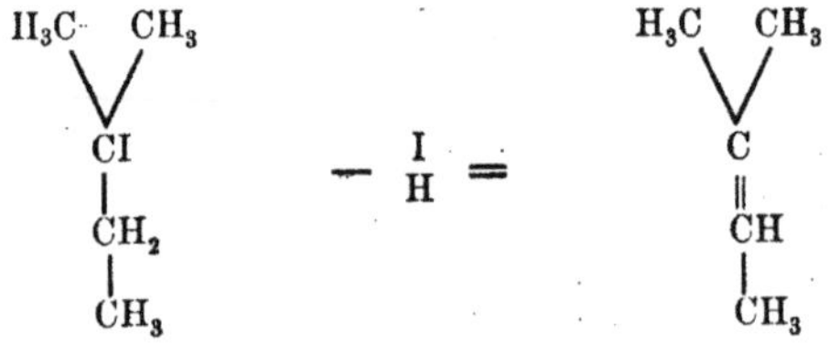

It boils at 35° C. (95° F.), unites with bromine, and with fuming hydriodic acid at 100°, reproducing the tertiary iodide.†

The higher olefines have been but little studied. They are obtained from the corresponding alcohols and iodides by the methods above described.

Hexene, or **Hexylene**, C_6H_{12}. Two hydrocarbons of this composition have been obtained, one from secondary, the other from tertiary hexyl alcohol. The former has a specific gravity of 0.699 at 0°, and boils at 65–66° C. (149–150.8° F.); the latter, which occurs in the light oils from Boghead and Cannel tar, boils at 70° C. (158° F.).

Heptene, or **Heptylene**, C_7H_{14}, also called *œnanthylene*, is formed from secondary heptyl chloride; also, by treating heptyl aldehyde (œnanthol), $C_7H_{14}O$, with phosphorus pentachloride, and decomposing the resulting chloride, $C_7H_{14}Cl_2$, with sodium. It boils at 100°. The same hydrocarbon occurs in the light oils from Boghead and Cannel tar.

Octene, or **Octylene**, C_8H_{16}, occurs in the same oils, and is easily prepared by distilling secondary octyl alcohol with zinc chloride. It boils at 125° C. (257° F.).

Cetene, $C_{16}H_{32}$, **Cerotene**, $C_{27}H_{54}$, and **Melene**, $C_{30}H_{60}$, are formed by destructive distillation of the corresponding alcohols, or sub-

* Flavitzky, Liebig's Annalen, clxix. 205.
† Ermolaiew, Zeitschrift für Chemie, 1871, 275.

stances containing them ; cetene from spermaceti, cerotene from Chinese wax, melene from beeswax.

Cetene is an oily liquid, boiling at 275° C. (527° F.); cerotene, a crystalline solid ; melene crystallizes from hot alcohol in scales, which melt at 62° C. (143.6° F.).

THIRD SERIES.—C_nH_{2n-2}.

Ethine or *Acetylene Series.*

These hydrocarbons may be arranged in two collateral series differing from one another in structure and in properties. The following are known :

a Series.				Boiling Point.		
Ethine	or	Acetylene,	C_2H_2	...		...
Propine	"	Allylene,	C_3H_4	...		...
Pentine	"	*a* Valerylene,	C_5H_8	50° C.	(122	° F.)
Hexine	"	Hexoylene,	C_6H_{10}	80° C.	(176	° F.)
Heptine	. . .		C_7H_{12}	107° C.	(224.6°	F.)
Octine	. . .		C_8H_{14}	133° C.	(271.4°	F.)
Decine	"	Decenylene,	$C_{10}H_{18}$	165° C.	(329	° F.)
Pentadecine	"	Benylene,	$C_{15}H_{28}$	225° C.	(437	° F.)
Hexdecine	"	Cetenylene,	$C_{16}H_{30}$	280° C.	(536	° F.)

β Series.		Boiling Point.		
β Allylene,	C_3H_4	...		
Crotonylene,	C_4H_6	18° C.	(64.4°	F.)
β Valerylene,	C_5H_8	45° C.	(113	° F.)
Diallyl,	C_6H_{10}	59° C.	(138.2°	F.)
Rutylene,	$C_{10}H_{18}$	150° C.	(302	° F.)

A general method of preparing the hydrocarbons of the *a* series, consists in heating the dibromides or di-iodides of the olefines with alcoholic potash, the reaction taking place by two stages, a monobrominated olefine being first formed, and then deprived, by the further action of the potash, of the elements of hydrobromic acid ; thus—

$$C_nH_{2n}Br_2 + KOH = KBr + H_2O + C_nH_{2n-1}Br,$$

and $\quad C_nH_{2n-1}Br + KOH = KBr + H_2O + C_nH_{2n-2}.$

The reaction amounts to the abstraction of 2HBr from the olefine dibromide, and shows that the hydrocarbons of this series have two of their carbon atoms united by three combining units ; thus—

$$
\begin{array}{ccccc}
CH_2Br & & & & CH \\
| & - & 2HBr & = & ||| \\
CH_2Br & & & & CH \\
\text{Ethene dibromide.} & & & & \text{Acetylene.}
\end{array}
$$

$$
\begin{array}{ccccc}
CH_3 & & & & CH_3 \\
| & & & & | \\
CHBr & - & 2HBr & = & C \\
| & & & & ||| \\
CH_2Br & & & & CH \\
\text{Propene dibromide.} & & & & \text{Allylene.}
\end{array}
$$

23 *

Acetylene does not admit of any other modification, at least in the free state; but in allylene, and the higher members of the series, another mode of grouping is possible, as shown by the following formulæ:—

$$CH_2 \!\!=\!\! C \!\!=\!\! CH_2 \qquad\qquad CH_2 \!\!=\!\! CH \!\!-\!\! CH_2 \!\!-\!\! CH_2 \!\!-\!\! CH \!\!=\!\! CH_2$$

$\qquad$ β-Allylene. $\qquad\qquad\qquad\qquad\qquad$ Diallyl.

The hydrocarbons of this latter (β) subseries are formed by various reactions, β-allylene, for example, by the electrolysis of itaconic acid:—

$$
\begin{array}{lcl}
CH_2 \!\!-\!\! COOH & & CH_2 \\
\mid & & \parallel \\
\!\!-\!\!C\!\!-\!\! & = \; 2CO_2 \; + \; H_2 \; + & C \\
\mid & & \parallel \\
CH_2 \!\!-\!\! COOH & & CH_2 \\
\text{Itaconic acid.} & & \text{β-Allylene.}
\end{array}
$$

The hydrocarbons of the two subseries likewise differ in certain of their properties. Those of the *a* series, when treated with an ammoniacal solution of argentic or cuprous chloride, yield metallic derivatives, in the form of crystalline precipitates, which, when heated with hydrochloric acid, reproduce the original hydrocarbons. This reaction affords a convenient method of separating acetylene and allylene from other gases. The hydrocarbons of the β subseries do not yield these metallic derivatives.

When the hydrocarbons of the *a* subseries are subjected to the action of chlorine, bromine, and other powerful reagents, the connection between the trebly-linked carbon atoms becomes loosened—as in the case of the olefines—so that the molecule which was previously saturated becomes bivalent or quadrivalent: thus, in the case of acetylene,

$$
\begin{array}{ccc}
C\!\!-\!\!H & \!\!-\!\!C\!\!-\!\!H & \!\!-\!\!\overset{\mid}{C}\!\!-\!\!H \\
\vert\vert\vert & \parallel & \mid \\
C\!\!-\!\!H & \!\!-\!\!C\!\!-\!\!H & \!\!-\!\!\underset{\mid}{C}\!\!-\!\!H
\end{array}
$$

$\quad$ Saturated. $\qquad\qquad$ Bivalent. $\qquad\qquad$ Quadrivalent.

When agitated with hydrobromic or hydriodic acid, they take up one or two molecules of these acids. The dihydrobromides and dihydriodides thus produced have the same composition as the dibromides of the olefines; thus:

$$C_nH_{2n-2}.2HBr = C_nH_{2n}Br_2.$$

The two classes of bodies are, however, isomeric, not identical.

Acetylene, or **Ethine,** C_2H_2.—This hydrocarbon is one of the constituents of coal gas.—It is produced—1. By synthesis from its elements. When an electric arc from a powerful voltaic battery passes between carbon poles in an atmosphere of hydrogen, the carbon and hydrogen unite in the proportion to form ethine.

2. By the action of heat upon ethene, or the vapor of alcohol, ether, or wood-spirit, or by passing induction-sparks through marsh-gas.

3. By passing the vapor of chloroform over ignited copper:

$$2CHCl_3 + Cu_6 = 3Cu_2Cl_2 + C_2H_2.$$

4. By the incomplete combustion of bodies containing carbon and hydrogen; for example:

$$4CH_4 + O_6 = 6H_2O + 2C_2H_2$$
$$2C_2H_4 + O_2 = 2H_2O + 2C_2H_2.$$

It is produced in considerable quantity in the imperfect combustion which takes place in a Bunsen's gas-burner, when the flame strikes down and the gas burns at the small orifice at the bottom of the tube; it may be collected and purified by aspirating the gas through an ammoniacal copper or silver solution.

5. By passing a mixture of marsh-gas and carbon monoxide through a red-hot tube:

$$CH_4 + CO = H_2O + C_2H_2.$$

6. By the action of alcoholic potash on monobromethene:

$$C_2H_3Br + HKO = KBr + H_2O + C_2H_2.$$

7. By the electrolysis of fumaric or maleic acid:

$$C_2H_2\!<\!^{CO_2H}_{CO_2H} = 2CO_2 + H_2 + C_2H_2.$$

The crude acetylene obtained by either of these processes is purified in the manner above mentioned.

Acetylene is a colorless gas of specific gravity 0.92, having a peculiar and unpleasant odor, moderately soluble in water, not condensed by cold or pressure. It burns with a very bright and smoky flame, one volume of the gas consuming $2\frac{1}{2}$ volumes of oxygen, and producing 2 volumes of carbon dioxide. When mixed with *chlorine*, it detonates almost instantly, even in diffused daylight, with separation of carbon.

Acetylene passed into an ammoniacal solution of *cuprous chloride* forms a red precipitate consisting of cuproso-vinyl oxide, $C_4(Cu_2)_2H_2O$, or $[C_2(Cu_2)H]_2O$, that is to say, vinyl-oxide, $(C_2H_3)_2O$, having four of its hydrogen-atoms replaced by four atoms of (apparently) univalent copper (see p. 372). The constitution of this compound may be understood from the following formulæ:

Vinyl oxide.   Cuproso-vinyl oxide.

Its formation from cuprous chloride and acetylene is represented by the equation:

$$2Cu_2Cl_2 + 2C_2H_2 + H_2O = 4HCl + C_4Cu_4H_2O.$$

On heating it with hydrochloric acid, the opposite reaction takes place, cuprous chloride and water being reproduced, and pure acetylene evolved as gas.

When this copper compound is heated with zinc and dilute ammonia, the nascent hydrogen thereby evolved unites with the elements of ethine, producing ethene:

$$C_4Cu_4H_2O + 2H_2 = Cu_4 + H_2O + 2C_2H_2$$

and

$$C_2H_2 + H_2 = C_2H_4.$$

Acetylene, briskly agitated with strong *sulphuric acid*, is absorbed, producing vinyl-sulphuric acid, $C_2H_4SO_2$:

$$C_2H_2 + H_2SO_4 = (C_2H_3)HSO_4;$$

and this acid, distilled with water, is resolved into sulphuric acid and vinyl alcohol :

$$(C_2H_3)HSO_4 \quad + \quad H_2O = H_2SO_4 \quad C_2H_3(OH)$$
$$\text{Vinyl-sulphuric} \qquad\qquad\qquad\qquad\qquad \text{Vinyl}$$
$$\text{acid.} \qquad\qquad\qquad\qquad\qquad\qquad\quad \text{alcohol.}$$

Acetylene unites with *bromine*, forming a dibromide, $C_2H_2Br_2$.

When a series of strong induction sparks is passed through a mixture of acetylene and *nitrogen*, the two gases unite and form h y d r o c y a n i c a c i d : $C_2H_2 + N_2 = 2CNH$.

Bromacetylene, C_2HBr, is produced by the action of alcoholic potash on dibromethene dibromide :

$$C_2H_2Br_2.Br_2 = HBr + Br_2 + C_2HBr.$$

It is a spontaneously inflammable gas, which liquefies under a pressure of three atmospheres, is soluble in water, and very soluble in dibromethine. It unites with bromine, forming the compound $C_2HBr.Br_2$, and when passed into an ammoniacal solution of cuprous chloride, yields a precipitate of cuproso-vinyl oxide.

Allylene or **Propine**, C_3H_4.—This compound is produced by the action of sodium ethylate on bromopropene :

$$C_3H_5Br + C_2H_5NaO = NaBr + C_2H_5(HO) + C_3H_4.$$

its formation being a particular case of the general reaction given on page 537. It is a colorless gas, having an unpleasant odor, burning with a smoky flame, and forming, with mercurous salts, a gray precipitate ; with silver salts, a white precipitate ; and with cuprous chloride, a yellow precipitate, analogous in composition to that formed by ethine. With *bromine* it forms the compounds $C_3H_4Br_2$ and $C_3H_4Br_4$, the former boiling at 130⁰ C. (266⁰ F.), the latter decomposing when distilled. With iodine it forms the compound $C_3H_4I_2$, which boils at 198⁰ C. (388.4⁰ F.). It unites with 2 molecules of hydriodic acid, forming the compound CH_3—CI_2—CH_3, and similarly with HBr and HCl.

ß *Allylene*, or *Allene*, $CH_2{=}C{=}CH_2$, is formed, as already stated, by the electrolysis of itaconic acid ; also by the action of sodium on the modification of dichloropropene obtained from dichlorhydrin (see GLYCERIN) ; probably also from allyl iodide. With bromine it forms a crystalline tetrabromide, $C_3H_4Br_4$.

Crotonylene, C_4H_6 or CH_3—$CH{=}C{=}CH_2$, is formed by abstraction of BrH from monobromobutene, CH_3—CH_2—$CBr{=}CH_2$; also by distilling erythrite, $C_4H_{10}O_4$, with formic acid. It is a liquid boiling at 20–25⁰ C. (68–77⁰ F.), and forming a tetrabromide which crystallizes in shining rhombic plates, melting at 116⁰ C. (240.8⁰ F.). It does not form any metallic derivative with copper or silver, whence, and from its mode of formation, it may be referred to the ß series.

Valerylene or **Pentine**, C_5H_8.—Of this hydrocarbon two modifications are known. *a Valerylene*, probably $HC{\equiv}C$—C_3H_7, is formed by abstraction of H_2Cl_2 from methylpropyl-ketonic chloride, H_3C—CCl_2—C_3H_7. It boils at 50⁰ C. (122⁰ F.), and yields metallic derivatives with silver and copper.

ß *Valerylene*, probably $\begin{smallmatrix}H_3C\\H_3C\end{smallmatrix}{>}C{=}C{=}CH_2$, is formed by abstraction of HBr from bromisopentene, $\begin{smallmatrix}H_3C\\H_3C\end{smallmatrix}{>}C{=}CH$—$CBr{=}CH_2$. It is a liquid having an alliaceous odor, a specific gravity of 0.700, and boiling at 45° C. (113° F.). It does not form compounds with silver and copper. It forms a dibromide which boils at 170° C. (338° F.), and a liquid tetrabromide.

Hexines, C_6H_{10}.—*Hexoylene,* probably CH_3—$(CH_2)_3$—$C≡CH$, obtained by abstraction of HBr from monobromhexene, boils at 76–80° C. (168.8–176° F.).

Diallyl, $H_2C=CH$—CH_2—CH_2—$CH=CH_2$, is formed by the action of sodium or silver on allyl iodide. It is a volatile, pungent liquid, boiling at 59° C. (138.2° F.), and forming a crystalline tetrabromide and tetriodide, the former melting at 63° C. (145.4° F.). When shaken with aqueous hypochlorous acid, it forms the compound $C_6H_{10}Cl_2(OH)_2$. Heated with concentrated hydriodic acid it yields the compound $C_6H_{12}I_2$ or H_3C—CHI—CH_2—CH_2—CHI—CH_3, as a heavy oily liquid ; and on heating this compound with silver acetate, it is converted into the corresponding diacetate, which, when heated with caustic potash, yields the corresponding hexene glycol,

$$H_3C—CHOH—CH_2—CH_2—CHOH—CH_3.$$

The higher members of the series C_nH_{2n-2} are obtained either from the dibromides of the corresponding olefines, or from substitution-products isomeric therewith. Thus h e p t i n e, or œ n a n t h i d e n e, is formed by the action of potash on œnanthidene dibromide (obtained by treating œnanthyl with phosphorus pentachloride) ; r u t y l e n e is prepared from d i a m y l e n e ; and the isomeric body d e c i n e or d e c e n y l e n e from the dibrominated derivative of the hydrocarbon $C_{10}H_{22}$, which exists in petroleum.

SERIES C_nH_{2n-4}.

The only known member of the fatty group belonging to this series is v a l y l e n e or p e n t o n e, C_5H_6, which is formed by the action of alcoholic potash on valerylene dibromide, $C_5H_8Br_2$. It is a light oil, smelling like garlic, and boiling at 50°. It forms precipitates with ammoniacal copper and silver solutions. With bromine in a freezing mixture, it yields a crystalline mass consisting of $C_5H_6Br_6$, saturated with a thick liquid which is a mixture of the same compound with $C_5H_6Br_4$, and probably $C_5H_6Br_2$.

The t e r p e n e s, $C_{10}H_{16}$, also come under the formula C_nH_{2n-4}, but they belong rather to the aromatic group.

SERIES C_nH_{2n-6}.

Dipropargyl, C_6H_6 or $HC≡C$—CH_2—CH_2—$C≡CH$, the only known hydrocarbon of the fatty group belonging to this series, is isomeric with benzene, but differs greatly from that body in its properties. It is prepared by distilling diallyl tetrabromide, $C_6H_{10}Br_4$, with a large excess of caustic potash, whereby it is converted into dibromodiallyl, $C_6H_8Br_2$, and boiling the latter with alcoholic potash :

$$\begin{array}{c} H_2C—CH=CHBr \\ | \\ H_2C—CH=CHBr \\ \text{Dibromodiallyl.} \end{array} \quad - \quad 2HBr \quad = \quad \begin{array}{c} H_2C—C≡CH \\ | \\ H_2C—C≡CH \\ \text{Dipropargyl.} \end{array}$$

Dipropargyl is a mobile, limpid, highly refractive liquid, having an intensely pungent odor, a specific gravity of 0.81 at 18° C. (64.4° F.), and

boiling at 85° C. (185° F.).   With ammoniacal cuprous chloride it forms
a greenish-yellow precipitate, $C_6H_4Cu_2 + 2H_2O$, and with silver nitrate, a
white precipitate, $C_6H_4Ag_2 + 2H_2O$, which blackens on exposure to light,
and explodes at 100°, leaving carbon and silver.

Dipropargyl is easily distinguished from benzene by its property of combining explosively with bromine, forming the tetrabromide $C_6H_6Br_4$,
a viscid liquid, of specific gravity 2.460 at 19° C. (66.2° F.), which combines in the dark with more bromine, forming the octobromide $C_6H_6Br_8$,
which boils at 140° C. (284° F.).

HALOGEN DERIVATIVES OF THE HYDROCARBONS.
HALOÏD ETHERS.

These compounds are formed, as already observed, by substitution of a
halogen element (Br, Cl, I) for an equivalent quantity of hydrogen in a
hydrocarbon, or of hydroxyl in the corresponding alcohol.   They are
monatomic, diatomic, triatomic, etc., according as 1, 2, 3, or more atoms
of the halogen are thus introduced.

Monatomic Haloïd Ethers, $C_nH_{2n+1}X$.—These ethers are derived
from the paraffins.   Chlorine and bromine act directly on these hydrocarbons, the action of the former being accelerated by light, that of the latter
by heat.   Iodine likewise acts on the paraffins, removing an atom of hydrogen, and taking its place; but the iodoparaffin thus formed is for the
most part reconverted into the original hydrocarbon by the action of the
hydriodic acid formed at the same time : thus,

$$C_3H_7I + HI = C_3H_8 + I_2.$$

This reverse action may, however, be prevented by the addition of a substance, like mercuric oxide, which decomposes the hydriodic acid as fast
as it is formed, and in that case an iodoparaffin or alcoholic iodide is obtained : e. g.,

$$2C_3H_8 + 2I_2 + HgO = 2C_3H_7I + H_2O + HgI_2.$$

These monatomic haloïd ethers are also produced by the action of the
haloïd compounds of hydrogen and of phosphorus on the corresponding
alcohols (*q. v.*), and by addition of a haloïd acid to an unsaturated hydrocarbon, *i. e.*, a hydrocarbon in which a pair of carbon-atoms are joined
together by two combining units (p. 532) : *e. g.*,

$$
\begin{array}{ccccc}
CH_3 & & & & CH_3 \\
| & & & & | \\
CH & + & I & = & CHI \\
\| & & & & | \\
CH_2 & & H & & CH_3 \\
\text{Propene.} & & & & \text{Isopropyl} \\
& & & & \text{iodide.}
\end{array}
$$

These ethers are converted—

α. Into the original paraffins by the action of *nascent hydrogen*
(evolved by hydrochloric acid and zinc, or by sodium amalgam) :

β. Into the corresponding alcohols by the action of moist *silver oxide :*

$$C_2H_5I + AgOH = AgI + C_2H_5OH.$$

Potassium hydroxide acts in a similar manner, but the resulting .alc°ho¹
(especially if a secondary alcohol) is partly converted by dehydration into
the corresponding olefine, *e. g.*, $C_3H_8O - H_2O = C_3H_6$.

γ. Into o x y g e n - e t h e r s by the action of the sodium or potassium derivatives of the alcohols ; thus,

$$C_2H_5I \quad + \quad NaOC_2H_5 \quad = \quad NaI \quad + \quad C_2H_5.O.C_2H_5$$

| Ethyl iodide. | Sodium ethylate. | | Ethyl oxide. |

δ. Into c o m p o u n d e t h e r s by the silver salts of the corresponding acids :

$$C_2H_5I \quad + \quad AgC_2H_3O_2 \quad = \quad AgI \quad + \quad C_2H_5.C_2H_3O_2$$

| Ethyl iodide. | Silver acetate. | | Ethyl acetate. |

The individual ethers of this group will be described in connection with the corresponding alcohols.

Monatomic Haloïd Ethers, $C_nH_{2n-1}X$.

These ethers are derived from the olefines, C_nH_{2n}, by substitution of 1 atom of chlorine, etc., for hydrogen. They cannot, however, be formed by direct substitution, as the action of chlorine, bromine, or iodine on the olefines gives rise to addition-products, $C_nH_{2n}Cl_2$, etc. ; but they may be obtained by treating these addition-products with alcoholic potash or silver oxide ; thus,

$$C_2H_4Cl_2 \quad + \quad KOH \quad = \quad KCl \quad + \quad H_2O \quad + \quad C_2H_3Cl.$$

These ethers, like the olefines from which they are derived, can take up 2 atoms of chlorine, bromine, or iodine, forming compounds which can likewise give up hydrochloric, hydrobromic, or hydriodic acid, under the influence of alcoholic potash ; the body thus formed can take up 2 atoms of chlorine, bromine, or iodine, then give up HCl, HBr, or HI ; and thus, by a series of perfectly similar reactions, we at length arrive at bodies consisting of the primitive olefine with all its hydrogen replaced by chlorine, bromine, or iodine, and the dichlorides, dibromides, and di-iodides of these last-mentioned bodies : thus, from ethene may be derived the following series of brominated compounds :

Ethene	C_2H_4	Ethene bromide	$C_2H_4Br_2$
Bromethene	C_2H_3Br	Bromethene bromide	$C_2H_3Br.Br_2$
Dibromethene	$C_2H_2Br_2$	Dibromethene bromide	$C_2H_2Br_2.Br_2$
Tribromethene	C_2HBr_3	Tribromethene bromide	$C_2HBr_3.Br_2$
Tetrabromethene	C_2Br_4	Tetrabromethene bromide	$C_2Br_4.Br_2$

Further, a monochlorinated or monobrominated olefine may give up the atom of chlorine or bromine which it contains, in the form of hydrochloric or hydrobromic acid, whereby it is reduced to a hydrocarbon of the series C_nH_{2n-2}. This reaction may take place at 130°–150° C. (266°–302° F.), under the influence of alcoholic potash, or, better, of sodium ethylate (obtained by dissolving sodium in anhydrous alcohol) ; thus :

$$C_2H_3Br \quad + \quad NaOC_2H_5 \quad = \quad NaBr \quad + \quad C_2H_6O \quad + \quad C_2H_2$$

| Bromethene. | Sodium ethylate. | | Ethyl alcohol. | Acetylene. |

Chlorethene or *Vinyl Chloride*, $C_2H_3Cl = CH_2{=}CHCl$ (the univalent radicle $CH_2{=}CH-$ being called *vinyl*), is a gas having an alliaceous odor, and liquefying at 18° C. (64.4° F.).

Bromethene or *Vinyl Bromide*, C_2H_3Br, smells like the chloride, has a specific gravity of 1.52, and boils at 23° C. (73.4° F.). When kept in sealed tubes, it sometimes solidifies to a white amorphous mass, having the appearance of porcelain.

Iodethene or *Vinyl iodide*, C_2H_3I, has a specific gravity of 1.98, and boils at 55° C. (131° F.).

Propene yields three series of monatomic haloïd ethers: *e. g.*,

(1.)	(2.)	(3.)
CH_3	CH_3	CH_2Br
$\vert$	$\vert$	$\vert$
CH	CBr	CH
$\Vert$	$\Vert$	$\Vert$
$CHBr$	CH_2	CH_2

1. The only known ether of the first group is the **bromide**, which is a liquid boiling at 48○ C. (118.4○ F.), produced by addition of hydrobromic acid to allylene:

$$CH_3—C≡CH + HBr = CH_3—CH≔CHBr.$$

2. The ethers of the second group, called **chloropropene**, etc., are formed by the action of alcoholic potash on the chloride, etc., of propene, the action consisting in the abstraction of HCl, etc. (p. 543):

$$CH_3—CHCl—CH_2Cl - HCl = CH_3—CCl≔CH_2 ;$$

also from acetone chloride (p. 547);

$$CH_3—CCl_2—CH_3 - HCl = CH_3—CCl≔CH_2.$$

Chloropropene has a specific gravity of 0.918 at 9○ C. (48.2○ F.), and boils at 23○ C. (73.4○ F.). *Bromopropene* has a specific gravity of 1.411 at 15○ C. (59○ F.), and boils at 57.6○ C. (135.7○ F.). It unites directly with hydrogen bromide, forming **propene bromide**, $CH_3—CHBr—CH_2Br$, and **acetone bromide**, $CH_3—CBr_2—CH_3$.

3. The ethers of the third group, called **allyl ethers**, are analogous in constitution to allyl alcohol, $CH_2OH—CH≔CH_2$, and will be described in connection therewith.

Diatomic Haloïd Ethers, $C_nH_{2n}X_2$.

These ethers are formed: 1. From the paraffins, by direct substitution of Cl_2Br_2, etc., for H_2. 2. From the olefines, C_nH_{2n}, by direct addition of Cl_2, etc. 3. By the action of the chlorides, bromides, and iodides of phosphorus on aldehydes and ketones; thus,

C_2H_4O	$+$	PCl_5	$=$	$C_2H_4Cl_2$	$+$	$POCl_3$
Aldehyde.				Dichlorethane.		

C_3H_6O	$+$	PCl_5	$=$	$C_3H_6Cl_2$	$+$	$POCl_3$
Acetone.				Dichloro-propane.		

Methane Derivatives.—*Dichloromethane*, CH_2Cl_2, is formed by chlorination of methyl chloride, CH_3Cl; and by the action of chlorine on di-iodomethane or on methyl iodide:

$$CH_3I + Cl_2 = CH_2Cl_2 + HI.$$

It is a colorless liquid, having a specific gravity of 1.36 at 0○, and boiling at 40○ C. (104○ F.).

Di-iodomethane, CH_2I_2, is produced by heating trichloro- or tri-iodo-methane to 130○ C. (266○ F.) with fuming hydriodic acid:

$$CHCl_3 + 4HI = CH_2I_2 + I_2 + 3HCl.$$

It crystallizes in colorless shining laminæ of specific gravity 3.34 at 0°, melts at 6° C. (42.8 F.), and boils at 182° C. (359.6^O F.).

Ethane Derivatives, $C_2H_4X_2$.—These ethers admit of two modifications, accordingly as the two atoms of the chlorous radicle X are attached to different atoms, ór to the same atom of carbon; thus

$$\begin{array}{cc} CH_2X & CH_3 \\ | & | \\ CH_2X & CHX_2. \end{array}$$

The ethers of the first modification are formed by direct addition of chlorine, bromine, or iodine, to ethene; the double linking of the two car-bon-atoms in the molecule of free ethene being loosened by the entrance of the two chlorine-atoms into the group, and the ethene then becoming a bivalent radicle; thus,

$$\begin{array}{ccccc} CH_2 & & & & CH_2Cl \\ \| & + & Cl_2 & = & | \\ CH_2 & & & & CH_2Cl. \end{array}$$

These ethers are accordingly regarded as chloride, bromide, or iodide of ethene.

The ethers of the second modification are produced by the action of the perchloride, etc., of phosphorus on aldehyde, the reaction consisting in the replacement of the oxygen-atom of this compound by 2 atoms of chlorine, etc., *e. g.*,

$$\begin{array}{ccccccc} CH_3 & & & & CH_3 & & \\ | & + & PCl_3.Cl_2 & = & | & + & PCl_3O. \\ CHO & & & & CHCl_2 & & \end{array}$$

They may be supposed to contain the bivalent radicle e t h i d e n e, CH_3—CH⁼; but this radicle has not been isolated.

Ethene Chloride, $\begin{array}{c} CH_2Cl \\ | \\ CH_2Cl \end{array}$.—This compound has

Fig. 163.

long been known by the name of *Dutch liquid*, having been discovered by four Dutch chemists in 1795. When equal measures of ethene gas and chlorine are mixed over water, absorption of the mixture takes place, and a yellowish oily liquid is produced, which collects upon the surface of the water, and ultimately sinks to the bottom in drops. It may be easily prepared, in quantity, by causing the two gases to combine in a glass globe (fig. 163) having a narrow neck at the lower part, dipping into a small bottle, to receive the product. The two gases are conveyed by separate tubes, and allowed to mix in the globe, the ethene gas being kept a little in ex-cess. The chlorine should be washed with water, and the ethene passed through strong oil of vitriol, to remove vapor of ether: the presence of sulphu-rous and carbonic acids is not injurious. Combina-tion takes place very rapidly, and the liquid product trickles down the sides of the globe into the receiver. When a considerable quantity has been collected, it is agitated, first with water, and afterwards with concentrated sulphuric acid, and, lastly, purified by distillation.

Pure ethene chloride is a thin, colorless liquid, of fragrant odor and sweet taste : it is slightly soluble in water, and readily so in alcohol and ether. It has a specific gravity 1.271 at 0°, and boils at 85° C. (185° F.). It is unaffected by oil of vitriol or solid potassium hydrate. When inflamed it burns with a greenish, smoky light. When treated with an alcoholic solution of potash, it is slowly resolved into potassium chloride, which separates, and chlorethene, C_2H_3Cl, whose vapor requires to be cooled down to —18° C. (—0.4° F.) before it condenses ; at this temperature it forms a limpid, colorless liquid. Chlorine is absorbed by this latter substance, and a compound is produced, which contains $C_2H_3Cl_3$; and this is in turn decomposed by an alcoholic solution of potash into potassium chloride and another volatile liquid, $C_2H_2Cl_2$. This series of reactions is analogous to that already noticed in the case of the bromine compounds (p. 543).

Ethidene Chloride, CH_3—$CHCl_2$, is best prepared by the action of phosphoric chloride on aldehyde : it is also formed by the action of chlorine on ethyl chloride. It is a colorless liquid smelling like chloroform, having a specific gravity of 1.198, and boiling at 58°–59° C. (136.4–138.2° F.). When subjected to the action of chlorine, it yields the compounds $CH_3.CCl_3$ and $CH_2Cl.CHCl_2$, together with other products.

Ethene Bromide, CH_2Br—CH_2Br, obtained by saturating bromine with olefiant gas, is an oily fragrant liquid, of specific gravity 2.163 at 21° C. (69.8° F.), boiling at 129° C. (264.2° F.), and solidifying at 0° to a crystalline mass which melts at 9° C. (48.2° F.).

Ethidene Bromide, CH_3—$CHBr_2$, obtained similarly to the chloride, is a liquid boiling at 110–114° C. (230–237.2° F.), and decomposing when heated with water.

These two bromides are also formed by heating monobromethene, CH_2—$CHBr$, with hydrobromic acid, the ethene compound when the acid is highly concentrated, the ethidene compound when it is more dilute.

Ethene Iodide, CH_2I—CH_2I, is produced by the direct combination of iodine with ethene in sunshine or under the influence of heat, also by passing ethene gas into an alcoholic solution of iodine. It crystallizes from alcohol in shining needles, melts at 73° C. (163.4° F.), and decomposes at higher temperatures into ethene and iodine.

Ethidene Iodide, CH_3I—CHI_2, is obtained by heating the chloride with aluminium iodide :—

$$3C_2H_4Cl_2 + 2AlI_3 = 2AlCl_3 + 3C_2H_4I_2 ;$$

also by addition of hydriodic acid to acetylene : $C_2H_2 + 2HI = C_2H_4I_2$. It is a liquid having a specific gravity of 2.84 at 0°, and boiling at 178° C. (352.4° F.).

Propane Derivatives, $C_3H_6X_2$.—The diatomic ethers derived from propane admit of four modifications, as exhibited by the following formulæ of the bromide :—

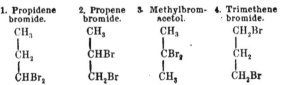

1. Propidene bromide.	2. Propene bromide.	3. Methylbrom- acetol.	4. Trimethene bromide.
CH_3	CH_3	CH_3	CH_2Br
CH_2	$CHBr$	CBr_2	CH_2
$CHBr_2$	CH_2Br	CH_3	CH_2Br

1. The p r o p i d e n e or p r o p y l i d e n e e t h e r s are formed by the action of phosphorus pentachloride, etc., on propylic aldehyde, CH_3—CH_2—CHO. The *chloride* is a liquid having an alliaceous odor, and boiling at 84-87° C. (183.2-188.6° F.). The bromide and iodide are not known.

2. The p r o p e n e or p r o p y l e n e e t h e r s are formed by direct combination of propene with chlorine, bromine, and iodine. When treated with silver oxide and water they are converted into propene alcohol or propylene glycol, CH_3—CHOH—CH_2OH.

The *chloride*, CH_3—CHCl—CH_2Cl, is also formed by the action of chlorine, in sunshine and in presence of iodine, on pseudopropyl chloride, CH_3—CHCl—CH_3. If iodine is not present, acetone chloride is formed at the same time. Propene chloride has a specific gravity of 1.165 at 14° C. (57.2° F.), and boils at 98° C. (208.4° F.). The *bromide*, formed by direct combination or by the action of bromine on pseudopropyl bromide, is a liquid of specific gravity 1.946 at 17° C. (62.6° F.), and boiling at 142° C. (287.6° F.). The *iodide*, produced by direct combination of propene with iodine at 50° C. (122° F.), is a colorless oil not distillable without decomposition.

The chloride or bromide of propene heated with water at 200° C. (392° F.) is converted into propionic aldehyde and acetone :—

$$2 \begin{matrix} CH_3 \\ | \\ CHCl \\ | \\ CH_2Cl \end{matrix} + 2H_2O = \begin{matrix} CH_3 \\ | \\ CH_2 \\ | \\ CHO \end{matrix} + \begin{matrix} CH_3 \\ | \\ CO \\ | \\ CH_3 \end{matrix} + 4HCl$$

Propene chloride.   Propionic aldehyde.   Acetone.

3. The propene derivatives of the form CH_3—CX_2—CH_3, are produced from acetone by substitution of Cl_2 or Br_2 for O, the substitution being effected by the action of PCl_5 or PBr_5.

Acetone chloride, or *Methylchloracetol*, CH_3—CCl_2—CH_3, is also produced, together with propene chloride, by addition of HCl to allylene, C_3H_4, thus:

$$2 \begin{matrix} CH_3 \\ | \\ C \\ ||| \\ CH \end{matrix} + 4HCl = \begin{matrix} CH_3 \\ | \\ CCl_2 \\ | \\ CH_3 \end{matrix} + \begin{matrix} CH_3 \\ | \\ CHCl \\ | \\ CH_2Cl \end{matrix} ;$$

and by chlorination of pseudopropyl chloride. It is a colorless liquid of specific gravity 1.827 at 16° C. (60.8° F.), and boiling at 70° C. (158° F.). Sodium reduces it to ordinary propene. With alcoholic potash it gives chloropropene, CH_3—$CCl\overline{}CH_2$, which is also formed in the same manner from propene dichloride. Heated to 150° C. (302° F.) with water, it is partially converted into acetone.

4. T r i m e t h e n e B r o m i d e, CH_2Br—CH_2—CH_2Br, is formed by heating allyl alcohol, CH_2—CH—CH_2OH, with hydrobromic acid ; also, together with propene bromide, by heating allyl bromide to 100° with hydrobromic acid. It is a colorless liquid, of specific gravity 1.93 at 19° C. (66.2° F.), and boiling at 162° C. (323.6° F.); converted by alcoholic potash into allyl bromide and allyl-ethyl oxide, CH_2—CH—CH_3—O—C_2H_5.

T r i m e t h e n e c h l o r i d e, CH_2Cl—CH_2—CH_2Cl, formed by heating the bromide to 169° C. (336.2° F.) with mercuric chloride, is a fragrant liquid having a specific gravity of 1.201 at 15° C. (59° F.), boiling at 117° C. 242.6° F.); converted by alcoholic potash into allyl-ethyl oxide.

Triatomic Haloïd Ethers, $C_nH_{2n-1}X_3$.

Methenyl Chloride or **Chloroform,** $CHCl_3$.—This compound is produced: 1. Together with methane chloride, CH_2Cl_2, when a mixture of chlorine and gaseous methyl chloride is exposed to the sun's rays. 2. By the action of alkalies on chloral (trichloraldehyde):—

$$C_2HCl_3O + KOH = CHCl_3 + CHKO_2$$
Chloral.　　　　　　　　　Chloroform.　Potassium
　　　　　　　　　　　　　　　　　　　　formate.

3. By boiling trichloracetic acid with aqueous alkalies:—

$$C_2HCl_3O_2 + 2KOH = CHCl_3 + K_2CO_3 + H_2O.$$

4. By the action of nascent hydrogen on carbon tetrachloride:—

$$CCl_4 + H_2 = HCl + CHCl_3.$$

5. By the action of hypochlorites, or of chlorine in presence of alkalies, on various organic substances, as methyl-, ethyl-, and amyl-alcohols, acetic acid, acetone, etc. The reaction is complicated, giving rise to several other products ; with common alcohol and calcium hypochlorite the principal reaction appears to be—

$$2C_2H_6O + 5CaCl_2O_2 = 2CHCl_3 + 2CaCO_3 + 2CaCl_2 + CaH_2O_2 + 4H_2O.$$

Chloroform is prepared on the large scale by cautiously distilling together good commercial chloride of lime, water, and alcohol. The whole product distils over with the first portions of water, so that the operation may be soon interrupted with advantage. The chloroform, which constitutes the oily portion of the distillate, is purified by agitation with water, desiccation with calcium chloride, and distillation in a water-bath.

Chloroform is a thin colorless liquid of agreeable ethereal odor, much resembling that of Dutch liquid, and of a sweetish taste. Its density is 1.525 at 0°, and it boils at 61° C. (141.8 F.): the density of its vapor (compared with air) is 4.20. Chloroform is difficult to kindle, and burns with a greenish flame. It is nearly insoluble in water, and is not affected by concentrated sulphuric acid. When boiled with aqueous potash in a closed tube, it is converted into potassium chloride and formate :—

$$CHCl_3 + 4HOK = 3KCl + CHKO_2 + 2H_2O.$$

Chloroform is well known for its remarkable effects upon the animal system n producing temporary insensibility to pain when its vapor is inhaled.

Bromoform, $CHBr_3$, is a heavy, volatile liquid, prepared by the simultaneous action of bromine and aqueous alkalies on alcohol, wood-spirit, and acetone. It has a density of 2.9 at 12° C. (53.6^o F.), solidifies at $—9^o$ C. (15.8^o F.), and boils at $150–152^o$ C. ($302–305.6^o$ F.). It is converted by caustic potash into potassium bromide and formate.

Iodoform, CHI_3, is a solid, yellow, crystallizable substance, easily obtained by adding alcoholic solution of potash to tincture of iodine, avoiding excess, evaporating the whole to dryness, and treating the residue with water. It melts at 119° C. (246.2^o F.), and distils with vapor of water. It is nearly insoluble in water, but dissolves in alcohol, and is decomposed by alkalies in the same manner as the preceding compounds. Bromine converts it into b r o m i o d o f o r m, $CHBr_2I$, a colorless liquid which solidifies at 0°. Iodoform distilled with phosphorus pentachloride or mercuric chloride, is converted into chloriodoform, $CHCl_2I$, a colorless liquid of specific gravity 1.96, which does not solidify at any temperature.

Trichlorethane, $C_2H_3Cl_3$, admits of two modifications, viz. : (1) CH_3—CCl_3, formed by the action of chlorine in sunshine on ethyl chloride. It is a liquid smelling like chloroform, boiling at 74.5○ C. (167.9○ F.), and converted by alcoholic potash into acetic acid :—

$$CH_3.CCl_3 + 4KOH = CH_3.CO.OK + 3KCl + 2H_2O.$$

(2) CH_2Cl—$CHCl_2$, obtained by the action of chlorine on vinyl chloride, $H_2C{=}CHCl$ (see VINYL ALCOHOL). It boils at 115○ C. (239○ F.), and has a specific gravity of 1.422 at 0○. Alcoholic potash solution converts it into dichlorethene, $C_2H_2Cl_2$.

Tribromethane, $C_2H_3Br_3$, is obtained by adding bromine to bromethene, C_2H_3Br, cooled by a freezing mixture. It is a colorless liquid, which smells like chloroform, has a specific gravity of 2.620 at 23○ C. (73.4○ F.), and boils at 186.5○ C. (367.7○ F.). Alcoholic potash abstracts HBr, converting it into dibromethene, $C_2H_2Br_2$.

Trichloropropane, $C_3H_5Cl_3$, admits of four modifications, represented by the following formulæ :—

(1) CH_3—CH_2—CCl_3.   (3) CH_3—CCl_2—CH_2Cl.
(2) CH_3—$CHCl$—$CHCl_2$.   (4) CH_2Cl—$CHCl$—CH_2Cl.

The last of these, which is the most important, is analogous in composition to glycerin, $C_3H_5(OH)_3$, and will be described, together with the corresponding bromine- and iodine-derivatives, in connection with that compound.

Haloïd Ethers of Higher Orders.

Tetrachloromethane, or **Carbon Tetrachloride,** CCl_4, is formed by passing the vapor of carbon bisulphide, together with chlorine, through a red-hot porcelain tube. A mixture of sulphur chloride and carbon tetrachloride is formed, which is distilled with potash, whereby the chloride of sulphur is decomposed, and pure carbon tetrachloride passes over. It is a colorless liquid of 1.56 specific gravity, and boils at 77○ C. (170.6○F.). The same compound is formed by exhausting the action of chlorine upon marsh-gas or methyl chloride in sunshine. An alcoholic solution of potash converts it into a mixture of potassium chloride and carbonate.

Tetrabromomethane, or **Carbon Tetrabromide,** CBr_4, is formed by the action of bromine, in presence of iodine bromide or antimonious bromide, on carbon sulphide, bromopicrin, bromoform, or chloroform. It is a white crystalline substance, having a specific gravity of 3.42 at 14○ C. (57.2○ F.), melts at 91○ C. (195.8 F.), and boils with slight decomposition at 189.5○ C. (373.1○ F.). Heated to 100○ with alcohol in a sealed tube, it yields hydrogen bromide, aldehyde, and bromoform :—

$$CB_4 + C_2H_6O = HBr + C_2H_4O + CHBr_3.$$

Tetra-iodomethane, or **Carbon Tetra-iodide,** CI_4, is produced by heating tetrachlorethane with iodide of aluminium. It crystallizes from ether in dark-red regular octohedrons of specific gravity 4.32 at 20° C. (68° F.), and decomposes on exposure to the air, especially at high temper atures, yielding iodine and carbon dioxide.

Tetrabromethane, $C_2H_2Br_4$, formed by direct combination of bromine with dibromethane, $C_2H_2Br_2$, or with acetylene, C_2H_2, is a liquid having a specific gravity of 2.88 at 22° C. (71.6° F.), boiling at 200° C. (392° F.), and solidifying in a freezing mixture to a white crystalline mass. By heating with alcoholic potash it is converted into tribromethene, $CHBr_3$.

Tetrabromethene, or **Carbon Dibromide,** C_2Br_4, is formed by the action of alcoholic potash on pentabromethane :—

$$C_2HBr_5 + KOH = C_2Br_4 + KBr + H_2O,$$

or by treating common alcohol or ether with bromine :—

$$\underset{\text{Alcohol.}}{C_2H_6O} + 8Br = C_2Br_4 + 4HBr + H_2O$$

$$\underset{\text{Ether.}}{C_4H_{10}O} + 16Br = 2C_2Br_4 + 8HBr + H_2O.$$

It crystallizes in white plates, melts at 50° C. (120.2° F.), and sublimes without alteration.

Tetrachlorethene, or **Carbon Dichloride,** C_2Cl_4, obtained by passing the vapor of the trichloride or tetrachloride through a red-hot tube, either alone or mixed with hydrogen, or by the action of nascent hydrogen (zinc and dilute sulphuric acid) on the trichloride, is a mobile liquid of specific gravity 1.629, boiling at 117° C. (242.6° F.). When heated to 200° C. (392° F.) with potassium hydroxide, it is completely converted into potassium chloride and oxalate, with evolution of hydrogen :

$$C_2Cl_4 + 6KOH = 4KCl + C_2K_2O_4 + 2H_2O + H_2.$$

It absorbs chlorine and bromine in sunshine, forming in the one case the trichloride, C_2Cl_6, and in the other the chlorobromide, $C_2Cl_4Br_2$.

Pentabromethene, C_2HBr_5, obtained by slowly distilling tribromethene, C_2HBr_3, with bromine, or by the action of bromine on acetylene, crystallizes from alcohol in prisms smelling like camphor, melting at 45–50° C. (120.2–122° F.), and decomposed by distillation.

Hexbromethene, or **Carbon Tribromide,** C_2Br_6, is formed by heating the dibromide with bromine in a sealed tube, or by heating either of the compounds, $C_2H_2Br_4$, C_2HBr_5, with bromine in a sealed tube :

$$C_2H_2Br_4 + Br_4 = 2HBr + C_2Br_6$$
$$C_2HBr_5 + Br_2 = HBr + C_2Br_6.$$

This compound dissolves sparingly in alcohol or ether, easily in carbon sulphide, from which it separates on evaporation in hard, rather thick rectangular prisms. At 200–210° C. (392–410° F.) it melts and decomposes, yielding carbon dibromide and free bromine.

Hexchlorethane, or **Carbon Trichloride,** $C_2Cl_6 = CCl_3—CCl_3$, is the final product of the action of chlorine in sunshine on ethyl chloride, C_2H_5Cl, or ethene chloride, $C_2H_4Cl_2$. It is a white, crystalline substance, of aromatic odor, insoluble in water, but easily dissolved by alcohol and ether : it melts at 160° C. (320° F.), and boils at 182° C. (359.6° F.). It burns with difficulty, and is not altered by distillation with aqueous or alcoholic potash. Its vapor, passed through a red-hot porcelain tube filled with fragments of glass or rock-crystal, is decomposed into free chlorine, and the dichloride, C_2Cl_4.

Tetrachlorodibromethane, or **Carbon Chlorobromide,** $C_2Cl_4Br_2$, formed by the action of bromine in sunshine on the dichloride, C_2Cl_4, is a white crystalline body resembling the trichloride.

Carbon monochloride, C_2Cl_2, analogous to ethine or acetylene, is obtained by passing the vapor of chloroform or of carbon-dichloride through a red-hot tube. It forms white needles, subliming between 175° and 200° C. (347–392° F.).

NITROPARAFFINS, $C_nH_{2n+1}(NO_2)$.

These compounds, isomeric with the nitrous ethers of the fatty series, are produced, together with the latter, by the action of silver nitrite on the iodoparaffins or alcoholic iodides,—nitromethane, for example, from methyl iodide :

$$H_3CI + AgNO_2 = AgI + H_3CNO_2.$$

They are also formed by the action of nitrogen tetroxide, or of fuming nitric acid, on the hydrocarbons of the ethene series ;

$$\underset{\text{Ethene.}}{C_2H_4} + N_2O_4 = \underset{\text{Dinitroethane.}}{C_2H_4(NO_2)_2}$$

They have their nitrogen-atom in direct union with a carbon-atom, and are consequently converted by nascent hydrogen (evolved from water by sodium-amalgam) into amidoparaffins or amines ; e. g.,

$$\underset{\text{Nitromethane.}}{C\begin{Bmatrix}H_3\\NO_2\end{Bmatrix}} + H_6 = 2H_2O + \underset{\text{Methylamine.}}{C\begin{Bmatrix}H_3\\NH_2\end{Bmatrix}}$$

The nitrous ethers, on the other hand, which are formed by the action of nitrous acid on the corresponding alcohols, have their nitrogen united with carbon, only through the medium of oxygen ; e. g.,

$$\underset{\substack{\text{Methyl}\\\text{alcohol.}}}{O\begin{Bmatrix}CH_3\\H\end{Bmatrix}} + \underset{\substack{\text{Nitrous}\\\text{acid.}}}{O\begin{Bmatrix}NO\\H\end{Bmatrix}} = H_2O + \underset{\substack{\text{Methyl}\\\text{nitrite.}}}{O\begin{Bmatrix}CH_3\\NO\end{Bmatrix}}$$

and are accordingly converted by nascent hydrogen into ammonia and the corresponding alcohols ; e. g.,

$$\underset{\text{Methyl nitrite.}}{O\begin{Bmatrix}CH_3\\NO\end{Bmatrix}} + H_6 = H_2O + NH_3 + \underset{\text{Methyl alcohol.}}{O\begin{Bmatrix}CH_3\\H\end{Bmatrix}}$$

The nitroparaffins are mostly decomposed, with explosion, when rapidly heated. They are not decomposed by potash or soda, whereas the nitrous ethers are quickly resolved thereby into nitrous acid and the corresponding alcohols.

Nitromethane, $CH_3.NO_2$, is formed, as above stated, by the action of silver nitrite on methyl iodide. The action is very violent, and the whole of the methyl iodide is converted into nitromethane, without a trace of the isomeric methyl nitrite. It is also formed by heating the potassium salt of monochloracetic acid with potassium nitrite:

$$2(CH_2Cl.CO_2K) \ + \ 2NO_2K \ + \ H_2O \ = \ 2(CH_3.NO_2)$$
$$+CO_3K_2 \ + \ 2KCl \ + \ CO_2$$

Nitromethane is a heavy oil, having a peculiar odor, and boiling at 99–101°. When treated with an alcoholic solution of soda, it yields transparent needles of the sodium derivative, $CH_2Na(NO_2)$, the aqueous solution of which gives characteristic precipitates with various metallic solutions.

Bromonitromethane, $CH_2Br.NO_2$, formed by the action of bromine on the dry sodium compound, is a heavy, pungent, strongly refracting liquid, boiling at 143–144° C. (289–291.2° F.).

Nitroethane, $C_2H_5NO_2 = CH_3—CH_2—NO_2$, prepared like the methane-derivative, is a colorless, strongly refracting liquid, having a fragrant ethereal odor, a specific gravity = 1.0582 at 13° C. (55.4° F.), and boiling at 111–113° C. (231.8–235.4° F.) : ethyl nitrite boils at 16° C. (60.8° F.). The vapor is inflammable, and burns with a pale yellow flame, but does not explode, even when heated above its boiling point. By nascent hydrogen it is converted into amidethane or ethylamine, $C_2H_5.NH_2$.

Bromonitroethane, $C_2H_4Br.NO_2$, and *dibromonitroethane,* $C_2H_3Br_2.(NO_2)$, are formed by the action of bromine on sodium-nitroethane. The former is a heavy acid oil, boiling at 145–148° C. (293–298.4° F.), the latter a mobile neutral liquid, boiling at 162–164° C. (323.6–327.2° F.).

Nitropropanes, $C_3H_7NO_2$.—There are two isomeric bodies having this composition, obtained by the action of silver nitrite on the iodides of normal propyl and pseudopropyl respectively.

Nitropropane, $CH_3—CH_2—CH_2—NO_2$, is a limpid mobile liquid, very little heavier than water, and boiling at 122–127° C. (251.6–260.6° F.).

Pseudo-nitropropane, $\frac{H_3C}{H_3C}{>}CH—NO_2$, boils at 112–117° C. (233.6–242.6° F.). Both form crystalline sodium compounds.

Nitro-isobutane, $C_4H_9NO_2 = \frac{H_3C}{H_3C}{>}CH—CH_2—NO_2$, from isobutyl iodide, is an oil smelling like peppermint, boiling at 135–140° C. (275–284° F.) and converted by nascent hydrogen into isobutylamine.

Nitro-isopentane, $C_5H_{11}NO_2 = \frac{H_3C}{H_3C}{>}CH—CH_2—CH_2—NO_2$, formed by the action of silver nitrite on ordinary amyl iodide, boils at 155–160° C. (311–320° F.).

Reactions of Nitroparaffins with Metallic Salts.—The following table exhibits the reactions of various metallic salts with the sodium compounds of nitromethane, nitroethane, and the two nitropropanes, which are sufficiently characteristic to be used for distinguishing these compounds one from the other.

	Sodium-derivative of			
	Nitromethane.	Nitro-ethane.	Nitropropane.	Pseudonitro-propane.
Mercuric chloride.	Light yellow precipitate; explosive.	White crystalline precipitate.	White crystalline precipitate.	White crystalline precipitate.
Mercurous nitrate.	Black flocculent precipitate.	Dirty-gray precipitate.	Black flocculent precipitate.	Black flocculent precipitate.
Ferric chloride.	Dark brownish-red precipitate.	Blood-red coloration.	Blood-red solution.	Blood-red solution.
Cupric sulphate.	Grass-green precipitate.	Deep-green solution.	Deep-green solution.	Deep-green solution.
Lead acetate.	White precipitate.	0	White precipitate.	0
Silver nitrate.	White precipitate, almost immediately turning black.	White precipitate, soon turning brown.	White precipitate, gradually turning brown.	Light yellow precipitate, soon turning black.

Action of Nitrous Acid on the Nitroparaffins.—The reactions of these bodies with nitrous acid (or better, with potassium nitrite and sulphuric acid) differ considerably, according as they contain primary, secondary, or tertiary alcohol-radicles (p. 524).

On treating the nitro-compounds of the p r i m a r y alcohol radicles, in which the NO_2-group is united to the group CH_2, with potassium nitrite and dilute sulphuric acid, the solution acquires a deep-red color, and compounds called n i t r o l i c a c i d s , having the general formula—

$$C_nH_{2n+1}-C\begin{smallmatrix}NOH\\NO_2\end{smallmatrix}$$

are produced by substitution of nitrosyl, NO, for one of the hydrogen-atoms in the group, CH_2; thus—

Nitropropane,
$C_2H_5-CH_2-NO_2$ } gives { Propyl-nitrolic acid,
$C_2H_5-C\begin{smallmatrix}N-OH\\NO_2\end{smallmatrix}$

The nitrolic acids are colorless, crystalline bodies, soluble in ether. Their alkali-salts have a deep-red color : hence the red tint produced at the commencement of the reaction above mentioned, and disappearing when the sulphuric acid is added in excess.

The nitro-compounds of the s e c o n d a r y alcohol-radicles, in which the NO_2 is linked to the group CH, give, when similarly treated, deep-blue solutions, which after a while deposit colorless compounds called p s e u d o - n i t r o l s ; e. g.—

Isonitropropane,
$\begin{smallmatrix}H_3C\\H_3C\end{smallmatrix}>CH-NO_2$ } gives { Isopropyl-pseudonitrol,
$\begin{smallmatrix}H_3C\\H_3C\end{smallmatrix}> C <\begin{smallmatrix}NO\\NO_2\end{smallmatrix}$

These compounds are colorless in the solid state, dark-blue in the fused state or in solution.

24

The nitro-compounds of the t e r t i a r y alcohol-radicles, in which the NO_2-group is associated only with carbon : *e. g.*, $(CH_3)_3C.NO_2$, are neither decomposed nor colored by nitrous acid.

These reactions afford a characteristic and delicate test for distinguishing the primary, secondary, and tertiary alcohol-radicles (in their iodides) from one another.

Nitroparaffins of Higher Orders.

Trinitromethane, or **Nitroform,** $CH(NO_2)_3$, is produced by the action of nitric acid on various organic compounds, but is most conveniently obtained as an ammonium salt, $C(NO_2)_3NH_4$, by the action of water or alcohol on trinitracetonitril or cyanotrinitro-methane : $C(NO_2)_3CN + 2H_2O = C(NO_2)_3NH_4 + CO_2$; and on treating the yellow crystalline salt thus formed with strong sulphuric acid, nitroform is obtained in the free state.

Nitroform, at temperatures above 15° C. (59° F.), is a colorless oil ; below that temperature it solidifies in colorless cubic crystals. It is moderately soluble in water, forming a dark yellow solution. It cannot be distilled, as it explodes with violence when heated.

The atom of hydrogen in nitroform may be replaced either by metals or by chlorous radicles, namely, bromine and nitryl. The metallic derivatives or *salts of nitroform*, are for the most part yellow and crystallizable : they explode when heated.

Bromonitroform, $C(NO_2)_3Br$, produced by exposing nitroform to the action of bromine for some days under the influence of direct sunshine, or more easily by treating an aqueous solution of mercuric nitroform, $C_2(NO_2)_6Hg$, with bromine, is a colorless liquid above 12° C. (53.6° F.), but solidifies below that temperature to a white crystalline mass. It is somewhat soluble in water, and may be distilled with aqueous vapor, or in a current of air. It decomposes at 140° C. (284° F.).

Tetranitromethane, $C(NO_2)_4$, is produced by treating nitroform with fuming nitric and sulphuric acids, heating the liquid to 100°, and passing air through it. A liquid then distils over, from which water throws down tetranitromethane as a heavy oil.

Tetranitromethane is liquid at ordinary temperatures, but solidifies at 13° C. (55.4° F.) to a white crystalline mass. It is insoluble in water, but soluble in alcohol and ether. It boils at 126° C. (258.8° F.), and unlike nitroform, may be distilled without decomposition. When quickly heated it decomposes, with evolution of nitrous vapors, but without explosion. It does not take fire by contact with flame, but a glowing coal on which it is poured burns with a bright light.

Nitrotrichloromethane, Nitrochloroform or **Chloropicrin,** $C(NO_2)Cl_3$, is produced by the action of nitric acid on various chlorinated organic compounds—chloral, for example ; also by that of chlorine or hypochlorites on nitro-compounds, such as fulminating silver, $C_2N_2O_2Ag_2$, and picric acid, $C_6H_3(NO_2)_3O$. To prepare it, 10 parts of freshly prepared bleaching powder, made into a thick paste with water, are introduced into a retort, and a saturated solution of 1 part of picric acid heated to 30° C (86° F.) is added. Reaction then generally takes place, without further heating, and the chloropicrin distils over with vapor of water.

Chloropicrin is a colorless liquid, having a specific gravity of 1.665, and boiling at 112° C. (233.6° F.). It has a very pungent odor, and explodes when suddenly heated. By the action of nascent hydrogen (evolved by the action of iron on acetic acid) it is reduced to a m i d o m e t h a n e or m e t h y l a m i n e :

$$CCl_3.NO_2 + 12H = CH_3.NH_2 + 3HCl + 2H_2O.$$

Bromopicrin, $C(NO_2)Br_3$, is formed in like manner by heating picric acid with calcium hypobromite (slaked lime and bromine). It closely resembles chloropicrin, solidifies below 10° C. (50° F.); and may be distilled in a vacuum without decomposition.

Dinitrodichloromethane, $C(NO_2)_2Cl_2$ (commonly called *Marignac's oil*), obtained by distilling chloronaphthalene with nitric acid, is very much like chloropicrin.

AMIDOPARAFFINS.

These are derivatives formed by the replacement of 1, 2, or more hydrogen-atoms in a paraffin by the monatomic radicle amidogen, NH_2; *e. g.*,

$$Amido\text{-}ethane, \quad C_2H_5NH_2$$
$$Diamido\text{-}ethane, \quad C_2H_4(NH_2)_2.$$

They are volatile and strongly basic compounds, exhibiting in their behavior with acids and other bodies the closest resemblance to ammonia : they may indeed be regarded as derived from one or more molecules of ammonia by the substitution of alcohol-radicles, mono- or polyatomic, for equivalent quantities of hydrogen : hence they are called a m i n e s ; *e. g.*,

$$Amidoethane = N \begin{cases} C_2H_5 \\ H \\ H \end{cases} \text{Ethylamine.}$$

$$Diamidoethane = N_2 \begin{cases} (C_2H_4)'' \\ H_4 \end{cases} \text{Ethene-diamine.}$$

The mono-derivatives are formed by the action of reducing agents on the mono-nitroparaffins : nitroethane, for example, treated with iron and acetic acid, is converted into ethylamine :—

$$C_2H_5(NO_2) + H_6 = 2H_2O + C_2H_5NH_2.$$

But they are more readily prepared by other methods, which, together with the properties and reactions of these bodies, will be discussed in a future chapter.

AZOPARAFFINS, NITRILS, or CYANIDES,

$$C_nH_{2n-1}N.$$

These compounds may be regarded as derived from the paraffins by substitution of trivalent nitrogen for 3 atoms of hydrogen, or as compounds of nitrogen with trivalent radicles, C_nH_{2n-1} (nitrils), or as compounds of the univalent radicle, $N\equiv C-$ (cyanogen) with a univalent radicle $C_{n-1}H_{2n-1}$ or C_pH_{2p+1};* *e. g.*,

* If $n-1=p$, $2n-1=2p+1$.

Azomethane, Methenyl Nitril, or Hydrogen Cyanide, $N{\equiv}C{-}H.$
Azoethane, Ethenyl Nitril, or Methyl Cyanide, $N{\equiv}C{-}CH_3.$
Azopropane, Propenyl Nitril, or Ethyl Cyanide, $N{\equiv}C{-}C_2H_5.$

There are also cyanides derived from hydrocarbons belonging to the aromatic group, *e. g.*, p h e n y l c y a n i d e, $C_7H_5N{=}C_6H_5.CN$, from toluene, C_7H_8.

C y a n o g e n is obtained in the free state, as already described (p.235) by the action of heat on mercuric cyanide. It is a colorless, inflammable gas, of specific gravity 26 (H=1.) Its molecular weight is therefore 52, and it is represented by the formula,

$$C_2N_2 = N{\equiv}C{-}C{\equiv}N,$$

which gives $2 \times 12 + 2 \times 14 = 52$.

The univalent radicle CN is often represented by the shorter formula Cy.

In the preparation of cyanogen from mercuric cyanide, a brownish or blackish substance, called p a r a c y a n o g e n, is always formed in small quantity. It is insoluble in water, and when calcined in a gas which does not act upon it, such as nitrogen or carbon dioxide, is completely volatilized in the form of cyanogen gas, without leaving any carbonaceous residue : hence it appears to be isomeric or polymeric with cyanogen.

H y d r o g e n C y a n i d e; H y d r o c y a n i c or P r u s s i c A c i d, HCN.—This very important compound, so remarkable for its poisonous properties, was discovered as early as 1782 by Scheele.

Pure anhydrous hydrocyanic acid may be obtained by passing dry sulphuretted hydrogen gas over dry mercuric cyanide, gently heated in a glass tube connected with a small receiver cooled by a freezing mixture. It is a thin, colorless, and exceedingly volatile liquid, which has a density of 0.7058 at 7.2° C. (45° F.), boils at 26.1° C. (80° F.), and solidifies when cooled to —18° C. (0.4° F.) : its odor is very powerful and most characteristic, much resembling that of peach-blossoms or bitter almond-oil ; it has a very feeble acid reaction, and mixes with water and alcohol in all proportions. In the anhydrous state this substance constitutes one of the most formidable poisons known, and even when largely diluted with water, its effects upon the animal system are exceedingly energetic: it is employed, however, in medicine, in very small doses. The inhalation of the vapor should be carefully avoided in all experiments in which hydrocyanic acid is concerned, as it produces headache, giddiness, and other disagreeable symptoms ; ammonia and chlorine are the best antidotes.

The acid in its pure form can scarcely be preserved ; even when enclosed in a carefully stoppered bottle, it is observed after a very short time to darken, and eventually to deposit a black substance containing carbon, nitrogen, and perhaps hydrogen : ammonia is formed at the same time, and many other products. Light favors this decomposition. Even in the dilute state it is apt to decompose, becoming brown and turbid, but not always with the same facility, some samples resisting change for a great length of time, and then solidifying in a few weeks to a brown pasty mass.

When hydrocyanic acid is mixed with concentrated mineral acids, hydrochloric acid, for example, the whole solidifies to a crystalline paste of sal· ammoniac and formic acid :—

$$CNH + 2H_2O = NH_3 + CH_2O_2.$$

On the other hand, when dry ammonium formate is heated to 200° C. (392° F.), it is almost entirely converted into hydrocyanic acid and water.

Aqueous solution of hydrocyanic acid may be prepared by various means. The most economical, and by far the best, where considerable quantities are wanted, is to decompose yellow potassium ferrocyanide at boiling heat with dilute sulphuric acid. 500 grains of the powdered ferrocyanide, K_4FeCy_6, are dissolved in four or five ounces of warm water, and introduced into a capacious flask or globe, connected by a perforated cork and wide bent tube with a Liebig's condenser well supplied with cold water; 300 grains of oil of vitriol are diluted with three or four times as much water and added to the contents of the flask ; and the distillation is carried on till about half the liquid has distilled over, after which the process may be interrupted. The residue is in the retort is a white or yellow mass, consisting of potassio-ferrous ferrocyanide, $K_2Fe_2Cy_6$ (see p. 562), mixed with potassium sulphate.

$$2K_4Fe''Cy_6 + 3H_2SO_4 = 6HCy + K_2Fe''_2Cy_6 + 3K_2SO_4.$$

When hydrocyanic acid is wanted for the purposes of pharmacy, it is best to prepare a strong solution in the manner above described, and then, having ascertained its exact strength, to dilute it with pure water to the standard of the Pharmacopœia, viz., 2 per cent. of real acid. This examination is best made by precipitating with excess of silver nitrate a known weight of the acid to be tried, collecting the insoluble silver cyanide upon a small filter previously weighed, then washing, drying, and lastly, re-weighing the whole. From the weight of the cyanide that of the hydrocyanic acid can be easily calculated, a molecule of the one ($CNAg = 134$), corresponding with a molecule of the other ($CNH = 27$); or the weight of the silver cyanide may be divided by 5, which will give a close approximation to the truth.

Another very good method for determining the amount of hydrocyanic acid in a liquid has been suggested by Liebig. It is based upon the property possessed by potassium cyanide of dissolving a quantity of silver cyanide sufficient to produce with it a double cyanide, $KCy.AgCy$. Hence a solution of hydrocyanic acid, which is supersaturated with potash, and mixed with a few drops of solution of common salt, will not yield a permanent precipitate with silver nitrate before the whole of the hydrocyanic acid is converted into the above double salt. If we know the amount of silver in a given volume of the nitrate solution, it is easy to calculate the quantity of hydrocyanic acid ; for this quantity will stand to the amount of silver in the nitrate consumed, as 2 molecules of hydrocyanic acid to 1 atom of silver, i. e.:

$$108 : 54 = \text{silver consumed} : x.$$

It is a common remark, that the hydrocyanic acid made from potassium ferrocyanide keeps better than that made by other means. The cause of this is ascribed to the presence of a trace of mineral acid. Everitt found that a few drops of hydrochloric acid, added to a large bulk of the pure dilute acid, preserved it from decomposition, while another portion, not so treated, became completely spoiled.

A very convenient process for the extemporaneous preparation of an acid of definite strength, is to decompose a known quantity of potassium cyanide with solution of tartaric acid : 100 grains of crystallized tartaric acid in powder, 44 grains of potassium cyanide, and 2 measured ounces of distilled water, shaken up in a phial for a few seconds, and then left at rest, in order that the precipitate may subside, will yield an acid of very nearly the required strength. A little alcohol may be added to complete the separation of the cream of tartar ; no filtration or other treatment need be employed.

Bitter almonds, the kernels of plums and peaches, the seeds of the apple, the leaves of the cherry-laurel, and various other parts of plants belonging to the great natural order *Rosaceæ*, yield on distillation with water a sweet-smelling liquid containing hydrocyanic acid. This is probably due in all cases to the decomposition of a substance called amygdalin under the influence of emulsin or synaptase, a nitrogenized ferment present in the organic structure (see GLUCOSIDES). The reaction is expressed by the equation :

$$C_{20}H_{27}NO_{11} \quad + \quad 2H_2O \quad = \quad CNH \quad + \quad C_7H_6O \quad + \quad 2C_6H_{12}O_6.$$

| Amygdalin. | | | Prussic acid. | Bitter almond oil. | Glucose. |

Hydrocyanic acid exists ready formed to a considerable extent in the juice of the bitter cassava.

The presence of hydrocyanic acid is detected with the utmost ease : its remarkable odor and high degree of volatility almost sufficiently characterize it. With solution of silver nitrate it gives a dense curdy white precipitate, much resembling the chloride, but differing from that substance in not blackening so readily by light, in being soluble in boiling nitric acid, and in suffering complete decomposition when heated in the dry state, metallic silver being left : the chloride, under the same circumstances, merely fuses, but undergoes no chemical change. The production of Prussian blue by "Scheele's test" is an excellent and most decisive experiment, which may be made with a very small quantity of the acid. The liquid to be examined is mixed with a few drops of solution of ferrous sulphate and an excess of caustic potash, and the whole exposed to the air for 10 or 15 minutes, with agitation, whereby the ferrous salt is partly converted into ferric salt : hydrochloric acid is then added in excess, which dissolves the precipitated iron oxide, and, if hydrocyanic acid is present, leaves Prussian blue as an insoluble powder. The reaction will be explained in connection with the ferrocyanides (p. 562).

Another very delicate test for hydrocyanic acid will be mentioned in connection with thiocyanic acid.

Metallic Cyanides.—The most important of the metallic cyanides are the following : they bear the most perfect analogy to the haloïd-salts.

Potassium Cyanide, CNK or KCy.—Potassium heated in cyanogen gas, takes fire and burns in a very beautiful manner, yielding potassium cyanide : the same substance is produced when potassium is heated in the vapor of hydrocyanic acid, hydrogen being liberated. When pure nitrogen is transmitted through a white-hot tube containing a mixture of potassium carbonate and charcoal, a small quantity of potassium cyanide is formed, which settles on the cooler portions of the tube as a white amorphous powder : carbon monoxide is at the same time evolved.* When azotized organic matter of any kind, capable of furnishing ammonia by destructive distillation, as horn-shavings, parings of hides, etc., is heated to redness with potassium carbonate in a close vessel, a very abundant production of potassium cyanide results, which cannot, however, be advantageously extracted by direct means, but in practice is always converted into ferrocyanide, which is a much more stable substance, and crystallizes better.

---

* According to the experiments of Margueritte and De Sourdeval, the formation of cyanide appears to be more abundant if the potash be replaced by baryta. If the barium cyanide thus formed be exposed to a stream of superheated steam at 300° C. (572° F.), the nitrogen of the salt is eliminated in the form of ammonia. Margueritte and De Sourdeval recommend this process as a method of preparing ammonia by means of atmospheric nitrogen.

, Potassium cyanide may be prepared by passing the vapor of hydrocyanic acid into a cold alcoholic solution of potash : the salt is then deposited
in the crystalline form, and may be separated from the liquid, pressed
and dried. But it is more generally made from the ferrocyanide, which,
when heated to whiteness in a nearly close vessel, evolves nitrogen and
other gases, and leaves a mixture of carbon, iron carbide, and potassium
cyanide, which latter salt is not decomposed unless the temperature is excessively high.

Liebig has given a very easy and excellent process for making potassium
cyanide, which does not, however, yield it pure, but mixed with potassium
cyanate. For most of the applications of potassium cyanide, electro-plat-
ing and gilding, for example, for which a considerable quantity is now
required, this impurity is of no consequence. Eight parts of potassium
ferrocyanide are rendered anhydrous by gentle heat, and intimately mixed
with three parts of dry potassium carbonate: this mixture is thrown into
a red-hot earthen crucible, and kept in fusion, with occasional stirring,
until gas ceases to be evolved, and the fluid portion of the mass becomes
colorless. The crucible is left at rest for a moment, and then the clear
salt is decanted from the heavy black sediment at the bottom, which is
principally metallic iron in a state of minute division. The reaction is
represented by the equation :—

$$\underset{\text{Ferrocyanide.}}{K_4FeCy_6} + \underset{\text{Carbonate.}}{K_2CO_3} = \underset{\text{Cyanide.}}{5KCy} + \underset{\text{Cyanate.}}{KCyO} + Fe + CO_2.$$

The product may be advantageously used, instead of potassium ferrocyanide, in the preparation of hydrocyanic acid, by distillation with
diluted oil of vitriol.

Potassium cyanide is often produced in considerable quantity in blast-
furnaces in which iron ores are smelted with coal or coke.

Potassium cyanide forms colorless, cubic or octohedral, anhydrous crystals, deliquescent in the air, and exceedingly soluble in water : it dissolves
in boiling alcohol, but separates in great measure on cooling. It is readily
fusible, and undergoes no change at a moderate red or even white heat,
when excluded from air; otherwise, oxygen is absorbed and the cyanide
becomes cyanate. Its solution always has an alkaline reaction, and when
exposed to the air exhales the odor of hydrocyanic acid : it is decomposed
by the weakest acids, even the carbonic acid of the air, and when boiled
in a retort is slowly converted into potassium formate, with separation of
ammonia. It is said to be as poisonous as hydrocyanic acid itself.

Sodium cyanide, NaCy, is a very soluble salt, corresponding closely with
the foregoing, and obtained by similar means.

Ammonium cyanide, NH_4Cy, is a colorless, crystallizable, and very volatile substance, prepared by distilling a mixture of potassium cyanide and
sal-ammoniac; or by mingling the vapor of anhydrous hydrocyanic acid
with ammoniacal gas ; or, lastly, by passing ammonia over red-hot charcoal. It is very soluble in water, subject to spontaneous decomposition,
and is slightly poisonous.

Mercuric cyanide, $Hg(CN)_2$, or $HgCy_2$.—One of the most remarkable properties of cyanogen is its powerful attraction for certain of the less oxidable
metals, as silver, and more particularly for mercury and palladium. Dilute
hydrocyanic acid dissolves finely-powdered mercuric oxide with the utmost
ease : the liquid loses all odor, and yields on evaporation crystals of mercuric cyanide. Potassium cyanide is in like manner decomposed by mercuric
oxide, potassium hydroxide being produced. Mercuric cyanide is generally
prepared from potassium ferrocyanide ; 2 parts of the salt are dissolved in
15 parts of hot water, and 3 parts of dry mercuric sulphate are added ;
the whole is boiled for 15 minutes, and filtered hot from the iron oxide,

which separates. The solution, on cooling, deposits the mercuric cyanide in crystals. Mercuric cyanide forms white, translucent, dimetric prisms, much resembling those of corrosive sublimate : it is soluble in 8 parts of cold water, and in a much smaller quantity at a higher temperature, also in alcohol. The solution has a disagreeable metallic taste, is very poisonous, and is not precipitated by alkalies. Mercuric cyanide is used in the laboratory as a source of cyanogen.

Silver cyanide, AgCy, has been already described (p. 558).—*Zinc cyanide*, $ZnCy_2$, is a white insoluble powder, prepared by mixing zinc acetate with hydrocyanic acid.—*Cobalt cyanide*, $CoCy_2$, is obtained by similar means : it is dirty-white, and insoluble.—*Palladium cyanide*, $PdCy_2$, forms a yellowish-white precipitate when the chloride of that metal is mixed with a soluble cyanide.—*Auric cyanide*, $AuCy_3$, is yellowish-white and insoluble, but freely dissolved by solution of potassium cyanide.

Iron Cyanides.—These compounds are scarcely known in the separate state, on account of their great tendency to form double salts. On adding potassium cyanide to a ferrous salt, a yellowish-red flocculent precipitate is formed, consisting chiefly of ferrous cyanide, $FeCy_2$, but always containing a certain quantity of potassium cyanide, and dissolved as ferrocyanide by excess of that salt. Ferric cyanide, Fe_2Cy_6, is known only in solution. Pelouze obtained an insoluble green compound containing Fe_3Cy_8, or $FeCy_2.Fe_2Cy_6$, by passing chlorine gas into a boiling solution of potassium ferrocyanide.

The iron cyanides unite with other metallic cyanides, forming two very important groups of compounds, called ferrocyanides and ferricyanides, the composition of which may be illustrated by the respective potassium-salts :—

Ferrocyanide . . . . $K_4Fe''Cy_6$, or $4KCy.Fe''Cy_2$
Ferricyanide . . . . $K_3Fe'''Cy_6$, or $3KCy.Fe'''Cy_3$.

It will be seen from these formulæ, that ferro- and ferricyanides containing the same quantity of cyanogen, differ from one another only by one atom of univalent metal, and, accordingly, it is found that the former may be converted into the latter, by the action of oxidizing (metal-abstracting) agents, and the latter into the former by the action of reducing (metal-adding) agents. Thus potassium ferrocyanide is easily converted into the ferricyanide by the action of chlorine, and many double ferrocyanides may be formed from ferricyanides by the action of alkalies in presence of a reducing agent ; thus potassium ferricyande, $K_3Fe'''Cy_6$, is easily converted into ammonio-tripotassic ferrocyanide, $(NH_4)K_3Fe''Cy_6$, by the action of ammonia in presence of glucose.*

Ferrocyanides.

Potassium Ferrocyanide, K_4FeCy_6, or $4KCy.FeCy_2$, commonly called *yellow prussiate of potash.*—This important salt is formed—1. By digesting precipitated ferrous cyanide in aqueous solution of potassium cyanide.

* The ferrocyanides and ferricyanides are sometimes regarded as salts of peculiar compound radicles containing iron, viz., *ferrocyanogen*, $Fe''Cy_6$, and *ferricyanogen*, $Fe'''Cy_6$, the first being quadrivalent, the second trivalent; but there is nothing gained by this assumption. For a discussion of the formulæ of these salts, and of the double cyanides in general, see Watts's Dictionary of Chemistry, vol. ii. p. 201.

2. By digesting ferrous hydrate with potassium cyanide, potash being formed at the same time :

$$6KCy + FeH_2O_2 = 2KHO + K_4FeCy_6.$$

3. Ferrous cyanide with aqueous potash :

$$3FeCy_2 + 4KHO = 2FeH_2O_2 + K_4FeCy_6.$$

4. Aqueous potassium cyanide with metallic iron : if the air be excluded, hydrogen is evolved :

$$6KCy + Fe + 2H_2O = K_4FeCy_6 + 2KHO + H_2;$$

but if the air has access to the liquid, oxygen is absorbed, and no hydrogen is evolved :

$$6KCy + Fe + H_2O + O = K_4FeCy_6 + 2KHO.$$

5. Ferrous sulphide with aqueous potassium cyanide :

$$6KCy + FeS = K_2S + K_4FeCy_6.$$

6. Any soluble ferrous salt with potassium cyanide ; e. g. :

$$6KCy + FeSO_4 = K_2SO_4 + K_4FeCy_6.$$

Potassium ferrocyanide is manufactured on the large scale by the following process : Dry refuse animal matter of any kind is fused at a red-heat with impure potassium carbonate and iron filings, in a large iron vessel from which the air should be excluded as much as possible; potassium cyanide is generated in large quantity. The melted mass is afterwards treated with hot water, which dissolves out the cyanide and other salts, the cyanide being quickly converted by the oxide or sulphide* of iron into ferrocyanide. The filtered solution is evaporated, and the first-formed crystals are purified by re-solution. If a sufficient quantity of iron be not present, great loss is incurred by the decomposition of the cyanide into potassium carbonate and ammonia.

A new process for the preparation of potassium ferrocyanide has lately been proposed by Gélis. It consists in converting carbon bisulphide into ammonium thiocarbonate by agitating it with ammonium sulphide : $CS_2 + (NH_4)_2S = (NH_4)_2CS_3$, and heating the product thus obtained with potassium sulphide, whereby potassium thiocyanate (p. 572) is formed, with evolution of ammonium sulphide and hydrogen sulphide :

$$2(NH_4)_2CS_3 + K_2S = 2KCNS + 2(NH_4)HS + 3H_2S.$$

The potassium thiocyanate is dried, mixed with finely divided metallic iron, and heated for a short time in a closed iron vessel to dull redness, whereby the mixture is converted into potassium ferrocyanide, potassium sulphide, and iron sulphide :

$$6KCyS + Fe_6 = K_4FeCy_6 + 5FeS + K_2S.$$

By treatment with water, the sulphide and ferrocyanide of potassium are dissolved, and on evaporation the ferrocyanide is obtained in crystals. It remains to be seen whether this ingenious process is capable of being carried out upon a large scale.

Potassium ferrocyanide forms large, transparent, yellow crystals,

---

* The sulphur is derived from the reduced sulphate of the crude pearl-ashes and the animal substances used in the manufacture.

24 *

$K_4FeCy_6 + 3Aq$., derived from an octohedron with a square base : they cleave with facility in a direction parallel to the base of the octohedron, and are tough and difficult to powder. They dissolve in 4 parts of cold and 2 parts of boiling water, and are insoluble in alcohol. They are permanent in the air, and have a mild saline taste. The salt has no poisonous properties, and, in small doses at least, is merely purgative. Exposed to a gentle heat, it loses 3 molecules of water, and becomes anhydrous : at a high temperature it yields potassium cyanide, iron carbide, and various gaseous products ; if air be admitted, the cyanide becomes cyanate.

Potassium ferrocyanide is a chemical reagent of great value : when mixed in solution with neutral or slightly acid salts of the heavy metals, it gives rise to precipitates which very frequently present highly characteristic colors. In most of these compounds the potassium is simply displaced by the new metal : the beautiful brown ferrocyanide of copper contains, for example, Cu_2FeCy_6, or $2CuCy_2.FeCy_2$, and that of lead, Pb_2FeCy_6.

With *ferrous salts*, potassium ferrocyanide gives a precipitate which is perfectly white, if the air be excluded and the solution is quite free from ferric salt, but quickly turns blue on exposure to the air. It consists of potassio-ferrous ferrocyanide, $K_2Fe_2Cy_6$, or potassium ferrocyanide having half the potassium replaced by iron. The same salt is produced in the preparation of hydrocyanic acid by distilling potassium ferrocyanide with dilute sulphuric acid (p. 557).

When a soluble ferrocyanide is added to the solution of a *ferric salt*, a deep blue precipitate is formed, consisting of ferric ferrocyanide, Fe_7Cy_{18}, or $Fe'''_4Fe''_3Cy_{18}$, or $2(Fe_2)^{vi}Cy_6.FeCy_2$, which in combination with 18 molecules of water constitutes ordinary Prussian blue. This beautiful pigment is best prepared by adding potassium ferrocyanide to ferric nitrate or chloride :

$$3K_4FeCy_6 + 2(Fe_2)^{vi}Cl_6 = 12KCl + Fe_7Cy_{18}.$$

It is also formed by precipitating a mixture of ferrous and ferric salts with potassium cyanide :

$$18KCy + 3FeCl_2 + 2(Fe_2)^{vi}Cl_6 = 18KCl + Fe_7Cy_{18}.$$

This reaction explains Scheele's test for prussic acid (p. 558). Prussian blue is also formed by the action of air, chlorine-water, and other oxidizing agents, on potassio-ferrous ferrocyanide ; probably thus :

$$6K_2Fe_2Cy_6 + O_3 = Fe_7Cy_{18} + 3K_4FeCy_6 + Fe_2O_3.$$

It is chiefly by this last reaction that Prussian blue is prepared on the large scale, potassium ferrocyanide being first precipitated by ferrous sulphate, and the resulting white or light blue precipitate either left to oxidize by contact with the air, or subjected to the action of nitric acid, chlorine, hypochlorites, chromic acid, etc. The product, however, is not pure ferric ferrocyanide : for it is certain that another and simpler reaction takes place at the same time, by which the potassio-ferrous ferrocyanide $(K_2Fe'')Fe''Cy_6$, is converted, by abstraction of an atom of potassium, into potassio-ferrous ferricyanide $(KFe'')Fe'''Cy_6$, which also possesses a fine deep-blue color. Commercial Prussian blue is, therefore, generally a mixture of this compound with ferric ferrocyanide, $Fe'''_4Fe''_3Cy_{18}$, the one or the o   predominating according to the manner in which the process is conducted.

Prussian blue in the moist state forms a bulky precipitate, which shrinks to a comparatively small compass when well washed and dried by a gentle heat. In the dry state it is hard and brittle, much resembling in appearance the best indigo : the freshly fractured surfaces have a beautiful copper-red lustre, similar to that produced by rubbing indigo with a hard body. Prussian blue is quite insoluble in water and dilute acids, with the exception of oxalic acid, in a solution of which it dissolves, forming a deep-blue liquid, which is sometimes used as ink : concentrated oil of vitriol converts it into a white, pasty mass, which again becomes blue on addition of water. Alkalies destroy the color instantly : they dissolve out a ferrocyanide, and leave ferric oxide. Boiled with water and mercuric oxide, it yields mercuric cyanide and ferric oxide. Heated in the air, Prussian blue burns like tinder, leaving a residue of ferric oxide. Exposed to a high temperature in a close vessel, it gives off water, ammonium cyanide, and ammonium carbonate, and leaves carbide of iron. It forms a very beautiful pigment, both as oil and water color, but has little permanency.

Common or basic Prussian blue is an inferior article, prepared by precipitating a mixture of ferrous sulphate and alum with potassium ferrocyanide, and exposing the precipitate to the air. It contains alumina, which impairs the color, but adds to the weight.

Soluble Prussian blue is obtained by adding ferric chloride to an excess of potassium ferrocyanide; it is insoluble in the saline liquor, but soluble in pure water. It has a deep-blue color, and probably consists of potassio-ferrous ferricyanide.

Hydrogen Ferrocyanide, *Hydroferrocyanic Acid*, H_4FeCy_6, is prepared by decomposing ferrocyanide of lead or copper suspended in water by a stream of sulphuretted hydrogen gas, and separates on evaporating the filtered solution in a vacuum over oil of vitriol. Its solution in water has a powerfully acid taste and reaction, and decomposes alkaline carbonates with effervescence: it does not dissolve mercuric oxide in the cold, but when heat is applied, undergoes decomposition, forming mercuric cyanide and ferrous cyanide: $H_4FeCy_6 + 2HgO = 2HgCy_2 + FeCy_2 + 2H_2O$; but the ferrous cyanide is immediately oxidized by the excess of mercuric oxide, with separation of metallic mercury. In the dry state the acid is very permanent, but when long exposed to the air in contact with water, it is entirely converted into Prussian blue.

Sodium Ferrocyanide, $Na_4FeCy_6.12Aq.$, crystallizes in yellow four-sided prisms, which are efflorescent in the air, and very soluble.

Ammonium Ferrocyanide, $(NH_4)_4FeCy_6.3Aq.$, is isomorphous with potassium ferrocyanide: it is easily soluble, and is decomposed by ebullition. *Barium Ferrocyanide*, Ba_2FeCy_6, prepared by boiling potassium ferrocyanide with a large excess of barium chloride, or Prussian blue with baryta-water, forms minute yellow, anhydrous crystals, which have but a small degree of solubility even in boiling water. The corresponding compounds of *strontium*, *calcium*, and *magnesium* are more freely soluble. The ferrocyanides of *silver*, *lead*, *zinc*, *manganese*, and *bismuth* are white and insoluble ; those of *nickel* and *cobalt* are pale-green and insoluble ; and lastly, that of *copper* has a beautiful reddish-brown tint.

There are also several double ferrocyanides. When, for example, concentrated solutions of calcium chloride and potassium ferrocyanide are mixed, a sparingly soluble crystalline precipitate falls, containing $K_2CaFeCy_6$.

Ferricyanides.

These salts are formed, as already observed, by abstraction of metal from the ferrocyanides ; in other words, by the action of oxidizing agents.

Potassium Ferricyanide, $K_3Fe'''Cy_6$, or $K_6(Fe_2)^{vi}Cy_{12}$, often called *red prussiate of potash*, is prepared by slowly passing chlorine, with agitation, into a somewhat dilute and cold solution of potassium ferrocyanide, until the liquid acquires a deep reddish-green color, and ceases to precipitate a ferric salt. The solution is evaporated until a skin begins to form upon the surface, then filtered and left to cool ; and the salt is purified by re-crystallization. It forms prismatic, or sometimes tabular crystals, belonging to the monoclinic system, of a beautiful ruby-red tint, permanent in the air, and soluble in 4 parts of cold water : the solution has a dark-greenish color. The crystals burn and emit sparks when introduced into the flame of a candle. The salt is decomposed by excess of chlorine, and by deoxidizing agents, as sulphuretted hydrogen.

Hydrogen ferricyanide is obtained in the form of a reddish-brown acid liquid, by decomposing lead ferricyanide with sulphuric acid : it is very unstable, and is resolved by boiling into hydrated ferric cyanide, an insoluble dark-green powder containing $Fe_2Cy_6.3Aq.$, and hydrocyanic acid. The ferricyanides of *sodium,* **ammonium,** and of the *alkaline earth-metals,* are soluble ; those of most of the other metals are insoluble. Potassium ferricyanide added to a ferric salt occasions no precipitate, but merely a darkening of the reddish-brown color of the solution ; with *ferrous salts,* on the other hand, it gives a deep-blue precipitate, consisting of ferrous ferricyanide, $Fe_5Cy_{12}+xAq.$, or $Fe''_3(Fe_2)^{vi}Cy_{12}+xAq.$, which, when dry, has a brighter tint than Prussian blue : it is known under the name of *Turnbull's blue.* Hence, potassium ferricyanide is as delicate a test for ferrous salts as the yellow ferrocyanide is for ferric salts.

CobaltICYANIDES.—This name is applied to a series of compounds analogous to the preceding, containing cobalt in place of iron ; a hydrogen-acid has been obtained, and a number of salts, which much resemble the ferricyanides. Several other metals of the same isomorphous family are found capable of replacing iron in these compounds.

Nitroprussides.—These are salts produced by the action of nitric acid upon ferrocyanides and ferricyanides. The general formula of these salts appears to be $M_2(NO)Fe''Cy_5$, which exhibits a close relation to those of the ferro- and ferricyanides.

The formation of the nitroprussides appears to consist in the reduction of the nitric acid to the state of nitrogen dioxide or nitrosyl, NO, which replaces 1 molecule of metallic cyanide, MCy, in a molecule of ferricyanide $M_3F'''Cy_6$. The formation of these salts is attended with the production of a variety of secondary products, such as cyanogen, oxamide, hydrocyanic acid, nitrogen, carbonic acid, etc. One of the finest compounds of this series is the nitroprusside of sodium, $Na_2(NO)Fe''Cy_5+2Aq.$, which is readily obtained by treating 2 parts of powdered potassium ferrocyanide with 5 parts of common nitric acid previously diluted with its own volume of water. The solution, after the evolution of gas has ceased, is digested on the water-bath, until ferrous salts no longer yield a blue, but a slate-colored precipitate. The liquid is now allowed to cool, when much potassium nitrate, and occasionally oxamide, is deposited : it is filtered and neutralized with sodium carbonate, which yields a green or brown precipitate, and a ruby-colored filtrate. This, on evaporation, gives a crystallization of the nitrates of potassium and sodium, together with the nitroprusside. The crystals of the latter are selected and purified by crystallization ; they are rhombic and of a splendid ruby color. The soluble nitroprussides strike a most beautiful violet tint with soluble sulphides, affording an extremely delicate test for alkaline sulphides.

ALCOHOLIC CYANIDES OR HYDROCYANIC ETHERS.

These compounds play an important part in organic chemistry : for example, in the conversion of alcohols into acids containing a greater number of carbon-atoms.

The cyanides of univalent alcohol-radicles may also be regarded as compounds of nitrogen with trivalent radicles : hence, as already observed (p. 555), they are often called n i t r i l s.

These alcoholic cyanides are produced :—

1. By distilling a mixture of potassium cyanide and the potassium-salt of ethylsulphuric acid, $\left.\begin{array}{c}C_2H_5 \\ H\end{array}\right\} SO_4$, or a similar acid :—

$$KCN + (C_2H_5)KSO_4 = K_2SO_4 + C_2H_5.CN.$$

2. By the dehydrating action of phosphoric oxide on the ammonium-salts of the monobasic acids, $C_nH_{2n}O_2$ and $C_nH_{2n-6}O_2$, homologous with acetic and benzoic acid respectively, thus :—

$$\underset{\substack{\text{Ammonium} \\ \text{acetate.}}}{C_2H_3O_2.NH_4} - 2H_2O = \underset{\substack{\text{Ethenyl} \\ \text{nitril.}}}{C_2H_3N}$$

$$\underset{\substack{\text{Ammonium} \\ \text{benzoate.}}}{C_7H_5O_2.NH_4} - 2H_2O = \underset{\substack{\text{Benzonitril.}}}{C_7H_5N}$$

The bodies obtained by these two processes are oily liquids, exhibiting the same properties whether prepared by the first or the second method, excepting that those obtained by the latter have an aromatic fragrant odor, whereas those prepared by the former have a pungent and repulsive odor, due to the presence of certain isomeric compounds, to be noticed further on. *Methyl cyanide, Ethenyl-nitril,* or *Acetonitril,* boils at 77° C. (170.6° F.); *Ethyl cyanide,* or *Propenylnitril,* at 82° C. (179.6° F.); *Butyl cyanide,* or *Valeronitril,* at 125–128° C. (257–262.4° F.); *Isopentyl cyanide, Amyl cyanide,* or *Capronitril,* at 146° C. (294.8° F.); *Phenyl cyanide,* or *Benzonitril,* at 190.6° C. (375° F.).

All these cyanides, when heated with fuming sulphuric acid or sulphuric oxide, are converted into sulpho-acids ; thus :—

$$\underset{\substack{\text{Methyl} \\ \text{cyanide.}}}{CH_3.CN} + H_2O + 2SO_4H_2 = SO_4H(NH_4) + \underset{\substack{\text{Sulphacetic} \\ \text{acid.}}}{C_2H_4SO_5}$$

$$\underset{\substack{\text{Methyl} \\ \text{cyanide.}}}{CH_3.CN} + 3H_2SO_4 = SO_4H(NH_4) + CO_2 + \underset{\substack{\text{Disulpho-} \\ \text{metholic acid.}}}{CH_4S_2O_6}$$

By heating with caustic potash or soda, they are resolved into ammonia and the corresponding fatty or aromatic acid, just as hydrocyanic acid similarly treated is resolved into ammonia and formic acid ; thus :—

$$\underset{\substack{\text{Hydrogen} \\ \text{cyanide.}}}{HCN} + 2H_2O = NH_3 + \underset{\substack{\text{Formic} \\ \text{acid.}}}{CH_2O_2}$$

$$\underset{\substack{\text{Ethyl} \\ \text{cyanide.}}}{C_2H_5CN} + 2H_2O = NH_3 + \underset{\substack{\text{Propionic} \\ \text{acid.}}}{C_3H_6O_2}$$

$$\underset{\substack{\text{Phenyl} \\ \text{cyanide.}}}{C_6H_5CN} + 2H_2O = NH_3 + \underset{\substack{\text{Benzoic acid.}}}{C_7H_6O_2}$$

The alcoholic cyanides or nitrils, treated with nascent hydrogen, are converted into the corresponding amine-bases, e. g. :—

$$CNCH_3 \ + \ 2H_2 \ = \ NH_2(CH_2CH_3) \text{ or } C_2H_7N$$
$$\text{Methyl} \qquad\qquad\qquad \text{Ethylamine.}$$
$$\text{cyanide.}$$

Ethene Cyanide, $(C_2H_4)''(CN)_2$, is obtained by distilling potassium cyanide with ethene bomide :—

$$C_2H_4Br_2 \ + \ 2KCN \ = \ 2KBr \ + \ C_2H_4(CN)_2.$$

It is a crystalline body, melting at 50° C. (140° F.), and converted by alcoholic potash into ammonia and succinic acid :—

$$C_2H_4(CN)_2 \ + \ 4H_2O \ = \ 2NH_3 \ + \ C_4H_6O_4.$$

Isocyanides, or Carbamines.—On examining the equations just given for the decomposition of the alcoholic cyanides under the influence of alkalies, it is easy to see that the reaction might be supposed to take place in a different way, each cyanide yielding, not ammonia and an acid containing the same number of carbon-atoms as itself, but an alcoholic ammonia or amine, and formic acid ; thus :—

$$C_2H_5{\cdot}CN \ + \ 2H_2O \ = \ NH_2C_2H_5 \ + \ CH_2O_2$$
$$\text{Ethyl} \qquad\qquad\qquad \text{Ethyl-} \qquad \text{Formic}$$
$$\text{cyanide.} \qquad\qquad\qquad \text{amine.} \qquad \text{acid.}$$

In the one case the alcohol-radicle remains united with the carbon, producing a homologue of formic acid, together with ammonia ; in the other it remains united with the nitrogen, producing a homologue of ammonia, together with formic acid.

A class of cyanides exhibiting the second of these reactions has been discovered by Dr. Hofmann.* They are obtained by distilling a mixture of an alcoholic ammonia-base and chloroform with alcoholic potash : for example :—

$$C_6H_7N \ + \ CHCl_3 \ = \ 3HCl \ + \ C_7H_5N$$
$$\text{Aniline.} \qquad \text{Chloro-} \qquad\qquad\qquad \text{Phenyl-}$$
$$\text{form.} \qquad\qquad\qquad\qquad \text{isocyanide.}$$

The potash serves to neutralize the hydrochloric acid produced, which would otherwise quickly decompose the isocyanide. Phenyl-isocyanide, or phenyl-carbamine, when freed from excess of aniline by oxalic acid, then dried with caustic potash and rectified, is an oily liquid, green by transmitted, blue by reflected light, and having an intolerably pungent and suffocating odor. It is isomeric with benzonitril, and is resolved by boiling with dilute acids into formic acid and aniline :—

$$C_7H_5N \ + \ 2H_2O \ = \ CH_2O_2 \ + \ C_6H_7N.$$

It is a remarkable fact that, whereas the normal alcoholic cyanides are easily decomposed by boiling alkaline solutions, the isocyanides are scarcely altered by alkalies, but are easily hydrated under the influence of acids.

The isocyanides of ethyl and amyl have been obtained by similar processes ; namely, by distilling methylamine and ethylamine respectively with chloroform, also by the action of ethylic and amylic iodides on silver cyanide. They resemble the phenyl compound in their reactions, and are also characterized by extremely powerful odors. The repulsive odor possessed by the normal alcoholic cyanides when prepared by distilling potas-

* Proceedings of the Royal Society, xvi. 144, 148, 150.

sium cyanide with the ethyl-sulphates or homologous salts, appears to be due to the presence of small quantities of these isocyanides.

The difference of constitution between the normal cyanides and the isocyanides may be represented by the following formulæ, taking the methyl compounds for example:—

$$N\equiv C—CH_3 \text{ or } C^{iv}\begin{cases} N''' \\ CH_3 \end{cases} \qquad C\equiv N—CH_3 \text{ or } N^v\begin{cases} C^{iv} \\ CH_3 \end{cases}$$

Cyanide.   Isocyanide.

In the isocyanide the carbon belonging to the alcohol-radicle is united directly with the nitrogen ; in the cyanide, only through the medium of the carbon belonging to the cyanogen.

This difference of structure may perhaps account for the difference in reactions of the cyanides and isocyanides, under the influence of hydrating agents, thus :

$$C\begin{cases} N \\ CH_3 \end{cases} + \begin{matrix} HHO \\ HHO \end{matrix} = NH_3 + C\begin{cases} CH_3 \\ O'' \\ OH \end{cases}$$

Methyl   Water   Ammonia.   Acetic
cyanide.   (2 mol.)   acid.

$$N\begin{cases} C \\ CH_3 \end{cases} + \begin{matrix} HHO \\ HHO \end{matrix} = N\begin{cases} C \\ CH_3 \end{cases} + C\begin{cases} H \\ O'' \\ OH \end{cases}$$

Methol   Water   Methylamine.   Formic
isocyanide. (2 mol.)   acid.

The isocyanides of methyl and ethyl (methyl- and ethyl-carbamines) unite with acids, forming crystallizable salts.

OXYGEN- AND SULPHUR-COMPOUNDS OF CYANOGEN.

Cyanic Acid, CHNO.—Of this acid there are two possible modifications represented by the formulæ :

$$N\equiv C—OH \text{ and } CO\equiv N—H .$$

These modifications are actually exhibited in the metallic cyanates and the cyanic ethers ; but the acid itself is known in one modification only, the particular constitution of which has not yet been determined. It is produced when cyanuric acid, deprived of its water of crystallization, is heated to dull redness in a hard glass retort connected with a receiver cooled by ice. The cyanuric acid is resolved, without any other product, into cyanic acid, which condenses in the receiver to a limpid, colorless liquid, of exceedingly pungent and penetrating odor, like that of the strongest acetic acid : it even blisters the skin. When mixed with water it decomposes almost immediately, giving rise to ammonium bicarbonate :.

$$CHNO + H_2O = CO_2 + NH_3.$$

In consequence of this decomposition, cyanic acid cannot be separated from a cyanate by a stronger acid. A trace of it, however, always escapes decomposition, and communicates to the carbon dioxide evolved a pungent smell similar to that of sulphurous acid. The cyanates may be easily distinguished by this smell, and by the simultaneous formation of an ammonia-salt, which remains behind.

~ Pure cyanic acid cannot be preserved : shortly after its preparation it changes spontaneously, with sudden rise of temperature, into a solid, white, opaque, amorphous substance, called *cyamelide*. This body has the same composition as cyanic acid : it is insoluble in water, alcohol, ether, and dilute acids : it dissolves in strong oil of vitriol by the aid of heat, with evolution of carbon dioxide and production of ammonia ; boiled with a solution of caustic alkali, it dissolves, ammonia being disengaged, and a mixture of cyanate and cyanurate of the base generated. By dry distillation it is reconverted into cyanic acid.

Potassium Cyanate, CNKO.—Of this salt there are two modifications, viz.,

$$N\equiv C—OK \qquad \text{and} \qquad CO\!=\!NK$$
Normal cyanate.   Isocyanate.

The normal cyanate, formed by passing gaseous cyanogen chloride (p. 572) into cold aqueous potash, crystallizes in long needles, and is converted by fusion into the isocyanate.

The isocyanate (ordinary potassium cyanate) is best prepared by oxidizing potassium cyanide with litharge. The cyanide, already containing a portion of cyanate, described at page 559, is remelted in an earthen crucible, and finely powdered lead oxide added by small portions ; the oxide is instantaneously reduced, and the metal, at first in a state of minute division, ultimately collects to a fused globule at the bottom of the crucible. The salt is poured out, and, when cold, powdered and boiled with alcohol ; the hot filtered solution deposits crystals of potassium isocyanate on cooling. The great deoxidizing power exerted by potassium cyanide at a high temperature renders it a valuable agent in many of the finer metallurgic operations.

Another method of preparing the isocyanate is to mix dried and finely powdered potassium ferrocyanide with half its weight of equally dry manganese dioxide ; heat this mixture in a shallow iron ladle, with free exposure to air and frequent stirring, until the tinder-like combustion is at an end ; and boil the residue in alcohol, which extracts the isocyanate.

The salt crystallizes from alcohol in thin, colorless, transparent plates, which suffer no change in dry air, but on exposure to moisture are gradually converted, without much alteration of appearance, into potassium bicarbonate, ammonia being at the same time given off. Water dissolves potassium isocyanate in large quantity ; the solution is slowly decomposed in the cold, and rapidly at a boiling heat, into potassium bicarbonate and ammonia. When a concentrated solution is mixed with a small quantity of dilute mineral acid, a precipitate falls, consisting of acid potassium cyanurate. Potassium isocyanate is reduced to cyanide by ignition with charcoal in a covered crucible. Mixed with solutions of lead and silver, it gives rise to white insoluble isocyanates of those metals.

Ammonium Cyanate (probably *iso*), CN_2H_4O, or NH_4CNO.—When the vapor of cyanic acid is mixed with excess of ammoniacal gas, a white, crystalline, solid substance is produced, which has all the characters of a true, although not neutral ammonium cyanate. It dissolves in water, and if mixed with an acid, evolves carbon dioxide : with an alkali, it yields ammonia. But if the solution be heated, or if the crystals be merely exposed for a certain time to the air, a portion of ammonia is dissipated, and the properties of the compouud are completely changed. It may now be mixed with acids without the least sign of decomposition, and does not evolve the smallest trace of ammonia when treated with cold caustic alkali. The result of this transformation, as already observed (p. 488), is u r e a.

Cyanuric Acid, $C_3N_3H_3O_3$.—This substance may be prepared by heating dry and pure urea in a flask or retort ; the urea melts, boils, gives off ammonia in large quantity, and at length becomes converted into a dirty-white, solid, amorphous mass, which is impure cyanuric acid. This is dissolved by the aid of heat in strong oil of vitriol, and nitric acid added by small portions till the liquid becomes nearly colorless ; it is then mixed with water, and left to cool, whereupon the cyanuric acid separates. The urea may likewise be decomposed very conveniently by gently heating it in a tube, while dry chlorine or hydrochloric acid gas passes over ·it. A mixture of cyanuric acid and sal-ammoniac results, which is separated by dissolving the latter in water. The reaction with chlorine is represented by the equation :

$$3CON_2H_4 + Cl_2 = C_3N_3H_3O_3 + 2NH_4Cl + HCl + N.$$

Cyanuric acid forms colorless efflorescent crystals, seldom of large size, derived from an oblique rhombic prism. It is very little soluble in cold water, and requires 24 parts for solution at a boiling heat : it reddens litmus feebly, has no odor, and but little taste. The acid is tribasic: the crystals contain $C_3N_3H_3O_3.2Aq$, and are easily deprived of their water of crystallization. In point of stability, cyanuric acid offers a most remarkable contrast to its isomeride, cyanic acid ; it dissolves, as above indicated, in hot oil of vitriol, and even in strong nitric acid, without decomposition, and, in fact, crystallizes from the latter in the anhydrous state. Long-continued boiling with these powerful agents resolves it into ammonia and carbonic acid.

The connection between cyanic acid, urea, and cyanuric acid, may be thus recapitulated :

Ammonium cyanate is converted by heat into urea.

Urea is decomposed by the same means into cyanuric acid and ammonia.

Cyanuric acid is changed by a very high temperature into cyanic acid, one molecule of cyanuric acid splitting into three molecules of cyanic acid.

Cyanic and Cyanuric Ethers.

Of each of these ethers there are two series, analogous to the alcoholic cyanides and isocyanides. The difference of their structure is exhibited by the following formulæ, taking the methyl-compounds as examples :

$$N{\equiv}C{-}O{-}CH_3 \text{ or } C\begin{cases} N''' \\ OCH_3 \end{cases} \qquad O{=}C{=}N{-}CH_3 \text{ or } N\begin{cases} (CO)'' \\ CH_3 \end{cases}$$

Normal cyanate.   Isocyanate.

The corresponding cyanuric ethers are represented by the formulæ,

$$C_3\begin{cases} N_3 \\ (OCH_3)_3 \end{cases} \text{ and } N_3\begin{cases} (CO)_3 \\ (CH_3)_3 \end{cases}.$$

The normal cyanic ethers, discovered by Cloez, and the normal cyanuric ethers, discovered by Hofmann,[*] are produced simultaneously by the action of gaseous cyanogen chloride on the sodium alcohols : normal ethyl cyanate, for example, from cyanogen chloride and sodium ethylate; thus :

$$\begin{cases} N \\ Cl \end{cases} + NaOC_2H_5 = NaCl + C\begin{cases} N \\ OC_2H_5 \end{cases}.$$

[*] The isocyanic and isocyanuric ethers having been discovered first (by Wurtz in 1848), were originally called cyanic and cyanuric ethers.

They are decomposed by water, assisted by acids or bases, into cyanic or cyanuric acid and an alcohol : *e. g.*:

$$C\left\{\begin{array}{l} N \\ OC_2H_5 \end{array}\right. + HOH = HOC_2H_5 + C\left\{\begin{array}{l} N \\ OH \end{array}\right.$$

The cyanates of methyl, ethyl, and amyl are colorless, oily liquids, decomposed by heat into a volatile portion and a solid residue. The corresponding cyanurates are crystalline solids.

The isocyanic and isocyanuric ethers, or alcoholic carbimides, are produced simultaneously by distilling a dry mixture of potassium isocyanate and methylsulphate, ethylsulphate, etc., *e. g.*:

$$N\left\{\begin{array}{l} (CO) \\ K \end{array}\right. + \left.\begin{array}{l} C_2H_5 \\ K \end{array}\right\} SO_4 = K_2SO_4 + N\left\{\begin{array}{l} (CO) \\ C_2H_5 \end{array}\right.$$

Ethylic isocyanate and isocyanurate thus obtained are easily separated by distillation, the former boiling at 60° C. (140° F.), the latter at 276° C. (528.8° F.). The former is a mobile liquid, the latter a crystalline solid, melting at 85° C. (185° F.). The isocyanurate may likewise be obtained by distilling a mixture of potassium cyanurate and ethylsulphate.

The ethers of this class, when heated with a strong solution of caustic alkali, are resolved into carbon dioxide and an alcoholic ammonia or amine, *e. g.*:

$$N\left\{\begin{array}{l} (CO) \\ CH_3 \end{array}\right. + H_2O = CO_2 + N\left\{\begin{array}{l} H_2 \\ CH_3 \end{array}\right.$$
Methyl isocyanate.   Methylamine.

Pulminic Acid, $C_2N_2H_2O_2$.—This compound, polymeric with cyanic and cyanuric acids, is one of the products formed by the action of nitrous acid upon alcohol in presence of a salt of silver or mercury. The acid itself, or hydrogen fulminate, has not been obtained.

Silver fulminate is prepared by dissolving 40 or 50 grains of silver, which need not be pure, in about $\frac{3}{4}$ oz. by measure of nitric acid of sp. gr. 1.37, with the aid of a little heat. To the highly acid solution, while still hot, 2 measured ounces of alcohol are added, and heat is applied until reaction commences. The nitric acid oxidizes part of the alcohol to aldehyde and oxalic acid, becoming itself reduced to nitrous acid, which, in turn, acts upon the alcohol in such a manner as to form nitrous ether, fulminic acid, and water, 1 molecule of nitrous ether and 1 molecule of nitrous acid containing the elements of 1 molecule of fulminic acid and 2 molecules of water:

$$C_2H_5NO_2 + HNO_2 = C_2N_2H_2O_2 + 2H_2O.$$

The silver fulminate slowly separates from the hot liquid, in the form of small, brilliant, white, crystalline plates, which may be washed with a little cold water, distributed upon separate pieces of filter-paper in portions not exceeding a grain or two each, and left to dry in a warm place. When dry, the papers are folded up and preserved in a box. The only perfectly safe method of keeping the salt is by immersing it in water. Silver fulminate is soluble in 36 parts of boiling water, but the greater part crystallizes out on cooling : it is one of the most dangerous substances known, exploding with fearful violence when strongly heated, or when rubbed or struck with a hard body, or when touched with concentrated sulphuric acid : the metal is reduced, and a large volume of gaseous matter suddenly liberated. Nevertheless, when very cautiously mixed with copper oxide, it may be burned in a tube with as much facility as any other organic substance. Its composition thus determined is expressed by the formula $Ag_2C_2N_2O_2$.

Fulminic acid is bibasic : when silver fulminate is digested with caustic potash, one-half of the silver is precipitated as oxide, and a *silver-potassium fulminate*, $AgKC_2N_2O_2$, is produced, which resembles the neutral silver-salt, and detonates by a blow. Corresponding compounds containing sodium or ammonium exist : but a pure fulminate of an alkali-metal has never been formed. If silver fulminate be digested with water and copper, or zinc, the silver is entirely displaced, and a fulminate of the other metal produced. The zinc-salt mixed with baryta-water gives rise to a precipitate of zinc oxide, while *zinco-baric fulminate*, $ZnBa(C_2N_2O_2)_2$, remains in solution. *Mercuric fulminate*, $HgC_2N_2O_2$, is prepared by a process very similar to that by which the silver-salt is obtained. One part of mercury is dissolved in 12 parts of nitric acid; the solution is mixed with an equal quantity of alcohol ; and gentle heat is applied, the reaction, if too violent, being moderated by adding more spirit from time to time. Much carbonic acid, nitrogen, and red vapors are disengaged, together with a large quantity of nitrous ether and aldehyde ; these are sometimes condensed and collected for sale, but are said to contain hydrocyanic acid. The mercuric fulminate separates from the hot liquid, and after cooling may be purified from an admixture of reduced metal by solution in boiling water and recrystallization. It much resembles the silver-salt in appearance, properties, and degree of solubility. It explodes violently by friction or percussion, but unlike the silver-compound, merely burns with a sudden and almost noiseless flash when kindled in the open air. It is manufactured on a large scale for the purpose of charging *percussion-caps;* sulphur and potassium chlorate, or more frequently nitre, are added, and the powder, pressed into the cap, is secured by a drop of varnish.

The relation of composition between the three isomeric acids is shown by comparison of their silver-salts : the first acid is monobasic, the second bibasic, and the third tribasic :—

Silver cyanate	. . . . .	$Ag\,C\,N\,O$
Silver fulminate	. . . . .	$Ag_2C_2N_2O_2$
Silver cyanurate	. . . .	$Ag_3C_3N_3O_3.$

Fulminic, as well as cyanic acid, may be converted into urea. Dr. Gladstone has shown that, when a solution of copper fulminate is mixed with excess of ammonia, filtered, treated with sulphuretted hydrogen in excess, and again filtered from the insoluble copper sulphide, the liquid obtained is a mixed solution of urea and ammonium thiocyanate.

Another view regarding the constitution of fulminic acid was proposed by Gerhardt. The fulminates may be considered as methyl cyanide (acetonitril), in which one atom of hydrogen is replaced by NO_2 and 2 atoms of hydrogen by mercury or silver :—

$C\ H\ H\ H\ CN$	. . . .	Methyl cyanide.
$C(NO_2)AgAgCN$	. . . .	Silver fulminate.
$C(NO_2)\ Hg''CN$	. . . .	Mercuric fulminate.

This view has received some support by the interesting observation, made by Kekulé, that the action of chlorine upon mercuric fulminate, gives rise to the formation of chloropicrin, $C(NO_2)Cl_3$ (p. 554). The connection of fulminic acid with the methyl series is thus established.

Fulminuric Acid, $C_3N_3H_3O_3$.—This acid, isomeric with cyanuric acid, was discovered simultaneously by Liebig and by Schischkoff. It is obtained by the action of a soluble chloride upon mercuric fulminate. On boiling mercuric fulminate with an aqueous solution of potassium chloride, the mercury salt gradually dissolves, and the clear solution, after some time, becomes turbid, in consequence of a separation of mercuric oxide; it then contains potassium fulminurate :—

$$3HgC_2N_2O_2 + 8KCl + H_2O = 4KCl + 2HgCl_2 + HgO + 2C_3N_3HK_2O_3.$$

If, instead of potassium chloride, sodium or ammonium chloride be employed, the corresponding sodium and ammonium-compounds are obtained. The fulminurates crystallize with great facility : they are not explosive.

Fulminuric acid has the same composition as cyanuric acid, but it is bibasic, whereas cyanuric acid is tribasic.

Cyanogen Chlorides.—Chlorine forms with cyanogen, or its elements, two compounds, which are polymeric, and analogous to cyanic and cyanuric acids. *Gaseous cyanogen chloride*, CyCl, is formed by passing chlorine gas into anhydrous hydrocyanic acid, or by passing chlorine over moist mercuric cyanide contained in a tube sheltered from the light. It is a permanent and colorless gas at the temperature of the air, of insupportable pungency, and soluble to a very considerable extent in water, alcohol, and ether. At —18° C. (0.4° F.) it congeals to a mass of colorless crystals, which at —15° C. (5° F.) melt to a liquid whose boiling point is —11.6° C. (12.3° F.). At the temperature of the air it is condensed to the liquid form under a pressure of four atmospheres, and when long preserved in this state in hermetically sealed tubes, gradually passes into the solid modification.

On passing gaseous cyanogen chloride into a solution of ammonia in anhydrous ether, c y a n a m i d e, CN_2H_2, is formed together with sal-ammoniac.

Solid cyanogen chloride, $C_3N_3Cl_3$, or Cy_3Cl_3, is generated when anhydrous hydrocyanic acid is put into a vessel of chlorine gas, and the whole exposed to the sun : hydrochloric acid is formed at the same time. It forms long colorless needles, which exhale a powerful and offensive odor, compared by some to that of the excrement of mice ; it melts at 140° C. (284° F.), and sublimes unchanged at a higher temperature. When heated in contact with water, it is decomposed into cyanuric and hydrochloric acid. It dissolves in alcohol and ether without decomposition.

Cyanogen Bromide and *Iodide* correspond with the first of the preceding compounds, and are prepared by distilling bromine or iodine with mercuric cyanide. They are colorless, volatile, solid substances, of powerful odor.

Cyanogen Sulphide, C_2N_2S, or Cy_2S, recently obtained by Linnemann by the action of cyanogen iodide upon silver thiocyanate, crystallizes in transparent, volatile, rhombic plates, having an odor similar to that of cyanogen iodide. It melts at 60° C. (140° F.), but decomposes rapidly at a higher temperature ; dissolves in ether, alcohol, and water, and separates from hot concentrated solutions, on cooling, in the crystalline form.

Thiocyanic Acid, CNHS, also called *Sulphocyanic acid.*—This acid is the sulphur analogue of cyanic acid, and, like the latter, is monobasic, the thiocyanates of monad metals being represented by the formula MCNS.

Potassium Thiocyanate, CNKS.—To prepare this salt, yellow potassium ferrocyanide, deprived of its water of crystallization, is intimately mixed with half its weight of sulphur, and the whole heated to tranquil fusion in an iron pot, and kept for some time in that condition. When cold, the melted mass is boiled with water, which dissolves out a mixture of potassium thiocyanate and iron thiocyanate, leaving little behind but the excess of sulphur. This solution, which becomes red on exposure to the air, from oxidation of the iron, is mixed with potassium carbonate, by which the iron is precipitated, and potassium substituted : an excess of the carbonate must be, as far as possible, avoided. The filtered liquid is concentrated, by evaporation over an open fire, to a small bulk, and left to cool and crystallize. The crystals are drained, purified by re-solution, if necessary, or dried by inclosing them, spread on filter-paper, over a surface of oil of vitriol covered with a bell-jar.

The reaction between the sulphur and the potassium ferrocyanide is represented by the equation :—

$$K_4FeC_6N_6 + S_6 = 4KCNS + Fe(CNS)_2.$$

Another, and even better process, consists in gradually heating to low redness in a covered vessel a mixture of 46 parts of dried potassium ferro-cyanide, 32 of sulphur, and 17 of pure potassium carbonate. The mass is exhausted with water, the aqueous solution is evaporated to dryness, and the residue is exhausted with alcohol. The alcoholic liquid deposits splendid crystals on cooling or evaporation.

Potassium thiocyanate crystallizes in long, slender, colorless prisms, or plates, which are anhydrous : it has a bitter saline taste, and is destitute of poisonous properties : it is very soluble in water and alcohol, and deliquesces when exposed to a moist atmosphere. When heated, it melts to a colorless liquid, at a temperature far below that of ignition.

Chlorine, passed into a strong solution of potassium thiocyanate, throws down a large quantity of a bulky, deep-yellow, insoluble substance, formerly called s u l p h o c y a n o g e n, from its supposed identity with the radicle of the sulphocyanates : it is, however, invariably found to contain hydrogen, and is represented by the formula $C_3N_3HS_3$. This yellow substance, now generally called p e r s u l p h o c y a n o g e n, is quite insoluble in water, alcohol, and ether. When heated in the dry state, it evolves sulphur and carbon bisulphide, and leaves a pale, straw-yellow substance called h y d r o m e l l o n e, $C_6N_9H_3$, the decomposition being represented by the equation :—

$$3C_3N_3HS_3 = 3CS_2 + S_3 + C_6N_9H_3.$$

Hydrogen Thiocyanate, or *Thiocyanic Acid*, HCNS, is obtained by decomposing lead thiocyanate, suspended in water, with sulphuretted hydrogen. The filtered solution is colorless, very acid, and not poisonous : it is easily decomposed, in a very complex manner, by ebullition, and by exposure to the air. By neutralizing the liquid with ammonia, and evaporating very gently to dryness, *ammonium thiocyanate*, NH_4CNS, is obtained as a deliquescent, saline mass. The salt may be conveniently prepared by digesting hydrocyanic acid with yellow ammonium sulphide (containing excess of sulphur), and boiling off the excess of the latter :—

$$2HCN + (NH_4)_2S + S_2 = H_2S + 2(NH_4)CNS.$$

The thiocyanates of *sodium, barium, strontium, calcium, manganese*, and *ferrous thiocyanate*, are colorless and very soluble ; those of *lead* and *silver* are white and insoluble. A soluble thiocyanate mixed with a ferric salt gives no precipitate, but causes the liquid to assume a blood-red tint : hence the use of potassium thiocyanate as a test for iron in the state of ferric salt. The red color produced by thiocyanates in ferric solutions is exactly like that caused under similar circumstances by meconic acid. The two substances may, however, be readily distinguished by the addition of a solution of gold chloride, which destroys the color produced by thiocyanates. The ferric meconate may also be distinguished from the thiocyanate by an addition of corrosive sublimate, which bleaches the thiocyanate, but has little effect upon the meconate. This is a point of considerable practical importance, as in medico-legal inquiries, in which evidence of the presence of opium is sought for in complex organic mixtures, the detection of meconic acid is usually the object of the chemist ; and since traces of alkaline thiocyanate are to be found in the saliva, it becomes very desirable to remove that source of error and ambiguity.

The great facility with which hydrocyanic acid may be converted into ammonium thiocyanate enables us to ascertain its presence by the iron test just described. The cyanide to be examined is mixed in a watch-glass with some hydrochloric acid and covered with another watch-glass, to which a few drops of yellow ammonium sulphide adhere. On heating the mixture, hydrocyanic acid is disengaged, which combines with the ammonium sulphide, and produces ammonium thiocyanate : this, after expulsion of the excess of sulphide, yields the red color with solution of ferric chloride.

Thiocyanic Ethers.—These ethers exhibit isomeric modifications analogous to those of the alcoholic cyanates and isocyanates (p. 570). The normal thiocyanates of methyl and its homologues were discovered by Cahours ; and Hofmann has obtained the corresponding isothiocyanates. The same chemist some years ago obtained phenyl isothiocyanate. Allyl thiocyanate has long been known as a natural product.

Normal Ethyl Thiocyanate, $C\left\{{N \atop S_2CH_5}\right.$, is obtained by saturating a concentrated solution of potassium thiocyanate with ethyl chloride :

$$C\left\{{N \atop SK}\right. \ + \ C_2H_5Cl \ = \ KCl \ + \ C\left\{{N \atop SC_2H_5}\right. ;$$

also by distilling a mixture of calcium ethylsulphate and potassium thiocyanate. It is a mobile, colorless, strongly refracting liquid, having a somewhat pungent odor like that of mercaptan. It boils at 146° C. (294.8° F.). With ammonia it does not combine directly, but yields products of decomposition.

The methyl and amyl thiocyanic ethers resemble the ethyl compound, and are obtained by similar processes. The methyl ether boils at about 132° C. (269.6° F.) ; the amyl ether at 197° C. (386.6° F.).

Ethyl Isothiocyanate, or *Ethylic Thiocarbimide,* $N\left\{{(CS)'' \atop C_2H_5}\right.$, is produced by distilling diethyl-thiocarbamide with phosphoric oxide, which abstracts ethylamine :

$$N_2\left\{{(CS)'' \atop (C_2H_5)_2 \atop H_2}\right. \ - \ N\left\{{C_2H_5 \atop H_2}\right. \ = \ N\left\{{(CS)'' \atop C_2H_5}\right.$$

Diethyl-thiocarbamide.　　Ethylamine.　　Ethyl-isothiocyanate.

This ether differs essentially in all its properties from ethyl thiocyanate. It boils at 134° C. (273.2° F.), and has a powerfully irritating odor, like that of mustard-oil, and quite different from that of normal ethyl sulphocyanate. It unites directly with ammonia in alcoholic solution, forming ethylthiocarbamide, $N_2(CS)''(C_2H_5)H_3$, and forms similar compounds with methylamine and ethylamine. The pungent odor, and the direct combination with ammonia and amines, are characteristic of all the ethers of this group.

Phenyl Isothiocyanate, $N(CS)''(C_6H_5)$, is obtained by distilling phenylsulphocarbamide, $N_2(CS)''(C_6H_5)H_3$, with phosphoric oxide ; *naphthyl isothio-*

cyanate, $N(CS)''(C_{10}H_7)$, in like manner from dinaphthylsulphocarbamide. The former boils at 220° C. (428° F.).

Allyl Isothiocyanate, or *Allylic Thiocarbimide*, $N \begin{cases} (CS)'' \\ C_3H_5 \end{cases}$.—This is the in-tensely pungent volatile oil obtained by distilling the seeds of black mustard with water. It does not exist ready-formed in the seeds, but is produced by the decomposition of m y r o n i c a c i d under the influence ot m y r o s i n, an albuminous substance analogous to the synaptase of bitter almonds. The same compound, or perhaps its isomeride, normal allyl thiocyanate, is produced by the action of potassium thiocyanate or silver thiocyanate on allyl iodide or allyl oxide.

Oil of mustard is a transparent, colorless, strongly refracting oil, pos-sessing in the highest degree the sharp penetrating odor of black mustard. The smallest quantity of the vapor excites tears, and is apt to produce inflammation of the eyes. It has a burning taste, and rapidly blisters the skin. Its specific gravity is 1.009 at 15° C. (59° F.). It boils at 148° C. (298.4° F.). It is sparingly soluble in water, easily soluble in alcohol and ether ; dissolves sulphur and phosphorus when heated, and deposits them in the crystalline state on cooling. It is violently oxidized by nitric and by nitromuriatic acid. Heated in a sealed tube with potassium monosul-phide, it yields potassium thiocyanate and a l l y l s u l p h i d e (volatile oil of garlic).

$$2(C_3H_5)NCS + K_2S = 2KCNS + (C_3H_5)_2S.$$

It likewise yields garlic oil when decomposed by potassium. Heated to 120° C. (248° F.) in a sealed tube with pulverized soda-lime, it yields sodium thiocyanate and a l l y l o x i d e, the oxidized constituent of garlic oil :

$$2(C_3H_5)NCS + Na_2O = 2NaCNS + (C_3H_5)_2O.$$

Aqueous potash, soda, baryta, and the oxides of lead, silver, and mercury, in presence of water, convert oil of mustard into s i n a p o l i n e, $C_7H_{12}N_2O$, with formation of metallic sulphide and carbonate ; thus :

$$2(C_3H_5)NCS + 3PbO + H_2O = 2PbS + PbCO_3 + C_7H_{12}N_2O.$$

Sinapoline is a basic substance, which crystallizes in colorless plates, soluble in water and alcohol, and having a distinct alkaline reaction.

Oil of mustard readily unites with ammonia, forming t h i o s i n a m i n e, $C_4H_8NS.NH_3$, or a l l y l - t h i o c a r b a m i d e, $N_2 \begin{cases} (CS)'' \\ C_3H_5, \\ H_3 \end{cases}$ which is also a basic compound, forming colorless prismatic crystals, having a bitter taste, and soluble in water. The solution does not affect test-paper. Thiosina-mine melts when heated, but cannot be sublimed. Acids combine with it, but do not form crystallizable salts ; the double salts of the hydrochloride with platinic and mercuric chlorides are the most definite.

Thiosinamine is decomposed by metallic oxides, as lead oxide or mercuric oxide, with production of a metallic sulphide and s i n a m i n e, $C_4H_6N_2$, a basic compound which crystallizes very slowly from a concentrated aqueous solution, in brilliant, colorless crystals containing water. It has a power-fully bitter taste, is strongly alkaline to test-paper, and decomposes ammo-nium salts at the boiling heat. Its oxalate is crystallizable. The formation of sinamine from thiosinamine by the action of mercuric oxide is represented by the equation, $C_4H_8N_2S + HgO = HgS + H_2O + C_4H_6N_2$.

Seleniocyanates.—A series of salts containing selenium, and corre-sponding in composition and properties with the thiocyanates, have been discovered and examined by Mr. Crookes.*

* Journal of the Chemical Society, iv. 12.

Cyanamide, $CN.NH_2$, is formed by the action of chloride or bromide of cyanogen on ammonia dissolved in ether :

$$CNCl + NH_3 = HCl + CN.NH_2,$$

also by the action of carbon dioxide on sodamide, NH_2Na :

$$CO_2 + 2NH_2Na = 2NaOH + CN.NH_2,$$

and by desulphurizing thiocarbamide with oxide of lead or mercury :

$$CS(NH_2)_2 + O = CN_2H_2 + H_2O + S.$$

Cyanamide forms colorless crystals, easily soluble in water, alcohol, and ether, melting at 40° C. (104° F.). The solutions give with ammoniacal silver nitrate a yellow precipitate of argentocyanamide, CN_2Ag_2, and with cupric sulphate a black precipitate of cuprocyanamide, CN_2Cu. By nitric, sulphuric, or phosphoric acid cyanamide is converted into c a r b a m i d e, $CO(NH_2)_2$. In this and most of its reactions it behaves as if it had the structure of c a r b o d i i m i d e, $C\diagup{NH}\diagdown{NH}$; thus :

$$C{\diagup NH \diagdown NH} + H_2O = CO\diagup{NH_2}\diagdown{NH_2}.$$

Hydrogen sulphide converts it, in like manner, into t h i o c a r b a m i d e, $CS(NH_2)_2$.

Alcoholic derivatives of cyanamide are formed by the action of cyanogen chloride on primary amines dissolved in ether :

$$C_2H_5.NH_2 + CNCl = HCl + CN.NH(C_2H_5) ;$$
Ethylamine.   Ethylcyanamide.

also by heating the corresponding thiocarbamides with mercuric oxide and water :

$$CS\diagup{NH(C_2H_5)}\diagdown{NH_2} + HgO = HgS + H_2O + CN.NH(CH_5).$$

Methyl-cyanamide and *Ethyl-cyanamide* are thick uncrystallizable syrups, having a neutral reaction, and easily converted into polymeric modifications.

Allyl-cyanamide, $CN.NH(C_3H_5)$, prepared from allyl-thiocarbamide, easily changes into the polymeric compound, t r i a l l y m e l a m i n e, $C_3N_6H_3(C_3H_5)_3$.

D i c y a n i m i d e, $NH(CN)_2$, is formed by the action of potash on normal potassium cyanate :

$$3(CN.OK) + KOH = NH(CN) + CO_3K_2 + K_2O.$$

D i c y a n o d i a m i d e, $C_2N_4H_4 = C_2N_2(NH_2)_2$ (*Param*), is formed, by polymerization of cyanamide, when the aqueous solution of the latter is left to itself, or evaporated ; also when thiocarbamide is boiled with mercuric oxide or silver oxide. It is easily soluble in water and in alcohol, and crystallizes in rhombic plates, melting at 205° C. (401° F.). On adding silver nitrate to its aqueous solution, the compound $C_2N_4H_3Ag+NO_3H$ crystallizes out in silky needles.

Cyanuramides.—From cyanuric acid, $C_3N_3(OH)_3$, may be derived three amides, viz. :

$$C_3N_3 \begin{cases} (OH)_2 \\ NH_2 \end{cases} \qquad C_3N_3 \begin{cases} OH \\ (NH_2)_2 \end{cases} \qquad C_3N_3(NH_2)_3$$
$$\text{Ammelide.} \qquad\qquad \text{Ammeline.} \qquad\qquad \text{Melamine.}$$

Melamine or *Cyanuramide*, $C_3N_6H_6 = N_3 \begin{cases} (C_3N_3) \\ H_6 \end{cases}$, is formed by poly-merization of cyanamide when the latter is heated to 150°; it is, however, more easily prepared by heating melam with potash-lye, the melam then being resolved into melamine and ammeline.

Melamine crystallizes in shining rhombic octohedrons, easily soluble in water, but insoluble in alcohol and ether. It forms well-crystallized salts containing 1 equivalent of acid, *e. g.*, $C_3N_6H_6.HCl$. When boiled with acids or alkalies, it is converted successively, by assumption of H_2O and elimination of NH_3, into ammeline, ammelide, and cyanuric acid.

Triethyl- and *trimethyl-melamine* are formed by polymerization of ethyl- and methyl-cyanamide, when the aqueous solutions of those bodies are evaporated. They are crystalline, strongly alkaline bodies, which are converted, by boiling with hydrochloric acid, into the corresponding deriva-tives of ammeline, and ultimately into isocyanuric ethers.

Ammeline, $C_3N_5H_5O = C_3N_3 \begin{cases} OH \\ (NH_2)_2 \end{cases}$, is a white powder insoluble in water, alcohol, and ether, but soluble in acids and alkalies, and forming crystalline salts, which are decomposed by water.

Ammelide, $C_3N_4H_4O_2 = C_3N_3 \begin{cases} (OH)_2 \\ NH_2 \end{cases}$, is a white powder which dis-solves in alkalies and in concentrated acids, but does not form definite salts with the latter. Its ammoniacal solution gives with silver nitrate a white precipitate having the composition

$$C_3N_3 \begin{cases} OH \\ OAg \\ NH_2 \end{cases}.$$

Melam, $C_6H_9N_{11}$, is a buff-colored, insoluble, amorphous substance, ob-tained by the distillation of ammonium thiocyanate at a high temperature. It may be prepared in large quantity by intimately mixing 1 part of per-fectly dry potassium thiocyanate with 2 parts of powdered sal-ammoniac, and heating the mixture for some time in a retort or flask : carbon bisul-phide, ammonium sulphide, and sulphuretted hydrogen, are disengaged and volatilized, while a mixture of melam, potassium chloride, and sal-ammoniac remains ; the two latter substances are removed by washing with hot water. Melam dissolves in concentrated sulphuric acid, and gives, by dilution with water and long boiling, cyanuric acid. The same substance is produced, with disengagement of ammonia, when melam is fused with potassium hydrate. When strongly heated, melam is resolved into mellone and ammonia :—

$$3C_6N_{11}H_9 = 2C_9N_{12} + 9NH_3 ;$$

and by prolonged boiling with moderation strong caustic potash, it is re-solved into melamine and ammeline :—

$$C_6N_{11}H_9 + H_2O = C_3N_6H_6 + C_3N_5H_5O.$$

25

Mellone and **Mellonides.**—The name mellone was given by Liebig to a yellow insoluble substance obtained as a residue in the decomposition of perthiocyanogen, or of melam, at a low red heat. The composition of the product thus obtained varies with the duration of the heating. If the decomposition be stopped at a certain point, the product has the composition of d i c y a n u r a m i d e , $N_3 \begin{cases} (C_3N_3)''' \\ (C_3N_3)''' \\ H_3 \end{cases}$; but this, when further heated still, gives off ammonia, and becomes continually richer in carbon, approaching, in fact, continually nearer to the composition of t r i c y a n u - r a m i d e , $C_9N_{12} = N_3(C_3N_3)_3$, which is probably the ultimate product of the decomposition, though it has never been actually attained, the product always containing a small quantity of hydrogen, however long the heating may be continued.

Mellone is also produced by ignition of ammonium thiocyanate and mercury thiocyanate. When boiled with nitric acid, it is resolved into ammonia and cyanuric acid.

M e l l o n i d e s , $C_9N_{13}M_3$ (M denoting a monatomic positive radicle).— *Tripotassic mellanide,* $C_9N_{13}K_3$, is produced by ignition of potassium thiocyanate, the preparation being greatly facilitated by the presence of a metallic compound, such as trichloride of antimony or bismuth, capable of taking up a portion of the sulphur which is set free as carbon sulphide. It may also be prepared by fusing potassium thiocyanate with crude mellone, or the ferrocyanide with half its weight of sulphur. The fused mass obtained by either process is dissolved in boiling water, from which the tripotassic mellonide crystallizes on cooling in silky needles, containing $C_9N_{13}K_3 +$ $5H_2O$, insoluble in alcohol and in ether. Acetic acid converts this salt into *dipotassic mellonide,* $C_9H_{13}K_2H$, which is also soluble. Hydrochloric acid produces the *monopotassic salt,* $C_9N_{13}KH_2$, which is insoluble. These three salts stand to each other in the same relation as the several salts of phosphoric and cyanuric acids. Tripotassic mellonide produces, with soluble silver salts, a white precipitate, $C_9N_{13}Ag_3$; with lead salts and mercury salts, precipitates containing respectively $(C_9N_{13})_2Pb_3$ and $(C_9N_{13})_2Hg_3$. The latter, dissolved in hydrocyanic acid and treated with sulphuretted hydrogen, yields *hydromellonic* acid, $C_9N_{13}H_3$, which is known only in solution, has an acid taste, and is decomposed by evaporation into ammonia and mellone.

Cyameluric Acid, $C_6N_7O_3H_3$.—The potassium salt of this acid is formed by boiling tripotassic mellonide with strong potash-lye, and the acid itself may be separated from the solution of the salt by a mineral acid in the form of a crystalline precipitate. It is a strong tribasic acid, converted by heating with mineral acids into cyanuric acid.

ALCOHOLS AND ETHERS.

The term alcohol, originally limited to one substance, viz., spirit of wine, is now applied to a large number of organic compounds, many of which, in their external characters, exhibit but little resemblance to common alcohol. They are all, however, analogously constituted, having the composition of hydrocarbons in which one or more of the hydrogen-atoms are replaced by hydroxyl; they may therefore be regarded as compounds of hydroxyl with univalent or multivalent hydrocarbon-radicles, hence called alcohol-radicles. Thus from propane C_3H_8, are derived the three alcohols:—

$$C_3H_7(OH) \qquad C_3H_6(OH)_2 \qquad C_3H_5(OH)_3$$

Propyl alcohol.   Propene alcohol.   Propenyl alcohol.

Alcohols are accordingly classed as monatomic, diatomic, triatomic, etc., or generally as monatomic and polyatomic, according to the number of hydroxyl-groups which they contain, or according to the equivalent values of their hydrocarbon-radicles.

The replacement, partial or total, of the hydroxyl in an alcohol by chlorine, bromine, iodine, or fluorine, gives rise to haloïd ethers; thus:—

From $C_3H_7(OH)$ are derived C_3H_7Cl, C_3H_7Br, etc.
" $C_3H_6(OH)_2$ " $C_3H_6Cl(OH)$, $C_3H_6Cl_2$, etc.
" $C_3H_5(OH)_3$ " $C_3H_5Cl(OH)_2$, $C_3H_5Cl_2(OH)$, $C_3H_6Cl_3$, etc.

These substitutions are effected by treating the alcohols with the chlorides, bromides, and iodides of hydrogen or phosphorus, as in the following equations, which represent the formation of ethyl chloride from common alcohol:—

$$C_2H_5(OH) \quad + \quad HCl \quad = \quad H(OH) \quad + \quad C_2H_5Cl$$
$$3C_2H_5(OH) \quad + \quad PCl_3 \quad = \quad P(OH)_3 \quad + \quad 3C_2H_5Cl$$
$$3C_2H_5(OH) \quad + \quad POCl_3 \quad = \quad PO(OH)_3 \quad + \quad 3C_2H_5Cl$$

Instead of the bromides and iodides of phosphorus, the elements phosphorus and bromine or iodine, in the proportions required to form them, are often used in these processes.

These haloïd ethers are also formed in many instances by direct substitution of chlorine, bromine, etc., for hydrogen in saturated hydrocarbons, as explained in the preceding pages.

The treatment of the haloïd ethers with caustic aqueous alkalies gives rise to a substitution opposite to that exhibited in the above equations, reconverting the ethers into alcohols, e. g.:—

$$C_2H_5Cl \quad + \quad KOH \quad = \quad KCl \quad + \quad C_2H_5(OH).$$

A considerable portion of the alcohol thus produced is, however, converted, by dehydration, into the corresponding olefine; e. g.: $C_3H_8O - H_2O = C_3H_6$. A better result is obtained by heating the haloïd ether with moist silver oxide, which acts like a hydroxide, AgOH. A still better method is to convert the alcoholic chloride, etc., into an acetate, by heating it with acetate of silver or potassium, and boil the resulting acetic ether with caustic potash or soda; thus:—

$$C_2H_5Cl \quad + \quad \underset{\substack{\text{Potassium} \\ \text{acetate.}}}{KC_2H_3O_2} \quad = \quad KCl \quad + \quad \underset{\text{Ethyl acetate.}}{C_2H_5.C_2H_3O_2}$$

and

$$\underset{\text{Ethyl acetate.}}{C_2H_5.C_2H_3O_2} \quad + \quad KOH \quad = \quad \underset{\text{Ethyl alcohol.}}{C_2H_5OH} \quad + \quad \underset{\substack{\text{Potassium} \\ \text{acetate.}}}{K.C_2H_3O_2}$$

The replacement of the hydroxyl in an alcohol by the corresponding radicles, potassoxyl, OK, methoxyl, OCH_3, ethoxyl, OC_2H_5, etc. (p. 255),— or of the hydrogen in the hydroxyl by potassium, methyl, ethyl, etc.,— gives rise to o x y g e n - e t h e r s ; thus :—

$$\underset{\substack{\text{Ethyl} \\ \text{alcohol.}}}{C_2H_5(OH)} \quad \text{yields} \quad \underset{\substack{\text{Potassium} \\ \text{ethylate.}}}{C_2H_5(OK)} \quad \underset{\substack{\text{Methyl} \\ \text{ethylate.}}}{C_2H_5(OCH_3)} \quad \underset{\substack{\text{Ethyl} \\ \text{ethylate.}}}{C_2H_5(OC_2H_5)}$$

$$\underset{\substack{\text{Ethene} \\ \text{alcohol.}}}{C_2H_4(OH)_2} \quad \text{''} \quad \underset{\substack{\text{Monethylic} \\ \text{ethenate.}}}{C_2H_4(OH)(OC_2H_5)} \quad \underset{\substack{\text{Diethylic} \\ \text{ethenate.}}}{C_2H_4(OC_2H_5)_2}$$

These substitutions may be effected in various ways. The simplest is to replace an atom of hydrogen in the alcohol by potassium or sodium, and act on the resulting compound with a haloïd ether ; thus :—

$$\underset{\substack{\text{Ethene} \\ \text{alcohol.}}}{2C_2H_4(OH)_2} \quad + \quad Na_2 \quad = \quad \underset{\substack{\text{Sodium} \\ \text{ethenate.}}}{2C_2H_4(OH)(ONa)} \quad + \quad H_2,$$

$$\underset{\substack{\text{Sodium} \\ \text{ethenate.}}}{C_2H_4(OH)(ONa)} \quad + \quad \underset{\substack{\text{Ethyl} \\ \text{iodide.}}}{C_2H_5I} \quad = \quad \underset{\substack{\text{Sodium} \\ \text{iodide.}}}{NaI} \quad + \quad \underset{\substack{\text{Monethylic} \\ \text{ethenate.}}}{C_2H_4(OH)(OC_2H_5)}$$

In the polyatomic alcohols, two hydroxyl groups may also be replaced by one atom of oxygen, giving rise to another class of óxygen ethers ; thus, from ethene alcohol, $C_2H_4(OH)_2$, is derived e t h e n e o x i d e, C_2H_4O.

The replacement of the hydrogen of the hydroxyl in an alcohol by acid radicles (p. 509), produces e t h e r e a l s a l t s or c o m p o u n d e t h e r s : thus, from methyl alcohol, $CH_3(OH)$, are derived :—

Methyl nitrate, $CH_3(ONO_2)$, or

$$H-\underset{\underset{H}{|}}{\overset{\overset{H}{|}}{C}}-O-\underset{\underset{O}{\|}}{\overset{\overset{O}{\|}}{N}}$$

Methyl acetate, $CH_3(OC_2H_3O)$, or

$$H-\underset{\underset{H}{|}}{\overset{\overset{H}{|}}{C}}-O-\underset{}{\overset{\overset{O}{\|}}{C}}-\underset{\underset{H}{|}}{\overset{\overset{H}{|}}{C}}-H$$

Acid methyl sulphate, $CH_3(OSO_3H)$, or

$$H-\underset{\underset{H}{|}}{\overset{\overset{H}{|}}{C}}-O-\underset{\underset{O}{\|}}{\overset{\overset{O}{\|}}{S}}-O-H$$

Neutral methyl sulphate, $CH_3(OSO_3CH_3)$, or

$$H-\underset{\underset{H}{|}}{\overset{\overset{H}{|}}{C}}-O-\underset{\underset{O}{\|}}{\overset{\overset{O}{\|}}{S}}-O-\underset{\underset{H}{|}}{\overset{\overset{H}{|}}{C}}-H$$

These ethereal salts may also be derived from the corresponding acids by substitution of alcohol-radicles for hydrogen, being in fact related to the alcohols in the same manner as metallic salts to metallic hydroxides. When distilled with alkalies, they are resolved into an acid and an alcohol ; *e. g.* :

$$\underset{\text{Ethyl acetate.}}{C_2H_5(OC_2H_3O)} + \underset{\substack{\text{Potassium} \\ \text{hydroxide.}}}{K(OH)} = \underset{\substack{\text{Potassium} \\ \text{acetate.}}}{K(OC_2H_3O)} + \underset{\substack{\text{Ethyl} \\ \text{alcohol.}}}{C_2H_5(OH)}$$

The number of compound ethers that can be formed by a given acid and alcohol depends upon the number of hydroxyl-groups which each of them contains, in other words, on their atomicity : thus, ethyl alcohol, C_2H_5OH, and acetic acid, $C_2H_3O.OH$, both of which are monatomic, yield but one ether, viz., ethyl acetate, $C_2H_3O.O.C_2H_5$; but ethene alcohol or glycol, $C_2H_4.(OH)_2$, which is diatomic, forms two ethers with acetic acid, viz. :

$$\underset{\text{Mono-acetate.}}{C_2H_4{<}^{OH}_{O.C_2H_3O}} \quad \text{and} \quad \underset{\text{Diacetate.}}{C_2H_4{<}^{O.C_2H_3O}_{O.C_2H_3O}}$$

and glycerin, $C_3H_5(OH)_3$, which is triatomic, forms three acetic ethers, viz. :

$$\underset{\text{Acetin.}}{C_3H_5{<}^{OH}_{\substack{OH \\ O.C_2H_3O}}} \quad \underset{\text{Diacetin.}}{C_3H_5{<}^{OH}_{\substack{O.C_2H_3O \\ O.C_2H_3O}}} \quad \underset{\text{Triacetin.}}{C_3H_5{<}^{O.C_2H_3O}_{\substack{O.C_2H_3O \\ O.C_2H_3O}}}$$

On the other hand, nitric acid, $NO_2.OH$, which is monatomic, forms but one ethylic ether, viz., ethyl nitrate, $NO_2.O.C_2H_5$; but sulphuric acid, $SO_2(OH)_2$, which is diatomic, forms two ethylic ethers, viz. :

$$\underset{\substack{\text{Ethyl-sulphuric} \\ \text{acid.}}}{SO_2{<}^{OH}_{O.C_2H_5}} \quad \underset{\substack{\text{Diethylic} \\ \text{sulphate.}}}{SO_2{<}^{O.C_2H_5}_{O.C_2H_5}}$$

Compound ethers are formed : 1. By the action of the acids upon alcohols ; *e. g.* :

$$\underset{\text{Ethyl alcohol.}}{C_2H_5.OH} + \underset{\text{Nitrous acid.}}{NO.OH} = H_2O + \underset{\text{Ethyl nitrite.}}{NO.O.C_2H_5}$$

This action takes place slowly at ordinary, more quickly at high temperatures ; but the etherification is never complete, as the water separated in the process always exerts a reverse action on the ether, reconverting part of it into alcohol and acid.

Polybasic acids, acting upon alcohols, produce for the most part acid ethers.

2. By the action of chlorides of acid radicles on alcohols or their sodium derivatives :

$$\underset{\substack{\text{Sulphuric} \\ \text{chloride.}}}{SO_2Cl_2} + \underset{\text{Ethyl alcohol.}}{2(C_2H_5.OH)} = 2HCl + \underset{\text{Ethyl sulphate.}}{SO_2(O.C_2H_5)_2}$$

3. By the action of haloïd ethers on the potassium or silver salts of the corresponding acids :

$$C_2H_3O.OAg + C_2H_5I = AgI + C_2H_3O.O.C_2H_5.$$

Neutral compound ethers are mostly volatile : the acid ethers are non-volatile, and as they still contain unreplaced hydrogen belonging to one or more HO groups, they act as acids, exchanging this hydrogen for metals or for alcohol-radicles.

All compound ethers, but especially the acid ethers, are resolved into their components, alcohol and acid, by heating them with water, or more easily with potash or soda in aqueous or alcoholic solution :

$$C_2H_3O.O.C_2H_5 \ + \ KOH \ = \ C_2H_3O.O.K \ + \ C_2H_5.OH$$
Ethyl acetate.   Potassium   Ethyl
                        acetate.   alcohol.

This process is called s a p o n i f i c a t i o n, a term originally applied to the formation of soaps by boiling neutral fats (glycerin-ethers) with caustic alkali, but now extended to all similar decompositions.

Isomerism in the Compound Ethers.—The ethers of polybasic organic acids exhibit isomeric modifications, depending on the structure of the acid itself. The most important case of this kind of isomerism is presented by the s u l p h u r o u s e t h e r s.

Sulphurous acid is susceptible of two modifications, viz. :

$$\overset{O}{\underset{\parallel}{HO{-}S{-}OH}} \text{ or } SO\overset{OH}{\underset{OH}{<}} \qquad \overset{O}{\underset{\underset{O}{\parallel}}{\overset{\parallel}{H{-}S{-}OH}}} \text{ or } SO_2\overset{H}{\underset{OH}{<}}$$

          Symmetrical.   Unsymmetrical.

Both these modifications yield neutral alcoholic derivatives or sulphurous ethers, but only the unsymmetrical acid appears to form acid ethers, and these acid ethers are called s u l p h o n i c a c i d s, *e. g.*:

Symmetrical :

Dimethylic sulphite, $SO\overset{OCH_3}{\underset{OCH_3}{<}}$

Unsymmetrical :

Methysulphonic acid, $SO_3\overset{CH_3}{\underset{OH}{<}}$

Methylic methysulphonate, $SO_2\overset{CH_3}{\underset{OCH_3}{<}}$

The sulphonic acids are formed by the following general reactions :—

1. By heating the haloïd ethers to 120–150° C. (248°–302° F.) with a concentrated solution of potassium or sodium sulphite ; thus :

$$C_2H_5I \ + \ SO_2\overset{K}{\underset{OK}{<}} \ = \ SO_2\overset{C_2H_5}{\underset{OK}{<}} \ + \ KI$$
                              Potassium ethyl-
                              sulphonate.

$$C_2H_4Br_2 \ + \ 2SO_2\overset{K}{\underset{OK}{<}} \ = \ \begin{matrix} SO_2\overset{OK}{<} \\ \qquad C_2H_4 \\ SO_2\underset{OK}{<} \end{matrix} \ + \ 2KBr$$
                              Potassium ethene-
                              sulphonate.

2. By oxidation of the hydrosulphides (mercaptans) and disulphides (also of the thiocyanates) of the alcohol radicles with nitric acid :

$$S\overset{C_2H_5}{\underset{H}{<}} \ + \ O_3 \ = \ SO_2\overset{C_2H_5}{\underset{OH}{<}}$$

$$(C_2H_5)S_2 \ + \ O_5 \ + \ H_2O \ = \ C_4H_5.2(SO_2.OH).$$

This mode of formation shows that the sulphur-atom of a sulphonic acid is directly united to the alcohol-radicle (and therefore to an atom of carbon), and hence it may be inferred, from the first reaction, that in the metallic sulphites an atom of metal is also directly united to the sulphur.

The sulphonic acids are very stable compounds, not decomposed by boiling with caustic alkaline solutions. When fused with potassium hydroxide, however, they are resolved, into an alcohol and sulphurous acid:

$$C_2H_5.SO_2.OK + KOH = C_2H_5.OH + K.SO_3.OK.$$

By the action of phosphorus pentachlorides they are converted into acid chlorides or chloranhydrides, which, by the action of nascent hydrogen, may be converted into mercaptans:

$$C_2H_5.SO_2.OH + PCl_5 = PCl_3O + HCl + C_2H_5.SO_2.Cl$$
$$\text{and } C_2H_5.SO_2.Cl + 3H_2 = HCl + 2H_2O + C_2H_5.SH.$$

The ethers of the sulphonic acids are formed by the action of acid chlorides on the sodium-alcohols:

$$C_2H_5.SO_2.Cl + Na.OC_2H_5 = NaCl + C_2H_5.SO_2.OC_2H_5.$$

They are identical with the neutral ethers of unsymmetrical sulphurous acid, and may accordingly be also produced by the action of silver sulphite on alcoholic iodides:

$$Ag.SO_2.OAg + 2C_2H_5I = 2AgI + C_2H_5.SO_2.O.C_2H_5.$$

Analogous to the sulphonic acids are the sulphinic acids, or acid ethers of hyposulphurous acid, $SO<^H_{OH}$, the zinc salts of which are formed by the action of sulphur dioxide on the zinc compounds of the alcohol-radicles; e. g:

$$Zn<^{CH_3}_{CH_3} + 2SO_2 = SO<^{CH_3}_{O-Zn-O}>^{H_3C}SO.$$

By converting the zinc salts into barium salts, and decomposing the latter with sulphuric acid, the free sulphinic acids are obtained as strongly acid liquids which decompose when heated, e. g:

$$SO<^{CH_3}_{OH}.$$

The phosphorous ethers likewise exhibit isomeric modifications, derivable from symmetrical and unsymmetrical phosphorous acid $P(OH)_3$ and $HPO(OH)_2$. The neutral ethers of symmetrical phosphorous acid are formed by the action of phosphorous trichloride on the alcohols. The ethylic ether $P(O.C_2H_5)_3$, is a liquid boiling at 191° C. (275.8° F.).

Unsymmetrical phosphorous acid yields acid ethers called phosphonic acids, e. g., $(C_2H_5)PO<^{OH}_{OH}$, which are formed by the action of nitric acid on the primary phosphines:

$$\underset{\substack{\text{Primary methyl}\\\text{phosphine.}}}{CH_3.PH_2} + O_3 = \underset{\substack{\text{Methylphosphonic}\\\text{acid.}}}{CH_3.PO(OH)_2}$$

They are crystalline bodies resembling spermaceti, dissolve in water, have a strong acid reaction, and act as bibasic acids, forming acid and neutral salts.

Hypophosphorous acid, $H_2.PO.OH$, yields analogous acid ethers, called p h o s p h i n i c   a c i d s , which are formed by the action of nitric acid on the secondary phosphines; *e. g :*

$$(CH_3)_2PH + O_2 = (CH_3)_2PO.OH.$$

A r s e n i o u s   a c i d , AsH_3O_3, forms :

(*a*) *Symmetrical ethers*, like $As(O.CH_3)_3$, which are produced by the action of arsenious bromide, $AsBr_3$, on the sodium alcohols; they are volatile without decomposition, and are resolved by water into arsenious acid and alcohols.

(*ß*) *Unsymmetrical ethers*, called a r s o n i c   a c i d s , analogous to the phosphonic acids, *e. g.*, m e t h y l a r s o n i c   a c i d , $CH_3.AsO(OH)_2$; and lastly, there is a d i m e t h y l a r s i n i c   a c i d $(CH_3)_2AsO.OH$ (cacodylic acid), analogous to the phosphinic acids (see ARSENIC BASES.)

The n i t r o u s   e t h e r s , as already observed, are isomeric with the nitroparaffins (p. 551).

The action of haloïd ethers, or of certain ethereal salts, on the hydrosulphides and sulphides of the alkali-metals, gives rise to a l c o h o l i c h y d r o s u l p h i d e s   and   s u l p h i d e s , that is to say, alcohols and ethers containing sulphur in place of oxygen ; thus :—

$$C_2H_5Cl + KSH = KCl + C_2H_5SH$$

Ethyl chloride.   Ethyl hydrosulphide.

$$2C_2H_5OSO_3K + KSK = 2KOSO_3K + C_2H_5SC_2H_5$$

Potassium ethyl sulphate.   Potassium sulphide.   Potassium sulphate.   Ethyl sulphide.

The alcoholic hydrosulphides, or thio-alcohols, are also called m e r c a p t a n s , from their property of readily combining with mercury (*corpora mercurio apta*). Their reactions are closely analogous to those of the oxygen-alcohols.

MONATOMIC ALCOHOLS AND ETHERS.

1. Containing the Radicles, C_nH_{2n+1}, homologous with Methyl.

The alcohols of this series are the best known and most important of all this class of bodies. They may be formed from the corresponding haloïd ethers by the action of alkalies, and several of them are produced by the fermentation of sugar. There are also synthetical processes by which these alcohols may be built up in regular order, from the lowest upwards ; but these will be better understood further on.

The names and formulæ of the known alcohols of this series are as follows :—

Methyl alcohol	CH_4O
Ethyl alcohol	C_2H_6O
Propyl alcohol	C_3H_8O
Butyl alcohol	$C_4H_{10}O$
Pentyl or Amyl alcohol	$C_5H_{12}O$
Hexyl alcohol	$C_6H_{14}O$
Heptyl alcohol	$C_7H_{16}O$
Octyl alcohol	$C_8H_{18}O$
Nonyl alcohol	$C_9H_{20}O$
Hexdecyl or Cetyl alcohol	$C_{16}H_{34}O$
Ceryl alcohol	$C_{27}H_{56}O$
Melissyl alcohol	$C_{30}H_{62}O$

The formula of methyl alcohol is that of methane or marsh-gas having one atom of hydrogen replaced by hydroxyl; and the rest may be derived from it by replacement of one or more of the other hydrogen-atoms by methyl and its homologues. If we replace only one atom of hydrogen in this manner we obtain the series :—

Methyl alcohol.	Ethyl alcohol.	Propyl alcohol.	Butyl alcohol.	Pentyl alcohol.
$C\begin{cases}H\\H\\H\\OH\end{cases}$	$C\begin{cases}CH_3\\H\\H\\OH\end{cases}$	$C\begin{cases}C_2H_5\\H\\H\\OH\end{cases}$	$C\begin{cases}C_3H_7\\H\\H\\OH\end{cases}$	$C\begin{cases}C_4H_9\\H\\H\\OH\end{cases}$

Now it is clear that, so long as the type of an alcohol is preserved—that is, of a hydrocarbon having at least one hydrogen-atom replaced by hydroxyl—the first two alcohols of this series do not admit of any other mode of formulation: in other words, they are not susceptible of isomeric modifications. But the higher members of the series admit of isomeric modifications analogous to those of the haloïd ethers already mentioned (p. 524), and distinguished by similar names, an alcohol being designated as p r i m a r y, s e c o n d a r y, or t e r t i a r y, according as the carbon-atom which is in combination with hydroxyl, is likewise directly united to one, two, or three other carbon-atoms. Moreover, the four-carbon-alcohol and all above it in the series, admit of further modifications, according to the structure of the radicles (*normal* or *iso*) contained in them—all these modifications, like those of the haloïd ethers, depending upon the struc-ture of the paraffins from which the alcohols are derived; thus—

From :

Butane.	Normal Primary Butyl Alcohol.		Secondary Butyl Alcohol.	
CH_3	CH_2OH		CH_3	
CH_2 are derived	CH_2 or $H_2C\begin{cases}C_2H_4OH\\CH_3\end{cases}$ and		CH_2 or $HO.HC\begin{cases}C_2H_5\\CH_3\end{cases}$	
CH_2	CH_2		$CHOH$	
CH_3	CH_3		CH_3	

and rom

Isobutane.	Isoprimary Butyl Alcohol.		Tertiary Butyl Alcohol.	
$H_3C\ \ CH_3$	$H_3C\ \ CH_3$		$H_3C\ \ CH_3$	
$\vee$	$\vee$		$\vee$	
CH are derived	CH or $HC\begin{cases}CH_3\\CH_3\\CH_2OH\end{cases}$ and		COH or $HO.C\begin{cases}CH_3\\CH_3\\CH_3\end{cases}$	
CH_3	CH_2OH		CH_3	

The primary alcohols are formed by substitution of OH for H in a methyl-group CH_3; the secondary and tertiary alcohols by similar substi-tution in a methane group CH_2, and a methenyl-group CH respectively.

A very convenient nomenclature for these isomeric alcohols has been proposed by Kolbe. Methyl alcohol, $CH_3(OH)$, is called c a r b i n o l ; and the alcohols formed from it by successive substitution of methyl, ethyl, etc., for an atom of hydrogen, are named according to the radicles which they contain ;* thus—

* Odling modifies this nomenclature by restricting the term c a r b i n o l to the tertiary alcohols, designating the secondary alcohols as p s e u d o - a l c o h o l s or p s e u d o l s, and employing the term a l c o h o l especially to designate the

25 *

Carbinol, or Methyl alcohol . . . . . $C(OH)H_3$
Methyl carbinol, or Ethyl alcohol . . . . $C(OH)H_2CH_3$
Ethyl carbinol, or Propyl alcohol . . . . $C(OH)H_2C_2H_5$
Dimethyl carbinol, or Pseudopropyl alcohol . . $C(OH)H(CH_3)_2$
Propyl carbinol, or Butyl alcohol . . . . $C(OH)H_2(C_3H_7)$
Isopropyl carbinol, or Isobutyl alcohol . . . $C(OH)H_2CH(CH_3)_2$
Methyl-ethyl carbinol, or Secondary Butyl alcohol $C(OH)HCH_3C_2H_5$
Trimethyl carbinol, or Tertiary Butyl alcohol . $C(OH)(CH_3)_3.$

Primary, secondary, and tertiary alcohols are distinguished from one another by the products which they yield by oxidation. Primary alcohols of the series $C_nH_{2n+2}O$, containing the group CH_2OH, are converted by oxidation with chromic acid, first into the corresponding a l d e h y d e s $C_nH_{2n}O$ by removal of H_2, or conversion of the group CH_2OH into COH, and then by further oxidation into the corresponding a c i d s $C_nH_{2n}O_2$ (fatty acids) ; thus :—

$$CH_3\text{—}CH_2\text{—}CH_2OH \ + \ O \ = \ H_2O \ + \ CH_3\text{—}CH_2\text{—}COH$$
Normal propyl alcohol.   Propyl aldehyde.

$$CH_3\text{—}CH_2\text{—}CH_2OH \ + \ O_2 \ = \ 2H_2O \ + \ CH_3\text{—}CH_2\text{—}COOH$$
Normal propyl alcohol.   Proprionic acid.

A secondary alcohol, on the other hand, which contains two alcohol-radicles united by the group CHOH, is converted, by removal of H_2 from this group, into a k e t o n e, which is a compound consisting of two alcohol-radicles joined by the group CO ; thus :—

$$CH_3\text{—}CHOH\text{—}CH_3 \ + \ O \ = \ H_2O \ + \ CH_3\text{—}CO\text{—}CH_3$$
Secondary propyl alcohol.   Dimethyl ketone.

Conversely, the aldehydes treated with nascent hydrogen (action of so-dium amalgam) are converted into primary alcohols, and the ketones by similar treatment into secondary alcohols.

Tertiary alcohols do not yield by oxidation either aldehydes, ketones, or acids containing the same number of carbon-atoms as the alcohols them-selves, but are split up into compounds containing a smaller number of carbon-atoms ; tertiary butyl alcohol, for example, into formic and pro-prionic acid :—

$$(CH_3)_3.COH \ + \ O_4 \ = \ CH_2O_2 \ + \ C_3H_6O_2 \ + \ H_2O.$$

METHYL ALCOHOL AND ETHERS.

Methyl Alcohol, Hydroxymethane, Carbinol, Me-thol, CH_4O or $CH_3(OH)$.—This is the simplest member of the series. It is produced :—

alcohols, proper or primary alcohols, which may also be called m e t h o l s ; thus—

		Propyl Alcohol or Ethyl-Methol.	Isobutyl Alcohol or Isopropyl-Methol.
Primary	. . . . .	$H_2O\begin{cases}CH_2OH\\CH_3\end{cases}$	$HO\begin{cases}CH_2OH\\CH_3\\CH_3\end{cases}$
		Dimethyl Pseudol.	Ethyl-Methyl Pseudol.
Secondary	. . . . .	$HO\,CH\begin{cases}CH_3\\CH_3\end{cases}$	$HO.HO\begin{cases}C_2H_5\\CH_3\end{cases}$
		Trimethyl Carbinol.	Ethyl-Dimethyl Carbinol.
Tertiary	. . . .	$HO.C\begin{cases}CH_3\\CH_3\\CH_3\end{cases}$	$HO.C\begin{cases}C_2H_5\\CH_3\\CH_3\end{cases}$

1. From marsh-gas, by subjecting that compound to the action of chlorine in sunshine, whereby chloromethane, or methyl chloride, CH_3Cl, is produced, and distilling with potash.

2. From wintergreen oil, which consist chiefly of acid methyl salicylate, $C_7H_4O_3.H.CH_3$, by distillation with potash, whereby potassium salicylate is formed, and methyl alcohol distils over :—

$$C_7H_4O_3.H.CH_3 \;+\; KOH \;=\; C_7H_4O_3.HK \;+\; CH_3(OH).$$

This reaction, which consists in the interchange of methyl and potassium, yields very pure methyl alcohol.

3. From crude wood-vinegar, the watery liquid obtained by the destructive distillation of wood : it was in this liquid that methyl alcohol was first discovered by P. Taylor, in 1812 : hence it is often called *wood-spirit*. Crude wood-vinegar probably contains about $\frac{1}{100}$ part of methyl alcohol, which is separated from the great bulk of the liquid by distilling it, and collecting apart the first portions which pass over. The acid solution thus obtained is neutralized with slaked lime, and the clear liquid, separated from the oil which floats on the surface, and from the sediment at the bottom is again distilled. A volatile liquid is thus obtained, which burns like weak spirit ; this may be strengthened by rectification, and ultimately rendered pure and anhydrous by careful distillation from quicklime at the heat of a water-bath.

Pure methyl alcohol is a thin, colorless liquid, very similar in smell and taste to ethyl alcohol : crude wood-spirit, on the other hand, which contains many impurities, has an offensive odor and a nauseous, burning taste. Methyl alcohol boils at 66.6° C. (151.9° F.), and has a density of 0.798 at 20° C. (68° F.). Vapor-density (referred to hydrogen) = 16. Methyl alcohol when pure mixes in all proportions with water : it dissolves resins and volatile oils as freely as ethyl alcohol, and is often substituted for ethyl alcohol in various processes in the arts. It may be burnt instead of ordinary spirit in lamps : the flame is pale-colored, like that of ethyl alcohol, and deposits no soot. Methyl alcohol dissolves caustic baryta : the solution deposits, by evaporation in a vacuum, acicular crystals, containing $BaO.2CH_4O$. It dissolves calcium chloride in large quantity, and gives rise to a crystalline compound containing $CaCl_2.2CH_4O$.

Potassium and *sodium* dissolve in it, with evolution of hydrogen yielding potassium and sodium methylates, CH_3OK, and CH_3ONa.

By *oxidation*, as by exposure to the air in contact with platinum black, it is converted into formic acid, CH_2O_2, which is derived from it by substitution of 1 atom of oxygen for 2 atoms of hydrogen :

$$CH_4O \;+\; O_2 \;=\; H_2O \;+\; CH_2O_2.$$

Methyl Chloride, or **Chloromethane,** CH_3Cl, is formed when a mixture of equal volumes of methane (marsh-gas) and chlorine is exposed to reflected sunlight. It is more easily prepared, however, by heating a mixture of 2 parts of common salt, 1 part of wood-spirit, and 3 parts of concentrated sulphuric acid. It is a gaseous body, which may be conveniently collected over water, as it is but slightly soluble in that liquid. It is colorless ; has a peculiar odor and sweetish taste, and burns, when kindled, with a pale flame, greenish towards the edges, like most combustible chlorine-compounds. Its density, referred to hydrogen as unity, is 25.25 ; it is not liquefied at —18° C. (0.4° F.). The gas is decomposed by transmission through a red-hot tube, with slight deposition of carbon, into hydrochloric acid gas and a hydrocarbon which has been but little examined. By the action of chlorine in sunshine it is successively converted into *methene chloride,* or *dichloromethane,* CH_2Cl_2, a liquid boiling at

30.5° C. (86.9° F.) ; *methenyl chloride, trichloromethane,* or *chloroform,* $CHCl_3$; and *carbon tetrachloride,* CCl_4.

Methyl Iodide, or **Iodomethane,** CH_3I, is a colorless and feebly combustible liquid, obtained by distilling together 1 part of phosphorus, 8 of iodine, and 12 or 15 of methyl-alcohol. It is insoluble in water, has a density of 2.237, and boils at 44° C. (111.2° F.). The density of its vapor, referred to hydrogen as unity, is 71. When digested in sealed tubes with *zinc,* it yields a colorless gaseous mixture containing ethane, or dimethyl, C_2H_6, and the residue contains zinc iodide, together with zinc methide, $Zn(CH_3)_2$:

$$2CH_3I + Zn = ZnI_2 + C_2H_6$$
$$2CH_3I + Zn_2 = ZnI_2 + Zn(CH_3)_2.$$

Methyl Ether, Methyl Oxide, or **Methoxyl-methane,** $C_2H_6O =$ $(CH_3)_2O = C \begin{cases} H_3 \\ OCH_3 \end{cases}$.—This compound, which bears the same relation to methyl alcohol that anhydrous potassium oxide bears to potassium hydroxide, is produced by abstraction of the elements of water from methyl alcohol: $2CH_4O - H_2O = C_2H_6O$.

It may be prepared by heating 1 part of methyl-alcohol and 4 parts of concentrated sulphuric acid, and passes over as a colorless gas, which may be collected over mercury. It does not liquefy at —16° C. (3.2° F.). It has an ethereal odour, and burns with a pale and feebly-luminous flame. Its specific gravity is 1.617 referred to air, or 23 referred to hydrogen as unity. Cold water dissolves about 33 times its volume of this gas, acquiring thereby its characteristic taste and odor: on boiling the solution, the gas is again liberated. Alcohol, wood-spirit, and concentrated sulphuric acid dissolve it in still larger quantity.

Methyl Nitrate, $CH_3.NO_3$, or $CH_3.O.NO_2$. This ether is obtained by distilling 50 grams of pounded nitre with 50 grams of methyl alcohol and 100 grams of sulphuric acid, in a retort without external heating. It is a colorless liquid of sp. gr. 1.182 at 20° C. (68° F.) ; boils at 60° C. (140° F.) ; has a faint ethereal odor. Its vapor detonates violently when heated to 150° C. (302° F.). Heated with *alcoholic ammonia,* it yields methylamine nitrate, $CH_5N.NO_3H$. Distilled with aqueous *potash,* it yields methyl ether.

Methyl Nitrite, $CH_3.O.NO$, isomeric with nitromethane, $CH_3.NO_2$ (p. 552), is produced by the action of nitrous acid on methyl alcohol. It is a gas having a pleasant odor, and condensing at very low temperatures to a yellowish liquid which boils at —12° C. (10.4° F.).

Methyl Sulphates.—Sulphuric acid, being a bibasic acid, yields two methyl ethers—one acid, the other neutral.

Acid Methyl sulphate, Methyl and *Hydrogen sulphate, Methylsulphuric acid,* or *Sulphomethylic acid,* $CH_3.H.SO_4$, or $SO_2 {<}{OCH_3 \atop OH}$.—To prepare this acid ether, 1 part of methyl alcohol is slowly mixed with two parts of concentrated sulphuric acid, and the whole is heated to boiling, and left to cool, after which it is diluted with water, and neutralized with barium carbonate. The solution is filtered from the insoluble sulphate, and evaporated, first in a water-bath, and afterwards in a vacuum to the proper degree of concentration. The salt crystallizes in beautiful, square, colorless tables, containing $(CH_3)_2Ba(SO_4)_2.2H_2O$, which effloresce in dry air, and are very soluble in water. By exactly precipitating the base from this substance with dilute sulphuric acid, and leaving the filtered liquid to evaporate in the air, methylsulphuric acid may be procured in the form of a sour, syrupy liquid, or in minute acicular crystals, very soluble in water and alcohol. It is very instable, being easily decomposed by heat. *Potassium methylsulphate*, CH_3KSO_4, crystallizes in small, nacreous, deliquescent rhombic tables. The *lead salt* is also very soluble.

Neutral Methyl sulphate, or *Dimethylic sulphate* $(CH_3)_2SO_4$, or $SO_2(OCH_3)_2$.— This ether is prepared by distilling 1 part of methyl alcohol with 8 or 10 parts of strong sulphuric acid: the distillation may be carried nearly to dryness. The oleaginous liquid found in the receiver is agitated with water, and purified by rectification from powdered anhydrous baryta. The product is a colorless, oily liquid, of alliaceous odor, having a density of 1.324, and boiling at 188° C. (370.4° F.). It is neutral to test-paper, and insoluble in water, but decomposed by that liquid, slowly in the cold, rapidly and with violence at a boiling temperature, into methylsulphuric acid and and methyl alcohol. Anhydrous lime and baryta have no action on this ether : their hydrates, however, and those of potassium and sodium, decompose it instantly, with production of a methylsulphate of the base, and methyl alcohol. When neutral methylsulphate is heated with common salt, it yields sodium sulphate and methyl chloride; with mercuric cyanide, or potassium cyanide, it gives a sulphate of the base and methyl cyanide; with dry sodium formate, it yields sodium sulphate and methyl formate.

Methyl Sulphite (symmetrical), $SO(O.CH_3)_2$, formed by the action of sulphur dichloride, S_2Cl_2, on methyl alcohol, as a fragrant liquid having a specific gravity of 1.045, and boiling at 121° C. (249.8° F.).

Methylsulphonic Acid, $CH_3.SO_3H$, is prepared by heating methyl iodide with a concentrated solution of potassium or sodium sulphite, converting the resulting methylsulphonate into a lead-salt, decomposing the latter with hydrogen sulphide, and evaporating the filtered solution. The acid then remains as a viscid uncrystallizable liquid, soluble in water. Its salts are easily soluble in water, and crystallize well ; the barium salt, $(CH_3.SO_3)_2.Ba$, in rhombic plates.

Methylsulphonic Chloride, $CH_3.SO_2.Cl$, boils at 153° C. (307.4° F.), and is decomposed by water into hydrochloric and methylsulphonic acids.

Trichloromethylsulphonic Chloride, $CCl_3.SO_2.Cl$, is formed by the action of moist chlorine on carbon bisulphide. To prepare it, a mixture of 500 grams of hydrochloric acid, 300 grams of coarsely pounded potassium dichromate, 200 grams of nitric acid, and 30 grams of carbon bisulphide is left to itself in an open flask for about a week, water is added, and the crystals of the compound, CCl_4SO_2, are separated from the saline solution by filtration.

The chloride or chloranhydride thus formed is a colorless crystalline body, which melts at 135° C. (275° F.), and boils at 170° C. (338° F.). It has a camphorous tear-exciting odor, dissolves in alcohol and ether, but is insoluble in water.

Trichloromethylsulphonic Acid, $CCl_3.SO_3H$, is obtained by boiling the chloride just described with baryta-water, and decomposing the resulting barium salt with sulphuric acid. It crystallizes in deliquescent prisms ; the barium salt, $(CCl_3.SO_3)_2Ba+H_2O$, in laminæ.

This trichlorinated acid heated in aqueous solution with sodium-amalgam is converted successively into the acids, $CHCl_2.SO_3H$, $CH_2Cl.SO_3H$, and finally into m e t h y l s u l p h o n i c a c i d, $CH_3.SO_3H$. This series of reactions, discovered by Kolbe in 1845, afforded one of the earliest instances of the formation of an organic compound from inorganic materials.

Methyl Borate, $(CH_3)_3BO_3 = B(OCH_3)_3$, is formed by the action of gaseous boron chloride on anhydrous methyl alcohol. It is a limpid liquid, of specific gravity 0.9551 at 0°, boiling at 72° C. (161.6^O F.). Water decomposes it into boric acid and methyl alcohol.

Methyl Phosphates.—Two methyl phosphates, viz., methylphosphoric acid, $PO(OH)_2(OCH_3)$, and dimethylphosphoric acid, $PO(OH)(OCH_3)_2$, are formed by the action of phosphorus oxychloride on methyl alcohol under different circumstances.

Methyl-phosphonic Acid, $(CH_3)PO.(OH)_2$ (p. 583), is a crystalline body melting at 105° C. (221° F.), and converted by phosphorus trichloride into methylphosphonic chloride, $(CH_3)PO.Cl_2$, which melts at 32° C. (89.6^O F.), boils at 163° C. (325.4^O F.), and is reconverted into the acid by the action of water.

Dimethylphosphinic acid, $(CH_3)_2PO.OH$, is a mass resembling paraffin, melting at 76° C. (168.8^O F.), and volatilizing without decomposition.

On *Methyl-arsonic* and *Methyl-arsinic Acids*, see ARSENIC BASES.

Methyl Silicate, $Si(OCH_3)_4$, is obtained by acting upon perfectly pure and dry methyl alcohol with silicium tetrachloride, and distilling the product. It is a colorless liquid, of pleasant, ethereal odor, specific gravity 1.0589 at 0°, distilling between 121° and 126° C. (249.8^O and 258.8^O F.). It dissolves with moderate facility in water, and the solution does not become turbid, from separation of silica, for some weeks. Its observed vapor-density is 5.38 referred to air, or 77.6 referred to hydrogen, the calculated number being 76.

Methyl Hydrosulphide, CH_3SH, also called **Methyl Mercaptan.**— This compound, which has the composition of methyl alcohol with the oxygen replaced by sulphur, is formed by distilling in a water-bath, with efficient condensation, a mixture of calcium methylsulphate and potassium hydrosulphide :

$$Ca(CH_3)_2(SO_4)_2 + 2KSH = K_2SO_4 + CaSO_4 + 2CH_3SH.$$

It is a liquid lighter than water, and having an extremely offensive odor. It forms with lead-acetate a yellow precipitate, and with mercuric oxide a white compound, $(CH_3)_2S_2Hg$, which crystallizes from alcohol in shining laminæ.

Methyl Sulphide, $(CH_3)_2S$, or $H_3C—S—CH_3$, is obtained by passing gaseous methyl chloride into a solution of potassium monosulphide in methyl alcohol. It is a colorless, mobile, fetid liquid, of specific gravity 0.845 at 21° C. (69.8^O F.), boiling at 41° C. (105.8^O F.). It forms several substitution-products with chloride.

Methyl Bisulphide, $(CH_3)_2S_2$, is prepared by passing gaseous methyl chloride through an alcoholic solution of potassium bisulphide. It is a limpid, strongly refracting liquid, having a specific gravity of 1.046 at 18° C. (64.4^O F.), and an intolerable odor of onions ; boils between 116° and 118° C. (244.4^O F.). It forms substitution-products with bromine and chlorine.

By substituting pentasulphide for bisulphide of potassium in the preceding preparation, a *trisulphide of methyl*, $(CH_3)_2S_3$ is obtained, boiling at about 200° C. (392° F.).

Methyl Telluride, or **Telluro-methyl**, $(CH_3)_2Te$, obtained by distilling potassium telluride with potassium methylsulphate, is an oily fetid liquid, resembling ethyl telluride, which will be described hereafter. The corresponding selenium-compound has also been obtained.

ETHYL ALCOHOL AND ETHERS.

Ethyl Alcohol, Hydroxyl-ethane, or Methyl Carbonol,

$$C_2H_6O \ = \ C_2H_5(OH) \ = \ \begin{matrix} CH_3 \\ | \\ CH_2(OH) \end{matrix} \ = \ C \left\{ \begin{matrix} CH_3 \\ H \\ H \\ OH \end{matrix} \right.$$

This important compound, the oldest and best known of the whole group of alcohols, and generally designated by the simple name "alcohol," is produced:—

1. From e t h e n e, C_2H_4, by addition of the elements of water. When ethene gas and strong sulphuric acid are violently agitated together for a long time, the gas is absorbed, and ethylsulphuric acid, $C_2H_6SO_4$, is produced; and this compound, distilled with water, yields sulphuric acid and ethyl alcohol:—

$$C_2H_6SO_4 \ + \ H_2O \ = \ H_2SO_4 \ + \ C_2H_6O.$$

Now we have seen that ethene can be formed by addition of hydrogen to acetylene, C_2H_2, which is itself formed by direct combination of carbon and hydrogen. It follows, therefore, that alcohol can be produced synthetically from its elements.

2. From e t h y l c h l o r i d e, b r o m i d e or i o d i d e by the reactions already mentioned (p. 543).

3. By the fermentation of certain kinds of s u g a r. When a moderately warm solution of cane-sugar or grape-sugar (glucose) is mixed with certain albuminous matters, as blood, white of egg, flour-paste, and especially beer-yeast, in a state of decomposition, a peculiar process, called *fermentation*, is set up, by which the sugar is resolved into ethyl alcohol and carbon dioxide. In the case of glucose, $C_6H_{12}O_6$, these products result from a simple splitting up of the molecule:—

$$C_6H_{12}O_6 \ = \ 2CO_2 \ + \ 2C_2H_6O.$$

Cane-sugar, $C_{12}H_{22}O_{11}$, is first converted into glucose by assumption of water, $(C_{12}H_{22}O_{11} + H_2O = 2C_6H_{12}O_6)$, and the latter is then decomposed as above.[*]

If ordinary cane-sugar be dissolved in a large quantity of water, a due proportion of active yeast added, and the whole maintained at a temperature of $21^o\text{--}26^o$ C. ($70^o\text{--}80^o$ F.), the change will go on with great rapidity. The gas disengaged is nearly pure carbon dioxide: it is easily collected and examined, as the fermentation, once commenced, proceeds perfectly well in a close vessel, such as a large bottle or flask fitted with a cork and a conducting tube. When the effervescence is at an end, and the liquid has become clear, it will yield alcohol by distillation.

[*] Side by side with this principal decomposition, a variety of other changes are simultaneously accomplished. According to Pasteur, glycerin, succinic acid, cellulose, fats, and occasionally lactic acid, are observed among the products of alcoholic fermentation. Some of the homologues of ethyl alcohol are also found among the products.

The spirit first obtained by distilling a fermented saccharine liquid is very weak, being diluted with a large quantity of water. By a second distillation, in which the first portions of the distilled liquid are collected apart, it may be greatly strengthened: the whole of the water cannot, however, be thus removed. The strongest rectified spirit of commerce has a density of about 0.835, and yet contains 13 or 14 per cent. of water. Pure or *absolute* alcohol may be obtained from it by redistilling it with half its weight of fresh quicklime. The lime is reduced to coarse powder, and put into a retort; the alcohol is added, and the whole mixed by agitation. The neck of the retort is securely stopped with a cork, and the mixture left for several days. The alcohol is distilled off by the heat of a water-bath.

Pure alcohol is a colorless, limpid liquid, of pungent and agreeable taste and odor; its specific gravity, at 15.5° C. (60° F.), is 0.7938, and that of its vapor, referred to air, 1.613. It is very inflammable, burning with a pale bluish-flame, free from smoke; it has never been frozen. Alcohol boils at 78.4° C. (173° F.) when in the anhydrous state; in a diluted state the boiling point is higher, being progressively raised by each addition of water. In the act of dilution a contraction of volume occurs, and the temperature of the mixture rises many degrees: this takes place not only with pure alcohol, but also with rectified spirit. Alcohol is miscible with water in all proportions, and, indeed, has a great attraction for the latter, absorbing its vapor from the air, and abstracting the moisture from membranes and other similar substances immersed in it. The solvent powers of alcohol are very extensive; it dissolves a great number of saline compounds, and likewise a considerable proportion of potash. With some salts it forms definite crystalline compounds, called a l c o h o l a t e s: with *zinc chloride*, $ZnCl_2.2C_2H_6O$; with *calcium chloride*, $CaCl_2.4C_2H_6O$; with *magnesium nitrate*, $Mg(NO_3)_2.6C_2H_6O$. Alcohol dissolves, moreover, many organic substances, as the vegeto-alkalies, resins, essential oils, and various other bodies: hence its great use in chemical investigations and in several of the arts.

Potassium and *sodium* dissolve in ethyl alcohol in the same manner as in methyl alcohol, forming the compounds C_2H_5KO and C_3H_5NaO.

Alcohol, passed through a red-hot tube, is resolved into marsh-gas, hydrogen, and carbon monoxide.

$$C_2H_6O = CH_4 + H_2 + CO.$$

Small quantities of ethene, benzene, and naphthaline are, however, formed at the same time by the mutual action of these primary products, and carbon is deposited.

By *oxidation*, alcohol is converted first into aldehyde, C_2H_4O, then into acetic acid, $C_2H_4O_2$:

$$C_2H_6O + O = H_2O + C_2H_4O,$$
and $$C_2H_4O + O = C_2H_4O_2.$$

Chlorine gas is rapidly absorbed by anhydrous alcohol, turning it yellow, and causing considerable rise of temperature. At the same time it rapidly abstracts hydrogen, which is partly replaced by the chlorine, producing hydrochloric acid, aldehyde, acetic acid, ethyl acetate, ethyl chloride, and chloral. The mixture of these substances, freed by water from the soluble constituents, was formerly called *heavy muriatic ether*. The formation of the several products is represented by the following equations:—

$$C_2H_6O + Cl_2 = 2HCl + C_2H_4O,$$
Alcohol.   Aldehyde.

$$C_2H_6O + 4Cl_2 = 5HCl + C_2HCl_3O,$$
Alcohol.   Chloral.

$$C_2H_6O \quad + \quad HCl \quad = \quad H_2O \quad + \quad C_2H_5Cl,$$
Alcohol. · · · · Ethyl chloride.

$$C_2H_6O \quad + H_2O + 2Cl_2 = \quad 4HCl \quad + \quad C_2H_4O_2,$$
Alcohol. · · · · Acetic acid.

$$C_2H_6O \quad + \quad C_2H_4O_2 \quad = \quad H_2O \quad + \quad C_2H_5.C_2H_3O_2.$$
Alcohol. · · Acetic acid. · · · · Ethyl acetate.

When the action of the chlorine is continued for a long time, c h l o r a l is always the principal product. This compound is a heavy oily liquid, having the composition of aldehyde with 3 atoms of hydrogen replaced by chlorine; but it cannot be formed by the direct action of chlorine upon aldehyde. When alcohol containing water is used, scarcely any chloral is obtained, the chief product being aldehyde.

Chlorine, in presence of *alkalies*, converts alcohol into chloroform, $CHCl_3$, and carbon dioxide :—

$$C_2H_6O + 5Cl_2 + H_2O = CO_2 + 7HCl + CHCl_3.$$

The same products are formed by distilling dilute alcohol with bleaching powder.

Aqueous alcohol heated with strong *sulphuric acid* is converted into ethyl-sulphuric acid, $C_2H_6SO_4$; but when anhydrous alcohol is exposed to the vapor of sulphuric oxide, SO_3, a white crystalline substance is formed, called e t h i o n i c o x i d e, formerly *sulphate of carbyl*, $C_2H_4S_2O_6$. This, when dissolved in water or in aqueous alcohol, is converted into e t h i o n i c a c i d, $C_2H_6S_2O_7$, a bibasic acid, which forms a soluble barium salt. Lastly, a solution of ethionic acid, when boiled, is resolved into sulphuric acid and i s e t h i o n i c a c i d, an acid isomeric with ethyl-sulphuric acid.

Commercial Spirit, Wine, Beer, etc. Vinous Fermentation.—The strength of commercial spirit, when free from sugar and other substances added subsequently to distillation, is inferred from its density : a table exhibiting the proportions of real alcohol and water in spirits of different densities will be found at the end of the volume. The excise *proof spirit* has a specific gravity of 0.9198 at 60° F., and contains 49¼ per cent. by weight of real alcohol.

The high duty on spirits of wine in this country has hitherto interfered with the development of many branches of industry, which are dependent on the free use of this important liquid. The labors of the scientific chemist have been likewise often checked by this inconvenience. A remedy for the evil has been supplied by a very important measure, proposed and carried out by the late Mr. John Wood, Chairman of the Board of Inland Revenue. This measure consists in issuing, for manufacturing and scientific purposes, duty free, a mixture of 90 per cent. of spirits of wine of strength not less than corresponds with a density of 0.830, and 10 per cent. of partially purified wood-spirit, which is now sold by licensed dealers under the name of *Methylated Spirit*. It appears that a mixture of this kind is rendered permanently unfit for human consumption, the separation of the two substances, in consequence of their close analogy, being not only difficult, but to all appearance impossible: at the same time, and for the same reasons, this mixture is not materially impaired for the greater number of the more valuable purposes in the arts for which spirits are usually employed. Methylated spirit may be used, instead of pure spirit, as a solvent of resinous substances, and of many chemical preparations, especially of the alkaloids and other organic products. It may be used for the production of fulminating mercury, ether, chloroform, iodoform, olefiant gas, and all its derivatives—in fact, for an endless number of laboratory purposes. Methylated spirit may also be substituted for pure spirit of wine in the preservation of anatomical preparations. The introduction of this spirit has already exerted a very beneficial effect upon the development of organic chemistry in England.*

Wine, Beer, etc., owe their intoxicating properties to the alcohol they contain, the quantity of which varies very much. Port and sherry, and some other strong wines, contain from 19 to 25 per cent. of alcohol, while in the lighter wines of France and Germany it sometimes falls as low as 12 per cent. Strong ale contains about 10 per cent.; ordinary spirits, as brandy, gin, whiskey, 40 to 50 per cent., or occasionally more. These latter owe their characteristic flavors to certain essential oils and compound ethers, present in very small quantity, either generated in the act of fermentation or purposely added.

In making wine, the expressed juice of the grape is simply set aside in large vats, where it undergoes spontaneously the necessary change. The vegetable albumin of the juice absorbs oxygen from the air, runs into decomposition, and in that state becomes a ferment to the sugar, which is gradually converted into alcohol. If the sugar be in excess, and the azotized matter deficient, the resulting wine remains sweet; but if, on the other hand, the proportion of sugar be small and that of albumin large, a *dry* wine is produced. When the fermentation stops, and the liquor becomes clear, it is drawn off from the lees, and transferred to casks, to ripen and improve.

The color of red wine is derived from the skins of the grapes, which in such cases are left in the fermenting liquid. Effervescent wines, as champagne, are bottled before the fermentation is complete; the carbonic acid is disengaged under pressure, and retained in solution in the liquid. A certain quantity of sugar is frequently added. The process requires much delicate management.

During the fermentation of the grape-juice, or *must,* a crystalline, stony matter, called *argol,* is deposited. This consists chiefly of acid potassium tartrate with a little coloring matter, and is the source of all the tartaric acid met with in commerce. The salt in question exists in the juice in considerable quantity; it is but sparingly soluble in water, but still less so in dilute alcohol: hence, as the fermentation proceeds, and the quantity of spirit increases, it is slowly deposited. The acid of the juice is thus removed as the sugar disappears. It is this circumstance which renders grape-juice alone fit for making good wine; when that of gooseberries or currants is employed as a substitute, the malic and citric acids which these fruits contain cannot be thus withdrawn. There is then no other resource but to add sugar in sufficient quantity to mask and conceal the natural acidity of the liquor. Such wines are necessarily ascescent, prone to a second fermentation, and, to many persons at least, very unwholesome.

Beer is a well-known liquor, of great antiquity, prepared from germinated grain, generally barley, and is used in countries where the vine does not flourish. The operation of *malting* is performed by steeping the barley in water until the grains become swollen and soft, then piling it in a heap or *couch,* to favor the rise of temperature caused by the absorption of oxygen from the air, and afterwards spreading it upon a floor, and turning it over from time to time to prevent unequal heating. When germination has proceeded far enough, the vitality of the seed is destroyed by kiln-drying.

* See Report on the Supply of Spirits of Wine, free from duty, for use in the Arts and Manufactures, addressed to the Chairman of Inland Revenue by Professors Graham, Hofmann, and Redwood. (Quarterly Journal of Chemical Society, vol. viii. p. 120.)

During this process, a peculiar nitrogenous substance called *diastase* is produced, which acts as a ferment on the starch of the grain, converting a portion of it into sugar and rendering it soluble.

In brewing, the crushed malt is infused in water at about 77° C. (170° F.), and the mixture is left to stand for two hours or more. The easily soluble diastase has thus an opportunity of acting upon the unaltered starch of the grain, and changing it into dextrin and sugar. The clear liquor, or *wort*, strained from the exhausted malt, is next pumped up into a copper boiler, and boiled with the requisite quantity of hops, to communicate a pleasant bitter flavor, and confer on the beer the property of keeping without injury. The flowers of the hop contain a bitter, resinous principle, called *lupulin*, and an essential oil.

When the wort has been sufficiently boiled, it is drawn from the copper, and cooled as rapidly as possible, to near the ordinary temperature of the air, in order to avoid an irregular acid fermentation, to which it would otherwise be liable. It is then transferred to the fermenting vessels, which in large breweries are of great capacity, and mixed with a quantity of yeast, the product of a preceding operation, by which the change is speedily induced. This is the most critical part of the whole operation, and one in which the skill and judgment of the brewer are most called into play. The process is in some measure under control by attention to the temperature of the liquid ; and the extent to which the change has been carried is easily known by the diminished density, or *attenuation* of the wort. The fermentation is never suffered to run its full course, but is always stopped at a particular point, by separating the yeast, and drawing off the beer into casks. A slow and almost insensible fermentation succeeds, which in time renders the beer stronger and less sweet than when new, and charges it with carbonic acid.

Highly colored beer is made by adding to the malt a small quantity of strongly dried or charred malt, the sugar of which has been changed to caramel : porter and stout are so prepared.

The yeast of beer is a very remarkable substance. To the naked eye it is a greenish-yellow soft solid, nearly insoluble in water, and dries up to a pale-brownish mass, which readily putrefies when moistened, and becomes offensive. Under the microscope it exhibits a kind of organized appearance, being made up of little transparent globules, which sometimes cohere in clusters or strings, like some of the lowest members of the vegetable kingdom. Whatever may be the real nature of the substance, no doubt can exist that it is formed from the soluble azotized portion of the grain during the fermentative process. No yeast is ever produced in liquids free from azotized matter ; that added for the purpose of exciting fermentation in pure sugar is destroyed, and rendered inert thereby. When yeast is deprived, by straining and strong pressure, of as much water as possible, it may be kept in a cool place, with unaltered properties for a long time ; otherwise it quickly spoils.

The distiller, who prepares spirits from grain, makes his wort, or *wash*, much in the same manner as the brewer ; he uses, however, with the malt a large quantity of raw grain, the starch of which suffers conversion into sugar by the diastase of the malt, which is sufficient for his purpose. He does not boil his infusion with hops, but proceeds at once to the fermentation, which he pushes as far as possible by large and repeated doses of yeast. Alcohol is manufactured in many cases from potatoes. The potatoes are ground to pulp, mixed with hot water and a little malt, to furnish diastase, made to ferment, and then the fluid portion is distilled. The potato-spirit is contaminated by a very offensive volatile oil, again to be mentioned : the crude product from corn contains a substance of a similar kind. The business of the rectifier consists in removing or modifying these volatile oils, and in replacing them by others of a more agreeable character.

In making *bread*, the vinous fermentation plays an important part: the yeast added to the dough converts the small portion of sugar the meal naturally contains into alcohol and carbonic acid. The gas thus disengaged forces the tough and adhesive materials into bubbles, which are still further expanded by the heat of the oven, which at the same time dissipates the alcohol: hence the light and spongy texture of all good bread. The use of *leaven* is of great antiquity: this is merely dough in a state of incipient putrefaction. When mixed with a large quantity of fresh dough, it excites in the latter the alcoholic fermentation, in the same manner as yeast, but less perfectly; it is apt to communicate a disagreeable sour taste and odor. Sometimes carbonate of ammonium is employed to lighten the dough, being completely volatilized by the high temperature of the oven. Bread is now sometimes made by mixing a little hydrochloric acid and sodium carbonate in the dough; if proper proportions be taken and the whole thoroughly mixed, the operation appears to be very successful.

Another mode of bread-making is that invented by the late Dr. Dauglish, which consists in agitating the dough in a strong vessel with water saturated under pressure with carbonic acid gas. When the dough thus treated is subsequently released from this pressure and exposed to the air, the gas escapes in bubbles, and lightens the mass as effectually as that evolved within its substance by fermentation. The bread thus made, called "aërated bread," is of excellent quality, not being subject to the deterioration which so frequently takes place in ordinary bread, when the fermentation is allowed to go too far.

Vinous fermentation, that is to say, the conversion of sugar into alcohol and carbon dioxide, never takes place except in presence of some nitrogenous body of the albuminoid class in a state of decomposition, and it is always accompanied by the development of certain minute living organisms—fungi and infusoria—like those already mentioned as existing in yeast. So constantly indeed is this the case that many chemists and physiologists regard these organisms as the exciting cause of fermentation and putrefaction; and this view appears to be corroborated by the fact that each particular kind of fermentation takes place most readily in contact with a certain living organism, or at least with nitrogenous matter containing it; thus beer-yeast contains two species of fungus, called *Torvula cerevisiæ* and *Penicillium glaucum*, the cells of which are of very different sizes, so that they may be separated by filtering an infusion of the yeast, the larger cells of the *Torvula* remaining on the filter, while those of the *Penicillium*, which are much smaller, pass through with the liquid. Now, it is found that the residue on the filter brings a solution of sugar into the state of vinous fermentation, whereas the filtered liquid induces lactous fermentation; but whether this effect is due to the fungi themselves, or to the peculiar state of the albuminous matter in which they occur, is a question not yet decided. The investigation is attended with peculiar difficulties, arising chiefly from the universal diffusion of the germs of these minute organisms, which are present not only in all decaying albuminous matter, and on the skins of fruits, leaves, and other parts of plants, but are likewise diffused through the air; so that in experiments made for the purpose of ascertaining whether fermentation can take place without them, it is extremely difficult to ensure their complete exclusion from the substances under examination.*

* See the article "Fermentation," in Watts's Dictionary of Chemistry, vol. ii. p. 623.

Ethyl Chloride, or **Chlorethane**, C_2H_5Cl, or $H_3C—CH_2Cl$, often called *Hydrochloric ether*.—To prepare this compound, rectified spirit of wine is saturated with dry hydrochloric acid gas, and the product distilled at a very gentle heat; or a mixture of 3 parts oil of vitriol and 2 parts of alcohol is poured upon 4 parts of dry common salt in a retort, and heat applied; in either case the vapor of the hydrochloric ether should be conducted through a little tepid water in a wash-bottle, and thence into a small receiver surrounded by ice and salt. It is purified from adhering water by contact with a few fragments of fused calcium chloride.

Ethyl chloride is a thin, colorless, and excessively volatile liquid, of a penetrating, aromatic, and somewhat alliaceous odor. At the freezing point of water, its sp. gr. is 0.921, and it boils at 12.5° C. (54.5° F.). It is soluble in 10 parts of water, is but incompletely decomposed by solution of silver nitrate when the two are heated together in a sealed tube, but is quickly resolved into potassium chloride and ethyl alcohol by a hot aqueous solution of caustic potash:

$$C_2H_5Cl \ + \ KOH \ = \ KCl \ + \ C_2H_5OH.$$

With alcoholic potash, on the other hand, or potassium ethylate, it yields ethyl oxide, or common ether:

$$C_2H_5Cl \ + \ C_2H_5OK \ = \ KCl \ + \ (C_2H_5)_2O.$$

Heated with soda-lime, it yields ethene or olefiant gas:

$$2C_2H_5Cl \ + \ Na_2O \ = \ 2NaCl \ + \ H_2O \ + \ 2C_2H_4.$$

When vapor of ethyl chloride is mixed with chlorine gas in a vessel exposed, first to diffused daylight, and afterwards to direct sunshine, hydrochloric acid is formed, and the chlorine displaces one atom of hydrogen in the ethyl chloride, producing monochlorinated ethyl chloride, ethene chloride or dichlorethane, $C_2H_2Cl_4$ (p. 517). By the prolonged action of chlorine in excess, the compounds $C_2H_3Cl_3$, $C_2H_2Cl_4$, C_2HCl_5, and C_2Cl_6, are produced (pp. 549, 550).

Ethyl Bromide, or **Bromethane**, C_2H_5Br, also called *Hydrobromic ether*, is prepared by distilling a mixture of 8 parts bromine, 1 part phosphorus, and 39 parts alcohol. It is a very volatile liquid, heavier than water, having a penetrating taste and odor, boiling at 41° C. (105.8° F.).

Ethyl Iodide, or **Iodethane**, C_2H_5I, also called *Hydriodic ether*, may be conveniently prepared with 5 parts of phosphorus, 70 parts of alcohol (of 0.84 sp. gr.), and 100 parts of iodine. The phosphorus is introduced into a tubulated retort, covered with part of the alcohol, and heated to fusion. The rest of the alcohol is poured upon the iodine, and the solution thus obtained is allowed to flow gradually through a tap-funnel into the retort. The brown liquid is at once decolorized, and ethyl iodide distils over, which is condensed by a good cooling apparatus. The distillate, consisting of alcohol and ethyl iodide, is again poured on the residuary iodine, which is thus rapidly dissolved, introduced into the retort, and ultimately entirely converted into ethyl iodide. The latter is washed with water to remove adhering alcohol, separated from this water by a tap-funnel, digested with calcium chloride, and rectified in the water-bath. Ethyl iodide may also be formed by heating in a sealed glass vessel a mix-

ture of hydriodic acid and olefiant gas. It is a colorless liquid, of penetrating ethereal odor, having a density of 1.92, and boiling at ',2ºC. (161.6º
F.). It becomes red by exposure to light, from the commencement of decomposition. This substance has become highly important as a source of
ethyl, and from its remarkable deportment with ammonia, which will be
discussed in the Section on Amines.

Ethyl Oxide, or **Ethylic ether,** $C_4H_{10}O = C_2H_5O.C_2H_5 = (C_2H_5)_2O.$ —
This compound, also called *common ether*, or simply *either*, contains the elements of 2 molecules of alcohol *minus* 1 molecule of water :—

$$2C_2H_6O \quad - \quad H_2O \quad = \quad C_4H_{10}O ;$$

and it is in fact produced by the action of various dehydrating agents, such
as zinc chloride, phosphoric oxide, and strong sulphuric acid, upon alcohol.
The process does not appear, however, to be one of direct dehydration, at
least in the case of sulphuric acid ; for when that acid is heated with alcohol to a certain temperature it does not become weaker by taking water
from the alcohol, but ether and water distil over together, and the sulphuric acid remains in its original state, ready to act in the same manner
on a fresh portion of alcohol. The reaction is in fact one of substitution,
the ultimate result being the conversion of alcohol, $C_2H_5(OH)$, into ether,
$C_2H_5(OC_2H_5)$, by the substitution of ethyl for hydrogen. The manner in
which this takes place will be better understood when another mode of
the formation of ether has been explained.

When a solution of sodium ethylate, $NaOC_2H_5$, in anhydrous alcohol,
obtained by dissolving sodium to saturation in that liquid, is mixed with
ethyl iodide, double decomposition takes place, resulting in the formation
of sodium iodide and ethyl oxide :—

$$NaOC_2H_5 \quad + \quad C_2H_5I \quad = \quad NaI \quad + \quad C_2H_5OC_2H_5.$$

The result would be the same if chloride or bromide of ethyl were substituted for the iodide: moreover, when methyl iodide is added, instead of
the ethyl iodide, an oxygen-ether is formed containing both ethyl and
methyl :—

$$\underset{\text{Sodium ethylate.}}{NaOC_2H_7} \quad + \quad \underset{\substack{\text{Methyl}\\\text{iodide.}}}{CH_3I} \quad = \quad NaI \quad + \quad \underset{\substack{\text{Ethyl-methyl}\\\text{ether.}}}{C_2H_5OCH_3.}$$

In each case the reaction consists in an interchange between the sodium
and the alcohol-radicle.

Now, when alcohol is heated with strong sulphuric acid, the first result
is the formation of ethylsulphuric acid, $SO_2(OC_2H_5)OH$, by substitution of
ethyl for hydrogen in the acid :—

$$\underset{\substack{\text{Sulphuric}\\\text{acid.}}}{SO_2(OH)(OH)} \quad + \quad \underset{\text{Alcohol.}}{C_2H_5(OH)} \quad = \quad \underset{\text{Water.}}{H(OH)} \quad + \quad \underset{\substack{\text{Ethylsulphuric}\\\text{acid.}}}{SO_2(OC_2H_4)(OH);}$$

and when the ethylsulphuric acid thus formed is brought in contact, at a
certain temperature, with a fresh portion of alcohol, the reverse substitution takes place, resulting in the formation of ethyl oxide and sulphuric
acid :—

$$\underset{\substack{\text{Ethylsulphuric}\\\text{acid.}}}{SO_2(OC_2H_5)(OH)} \quad + \quad \underset{\text{Alcohol.}}{C_2H_5(OH)} \quad = \quad \underset{\text{Ether.}}{C_2H_5OC_2H_5} \quad + \quad \underset{\substack{\text{Sulphuric}\\\text{acid.}}}{SO_2(OH)_2.}$$

The sulphuric acid is thus reproduced in its original state, and if the supply of alcohol be kept up, and the temperature maintained within certain

limits, the same series of actions is continually repeated, and ether and water distil over together.

The most favorable temperature for etherification is between 127⁰ and 154⁰ C. (260⁰ and 310⁰ F.); below 127⁰ very little ether is produced, and above 154⁰ a different reaction takes place, resulting in the formation of olefiant gas. The maintenance of the temperature within the ether-producing limits is best effected by boiling the mixture of sulphuric acid and alcohol in a flask into which a further quantity of alcohol is supplied in a continuous and regulated stream. This is called the *continuous* ether process.

A wide-necked flask *a* (fig. 164), is fitted with a sound cork perforated by three apertures, one of which is destined to receive a thermometer with the graduation on the stem; a second, the vertical portion of a long, narrow tube, terminating in an orifice of about $\frac{1}{20}$ of an inch in diameter;

Fig. 164.

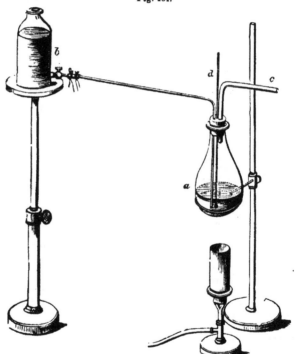

and the third, a wide bent tube, *c*, connected with the condenser, to carry off the volatilized products. A mixture is made of 8 parts by weight of concentrated sulphuric acid, and 5 parts of rectified spirit of wine, of about 0.834 sp. gr. This is introduced into the flask, and heated by a lamp. The liquid soon boils, and the thermometer very shortly indicates a temperature of 140⁰ C. (284⁰ F.). When this happens, alcohol of the above density is suffered slowly to enter by the narrow tube, which is put into communication with a reservoir, *b*, of that liquid, consisting of a large bottle perforated by a hole near the bottom, and fc nished with a small brass stopcock fitted by a cork : the stopcock is secv,ed to the end of the long tube by a caoutchouc connector. As the tub€ passes nearly to the bottom of the flask, the alcohol gets thoroughly mixel with the acid liquid, the pressure of the fluid column being sufficient to insure the regularity of the flow ; the quantity is easily adjusted by the aid of the stopcock. For condensation a Liebig's condenser may be used, supplied with ice water.

The degree of heat, and the supply of alcohol, must be so adjusted that the thermometer may remain at 140°, or as near that temperature as possible, while the contents of the flask are maintained in a state of *rapid and violent ebullition*—a point of essential importance. Ether and water distil over together, and collect in the receiver, forming two distinct strata : the mixture slowly blackens, from some slight secondary action of the acid upon the spirit, or upon the impurities in the latter, but retains, after many hours' ebullition, its etherifying powers unimpaired. The acid, however, slowly volatilizes, partly in the state of *oil of wine*, and the quantity of liquid in the flask is found, after the lapse of a considerable interval, sensibly diminished. This loss of acid constitutes the only limit to to the duration of the process, which might otherwise be continued indefinitely.

On the large scale, the flask may be replaced by a vessel of lead, the tubes being also of the same metal : the stem of the thermometer may be made to pass air-tight through the cover, and heat may perhaps be advantageously applied by high-pressure steam, or hot oil, circulating in a spiral tube of metal immersed in the mixture of acid and spirit.

The crude ether is to be separated from the water on which it floats, agitated with a little solution of caustic potash, and redistilled by the heat of warm water. The aqueous portion, treated with an alkaline solution, and distilled, yields alcohol containing a little ether. Sometimes the spontaneous separation before mentioned does not occur, from the accidental presence of a larger quantity than usual of undecomposed alcohol ; the addition of a little water, however, always suffices to determine it.

Pure ethylic ether is a colorless, transparent, fragrant liquid, very thin and mobile. Its specific gravity at 15.5° C. (59.9° F.) is about 0.720 ; it boils at 35.6° C. (96° F.) under the pressure of the atmosphere, and bears without freezing the severest cold. When dropped on the hand it occasions a sharp sensation of cold, from its rapid volatilization. Ether is very combustible, and burns with a white flame, generating water and carbon dioxide. Although the substance itself is one of the lightest liquids, its vapor is very heavy, having a density of 2.586 (referred to air). Mixed with oxygen gas, and fired by the electric spark, or otherwise, it explodes with the utmost violence. Preserved in an imperfectly stopped vessel, ether absorbs oxygen, and becomes acid from the production of acetic acid : this attraction for oxygen is increased by elevation of temperature. It is decomposed by transmission through a red-hot tube into ethene, methane, aldehyde, and acetylene, and two substances yet to be described.

Ether is miscible with alcohol in all proportions, but not with water ; it dissolves to a small extent in that liquid, 10 parts of water taking up about 1 part of ether. It may be separated from alcohol, provided the quantity of the latter is not excessive, by addition of water, and in this manner samples of commercial ether may be conveniently examined. Ether dissolves oily and fatty substances generally, and phosphorus to a small extent, also a few saline compounds and some organic principles ; but its powers in this respect are much more limited than those of alcohol or water.

Anhydrous ether, subject to the action of chlorine, yields the three substitution-products, $C_4H_8Cl_2O$, $C_4H_6Cl_4O$, and $C_4Cl_{10}O$, the first two of which . are liquids, while the third, produced by the prolonged action of chlorine on ether in sunshine, is a crystalline solid. The second chlorine compound is converted by hydrogen sulphide into the two crystalline compounds $C_4H_6Cl_2SO$ and $C_4H_6S_2O$.

Ethyl-methyl oxide, Ethyl-methyl ether, Ethyl methylate, or Methyl ethylate, $C_3H_8O = C_2H_5OCH_3$, is produced, as already mentioned, by the action of methyl iodide on potassium ethylate, or of ethyl iodide on potassium methylate. It is a very inflammable liquid, boiling at 11° C. (51.8^O F.).

Ethyl Nitrate, $C_2H_5NO_3$, or $C_2H_5ONO_2$.—*Nitric ether.*—When nitric acid is heated with alcohol alone, part of the alcohol is oxidized, and the nitric acid is reduced to nitrous acid, which, with the remainder of the alcohol, forms ethyl nitrite, $C_2H_5NO_2$, together with other products; but by adding urea to the liquid, which decomposes the nitrous acid as fast as it is formed, this action may be prevented, and the alcohol and nitric acid then form ethyl nitrate. The experiment is most safely conducted on a small scale, and the distillation must be stopped when seven-eighths of the whole have passed over; a little water added to the distilled product separates the nitric acid. Nitric ether boils at 85° or 86° C. (185°–186.8^O F.), and has a density of 1.112 at 15° C. (59° F.) ; it is insoluble in water, has an agreeable sweet taste and odor, and is not decomposed by an aqueous solution of caustic potash, although that substance dissolved in alcohol attacks it even in the cold, with production of potassium nitrate. Its vapor is apt to explode when strongly heated.

Ethyl Nitrite, $C_2H_5NO_2 = C_2H_5$—O—NO.—*Nitrous ether.*—This compound, isomeric with nitro-ethane, C_2H_5—NO_2, can be obtained pure only by the direct action of the acid itself upon alcohol. One part of starch and 10 parts of nitric acid are gently heated in a capacious retort or flask, and the vapor of nitrous acid thereby evolved is conducted into alcohol mixed with half its weight of water, contained in a two-necked bottle, which is to be plunged into cold water, and connected with a good condensing arrangement. Rise of temperature must be carefully avoided. The product of this operation is a pale-yellow volatile liquid, having an exceedingly agreeable odor of apples : it boils at 16.4^O C. (61.5^O F.), and has a density of 0.947 at 15° C. (59° F.). It is decomposed by potash, without darkening, into potassium nitrite and alcohol.

The *sweet spirits of nitre* of pharmacy, prepared by distilling three pounds of alcohol with four ounces of nitric acid, is a solution of nitrous ether, aldehyde, and several other substances, in spirit of wine.

Ethyl Sulphates.—There are two of these ethers, analogous to the methyl sulphates.

Acid Ethyl Sulphate, Ethylsulphuric acid, or *Sulphovinic acid,* $C_2H_6SO_4 = C_2H_5.O.SO_3H = SO_2(OC_2H_5)(OH) = SO_4(C_2H_5)H$, which has the composition of sulphuric acid, SO_4H_2, with half the hydrogen replaced by ethyl, is formed by the action of sulphuric acid upon alcohol. To prepare it, strong rectified spirit of wine is mixed with twice its weight of concentrated sulphuric acid : the mixture is heated to its boiling point, and then left to cool. When cold it is diluted with a large quantity of water, and neutralized with chalk, whereby much calcium sulphate is produced. The mass is placed upon a cloth filter, drained, and pressed ; and the clear solution is evaporated to a small bulk by the heat of a water-bath, filtered from a little sulphate, and left to crystallize : the product is *calcium ethylsulphate,*

26

in beautiful, colorless, transparent crystals, containing $Ca(C_2H_5)_2(SO_4)_2$. $2H_2O$. They dissolve in an equal weight of cold water, and effloresce in a dry atmosphere.

Barium Ethylsulphate, $Ba(C_2H_5)_2SO_4.2H_2O$, equally soluble, and still more beautiful, may be produced by substituting, in the above process, barium carbonate for chalk. From this salt the acid may be procured by exactly precipitating the base with dilute sulphuric acid, and evaporating the filtered solution in a vacuum at the temperature of the air. It forms a sour, syrupy liquid, in which sulphuric acid cannot be recognized by the ordinary reagents, and is very easily decomposed by heat, and even by long exposure in the vacuum of the air-pump. All the ethylsulphates are soluble ; the solutions are decomposed by boiling. The *lead-salt* resembles the barium-compound. The *potassium salt*, $K(C_2H_5)SO_4$,—easily made by decomposing calcium ethylsulphate with potassium carbonate—is aubydrous, permanent in the air, very soluble, and crystallizes well.

Potassium ethylsulphate distilled with strong sulphuric acid, yields ethyl oxide with dilute sulphuric acid, alcohol ; and with strong acetic acid, ethyl acetate.

Isethionic acid, isomeric with ethylsulphuric acid, will be described amongst ethenic ethers.

Neutral Ethyl sulphate $(C_2H_5)_2SO_4$, or $SO_2(OC_2H_5)_2$, is formed by passing the vapor of sulphuric oxide into perfectly anhydrous ether. A syrupy liquid is produced, which, when shaken with 4 volumes of water and 1 volume of ether, separates into two layers, the lower containing ethylsulphuric acid and various other compounds, while the upper layer consists of an ethereal solution of neutral ethyl sulphate. At a gentle heat the ether is volatilized, and the ethyl sulphate remains as a colorless liquid. It cannot be distilled without decomposition.

Ethyl Sulphites.—*The symmetric neutral sulphite*, $SO{<}{OC_2H_5 \atop OC_2H_5}$, is formed by the action of thionyl chloride, $SOCl_2$, or of sulphur dichloride, S_2Cl_2, on absolute alcohol :—

$$SOCl_2 + 2(C_2H_5.OH) = 2HCl + SO(OC_2H_5)_2$$
$$S_2Cl_2 + 3(C_2H_5.OH) = C_2H_5.SH + 2HCl + SO(OC_2H_5)_2.$$

The mercaptan, likewise formed in the last reaction, suffers further decomposition.

Ethyl sulphite is a limpid, strong-smelling liquid, having a specific gravity of 1.085 at 0°, boiling at 161° C. (321.8° F.), decomposed by water into alcohol and sulphurous acid. With *phosphorus pentachloride* it forms the chloride, $SO{<}{OC_2H_5 \atop Cl}$ (isomeric with ethylsulphonic chloride), which boils at 122° C. (251.6° F.), and is decomposed by water into alcohol hydrochloric acid and sulphurous acid.

When ethyl sulphite is mixed with an equivalent quantity of potassium hydroxide in dilute solution, crystalline scales are formed, consisting of the salt, $SO{<}{O.C_2H_5 \atop OK}$, which may be regarded as the potassium-salt of ethylsulphurous acid, isomeric with ethylsulphonic acid. The aqueous solution of this salt easily splits up into potassium sulphite and ethyl alcohol.

Ethylsulphonic acid, $C_2H_5.SO_2.OH$, formed by the action of ethyl iodide on potassium or sodium sulphite (p. 582), is a thick crystallizable liquid, which is oxidized by nitric acid to ethylsulphuric acid, $SO_4(C_2H_5)H$. Its *lead-salt*, $(C_2H_5.SO_3)_2Pb$, crystallizes in easily soluble laminæ. The *ethylic ether*, $C_2H_5.SO_2.OC_2H_5$, likewise obtained by heating silver sulphite with ethyl iodide, is a liquid having a specific gravity of 1.17 at 0°, and boiling at 208° C. (406.4° F.).

The *chloride*, $C_2H_5.SO_2Cl$, is a liquid boiling at 173° C. (343.4° F.).

Chlorethylsulphonic acid, $C_2H_4Cl.SO_2.OH$, is produced by the action of phosphorus pentachloride on isethionic acid, $C_2H_4(OH).SO_3H$.

Ethylsulphinic Acid, $C_2H_5.SO.OH$.—The zinc salt, $(C_2H_5.SO_2)_2Zn$, formed by the action of sulphur dioxide on zinc ethide (p. 583), crystallizes in shining laminæ. By oxidation with nitric acid, the acid is converted into ethylsulphonic acid.

Ethyl Phosphates.—Three ethyl orthophosphates have been obtained, two acid and one neutral, analogous in composition to the sodium phosphates ; also a neutral pyrophosphate.

Monethylic Phosphate, or Ethylphosphoric acid, $(C_2H_5)H_2PO_4$, or $(PO)'''(OC_2H_5)$ $(OH)_2$, also called *Phosphovinic acid*.—This acid is bibasic. Its barium salt is prepared by heating to 82° C. (179.6° F.) a mixture of equal weights of strong alcohol and syrupy phosphoric acid, diluting this mixture, after a lapse of 24 hours, with water, and neutralizing with barium carbonate. The solution of ethylphosphate, separated by filtration from the insoluble phosphate, is evaporated at a moderate temperature. The salt crystallizes in brilliant hexagonal plates, which have a pearly lustre, and are more soluble in cold than in hot water ; it dissolves in 15 parts of water at 20° C. (68° F.). The crystals contain $(C_2H_5)BaPO_4.6H_2O$. From this salt the acid may be obtained by precipitating the barium with dilute sulphuric acid, and evaporating the filtered liquid in the vacuum of the air-pump : it forms a colorless, syrupy liquid, of intensely sour taste, sometimes exhibiting appearances of crystallization. It is very soluble in water, alcohol, and ether, and easily decomposed by heat when in a concentrated state. The ethylphosphates of calcium, silver, and lead are but slightly soluble ; those of the alkali-metals, magnesium, and strontium, are freely soluble.

Diethylic Phosphate, or *Diethylphosphoric acid*, $(C_2H_5)_2.HPO_4$, or $(PO)'''$ $(OC_2H_5)_2(OH)$, is a monobasic acid, obtained, together with the preceding, by the action of syrupy phosphoric acid upon alcohol. Its barium, silver, and lead salts are more soluble than the methyl phosphates. The calcium salt $(C_2H_5)_4Ca(PO_4)_2$, and the lead salt, $(C_2H_5)_2Pb(PO_4)_2$, are anhydrous.

Triethylic Phosphate, $(C_2H_5)_3PO_4$, or $(PO)'''(OC_2H_5)_3$, is obtained in small quantity by heating the lead salt of diethylphosphoric acid to 100° ; more easily by the action of ethyl iodide on triargentic phosphate, or of phosphorus oxychloride on sodium ethylate :—

$$3C_2H_5ONa + (PO)Cl_3 = 3NaCl + (PO)(OC_2H_5)_3 .$$

It is a limpid liquid of specific gravity 1.072 at 12° C. (53.6° F.), boiling at 215° C. (419° F.), soluble in alcohol and ether, and also in water, by which, however, it is slowly decomposed.

Tetrethylic Pyrophosphate, $(C_2H_5)_4.P_2O_7$, produced by the action of ethyl iodide on argentic pyrophosphate, is a viscid liquid of specific gravity 1.172 at 17° C. (62.6° F.), decomposed by potash, with formation of potassium diethyl-phosphate.

Ethyl Phosphites.—*Symmetrical Triethyl Phosphite*, $P(O.C_2H_5)_3$, formed by the action of phosphorus trichloride on ethyl alcohol, boils at 191°.

Ethylphosphonic acid, $(C_2H_5)PO(OH)_2$, formed by oxidation of primary ethylphosphine with nitric acid, is a solid body melting at 44° C. (111.2° F.) (p. 583).

Ethyl Borates.—*Triethylic Borate*, $(C_2H_5)_3BO_3$, is formed by the action of boron trichloride on alcohol :—

$$3C_2H_5(OH) + BCl_3 = 3HCl + (C_2H_5)_3BO_3 ;$$

also, together with monethylic borate, by heating boric oxide with abso. lute alcohol.

It is a thin limpid liquid, of agreeable odor, specific gravity 0.885, boiling at 119° C. (246.2° F.), decomposed by water. Its alcoholic solu tion burns with a green flame, throwing off a thick smoke of boric acid. Treated with zinc-ethyl, it yields b o r e t h y l, $B(C_2H_5)$, a colorless, mobile, pungent, spontaneously inflammable liquid, having a specific gravity of 0.696, and boiling at 95° C. (203° F.).

Monethylic Borate, $C_2H_5BO_2$, is a dense inodorous liquid, having at 120° the consistence of oil of vitriol. It cannot be distilled without decompo. sition, being resolved at high temperatures into triethylic borate and monethylic triborate, $C_2H_5B_3O_5$, or $C_2H_5BO_2.B_2O_3$:

$$4(C_2H_5)BO_2 = (C_2H_5)_3BO_3 + (C_2H_5)B_3O_5.$$

The latter remains as a mass resembling gum-arabic : it attracts moisture from the air, and becomes covered with a crust of boric acid.

Ethyl Silicates.—*Tetrethylic Silicate*, $(C_2H_5)_4SiO_4$, or $Si(OC_2H_5)_4$, is produced by treating silicic chloride with a small quantity of anhydrous alcohol :

$$4C_2H_5OH + SiCl_4 = 4HCl + Si(OC_2H_5)_4.$$

It is a colorless liquid, having a rather pleasant ethereal odor, and strong peppery taste ; specific gravity 0.933 at 20° C. (68° F.). It boils without decomposition between 165° and 166° C. (329° and 330.8° F.), and when set on fire burns with a dazzling flame, diffusing a white smoke of finely divided silica. It is decomposed slowly by water, quickly by ammonia and the fixed alkalies.

Silicic ethers containing ethyl and methyl, and ethyl and amyl, have likewise been obtained.

Ethylic Thio-alcohol and Ethers.

Ethyl Hydrosulphide, Ethylic Thio-alcohol, or **Mercaptan,** C_2H_5SH.—This compound, the sulphur analogue of ethyl alcohol, is pro- duced, analogously to methyl hydrosulphide (p. 590), by the action of potassium hydrosulphide on calcium ethylsulphate. A solution of caustic potash of specific gravity 1.28 or 1.3, is saturated with sulphuretted hy- drogen, and mixed in a retort with an equal volume of solution of calcium ethylsulphate of the same density. The retort is connected with a good condenser, and heat is applied by means of a bath of salt and water. Mercaptan and water distil over together, and are easily separated by a tap-funnel. The product thus obtained is a colorless, limpid liquid, of specific gravity 0.8325 at 21° C. (69.8° F.), slightly soluble in water, easily miscible with alcohol. It boils at 36° C. (96.8° F.). The vapor of mercaptan has a most intolerable odor of onions, which adheres to the clothes and person with great obstinacy : it is very inflammable, and burns with a blue flame.

When mercaptan is brought in contact with mercuric oxide, even in the cold, violent reaction ensues, water is formed, and a white substance is produced, soluble in alcohol, and separating from that liquid in distinct crystals which contain $Hg(SC_2H_5)_2$. This compound is decomposed by sulphuretted hydrogen, mercuric sulphide being thrown down, and mercaptan reproduced. By adding solutions of lead, copper, silver, and gold to an alcoholic solution of mercaptan, corresponding compounds containing those metals are formed. Caustic potash produces no effect upon mercaptan, but potassium displaces hydrogen, and gives rise to a crystallizable compound, C_2H_5SK, soluble in water. Sodium acts in a similar manner.

Ethyl Sulphides.—Three of these compounds have been obtained, analogous in composition to the methyl sulphides, and produced by similar reactions. The *monosulphide*, $(C_2H_5)_2S$, or $C_2H_5SC_2H_5$, is a colorless oily liquid, having a very pungent alliaceous odor, a specific gravity of 0.825 at 20° C. (68° F.), and boiling at 91° C. (195.8° F.). It is very inflammable, and burns with a blue flame. When poured into chlorine gas, it takes fire; but when dry chlorine is passed into a flask containing it, not at first into the liquid, the vessel being kept cool and in the shade, substitution-products are formed and hydrochloric acid is copiously evolved. The product consists chiefly of d i c h l o r e t h y l i c s u l p h i d e, $(C_2H_4Cl)_2S$. If the action takes place in diffused daylight, and without external cooling, the compounds $(C_2H_2Cl_3)_2S$ and $(C_2HCl_4)_2S$ are obtained, which may be separated by fractional distillation, the first boiling between 189° and 192° C. (372.2–377.6° F.), the second between 217° and 222° C. (422.6–431.6° F.). The action of chlorine on ethyl sulphide in sunshine yields a more highly chlorinated compound, probably $(C_2Cl_5)_2S$.

Ethyl bisulphide, $(C_2H_5)_2S_2$, obtained by distilling potassium bisulphide with potassium ethylsulphate or with ethyl oxalate, is a colorless oily liquid, very inflammable, boiling at 151° C. (303.8° F.). The *trisulphide*, $(C_2H_5)_2S_3$, is a heavy oily liquid, obtained by acting in like manner on potassium pentasulphide.

Triethylsulphurous Compounds.—When ethyl monosulphide and ethyl iodide are heated together, they unite and form s u l p h u r o u s i o d o t r i e t h i d e, $(C_2H_5)S.C_2H_5I$, or $S^{iv}(C_2H_5)_3I$, which crystallizes in needles. The same compound is formed by the action of ethyl iodide on ethyl hydrosulphide:

$$2C_2H_5I + C_2H_5SH = HI + S(C_2H_5)_3I,$$

or of hydrogen iodide on ethyl monosulphide:

$$HI + 2(C_2H_5)_2S = C_2H_5SH + S(C_2H_5)_3I.$$

Sulphurous iodotriethide is insoluble in ether, slightly soluble in alcohol, and crystallizes from the solution in white deliquescent needles belonging to the monoclinic system. It unites with metallic chlorides.

Ethyl chloride and ethyl bromide unite in like manner, but less readily, with ethyl sulphide, forming the compounds $S(C_2H_5)_3Cl$ and $S(C_2H_5)_3Br$, both of which crystallize in needles.

By treating the iodine-compound with recently precipitated silver oxide, a strongly alkaline solution is obtained, which dries up over oil of vitriol to a crystalline deliquescent mass, consisting of s u l p h u r o u s t r i e t h y l-h y d r o x y l a t e, $(C_2H_5)_3S(OH)$. The solution of this substance dissolves the skin like caustic potash, and forms similar precipitates with various metallic salts. It neutralizes acids, forming definite crystallizable salts, *e. g.*, the *nitrate*, $(C_2H_5)_3S(ONO_2)$, the *acetate*, $(C_2H_5)_3S(OC_2H_3O)$, etc.

The function of the sulphur in these compounds is analogous to that of nitrogen in the ammonium-compounds. Nitrogen (quinquivalent) forms, with 4 atoms of hydrogen or of an alcohol-radicle, univalent radicles, NH_4, $N(C_2H_5)_4$, etc., which combine, like the alkali-metals, with chlorine, iodine, hydroxyl, etc. ; and in like manner sulphur (quadrivalent) forms with 3 atoms of methyl, ethyl, etc., univalent radicles like $S(C_2H_5)_3$ (triethylsulphine), which also combine with chlorine, hydroxyl, etc., forming the s u l p h o n i u m c o m p o u n d s $S^{iv}(C_2H_5)_3Cl$, $S^{iv}(C_2H_5)_3OH$, etc., the hydroxides being strong alkaline bases, like tetrethylammonium hydroxide (see p. 508; also AMINES).

Similar compounds, containing sexvalent sulphur, are obtained by combining ethyl sulphide and ethene sulphide, $S(C_2H_4)$, with ethene-dibromide; thus :—

$$S(C_2H_4) \ + \ C_2H_4Br_2 \ = \ \begin{matrix} Br \\ Br \end{matrix}\!\!>\!\!S\!\!<\!\!\begin{matrix} C_2H_4 \\ C_2H_4 \end{matrix}$$
Diethene-sulphonium dibromide.

$$S(C_2H_5)_2 \ + \ C_2H_4Br_2 \ + \ \begin{matrix} Br \\ Br \end{matrix}\!\!>\!\!\overset{vi}{S}\!\!=\!\!C\!\!<\!\!\begin{matrix} C_2H_5 \\ C_2H_4 \\ C_2H_5 \end{matrix}$$
Ethene-diethyl-sulphonium dibromide.

Analogous compounds are also formed by selenium, *e. g.*, *Trimethy-selenonium iodide*, $Se(CH_3)_3I$.

Ethyl Telluride, Telluric Ethide or Tellurethyl, $Te(C_2H_5)_2$.—This compound is obtained by distilling potassium telluride with potassium ethylsulphate :—

$$TeK_2 \ + \ 2K(C_2H_5)SO_4 \ = \ 2K_2SO_4 \ + \ Te(C_2H_5)_2.$$

It is a heavy, oily, yellowish-red liquid, very inflammable, and having a most intolerable odor. It acts as a bivalent radicle, uniting with chlorine, bromine, oxygen, etc., to form compounds in which the tellurium enters as a tetrad, *e. g.*, $Te(C_2H_5)_2Cl_2$, $Te(C_2H_5)_2O$, etc. The *nitrate* is obtained by treating tellurethyl with nitric acid; the other salts by double decomposition; the *chloride*, for example, settles down as a heavy oil, on adding hydrochloric acid to a solution of the nitrate. The *oxide* is best prepared by treating the chloride with water and silver oxide; it dissolves in water, forming a slightly alkaline liquid.

Selenic Ethide, or Selenethyl, $Se(C_2H_5)_2$, prepared like tellurethyl, is also a fetid liquid, exactly resembling the tellurium compound in its chemical relations.

PROPYL ALCOHOLS AND ETHERS.

It has already been observed that the three-carbon alcohol, C_3H_8O, is susceptible of two isomeric modifications; namely :—

$$\begin{matrix} CH_3 \\ | \\ CH_2 \\ | \\ CH_2OH \end{matrix} \quad or \quad C\begin{cases} CH_2CH_3 \\ H_2 \\ OH \end{cases} \qquad \begin{matrix} CH_3 \\ | \\ HCOH \\ | \\ CH_3 \end{matrix} \quad or \quad C\begin{cases} CH_3 \\ CH_3 \\ H \\ OH \end{cases}$$

Normal propyl alcohol.   Pseudopropyl alcohol.

each of which may give rise to a corresponding set of ethers.

Normal Propyl Alcohol was discovered by Chancel in 1853, in the fusel oil of the residues left in the distillation of brandy from wine. It may be obtained synthetically from ethyl alcohol by the following series of processes :—

1. Ethyl cyanide, C_3H_5N, or $CH_2CH_3.CN$ (prepared by distilling a mixture of potassium cyanide and potassium ethylsulphate, p. 555), is converted into propionic acid, $CH_2CH_3.COOH$, by boiling with strong caustic potash.

2. A mixture of the calcium salts of propionic and formic acids is subjected to dry distillation, whereby propionic aldehyde is obtained :—

$$CH_2CH_3.COOCa' \ + \ H.COOCa'* \ = \ Ca_2CO_3 \ + \ CH_2CH_3.COH.$$

Calcium propionate.   Calcium formate.   Calcium carbonate.   Propionic aldehyde.

3. The propionic aldehyde, treated with water and sodium amalgam, takes up 2 atoms of hydrogen, and is converted into normal propyl alcohol :—

$$CH_2CH_3.COH \ + \ H_2 \ = \ CH_2CH_3.CH_2OH.$$

The series of processes just described affords a general method of building up the normal primary alcohols of the fatty group, one from the other. It has not, however, been actually carried out higher than the six-carbon or hexyl alcohol.

Another method of passing from ethyl-alcohol to propyl alcohol—also generallly applicable as a method of synthesis of primary alcohols—is to convert ethyl cyanide or propionitril into propylamine by the action of nascent hydrogen (water and sodium amalgam) and the propylamine, by the action of nitrous acid, into propyl nitrite, which may then be converted into the alcohol by distillation with an alkali :—

$$CH_3.CH_2.CN \ + \ 2H_2 \ = \ CH_3.CH_2.CH_2.NH_2$$

Ethyl cyanide.   Propylamine.

$$C_3H_7.NH_2 \ + \ 2NO_2H \ = \ 2H_2O \ + \ N_2 \ + \ C_3H_7NO_2$$

Propylamine.   Propyl nitrite.

Normal propyl alcohol is an oily liquid, boiling at 96° C. (204.8° F.), and having a specific gravity of 0.8205 at 0°. By oxidization with a mixture of sulphuric acid and potassium dichromate, it is converted into propionic acid.

Normal propyl chloride, C_3H_7Cl, boils at 46.5° C. (115.7° F.); the *bromide* at 70°–71° C. (158°–159.8° F.); the *iodide* at 102° C. (215.6° F.); the *oxide*, $(C_3H_7)_2O$, at 85°–86° C. (185°–186.8° F.).

Pseudopropyl Alcohol, or **Secondary Propyl Alcohol,** $CH(CH_3)_2OH$.—This alcohol is prepared :

1. From *acetone*, $(CO)(CH_3)_2$, by direct addition of hydrogen, evolved by the action of water on sodium amalgam :

$$\begin{array}{ccc} H_3C\ CH_3 & & H_3C\ CH_3 \\ \diagdown\diagup & + \ H_2 \ = & \diagdown\diagup \\ CO & & HCHO \end{array}$$

This mode of synthesis affords direct proof of the constitution of psendopropylic alcohol, the addition of the two hydrogen-atoms being tantamount to the replacement of the bivalent radicle oxygen by the two monad radicles, hydrogen and hydroxyl.

* For the sake of simplicity, the equivalent (20) of calcium is used in this equation, instead of the atomic weight.

2. Pseudopropyl iodide is prepared by the action of iodine and phosphorus on *glycerin;* this iodide is easily converted into the oxalate or acetate by treatment with silver oxalate or acetate ; and from either of these ethers the alcohol may be obtained by distillation with potash or soda.

Pseudopropyl alcohol is a colorless, not very mobile liquid, having a peculiar odor, a specific gravity of 0.791 at 15° C. (59° F.), boiling at 83° to 84° C. (181.4–183.2° F.) under a barometric pressure of 739 millimetres, not freezing at —20° C. (4° F.). It does not act on polarized light. It is very difficult to dry, as it mixes with water in all proportions, and forms with it three definite and very stable hydrates, viz., $3C_3H_8O.2H_2O$, boiling at 78°–80° C. (172.4°–176° F.) ; $2C_3H_8O.H_2O$, boiling at 80° C. (176° F.); and $3C_3H_8O.H_2O$, boiling at 81° C. (177.8° F.). The second of these hydrates exhibits a very close resemblance to ethyl alcohol, has the same percentage composition, boils at nearly the same temperature, and likewise yields acetic acid by oxidation (see below) ; moreover it retains its water of hydration so obstinately, that it does not even change the white color of anhydrous cupric sulphate to blue. The readiest mode of distinguishing between this hydrate and ethyl alcohol is to submit them to the action of iodine and phosphorus, whereby the former is converted into pseudopropyl iodide, the latter into ethyl iodide.

The characteristic property of pseudopropyl alcohol is that it yields acetone by oxidation with dilute chromic acid, this transformation being the reverse of that by which it is produced :

$$\underset{\text{HCOH}}{\overset{\text{H}_3\text{C CH}_3}{\diagdown\diagup}} \;+\; \text{O} \;=\; \underset{\text{CO}}{\overset{\text{H}_3\text{C CH}_3}{\diagdown\diagup}} \;+\; \text{H}_2\text{O}\,.$$

On pushing the oxidation further, the acetone breaks up into acetic acid, carbon dioxide, and water:

$$\underset{\text{Acetone.}}{CO(CH_3)_2} \;+\; O_4 \;=\; \underset{\text{Acetic acid.}}{CO(CH_3)OH} \;+\; CO_2 \;+\; H_2O.$$

The evolution of carbon dioxide in this reaction affords a further distinction between hydrated pseudopropyl alcohol and ethyl alcohol.

The formation of a ketone by oxidation is, as already observed, the essential characteristic of a secondary alcohol, and is an immediate consequence of its structure (p. 586).

Pseudopropyl alcohol, heated with acetic acid, or with potassium acetate and sulphuric acid, is converted into *pseudopropyl acetate,* $CH(CH_3)_2OC_2H_3O$.

PSEUDOPROPYL IODIDE, $CH(CH_3)_2$, is most conveniently prepared by the action of hydriodic acid, concentrated and in large excess, on glycerin (propenyl alcohol) $C_3H_8O_3$:

$$C_3H_8O_3 \;+\; 5HI \;=\; C_3H_7I \;+\; 3H_2O \;+\; 2I_2.$$

The iodine, as fast as it is set free by the reaction, may be reconverted into hydriodic acid by means of phosphorus, and will then be ready to act upon another portion of glycerin. The iodide may also be produced by the action of hydriodic acid on pseudopropyl alcohol, allyl iodide, C_3H_5I, propene, or propene alcohol.

Pseudopropyl iodide is an oil boiling at 80°–90° C. (176°–194° F.), and having a specific gravity of 1.70. With *sodium* in presence of ether it yields propene, propane, and di-isopropyl, C_6H_{14} (p. 528). *Bromine* expels the iodine, and forms pseudopropyl bromide.

By treatment with zinc and hydrochloric acid, which evolves hydrogen, pseudopropyl iodide is converted into propane: $C_3H_7I + H_2 = HI + C_3H_8$; the propane, exposed to the action of chlorine in diffused daylight, is partly converted into normal propyl chloride; this compound, heated with potassium acetate and strong acetic acid, yields normal propyl acetate; and the latter, heated with potash-lye in sealed tubes, yields n o r m a l p r o p y l a l c o h o l. This series of reactions affords a general method of converting a secondary alcohol into the corresponding normal primary alcohol.

TETRYL OR BUTYL ALCOHOLS AND ETHERS.

Theory indicates the existence of four alcohols included in the formula $C_4H_{10}O$, two primary, one secondary, and one tertiary; thus,

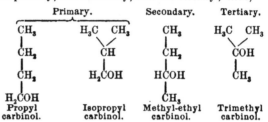

Primary.		Secondary.	Tertiary.

Propyl carbinol. — Isopropyl carbinol. — Methyl-ethyl carbinol. — Trimethyl carbinol.

Propyl Carbinol, or Normal Butyl Alcohol, $C \begin{cases} CH_2CH_2CH_3 \\ H_2 \\ OH \end{cases}$.—This alcohol is obtained from butyl chloride, C_4H_9Cl (produced by the action of chlorine on butane or diethyl, C_4H_{10}), by heating that chloride with potassium acetate and strong acetic acid, whereby it is converted into butyl acetate, and treating that compound with barium hydrate. It may also be prepared from normal propyl alcohol, in the same manner as the latter is obtained from ethyl alcohol, viz., by successive conversion into propyl cyanide or butyronitril, C_4H_7N or $CH_2CH_2CH_3.CN$, normal butyric acid, $CH_2CH_2CH_3.COOH$, butyric aldehyde, $CH_2CH_2CH_3.COH$ (prepared by heating a mixture of the calcium salts of butyric and formic acids), and finally into the alcohol, $CH_2CH_2CH_3.CH_2OH$, by the action of nascent hydrogen on the aldehyde; or, lastly, by converting the butyronitril into butylamine, $C_4H_{11}N$, the latter into butyl nitrite by the action of nitrous acid, and distilling the nitrite with an alkali.

Normal butyl alcohol boils at 115° C. (239° F.), smells like isobutyl alcohol, is much lighter than water, slightly soluble therein; with iodine and phosphorus it yields normal butyl iodide, boiling at 127° C. (260.6° F.). Oxidation with dilute chromic acid converts it into normal butyric acid $CH_2CH_2CH_3.COOH$.

Isopropyl Carbinol, or Isobutyl Alcohol, $C \begin{cases} CH(CH_3)_2 \\ H_2 \\ OH \end{cases}$.—This variety of primary butyl-alcohol was found by W u r t z in the fusel-oil obtained by fermenting the molasses of beet-root sugar. To separate it, this oil is submitted to fractional distillation, and the liquid boiling between 108° C. (226.4° F.) and 118° C. (244.4° F.) is repeatedly rectified over potassium hydroxide, till it boils constantly at 110° C. (230° F.).

Pure isobutyl alcohol is a colorless liquid, having an odor somewhat like

26 *

that of amyl alcohol, but less pungent, and more vinous: specific gravity $= 0.8032$ at 18.5° C. (65.3° F.). It dissolves in $10\frac{1}{2}$ times its weight of water, and is separated therefrom, as an oil, by calcium chloride, sodium chloride, and other soluble salts. By oxidation it is converted into isobutyric acid, $CH(CH_3)_2.COOH$.

Isobutyl alcohol is acted upon by acids and other chemical reagents much in the same manner as common alcohol. With strong *sulphuric acid* it yields i s o b u t y l - s u l p h u r i c a c i d, $(C_4H_9)HSO_4$, if the mixture is kept cool, but on heating the liquid, isobutene, $\frac{H_3C}{H_3C}{>}C{\equiv}CH_2$, is given off mixed with sulphurous oxide and carbon dioxide. Heated with *hydrochloric acid* in a sealed tube, or treated with *phosphorus pentachloride* or *oxychloride,* it is converted into i s o b u t y l c h l o r i d e or c h l o r o - i s o b u t a n e, $(CH_3)_2CH—CH_2Cl$, an ethereal liquid, having a pungent odor, and boiling at 70° C. (158° F.). Isobutyl bromide, C_4H_9Br, obtained in like manner, boils at 89° C. (192.2° F.), the i o d i d e, C_4H_9I, at 121° C. (249.8° F.). The iodide is decomposed by potassium or sodium, yielding i s o - o c -

$$(CH_3)_2CH—CH_3,$$
t a n e or i s o - d i b u t y l, | a limpid liquid, lighter than
$$(CH_3)_2CH—CH_2,$$
water, and boiling at 105° C. (221° F.). The same hydrocarbon is obtained by the electrolysis of ordinary valeric acid, $C_5H_{10}O_2$.

Methyl-ethyl Carbinol, or Secondary Butyl Alcohol, $C \begin{cases} CH_3 \\ C_2H_5. \\ H \\ OH \end{cases}$—

This alcohol is obtained from erythrite, a saccharine substance having the composition of a tetratomic alcohol, $C_4H_{10}O_4$, or $C_4H_6(OH)_4$. The erythrite, distilled with fuming hydriodic acid, yields methyl-ethyl-iodomethane, or secondary butyl iodide, $C(CH_3)(C_2H_5)HI$, and this liquid, treated with moist silver oxide is converted into methyl-ethyl carbinol :

$$C(CH_3)(C_2H_5)HI + AgOH = AgI + C(CH_3)(C_2H_5)HOH.$$

Methyl-ethyl carbinol is a colorless oily liquid, having a strong odor and burning taste, a specific gravity of 0.85 at 0°, and boiling at $95^{\circ}–98^{\circ}$ C. ($203^{\circ}–208.4^{\circ}$ F.) (about 20° C. (36° F.) lower than the normal primary alcohol). When heated at 250° C. (482° F.), it is for the most part resolved into water and butene: $C_4H_{10}O = H_2O + C_4H_8$.

Methyl-ethyl-iodomethane, or *Secondary Butyl iodide,* prepared as above, or by the action of strong hydriodic acid on the alcohol, is a liquid having a pleasant ethereal odor, a specific gravity of 1.632 at 0°, 1.600 at 20° C. (68° F.), and 1.584 at 30° C. (86° F.). It boils at 118° C. (244.4° F.). Bromine decomposes it, expelling the iodine and forming butene dibromide, $C_4H_8Br_2$. When distilled with alcoholic potash, it gives off butene. This tendency to give off the corresponding olefine is characteristic of all the secondary alcohols and ethers, as will be further noticed in connection with the five-carbon compounds.

Trimethyl Carbinol, or Tertiary Butyl Alcohol, $C \begin{cases} (CH_3)_3 \\ OH \end{cases}$, is produced by treating zinc methide with carbonyl chloride (phosgene gas) or acetyl chloride, and submitting the product to the action of water.

$$\underset{\substack{\text{Carbonyl} \\ \text{chloride.}}}{2COCl_2} + \underset{\substack{\text{Zinc} \\ \text{methide.}}}{Zn(CH_3)_2} = \underset{\substack{\text{Zinc} \\ \text{chloride.}}}{ZnCl_2} + \underset{\substack{\text{Acetyl} \\ \text{chloride.}}}{2COCH_3Cl}$$

$$COCH_3Cl \quad + \quad Zn(CH_3)_2 \quad = \quad ZnO \quad + \quad C\begin{cases}(CH_3)_3 \\ Cl\end{cases}$$

Acetyl   Zinc   Zinc   Trimethyl
chloride.   methide.   oxide.   chloromethane.

$$C\begin{cases}(CH_3)_3 \\ Cl\end{cases} + \quad HOH \quad = \quad HCl \quad + \quad C\begin{cases}(CH_3)_3 \\ OH\end{cases}$$

Trimethyl   Water.   Trimethyl
chloromethane.   carbinol.

It may also be formed from the primary isoalcohol by the following series of processes : (1) The alcohol treated with sulphuric acid, or other dehydrating agent, yields isobutene, C_4H_8.—(2) This olefine, treated with strong hydriodic acid, yields tertiary butyl iodide, the iodine attaching itself to the carbon atom which is not in direct combination with hydrogen.—(3) The iodide treated with silver oxide is converted into the tertiary alcohol. The series of transformations is represented by the following formulæ :—

$$(CH_3)_2CH—CH_2OH \qquad\qquad (CH_3)_2C{=\!=}CH_2$$
Isobutyl alcohol.   Isobutene.

$$(CH_3)_2CI—CH_3 \qquad\qquad (CH_3)_2C(OH)—CH_3$$
Tertiary Butyl iodide.   Tertiary Butyl alcohol.

Trimethyl carbinol, when perfectly anhydrous, crystallizes in rhombic prisms or tables, melting at $25.5°$ C. ($77.9°$ F.). In the liquid state it has a specific gravity of 0.7788 at $30°$ C. ($86°$ F.), and boils at $82.5°$ C. ($180.5°$ F.). It mixes with water in all proportions, and likewise forms a definite hydrate, $2C_4H_{10}O.H_2O$, which crystallizes in a freezing mixture, and boils without decomposition at $80°$ C. ($176°$ F.). By oxidation with chromic acid, trimethyl carbinol is resolved into acetic acid and acetone, together with a small quantity of isobutyric acid.

PENTYL ALCOHOLS AND ETHERS.

The formula $C_5H_{12}O$ may include eight different alcohols : four primary, three secondary, and one tertiary, viz. :—

Primary, $C\begin{cases}CH_2CH_2CH_2CH_3 \\ H \\ H \\ OH\end{cases}$ $C\begin{cases}CH_2CH(CH_3)_2 \\ H \\ H \\ OH\end{cases}$ $C\begin{cases}OH\begin{cases}CH_2CH_3 \\ CH_3\end{cases} \\ H \\ H \\ OH\end{cases}$ $C\begin{cases}C(CH_3)_3 \\ H \\ H \\ OH\end{cases}$

        Butyl carbinol.   Isobutyl carbinol.

Secondary, $C\begin{cases}CH_2CH_2CH_3 \\ CH_3 \\ H \\ OH\end{cases}$ $C\begin{cases}CH(CH_3)_2 \\ CH_3 \\ H \\ OH\end{cases}$ and $C\begin{cases}CH_2CH_3 \\ CH_2CH_3 \\ H \\ OH\end{cases}$

  Methyl-propyl   Methyl isopropyl   Diethyl
    carbinol.   carbinol.   carbinol.

Tertiary, $C\begin{cases}CH_2CH_3 \\ CH_3 \\ CH_3 \\ OH\end{cases}$ Dimethyl-ethyl carbinol.

Six of these are known, viz., the first, second, fifth, sixth, seventh, and eighth.

Butyl Carbinol, or Normal Primary Pentyl Alcohol,
$H_2C\begin{cases}C_3H_6OH \\ CH_3\end{cases}$, is prepared from normal butyl alcohol in the same manner

as the latter from normal propyl alcohol, viz., by successive conversion into butyl cyanide, $CH_2CH_2CH_2CH_3.CN$, normal valeric acid, $CH_2CH_2CH_2CH_3.$ COOH, valeric aldehyde, $CH_2CH_2CH_2CH_3.COH$, and the alcohol, $CH_2CH_2CH_2$ $CH_3.CH_2OH$. It is a liquid boiling at 135° C. (275° F.), i. e., 3 degrees higher than isopentyl alcohol. The chloride, bromide, iodide, and acetate obtained from it boil at higher temperatures than the corresponding iso-pentyl compounds. By oxidation it yields normal valeric acid.

Isobutyl Carbinol, Isopentyl Alcohol, or Amyl Alcohol,

$CH \left\{ \begin{matrix} C_2H_4OH \\ (CH_3)_2 \end{matrix} \right.$. — This is the ordinary amyl alcohol produced by fermen-tation. In the manufacture of brandy from corn, potatoes, or the must of grapes, the ethyl alcohol is found to be accompanied by an acrid oily liquid called *fusel-oil*, which is very difficult to separate completely from the ethyl alcohol. It passes over, however, in considerable quantity towards the end of the distillation, and may be collected apart, washed by agitation with several successive portions of water to free it from ethyl alcohol, and re-distilled. The liquid thus obtained consists chiefly of amyl alcohol, sometimes mixed with propylic, butylic, and other alcohols. The amyl alcohol may be obtained pure by fractional distillation, the por-tion which passes over between 128° and 132° C. (262.4^O and 269.6^O F.) being collected apart. Potato fusel-oil consists almost wholly of ethyl and amyl alcohols, the latter constituting the greater quantity.

Amyl alcohol is an oily, colorless, mobile liquid, having a peculiar odor, and a burning acrid taste. Its vapor when inhaled produces coughing and oppression of the chest. It has a specific gravity of 0.825 at 0°, and boils at 130° C. (266° F.). When dropped on paper it forms a greasy stain, which, however, disappears after a while. It is not perceptibly soluble in water, but floats on the surface of that liquid like an oil ; com-mon alcohol, ether, and various essential oils dissolve it readily.

Amyl alcohol usually exerts a rotatory action on polarized light, but the rotatory power varies considerably in different samples. Pasteur, indeed, has shown that ordinary amyl alcohol is a mixture of two iso-meric alcohols, having the same vapor-density, but differing in their optical properties, one of them turning the plane of polarization to the right, whereas the other is optically inactive. They are separated by converting the crude amyl alcohol into amylsulphuric acid, saturating with barium carbonate, and crystallizing the barium amylsulphate thus formed. The salt obtained from the active amyl alcohol is $2\frac{1}{2}$ times more soluble than that obtained from the inactive alcohol, and consequently the latter crystallizes out first ; and by precipitating the barium from the solu-tion of either salt with sulphuric acid, and distilling the amylsulphuric acid thus separated with water, the corresponding amyl alcohol is ob-tained. The difference of optical character between the two alcohols— which is traceable through many of their derivatives—has not been satis-factorily explained ; but it perhaps depends upon the arrangement of the molecules, rather than upon that of the atoms within the molecule. On the other hand, it is possible that the active and inactive alcohols may contain different radicles, as indicated by the second, third, and fourth formulæ of primary amyl alcohols above given.

Vapor of amyl alcohol passed through a red-hot tube, yields a mixture of ethene, propene, butene, and isopentene or amylene.

Amyl alcohol takes fire easily and burns with a blue flame. When ex-posed to the air in contact with platinum black, or treated with a mixture of potassium chromate and dilute sulphuric acid, it is oxidized to isovaleric acid, $CH_2CH(CH_3)_2.COOH$.

Amyl alcohol, heated to 220° C. (428° F.) with a mixture of *potassium hydroxide* and *lime*, is converted into potassium valerate, with evolution of hydrogen :

$$C_5H_{12}O + KHO = C_5H_9KO_2 + 2H_2.$$

Potassium and *sodium* dissolve in amyl alcohol as in ethyl alcohol, yielding the compounds, $C_5H_{11}KO$, and $C_5H_{11}NaO$, which, when treated with amyl iodide, yield a m y l o x i d e or a m y l e t h e r, $(C_5H_{11})_2O$, and with ethyl iodide, e t h y l - a m y l o x i d e, $(C_2H_5)(C_5H_{11})O$.

Chlorine acts upon amyl alcohol as upon ethyl alcohol (p. 592), excepting that it finally removes only four atoms of hydrogen, instead of five :

$$\underset{\text{Amyl alcohol.}}{C_5H_{12}O} + 3Cl_2 = 4HCl + \underset{\text{Chloramylal.}}{C_5H_8Cl_2O}.$$

Amyl alcohol is acted upon by acids, like common alcohol, yielding ethers. When mixed with strong *sulphuric acid*, it is converted into amylsulphuric acid, $(C_5H_{11})HSO_4$; and, on distilling the mixture, amyl oxide, $(C_5H_{11})_2O$, passes over, together with amylene, and several other hydrocarbons.

ISOPENTYL CHLORIDE, or AMYL CHLORIDE, $C_5H_{11}Cl$, or $HC\begin{cases} C_2H_4Cl \\ (CH_3)_2 \end{cases}$, is prepared by distilling equal weights of amyl alcohol and phosphorus pentachloride, washing the product repeatedly with alkaline water, and rectifying it from calcium chloride. Less pure it may be obtained by saturating amyl alcohol with hydrochloric acid. It is a colorless liquid, of agreeable aromatic oder, insoluble in water, and neutral to test-paper: it boils at 102° C. (215.6° F.), and ignites readily, burning with a flame green at the edges. By the long continued action of chlorine, aided by powerful sunshine, it is converted into o c t o c h l o r i n a t e d a m y l c h l o r i d e, or n o n o c h l o r o p e n t a n e, $C_5H_3Cl_9$, a volatile, colorless liquid, smelling like camphor : the whole of the hydrogen has not yet, however, been removed. The *bromide*, $C_5H_{11}Br$, is a volatile, colorless liquid, heavier than water, boiling at 119° C. (246.2° F.). It is obtained by distilling amyl alcohol with bromine and phosphorus (see Ethyl-bromide, p. 597). Its odor is penetrating and alliaceous. The bromide is decomposed by an alcoholic solution of potash, with reproduction of the alcohol and formation of potassium bromide. The *iodide*, $C_5H_{11}I$, is prepared by distilling a mixture of 15 parts of amyl alcohol, 8 of iodine, and 1 of phosphorus. It is colorless when pure, heavier than water, volatile without decomposition at 146° C. (294.8° F.), and in other respects resembles the bromide : it is partly decomposed by exposure to light. Heated to 290° C. (554° F.) in sealed tubes, with zinc, it yields d e c a n e or d i a m y l, $C_{10}H_{22}$, or $C_5H_{11}.C_5H_{11}$, a colorless ethereal liquid boiling at 155° C. (311° F.). At the same time there is formed a compound of zinc iodide with zinc amylide, $Zn(C_5H_{11})_2$, which is decomposed by contact with water, yielding zinc oxide and pentane or amyl hydride (p. 499) :

$$Zn(C_5H_{11})_2 + H_2O = ZnO + 2C_5H_{12}.$$

AMYL OXIDE, $(C_5H_{11})_2O$, obtained by the processes already mentioned, is a colorless oily liquid, of specific gravity 0.779, and boiling at 176° C. (348.8° F.).

AMYLSULPHURIC, or SULPHAMYLIC ACID, $(C_5H_{11})HSO_4$, or $C_5H_{11}(OSO_3H)$. The barium salt of this acid, $(C_5H_{11})_2Ba(SO_4)_2 2Aq.$, prepared like the ethylsulphate (p. 601), crystallizes, on evaporating the solution, in small brilliant pearly plates ; the difference of solubility of the salts prepared from optically active and optically inactive amyl alcohol has already been mentioned. The barium may be precipitated from the salt by dilute sulphuric acid, and the sulphamylic acid concentrated by spontaneous evaporation to a syrupy, or even crystalline state: it has an acid and bitter taste, strongly reddens litmus-paper, and is decomposed by ebullition into amyl alcohol and sulphuric acid. The potassium salt forms groups of small radiating needles, very soluble in water. The sulphamylates of calcium and lead are also soluble and crystallizable.

Amyl hydrosulphide, $C_5H_{11}SH$, and Amyl sulphide $(C_5H_{11})_2S$, resemble the ethyl-compounds in their properties and reactions.

Fusel-oil of Grain-spirit.—The fusel-oil, separated in large quantities from grain-spirit by the London rectifiers, consists chiefly of amyl alcohol mixed with ethyl alcohol and water. Sometimes it contains in addition more or less of the ethyl- or amyl-compounds of certain fatty acids thought to be identical with œnanthylic and palmitic acids. These last-named substances form the principal part of the nearly solid fat produced in this manner in whiskey distilleries conducted on the old plan. Mulder has described, under the name of *corn-oil*, another constituent of the crude fusel-oil of Holland: it has a very powerful odor, resembling that of some of the umbelliferous plants, and is unaffected by solution of caustic potash. According to Mr. Rowney, the fusel-oil of the Scotch distilleries contains in addition a certain quantity of capric acid, $C_{10}H_{20}O_2$. Amyl alcohol, in addition to isobutyl alcohol, has been separated from the spirit distilled from beet-molasses and from artificial grape-sugar made by the aid of sulphuric acid.

Propyl-methyl Carbinol, $HO.HC{<}^{CH_2CH_2CH_3}_{CH_3}$, or, C_3H_7—$CHOH$—CH_3.—This secondary alcohol is produced by decomposing propyl-methyl ketone (obtained by distilling a mixture of calcium butyrate and acetate) with water and sodium amalgam:

$$C_3H_7—CO—CH_3 + H_2 = C_3H_7—CHOH—CH_3.$$

It is a liquid smelling like ordinary amyl alcohol, but less pungent; boils at 120° C. (248° F.); has a specific gravity of 0.825 at 0°; is oxidized by potassium permanganate to propyl-methyl ketone.

Isopropyl-methyl Carbinol, $HO.HC{<}^{CH(CH_3)_2}_{CH_3}$, also called **Amylene hydrate,** $(C_5H_{10})\left\{^H_{OH}\right.$.—This is a secondary alcohol produced from amylene, C_5H_{10}, by combining that substance with hydriodic acid, and decomposing the resulting hydriodide, $C_5H_{10}.HI$, with moist silver oxide, whereby silver iodide and amylene hydrate are obtained:

$$\begin{matrix} CH(CH_3)_2 \\ | \\ CH \\ \| \\ CH_2 \end{matrix} \quad + \quad \left.\begin{matrix} I \\ H \end{matrix}\right\} \quad = \quad \begin{matrix} CH(CH_3)_2 \\ | \\ CHI \\ | \\ CH_3 \end{matrix}$$

and $\begin{matrix} CH(CH_3)_2 \\ | \\ CHI \\ | \\ CH_3 \end{matrix} \quad + \quad AgOH \quad = \quad AgI \quad + \quad \begin{matrix} CH(CH_3)_2 \\ | \\ CHOH \\ | \\ CH_3 \end{matrix}$

A portion of the hydriodide is at the same time resolved, by the heat evolved in the reaction, into hydriodic acid and amylene; and, on submitting the resulting liquid to fractional distillation, the amylene passes over first, and then, between 105° and 108° C. (221–226.4° F.), the amylene hydrate or isopropylmethyl carbinol.

This alcohol is a liquid having a specific gravity of 0.829 at 0°, and a pungent ethereal odor, quite distinct from that of ordinary amyl alcohol. Heated with strong *sulphuric acid*, it is converted, not into amylsulphuric acid, but into hydrocarbons polymeric with amylene, viz., diamylene, or decene, $C_{10}H_{20}$, and triamylene, or pentadecene, $C_{15}H_{30}$ (p. 533). *Hydriodic acid* converts it, at ordinary temperatures, into amylene hydriodide, C_5H_{10}. HI, boiling at 130° C. (266° F.), amyl iodide at 146° C. (294.8° F.). *Hydrochloric acid* converts it (even at 0°) into amylene hydrochloride, C_5H_{10}. HCl, having a boiling point 10° below that of amyl chloride. On mixing it with two atoms of *bromine* at a very low temperature, a red liquid is formed, which, as soon as it attains the ordinary temperature of the air, is resolved into water and amylene bromide. Heated for some time to 100° with strong *acetic acid*, it yields amylene, together with a small quantity of amylene acetate. *Sodium* dissolves in amalene hydrate with evolution of hydrogen, forming a colorless translucent mass, which has the compositiou $C_5H_{10}NaOH$, and is decomposed by amylene hydriodide in the manner shown by the equation:

$$C_5H_{10}NaOH \ + \ C_5H_{10}HI \ = \ C_5H_{10} \ + \ C_5H_{10}H(OH) \ + \ NaI.$$

Sodium compound.　　Amylene　　Amylene.　　Amylene
　　　　　　　　　hydriodide.　　　　　　　　hydrate.

From these reactions it is apparent that amylene hydrate or isopropylmethyl carbinol is especially distinguished from the primary amyl alcohols by the facility with which it gives up the corresponding olefine. This peculiarity is exhibited also by all the secondary alcohols of the series. These alcohols may indeed be regarded as intermediate links between the primary monatomic alcohols and the diatomic alcohols or glycols, *e. g.*,

$$C_5H_{11}(OH) \qquad C_5H_{10}\begin{Bmatrix} H \\ OH \end{Bmatrix} \qquad C_5H_{10}\begin{Bmatrix} OH \\ OH \end{Bmatrix}$$

Amyl alcohol.　　　　　Amylene hydrate.　　　Amylene glycol.

Diethyl Carbinol, C_2H_5—CHOH—C_2H_5, is produced by heating ethyl formate with ethyl iodide and granulated zinc, and decomposing the product with ice-cold water. The action of the zinc on the ethyl iodide produces zinc-ethyl, and this reacts with the ethyl formate according to the following equation:

$$HCO.OC_2H_5 + 2Zn(C_2H_5)_2 = HC\begin{Bmatrix}(C_2H_5)_2 \\ ZnOC_2H_5\end{Bmatrix} + Zn\begin{Bmatrix}C_2H_5 \\ OC_2H_5\end{Bmatrix};$$

and on treating this product with water, the compound, $HC\begin{Bmatrix}(C_2H_5)_2 \\ ZnOC_2H_5\end{Bmatrix}$ is decomposed, yielding diethyl carbinol, together with zinc hydroxide and ethene,

$$HC\begin{Bmatrix}(C_2H_5)_2 \\ ZnOC_2H_5\end{Bmatrix} + 2H_2O = HOCH\begin{Bmatrix}C_2H_5 \\ C_2H_5\end{Bmatrix} + Zn(OH_2) + C_2H_6.$$

The final result is the replacement of the oxygen-atom of the group HO, in formic acid, HCO.OH, by 2 atoms of ethyl.

Diethyl carbinol is a liquid which smells like amyl alcohol, has a specific gravity of 0.832 at 0°, and boils at 116–117° C. (240.8–242.6° F.). By oxidation with chromic acid, it is converted with diethyl ketone, $CO(C_2H_5)_2$. The corresponding iodide boils at 145°; the acetate at 132°.*

---

* W a g n e r, Liebig's Annalen, clxxv. 351; Chem. Soc. Journ. 1875, p. 627.

Ethyl-dimethyl Carbinol, or **Tertiary Amyl Alcohol,**

$$C \begin{cases} C_2H_5 \\ (CH_3)_2 \\ OH \end{cases}$$, is prepared like tertiary butyl alcohol, by treating zinc-me-thide with propionyl chloride, C_3H_5OCl, and decomposing the product with water. It smells very much like tertiary butyl alcohol, has a specific gravity of 0.828 at 0°, solidifies to a crystalline mass at —30° C. (—22° F.), boils between 98.5° and 100° C. (209.3° and 212° F.), and does not solidify at —17° C. (1.4° F.), but merely becomes viscid. By oxidation with dilute chromic acid, it yields nothing but acetic acid.

The boiling points of the six known pentyl alcohols become gradually lower, from the normal primary to the tertiary, as their structure becomes more complex; thus—

		Boiling Point.
Primary,	Butyl carbinol . . .	137°
	Isobutyl carbinol . . .	128°–132°
Secondary,	Propyl-methyl carbinol . .	120°–123°
	Diethyl carbinol . . .	116°–117°
	Isopropyl-methyl carbinol .	104°–108°
Tertiary,	Ethyl-dimethyl carbinol .	98.5°–100°

HEXYL ALCOHOLS AND ETHERS.

The number of possible modifications of an alcohol increases with the number of carbon-atoms in its molecular formula. Thus, we have seen that there may be two propyl alcohols, C_3H_8O, four butyl alcohols, $C_4H_{10}O$, and eight amyl alcohols, $C_5H_{12}O$. The six-carbon formula, $C_6H_{14}O$, will in like manner be found to include seventeen isomeric alcohols—eight primary, six secondary, and three tertiary; but as the manner in which these modifications arise has been sufficiently explained in the preceding pages, the further development of the theoretical formulæ may be left as an exercise for the student.

The number of modifications of the six-carbon alcohol actually known is eight: of which two are primary, three secondary, and the remaining three tertiary.

Primary Hexyl Alcohols, $C_6H_{13}(OH)$.—The normal alcohol, $CH_3.CH_2.CH_2.CH_2.CH_2.CH_2OH$, or $H_2C \begin{cases} C_4H_8OH \\ CH_3 \end{cases}$, is obtained from the essential oil of *Heracleum giganteum*, which is a mixture of hexyl butyrate and octyl acetate. The hexyl and octyl alcohols are isolated by decomposing the oil with alcoholic potash, and separated by fractional distillation. The hexyl alcohol thus obtained has a strong aromatic odor, a specific gravity of 0.819 at 23° C. (73.4° F.), and boils at 156.6° C. (313.8° F.). By oxidation it yields a caproic acid, $C_6H_{12}O_2$, having the same boiling point (204.5° C., 400.1° F.), as normal caproic acid. The corresponding *iodide* is a heavy colorless liquid, boiling at 179.5° C. (355.1° F.). The *acetate*, $C_6H_{13}.C_2H_3O_2$, has a pleasant fruity odor, and boils at 169° C. (336.2° F.).

The same alcohol is obtained, together with butyl-methyl carbinol, by treating normal hexane, $CH_3.(CH_2)_4.CH_3$, from American petroleum with chlorine, converting the resulting hexyl chloride into the acetate by treatment with silver acetate, and distilling this acetate with potash. The mixture of alcohols thus obtained cannot be completely separated by fractional distillation, but it yields by oxidation the corresponding products, viz., normal caproic acid and methyl-butyl ketone.

Lastly, normal hexyl alcohol is obtained, according to Rossi, by the action of sodium-amalgam and water on normal caproic aldehyde.

Another primary hexyl alcohol, boiling at about $150°$ C. $(302°$ F.), and yielding caproic acid by oxidation, was found by Faget in fusel-oil. The statements respecting it are not very exact, but as it is produced by fermentation, it is probably the i s o p r i m a r y alcohol, $HC \begin{cases} C_3H_6OH \\ CH_3 \\ CH_3 \end{cases}$.

Secondary Hexyl Alcohols.—1. *Methyl-butyl Carbinol,* $HO.HC {<}^{C_4H_9}_{CH_3}$, discovered by Wanklyn and Erlenmeyer, is produced by treating mannite, $C_6H_{14}O_6$ (a saccharine body obtained from manna), with a large excess of very strong hydriodic acid, whereby it is converted into secondary hexyl iodide :—

$$C_6H_{14}O_6 \ + \ 11HI \ = \ C_6H_{13}I \ + \ 6H_2O \ + \ I_{10};$$

and digesting this compound with silver oxide and water :—

$$C_6H_{13}I \ + \ AgHO \ = \ AgI \ + \ C_6H_{14}O.$$

Methyl-butyl carbinol is a viscid liquid, having a pleasant, refreshing odor ; it boils at $137°$ C. $(278.6°$ F.); has a sp. gr. of 0.8327 at $0°$, 0.8209 at $16°$ C. $(60.8°$ F.), and 0.7422 at $99°$ C. $(210.2°$ F.), so that it expands somewhat rapidly by heat. Strong hydrochloric acid converts it into the corresponding chloride, $ClHC {<}^{C_4H_9}_{CH_3}$, which boils at $120°$ C. $(248°$ F.), and yields hexene when digested at $100°$ with alcoholic potash. The iodide boils at $167–168°$ C. $(332.6–334.4°$ F.).

The alcohol is converted by oxidation with a mixture of potassium dichromate and sulphuric acid into m e t h y l - b u t y l k e t o n e, $CO {<}^{C_4H_9}_{CH_3}$, which, when further treated with the oxidizing mixture, yields acetic, carbonic, and normal butyric acids.

2. M e t h y l - k a t a b u t y l C a r b i n o l, $HO.HC {<}^{C(CH_3)_3}_{CH_3}$, also called *Pinacolyl alcohol,* is formed by the action of nascent hydrogen on pinacolin (*q. v.*). It solidifies at low temperatures to crystals which melt at $+4°$ C. $(39.2°$ F.), boils at $120°$ C. $(248°$ F.), and has a specific gravity of 0.834. When oxidized by potassium dichromate and sulphuric acid, it is converted into m e t h y l - k a t a b u t y l k e t o n e, $CH_3—CO—C(CH_3)_3$, which splits up on further oxidation into carbon dioxide and trimethylacetic acid.

3. *Ethyl-propyl Carbinol,* $HO.HC {<}^{CH_2CH_2CH_3}_{CH_2CH_3}$, also called *Hexene hydrate.*—The iodide analogous to this alcohol is prepared from dichlorethyl oxide, $O {<}^{C_2H_3Cl_2}_{C_2H_5}$, by successive treatment with zinc-ethyl and hydriodic acid ; thus :

$$
\begin{array}{lll}
CH_2Cl & & CH_2.C_2H_5 \\
| & & | \\
CHCl \ + \ 2Zn {<}^{C_2H_5}_{C_2H_5} \ = \ 2Zn {<}^{C_2H_5}_{Cl} \ + & CH.C_2H_5 \\
| & & | \\
OC_2H_5 & & OC_2H_5 \\
\text{Dichlorethyl} \quad \text{Zinc-ethyl.} \quad \text{Zinc-ethylo-} & \text{Diethylated} \\
\text{oxide.} \qquad\qquad\qquad \text{chloride.} & \text{ethyl-oxide.}
\end{array}
$$

$$CH_2.C_2H_5 \atop CH.C_2H_5 \atop OC_2H_5 \quad + \quad 2HI \quad = \quad IHC{<}^{CH_2.C_2H_5}_{C_2H_5} \quad + \quad H_2O \quad + \quad C_2H_5I;$$

Diethylated
ethyl oxide.

Ethyl-propyl-
carbinyl iodide.

and the alcohol is obtained by converting the iodide into the acetate, and distilling the latter with potash.

Tertiary Hexyl Alcohols.—Three of these alcohols are possible, and have been obtained, viz.:

$$\text{Methyl-diethyl carbinol,} \quad C{\left\{{CH_3 \atop (C_2H_5)_2 \atop OH}\right.} \quad \text{or} \quad HO.C{\left\{{CH_3 \atop C_2H_5 \atop C_2H_5}\right.}$$

$$\text{Propyl-dimethyl carbinol,} \quad C{\left\{{CH_2CH_2CH_3 \atop (CH_3)_2 \atop OH}\right.} \quad \text{or} \quad HO.C{\left\{{C_3H_7 \atop CH_3 \atop CH_3}\right.}$$

$$\text{Isopropyl-dimethyl carbinol,} \quad C{\left\{{CH(CH_3)_2 \atop (CH_3)_2 \atop OH}\right.} \quad \text{or} \quad HO.C{\left\{{CH(CH_3)_2 \atop CH_3 \atop CH_3}\right.}$$

The first is prepared by treating acetyl chloride, $COCH_3Cl$, with zinc-ethyl, and decomposing the resulting methyl-diethyl-chloromethane, $C(CH_3)(C_2H_5)_2Cl$, with water. It boils at 120° C. (248° F.), and yields by oxidation nothing but acetic acid.

The second, obtained in like manner from butyryl chloride, $CO(C_3H_7)Cl$, and zinc-methyl, boils at 115° C. (239° F.), and is resolved by oxidation into acetic acid, propionic acid, and carbon dioxide.

The third, obtained from isobutyryl chloride and zinc-methyl, is a liquid which solidifies at —35° C. (—31° F.), boils at 112° C. (233.6° F.), and yields by oxidation acetone, acetic acid, and carbon dioxide.

HEPTYL ALCOHOLS, $C_7H_{15}OH$.

Of these alcohols six have been obtained, two primary, one secondary, and three tertiary.

Normal Heptyl Alcohol, $CH_3(CH_2)_5.CH_2OH$, or $H_2C{<}^{C_5H_{10}OH}_{CH_3}$, is prepared, either by the action of nascent hydrogen (evolved by the action of sodium-amalgam on water), on heptyl aldehyde (œnanthol):

$$C_7H_{14}O \quad + \quad H_2 \quad = \quad C_7H_{16}O;$$
Aldehyde.   Alcohol.

or from normal heptane, C_7H_{16}, in the same manner as normal hexyl alcohol from hexane (p. 616). It is a colorless, oily liquid, insoluble in water, boiling at 177° C. (350.6° F.), and converted by oxidation into normal œnanthylic acid.

Another heptyl alcohol, probably the *isoprimary*, was separated by Faget from fusel-oil.

Secondary Heptyl Alcohol, or Dipropyl Carbinol, C_3H_7—CHOH —C_3H_7, prepared by hydrogenation of dipropyl ketone. boils at 150° C. (302° F.), and is reconverted into the ketone by oxidation.

Tertiary Heptyl Alcohols.—*Triethyl Carbinol*, $(C_2H_5)_3COH$, is obtained by treating propionyl chloride, C_3H_5OCl, with zinc-methyl, and the product with water. It remains liquid at —20⁰ C. (—4⁰ F.), boils at 140⁰–142⁰ C. (284⁰–287.6⁰ F.), is slightly soluble in water, and has a specific gravity of 0.8593 at 0⁰. By oxidation with chromic acid it yields heptene, C_7H_{14}, together with carbon dioxide, and apparently also acetic and propionic acids.

Dimethyl-isobutyl Carbinol, $\left.\begin{matrix} C_4H_9 \\ (CH_3)_2 \end{matrix}\right\}COH$, obtained by treating isovaleric chloride, $CH(CH_3)_2$—CH_2—$COCl$, with zinc-methyl, and decomposing the product with water, boils at 129⁰–131⁰ C. (264.2⁰–267.8⁰ F.), and is converted by oxidation into acetic and isobutyric acids.

Dimethyl-katabutyl Carbinol, or *Pentamethylated Ethyl Alcohol*, $\left.\begin{matrix} C(CH_3)_3 \\ (CH_3)_2 \end{matrix}\right\}COH$, prepared by treating trimethyl-acetyl chloride, $C(CH)_3$. $COCl$, with zinc-methyl, and the product with water, melts at 17⁰ C. (62.6⁰ F.), boils at 131⁰–132⁰ C. (267.8⁰–269.6⁰ F.), and forms with water a crystalline hydrate, $2C_7H_{16}O.H_2O$, which melts at 83⁰ C. (181.4⁰ F.).

OCTYL ALCOHOLS, $C_8H_{17}OH$.

Five of these alcohols are known—one primary, three secondary, and one tertiary.

Primary Octyl Alcohol or **Heptyl Carbinol**, $C_7H_{15}.CH_2OH$, is contained, together with the corresponding acetate, $C_8H_{17}.C_2H_3O_2$, in the volatile oil obtained from the seed of the cow-parsnep (*Heracleum sphondylium*) ; also as a butyric ether in the seeds of the common parsnep (*Pastinaca sativa*) ; and, together with hexyl butyrate, in the oil of *Heracleum giganteum*.

The comparatively small portion of cow-parsnep oil, which boils between 190⁰ and 195⁰, consists mainly of the alcohol ; but by far the greater portion passes over between 200⁰ and 212⁰ C. (392⁰ and 413.6⁰ F.), and this, by continued fractionation, yields primary octyl acetate, $C_{10}H_{20}O_2$, boiling between 206⁰ and 208⁰ C. (402.8–406.4⁰ F.). This compound is insoluble in water, easily soluble in alcohol and ether, and has a sp. gr. of 0.8717 at 16⁰ C. (60.8⁰ F.). Heated with alcoholic potash, it yields potassium acetate and primary octyl alcohol.

This alcohol is a colorless oily liquid, having a sp. gr. of 0.830 at 16⁰ C. (60.8⁰ F.), boiling between 190⁰ C. (374⁰ F.) and 192⁰ C. (377.6⁰ F.), nearly insoluble in water, miscible with alcohol and ether ; it has a peculiarly pungent aromatic odor, and tastes sweetish at first, afterwards burning and sharp. By boiling with potassium dichromate and dilute sulphuric acid, it is converted into an acid, $C_8H_{16}O_2$, melting at 16⁰ to 17⁰ C. (62.6⁰ F.), and isomeric or identical with the caprylic acid of natural fats, together with the corresponding octylic ether, $C_8H_{17}.C_8H_{15}O_2$. This reaction shows it to be a primary alcohol ; and from the boiling point of the octane obtained from it (122⁰–125⁰ C.) (251.6⁰–257⁰ F.),[*] Schorlemmer infers

---

[*] Normal pentane, C_5H_{12}, boils at 38° C. (100.4° F.), and the average difference of boiling point between any two consecutive members of the normal paraffin series is about 31° C. (87.8° F.) (p. 529) : hence the boiling point of normal octane should be $38 + 3 \times 31 = 131°$ C. (267.8° F.). On the other hand, isopentane boils at 30° C. (86° F.), and the difference of boiling point between two consecutive isoparaffins is also about 31° C. (87.8° F.) : hence the boiling point of iso-octane should be $30 + 3 \times 31 = 123°$. (S c h o r l e m m e r, Proceedings of the Royal Society, xvi. 376 ; see also Watts's Dictionary of Chemistry, First Supplement, p. 879.

that it is an iso-alcohol, $HC \begin{cases} C_6H_{10}OH \\ (CH_3)_2 \end{cases}$

The haloïd octyl ethers obtained from this alcohol exhibit the following properties:

	Boiling Point.	Sp. gr.
Chloride, $C_8H_{17}Cl$.	179.5°–180° C. (355.1°–356° F.)	0.8802 at 16° C. (60.8°F.)
Bromide, $C_8H_{17}Br$:	198°–200° C. (388.4°–392° F.)	1.1116 " "
Iodide, $C_8H_{17}I$.	220°–222° C. (428°–431.6° F.)	1.1338 " "

Primary octyl alcohols, convertible by oxidation into caprylic acids, are also obtained from octane; they differ from the alcohol obtained from heracleum oil, but their exact structure has not been ascertained.

Secondary Octyl Alcohols.—1. *Methyl-hexyl Carbinol*,

$HO.HC \begin{cases} C_6H_{13} \\ CH_3 \end{cases}$, is produced by heating castor-oil with excess of solid potassium hydroxide. Castor-oil contains ricinoleic acid, $C_{18}H_{34}O_3$, and this acid, when heated with potash, yields free hydrogen, a distillate containing methyl-hexyl carbinol, together with products of its decomposition, and a residue of potassium sebate, $C_{10}H_{16}K_2O_4$:

$$C_{18}H_{34}O_3 + 2KOH = C_8H_{18}O + C_{10}H_{16}K_2O_4 + H_2.$$

To separate the alcohol, the distillate is repeatedly rectified over fused potash, the portion boiling below 200° C. (392° F.) only being collected: this liquid, subjected to fractional distillation yields a portion boiling at 181° C. (357.8° F.), which is the pure secondary octyl alcohol. The portions of the original distillate having a lower boiling point, consist of olefines, amongst which octene, C_8H_{16}, boiling at 150° C. (302° F.), preponderates.

The same alcohol is obtained from the octane of American petroleum, by converting this hydrocarbon into octyl chloride, then into the acetate, and heating the latter with alcoholic potash.

Methyl-hexyl carbinol is a limpid oily liquid, having a strong aromatic odor, and making grease-spots on paper. It has no action on polarized light. It has a specific gravity of 0.823 at 17° C. (62.6° F.), and boils at 181° C. (357.8° F.). It is insoluble in water, but dissolves in alcohol, ether, wood-spirit, and acetic acid. It mixes with sulphuric acid, forming octyl-sulphuric acid, $C_8H_{17}HSO_4$, generally also octene and neutral octyl sulphate. Fused zinc chloride converts it into octene. With potassium and sodium it yields substitution-products.

Methyl-hexyl carbinol, oxidized with potassium dichromate and sulphuric acid, yields the corresponding ketone, viz., m e t h y l - h e x y l k e t o n e, CH_3—CO—C_6H_{13}, and by the prolonged action of the oxidizing mixture, this ketone is further oxidized to caproic and acetic acids:

$$C_8H_{18}O + O_4 = C_6H_{12}O_2 + C_2H_4O_2 + H_2O.$$

These reactions show that the alcohol produced from castor-oil is a secondary alcohol; and from considerations similar to those above adduced with respect to the primary alcohol, it is inferred to be a secondary iso-alcohol, represented by the formula:

$$HO.HC \begin{cases} (CH_2)_3CH(CH_3)_2 \\ CH_3. \end{cases}$$

The Chloride, $C_8H_{17}Cl$, produced by the action of phosphorus pentachloride on this alcohol, has an odor of oranges, a specific gravity of 0.892 at 18° C. (64.4° F.), and boils at 175° C. (347° F.). Heated with alcoholic potash, it yields octene, C_8H_{16}; by alcohol and potassium acetate, it is converted into octene and octyl-acetate.

2. *Ethyl-isopentyl Carbinol,* $HO.CH \begin{cases} CH_2CH_3 \\ (CH_2)_2CH(CH_3)_2 \end{cases}$, is obtained, together with the primary alcohol, from the octane produced by the action of zinc and hydrochloric acid on the secondary octyl iodide obtained from the alcohol last described. The octyl chloride prepared from this octane smells faintly of oranges, and the acetate prepared from it yields, when heated with alcoholic potash, an octyl alcohol, which boils at 182–186° C. (359.6–366.8° F.), and is converted by oxidation into a ketone, $C_8H_{16}O$, isomeric with methyl-hexyl ketone, but differing from it by yielding, when further oxidized, not caproic and acetic, but propionic and ordinary valeric acid : hence, it consists of ethyl-isopentyl ketone (see KETONES):

$$\begin{array}{c} C_2H_5 \\ | \\ CO \\ | \\ CH_2CH_2CH(CH_3)_2 \end{array} + O_4 = \begin{array}{c} C_2H_5 \\ | \\ COOH \end{array} + \begin{array}{c} CH_2CH(CH_3)_2 \\ | \\ COOH \end{array} + H_2O \ ;$$

Ethyl-isopentyl ketone.　　Propionic acid.　　Valeric acid.

and the alcohol from which it is obtained is e t h y l - i s o p e n t y l c a r b i n o l.

3. A secondary octyl alcohol, different from both the preceding, has been obtained from octene, by heating this hydrocarbon in a sealed tube with hydriodic acid, converting the resulting octyl iodide, or octene hydriodide, $C_8H_{16}.HI$, into the acetate, and distilling the latter with finely pulverized potassium hydroxide. This alcohol has an aromatic odor, a specific gravity of 0.811 at 0° C. (32° F.), and boils at 174–178° C. (345.2–352.4° F.). By oxidation it yields the same products as the castor-oil alcohol, and must therefore likewise consist of methyl-hexyl carbinol ; but as it boils at a lower temperature than the latter, it probably contains a different modification of hexyl.

Tertiary Octyl Alcohol, or Propyl-diethyl Carbinol,

$HO.C \begin{cases} C_3H_7 \\ C_2H_5 \\ C_2H_5 \end{cases}$, is formed by treating butyryl chloride, C_3H_7COCl, with zinc-ethyl, and decomposing the product with water. It is a somewhat viscid liquid, lighter than water, and insoluble therein, and does not solidify in a freezing mixture. With phosphorus pentachloride it yields an octyl chloride boiling at 155° C. (311° F.). By oxidation with chromic acid mixture it yields propionic and acetic acids.

Nonyl Alcohol, $C_9H_{19}OH$, is obtained by the series of reactions above described from the nonane of American petroleum, and likewise occurs, together with the nonene, C_9H_{18}, in that portion of the liquid obtained by distilling amyl alcohol with zinc chloride, which boils between 134° C. (273.2° F.) and 150° C. (302° F.). Nonyl alcohol boils at about 200° C. (392° F.). Nonyl chloride, $C_9H_{19}Cl$, has a specific gravity of 0.899 at 16° C. (60.8 F.), and boils at 196° C. (384.8° F.).

Decyl Alcohol, $C_{10}H_{21}OH$, from petroleum decane, boils at 210–215° C. (410–419° F.). An isomeric alcohol, probably tetra-ethylated ethyl alcohol, $CH(C_2H_5)_2.—C(C_2H_5)_2OH$, is formed by the action of dibromacetyl bromide, $CHBr.COBr$, on zinc-ethyl ; it boils at 155–157° C. (311–314.6° F.).

The alcohols of the series $C_nH_{2n+2}O$ containing 11 to 15 carbon-atoms, are not known, but compound ethers containing 12 and 14 carbon-atoms appear to occur in spermaceti.

Hexdecyl, or **Cetyl Alcohol**, $C_{16}H_{34}O = C_{16}H_{33}(OH)$, also called *Ethal*, is obtained from spermaceti, a crystalline fatty substance found in peculiar cavities in the head of the sperm whale (*Physeter macrocephalus*). This substance consists of cetyl palmitate, $C_{32}H_{64}O_2$, or $C_{16}H_{33}.C_{16}H_{31}O_2$, and when heated for some time with solid potash, is resolved into potassium palmitate and cetyl alcohol :—

$$C_{16}H_{33}.C_{16}H_{31}O_2 \ + \ KOH \ = \ KC_{16}H_{31}O_2 \ + \ C_{16}H_{33}(OH).$$

The cetyl alcohol is dissolved out from the fused mass by alcohol and ether, and purified by several crystallizations from ether.

Cetyl alcohol, or ethal, is a white crystalline mass, which melts at about 50° C. (122° F.), and crystallizes by slowly cooling in shining laminæ. It has neither taste nor smell, is insoluble in water, but dissolves in all proportions in alcohol and ether. When heated it distils without decomposition. With sodium it gives off hydrogen and yields sodium cetylate, $C_{16}H_{33}NaO$. It is not dissolved by aqueous alkalies ; but when heated with a mixture of potash and lime, it gives off hydrogen, and is converted into palmitic acid :—

$$C_{16}H_{34}O \ + \ KOH \ = \ KC_{16}H_{31}O_2 \ + \ 2H_2.$$

Distilled with phosphorus pentachloride it yields cetyl chloride, $C_{16}H_{33}Cl$, a limpid oily liquid, having a specific gravity of 0.8412 at 12° C. (53.6° F.), and distilling with partial decomposition at a temperature above 200° C. (392° F.). Cetyl iodide, $C_{16}H_{33}I$, obtained by treating the alcohol with iodine and phosphorus, is a solid substance which melts at 22° C. (71.6° F.), dissolves in alcohol and ether, and crystallizes from alcohol in interlaced laminæ.

According to Heintz, cetyl alcohol, or ethal, prepared as above, is not a definite compound, but a mixture of hexdecyl alcohol, $C_{16}H_{34}O_2$, with small quantities of three other alcohols of the same series, containing respectively 12, 14, and 18 atoms of carbon, inasmuch as, when fused with potash-lime, it yields the corresponding fatty acids, $C_nH_mO_2$.

Ceryl Alcohol, $C_{27}H_{56}O = C_{27}H_{55}(OH)$; also called *Cerotic alcohol* and *Cerotin*.—This alcohol is obtained from Chinese wax or Pela, a secretion enveloping the branches of certain trees in China, and supposed to be produced by the puncture of an insect. This wax consists mainly of ceryl cerotate, $C_{27}H_{55}.C_{27}H_{53}O_2$, and is decomposed by fused potash in the same manner as spermaceti, yielding potassium cerotate and ceryl alcohol :—

$$C_{27}H_{55}.C_{27}H_{53}O_2 \ + \ KOH \ = \ KC_{27}H_{53}O_2 \ + \ C_{27}H_{55}(OH).$$

On digesting the fused mass with boiling water, a solution of potassium cerotate is obtained, holding ceryl alcohol in suspension ; and on precipitating the cerotic acid with barium chloride and treating the resulting precipitate with alcohol, the ceryl alcohol dissolves, and may be purified by repeated crystallization from alcohol or ether. It then forms a waxy substance, melting at 97° C. (206.6° F.). Heated with potash-lime, it gives off hydrogen, and is converted into potassium cerotate. At very high temperatures it distils, partly undecomposed, partly resolved into water and cerotene, $C_{27}H_{54}$; by this character it would appear to be related to the secondary alcohols. With sulphuric acid in excess, it forms hydrated neutral ceryl sulphate, $(C_{27}H_{55})_2SO_4.H_2O$.

Myricyl Alcohol.—$C_{30}H_{62}O = C_{30}HO_{61}(OH)$.—This alcohol, the highest known member of the series, $C_nH_{2n+2}O$, is obtained from myricin, the portion of common bees' wax which is insoluble in boiling alcohol. Myricin consists of myricyl palmitate, $C_{30}H_{61}.C_{16}H_{31}O_2$, and when heated with potash is decomposed in the same manner as spermaceti and Chinese wax, yielding potassium palmitate and myricyl alcohol. On dissolving the product in water, precipitating with barium chloride, exhausting the precipitate with boiling alcohol, and dissolving the substance deposited from the alcohol in light petroleum, pure myricyl alcohol separates as a crystalline substance, having a silky lustre. When heated, it partly sublimes unaltered, and is partly resolved (like ceryl alcohol) into water and melene, $C_{30}H_{60}$. With strong *sulphuric acid* it yields myricyl sulphate. Heated with *potash-lime*, it gives off hydrogen, and is converted into potassium melissate :

$$C_{30}H_{62}O \;+\; KOH \;=\; KC_{30}H_{59}O_2 \;+\; 2H_2.$$

The mother-liquor from which the myricyl alcohol has crystallized out, as above mentioned, retains a small quantity of an isomeric alcohol, which melts at 72° C. (161.6° F.), and when treated with potash-lime yields an acid containing a smaller proportion of carbon.

---

2. Monatomic Alcohols, $C_nH_{2n}O$, or $C_nH_{2n-1}OH$.

Two alcohols of this series are known, viz. :

$$\text{Vinyl alcohol, } C_2H_4O = C_2H_3(OH)$$
$$\text{Allyl alcohol, } C_3H_6O = C_3H_5(OH).$$

The first, discovered by Berthelot in 1860, is produced by combining ethine or acetylene with sulphuric acid, whereby vinyl sulphuric acid $(C_2H_3)HSO_4$ is formed, and distilling the product with water, just as in the preparation of ethyl alcohol from ethene :

$$HHSO_4 \;+\; C_2H_2 \;=\; (C_2H_3)HSO_4$$
$$(C_2H_3)HSO_4 \;+\; HOH \;=\; HHSO_4 \;+\; C_2H_3(OH).$$

It is an easily decomposible liquid, having a highly pungent odor, somewhat more volatile than water, soluble in 10 to 15 parts of that liquid, and precipitated from the solution by potassium carbonate. It is isomeric with acetic aldehyde and ethene oxide (p. 648). The univalent radicle vinyl, C_2H_3, which may be supposed to exist in it, is related to the trivalent radicle ethenyl (p. 507) in the same manner as allyl to propenyl (see below).

Allyl Alcohol, $C_3H_6O = C_3H_5(OH) = \begin{array}{c} CH_2 \\ \parallel \\ CH \\ \mid \\ CH_2OH \end{array}$.—This alcohol, discovered by Cahours and Hofmann in 1856, may be supposed to contain the univalent radicle allyl, C_3H_5, derived from a saturated hydrocarbon by abstraction of one atom of hydrogen, and isomeric with the trivalent radicle propenyl, (C_3H_5), derived from propane, CH_3—CH_2—CH_3, by abstraction of three atoms of hydrogen. Allyl and propenyl compounds, indeed, are easily converted one into the other by addition or subtraction of two atoms of a monad element or radicle.

To obtain the alcohol, allyl iodide is first prepared by the action of phosphorus tetriodide on propenyl alcohol (glycerin), $(C_3H_5)(OH)_3$:

$$2(C_3H_5)(OH)_3 + P_2I_4 = 2C_3H_5I + 2P(OH)_3 + I_2.$$

The allyl iodide is next decomposed by silver oxalate, yielding allyl oxalate :

$$2C_3H_5I + Ag_2C_2O_4 = 2AgI + (C_3H_5)_2C_2O_4;$$

and the allyl oxalate is decomposed by ammonia, yielding oxamide, $C_2O_2(NH_2)_2$, and allyl alcohol :

$$(C_3H_5)_2C_2O_4 + 2NH_3 = (C_2O_2)(NH_2)_2 + 2C_3H_5(OH).$$

Allyl alcohol is a colorless liquid, having a pungent odor and a spirituous burning taste. It mixes in all proportions with water, common alcohol, and ether ; boils at 103° C. (217.4° F.); burns with a brighter flame than common alcohol.

Allyl alcohol is a primary alcohol, similar in all its ordinary reactions to ethyl alcohol. By oxidation in contact with platinum black, or more quickly by treatment with potassium dichromate and sulphuric acid, it is converted into acrylic aldehyde (acrolein), C_3H_4O, and acrylic acid, $C_3H_4O_2$, compounds related to it in the same manner as common aldehyde and acetic acid to ethyl alcohol. Heated with phosphoric oxide, it yields allylene, C_3H_4. With potassium and sodium it yields substitution-products. Strong sulphuric acid converts it into allyl-sulphuric acid. With the bromides and chlorides of phosphorus it yields allyl bromide, C_3H_5Br, and allyl chloride, C_3H_5Cl.

ALLYL BROMIDES.—The *monobromide*, C_3H_5Br, prepared as just mentioned, or by distilling propene bromide, $C_3H_6Br_2$, with alcoholic potash, is a liquid of specific gravity 1.47, and boiling at 62° C. (143.6° F.). A *tribromide of allyl*, $C_3H_5Br_3$, is obtained by adding bromine to the mono-iodide in a vessel surrounded by a freezing mixture. It is a liquid of specific gravity 1.436 at 23° C. (73.4° F.), boiling at 217° C. (422.6° F.), and solidifying when cooled below 10° C. (50° F.). It is isomeric with propenyl bromide or tribromhydrin, obtained by the action of phosphorus pentabromide on glycerin.

A *diallyl tetrabromide*, $C_6H_{10}Br_4$, is formed by the direct combination of diallyl (p. 541), with bromine ; it is a crystalline body, melting at 37° C. (98.6° F.).

ALLYL IODIDES.—The *mono-iodide*, C_3H_5I, obtained, as above described, by distilling glycerin with phosphorus tetriodide, is a liquid of specific gravity 1.780 at 16° C. (60.8° F.), and boiling at 100° C. (212° F.). It is decomposed by sodium, with formation of diallyl, C_6H_{10}. By the action of zinc or mercury and hydrochloric or dilute sulphuric acid, it is converted into propene (or allyl hydride):—

$$2C_3H_5I + Zn_2 + 2HCl = ZnCl_2 + ZnI_2 + 2C_3H_6.$$

Diallyl tetriodide, $C_6H_{10}I_4$, is a crystalline body obtained by dissolving iodine in diallyl at a gentle heat.

ALLYL-SULPHURIC ACID, $(C_3H_5)HSO_4$, is produced by adding allyl alcohol to strong sulphuric acid. The solution, diluted with water, and neutralized with barium carbonate, yields barium allyl-sulphate, $(C_3H_5)_2Ba(SO_4)_2$.

ALLYL OXIDE, $(C_3H_5)_2O$, is produced by the action of allyl iodide on potassium allyate (the gelatinous mass obtained by dissolving potassium in allyl alcohol):—

$$C_3H_5OK + C_3H_5I = KI + (C_3H_5)_2O.$$

It is a colorless liquid, boiling at 82° C. (179.6° F.).

ALLYL SULPHIDE, $(C_3H_5)_2S$.—This compound exists, together with a small quantity of allyl oxide, in volatile oil of garlic, and is formed artificially by distilling allyl iodide with potassium monosulphide :—

$$2C_3H_5I \; + \; K_2S \; = \; 2KI \; + \; (C_3H_5)_2S \;.$$

To prepare it from garlic, the sliced bulbs are distilled with water, and the crude oil thus obtained—which is a mixture of the sulphide and oxide of allyl—is subjected to the action of metallic potassium, renewed until it is no longer tarnished, whereby the allyl oxide is decomposed, after which the sulphide may be obtained pure by re-distillation. In this state it forms a colorless liquid, lighter than water, of high refractive power, possessing in a high degree the peculiar odor of the plant, and distilling without decomposition. Allyl sulphide, dissolved in alcohol and mixed with solutions of platinum, silver, and mercury, gives rise to crystalline compounds, consisting of a double sulphide of allyl and the metal, either alone or mixed with a double chloride.

Volatile Oil of Mustard, consisting essentially of allyl isothiocyanate, $C_3H_5.CNS$, is described in connection with the thiocyanic ethers (p. 575).

Allyl Hydrosulphide, or *Allyl Mercaptan*, $C_3H_5(SH)$, obtained by distilling allyl iodide with potassium hydrosulphide, is a volatile oily liquid, having an odor like that of garlic oil, but more ethereal, boiling at 90° C. (194° F.). It attacks mercuric oxide like ethyl mercaptan, forming the compound $Hg(C_3H_5)_2S_2$.

DIATOMIC ALCOHOLS AND ETHERS.

The diatomic alcohols are derived from saturated hydrocarbons by substitution of two equivalents of hydroxyl for two atoms of hydrogen, and may therefore be regarded as compounds of bivalent radicles with two equivalents of hydroxyl : ethene alcohol, for example, may be represented by either of the formulæ :—

$$\begin{matrix} CH_2OH \\ | \\ CH_2OH \end{matrix} \qquad\qquad C_2H_4{<}^{OH}_{OH}$$

the first representing it as a derivative of ethane, $\begin{matrix} CH_3 \\ | \\ CH_3 \end{matrix}$; the second as a compound of ethene, C_2H_4, with hydroxyl, or as derived from a double molecule of water, $H_2(OH)_2$, by substitution of ethene for two atoms of hydrogen.

The diatomic alcohols of the fatty group, called g l y c o l s, are represented by the general formula—

$$C_nH_{2n+2}O_2 \quad \text{or} \quad (C_nH_{2n})(OH)_2 \;.$$

They may be regarded as compounds of olefines with two equivalents of hydroxyl. The following are known :—

Ethene alcohol	$C_2H_6O_2$ = $C_2H_4(OH)_2$
Propene alcohol	$C_3H_8O_2$ = $C_3H_6(OH)_2$
Tetrene or Butene alcohol	$C_4H_{10}O_2$ = $C_4H_8(OH)_2$
Pentene or Amylene alcohol	$C_5H_{12}O_2$ = $C_5H_{10}(OH)_2$
Octene alcohol	$C_8H_{18}O_2$ = $C_8H_{16}(OH)_2$

27

The glycols are formed by the following processes :—

1. By combining an olefine with bromine ; treating the resulting dibromide with an alcoholic solution of potassium acetate, or with silver acetate, whereby it is converted into a diacetate of the olefine ; and decomposing this compound with solid potassium hydroxide, whereby potassium acetate and a diatomic alcohol are formed, the latter of which may be distilled off :—

$$\begin{array}{l} CH_2Br \\ | \\ CH_2Br \end{array} + 2AgOC_2H_3O = 2AgBr + \begin{array}{l} CH_2OC_2H_3O \\ | \\ CH_2OC_2H_3O \end{array}$$

Ethene bromide.　Silver acetate.　　　　　　　Ethene diacetate.

$$\begin{array}{l} CH_2OC_2H_3O \\ | \\ CH_2OC_2H_3O \end{array} + 2KOH = 2KOC_2H_3O + \begin{array}{l} CH_2OH \\ | \\ CH_2OH \end{array}$$

Ethene diacetate.　Potassium　　Potassium　　Ethene
　　　　　　　hydroxide.　　acetate.　　alcohol.

2. By combining an olefine with hypochlorous acid, and treating the resulting compound (a chlorhydrin) with moist silver oxide :

$$\begin{array}{l} CH_2 \\ | \\ CH_2 \end{array} + ClOH = \begin{array}{l} CH_2Cl \\ | \\ CH_2OH \end{array}$$

Ethene.　　　　　　　　　　　Ethene
　　　　　　　　　　　chlorhydrin.

$$\begin{array}{l} CH_2Cl \\ | \\ CH_2OH \end{array} + AgOH = AgCl + \begin{array}{l} CH_2OH \\ | \\ CH_2OH \end{array}$$

Properties.—The glycols are colorless, inodorous, more or less viscid liquids, having a sweetish taste, freely soluble in water and alcohol ; ethene alcohol is but sparingly soluble in ether; the rest dissolve easily in that liquid.

The chemical reactions of the glycols have been studied chiefly in the case of ethene alcohol. They are, for the most part, similar to those of the monatomic alcohols : but inasmuch as the glycols contain two atoms of replaceable hydrogen, or of hydroxyl, the reactions generally take place by two stages, yielding two series of products.

1. Ethene alcohol treated with *nitric acid* gives up 2 or 4 atoms of hydrogen in exchange for oxygen, and is converted into g l y c o l l i c   a c i d, $C_2H_4O_3$, or o x a l i c   a c i d, $C_2H_2O_4$, according as the action takes place at ordinary or at higher temperatures ; thus :

$$\begin{array}{l} CH_2OH \\ | \\ CH_2OH \end{array} + O_2 = H_2O + \begin{array}{l} CH_2OH \\ | \\ COOH \end{array}$$

and

$$\begin{array}{l} CH_2OH \\ | \\ CH_2OH \end{array} + O_4 = 2H_2O + \begin{array}{l} COOH \\ | \\ COOH \end{array}$$

Under certain circumstances the corresponding aldehydes are also produced, as g l y o x a l, $\begin{array}{l} COH \\ | \\ COH \end{array}$, from ethene alcohol, by removal of four hydrogen-atoms without substitution.

2. *Potassium* and *sodium* eliminate one or two atoms of hydrogen from the glycols, and form substitution-products. Ethene alcohol is strongly at-

tacked by sodium, yielding s o d i u m e t h e n a t e, $C_2H_5NaO_2$; and this compound, fused with excess of sodium, is converted into d i s o d i u m e t h e n a t e, $C_2H_4Na_2O_2$. These compounds, treated with monatomic alcoholic iodides, yield the alcoholic ethers of the glycols ; thus :

$$\begin{matrix} CH_2ONa \\ | \\ CH_2OH \\ \text{Sodium} \\ \text{ethenate.} \end{matrix} \quad + \quad C_2H_5I \quad = \quad NaI \quad + \quad \begin{matrix} CH_2OC_2H_5 \\ | \\ CH_2OH \\ \text{Ethyl} \\ \text{ethenate.} \end{matrix}$$

$$\begin{matrix} CH_2ONa \\ | \\ CH_2ONa \\ \text{Disodium} \\ \text{ethenate.} \end{matrix} \quad + \quad 2C_2H_5I \quad = \quad 2NaI \quad + \quad \begin{matrix} CH_2OC_2H_5 \\ | \\ CH_2OC_2H_5 \\ \text{Diethyl} \\ \text{ethenate.} \end{matrix}$$

3. *Oxygen acids*, heated with glycols in closed vessels, act upon them in the same manner as upon the monatomic alcohols, converting them into e t h e r e a l s a l t s or c o m p o u n d e t h e r s, mono-acid or di-acid, according to the proportions used. In the di-acid glycol-ethers, the two radicles by which the hydrogen is replaced may belong either to the same or to different acids, *e. g.*,

$$\begin{matrix} CH_2OH \\ | \\ CH_2OH \\ \text{Ethene} \\ \text{alcohol.} \end{matrix} \quad + \quad \begin{matrix} HOC_2H_3O \\ \text{Acetic} \\ \text{acid.} \end{matrix} \quad = \quad H_2O \quad + \quad \begin{matrix} CH_2OH \\ | \\ CH_2OC_2H_3O \\ \text{Ethene mono-} \\ \text{acetate.} \end{matrix}$$

$$\begin{matrix} CH_2OH \\ | \\ CH_2OH \\ \text{Ethene} \\ \text{alcohol.} \end{matrix} \quad + \quad \begin{matrix} 2HOC_2H_3O \\ \text{Acetic} \\ \text{acid.} \end{matrix} \quad = \quad 2H_2O \quad + \quad \begin{matrix} CH_2OC_2H_3O \\ | \\ CH_2OC_2H_3O \\ \text{Ethene} \\ \text{diacetate.} \end{matrix}$$

$$\begin{matrix} CH_2OH \\ | \\ CH_2OC_2H_3O \\ \text{Ethene} \\ \text{mono-acetate.} \end{matrix} \quad + \quad \begin{matrix} HOC_4H_7O \\ \text{Butyric} \\ \text{acid.} \end{matrix} \quad = \quad H_2O \quad + \quad \begin{matrix} CH_2OC_4H_7O \\ | \\ CH_2OC_2H_3O \\ \text{Ethene} \\ \text{butyracetate.} \end{matrix}$$

The *haloïd acids* (HCl and HBr) act in the same manner as oxygen-acids, excepting that the reaction never goes beyond the first stage ; *e. g.*,

$$\begin{matrix} CH_2OH \\ | \\ CH_2OH \\ \text{Ethene} \\ \text{alcohol.} \end{matrix} \quad + \quad HCl \quad = \quad H_2O \quad + \quad \begin{matrix} CH_2Cl \\ | \\ CH_2OH \\ \text{Ethene} \\ \text{chlorhydrin.} \end{matrix}$$

With *hydriodic acid*, a further action generally takes place, resulting in the separation of iodine, and the formation of an olefine ; thus :

$$CH_2I—CH_2OH \quad + \quad IH \quad = \quad C_2H_4 \quad + \quad H_2O \quad + \quad I_2.$$

The m o n o c h l o r h y d r i n s, etc., of the glycols (haloïd hydrins), are also produced, as above mentioned, by direct addition of hypochlorous acid to the olefines. When treated with nascent hydrogen, they are converted into monatomic alcohols ; *e. g.*,

$$C_2H_4Cl(OH) \quad + \quad H_2 \quad = \quad HCl \quad + \quad C_2H_5(OH).$$

When heated with metallic salts they form mono-acid compound ethers

$$\begin{matrix} CH_2Cl \\ | \\ CH_2.OH \end{matrix} \quad + \quad C_2H_3O.OK \quad = \quad KCl \quad + \quad \begin{matrix} CH_2OC_2H_3O \\ | \\ CH_2.OH \end{matrix}$$

Ethene   Potassium   Ethene
chlorhydrin.   acetate.   monoacetate.

The haloïd hydrins treated with alkalies are converted into the oxygen-ethers or anhydrides of the glycols, $C_nH_{2n}O$; thus :

$$\begin{matrix} CH_2Cl \\ | \\ CH_2OH \end{matrix} \quad + \quad KOH \quad = \quad KCl \quad + \quad H_2O \quad + \quad O{\Large<}\begin{matrix} CH_2 \\ \\ CH_2 \end{matrix}.$$

These oxides are isomeric with the aldehydes and ketones.

The dichlorinated ethers, etc. (haloïd dihydrins), produced, as already observed, by direct combination of chlorine, bromine, and iodine, with the olefines (p. 544), may also be formed by the action of the chlorides, bromides, and iodides of phosphorus on the glycols ; *e. g.*,

$$C_2H_4(OH)_2 \quad + \quad 2PCl_5 \quad = \quad 2POCl_3 \quad + \quad 2HCl \quad + \quad C_2H_4Cl_2.$$

Methene Glycol, $CH_2(OH)_2$ is not known, and in all probability can-not exist. It appears indeed that a single carbon-atom is not capable of attracting to itself more than one hydroxyl group ; and in reactions, where such dihydroxyl-compounds might be expected to arise, water is separated, and the corresponding oxides (*i. e.*, aldehydes) are actually produced. Thus, on heating ethidene dichloride, CH_3—$CHCl_2$, with silver acetate, and distilling the resulting acetic ether with potash, the product obtained is not ethidene glycol, CH_3—$CH(OH)_2$, but the products of its decomposition, namely, water and aldehyde, CH_3—CHO.

Ethers of methene, or of methene-glycol, have, however, been produced. The haloïd ethers, CH_2I_2, etc., have already been described (p. 544).

Methene Dimethylate, $CH_2(OCH_3)_2$, also called *Methylal* and *Formal*, is formed by the oxidation of methyl alcohol with manganese dioxide and sulphuric acid. It is an ethereal liquid, having a specific gravity of 0.855, and boiling at 42° C. (107.6° F.). It mixes readily with alcohol and ether, and dissolves in three parts of water.

Methene Diacetate, $CH_2(OC_2H_3O)_2$, formed by the action of methene iodide on silver acetate, is an oily liquid, heavier than water, and boiling at 170° C. (338° F.). It is insoluble in water, and, when boiled with alkalies, is converted into methene oxide.

Methene Oxide, CH_2O, isomeric with formic aldehyde, H.CHO, is produced by the action of oxide or of oxalate of silver on methene iodide, in the latter case with evolution of carbon monoxide and dioxide :

$$CH_2I_2 \quad + \quad C_2O_4Ag_2 \quad = \quad CH_2O \quad + \quad CO_2 \quad + \quad CO \quad + \quad 2AgI.$$

The distillate, however, consists, not of CH_2O, but of the polymeric com-pound, $C_3H_6O_3$, which collects in the receiver as a crystalline mass, solu-ble in water, alcohol, and ether, melting at 152° C. (305.6° F.), and sub-liming below 100°. The density of its vapor is 1.06 referred to air, or 15 referred to hydrogen as unity ; and as this is half the molecular weight represented by the formula CH_2O ($12 + 2 + 16 = 30$), it follows that this formula correctly represents the molecule of the compound in the state of vapor. On the other hand, the solid compound, when treated with hydrogen sulphide, is converted into m e t h e n e s u l p h i d e, a body which melts at 218○ C. (424○ F.), sublimes easily, and yields a vapor whose density (69 referred to hydrogen), shows that the molecule of this compound is not CH_2S, but $C_3H_6S_3$. $\left(\dfrac{36+6+96}{2}=69\right)$. Hence it is inferred that the oxide from which it is formed has also, in the solid state, the constitution represented by the formula $C_3H_6O_3$, or $(CH_2)_3O_3$.

Methene-disulphonic or **Methionic acid,** $CH_2{<}^{SO_3H}_{SO_3H}$, formed by the action of fuming sulphuric acid on acetamide or acetonitril (methyl cyanide, $CH_3.CN$; see below), crystallizes in long deliquescent needles. It is very stable, not being altered by boiling with nitric acid. The barium-salt, $CH_2(SO_3)_2Ba + 2H_2O$, forms nacreous laminæ, sparingly soluble in water ; it is precipitated from the aqueous solution of the acid by barium chloride.

Methene-hydrinsulphonic acid, $CH_2{<}^{OH}_{SO_3H}$, or *Oxymethyl-sulphonic acid*, is obtained by acting on methyl alcohol with sulphuric anhydride, and boiling the product with water. It crystallizes with difficulty, and is moderately stable. Its barium salt crystallizes in small anhydrous tables.

Sulphacetic acid, $CH_2{<}^{SO_3H}_{CO_2H}$ (Methene carbonyl-sulphonic acid), is the first product of the action of sulphuric acid on acetamide or acetonitril :

$$\underset{\substack{\text{Acetamide.}}}{\overset{\substack{CH_3 \\ | \\ CONH_2}}{}} + 2SO_4H_2 = \underset{\substack{\text{Acid ammonium} \\ \text{sulphate.}}}{(NH_4)HSO_4} + \underset{\substack{\text{Sulphacetic} \\ \text{acid.}}}{CH_2{<}^{SO_3H}_{CO_2H}}$$

It is also produced by heating a mixture of glacial acetic acid and sulphuric anhydride, and as a sodium salt by boiling a solution of sodium sulphite with sodium monochloracetate :

$$\overset{\substack{CH_2Cl \\ | \\ CO.ONa}}{} + SO_3{<}^{Na}_{Na} = NaCl + CH_2{<}^{SO_3Na}_{CO.ONa.}$$

By neutralizing the solution obtained by the second process with lead carbonate, and decomposing the solution of the resulting lead-salt with hydrogen sulphide, a solution is obtained which, when concentrated, yields sulphacetic acid in transparent prisms.

Sulphacetic acid is a strong bibasic acid. When heated with sulphuric anhydride, it is converted into methene-disulphonic or methionic acid :

$$CH_2(SO_3H).CO_2H + SO_3 = CH_2(SO_3H)_2 + CO_2.$$

Hence, also, methionic acid is obtained as the ultimate product of the action of fuming sulphuric acid on acetamide or acetonitril.

Ethene Glycol, $C_2H_6O_2 = C_2H_4(OH)_2 = {\overset{CH_2OH}{\underset{CH_2OH}{|}}}$, prepared by distilling the monoacetate or diacetate of ethene with caustic potash, is a colorless viscid liquid, having a specific gravity of 1.125 at 0○, and boiling at 197.5○ C. (387.5○ F.). It mixes freely with water and alcohol, but is only slightly soluble in ether. Its reaction with sodium, and the formation

53 *

of ethylic ethers by treating the resulting sodium derivatives with ethyl iodide, have been already described (pp. 626, 627).

Ethylic Ethenate, $C_2H_4(OH)(OC_2H_5)$, likewise formed by direct combination of ethene oxide with ethyl alcohol, is a liquid having a fragrant odor, and boiling at 127° C. (260.6° F.). The *diethylic ether,* $C_2H_4(OC_2H_5)_2$, boils at 123° C. (253.4° F.).

Glycol heated with *zinc chloride* yields **aldehyde**;

$$
\begin{matrix} CH_2OH \\ | \\ CH_2OH \end{matrix} \quad - \quad H_2O \quad = \quad \begin{matrix} CH_3 \\ | \\ CHO \end{matrix} .
$$

With *phosphorus pentachloride* it forms **ethene dichloride**:

$$C_2H_4(OH)_2 \; + \; PCl_5 \; = \; PCl_3O \; + \; H_2O \; + \; C_2H_4Cl_2.$$

Hydriodic acid reduces it to **ethyl iodide**:

$$C_2H_4(OH)_2 \; + \; 3HI \; = \; C_2H_5I \; + \; 2H_2O \; + \; I_2.$$

Ethene Chlorhydrate or **Ethene Chlorhydrin,** $C_2H_4{<}^{OH}_{Cl}$, is formed by heating glycol with hydrochloric acid, or by agitating ethene with aqueous hypochlorous acid. It is a colorless liquid, miscible with water, and boiling at 128° C. (262.4° F.). By oxidation with chromic acid mixture it is converted into monochloracetic acid:

$$C_2H_4(OH)Cl \; + \; O_2 \; = \; H_2O \; + \; C_2H_3ClO_2.$$

By heating with potassium iodide it is converted into **ethene iodhydrin**, $C_2H_4(HO)I$, a viscid liquid, which decomposes when distilled.

Ethene Nitrate, $C_2H_4(NO_3)_2$, produced by heating ethene iodide with silver nitrate in alcoholic solution, or by dissolving glycol in a mixture of strong nitric and sulphuric acid, is a yellowish liquid, of specific gravity 1.483 at 8° C. (46.4° F.), insoluble in water, exploding when heated. By alkalies is resolved into glycol and nitric acid.

Ethene-sulphuric Acid, $C_2H_4.SO_4H_2 = SO_2{<}^{OH}_{OC_2H_5}$, is produced by heating glycol with sulphuric acid. Its barium salt, obtained by neutralizing the cooled acid solution with barium carbonate, is very soluble, deliquescent, and decomposes when heated with baryta-water into ethene and barium sulphate.

Ethene Oxide, C_2H_4O, or $O{<}^{CH_2}_{CH_2}$, isomeric with acetic aldehyde, is formed by the action of caustic potash on ethene chlorhydrin. The action is violent, and the ethene oxide is given off in vapor, which may be condensed in a receiver surrounded by a freezing mixture, and containing a few lumps of calcium chloride, over which the product may be afterwards dried. It is an ethereal mobile liquid, having a specific gravity of 0.898 at 0°, boiling at 13.5° C. (56.3° F.) (aldehyde boils at 21° C., 69.8° F.), and miscible in all proportions with water and with alcohol. When the aqueous solution is treated with *sodium amalgam,* in a vessel surrounded with a freezing mixture, the ethene oxide takes up hydrogen, and is converted into **ethyl alcohol**:

$$C_2H_4O \; + \; H_2 \; = \; C_2H_6O.$$

Ethene oxide unites with *ammonia* in several proportions, forming the following basic compounds, all of which are syrupy liquids :

Monoxethenamine	.	.	.	.	$C_2H_4O.NH_3$
Dioxethenamine	.	.	.	.	$(C_2H_4O)_2.NH_3$
Trioxethenamine	.	.	.	.	$(C_2H_4O)_3.NH_3$
Tetroxethenamine	.	.	.	.	$(C_2H_4O)_4.NH_3$.

This character distinguishes ethene oxide from aldehyde, which forms with ammonia a crystalline compound not possessing basic properties. A further distinction between these two isomeric bodies is, that aldehyde forms crystalline compounds with the acid sulphites of the alkali-metals, a property not possessed by ethene oxide.

Ethene oxide is a powerful base, uniting directly with *acids*, precipitating magnesia from a solution of magnesium chloride at ordinary temperatures, and ferric oxide and alumina from their saline solutions at 100° C.

(212° F.). With *hydrochloric acid*, it forms ethene chlorhydrin, $C_2H_4{<}^{Cl}_{OH}$,

and with *acetic acid*, ethene acetohydrin, $C_2H_4{<}^{OC_2H_3O}_{OH}$. It also unites with *water* in several proportions, forming glycol and the following compounds.

Polyethenic Alcohols.—These are bodies which contain the elements of two or more molecules of ethene oxide combined with one molecule of water, and may be regarded as formed by the union of two or more molecules of glycol (mono-ethenic alcohol), with elimination of a number of water-molecules less by one than the number of glycol-molecules which enter into combination ; or as derived from three or more molecules of water, by substitution of ethene for the whole of the hydrogen except two atoms ; thus :

$C_2H_6O_2$ or $(C_2H_4)H_2O_2$
Monethenic alcohol
(glycol).
$= C_2H_4O.H_2O$
Ethene
oxide.

$C_4H_{10}O_3$ or $(C_2H_4)_2H_2O_3$
Diethenic alcohol.
$= 2C_2H_4O.H_2O = 2C_2H_6O_2{-}H_2O$
Ethene
oxide.
Glycol.

$C_6H_{14}O_4$ or $(C_2H_4)_3H_2O_4$
Triethenic alcohol.
$= 3C_2H_4O.H_2O = 3C_2H_6O_2{-}2H_2O$
Ethene
oxide.
Glycol.

$C_8H_{18}O_5$ or $(C_2H_4)_4H_2O_5$
Tetrethenic alcohol.
$= 4C_2H_4O.H_2O = 4C_2H_6O_2{-}3H_2O$
Ethene
oxide.
Glycol.

$C_{2n}H_{4n+2}O_{n+1}$ or $(C_2H_4)_nH_2O_{n+1} = nC_2H_4O.H_2O = nC_2H_6O_2{-}(n{-}1)H_2O$
n-ethenic alcohol.
Ethene
oxide.
Glycol.

The polyethenic alcohols are formed—1. By heating ethene oxide with water in sealed tubes. In this manner Wurtz obtained diethenic alcohol together with monethenic, and a small quantity of tri-ethenic alcohol. 2. By heating ethene oxide with glycol in sealed tubes : this process yields the di- and tri-ethenic alcohols. 3. By heating glycol with ethene bromide in sealed tubes to 100°–120° C. (212°–248° F.). The first products of this reaction are diethenic alcohol, ethene bromhydrin and water :

$3(C_2H_4)H_2O_2 + C_2H_4Br_2 = (C_2H_4)_2H_2O_3 + 2(C_2H_4)Br(OH) + H_2O$;
Monethenic
alcohol.
Ethene
bromide.
Diethenic
alcohol.
Ethene brom-
hydrin.

and the other polyethenic alcohols are formed, each from the one next below it in the series, by the action of ethene bromhydrin, according to the general equation :

$$(C_2H_4)_nH_2O_{n+1} \ + \ (C_2H_4)Br(OH) \ = \ (C_2H_4)_{n+1}H_2O_{n+2} \ + \ HBr.$$

The hydrobromic acid thus formed then acts on the excess of glycol present, reproducing ethene bromhydrin, and thus the action is continued. By this process, the 2-, 3-, 4-, 5-, and 6-ethenic alcohols have been obtained and separated by fractional distillation ; and when a sufficient excess of glycol is present, the temperature being kept between 110° C. (230° F.) and 120° C. (248° F.), still higher members of the series are produced.

The polyethenic alcohols are syrupy liquids, becoming more viscid as their molecular weight increases : their boiling point rises by about 45° C. (81° F.) for each addition of C_2H_4O.

Diethenic alcohol, $C_4H_{10}O_3$ or $(C_2H_4)_2H_2O_3$, boils at about 245° C. (573° F.) ; the density of its vapor is 3.78 (air $= 1$); by calculation it should be 3.67, so that it exhibits the normal condensation to two volumes. By contact with platinum black, or by treatment with nitric acid, it is oxidized to d i g l y c o l l i c   a c i d, $C_4H_6O_5$, an acid isomeric with malíc acid, and formed from diethenic alcohol by substitution of O for H_2, just as glycollic acid, $C_2H_4O_3$, is formed from monethenic alcohol, $C_2H_6O_2$. *Triethenic alcohol*, $C_6H_{14}O_4$, or $(C_2H_4)_3H_2O_4$, is oxidized in like manner to e t h e n e - d i g l y - c o l l i c   a c i d, $C_6H_{12}O_5$.

Ethene Hydrosulphide or Thiohydrate.—*Ethenic* or *glycolic mercaptan*, $C_2H_4(SH)_2$, formed by the action of ethene bromide on an alcoholic solution of potassium hydrosulphide, is a colorless oil, of specific gravity 1.12, boiling at 146° C. (294.8° F.), insoluble in water, soluble in alcohol and ether. With *lead acetate* it forms a yellow precipitate consisting of $C_2H_4S_2Pb$, and similar compounds with other metallic salts.

Ethene Sulphide, C_2H_4S, formed by the action of ethene bromide on potassium sulphide in alcoholic solution, is a crystalline body, melting at 110° C. (230° F.), and boiling at 200° C. (392° F.). Its vapor-density ($H = 1$) is 60, showing that its molecular formula is $(C_2H_4)_2S_2$. Its constitution is therefore $C_2H_4{<}^S_S{>}C_2H_4$.

Ethene Hydroxysulphide or Monothio-hydrate, $C_2H_4(OH)(SH)$, formed by the action of ethene chlorhydrin on potassium hydrosulphide, is a liquid of similar properties, and forms salts in which half the hydrogen is replaced by a metal. . Nitric acid oxidizes it to i s e t h i o n i c   a c i d, $C_2H_6SO_4$.

Ethene-sulphonic Acids.—*Ethene-disulphonic acid*, $C_2H_4{<}^{SO_3H}_{SO_3H}$, is formed by oxidation of ethene hydrosulphide, $C_2H_4(SH)_2$, and ethene thiocyanate, $C_2H_4(CNS)_2$; by the action of fuming sulphuric acid on alcohol or ether ; and by boiling ethene bromide with a strong solution of potassium sulphite :—

$$C_2H_4Br_2 \ + \ 2(KSO_2.OK) \ = \ 2KBr \ + \ C_2H_4{<}^{SO_2.OK}_{SO_2.OK} \ .$$

It forms a thick liquid, very soluble in water, difficult to crystallize ; the crystallized acid melts at 94°C. (201.2° F.). The barium salt, $C_2H_4(SO_3)_2Ba$, crystallizes from water in six-sided tables.

Ethene-hydrinsulphonic acid, or *Isethionic acid*, $C_2H_4<^{OH}_{SO_3H_5}$, isomeric with ethylsulphuric acid, $SO_2<^{OH}_{OC_2H_5}$, is produced by the oxidation of ethene mono-thiohydrate, $C_2H_4<^{OH}_{SH}$, with nitric acid; by the action of nitrous acid on taurine :—

$$C_2H_4<^{NH_2}_{SO_3H} + NO_2H = N_2 + H_2O + C_2H_4<^{OH}_{SO_3H};$$

by heating ethene-chlorhydrin with potassium sulphite :—

$$C_2H_4<^{OH}_{Cl} + KSO_3K = KCl + C_2H_4<^{OH}_{SO_3K};$$

and by boiling ethionic acid with water (p. 634).

When vapor of sulphuric anhydride is passed into well-cooled alcohol or ether, and the product is boiled with water for several hours, a solution is formed, containing ethionic, sulphuric, and a small quantity of methionic acid : and on saturating this liquid with barium carbonate, filtering from barium sulphate, and leaving the filtrate to cool, methionate of barium crystallizes out first, and afterwards the isethionate.

Isethionic acid is a viscid liquid, which does not easily crystallize : it is not decomposed by boiling with water. Its salts are stable, and crystallize well. The *ammonium salt* crystallizes in rhombic plates melting at 190° C. (374° F.); when heated to 210–220° C. (410–428° F.), it is converted into a m i d e t h y l s u l p h o n i c a c i d or t a u r i n e :—

$$C_2H_4<^{OH}_{SO_3.NH_4} = H_2O + C_2H_4<^{NH_2}_{SO_3H}.$$

By the action of phosphorus pentachloride on isethionic acid or its salts, the chloride, $C_2H_4Cl.SO_2Cl$, is formed, as a liquid which boils at 200° C. (392° F.), and is converted by boiling with water into c h l o r e t h y l s u l - p h o n i c a c i d, $C_2H_4Cl.SO_3H$.

Taurine, $C_2H_4(NH_2).SO_3H$, occurs in combination with cholic acid (as taurocholic acid), in the bile of oxen and other animals, and in various other animal secretions. It may be prepared by boiling taurocholic acid with an alkali :—

$$\underset{\text{Taurocholic acid.}}{C_{26}H_{45}NSO_7} + H_2O = \underset{\text{Cholic acid.}}{C_{24}H_{40}O_5} + \underset{\text{Taurine.}}{C_2H_7NSO_3.}$$

It is formed artificially, as already observed, by heating ammonium isethionate to 230° C. (446° F.); also by heating chlorethylsulphonic acid with aqueous ammonia.

Taurine crystallizes in large monoclinic prisms, easily soluble in hot water, insoluble in alcohol, melting and decomposing at about 240° C. (464° F.). It forms salts with alkalies, and dissolves in acids, but separates from the solution unaltered.

Taurine is not decomposed by boiling with acids or with alkalies, but by fusion with potassium hydroxide it is decomposed in the manner shown by the equation :—

$$C_2H_4(NH_2).SO_3K + 2KOH = C_2H_3KO_2 + SO_3K_2 + NH_3 + H_2.$$

Nitrous acid converts it into isethionic acid.

Ethionic acid and *Anhydride*.—The anhydride, $C_2H_4S_2O_6$, formerly called *sulphate of carbyl*, is formed by passing the vapor of sulphur trioxide into anhydrous alcohol; also by direct union of ethene with sulphur trioxide. It is a very deliquescent crystalline mass, which melts at 80° C. (176° F.). It readily takes up the elements of water, and is converted into ethionic acid, $C_2H_6S_2O_7$:—

$$C_2H_4\!\!<^{O.SO_2}_{SO_2}\!\!>O \;+\; HOH \;=\; C_2H_4\!\!<^{O.SO_2.OH}_{SO_2.OH}\;.$$
$$\text{Anhydride.} \qquad\qquad\qquad\qquad \text{Acid.}$$

Ethionic acid, having one of its sulphur-atoms connected with a carbon-atom directly, the other only through the medium of oxygen, acts both as a sulphonic acid and as an acid ether of sulphuric acid; it is therefore bibasic. Both the free acid and its salts are resolved by boiling with water into sulphuric and isethionic acids :—

$$C_2H_4\!\!<^{OSO_3H}_{SO_3H} \;+\; H_2O \;=\; SO_4H_2 \;+\; C_2H_4\!\!<^{OH}_{SO_3H}\;.$$

ETHIDENE COMPOUNDS.

Ethidene dichloride, CH_3—$CHCl_2$, has already been described (p. 546). The *oxide*, CH_3—CHO, is ordinary aldehyde (see ALDEHYDES).

Ethidene Dimethylate, CH_3—$CH(OCH_3)_2$, occurs in crude wood-spirit, and is formed in the oxidation of a mixture of methylic and ethylic alcohols; also by heating aldehyde with methyl alcohol. It is a colorless ethereal liquid, having a specific gravity of 0.8555, and boiling at 65° C. (149 F.).

Ethidene Methyl-ethylate, CH_3—$CH<^{OCH_3}_{OC_2H_5}$, formed simultaneously with the preceding by the oxidation of a mixture of methyl and ethyl alcohol, boils at 85° C. (185° F.).

Ethidene-Diethylate, or *Acetal*, CH_3—$CH(OC_2H_5)_2$, isomeric with ethene diethylate, is formed by oxidation of ethyl alcohol, and is found among the first portions of the distillate obtained in the preparation of ordinary spirit. It is formed also by the action of sodium ethylate on ethidene dichloride, and by heating aldehyde with alcohol to 100° in sealed tubes :—

$$CH_3\text{—}CHO \;+\; 2(C_2H_5OH) \;=\; H_2O \;+\; CH_3\text{—}CH(OC_2H_5)_2\;.$$

It is a liquid smelling like alcohol, having a specific gravity of 0.821 at 22° C. (71.6° F.), boiling at 104° C. (219.2° F.). With *chlorine* it yields mono-, di-, and trichloracetal.

Ethidene-sulphonic acids.—The relation of these somewhat unstable acids to their isomerides, the ethene-sulphonic acids, is shown by the following formulæ :—

$$
\begin{array}{ll}
\begin{array}{c}
CH_2.OH \\
| \\
CH_2.SO_3H \\
\text{Isethionic acid.}
\end{array}
&
\begin{array}{c}
CH_3.CH<^{OH}_{SO_3H} \\
\text{Ethidene-oxysulphonic} \\
\text{acid.}
\end{array}
\\[4ex]
\begin{array}{c}
CH_2.SO_2H \\
| \\
CH_2.SO_2H \\
\text{Ethene-disulphonic} \\
\text{acid.}
\end{array}
&
\begin{array}{c}
CH_3.CH<^{SO_3H}_{SO_3H} \\
\text{Ethidene-disulphonic} \\
\text{acid.}
\end{array}
\end{array}
$$

Ethidene-disulphonic acid, $CH_3.CH(SO_3H)_2$, and *Ethidene-chlorosulphonic acid,*
$CH_3.CH{<}^{Cl}_{SO_3H}$, are formed by heating ethidene dichloride (p. 546) with an aqueous solution of neutral sodium sulphite to about 140° C. (284° F.). The former is very unstable; the latter is moderately stable, and forms well-crystallized salts: its sodium salt crystallizes in nacreous laminæ.

Propene Glycols, $C_3H_8O_2 = C_3H_6(OH)_2$.—There are two of these diatomic alcohols, represented by the following formulæ :—

$$
\begin{array}{cc}
CH_3 & CH_2.OH \\
| & | \\
CHOH & CH_2 \\
| & | \\
CH_2OH & CH_2.OH \\
\text{Propene glycol.} & \text{Trimethene glycol.}
\end{array}
$$

They are analogous to the second and fourth modifications of the diatomic haloïd derivatives of propene, already described (p. 546). Glycols analogous to the first and third of these derivatives, viz., $CH_3—CH_2—CX_2$ and $CH_3—CX_2—CH_3$, are not known, and probably cannot exist, since it appears, as already noticed, that two hydroxyl groups cannot be attached to one carbon-atom.

Propene glycol is formed by heating the corresponding bromide, $CH_3—CHBr—CH_2Br$, with silver acetate, and the resulting acetic ether with potash; also by heating propene chloride with water and lead oxide. It is a colorless oily liquid, having a sweet taste, a specific gravity of 1.051 at 0°, and boiling at 188°–189° C. (370.4°–372.2° F.). In contact with platinum black it is oxidized to lactic acid:—

$$CH_3—CHOH—CH_2OH + O_2 = H_2O + CH_3—CHOH—COOH.$$

Heated to 100° with strong hydriodic acid, it gives pseudopropyl iodide, $CH_3—CHI—CH_3$. Phosphorus pentachloride converts it into propene dichloride, $CH_3—CHCl—CH_2Cl$.

Propene chlorhydrin, $CH_3—CHOH—CH_2Cl$, is formed by the action of gaseous hydrogen chloride on the glycol, and by that of hypochlorous acid on propene. It has a specific gravity of 1.302 at 0°, and boils at 127° C. (260.6° F.).

The corresponding *bromhydrin* boils at about 146° C. (294.8° F.).

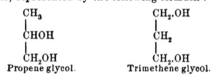

Propene oxide, $CH_3—CH—CH_2$, formed by the action of aqueous potash on propene chlorhydrin, is a volatile liquid, boiling at 35° C. (95° F.), having a specific gravity of 0.859 at 0°, easily soluble in water. Nascent hydrogen converts it into secondary propyl alcohol, $CH_3—CHOH—CH_3$.

Trimethene glycol, prepared from trimethene bromide (p. 547), is a thick saccharine liquid, boiling at 212° C. (413.6° F.).

Butene Glycols, $C_4H_{10}O_2 = C_4H_8(OH)_2$.—Six of these compounds are theoretically possible, four derived from normal butane, and two from isobutane. Only two of them, however, are at present known, viz :—

(1) *Butene glycol,* $CH_3—CHOH—CH_2—CH_2OH$, is formed in small quantity by the action of sodium amalgam on acetic aldehyde in aqueous

solution ; but is best prepared by leaving a cold mixture of acetic aldehyde and dilute hydrochloric acid to itself for two or three days, whereby
the acetic aldehyde is converted into the aldehyde of butene glycol, called
aldol, and treating the latter with sodium amalgam :—

$$CH_3\text{—}CHO + CH_3\text{—}CHO = CH_3\text{—}CHOH\text{—}CH_2\text{—}CHO$$
Acetic aldehyde (2 mol.).   Aldol.

and

$$CH_3\text{—}CHOH\text{—}CH_2\text{—}CHO + H_2 = CH_3\text{—}CHOH\text{—}CH_2\text{—}CH_2OH .$$
Aldol.   Butene glycol.

Butene glycol is a thick liquid, boiling at 204° C. (399.2° F.), miscible
with water and with alcohol. By oxidation with nitric acid or with obro-
mic acid mixture it is resolved into acetic and oxalic acids, together with
a small quantity of butyric aldehyde.

(2) *Isobuteneglycol*, $(CH_3)_2\text{=}COH\text{—}CH_2OH$, prepared from the corresponding bromide, has a specific gravity of 1.048 at 0°, and boils at 183°–
184° C. (361.4°–363.2° F.). By oxidation with nitric acid it is converted
into oxyisobutyric acid, $(CH_3)_2\text{=}COH\text{—}COOH$.

Isopentene or **Amylene Glycol,** $C_5H_{10}(OH)_2 = (CH_3)_2\text{=}$
$CH\text{—}CHOH\text{—}CH_2OH$, the only 5-carbon glycol known, is prepared by distilling amylene diacetate with potash, or by addition of hydrogen dioxide
to amylene. It boils at 177° C. (350.6° F.), has a specific gravity of 0.987
at 0°, and is converted by oxidation with nitric acid into isovaleric acid.
The corresponding oxide, $C_5H_{10}O$, is a liquid insoluble in water, having a
specific gravity of 0.824 at 0°, and boiling at 95° C. (203° F.).

Hexene Glycols, $C_6H_{12}(OH)_2$. — *Normal hexene glycol*,
$CH_2OH\text{—}(CH_2)_4\text{—}CH_2OH$, from hexene bromide (obtained from mannite,
or from the hexyl chloride of American petroleum), has a specific gravity
of 0.967 at 0°, and boils at 207° C. (404.6° F.).

Mesohexene Glycol, $\dfrac{HO.H_2C}{H_3C}{>}CH\text{—}HC{<}\dfrac{CH_2OH}{CH_3}$, or *Diallyl*

Dihydrate, $C_6H_{10}\begin{Bmatrix} H_2 \\ (OH)_2 \end{Bmatrix}$, is prepared by heating diallyl in a sealed
flask with strong hydriodic acid, converting the resulting diallyl hydriodide,
$C_6H_{10}\begin{Bmatrix} H_2 \\ I_2 \end{Bmatrix}$, into the diacetate, $C_6H_{10}\begin{Bmatrix} H_2 \\ (OC_2H_3O)_2 \end{Bmatrix}$, by treating it with silver
acetate suspended in ether, and decomposing the diacetate with potash.
It is a thick, colorless syrup, having a specific gravity of 0.9638 at 0° and
0.9202 at 65° C. (149° F.) ; boils between 212° C. (413.6° F.) and 215° C.
(419° F.).

Pinacone, $(CH_3)_2COH\text{—}COH(CH_3)_2$, is formed, together with pseudopropyl alcohol, by the action of sodium or sodium-amalgam on acetone in
aqueous solution :

$$(CH_3)_2CO + CO(CH_3)_2 + H_2 = (CH_3)_2COH\text{—}COH(CH_3)_2.$$

It crystallizes from the concentrated aqueous solution as a hydrate, $C_6H_{14}O_2$
$+ 6H_2O$, in large square tables which melt at 42° (107.6° F.), and gradually
effloresce in contact with the air. The hydrate gives up its water when
heated, and at 171°–172° C. (339.8°–341.6° F.) yields a distillate of auhy-
drous pinacone, which solidifies to an indistinctly crystalline mass, melting
at 38° C. (100.4° F.), and reconverted into the hydrate by solution in water.
Pinacone, when heated with acids, is converted into pinacolin, $C_6H_{12}O$
(see ALDEHYDES).

TRIATOMIC ALCOHOLS AND ETHERS.

Triatomic alcohols may be derived from saturated hydrocarbons by sub-stitution of three atoms of hydroxyl for three atoms of hydrogen, and may accordingly be regarded as compounds of trivalent alcohol-radicles with three atoms of hydroxyl, or as compounds derived from a triple molecule of water, H_6O_3, by substitution of a trivalent alcohol-radicle for three atoms of hydrogen. The hydrocarbons of the series $C_nH_{2n}+_2$ should accordingly yield a series of triatomic alcohols of the form $(C_nH_{2n}-_1)(OH)_3$, viz. :—

Methenyl alcohol	$CH(OH)_3$
Ethenyl alcohol	$C_2H_3(OH)_3$
Propenyl alcohol	$C_3H_5(OH)_3$
Tetrenyl alcohol	$C_4H_7(OH)_3$
Pentenyl alcohol	$C_5H_9(OH)_3$
etc.	etc.

Of these, however, only two are known, viz., p r o p e n y l a l c o h o l, or g l y c e r i n, and p e n t e n y l a l c o h o l or a m y l g l y c e r i n.

Each triatomic alcohol subjected to the action of acids, or of the chlorides, bromides, or iodides of phosphorus, may yield three classes of ethers, derived from it by substitution of a halogen element, or acid radicle, for part or the whole of the hydroxyl ; thus, from glycerin may be obtained the three hydrochloric ethers, $C_3H_5Cl(OH)_2$, $C_3H_5Cl_2OH$, $C_3H_5Cl_3$, and the three acetic ethers, $C_3H_5(OC_2H_3O)(OH)_2$, $C_3H_5(OC_2H_3O)_2OH$, and $C_3H_5(OC_2H_3O)_3$.

Propenyl Alcohol or Glycerin,

$$C_3H_8O_3 = (C_3H_5)\begin{cases}OH\\OH\\OH\end{cases} \quad or \quad \begin{array}{l}CH_2OH\\ |\\CHOH\\ |\\CH_2OH\end{array}$$

This compound is obtained by the action of alkalies on natural fats, which are, in fact, the propenylic ethers of certain fatty acids ; thus stearin, one of the constituents of mutton suet, consists of *propenyl tristearate*, (C_3H_5) $(OC_{18}H_{35}O)_3$, a compound derivable from glycerin itself, by substitution of stearyl, $C_{18}H_{35}O$, for hydrogen. Now, when stearin is boiled with a caustic alkali, it is converted into a stearate of the alkali-metal and glycerin ; thus :

$$\underset{\text{Stearin.}}{C_3H_5(OC_{18}H_{35}O)_3} + 3KHO = \underset{\substack{\text{Potassium.}\\ \text{stearate.}}}{3KOC_{18}H_{35}O} + \underset{\text{Glycerin.}}{C_3H_5(OH)_3}$$

A similar reaction takes place when any other similarly constituted fat is treated with a caustic alkali. The metallic salts of the fatty acids thus obtained are the well-known bodies called s o a p s , and the process is called s a p o n i f i c a t i o n ; this term, originally restricted to actual soap-making, has been extended to all cases of the resolution of a compound ether into an acid and an alcohol, such, for example, as the conversion of ethyl acetate into acetic acid and ethyl alcohol by the action of alcoholic potash.

Glycerin was originally obtained by heating together olive or other suitable oil, lead oxide, and water, as in the manufacture of common *lead-plaster ;* an insoluble soap of lead is thereby formed, while the glycerin remains in the aqueous liquid. The latter is treated with sulphuretted hydrogen, digested with animal charcoal, filtered, and evaporated in a vacuum at the temperature of air. Glycerin is now produced in very large

54

quantity and perfect purity in the decomposition of fatty substances by means of over-heated steam, a process which Mr. George Wilson has introduced into the manufacture of candles. In this reaction a fatty acid and glycerin are produced by assimilation of the elements of water; they are carried over by the excess of steam in a state of mechanical mixture, which rapidly separates into two layers in the receiver. The reaction is exactly similar to that which takes place when a caustic alkali is used to effect the saponification, e. g. :

$$C_3H_5(OC_{18}H_{35}O)_3 \ + \ 3H_2O \ = \ 3HOC_{18}H_{35}O \ + \ C_3H_5(OH)_3$$
Stearin.   Stearic acid.   Glycerin.

Glycerin may also be produced from propenyl bromide, $(C_3H_5)Br_3$, a compound formed, as already observed, by the action of bromine on allyl iodide, C_3H_5I. The process consists in converting the propenyl bromide into propenyl triacetate, $(C_3H_5)(OC_2H_3O)_3$, by the action of silver acetate, and decomposing this ether with potash. This mode of formation must not, however, be regarded as an actual synthesis of glycerin from compounds of simpler constitution : for the allyl-compounds are themselves prepared from glycerin (p. 624).

Glycerin is a nearly colorless and very viscid liquid, of specific gravity 1.27, and boiling at 290° C. (554° F.). When quite pure and anhydrous, it crystallizes on exposure to a very low temperature, especially if agitated, as in railway transport. The crystals are monoclinic, perfectly colorless, and melt at 15.6° C. (60° F.).* Glycerin has an intensely sweet taste, and mixes with water in all proportions : its solution does not undergo the alcoholic fermentation, but when mixed with yeast and kept in a warm place, it is gradually converted into propionic acid. Glycerin has no action on vegetable colors. Exposed to heat, it volatilizes in part, darkens, and decomposes, giving off, amongst other products, a substance called *acrolein*, C_3H_4O, having an intensely pungent odor.

Concentrated *nitric acid* converts glycerin into g l y c e r i c a c i d, $C_3H_6O_4$, an acid related to glycerin in the same manner as glycollic acid to glycol, and acetic acid to ethyl alcohol ; being formed from it by substitution of oxygen for two atoms of hydrogen in immediate relation to hydroxyl ; thus :

$$\begin{array}{llll}
CH_2OH & & & CH_2OH \\
| & & & | \\
CHOH & + \ O_2 \ = \ H_2O \ + & & CHOH \\
| & & & | \\
CH_2OH & & & COOH \\
\text{Glycerin.} & & & \text{Glyceric acid.}
\end{array}$$

The formula of glycerin indicates the possibility of effecting a second substitution of the same kind, which would yield diglyceric acid, $C_3H_4O_5$, but this acid has not been actually obtained.

Glycerin, treated with a mixture of strong nitric and sulphuric acids, forms n i t r o g l y c e r i n, $C_3H_5(NO_2)_3O_3$, a heavy oily liquid which explodes powerfully by percussion. It is much used for blasting in mines and quarries, but is very dangerous to handle, and has given rise to several fatal accidents.

Glycerin combines with the elements of sulphuric acid, forming a s u l-p h o g l y c e r i c a c i d, $C_3H_8O_3SO_3$, which gives soluble salts with lime, baryta, and lead oxide.

Monatomic *oxygen-acids* (acetic, benzoic, stearic, etc.), heated in sealed tubes with glycerin, yield compound ethers in which 1, 2, or 3 hydrogen-atoms of the glycerin are replaced by an equivalent quantity of the acid

* Roos, Chem. Soc. J. 1876, i. 651.

radicle, according to the proportions employed. The resulting compound ethers are denoted by names ending in *in*; thus:

$$C_3H_5(OH)_3 \quad + \quad HOC_2H_3O \quad = \quad C_3H_5(OH)_2OC_2H_3O \quad + \quad H_2O$$
Glycerin.   Acetic acid.   Mono-acetin.

$$C_3H_5(OH)_3 \quad + \quad 2HOC_2H_3O \quad = \quad C_3H_5(OH)(OC_2H_3O)_2 \quad + \quad 2H_2O$$
Glycerin.   Acetic acid.   Diacetin.

$$C_3H_5(OH)_3 \quad + \quad 3HOC_2H_3O \quad = \quad C_3H_5(OC_2H_3O)_3 \quad + \quad 3H_2O \, .$$
Glycerin.   Acetic acid.   Triacetin.

The glyceric ethers or g l y c e r i d e s thus produced are, for the most part, oily liquids, increasing in viscidity as the acid from which they are formed has a higher molecular weight; those formed from the higher members of the fatty acid series, $C_nH_{2n}O_2$ (such as palmitic and stearic acids), are solid fats. Some of the triacid glycerides, produced artificially in the way just mentioned, are identical with natural fats occurring in the bodies of plants and animals; thus tristearin is identical with the stearin of beef and mutton suet: triolein with the olein of olive oil, etc.

Hydrochloric and *hydrobromic acids* act upon glycerin in the same manner as oxygen-acids, excepting that the reaction always stops at the second stage (just as in the action of these acids on the glycols it stops at the first stage). The ethers thus formed are called c h l o h y d r i n s, b r o m h y - d r i u s, etc., *e. g.*:

$$C_3H_5(OH)_3 \quad + \quad HCl \quad = \quad C_3H_5(OH)_2Cl \quad + \quad H_2O$$
Glycerin.   Chlorhydrin.

$$C_3H_5(OH)_3 \quad + \quad 2HCl \quad = \quad C_3H_5(OH)Cl_2 \quad + \quad 2H_2O$$
Glycerin.   Dichlorhydrin.

Hydriodic acid acts somewhat differently, producing an ether, $C_6H_{11}IO_3$, which may be regarded as a double molecule of glycerin, having four equivalents of hydroxyl replaced by two atoms of oxygen, and a fifth by iodine, $C_6H_{10}O_2(OH)I$.

The *chlorides* and *bromides of phosphorus* act upon glycerin in the same manner as hydrochloric and hydrobromic acid, but their action goes on to the third stage, producing trichlorhydrin, or propenyl chloride, and the corresponding bromine-compound:

$$C_3H_5(OH)Cl_2 \quad + \quad PCl_5 \quad = \quad PCl_3O \quad + \quad HCl \quad + \quad C_3H_5Cl_3$$
Dichlorhydrin.   Trichlor- hydrin.

Iodide of phosphorus acts on glycerin in a totally different manner, yielding iodopropene or allyl iodide, C_3H_5I (p. 624).

Monochlorhydrins, $C_3H_5 \left\{ \begin{array}{l} (OH)_2 \\ Cl \end{array} \right.$—Of these compounds there are two modifications, viz.:

$$
\begin{array}{ll}
CH_2.Cl & CH_2.OH \\
| & | \\
CH.OH & CHCl \\
| & | \\
CH_2.OH & CH_2.OH \\
\text{Unsymmetrical.} & \text{Symmetrical.}
\end{array}
$$

The first is the chief product obtained by saturating glycerin with hydrochloric acid gas, and heating the liquid for some time over the water-bath. To purify it, the acid liquid is saturated with sodium carbonate, then agitated with ether, the ethereal solution is evaporated, and the residual liquid subjected to fractional distillation.

Unsymmetrical chlorhydrin is a viscid liquid, easily soluble in water, alcohol, and ether, having a specific gravity of 1.31, and boiling at 220°–227° C. (428°–440.6° F.); converted by sodium amalgam into ordinary propene-glycol, CH_3—$CH(OH)$—CH_2OH.

Symmetrical chlorhydrin is obtained by agitating allyl alcohol with aqueous hypochlorous acid:

$$\begin{array}{ccc} CH_2 & & CH_2OH \\ \parallel & & \mid \\ CH & + \quad ClOH \quad = & CHCl \\ \mid & & \mid \\ CH_2OH & & CH_2OH \end{array}$$

It has a specific gravity of 1.4 at 13° C. (55.4° F.), and boils at 230°–235° C. (446°–455° F.).

Dichlorhydrins, $C_3H_5\left\{\begin{array}{l} OH \\ Cl_2 \end{array}\right.$ —Of these also there are two modifications, viz.:

$$\begin{array}{cc} CH_2Cl & CH_2.OH \\ \mid & \mid \\ CH.OH & CHCl \\ \mid & \mid \\ CH_2.Cl & CH_2Cl \\ \textbf{Symmetrical.} & \textbf{Unsymmetrical.} \end{array}$$

Both are formed by heating glycerin with strong hydrochloric acid, the first, however, predominating. This, which is the ordinary modification, is best prepared by saturating a mixture of equal volumes of glycerin and glacial acetic acid with hydrochloric acid gas at 100° C., neutralizing the product with sodium carbonate, and subjecting it to fractional distillation. It may also be prepared by distilling glycerin with S_2Cl_2. Lastly, it may be obtained quite pure by mixing epichlorhydrin, C_3H_5OCl, with fuming hydrochloric acid.

Symmetrical dichlorhydrin, or *dichlorinated pseudopropyl alcohol*, is a liquid having an ethereal odor, a specific gravity of 1.383 at 19° C. (66.2° F.), boiling at 174° C. (345.2° F.), easily soluble in alcohol and ether, sparingly in water. By *sodium amalgam* it is converted into pseudopropyl alcohol, CH_3—$CHOH$—CH_3, and by *oxidation* with dilute chromic acid mixture, into acetone dichloride, CH_3—Cl_2—CH_3. With *potassium cyanide* it forms dicyanhydrin, $CH_2.CN$—$CH(OH)$—$CH_2.CN$.

Unsymmetrical Dichlorhydrin (commonly called *chloride of allyl alcohol*) is formed by addition of chlorine to allyl alcohol, CH_2=CH—$CH_2(OH)$, or of hypochlorous acid to allyl chloride:

$$\begin{array}{ccc} CH_2 & & CH_2.OH \\ \parallel & & \mid \\ CH & + \quad ClOH \quad = & CHCl \\ \mid & & \mid \\ CH_2Cl & & CH_2Cl \end{array}$$

It has a specific gravity of 1.379 at 0°; boils at 182°–183° C. (359.6°–361.4° F.); is converted by metallic sodium into allyl alcohol.

By the action of caustic potash or soda, both dichlorhydrins are converted, by abstraction of HCl, into epichlorhydrin.

Trichlorhydrin, *Glyceryl trichloride,* or *Allyl trichloride,* $C_3H_5Cl_3 =$ CH_2Cl—$CHCl$—CH_2Cl, is formed by the action of phosphorus pentachloride on either of the dichlorhydrins ; also (together with $CH_3.CHCl.CHCl_2$) by the action of iodine chloride on propene chloride, $CH_3.CHCl.CH_2Cl$; and by chlorination of propene and of pseudopropyl iodide. It is a liquid which smells like chloroform, has a specific gravity of 1.417 at 15° C. (59° F.), and boils at 158° C. (316.4° F.). Heated with solid potassium or sodium hydroxide, it forms d i c h l o r o g l y c i d e, $CH_3\!=\!CCl\!-\!CH_2Cl$, boiling at 94° C. (201.2° F.). Heated with sulphuric acid, it forms m o n o - c h l o r a c e t o n e, $CH_3\!-\!CO\!-\!CH_2Cl$.

Bromhydrins. — $Monobromhydrin$, $C_3H_5(OH)_2Br = CH_2.Br\!-\!CH.$ $OH\!-\!CH_2.OH$, obtained by the action of hydrobromic acid on glycerin, is an oily liquid, boiling in a vacuum at 180° C. (356° F.).

$Symmetrical\ Dibromhydrin$, $CH_2Br\!-\!CH.OH\!-\!CH_2Br$, prepared from glycerin and hydrobromic acid, is a liquid having an ethereal odor, a specific gravity of 2.11 at 18° C. (64.4° F.), and boiling at 219° C. (426.2° F.). $Unsymmetrical\ dibromhydrin$, $CH_2Br\!-\!CHBr\!-\!CH_2OH$, from allyl alcohol and bromine, boils at 212-214° C. (413.6-417.2° F.).

$Tribromhydrin$, $Glyceryl\ tribromide$, or $Allyl\ tribromide$, $C_3H_5Br_3\!=\!CH_2Br\!-\!CHBr\!-\!CH_2Br$, is obtained by the action of bromine on allyl iodide :

$$C_3H_5I\ +\ 2Br_2\ =\ IBr\ +\ C_3H_5Br_3.$$

It crystallizes in colorless, shining prisms, melts at 16° C. (60.8° F.), boils at 219°-220° C. (426.2°-428° F.) : is converted by caustic potash or soda into d i b r o m o g l y c i d e, $CH_2\!=\!CBr\!-\!CH_2Br$, boiling at 151°-152° C. 303.8°-305.6° F.).

Iodhydrins. — $Monoiodhydrin$, $C_3H_5\left\{\begin{smallmatrix}(OH)_2\\I\end{smallmatrix}\right.$, obtained by heating glycerin to 100° C. with hydriodic acid, is a viscid liquid, having a specific gravity of 1.783.

$Di\text{-}iodhydrin$, $C_3H(OH)I_2\!=\!CH_2I.CH(OH).CH_2I$, obtained by heating symmetrical dichlorhydrin with aqueous potassium iodide, is a thick oil, having a specific gravity of 2.4, and solidifying to a crystalline mass at —15° C. (5° F.).

$Tri\text{-}iodhydrin$ or $Glyceryl\ tri\text{-}iodide$, $C_3H_5I_2$, does not appear to be capable of existing, inasmuch as the action of hydriodic acid, or of iodine and phosphorus on glycerin yields nothing but allyl iodide and free iodine.

Glycide compounds.—When dichlorhydrin is treated with potash, it gives up a molecule of hydrochloric acid, and is converted into a compound, C_3H_5OCl, called e p i c h l o r h y d r i n :

$$C_3H_5(OH)Cl_2\ -\ HCl\ =\ C_3H_5OCl.$$

This compound may be regarded as the hydrochloric ether of an alcohol, $C_3H_5O(OH)$, called $glycide$, formed from glycerin by abstraction of H_2O. Dibromhydrin, $C_3H_5(OH)Br_2$, treated in the same manner, yields e p i - b r o m h y d r i n, or the hydrobromic ether of glycide, C_3H_5OBr. Epichlorhydrin heated with dry potassium iodide is converted into e p i - i o d h y - d r i n, C_3H_5OI :

$$C_3H_5OCl\ +\ KI\ =\ KCl\ +\ C_3H_5OI.$$

These glycidic ethers are easily reconverted into bodies of the glycerin

54 *

type. Thus epichlorhydrin combines with acetic acid, forming g l y c e r i c
a c e t o c h l o r h y d r i n:

$$(C_3H_5)ClO + HOC_2H_3O = (C_3H_5)Cl(OH)(OC_2H_3O);$$

and with alcohol in like manner, forming g l y c e r i c e t h y l c h l o r h y-
d r i n, $(C_3H_5)Cl(OH)(OC_2H_5)$.

Epichlorhydrin unites directly with *water*, forming g l y c e r i c m o n o-
c h l o r h y d r i n, $C_3H_5(OH)_2Cl$.

Epichlorhydrin or *Glycidic Hydrochloride*, isomeric with
monochloracetone, may be obtained by agitating either of the dichlorhydrins with strong potash-lye:

$$
\begin{array}{c}
CH_2Cl \\
| \\
CH.OH \\
| \\
CH_2Cl
\end{array}
+ KOH = KCl + H_2O +
\begin{array}{c}
\quad CH_2 \\
O{<}| \\
\quad CH \\
\quad \| \\
\quad CH_2Cl.
\end{array}
$$

It is a mobile liquid, having a specific gravity of 1.194 at 0°, boiling at
119° C. (246.2° F.) ; insoluble in water ; easily soluble in alcohol and ether.
It smells like chloroform, and has a burning sweetish taste. When heated
with fuming *hydrochloric acid*, it is converted into s y m m e t r i c a l d i-
c h o r h y d r i n:

$$
\begin{array}{c}
\quad CH_2 \\
O{<}\| \\
\quad CH \\
\quad | \\
\quad CH_2Cl
\end{array}
+ HCl =
\begin{array}{c}
CH_2Cl \\
| \\
CH.OH ; \\
| \\
CH_2Cl
\end{array}
$$

and by prolonged heating with water, into m o n o c h l o r h y d r i n. Heated
with *alcohols*, it yields e t h e r s of c h l o r h y d r i n, such as $C_3H_5\begin{cases} Cl \\ OH \\ O.C_2H_5 \end{cases}$,
which, when distilled with potassium hydroxide, are converted into a l c o-
h o l i c g l y c i d e s or g l y c i d i c e t h e r s; thus:

$$
\begin{array}{c}
CH_2Cl \\
| \\
CH.OH \\
| \\
CH_2.O.C_2H_5
\end{array}
+ KOH = KCl + H_2O +
\begin{array}{c}
\quad CH_2 \\
O{<}| \\
\quad CH \\
\quad | \\
\quad CH_2.O.C_2H_5
\end{array}
$$
Ethylic chlor-
hydrin.   Ethyl
glycide.

Ethyl-glycide, $C_3H_5.O.OC_2H_5$ (or epiethylin), boils at 126°–130°C. (258.8°
F.) ; *amyl-glycide*, $C_3H_5.O.O.C_5H_{11}$, at 188° C. (370.4° F.).

Epibromhydrin, C_3H_5OBr, from either of the two bromhydrins, boils
at 139°–140° C. (282.2°–284° F.); *epi-iodhydrin*, C_3H_5OI, obtained by
heating epichlorhydrin with solution of potassium iodide, boils at about
160° C. (320° F.).

Trichlorhydrin, heated with potassium hydroxide, which abstracts HCl,
yields d i c h l o r o g l y c i d e, $C_3H_4Cl_2 = CH_2{=}CCl{-}CH_2Cl.$

Alcoholic Ethers of Glycerin.—Derivatives of glycerin containing alcohol-radicles are formed by heating the mono- and dichlorhydrins
with sodium alcohol, thus:—

$$C_3H_5\begin{cases} OH \\ Cl_2 \end{cases} + 2(C_2H_5.O.Na) = 2NaCl + C_3H_5\begin{cases} OH \\ (O.C_2H_5)_2 \end{cases}.$$
Dichlorhydrin.   Diethylin.

Mono-ethylin, $C_3H_5 \begin{Bmatrix} (OH)_2 \\ OC_2H_5 \end{Bmatrix}$, is soluble in water, and boils at 230°
C. (446° F.). Diethylin is sparingly soluble in water, smells like
peppermint, has a specific gravity of 0.92, and boils at 191° C. (375.8° F.).
Triethylin, $C_3H_5(O.C_2H_5)_3$, formed by the action of ethyl iodide on the
sodium-derivative of diethylin, is a liquid insoluble in water, boiling at
185° C. (365° F.).

Allylin, $C_3H_5 \begin{Bmatrix} (OH)_2 \\ O.C_3H_5 \end{Bmatrix}$, is produced by heating glycerin with oxalic
acid, and occurs (together with glyceric oxide) in the residue left after
heating the product to 260° C. (500° F.). It is a viscid liquid, boiling at
225°–240° C. (437°–464° F.).

Glyceric oxide or glycerin ether, $(C_3H_5)_2O_3$, occurring together
with allylin, as just mentioned, boils at 169°–172° C. (336.2°–341.6° F.).

The ethers of glycerin containing acid radicles, including the natural
oils and fats, will be described under the respective acids.

Polyglycerins.—Two, three, or more molecules of glycerin can unite
into a single molecule, with elimination of a number of water-molecules
less by one than the number of glycerin molecules which combine together; thus :—

$$2C_3H_5(OH)_3 \quad - \quad H_2O \quad = \quad (C_3H_5)_2 \begin{Bmatrix} O \\ (OH)_4 \end{Bmatrix}$$
Glycerin.   Diglycerin.

$$3C_3H_5(OH)_3 \quad - \quad 2H_2O \quad = \quad (C_3H_5)_3 \begin{Bmatrix} O_2 \\ (OH)_5 \end{Bmatrix}$$
Glycerin.   Triglycerin.

Generally :—

$$nC_3H_5(OH)_3 \quad - \quad (n-1)H_2O \quad = \quad (C_3H_5)_n \begin{Bmatrix} O_{n-1} \\ (OH)_{n+2} \end{Bmatrix}.$$

The product is a polyglycerin whose atomicity (determined by the
number of equivalents of hydroxyl contained in it) is $n + 2$.

The mode of preparing the polyglycerins is similar to that of the polyethenic alcohols (p. 631), and consists in heating glycerin with monochlor-
hydrin, whereby diglycerin and hydrochloric acid are formed :—

$$C_3H_5Cl(OH)_2 \quad + \quad C_3H_5(OH)_3 \quad = \quad (C_3H_5)_2O(OH)_4 \quad + \quad HCl.$$

The hydrochloric acid thus formed converts a fresh quantity of glycerin
into chlorhydrin, which then acts in a similar manner on the diglycerin,
and converts it into triglycerin; and in this manner the process is con-
tinned. The polyglycerins may then be separated by fractional distillation. Their properties are but little known.

Isopentenyl Alcohol, or **Amyl Glycerin,** $C_5H_{12}O_3 = (C_5H_9)(OH)_3$.
—This compound is formed from bromo-isopentene dibromide, $C_5H_9Br.Br_2$,
or isopentenyl bromide, $C_5H_9Br_3$, by the series of processes represented in
the following equations :—

$$C_5H_9Br_3 \quad + \quad 2AgOC_2H_3O \quad = \quad 2AgBr \quad + \quad C_5H_9 \begin{Bmatrix} (OC_2H_3O)_2 \\ Br \end{Bmatrix}$$
Isopentenyl   Silver   Silver   Isopentenyl diaceto-
bromide.   acetate.   bromide.   bromhydrin.

$$C_5H_9 \begin{Bmatrix} (OC_2H_3O)_2 \\ Br \end{Bmatrix} \quad + \quad 2KOH \quad = \quad 2KOC_2H_3O \quad + \quad C_5H_9 \begin{Bmatrix} (OH)_2 \\ Br \end{Bmatrix}$$
Isopentenyl diaceto-   Potassium   Potassium   Isopentenyl
bromhydrin.   hydrate.   acetate.   bromhydrin.

$$C_5H_9 \left\{ \begin{matrix} (OH)_2 \\ Br \end{matrix} \right. + KOH = KBr + (C_5H_9)(OH)_3$$

Isopentenyl
bromhydrin.

Isopentenyl
alcohol.

Amyl glycerin is a thick colorless liquid, having a sweet aromatic taste, and soluble in water.

TETRATOMIC ALCOHOLS AND ETHERS.

The only tetratomic alcohol at present known is:

Erythrite, $C_4H_{10}O_4 = (C_4H_6)(OH)_4$, also called *Erythromannite, Erythroglucin*, and *Phycite*, which is the tetratomic alcohol analogous to butyl alcohol, $C_4H_{10}O$, and butyl glycol, $C_4H_{10}O_2$; the corresponding glycerin is not known.

Erythrite is a saccharine substance, existing ready formed in *Protococcus vulgaris*. It was originally discovered by Dr. Stenhouse among the products of decomposition of erythric acid.* It crystallizes in large transparent prisms, is readily soluble in water, sparingly soluble in alcohol, insoluble in ether: not fermentable. Heated with *hydriodic acid*, it yields secondary butyl iodide, C_4H_9I (p. 582):—

$$C_4H_{10}O_4 + 7HI = C_4H_9I + 4H_2O + 3I_2.$$

Heated with *oxygen acids*, it forms compound ethers, in the manner of alcohols in general; thus, with *benzoic acid*, $C_7H_6O_2$, or HOC_7H_5O, it forms a dibenzoate, $(C_4H_6)(OH)_2(OC_7H_5O)_2$, and a hexbenzoate, $(C_4H_6)(OC_7H_5)(O)_4$. $2C_7H_2O_2$, consisting of neutral benzoyl-erythrite, united with two molecules of benzoic acid.

A tetratomic ether, viz., Ethylic Orthocarbonate or Orthocarbonic Ether, $C(OC_2H_5)_4$, analogous to carbon tetrachloride, CCl_4, is formed by the action of sodium ethylate on chloropicrin (p. 554):

$$CCl_3(NO_2) + 4NaOC_2H_5 = 3NaCl + NaNO_2 + C(OC_2H_5)_4.$$

It is a liquid having an ethereal odor, and boiling at 158°–159° C. (316.4–318.2° F.). Heated with ammonia, it yields guanidine, CN_3H_5, and ethyl alcohol:

$$C(OC_2H_5)_4 + 3NH_3 = CN_3H_5 + 4C_2H_5OH.$$

The corresponding alcohol, $C(OH)_4$, has probably no existence, inasmuch as one carbon-atom appears to be incapable of linking together two or more hydroxyl groups.

HEXATOMIC ALCOHOLS AND ETHERS.

This class of compounds includes most of the saccharine substances found in plants, and others produced from them by artificial transformation. Three of the natural sugars, mannite, dulcite, and sorbite, having the composition $C_6H_{14}O_6$, or $C_6H_8(OH)_6$, are saturated hexatomic alcohols derived from the saturated hydrocarbon, C_6H_{14}. Several others, called glucoses, contain $C_6H_{12}O_6$, that is to say, two atoms of hydrogen

---

* See the Chapter on Coloring Matters.

less than mannite and dulcite, and may therefore be regarded—so far as composition is concerned—as the aldehydes of these alcohols; moreover, ordinary glucose (grape-sugar) is converted into mannite by the action of nascent hydrogen, just as acetic aldehyde, C_2H_4O, is converted into common alcohol, C_2H_6O. Further, there are diglucosic alcohols, or saccharoses, $C_{12}H_{22}O_{11}(=2C_6H_{12}O_6-H_2O)$, related to glucoses in the same manner as diethenic alcohol to glycol, or diglycerin to glycerin: the most important of these are cane-sugar and milk-sugar; and, lastly, there are certain vegetable products called amyloses—viz., starch, cellulose, and a few others, represented by the formula $C_6H_{10}O_5$, or multiples thereof, which may be regarded as the oxygen-ethers or anbydrides of the glucoses, or of the diglucosic alcohols, inasmuch as they differ therefrom by a molecule of water.

These three groups of compounds, which consist of carbon united with hydrogen and oxygen in the proportion to form water, are included under the general name of carbohydrates.

Saturated Hexatomic Alcohols.

Mannite, $C_6H_{14}O_6 = (C_6H_8)(OH)_6$.—This is the chief component of *manna*, an exudation from a species of ash; it is also found in the juice of certain other plants, in several sea-weeds, and in mushrooms. It is best prepared by treating manna with boiling alcohol, and filtering the solution whilst hot; it then crystallizes on cooling in tufts of slender needles. Mannite may be produced artificially by treating a solution of glucose with sodium-amalgam, the glucose then taking up 2 atoms of hydrogen:

$$C_6H_{12}O_6 + H_2 = C_6H_{14}O_6.$$

The same transformation of glucose sometimes takes place under the action of certain ferments.

Mannite crystallizes in thin four-sided prisms, easily soluble in water and in hot alcohol, insoluble in ether. It is slightly sweet, melts at 166° C. (330.8° F.), has no action on polarized light, and is not fermentable, except under very unusual conditions.

By oxidation in contact with *platinum black*, mannite is converted into mannitic acid, $C_6H_{12}O_7$, and mannitose, $C_6H_{12}O_6$, a kind of sugar isomeric with glucose. By oxidation with *nitric acid* it yields saccharic acid, $C_6H_{10}O_8$, and ultimately oxalic acid. Mannitic acid and saccharic acid are related to mannite in the same manner as glycollic acid and oxalic acid to glycol; the relation between the three compounds is shown by the following formulæ:

CH$_2$OH	COOH	COOH
CHOH	CHOH	CHOH
CHOH	CHOH	CHOH
CHOH	CHOH	CHOH
CHOH	CHOH	CHOH
CH$_2$OH	CH$_2$OH	COOH
Mannite.	Mannitic acid.	Saccharic acid.

By *fuming nitric acid*, or more easily by a mixture of *nitric* and *sulphuric acids*, mannite is converted into n i t r o m a n n i t e , $C_6H_8(NO_2)_6O_6$, a crystalline body which explodes violently by percussion or when suddenly heated, and is reconverted into mannite by ammonium sulphide. With sulphuric acid mannite forms s u l p h o m a n n i t i c   a c i d , $C_6H_{14}O_6$.$3SO_3$.

Mannite, treated with *hydriodic acid*, is converted into secondary hexyl iodide, or hexene hydriodide.

$$C_6H_{14}O_6 \;+\; 11HI \;=\; C_6H_{13}I \;+\; 6H_2O \;+\; 5I_2.$$

Mannite, heated with *organic acids*, forms compound ethers, after the manner of alcohols in general, the elements of the mannite and the acid uniting together, with elimination of one or more molecules of water. The resulting compounds, called m a n n i t a n i d e s , bear a considerable resemblance to the fats ; but their composition has not been very exactly determined.

These ethers, when saponified with alkalies, yield, not mannite, but m a n n i t a n , $C_6H_{12}O_5$, a compound differing from mannite by one molecule of water. The same compound is obtained in small quantity by heating mannite to 200° C. (392° F.), and more easily by prolonged boiling of mannite with strong hydrochloric acid. It is a syrupy liquid, which volatilizes slowly at 140° C. (284° F.), and dissolves easily in water and in cold absolute alcohol : this last property affords the means of separating it from mannite. When exposed to the air, it slowly absorbs water, and is reconverted into mannite ; the change is accelerated by boiling with acids or with alkalies.

Mannite, boiled with *butyric acid*, gives up two molecules of water, and is converted into m a n n i d e , $C_6H_{10}O_4$, which is also a syrupy liquid, but differs from mannitan in being much more volatile, evaporating rapidly at 140° C. (284° F.), and in being quickly reconverted into mannite by exposure to moist air. It dissolves easily in water, and in absolute alcohol.

The two anhydrides of mannite may be represented by the following formulæ —   :

$$C_6H_8 \begin{cases} O \\ (OH)_4 \end{cases}$$
Mannitan.

$$C_6H_8 \begin{cases} O \\ (OH)_2 \\ O \end{cases}$$
Mannide.

Dulcite, $C_6H_{14}O_6$, also called *Dulcin, Dulcose*, and *Melampyrite*.—This sugar, isomeric with mannite, is obtained from a crystalline substance of unknown origin, imported from Madagascar : it is extracted therefrom by boiling with water, and crystallizes from the filtered solution. Dulcite is likewise obtained from *Melampyrum nemorosum*, by mixing the aqueous decoction of the plant with lime, concentrating, adding hydrochloric acid in slight excess, and evaporating a little ; it then separates in crystals as the liquid cools.

Dulcite is a sweet substance resembling mannite in most of its properties, but differing from it in its crystalline form, which is that of a monoclinic prism, whereas the crystals of mannite are trimetric ; also in its melting point, dulcite melting at 182° C. (359.6° F.), mannite at 166° C. (330.8° F.), and by yielding, when oxidized with nitric acid, not saccharic acid, but mucic acid, which is isomeric therewith. Dulcite, heated with hydriodic acid, yields a secondary hexyl alcohol, identical with that obtained from mannite. Heated with organic acids, it forms ethers called d u l c i t a n i d e s , analogous to the mannitanides, and yielding by saponification, not dulcite, but d u l c i t a n , $C_6H_{12}O_5$, which may likewise be obtained by heating dulcite, or by boiling it with hydrochloric acid. *Hexacetodulcite*, $C_6H_8(OC_2H_3O)_6$, melts at 171° C. (339.8° F.).

Isodulcite, $C_6H_{14}O_6$, or $C_6H_{12}O_5.H_2O$, a saccharine substance isomeric with mannite and dulcite, is produced by the action of dilute acids on quercitrin, a yellow substance occurring in quercitron bark. It forms large transparent, regularly developed crystals, resembling those of cane-sugar : it is sweeter than grape-sugar, not fermentable, dissolves in 2.09 parts of water at 18° C. (64.4° F.), and easily in absolute alcohol. The solutions turn the plane of polarization to the right. Isodulcite melts, with loss of water, between 105° C. (221° F.) and 110° C. (230° F.), is colored yellow or brown by strong sulphuric acid and caustic alkalies, and reduces cupric oxide. By a mixture of nitric and sulphuric acids, it is converted into a slightly explosive nitro-compound, $C_6H_9(NO_2)_3O_6$.

Sorbite, $C_6H_{14}O_6 + 1\frac{1}{2}H_2O$, occurs in the berries of the mountain-ash in small crystals, easily soluble in water. It gives off its water when heated, and melts at 110° C. (230° F.).

Mannite, dulcite, and sorbite are distinguished from the true sugars (glucoses and saccharoses) by not fermenting in contact with yeast, and by not reducing an alkaline solution of cupric oxide.

Pinite and **Quercite** are naturally occurring saccharine bodies, isomeric with mannitan and dulcitan.

Pinite is contained in the sap of a Californian pine (*Pinus Lambertiana*), and is deposited from the aqueous extract of the hardened juice, in hard white crystalline nodules, as sweet as cane-sugar, very soluble in water, nearly insoluble in alcohol. It turns the plane of polarization of a luminous ray to the right; it is not fermentable. With *benzoic acid*, it forms dibenzopinite, $C_6H_8\begin{cases}(OC_7H_5O)_2\\O\\(OH)_2\end{cases}$, and tetrabenzopinite, $C_6H_8\begin{cases}O\\(OC_7H_5O)_4\end{cases}$; and similar compounds with *stearic acid*.

Quercite is a saccharine substance extracted from acorns, by treating the aqueous infusion with milk of lime to remove tannic acid, leaving the liquid to ferment with yeast to remove fermentable sugar, evaporating the filtrate to a syrup, and leaving it to crystallize. It forms hard monoclinic crystals, which grate between the teeth, and are soluble in water and in hot dilute alcohol. Heated in a sealed tube with *benzoic acid*, it forms dibenzoquercite, having the same composition as dibenzopinite.

CARBOHYDRATES.

This name, as already observed, is given to a class of compounds containing 6 or 12 atoms of carbon united with hydrogen and oxygen in the proportion to form water. They may be regarded as derivatives of the hexatomic alcohol, $C_6H_{14}O_6$, and are divided into the three following groups :—

Glucoses, $C_6H_{12}O_6$.	*Saccharoses,* $C_{12}H_{22}O_{11}$.	*Amyloses,* $C_6H_{10}O_5$.
+ Grape-sugar*	+ Cane-sugar	+ Starch.
or Dextrose.	or Saccharose.	+ Dextrin.
— Fruit-sugar	+ Parasaccharose.	+ Glycogen.
or Levulose.	+ Melitose.	— Inulin.
Mannitose.	+ Melezitose.	Cellulose.
+ Galactose.	+ Trehalose.	Tunicin.
Inosite.	+ Mycose.	Gum.
Sorbin.	Synanthrose.	
+ Eucalyn.	+ Milk-sugar	
Dambose.	or Lactose.	
	+ Maltose.	

Most of these compounds occur in the vegetable organism ; a few also, as dextrose, lactose, glycogen, and inosite, in the animal organism.

Glucoses, $C_6H_{12}O_6$.

The sugars included in this formula may be regarded as aldehydes of the saturated alcohols, $C_6H_{14}O_6$. Ordinary glucose (grape-sugar) is converted into mannite by the action of nascent hydrogen (p. 645), and, on the other hand, mannite, when slowly oxidized in contact with platinum black, is partly converted into mannitose. Nevertheless, the glucoses still exhibit the characteristic property of alcohols, namely, that of forming ethers by combination with acids and elimination of water. The formula of a glucose may indeed be derived from that of mannite given on page 645, by removing two hydrogen-atoms from one of the groups, CH_2OH, the other groups remaining as before ; the glucoses may therefore be expected to act as pentatomic alcohols. Bodies thus constituted may be called al- coholic aldehydes.

The following varieties of glucose are known :—

1. *Ordinary glucose*, produced by hydration of starch under the influence of dilute acids or of diastase, and existing ready-formed, together with other kinds of sugar, in honey and various fruits, especially in grapes, and alone in diabetic urine.

2. *Levulose*, existing in cane-sugar which has been acted upon by acids, and obtained pure by the action of dilute acids upon a variety of starch called inulin.

3. *Mannitose*, produced by oxidation of mannite.

4. *Galactose*, formed by the action of acids on milk-sugar.

5. *Inosite*, existing in muscular flesh.

6. *Sorbin*, obtained from mountain-ash berries.

7. *Eucalyn*, existing, together with another kind of sugar, in the so-called Australian manna.

8. *Dambose*, obtained from a saccharine substance existing in African caoutchouc.

The first three of these glucoses exhibit but very slight diversity in their chemical properties, differing chiefly indeed in their action on polarized light, and a few other physical properties. They all yield saccharic acid by oxidation. Galactose differs from them in yielding mucic acid when oxidized. Inosite, sorbin, and eucalyn exhibit still greater differences in their chemical properties, especially in not being fermentable, except under very peculiar circumstances, whereas the four other glucoses

---

* The + sign indicates that the substance turns the plane of polarization to the right; the — sign indicates rotation to the left.

undergo vinous fermentation when placed, under certain conditions, in contact with yeast.

All the glucoses, except inosite, are decomposed by boiling with aqueous alkalies; this property distinguishes them from mannite and dulcite. They are not carbonized by strong sulphuric acid at ordinary tempera. tures. When boiled with a solution of potassio-cupric tartrate, they throw down the copper in the form of red cuprous oxide.

1. Ordinary Glucose, Dextro-glucose, Dextrose, $C_6H_{12}O_6$.—This variety of sugar is very abundantly diffused through the vegetable kingdom : it may be extracted in large quantity from the juice of sweet grapes (whence it is often called *grape-sugar*), also from honey, of which it forms the solid crystalline portion, by washing with cold alcohol, which dissolves the fluid syrup. The appearance of this substance, to an enormous extent, in the urine, is the most characteristic feature of the disease called *diabetes*. It exists in diabetic urine unmixed with any other kind of sugar, and is easily obtained by concentrating the liquid till it crystallizes, washing the crystals with cold alcohol, dissolving them in water, and re-crystallizing. It may also be prepared from starch by the action of diastase, a peculiar ferment existing in germinating barley, or by boiling with dilute sulphuric acid. In these reactions the starch takes up the elements of water, and is resolved into glucose and dextrin, a compound isomeric with starch itself.

$$3C_6H_{10}O_5 \quad + \quad H_2O \quad = \quad C_6H_{12}O_6 \quad + \quad 2C_6H_{10}O_5$$
$$\text{Starch.} \qquad\qquad\qquad\qquad \text{Glucose.} \qquad \text{Dextrin.}$$

Glucose, when required in considerable quantity, is always prepared from starch. Cellulose is likewise converted into glucose by the action of acids. Lastly, glucose is produced by boiling natural glucosides with dilute acids.

Glucose is much less sweet than cane-sugar, and less soluble in water and in alcohol ; anhydrous glucose dissolves in 1.224 parts of water at 15° C. (59° F.). From its aqueous solution it separates in granular warty masses, consisting of a hydrate, $C_6H_{12}O_6.H_2O$, which leaves anhydrous glucose as a fused transparent mass when heated to 100° C. (212° F.), or as a dry white powder when heated to 55° C. (131° F.) or 60° C. (140° F.) in a stream of dry air. The alcoholic solution deposits anhydrous glucose in microscopic needles which melt at 140° C. (284° F.). In the state of solution glucose turns the plane of polarization of a ray of light to the right (hence the names *dextroglucose* and *dextrose*) : its specific or molecular rotatory power is $+56°$, and does not vary with the temperature (pp. 518-9).

Glucose may be heated to 120° C. (248° F.) or even 130° C. (266° F.) without alteration, but at 170° C. (338° F.) it gives off water, and is converted into g l u c o s a n, $C_6H_{10}O_5$, which, when freed from caramel (p. 654) by means of charcoal, and from glucose by fermentation, forms a colorless mass, scarcely sweet to the taste, and having somewhat less dextro-rotatory power than glucose. At higher temperatures glucose blackens and suffers complete decomposition. Glucose, boiled for some time with *dilute sulphuric* or *hydrochloric acid*, is converted into brown substances called ulmin, ulmic acid, etc.—*Strong sulphuric acid* converts it at ordinary temperatures into sulpho-saccharic acid, $C_6H_{12}O_5SO_3$, which forms a soluble barium salt.

Lime, baryta, and *lead oxide* dissolve slowly in aqueous solution of glucose, and on adding alcohol to the liquid, compounds of these oxides with glucose are precipitated. The barium compound is said to contain $(C_6H_{12}O_6)_2$ $(BaO)_3.2H_2O$; the calcium compound $(C_6H_{12}O_6)_2(CaO)_3.2H_2O$; the lead compound $(C_6H_{12}O_6)_2(PbO)_3(OH)_6$. These compounds are, however, very

28

unstable, being decomposed at the heat of boiling water. Glucose also combines with *sodium chloride*, forming the compound $(C_6H_{12}O_6)_2NaCl.H_2O$.

Glucose, boiled with a *cupric salt* in presence of *alkalies*, easily reduces the cupric oxide to cuprous oxide: by this character it is easily distinguished from cane-sugar.

When solutions of cane-sugar and glucose are mixed with two separate portions of solution of cupric sulphate, and caustic potash is added in excess to each, deep-blue liquids are obtained, which, on being heated, exhibit different characters: the one containing cane-sugar is at first but little altered; a small quantity of red powder falls after a time, but the liquid long retains its blue tint; with the glucose, on the other hand, the first application of heat throws down a copious greenish precipitate, which rapidly changes to scarlet, and eventually to dark-red cuprous oxide, leaving a nearly colorless solution. If only small quantities of material are available, a mixture of cupric sulphate and tartaric acid, to which an excess of potash has been added, may be used with advantage. This solution, called *potassio-cupric tartrate*, is an excellent test for distinguishing the two varieties of sugar, or discovering an admixture of glucose with cane-sugar.

Glucose mixed in dilute solution with *yeast* and exposed to a temperature of 21º–26º C. (69.8º–78.8º F.), easily undergoes vinous fermentation (p. 591).

2. **Levulose**, $C_6H_{12}O_6$.—This sugar, distinguished from dextrose by turning the plane of polarization to the left, occurs, together with dextrose, in honey, in many fruits, and in other saccharine substances. The mixture of these two sugars in equivalent quantities constitutes f r u i t - s u g a r or i n v e r t s u g a r, which is itself levorotatory, because the specific rotatory power of levulose is, at ordinary temperatures, greater than that of dextrose.

Cane-sugar may be *inverted*, that is, transformed into a mixture of equal parts of dextrose and levulose, by warming with dilute acids:

$$C_{12}H_{22}O_{11} \ + \ H_2O \ = \ C_6H_{12}O_6 \ + \ C_6H_{12}O_6.$$

The same change is brought about by contact with yeast, or with pectase, the peculiar ferment of fruits; it likewise takes place slowly when a solution of cane-sugar is left to itself.

To separate the levulose, the invert sugar obtained from 10 grams of cane-sugar is mixed with 6 grams of slaked lime and 100 grams of water, whereby a solid calcium-compound of levulose is formed, while the whole of the dextrose remains in solution, and may be separated from the precipitate by pressure. The calcium salt of levulose, suspended in water and decomposed by carbon dioxide, yields a solution of pure levulose, which may be filtered and concentrated by evaporation. Levulose may be at once obtained in the pure state by the action of dilute acids on inulin.

Levulose is a colorless uncrystallizable syrup, as sweet as cane-sugar, more soluble in alcohol than dextrose. Its rotatory power is much greater than that of dextrose at ordinary temperatures, but diminishes as the temperature rises. For the transition tint between the blue and the purple $[a] = -106º$ at 14º C. (57.2º F.); $= -79.5º$ at 52º C. (125º F.); $= -53º$ at 90º C. (194º F.). Now, the rotatory power of dextrose is the same at all temperatures, and equal to $+56º$; consequently that of invert sugar, which is $-25º$ at 15º C. (59º F.), diminishes by about one-half at 52º C. (125º F.), becomes nothing at 90º C. (194º F.), and changes sign above that temperature.

Levulose exhibits, for the most part, the same chemical reactions as dextrose, but is more easily altered by heat or by acids, and on the contrary offers greater resistance to the action of alkalies or of ferments.

Levulosan, $C_6H_{10}O_5$, the oxygen-ether or anhydride of levulose, is produced, together with dextrose, by melting cane-sugar for some time at 160° C. (320° F.) :

$$C_{12}H_{22}O_{11} = C_6H_{10}O_5 + C_6H_{12}O_6.$$

The dextrose may be removed from the liquid by fermentation, and the levulosan, which is unfermentable, may be obtained by evaporation as an uncrystallizable syrup. By boiling with water or dilute acids, it is converted into a fermentable levorotatory sugar, probably levulose.

3. **Mannitose**, $C_6H_{12}O_6$.—This is the sugar produced, together with mannitic acid, by the oxidation of mannite in contact with platinum black. It may be separated by saturating the liquid with lime, precipitating the calcium mannitate with alcohol, evaporating the filtrate to a syrup, adding alcohol, again filtering, and evaporating to dryness.

Mannitose is syrupy, uncrystallizable, fermentable, inactive to polarized light, and resembles the other glucoses in its chemical reaction.

4. **Galactose**, $C_6H_{12}O_6$, is produced by boiling milk-sugar with dilute acids. It is soluble in water, sparingly soluble in cold alcohol, crystallizes more readily than ordinary glucose; has a dextro-rotatory power of 83.3°; and is very easily fermentable. It resembles dextrose in most of its reactions, but is distinguished from all the four glucoses above described by yielding mucic instead of saccharic acid, when oxidized by nitric acid.

5. **Inosite**, or **Phaseomannite**, $C_6H_{12}O_6$, is a variety of glucose occurring in the muscular substance of the heart and other organs of the animal body; also in green kidney beans, the unripe fruit of *Phaseolus vulgaris*, and in many other plants. It forms prismatic crystals, resembling gypsum, soluble in water, but insoluble in alcohol and ether. It may be boiled with a strong aqueous potash or baryta without alteration or coloration. If this sugar be evaporated with nitric acid nearly to dryness, the residue mixed with a little ammonia and calcium chloride, and again evaporated, a beautiful and characteristic rose tint is produced.

Inosite does not ferment with yeast, but in contact with cheese, flesh, or decaying membrane and chalk, it undergoes lactous fermentation, producing lactic, butyric, and carbonic acids.

6. **Sorbin**, $C_6H_{12}O_6$, is a crystallizable sugar existing in the juice of ripe mountain-ash berries (*Sorbus aucuparia*). The juice when allowed to stand for some time in open vessels, deposits a brown crystalline matter, which may be obtained in transparent colorless crystals belonging to the trimetric system. This substance is almost insoluble in alcohol, but easily soluble in water, to which it imparts an exceedingly sweet taste. A solution of sorbin, mixed with ammonia and lead acetate, yields a white flocculent precipitate, containing $C_{12}H_{18}Pb_3O_{12}.PbO_6$. With *sodium chloride* it forms a compound which crystallizes in cubes.

Sorbin is converted by *nitric acid* into tartaric, racemic, and aposorbic acids. It does not ferment with yeast, but in contact with cheese and chalk, at 40° C. (104° F.), it undergoes lactous fermentation, yielding a large quantity of lactic acid, together with alcohol and butyric acid.

7. **Eucalyn,** $C_6H_{12}O_6$, is an unfermentable sugar, separated in the fermentation of melitose (the sugar of the *Eucalyptus* of Tasmania), in consequence of the destruction of a fermentable kind of sugar, which, in combination with eucalyn, constitutes melitose:

$$C_{12}H_{22}O_{11} \quad + \quad H_2O \quad = \quad 2CO_2 \quad + \quad 2C_2H_6O \quad + \quad C_6H_{12}O_6$$
$$\text{Melitose.} \qquad\qquad\qquad\qquad\qquad \text{Alcohol.} \qquad \text{Eucalyn.}$$

On evaporating the liquid, the eucalyn remains as an uncrystallizable syrup, having a specific rotatory power of + 165° nearly. It is not rendered fermentable by the action of sulphuric acid.

8. **Dambose,** $C_6H_{12}O_6$, obtained by heating dambonite with hydriodic acid, crystallizes in six-sided prisms, melting at 212° C. (413.6° F.).

Dambonite, its methylic ether, $C_6H_{10}(CH_3)_2O_6$, occurs in a peculiar kind of caoutchouc from the Gaboon in West Africa, from which it may be extracted by alcohol. It crystallizes from alcohol in six-sided prisms, dissolves easily in water, melts at 190° C. (374° F.), and sublimes at 200°–210° C. (392°–410° F.) in slender shining needles.

GLUCOSIDES.—When ordinary glucose is heated to 100°–120° C. (212–248° F.) for fifty or sixty hours with acetic, butyric, stearic, benzoic, and other organic acids, the two unite, with elimination of water, and compound ethers called glucosides are formed, analogous to the mannitanides. A number of these artificial glucosides have been prepared by Berthelot, who regards them as derivatives of *glucosan*, $C_6H_{10}O_5$, because when heated with alkalies they yield glucosan, not glucose. Thus, there is a glucoso-butyric ether to which Berthelot assigns the formula $C_6H_8(C_4H_7O)_2O_5$, and an acetic ether, which he regards as hexaceto-glucosan, $C_6H_4(C_2H_3O)_6O_5$; but they are merely oily liquids, which are very difficult to obtain pure, and therefore their analyses are not much to be depended on.

A considerable number of bodies of similar constitution exist ready formed in plants, many of them constituting the bitter principles of the vegetable kingdom. None of these natural glucosides have been produced artificially, but they are all resolved by boiling with dilute acids into glucose and some other compound. The most important of them will be described in a future chapter.

POLYGLUCOSIC ALCOHOLS.

The compounds of this group, including cane-sugar and other bodies more or less resembling it, may be regarded as formed by the combination of two or more molecules of glucose, with elimination of a number of molecules of water, less by one than the number of glucose molecules which enter in the combination:

$$2C_6H_{12}O_6 \quad - \quad H_2O \quad = \quad C_{12}H_{22}O_{11}, \text{ Diglucosic alcohol.}$$
$$3C_6H_{12}O_6 \quad - \quad 2H_2O \quad = \quad C_{18}H_{32}O_{16}, \text{ Triglucosic alcohol.}$$
$$nC_6H_{12}O_6 \quad - \quad (n-1)H_2O \quad = \quad C_{6n}H_{10n+2}O_{5n+1}.$$

The only known alcohols of this class are diglucosic alcohols, $C_{12}H_{22}O_{11}$; but starch, cellulose, and other plant constituents, appear to be the oxygen-ethers, or anhydrides, of polyglucosic alcohols of higher orders.

Cane-sugar or **Saccharose,** $C_{12}H_{22}O_{11}$.—This most useful substance is found in the juice of many of the grasses, in the sap of several forest trees, in the root of the beet and the mallow, and in several other plants. Most sweet fruits contain cane-sugar, together with invert sugar (p. 650); some, as walnuts, hazel-nuts, almonds, coffee-beans, and St. John's-bread (the fruit of *Ceratonia siliqua*), contain only cane-sugar. Honey and the nectars of flowers contain cane-sugar together with invert sugar; the sugar in the nectary of cactuses is almost wholly cane-sugar.

Sugar is extracted most easily and in greatest abundance from the sugar-cane (*Saccharum officinarum*), cultivated for the purpose in many tropical countries. The canes are crushed between rollers, and the expressed juice is suffered to flow into a large vessel, where it is slowly heated nearly to its boiling point. A small quantity of slaked lime mixed with water is then added, which occasions the separation of a coagulum consisting chiefly of earthy phosphates, waxy matter, a peculiar albuminous principle, and mechanical impurities. The clear liquid separated from the coagulum is rapidly evaporated in open pans, heated by a strong fire made with the crushed canes of the preceding year, which have been dried in the sun, and preserved for the purpose. When sufficiently concentrated, the syrup is transferred to a shallow vessel, and left to crystallize, during which time it is frequently agitated in order to hasten the change and hinder the formation of large crystals. It is lastly drained from the dark uncrystallizable syrup, or *molasses*, and sent into commerce, under the name of *raw* or *Muscovado* sugar. The refining of this crude product is effected by redissolving it in water, adding a quantity of albumin in the shape of serum of blood or white of egg, and sometimes a little lime-water, and heating the whole to the boiling point : the albumin coagulates, and forms a kind of network of fibres, which inclose and separate from the liquid all mechanically suspended impurities. The solution is decolorized by filtration through animal charcoal, evaporated to the crystallizing point, and put into conical earthen moulds, where it solidifies, after some time, to a confusedly crystalline mass, which is drained, washed with a little clean syrup, and dried in a stove : the product is ordinary *loaf-sugar*. When the crystallization is allowed to take place quietly and slowly, *sugar candy* results, the crystals under these circumstances acquiring large volume and regular form. The evaporation of the decolorized syrup is best conducted in strong close boilers exhausted of air : the boiling point of the syrup is reduced in consequence from 110° C. (230° F.) to 65.5° C. (150° F.), or below, and the injurious action of the heat upon the sugar is in great measure prevented. Indeed, the production of molasses in the rude colonial manufacture is chiefly the result of the high and long-continued heat applied to the cane-juice, and might be almost entirely prevented by the use of vacuum-pans, the product of sugar being thereby greatly increased in quantity, and so far improved in quality as to become almost equal to the refined article.

In many parts of the continent of Europe, sugar is manufactured on a large scale from beet-root, which contains about 8 per cent. of that substance. The process is far more complicated and troublesome than that just described, and the raw product much inferior. When refined, however, it is scarcely to be distinguished from the preceding. In the Western States of America, sugar is prepared in considerable quantity from the sap of the sugar maple (*Acer saccharinum*), which is common in those parts. The tree is tapped in the spring by boring a hole a little way into the wood, and inserting a small spout to convey the liquid into a vessel placed for its reception. This is boiled down in an iron pot, and furnishes a coarse sugar, which is almost wholly employed for domestic purposes, but little finding its way into commerce.

Pure sugar slowly separates from a strong solution in large, transparent colorless crystals, having the figure of a modified monoclinic prism. The crystals have a specific gravity of 1.6, and are unchangeable in the air. Sugar has a pure sweet taste, is very soluble in water, requiring for solution only one-third of its weight in the cold, and is also dissolved by alcohol, but less easily. When moderately heated it melts, and solidifies on cool-ing to a glassy amorphous mass, familiar as *barley-sugar*.

Cane-sugar, heated a little above 160° C. (320° F.), is converted, with-out loss of weight, into a mixture of dextrose and levolusan (p. 651).

$$C_{12}H_{22}O_{11} \;=\; C_6H_{12}O_6 \;+\; C_6H_{10}O_5 \,.$$

At a higher temperature, water is given off, the dextrose being probably converted into glucosan (p. 650): afterwards, at about 210° C. (410° F.), more water goes off, and a brown substance called c a r a m e l remains, consisting of a mixture of several compounds, all formed from sugar by elimination of water. At a still higher temperature, an inflammable gaseous mixture is given off, consisting of carbon monoxide, marsh-gas, and carbon dioxide ; a distillate is obtained, consisting of brown oils, acetic acid, acetone, and aldehyde ; and a considerable quantity of charcoal remains behind. The brown oils contain a small quantity of f u r f u r o l, and a bitter substance called a s s a m a r.

By prolonged boiling with *water*, cane-sugar is converted into i n v e r t s u g a r. This transformation is accelerated by the presence of acids, and apparently also of certain salts. Different acids act with various degrees of rapidity—mineral more quickly than organic acids, sulphuric acid most quickly of all. By prolonged boiling even with very dilute acids, sugar is decomposed, yielding a number of brown amorphous products called u l m i n, u l m i c a c i d,* etc.; if the air has access to the liquid, formic acid is likewise produced. Concentrated hydrochloric acid decomposes sugar very quickly.

Strong sulphuric acid decomposes dry sugar when heated, and a con-centrated solution, even at ordinary temperatures, with copious evolution of sulphurous oxide, and formation of a large quantity of black carbona-ceous matter. By this reaction cane-sugar may be distinguished from glucose.

Cane-sugar is very easily oxidized. It reduces silver and mercury salts when heated with them, and precipitates gold from the chloride. Pure cupric hydrate is but slowly reduced by it, even at the boiling heat ; in presence of alkali, however, a blue solution is formed, and on boiling the liquid, cuprous oxide is slowly precipitated (P. 650). Cane-sugar takes fire when triturated with 8 parts of lead dioxide, and forms with potas-sium chlorate a mixture which detonates on percussion, and burns vividly when a drop of oil of vitriol is let fall upon it. Distilled with a mixture of sulphuric acid and manganese dioxide, it yields formic acid. Heated with dilute nitric acid, it yields saccharic and oxalic acids. 1 part sugar mixed with 3 parts nitric acid, of specific gravity 1.25 to 1.30, and heated to 50° C. (122° F.), is wholly converted into saccharic acid :

$$C_{12}H_{22}O_{11} \;+\; O_6 \;=\; 2C_6H_{10}O_8 \;+\; H_2O \,.$$

---

* Under the names *ulmin* and *ulmic acid* have been confounded a number or brown or black uncrystallizable substances produced by the action of powerfu chemical agents upon sugar, lignin, etc., or generated by the putrefactive deca) of vegetable fibre. Common garden mould, for example, treated with dilute, boil ing solution of caustic potash, yields a deep-brown solution, from which acids precipitate a flocculent, brown substance, having but a slight degree of solu-bility in water. This is generally called *ulmic* or *humic* acid, and its origin is ascribed to the reaction of the alkali on the *ulmin* or *humus* of the soil. These bodies differ exceedingly in composition, and are too indefinite to admit of ready investigation.

At the boiling heat, the product consists chiefly of oxalic acid. Very strong nitric acid, or a mixture of strong nitric and sulphuric acids, converts sugar into nitrosaccharose, probably $C_{12}H_{18}(NO_2)_4O_{11}$. Sugar is likewise oxidized by chloride of lime, but the products have not been examined.

Cane-sugar does not turn brown when triturated with alkalies, a character by which it is distinguished from glucose: it combines with them, however, forming compounds called sucratés. By boiling with potash-lye it is decomposed, but much more slowly than the glucoses.

Potassium- and *Sodium-compounds* of cane-sugar, $C_{12}H_{21}KO_{11}$ and $C_{12}H_{21}NaO_{11}$, are formed, as gelatinous precipitates, on mixing an alcoholic solution of cane-sugar with potash- or soda-lye.

A *Barium-compound*, $C_{12}H_{20}BaO_{11}.H_2O$, or $C_{12}H_{22}O_{11}.BaO$, is obtained, as a crystalline precipitate, on adding hydrate or sulphide of barium to an aqueous solution of sugar. It may be crystallized from boiling water, but is insoluble in alcohol.

Calcium-compounds.—Lime dissolves in sugar-water much more rapidly than in pure water. The solution has a bitter taste, and is completely but slowly precipitated by carbonic acid. There are three or four of these compounds, which may be approximately represented by the following formulæ :—

1. $C_{12}H_{22}O_{11}.CaO.$
2. $2C_{12}H_{22}O_{11}.3CaO$ (?)
3. $C_{12}H_{32}O_{11}.2CaO.2H_2O.$
4. $C_{12}H_{22}O_{11}.3CaO.$

Magnesia and *lead oxide* are also dissolved by sugar-water. A crystalline lead-compound, $C_{12}H_{18}Pb_2O_{11}$, is precipitated on mixing sugar-water with neutral lead-acetate and ammonia.

Sugar also forms crystalline compounds with *sodium chloride.*

Cane-sugar is not directly fermentable, but when its dilute aqueous solution is mixed with yeast, and exposed to a warm atmosphere, it is first resolved into a mixture of dextrose and levulose, which then enter into fermentation, yielding alcohol and carbon dioxide.

Parasaccharose, $C_{12}H_{22}O_{11}$.—This is an isomeride of cane-sugar, produced, according to Jodin,* by spontaneous fermentation. An aqueous solution of cane-sugar containing ammonium phosphate, left to itself for three months in summer, yielded, under circumstances not further specified, a crystalline sugar, isomeric with saccharose, together with an amorphous sugar having the composition of a glucose, both dextro-rotatory. Parasaccharose is very soluble in water, nearly insoluble in alcohol of 90 per cent. Its specific rotatory power at 10° C. (50° F.) $= + 108^O$, appearing to increase a little with rise of temperature. It does not melt at 100°, but becomes colored, and appears to decompose. It reduces an alkaline cupric solution, but only half as strongly as dextro-glucose. It is not perceptibly altered by dilute sulphuric acid, even at 100°; hydrochloric acid weakens its rotatory power, turns the solution brown, and heightens its reducing power for cupric oxide.

Melitose, $C_{12}H_{22}O_{11}$.—A kind of sugar obtained from the manna which falls in opaque drops from various species of *Eucalyptus* growing in Tasmania. It is extracted by water, and crystallizes in extremely thin interlaced needles, having a slightly saccharine taste.

The crystals of melitose are hydrated, containing $C_{12}H_{22}O_{11}.3H_2O$. They

* Comptes Rendus, liii. 1252; liv. 720.

give off 2 molecules of water at 100° C., and become anhydrous at 130° C. (266° F.). They dissolve in 9 parts of cold water, very easily in boiling water, and dissolves also in boiling alcohol more freely than mannite. The alcoholic solution yields small but well-developed crystals. The aqueous solution turns the plane of polarization to the right : for the transition tint $[a] = + 102°$.

Melitose, heated with dilute sulphuric acid, is resolved into a fermentable sugar (probably dextrose), and non-fermentable eucalyn (p. 652). Melitose ferments in contact with yeast, but is resolved, in the first instance, into glucose and eucalyn. It does not reduce an alkaline cupric solution, and is not altered by boiling with dilute alkalies or with barytawater. It is oxidized by nitric acid, yielding a certain quantity of mucic acid, together with a large quantity of oxalic acid.

Melezitose, $C_{12}H_{22}O_{11}$.—This variety of sugar is found in the so-called manna of Briançon, which exudes from the young shoots of the larch (*Larix europœa*). The manna is exhausted with alcohol, which, when evaporated, yields melezitose in very small, hard, shining, efflorescent crystals, which give off 4 per cent. of water when heated, and melt below 140° C. (284° F.) without further alteration, forming a liquid which solidifies to a glass on cooling. Melezitose is dextro-rotatory ; $[a] = + 94.1°$. It dissolves easily in water, is nearly insoluble in cold, slightly soluble in boiling alcohol.

Melezitose decomposes at about 200° C. (392° F.). It is carbonized by cold strong sulphuric acid, quickly turns brown with boiling hydrochloric acid, and forms oxalic acid with nitric acid. By an hour's boiling with dilute sulphuric acid, it is converted into glucose. In contact with yeast, it passes slowly, or sometimes not at all, into vinous fermentation. It is not altered at 100° C. by aqueous alkalies, and scarcely by potassio-cupric tartrate.

Trehalose, $C_{12}H_{22}O_{11}.2H_2O$, is obtained from *Trehala manna*, the produce of a species of Echinops growing in the East, by extraction with boiling alcohol. It forms shining rhombic crystals, containing $C_{12}H_{22}O_{11}.2H_2O$, which melt when quickly heated to 109° C. (228.2° F.) ; but if slowly heated give off their water even below 100°. It has a strongly saccharine taste, dissolves easily in water and in boiling alcohol, but is insoluble in ether. The aqueous solution is dextro-rotatory ; $[a] = + 199°$.

By several hours' boiling with dilute sulphuric acid, it is converted into dextrose. With strong nitric acid it forms a detonating nitro-compound ; heated with dilute nitric acid it yields oxalic acid. In contact with yeast it passes slowly and imperfectly into alcoholic fermentation. It is not altered by boiling with alkalies, and does not reduce cuprous oxide from alkaline cupric solutions. Heated with acetic or butyric acid, it yields compounds not distinguishable from those which are formed in like manner from dextrose (p. 649).

Mycose, $C_{12}H_{22}O_{11}.2H_2O$, is a kind of sugar, very much like trehalose, obtained from ergot of rye by precipitating the aqueous extract of the fungus with basic lead acetate, removing the lead from the filtrate by sulphydric acid, evaporating to a syrup, and leaving the liquid to crystallize. It differs from trehalose only in possessing a somewhat feebler rotatory power ; $[a] = + 192.5°$, and in not being completely dehydrated at 100°.

Synanthrose, $C_{12}H_{22}O_{11}$, occurs in the tubers of *Dahlia variabilis*, of the Jerusalem artichoke (*Helianthus tuberosus*), and other plants of the compo-

site or synanthraceous order. It is a light amorphous powder, very deliquescent, not sweet, and without action on polarized light. It is resolved
by dilute acids or by yeast into dextrose and levulose, and yields saccharic
acid by oxidation.

Milk-sugar, Lactin, or **Lactose,** $C_{12}H_{22}O_{11}.H_2O$.—This kind
of sugar is an important constituent of milk : it is obtained in large quantities by evaporating *whey* to a syrupy state, and purifying the lactose,
which slowly crystallizes out, with animal charcoal. It forms white,
translucent, four-sided, trimetric prisms, of great hardness. It is slow and
difficult of solution in cold water, requiring for that purpose five or six times
its weight. Its specific rotatory power in aqueous solution is $[a] = + 58.3^{\circ}$.
It has a faint, sweet taste, and in the solid state feels gritty between the
teeth. When heated, it loses water, and at a high temperature blackens
and decomposes. Milk-sugar combines with bases, forming compounds
which have an alkaline reaction, and are easily decomposed. Dilute acids
convert it into galactose (p. 641).

Milk-sugar, when distilled with oxidizing mixtures, such as sulphuric
acid and manganese dioxide, yields formic acid. With nitric acid, it forms
mucic, saccharic, tartaric, and a small quantity of racemic acid, and finally
oxalic acid. Very strong nitric acid, or a mixture of nitric and sulphuric
acids, converts milk-sugar into a crystalline substitution-product called
nitrolactin.

Milk-sugar is not brought immediately by yeast into the state of alcoholic
fermentation ; but when it is left for some time in contact with yeast, fermentation gradually sets in. When cheese or gluten is used as the ferment,
the milk-sugar is converted into lactic acid. Alcohol is, however, always
formed at the same time, especially if no chalk is added to neutralize the
acid as it forms ; the quantity of alcohol formed is greater also as the
solution is more dilute.

Maltose.—This is a sugar isomeric with lactose, produced by the
action of malt-extract on starch. It is less soluble in alcohol than dextrose, has a reducing power about two-thirds as great as that of dextrose,
and a specific rotatory power of $+ 56^{\circ}$. By the prolonged action of acids
it is converted into dextrose.

OXYGEN-ETHERS OR ANHYDRIDES OF THE POLY-GLUCOSIC ALCOHOLS—
AMYLOSES.

These compounds, which are important constituents of the vegetable
organism, may be derived from glucose and the poly-glucosic alcohols by
abstraction of a molecule of water:

$$C_6H_{12}O_6 \quad - \quad H_2O \quad = \quad C_6H_{10}O_5,$$
Glucose.

$$C_{12}H_{22}O_{11} \quad - \quad H_2O \quad = \quad C_{12}H_{20}G_{10}, \text{ or } 2C_6H_{10}O_5,$$
Diglucosic
alcohol.

$$C_{18}H_{32}O_{16} \quad - \quad H_2O \quad = \quad C_{18}H_{30}O_{15}, \text{ or } 3C_6H_{10}O_5,$$
Triglucosic
alcohol.

$$\vdots$$

$$C_6H_{10n+2}O_{5n+1} \quad - \quad H_2O \quad = \quad C_{6n}H_{10n}O_{5n}, \text{ or } nC_6H_{10}O_5.$$

28 *

All these bodies are therefore isomeric or polymeric one with the other. Their compounds with metallic oxides, etc., have not been sufficiently investigated to fix their exact molecular weight, or to determine in each case the value of n; but from the mode of conversion of starch into glucose, and the constitution of certain substitution-products obtained by the action of nitric acid on cellulose, it appears most probable that in these bodies $n = 3$.

Starch, $nC_6H_{10}O_5$, probably $C_{18}H_{30}O_{15}$, also called *Fecula* and *Amidin*. —This is one of the most important and widely diffused of the vegetable proximate principles, being found to a greater or less extent in every plant. It is most abundant in certain roots and tubers, and in soft stems : seeds often contain it in large quantity. From these sources the starch can be obtained by rasping or grinding the vegetable structures to pulp, and washing the mass upon a sieve, by which the torn cellular tissue is retained, while the starch passes through with the liquid, and eventually settles down from the latter, as a soft, white, insoluble powder, which may be washed with cold water, and dried at a very gentle heat. Potatoes treated in this manner yield a large proportion of starch. Starch from grain may be prepared in the same manner, by mixing the meal with water to a paste, and washing the mass upon a sieve : a nearly white insoluble substance called *gluten* is then left, containing a large proportion of nitrogen. The gluten of wheat flour is extremely tenacious and elastic. The value of meal as an article of food greatly depends upon this substance. Starch from grain is commonly manufactured on the large scale by steeping the material in water for a considerable time, when the lactic acid, always developed under such circumstances from the sugar of the seed, disintegrates, and in part dissolves the azotized matter, thereby greatly facilitating the mechanical separation of that which remains. A still more easy and successful process has lately been introduced, in which a very dilute solution of caustic soda, containing about 200 grains of alkali to a gallon of liquid, is employed with the same view. Excellent starch is thus prepared from rice. Starch is insoluble in cold water, as indeed its mode of preparation sufficiently shows ; it is equally insoluble in alcohol and other liquids, which do not effect its decomposition. To the naked eye it presents the appearance of a soft, white, and often glistening powder : under the microscope it is seen to be altogether destitute of crystalline structure, but to possess, on the contrary, a kind of organization, being made up of multitudes of little rounded transparent bodies, upon each of which a series of depressed parallel rings, surrounding a central spot or hilum, may often be traced. The starch-granules from different plants vary both in magnitude and form : those from the *Canna coccinea*, or *tous les mois*, and potato being the largest ; and those from wheat, and the cereals in general, very much smaller. Figure 165 represents granules of potato-starch, highly magnified.

Fig. 165.

When a mixture of starch and water is heated to near the boiling point of the latter, the granules burst and disappear, producing, if the proportion of starch is considerable, a thick gelatinous mass, very slightly opalescent, from the shreds of fine membrane, the envelope of each separate granule. By the addition of a large quantity of water, this gelatinous starch, or *amidin*, may be so far diluted as to pass in great measure through filter-paper. It is very doubtful, however, how far the substance itself is really soluble in water, at least when cold; it is more likely to be merely suspended in the liquid in the form of a swollen, transparent, and insoluble jelly, of extreme tenuity. Gelatinous starch, exposed in a thin layer to a dry atmosphere, becomes converted into a yellowish, horny subtance, like gum, which, when put into water, again softens and swells.

Thin gelatinous starch is precipitated by many of the metallic oxides, as lime, baryta, and lead oxide; also by a large addition of alcohol. *Infusion of galls* throws down a copious yellowish precipitate containing tannic acid, which re-dissolves when the solution is heated. By far the most characteristic reaction, however, is that with free *iodine*, which forms with starch a deep indigo-blue compound, which appears to dissolve in pure water, although it is insoluble in solutions containing free acid or saline matter. The color of the blue liquid is destroyed by heat, temporarily if the heat be quickly withdrawn, and permanently if the boiling be long continued, in which case the compound is decomposed, and the iodine volatilized. Dry starch, put into iodine-water, acquires a purplish-black color.

The unaltered and the gelatinous starch, when dry, have the same empirical formula, $C_6H_{10}O_5$. A compound of starch and lead oxide was found to contain, when dried at 100°, $C_6H_{10}O_5.PbO$, or $C_{18}H_{30}O_{15}.3PbO$.

Starch is an important article of food, especially when associated, as in ordinary meal, with albuminous substances. Arrow-root, and the fecula of the *Canna coccinea*, are very pure varieties, employed as articles of diet; arrowroot is obtained from the *Maranta arundinacea*, cultivated in the West Indies; it is with difficulty distinguished from potato-starch.—*Tapioca* is prepared from the root of the *Jatropha Manihot*, being thoroughly purified from its poisonous juice.—*Cassava* is the same substance modified while moist by heat.—*Sago* is made from the soft central portion of the stem of a palm; and *Salep* from the fleshy root of the *Orchis mascula*.

STARCH FROM ICELAND MOSS.—The lichen called *Cetraria Islandica*, purified by a little cold solution of potash from a bitter principle, yields, when boiled in water, a slimy and nearly colorless liquid, which gelatinizes on cooling, and dries up to a yellowish amorphous mass, which does not dissolve in cold water, but merely softens and swells. A solution of this substance in warm water is not affected by iodine, although the jelly is turned blue. It is precipitated by alcohol, lead acetate, and infusion of galls, and is converted into glucose by boiling with dilute sulphuric acid. According to Mulder, it contains $C_6H_{10}O_5$. The jelly from certain *algæ*, as that of Ceylon, and the so-called *Carragheen moss*, closely resembles the above.

INULIN.—This substance which differs from common starch in some important particulars, is found in the root of *Inula Helenium, Helianthus tuberosus, Dahlia*, and several other plants: it may be easily obtained by washing the rasped root on a sieve, and allowing the inulin to settle down from the liquid; or by cutting the root into thin slices, boiling these in water, and filtering while hot; the inulin separates as the solution cools. It is a white, amorphous, tasteless substance, nearly insoluble in cold water, but freely dissolved by the aid of heat; the solution is precipitated by alcohol, but not by acetate of lead or infusion of galls. Iodine colors it brown. Inulin has the same percentage composition as common starch. By boiling with dilute acids, it is completely converted into levulose (p. 650).

Glycogen, $nC_6H_{10}O_5$, was obtained by Bernard from the liver of several animals (calf or pig) by exhaustion with water and precipitating with boiling alcohol. The precipitate is purified by boiling with dilute potash, repeatedly dissolving in strong acetic acid, and precipitating by alcohol. Glycogen also enters largely into the composition of most of the tissues of the embryo. The muscles of fœtal calves of three to seven months have been found to yield from 20 to 50 per cent. of it.

Glycogen is a white, amorphous, starch-like substance, without odor or taste, yielding an opalescent solution with water, but insoluble in alcohol. It does not reduce an alkaline solution of copper. This substance does not ferment with yeast, but is converted into glucose by boiling with dilute acids, or by contact with diastase, pancreatic juice, saliva, or blood.

Dextrin, $C_6H_{10}O_5$. When gelatinous starch is boiled with a small quantity of dilute sulphuric, hydrochloric, or, indeed, almost any acid, it speedily loses its consistency, and becomes thin and limpid, from having suffered conversion into a soluble gum-like substance, called dextrin, on account of its dextro-rotatory action on polarized light. The experiment is most conveniently made with sulphuric acid, which may be afterwards withdrawn by saturation with chalk. The liquid filtered from the nearly insoluble gypsum may then be evaporated to dryness on a water-bath. The result is a gum-like mass, destitute of crystalline structure, soluble in cold water, precipitable from its solution by alcohol, and capable of combining with lead oxide.

When the boiling with the dilute acid is continued for a considerable time, the dextrin first formed undergoes a further change, and becomes converted into dextro-glucose, which can be thus artificially produced with the greatest facility. The length of time required for this change depends upon the quantity of acid present; if the latter be very small, it is necessary to continue the boiling many successive hours, replacing the water which evaporates. With a larger proportion of acid, the conversion is much more speedy. A mixture of 15 parts of potato-starch, 60 parts water, and 6 parts sulphuric acid, may be kept boiling for about four hours, the liquid neutralized with chalk, filtered, and rapidly evaporated to a small bulk. By digestion with animal charcoal and a second filtration, much of the color will be removed, after which the solution may be boiled down to a thin syrup, and left to crystallize: in the course of a few days it solidifies to a mass of glucose.

There is another method of preparing this substance from starch which deserves particular notice. Germinating seeds, and buds in the act of development, are found to contain a small quantity of a peculiar azotized substance, called d i a s t a s e, formed at this particular period from the gluten of vegetable albuminous matter. This substance likewise converts starch into dextrin and glucose, and at a temperature much below the boiling point. When a little infusion of malt, or germinated barley, in tepid water, is mixed with a large quantity of thick gelatinous starch, and the whole kept at about 71° C. (160° F.), complete liquefaction takes place in the space of a few minutes, from the production of dextrin and glucose. If a greater degree of heat be employed, the diastase is coagulated and rendered insoluble and inactive. Very little is known respecting diastase itself; it seems very much to resemble vegetable albumin, but has never been obtained in a state of purity.

The change of starch or dextrin into sugar, whether produced by the action of dilute acids or by diastase, takes place quite independently of the oxygen of the air, and is unaccompanied by any secondary product. The acid takes no direct part in the reaction; it may, if not volatile, be all withdrawn without loss after the experiment. The whole reaction lies between the starch and the elements of water, a fixation of the latter occurring in the new product, as will be seen on comparing the composition of starch and glucose. Dextrin itself has exactly the same composition as the original starch.

It was formerly supposed that, in the action of acids [or of diastase] upon starch, the starch is first converted into dextrin by a mere alteration of physical structure, and that the dextrin then takes up the elements of water, and is converted into glucose, this second stage of the process occupying a much longer time than the first ; but from the experiments of Musculus* it appears that when the conversion is effected by a dilute acid, both dextrin and glucose are produced at the very commencement of the reaction, and always in the proportion of 1 molecule of glucose to 2 molecules of dextrin : whence it may be inferred that the molecule of starch contains $C_{18}H_{30}O_{15}$, and that it is resolved into glucose and dextrin by taking up a molecule of water :

$$\underset{\text{Starch.}}{C_{18}H_{30}O_{15}} + H_2O = \underset{\text{Glucose.}}{C_6H_{12}O_6} + \underset{\text{Dextrin.}}{2C_6H_{10}O_5}$$

and that the dextrin, after several hours' boiling, is completely converted into glucose, which is therefore the sole ultimate product of the reaction. When malt extract is used as the converting agent the starch is first resolved into dextrin and maltose (p. 657), in various proportions according to the temperature and other conditions of the reaction ; and the dextrin is afterwards very gradually converted into maltose.†

Dextrin is used in the arts as a substitute for gum ; it is sometimes made in the manner above described, but more frequently by heating dry potato-starch to 400° C. (752° F.), by which it acquires a yellowish tint and becomes soluble in cold water. It is sold in this state under the name of *British Gum*.

Cellulose, $nC_6H_{10}O_5$, probably $C_{18}H_{30}O_{15}$; also called *Lignin.*—This substance constitues the fundamental material of the structure of plants : it is employed in the organization of cells and vessels of all kinds, and forms a large proportion of the solid parts of every vegetable. It must not be confounded with *ligneous* or *woody tissue*, which is in reality cellulose with other substances superadded, encrusting the walls of the original membraneous cells, and conferring stiffness and inflexibility. Pure cellulose, on the other hand, has the same percentage composition as starch ; but woody tissue, even when freed as much as possible from coloring matter and resin by repeated boiling with water and alcohol, yields, on analysis, a result indicating an excess of hydrogen above that required to form water with the oxygen, besides traces of nitrogen.

The properties of cellulose may be conveniently studied in fine linen and cotton, which are almost entirely composed of it, the associated vegetable principles having been removed or destroyed by the variety of treatment to which the fibre has been subjected. Pure cellulose is tasteless, insoluble in water and alcohol, and absolutely innutritious : it is not sensibly affected by boiling water, unless it happens to have been derived from a soft or imperfectly developed portion of the plant, in which case it is disintegrated and rendered pulpy. Dilute acids and alkalies exert but little action on the cellulose, even at a boiling temperature ; strong oil of vitriol converts it, in the cold, into a nearly colorless, adhesive substance, which dissolves in water, and presents the characters of dextrin. This transformation may be conveniently effected by very slowly adding concentrated sulphuric acid to half its weight of lint, or linen cut into small shreds, taking care to avoid any rise of temperature, which would be

* Comptes Rendus, l. 785 ; liv. 194 ; Ann. Ch. Phys. [3], lx 208 ; [4], v 177
† O'Sullivan, Chem. Soc. J. 1876, ii. 125.

attended with charring or blackening. The mixing is completed by tritu‑ration in a mortar, and the whole left to stand a few hours; after which it is rubbed up with water, warmed, and filtered from a little insoluble matter. The solution may then be neutralized with chalk, and again fil‑tered. The gummy liquid retains lime, partly in the state of sulphate, and partly in combination with sulpholignic acid, an acid composed of the elements of sulphuric acid in union with those of cellulose. If the liquid, previous to the neutralization, be boiled during three or four hours, and the water replaced as it evaporates, the dextrin becomes entirely changed into glucose. Linen rags may, by these means, be made to furnish more than their own weight of that substance.

If a piece of unsized paper be dipped for a few seconds into a mixture of 2 volumes of concentrated sulphuric acid and 1 volume of water, and then thoroughly washed with water and dilute ammonia, a substance is obtained which resembles parchment, and has the same composition as cellulose; it occurs in commerce under the name of parchment paper (papyrin). An excellent application of this substance in diffusion experi‑ments is mentioned on p. 161.

Cellulose dissolves in an ammoniacal solution of cupric oxide (prepared by dissolving basic cupric carbonate in strong ammonia), from which it is precipitated by acids in colorless flakes.

Cellulose is not colored by iodine.

XYLOÏDIN and PYROXYLIN.—When the starch is mixed with nitric acid of specific gravity 1.5, it is converted, without disengagement of gas, into a transparent, colorless jelly, which, when put into water, yields a white, curdy, insoluble substance: this is *xyloïdin*. When dry, it is white and tasteless, insoluble even in boiling water, but freely dissolved by dilute nitric acid, and the solution yields oxalic acid when boiled. Other sub‑stances belonging to the same class also yield xyloïdin; paper dipped into the strongest nitric acid, quickly plunged into water, and afterwards dried, becomes in great part so changed: it assumes the appearance of parch‑ment, and becomes highly combustible.

If pure finely divided cellulose, such as cotton-wool, be steeped for a few minutes into a mixture of nitric acid of sp. gr. 1.5 and concentrated sul‑phuric acid, then squeezed, thoroughly washed, and dried by very gentle heat, it will be found to have increased in weight about 70 per cent., and to have become highly explosive, taking fire at a temperature not much above $149°$ C. ($300°$ F.), and burning without smoke or residue. This is *pyroxylin*, the *gun-cotton* of Schönbein.

Xyloïdin and pyroxylin are substitution-products consisting of starch and cellulose in which the hydrogen is more or less replaced by nitryl, NO_2. Xyloïdin consists of $C_6H_9(NO_2)O_5$, or $C_{18}H_{27}(NO_2)_3O_{15}$. Of pyroxylin several varieties are known, distinguished by their different degrees of stability and solubility in alcohol, ether, and other liquids. According to Hadow,* the three principal varieties are:—

α.—$C_{18}H_{21}(NO_2)_9O_{15}$, or $C_6H_7(NO_2)_3O_5$, insoluble in a mixture of ether and alcohol, but soluble in ethylic acetate. It is produced by repeated immersion of cotton-wool in a mixture of 2 molecules of nitric acid, HNO_3, 2 molecules of oil of vitriol, H_2SO_4, and 3 molecules of water.

β.—$C_{18}H_{22}(NO_2)_8O_{15}$, soluble in ether-alcohol, insoluble in glacial acetic acid. Produced when the acid mixture contains half a molecule more water than in α.

γ.—$C_{18}H_{23}(NO_2)_7O_{15}$ (Gladstone's *cotton-xyloïdin*), soluble in ether and in

* Chem. Soc. Journal, vii. 201.—A series of elaborate and valuable researches on gun-cotton has been published by Abel (Proceed. Royal Soc.) xv. 182; Chem. Soc. Journ. [2], xv. 310.

glacial acetic acid. Produced when the acid mixture contains one molecule more water than in *a*.

The first of these, which consist of *trinitrocellulose*, is the most explosive of the three, and the least liable to spontaneous decomposition. It is the only one adapted for use as an explosive agent, and is especially distinguished as "gun-cotton." From the experiments of General von Lenk, of the Austrian service, it appears that, to insure the uniform production of this particular compound, the following precautions are necessary:—

1. The cleansing and perfect desiccation of the cotton previously to its immersion in the mixed acids. 2. The employment of the strongest acids procurable in commerce. 3. The steeping of the cotton in a fresh strong mixture of acids after the first immersion and partial conversion into gun-cotton. 4. The continuance of the steeping for forty-eight hours. 5. The thorough purification of the gun-cotton thus produced from every trace of free acid, by washing the product in a stream of water for several weeks; subsequently a weak solution of potash may be used, but this is not essential.

The solution of the less highly nitrated compounds in alcohol and ether is called collodion.. This solution, when left to evaporate, dries up quickly to a thin, transparent, adhesive membrane: it is employed with great advantage in surgery as an air-tight covering for wounds and burns. It is also largely used in photography (p. 103).

Tunicin, $C_6H_{10}O_5$, is a substance closely resembling vegetable cellulose, and perhaps identical therewith, occurring in the mantle of Ascidians (*Cynthia, Pallusia*, etc.), from which it may be separated by boiling the mantles, first with hydrochloric and then with strong potash-solution, and washing with water. When treated with strong sulphuric acid, it deliquesces to a colorless liquid, which, after boiling with water, is found to contain a glucose.

Gum.—*Gum-arabic,* which is the produce of several species of acacia growing in Egypt and Arabia, may be taken as the type of this class of bodies. When pure, it forms white or slightly yellowish irregular masses, which are destitute of crystalline structure, and break with a smooth conchoïdal fracture. It is soluble in cold water, forming a viscid, adhesive, tasteless solution. It consists of the potassium and calcium salts of a r a-b i n or a r a b i c a c i d, which may be precipitated from the aqueous solution by addition of hydrochloric acid and alcohol, as a white amorphous mass. Arabin, when dried at 100° C., has the composition $C_6H_{10}O_5 + \frac{1}{2}H_2O$, or $2C_6H_{10}O_5 + H_2O$, and is therefore isomeric with cane-sugar; at 150° C. (302° F.), it gives off all its water, leaving a residue having the composition $C_6H_{10}O_5$.

Gum-arabic contains 70.4 per cent. arabin and 17.6 water, the remaining 12 per cent. consisting of metallic salts, silica, iron oxide, etc. Its aqueous solution turns the plane of polarization to the left, and gives precipitates with basic lead acetate, copper salts, etc. Arabin is oxidized by nitric acid to mucic and saccharic acids. Heated with dilute sulphuric acid, it is converted into a non-fermentable glucose called a r a b i n o s e, which crystallizes in rhombic needles, and exhibits strong dextro-rotation.

Gum Senegal, obtained from a species of acacia growing in Senegal, is very much like gum-arabic, but usually occurs in larger lumps. It contains 81.1 per cent. arabin, 1.6 water, and 2 or 3 per cent. of saline matters. It forms a somewhat stronger mucilage than gum-arabic, and is much used in calico-printing for thickening colors and mordants.

Mucilage, abundant in linseed, in the roots of the mallow, in *salep,* the fleshy root of *Orchis mascula,* and in other plants, differs in some respects from gum-arabic, although it agrees in the property of dissolving in cold water. The solution is less transparent than that of gum, and is precipitated by neutral lead acetate. *Gum-tragacanth*, from *Astragalus verus*, a tree growing in Armenia and the north of Persia, is chiefly composed of a kind mucilage called *bassorin;* it does not dissolve in water, but merely softens and assumes a gelatinous aspect. It is dissolved by caustic alkali. *Cerasin* is the insoluble portion of the gum of the cherry tree ; it resembles bassorin. The composition of these various substances agrees closely with that of starch. Mucilage treated with acids yields dextrose.

AMINES.

These are compounds which may be derived from hydrocarbons by substitution of the univalent radicle amidogen, NH_2, for an equivalent quantity of hydrogen, or from the alcohols by similar substitution of NH_2 for OH : they are called monamines, diamines, triamines, etc., according to the number of amidogen groups thus introduced ; thus from ethane and the corresponding alcohols are derived the following monamine and diamine :

$$
\begin{array}{ccc}
CH_3 & CH_3 & CH_3 \\
| & | & | \\
CH_3 & CH_2(NH_2) & CH_2(OH) \\
\text{Ethane.} & \text{Amidethane or} & \text{Ethyl alcohol.} \\
& \text{Ethylamine.} &
\end{array}
$$

$$
\begin{array}{ccc}
CH_3 & CH_2.NH_2 & CH_2.OH \\
| & | & | \\
CH_3 & CH_2.NH_2 & CH_2.OH \\
\text{Ethane.} & \text{Diamidethane or} & \text{Ethene-alcohol.} \\
& \text{Ethane-diamine.} &
\end{array}
$$

The amines are basic compounds, capable of uniting with acids **and** forming salts which bear a close resemblance to the salts of ammonia ; the amines themselves in the free state are also very much like ammonia, being volatile bodies having a more or less ammoniacal odor and alkaline reaction. They may, in fact, be regarded as derivatives of ammonia, formed by substitution of alcohol radicles, univalent or multivalent, for an equivalent quantity of hydrogen, in a single, double, or triple molecule of ammonia, NH_3, N_2H_6, N_3H_9, etc., *e. g.* :

$$
\text{Ethylamine,} \quad
\begin{array}{c}
CH_3 \\
| \\
CH_2.NH_2
\end{array}
= N
\begin{cases}
H \\
H \\
C_2H_5
\end{cases}
$$

$$
\text{Ethene-diamine,} \quad
\begin{array}{c}
CH_2.NH_2 \\
| \\
CH_2.NH_2
\end{array}
= N_2
\begin{cases}
H_2 \\
H_2 \\
(C_2H_4)''
\end{cases}
$$

$$
\text{Diethene-triamine,} \quad
\begin{array}{c}
C_2H_3(NH_2)_2 \\
| \\
C_2H_4.NH_2
\end{array}
= N_3
\begin{cases}
H_3 \\
(C_2H_4)'' \\
(C_2H_4)''
\end{cases}
$$

MONAMINES.

These bases are derived from a single molecule of ammonia, NH_3. Now this molecule may give up one, two, or all three of its hydrogen-atoms in exchange for univalent alcohol-radicles,—methyl and its homologous, for example,—producing p r i m a r y , s e c o n d a r y , and t e r t i a r y amines. If A, B, C denote three such alcohol-radicles, the amines formed by substituting them for hydrogen in ammonia will be represented by the general formulæ:

$$N\begin{Bmatrix} A \\ H \\ H \end{Bmatrix} \qquad N\begin{Bmatrix} A \\ B \\ H \end{Bmatrix} \qquad N\begin{Bmatrix} A \\ B \\ C \end{Bmatrix}$$

Primary.   Secondary.   Tertiary.

In the secondary and tertiary amines the alcohol-radicles denoted by A, B, C may be either the same or different; for example:

Secondary.   Tertiary.

$$N\begin{Bmatrix} CH_3 \\ CH_3 \\ H \end{Bmatrix} \quad N\begin{Bmatrix} CH_3 \\ C_2H_5 \\ H \end{Bmatrix} \quad N\begin{Bmatrix} CH_3 \\ CH_3 \\ CH_3 \end{Bmatrix} \quad N\begin{Bmatrix} CH_3 \\ CH_3 \\ C_2H_5 \end{Bmatrix} \quad N\begin{Bmatrix} CH_3 \\ C_2H_5 \\ C_5H_{11} \end{Bmatrix}$$

Dimethyl-   Methyl-   Trimethyl-   Dimethyl-   Methyl-ethylamine.   ethylamine.   amine.   ethylamine.   amylamine.

The salts of these amines are analogous in composition to the ammonium-salts, and, like the latter, may be regarded either as compounds of ammonia-molecules with acids, or of ammonium-molecules with halogen elements and acid radicles analogous thereto (see p. 508) ; thus :

NH_3 $+$ HCl $=$ $NH_4.Cl$,   Ammonium chloride.
Ammonia.

$NH_2(C_2H_5)$ $+$ HCl $=$ $NH_3(C_2H_5).Cl$,   Ethylammonium chloride.
Ethylamine.

$NH(C_2H_5)_2$ $+$ HCl $=$ $NH_2(C_2H_5)_2.Cl$,   Diethylammonium chloride.
Diethylamine.

$N(C_2H_5)_3$ $+$ HCl $=$ $NH(C_2H_5)_3.Cl$,   Triethylammonium chloride.
Triethylamine.

$2N(C_2H_5)_3$ $+$ H_2SO_4 $=$ $[NH(C_2H_5)_3]_2SO_4$, Triethylammonium sulphate.
Triethylamine.

All these salts when heated with potash, give off the amine, just as ammonia-salts give off ammonia.

The tertiary amines can unite with the chlorides, etc., of alcohol-radicles in the same manner as with acids : thus triethylamine, $N(C_2H_5)_3$, unites directly with ethyl iodide, C_2H_5I, forming a compound which may be regarded either as *triethylamine ethiodide*, $N(C_2H_2)_3.C_2H_5I$, or as *tetrethylammonium iodide*, $N(C_2H_5)_4I$. Now this iodide, when heated with potash, does not give off ammonia or a volatile ammonia-base ; but when heated with silver oxide and water, it is converted, by exchange of iodine for hydroxyl, into a strongly alkaline base, called *tetrethylammonium hydroxide*, which may be obtained in the solid state, and exhibits reactions closely analogous to those of the fixed caustic alkalies. Its formation is represented by the equation :

$$N(C_2H_5)_4I \quad + \quad AgOH \quad = \quad AgI \quad + \quad N(C_2H_5)_4(OH).$$

Moreover, this base can exchange its hydroxyl for chlorine, bromine, and other acid radicles, just like potash or soda, forming solid crystallizable salts like the iodide above mentioned. These compounds, containing four equivalents of alcohol-radicle, are, in fact, analogous in every respect to ammonium-salts, excepting that the corresponding hydroxides are capable of existing in the solid state, whereas ammonium hydroxide, $NH_4(OH)$, splits up, as soon as formed, into ammonia and water. The radicles $N(C_2H_5)_4$, etc., corresponding with ammonium, are not known in the free state.

The monamines containing more than one carbon-atom are susceptible of isomeric modifications similar to those of the alcohols, and depending on the number of alcohol-radicles in the molecule: thus ethylamine, $CH_2(CH_2H_5)$, is isomeric with dimethylamine, $NH(CH_3)_2$; propylamine, $NH_2(C_3H_7)$, is isomeric with methyl-ethylamine, $NH(CH_3)(C_2H_5)$, and with trimethylamine, $N(CH_3)_3$, etc. etc., the number of possible modifications of course increasing with the complexity of the molecules. Moreover, a monamine, either primary, secondary, or tertiary, may admit of modification in the alcohol-radicle itself; thus the primary monamine, $NH_2(C_3H_7)$, may exhibit the two following modifications :—

$$N\begin{cases}CH_2CH_2CH_3\\H\\H\end{cases} \qquad N\begin{cases}CH(CH_3)_2\\H\\H\end{cases}$$
$$\text{Propylamine.} \qquad\qquad \text{Isopropylamine.}$$

General Modes of Formation.—1. By heating the iodides or bromides of the alcohol-radicles to 100° in sealed tubes with alcoholic ammonia. The hydrogen of the ammonia is then replaced by the alcohol-radicle, forming a mixture of primary, secondary, and tertiary amines, which unite with the halogen-acid, produced at the same time, to form ammonium salts, thus :—

$$NH_3 \quad + \quad \underset{\text{Ethyl iodide.}}{C_2H_4I} \quad = \quad \underset{\substack{\text{Ethylammonium}\\\text{iodide.}}}{NH_2(C_2H_5).HI}$$

$$NH_3 \quad + \quad 2C_2H_5I \quad = \quad \underset{\substack{\text{Diethylammonium}\\\text{iodide.}}}{NH(C_2H_5)_2.HI} \quad + \quad HI$$

$$NH_3 \quad + \quad 3C_2H_5I \quad = \quad \underset{\substack{\text{Triethylammonium}\\\text{iodide.}}}{N(C_2H_5)_3.HI} \quad + \quad 2HI$$

From the mixture of ammonium salts thus obtained, the three amines, being volatile, may be at once separated by distillation with aqueous potash or soda :—

$$NH_2(C_2H_5).HI \quad + \quad KOH \quad = \quad KI \quad + \quad H_2O \quad + \quad \underset{\text{Ethylamine.}}{NH_2(C_2H_5)}$$

$$NH(C_2H_5)_2.HI \quad + \quad .KOH \quad = \quad KI \quad + \quad H_2O \quad + \quad \underset{\text{Diethylamine.}}{NH(C_2H_5)_2}$$

$$N(C_2H_5)_3.HI \quad + \quad KOH \quad = \quad KI \quad + \quad H_2O \quad + \quad \underset{\text{Triethylamine.}}{N(C_2H_5)_3}$$

while the tetrethylammonium iodide remains unaltered, but may be converted into the corresponding hydroxide, $N(C_2H_5)_4OH$, by the action of moist silver oxide in the manner already explained.

The primary, secondary, and tertiary amines cannot be separated by fractional distillation, but their separation may be effected by the following process, devised by Dr. Hofmann. The mixture is treated with diethylic oxalate (oxalic ether), whereby the primary amine is converted into diethyloxamide (see AMIDES), which is easily soluble in water, while the diethylamine is converted into the ethylic ether of diethyloxamic acid, which is insoluble in water, and the triethylamine remains unaltered :—

$$\underset{\text{Ethylamine.}}{2NH_2(C_2H_5)} \quad + \quad \underset{\text{Diethylicoxalate.}}{C_2O_2(OC_2H_5)_2} \quad = \quad \underset{\text{Diethyloxamide.}}{C_2O_2(NH.C_2H_5)_2} \quad + \quad \underset{\text{Alcohol.}}{2C_2H_5OH}$$

$$\underset{\text{Diethylamine.}}{NH(C_2H_5)_2} \quad + \quad C_2O_2(OC_2H_5)_2 \quad = \quad \underset{\substack{\text{Ethylic diethyl-}\\\text{oxamate.}}}{C_2O_2\!<\!\genfrac{}{}{0pt}{}{N(C_2H_5)_2}{OC_2H_5}} \quad + \quad C_2H_5OH$$

On distilling the product of this reaction, the unaltered t r i e t h y l-a m i n e passes over. From the residue, water extracts the diethylox-amide, which is resolved by boiling with potash into oxalic acid and e t h y l a m i n e :—

$$C_2O_2(NH.C_2H_5)_2 + 2KOH = C_2O_4K_2 + 2NH_2(C_2H_5),$$

and the diethyloxamic ether, which is not dissolved by the water, yields, by distillation with potash, d i e t h y l a m i n e :—

$$C_2O_2{<}{}^{N(C_2H_5)_2}_{OC_2H_5} + 2KOH = C_2O_4K_2 + C_2H_5(OH) + NH(C_2H_5)_2 .$$

The primary amines may also be obtained by the following processes :—
2. By the action of nascent hydrogen on the nitroparaffins :—

$$\underset{\text{Nitromethane.}}{CH_3.NO_2} + 3H_2 = 2H_2O + \underset{\substack{\text{Amidomethane,}\\\text{or Methylamine.}}}{CH_3.NH_2}$$

3. By the action of nascent hydrogen on the nitrils (azoparaffins, p. 555):—

$$\underset{\text{Formonitril.}}{HCN} + 2H_2 = \underset{\text{Methylamine.}}{CH_3.NH_2}$$

$$\underset{\text{Acetonitril.}}{CH_3.CN} + 2H_2 = \underset{\text{Ethylamine.}}{CH_3.CH_2.NH_2}$$

4. By distilling an isocyanic or isocyanuric ether with potash-lye :—

$$\underset{\text{Methyl isocyanate.}}{CO{=}N{-}CH_3} + 2KOH = CO_3K_2 + \underset{\text{Methylamine.}}{{}^{H}_{H}{>}N{-}CH_3}$$

This reaction, which is exactly analogous to the formation of ammonia from cyanic acid, $(CO{=}NH + 2KOH = CO_3K_2 + NH_3)$, is that by which the primary amines were first obtained by Wurtz.

Amines are also produced by special reactions, as by the decomposition of more complex nitrogen-compounds and of amido-acids.

Properties and Reactions.—The amines, as already observed, bear a strong resemblance to ammonia. The lower members of the group are gases, easily soluble in water, and having a strong ammoniacal odor ; they are distinguished from ammonia by their ready combustibility. The higher members are liquids, more or less oily. Amines expel ammonia from its salts when heated with them, and in like manner the monamines are re-placed by the diamines, and these by the triamines : hence it might be inferred that the basic power of an amine increases with the number of alcohol-radicles which it contains ; but in this, as in many similar cases, the expulsion of one base or acid by another is a question of relative vola-tility as well as of strength of combination.

Amines form double salts with *platinic chloride, auric chloride,* and other metallic haloïd salts, exactly similar to those of ammonia, *e. g.:*

Ethylammonium Aurochloride   .   .   . $NH_3(C_2H_5)Cl.AuCl_3$
Tetramethylammonium Platinochloride . . $2N(C_2H_5)_4Cl.PtCl_4.$

They can also replace ammonia in the alums, and in the salts of platin-ammonium, cuprammonium, etc. Examples of such compounds are :

Ethylammonium alum   .   .   .   . $(SO_4)_2Al(NH_3.C_2H_5) + 12H_2O$

Platosethylammonium chloride .   .   . $Pt{<}{}^{NH_2(C_2H_5)Cl}_{NH_2(C_2H_5)Cl}$

Platosodiethyldiammonium chloride .   . $Pt\underset{NH_2.NH_3(C_2H_5)Cl}{\overset{NH_2.NH_3(C_2H_5)Cl}{<}}$.

Primary amines treated with *nitrous acid* (or their haloïd salts with potassium nitrite) are converted into the corresponding alcohols; thus:

$$C_2H_5.NH_2 + NO_2H = N_2 + H_2O + C_2H_5.OH.$$

This reaction—analogous to the resolution of ammonium nitrite into ammonia and water ($NH_3 + NO_2H = N_2 + 2H_2O$)—affords, as already observed (p. 607), the means of passing from one alcohol of a series to the next highest: thus methyl alcohol yields methyl cyanide or formonitril, C_2H_3N; this may be converted by nascent hydrogen into ethylamine, C_2H_7N; and this base, as above, into ethyl alcohol.

Secondary amines are converted by nitrous acid into nitroso-compounds:

$$NH(CH_3)_2 + NO.OH = H_2O + N\begin{cases}NO \\ CH_3, \\ CH_3\end{cases}$$

Dimethylamine.   Nitroso-
                                                  dimethylamine.

and tertiary amines are but slightly attacked by nitrous acid.

METHYL BASES.

Methylamine or **Amidomethane**, $CH_5N = H_3C—NH_2 = N\begin{cases}H_2 \\ CH_3\end{cases}$. This base may be formed by either of the general reactions above given, also by the decomposition of certain natural alkaloïds, as morphine, narcotine, and theine; but it is best prepared by heating methyl isocyanate with potash in a retort attached to a receiver cooled by a freezing mixture. The distillate, which is an aqueous solution of methylamine, is saturated with hydrochloric acid, and evaporated to dryness, whereby a crystalline residue is obtained consisting of methylammonium chloride; and this when distilled with dry lime yields methylamine in the form of a gas, which must be collected over mercury.

Methylamine is a colorless gas, having an ammoniacal and slightly fishy odor, a specific gravity of 1.08, and condensing to a liquid at —18° C. (0.4° F.). It is the most soluble of all gases, one volume of water at 12° C. (53.6° F.) absorbing 1040 volumes of the gas: it is likewise very readily absorbed by charcoal. It is distinguished from ammonia by its odor, and by the facility with which it burns. In its deportment with acids and other substances, however, it bears the closest resemblance to ammonia. Its aqueous solution also possesses all the properties of aqueous ammonia, excepting that it does not dissolve the oxides of nickel, cobalt, or cadmium. *Iodine* added to the aqueous solution, throws down a dark-red precipitate of di-iodomethylamine, while methylammonium iodide remains in solution:

$$3(CH_3.NH_2) + 2I_2 = CH_3.NI_2 + 2(CH_3.NH_2.HI).$$

Bromine acts in a similar manner. The gas passed over heated potassium is resolved into cyanogen and hydrogen:

$$CH_3.NH_2 + K = CNK + 5H.$$

The salts of methylamine are easily soluble in water. The *hydrochloride*, $CH_5N.HCl$, crystallizes in large deliquescent laminæ, which melt at 100° C., and distil without decomposition. With platinum chloride it forms a yellow crystalline precipitate of the *platinochloride*, $2(CH_5N.HCl).PtCl_4$, and with auric chloride the double salt, $CH_5N.HCl.AuCl_3$, which crystallizes in needles. Methylamine also forms platinum-bases analogous to the ammonio-platinum compounds (p. 468).

Dimethylamine, $C_2H_7N = NH(CH_3)_2$, isomeric with ethylamine, is formed by heating ammonia with methyl iodide, and separated from simultaneously formed methylamine and trimethylamine by means of oxalic ether (p. 666). It is a gas which dissolves easily in water, and condenses to a liquid below $+ 8^{\circ}$ C. (46.4° F.). Its platinochloride crystallizes in large needles.

Trimethylamine, $C_3H_9N = N(CH_3)_3$, isomeric with propylamine and methyl-ethylamine, is obtained in perfect purity by the action of heat on tetramethyl-ammonium hydroxide, which is thereby completely resolved into trimethylamine and methyl alcohol :

$$N(CH_3)_4OH = N(CH_3)_3 + CH_3(OH).$$

It is also produced by distilling codeine and narcotine with potash, and is contained in large quantity in herring-pickle, the peculiar odor of which is due to its presence. It is a liquid easily soluble in water, and boiling at 9.8° C. (49.6° F.). Its salts are soluble ; the hydrochloride very deliquescent.

Tetramethyl-ammonium compounds.—The *iodide*, $N(CH_3)_4I$, may be obtained by adding methyl iodide to trimethylamine. The two substances unite with a sort of explosion. It is more easily prepared, however, by digesting methyl iodide with an alcoholic solution of ammonia. In this reaction a mixture of the iodides of ammonium, methyl-ammonium, dimethyl-ammonium, trimethyl-ammonium, and tetramethyl-ammonium is produced. The first and last compounds are formed in largest quantity, and may be separated by crystallization, the iodide of tetramethyl-ammonium being but sparingly soluble in water. It crystallizes in hard, flat needles of dazzling whiteness. The *hydroxide*, $N(CH_3)_4OH$, resembles the corresponding ethyl-compound (*infra*), and is decomposed by heat, as above mentioned, into methyl alcohol and trimethylamine.

ETHYL BASES.

Ethylamine or **Amido-ethane,** $C_2H_7N = H_3C—CH_2(NH_2) = N \begin{cases} H_2 \\ C_2H_5 \end{cases}$, is prepared either by heating ethyl isocyanate with potash, or by digesting ethyl bromide or iodide with alcoholic ammonia, and treating the mixture of salts thereby produced with oxalic ether in the manner already described. It is a very mobile liquid, having a specific gravity of 0.6964 at 8° C. (46.4° F.), boiling at 19° C. (66.2° F.), and giving off an inflammable vapor which has a specific gravity of 1.57. It has a most powerful ammoniacal odor, and restores the blue color to reddened litmus-paper. It produces white clouds with hydrochloric acid, is absorbed by water with great avidity, and reacts like ammonia with metallic salts, precipitating the salts of magnesium, aluminium, iron, manganese, bismuth, chromium, uranium, tin, lead, and mercury ; and forming with zinc-salts a white precipitate, which is soluble in excess. It dissolves silver chloride, and yields with copper-salts a blue precipitate, which is soluble in excess of ethylamine.

On adding ethylamine to *oxalic ether*, a white precipitate of d i e t h y l o x - a m i d e, $C_2O_2(NH.C_2H_5)_2$, is produced. Treated with *chlorine*, it yields ethylammonium chloride and d i c h l o r e t h y l a m i n e, $NCl_2C_2H_6$, a yellow liquid which has a penetrating, tear-exciting odor, and when treated with potash is converted into ammonia, potassium acetate, and potassium chloride :

$$NCl_2(C_2H_5) + 3KHO = C_2H_3KO_2 + 2KCl + NH_3 + H_2O.$$

Ethylamine is decomposed by *nitrous acid*, with formation of ethyl nitrite and evolution of nitrogen :

$$C_2H_7N + 2HNO_2 = C_2H_5NO_2 + 2H_2O + N_2.$$

On passing the vapor of *cyanic acid* into a solution of ethylamine, the liquid becomes hot, and deposits, after evaporation, crystals of e t h y l - u r e a : $C_2H_7N + CNHO = C_3H_8N_2O$ or $CH_3(C_2H_3)N_2O$.

The salts of ethylamine resemble those of ammonia and of methylamine. The *hydrochloride*, $NH_2(C_2H_5).HCl$, crystallizes in large deliquescent plates, melting at 80° C. (176° F.) : the *platinochloride*, $2NH_3(C_2H_5)Cl.PtCl_4$, in orange-yellow scales.

Diethylamine, $C_4H_{11}N = NH(C_2H_5)_2$.—A mixture of the solutions of ethylamine and ethyl bromide, heated in a sealed tube for several hours, solidifies to a crystalline mass of diethylammonium bromide : $NH_2(C_2H_5) + C_2H_5Br = NH_2(C_2H_5)_2Br$. This bromide, distilled with potash, yields diethylamine as a colorless liquid, still very alkaline, and soluble in water, but less so than ethylamine. This compound boils at 57° C. (134.6° F.). It forms beautifully crystallizable salts with acids. A solution of diethylammonium chloride forms with platinic chloride a very soluble double salt, $2NH_2(C_2H_5)_2Cl.PtCl_4$, crystallizing in orange-red grains, very different from the orange-yellow leaves of the corresponding ethyl-ammonium salt.

Diethylamine behaves with cyanic acid like ammonia and ethylamine, giving rise to d i e t h y l - u r e a, $CH_3(C_2H_5)_2N_2O$.

Triethylamine, $C_6H_{16}N = N(C_2H_5)_3$.—The formation of this body is perfectly analogous to that of ethylamine and of diethylamine. On heating for a short time a mixture of diethylamine with ethyl bromide in a sealed glass tube, a beautiful fibrous mass of triethylammonium bromide is obtained, from which the triethylamine may be separated by potash. Triethylamine is a colorless, powerfully alkaline liquid, boiling at 89° C. (192.2° F.). Its salts crystallize remarkably well. With platinic chloride it forms a very soluble double salt, $2NH(C_2H_5)_3Cl.PtCl_4$, which crystallizes in magnificent, large, orange-red rhombs.

Tetrethylammonium compounds.—When anhydrous triethylamine is mixed with dry ethyl iodide, a powerful reaction ensues, the mixture boils, and solidifies on cooling to a white crystalline mass of t e t r e t h y l - a m m o n i u m   i o d i d e : $N(C_2H_5)_3 + C_2H_5I = N(C_2H_5)_4I$. This iodide is readily soluble in hot water, from which it separates on cooling in beautiful crystals of considerable size. It is not decomposed by potash, but may be boiled with the alkali for hours without yielding a trace of volatile base. The iodine may, however, be readily removed by treating the solution with silver-salts. If in this case silver sulphate or nitrate be used, we obtain, together with silver iodide, the sulphate or nitrate of tetrethyl-ammonium, which crystallizes on evaporation ; on the other hand, if the iodide be treated with freshly precipitated silver oxide, the h y d r o x i d e of t e t r e t h y l a m m o n i u m, $N(C_2H_5)_4OH$, is separated. On filtering off the silver precipitate, a clear colorless liquid is obtained, which contains the isolated base in solution. It has a strongly alkaline reaction, and intensely bitter taste. The solution of tetrethyl-ammonium hydroxide has a remarkable analogy to potash and soda. Like these substances, it destroys the epidermis and saponifies fatty substances, with formation of true soaps. With metallic salts it exhibits exactly the same reactions as potash. On evaporating a solution of the base in a vacuum, long slender needles are deposited, consisting of the hydroxide with a certain amount of crystallization-water. After some time these needles disappear, and a semisolid mass is left, which is the anhydrous base. A concentrated solution of this compound in water may be boiled without decomposition, but on heating the dry substance, it is decomposed into triethylamine, water, and ethene gas :

$$N(C_2H_5)_4(OH) = H_2O + N(C_2H_5)_3 + C_2H_4.$$

The other salts of tetrethylammonium are obtained by treating the hydroxide with acids : several of them form beautiful crystals. The *platinochloride*, $[N(C_2H_5)_4Cl]_2.PtCl_4$, crystallizes in orange-yellow octohedrons, slightly soluble in water, less soluble in alcohol and ether.

The amines of the series $C_nH_{2n+3}N$, which contain more than three atoms of carbon, admit, as already observed, of isomeric modifications, depend on the structure of their alcohol-radicles.

Propylamine (normal), $C_3H_9N = N \begin{cases} CH_2CH_2CH_3 \\ H_2 \end{cases}$, produced by hydrogenation of proprionitril (ethyl cyanide), C_3H_5N, or by the action of boiling potash on the mixture of propyl isocyanate and isocyanurate obtained by distilling normal propyl iodide with silver cyanate, is a strongly alkaline liquid, boiling at 50° C. (122° F.), and having a specific gravity of 0.7283 at 0° C. The *platinochloride*, $2(C_3H_9N.HCl).PtCl_4$, crystallizes in orange-yellow monoclinic prisms.

The *nitrite*, $C_3H_9N.NO_2$, is resolved by heat into pseudopropyl alcohol, nitrogen, and water, $CH_3—CH_2—CH_2(NH_2) + NO_2H \rightleftharpoons CH_3—CHOH—CH_3 + N_2 + H_2O.$

Isopropylamine, $CH_3—CH.NH_2—CH_3$, or $N \begin{cases} CH(CH_3)_2 \\ H_2 \end{cases}$, is produced as a formate by the action of hydrochloric acid on isopropyl isocyanide :—

$$C{\equiv}N—C_3H_7 + 2H_2O = NH_2(C_3H_7).CH_2O_2.$$

The base liberated from this salt by successive treatment with hydrochloric acid and potash, is a sweetish ammoniacal liquid, boiling at 32° C. (89.6° F.). Its platinochloride crystallizes in golden-yellow scales or flattened needles.

Butylamines.—Of primary butylamine, $C_4H_{11}N$, there are three modifications, viz. :—

1. **Normal Butylamine**, $CH_3.(CH_2)_3.NH_2$, obtained by the action of nascent hydrogen on normal propyl cyanide or butyro-nitril, $CH_3(CH_2)_2.CN$. It is a colorless, hygroscopic, fuming liquid, boiling at 75.5° C. (167.9° F.), and having a specific gravity of 0.755 at 0° C. Nitrous acid converts it into isobutyl alcohol, $CH(CH_3)_2—CH_2OH.$

2. **Isobutylamine**, $CH(CH_3)_2.CH_2.NH_2$, is obtained by distilling a mixture of potassium isobutylsulphate with potassium cyanate, and treating the resulting isobutyl isocyanate with potash. It boils at 68° C. (154.4°

F.), and has a specific gravity of 0.7357 at 15° C. (59° F.). With nitrous acid it yields tertiary butyl alcohol or trimethyl carbinol, $(CH_3)_3.COH$. Now it has been already stated that normal butylamine similarly treated is converted into isobutyl alcohol, and normal propylamine into isopropyl alcohol. Hence it appears that the action of nitrous acid on primary amines gives rise to the alcohol of a radicle containing one or more methyl-group than the original amines :—

$$CH_3.CH_2.CH_2.CH_2NH_2 \ + \ NO_2H \ = \ N_2 \ + \ H_2O \ + \ CH(CH_3)_2.CH_2OH$$
Normal Butylamine.   Isobutyl alcohol.

$$CH(CH_3)_2.CH_2.NH_2 \ + \ NO_2H \ = \ N_2 \ + \ H_2O \ + \ (CH_3)_3C.COH$$
Isobutylamine.   Tertiary Butyl
                                                        alcohol.

3. **Tertiary Butylamine**, or **Katabutylamine**, $(CH_3)_3C.NH_2$, is prepared by heating isobutyl iodide with dry silver cyanate, mixing the resulting solid compound of silver iodide and isobutyl cyanate with finely pulverized potassium hydroxide, and distilling the mixture by small quantities at a time. It is a liquid - boiling at 45°–46° C. (113–114.8° F.), and having a specific gravity of 0.6987 at 15° C. (59° F.).

Dibutyalamine, $NH(C_4H_9)_2$, and **Tributylamine**, $N(C_4H_9)_3$, are obtained as bye-products in the preparation of normal valeric acid by the action of alcoholic potash on normal butyl cyanide. The former boils at 160° C. (320° F.), the latter between 211° C. (411.8° F.) and 215° C. (419° F.).

Pentylamines.—Of primary pentylamines, $C_5H_{13}N$, there are two known modifications, viz., i s o p e n t y l a m i n e or a m y l a m i n e, derived from the ordinary amyl alcohol of fermentation, and p s e u d o p e n t y l - a m i n e, derived from isopropyl-methyl carbinol (p. 614). They are represented by the following formula :—

$$C \left\{ \begin{matrix} CH_2CH(CH_3)_2 \\ H_2 \\ NH_2 \end{matrix} \right. \text{or N} \left\{ \begin{matrix} CH_2CH_2CH(CH_3)_2 \\ H \\ H \end{matrix} \right. \qquad C \left\{ \begin{matrix} CH(CH_3)_2 \\ CH_3 \\ H \\ NH_2 \end{matrix} \right. \text{or N} \left\{ \begin{matrix} CH(CH_3)[CH(CH_3)_2] \\ H \\ H \end{matrix} \right.$$

Isopentylamine.   Pseudopentylamine.

Isopentylamine or **Ordinary Amylamine**, is obtained by distilling isopentyl isocyanate or isocyanurate, or primary isopentyl-carbamide, $CON_2H_3(C_5H_{11})$, with potash ; pseudo-pentylamine in like manner from secondary isopentyl-carbamide, $CON_2H_2(CH_3)(C_4H_9)$. Both are colorless liquids, of penetrating ammoniacal odor, and slightly soluble in water, to which they impart an alkaline reaction. Amylamine boils at 95° C. (203° F.), and has a specific gravity of 0.7503 at 18° C. (64.4° F.); pseudopentylamine boils at 78.5° C. (173.3° F.), and has a specific gravity of 0.755 at 0° C. Platinochloride of amylamine crystallizes in golden-yellow scales ; that of pseudopentylamine in fine crystals derived from a monoclinic prism. The aurochloride of pseudopentyl, $C_5H_{14}N.AuCl_3$, forms large yellow monoclinic crystals.

Diamylamine, $C_{10}H_{23}N = NH(C_5H_{11})_2$.—An aromatic liquid, less soluble in water, and less alkaline than amylamine. It boils at about 170° C. (338° F.).

Triamylamine, $C_{15}H_{33}N = N(C_5H_{11})_3$.—A colorless liquid, having properties similar to those of the two preceding bases, but boiling at 257° C. (494° F.). The salts of triamylamine are very sparingly soluble in water, and melt, when heated, to colorless liquids, floating upon water.

Tetramylammonium Hydroxide, $C_{20}H_{45}NO = N(C_5H_{11})_4OH$.—This substance is far less soluble than the corresponding bases of the methyl and ethyl series, and separates as an oily layer on adding potash to the aqueous solution. On evaporating the solution in an atmosphere free from carbonic acid, the alkali may be obtained in splendid crystals of considerable size. When submitted to distillation it splits into water, triamylamine, and amylene :

$$N(C_5H_{11})_4OH = H_2O + N(C_5H_{11})_3 + C_5H_{10}.$$

Bromethyl- and **Ethenyl-** or **Vinyl-bases.**—Tertiary monamines unite directly with ethene bromide, giving rise to brominated ammonium-bases : *e. g.,*

$$(C_2H_5)_3N + C_2H_4Br_2 = \left.\begin{matrix}(C_2H_5)_3\\ C_2H_4Br\end{matrix}\right\} N.Br.$$

In these compounds, the bromine directly attached to the nitrogen is easily replaceable by double decomposition, by the action of silver nitrate, for example, whereby the bromide is converted into the nitrate $(C_2H_5)_3$ $(C_2H_4Br)N.NO_3$.

The other bromine-atom is more intimately combined, but this also may be removed by the action of moist silver oxide, the group C_2H_4Br or CH_2Br—CH_2 being converted, by elimination of HBr, into ethenyl or vinyl CH_2=CH, and the bromide of triethyl-bromethyl-ammonium being converted into triethyl-vinyl-ammonium hydroxide :

$$\left.\begin{matrix}(C_2H_5)_3\\ C_2H_4Br\end{matrix}\right\} N.Br + 2AgOH = 2AgBr + H_2O + \left.\begin{matrix}(C_2H_5)_3\\ C_2H_3\end{matrix}\right\} N.OH.$$

Oxethene-bases or **Hydramines.**—When etheneoxide, C_2H_4O or CH_2—O—CH_2, is treated with aqueous ammonia, 1, 2, or 3 molecules of the oxide unite with one mol. ammonia, producing the following bases :

$CH_2(OH)CH_2.NH_2$,	Ethene-hydramine.
$\begin{matrix}CH_2(OH)CH_2\\ CH_2(OH)CH_2\end{matrix}> NH$,	Diethene-hydramine.
$[CH_2(OH)CH_2]_3N$,	Triethene-hydramine.

The hydrochlorides of these bases are also formed by the action of ammonia on ethene chlorhydrin, $C_2H_4Cl(OH)$. The bases are separated from one another by fractional crystallization of their hydrochlorides or platinochlorides. They are viscid, alkaline liquids, decomposed by distillation.

Choline, $C_5H_{15}NO_2$.—This base, which has the composition of trimethylethenehydrinammonium hydroxide, $C_2H_4 {<}{\overset{OH}{\underset{N(CH_3)_3OH}{}}}$, is formed synthetically by heating ethene oxide or ethene-chlorhydrin with trimethylamine in aqueous solution :

$$(CH_3)_3N + C_2H_4O + H_2O = C_2H_4 {<}{\overset{OH}{\underset{N(CH_3)_3OH}{}}}.$$

It exists ready-formed in the animal organism, namely in bile, from which it was first obtained (hence its name) ; also in the brain, and in white of egg, in both of which it occurs in the form of lecithin, a compound of choline with glycerophosphoric acid and fatty acids. It is also produced from sinapine, the alkaloïd of white mustard, by boiling with alkalies, and is thence also called sincaline.

Choline is a deliquescent substance difficult to crystallize, having a strong

29

acid reaction, and absorbing carbonic acid from the air. Its *platinochloride*, $2(C_5H_{15}NO_2.HCl).PtCl_4$, crystallizes in fine red-yellow tables, insoluble in alcohol.

Choline heated with hydriodic acid yields the compound $(CH_3)_3N\diagdown^{C_2H_3I}_I$, which, by action of moist silver oxide is converted into t r i m e t h y l - v i n y l - a m m o n i u m h y d r o x i d e :

$$(CH_3)_3N\diagdown^{C_2H_3}_{OH} = C_5H_{13}NO.$$

This base, called n e u r i n e, is very much like choline—from which it differs by H_2O—and likewise exists in the brain.

Betaine, $C_5H_{11}NO_2$, the alkaloïd of beet-juice, which has the constitution of trimethyl-glycocine, $(CH_3)_3N\diagdown^{CH_2.CO}_O$ (see AMIDES), is also nearly related to choline, and is formed by oxidation of choline hydrochloride, whence it is also called o x y c h o l i n e, and o x y n e u r i n e. It is formed synthetically, as a hydrochloride, by heating trimethylamine with monochloracetic acid :

$$(CH_3)_3N + CH_2Cl.CO.OH = (CH_3)_3N\diagdown^{CH_2CO}_O .HCl.$$

Betaine crystallizes from alcohol in shining deliquescent crystals containing one molecule of water. It has a neutral reaction, a sweetish taste, and is decomposed by boiling alkalies, with evolution of trimethylamine.

DIAMINES.

These are bases derived from a double molecule of ammonia, N_2H_6, by substitution of bivalent alcohol-radicles for a part or the whole of the hydrogen, or from the paraffins by substitution of two NH_2-groups for 2 atoms of hydrogen.

Diamines are formed by the action of ammonia on the chlorides, bromides, and iodides of diatomic alcohol-radicles (olefines).

E t h e n e - d i a m i n e s.—By the action of ammonia on ethene dibromide, a number of compounds are produced, among which are the hydrobromides of three bases derived from two molecules of ammonia, by substitution of 1, 2, or 3 molecules of ethene, C_2H_4, for equivalent quantities of hydrogen ; thus :

$$2NH_3 + C_2H_4Br_2 = N\diagdown^{H\ H}_{H\ H}\diagdown^{C_2H_4}N.2HBr$$

Ethene-diamine.

$$4NH_3 + 2C_2H_4Br_2 = N\diagdown^{C_2H_4}_{H\ H}\diagdown^{C_2H_4}N.2HBr + 2NH_4Br$$

Diethene-diamine.

$$6NH_3 + 3C_2H_4Br_2 = N\diagdown^{C_2H_4}_{C_2H_4}\diagdown^{C_2H_4}N.2HBr + 4NH_4Br$$

Triethene-diamine.

The mixture of hydrobromides thus obtained is decomposed by distillation with potash, and the three diamines which pass over are separated by fractional distillation.

Ethene-diamine, $C_2H_8N_2 = C_2H_4(NH_2)_2 = N_2\begin{cases} C_2H_4 \\ H_4 \end{cases}$, is also produced by the action of tin and hydrochloric acid on dicyanogen: $C_2N_2 + 4H_2 = C_2H_8N_2$. It is a colorless liquid, boiling at 123° C. (253.4° F.), having a strong alkaline reaction, and ammoniacal odor. Nitrous acid converts it into ethene oxide:

$$C_2H_4(NH_2)_2 + N_2O_3 = 2H_2O + 2N_2 + C_2H_4O.$$

Diethene-diamine, $N_2\begin{cases} C_2H_4 \\ C_2H_4 \\ H_2 \end{cases}$, boils at 170° C. (338° F.); *tri-ethene-diamine*, $N_2(C_2H_4)_3$, at 220° C. (428° F.).

These diamines are bi-acid bases, capable of uniting directly with 2 equivalents of an acid. They can also unite with the bromides and iodides of the alcohol-radicles, producing iodides of ammonium bases, thus:

from Ethene-diamine are obtained:

Iodide of Ethene-diammonium . $[N_2H_6(C_2H_4)]I_2$.
Iodide of Diethyl-ethene-diammonium . $[N_2H_4(C_2H_4)(C_2H_5)_2]I_2$.
Iodide of Tetrethyl-ethene-diammonium . $[N_2H_2(C_2H_4)(C_2H_5)_4]I_2$.
Iodide of Pentethyl-ethene-diammonium . $[N_2H(C_2H_4)(C_2H_5)_5]I_2$.
Iodide of Hexethyl-ethene-diammonium . $[N_2(C_2H_4)(C_2H_5)_6]I_2$.

from Diethene-diamine;

Iodide of Diethene-diammonium . $[N_2H_4(C_2H_4)_2]I_2$.
Iodide of Diethyl-diethene-diammonium . $[N_2H_2(C_2H_4)_2(C_2H_5)_2]I_2$.
Iodide of Triethyl-diethene-diammonium. . $[N_2H(C_2H_4)_2(C_2H_5)_3]I_2$.
Iodide of Tetrethyl-diethene-diammonium. $[N_2(C_2H_4)_2(C_2H_5)_4]I_2$.

from Triethene-diamine:

Iodide of Diethyl-triethene-diammonium . $[N_2(C_2H_4)_3(C_2H_5)_2]I_2$.

The diamines also unite directly with water to form ammonium oxides; thus:

$$C_2H_4\begin{matrix} NH_2 \\ NH_2 \end{matrix} + H_2O = C_2H_4\begin{matrix} NH_3 \\ NH_3 \end{matrix}O.$$

These oxides are very unstable, and give off the water again on mere distillation over potassium hydroxide; when treated with acids, they form diamine-salts, with separation of water.

Methenyl-diamine, $CN_2H_4 = N_2\begin{cases} (CH)''' \\ H_3 \end{cases}$, or *Formylimidamide*,

$CH\begin{matrix} NH \\ NH_2 \end{matrix}$. This base, nearly related to cyanimide or carbodiimide,

$C\begin{matrix} NH_2 \\ NH_2 \end{matrix}$, is known only in combination with acids. The *hydrochloride*, $CN_2H_4.HCl$, is produced together with ethyl chloride and ethyl formate, by heating the compound $CNH.HCl$, formed by direct combination of anhydrous hydrocyanic acid and gaseous hydrochloric acid, with alcohol:

$$2(CNH.HCl) + 2(C_2H_5.OH) = CN_2H_4.HCl + C_2H_5Cl + C_2H_5.CHO_2.$$

It crystallizes in needles melting at 81° C. (177.8° F.), and is resolved by alkalies into ammonia and formic acid :

$$CN_2H_4 + 2H_2O = 2NH_3 + CHO.OH.$$

Methyl-methenyl-diamine, Ethenyl-diamine, or Acediamine,
$C_2N_2H_6 = (H_3C)C{<}{\stackrel{NH}{NH_2}}$. The hydrochloride of this base, $C_2N_2H_6.HCl$, together with acetic acid, is formed by heating acetamide in dry hydrochloric acid gas :

$$2C_2H_5NO_4 + HCl = C_2H_6N_2.HCl + C_2H_4O_2 .$$

It crystallizes in long needles, soluble in water and alcohol. The free base is very unstable, and when separated from the hydrochloride by an alkali, is quickly resolved into ammonia and acetic acid.

Hydroxyl-methenyl-diamine or Isuret, $CON_2H_4 = CH{<}{\stackrel{NH}{NH.OH}}.{—}$
This base, isomeric with carbamide, is formed by evaporating an alcoholic solution of hydroxylamine (p. 192) and hydrocyanic acid :

$$CHN + NH_2.OH = CH{<}{\stackrel{NH}{NH.OH}}.$$

It crystallizes in rhombic prisms resembling carbamide (urea); melts with partial decomposition at 104°–105° C. (219.2°–221° F.); has an alkaline reaction, and unites with 1 equiv. of acids, forming crystalline salts, which are resolved, on heating their solutions, into formic acid, ammonia, and hydroxylamine. The base itself is also decomposed by boiling its aqueous solution yielding nitrogen, carbon dioxide, ammonia, guanidine, carbamide, and biuret.

TRIAMINES.

Diethene-triamine, $N_4{\left\{{\stackrel{\textstyle C_2H_4}{\stackrel{\textstyle C_2H_4}{H_8}}}\right.}$, and **Triethene-triamine,** $N_3{\left\{{\stackrel{\textstyle (C_2H_4)_3}{H_3}}\right.}$
—The hydrobromides of these bases are found amongst the products obtained by heating ethene bromide with ammonia. Their formation is represented by the equations,

$$4NH_3 + 2C_2H_4Br_2 = NH_4Br + \underset{\text{Diethene-triamine.}}{N_3(C_2H_4)_2H_53HBr}$$

$$6NH_3 + 3C_2H_4Br_2 = 3NH_4Br + \underset{\text{Triethene-triamine.}}{N_3(C_2H_4)_3H_3.3HBr}$$

These bases cannot be separated by distillation, on account of the proximity of their boiling points, but their separation is effected by converting them into platino-chlorides, which crystallize with great facility.

Both these triamines form well-defined crystalline salts, containing 1, 2, and 3 equivalents of acids ; and the aqueous solutions of these salts, treated with solid potassium hydroxides, yield the free bases in the form of strongly alkaline oily liquids, soluble in every proportion of water and alcohol, nearly insoluble in ether. Diethene-triamine boils at 208° C. (406.4° F.), triethene-triamine at 216° C. (420.8° F.).

Triethene Tetramine, $N_4{\left\{{\stackrel{(C_2H_4)_3}{H_6}}\right.}$.—The hydrobromide of this base is also one of the products of the action of ammonia on ethene bromide, but it is more easily obtained by submitting ethene-diamine to the action of ethene bromide :

$$2N_3(C_2H_4)H_4 + C_2H_4Br_2 = N_4(C_2H_4)_3H_6.2HBr.$$

The base, separated from the bromide by silver oxide, is a strongly alkaline liquid, which dries up to a non-crystallizing syrup. The *platinochloride*, $N_4(C_2H_4)_3H_6.4HCl.2PtCl_4$, is a pale yellow amorphous, almost insoluble salt.*

Guanidine, CH_5N_3, = **Carbotriamine,** $N_3 \begin{Bmatrix} C^{iv} \\ H_5 \end{Bmatrix}$, or **Carbimido-**

diamide, $C{\equiv}N \begin{matrix} {\diagup}NH_2 \\ {\diagdown}NH_2 \end{matrix}$.—This base is produced: 1. By the action of aqueous

ammonia at 150° C. (302° F.) on chloropicrin, or on ethyl orthocarbonate:

$$CCl_3(NO_2) + 3NH_3 = CN_3H_5.HCl + 2HCl + NO_2H.$$
Chloropicrin.

$$C(OC_2H_5)_4 + 3NH_3 + H_2O = CN_3H_5 + H_2O + 4C_2H_6O.$$
Ethyl Ortho-
carbonate.

2. By heating cyanamide in alcoholic solution with ammonium chloride:

$$CH_2N_2 + NH_4Cl = CN_3H_5.HCl.$$

3. Together with parabanic acid, by heating guanine (an alkaloid obtained from guano), with a mixture of hydrochloric acid and potassium chlorate, which exerts an oxidizing action:

$$C_5H_5N_5O + O_3 + H_2O = CN_3H_5 + C_3N_2H_2O_3 + CO_2.$$

Guanidine, separated from its sulphate by baryta-water, forms colorless crystals, easily soluble in water and alcohol: the solution has a powerful alkaline reaction, and absorbs carbonic acid from the air, forming a carbonate, $2CH_5N_3.H_2CO_3$, which is also alkaline, and crystallizes in square prisms.

The *nitrate*, $CH_5N_3.NO_3H$, forms large laminæ easily soluble in water. The hydrochloride, $CH_5N_3.HCl$, yields a platinochloride, crystallizing in yellow needles.

Alcoholic derivatives of guanidine are formed by reactions analogous to those which yield guanidine itself, especially by heating cyanamide with the hydrochlorides of primary amines: *e. g.*,

$$CN_2H_2 + NH_2(CH_3).HCl = CN_3H_4(CH_3).HCl.$$

Methyl-guanidine, $CN_3H_4(CH_3)$, is also formed by boiling creatine with water and mercuric oxide. When separated from its hydrochloride by silver oxide it forms a deliquescent crystalline mass, which unites with 1 eq. of acid, forming well-crystallized salts.

Triethyl-guanidine, $CN_3H_2(C_2H_5)_3$, is formed by boiling diethyl-thiocarbamide and ethylamine in alcoholic solution with mercuric oxide:

$$CS{\diagdown}{\diagup} \begin{matrix} NH.C_2H_5 \\ NH.C_2H_5 \end{matrix} + NH_2.C_2H_5 + HgO =$$

$$HgS + H_2O + (C_2H_5)N{=}C{\diagdown}{\diagup} \begin{matrix} NH.C_2H_5 \\ NH.C_2H_5 \end{matrix}.$$

The phenyl and tolyl-derivations of guanidine will be described amongst the compounds of the aromatic group; the derivatives containing acid radicles amongst the derivatives of uric acid.

* For further details respecting these higher ethene-bases, see Watts's Dictionary of Chemistry, vol. ii. pp. 588-593.

Phosphorus, Arsenic, and Antimony Bases.

Phosphorus, arsenic, and antimony, being like nitrogen, either trivalent or quinquivalent, are capable of forming compounds analogous to the amines and the ammonium salts.

PHOSPHORUS BASES OR PHOSPHINES.

Tertiary phosphines and phosphonium bases are formed by the action of alcoholic iodides on phosphine, PH_3: thus,

$$PH_3 \quad + \quad 3C_2H_5I \quad = \quad 2HI \quad + \quad \underset{\substack{\text{Triethyl-phosphine}\\\text{hydriodide.}}}{P(C_2H_5)_3.HI.}$$

$$P(C_2H_5)_3 \quad + \quad C_2H_5I \quad = \quad \underset{\substack{\text{Triethyl-phosphonium}\\\text{iodide.}}}{P(C_2H_5)_4I}$$

or more readily by heating phosphonium iodide, PH_4I (p. 207), to 150°–180° C. (302°–356° F.) with alcoholic iodides :

$$PH_4I \quad + \quad 3C_2H_5I \quad = \quad 3HI \quad + \quad P(C_2H_5)_3.HI;$$
$$\text{and } P(C_2H_5)_3HI \quad + \quad C_2H_5I \quad = \quad HI \quad + \quad P(C_2H_5)_4I.$$

On heating the product with potash-lye, the hydriodide of the tertiary phosphine is decomposed and the base set free, while the iodide of the phosphonium-base remains unaltered.

The tertiary phosphines are also formed by distilling the zinc-compounds of the alcohol-radicles with phosphorus trichloride in an atmosphere of hydrogen :

$$3Zn(C_2H_5)_2 \quad + \quad 2PCl_3 \quad = \quad 3ZnCl_2 \quad + \quad 2P(C_2H_5)_3.$$

The primary and secondary phosphines are formed by heating the alcoholic iodides with phosphonium iodide to about 150° C. (302° F.) in presence of certain metallic oxides, such as zinc oxide (ordinary zinc-white of commerce), the product being a mixture of the primary and secondary phosphines :

$$2PH_4I \quad + \quad 2C_2H_5I \quad + \quad ZnO \quad = \quad ZnI_2 \quad + \quad H_2O \quad + \quad 2P(C_2H_5)H_3I$$
$$\text{and } PH_4I \quad + \quad 2C_2H_5I \quad + \quad ZnO \quad = \quad ZnI_2 \quad + \quad H_2O \quad + \quad P(C_2H_5)_2H_2I.$$

On treating the crystalline product with *water*, monethyl-phosphine is set free :

$$P(C_2H_5)H_3I \quad + \quad H_2O \quad = \quad HI \quad + \quad H_2O \quad + \quad P(C_2H_5)H_2,$$

while the diethylphosphine salt remains unaltered, but may be decomposed by boiling it with caustic soda, the diethylphosphine being then set free.

The phosphines are colorless, strongly refracting liquids, having an extremely pungent intoxicating odor. They are nearly insoluble in water. In contact with the air, they oxidize rapidly and mostly take fire : hence their preparation must be conducted in an atmosphere of hydrogen. The salts of the primary phosphines are easily decomposed by water, those of the secondary and tertiary phosphines are not.

The primary phosphines are converted by oxidation into phosphonic acids, those of the secondary phosphines into phosphinic acids (p. 588) : *e. g.*,

$$P(CH_3)_2H_2 + O_3 = H_3C-P{\displaystyle \mathop{=}^{OH}_{OH}}O , \text{ Methylphosphonic acid.}$$

$$P(CH_3)_2 + O_2 = {H_3C \atop H_3C}{>}P{<}{O \atop OH}, \text{ Dimethyl-phosphinic acid.}$$

The tertiary phosphines unite directly with 1 atom of oxygen, forming neutral oxides, like $(CH_3)_3PO$, and similarly with 1 atom of sulphur, 2 atoms of bromine, iodine, etc., and also with carbon bisulphide.

The following table exhibits the chief distinguishing characters of the methyl- and ethyl-phosphines :—

	Boiling Point.	Sp. Gravity.	Hydriodide.	Platino-chloride.
Methyl phosphines :—				
$P(CH_3)H_2$	Gas liquefying at —14° C. (6.8° F.)	..	Thick crystals.	
$P(CH_3)_2H$	25° C. (77° F.)	Lighter than water.	..	Crystallizes well.
$P(CH_3)_3$	40° C. (104° F.)	Lighter than water.		
Ethyl-phosphines :—				
$P(C_2H_5)H_2$ . . .	25° C. (77° F.)	Lighter than water.	Quadrilateral plates.	
$P(C_2H_5)_2H$	85° C. (185° F.)	Lighter than water.	..	Orange-yellow prisms.
$P(C_2H_5)_3$	127° C. (260.6° F.)	0.812 at 12° C. (53.6° F.)		Red needles.

The boiling point of the isopropyl, isobutyl, and isopentyl or amyl phosphines are as follows :—

Isopropyl-phosphines, B. P.	$P(C_3H_7)H_2$ 41°	$P(C_3H_7)_2H$ 118°	$P(C_3H_7)_3$ —
Isobutyl-phosphines, B. P.	$P(C_4H_9)H_2$ 62°	$P(C_4H_9)_2H$ 153°	$P(C_4H_9)_3$ 215°
Amyl-phosphines, B. P.	$P(C_5H_{11})H_2$ 106–107°	$P(C_5H_{11})_2H$ 210–215°	$P(C_5H_{11})_3$ 300°

ARSENIC BASES.

Triethylarsine, $As(C_2H_5)_3$, is produced by distilling an alloy of arsenic and sodium with ethyl iodide. At the same time, also, there is formed another body, containing $As_2(C_2H_5)_4$, analogous to arsendimethyl or cacodyl. Both compounds are liquids of powerful odor; they may be separated by distillation in an atmosphere of carbon dioxide, the triethylarsine passing over last.

Triethylarsine may be obtained pure by a process analogous to that employed for the preparation of triethylphosphine, namely, by distilling arsenious chloride, $AsCl_3$, with zinc-ethyl. It is a colorless liquid of most disagreeable odor, similar to that of arseniëtted hydrogen, soluble in water, alcohol, and ether, and boiling at 140° C. (284° F.). Triethylarsine combines directly with oxygen, sulphur, bromine, and iodine, giving rise to a series of compounds containing 2 atoms of bromine or iodine, 1 atom of sulphur or oxygen, and analogous to the corresponding compounds of triethylstibine.

Triethylarsine submitted to the action of ethyl iodide yields a crystalline compound, $As(C_2H_5)_4I$, from which freshly precipitated silver oxide separates the corresponding hydroxide, $As(C_2H_5)_4OH$, a powerfully alkaline substance, similar to the corresponding nitrogen-, phosphorus-, and anti-mony-compounds.

Analogous substances exist in the methyl series. *Trimethylarsine*, $As(CH_3)_3$, is formed, together with arsendimethyl or cacodyl, $As_2(CH_3)_4$, when an alloy of arsenic and sodium is submitted to the action of methyl iodide. It unites with methyl iodide, producing *tetramethylarsonium iodide*, $As(CH_3)_4I$, from which silver oxide separates the hydrate, $As(CH_3)_4OH$. The iodide just mentioned is formed, together with iodide of cacodyl, when cacodyl is acted upon by methyl iodide:

$$As_2(CH_3)_4 \ + \ 2CH_3I \ = \ As(CH_3)_4I \ + \ As(CH_3)_2I.$$

By substituting ethyl iodide for methyl iodide in this reaction, the com-pound $As(CH_3)_2(C_2H_5)_2I$ is formed. All these iodides, treated with moist silver oxide, yield the corresponding hydroxides.

Arsendimethyl or **Cacodyl,** $As_2(CH_3)_4$ or $\begin{vmatrix} As(CH_3)_2 \\ As(CH_3)_2 \end{vmatrix}$. — The arsenic in this compound is trivalent, as in those just described, one unit of equivalence of each of the arsenic atoms being satisfied by combination with the other, just as in the solid hydrogen arsenide, As_2H_4 (p. 213). When, however, the arsendimethyl combines with chlorine or other mon-atomic radicles, the molecule splits into two; thus:

$$As_2(CH_3)_4 \ + \ Cl_2 \ = \ 2As(CH_3)_2Cl.$$

Cacodyl, so called from its repulsive odor, constitutes, together with its products of oxidation, the spontaneously inflammable liquid known as *Cadet's fuming liquid*, or *Alkarsin*. This liquid is prepared by distilling equal weights of potassium acetate and arsenious oxide in a glass retort connected with a condenser and tubulated receiver cooled by ice, a tube being attached to the receiver to carry away the permanently gaseous products to some distance from the experimenter. At the close of the ope-ration, the receiver is found to contain two liquids, besides a quantity of reduced arsenic: the heavier of these is the crude cacodyl; the other con-sists chiefly of water, acetic acid, and acetone. The gas given off during the distillation is principally carbon dioxide. The crude cacodyl is re-peatedly washed by agitation with water previously freed from air by boiling, and afterwards redistilled from potassium hydrate in a vessel filled with pure hydrogen gas. All these operations must be conducted in the open air.

Pure cacodyl is obtained by decomposing the chloride with metallic zinc, dissolving out the zinc chloride with water, and dehydrating the oily liquid with calcium chloride. The strong tendency of cacodyl to take fire in the air, and the extremely poisonous character of its vapor, render it necessary to perform all the distillations in sealed vessels filled with carbon dioxide.

Cacodyl is a colorless, transparent liquid, boiling at 170° C. (338° F.), and crystallizing at 6° C. (42.8° F.) in large transparent prisms. It smells like alkarsin, and is even more inflammable. At a temperature below redness it is resolved into metallic arsenic and a mixture of 2 vols. methane and 1 vol. ethene: $2As_2C_2H_6 = As_4 + 2CH_4 + C_2H_4$.

Cacodyl instantly takes fire when poured out into air or oxygen or chlo-rine. With very limited access of air it throws off white fumes, passing into oxide and ultimately into cacodylic acid: it combines also directly with sulphur. It can take up 2 atoms of a monad or 1 atom of a dyad element, forming compounds like the chloride, $As_2(CH_3)_4Cl_2 = 2As(CH_3)_2Cl,$

and the oxide, $As_2(CH_3)_4O$, in which the arsenic is trivalent; or again, 6 atoms of a monad or 3 atoms of a dyad element, forming compounds like the trichloride, $As_2(CH_3)_4Cl_6 = 2As(CH_3)_2Cl_3$, in which the arsenic is quinquivalent. These last-mentioned bodies are the most stable of all the cacodyl compounds.

Cacodyl Chloride or *Arsen-chlorodimethide*, $As(CH_3)_2Cl$, is obtained by distilling alkarsin with strong hydrochloric acid, or better, by mixing the dilute alcoholic solutions of alkarsin and mercuric chloride, distilling the resulting precipitate of cacodylic chloromercurate, $As_2(CH_3)_4O.2HgCl_2$, with very strong hydrochloric acid, and digesting the distillate for several days in a sealed bulb apparatus with calcium chloride and quicklime, and finally distilling it in an atmosphere of carbon dioxide.

Cacodyl chloride is a colorless liquid which does not fume in the air, but emits an intensely poisonous vapor. It is heavier than water, insoluble in that liquid and in ether, but easily soluble in alcohol. The boiling point of this compound is a little above 100° C.; its vapor is colorless, spontaneously inflammable in the air, and has a density of 4.56. Dilute nitric acid dissolves the chloride without change; with the concentrated acid ignition and explosion occur. Cacodyl chloride combines with cuprous chloride, forming a white, insoluble, crystalline double salt, containing $As_2(CH_3)_4Cl_2.Cu_2Cl_2$; also with cacodyl oxide. It forms a thick, viscid hydrate, easily dehydrated by calcium chloride. *Cacodyl trichloride*, $As(CH_3)_2Cl_3$, is produced by the action of phosphorus pentachloride on cacodylic acid:

$$As(CH_3)_2O(OH) + 2PCl_5 = As(CH_3)_2Cl_3 + 2POCl_3 + HCl;$$

also by the action of chlorine gas on the monochloride. Prepared by the first method, it forms splendid large prismatic crystals, which are instantly decomposed, between 40° and 50° C. (104–127° F.), into methyl chloride and arsen-monomethyl chloride:

$$As(CH_3)_2Cl_3 = CH_3Cl + As(CH_3)Cl_2.$$

Cacodyl Iodide, $As(CH_3)_2I$, is a thin, yellowish, heavy liquid, of offensive odor, prepared by distilling alkarsin with strong solution of hydriodic acid. A yellow crystalline oxyiodide is formed at the same time. *Cacodyl bromide* and *fluoride* have also been obtained.

Cacodyl Cyanide, $As(CH_3)_2CN$, is easily formed by distilling alkarsin with strong hydrocyanic acid, or mercuric cyanide. Above 33° C. (91.4° F.), it is a colorless, ethereal liquid, but below that temperature it crystallizes in colorless four-sided prisms, of beautiful diamond lustre. It boils at about 140° C. (284° F.), and is but slightly soluble in water. It takes fire only when heated. The vapor of this substance is most fearfully poisonous: the atmosphere of a room is said to be so far contaminated by the evaporation of a few grains of it as to cause instantaneous numbness of the hands and feet, vertigo, and even unconsciousness.

Cacodyl Oxide, $As_2(CH_3)_4O$, is formed by the slow oxidation of cacodyl. When air is allowed access to an aqueous solution of alkarsin, so slowly that no sensible rise of temperature follows, that body is gradually converted into a thick, syrupy liquid, full of crystals of cacodylic acid. On dissolving this mass in water, and distilling, water having the odor of alkarsin passes over, and afterwards an oily liquid, which is the cacodyl oxide. Impure cacodylic acid remains in the retort.

Cacodyl oxide, purified by rectification from caustic baryta, is a colorless, oily liquid, having a pungent odor, sparingly soluble in water, solidifying at —25° C. (—13° F.), and boiling at 150° C. (302° F.). It strongly resembles alkarsin in odor, in its relations to solvents, and in the greater

29 *

number of its reactions; but it neither fumes in the air, nor takes fire at
common temperatures : its vapor mixed with air, and heated to about 88⁰ C.
(190.4⁰ F.), explodes with violence. It dissolves in hydrochloric, hydrobro
mic, and hydriodic acids, forming chloride, bromide, and iodide of cacodyl
   Cacodyl dioxide, $As_2(CH_3)_4O_2$, is the thick syrupy liquid produced by the
slow oxidation of cacodyl or of alkarsin. It is decomposed by water, and
then yields a distillate of cacodyl monoxide, with a residue of cacodylic
acid :

$$2As_2(CH_3)_4O_2 \ + \ H_2O \ = \ As_2(CH_3)_4O \ + \ 2As(CH_3)_2O(OH).$$

   Cacodylic or *Dimethyl-arsenic Acid*, $\begin{matrix} H_3C \\ H_3C \end{matrix} > As \begin{matrix} O \\ OH \end{matrix}$, also called *Alkargen.*
—This is the ultimate product of the action of oxygen at a low tempera-
ture upon cacodyl or alkarsin in presence of water : it is best prepared by
adding mercuric oxide to alkarsin, covered with a layer of water and artificially cooled, until the mixture loses all odor, and afterwards decomposing any mercuric cacodylate that may have been formed, by the cautious
addition of more alkarsin. The liquid yields, by evaporation to dryness and
solution in alcohol, crystals of cacodylic acid. The sulphide and other
compounds of cacodyl yield the same substance on exposure to air. Cacodylic acid forms brilliant, colorless, brittle crystals, which have the form
of a modified square prism : it is permanent in dry air, but deliquescent
in a moist atmosphere. It is not at all poisonous, though it contains more
than 50 per cent. of arsenic. It is very soluble in water and in alcohol,
but not in ether : the solution has an acid reaction. When mixed with
alkalies and evaporated, it leaves a gummy amorphous mass. With the
oxides of silver and mercury, on the other hand, it yields crystallizable
compounds. It unites with cacodyl oxide, and forms a variety of combinations with metallic salts. Cacodylic acid is exceedingly stable : it is
not affected by red fuming nitric acid, nitromuriatic acid, or even chromic
acid in solution : it may be boiled with these substances without the least
change. It is deoxidized, however, by phosphorous acid and stannous
chloride, yielding cacodyl oxide. Dry hydriodic acid gas decomposes it,
with production of water, cacodyl iodide, and free iodine. With dry hydrochloric acid gas, or with the concentrated aqueous acid, cacodylic acid
unites directly, forming the compound $As(CH_3)_2O_2H.HCl$. But by exposing cacodylic acid for a long time to a stream of hydrochloric acid gas,
a r s e n - m o n o m e t h y l d i c l o r i d e is obtained, together with water and
methyl chloride :

$$As(CH_3)_2O_2H \ + \ 3HCl \ = \ As(CH_3)Cl_2 \ + \ 2H_2O \ + \ CH_3Cl.$$

Phosphorus pentachloride converts cacodylic acid into cacodylic trichloride
(p. 681).
   Cacodyl Sulphide, $As_2(CH_3)_4S$, is formed by adding barium sulphide to
crude cacodyl, or by distilling barium sulphydrate with cacodyl chloride.
It is a transparent liquid which retains its fluidity at —40⁰, and boils at
a temperature considerably above 100⁰ C.
   Cacodyl disulphide, $As_2(CH_3)_4S_2$, is formed by the action of sulphur on
cacodyl or the monosulphide, or by treating cacodylic acid with sulphuretted hydrogen in a vessel externally cooled. It separates from the solution in large rhombic crystals. Its alcoholic solution yields with various
metallic solutions, precipitates consisting of salts of t h i o c a c o d y l i c
a c i d , $As(CH_3)_2S(SH)$, analogous to cacodylic acid. The lead salt,
$As_2(CH_3)_4S_4Pb$, forms small white crystals.

   Arsenmonomethyl, $As(CH_3)$.—This radicle, which is not known in
the separate state, is either bivalent or quadrivalent. Its *dichloride,*

$As(CH_3)Cl_2$, is produced either by the decomposition of cacodyl trichloride by heat : $As(CH_3)_2Cl_3 = As(CH_3)Cl_2 + CH_3Cl$; or by the prolonged action of hydrochloric acid on cacodylic acid. It is a colorless, heavy mobile liquid, having a strong reducing power ; it boils at 133° C. (271.4° F.). Its vapor exerts a most violent action on the mucous membranes ; on smelling it, the eyes, nose, and whole face swell up, and a peculiar lancinating pain is felt, extending down to the throat. The *tetrachloride*, $As(CH_3)Cl_4$, is obtained in large crystals by passing chlorine over a mixture of the dichloride and carbon bisulphide cooled to -10° C. (14° F.). It is very unstable, decomposing even near 0° into methyl chloride and arsenious chloride, $AsCl_3$. There is also a chlorobromide, $As(CH_3)ClBr$, and a di-iodide, $As(CH_3)I_2$.

The *oxide*, $As(CH_3)O$, obtained by decomposing the dichloride with potassium carbonate, forms large cubical crystals, soluble in water, alcohol, and ether, and resolved by distillation with potash into arsenious oxide and cacodyl oxide : $4As(CH_3)O = As_2O_3 + As_2(CH_3)_4O$.

Arsenmethylic Acid, $As(CH_3)O(OH)_2$, is obtained as a silver-salt by decomposing arsenmethyl dichloride with a slight excess of silver-oxide ; and this salt, decomposed by sulphuric acid, yields the acid, which remains on evaporation in the form of a laminated mass. It is bibasic.

Arsenmethyl sulphide, $As(CH_3)S$, is obtained as a white mass by passing hydrogen sulphide over the dichloride.

On comparing the combining or equivalent values of the several arsenides of methyl, it will be seen that they all unite with elementary bodies and compound radicles, in such proportions as to form compounds in which the arsenic is either trivalent or quinquivalent, the last-mentioned compounds being by far the most stable. Thus :

Arsenmonomethyl, $As(CH_3)$, is bi- and quadri-valent, forming the chlorides $As(CH_3)Cl_2$ and $As(CH_3)Cl_4$.

Arsendimethyl, $As(CH_3)_2$, is uni- and tri-valent, forming the chlorides $As(CH_3)_2Cl$ and $As(CH_3)_2Cl_3$.

Arsentrimethyl, $As(CH_3)_3$, is bivalent only, and forms the chloride $As(CH_3)_3Cl_2$.

Arsentrimethylium, or Tetramethylarsonium, $As(CH_3)_4$, is univalent, forming the chloride $As(CH_3)_4Cl$.

ANTIMONY-BASES OR STIBINES.

Of these only the tertiary bases and the corresponding stibonium-compounds are known.

Triethylstibine or **Stibethyl**, $Sb(C_2H_5)_3$, is obtained by distilling ethyl iodide with an alloy of antimony and potassium. It is a transparent, very mobile liquid, having a penetrating odor of onions. It boils at 158° C. (136.4° F.). In contact with atmospheric air, it emits a dense white fume, and frequently even takes fire, burning with a white brilliant flame. It is analogous in many of its reactions to triethylamine, but has much more powerful combining tendencies, uniting readily with 2 atoms of chlorine, bromine, or iodine, and 1 atom of oxygen or sulphur, thereby forming compounds in which the antimony is quinquivalent, such as $Sb(C_2H_5)_3Cl_2$, $Sb(C_2H_5)_3O$, etc. The same tendency to act as a bivalent-radicle is, however, exhibited by triethylamine, which, though it does not unite directly with elementary bodies, can nevertheless take up a molecule of hydrogen chloride, ethyl iodide, etc., likewise producing compounds in which the nitrogen is quinquivalent, *e. g.*,

$$N(C_2H_5)_3HCl, \quad N(C_2H_5)_3(C_2H_5)I, \text{ etc.}$$

Stibethyl oxide, $Sb(C_2H_5)_3O$, forms a viscid, transparent, bitter, non-poisonous mass, soluble in water and alcohol ; not volatile without decomposition. Treated with acids, it forms crystallizable salts containing 2 molecules of a monatomic or 1 molecule of a diatomic acid radicle, *e. g. :* $Sb(C_2H_5)_3(NO_3)_2$, $Sb(C_2H_5)_3SO_4$, etc. The *sulphide,* $Sb(C_2H_5)_3S$, forms beautiful crystals of silvery lustre, soluble in water and alcohol. Their taste is bitter, and their odor similar to that of mercaptan. The solution of this compound exhibits the deportment of an alkaline sulphide : it precipitates metals from their solutions as sulphides, a soluble salt of stibethyl being formed at the same time. This deportment, indeed, affords the simplest means of preparing the salts of stibethyl. The *chloride,* $Sb(C_2H_5)_3Cl_2$, is a colorless liquid, having the odor of turpentine oil. The *iodide,* $Sb(C_2H_5)_3I_2$, forms colorless needles of intensely bitter taste.

The analogy of triethylstibine to triethylamine is best exhibited in its deportment with ethyl iodide. The two substances combine, forming tetrethylstibonium iodide, $Sb(C_2H_5)_4I$, from which silver oxide separates a powerful alkaline base analogous to tetrethylammonium hydroxide :

$$N(C_2H_5)_4(OH) \qquad\qquad Sb(C_2H_5)_4OH.$$

A series of analogous substances exists in the methyl series. They have been examined by Landolt, who has described several of their compounds, and separated the methyl-antimony-base analogous to tetramethylammonium hydrate.

The *iodide,* $Sb(CH_3)_4I$, produced by the action of methyl iodide upon trimethylstibine, $Sb(CH_3)_3$, crystallizes in white six-sided tables, which are easily soluble in water and alcohol, and slightly soluble in ether. It has a very bitter taste, and is decomposed by the action of heat. When treated with silver oxide, it yields a powerfully alkaline solution, exhibiting all the properties of potash, from which, on evaporation, a white crystalline mass, the *hydroxide of tetramethylstibonium,* $Sb(CH_3)_4(OH)$, crystallizes. This compound forms an acid sulphate, $Sb(CH_3)_4SO_4H$, which crystallizes in tables.

Bismethyl or **Triethylbismuthine,** $Bi(C_2H_5)_3$, analogous in composition to triethylstibine and triethylarsine, is formed by the action of ethyl iodide on an alloy of bismuth and potassium, and is extracted from the residue by ether. It is a yellow liquid of specific gravity 1.82, has a most nauseous odor, and emits vapors which take fire in contact with the air. It unites with oxygen, chlorine, bromine, iodine, and nitric acid.

––––––––––

DIATOMIC BASES OF THE PHOSPHORUS AND ARSENIC SERIES.

The action of ethene bromide on triethylphosphine gives rise to the formation of two crystalline bromides, according to the proportions in which the substances are brought in contact. These bromides are $C_8H_{19}PBr_2 = (C_2H_5)_3P + C_2H_4Br_2$ and $C_{14}H_{34}P_2Br_2 = 2(C_2H_5)_3P + C_2H_4Br_2$. The first of these compounds is the bromide of a phosphonium in which 3 atoms of hydrogen are replaced by ethyl and 1 atom by the univalent radicle bromethyl, C_2H_4Br, namely, $[(C_2H_4Br)(C_2H_5)_3P]Br$. Half the bromine in this salt is unaffected by the action of silver-salts ; it may accordingly be designated as *bromide of bromethyl-triethyl-phosphonium.* Numerous salts of this compound are known, but the free base cannot be obtained, since silver oxide eliminates the latent bromine, giving rise to the formation of a base containing $[C_2H_5O)(C_2H_5)_3P]OH$. The second compound is the dibromide of *ethene-hexethyl-diphosphonium* $[(C_2H_4)(C_2H_5)_6P_2]Br_2$. This radicle, analogous to diammonium, N_2H_8, forms a series of very stable and beautiful salts, especially an iodide, which is difficultly soluble in water. In all these salts the base, which is composed of 1 atom of ethene, 6 atoms of ethyl, and 2 atoms of phosphorus, is united with 2 equivalents of univalent acid radicle; the platinum-salt contains $(C_2H_4)(C_2H_5)_6P_2Br_2PtCl_4$. The very caustic and stable base has the composition $[(C_2H_4)(C_2H_5)_6P_2]$ $(OH)_2$.

The dibromide of ethene-hexethyl-diphosphonium may be formed by the action of triethylphosphine upon the brominated bromide which has been mentioned as the first product of the action of ethene dibromide upon triethylphosphine: $C_8H_{19}PBr_2.(C_2H_5)_3P = C_{14}H_{34}P_2Br_2$. If the triethylphosphine be replaced in this process by ammonia or by monamines in general, or by monarsines, an almost unlimited series of diatomic salts may be formed, in which phosphorus and nitrogen or phosphorus and arsenic are associated.

Thus the action of ammonia, of ethylamine, and triethylarsine, gives rise respectively to the following compounds:

Dibromide of Ethene-triethyl-phosphammonium . . .	$[(C_2H_4)(C_2H_5)_3H_3PN]Br_2$.
Dibromide of Ethene-tetrethyl-phosphammonium . . .	$[(C_2H_4)(C_2H_5)_4H_2PN]Br_2$.
Dibromide of Ethene-hexethyl-phospharsonium . . .	$[(C_2H_4)(C_2H_5)_6PAs]Br_2$.

Treated with silver oxide those bromides yield the very caustic diatomic bases—

Hydroxide of Ethene-triethyl-phosphammonium . . .	$[(C_2H_4)(C_2H_5)_3H_3PN](OH)_2$.
Hydroxide of Ethene-tetrethyl-phosphammonium . . .	$[(C_2H_4)(C_2H_5)_4H_2PN](OH)_2$.
Hydroxide of Ethene-hexethyl-phospharsonium	$[(C_2H_4)(C_2H_5)_6PAs](OH)_2$.

The arsenic bases, when submitted to the action of ethene dibromide, give rise to perfectly analogous results. The limits of this Manual will not permit us to examine these compounds in detail.

ORGANO-BORON COMPOUNDS.

Boric Triethide, or **Borethyl,** $B(C_2H_5)_3$, discovered by Frankland, is obtained, together with zinc-ethoxide, by treating triethylic borate with zinc-ethide:—

$$B(OC_2H_5)_3 + 3Zn(C_2H_5)_2 = 3Zn(OC_2H_5)_2 + B(C_2H_5)_3.$$

It is a colorless, mobile liquid, having a pungent odor, irritating the eyes, of sp. gr. 0.696 at 23○ C. (73.4○ F.), and boiling at 95○ C.(203○ F.). Borethyl is insoluble in water, but very slowly decomposed when left in prolonged contact with it. When exposed to the air it takes fire, burning with a beautiful green and somewhat smoky flame. It combines with ammonia, forming the compound $NH_3.B(C_2H_5)_3$. By the gradual action of dry air, and ultimately of dry oxygen, borethyl is converted into the diethylic ether of ethylboric acid, $(C_2H_5)B{<}^{OC_2H_5}_{OC_2H_5}$, a liquid boiling with partial decomposition, between 95° and 115° C. (203° and 239° F.), but distilling unchanged under reduced pressure. In contact with water this liquid is immediately transformed into ethylboric acid, $(C_2H_5)B(OH)_2$.

*Diboric Ethopentethoxide** (or Ethopentethylate), $B_2(C_2H_5)(OC_2H_5)_5$, is formed by heating 2 molecules of boric ether with 1 molecule of zinc ethide :—

$$2B(OC_2H_5)_3 + Zn(C_2H_5)_2 = Zn{<}^{C_2H_5}_{OC_2H_5} + B_2(C_2H_5)(OC_2H_5)_5 \ .$$

It is a colorless, mobile liquid, boiling at about 120° C. (248° F.), and condensing unchanged. Its vapor-density, taken between 114° and 120° C. (237.2° and 248° F.), is 69 (H $=$1), which represents a four-volume condensation, indicating that the compound, in passing from the liquid to the gaseous state, is broken up into diethylic ethylborate and boric ether :—

$$B_2(C_2H_5)(OC_2H_5)_5 = (C_2H_5)B(OC_2H_5)_2 = B(OC_2H_5)_3 \ ,$$

just as sal-ammoniac, NH_4Cl, is resolved by heat into NH_3 and HCl, which together occupy four volumes of vapor (p. 246). A similar deportment is exhibited by ammonioboric methide, $NH_3.B(CH_3)_3$, which is resolved by heat into NH_3 and $B(CH_3)_3$, the mixed vapor resulting from the decompositiou having also a four-volume condensation. These results render it probable that the compounds under consideration have a constitution analogous to that of the ammonium salts, and that boron, though generally triadic, may, like nitrogen, enter into combination also as a pentad. On this view the constitution of ammonioboric methide and diboric ethopentethoxide may be represented by the following formulæ :—

$$NH_3 \qquad\qquad (C_2H_5)B(OC_2H_5)_2$$
$$\| \qquad\qquad\qquad \|$$
$$B(CH_3)_3 \qquad\qquad B(OC_2H_5)_3$$

Ammonioboric   Diboric
methide.   Ethopentethoxide.

In contact with water, diboric ethopentethoxide is immediately decomposed into boric acid, ethyl-boric acid, and alcohol :—

$$B_2(C_2H_5)(OC_2H_5)_5 + 5H_2O = B(OH)_3 + (C_2H_5)B(OH)_2 + 5C_2H_5(OH).$$

Boric Diethylethoxide, $(C_2H_5)_2B(OC_2H_5)$, is produced by boiling boric ether with 2 molecules of zinc ethide :—

$$B(OC_2H_5)_3 + 2Zn(C_2H_5)_2 = 2Zn{<}^{C_2H_5}_{OC_2H_5} + (C_2H_5)_2B(OC_2H_5).$$

When purified by distillation in an atmosphere of carbon dioxide, it is a colorless, mobile pungent liquid boiling at 102° C. (215.6° F.), and having at 135.5° C. (275.9° F.) a vapor-density of 56.5 (H $=$ 1), indicating a normal two-volume condensation. It takes fire in the air, and burns with a green flame. Exposed to dry air and then to oxygen, it oxidizes to boric ethylodiethoxide, $(C_2H_5)B(OC_2H_5)_2$.

Boric Diethylhydroxide, $(C_2H_5)_2B(OH)$, formed by agitating boric diethyl-ethoxide with water, is a spontaneously inflammable ethereal liquid, resembling boric ethide, and decomposing when distilled.

Boric Ethyl-hydroxethoxide, $(C_2H_5)B{<}^{OC_2H_5}_{OH}$, is formed by exposing the last compound in a cooled vessel to a slow current of dry air. It is liquid at ordinary temperatures, but solidifies below 8° C. (46.4° F.)

* This and the following compounds have quite recently been discovered by Frankland.(Proc. Roy. Soc. 1876, vol. xxv. p. 165).

to a white crystalline body smelling like borethide, and having a pungent taste. It is rapidly decomposed by water into ethylboric acid and alcohol:

$$(C_2H_5)B(OH)(OC_2H_5) + H_2O = (C_2H_5)B(OH)_2 + C_2H_5OH.$$

It is not spontaneously inflammable, and cannot be distilled under ordinary atmospheric pressure without decomposition.

Boric Trimethide, or **Bor-methyl,** $B(CH_3)_3$, obtained like the ethyl-compound, is a colorless gas, which condenses to a liquid at low temperatures. It unites with ammonia, forming the compound $H_3N\!=\!\!B(CH_3)_3$.

ORGANO-SILICON COMPOUNDS.

Silicon is a tetrad element, and forms with alcohol-radicles compounds bearing a very close analogy to the hydrocarbons.

Silicio-Tetramethide, $Si(CH_3)_4$, is produced by heating silicon tetrachloride with zinc methide :—

$$SiCl_4 + 2Zn(CH_3)_2 = 2ZnCl_2 + Si(CH_3)_4.$$

It is a mobile liquid, which boils at 30° C. (86° F.), is not decomposed by water, and reacts generally like a hydrocarbon, viz., tetramethyl-methane.

Silicio Tetrethide, $Si(C_2H_5)_4$, prepared like the methyl-compound, is a liquid boiling at 153° C. (307.4° F.). With chlorine it yields the monochlorinated derivative, $Si\begin{Bmatrix}(C_2H_5)_3\\C_2H_4Cl\end{Bmatrix}$, which boils at 185° C. (365° F.) and reacts exactly like the chloride of a hydrocarbon, being converted by potassium acetate into the acetic ether, $Si\begin{Bmatrix}(C_2H_5)_3\\C_2H_4(OC_2H_3O)\end{Bmatrix}$, which boils at 211° C. (411.8° F.), and is decomposed by alkalies into acetic acid, and an alcohol having the composition $Si\begin{Bmatrix}(C_2H_5)_3\\C_2H_4.OH\end{Bmatrix}$. This alcohol is a liquid insoluble in water, and boiling at 190° C. (374° F.).

Silicic tetrethide and its derivatives may be regarded as *nonyl-compounds* in which 1 atom of carbon is replaced by silicon ; thus :

$Si(C_2H_5)_4$	$= SiC_8H_{20}$	Silicononane,
$Si\begin{Bmatrix}(C_2H_5)_3\\C_2H_4Cl\end{Bmatrix}$	$= SiC_8H_{19}Cl$	Silicononyl chloride,
$Si\begin{Bmatrix}(C_2H_5)_3\\C_2H_4.O.C_2H_3O\end{Bmatrix}$	$= SiC_8H_{19}(O.C_2H_3O)$.	" acetate,
$Si\begin{Bmatrix}(C_2H_5)_3\\C_2H_4(OH)\end{Bmatrix}$	$= SiC_8H_{19}(OH)$ . . .	" alcohol.

Ethylated Silicic Ethers.—When normal silicic ether, $Si(OC_2H_5)_4$, is heated with zinc-ethyl and sodium, one or all of the ethoxyl-groups are replaced by ethyl, the product being a mixture of mono-, di-, and triethylated silicic ethers, and silicic tetrethide, which may be separated by fractional distillation.

Triethylsilicic ether, $(C_2H_5)_3Si.(OC_2H_5)$, or *Silicoheptyl-ethyl oxide,* $SiC_6H_{15}.O.C_2H_5$, is a liquid, of specific gravity 0.841 at 0° C., boiling at

153° C. (307.4° F.), insoluble in water. Treated with acetic anhydride, it yields an acetic ether, convertible by the action of potash into t r i e t h y l silicon-h y d r o x i d e, $(C_2H_5)_3Si(OH)$, which is a tertiary silicic alcohol, analogous to triethyl-carbinol, and is hence called t r i e t h y l-s i l i c o l. It is a colorless, oily liquid, smelling like camphor, insoluble in water, boiling at 154° C. (309.2° F.).

Diethylsilicic ether, $(C_2H_5)_2Si(OC_2H_5)_2$, is a fragrant liquid, insoluble in water, boiling at 155.8° C.(312.4° F.), and having a specific gravity of 0.875 at 0° C. When heated with acetyl chloride, it yields the compounds $(C_2H_5)_2Si(OC_2H_5)Cl$ and $(C_2H_5)_2SiCl_2$. The latter is a fuming liquid, boiling at 148° C. (298.4° F.), and converted by water into d i-e t h y l s i l i c i c o x i d e $(C_2H_5)_2SiO$, analogous to diethyl ketone, $(C_2H_5)_2CO$.

Monethyl silicic ether, $(C_2H_5)Si(OC_2H_5)_3$, is a liquid smelling like camphor, boiling at 159° C. (318.2° F.), slowly decomposed by water. Heated with acetyl chloride, it yields e t h y l s i l i c o n t r i c h l o r i d e, $(C_2H_5)SiCl_3$, a strongly fuming liquid which boils at 100°, and is converted by water into e t h y l s i l i c i c or s i l i c o p r o p i o n i c a c i d, C_2H_5.$SiO.OH$. This acid is a white amorphous powder, which smoulders away when heated in the air, and dissolves in aqueous potash and soda, forming **silico-propionates**.

ORGANO-METALLIC COMPOUNDS.

The name is especially applied to compounds of alcohol-radicles with metals not belonging to the pentad group. Those containing the more basic metals, such as zinc, are formed by the direct action of the metals, or better, of their sodium alloys, on the alcoholic iodides : *e. g.*,

$$\underset{\text{Zinc-sodium.}}{ZnNa_2} + 2C_2H_5I = 2NaI + \underset{\text{Zinc ethide.}}{Zn(C_2H_5)_2}$$

Those containing the less basic metals (tin, lead, etc.) are produced by the action of organo-zinc or mercury-compounds on metallic chlorides : *e. g.*,

$$SnCl_4 + 2Zn(C_2H_5)_2 = 2ZnCl_2 + Sn(C_2H_5)_4.$$

Potassium Ethide, C_2H_5K, and **Sodium Ethide**, C_2H_5Na, are known only in combination or mixture with zinc ethide, in which state they are obtained by the action of potassium or sodium on zinc ethide. These compounds and their homologues, discovered by Wanklyn, have played an important part in chemical synthesis. They absorb carbon-dioxide, producing s a l t s of the fatty acids:

$$\underset{\text{Sodium propionate.}}{C_2H_5Na + CO_2 = C_2H_5.CO_2Na.}$$

By the action of *carbon monoxide* they are converted into k e t o n e s:

$$2C_2H_5Na + CO = Na_2 + (C_2H_5)_2CO.$$

Magnesium Ethide, $Mg(C_2H_5)_2$.—When ethyl iodide is heated with magnesium filings in a vessel from which the air is excluded, magnesium ethiodide is formed in the first instance according to the equation:

$$Mg + C_2H_5I = Mg{<}^{C_2H_5}_{I};$$

and this compound, when heated, is resolved into iodide and ethide :

$$2Mg{<}_I^{C_2H_5} = MgI_2 + Mg(C_2H_5)_2.$$

Magnesium ethide is a liquid which takes fire spontaneously in the air, and is rapidly decomposed by water, with formation of ethane :

$$Mg(C_2H_5)_2 + H_2O = 2C_2H_6 + MgO.$$

Zinc Ethide or **Zinc-ethyl**, $Zn(C_2H_5)_2$, is formed by heating ethyl iodide with zinc in a sealed glass tube, or for larger quantities, in a strong and well-closed copper cylinder. The reaction takes place in the same manner as with magnesium, zinc ethiodide, $Zn(C_2H_5)I$, being formed in the first instance as a white crystalline mass, which, when distilled in an atmosphere of hydrogen, is resolved into zinc-iodide and zinc-ethide, the latter distilling over.

Zinc-ethide is a mobile and very volatile disagreeable-smelling liquid, having a specific gravity of 1.182, and boiling at 118° C. (244.4° F.). It takes fire instantly on coming in contact with the air, diffusing white fumes of zinc oxide. *Water* decomposes it violently, with formation of zinc hydroxide, and evolution of ethane :

$$Zn(C_2H_5)_2 + 2H_2O = ZnH_2O_2 + 2C_2H_6.$$

When gradually mixed with dry *oxygen*, it passes through two stages of oxidation, yielding first zinc-ethyl-ethoxide, $Zn{\{}_{OC_2H_5}^{C_2H_5}$, and finally zinc-ethoxide, $Zn(OC_2H_5)_2$. With *iodine* and other halogens, the reaction also takes place by two stages, but consists in the successive substitution of the halogen for the ethyl; thus :

$$Zn(C_2H_5)_2 + I_2 = C_2H_5I + Zn(C_2H_5)I,$$

and

$$Zn(C_2H_5)I + I_2 = C_2H_5I + ZnI_2.$$

Zinc Methide, $Zn(CH_3)_2$, prepared in like manner, is a mobile spontaneously inflammable liquid, which boils at 46° C. (114.8° F.), and resembles the ethyl compound in all its reactions.

Zinc Isopropide, $Zn(C_3H_7)_2$, boils at 146° C. (294.8° F.). **Zinc isopentide**, or **Zinc amylide**, $Zn(C_5H_{11})_2$, boils at 220° C. (428° F.), and fumes strongly in the air, but does not take fire spontaneously.

These organo-zinc compounds, discovered by Frankland, are very important reagents in organic synthesis, serving to effect the substitution of the positive radicle ethyl, etc., for chlorine, iodine, and other negative elements, and thus enabling us to build up carbon-compounds from others lower in the scale. With *carbon oxychloride* (phosgene) they form k e t o n e s :

$$COCl_2 + Zn(CH_3)_2 = ZnCl_2 + \underset{\text{Acetone.}}{CO(CH_3)_2};$$

also with the *chlorides* of *acid radicles* at ordinary temperatures : *e. g.*,

$$2(CH_3.COCl) + Zn(C_2H_5)_2 = ZnCl_2 + 2CO{<}_{C_2H_5}^{CH_3}$$
$$\text{Methyl-ethyl ketone.}$$

They also serve, as already mentioned, for the preparation of other organo-metallic bodies : *e. g.*,

$$Zn(C_2H_5)_2 + HgCl_2 = ZnCl_2 + Hg(C_2H_5)_2$$
$$2Zn(C_2H_5)_2 + SnCl_4 = 2ZnCl_2 + Sn(C_2H_5)_4$$
$$3Zn(C_2H_5)_2 + 2AsCl_3 = 3ZnCl_2 + 2As(C_2H_5)_3.$$

Sulphur dioxide is absorbed by these zinc-compounds, with formation of the zinc salts of m e t h y l - and e t h y l - d i t h i o n i c or s u l p h i n i c a c i d s (p. 582). *Nitrogen dioxide* dissolves in zinc-ethide, forming a crystalline compound, which, by the action of water and carbon dioxide, is converted into the zinc salt of d i n i t r o e t h y l i c a c i d , $C_2H_5.N_2O_2H$, the structure of which has not yet been satisfactorily made out.

M e r c u r i c E t h i d e , $Hg(C_2H_5)_2$.—This compound is formed by the action of mercuric chloride on zinc ethide, but it is more easily prepared by the action of sodium-amalgam on ethyl iodide in presence of acetic ether :

$$2C_2H_5I \ + \ Na_2 \ + \ Hg \ = \ 2NaI \ + \ Hg(C_2H_5)_2.$$

The acetic ether takes no part in the reaction ; nevertheless its presence appears to be essential.

Mercuric ethide is a transparent colorless liquid, boiling at 159° C. (318.2° F.), and having a sp. gr. of 2.44. It burns with a smoky flame, giving off a large quantity of mercurial vapor. Chlorine, bromine, and iodine remove one equivalent of ethyl from this compound, and take its place, forming m e r c u r i c c h l o r e t h i d e , etc. ; thus :

$$Hg(C_2H_5)_2 \ + \ Cl_2 \ = \ C_2H_5Cl \ + \ Hg(C_2H_5)Cl.$$

A similar action is exerted by acids, *e. g.*, by hydrobromic acid, the products being ethane and mercuric bromethide :

$$Hg(C_2H_5)_2 \ + \ HBr \ = \ C_2H_6 \ + \ Hg(C_2H_5)Br.$$

The chlorethide or bromethide is converted by water into mercuric ethylhydrate, $Hg(C_2H_5)(OH)$. Mercuric ethide serves for the preparation of several other organo-metallic bodies.

Mercuric Methide, $Hg(CH_3)_2$, prepared by similar processes, is a liquid slightly soluble in water, boiling at 95° C. (203° F.), and having a specific gravity of 3.069.

Mercuric Isopentide or **Amylide,** $Hg(C_5H_{11})_2$ is a thick liquid, of specific gravity 1.66, decomposed by distillation.

Mercuric Allyliodide, $Hg{<}^{C_3H_5}_{I}$, produced by agitating allyl iodide with mercury, crystallizes from alcohol in silvery laminæ, melting at 135° C. (275° F.). With hydrogen iodide, it yields mercuric iodide and p r o - p e n e :

$$Hg(C_3H_5)I \ + \ HI \ = \ HgI_2 \ + \ C_3H_6.$$

A l u m i n i u m M e t h i d e , $Al(CH_3)_3$, or $Al_2(CH_3)_6$. This compound, discovered by Buckton and Odling, is formed by heating mercuric methide with aluminium. It is a mobile liquid, which crystallizes at a little above 0°, and boils at 130° C. (266° F.). At and above 220° C. (428° F.) the density of its vapor, compared with that of air, is 2.8, which is near to the theoretical density calculated for the formula $Al(CH_3)_3$, namely, 2.5. This seems to show that the true formula of the compound is $Al(CH_3)_3$, and not $Al_2(CH_3)_6$, and, consequently, that aluminium is a triad, not a tetrad (p. 257). At temperatures near the boiling point, however, the vapor-density becomes 4.4, approximating to the theoretical density calculated for the formula, $Al_2(CH_3)_6$.

Aluminium Ethide, $Al(C_2H_5)_2$, or $Al_2(C_2H_5)_6$, resembles the methyl compound. It boils at 194° C. (381.2° F.), and its vapor likewise exhibits, at temperatures considerably above its boiling point, a density nearly equal to that required by the formula $Al(C_2H_5)_3$, for a two-volume condensation.

Tin Compounds.—Tin forms two ethyl-compounds, $Sn(C_2H_5)_2$ and $Sn(C_2H_5)_4$, analogous to stannous and stannic chloride; also a stannoso-stannic ethide, $Sn_2(C_2H_5)_6$, analogous in constitution to ethane, C_2H_6. Stannic ethide is a saturated compound, but the other two are unsaturated bodies, capable of uniting with chlorine, bromine, oxygen, and acid radicles, and being thereby converted into compounds of the stannic type.

STANNOUS ETHIDE, $Sn(C_2H_5)_2$.—When ethyl iodide and tinfoil are heated together in a sealed glass tube to about 150° or 180° C. (302–356° F.), stannic iodethide, $Sn(C_2H_5)_2I_2$, is produced, crystallizing in colorless needles. The same compound is obtained when tin and ethyl iodide are exposed to the rays of the sun concentrated by a concave mirror. The reaction is considerably facilitated if the tin be alloyed with one-tenth of its weight of sodium. This iodide is decomposed by sodium or zinc, which abstracts the iodine, and leaves stannous ethide in the form of a thick, oily liquid, insoluble in water, having the sp. gr. 1.558, and decomposed by distillation. Stannous ethide combines directly with 2 atoms of chlorine, iodine, and bromine, forming stannic chlorethide, $Sn^{iv}(C_2H_5)_2Cl_2$, etc. Exposed to the air, it absorbs oxygen, and is converted into stannic oxethide, $Sn^{iv}(C_2H_5)_2O$, a whitish, tasteless, inodorous powder, which when treated with oxygen acids, yields well-crystallized stannic salts, such as $Sn(C_2H_5)_2(NO_3)_2$, $Sn(C_2H_5)_2SO_4$, etc.

STANNOSO-STANNIC ETHIDE, $Sn_2(C_2H_5)_6$, is always produced in small quantity when stannous ethide is prepared by the methods above mentioned. It is readily obtained in the free state by digesting an alloy of 1 part of sodium and 5 parts of tin with ethyl iodide, exhausting the mass with ether, evaporating the ethereal solution, and exhausting the residue with alcohol. The stannoso-stannic ethide, being insoluble in that liquid, then remains behind. It is a yellow oil, boiling at 180° C. (356° F.), combining directly with chlorine, bromine, and iodine to form two molecules of a stannic compound; e. g.:

$$Sn_2(C_2H_5)_6 \ + \ Cl_2 \ = \ 2Sn^{iv}(C_2H_5)_3Cl ;$$
Stannic chloro-triethide.

also with oxygen, forming distannic oxy-hexethide, $Sn^{iv}_2(C_2H_5)_6O$. This oxide is, however, best obtained by distilling stannic oxy-diethide, $Sn^{iv}(C_2H_5)_2O$ (above described), with potash. It is an oily liquid, soluble in alcohol, ether, and water; the aqueous solution has a strong alkaline reaction. It is easily acted upon by oxygen-acids, yielding the corresponding sulphate, $Sn_2(C_2H_5)_6SO_4$, etc.

STANNIC ETHIDE, $Sn^{iv}(C_2H_5)_4$, is produced by the action of zinc ethide on stannic chloride; also by the distillation of stannous ethide: $2Sn(C_2H_5)_2$ $= Sn + Sn(C_2H_5)_4$. It is a colorless, nearly odorless liquid, of sp. gr. 1.19, boiling at 181° C. (257.8° F.), and very inflammable, burning with a highly luminous flame. When treated with chlorine, bromine, etc., or with acids, it forms substitution-products: thus, with iodine, it splits up into ethyl iodide and stannic iodotriethide:

$$Sn(C_2H_5)_4 \ + \ I_2 \ = \ C_2H_5I \ + \ Sn(C_2H_5)_3I,$$

with strong hydrochloric acid, it yields ethane and stannic chlorotriethide, $Sn(C_2H_5)_4 + HCl = C_2H_6 + Sn(C_2H_5)_3Cl.$

Stannous Methide, $Sn(CH_3)_2$, and **Stannic Methide**, $Sn(CH_3)_4$, resemble the corresponding ethyl compounds, and are obtained by similar reactions.

Plumbic Ethide, $Pb(C_2H_5)_4$, is produced by the action of plumbic chloride on zinc ethide:

$$2Zn(C_2H_5)_2 \;+\; 2PbCl_2 \;=\; 2ZnCl_2 \;+\; Pb \;+\; Pb(C_2H_5)_4.$$

It is a colorless limpid liquid, soluble in ether but not in water. When protected from the air, it boils, with partial decomposition, at about 200° C. $(392^\circ$ F.). It is not acted upon by oxygen at ordinary temperatures, but chlorine, bromine, and iodine act violently upon it in the same manner as on stannic ethide, forming plumbic chloro-triethide, $Pb(C_2H_5)_3Cl$, etc. Plumbic ethide is interesting, as affording a proof that lead is really a tetrad.

ALDEHYDES.

THESE are bodies containing the bivalent group CO, associated, on the one hand, with a monatomic alcohol radicle, and on the other with hydrogen; e. g.,

$$\text{H—CO—CH}_3 \qquad\qquad \text{H—CO—C}_4\text{H}_9$$
Acetic aldehyde.   Valeric aldehyde.

They are derived from primary alcohols by elimination of one or more molecules of hydrogen (H_2), without introduction of an equivalent quantity of oxygen, so that they hold a position intermediate between the alcohols and the acids; thus:

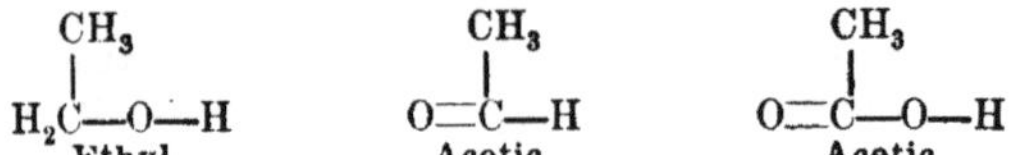

$$\begin{array}{ccc}
\text{CH}_3 & \text{CH}_3 & \text{CH}_3 \\
| & | & | \\
\text{H}_2\text{C—O—H} & \text{O}{=}\text{C—H} & \text{O}{=}\text{C—O—H} \\
\text{Ethyl} & \text{Acetic} & \text{Acetic} \\
\text{alcohol.} & \text{aldehyde.} & \text{acid.}
\end{array}$$

The hydrogen eliminated in the conversion of a primary alcohol into an aldehyde is that which belongs to the group CH_2OH; consequently a monatomic alcohol can yield but one aldehyde; but a diatomic alcohol can yield two, by removal of H_2 and of $2H_2$; a triatomic alcohol three, and so on. At present, however, we are acquainted only with aldehydes derived from monatomic and diatomic alcohols.

Aldehydes derived from Monatomic Alcohols.

Of these aldehydes there are two series belonging to the fatty group, viz.:—

1. *Aldehydes*, $C_nH_{2n}O$, *corresponding with the Fatty acids.*

Formic aldehyde	. .	CH_2O	Caproic aldehyde . .	$C_6H_{12}O$
Acetic aldehyde	. .	C_2H_4O	Œnanthylic aldehyde .	$C_7H_{14}O$
Propionic aldehyde	. .	C_3H_6O	Caprylic aldehyde . .	$C_8H_{16}O$
Butyric aldehyde	. .	C_4H_8O	Euodic aldehyde . .	$C_{11}H_{22}O.$
Valeric aldehyde	. .	$C_5H_{10}O$		

2. *Aldehydes*, $C_nH_{2n-2}O$, *corresponding with the Acrylic acids.*

Acrylic aldehyde, or Acrolein . . . C_3H_4O
Crotonic aldehyde . . . . . C_4H_6O.

All these aldehydes contain two atoms of hydrogen less than the cor. responding alcohols, and one atom of oxygen less than the corresponding acids.

The aldehydes of the fatty groups are produced : 1. By oxidation of primary alcohols, either by the action of atmospheric oxygen, or by that of a mixture of dilute sulphuric acid and potassium dichromate or manganese dioxide, or by the action of chlorine on the alcohol diluted with water, the chlorine in this case decomposing the water, and thus acting as an oxidizing agent ; *e. g.* :—

$$\underset{\text{Ethyl alcohol.}}{\overset{CH_3}{\underset{|}{CH_2OH}}} + O = H_2O + \underset{\text{Acetic aldehyde.}}{\overset{CH_3}{\underset{|}{COH}}}$$

2. By distilling an intimate mixture of the calcium salt of the corresponding acid with calcium formate ; *e. g.* :—

$$\underset{\text{Acetic acid.}}{CH_3.CO.OH} + \underset{\text{Formic acid.}}{H.CO.OH} = CO_2 + H_2O + \underset{\text{Acetic aldehyde.}}{CH_3.CO.H}$$

3. By the action of nascent hydrogen on the anhydrides and chlorides (chloranhydrides) of the fatty acids :—

$$\underset{\text{Acetyl chloride.}}{CH_3.COCl} + H_2 = HCl + \underset{\text{Acetic aldehyde.}}{CH_3.COH}$$

$$\underset{\text{Acetic anhydride.}}{\overset{CH_3.CO}{\underset{CH_3.CO}{>}}O} + 2H_2 = H_2O + 2(CH_3.COH).$$

4. From the corresponding di-halogen derivatives of the hydrocarbons, by heating with water, or better, with lead oxide :—

$$\underset{\text{Ethidene chloride.}}{CH_3.CHCl_2} + PbO = PbCl_2 + CH_3.CHO .$$

5. By the slow oxidation of albuminous substances by means of manganese dioxide and sulphuric acid.

Some of the aldehydes occur in the essential oils of plants.

The aldehydes are colorless, neutral, volatile liquids, having more or less pungent odors. Only the lower aldehydes are soluble in water.

Reactions.—1. Aldehydes are easily converted by *oxidation* into the corresponding a c i d s , either on exposure to the air or by contact with moist silver oxide, in the latter case with reduction of metallic silver :—

$$\underset{\substack{\text{Acetic}\\\text{aldehyde.}}}{CH_3COH} + 3AgOH = \underset{\text{Silver acetate.}}{CH_3.CO.OAg} + 2H_2O + Ag_2 .$$

This oxidation takes place with peculiar facility with ammoniacal silver solution, the silver being frequently deposited in a bright specular film.

2. Aldehydes fused with *potash* are converted into the corresponding acids, with evolution of hydrogen ; *e. g.*,

$$\underset{\substack{\text{Valeric}\\\text{aldehyde}}}{C_5H_{10}O} + KOH = \underset{\text{Valeric acid.}}{C_5H_9KO_2} + H_2.$$

3. *Nascent hydrogen*, evolved by the action of water on sodium amalgam, converts them into the corresponding alcohols; *e. g.*, $C_2H_4O + H_2 = C_2H_6O$. If, however, the aldehyde belongs to a non-saturated series, the action goes further, an additional quantity of hydrogen being then taken up, whereby the alcohol first formed is converted into a saturated alcohol belonging to another series ; thus :—

$$C_3H_4O \ + \ H_2 \ = \ C_3H_6O; \quad \text{and} \quad C_3H_6O \ + \ H_2 \ = \ C_3H_8O \ .$$

Acrylic aldehyde.   Allyl alcohol.   Allyl alcohol.   Propyl alcohol.

Nascent hydrogen evolved by the action of zinc on sulphuric acid does not appear to unite with aldehydes.

4. *Phosphorus pentachloride* converts aldehydes into chloraldehydes, compounds derived from aldehydes by substitution of Cl_2 for O; thus :—

$$\begin{array}{c}CH_3\\ |\\ CHO\end{array} \ + \ PCl_5 \ = \ PCl_3O \ + \ \begin{array}{c}CH_3\\ |\\ CHCl_2\end{array}$$

Aldehyde.   Chloraldehyde.

The compounds thus produced are isomeric with the chlorides of the olefines ; *e. g.*, acetic chloraldehyde, $CH_3.CHCl_2$, or ethidene chloride, with ethene chloride, $C_2H_4.Cl_2$ (p. 517).

5. *Chlorine* and *bromine* convert aldehydes into chlorides and bromides of acid radicles :

$$C_2H_4O \ + \ Cl_2 \ = \ HCl \ + \ C_2H_3O.Cl$$
Aldehyde.   Acetyl chloride.

$$C_2H_4O \ + \ 2Cl_2 \ = \ 2HCl \ + \ C_2H_2ClO.Cl$$
Aldehyde.   Chloracetyl chloride.

6. The *alkali-metals* dissolve in aldehydes, eliminating an equivalent quantity of hydrogen :

$$2C_2H_4O \ + \ K_2 \ = \ H_2 \ + \ 2C_2H_3KO.$$

7. Aldehydes heated with *hydrocyanic* and *hydrochloric acid* are converted, first into cyanhydrins, which then, by the action of the hydrochloric acid, or by that of alkalies, are converted into oxy-acids ; thus :

$$CH_3.CHO \ + \ CNH \ = \ CH_3.CH{<}^{OH}_{CN} \ .$$

Aldehyde.   Ethidene-cyanhydrin.

$$CH_3.CH{<}^{OH}_{CN} \ +2H_2O \ = \ NH_3 \ + \ CH_3.CH{<}^{OH}_{CO.OH}$$

Lactic acid.

8. Aldehyde likewise unites with other carbon-compounds, as with the anhydrides and chlorides of the fatty acids, forming compound ethers :

$$CH_3.CHO \ + \ C_2H_3OCl \ = \ CH_3.CH{<}^{Cl}_{O.C_2H_3O}$$

Acetyl chloride.   Ethidene chloro-acetate.

$$CH_3.CHO \ + \ {C_2H_3O \atop C_2H_3O}{>}O \ = \ CH_3.CH{<}^{O.C_2H_3O}_{O.C_2H_3O}$$

Acetic anhydride.   Ethidene diacetate.

9. Aldehydes combine with ammonia, forming compounds called alde-hyde-ammonias, the mode of formation of which is shown in the following equation :—

$$CH_3.CHO \ + \ NH_3 \ = \ CH_3.CH{<}^{OH}_{NH_2} \ .$$

These compounds are easily soluble in water, but insoluble in ether, and are therefore precipitated by ammonia from the ethereal solutions of the aldehydes. They are rather unstable, and are easily resolved by acids into ammonia and the aldehydes.

10. Aldehydes also unite with *aniline*, water being eliminated, and form bases derived from a double molecule of aniline, $(C_6H_7N)_2$, by substitution of two equivalents of a diatomic radicle for four atoms of hydrogen ; *e. g.*,

$$2C_2H_4O \quad + \quad 2C_6H_7N \quad = \quad 2H_2O \quad + \quad C_{12}H_{10}(C_2H_4)_2N_2$$

$$\text{Acetic} \qquad\qquad \text{Aniline.} \qquad\qquad\qquad\qquad \text{Diethidene-}$$
$$\text{aldehyde.} \qquad\qquad\qquad\qquad\qquad\qquad\qquad \text{dianiline.}$$

11. All aldehydes unite directly with the *acid sulphites of the alkali-metals*, forming crystalline compounds, which may be regarded as salts of ethidene-oxysulphonic acid (p. 634), as shown by the following equation :

$$CH_3.CHO \quad + \quad SO_3HNa \quad = \quad CH_3.CH\!\!<\!\!^{OH}_{SO_3Na}.$$

From these salts, the aldehydes may be separated by distillation with dilute sulphuric acid or solution of sodium carbonate. The reaction affords a ready means of purifying the aldehydes, and of detecting their presence in mixtures.

12. Under the influence of certain reagents (small quantities of acids, alkalies, and salts) two or more molecules of an aldehyde may unite together, forming a polymeride ; *e. g.*,

$$CH_3.CHO \quad + \quad {OCH.CH_3 \atop OCH.CH_3} \quad = \quad CH_3.CH\!\!<\!\!{^{O-C}_{O-C}}\!\!>\!\!{^{H.CH_3}_{O}}_{H.CH_3}$$

$$\text{Aldehyde (3 mol.).} \qquad\qquad\qquad\qquad \text{Paraldehyde.}$$

ALDEHYDES BELONGING TO THE SERIES $C_nH_{2n}O$.

Formic Aldehyde, CH_2O or $H.CHO$, also called *Methylic aldehyde.*—This compound, discovered by Hofmann, is produced when a current of air charged with vapor of methyl alcohol is directed upon an incandescent spiral of platinum wire ; and by suitable condensing arrangements, a liquid may be obtained consisting of a solution of the aldehyde in methyl alcohol. This liquid, rendered slightly alkaline by ammonia, and gently warmed with silver nitrate, yields a beautiful specular deposit of silver, with greater ease even than ordinary acetic aldehyde. The same solution, heated with a few drops of caustic potash, deposits drops of a brownish oil, having the odor of the resin of acetic aldehyde.

Formic aldehyde is likewise obtained by the dry distillation of calcium formate : $Ca(CHO_2)_2 = CH_2O + CaCO_3$. It has not yet been obtained in the pure state ; but on heating the distillate obtained as above, part of the formic aldehyde escapes, and another portion remains as a solid polymeric modification, viz.,

Paraformaldehyde or *Trioxymethene*, $C_3H_6O_3$ or $(CH_2)_3O_3$, a compound likewise formed by the action of oxide or oxalate of silver on methene iodide, CH_2I_2, or best by distilling glycollic acid with a small quantity of sulphuric acid. It is a crystalline mass, insoluble in water, alcohol, and ether, melting at $152°$ C. $(305.6°$ F.), subliming below $100°$. Its vapor-density, 1.06, corresponds with the formula CH_2O, showing that the molecule $C_3H_6C_3$ is split up when heated into three molecules of formaldehyde. On cooling, however, the triple molecule is reproduced. $C_3H_6O_3$ is also converted into CH_2O when heated with water to $130°$ C. $(266°$ F.).

When ammonia is passed over paraformaldehyde, h e x m e t h e n a m i n e, $C_6H_{12}N_4 = (CH_2)_6N_4$, is produced, a base which crystallizes from alcohol in shining rhombohedrons, and sublimes without alteration when cautiously heated. It is a monoacid base, forming crystalline salts, the hydrochloride, $C_6H_{12}N_4$, HCl, for example. By boiling with dilute acids it is reconverted into formaldehyde.

Parathioformaldehyde or *Trithiomethene*, $(CH_2)_3S_3$, is produced by passing hydrogen sulphide into the aqueous solution of formaldehyde, also by the action of zinc and hydrochloric acid on carbon bisulphide, and by heating methene iodide with potassium sulphide in alcoholic solution. It is a solid body having an alliaceous odor, is insoluble in water, but dissolves in alcohol, and crystallizes therefrom in slender needles, melting at 218º, and easily subliming. The vapor-density agrees with the formula $C_3H_6S_3$, whence it is inferred that solid paraformaldehyde is correctly represented by the formula $C_3H_6O_3$. The thioaldehyde, $C_3H_6S_3$, heated to 170º C. (338º F.) with silver sulphate, is converted into $C_3H_4O_3$.

Acetic Aldehyde, or **Acetaldehyde,** $C_2H_4O = CH_3.CHO = C_2H_3O.H$, also called *Ethyl Aldehyde*, but more generally by the simple name a l d e h y d e.*—This substance is formed by oxidation of ethyl-alcohol; also among other products, when the vapor of ether or alcohol is transmitted through a red-hot tube; by the action of chlorine on weak alcohol; and by the other general reactions above mentioned. It is best prepared by the following process: 6 parts of oil of vitriol are mixed with 4 parts of rectified spirit of wine, and 4 parts of water; this mixture is poured upon 6 parts of powdered manganese dioxide contained in a capacious retort, in connection with a condenser cooled by ice-cold water; gentle heat is applied, and the process is interrupted when 6 parts of liquid have passed over. The distilled product is put into a small retort, with its own weight of calcium chloride, and redistilled; and this operation is repeated. The aldehyde, still retaining alcohol and other impurities, is mixed with twice its volume of ether, and saturated with dry ammoniacal gas; a crystalline compound of aldehyde and ammonia then separates, which may be washed with a little ether, and dried in the air. From this substance the aldehyde may be separated by distillation in a water-bath with sulphuric acid diluted with an equal quantity of water; by careful rectification from calcium chloride, at a temperature not exceeding 30.5º C. (86.9º F.), it is obtained pure and anhydrous.

Considerable quantities of aldehyde occur in the "first runnings" obtained in the manufacture of alcohol from sugar-beet and from potatoes, being probably formed by oxidation of the spirit during the filtration through charcoal, to which it is subjected for the removal of fusel-oil and other impurities.

Aldehyde is a limpid, colorless liquid, of characteristic ethereal odor, which, when strong, is exceedingly suffocating. It has a density of 0.807 at 0º, boils at 21º-22º C. (69.8º-71.6º F.), and mixes in all proportions with water, alcohol, and ether: it is neutral to test-paper, but becomes acid on exposure to air, from production of acetic acid: under the influence of platinum-black this change is very speedy. When a solution of aldehyde is heated with caustic potash, a brown resin-like substance is produced, the so-called a l d e h y d e - r e s i n. It reduces silver oxide at a gentle heat, without evolution of gas, the metal being deposited on the inner surface of the vessel as a brilliant and uniform film; the liquid contains silver acetate.

---

* Alcohol dehydrogenatum.

The principal reactions of aldehyde have been already mentioned (p. 693). It is converted by nascent hydrogen into ethyl alcohol, by oxidation into acetic acid ; by phosphorus pentachloride into chloraldehyde or ethidene dichloride, CH_3—$CHCl_2$. It unites with the acid sulphites of the alkali-metals, forming compounds like the ammonium salt, $CH_3.CH{<}^{OH}_{O.SO_3(NH_4)}$, which are resolved at 100° into aldehyde, sulphurous oxide, and neutral sulphites ; thus :

$$2[CH_3.CH{<}^{OH}_{SO_3K}] = 2(CH_3.CHO) + SO_3K_2 + SO_2 \ H_2O.$$

Aldehyde unites also with *acetic oxide*, forming the compound C_2H_4O. $(C_2H_3O)_2O$ or $CH_3.CH{<}^{OC_2H_3O}_{OC_2H_3O}$, and with *ethyl oxide*, forming e t h i d e n e

d i e t h y l a t e or a c e t a l, $CH_3.CH{<}^{OC_2H_5}_{OC_2H_5}$; in like manner with *methyl oxide*, forming e t h i d e n e d i m e t h y l a t e, $CH_3.CH(OCH_3)_2$ (p. 634).

With dry *hydrogen cyanide*, aldehyde forms the compound $CH_3.CH{<}^{OH}_{CN}$, a liquid soluble in water and in alcohol, boiling at 183° C. (361.4° F.), with partial decomposition into aldehyde and hydrogen cyanide, and converted, by heating with strong hydrochloric acid or with alkalies, into α - l a c t i c a c i d, $CH_3.CH{<}^{OH}_{CO_2H}.$

A l d e h y d e - a m m o n i a, $C_2H_4O.NH_3 = CH_3.CH{<}^{OH}_{NH_2}$, obtained by passing dry ammonia gas into an ethereal solution of aldehyde, crystallizes in large shining rhombohedrons ; it has a mixed odor of ammonia and turpentine, dissolves very easily in water, with less facility in alcohol, and with difficulty in ether ; melts at about 76° C. (168.8° F.), and distils unchanged at 100°. Acids decompose it, with production of ammoniacal salts and separation of aldehyde. Hydrogen sulphide converts it into t h i a l-d i n e, $C_6H_{13}NS_2$. Sulphurous oxide gas is rapidly absorbed by a solution of aldehyde-ammonia, forming the crystalline compound $C_2H_4ONH_3.SO_2$, isomeric with taurine (p. 633), and with the compound formed by direct combination of aldehyde with acid ammonium sulphite.

POLYMERIC MODIFICATIONS OF ALDEHYDE.—Pure anhydrous aldehyde, treated at ordinary temperatures with a small quantity of gaseous hydrochloric acid, sulphurous oxide, or carbonyl chloride (phosgene), or of zinc chloride or strong sulphuric acid, is converted into a polymeride, called p a r a l d e h y d e, which crystallizes on cooling the liquid to 0° in large transparent prisms, melting at 10.5° C. (50.9° F.) to a liquid which boils at 124° C. (355.2° F.), has a specific gravity of 0.998 at 15° C. (59° F.), and a vapor-density three times as great as that of aldehyde itself: hence its molecular formula is $C_6H_{12}O_3$. When distilled in contact with either of the reagents above mentioned, it is reconverted into ordinary aldehyde, but it does not undergo this change when heated alone.

When either of the same reagents acts on aldehyde cooled by a freezing mixture, another polymeric modification is formed, called m e t a l d e-h y d e, which also crystallizes in transparent prisms or in white needles, but differs from paraldehyde in subliming when heated without previous fusion, and in being at the same time partly reconverted into aldehyde. In consequence of this transformation, its vapor-density, and therefore its molecular formula, have not been determined.

When aldehyde is left for some time in contact with dilute hydrochloric

30

acid, two molecules of it unite in such a manner as to form the aldehyde of butene glycol, called a l d o l :—

$$CH_3.CHO \ + \ CH_3.CHO \ = \ CH_3.CH(OH).CH_2.CHO .$$

Under the influence of certain other reagents, condensed products are formed by the union of two or more molecules of aldehyde, and elimination of the elements of water; thus when aldehyde is heated with zinc chloride, crotonic aldehyde is produced :—

$$CH_3\!-\!CHO + CH_3\!-\!CHO = H_2O + CH_3\!-\!CH\!=\!CH\!-\!CHO .$$

Halogen Derivatives of Aldehyde.

Monochloraldehyde, $CH_2Cl.CHO$, formed by the action of sulphuric acid on chloracetal, $CH_2Cl\!-\!CH(OC_2H_5)_2$, is known only in aqueous solution. By exposure to the air, and by the action of silver oxide, it is converted into chloracetic acid.

Dichloraldehyde, $CHCl_2.CHO$, produced by distillation of dichloracetal with strong sulphuric acid, is a liquid insoluble in water, and boiling at $88°\!-\!90°$ C. ($190.4°\!-\!194°$ F.). By keeping, it is gradually converted into two solid polymeric modifications, one of which is amorphous, and is reconverted into liquid dichloraldehyde at $120°$ C. ($248°$ F.), while the other crystallizes from alcohol in large prisms, melts at $130°$ C. ($266°$ F.), and does not pass into the liquid modification till heated to $240°$ C. ($464°$ F.).

Trichloraldehyde, or **Chloral,** $CCl_3.CHO$, is formed by the prolonged action of chlorine on ethyl alcohol. To prepare it, chlorine is passed into absolute alcohol as long as hydrochloric acid continues to be evolved, and the product is agitated with three times its volume of strong sulphuric acid. On gently warming the mixture in a water-bath, the impure chloral separates as an oily liquid, which floats on the surface of the acid; it is purified by distillation from fresh oil of vitriol, and afterwards from a small quantity of quicklime, which must be kept completely covered by the liquid until the end of the operation. Chloral has also been obtained from starch, by distillation with hydrochloric acid and manganese dioxide.

Chloral is a thin, oily, colorless liquid, of peculiar, pungent, tear-exciting odor : it has but little taste. When dropped upon paper it leaves a greasy stain, which is not, however, permanent. It has a density of 1.502, and boils at $94°$ C. ($201.2°$ F.), and changes on prolonged keeping into a solid polymeride. Chloral is freely soluble in water, alcohol, and ether; the aqueous solution is not affected by silver nitrate. Caustic baryta and lime decompose the vapor of chloral when heated in it, with appearance of ignition; the oxide is converted into chloride, carbon is deposited, and carbon monoxide is set free. Solutions of caustic alkalies also decompose it, with production of a formate and chloroform :—

$$CCl_3.CHO \ + \ KOH \ = \ CHCl_3 \ + \ CHO.OK.$$

By oxidation with nitric acid, it is converted into trichloracetic acid.

With a small quantity of *water*, chloral forms a h y d r a t e, $C_2HCl_3O.H_2O = CCl_3.CH{<}^{OH}_{OH}$, which forms large monoclinic crystals, melting at $46°$ C. ($114.8°$ F.), and distilling at $96°\!-\!98°$ C. ($204.8°\!-\!208.4°$ F.). It is easily soluble in water, has a peculiar odor, and sharp scratching taste, and produces sleep when taken internally or introduced under the skin. Strong sulphuric acid decomposes it into water and alcohol.

With *hydrogen sulphide*, chloral forms, in like manner, a crystalline sulphydrate, $CCl_3.CH\diagdown_{SH}^{OH}$, which melts at 77° C. (170.6° F.), and boils at 123° C. (251.6° F.).

With *alcohol*, chloral forms an alcoholate, $CCl_3.CH\diagdown_{OH}^{OC_2H_5}$, which is a crystalline body, melting at 56° C. (132.8° F.), and boiling at 114°–115° C. (237.2°–239° F.), and decomposed by strong sulphuric acid, with reproduction of chloral. Treated with *acetyl chloride*, it is converted into the acetic ether, $CCl_3.CH\diagdown_{O.C_2H_3O}^{O.C_2H_5}$, which boils at 198° C. (388.4° F.).

Dibromaldehyde, $CHBr_2.CHO$, formed by direct bromination of aldehyde, is a liquid which boils at 140°–142° C. (284°–287.6° F.), is converted on standing into a solid polymeride, forms a crystalline hydrate, and unites with hydrogen cyanide, forming the compound, $CHBr_2.CH\diagdown_{OH}^{CN}$, convertible into dibromolactic acid.

Tribromaldehyde or **Bromal,** $CBr_3.CHO$, formed by the action of bromine on alcohol, is a liquid very much like chloral, boiling at 172°–173° C. (341.6°–343.4° F.), and decomposed by alkalies into formic acid and bromoform. It forms a solid hydrate, melting at 53° C. (127.4° F.), an alcoholate which melts at 44° C. (111.2° F.), and decomposes at 100°, and with hydrogen cyanide the compound, $CBr_3.CH\diagdown_{OH}^{CN}$, convertible into tribromolactic acid.

Sulphur-derivatives of Aldehyde.—When hydrogen sulphide is passed through aqueous aldehyde, a compound of aldehyde and thioaldehyde is produced, in the form of an oil, having an offensive odor, solidifying at —8° C. (17.6° F.), and converted by contact with hydrochloric acid into parathioaldehyde, $(C_2H_4S)_3$. This compound, analogous to paraldehyde, forms white needles, insoluble in water, and subliming at 45° C. (113° F.). Its vapor density agrees with the formula, $C_6H_{12}S_3$.

On passing hydrogen sulphide into an aqueous solution of aldehyde-ammonia, thialdine, $C_6H_{13}NS_2$, separates in large colorless crystals, melting at 43° C. (109.4° F.), slightly soluble in water, easily in alcohol and ether, and having an offensive odor. It is a strong base; its hydrochloride has the composition $C_6H_{13}NS_2.HCl$.

On mixing an alcoholic solution of aldehyde-ammonia with *carbon bisulphide*, carbothialdine, $C_5H_{10}N_2S_2$, is formed in large shining crystals, separable by boiling with acids into aldehyde, carbon bisulphide, and ammonia.

Propionic Aldehyde, $C_3H_6O = CH_3.CH_2.CHO$, prepared by oxidation of normal propyl alcohol, or by distilling a mixture of calcium propionate and formate, is very much like acetaldehyde, has a specific gravity of 0.804 at 20° C. (68° F.), boils at 49° C. (120.2° F.), and dissolves in 5 volumes of water. With phosphorus pentachloride it forms propidene chloride, $CH_3.CH_2.CHCl_2$.

The higher aldehydes of the series are susceptible of isomeric modifications analogous to those of the primary alcohols. Only a few of those are, however, known : thus, of the four possible 5-carbon aldehydes, analogous to the four primary pentyl alcohols, two only have yet been obtained.

The following table exhibits the specific gravities and boiling points of these higher aldehydes. All of them are liquid at ordinary temperatures, except palmitic aldehyde, formed by oxidation of cetyl alcohol, which is a white crystalline mass :—

Aldehydes.		Boiling Point.	Specific gravity.
Butyric	Normal, $CH_3.CH_2.CH_2.CHO$	75° C. (167° F.)	0.834 at 0°
	Iso, $(CH_3)_2CH CHO$	61° C. (141.8° F.)	0.822 at 0°
Valeric	Normal, $CH_3.(CH_2)_3.CHO$	102° C. (215.6° F.)	
	Iso, $(CH_3)_2.CH.CH_2.CHO$	92-93°C.(197.6-199.4° F.)	0.768 at 12°C. (53.6° F.)
Caproic (Iso), $(CH_3)_2CH.(CH_2)_2CHO$		121° C. (249.8° F.)	•
Œnanthylic, $CH_3.(CH_2)_5.CHO$		154° C. (309.2° F.)	0.827
Palmitic, $C_{16}H_{32}O$.		melts at 50° C. (122° F.)	

Isobutyric aldehyde treated with a small quantity of strong sulphuric acid, is converted into p a r a i s o b u t y r i c a l d e h y d e, $(C_4H_8O)_3$, which crystallizes in shining needles, melting at 60° C. (140.2° F.), and boiling at 194° C. (381.2° F.).

Normal butyric aldehyde, heated with alcoholic ammonia to a temperature not exceeding 100°, yields two bases, d i b u t y r a l d i n e, $C_8H_{17}NO$, and t e t r a b u t y r a l d i n e, $C_{16}H_{29}NO$, the former of which, when subjected to dry distillation, gives off H_2O, and is converted into p a r a c o n i n e, $C_8H_{15}N$, an oily base, very much like the conine of hemlock.

Trichlorobutyric aldehyde or *Butyric chloral,* $C_4H_5Cl_3O$ (formerly supposed to be crotonic chloral, $C_4H_3Cl_3O$), is formed, together with several other products, by the action of chlorine on acetaldehyde. It is a liquid boiling at 155°–165° C. (311°–329° F.).

Isovaleric aldehyde or *Valeral,* heated with alcoholic ammonia, yields two bases, v a l e r i d i n e, $C_{10}H_{19}N$, and v a l e r i t r i n e, $C_{15}H_{27}N$, the latter of which is a liquid, boiling at 250°–260° C. (482°–500° F.), and smelling like conine.

Œnanthylic Aldehyde or *Œnanthol,* is formed in the dry distillation of castor-oil. It oxidizes in the air to œnanthylic acid, and is converted by nascent hydrogen into normal heptyl alcohol.

ALDEHYDES OF THE SERIES, $C_nH_{2n-2}O$.

These aldehydes are related to the alcohols of the allylic series in the same manner as the aldehydes, $C_nH_{2n}O$, are related to the fatty alcohols. They are unsaturated compounds, and capable of taking up 2 atoms of a monotomic element or radicle.

A c r y l i c A l d e h y d e or A c r o l e i n, $C_3H_4O=CH_2\text{=}CH\text{—}CHO$, the lowest member of the series, is produced :

1. By the oxidation of allyl alcohol.

2. By the dehydration of glycerin, when that substance is heated with phosphoric oxide, strong sulphuric acid, or acid potassium sulphate : $C_3H_8O_3—2H_2O = C_3H_4O$.

It is always produced in the destructive distillation of neutral fats containing glycerin, and is the cause of the intolerably pungent odor attending that process.

Pure acrolein is a thin, colorless, highly volatile liquid, lighter than water, and boiling at 52.2° C. (126° F.). Its vapor is intensely irritating. It is sparingly soluble in water, freely in alcohol and ether.

Acrolein, by keeping, undergoes partial decomposition, yielding a white, flocculent, indifferent body, d i s a c r y l ; the same substance is sometimes produced, together with acrylic acid, by exposure to the air. In contact with *alkalies*, acrolein suffers violent decomposition, producing, like aldehyde, a resinous body. When exposed for some time to the air, or mixed with silver oxide, it is rapidly oxidized into a c r y l i c   a c i d, $C_3H_4O_2$.

Nascent hydrogen converts it into a l l y l   a l c o h o l. With *phosphorous pentachloride* it yields the compound $CH_2=CH-CHCl_2$, a liquid boiling at 84° C. (183.2° F.). With *hydrochloric acid* acrolein forms a crystalline compound, $C_3H_4O.HCl$, which melts at 32° C. (89.6° F.), and is resolved by distillation into hydrochloric acid and acrolein.

Acrolein mixed with alcoholic ammonia is immediately converted into a c r o l e i n - a m m o n i a, C_6H_9NO:

$$2C_3H_4O \ + \ NH_3 \ = \ C_6H_9NO \ + \ H_2O.$$

This compound is a yellowish mass, which turns brown on drying. It forms amorphous salts, and yields picoline, C_6H_7N, by distillation.

C r o t o n i c   A l d e h y d e, $C_4H_6O = CH_3-CH=CH-CHO$, is produced by heating acetaldehyde to 100° with dilute hydrochloric acid, or with zinc chloride, and a little water:

$$CH_3.CHO \ + \ CH_3.CHO \ = \ H_2O \ + \ CH_3.CH=CH.CHO,$$

aldol being probably formed as an intermediate product.

Crotonic aldehyde is a colorless liquid, having an extremely pungent odor, and boiling at 104°–105° C. (219.2°–221° F.). In contact with the air, or with moist silver oxide, it is easily oxidized to crotonic acid. It unites with hydrogen chloride, forming monochlorobutyric aldehyde:

$$CH_3.CH=CH.CHO \ + \ HCl \ = \ CH_3.CH_2.CHCl.CHO.$$

Aldehydes derived from Diatomic Alcohols.

Diatomic alcohols can yield by oxidation two classes of aldehydes, according as one or two molecules of hydrogen (H_2) are removed. Propene glycol, $C_3H_8O_2$, for example, might yield the two aldehydes, $C_3H_6O_2$ and $C_3H_4O_2$. Only a few of these compounds have, however, been obtained.

Of aldehydes derived from the glycols, $C_nH_{2n+2}O_2$, only three are at present known, viz., g l y o x a l, $C_2H_2O_2$, which is the second aldehyde of ordinary glycol, $C_2H_6O_2$; a l d o l, $C_4H_8O_2$, which is the first, and succinic aldehyde, $C_4H_6O_2$, which is the second aldehyde of butene glycol.

G l y o x a l, $C_2H_2O_2$, and the aldehydic acid, g l y o x a l i c   a c i d, $C_2H_2O_3$, are intermediate in composition between glycol and oxalic acid:

CH_2OH	COH	COH	$CO.OH$
CH_2OH	COH	$CO.OH$	$CO.OH$
Glycol.	Glyoxal.	Glyoxylic acid.	Oxalic acid.

Both are formed as intermediate products in the oxidation of glycol, and are converted by further oxidation into oxalic acid. They are, however, most easily obtained by oxidizing ethyl alcohol with nitric acid.

Alcohol (2 parts) is introduced into a tall glass cylinder ; 1 part of water is poured into the vessel through a funnel, the tube of which reaches to the bottom, and afterwards 1 part of fuming nitric acid is introduced in the same way, so as to form three distinct layers of liquid. The mixture is left to itself at a temperature of 20°–22° C. (68°–71.6° F.) till the three liquids have become mixed ; after which the mixture is evaporated in the water-bath, diluted with water, saturated with chalk at the boiling heat, then filtered, and the filtrate is mixed with alcohol, whereupon glycollate and glyoxylate of calcium are precipitated, and glyoxal remains in solution. On recrystallizing the calcium salts, the glyoxylate crystallizes out first, and afterwards the glycollate. The glyoxal is separated from the solution by agitation with acid sodium sulphite, with which it forms a crystalline compound; this sodium compound treated with barium chloride yields the corresponding barium compound ; and from this the glyoxal may be separated by dilute sulphuric acid.

Glyoxal, $C_2H_2O_2 = \begin{matrix} COH \\ | \\ COH \end{matrix}$ ' isomeric with glycollide, $\begin{matrix} CH_2 \\ | \\ CO \end{matrix} \!\!>\!\! O$, remains, on evaporating the solution obtained in the manner just described, as a transparent, amorphous, deliquescent mass, very soluble in water, alcohol, and ether. It reduces an ammoniacal solution of silver nitrate, forming a silver speculum. Nitric acid oxidizes it to glycollic and oxalic acids. Being a double aldehyde (containing the group COH twice) it unites with two molecules of acid potassium sulphite, forming the compound $C_2H_2O_2$ $(SO_3KH)_2$. When its solution in acetic acid is left in contact with a small quantity of hydrochloric acid, a condensation-product is formed, perhaps $C_{12}H_{12}O_{12}$.

Fixed caustic alkalies and alkaline earths convert glyoxal into salts of glycollic acid, e. g., $C_2H_2O_2 + KOH = C_2H_3KO_3$. A syrupy solution of glyoxal heated with a strong solution of aqueous ammonia yields two crystalline bases, g l y o x a l i n e and g l y c o s i n e:

$$2C_2H_2O_2 \ + \ 2NH_3 \ = \ C_3H_4N_2 \ + \ CH_2O_2 \ + \ 2H_2O$$
Glyoxal.   Glyoxaline.   Formic
                                                       acid.

$$3C_2H_2O_2 \ + \ 4NH_3 \ = \ C_6H_6N_4 \ + \ 6H_2O.$$
Glyoxal.   Glycosine.

Glyoxylic or **Glyoxalic Acid,** $C_2H_2O_3 = CHO—CO_2H$, is formed by oxidation of alcohol as above described ; also by reducing oxalic acid with zinc and hydrochloric acid :

$$\begin{matrix} CO.OH \\ | \\ CO.OH \end{matrix} \ + \ H_2 \ = \ \begin{matrix} COH \\ | \\ CO.OH \end{matrix} \ + \ H_2O,$$

and by heating dichloracetic acid with silver oxide, or ethylic dichloracetate with water to 120° C. (248° F.) :

$$\begin{matrix} CHCl_2 \\ | \\ CO.OH \end{matrix} \ + \ Ag_2O \ = \ \begin{matrix} CHO \\ | \\ COOH \end{matrix} \ + \ 2AgCl.$$

When concentrated, it forms a thick syrup easily soluble in water, and distils undecomposed with aqueous vapor.

The glyoxylates contain 1 molecule of water very closely combined, excepting the *ammonium salt*, which is anhydrous.

This salt, $C_2HO_3.NH_4$, crystallizes in needles, and dissolves easily in water. The *silver salt*, $C_2HO_3Ag + H_2O$, is a white crystalline precipitate; the *calcium salt*, $(C_2HO_3)_2Ca + 2H_2O$, forms hard prisms, sparingly soluble in cold water.

Glyoxylic acid is readily oxidized to oxalic acid. By boiling its calcium salt with water, it is resolved into glycollic and oxalic acids:

$$2C_2H_2O_3 + H_2O = C_2H_4O_3 + C_2H_2O_4.$$

By nascent hydrogen (zinc and hydrochloric acid), it is converted into glycollic acid.

Glyoxylic acid is both acid and aldehyde. Its aldehydic nature is shown by its power of reducing an ammoniacal silver solution, and of forming double salts with alkaline bisulphites.

Aldol, $C_4H_8O_2$, the aldehyde of butene glycol, is polymeric with acetaldehyde, and is formed by combination of two molecules of that body, under the influence of dilute hydrochloric acid; thus:

$$CH_3.CHO + CH_3.CHO = CH_3.CH(OH).CH_2.CHO.$$

A cold mixture of acetaldehyde and dilute hydrochloric acid is left for two days till it acquires a yellow color. The liquid is then neutralized with sodium carbonate, and shaken up with ether; the ether is evaporated; and the aldol which remains is evaporated in a vacuum.

Aldol is a colorless inodorous liquid, of specific gravity 1.120 at $0°$, not miscible with water. On standing it changes to a viscid liquid, which cannot be poured out of the containing vessel. In a vacuum it distils between $95°$ C. ($203°$ F.) and $105°$ C. ($221°$ F.), but when heated under ordinary pressure it gives off water, and is converted into c r o t o n i c a l d e h y d e :

$$
\begin{array}{ccccc}
CH_3 & & & & CH_2 \\
| & & & & | \\
CH(OH) & & & & CH \\
| & = & H_2O & + & || \\
CH_2 & & & & CH \\
| & & & & | \\
CHO & & & & CHO.
\end{array}
$$

It reduces silver from the ammonio-nitrate, and when heated with water and silver oxide, is converted into $β$-o x y b u t y r i c a c i d :

$$CH_3.CH(OH).CH_2.CO(OH).$$

Succinic Aldehyde, $C_4H_6O_2 = \begin{array}{c} CH_2.COH \\ | \\ CH_2.COH \end{array}$, is a liquid soluble in water, and boiling at $201°$–$203°$ C. ($393.8°$–$397.4°$ F.). By oxidation it is converted into normal oxyisobutric acid and succinic acid.

KETONES.

These bodies contain the group CO associated with two monatomic alcohol-radicles, which may either be the same or different, *e. g.*,

$$
CO{<}^{CH_3}_{CH_3} \qquad\qquad CO{<}^{CH_3}_{C_2H_5}
$$

Dimethyl ketone or acetone.   Methyl-ethyl ketone.

They may be regarded as derived from aldehydes by substitution of an alcohol-radicle for the hydrogen atom attached to the group CO ; thus

Acetic aldehyde, $\quad\quad$ $CH_3.CO.H$
Acetic ketone or acetone, $\quad$ $CH_3.CO.CH_3$,

or as compounds of an alcohol-radicle with an acid-radicle—acetone, for example, as methyl-acetyl.

The only bodies of this class that have been carefully studied are those which contain the alcohol-radicles $C_nH_{2n}+_1$, and are analogous to the aldehydes $C_nH_{2n}O$, and the fatty acids $C_nH_{2n}O_2$.

The ketones of this group, containing two equivalents of the same alcohol-radicle, are produced :—

1. By the action of carbon monoxide on sodium ethide and its homologues :—

$$CO \quad + \quad 2NaC_nH_{2n}+_1 \quad = \quad Na_2 \quad + \quad CO(C_nH_{2n}+_1)_2 .$$

For example :—

$$CO \quad + \quad 2NaC_2H_5 \quad = \quad Na_2 \quad + \quad CO(C_2H_5)_2 .$$
Carbon $\quad\quad$ Sodium-ethyl. $\quad\quad\quad\quad\quad\quad\quad$ Propione.
monoxide.

2. By the action of zinc-methyl and its homologues on the acid chlorides, $C_nH_{2n-1}OCl$; e. g. :—

$$Zn(CH_3)_2 \quad + \quad 2COCH_3Cl \quad = \quad ZnCl_2 \quad + \quad 2CO(CH_3)_2 .$$
Zinc methide. $\quad\quad$ Acetic chloride. $\quad\quad\quad\quad\quad\quad\quad\quad$ Acetone.

3. By the oxidation of the secondary alcohols ; thus :—

$$CH(CH_3)_2.OH \quad + \quad O \quad = \quad H_2O \quad + \quad CO(CH_3)_2 .$$
Isopropyl alcohol. $\quad\quad\quad\quad\quad\quad\quad\quad\quad\quad\quad\quad$ Acetone.

4. By the dry distillation of the calcium-salts of the fatty acids ; e.g. :—

$$Ca \begin{cases} O(COCH_3) \\ O(COCH_3) \end{cases} = COCaO_2 \quad + \quad CO(CH_3)_2 .$$
Calcium acetate. $\quad\quad\quad\quad$ Calcium $\quad\quad$ Acetone.
$\quad\quad\quad\quad\quad\quad\quad\quad\quad\quad$ carbonate.

The ketones formed in this manner from the successive members of the fatty acid series differ from one another by twice CH_2 ; thus :—

Acetic acid	. .	$C_2H_4O_2$,	yields Acetone	. .	C_3H_6O	.
Propionic acid	. .	$C_3H_6O_2$,	" Propione	. .	$C_5H_{10}O$	.
Butyric acid	. .	$C_4H_8O_2$,	" Butyrone	. .	$C_7H_{14}O$	.
Valeric acid	. .	$C_5H_{10}O_2$,	" Valerone	. .	$C_9H_{18}O$	.

The intervals are filled up by ketones containing different alcohol-radicles ; thus ethyl-methyl ketone, C_4H_8O, or $CO(CH_3)(C_2H_5)$, is intermediate between acetone and propione.

The ketones containing two different alcohol-radicles may be obtained by the second of the processes above given ; e. g. :—

$$2COCH_3Cl \quad + \quad Zn(C_2H_5)_2 \quad = \quad ZnCl_2 \quad + \quad 2CO(CH_3)(C_2H_5).$$
Acetic $\quad\quad\quad\quad$ Zinc ethyl. $\quad\quad\quad\quad\quad\quad\quad\quad$ Ethyl-methyl
chloride. $\quad\quad\quad\quad\quad\quad\quad\quad\quad\quad\quad\quad\quad\quad\quad$ ketone.

Or by distilling a mixture of calcium-salts of two different fatty acids ; thus :—

$$Ca(CO_2C_4H_9)_2 \quad + \quad Ca(CO_2CH_3)_2 \quad = \quad 2CaCO_3 \quad + \quad 2CO(CH_3)(C_4H_9).$$
Valerate. $\quad\quad\quad\quad\quad$ Acetate. $\quad\quad\quad\quad\quad\quad\quad\quad\quad\quad$ Methyl-butyl ketone.

The formation of aldehydes by distilling a mixture of a formate with the salt of another fatty acid (p. 693), is a particular case of this last reaction.

Ketones are also formed : 5. By the gradual oxidation of the acids of the lactic series, $C_nH_{2n}O_3$, with chromic acid mixture :—

$$(CH_3)_2.C(OH).CO_2H \ + \ O \ = \ CO_2 \ + \ H_2O \ + \ (CH_3)_2CO \ .$$
Oxyisobutyric acid.   Dimethyl ketone.

6. By the action of alkalies on the carbo-ketonic acids (*q. v.*).

7. By the dry distillation of wood, sugar, and many other carbon compounds.

Every ketone is isomeric with an aldehyde belonging to the same series; thus acetone is isomeric with propionic aldehyde, $C_2H_5.COH$; butyrone, $CO(CO_3H_7)_2$, with œnanthylic aldehyde, $C_6H_{13}.COH$, etc. Formic acetone, COH_2, is identical with formic aldehyde.

Ketones are, for the most part, volatile liquids, insoluble in water, and not capable of reducing an ammoniacal silver solution. Those in which the group CO is associated with a methyl group, resemble the aldehydes in forming crystalline compounds with alkaline bisulphites, from which the ketone may be liberated by distillation with an alkali. Ketones differ from aldehydes : 1. In not being converted by oxidation into the corresponding acids.—2. In being converted by nascent hydrogen into secondary alcohols, whereas the aldehydes are converted into primary alcohols. —3. In not combining with aniline.

The formation of secondary alcohols by hydrogenation of ketones is attended, to a small amount, with a condensation of two ketone molecules, resulting in the formation of a diatomic alcohol (a pinacone); *e. g.* :—

$$2(CH_3)_2CO \ + \ H_2 \ = \ \begin{matrix} (CH_3)_2.C.OH \\ | \\ (CH_3)_2.C.OH \end{matrix}$$
Acetone.   Pinacone.

Ketones, like aldehydes, unite with *hydrogen cyanide*, forming c y a n - h y d r i n s , acetone, for example, yielding $(CH_3)_2C{<}^{OH}_{CN}$, which, by the action of acids or alkalies, may be converted into the corresponding o x y - a c i d s , $C_nH_{2n}O_3$.

Ketones are not convertible by oxidation into acids containing the same number of carbon-atoms as themselves. When boiled with chromic acid mixture, they decompose in such a manner that the group CO separates in combination with the lower alcohol-radicle, and forms an acid ; while, if the other radicle is a primary radicle, it is likewise oxidized to an acid, normal or iso, according to the constitution of the radicle ; thus :—

$$CH_3(CH_2)_3{>}^{CH_3}CO \ + \ O_3 \ = \ CH_3.CO.OH \ + \ CH_3.(CH_2)_2.COOH \ .$$
Methyl-butyl ketone.   Acetic acid.   Butyric acid.

$$(CH_3)_2CH.CH_2{>}^{CH_3}CO \ + \ O_3 \ = \ CH_3.CO.OH \ + \ (CH_3)_2CH.COOH \ .$$
Methyl-isobutyl ketone.   Acetic acid.   Isobutyric acid.

If, on the other hand, the higher radicle is secondary, it will be oxidized to a ketone, which will be further split up by oxidation :

$$(CH_3)_2CH{>}^{CH_3}CO \ + \ O_2 \ = \ CH_3.CO.OH \ + \ (CH_3)_2CO ;$$
Methyl-iso- propyl ketone.   Acetic acid.   Acetone.

30 *

and if the higher radicle is tertiary, such as $C(CH_3)_3$, it will be immediately split up.*

Dimethyl Ketone.—Acetone, $CO(CH_3)_2$.—This compound is formed: 1. By the dry distillation of acetates.—2. By oxidation of pseudopropyl alcohol.—3. By the action of water at 180°–200° C. (356°–392° F.) on acetone chloride or bromide (p. 547):

$$CH_3.CCl_2.CH_3 \ + \ H_2O \ = \ 2HCl \ + \ CH_3.CO.CH_3 ;$$

or on propene chloride (or bromide), $CH_3.CHCl.CH_2Cl$ (in which case a molecular transposition of H and Cl must be supposed to take place in the first instance), or on monochloro- or monobromopropene:

$$CH_3\!\!-\!\!CCl\!\!=\!\!CH_2 \ + \ H_2O \ = \ HCl \ + \ CH_3.CO.CH_3 .$$

4. Together with other products, by passing the vapor of strong acetic acid through an iron tube heated to dull redness, and by the dry distillation of citric acid, tartaric acid, sugar, starch, gum, and wood: hence it occurs in crude wood-spirit.

Acetone is best prepared by the dry distillation of acetates, the calcium or the lead salt being the most convenient for the purpose. The crude distillate is saturated with potassium carbonate, and afterwards rectified in a water-bath from calcium chloride.

Pure acetone is a colorless limpid liquid, of peculiar odor: it has a density of 0.792, and boils at 55.5° C. (131.9° F.): the density of its vapor (referred to air) is 2.022. Acetone is very inflammable, and burns with a bright flame: it is miscible in all proportions with water, alcohol, and ether.

Nascent hydrogen converts it into pseudopropyl alcohol (p. 607); but at the same time a portion of the acetone doubles its molecule, and likewise takes up hydrogen, being thereby converted into a crystalline substance, pinacone, $C_6H_{14}O_2 = 2C_3H_6O + H_2$.

By *oxidation* with chromic acid mixture, acetone is converted into acetic and formic acids, the latter being for the most part further oxidized to H_2O and CO_2.

$$\underset{\text{Acetone.}}{CH_3.CO.CH_3} \ + \ O_3 \ = \ \underset{\text{Acetic acid.}}{CH_3.CO.OH} \ + \ \underset{\text{Formic acid.}}{H.CO.OH}$$

Acetone treated with *hydrocyanic acid*, water, and hydrochloric acid is converted into oxyisobutyric (acetonic) acid:

$$(CH_3)_2CO \ + \ CHN \ + \ 2H_2O \ + \ HCl \ = \ NH_4Cl \ + \\ (CH_3)_2COH.COOH.$$

By *phosphorus pentachloride*, acetone is converted into acetone chloride or methylchloracetol, $CH_3.CCl_2.CH_3$; similarly with PBr_5. With P_2S_5, it forms thioacetone, $CH_3.CS.CH_3$, a yellowish offensive-smelling liquid, boiling at 183°–185° C. (361.4°–365° F.), insoluble in water.

Chlor- and *Brom-acetones.*—These compounds are formed by the direct action of chlorine and bromine on acetone, and in other ways.

Monochloracetone, $CH_3.CO.CH_2Cl$, is formed by passing chlorine into cold acetone, and by the action of hypochlorous acid on monochloro- or monobromopropene:

$$CH_3\!\!-\!\!CBr\!\!=\!\!CH_2 \ + \ ClOH \ = \ HBr \ + \ CH_3.CO.CH_2Cl .$$

* See Watts's Dictionary of Chemistry, 2d Supplement, 711.

It is a liquid, insoluble in water, boiling at 119° C. (246.2° F.), and emitting vapors which excite a copious flow of tears.

Dichloracetones, $C_3H_4Cl_2O$.—Of these there are two, viz.:

$$CH_3.CO.CHCl_2 \qquad \text{and} \qquad CH_2Cl.CO.CH_2Cl.$$
$$\text{Unsymmetrical.} \qquad\qquad\qquad \text{Symmetrical.}$$

The former, produced by direct chlorination of acetone, is an oily liquid, having a density of 1.236 at 21° C. (69.8° F.), and boiling at 120° C. (248° F.).

The symmetrical modification, obtained by oxidation of symmetrical dichlorhydrin, $CH_2Cl.CHOH.CH_2Cl$, crystallizes in rhombic plates, melts at 43° C. (109.4° F.), and boils at 172°–174° C. (341.6°–345.2° F.).

Condensation-products of Acetone.—When acetone is subjected to the action of dehydrating agents, such as sulphuric acid, quick-lime, zinc chloride, or hydrochloric acid, two or more molecules of it unite together, with separation of water, to form a condensation-product; thus:

$$2C_3H_6O \;-\; H_2O \;=\; \underset{\text{Mesityl oxide.}}{C_6H_{10}O}$$
$$3C_3H_6O \;-\; 2H_2O \;=\; \underset{\text{Phorone.}}{C_9H_{14}O}$$

These bodies are best prepared by saturating acetone with gaseous hydrogen chloride, leaving the liquid to itself for some time, and treating the product with alcoholic potash. On dilution with water an oily liquid is obtained, consisting of mesityl oxide and phorone, which may be separated by fractional distillation.

Mesityl Oxide, $\genfrac{}{}{0pt}{}{H_3C}{H_3C}{>}C{=}CH.CO.CH_3$, is a colorless oil, smelling like peppermint, and boiling at 130° C. (266° F.). Treated with phosphorus pentachloride, it is converted into the chloride, $C_6H_{10}Cl_2$. By boiling with dilute sulphuric acid, it is resolved into two molecules of acetone.

An isomeric compound, called **M e t a c e t o n e**, is formed by distilling sugar with lime; it is a liquid boiling at 84° C. (183.2° F.).

Phorone, $C_9H_{14}O$, probably $\genfrac{}{}{0pt}{}{H_3C}{H_3C}{>}C{=}CH - C\genfrac{}{}{0pt}{}{\diagup CH-CO-CH_3}{\diagdown CH_3}$.—This body crystallizes in large yellowish prisms, melts at 28° C. (82.4° F.), boils at 196° C. (384.8° F.), and when boiled with dilute sulphuric acid takes up water, and is resolved into 3 molecules of acetone:

$$C_9H_{14}O \;+\; 2H_2O \;=\; 3C_3H_6O.$$

Acetophorone, produced by heating acetone with quick lime, and camphophorone, obtained from camphoric acid, appear to be isomeric with the phorone obtained as above.

Acetone heated with strong *sulphuric acid* yields a distillate of m e s i t y-l e n e or t r i m e t h y l b e n z e n e, a body belonging to the aromatic group:

$$3C_3H_6O \;=\; 3H_2O \;+\; C_9H_{12}.$$

The same body is formed by the action of sulphuric acid on phorone. Other ketones heated with sulphuric acid likewise yield derivatives of benzene.

Acetonamines.*—Acetone, heated with ammonia, yields a mixture of three bases, the composition and mode of formation of which are indicated by the following formulæ:

Diacetonamine, $C_6H_{13}NO$ = $2C_3H_6O$ + NH_3 — H_2O.
Triacetonamine, $C_9H_{17}NO$ = $3C_3H_6O$ + NH_3 — $2H_2O$.
Dehydrotriacetonamine, $C_9H_{15}N$ = $C_9H_{17}NO$ — H_2O.

Diacetonamine is the chief product obtained at a moderate heat, and is best prepared by passing dry ammonia gas into a flask containing acetone in a state of gentle ebullition. It is a colorless liquid, having an ammoniacal odor and strong alkaline reaction; mixes in all proportions with water, alcohol, and ether, oxidizes and turns brown on exposure to the air; forms crystalline salts with hydrochloric, sulphuric, and oxalic acids; the *platinochloride*, $2(C_6H_{13}NO.HCl).PtCl_4$, crystallizes from water in orange-yellow monoclinic prisms containing 2 molecules H_2O.

Triacetonamine, $C_9H_{17}NO$, is the chief product obtained at higher temperatures, and is best prepared by boiling diacetonamine with acetone in a flask fitted with a reversed condenser. It crystallizes in large colorless square tablets or long needles, having a faint ammoniacal and camphorous odor, and melting at 39.6° C. (103.3° F.). It dissolves easily in water, alcohol, and ether; the aqueous solution has a strong alkaline reaction. It may be partly volatilized without decomposition at 100°, but decomposes at higher temperatures. The *hydrochloride*, $C_9H_{17}NO.HCl$, crystallizes from alcohol in small needles. The *platinochloride*, $2(C_9H_{17}NO.HCl).PtCl_4 + 3H_2O$, forms tufts of long golden-yellow needles, easily soluble in water, sparingly in alcohol.

An uncrystallizable modification of triacetonamine, called i s o t r i a c e t o n a m i n e, is contained in the mixture of bases formed by the action of ammonia on acetone. It forms a crystalline plátinochloride containing $2H_2O$.

D e h y d r o t r i a c e t o n a m i n e, $C_9H_{15}N$, is obtained by distilling the last mother-liquor of the same crude product with potash: it forms a platinum salt, $2(C_9H_{15}N.HCl).PtCl_4$, which crystallizes in oblique rhombic prisms.

D e h y d r o p e n t a c e t o n a m i n e, $C_{15}H_{23}N$, is obtained, amongst other products, by heating triacetonamine with hydrochloric acid:

$$5C_9H_{17}NO + 2HCl = 3C_{15}H_{23}N + 2NH_4Cl + 5H_2O.$$

The hydrochloride forms small colorless crystals, slightly soluble in water. The free base separates as an oil on adding an alkali to the solution of the hydrochloride.

Methyl-ethyl Ketone, $C_4H_8O = CO\diagup_{C_2H_5}^{CH_3}$, is formed: 1. By oxidation of secondary butyl alcohol. 2. By the action of zinc-ethyl on acetyl chloride, or of zinc-methyl on propionyl chloride. 3. By distilling a mixture of propionate and acetate of calcium. 4. By oxidation of methyl-ethyl-oxalic acid (see OXALIC ACID):

$$\frac{CH_3}{C_2H_5}\!\!>\!C\!<^{OH}_{CO_2H} + O = CO_2 + H_2O + \frac{CH_3}{C_2H_5}\!\!>\!CO.$$

* H e i n t z, Liebig's Annalen, clxxiv. 133; clxxviii. 305, 326; clxxxiii. 276, 283. S o k o l o f f and L a t s c h i n o f f.—*Deut. Chem. Ges. Ber.* vii. 1384.

5. By the action of alkalies on ethylic aceto-acetate (see CARBOKETONIC ACIDS):

$$CO\begin{cases} CH_3 \\ CH(CH_3).CO_2.C_2H_5 \end{cases} + 2KOH = CO\begin{cases} CH_3 \\ CH_2CH_3 \end{cases} + CO_3K_2 + C_2H_5(OH).$$

It is a fragrant liquid, of specific gravity 0.812 at 13⚬ C. (55.4⚬ F.), boiling at 81⚬ C. (177.8⚬ F.); unites with acid sulphites of alkali-metals; yields by oxidation two molecules of acetic acid.

The higher members of the series admit of isomeric modifications, the formula, $C_5H_{10}O$, including three, and $C_6H_{12}O$, six isomeric ketones.

The table (p. 710) exhibits the specific gravities, boiling points, and products of oxidation of the best known ketones of the series.

Methyl-katabutyl Ketone, CH_3—CO—$C(CH_3)_3$, is probably the compound called pinacolin, formed by heating pinacone, $C_6H_{14}O_2$ (p. 636), with hydrochloric or dilute sulphuric acid. It is likewise formed by the action of zinc-methyl on trimethacetyl chloride:

$$2CO\begin{cases} C(CH_3)_3 \\ Cl \end{cases} + Zn(CH_3)_2 \cdot = ZnCl_2 + 2CO\begin{cases} C(CH_3)_3 \\ CH_3 \end{cases}.$$

Nascent hydrogen converts it into pinacolyl alcohol or methyl-katabutyl carbinol, $C(CH_3)_3$—CHOH—CH_3 (p. 617).

Methyl-nonyl Ketone, CH_3—CO—C_9H_{19}, is the chief constituent of oil of rue, and may be extracted therefrom by agitation with acid sodium sulphite. It is formed artificially by distilling a mixture of calcium acetate and rutate:

$$Ca\begin{cases} O.CO.CH_3 \\ O.CO.CH_3 \end{cases} + Ca\begin{cases} O.CO.C_9H_{19} \\ O.CO.C_9H_{19} \end{cases} = 2CO_3Ca + 2CO\begin{cases} CH_3 \\ C_9H_{19} \end{cases}.$$

It is an oily liquid, with a bluish fluorescence, solidifying at low temperatures to a laminar mass.

KETONES.	Boiling point.	Specific gravity.	Products of oxidation.
Dimethyl ketone, $C_3H_6O = CO\diagdown^{CH_3}_{CH_3}$ .... }	58° C. (136.4° F.)	0.814 at 6° C. (42.8° F.)	Formic and acetic acids.
Methyl-ethyl ketone, $C_4H_8O = CO\diagdown^{CH_3}_{C_2H_5}$ .... }	81° C. (177.8° F.)	0.812 at 18° C. (55.4° F.)	Acetic acid (2 mol.)
Ketones, $C_5H_{10}O$:			
1. Methyl-propyl ketone, $CO\diagdown^{CH_3}_{(CH_2)(C_2H_5)}$ ...... }	99–101° C. (210.2–213.8° F.)	0.807 at 18° C. (64.4° F.)	Acetic and propionic acids.
2. Methyl-isopropyl ketone, $CO\diagdown^{CH_3}_{CH(CH_3)_2}$ ...... }	93 5° C. (200.3° F.)	0.810 at 13° C. (55.4° F.)	
3. Diethyl ketone or propione, $CO\diagdown^{C_2H_5}_{C_2H_5}$ ...... }	101° C. (213.8° F.)	0.813 at 20° C. (68° F.)	Acetic and propionic acids.
Ketones, $C_6H_{12}O$:			
1. Methyl-butyl ketone, $CO\diagdown^{CH_3}_{(CH_2)_3CH_3}$ ...... }	127° C. (260.6° F.)	0.829 at 0° C.	Acetic and butyric acids.
2. Methyl-isobutyl ketone, $CO\diagdown^{CH_3}_{CH_2.CH(CH_3)_2}$ .... }	114° C. (287.2° F.)	0.819 at 0° C.	Acetic and isobutyric acids.
3. Methyl-katabutyl ketone, $CO\diagdown^{CH_3}_{C(CH_3)_3}$ ...... }	106° C. (222.8° F.)	0.823 at 0° C.	Acetic and trimethylacetic acids
4. Ethyl-propyl ketone, $CO\diagdown^{C_2H_5}_{(CH_2)_2(CH_3)}$ ...... }	128° C. (262.4° F.)	0.833 at 0° C.	
Ketones, $C_7H_{14}O$:			
1. Dipropyl ketone, $CO\diagdown^{C_3H_7}_{C_3H_7}$ ...... }	144° C. (291.2° F.)	0.82 at 20° C. (68° F.)	Butyric and propionic acids.
2. Di-isopropyl ketone, $CO\diagdown^{C_3H_7}_{C_3H_7}$ ...... }	123–124° C. (253.4–255.2° F.)	....	Isobutyric acid, acetic acid, & CO_2.
3. Diethylacetone, $CO\diagdown^{CH_3}_{CH(C_2H_5)_2}$ ...... }	137–139° C. (278.6–282.2° F.)	0.817 at 22° C. (71 6° F.)	
4. Methyl-pentyl ketone, $CO\diagdown^{CH_3}_{C_5H_{11}}$ ...... }	155–156° C. (311–312.8° F.)	0.813 at 20° C. (68° F.)	Acetic and normal valeric acids.
Methyl-hexyl ketone, $C_8H_{16}O = CO\diagdown^{CH_3}_{C_6H_{13}}$ .... }	171° C. (339.8° F.)	0.818.	Acetic and caproic acids.
Di-isobutyl ketone or valerone, $C_9H_{18}O = CO(C_4H_9)_2$ .... }	182° C. (359.6° F.)	0.833 at 30° C. (86° F.)	
Ketones, $C_{11}H_{22}O$:			
1. Diamyl ketone or caprone, $CO(C_5H_{11})_2$ ...... }	220° C. (428° F.)		
2. Methyl-nonyl ketone, $CO\diagdown^{CH_3}_{C_9H_{19}}$ ...... }	225° C. (437° F.)	Melting point, 15° C. (59° F.)	Acetic and pelargonic acids.
Dihexyl ketone or œnanthone, $C_{13}H_{26}O = CO(C_6H_{13})_2$ .. }	255° C. (491° F.)	30° C. (86° F.)	
Diheptyl ketone, $C_{15}H_{30}O = CO(C_7H_{15})_2$ .. }	280° C. (536° F.)	40° C. (104° F.)	
Dinonyl ketone, $C_{19}H_{38}O = CO(C_9H_{19})_2$ .. }	350° C. (662° F.)	58° C. (136.4° F.)	

ORGANIC ACIDS.

Organic acids, or Carbon-acids, contain the univalent group, COOH or

$O\!=\!\overset{|}{C}\!-\!OH$ (called carboxyl or oxatyl), linked by its free combining unit with a hydrocarbon residue, and they may be regarded as derived from hydrocarbons, saturated or unsaturated, by the substitution of one or more of these univalent groups for an equal number of hydrogen-atoms; thus:

from CH$_4$ are derived CH$_3$—CO$_2$H and CH$_2\!\!<^{CO_2H}_{CO_2H}$
  Methane.   Acetic acid.   Malonic acid.

" C$_2$H$_6$ "   C$_2$H$_5$.CO$_2$H " C$_2$H$_4\!\!<^{CO_2H}_{CO_2H}$
  Ethane.   Propionic acid. Succinic acid.

" C$_3$H$_6$ "   C$_3$H$_5$.CO$_2$H " C$_3$H$_4\!\!<^{CO_2H}_{CO_2H}$
  Propene.   Crotonic acid. Citraconic acid.

Many acids are formed by oxidation from primary alcohols, the H$_2$ of the groups CH$_2$OH in these alcohols being replaced by O; thus:

Alcohols.   *Acids.*

  CH$_3$ CH$_3$
  CH$_2$ CH$_2$
  CH$_2$OH   COOH
Propyl alcohol.   ropionic acid.

  CH$_2$OH   CH$_2$OH   COOH
  CH$_2$ CH$_2$ CH$_2$
  CH$_2$OH   COOH   COOH
β Propene glycol. β Lactic acid. Malonic acid.

  CH$_2$OH   CH$_2$OH   COOH
  CHOH   CHOH   CHOH
  CH$_2$OH   COOH   COOH
  Glycerin.   Glyceric acid. Oxymalonic acid.

An acid may, however, contain one or more hydroxyl groups not directly connected with the group CO, and the hydrogen in these groups, as well as in the group CO$_2$H, is replaceable by alcohol-radicles (also by alkali-metals) producing acid ethers, or alcoholic acids, *e. g.*, ethyl-lactic acid, CH$_2$(OC$_2$H$_5$)—CH$_2$—CO.OH.

The total number of hydroxyl groups in an organic acid is the same as that of the alcohol from which it is derived, and this determines the atomicity of the acid, and the number of ethers which it is capable of forming with any given alcohol-radicles; thus β-lactic acid, though it contains only one atom of basic hydrogen, and therefore forms only one class of metallic salts, represented by the formula C$_3$H$_5$O$_3$M, is diatomic, like propene-glycol, and can form two ethylic ethers, viz., ethyl-β-lactic acid and diethyl-β-lactate or ethylic ethyl-lactate; thus:

  CH$_2$OH   CH$_2$OC$_2$H$_5$ CH$_2$OC$_2$H$_5$
  CH$_2$ CH$_2$ CH$_2$
  COOH   COOH   COOC$_2$H$_5$
β-Lactic acid Ethyl-β-lactic Diethylic
(monobasic). acid (mono-   β-lactate
      basic).   (neutral).

From these considerations it appears that monatomic acids must necessarily be monobasic; but diatomic acids may be either monobasic or bibasic; triatomic acids, either monobasic, bibasic, or tribasic; and so on.

Many of the most important acids are derived, in the manner above explained, from actually known alcohols; others, though they have no alcohols actually corresponding with them, are homologous with other acids derived from known alcohols; but there is also a considerable number of acids, especially those formed in the vegetable or animal organism, which cannot be regarded as derivatives of alcohols of any known series; but the number of these unclassified acids will doubtless diminish as their composition and reactions become more thoroughly known.

Acids may also be regarded as compounds of hydroxyl with oxygenated radicles (acid radicles) formed from the corresponding alcohol-radicles by substitution of O for H_2; e.g.,

$$C_2H_5.OH \qquad C_2H_3O.OH$$
Ethyl alcohol.   Acetic acid.

$$C_3H_6(OH)_2 \qquad C_3H_4O(OH)_2 \qquad C_3H_2O_2(OH)_2$$
Propene glycol.   Lactic acid.   Malonic acid.

The replacement of the hydroxyl in an acid by chlorine, bromine, or iodine, gives rise to acid chlorides, etc.; thus from acetic acid, $C_2H_3O(OH)$, is derived acetic chloride, C_2H_3OCl, etc. The replacement of the hydrogen within the radicle (radical hydrogen) by the same elements, or by the groups, CN, NO_2, NH_2, etc., gives rise to chlorinated, brominated, cyanated, nitrated, and amidated acids. Lastly, the replacement of the *extra-radical* or *typic* hydrogen by alcohol-radicles gives rise to ethereal salts of compound ethers; and its replacement by acid radicles yields acid oxides or anhydrides (p. 509). The derivatives of each acid will be described in connection with the acid itself.

Monatomic Acids.

These acids, being derived from monatomic alcohols by substitution of O for H_2, necessarily contain two atoms of oxygen. Each series of hydrocarbons yields a series of monatomic alcohols and a series of monatomic acids; thus:

Hydrocarbons.	Alcohols.	Acids.
C_nH_{2n+2}	$C_nH_{2n+2}O$	$C_nH_{2n}O_2$
C_nH_{2n}	$C_nH_{2n}O$	$C_nH_{2n-2}O_2$
C_nH_{2n-2}	$C_nH_{2n-2}O$	$C_nH_{2n-4}O_2$
C_nH_{2n-4}	$C_nH_{2n-4}O$	$C_nH_{2n-6}O_2$
etc.	etc.	etc.

The best known monatomic acids are those belonging to the series $C_nH_{2n}O_2$, $C_nH_{2n-2}O_2$, $C_nH_{2n-8}O_2$, and $C_nH_{2n-10}O_2$. The last two belong to the aromatic group. Of the other series only a few terms have hitherto been obtained.

1. Acids of the Fatty Series.

$$C_nH_{2n}O_2 \;=\; C_{n+1}H_{2n-1}.CO_2H \;=\; C_nH_{2n+1}O.OH.$$

These acids are called Fatty or Adipic Acids, because most of them are of an oily consistence, and the higher members of the series are solid fats. The following is a list of the known members of the series:

Formic acid	CH_2O_2	Lauric acid		$C_{12}H_{24}O_2$
Acetic acid	$C_2H_4O_2$	Myristic acid		$C_{14}H_{28}O_2$
Propionic acid	$C_3H_6O_2$	Palmitic acid		$C_{16}H_{32}O_2$
Butyric acid (normal)	$C_4H_8O_2$	Margaric acid		$C_{17}H_{34}O_2$
Valeric acid "	$C_5H_{10}O_2$	Stearic acid		$C_{18}H_{36}O_2$
Caproic acid "	$C_6H_{12}O_2$	Arachidic acid		$C_{20}H_{40}O_2$
Œnanthylic acid	$C_7H_{14}O_2$	Behenic acid		$C_{22}H_{44}O_2$
Caprylic acid	$C_8H_{16}O_2$	Cerotic acid		$C_{27}H_{54}O_2$
Pelargonic acid	$C_9H_{18}O_2$	Melissic acid		$C_{30}H_{60}O_2$
Rutic or Capric acid	$C_{10}H_{20}O_2$			

The higher members, from valeric acid upwards, are sometimes denoted by names indicating their number of carbon-atoms, and the alcohols from which they are derived, e. g., *pentoic*, *hexoic*, *heptoic*, etc.

The acid radicles in the formula $C_nH_{2n-1}(OH)$ may be regarded as compounds of carbonyl CO, with alcohol-radicles : $C_nH_{2n-1}O = CO(C_{n-1}H_{2n-1})$, and accordingly the several acids may be formulated as follows :—

$$COH.OH \qquad CO(CH_3).OH \qquad CO(C_2H_5)OH .$$
Formic.   Acetic.   Propionic.

All the acids of the series . containing more than three carbon-atoms admit of isomeric modifications, according to the constitution of the alcohol-radicles which they contain : butyric acid, $C_4H_8O_2$, for example, may exhibit the following modifications :—

$$CH_2CH_2CH_3 \qquad or \qquad CH(CH_3)_2$$
$$|\qquad\qquad\qquad\qquad\qquad |$$
$$COOH \qquad\qquad\qquad\qquad COOH$$
Normal butyric acid.   Isobutyric acid.

But none of these acids can exhibit modifications analogous to the secondary and tertiary alcohols : because in them the carbon-atom which is associated with hydroxyl has two of its other units of equivalence satisfied by an atom of bivalent oxygen, and therefore cannot unite directly with more than one other atom of carbon. Accordingly, it is found that the secondary and tertiary alcohols are not converted by oxidation into acids containing the same number of carbon-atoms as themselves.

Occurrence.—Most of the fatty acids are found in the bodies of plants or animals, some in the free state ; formic acid in ants and nettles ; valeric acid in valerian root ; pelargonic acid in the essential oil of *Pelargoneum roseum;* and cerotic acid in beeswax. Others occur as ethereal salts of monatomic or polyatomic alcohols : as cetyl palmitate in spermaceti ; ceryl cerotate in Chinese wax ; glyceric butyrate, palmitate, stearate, etc., in natural fats.

Formation.—1. By oxidation of the primary alcohols of the methyl series, as by exposure to the air in contact with platinum black, or by heating with aqueous chromic acid.—2. By the oxidation of aldehydes. In this case, an atom of oxygen is simply added ; e. g., C_2H_4O (aldehyde) $+ O = C_2H_4O_2$ (acetic acid).

3. By the action of carbon dioxide on the potassium or sodium compound of an alcohol-radicle of the methyl series ; thus :—

$$CO_2 \qquad + \qquad CH_3Na \qquad = \qquad CH_3.CO_2Na$$
Carbon   Sodium   Sodium
dioxide.   methide.   acetate.

4. By heating the ethylate of an alkali-metal in alcoholic solution with carbon monoxide under pressure ; e. g.,

$$C_2H_5OK \qquad + \qquad CO \qquad = \qquad C_2H_5.CO.OK$$
Potassium   Potassium
ethylate.   acetate.

This reaction, however, is slow, and yields but a small product.

5. By the action of alkalies or acids on the cyanides of the alcohol-radicles, C_nH_{2n+1}; thus :—

$$
\begin{matrix}
C_nH_{2n+1} \\
\mid \\
CN
\end{matrix}
\quad + \quad KOH \quad + \quad H_2O \quad = \quad
\begin{matrix}
C_nH_{2n+1} \\
\mid \\
CO_2K
\end{matrix}
\quad + \quad NH_3
$$

Alcoholic   Potassium   Water.   Potassium-salt Ammonia.
cyanide.   hydroxide.   of fatty acid.

and

$$C_nH_{2n+1}.CN \quad + \quad HCl \quad + \quad 2H_2O \quad = \quad C_nH_{2n+1}.CO_2H \quad + \quad NH_4Cl$$

Alcoholic   Hydrochloric   Water.   Potassium-salt.   Ammonium
cyanide.   acid.   chloride.

In this manner the cyanide of each alcohol-radicle yields the potassium-salt of the acid next higher in the series, that is, containing one atom of carbon more; methyl cyanide, for example, yielding acetic acid, ethyl cyanide yielding propionic acid, etc.; thus :—

$$CH_3.CN \quad + \quad KOH \quad + \quad H_2O \quad = \quad CH_3.CO_2K \quad + \quad NH_3$$

Methyl   Potassium
cyanide.   acetate.

6. By the action of water on the corresponding acid chlorides: e. g.,

$$C_2H_3OCl \quad + \quad HOH \quad = \quad HCl \quad + \quad C_2H_3O(OH)$$

Acetyl chloride.   Acetic acid.

Now, these acid chlorides can be produced, in some instances at least, by the action of carbonyl chloride (phosgene gas) on the corresponding paraffins; thus :—

$$CH_4 \quad + \quad COCl_2 \quad = \quad HCl \quad + \quad C_2H_3OCl$$

Methane.   Carbonyl   Acetyl
         chloride.   chloride.

$$C_4H_{10} \quad + \quad COCl_2 \quad = \quad HCl \quad + \quad C_5H_9OCl$$

Butane.   Carbonyl   Valeryl
        chloride.   chloride.

By these combined reactions, therefore, the paraffins may be converted into the corresponding fatty acids.

7. By the action of phosgene on the zinc-compounds of the alcohol-radicles, whereby acid chlorides are formed, to be subsequently decomposed by water :

$$Zn(CH_3)_2 \quad + \quad 2COCl_2 \quad = \quad ZnCl_2 \quad + \quad 2(CH_3.COCl)$$

Zinc-methyl.   Acetyl chloride.

8. By dissolving sodium in methylic or ethylic acetate, adding the iodide of an alcohol radicle, heating the mixture to 100° and distilling. The reaction, which is complex, and will be more fully explained hereafter (see CARBO-KETONIC ACIDS), may be viewed—so far as our present purpose is concerned—as consisting in the substitution of 1 or 2 atoms of sodium for hydrogen in the methyl-group of acetic acid, and the decomposition of the resulting compound by the alcoholic iodide: e. g.,

$$
\begin{matrix}
CH_2Na \\
\mid \\
CO_2C_2H_5
\end{matrix}
\quad + \quad CH_3I \quad = \quad NaI \quad + \quad
\begin{matrix}
CH_2(CH_3) \\
\mid \\
CO_2C_2H_5
\end{matrix}
$$

Ethylic sod-   Ethylic methyl
acetate.   acetate.

$$
\begin{matrix}
CHNa_2 \\
\mid \\
CO_2C_2H_5
\end{matrix}
\quad + \quad 2CH_3I \quad = \quad 2NaI \quad + \quad
\begin{matrix}
CH(CH_3)_2 \\
\mid \\
CO_2C_2H_5
\end{matrix}
$$

Ethylic di-   Ethylic dimethyl
sodacetate.   acetate.

The resulting ethers saponified with caustic potash yield the acids; thus:

$$\underset{\substack{\text{Ethylic dimethyl} \\ \text{acetate.}}}{\overset{\displaystyle CH(CH_3)_2}{\underset{\displaystyle CO_2C_2H_5}{\vert}}} + KOH = C_2H_5OH + \underset{\substack{\text{Dimethylacetic or} \\ \text{Isobutyric acid.}}}{\overset{\displaystyle CH(CH_3)_2}{\underset{\displaystyle CO_2K}{\vert}}}$$
$$\qquad\qquad\qquad\qquad\qquad\qquad\qquad \text{Alcohol.}$$

Ethyl-acetic or normal butyric acid, $CH_2(C_2H_5)$—CO_2H, and other members of the series, may be obtained in a similar manner.

The eight modes of formation above given are general, or capable of being made so. There are also special methods of producing particular acids of the series, but in most of these cases the reactions cannot be distinctly traced; thus formic, acetic, propionic, butyric, and valeric acids are produced by the oxidation of albumin, fibrin, casein, gelatin, and other similar substances; propionic and butyric acids, in certain kinds of fermentation; acetic acid by the destructive distillation of wood and other vegetable substances.

Properties.—Most of the fatty acids are, at ordinary temperatures, transparent and colorless liquids; formic and acetic acids are watery; propionic acid and the higher acids, up to pelargonic acid, are oily; rutic acid and those above it are solid at ordinary temperatures, most of them being crystalline fats; cerotic and melissic acids are of waxy consistence.

Reactions.—1. When the fatty acids are submitted to the action of *nascent oxygen* evolved by electrolysis, the carboxyl (COOH) contained in them is resolved into water and carbon dioxide, and the alcohol-radicle is set free (p. 520); thus:

$$\underset{\text{Valeric acid.}}{2(C_4H_9.CO_2H)} + O = H_2O + 2CO_2 + \underset{\text{dibutyl.}}{C_4H_9.C_4H_9}$$

2. When the ammonium salt of either of these acids is heated with *phosphoric oxide*, it gives up water, and is converted into the cyanide of the alcohol-radicle next below it; *e. g.*, ammonium acetate into methyl cyanide:

$$CH_3—COONH_4 - 2H_2O = CH_3—CN.$$

This reaction is the converse of the fifth mode of formation above given.

3. By distilling the potassium salt of a fatty acid with an equivalent quantity of *potassium formate*, the corresponding a l d e h y d e is obtained:

$$\underset{\text{Acetate.}}{CH_3.CO.OK} + \underset{\text{Formate.}}{H.CO.OK} = \underset{\text{Aldehyde.}}{CH_3.COH} + \underset{\text{Carbonate.}}{K_2CO_3};$$

and the aldehyde, treated with nascent hydrogen, is converted into a primary alcohol (pp. 586, 693).

4. By subjecting the barium or calcium salt of a fatty acid to dry distillation, a similar decomposition takes place, resulting in the formation of a k e t o n e; and the ketone, treated with nascent hydrogen, yields a secondary alcohol (pp. 586, 705).

By these reactions, the fatty acids may be converted into alcohols.

5. The fatty acids, heated with *alcohols* in sealed tubes, yield compound ethers, or e t h e r e a l s a l t s, water being eliminated:

$$\underset{\substack{\text{Butyric} \\ \text{acid.}}}{C_4H_7O(OH)} + \underset{\substack{\text{Ethyl} \\ \text{alcohol.}}}{HOC_2H_5} = H_2O + \underset{\substack{\text{Ethyl} \\ \text{butyrate.}}}{C_4H_7O(OC_2H_5)}$$

The conversion, however, is never complete, a portion, both of the acid

and of the alcohol, remaining unaltered, in whatever proportion they may
be mixed (p. 581).

The ethereal salts of the fatty acids are, for the most part, more easily
obtained by acting upon the alcohol with an acid chloride, or by passing
hydrochloric acid gas into a solution of the fatty acid in the alcohol; thus
butyric chloride and ethyl alcohol yield ethyl-butyrate:

$$C_4H_7OCl \;+\; HOC_2H_5 \;=\; HCl \;+\; C_4H_7O(OC_2H_5).$$

Another method very commonly adopted, is to distil a potassium salt of
the fatty acid with a mixture of the alcohol and strong sulphuric acid.
In this case an acid sulphuric ether is first formed (as ethyl-sulphuric acid
from ethyl alcohol, p. 601), and this acts upon the salt of the fatty acid
in the manner illustrated by the equation:

$$\underset{\substack{\text{Ethyl-sulphuric} \\ \text{acid.}}}{SO_2(HO)OC_2H_5} \;+\; \underset{\substack{\text{Potassium} \\ \text{butyrate}}}{C_4H_7O(OK)} \;=\; \underset{\substack{\text{Ethyl} \\ \text{butyrate.}}}{C_4H_7O(OC_2H_5)} \;+\; \underset{\substack{\text{Acid potassium} \\ \text{sulphate.}}}{SO_2(OH)(OK)}$$

The ethereal salts of the fatty acids are either volatile, oily or syrupy
liquids, or crystalline solids, for the most part insoluble in water, but
soluble in alcohol and in ether. When distilled with potash or soda, they
take up water and are saponified, that is to say, resolved into the alcohol
and acid; e. g., ethyl butyrate into butyric acid and ethyl alcohol:

$$C_4H_7O(OC_2H_5) \;+\; HOH \;=\; C_4H_7O(OH) \;+\; C_2H_5(OH).$$

6. The fatty acids are strongly acted upon by the *chlorides, bromides,
oxychlorides,* and *oxybromides of phosphorus,* yielding a c i d c h l o r i d e s and
b r o m i d e s, the phosphorus being at the same time converted into phosphorous or phosphoric acid; thus:

$$3C_2H_3O(OH) \;+\; PCl_3 \;\;=\; PO_3H_3 \;+\; 3C_2H_3OCl.$$
$$3C_2H_3O(OH) \;+\; PCi_3O \;=\; PO_4H_3 \;+\; 3C_2H_3OCl.$$
$$C_2H_3O(OH) \;+\; PCl_5 \;=\; PCl_3O \;+\; HCl \;+\; C_2H_3OCl.$$

These acid chlorides are, for the most part, oily liquids, having a pungent
acid odor; they are easily decomposed by water, yielding the fatty acid
and hydrochloric acid. This decomposition takes place also when they are
exposed to the air: hence they emit dense acid fumes. They react in an
exactly similar manner with alcohols, as above mentioned, yielding hydrochloric acid and a compound ether.

7. The chlorides of the acid radicles, $C_nH_{2n+1}O$, act violently on ammonia, forming ammonium chloride and the corresponding a m i d e s, acetic
chloride, for example, yielding acetamide:

$$C_2H_3OCl \;+\; 2NH_3 \;=\; NH_4Cl \;+\; NH_2(C_2H_3O).$$

8. The acid chlorides, distilled with a metallic salt of the corresponding
acid, yield a metallic chloride and the o x i d e or a n h y d r i d e corresponding with the acid; thus:

$$C_2H_3OCl \;+\; C_2H_3O(OK) \;=\; KCl \;+\; (C_2H_3O)_2O.$$

In like manner, when distilled with the potassium salt of another monatomic acid, they yield oxides or anhydrides containing two monatomic
acid radicles: e. g.,

$$\underset{\substack{\text{Acetic} \\ \text{chloride.}}}{C_2H_3OCl} \;+\; \underset{\substack{\text{Potassium} \\ \text{benzoate.}}}{C_7H_5O(OK)} \;=\; KCl \;+\; \underset{\substack{\text{Aceto-ben-} \\ \text{zoic oxide.}}}{\left.\begin{matrix} C_2H_3O \\ C_7H_5O \end{matrix}\right\}O}$$

The oxides of the fatty acid radicles may also be prepared by heating a dry lead-salt of the acid, in a sealed tube with carbon bisulphide : *e. g.*,

$$2Pb \begin{cases} OC_2H_3O \\ OC_2H_3O \end{cases} + \quad CS_2 \quad = \quad 2PbS \quad + \quad CO_2 \quad + \quad 2(C_2H_3O)_2O$$
Lead acetate.   Acetic oxide.

The oxides of the fatty acid radicles are gradually decomposed by water, quickly when heated, yielding two molecules of the corresponding acid :

$$(C_2H_3O)_2O \quad + \quad H_2O \quad = \quad 2C_2H_3O(OH).$$

Those containing two acid radicles yield one molecule of each of the corresponding acids.

In contact with *alcoholic oxides* (*oxygen-ethers*), the acid oxides are converted into ethereal salts :

$$(C_2H_3O)_2O \quad + \quad (C_2H_5)_2O \quad = \quad 2C_2H_3O(OC_2H_5)$$
Acetic oxide.   Ethyl oxide.   Ethyl acetate.

With *alcohols*, in like manner, they yield a mixture of a compound ether with the acid :

$$(C_2H_3O)_2O \quad + \quad C_2H_5(OH) \quad = \quad C_2H_3O(OC_2H_5) \quad + \quad C_2H_3O(OH)$$
Acetic oxide.   Ethyl alcohol.   Ethyl acetate.   Acetic acid.

The acid oxides are decomposed by *ammonia gas*, yielding a mixture of an ammonium-salt with an amide; *e. g.*,

$$(C_2H_3O)_2O \quad + \quad 2NH_3 \quad = \quad C_2H_3O(ONH_4) \quad + \quad NH_2C_2H_3O.$$

9. The fatty acids, subjected to the action of *chlorine* or *bromine*, give off hydrochloric or hydrobromic acid, and are converted into substitution-compounds containing one or more atoms of chlorine or bromine in place of hydrogen ; but it is only the hydrogen within the radicle that can be thus exchanged, not that belonging to the hydroxyl group (the so-called *typic* hydrogen), so that the number of chlorine or bromine-atoms introduced in place of hydrogen is always less by at least one than the number of hydrogen-atoms in the acid :

$$C_2H_3O(OH) \quad + \quad Cl_2 \quad = \quad HCl \quad + \quad C_2H_2ClO(OH)$$
Acetic acid.   Chloracetic acid.

$$C_2H_3O(OH) \quad + \quad 3Cl_2 \quad = \quad 3HCl \quad + \quad C_2Cl_3O(OH)$$
Acetic acid.   Trichloracetic acid.

The **iodated acids** of the same series (or rather their ethereal salts) are obtained by heating the corresponding bromine-compounds with potassium iodide :

$$C_2H_2BrO(OC_2H_5) \quad + \quad KI \quad = \quad KBr \quad + \quad C_2H_2IO(OC_2H_5) ;$$
Ethyl-brom-   Ethyl-iodacetate.
acetate.

and the ethers treated with potash yield potassium salts of the iodated acids, from which the acids may be obtained by decomposition with sulphuric acid.

10. The chlorinated and brominated fatty acids, boiled with *water* and *silver oxide*, exchange the whole of their chlorine or bromine for an equivalent quantity of hydroxyl, producing new acids, which differ from the primitive acids by a number of atoms of oxygen equal to the number of atoms of chlorine or bromine present ; *e. g.*,

$$2C_2H_3BrO_2 \quad + \quad Ag_2O \quad + \quad H_2O \quad = \quad 2AgBr \quad + \quad 2C_2H_4O_3$$
Bromacetic   Glycollic
acid.   acid.

$$C_4H_6Br_2O_2 \quad + \quad Ag_2O \quad + \quad H_2O \quad = \quad 2AgBr \quad + \quad C_4H_8O_4 \,.$$
Dibromo-   Dioxy-
butyric acid.   butyric acid.

Dichloracetic and trichloracetic acid are not sufficiently stable to exhibit this transformation, their molecules splitting up altogether when boiled with silver oxide.

11. The monochlorinated and monobrominated acids, subjected to the action of an alcoholic solution of *ammonia gas*, yield ammonium chloride and a new acid, in which the chlorine or bromine is replaced by amidogen. Thus monochloracetic acid yields a m i d a c e t i c a c i d, or g l y - c o c i n e :—

$$C_2H_3ClO_2 \quad + \quad 2NH_3 \quad = \quad NH_4Cl \quad + \quad C_2H_3(NH_2)O_2 \,.$$

There is another way of viewing these amidated acids, which will be con, sidered hereafter.

Formic Acid, $CH_2O_2 = CHO(OH) = H.CO.OH$.—This acid occurs in the concentrated state in the bodies of ants, in the hairs and other parts of certain caterpillars, and in stinging nettles. It may be produced by the first, second, and fourth of the above-mentioned general methods of forming the fatty acids—viz., by the slow oxidation of methyl alcohol, or of formic aldehyde, in contact with platinum black, and as a potassium salt by heating hydrocyanic acid (hydrogen cyanide) with an alcoholic solution of potash :—

$$HCN \quad + \quad KOH \quad + \quad H_2O \quad = \quad NH_3 \quad + \quad CHO(OK) \,.$$

It is also produced by certain special reactions—viz., 1. By passing carbon monoxide over moist potassium hydroxide, the gas being thereby absorbed, and producing potassium formate :—

$$CO \quad + \quad HOK \quad = \quad COH(OK) \,.$$

The absorption of the gas is accelerated by the presence of a considerable quantity of water, and still more by alcohol or ether.

2. By passing carbon dioxide and water-vapor over potassium at a moderate heat, acid potassium carbonate, $KHCO_3$, being formed at the same time :—

$$K_2 \quad + \quad 2CO_2 \quad + \quad H_2O \quad = \quad KHCO_3 \quad + \quad KCHO_2 \,.$$

3. By the action of sodium-amalgam on a strong solution of ammonium carbonate, and by boiling zinc-dust (a mixture of metallic zinc, oxide, and hydrate), or zinc carbonate with potash solution. In both these cases the production of formic acid is due to the nascent hydrogen, which, in presence of the alkali, unites with the CO_2 of the carbonate :—

$$CO_2 \quad + \quad KOH \quad + \quad H_2 \qquad HCO_2K \quad + \quad H_2O \,.$$

4. By distilling dry oxalic acid either alone or mixed with sand or pumice-stone, or better, with glycerin :—

$$C_2H_2O_4 \quad = \quad CO_2 \quad + \quad CH_2O_2$$

5. By boiling chloroform with alcoholic solution of potash :

$$HCCl_3 \quad + \quad 4KOH \quad = \quad 3KCl \quad + \quad 2H_2O \quad + \quad H.CO.OK \,.$$

6. By the oxidation of sugar, starch, gum, and organic substances in general. · This reaction affords a convenient method of preparing the acid : ·1 part of sugar, 3 parts of manganese dioxide, and 2 parts of water, are mixed in a very capacious retort, or large metal still; 3 parts of oil of vitriol, diluted with an equal weight of water, are then added, and when the first violent effervescence from the disengagement of carbon dioxide has subsided, heat is cautiously applied, and a considerable· quantity of liquid distilled over. This is very impure : it contains a volatile oily matter, and some substance which communicates a pungency not proper to formic acid in that dilute state. The acid liquid is neutralized with sodium carbonate, and the resulting formate purified by crystallization, and, if needful, by animal charcoal. From this or any other of its salts, solution of formic acid may be readily obtained by distillation with sulphuric acid.

The best mode of preparation, however, consists in heating oxalic acid in contact with glycerin. Very concentrated glycerin is added to crystallized oxalic acid, $C_2H_2O_4 + 2H_2O$, and the mixture is heated to 100°–110° C. (212°–230° F.), whereupon carbon dioxide escapes, and dilute formic acid distils over. As soon as the evolution of gas ceases, more oxalic acid is added, and the heating continued, whereupon a stronger formic acid distils over, and on further addition of oxalic acid, and heating, an acid of constant strength (56 per cent.) passes over. The course of the reaction is as follows : The crystallized oxalic acid, when heated, gives up its water, and the remainder forms with the glycerin, the monoformic ether of glycerin, or monoformin :—

$$C_3H_5(OH)_3 \ + \ C_2O_4H_2 \ = \ CO_2 \ + \ H_2O \ + \ C_3H_5 \left\{ {(OH)_2 \atop OCHO} \right. ;$$

and the oxalic acid afterwards added is likewise resolved into anhydrous acid and water, which decomposes the formin into glycerin and formic acid :—

$$C_3H_5 \left\{ {(OH)_2 \atop OCHO} \right. + \ H_2O \ = \ C_3H_5(OH)_3 \ + \ CHO.OH \ .$$

The regenerated glycerin reacts with the anhydrous oxalic acid, reproducing monoformin.

To obtain the acid in its most concentrated state, the dilute acid is saturated with lead oxide, the liquid is evaporated to complete dryness, and the dried lead formate, reduced to fine powder, is very gently heated in a glass tube connected with a condensing apparatus, through which a current of dry sulphuretted hydrogen gas is passed.

Pure anhydrous formic acid is a clear, colorless liquid, which fumes slightly in the air, has an exceedingly penetrating odor, boils at about 100°,* and crystallizes in large brilliant plates when cooled below 0°. The specific gravity of the acid is 1.233. Its vapor is inflammable, and burns with a blue flame. Concentrated formic acid is extremely corrosive, attacking the skin, and forming a blister or an ulcer, painful and difficult to heal.

Formic acid mixes in all proportions with water, alcohol, and ether. The aqueous acid has an odor and taste much resembling those of acetic acid : it reddens litmus strongly, and decomposes alkaline carbonates with effervescence.

Formic acid also unites with water in definite proportion, forming a hydrate, $CH_2O_2 + H_2O$, which remains liquid at low temperatures, and distils at 106° C. (222.8^o F.). This hydrate may be regarded as a triatomic acid, viz., orthoformic acid, $HC(OH)_3$.

Formic acid is a powerful reducing agent. It may be readily distin-

* At 98.5° (Liebig); 100° (Person); 101.1° (Roscoe); 105.5° (Kopp).

guished from acetic acid by heating it with solution of *silver nitrate;* the metal is thus reduced, sometimes in the pulverulent state, sometimes as a specular coating on the glass tube, and carbon dioxide is evolved. Mercuric chloride is reduced by formic acid to calomel.

Formic acid heated with *oil of vitriol* splits up into water and carbon monoxide, $CH_2O_2 = H_2O + CO$.

Chlorine converts it into hydrochloric acid and carbon dioxide:

$$CH_2O_2 + Cl_2 = 2HCl + CO_2.$$

Formic acid heated with strong *bases* is converted into oxalic acid, with disengagement of hydrogen ; *e. g.*,

$$2CH_2O_2 + BaO = C_2BaO_4 + H_2 + H_2O.$$

Formates.—The composition of these salts is expressed by the formulæ, $MCHO_2$, $M''(CHO_2)_2$, $M'''(CHO_2)_3$, etc., according to the equivalent value of the metal or other positive radicle contained in them. They are all soluble in water ; their solutions form dark-red mixtures with ferric salts. When distilled with strong sulphuric acid, they give off carbon monoxide and leave a residue of sulphate. The formates of the alkali-metals heated with the corresponding salts of other fatty acids, yield a carbonate and aldehyde (p. 693).

Sodium formate crystallizes in rhombic prisms containing $CHO_2Na.Aq.$ It reduces many metallic oxides when fused with them. *Potassium formate,* CHO_2K, is difficult to crystallize, on account of its great solubility. *Ammonium formate* crystallizes in square prisms : it is very soluble, and is decomposed at high temperatures into hydrocyanic acid and water, the elements of which it contains : $CHO_2NH_4 = 2H_2O + CNH.$ The formates of *barium, strontium, calcium,* and *magnesium* form small, prismatic, easily soluble crystals. *Lead formate* crystallizes in small, diverging, colorless needles, which require for solution 40 parts of cold water. The *manganous, ferrous, zinc, nickel,* and *cobalt formates* are also crystallizable. *Cupric formate* is very beautiful, crystallizing in bright blue rhombic prisms of considerable size. *Silver formate* is white, but slightly soluble, and decomposed by the least elevation of temperature.

Methyl formate, $CHO_2.CH_3$, metameric with acetic acid, is prepared by heating in a retort equal weights of neutral methyl sulphate and sodium formate. It is a very volatile liquid, lighter than water, boiling between 36° and 38° C. (96.8°–100.4° F.).

Ethyl formate, $CHO_2.C_2H_5$, metameric with methyl acetate and propionic acid (p. 512), is prepared by distilling a mixture of 7 parts of dry sodium formate, 10 of oil of vitriol, and 6 of strong alcohol. The formic ether, separated by the addition of water to the distilled product, is agitated with a little magnesia, and left for several days in contact with calcium chloride. Ethyl formate is colorless, has an aromatic odor, a density of 0.915, and boils at 56° C. (132.8° F.). Water dissolves it to a small extent.

Ethyl Orthoformate, $HC(OC_2H_5)_3$, is produced by heating chloroform with sodium ethylate in alcoholic solution :

$$HCCl_3 + 3C_2H_5ONa = 3NaCl + HC(OC_2H_5)_3.$$

It is a liquid having an aromatic odor, boiling at 146° C. (294.8° F.), insoluble in water. When heated with glacial acetic acid it is resolved into formate and acetate of ethyl.

Acetic acid, $C_2H_4O_2 = C_2H_3O(OH) = CH_3(COOH)$.—This acid is found in small quantities in the juices of plants and in animal fluids. It may be produced by either of the first seven general methods of formation given on pages 713–714, and in particular by the slow oxidation of alcohol. When spirit of wine is dropped upon platinum black, the oxygen condensed in the pores of the latter reacts so powerfully upon the alcohol as to cause its instant inflammation. When the spirit is mixed with a little water, and slowly dropped upon the finely divided metal, oxidation still takes place, but with less energy, and vapor of acetic acid is abundantly evolved. In all these modes of formation, the acetic acid is ultimately producible from inorganic materials. It is also formed by the action of nascent hydrogen on trichloracetic acid, which may itself be produced from inorganic materials. Lastly, acetic acid is obtained, together with many other products, in the destructive distillation of wood and other vegetable substances.

Preparation.—1. Dilute alcohol, mixed with a little yeast, or almost any azotized organic matter susceptible of putrefaction, and exposed to the air, speedily becomes oxidized to acetic acid. Acetic acid is thus manufactured in Germany, by suffering such a mixture to flow over wood shavings steeped in a little vinegar, contained in a large cylindrical vessel through which a current of air is made to pass. The greatly extended surface of the liquid expedites the change, which is completed in a few hours. No carbonic acid is produced in this reaction.

The best vinegar is made from wine by spontaneous acidification in a partially filled cask to which the air has access. Vinegar is first introduced into the empty vessel, and a quantity of wine added; after some days, a second portion of wine is poured in, and after similar intervals, a third and a fourth. When the whole has become vinegar, a quantity is drawn off equal to that of the wine employed, and the process is recommenced. The temperature of the building is kept up to 30° C. (86° F.). Such is the plan adopted at Orleans. In England vinegar is prepared from a kind of beer made for the purpose. The liquor is exposed to the air in half-empty casks, loosely stopped, until acidification is complete. Frequently a little sulphuric acid is afterwards added, with the view of checking further decomposition, or *mothering*, by which the product would be spoiled.

When dry, hard wood, as oak and beech, is subjected to destructive distillation at a red heat, acetic acid is found among the liquid condensable products of the operation. The distillation is conducted in a large iron cylinder, to which a worm or condenser is attached; a sour, watery liquid, a quantity of tar, and much inflammable gas pass over, while charcoal of excellent quality remains in the retort. The acid liquid is subjected to distillation, the first portion being collected apart for the preparation of wood-spirit. The remainder is saturated with lime, concentrated by evaporation, and mixed with the solution of sodium sulphate; calcium sulphate is thereby precipitated, while the acetic acid is transferred to the soda. The filtered solution is evaporated to its crystallizing point; and the crystals are drained as much as possible from the dark, tarry mother-liquor, and deprived by heat of their combined water. The dry salt is then cautiously fused, by which the last portions of tar are decomposed or expelled: it is then redissolved in water, and recrystallized. Pure sodium acetate, thus obtained, readily yields acetic acid by distillation with sulphuric acid.

The strongest acetic acid is prepared by distilling finely powdered anhydrous sodium acetate with three times its weight of strong sulphuric acid. The liquid is purified by rectification to free it from sodium sulphate accidentally thrown up, and exposed to a low temperature. Crystals of pure

31

acetic acid, $C_2H_4O_2$, then form in large quantity : they may be drained from the weaker fluid portion, and suffered to melt. Below 15.5° C. (59.9° F.) this substance, often called *glacial acetic acid*, forms large, colorless, transparent crystals, which above that temperature fuse to a thin, color-less liquid, of exceedingly pungent and well-known odor ; it raises blisters on the skin. It is miscible in all proportions with water, alcohol, and ether, and dissolves camphor and several resins. When diluted it has a pleasant acid taste. Glacial acetic acid in the liquid state has a density of 1.063, and boils at 120° C. (248° F.). Its vapor is inflammable, and exhibits the variations of density already noticed (p. 246). At 300° C. (572° F.), or above, it is 2.08 compared with air, or 30 compared with hy-drogen, agreeing exactly with the theoretical density, which is half the molecular weight ; but at temperatures near the boiling point it is consid-erably greater, being 2.90 at 140° C. (284° F.), and 3.20 at 125° C. (257° F.) (referred to air).

Dilute acetic acid, or distilled vinegar, used in pharmacy, should always be carefully examined for copper and lead ; these impurities are contracted from the metallic vessel or condenser sometimes employed in the process. The strength of any sample of acetic acid cannot be safely inferred from its density, but it is easily determined by observing the quantity of dry sodium carbonate necessary to saturate a known weight of the liquid. Common vinegar contains from 5 to 15 per cent. of the pure acid.

Acetic acid exhibits all the reactions of the fatty acids in general (pp. 715–717). The acid itself does not readily conduct the electric current, but a solution of potassium acetate is decomposed by electrolysis, with formation of dimethyl, or ethane, and potassium carbonate :

$$2(CH_3.CO.OK) + H_2O = C_2H_6 + H_2 + CO_2 + CO(OK)_2 .$$

Acetic acid is not attacked by nitric acid, but *periodic acid* converts it by oxidation into formic acid and carbon dioxide, being itself reduced to iodic acid or even to free iodine :

$$C_2H_4O_2 + O_3 = CH_2O_2 + CO_2 + H_2O.$$

Potassium acetate distilled with *arsenious oxide* gives off a highly inflam-mable and characteristically fetid oil, consisting chiefly of arsendimethyl or cacodyl, $As_2(CH_3)_4$.

Acetates.—Acetic acid forms a large number of highly important salts, represented by the formulæ, $MC_2H_3O_2$, $M''(C_2H_3O_2)_2$, or $M'''(C_2H_3O_2)_3$, according to the equivalent value of the metal contained in them. Being a monobasic acid, it cannot form any acid salts properly so called, that is, by replacement of a *part* of its typic hydrogen (p. 712); but the normal acetates of the alkali-metals can take up a molecule of acetic acid, just as they take up water of crystallization, forming salts called acid ace-tates or diacetates, $MC_2H_3O_2.C_2H_4O_2$. There are also basic acetates, formed by the union of a molecule of a normal acetate with a molecule of metallic oxide or hydrate.

POTASSIUM ACETATES.—The *normal salt*, $KC_2H_3O_2$, crystallizes with great difficulty : it is generally met with as a foliated, white, crystalline mass, obtained by neutralizing potassium carbonate with acetic acid, evaporating to dryness, and heating the salt to fusion. It is extremely deliquescent, and soluble in water and alcohol : the solution is usually alkaline, from a little loss of acid by the heat to which it has been subjected. From the alcoholic solution, potassium carbonate is thrown down by a stream of carbon dioxide.

The *acid salt*, $KC_2H_3O_2.C_2H_4O_2$, is formed by evaporating a solution of the neutral salt in excess of acetic acid, and crystallizes by slow evaporation in long flattened prisms. It is very deliquescent, and decomposes at 200° C. (392° F.), giving off crystallizable acetic acid.

SODIUM ACETATE, $NaC_2H_3O_2 + 3Aq.$—The mode of preparation of this salt on the large scale has been already described: it forms large, transparent, colorless crystals, derived from a rhombic prism, which are easily rendered anhydrous by heat, effloresce in dry air, and dissove in 3 parts of cold, and in an equal weight of hot water; it is also soluble in alcohol. The taste of this salt is cooling and saline. The dry salt melts at 288° C. (550.4° F.), and begins to decompose at 315° C. (599° F.).

AMMONIUM ACETATES.—The *neutral acetate*, $NH_4C_2H_3O_2$, is a white odorless salt, obtained by saturating glacial acetic acid with dry ammonia gas. It is very difficult to obtain in the crystalline form, for its aqueous solution, when evaporated, gives off ammonia, and leaves the acid salt. When distilled with phosphoric oxide, it loses 2 molecules of water, and gives off ethenyl nitril or acetonitril, $(C_2H_3)N = NH_4C_2H_3O_2 - 2H_2O$. The aqueous solution, known in the Pharmacopœia as *Spiritus Mindereri*, is prepared by saturating aqueous acetic acid with ammonia or ammonium carbonate.

The *acid salt*, $NH_4C_2H_3O_2.C_2H_4O_2$, is obtained as a crystalline sublimate by heating powdered sal-ammoniac with potassium or calcium acetate, ammonia being given off at the same time; also as a radiated crystalline mass by evaporating the aqueous solution of the neutral salt.

The acetates of *barium, strontium,* and *calcium* are very soluble, and can be procured in crystals : *magnesium acetate* crystallizes with difficulty.

ALUMINIUM ACETATES.—The neutral salt, $Al_2(C_2H_3O_2)_6$, is very soluble in water, and dries up in the vacuum of the air-pump to a gummy mass without trace of crystallization. If foreign salts are present, the solution of the acetate becomes turbid on heating, from the separation of a basic compound, which redissolves as the liquid cools. Aluminium acetate is much employed in calico printing : it is prepared by mixing solutions of lead acetate and alum, and filtering from the insoluble lead sulphate. The liquid is thickened with gum or other suitable material, and with it the design is impressed upon the cloth by a wood-block, or by other means. Exposure to a moderate degree of heat drives off the acetic acid, and leaves the alumina in a state capable of entering into combination with the dye-stuff.

Some very interesting researches on aluminium acetate have been published by the late Mr. Walter Crum.* The solution obtained by decomposing aluminium sulphate, $Al_2(SO_4)_3$, with lead acetate, may be supposed to contain neutral aluminium acetate, $(Al_2)(C_2H_3O_2)_6$, or $Al_2O_3.3C_4H_6O_3$. This salt cannot, however, be obtained in the dry state. If the solution be rapidly evaporated at low temperatures, by being spread in thin layers on glass or porcelain, a basic *soluble* acetate is obtained, having the composition $Al_2O_3.2C_4H_6O_3 + 4Aq.$; but if the solution be left to stand in the cold, or submitted to the action of heat, *insoluble* basic salts are precipitated, differing in composition from the former by containing in the first case five, and in the second two, molecules of water instead of four.

The soluble aluminium acetate, when exposed in a dilute solution to the temperature of boiling water for several days, undergoes a very remarkable change, the whole, or nearly the whole, of the acetic acid being expelled by the action of heat, and a peculiar soluble modification of

* Chem. Soc. Quar. Journ., vi. 216.

alumina (already described under ALUMINIUM, p. 398), remaining in solution.

Manganese acetate forms colorless, rhombic, prismatic crystals, permanent in the air. *Ferrous acetate* crystallizes in small, greenish-white needles, very prone to oxidation; both salts dissolve freely in water. *Ferric acetate* is a dark, brownish-red, uncrystallizable liquid, of powerful astringent taste. *Cobalt acetate* forms a violet-colored, crystalline, deliquescent mass. The *nickel salt* separates in green crystals, which dissolve in 6 parts of water.

LEAD ACETATES.—The *normal salt*, $Pb(C_2H_3O_2)_2 + 3Aq.$, is prepared on a large scale by dissolving litharge in acetic acid; it may be obtained in colorless, transparent, prismatic crystals, but is generally met with in commerce as a confusedly crystalline mass, somewhat resembling loaf-sugar. From this circumstance and from its sweet taste, it is often called *sugar of lead*. The crystals are soluble in about $1\frac{1}{4}$ parts of cold water, effloresce in dry air, and melt when gently heated in their water of crystallization; this water is easily driven off, and the anhydrous salt obtained, which melts, and afterwards decomposes, at a high temperature. Acetate of lead is soluble in alcohol. The aqueous solution has an intensely sweet, and at the same time, astringent taste, and is not precipitated by ammonia. It is an article of great value to the chemist.

Basic Acetates (Subacetates) of Lead.—A *sesquibasic acetate*, $2Pb(C_2H_3O_2)_2.PbO$, is produced when the neutral anhydrous salt is so far decomposed by heat as to become converted into a porous white mass, decomposable only at a much higher temperature. It is soluble in water, and separates from the solution evaporated to a syrupy consistence in the form of crystalline scales. A *triplumbic acetate*, $Pb(C_2H_3O_2)_2.2PbO$, is obtained by digesting, at a moderate heat, 7 parts of finely powdered litharge, 6 parts of lead acetate, and 30 parts of water; or, by mixing a cold saturated solution of neutral lead acetate with a fifth of its volume of caustic ammonia, and leaving the whole for some time in a covered vessel. The salt separates in minute needles containing one molecule of water. The solution of basic acetate prepared by the first method is known in pharmacy under the name of *Goulard water*. There is also a *sexplumbic acetate*, $Pb(C_2H_3O_2)_2.5PbO$, formed by adding a great excess of ammonia to a solution of normal lead acetate, or by digesting the normal salt with a large quantity of oxide. It is a white, slightly crystalline substance, insoluble in cold, and but little soluble in boiling water. The solutions of the basic lead acetates have a strong alkaline reaction, and absorb carbonic acid with the greatest avidity, becoming turbid from precipitation of basic carbonate.

CUPRIC ACETATES.—The *normal acetate*, $Cu(C_2H_3O_2)_2 + Aq.$, is prepared by dissolving *verdigris* in hot acetic acid, and leaving the filtered solution to cool. It forms beautiful dark-green crystals, which dissolve in 14 parts of cold and 5 parts of boiling water, and are also soluble in alcohol. A solution of this salt, mixed with sugar and heated, yields cupric oxide in the form of minute red octohedral crystals: the residual copper solution is not precipitated by an alkali. Cupric acetate yields, by destructive distillation, strong acetic acid containing acetone and contaminated with copper. The salt is sometimes called *distilled verdigris*, and is used as a pigment.

Basic Cupric Acetates.—Common verdigris, made by exposing plates of copper to the air for several weeks, in contact with acetic acid or the marc of grapes, is a mixture of several basic cupric acetates, which have a green or blue color. One of these, $2Cu(C_2H_3O_2)_2.CuO + 6Aq.$, is obtained by digesting the powdered verdigris in warm water, and leaving the soluble part to spontaneous evaporation. It forms a blue, crystalline mass, but little soluble in cold water. When boiled, it deposits a brown powder, which is a subsalt with large excess of base. The green insoluble residue of the verdigris contains $Cu(C_2H_3O_2)_2.2CuO + 3Aq.$; it may be formed by digesting normal cupric acetate with the hydrated oxide. By boiling with water it is resolved into normal acetate and the brown basic salt.

SILVER ACETATE, $AgC_2H_3O_2$, is obtained by mixing potassium acetate with silver nitrate, and washing the precipitate with cold water to remove the potassium nitrate. It crystallizes from a warm solution in small colorless needles, which have but little solubility in the cold.

Mercurous acetate forms small scaly crystals, which are as feebly soluble as those of acetate of silver. *Mercuric acetate* dissolves with facility.

METHYL ACETATE, $CH_3.C_2H_3O_2$, occurs in crude wood-spirit. It is prepared by distilling 2 parts of methyl alcohol with 1 part of glacial acetic acid and 1 part of sulphuric acid, or 1 part of methyl alcohol with 1 part of potassium acetate and 2 parts of sulphuric acid. When purified by rectification over calcium chloride and quick-lime, it forms a colorless fragrant liquid, of sp. gr. 0.9562 at 0°, boiling at 55° or 56° C. (131°–132.8° F.). It dissolves in water, and mixes in all proportions with alcohol and ether.

ETHYL ACETATE, $C_2H_5.C_2H_3O_2$, may be prepared by heating together in a retort, 3 parts of potassium acetate, 3 parts of strong alcohol, and 2 parts of oil of vitriol. The distilled product is mixed with water, to separate the alcohol, digested first with a little chalk, and afterwards with fused calcium chloride, and, lastly, rectified. The pure ether is an exceedingly fragrant limpid liquid, having a density of 0.9105 at 0°, 0.9068 at 15° C. (59° F.), and boiling at 77° C. (170.6° F.). Alkalies decompose it in the manner already mentioned (p. 716). When treated with ammonia, it yields *acetamide*, $NH_2C_2H_3O$.
On the reaction of ethyl acetate and homologous ethers with sodium and the iodides of alcohol-radicles, see p. 714; also CARBOKETONIC ACIDS.

AMYL ACETATE, $C_5H_{11}.C_2H_3O_2$, prepared in a similar manner from ordinary amyl alcohol, boils at 140° C. (284° F.). It possesses in a remarkable manner the odor of the Jargonelle pear, and is manufactured on a large scale for flavoring liquors and confectionery.

ETHENE ACETATES.—These compounds may be derived from ethene alcohol (glycol) by substitution of one or two equivalents of acetyl for hydrogen. The *monacetate*, $(C_2H_4) \begin{cases} OH \\ OC_2H_3O \end{cases}$, is produced by heating ethene dibromide with an alcoholic solution of potassium acetate. The product is distilled, the portion coming over at 182° C. (359.6° F.) being kept separate. It is a colorless, oily liquid, miscible in every proportion with water or alcohol. Hydrochloric acid gas passed into ethene monacetate converts it into ethene acetochloride, or glycollic chloracetin, $C_2H_4 \begin{cases} Cl \\ OC_2H_3O \end{cases}$, which is precipitated, on addition of water, as an oily liquid boiling at 145° C. (293° F.). Treatment with potash decomposes it into ethene oxide, potassium acetate, and potassium chloride.

Ethene diacetate, $C_2H_4 \begin{cases} OC_2H_3O \\ OC_2H_3O \end{cases}$, is prepared by digesting a mixture of ethene dibromide, silver acetate, and glacial acetic acid in the water-bath, and exhausting the digested mass with ether. On distilling the ethereal solution, the ether first passes over, then the acetic acid, and lastly, when the temperature has reached 187° C. (368.6° F.), ethene diacetate. It is a colorless, neutral liquid, of sp. gr. 1.128 at 0°, boiling at 186° C. (366.8° F.), soluble in 7 parts of water and in every proportion in alcohol and ether. By distillation with potash it yields ethene alcohol, or glycol, $C_2H_4(HO)_2$.

PROPENYL or GLYCERYL ACETATES ; ACETINS.—These ethers are derived from propenyl alcohol (glycerin) by substitution of 1, 2, or 3 equivalents of acetyl for hydrogen. The formula of glycerin being $(C_3H_5)(OH)_3$, those of the three acetins are :

Monoacetin	$(C_3H_5)(OH)_2(OC_2H_3O)$
Diacetin	$(C_3H_5)(OH)(OC_2H_3O)_2$
Triacetin	$(C_3H_5)(OC_2H_3O)_3$.

They are oily liquids, produced by heating glycerin and acetic acid together, in various proportions, in sealed tubes.

ACETIC CHLORIDE or ACETYL CHLORIDE, C_2H_3OCl.—This compound, which has the constitution of acetic acid with chlorine substituted for hydroxyl, is produced, as already observed (P. 716), by the action of phosphorus trichloride, pentachloride, or oxychloride on glacial acetic acid. The product heated with water and dilute soda-solution, to remove phosphorus oxychloride and hydrochloric acid, and then rectified, yields acetic chloride as a colorless liquid, having a suffocating odor, and emitting dense fumes of hydrochloric acid in contact with the air. It has a density of 1.1305 at 0°, boils at 55° C. (131° F.), and is decomposed by water and alkaline solutions, yielding hydrochloric and acetic acids.

ACETIC OXIDE or ANHYDRIDE, $C_4H_6O_3 = (C_2H_3O)_2O$, sometimes called *Anhydrous Acetic acid.*—This compound is obtained :
1. By the action of acetyl chloride on potassium or sodium acetate :

$$C_2H_3O(ONa) \; + \; C_2H_3OCl \; = \; NaCl \; + \; (C_2H_3O)_2O.$$

2. By heating sodium acetate with benzoyl chloride, C_7H_5OCl, whereby benzo-acetic oxide, $(C_2H_3O)(C_7H_5O)O$, is formed in the first instance, and subsequently resolved into acetic and benzoic oxides, the former distilling over, while the latter remains :

$$C_2H_3O(ONa) \; + \; C_7H_5OCl \; = \; NaCl \; + \; C_2H_3O.O.C_7H_5O$$

and :
$$2(C_2H_3O.O.C_7H_5O) \; = \; (C_2H_3O)_2O \; + \; (C_7H_5O)_2O.$$

Acetic oxide is a heavy oil which boils at 121° C. (249.8° F.), and dissolves slowly in water, being gradually converted into acetic acid :

$$(C_2H_3O)_2O \; + \; H_2O \; = \; 2C_2H_3O(OH).$$

Acetyl Peroxide, $(C_2H_3O)_2O_2$, obtained by adding barium dioxide to an ethereal solution of acetyl monoxide, is a viscid liquid, which explodes with violence when heated, and acts as a powerful oxidizing agent.

Acids derived from Acetic Acid by Substitution.

Chloracetic Acids.—The three acids, C_2H_3ClO, $C_2H_2Cl_2O_2$, and $C_2HCl_3O_2$, are produced by the action of chlorine on acetic acid in sunshine; the second, however, is formed in small quantity only, the first or the third being produced in greatest abundance according as the acetic acid or the chlorine is in excess.

Monochloracetic acid, $CH_2Cl.COOH$, is produced, according to R. Hoffmann, by the action of chlorine on boiling glacial acetic acid in sunshine. Dr. H. Müller finds that the formation of monochloracetic acid is facilitated by dissolving a little iodine in the hydrated acetic acid, and passing a stream of chlorine through the boiling solution. On submitting the products of this reaction to repeated distillation, monochloracetic acid is obtained as a liquid, boiling at 186° C. (366.8° F.), and solidifying to a crystalline mass, which melts at 62° C. (143.6° F.), and dissolves easily in water. Heated with potash, it is converted into potassium glycollate, $KC_2H_3O_3$ (p. 717).

$$C_2H_3ClO_2 \;+\; 2KHO \;=\; KCl \;+\; KC_2H_3O_3 \;+\; H_2O.$$

Dichloracetic acid, $CHCl_2COOH$, is produced, together with the preceding compound, by the action of chlorine and iodine on boiling acetic acid, and is found in the portion of the product which boils above 188° C. (370.4° F.). According to Maumené, it may be obtained by exposing monochloracetic acid in flasks to the action of dry chlorine (5 atoms of chlorine to 3 molecules of chloracetic acid)for twenty-four hours, warming the product to expel hydrochloric acid and then distilling. It is a liquid having a specific gravity of 1.5216 at 15° C. (59° F.), and boiling at 190–191° C. (374–375.8° F.). According to Müller, it remains liquid when cooled; but according to Maumené, it crystallizes in rhombohedral plates. It forms a soluble silver salt, $AgC_2HCl_2O_2$, which is decomposed when its solution is heated with silver oxide to 75° or 80° C. (167°–176° F.), giving off a mixture of carbon monoxide and dioxide:

$$2AgC_2HCl_2O_3 \;+\; 3Ag_2O \;=\; 2CO \;+\; 2CO_2 \;+\; 4AgCl \;+\; 2Ag_2 \;+\; H_2O.$$

Trichloracetic acid, $CCl_3.COOH$.—Discovered by Dumas. When a small quantity of crystallizable acetic acid is introduced into a bottle of dry chlorine gas, and the whole exposed to the direct solar rays for several hours, the interior of the vessel is found coated with a white crystalline substance, which is a mixture of trichloracetic acid with a small quantity of oxalic acid.

Trichloracetic acid may also be produced synthetically, viz., by the action of chlorine and water on carbon dichloride, C_2Cl_4, this compound first taking up 2 atoms of chlorine and forming carbon trichloride, C_2Cl_6, and the latter being converted by the water into hydrochloric and trichloracetic acids:

$$C_2Cl_6 \;+\; 2H_2O \;=\; 3HCl \;+\; C_2HCl_3O_2.$$

Trichloracetic acid is a colorless and extremely deliquescent substance: it has a faint odor, and sharp caustic taste, bleaching the tongue and destroying the skin; the solution is powerfully acid. It melts at 52° C. (125.6° F.), and boils without decomposition at 195–200° C. (383–392° F.), giving off a very irritating vapor. The density of the fused acid is 1.617.

Potassium trichloracetate, $2KC_2Cl_3O_2.Aq.$, crystallizes in fibrous silky needles, permanent in the air. The *ammonium-salt*, $2(NH_4)C_2Cl_3O + 5Aq.$, is also crystallizable and neutral. The *silver-salt*, $AgC_2Cl_3O_2$, is soluble, and crystallizes in small, grayish scales, easily altered by light.

Trichloracetic acid boiled with excess of ammonia yields ammonium carbonate and chloroform :

$$C_2HCl_3O_2 \; + \; 2NH_3 \; + \; H_2O \; = \; (NH_4)_2CO_3 \; + \; CHCl_3.$$

With caustic potash, it yields a smaller quantity of chloroform, together with potassium chloride, carbonate, and formate. The chloride and formate are secondary products of the reaction of the alkali upon the chloroform.

Nascent hydrogen reduces trichloracetic to acetic acid. When potassium or sodium amalgam is put into a strong aqueous solution of trichloracetic acid, the temperature of the liquid rises, without disengagement of gas, and the solution is found to contain acetate and chloride of potassium or sodium, together with caustic alkali.

Bromacetic Acids.—The *mono-* and *di*-brominated acids are formed by heating acetic acid or ethyl acetate with bromine in sealed tubes to 180° C. (356° F.). $CH_2Br.CO_2H$ crystallizes in deliquescent rhombohedrons, and boils at 208° C. (406.4° F.). Its *ethylic ether* is a liquid boiling at 159° C. (318.2° F.). The acid is converted by ammonia into glycerin (p. 718). $C_2HBr_2.CO_2H$ is a crystalline mass melting at 45°–50° C. (113°–122° F.), and boiling at 232°–234° C. (449.6°–453.2° F.). Heated with moist silver oxide it yields bromo-glycollic acid, $C_2H_3BrO_3$. Its ethylic ether boils at 192°–195° C. (377.6°–383° F.). *Tribromacetic Acid*, C_2Br_3.CO_2H, is produced by the action of water on tribromacetyl bromide, CBr_3.$COCl$, and by oxidation of bromal, $CBr_3.COH$, with nitric acid. It forms deliquescent crystals, melts at 130° C. (266° F.), and boils at 245° C. (473° F.).

Iodacetic Acid, *mono-* and *di-*, are obtained by heating the chlorinated or brominated acids with potassium iodide ; also by heating acetic aubydride with iodide and iodic acid. $C_2H_3I.CO_2H$ crystallizes in colorless laminæ, melting with partial decomposition at 82° C. (179.6° F.); reconverted into acetic acid by boiling with HI. Its ethylic ether boils at 178°–180° C. (352.4°–356° F.).

Cyanacetic Acid, $C_3H_3NO_2 = CH_2(CN).CO_2H$, formed by heating monochloracetic acid with potassium cyanide and water, is a crystalline mass melting at about 80° C. (176° F.), and decomposed at 165° C. (329° F.) into CO and acetonitril, $CH_3.CN$.

Thiacetic Acid, C_2H_4OS, or $CH_3.COSH$.—This acid, discovered by Kekulé, is formed by the action of phosphorus pentasulphide on glacial acetic acid :

$$5C_2H_3O(OH) \; + \; P_2S_5 \; = \; P_2O_5 \; = \; 5C_2H_3O(SH).$$

Thiacetic acid is a colorless liquid, boiling at 93° C. (199.4° F.); it smells like acetic acid and hydrogen sulphide. With solution of lead acetate, it forms a crystalline precipitate containing $(C_2H_3OS)_2Pb$.

––––––––––––––––

Propionic Acid, $C_3H_6O_2 = C_3H_5O(OH) = C_2H_5.CO.OH$.—This acid is produced : 1. As a potassium-salt, by the combination of carbon dioxide with potassium-ethyl : $CO_2 + C_2H_5K = CO(C_2H_5)OK$. 2. By the action of acids or alkalies on ethyl cyanide (p. 714). 3. By the simultaneous action of water and carbonyl chloride on ethane (p. 714). 4. By the oxidation of normal propyl alcohol (p. 617), and of propionic aldehyde, C_3H_6O. 5. Together with acetic acid, by oxidizing propione, $C_5H_{10}O$, with dilute chromic acid. This is the process by which it was first obtained. 6. From lactic acid and from glyceric acid by the action of hydriodic acid :

$$\underset{\text{Lactic acid.}}{C_3H_6O_3} + 2HI = C_3H_6O_2 + H_2O + I_2.$$

$$\underset{\text{Glyceric acid.}}{C_3H_6O_4} + 4HI = C_3H_6O_2 + 2H_2O + 2I_2.$$

7. Together with several other products, in the fermentation of glycerin, and likewise of sugar, by the action of putrid cheese in presence of calcium carbonate.

Propionic acid is usually prepared by the second of the above-mentioned processes. Ethyl cyanide is added by drops to a moderately strong solution of potash heated in a tubulated retort, the distillate being repeatedly poured back as long as it smells of ethyl cyanide. The residue in the retort, consisting of potassium propionate, is then evaporated down to dryness, and distilled with syrupy phosphoric acid.

Propionic acid is a colorless liquid, having a specific gravity of 0.992 at 18° C. $(64.4^O$ F.), and boiling at 140° C. $(284^O$ F.). It is soluble in water, and when the water is quite saturated with it, the excess of acid floats on the surface in the form of an oil.

The propionates are soluble in water. The *barium salt*, $(C_3H_5O_2)Ba + H_2O$, crystallizes in rhombic prisms, and yields propione by dry distillation. The *silver salt*, $C_3H_5O_2Ag$, crystallizes in slender needles.

Substitution-products.—The replacement of one hydrogen-atom in propionic acid by a haloïd element, X, yields two series of substitution-products distinguished as α and β ; viz. :

$$\underset{\alpha.}{CH_3-CHX-CO_2H} \qquad\qquad \underset{\beta.}{CH_2X-CH_2-CO_2H}$$

Chloropropionic acids, $C_3H_5ClO_2$.—The α-acid is formed by decomposing lactyl chloride with water :

$$CH_3-CHCl-CO_2Cl + H_2O = HCl + CH_3-CHCl-CO_2H.$$

It is a thick liquid of specific gravity 1.28, and boiling at 186° C. $(366.8^O$ F.). When heated with moist silver oxide, it is converted into α-lactic acid, $CH_3-CHOH-CO_2H$. Its ethylic ether, obtained by heating lactyl chloride with alcohol, boils at 144° C. $(291.2^O$ F.).

The β-acid is produced by the action of chlorine-water on β-iodopropionic acid, and by addition of hydrogen chloride to acrylic acid, $C_3H_4O_2$:

$$CH_2{=}CH-CO_2H + ClH = CH_2Cl-CH_2-CO_2H.$$

It is crystalline, and melts at 40.5^O C. $(104.9^O$ F.). Its ethylic ether boils at about 155° C. $(311^O$ F.).

Bromopropionic acids, $C_3H_5BrO_2$.—The α-acid is obtained by the action of bromine on propionic acid, and by heating α-lactic acid (ordinary lactic acid) with hydrobromic acid, is a liquid which solidifies at -17° C. $(1.4^O$ F.), and boils at 202° C. $(395.6^O$ F.). Its ethylic ether smells like camphor, and boils at 190° C. $(374^O$ F.). Alcoholic ammonia converts it into α-amidopropionic acid or alanine, $C_3H_5(NH_2)O_2$ (see AMIDES). The β-acid, formed by the action of bromine-water on β-iodopropionic acid, and

31 *

by addition of hydrobromic acid to acrylic acid, is crystalline, and melts at 61.5° C. (142.7° F.).

-Iodopropionic acid, $C_3H_5IO_2$, is a thick oil, obtained by the action of phosphorus iodide on *-lactic acid.

The *β*-acid is obtained by the action of phosphorus iodide and a small quantity of water on glyceric acid :

$$CH_2OH.CHOH.CO_2H + 3HI = CH_2I.CH_2.CO_2H + 2H_2O + I_2 ;$$

also by addition of hydriodic acid to acrylic acid. It crystallizes in large colorless six-sided plates, having a peculiar odor, and melting at 82° C. (179.6° F.) ; easily soluble in hot water. By heating with strong hydriodic acid, it is reduced to propionic acid.

The di-substitution products of propionic acid admit of three modifications, viz. :

$$CH_3—CX_2—CO_2H \qquad CH_2X—CHX—CO_2H \qquad CHX_2—CH_2—CO_2H$$
$$a. \qquad\qquad\qquad β. \qquad\qquad\qquad γ.$$

The first two modifications of the chlorine and bromine derivatives are known.

a-Dibromopropionic acid, $CH_3—CBr_2—CO_2H$, obtained by direct bromination of propionic or *a*-bromopropionic acid, melts at 61° C. (141.8° F.), and boils at 220° C. (428° F.). Its ethylic ether boils at 190° C. (374° F.).

β-Dibromopropionic acid, $CH_2Br—CHBr—CO_2H$, obtained by oxidation of dibromopropyl alcohol, $C_3H_6Br_2O$ (formed by addition of bromine to allyl alcohol), and by addition of Br_2 to acrylic acid, $CH_2=CH—CO_2H$, forms crystals easily soluble in water, melting at 65° C. (149° F.), decomposed by distillation. The ethylic ether boils at 212° C. (413.6° F.).

a-Dichloropropionic acid, $CH_3—CCl_2—CO_2H$. The ethylic ether of this acid is formed by the action of alcohol on the chloride, $CH_3—CO—COCl$, produced by the action of phosphorus pentachloride on pyroracemic acid, $CH_3—CO—COOH$. It is a colorless liquid, having a pleasant odor of apples, a specific gravity of 1.2493 at 0°, and boiling at 160° C. (320° F.). Heated with water to 150° C. (302° F.), it is converted into pyroracemic acid.

β-Dichloropropionic acid, $CH_2Cl—CHCl—CO_2H$, is formed by the action of water on the chloride obtained by heating glyceric acid with phosphorus pentachloride.

Butyric acid, $C_4H_8O_2 = C_3H_7.CO_2H$.—Of this acid there are two modifications, viz. :

$$\begin{array}{ll} CH_2CH_2CH_3 & CH(CH_3)_2 \\ | & | \\ COOH & COOH \\ \text{Normal butyric.} & \text{Isobutyric.} \end{array}$$

Normal Butyric acid, also called *Fermentation Butyric*, *Propyl-formic*, and *Ethyl-acetic acid*, occurs, either free or combined with bases, in tamarinds and a few other plants, in certain beetles, in flesh-juice, in human perspiration, and in various kinds of decomposing vegetable and animal matter. As a glyceride, it occurs in various animal and vegetable fats, especially in the butter of cows' and goats' milk ; as a hexyl ether in the oil of *Heracleum giganteum*, and as an octyl ether in oil of parsnep (*Pastinaca sativa*). It is formed by the oxidation of normal butyl alcohol ; by the action of nascent hydrogen on crotonic acid, $CH_3—CH=CH—CO_2H$; by heating normal propyl cyanide (butyronitril) with alkalies or acids ;

by the action of sodium and ethyl iodide on ethyl acetate (p. 715); and by the fermentation of sugar in contact with putrid cheese.

This last reaction affords the most convenient method of preparing the acid. The fermentation takes place, however, only in neutral liquids, so that it is necessary to add chalk, in order to neutralize the acid as fast as it is formed. The sugar is first converted into lactic and afterwards into butyric acid. The following proportions give a good result : 6 pounds of cane-sugar and half an ounce of tartaric acid are dissolved in 26 pounds of boiling water, and the solution is left for some time to allow the cane-sugar to pass into grape-sugar. To this solution about 4 ounces of decayed cheese diffused in 8 pounds of sour skim-milk, together with 3 pounds of chalk, are added, and the whole is kept for some weeks at a temperature of 30°–35° C. (86°–95° F.). The mixture is frequently stirred, and generally solidifies in ten or twelve days to a thick mass of calcium lactate. This, however, soon disappears, the mixture becoming liquid, and the lactate being converted into butyrate, with simultaneous evolution of carbon dioxide and hydrogen :—

$$2C_3H_6O_3 = C_4H_8O_2 + 2CO_2 + 2H_2 .$$

When fermentation is completed, a solution of 8 pounds of crystallized sodium carbonate is added, and the filtered liquid, after concentration, is decomposed by sulphuric acid. Butyric acid then separates as an oily layer, which is dried over calcium chloride and rectified.

Butyric acid is a viscid liquid having a rancid odor, solidifying at —12° C. (10.4° F.), and boiling at 163° C. (325.4° F.). Specific gravity 0.958 at 14° C. (57.2° F.). It dissolves easily in water and in alcohol, and separates from the aqueous solution on addition of salts.

The metallic butyrates are, for the most part, soluble in water and crystallizable. The *calcium salt*, $(C_4H_7O_2)Ca + H_2O$, forms shining laminæ, and is *more soluble in cold than in hot water*. The *silver salt*, $C_4H_7O_2Ag$, crystallizes in shining laminæ, soluble in 400 parts of water at 14° C. (57.2° F.).

Ethyl Butyrate, $C_4H_7O_2.C_2H_5$, is a liquid having a pleasant fruity odor, and boiling at 119° C. (246.2° F.).

The butyrates unite with acetates, forming double salts, which react like salts of butyracetic acid, $C_4H_8O_2.C_2H_4O_2$, isomeric with propionic acid. This acid, obtained in the free state by the fermentation of calcium tartrate, differs from propionic acid in being resolved by distillation into butyric and acetic acids.

A *monochlorobutyric acid*, $C_4H_7ClO_2$, crystallizing in slender needles and melling at 99° C. (210° F.), is formed by the action of chlorine on butyric acid in presence of iodine.

Out of five possible *monobromobutyric acids*, two, viz.,

$$CH_3—CH_2—CHBr—CO_2H \qquad\qquad CH_3—CHBr—CH_2—CO_2H$$
$$a , \qquad\qquad\qquad\qquad\qquad\qquad \beta ,$$

are formed by the addition of HBr to crotonic acid. The action of bromine on butyric acid yields chiefly the *a*-acid, the ethylic ether of which boils at 170°–172° C. (338°–341.6° F.). The two acids are converted by boiling with potash-solution or moist silver oxide into the corresponding oxybutyric acids, $C_4H_8O_3$.

A *dibromobutyric acid*, $C_4H_6Br_2O_2$, melting at 45°–48° C. (113°–118.4° F.), is formed by further bromination of butyric acid.

Isobutyric Acid, $HC(CH_3)_2$—CO_2H, also called *Isopropyl-formic* and *Di-methyl-acetic acid*, occurs in the free state in St. John's bread, the fruit of *Ceratonia siliqua*, and as an octylic ether in parsnep oil.   It is produced by oxidation of isobutyl alcohol, by the action of alkalies or acids on pseudo-propyl cyanide ; and by the action of sodium and methyl iodide on ethyl acetate (p. 715).

This acid is very much like normal butyric acid, but has a less disagreeable odor, and is not miscible with water.   It has a specific gravity of 0.960 at 0°, and boils at 154° C. (309.2° F.).   Its *calcium salt*, $(C_4H_7O_2)_2Ca$ + $5H_2O$, crystallizes in monoclinic prisms, *much more soluble in hot water than in cold*.   The silver salt dissolves in 110 parts of water at 16° C.(60.8° F.).   The *ethylic ether*, $C_4H_7O_2.C_2H_5$, has a specific gravity of 0.8893 at 0°, and boils at 110° C. (230° F.).

Bromisobutyric acid, $(CH_3)_2CBr.CO_2H$, formed by heating isobutyric acid with bromine to 140° C.(284° F.), forms crystals melting at 45° C. (113° F.), and is partly decomposed by boiling.   Its ethylic ether has a specific gravity of 1.328 at 0°, and boils at 158°–159° C. (316.4°–318.2° F.).   By heating with moist silver oxide or baryta-water, this acid is converted into the corresponding o x y i s o b u t y r i c   a c i d.

Pentoic or **Valeric Acids,** $C_5H_{10}O_2$ = $C_5H_9O(OH)$. — These acids admit of four metameric modifications, namely :—

$CH_2CH_2CH_2CH_3$	$CH_2CH(CH_3)_2$	$CH(CH_3)(C_2H_5)$	$C(CH_3)_3$
$\vert$	$\vert$	$\vert$	$\vert$
COOH	COOH	COOH	COOH
Propyl acetic.	Isopropyl acetic.	Methyl-ethyl acetic.	Trimethyl acetic.

The first and second are obtained by oxidation of normal pentylic and isopentylic or amylic alcohol (p. 601) respectively ; the third is not known ; the fourth is produced from tertiary butylic alcohol or trimethyl-carbinol.

Propyl-acetic or **Normal Valeric Acid,** $C_4H_9.COOH$, is an oily liquid, smelling like butyric acid, having a sp. gr. of 0.9577 at 0°, and boiling at 185° C. (365° F.).

Isopentoic, Isopropyl-acetic or **Isovaleric Acid** (ordinary valeric acid), occurs in valerian root, in angelica root, in the berries of the guelder rose (*Viburnum Opulus*), and probably in many other plants.   It is produced by the oxidation of ordinary amyl alcohol, either by absorption of atmospheric oxygen under the influence of platinum black, or by treatment with aqueous chromic acid, or by heating it with a mixture of caustic potash and quick-lime, the reaction, in this last case, being attended with evolution of hydrogen :

$$C_5H_{12}O + KOH = C_5H_9O_2K + 2H_2.$$

The potassium salt, distilled with sulphuric acid, yields the acid.

The most advantageous mode of preparing isovaleric acid, is to oxidize amyl alcohol with a mixture of sulphuric acid and potassium dichromate. 4 parts of the dichromate in powder, 6 parts of oil of vitriol, and 8 parts of water are mixed in a capacious retort, and 1 part of amyl alcohol is added by small portions, with strong agitation, the retort being plunged into cold water to moderate the violence of the reaction.   When the change appears complete, the deep-green liquid is distilled nearly to dryness, the product mixed with excess of caustic potash, and the aqueous solution separated mechanically from a pungent, colorless, oily liquid which floats upon it, consisting of amyl valerate.   The alkaline solution is then evaporated to a small bulk, and decomposed by dilute sulphuric acid in excess. The greater part of the valeric acid then separates as an oily liquid lighter than water : this is a hydrate consisting of $C_5H_{10}O_2.H_2O$. When distilled alone, it undergoes decomposition : water, with a little of the acid, first appears, and eventually the pure acid, $C_5H_{10}O_2$, in the form of a thin, mobile, colorless oil, having the persistent and characteristic odor of valerian root. It has a sharp and acid taste, reddens litmus strongly, bleaches the tongue, and burns when inflamed with a bright, yet smoky light. Isovaleric acid has a density of 0.947 at 0°, and boils at 175° C. (347° F.). Placed in contact with water, it absorbs a certain quantity, and is itself to a certain extent dissolved.

Isovaleric acid is active or inactive to polarized light, accordingly as it has been prepared from active or inactive amyl alcohol. That which has been prepared from the active alcohol produces a right-handed rotation of 43° in a tube 50 centimetres long. The difference between these two acids, like that of the alcohols from which they are derived, is probably due to difference rather of molecular than of atomic arrangement (p. 519).

The metallic isovalerates are greasy to the touch, and mostly crystallize with difficulty. Small fragments of them thrown on water, rotate on the surface as they dissolve. The *barium salt* $(C_5H_9O_2)_2Ba+H_2O$ usually crystallizes in thin laminæ, soluble in two parts of cold, and in a smaller quantity of warm water. The *zinc salt* $(C_5H_9O_2)_2Zn$ forms shining scales. The *silver salt* $C_5H_9O_2Ag$ forms scales soluble in 540 parts of water at 20° C. (68° F.). A solution of potassium valerate, subjected to electrolysis, yields dibutyl, C_8H_{18} (p. 715).

*Ethyl isovale*rate, $C_5H_9O_2.C_2H_5$, is obtained by passing hydrochloric acid gas into an alcoholic solution of valeric acid. Ammonia converts it into valeramide, $C_5H_9ONH_2$. It is decomposed by sodium in the same manner as ethyl acetate (p. 715), yielding sodium ethylate and the sodium salt of an acid, $C_{15}H_{29}O_3$:

$$4Na \ + \ 3(C_2H_5.C_5H_9O_2) \ = \ 3C_2H_5ONa \ + \ C_{15}H_{27}O_3Na \ .$$

CHLOROVALERIC ACIDS.—*Trichlorisovaleric acid*, $C_5H_7Cl_3O_2$, obtained by the prolonged action of chlorine on isovaleric acid in the dark, aided towards the end of the process by a gentle heat, is an oily liquid, becoming very viscid at 18° C. (64.4° F.), perfectly mobile at 30° C. (86° F.). In contact with water it forms a very viscid hydrate, which sinks to the bottom. It dissolves in aqueous alkalies, and is precipitated by acids in its original state.

Tetrachlorisovaleric acid, $C_5H_6Cl_4O_2$, is the ultimate product of the action of chlorine on the preceding compound, aided by exposure to the sun. It is a semi-fluid, colorless oil, destitute of odor, of powerfully pungent taste, and heavier than water. It can neither be solidified by cold, nor distilled without decomposition. In contact with water, it forms a hydrate containing $C_5H_6Cl_4O_2.H_2O$, which is slightly soluble in water, easily soluble in alcohol and ether.

Neopentoic or **Trimethyl-acetic Acid,** $(CH_3)_3C.COOH$, is obtained from tertiary butyl iodide, $(CH_3)_3CI$, by converting this ether into the cyanide, and distilling the latter with potash ; also by oxidation of pinacolin, $CH_3—CO—C(CH_3)_3$. It is a laminar crystalline mass, melting at 34°–35° C. (93.2°–95° F.), and boiling at 161° C. (321.8° F.). It smells like acetic acid, and dissolves in 40 parts of water at 20° C. (68° F.).

The *barium salt*, $(C_5H_9O_2)_2Ba + 5H_2O$, and the *calcium salt*, $(C_5H_9O_2)_2Ca + 4H_2O$, crystallize in needles or prisms.

The *silver salt*, $C_5H_9O_2Ag$, is obtained by precipitation in glittering spangles. The *ethylic ether*, $C_5H_9O_2.C_2H_5$, has a specific gravity of 0.8772 at 0°, and boils at 118.5° C. (245.3° F.).

Hexoic or **Caproic acids,** $C_6H_{12}O_2 = C_5H_{11}.CO_2H$.—There are eight possible forms of these acids, analogous to the eight pentyl alcohols, and five of them are known, viz. :

1. **Normal Caproic** or **Pentyl-formic acid,** $CH_3(CH_2)_4.CO_2H$.—This acid is formed synthetically by heating normal pentyl cyanide, $CH_3(CH_2)_4$. CN, with strong potash-solution.   It is also produced, together with other acids of the fatty series, by the oxidation of albuminous substances, and of fatty acids of higher atomic weight.   It occurs in the free state in perspiration, as an octylic ether, $C_6H_{11}O_2.C_8H_{17}$, in the volatile oil of cow-parsnep (p. 619), as a glyceride in the butter of cows' milk, and abundantly in cocoa-nut oil.   From this oil it may be prepared by saponifying with strong soda-lye, and distilling the soap with dilute sulphuric acid. The distillate contains caproic and caprylic acids, and, when neutralized with baryta and evaporated, yields crystals of barium caprylate, and afterwards verucose crystals of the caproate, which, when decomposed by sulphuric acid, yields caproic acid.

Normal caproic acid is a clear mobile oil, having a somewhat sudorific and pungent odor, a specific gravity of 0.945 at 0°, 0.895 at 99° C. (210.2° F.), and boiling at 205° C. (401° F.).   The barium salt, $(C_6H_{11}O_2)_2Ba +$ $3H_2O$, dissolves in 6–7 parts of water at 23° C. (73.4° F.).   The *ethylic ether*, $C_6H_{11}.O_2C_2H_5$, has a fruity smell, and boils at 167° C. (332.6° F.).

2. **Isocaproic** or **Isopentyl-formic acid,** $CH(CH_3)_2.CH_2.CH_2.CO_2H$, prepared from isopentyl (amyl) cyanide, has a specific gravity of 0.931 at 0°, and boils at 199°–200° C. (390.2°–392° F.).   Its odor resembles that of the normal acid, but is stronger and more disagreeable.   The *calcium salt*, $(C_6H_{11}O_2)_2Ca+ 3H_2O$, dissolves in 9 parts of water at 18° C. (64.4° F.). The *ethylic ether*, $C_6H_{11}O_2.C_2H_5$, has a specific gravity of 0.887 at 0°, and boils at 160.4° C. (320.7° F.).

3. **Methylisopropyl-acetic acid,** $CH(CH_3)_2\!\!>\!\!CH.CO_2H$, prepared from the iodide of methyl-isopropyl carbinol (amylene hydrate, p. 614), through the medium of the cyanide, is an oil having a fruity smell.   Its calcium salt crystallizes in scales, and is more soluble in cold than in hot water.

4. **Diethyl acetic acid,** $C_2H_5\!\!>\!\!CH.CO_2H$, prepared by the action of sodium and ethyl iodide on methyl acetate (p. 714), is an oil sparingly soluble in water.   Its *silver salt*, $C_6H_{11}O_2Ag$, crystallizes in shining needles. The *ethylic ether*, $C_6H_{11}O_2.C_2H_5$, boils at 151° C. (303.8° F.).

Dimethyl-ethyl acetic acid, $\left.{(CH_3)_2 \atop C_2H_5}\right\} C.CO_2H$, formed from tertiary butyl iodide, $(CH_3)_2(C_2H_5)CI$, through the medium of the cyanide, is a liquid which boils at 187° C. (368.6° F.), and solidifies in the crystalline form at 14° C. (57.2° F.).   Its *barium salt*, $(C_6H_{11}O_2)_2Ba+ 5H_2O$, crystallizes in large laminæ.

Heptoic acids, $C_7H_{14}O_2 = C_6H_{13}.CO_2H$.   Of the 17 possible modifications of these acids, one only is accurately known, viz. :

Normal Heptoic or *Œnanthylic acid,* $CH_3(CH_2)_5.CO_2H$, which is formed by the action of boiling potash on normal hexyl cyanide, $CH_3(CH_2)_5$. CN, by the oxidation of normal heptyl alcohol, and from œnanthol or heptoio aldehyde, $C_7H_{14}O$, (p. 700), by oxidation in the air, or with nitric acid, or with chromic acid; also by oxidation of castor-oil with nitric acid.

Œnanthylic acid is a transparant colorless oil, having an unpleasant odor like that of codfish. It boils at 223° C. (433.4° F.). It is insoluble in water, but soluble in alcohol and ether. When heated with baryta, it gives off hexane, C_6H_{14}, the baryta abstracting carbon dioxide: $C_7H_{14}O_2 = CO_2 + C_6H_{14}$. The potassium-salt subjected to electrolysis yields d i h e x y l, $C_{12}H_{26}$. The *calcium salt*, $(C_7H_{13}O_2)_2.Ca + H_2O$, crystallizes in flattened needles, soluble in 100 parts of water at 8° C. (46.4° F.). The *ethylic ether*, $C_7H_{13}O_2.C_2H_5$, has a fruity odor, a specific gravity of 0.873 at 16° C. (60.8° F.), and boils at 188° C. (370.4° F.).

Another heptoic acid has been obtained by the action of amyl iodide on ethylic sodacetate, $CH_2Na.CO_2C_2H_5$.

Caprylic Acid, $C_8H_{16}O_2 = C_7H_{15}.CO_2H$, occurs as a glyceride in the butter of cows' milk and in cocoa-nut oil; it is also found in several kinds of fusel-oil, partly free, partly as an ethylic or amylic ether. It is best prepared by saponification of cocoa-nut oil; its barium-salt, being very sparingly soluble, is easily separated from the barium-salt of caproic acid formed at the same time.

Caprylic acid has a faint but unpleasant odor, especially when warmed. It solidifies at 12° C. (53.6° F.), melts at 15° C. (59° F.), and boils at 236°-238° C. (456.8°-460.4° F.). When boiled with nitric acid, it is converted into nitrocaprylic acid, $C_8H_{15}(NO_2)O_2$.

Primary octyl alcohol, $C_8H_{18}O$, obtained from Heracleum oil (p. 619), yields by oxidation with chromic acid an acid, $C_8H_{16}O_2$, solidifying at 12° C. (53.6° F.), melting at 16°-17° C. (60.8°-62.6° F.), boiling at 234°-238° C. (453.2°-460.4° F.), and doubtless identical with the caprylic acid of natural fats: but the primary octyl alcohol obtained from the octane of American petroleum yields by oxidation an acid isomeric with the above, inasmuch as when floating on water it remains liquid at 0°.

As the primary octyl alcohol of Heracleum oil is an iso-alcohol (p. 619), it follows that the caprylic acid formed from it by oxidation, and therefore also that obtained from natural fats, must be an iso-acid, viz., iso-heptyl-formic acid, $\begin{matrix} H_3C \\ H_3C \end{matrix} > CH - (CH_2)_4 - CO_2H.$

Pelargonic or **Nonylic Acid**, $C_9H_{18}O_2 = C_8H_{17}.CO_2H$, was first obtained from the leaves of the geranium (*Pelargonium roseum*), in which it exists ready formed. It may be procured in large quantity by the action of nitric acid upon the essential oil of rue (which contains nonyl-methyl ketone, $C_9H_{19}.CO.CH_3$); also, together with several other acids of the fatty series, by the action of boiling nitric acid on oleic acid. It is also formed synthetically by the action of boiling alcoholic potash on primary octyl cyanide, $C_8H_{17}.CN$, prepared from the octyl alcohol of Heracleum oil. It is a liquid having a slightly unpleasant odor, solidifying at 12°-12.5° C. (53.6°-54.5° F.), and boiling at 253°-254° C. (487.4°-489.2° F.).

Ethyl Pelargonate, $C_9H_{17}O_2.C_2H_5$, may be easily produced by dissolving the acid in strong alcohol, and passing a current of hydrochloric acid through the solution. It is a liquid of specific gravity 0.862, and boiling at 250° C. (482° F.). It has a powerful and most intoxicating vinous odor.

The aroma possessed by certain wines appears to be due to the presence of the ether of pelargonic acid, which, in this case, is probably generated during fermentation. When such wines, or the residues of their fermentation, are distilled on the large scale, an oily liquid passes over towards the close of the operation, which consists, in a great measure, of the crude ether: it may be purified by agitation with solution of potassium carbon ate, freed from water by a few fragments of calcium chloride, and redis- tilled. The pelargonic ether obtained by this process was originally de- scribed as *œnanthic ether*, and the acid as *œnanthic acid*.

Rutic or **Capric Acid**, $C_{10}H_{20}O_2$.—This acid exists as a glyceride in ordinary butter and in cocoa-nut oil; it occurs also in several kinds of fusel-oil, and is formed by the oxidation of oleic acid and of oil of rue. It may be obtained pure and in tolerable quantity from the liquid which remains in the distillation of the fusel-oil of the Scotch distilleries (p. 614) after the amyl alcohol has been distilled off at 132° C. (269.6° F.). This residue consists chiefly of amyl rutate, $C_{10}H_{19}O_2.C_5H_{11}$, and when distilled with potash gives off amyl alcohol and leaves potassium rutate, from which the rutic acid may be obtained by distillation with sulphuric acid.

Rutic acid is a colorless crystalline body, having a slight odor of the goat, becoming stronger when the acid is warmed. It melts at 27–30° C. (80.6–86° F.), boils at 268–270° C. (514.4–518° F.), is very soluble in cold alcohol and ether, insoluble in cold water, slightly soluble in boiling water, and dissolves without alteration in strong nitric acid.

The metallic rutates are mostly sparingly soluble in water. The barium salt, $(C_{10}H_{19}O_2)_2Ba$, separates from solution in boiling water in needle-shaped or large prismatic crystals which float on the water if not moistened.

Lauric Acid, $C_{12}H_{24}O_2$, occurs as a glyceride (laurostearin) in the fat of the bay-tree (*Lauris nobilis*), and in the solid fat and volatile oil of pichurim beans (*Fabœ Pichurim maj.*). It is prepared by saponifying these fats with caustic alkali, and decomposing the resulting soap with tartaric or hydrochloric acid. It likewise occurs, together with other fatty acids, or their glycerides, in cocoa-nut oil and the oils or fats of several other plants, also in spermaceti; and is separated from the mixtures of fatty acids resulting from the saponification of these substances by a complicated process of fractional precipitation with barium and magnesium salts, into the details of which we cannot enter.*

Lauric acid is insoluble in water, but dissolves easily in alcohol and ether, and crystallizes from alcohol in white, silky needles, which melt at about 43° C. (109.4° F.).

The laurates of the alkali-metals and of barium are soluble in water; the other salts are insoluble or sparingly soluble. The *calcium salt*, $(C_{12}H_{23}O_2)_2Ca$, is resolved by distillation into calcium carbonate and lauro- stearone:

$$(C_{12}H_{23}O_2)_2Ca \;=\; CO_3Ca \;+\; C_{23}H_{46}O.$$

Myristic Acid, $C_{14}H_{28}O_2$, occurs as a glyceride in nutmeg-butter and Otoba fat; also, together with lauric acid, in Dika bread, the fruit of *Mangifera gabonensis*, an African tree; and, together with other fatty acids, in cocoa-nut oil and spermaceti. It may be produced from crude ethal (cetyl alcohol) by heating with a mixture of potash and lime, its formation being doubtless due to the presence of methal or myristic alcohol, $C_{14}H_{30}O$, in the crude ethal (p. 622):

$$C_{14}H_{30}O \;+\; KHO \;=\; C_{14}H_{27}O_2K \;+\; 2H_2.$$

Lauric acid is likewise produced by a similar process from crude ethal, doubtless because that substance also contains lethal or lauric alcohol· $C_{12}H_{26}O$.

Pure myristic acid is most easily obtained by saponification of Otoba fat

* See **Watts's** Dictionary of Chemistry, vol. iii. p. 474.

(from *Myristica Otoba*). It forms white, shining crystalline laminæ, melting at 53.8° C. (128.8° F.). It is quite insoluble in water and in ether, but dissolves easily in hot alcohol, and crystallizes therefrom on cooling.

The myristates of the alkali-metals, $C_{14}H_{27}O_2K$, etc., are soluble in water, and not decomposed thereby (like the stearates). The other myristates are insoluble or sparingly soluble, and are obtained by precipitation.

Myristin $(C_3H_5)(C_{14}H_{27}O_2)_3$, the glyceride of myristic acid, is obtained by pressing nutmegs between hot plates, exhausting the crude fat thus obtained with spirits of wine, and crystallizing the undissolved portion from boiling ether. It is a crystalline fat having a silky lustre.

Palmitic Acid, $C_{16}H_{32}O_2$, occurs as a glyceride (tripalmitin) in many natural fats, often associated with stearin. Palm-oil, the produce of *Elais guianensis*, Chinese tallow, the produce of the tallow-tree (*Stillingia sebifera*), and Japan wax, from *Rhus succedanea*, consists mainly of tripalmitin. Palmitic acid is easily prepared by saponifying palm-oil with caustic potash, decomposing the soap with sulphuric acid, and crystallizing the separated fatty acid several times from hot alcohol till it exhibits a constant melting-point. Chinese tallow may be saponified with alcoholic potash, and Japan wax by fusion with solid potassium hydrate, and the soap treated in a similar manner.

Palmitic acid exists also as cetyl palmitate (cetin), $C_{16}H_{33}.C_{16}H_{31}O_2$, in spermaceti, and as myricyl palmitate (melissin), $C_{30}H_{61}.C_{16}H_{31}O_2$, in bees'-wax. It is produced, together with acetic acid, by melting oleic acid, $C_{18}H_{34}O_2$, with potassium hydrate:

$$C_{18}H_{34}O_2 \;+\; 2KOH \;=\; C_{16}H_{31}O_2K \;+\; C_2H_3O_2K \;+\; H_2.$$

Palmitic acid is a colorless, solid body, without taste or smell, lighter than water. It is insoluble in water, but dissolves abundantly in boiling alcohol or ether. The solutions are acid, and when concentrated, solidify in a mass on cooling. When dilute they yield the acid in tufts of slender needles. It melts at 62° C. (143.6° F.), and solidifies on cooling in a mass of shining nacreous laminæ. When heated in a dish it boils and evaporates without residue, and may be distilled almost without change. When gently heated in the air, it is but slightly altered, but at higher temperatures it takes fire, and burns with a bright smoky flame like other fats. It is attacked by chlorine at 100°, giving off hydrochloric acid, and forming oily substitution-products. Heated with alcohols, it forms compound ethers.

Palmitic acid forms normal or neutral salts, having the composition $MC_{16}H_{31}O_2$ for univalent, and $M''(C_{16}H_{31}O_2)_2$ for bivalent metals, and with the alkali-metals also, acid salts analogous to the acid acetates. The normal palmitates of potassium and sodium are soluble in water and alcohol; the rest are insoluble, and are obtained by precipitating a metallic salt with an alcoholic solution of sodium or potassium palmitate.

Ethyl palmitate, $C_2H_5.C_{16}H_{31}O_2$, obtained by passing hydrochloric acid gas into a saturated alcoholic solution of palmitic acid, crystallizes in prisms, and melts at 24° C. (75.2° F.).

Glyceryl Palmitates or Palmitins.—There are three of these ethers—viz., *Monopalmitin*, $C_3H_5 \begin{Bmatrix} (OH)_2 \\ C_{16}H_{31}O_2 \end{Bmatrix}$, *dipalmitin*, $C_3H_5 \begin{Bmatrix} OH \\ (C_{16}H_{31}O_2)_2 \end{Bmatrix}$, and *tripalmitin*, $C_3H_5(C_{16}H_{31}O_2)_3$. The first and second are obtained by heating palmitic acid with glycerin in sealed tubes; the third by heating a mixture of 1 part of monopalmitin and 10 parts of palmitic acid to 250° C. (482° F.) for twenty-eight hours. They are all crystalline fats. Tripalmitin thus obtained melts at 46° C. (114.8° F.). Natural palmitin, obtained from palm-oil and other fats, has the composition of tripalmitin, but exhibits three isomeric (or rather allotropic) modifications (like those of stearin) melting respectively at 46° C. (114.8° F.), 61.7° C. (143.1° F.), and 62.8° C. (145° F.): the first appears to be identical with artificial tripalmitin.

Palm-oil comes chiefly from the coast of Africa. It has, when fresh, a deep orange-red tint and a very agreeable odor: the coloring matter—the nature of which is unknown—is easily destroyed by exposure to light, especially at a high temperature, and also by oxidizing agents. The oil melts at 27° C. (80.6° F.). By cautious pressure it may be separated into fluid olein and solid palmitin, which, when purified by crystallization from hot ether, is perfectly white. By keeping, palm-oil seems to suffer a change similar to that produced by saponification: in this state it is found to contain traces of glycerin and a considerable quantity of oleic acid, together with palmitic acid. The oil becomes harder and rancid, and its melting point is raised at the same time.

Margaric Acid, $C_{17}H_{34}O_2$.—This name was formerly applied to an acid, intermediate between stearic and palmitic acids, supposed to be produced, together with others, by the saponification of natural fats; but it is now restricted, for reasons to be presently mentioned, to an acid prepared by a definite reaction—viz., by the action of boiling alcoholic potash on cetyl cyanide:—

$$C_{16}H_{33}CN + KOH + H_2O = NH_3 + C_{17}H_{33}O_2K.$$

The solid potassium salt thus obtained is decomposed by boiling dilute hydrochloric acid, and the separated margaric acid is purified by precipitating its ammoniacal solution with barium chloride, decomposing the precipitate with hydrochloric acid and ether, separating the ethereal solution by means of a pipette; and distilling off the ether. It forms white crystals, melting at 59.9° C. (139.8° F.), and is intermediate in all its properties between palmitic and stearic acids.

The so-called margaric acid, obtained by the saponification of natural fats, and regarded by Chevreul[*] and many other chemists, as a distinct acid having the composition $C_{17}H_{34}O_2$, has been shown by Heintz[†] to be a mixture, resolvable into stearic acid and other fatty acids of lower melting point, chiefly palmitic acid. Such mixtures of solid fatty acids, or of the corresponding glycerides, cannot be completely resolved into their constituent fats by crystallization from alcohol, ether, or other solvents, which was the method of separation resorted to in the earlier investigations. The only effectual method of separation is to subject the alcoholic solution of the acids to a series of fractional precipitations with acetate of lead, barium, or magnesium, the stearate then separating out first.

Stearic Acid, $C_{18}H_{36}O_2$, was discovered by Chevreul as a constituent of the more solid fats of the animal kingdom. It is most abundant in these, especially in beef- and mutton-suet; but exists also, together with palmitic, myristic acid, etc., in the softer fats, such as the butter of cows' milk, human fat, that of the goose, of serpents, of cantharides, and in spermaceti. It occurs also in vegetable fats, especially those of cacaobeans, of the berries of *Cocculus indicus*, and in shea-butter, obtained from the nuts of *Bassia Parkii*, a tree growing in West Africa. In all these fats it occurs as a glyceride, but in that of cocculus grains also in the free state.

[*] *Recherches sur les corps gras d'origine animale.* Paris, 1823.
[†] For references to Heintz's memoirs, see Gmelin's Handbook, vol. xv. p. 343.

Stearic acid is prepared from b e e f or m u t t o n - s u e t, or better, from
c a c a o - f a t, by saponifying the fat with soda-lye, heating the soap-paste
with water and dilute sulphuric acid, removing the separated fatty acids
after cooling, washing them with water, and then dissolving them in as
small a quantity as possible of hot alcohol. On cooling, the greater part
of the solid acid separates out, while the oleic acid remains in solution,
and may be separated by subjecting the mass, after draining, to strong
pressure, redissolving the residue in a small quantity of alcohol, leaving
it to separate by cooling, and again pressing the solid mass. From the
mixture of solid fatty acids thus obtained, the stearic acid may be separated in a comparatively pure state, by repeated crystallization from considerable quantities of alcohol, only the portion which first separates being
each time collected. But to obtain pure stearic acid, it is better to dissolve the impure stearic acid (4 parts), melting at 60° C. (140° F.), in
such a quantity of hot alcohol that nothing will separate out on cooling,
even to 0°, and mix the hot liquid with a boiling alcoholic solution of
magnesium acetate (1 part). The magnesium-salt which separates on
cooling is pressed and boiled for some time with a large quantity of dilute
hydrochloric acid, and the stearic acid thereby separated is repeatedly
crystallized from alcohol, till it melts constantly at 69° to 70° C. (156.2°
to 158.6° F.).

Stearic acid is also easily prepared from the fat of cocculus-berries, which
consists mainly of stearin, by saponifying it with potash, etc. According
to Buff and Oudemanns, the best material for the preparation of stearic
acid is s h e a - b u t t e r, which contains about 30 per cent. oleic acid, and
70 per cent. stearic, but no other solid fatty acid.

On the large scale, impure stearic acid is prepared for the manufacture
of stearin-candles, by saponifying some of the harder fats, generally with
lime. The resulting lime-soap, decomposed by sulphuric acid, yields a
mixture of fatty acids, which are pressed, first in the cold, and afterwards
at a higher temperature, in order to separate the oleic acid from the less
fusible palmitic and stearic acids. Another method, applied chiefly to
palm-oil, consists in decomposing the fat with super-heated steam, as described under GLYCERIN (p. 638). A third method consists in treating the
fat with sulphuric acid, and distilling the product.

Pure stearic acid crystallizes from alcohol in nacreous laminæ or needles;
it is tasteless and inodorous, and has a distinct acid reaction. At low
temperatures it is heavier than water, having a specific gravity of 1.01 at
0°; but between 9° and 10° C. (48.2°–50° F.) its specific gravity is the
same as that of water. It melts at 69°–69.2° C. (156.2°–156.6° F.) to a
colorless oil, which on cooling solidifies to a white, fine, scaly, crystalline
mass, lamino-crystalline on the fractured surface. When heated, it distils,
for the most part, without alteration. *Chlorine* converts it into chloro-
stearic acid, $C_{18}H_{35}ClO_2$. Heated with *bromine* and water in a sealed tube,
it is converted into bromostearic acid, $C_{18}H_{35}BrO_2$, and dibromostearic acid,
$C_{18}H_{34}Br_2O_2$.

S t e a r a t e s.—Stearic acid dissolves in a cold aqueous solution of alkaline carbonate, probably from formation of acid carbonate, and does not
expel the carbonic acid and form a mono-acid salt, till heated to about
100°. On the other hand, the stearates are decomposed by most other
acids, the separated stearic acid rising to the surface as an oil when the
liquid is warm. The stearates have the consistence of hard soaps and
plasters, and are mostly insoluble in water. The *normal potassium-salt*,
$C_{18}H_{35}O_2K$, separates on cooling from a solution of 1 part stearic acid and 1
part potassium hydrate in 10 parts of water, in white opaque granules.
The *acid salt*, $C_{18}H_{35}O_2K.C_{18}H_{36}O$, is obtained by decomposing the normal
salt with 1000 parts or more of water, and separates in silvery scales from

solution in boiling alcohol. *Normal sodium stearate*, $C_{18}H_{35}O_2Na$, is very much like the potassium-salt, but harder. The *acid salt*, $C_{18}H_{35}O_2Na$. $C_{18}H_{36}O_2$, obtained by decomposing the normal salt with 2000 parts or more of water, separates from the hot solution in nacreous laminæ. The stearates of the earth-metals and heavy metals are insoluble in water, and are obtained by precipitation.

S o a p s consist of mixtures of the sodium or potassium salts of stearic, palmitic, oleic, and other fatty or oily acids, and are produced by saponifying tallow, olive oil, and other fats with caustic alkalies. The soda-soaps are called h a r d s o a p s : they separate from the alkaline liquor, on addition of common salt, in hard, unctuous masses, which are the soaps in common use : this mode of separation is called *salting out*. The potash-soaps, on the other hand, cannot be thus separated ; for, on adding salt to their solution, they are decomposed and converted into soda-soaps ; but they are obtained in a semi-solid state by evaporating the solution. The products, called s o f t s o a p s , always contain a considerable excess of alkali, and are used for cleansing and scouring when a powerful detergent is required.

S t e a r i c e t h e r s are formed by heating stearic acid with alcohols, monatomic or polyatomic. *Ethyl stearate*, $C_2H_5.C_{18}H_{35}O_2$, is most easily obtained by passing hydrochloric acid gas into an alcoholic solution of stearic acid. It resembles white wax, is inodorous and tasteless, melts at 30° C. (86° F.), and cannot be distilled without decomposition. It is readily decomposed by boiling with caustic alkalies. There are three g l y c e r y l s t e a r a t e s or s t e a r i n s , analogous in composition to the palmitins :

Monostearin, $C_3H_5 \begin{cases} (OH)_2 \\ C_{18}H_{35}O_2 \end{cases}$, prepared by heating a mixture of equal parts of stearic acid and glycerin to 200° in a sealed tube for 36 hours, forms very small white needles, melting at 61° C. (141.8° F.), and solidifying again at 60° C. (140° F.).—*Distearin*, $C_3H_5 \begin{cases} OH \\ (C_{18}H_{35}O_2)_2 \end{cases}$, obtained by heating monostearin with 3 parts of stearic acid to 260° for three hours, forms white microscopic laminæ, melts at 58° C. (136.4° F.), and solidifies at 55° C. (131° F.).—*Tristearin*, $C_3H_5(C_{18}H_{35}O_2)_3$, is prepared by heating monostearin with 15 to 20 times its weight of stearic acid to 270° C. (518° F.) for three hours in a sealed tube ; also from various solid natural fats by solution in ether, and repeated crystallization from the hot solution. It crystallizes in masses of white pearly laminæ or needles, inodorous, tasteless, neutral, and volatilizing without decomposition under reduced pressure. Both natural and artificial tristearin exhibit three isomeric or allotropic modifications. Stearin separated from ether melts at 69.7° C. (157.5° F.); but if heated to 73.7° C. (164.6° F.) or higher and then cooled, it does not solidify till cooled to 51.7° C. (125° F.). It then melts at 52° C. (125.6° F.), and if heated a few degrees higher, passes into a third modification, which does not melt below 64.2° C. (147.4° F.).

Arachidic Acid, $C_{20}H_{40}O_2$, is a fatty acid obtained by saponification of oil of earth-nut (*Arachis hypogœa*). It crystallizes in very small, shining scales, melts at 75° C. (167° F.), and solidifies again at 73.5° C. (164.3° F.) to a radiated crystalline mass. It is but slightly soluble in cold alcohol of ordinary strength, but dissolves easily in boiling absolute alcohol and in ether.

The *silver-salt*, $C_{20}H_{39}O_2Ag$, is a white precipitate, which separates from boiling alcohol in slightly lustrous prisms, not altered by exposure to light. *Ethyl arachidate*, $C_{20}H_{39}O_2.C_2H_5$, is a crystalline mass, melting at 52.5° C. (126.5° F.). Berthelot has obtained three *glyceryl arachidates* or *arachins*, analogous to the stearins, by heating the acid with glycerin in sealed tubes.

Benic or **Behenic Acid,** $C_{22}H_{44}O_2$, is obtained, together with other acids, by saponification of oil of ben, the oil expressed from the fruits of *Moringa Nux Behen*. It is a white crystalline fat, melting at 76° C. (168.8° F.), and solidifying at 70° C. (158° F.).

Cerotic Acid, $C_{27}H_{54}O_2$, is the essential constituent of *cerin*, the portion of bees'-wax which is soluble in boiling alcohol. It is prepared by heating the wax several times in succession with boiling alcohol, till the deposit which forms on cooling melts at 70° or 72° C. (158°–161.6° F.), and may be further purified by precipitating it from the boiling alcoholic solution with lead acetate, decomposing the precipitate with strong acetic acid, and crystallizing the separated acid from boiling alcohol. Cerotic acid is also produced by the dry distillation of Chinese wax, which consists of ceryl cerotate, $C_{27}H_{53}O_2.C_{27}H_{55}$, or by melting that substance with potash, and decomposing the resulting potassium-salt with an acid (p. 622).

Pure cerotic acid crystallizes in small grains, melting at 78° C. (172.4° F.), and distilling without alteration. Chlorine converts it into chlorocerotic acid, $C_{27}H_{42}Cl_{12}O_2$, a thick transparent gum of a pale yellow color.

Ceryl cerotate, or *Chinese wax,* is produced on certain trees in China by the puncture of a species of *coccus*. It is crystalline, of a dazzling whiteness, like spermaceti, melts at 82° C. (179.6° F.); dissolves in alcohol; yields cerotic acid and cerylene, $C_{27}H_{54}$, by dry distillation. It is used in China for making candles.

Melissic Acid, $C_{30}H_{60}O_2$, the highest known member of the fatty series, is obtained by heating myricyl alcohol (p. 623) with potash-lime:

$$C_{30}H_{62}O_2 + KOH = C_{30}H_{59}O_2K + 2H_2.$$

It bears considerable resemblance to cerotic acid, but melts at a higher temperature, viz., at 88° or 89° C. (190.4°–192.2° F.). The *silver-salt,* $C_{30}H_{59}O_2Ag$, is a white precipitate.

Monatomic Acids, $C_nH_{2n-2}O_2$.—Acrylic Series.

This series comprises three isomeric groups of acids, which may be represented by the following general formulæ, in which n and p denote any whole numbers from 0 upwards :*

* The most general formulæ by which the structure of these acids can be represented are the following :—

The first of these formulæ is reducible to $C_n+_3H_{2n}+_4O_2$ or $C_n+_3H_2(_p+_2)_2O_2$, and the second and third to $C_n+_2H_{2n}+_2O_2$ or $C_n+_2H_2(_n+_2)-_2O_2$, both of which are included under the general formula of the series $C_mH_{2m}-_2O_2$. In all the known acids of the first group, however, except perhaps angelic acid, the value of p = 0, and in the only known acid of the second group, viz., isocrotonic acid, p = 1, so that for these known acids the two formulæ are reduced to the simpler forms given in the text.

$$
\begin{array}{ccc}
\text{(1.)} & \text{(2.)} & \text{(3.)} \\
C_nH_{2n+1} & C_nH_{2n} & C_nH_{2n} \\
| & \parallel & \parallel \\
CH & CH & C-C_pH_{2p+1} \\
\parallel & | & | \\
CH & CH_2 & CO_2H \\
| & | & \\
CO_2H & CO_2H &
\end{array}
$$

Of the acids of the first group, called **normal acrylic acids**, some occur, mostly as glycerides, in vegetable and animal organisms, others are formed artificially by special processes. Most of them are oily liquids. The known acids of this group are:

Acrylic acid	$C_3H_4O_2$	Physetoleic acid ⎫		
Crotonic acid	$C_4H_6O_2$	Hypogæic acid ⎬	$C_{16}H_{30}O_2$	
Angelic acid	$C_5H_8O_2$	Gaïdic acid ⎭		
Pyroterebic acid	$C_6H_{10}O_2$	Oleic acid ⎫	$C_{18}H_{34}O_2$	
? Damaluric acid	$C_7H_{12}O_2$	Elaïdic acid ⎭		
? Damolic acid	$C_{13}H_{24}O_2$	Doeglic acid	$C_{19}H_{36}O_2$	
Moringic acid ⎫	$C_{15}H_{28}O_2$	Brassic acid ⎫	$C_{22}H_{42}O_2$	
Cimicic acid ⎭		Erucic acid ⎭		

Of the acids of the second group only one is at present known, viz., isocrotonic acid, $C_4H_6O_2 = CH_2{=}CH{-}CH_2{-}CO_2H$ [$n=1$].

The acrylic acids of the third group are formed by a general synthetical process, viz., by abstraction of the elements of water from certain acid ethers or alcoholic acids, having the composition of oxalic acid, $C_2H_2O_4$, in which 1 atom of oxygen is replaced by 2 atoms of a monatomic alcohol-radicle, C_nH_{2n+1}; thus:

$$
\begin{array}{cccc}
 & CH_3 & CH_2CH_3 & CH_2CH_3 \\
 & | & | & | \\
HO-C{=}O & HO-C-CH_3 & HO-C-CH_3 & HO-C-CH_2CH_3 \\
| & | & | & | \\
HO-C{=}O & HO-C{=}O & HO-C{=}O & HC-C{=}O \\
\text{Oxalic acid.} & \text{Dimethoxalic} & \text{Ethometh-} & \text{Diethoxalic} \\
 & \text{acid.} & \text{oxalic acid.} & \text{acid.}
\end{array}
$$

Now, when the ethylic ethers of these acids are treated with phosphoric oxide or phosphorus trichloride, they give up a molecule of water (H_2O), at the expense of one of the molecules of hydroxyl (OH) and of an atom of hydrogen abstracted from one of the monad alcohol-radicles, which is thereby converted into a dyad radicle (an olefine) capable of saturating the unit of equivalence of the carbon-atom set free by abstraction of the hydroxyl. The product is the ethylic ether of an acrylic acid of the third group; thus:—

$$
\begin{array}{ccccc}
CH_3 & & & & CH_2 \\
| & & & & \parallel \\
HO-C-CH_3 & & & & C-CH_3 \\
| & - & H_2O & = & | \\
H_5C_2O-C{:=}O & & & & H_5C_2O-C{:=}O \\
\text{Ethylic dimeth-} & & & & \text{Ethylic-methyl} \\
\text{oxalate.} & & & & \text{acrylate}
\end{array}
$$

The ethylic ether thus formed is converted into methacrylic acid by saponification with potash in the usual way. In this manner the following acids have been obtained:—

Methacrylic acid, $CH_2 = C \big\langle^{CH_3}_{COOH}$, isomeric with Crotonic acid.

Methylcrotonic acid, $C_2H_4 = C \big\langle^{CH_3}_{COOH}$, isomeric with Angelic acid.

Ethylcrotonic acid, $C_2H_4 = C \big\langle^{CH_2CH_3}_{COOH}$, isomeric with Pyroterebic acid.

There might, of course, be an ethacrylic acid isomeric with methylcrotonic acid, and a propyl-acrylic acid isomeric with ethylcrotonic acid, but these have not yet been obtained.

The individual acids of the three groups differ from one another according to the values of n and p, and those which contain the radicle C_3H_7 and its higher homologues, are susceptible of further modifications according to the structure of these radicles.

The acids of all three series, when fused with potash, are resolved, with evolution of hydrogen, into two acids of the fatty series :—

$$C_5H_8O_2 \; + \; 2H_2O \; = \; C_3H_6O_2 \; + \; C_2H_4O_2 \; + \; H_2$$
$$\text{Angelic.} \qquad\qquad\qquad \text{Propionic.} \qquad \text{Acetic.}$$

$$C_{18}H_{34}O_2 \; + \; 2H_2O \; = \; C_{16}H_{32}O_2 \; + \; C_2H_4O_2 \; + \; H_2$$
$$\text{Oleic.} \qquad\qquad\qquad \text{Palmitic.} \qquad \text{Acetic.}$$

generally :

$$C_nH_{2n-2}O_2 \; + \; 2H_2O \; = \; C_{n-p}H_2{}_{(n-p)}O_2 \; + \; C_pH_{2p}O_2 \; + \; H_2 \; .$$

The manner in which this splitting up of the molecule takes place differs, however, according to the group to which the acid belongs. In the decomposition of an acrylic acid of the first group, one of the products is always acetic acid ; in the second, one of the products is always propionic acid ; while in the third, the two products may be any members of the fatty series of acids.

The final result of the action above mentioned is, in fact, to add 2 atoms of hydrogen and 2 atoms of oxygen to the molecule $C_nH_{2n-2}O_2$. Now, if in an acid of the first group, represented by the formula,

C_nH_{2n+1}—CH$\vdots$CH—CO$_2$H, we suppose H$_2$ to be added to the right of the

dotted line, we get CH_3—CO_2H, which is the formula of acetic acid, while the addition of O_2 to the left of the dotted line gives C_nH_{2n+1}—CO_2H, which is also the formula of a fatty acid. Similarly the addition of H_2 on the

right of the formula of the second group, $C_nH_{2n}\vdots$CH—CH$_2$—CO$_2$H, gives

CH_3—CH_2—CO_2H, which is the formula of propionic acid, while the addition of O_2 on the left gives $C_nH_{2n}O_2$, which represents another fatty acid.

The acrylic acids are also converted into fatty acids by the action of nascent hydrogen, $e. g.$, crotonic acid, $C_4H_6O_2$, into butyric acid, $C_4H_8O_2$.

Acrylic Acid, $C_3H_4O_2 = CH_2 = CH$—CO_2H [$n = 0$ in formula 1]. —This acid is the lowest possible member of the series, and does not admit of isomeric modifications. It is produced—(1) By oxidizing its aldehyde (acrolein), C_3H_4O, with moist silver oxide ; (2) From a or b iodopropionic acid, $CH_3.CHI.CO_2H$, or $CH_2I.CH_2.CO_2H$, by heating alone, or with alcoholic potash, whereby HI is abstracted ; (3) By the action of nascent hydrogen (zinc and sulphuric acid) on b-dibromopropionic acid :—

$$CH_2Br—CHBr—CO_2H \; + \; H_2 \; = \; 2HBr \; + \; CH_2 = CH—CO_2H \; .$$

Acrylic acid is a colorless liquid, having a pungent, slightly aromatic

odor, and miscible in all proportions with water. It solidifies at low temperatures, melts at 7° C. (44.6^O F.), and boils at 139°–140° C. (282.2–
284° F.). Nascent hydrogen converts it into propionic acid. It unites
with bromine, forming β-dibromopropionic acid, and with the haloïd acids
to form the β-substitution products of propionic acid. By fusion with
potash it is resolved into formic and acetic acids.

The metallic acrylates, excepting the silver salt, are very soluble in
water; they decompose at 100°. The *silver salt*, $C_3H_3O_2Ag$, and the *lead
salt*, $(C_3H_3O_2)_2Pb$, crystallize in shining needles.

Ethyl Acrylate, CH_2=CH—$CO_2C_2H_5$, prepared from the ethylic ether of
β-dibromopropionic acid by the action of zinc and sulphuric acid, is a
pungent-smelling liquid, having a specific gravity of 0.925 at 0°, and
boiling at 102° C. (215.6^O F.).

Chloracrylic Acid, $C_3H_3ClO_2$, prepared by treating dichloropropionic
acid with baryta-water, crystallizes in slender needles, which melt at 65° C.
(149° F.).—*Bromacrylic Acid*, $C_3H_3BrO_2$. Of this acid there are two modifications, CH_2=CBr—CO_2H and CHBr=CH—CO_2H, obtained respectively
from a- and β-dibromopropionic acid. The former melts at 70° C. (158° F.),
and unites with hydrobromic acid, reproducing a-dibromopropionic acid.

Crotonic Acids, $C_4H_6O_2 = C_3H_5$—CO_2H.—Of these acids there are
three modifications, viz.:

(1) **Solid Crotonic Acid**, CH_3—CH=CH—CO_2H.—This acid is formed
—a. By oxidation of its aldehyde, CH_3—CH=CH—COH (p. 701); β. By
the dry distillation of β-oxybutyric acid, CH_3—CH(OH)—CH_2—CO_2H (abstraction of OHH); γ. By the action of alcoholic potash on a-bromobutyric
acid, CH_3—CH_2—CHBr—CO_2H; δ. From allyl iodide by conversion into the
cyanide, and distillation of the latter with potash. Now allyl iodide has
the structure CH_2=CH—CH_2I (p. 624), and therefore the cyanide might
be expected to be $\overset{..}{C}H_2$—CH_2CN, and the acid obtained from it, CH_2=
CH—CO_2OH, which is the structure of isocrotonic acid. But as the acid
obtained in this manner is identical in its properties with that which is
produced by the first three reactions, it must be inferred that, either in the
conversion of the allyl iodide into cyanide, or of the latter into the acid,
a transposition of atoms has taken place within the molecule.

Solid crotonic acid crystallizes in slender woolly needles, or in large
plates, melts at 72° C. (161.6^O F.), boils at 182° C. (359.6^O F.), and dissolves in 12 parts of water at 20°. Nascent hydrogen converts it into
normal butyric acid. It unites with hydrogen bromide, forming monobromobutyric acid, and with bromine to form dibromobutyric acid. By oxidation with nitric acid, it is resolved into acetic and oxalic acids, and by
fusion with potash, into two molecules of acetic acid:

$$CH_3\text{—}CH\text{=}CH\text{—}CO_2H \ + \ 2H_2O \ = \ 2(CH_3CO_2H) \ + \ H_2.$$

Monochlorocrotonic Acid, $C_4H_5ClO_2$, is produced by the action of zinc and
hydrochloric acid on trichlorocrotonic acid; also by that of phosphorus
pentachloride, and subsequently of water, on ethylic aceto-acetate, $(CH_3$—
CO—CH_2—$COOC_2H$—), whence it appears to have the structure CH_3—
CCl=CH—CO_2H:

$$\begin{matrix}CH_3 \\ CO \\ CH_2 \\ COOC_2H_5\end{matrix} \quad + \ 2PCl_4Cl_2 \ = \ 2PCl_3O \ + \ C_2H_5Cl \ + \ HCl \ + \ \begin{matrix}CH_3 \\ CCl \\ \| \\ CH \\ COCl\end{matrix}$$

and

$$CH_3.CCl\text{=}CH.COCl \ + \ HOH \ = \ HCl \ + \ CH_3.CCl\text{=}CH\text{—}COOH.$$

This acid dissolves in 35 parts of water at 19° C. (66.2° F.), melts at 94° C. (201.2° F.), and boils at 206°–210° C. (402.8°–410° F.). Sodium amalgam converts it into crotonic acid.

Trichlorocrotonic Acid, $C_4H_3Cl_3O_2$, produced by oxidation of the corresponding aldehyde with cold strong nitric acid, melts at 44° C. (111.2° F.).

(2) **Iosocrotonic Acid**, $CH_2{=}CH{-}CH_2{-}CO_2H$ (Geuther's *quartenylic acid*), formed from its chloro-derivative (*infra*) by the action of sodium amalgam, is a non-solidifying liquid, having a specific gravity of 1.018 at 25° C. (77° F.), and boiling at 172° C. (341.6° F.). When heated in a sealed tube to 170°–180° C. (338°–356° F.), it is converted into solid crotonic acid. When fused with potash, it is resolved, not, as might be expected, into propionic and formic acids, but, in consequence of the transformation by heat just mentioned, into two molecules of acetic acid.

Chlorisocrotonic acid, $CH_2{=}CCl{-}CH_2{-}CO_2H$, is produced, together with chlorocrotonic acid, by the action of phosphorus pentachloride and water on ethylic aceto-acetate. It forms crystals slightly soluble in water, sublimes at ordinary temperatures, melts at 59.5° C. (139.1° F.), and boils at 195° C. (383° F.).

(3) **Methacrylic Acid**, $CH_2{=}C{<}^{CH_3}_{CO_2H}$ [$n=1, p=1$ in formula 3, p. 742], is formed by the action of phosphoric oxide or phosphorus trichloride on ethylic dimethoxalate (p. 742), and by that of phosphorus trichloride on oxyisobutyric acids:

$$\frac{H_3C}{H_3C}{>}C{<}^{OH}_{CO_2H} \;=\; H_2O \;+\; \frac{H_3C}{H_3C}{\setminus}C{-}CO_2H.$$

It is a liquid which does not solidify at 0°. By fusion with potash it is resolved into formic and propionic acids.

Five-carbon Acrylic Acids, $C_5H_8O_2 = C_4H_7.CO_2H.$ —Of these acids, two only are known, namely, angelic and methyl-crotonic acid.

Angelic Acid exists in the root of the archangel (*Angelica Archangelica*), and in sumbul or moschus root, a drug imported from Asia Minor, and probably also belonging to an umbelliferous plant. It is obtained from archangel root, by boiling the root with lime and water, and distilling the strained and concentrated liquid with dilute sulphuric acid. It is also produced by heating the essential oil of chamomile, which consists of angelic aldehyde together with a hydro-carbon, with potassium hydroxide:

$$C_5H_8O \;+\; KOH \;=\; C_5H_7KO_2 \;+\; H_2.$$

Also, together with oreoselin, by treating peucedanin or imperatorin (a neutral substance contained in the root of *Imperatoria Ostruthium*, and some other umbelliferous plants) with alcoholic potash:

$$\underset{\text{Peucedanin.}}{C_{12}H_{12}O_3} \;+\; KOH \;=\; \underset{\substack{\text{Potassium} \\ \text{angelate.}}}{C_5H_7KO_2} \;+\; \underset{\text{Oreoselin.}}{C_7H_6O_2}$$

Angelic acid crystallizes in long prisms and needles, melts at 45° C. (113° F.), boils at 190° C. (374° F.), and distils without decomposition. It has an aromatic taste and odor, dissolves sparingly in cold, abundantly in hot water, also in alcohol and ether.

It unites with *bromine*, forming a dibromovaleric acid, $C_5H_8Br_2O_2$, which is reconverted by sodium-amalgam into angelic acid. It is not altered by

32

nascent hydrogen, but when heated with concentrated *hydriodic acid* and a little phosphorus, it is converted into normal valeric acid. By fusion with *potash* it is resolved into acetic and propionic acids, a mode of decomposition which is consistent with either of the two following structural formulæ of the acid :

$$C_2H_5—CH{=}CH—CO_2H^* \text{ or } CH_3—CH{=}CH—CH_2—CO_2H.†$$

The angelates of the alkali-metals are soluble in water and in alcohol. *Calcium angelate*, $Ca(C_5H_7O_2)_2 +$ Aq., forms shining, very soluble laminæ. The *lead-salt*, $Pb(C_5H_7O_2)_2$, is a white precipitate.

Potassium angelate treated with phosphorus oxychloride yields angelic oxide, or anhydride, $(C_5H_7O)_2O$, which is a viscid uncrystallizable oil, boiling at 240° C. (464° F.).

Methyl-orotonic Acid, $CH_3—CH{=}C{<}^{CH_3}_{CO_2H}$, formed by the action of phosphorus trichloride on ethylic methyl-ethyloxalate (p. 743), is very much like angelic acid, but melts at 62° C. (143.6° F.). By fusion with potash it is resolved into acetic and propionic acids.

Identical with this is the so-called t i g l i c a c i d, which is contained as a glyceride, together with those of butyric, valeric, and other fatty acids, in the oil of *Croton Tiglium*.

Six-carbon Acids, $C_6H_{10}O_2{=}C_5H_9.CO_2H$.—1. *Pyroterebic acid* is formed by dry distillation of terebic acid, $C_7H_{10}O_4$ (one of the products of the action of nitric acid on turpentine-oil). It is an oily liquid, smelling like butyric acid, having a specific gravity of 1.01, and boiling at 210° C. (410° F.). By fusion with potash it is resolved into acetic and isobutyric acids.

2. *Hydrosorbic acid* is formed by the action of sodium-amalgam on sorbic acid, $C_6H_8O_2$. It is a liquid having a sudorific odor, a specific gravity of 0.969 at 19° C. (66.2° F.), and boiling at 204.5° C. (400.1° F.). By fusion with potash it is resolved into acetic and normal butyric acids. This and the preceding acid are therefore represented by the following formulæ :—

$$CH_3—CH_2—CH_2—CH{=}CH—CO_2H$$
$$\text{Hydrosorbic.}$$
$$CH(CH_3)_2—CH{=}CH—CO_2H$$
$$\text{Pyroterebic.}$$

3. *Ethyl-crotonic acid*, $CH_3—CH—C{<}^{CH_2.CH_3}_{CO_2H}$, is formed from the ethylic ether of diethoxalic acid (p. 743), by the action of phosphorus trichloride, or by heating with hydrochloric acid to 130°–150° C. (266°–302° F.). It crystallizes in shining square prisms, has an aromatic odor, sublimes even at ordinary temperatures, and melts at 41.5° C. (106.7° F.). By fusion with potash it is resolved into acetic and normal butyric acids. Its salts decompose, even during the evaporation of their aqueous solutions.

Seven- to Fifteen-Carbon Acids.—D a m a l u r i c acid, $C_7H_{12}O_2$, and D a m o l i c a c i d, $C_{13}H_{24}O_2$, are volatile acids, said to exist in the urine of cows and horses.—M o r i n g i c a c i d, $C_{15}H_{29}O_2$, is an oily acid, obtained, together with palmitic, stearic, and benic acids, by the saponifi-

cation of oil of ben (p. 741).—C i m i c i c   a c i d, $C_{15}H_{28}O_9$, is a yellow crystallizable acid, having a rancid odor, extracted by alcohol and ether from
a kind of bug (*Rhaphigaster punctipennis*).

Hypogæic Acid, $C_{16}H_{30}O_2$, is contained as a glyceride, together
with palmitin and arachin, in oil of earth-nut (*Arachis hypogœa*). To
obtain it, the mixture of fatty acids obtained by saponifying the oil is
dissolved in alcohol; the palmitic and arachidic acids are precipitated by
ammonia and magnesium acetate; the filtrate is mixed with ammonia and
lead acetate; the lead precipitate is decomposed by hydrochloric acid; and
the separated hypogæic acid is dissolved out by ether. It is also produced by oxidation of axinic acid ($C_{18}H_{28}O_2$), an acid obtained by saponification of *age* or *axin*, a fatty substance contained in the Mexican plant
Coccus Axin.—Hypogæic acid crystallizes from ether in stellate groups of
needles, melting at 34○ or 35○ C. (93.2○–95○ F.), easily soluble in alcohol
and ether. Its *potassium* and *sodium salts* are soluble in water; the *barium
salt* is soluble in hot, insoluble in cold water; the *copper* and *silver salts*
are obtained by precipitation. The *ethlyic ether*, $C_{16}H_{29}O_2.C_2H_5$, is a yellow
oil, not volatile without decomposition.

Nitrous acid converts hypogæic acid into the isomeric or allotropic compound, G a ï d i o   a c i d, related to it in the same manner as elaïdic acid to
oleic acid. It forms a colorless crystalline mass which melts at 38○ C.
(100.4○ F.).

P h y s e t o l e i c   a c i d, a crystalline acid obtained from sperm-oil, is
isomeric, if not identical, with hypogæic acid; it melts at 30○ C. (80○ F.),
and solidifies at 28○ C. (82.4○ F.).

Oleic Acid, $C_{18}H_{34}O_2$.—This acid, the most important of the series,
is obtained by saponification of olein, the fluid constituent of most natural
fats and fixed oils.

To obtain pure oleic acid, olive or almond oil is saponified with potash;
the soap is decomposed by tartaric acid; and the separated fatty acid,
after being washed, is heated for some hours in the water-bath, with half
its weight of lead oxide previously reduced to fine powder. The mixture
is then well shaken up with about twice its bulk of ether, which dissolves
the oleate of lead, and leaves the stearate; the liquid, after standing for
some time, is decanted and mixed with hydrochloric acid; the oleic acid
thereby eliminated dissolves in the ether; and the ethereal solution which
rises to the surface of the water is decanted, mixed with water, and freed
from ether by distillation.

Large quantities of crude oleic acid are now obtained in the manufacture
of stearine candles, by treating with dilute sulphuric acid the lime-soap
resulting from the action of lime upon tallow. The fatty acids resulting
from the decomposition are washed with hot water, and solidify in a mass
on cooling; and this mass, when subjected to pressure, yields a liquid rich
in oleic acid, but still retaining a considerable quantity of stearic acid.
After remaining for some time in a cold place, it deposits a quantity of
solid matter, and the liquid decanted from this is sent into the market as
oleic acid or *red oil.* It may be purified by the process just described.

Oleic acid crystallizes from alcoholic solution in dazzling white needles,
melting at 14○ C. (57.2○ F.) to a colorless oil, which solidifies at 4○ C. (39.2○
F.) to a hard, white, crystalline mass, expanding considerably at the same
time. Specific gravity = 0.898 at 19○ C. (66.2○ F.). The acid volatilizes
in a vacuum without decomposition. It is tasteless and inodorous, and reacts neutral when unaltered (not oxidized), also in alcoholic solution. It
is insoluble in water, very soluble in alcohol, and dissolves in all proportions
in ether. Cold strong sulphuric acid dissolves it without decomposition.

It dissolves solid fats, stearic acid, palmitic acid, etc., and is dissolved by bile, with formation of a soap and strong acid reaction.

Oleic acid, in the solid state, oxidizes but slowly in the air; but when melted, it rapidly absorbs oxygen, acquiring a rancid taste and smell and a decided acid reaction. Its decomposition by fusion with potash has been already mentioned. *Chlorine* and *bromine*, in presence of water, convert it into dichloroleic and dibromoleic acids. Bromine, added by drops to fused oleic acid, forms tribromoleic acid, $C_{18}H_{31}Br_3O_2$.

Strong *nitric acid* attacks oleic acid with violence, giving off red nitrous vapors, and producing volatile acids of the series $C_nH_{2n}O_2$, viz., acetic, propionic, butyric, valeric, caproic, œnanthylic, caprylic, pelargonic, and rutic acids; also fixed acids of the series $C_nH_{2n-2}O_4$, viz., suberic, pimelic, adipic, lipic, and azelaic acids, the number and proportion of these products varying with the duration of the action.

Nitrous acid converts oleic acid into a solid isomeric or allotropic modification, called elaïdic acid.

Oleates.—The formula of the neutral oleates is $M'C_{18}H_{33}O_2$, or $M''(C_{18}H_{33}O_2)_2$, according to the quantivalence of the metal: there are likewise acid oleates. The neutral oleates of the alkali-metals are soluble in water, and not so completely precipitated from their solutions by the addition of another soluble salt, as the stearates and palmitates. The acid oleates are liquid and insoluble in water. The oleates dissolve in cold absolute alcohol and in ether, a property by which they may be distinguished and separated from the stearates and palmitates.

Oleins.—Oleic acid forms three glycerides, viz., monolein (C_3H_5) $(OH)_2(C_{18}H_{33}O_2)$; diolein ($C_3H_5)(OH)(C_{18}H_{33}O_2)_2$; and triolein ($C_3H_5$) $(C_{18}H_{33}O_2)_3$, which are produced by heating oleic acid and glycerin together in sealed tubes, in various proportions. The first two solidify at about 15° C. (59° F.).

The olein of animal fats, of olive oil, and of several other oils, both animal and vegetable, which do not dry up in the air by slow oxidation, but are converted into viscid masses having a rancid odor and acid reaction (non-drying oils), appears to be identical with triolein, but there is great difficulty in obtaining it pure. Olive oil, cooled to 4° C. (39° F.) or a lower temperature, deposits a large quantity of solid fat, consisting mainly of palmitin (originally called *margarin*, from its pearly lustre), and the oil filtered therefrom consists mainly of olein. A purer olein is obtained by treating olive oil with a cold strong solution of caustic soda, which saponifies the solid fats, and leaves the olein unaltered. Olein, subjected to dry distillation, yields gaseous products, liquid hydrocarbons, acrolein, and sebic acid.

Appendix to Oleic Acid.—Some non-drying oils contain the glycerides of acids homologous with oleic acid; such is the case, as already observed, with croton oil, earth-nut oil, and sperm oil. Doegling train-oil, obtained from the doegling or bottle-nosed whale (*Balæna rostrata*), yields doeglio acid, $C_{19}H_{36}O_2$. Colza-oil, obtained from the seeds of certain species of *Brassica*, especially the summer rape or colza, *Brassica campestris*, var. *oleifera*, yields brassic acid, $C_{22}H_{42}O_2$; and the oil of black mustard-seed yields a similar and probably identical acid, called erucic acid.

Drying oils, such as linseed, poppy, hemp, and nut oils, contain the glycerides of linoleic acid, $C_{16}H_{28}O_2$, which may be prepared by saponifying linseed oil with potash, precipitating the aqueous solution of the resulting potassium salt with calcium chloride, dissolving out the calcium linoleate with ether, and decomposing it with hydrochloric acid. Linoleic acid is a yellowish oil of specific gravity 0.921, not altered by nitrous acid.

Castor oil, which is a non-drying oil, contains the glyceride of r i c i n o - l i c a c i d, $C_{18}H_{34}O_3$, which, when separated, forms a colorless oil, solidi- fying at 0°, and converted by nitrous acid into a solid modification, r i c i n - e l a ï d i c a c i d, which melts at 50° C. (122° F.). By dry distillation it is resolved into œnanthylic acid and œnanthol, and when heated with excess of caustic potash or soda, it yields secondary octyl alcohol, together with sebic acid and free hydrogen (p. 620).

Monatomic Acids, $C_nH_{2n-4}O_2$.—The known acids of this series are:

Tetrolic acid,	$C_4 H_4 O_2$	Stearolic acid,	$C_{18}H_{32}O_2$
Sorbic acid,	$C_6 H_8 O_2$	Behenolic acid,	$C_{22}H_{40}O_2$.
Palmitolic acid,	$C_{16}H_{28}O_2$		

They are formed from the acids of the preceding series by abstraction of 2 atoms of hydrogen, which is effected by the action of alcoholic potash on the chlorine or bromine compounds or derivatives of those acids ; e. g.,

$$\underset{\substack{\text{Bromocrotonic}\\\text{acid.}}}{C_4H_5BrO_2} + KOH = KBr + H_2O + \underset{\substack{\text{Tetrolic}\\\text{acid.}}}{C_4H_4O_2}$$

$$\underset{\substack{\text{Bromide of cro-}\\\text{tonic acid.}}}{C_4H_6Br_2O_2} + 2KOH = 2KBr + 2H_2O + \underset{\substack{\text{Tetrolic}\\\text{acid.}}}{C_4H_4O_2}$$

They contain 4 atoms of hydrogen less than the corresponding fatty acids, and are therefore unsaturated compounds of the second order, capable of uniting with 2 or with 4 atoms of a halogen-element.

Tetrolic Acid, $C_4H_4O_2$, produced as above from monochlorocrotonic acid, crystallizes in deliquescent rhombic plates, easily soluble in alcohol and ether. It melts at 76.5° C. (169.7° F.), and boils at 203° C. (307.4° F.).

Sorbic Acid, $C_6H_8O_2$, occurs in the unripe berries of the mountain ash (*Sorbus aucuparia*), from which it may be obtained by mixing the juice with milk of lime and distilling. It then passes over as an oily liquid, which solidifies when boiled with strong hydrochloric acid, or when warmed with potash.

Sorbic acid crystallizes in long needles, melts at 134.5° C. (274.1° F.), decomposes when distilled alone, but is easily volatilized with vapor of water. It is nearly insoluble in cold water, but dissolves with moderate facility in hot water and in alcohol. *Nascent hydrogen* converts it into h y d r o s o r b i c a c i d, $C_6H_{10}O_2$. With *bromine* it forms the compounds $C_6H_8Br_2O_2$ and $C_6H_8Br_4O_2$, melting respectively at 95° C. (203° F.) and 183° C. (361.4° F.).

The metallic sorbates are crystallizable. The *ammonium salt* crystallizes in long needles ; the *barium salt*, $(C_6H_7O_2)_2Ba$, and the *calcium salt*, $(C_6H_7O_2)_2Ca$, form silvery scales. The *silver salt*, $C_6H_7O_2Ag$, is a crystal- line precipitate. The *ethylic ether*, $C_6H_7O_2.C_2H_5$, is an aromatic liquid, boil- ing at 195.5° C. (383.9 F.).

Palmitolic Acid, $C_{16}H_{28}O_2$, prepared from the dibromide of hypogæic acid, or of its isomeride, gaïdic acid, crystallizes in shining needles which melt at 42° C. (107.6° F.). It unites with 2 and 4 atoms of bromine, but is not altered by nascent hydrogen.

Stearolic Acid, prepared in like manner from the dibromide of oleic or elaïdic acid, crystallizes in long prisms, which melt at 48° C. (118.4° F.) and distil almost without decomposition. It is insoluble in water, but dissolves readily in alcohol and ether. It unites with 2 and 4 atoms of *bromine*, but is not altered by nascent hydrogen. By carefully regulated fusion with *potash*, it is resolved into acetic and hypogæic acids ; at higher temperatures myristic acid is produced.

Behenolic Acid, $C_{22}H_{40}O_2$, from dibromide of erucic or brassic acid (p. 749), forms shining needles, melting at 75° C. (167° F.).

The last three acids, heated with fuming nitric acid, are converted into monobasic acids of the form $C_nH_{2n-4}O_4$, viz. :—

Palmitoxylic.	Stearoxylic.	Behenoxylic.
$C_{16}H_{28}O_4$	$C_{18}H_{32}O_4$	$C_{22}H_{40}O_4$

melting at 67° C. (152.6° F.), 86° C. (186.8° F.), 90° C. (194° F.).

These acids crystallize in shining plates or scales.

CARBOKETONIC ACIDS.

These acids are derived from ethyl acetate and other compound ethers of the fatty series. When sodium is heated with ethyl acetate, it dissolves, with little or no evolution of hydrogen, and the whole solidifies on cooling to a crystalline mass of sodium ethylate and ethylic acetosodacetate, $C_6H_9NaO_3$. The reaction is either

$$2(C_2H_3O.OC_2H_5) + 2Na = C_6H_9NaO_3 + NaOC_2H_5 + H_2$$
$$\text{or} \quad 3(C_2H_3O.OC_2H_5) + 4Na = C_6H_9NaO_3 + 3NaOC_2H_5 .$$

The quantity of hydrogen evolved in this reaction varies considerably according to the temperature and pressure under which it takes place, and the proportions of the materials used ; sometimes no gas is evolved, showing that the reaction takes place according to the second equation, and under no circumstances yet observed is the quantity of hydrogen given off exactly equivalent to the sodium dissolved, as it should be if the reaction took place entirely according to the first equation. It is most probable, therefore, that the two reactions generally take place together.

Ethylic acetosodacetate (Wanklyn's *sodium-triacetyl*) crystallizes in shining scales. When treated with acetic acid, it is converted into the compound $C_6H_{10}O_3$ or $CO{<}^{CH_3}_{CH_2-CO-OC_2H_5}$, which is the ethylic ether of aceto-acetic acid, $CO{<}^{CH_3}_{CH_2-CO-OH}$, derived from acetic acid; $CO{<}^{CH_3}_{OH}$, by substitution of acetyl, $CH_3.CO$, for one of the hydrogen-atoms in the methyl group. · This acid is not known in the free state ; when separated from its ether by boiling with alkalies or acids, it is resolved into acetone and carbon dioxide,

$$CO{<}^{CH_3}_{CH_2-COOH} = CO(CH_3)_2 + CO_2 .$$

The ethylic ether, $C_6H_{10}O_3$, (also called ethyl-diacetic acid), is a liquid having an odor of strawberries, a specific gravity of 1.03, and boiling at 180.8° C. (357.5° F.). By the action of sodium or of sodium ethylate, it is converted into ethylic acetosodacetate, the compound already mentioned as a direct product of the action of sodium on acetic ether :—

$$CO\big<{}^{CH_3}_{CH_2.CO_2C_2H_5} + NaOC_2H_5 = CO\big<{}^{CH_3}_{CHNa.CO_2C_2H_5} + C_2H_5OH .$$

The *potassium salt* is obtained in a similar manner. The *barium salt*, $(C_6H_9O_3)_2Ba$, obtained by dissolving the ether in baryta-water, is easily soluble in water. From its solution the corresponding salts of the heavy metals may be obtained by double decomposition.

Methylic Aceto-acetate, $C_5H_8O_3 = CO\big<{}^{CH_3}_{CH_2.CO_2CH_3}$, obtained by the action of sodium on methyl acetate, etc., is a colorless, pungent-smelling liquid, which boils at 170° C. (338° F.), and has a specific gravity of 1.037 at 9° C. (48.2° F.). It has a faint acid reaction, dissolves sparingly in water, freely in alcohol and ether; produces a dark-red coloration with ferric chloride; is resolved by boiling with hydrochloric acid or with alkalies, into acetone, carbon dioxide, and methyl alcohol. Its *sodium salt*, $C_5H_7O_3Na$, crystallizes in shining needles. The *barium salt*, $(C_5H_7O_3)_2Ba$, is obtained by dissolving the ether in baryta-water. Its aqueous solution mixed with cupric acetate yields the *copper salt* $(C_5H_7O_3)_2Cu + 2H_2O$ in green crystals, slightly soluble in water.

By treating the sodium derivatives of these aceto-acetic ethers with the iodides of alcohol-radicles, new ethers are obtained, in which the sodium of the original compound is replaced by an alcohol-radicle, *e. g.*,

$$CO\big<{}^{CH_3}_{CHNa-CO_2C_2H_5} + CH_3I = NaI + CO\big<{}^{CH_3}_{CH(CH_3)-CO_2C_2H_5}$$
Ethylic   Ethylic
acetosodacetate.   methylaceto-acetate.

In these last ethers an atom of hydrogen may also be replaced by sodium, by the action of metallic sodium or of sodium ethylate, producing compounds like

Ethylic Sodio-methylaceto-acetate, $CO\big<{}^{CH_3}_{CNa(CH_3)-CO_2C_2H_5}$; and in these again the sodium may be replaced by alcohol-radicles, yielding, for example,

Ethylic Methylethyl-aceto-acetate, $CO\big<{}^{CH_3}_{C(CH_3)(C_2H_5)-CO_2C_2H_5}$.

Lastly, these mono- and di-substituted aceto-acetic ethers, when heated alone, or better, with sodium ethylate, or with ethylic sodio-aceto-acetate, $C_6H_9NaO_3$, are resolved into the group C_2H_2O or $CH_2.CO$ (which, by polymerization, yields dehydracetic acid, $C_8H_8O_4$), and substituted acetic ethers, that is to say, ethers of the higher fatty acids; *e. g.*,

$$CO\big<{}^{CH_3}_{CH(CH_3)-CO_2C_2H_5} = CH_3-CH_2-CO_2C_2H_5 + CH_2CO$$
Ethylic Methylaceto-acetate.   Ethylic Methylacetate
           or Propionate.

$$CO\big<{}^{CH_3}_{C(CH_3)_2-CO_2C_2H_5} = CH(CH_3)_2-CO_2C_2H_5 + CH_2CO$$
Ethylic Dimethyl-   Ethylic Dimethyl-acetate
aceto-acetate.   or Isobutyrate.

This reaction explains the direct production of the ethers of the higher fatty acids by the simultaneous action of sodium and alcoholic-radicles on acetic ether (p. 714).

The substituted ethers of aceto-acetic acid, heated with aqueous alkalies, are decomposed in the same manner as ethylic aceto-acetate (p. 750), yielding various ketones ; *e. g.*,

$$CO<^{CH_2}_{CH(CH_3)-CO_2C_2H_5} + 2KOH = CO_3K_2 + C_2H_5OH + CO<^{CH_3}_{CH_2CH_3}$$

Methyl-acetone.

$$CO<^{CH_3}_{C(C_2H_5)_2-CO_2C_2H_5} + 2KOH = CO_3K_2 + C_2H_5OH + CO<^{CH_3}_{CH(C_2H_5)_2}$$

Diethyl-acetone.

The following table exhibits the boiling points and specific gravities of these substituted acetacetic ethers, and their reactions with ferric chloride :

	Boiling Point.	Spec. Gravity.	Reaction with Ferric Chloride.
$CO<^{CH_3}_{CH(CH_3)-CO_2CH_3}$	177° C. (350.6° F.)	1.020 at 9° C. (48.2° F.)	Violet-red.
$CO<^{CH_3}_{CH(C_2H_5)-CO_2CH_3}$	190° C. (374° F.)	0.995 at 14° C. (57.2° F.)	Violet-red.
$CO<^{CH_3}_{CH(CH_3)-CO_2C_2H_5}$	186° C. (366.8° F.)		
$CO<^{CH_3}_{C(CH_3)_2-CO_2C_2H_5}$	184° C. (363.2° F.)	0.991 at 16° C. (60.8° F.)	
$CO<^{CH_3}_{CH(C_2H_5)-CO_2C_2H_5}$	195° C. (383° F.)	0.983 at 16° C. (60.8° F.)	Blue.
$CO<^{CH_3}_{C(C_2H_5)-CO_2C_2H_5}$	210–212° C. (410–413° F.)	0.974 at 0° C.	

Isopropylated acetacetic ethers are obtained by treating the acetacetic ethers with sodium and isopropyl iodide.

Diatomic Acids.

These acids contain two hydroxyl groups, and are monobasic or bibasic, according as one or both of these hydroxyls belongs to a carboxyl group, COOH ; *e. g.*,

CH_2OH
|
COOH
Glycollic acid.
1-basic.

COOH
|
COOH
Oxalic acid.
2-basic.

Some of them are formed by oxidation (exchange of H_2 for O), from diatomic alcohols (glycols), containing the same number of carbon-atoms, those glycols which contain the group CH_2OH twice, yielding in this manner two diatomic acids, one mono- and the other bibasic, while those which contain this group only once, yield only one acid, which is monobasic; thus :—

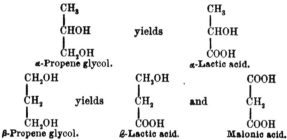

$$\begin{array}{c} CH_3 \\ | \\ CHOH \\ | \\ CH_2OH \\ \alpha\text{-Propene glycol.} \end{array} \quad \text{yields} \quad \begin{array}{c} CH_3 \\ | \\ CHOH \\ | \\ COOH \\ \alpha\text{-Lactic acid.} \end{array}$$

$$\begin{array}{c} CH_2OH \\ | \\ CH_2 \\ | \\ CH_2OH \\ \beta\text{-Propene glycol.} \end{array} \quad \text{yields} \quad \begin{array}{c} CH_2OH \\ | \\ CH_2 \\ | \\ COOH \\ \beta\text{-Lactic acid.} \end{array} \quad \text{and} \quad \begin{array}{c} COOH \\ | \\ CH_2 \\ | \\ COOH \\ \text{Malonic acid.} \end{array}$$

1. MONOBASIC ACIDS.

$$C_nH_{2n}O_3, \text{ or } C_nH_{2n} {\Large<}^{OH}_{COOH} \cdot$$

These acids are called l a c t i c a c i d s , after the most important member of the series, and o x y - f a t t y a c i d s , because they may be derived from the acids $C_nH_{2n}O_2$ by substitution of OH for H ; thus :—

$$\begin{array}{cc} CH_3.CO_2H & CH_2OH—CO_2H \\ \text{Acetic acid.} & \text{Oxyacetic acid.} \end{array}$$

They contain two hydroxyl groups, one a l o o h o l i c (connected with the radicle C_nH_{2n}), and having its hydrogen replaceable by monatomic alcohol-radicles, acid radicles, and alkali-metals ; the other, b a s i c (connected with CO), and having its hydrogen replaceable by metals in general to form salts. Each of them can, therefore, form three ethers, containing monatomic alcohol-radicles, but only one series of metallic salts, except with the alkali-metals, with each of which they can form two salts. Of the ethers, one is acid and the other two are neutral ; thus from glycollic acid, $CH_2OH.COOH$, are formed :—

$$\begin{array}{ccc} CH_2OCH_3 & CH_2OH & CH_2OCH_3 \\ | & | & | \\ COOH & COOCH_3 & COOCH_3 \\ \text{Methyl-glycollic} & \text{Monomethylic} & \text{Dimethylic} \\ \text{acid.} & \text{glycollate.} & \text{glycollate.} \end{array}$$

The known members of the series are the following :—

Carbonic acid . . . . $CH_2O_3 \ = \ CO{\Large<}^{OH}_{OH}$

Glycollic or Oxyacetic acid . $C_2H_4O_3 \ = \ CH_2{\Large<}^{OH}_{CO_2H}$

Lactic or Oxypropionic acid . $C_3H_6O_3 \ = \ C_2H_4{\Large<}^{OH}_{CO_2H}$

Oxybutyric acid . . . $C_4H_8O_3 \ = \ C_3H_6{\Large<}^{OH}_{CO_2H}$

Oxyvaleric acid . . . $C_5H_{10}O_3 \ = \ C_4H_8{\Large<}^{OH}_{CO_2H}$

Oxycaproic acid . . . $C_6H_{12}O_3 \ = \ C_5H_{10}{\Large<}^{OH}_{CO_2H}$

and others containing 7, 9, and 12 atoms of carbon.

32 *

· All those acids of the series which contain more than two atoms of carbon admit of isomeric modifications. Of these some are analogous in structure to the primary or secondary alcohols, and are called *normal* or *iso*-acids, according to the structure of the alcohol-radicles contained in them. Others are analogous in structure to the tertiary alcohols, and are formed from oxalic acid or its ethers by a series of transformations to be presently explained.

They are all monobasic, as already observed, except carbonic acid, which is bibasic, its two hydroxyl groups being similarly related to the group CO. This acid will be considered by itself.

The *normal* and *iso-lactic* acids are formed by the following general processes :—

1. By slow oxidation of the glycols in contact with platinum black, or by the action of dilute nitric acid ; *e. g.*,

$$H_3C{<}^{CH_2OH}_{CH_2OH} + O_2 = H_2O + H_3C{<}^{CH_2OH}_{COOH}$$

β-Propene glycol.　　　　　　　　　β-Lactic acid.

$$HO.HC{<}^{CH_3}_{CH_2OH} + O_2 = H_2O + HO.HC{<}^{CH_3}_{COOH}$$

α-Propene glycol.　　　　　　　　　α-Lactic acid.

2. By the action of moist silver oxide on the monochlorinated or monobrominated derivatives of the fatty acids (p. 717); *e. g.*,

$$CH_2Cl{-}CO_2H + AgOH = AgCl + CH_2OH{-}CO_2H$$

Chloracetic acid.　　　　　　　　　Glycollic acid.

3. From the glycollic chlorhydrins (p. 627), by converting them into cyanides by the action of potassium cyanide, and boiling the resulting cyanides with alkalies or acids :

$$H_2C{<}^{Cl}_{CH_2OH} + CNK = KCl + H_2C{<}^{CN}_{CH_2OH}$$

Chlorhydrin.　　　　　　　　　Cyanhydrin.

$$H_2C{<}^{CN}_{CH_2OH} + 2H_2O = NH_3 + H_2C{<}^{CH_2OH}_{COOH}$$

Cyanhydrin.　　　　　　　　　β-Lactic acid.

4. By heating an aldehyde or a ketone with hydrocyanic and hydrochloric acid, whereby a cyanhydrin is produced, and treating this compound with acids or alkalies as above :

$$CH_3{-}CHO + CNH = CH_3{-}CH{<}^{OH}_{CN}$$

Aldehyde.

$$CH_3{-}CH{<}^{OH}_{CN} + 2H_2O = NH_3 + CH_3{-}CH{<}^{OH}_{CO_2H}$$

α-Lactic acid.

5. By the action of nitrous acid on the amidated derivatives of the fatty acids :

$$H_2C{<}^{NH_2}_{CO_2H} + NO_2H = H_2O + N_2 + H_2C{<}^{OH}_{CO_2H}$$

Amidacetic acid.　　　　　　　　　Glycollic acid.

The *tertiary lactic acids* are represented by the general formula :

$${}^{C_nH_{2n}+1}_{C_nH_{2n}+1}{>}C{<}^{OH}_{COOH}\ .$$

They are obtained in the form of ethers by the action of the zinc-compound of an alcohol-radicle, C_nH_{2n+1}, on a neutral ether of oxalic acid containing a radicle of the same series, such as diethylic oxalate. The reaction consists in the replacement of an atom of oxygen in the oxalic ether by two equivalents of alcohol-radicle, and the simultaneous replacement of an equivalent of ethyl, methyl, etc., in the oxalic ether by an equivalent* of zinc, whereby an ether of zinc-diethoxalic acid, etc., is produced, which by certain obvious transformations may be converted into the required acid ; thus :

$$
\begin{array}{c}
\text{O}=\text{C—OCH}_3 \\
| \\
\text{O}=\text{C—OCH}_3 \\
\textbf{Dimethylic} \\
\textbf{oxalate.}
\end{array}
+ \; 2\text{Zn}'\text{C}_2\text{H}_5 \; = \; \text{Zn}'\text{OCH}_3 \; + \;
\begin{array}{c}
(\text{C}_2\text{H}_5)_2\text{C—OZn}' \\
| \\
\text{O}=\text{C—OCH}_3 \\
\textbf{Methylic} \\
\textbf{zinco-diethoxalate.}
\end{array}
$$

$$
\begin{array}{c}
(\text{C}_2\text{H}_5)_2\text{C—OZn}' \\
| \\
\text{O}=\text{C—OCH}_3 \\
\textbf{Methyl zinco-} \\
\textbf{diethoxalate.}
\end{array}
+ \; \text{HOH} \; = \; \text{Zn}'\text{OH} \; + \;
\begin{array}{c}
(\text{C}_2\text{H}_5)_2\text{C—OH} \\
| \\
\text{O}=\text{C—OCH}_3 \\
\textbf{Methylic} \\
\textbf{diethoxalate.}
\end{array}
$$

The methylic diethoxalate is easily decomposed by baryta-water, yielding methyl alcohol and barium diethoxalate :

$$
\begin{array}{c}
(\text{C}_2\text{H}_5)_2\text{C—OH} \\
| \\
\text{CO}_2\text{CH}_3
\end{array}
+ \; \text{Ba}'\text{OH} \; = \; (\text{CH}_3)\text{OH} \; + \;
\begin{array}{c}
(\text{C}_2\text{H}_5)_2\text{C—OH} \\
| \\
\text{CO}_2\text{Ba}'
\end{array}
$$

and this salt decomposed by sulphuric acid yields diethoxalic acid, $\begin{smallmatrix}C_2H_5 \\ C_2H_5\end{smallmatrix}\!\!>\!\!C\!\!<\begin{smallmatrix}OH \\ CO_2H\end{smallmatrix}$, isomeric with leucic acid.

In the first stage of the process, it is found better to use a mixture of ethyl iodide with metallic zinc, which produces zinc ethide, instead of the latter compound previously prepared. The other tertiary lactic acids are prepared in a similar manner.

Reactions.—1. All the acids of the series, $C_nH_{2n}O_3$, are reduced by *hydriodic acid* to the corresponding monatomic acids, $C_nH_{2n}O_2$, *e. g.*, lactic to propionic acid :

$$ C_3H_6O_3 \; + \; 2HI \; = \; H_2O \; + \; I_2 \; + \; C_3H_6O_2 . $$

By the action of *phosphorus pentachloride*, both their hydroxyl groups are replaced by chlorine ; thus glycollic acid yields glycollyl chloride, or chloracetyl chloride, $CH_2Cl.CO.Cl$:

$$
\text{CH}_2\!\!<\!\!\begin{smallmatrix}\text{OH} \\ \text{CO}_2\text{H}\end{smallmatrix}
+ \; 2\text{PCl}_5 \; = \; 2\text{POCl}_3 \; + \; 2\text{HCl} \; + \;
\text{CH}_2\!\!<\!\!\begin{smallmatrix}\text{Cl} \\ \text{COCl}\end{smallmatrix}
$$
Glycollic acid.   Glycollyl chloride.

The chlorine in the group COCl of the resulting chlorides is easily attacked by water and alcohols, forming acids or ethers : chloracetyl chloride, for example, yielding chloracetic acid, $CH_2Cl.CO.OH$, or its ethers.

2. The ethers of the tertiary lactic acids, when treated with PCl_3 or P_2O_5, are converted (by abstraction of H_2O) into ethers of the acrylic series

---

* To simplify the equations, the *equivalent* of zinc ($Zn'=32.5$) is used instead of the atom (65).

(p. 742), *e. g.*, ethylic dimethoxalate, $(CH_3)_2COH—CO_2C_2H_5$, into ethylic methylacrylate,

$$\begin{array}{c} H_2C \\ H_3C \end{array}\!\!\!\diagdown C—CO_2C_2H_5 \,.$$

3. The isolactic acids, when deprived of the elements of water by the action of heat or otherwise, are converted into **anhydrides**, which may be of three kinds. If the elements of 1 mol. H_2O, are removed partly from one and partly from another molecule of the acid, and the two residues unite, the result is an anhydride of the first order, differing in constitution and properties according as the two hydrogen-atoms are taken from the carboxyl-groups, or from the alcoholic hydroxyls; in the former case, the product is an **acid anhydride**; in the latter, an **alcoholic anhydride**, or **anhydro-acid**; thus:

$$H_2C\!\!\diagup^{OH}_{CO.OH} \quad\quad H_2C\!\!\diagup^{OH}_{CO}\!\!\diagdown O$$
$$H_2C\!\!\diagup^{CO.OH}_{OH} \quad - \quad H_2O \quad = \quad H_2C\!\!\diagup^{CO}_{OH}$$

Glycollic acid (2 mol.).   Glycollic anhydride.

$$H_2C\!\!\diagup^{CO.OH}_{OH} \quad\quad\quad H\ \diagup^{COOH}_{O}$$
$$H_2C\!\!\diagup^{OH}_{CO.OH} \quad - \quad H_2O \quad = \quad H_2C\!\!\diagdown_{COOH}$$

Diglicollic acid.

If, on the other hand, one molecule of water is removed from one molecule of the acid, the remainder constitutes an **anhydride of the second order**; *e. g.*:

$$\begin{array}{c} CH_2.OH \\ | \\ CO.OH \end{array} \quad - \quad H_2O \quad = \quad \begin{array}{c} CH_2 \\ | \quad\diagdown O \\ CO \diagup \end{array}$$

Glycollic acid.   Glycollide.

Glycollic Acid, $C_2H_4O_3 = CH_2\!\!\diagup^{OH}_{CO_2H}$.—This acid is produced in a variety of reactions, some of which have been already mentioned, viz., the oxidation of glycol by contact with platinum black or by treatment with dilute nitric acid; the decomposition of amidacetic acid (glycocine) by nitrous acid; the action of water or alkalies on bromacetic and chloracetic acid, or their salts (p. 718), *e. g.*, by boiling silver bromacetate with water:

$$C_2H_2BrAgO_2 \quad + \quad H_2O \quad = \quad AgBr \quad + \quad C_2H_4O_3.$$

It is also produced: *a.* By the action of alkalies on glyoxal and glyoxylic acid.

$$\underset{\text{Glyoxal.}}{C_2H_2O_2} \quad + \quad H_2O \quad = \quad \underset{\text{Glycollic acid.}}{C_2H_4O_3}$$

$$\underset{\substack{\text{Glyoxyllic} \\ \text{acid.}}}{2C_2H_2O_3} \quad + \quad H_2O \quad = \quad \underset{\substack{\text{Oxalic} \\ \text{acid.}}}{C_2H_2O_4} \quad + \quad \underset{\substack{\text{Glycollic} \\ \text{acid.}}}{C_2H_4O_3}$$

ß. Together with glyoxal, glyoxylic acid, and other products, by the action of nitric acid upon alcohol.

γ. By the action of nascent hydrogen (evolved by zinc and sulphuric acid) upon oxalic acid:

$$C_2H_2O_4 \quad + \quad 2H_2 \quad = \quad H_2O \quad + \quad C_2H_4O_3.$$

Glycollic acid differs somewhat in its properties, according to the manner in which it is prepared, being sometimes syrupy and uncrystallizable, sometimes separating from its solution in ether in large regular crystals. It has a very sour taste, dissolves easily in water, alcohol, and ether ; melts at 78° or 79° C. (172.4°–174.2° F.); begins to boil at 100°; decomposes when heated to above 150° C. (302° F.). All the glycollates are more or less soluble and crystallizable.

The *calcium salt*, $(C_2H_3O_3)_2Ca + 2H_2O$, is slightly soluble in cold water, and crystallizes in needles. The *silver salt*, $C_2H_3O_3Ag + \frac{1}{2}H_2O$, is also sparingly soluble. The *ethylic ether*, $C_2H_3O_3.C_2H_5$, is a liquid having a specific gravity of 1.03, and boiling at 150° C. (302° F.).

Acid and Alcoholic Derivatives of Glycollic Acid.—The alcoholic hydrogen of glycollic acid may be replaced either by acid or by alcoholic radicles. The acid derivatives are formed : *a.* By heating glycollic acid with monatomic acids :

$$CH_2OH.CO_2H \; + \; C_2H_3O.OH \; = \; H_2O \; + \; CH_2\!\!<\!\!{}^{OC_2H_3}_{CO_2H}$$
Glycollic acid.   Acetic acid.   Acetoglycollic acid.

ß. By the action of the alkali-salts of monatomic acids on monochloracetic acid :

$$CH_2Cl\!-\!CO_2H \; + \; C_7H_5O.OK \; = \; KCl \; + \; CH_2\!\!<\!\!{}^{OC_7H_5O}_{CO_2H}$$
Chloracetic acid.   Potassium benzoate.   Benzoglycollic acid.

The alcoholic derivatives are formed by the action of sodium alcoholates on chloracetic acid :

$$CH_2Cl\!-\!CO_2Na \; + \; C_2H_5ONa \; = \; NaCl \; + \; CH_2\!\!<\!\!{}^{OC_2H_5}_{CO_2Na}$$
Chloracetate.   Ethylate.   Ethyl-glycollate.

Methyl-glycollic acid, $CH_3OCH_3.CO_2H$, boils at 198° C. (384.4° F.) ; *ethyl-glycollic acid*, $CH_2OC_2H_5.CO_2H$, at 206° C. (402.8° F.). Both are permanent in the air, and are not decomposed by boiling with alkalies.

Anhydrides. — Diglycollic acid, $C_4H_6O_5 = 2C_2H_4O_3 - H_2O =$ $O\!\!<\!\!{}^{CH_2-COOH}_{CH_2-COOH}$, also called *Paramalic acid.*—This acid, isomeric with malic acid, and related to glycollic acid in the same manner as diethenic alcohol to glycol, is produced by the dehydration of glycollic acid, and by the oxidation of diethenic or triethenic alcohol. It is also formed in the preparation of glycollic acid by heating sodium chloracetate with caustic soda, which in fact is the process by which it was first obtained :

$$C_2H_3ClO_2 \; + \; 2NaHO \; = \; NaCl \; + \; H_2O \; + \; C_2H_3NaO_3$$
Chloracetic acid.   Sodium glycollate.

$$C_2H_3ClO_2 \; + \; C_2H_3NaO_3 \; = \; NaCl \; + \; C_4H_6O_5$$
Chloracetic acid.   Sodium. glycollate.   Diglycollic acid.

Diglycollic acid crystallizes in large rhombic prisms, melting at 150° C. (302° F.), and distilling with slight decomposition at 250° C. (482° F.); easily soluble in water and in alcohol ; not decomposed by boiling with alkalies. It is bibasic, forming with univalent metals, acid salts containing $C_4H_5M'O_5$, and normal salts, $C_4H_4M_2O_5$; with bivalent metals it forms only normal salts, $C_4H_4M''O_5$. The *calcium salt* is very soluble.

Glycollic Anhydride, $C_4H_6O_5 = O<^{CO-CH_2OH}_{CO-CH_2OH}$, obtained by exposing the acid for a few days to the vapors of sulphuric anhydride, is a lustreless powder, melting at 128°–130° C. (262.4°–266° F.); insoluble in alcohol, ether, and cold water; boiling water reconverts it into glycollic acid and dissolves it.

Glycollide, $C_2H_2O_2 = O<^{CH_2}_{\underset{CO}{|}}$, is formed by heating glycollic acid to 150° C. (302° F.). It is a white amorphous powder, melting at 180° C. (356° F.), insoluble in cold water, reconverted into glycollic acid by boiling with water. Heated with ammonia it yields glycollamide, $CH_2OH-CONH_2$.

Lactic or Oxypropionic Acids, $C_3H_6O_3 = C_2H_4<^{OH}_{CO.OH}$ ·

—Of these acids there are two isomeric modifications, viz. :—

$$\underset{\substack{\alpha\text{-Lactic or} \\ \text{Ethidene-lactic acid.}}}{\overset{CH_3}{\underset{|}{\overset{|}{CHOH}}}\;\;\overset{|}{\underset{|}{COOH}}}
\qquad
\underset{\substack{\beta\text{-Lactic or} \\ \text{Ethene-lactic acid.}}}{\overset{CH_2OH}{\underset{|}{\overset{|}{CH_2}}}\;\;\overset{|}{\underset{|}{CO.OH}}}$$

1. **Ethidene-lactic**, or **Isolactic Acid**, $HOHC<^{CH_3}_{CO_2H}$, is the ordinary lactic acid produced by a peculiar fermentation of sugar and other carbohydrates, hence called *fermentation lactic acid*. It is also produced by the first, second, fourth, and fifth of the general reactions already mentioned (p. 754), viz., by oxidation of a-propene-glycol; by the action of moist silver oxide on a-chloropropionic acid; by the action of hydrocyanic and hydrochloric acid on aldehyde; and by that of nitrous acid on a-amidopropionic acid (alanine); and lastly, by the action of nascent hydrogen on pyruvic acid :—

$$O<^{CH_2}_{\underset{CO_2H}{\underset{|}{CH}}} \;+\; H_2 \;=\; \overset{CH_3}{\underset{CO_2H}{\underset{|}{CH.OH}}}$$

Preparation of Ordinary Lactic Acid by Fermentation.—Various kinds of sugar and dextrin, when subjected to the action of particular ferments, are converted into lactic acid, the change consisting in a resolution of the molecule, preceded in some cases by the assumption of the elements of water :—

$$\underset{\text{Glucose.}}{C_6H_{12}O_6} \qquad = \qquad \underset{\text{Lactic acid.}}{2C_3H_6O_3}$$

$$\underset{\text{Milk sugar.}}{C_{12}H_{22}O_{11}} \;+\; H_2O \;=\; \underset{\text{Lactic acid.}}{4C_3H_6O_3}$$

This lactous fermentation requires a temperature between 20° and 40° C. (68° and 104° F.), and the presence of water and certain ferments—viz., albuminous substances in a peculiar state of decomposition, such as casein, glutin, or animal membranes, especially the coating of the stomach of the calf (rennet), or of the dog, or bladder. According to Pasteur and others, it depends upon the presence of a peculiar fungus, *Penicillium glaucum* (p.

596). The following is a good method for preparing the acid in considerable quantity : 2 gallons of milk are mixed with 6 pounds of raw sugar, 12 pints of water, 8 ounces of putrid cheese, and 4 pounds of chalk, which should be mixed up to a creamy consistence with some of the liquid. This mixture is exposed in a loosely covered jar to a temperature of about 30° C. (86° F.), with occasional stirring. The use of the chalk is to neutralize the lactic acid, which would otherwise coagulate the casein, render it insoluble, and thereby put a stop to the process. At the end of two or three weeks it will be found converted into a semi-solid mass of calcium lactate, which may be drained, pressed, and purified by re-crystallization from water. The lactate may be decomposed by the necessary quantity of pure oxalic acid, the filtered liquor neutralized with zinc carbonate, and, after a second filtration, evaporated until the zinc-salt crystallizes out on cooling. If, in the first part of the process, the solid calcium lactate be not removed at the proper time from the fermenting liquid, it will gradually redissolve and disappear, being converted into soluble butyrate (p. 731). An important modification of this process consists in employing commercial zinc-white instead of powdered chalk, which yields at once difficultly soluble zinc lactate, easily purified by recrystallization. The zinc lactate may, lastly, be redissolved in water and decomposed by sulphuretted hydrogen, in order to obtain the free acid. Together with the lactic acid a certain quantity of mannite is invariably formed. This is separated by agitating the concentrated aqueous solution with ether, in which the lactic acid alone is soluble.

Lactic acid may be extracted from a great variety of liquids containing decomposing organic matter, as *sauerkraut*, a preparation of white cabbage, the sour liquor of the starch-maker, etc.

Solution of lactic acid may be concentrated in the vacuum of the air-pump, over a surface of oil of vitriol, until it appears as a colorless, syrupy liquid, of sp. gr. 1.215. It has an intensely sour taste and acid reaction : it is hygroscopic, and very soluble in water, alcohol, and ether.

By prolonged evaporation over sulphuric acid it is partly resolved into water and anhydride, and by distillation it splits up into lactide, $C_3H_4O_2$, aldehyde, carbon monoxide, and water.

By oxidation with chromic acid mixture lactic acid yields formic and acetic acids. Boiled with dilute *nitric acid*, or with dioxide of lead or barium, it is converted into oxalic acid. Distilled with dilute sulphuric acid and dioxide of lead or manganese, it yields a large quantity of aldehyde, together with carbon dioxide. Hydriodic acid, or a mixture of phosphorus tetriodide and water, reduces it to propionic acid, with liberation of iodine :

$$C_3H_6O_3 + 2HI = C_3H_6O_2 + H_2O + I_2 .$$

By heating with hydrobromic acid it is converted into *α*-bromo-propionic acid :

$$CH_3—CH(OH)—CO_2H + HBr = H_2O + CH_3—CHBr—CO_2H.$$

Lactates.—The best defined of these salts are represented by the formulæ, $M'C_3H_5O_3$ and $M''(C_3H_5O_3)_2$. Barium and calcium also form acid lactates, *e. g.*, $Ca(C_3H_5O_3)_2.2C_3H_6O_3$. The lactates are, for the most part, sparingly soluble in cold water, and effloresce rapidly from their solutions : they are all insoluble in ether. When heated with excess of strong sulphuric acid, they give off a large quantity of pure carbon monoxide.

Sodium lactate, $C_3H_5O_3Na = CH_3—CH{<}^{OH}_{CO_2Na}$, obtained by neutralizing the acid with sodium carbonate, is an amorphous mass, which, when heated with metallic sodium is converted into disodic lactate, $CH_3 \big\langle {}^{ONa}_{CO_2Na}$, the alcohol hydrogen being likewise replaced by sodium.

Calcium lactate, $(C_3H_5O_3)_2Ca + 5H_2O$, is obtained in the fermentation process above described, or by boiling aqueous lactic acid with calcium carbonate. It dissolves in 9.5 parts of water at ordinary temperatures.— *Zinc lactate*, $(C_3H_5O_3)_2Zn + 3H_2O$, gives off its water quickly at 100°, dissolves in 6 parts of boiling water, in 5.8 parts of cold water, and is nearly insoluble in alcohol.—*Ferrous lactate* is precipitated in small yellowish needles on mixing ammonium lactate with ferrous chloride or sulphate.— *Ferric lactate* is a brown deliquescent mass.

Lactic Ethers.—Lactic acid, like the other members of the group, can form three different ethers containing the same univalent alcohol-radicle, according as the alcoholic or the basic hydrogen-atom, or both, are replaced ; thus :

$$
\begin{array}{llll}
C_2H_4OH & C_2H_4OC_2H_5 & C_2H_4OH & C_2H_4OC_2H_5 \\
| & | & | & | \\
COOH & COOH & COOC_2H_5 & COOC_2H_5 \\
\text{Lactic acid.} & \text{Ethyl-lactic} & \text{Monethylic} & \text{Diethylic lactate,} \\
 & \text{. acid.} & \text{lactate.} & \text{or ethylic ethyl-} \\
 & & & \text{lactate.}
\end{array}
$$

Monethylic lactate, $C_3H_5O_3.C_2H_5$, is produced by distilling potassium or sodium lactate with potassium ethylsulphate. It is a syrupy liquid, boiling at 176° C. (348.8° F.). Potassium dissolves in it, wit evolution of hydrogen, forming ethylic potassiolactate, $\begin{array}{l} C_2H_4OK h \\ | \\ COOC_2H_5 \end{array}$.—*Ethyl-lactic* acid, $C_2H_4(OC_2H_5)$—CO_2H, is obtained as a potassium or calcium salt by decomposing diethylic lactate with potash or milk of lime. When separated from these salts by sulphuric acid, it forms a viscid liquid, boiling with partial decomposition between 195° and 198°.—*Diethylic lactate*, $C_2H_4(OC_2H_5)$—$CO_2C_2H_5$, is produced by the action of ethyl-iodide on ethyl potassiolactate, and by that of sodium ethylate on ethyl chloropropionate :

$$
\begin{array}{ccccccc}
C_2H_4Cl\text{—}COC_2H & + & NaOC_2H_5 & = & NaCl & + & C_2H_4(OC_2H_5)\text{—}CO_2C_2H_5 \\
\text{Ethyl chloropro-}^5 & & \text{Sodium} & & & & \text{Diethylic} \\
\text{pionate.} & & \text{ethylate.} & & & & \text{lactate.}
\end{array}
$$

Methyl-lactic acid, $C_2H_5(CH_3)O_3$, and its zinc and silver salts have also been obtained.

The alcoholic hydrogen of lactic acid may also be replaced by acid radicles, forming such compounds as acetolactic acid, $C_2H_4 \big\langle {}^{OC_2H_3O}_{CO_2H}$.

Lactyl Chloride, $C_3H_4OCl_2$, or *a-Chloropropionyl Chloride*, C_2H_4Cl—$COCl$, is obtained, together with phosphorus oxychloride, by gently heating a mixture of calcium lactate with phosphorus pentachloride. It is a colorless liquid, boiling above 100°, and decomposed by water, forming hydrochloric and chloropropionic acids. With alcohol it forms ethylic a-chloropropionate. By heating with alkalies it is converted into lactic acid.

Lactic Anhydrides.—1. Dilactic acid, $C_6H_{10}O_5 = 2C_3H_6O_3 - H_2O =$ $\begin{array}{l} H_3C\text{—}CH\text{—}CO_2H \\ \quad\quad O \\ H_3C\text{—}CH\text{—}CO_2H \end{array}$, formed by heating lactic acid to 130°–200° C. (266°– 392° F.), is an amorphous mass, very bitter, and nearly insoluble in water, but reconverted into lactic acid by prolonged boiling with water.

2. L a c t i d e , $C_3H_4O_2 = HC—CH_3$, the second anhydride of lactic acid,

$$O—CO$$

is obtained by distillation of lactic or dilactic acid, and may be obtained pure by evaporating the distillate, washing the residue with cold alcohol, and crystallizing it from hot alcohol. Lactide thus purified crystallizes in rhombic plates which melt at 107° C. (224.6^O F.) ; it dissolves gradually in water.

Paralactic or *Sarcolactic acid* is a modification of lactic acid, existing in the animal organism, especially in muscular flesh, from which it may be extracted by cold water or dilute alcohol. It is most easily prepared from Liebig's extract of meat. It agrees in all its reactions with fermentation lactic acid, and must therefore have the same chemical structure—that is to say, it must be an ethidene-lactic acid—but it differs from ordinary lactic acid in its relation to polarized light, inasmuch as it turns the plane of polarization to the left, whereas ordinary lactic acid is optically inactive. Hence it is inferred that the two modifications, though chemically identical, differ from one another in physical structure; in other words, that the difference between them consists, not in the arrangement of the atoms within the molecules, but in the arrangement or modification of the molecules amongst themselves. There are other cases of difference in physical character coexisting with chemical identity, which are attributed to a similar difference of physical structure, as in the several modifications of turpentine oil.

Some of the paralactates differ from the ordinary lactates in solubility and other characters ; thus :

	Lactate.	Paralactate.
Calcium Salts.	$(C_3H_5O_3)_2Ca + 5H_2O$; nodular groups of needles, soluble in 9.5 parts of cold water.	$+4H_2O$; soluble in 12 parts of cold water.
Zinc Salts.	$(C_3H_5O_3)_2Zn + 3H_2O$; slender needles, soluble in 58 parts of cold water.	$+2H_2O$; thick shining prisms, soluble in 17 parts of water at 15° C. (59° F.).

Paralactic acid heated to 130° C. (266° F.) yields dilactic acid, convertible by water into ordinary lactic acid.

2. Ethene-lactic or **Hydracrylic Acid,** $\begin{matrix} CH_2—OH \\ | \\ CH_2—CO_2H \end{matrix}$.—This acid is produced by heating β-iodopropionic acid with moist silver oxide ; $CH_2I—CH_2—CO_2H + AgOH = AgI + CH_2OH—CH_2—CO_2H$. It is a thick, uncrystallizable syrup, which, when heated alone or boiled with sulphuric acid diluted with an equal weight of water, is converted by dehydration into acrylic acid :

$$\begin{matrix} CH_2—OH \\ | \\ CH_2—CO_2H \end{matrix} \quad - \quad H_2O \quad = \quad \begin{matrix} CH_2 \\ || \\ CH—CO_2H \end{matrix} .$$

By heating with hydriodic acid, it is reconverted into β-iodopropionic acid; by oxidation with chromic acid mixture, or nitric acid, it yields oxalic acid and carbon dioxide. When heated with silver oxide it is converted into carbacetoxylic acid, $C_3H_4O_4$.

The metallic hydracrylates are crystallizable. The *sodium salt*, $C_3H_5O_3Na$, crystallizes in flat prisms, which melt without loss of weight at 142°–148° C. (287.6°–298.4° F.), but give off water at 150° C. (302° F.), leaving sodium acrylate, $C_3H_3O_2Na$. The *calcium salt* $(C_3H_5O_3)_2Ca + 2H_2O$, forms large rhombic prisms, which give off their water of crystallization at 100°. The dehydrated salt unites without alteration at 140°–145° C. (284–293° F.), but at 190° C. (374° F.) it gives off water, and is converted into acrylate $(C_3H_5O_3)_2Ca - 2H_2O = (C_3H_3O_2)_2Ca$. The *zinc salt*, $(C_3H_5O_3)_2Zn + 4H_2O$, crystallizes from a moderately strong solution in large shining prisms, soluble in an equal weight of water at 15° C. (59° F.).*

Oxybutyric Acids, $C_4H_8O_3 = C_3H_6\underset{CO_2H}{\overset{OH}{<}}$.—Five of these acids are theoretically possible, and four are known.

1. *α-Oxybutyric acid*, $CH_3—CH_2—CHOH—CO_2H$, is formed by heating α-bromobutyric acid with moist silver oxide, or with aqueous potash. It is crystalline, very deliquescent, and melts at 43°–44° C. (109.4°–111.2° F.). Its salts are crystalline, the zinc-salt, $(C_4H_7O_3)_2Zn$, forming white laminæ, slightly soluble in cold water.

2. *β-Oxybutyric acid*, $CH_3—CHOH—CH_2—CO_2H$, is produced by boiling β-bromobutyric acid with silver oxide: by the action of sodium amalgam on ethylic aceto-acetate:

$$CO\underset{CH_2.CO_2.C_2H_5}{\overset{CH_3}{<}} + H_2 = CHOH\underset{CH_2.CO_2C_2H_5}{\overset{CH_3}{<}}$$

by the oxidation of aldol, $CH_3—CHOH—CH_2—CHO$ (p. 675); and apparently also by heating propene-chlorhydrin, $CH_3.CHOH.CH_2Cl$, with potassium cyanide, whereby it is converted into the corresponding cyanhydrin, and saponifying the latter with potash. The free acid is an uncrystallizable syrup; the calcium salt, $(C_4H_7O_3)_2Ca$, crystallizes with difficulty. The acid obtained from propene-chlorhydrin is resolved by distillation into water and solid crotonic acid, $CH_3—CH=CH—CO_2H$.

* According to Wislicenus (Liebig's *Annalen*, clxv. 6), hydracrylic acid prepared from β-iodopropionic acid is not identical with ethene-lactic acid prepared by combining ethene with carbonyl chloride (phosgene), and decomposing the resulting compound with water:

$$\underset{CH_2}{\overset{CH_2}{\|}} + COCl_2 = \underset{CH_2\,COCl}{\overset{CH_2Cl}{|}}$$

$$\text{and} \quad \underset{CH_2.COCl}{\overset{CH_2Cl}{|}} + 2H_2O = 2HCl + \underset{CH_2.COOH}{\overset{CH_2.OH}{|}}$$

The ethene-lactic acid thus obtained is converted by oxidation with chromic acid or nitric acid into malonic acid, $COOH—CH_2—COOH$, whereas hydracrylic acid, as above stated, is resolved thereby into oxalic acid and carbon dioxide. Moreover, the salts of hydracrylic acid are crystalline, whereas those of ethene-lactic acid are amorphous. To account for these differences, Wislicenus assigns to hydracrylic acid the formula, $CH_2OH—COH—OHOH$, and corresponding formulæ to

$$|—O—|$$

β-iodopropionic acid and acrylic acid. These formulæ, however, are somewhat improbable, as they do not contain the group COOH; moreover, the formation of ethene-lactic acid from ethene in the manner described by Wislicenus does not appear to be well established. (See Watts's Dictionary of Chemistry, 2d Suppl., 718.)

3. γ-*Oxybutyric acid*, or *Normal Oxybutyric acid*, CH_2OH—CH_2—CH_2—CO_2H, is prepared from succinic acid, CHO—CH_2—CH_2—CHO.

4. *Oxyisobutyric acid*, $\dfrac{H_3C}{H_3C}{>}COH$—CO_2H, is produced :

a. By the action of hydrocyanic and hydrochloric acids on acetone; hence called acetonic acid:

$$\dfrac{H_3C}{H_3C}{>}CO \quad + \quad HCN \quad = \quad \dfrac{H_3C}{H_3C}{>}COH\text{—}CN .$$

$$\dfrac{H_3C}{H_3C}{>}COH\text{—}CN \quad + \quad 2H_2O \quad = \quad NH_3 \quad + \quad \dfrac{H_3C}{H_3C}{>}COH\text{—}COOH.$$

β. From ethyl oxalate by the action of methyl iodide and zinc (p. 742) : hence called d i m e t h y l o x a l i c· acid. γ. From bromisobutyric acid, $(CH_3)_2CBr$—CO_2H, by boiling with baryta-water. δ. By oxidation of isopentene glycol (p. 636).

Oxyisobutyric acid crystallizes in slender prisms, soluble in water and in ether. It sublimes at 50º C. (122º F.), melts at 79º C. (174.2º F.), and distils at 212º C. (413.6º F.). Its ethylic ether, treated with phosphorus trichloride, is converted into ethylic methylacrylate :

$$\dfrac{H_3C}{H_3C}{>}COH\text{—}CO_2C_2H_5 \quad - \quad H_2O \quad = \quad \dfrac{H_2C}{H_3C}{>}C\text{—}CO_2C_2H_5.$$

By oxidation with chromic acid mixture, oxyisobutyric acid is resolved into carbon dioxide and acetone. Its *barium salt*, $(C_4H_7O_3)_2Ba$, forms shining needles, easily soluble in water. The *zinc salt*, $(C_4H_7O_3)_2Zn + 2H_2O$, crystallizes in shining six-sided prisms or tables, sparingly soluble in cold water.

Oxyvaleric Acids, $C_5H_{10}O_3 = C_4H_8{<}^{OH}_{CO_2H}$.—1. *a - Oxyisovaleric* or *Isopropyl-hydroxalic acid*, $\dfrac{H_3C}{H_3C}{>}CH\text{—}HC{<}^{OH}_{CO_2H}$.—This acid, prepared from the corresponding bromovaleric acid, forms large tabular crystals, easily soluble in water, melting at 80º C. (176º F.), and volatilizing at about 100º. When oxidized with chromic acid mixture, it yields carbon dioxide and isobutyric acid.

The ethylic ether of this acid is formed, together with that of oxyisocaprylic acid (*infra*), by the action of isopropyl iodide and zinc on ethylic oxalate, and may be obtained by evaporating the potash-solution used in the preparation of oxyisocaprylic acid, acidulating with sulphuric acid, agitating with ether and evaporating the ethereal solution. The oxyisovaleric acid then remains as a thick liquid which solidifies in the exsiccator.

2. *Methyl-ethyloxalic acid*, $\dfrac{H_3C}{H_5C_2}{>}C{<}^{OH}_{CO_2H}$, obtained by the action of a mixture of methylic and ethylic iodide and zinc on ethyl oxalate, forms white crystals, easily soluble in water, and melting at 63º C. (145.4º F.). Its ethylic ether is soluble in water, boils at 165º C. (329º F.), and is converted by phosphorus trichloride into the ethylic ether of m e t h y l - c r o t o n i c acid.

Oxycaproic Acids, $C_6H_{12}O_3 = C_5H_{10}{<}^{OH}_{CO_2H}$.—1. *Leucic acid*, probably *a-Oxycaproic acid*, C_4H_9—$CHOH$—CO_2H, is produced by the action of nitrous acid on leucine or amidocaproic acid (see AMIDES). It forms needles or monoclinic prisms, soluble in water, alcohol, and ether, melting at about

73º C. (163.4º F.), and volatilizing at 100º. When heated for some time at that temperature, it gives off water, and leaves a syrupy oxide or anhydride. It forms crystallizable salts analogous to the lactates ; the zinc salt crystallizes in shining scales.

2. *Diethyloxalic acid*, $(C_2H_5)_2C{<}^{OH}_{CO_2H}$, formed by the action of ethyl iodide and zinc an ethylic oxalate, is crystalline, easily soluble in water and ether, melts at 74.5º C. (166.1º F.), and sublimes at about 50º C. (122º F.). Its methylic ether boils at 165º C. (320º F.), and is converted by the action of phosphorus trichloride into methylic ethylcrotonate (p. 746). The free acid, oxidized by chromic acid mixture, yields diethyl-ketone, $CO(C_2H_5)$.

Oxyisoheptoic or Amylhydroxalic Acid, $C_7H_{14}O_3 =$

$^{C_5H_{11}}_{H}{>}C{<}^{OH}_{CO_2H}$, and **Diamyl-oxalic Acid**, $C_{12}H_{24}O_3 =$

$^{C_5H_{11}}_{C_5H_{11}}{>}C{<}^{OH}_{CO_2H}$, are formed by the action of zinc and isopentyl iodide on ethylic oxalate. The former is a thick syrup ; its ethylic ether boils at 203º C. (397.4º F.). The latter crystallizes in silky needles, is nearly insoluble in water, and melts at 122º C. (251.6º F.). Its ethylic ether boils at 262º C. (503.6º F.).

Oxyisocaprylic or Di-isopropyloxalic Acid,

$C_8H_{16}O_3 = (C_3H_7)_2C{<}^{OH}_{CO_2H}$, obtained by the action of isopropyl iodide and zinc on ethylic oxalate, forms slender needles, slightly soluble in water, melts at 110º–111º C.(230º–231.8º F.), and sublimes at a moderate heat.

Carbonic Acid, $CH_2O_3 = CO{<}^{OH}_{OH}$.—This acid belongs to the lactic series, so far as its constitution is concerned, being derived from the unknown methene glycol, $CH_2{<}^{OH}_{OH}$, by substitution of O for H_2 ; but it differs from all the other acids of the series in being bibasic, both the hydroxyl groups contained in it being immediately connected with an atom of oxygen, so that either of the hydrogen-atoms may be regarded as belonging to the group CO_2H.

Carbonic acid itself, or hydrogen carbonate, is not known, inasmuch as when a metallic carbonate is decomposed by a stronger acid, the hydrogen carbonate, CH_2O_3, always splits up into water and carbon dioxide, which escapes as gas. The corresponding sulphur compound, CH_2S_3, is, however, obtained as an oily liquid when a metallic thio-carbonate is decomposed by an acid.

With the alkali-metals carbonic acid forms acid and normal or neutral salts, according as one or both of the hydrogen-atoms are replaced ; *e. g.*,

Acid sodium carbonate, $CHNaO_3$, or $CO{<}^{OH}_{ONa}$

Normal sodium carbonate, CNa_2O_3, or $CO(ONa)_2$.

With dyad-metals, carbonic acid forms only normal salts, $CM''O_3$, and basic salts ; the so-called acid carbonates of barium, calcium, etc., are known only in solution, and are, in fact, merely solutions of neutral carbonates in aqueous carbonic acid, which give off carbon dioxide on boiling. The basic carbonates of dyad metals may be viewed as compounds of normal carbonates with metallic oxides or hydrates ; for example, slaked lime, produced by exposing quicklime to moist air, has the composition of a dicalcic carbonate, $CaO.CaCO_3.Aq.$; and native green copper carbonate, or malachite, consists of $CuO.CuCO_3 + H_2O$. These basic carbonates may, however, be viewed in another way, namely, as derived from a tetratomic carbonic acid, or ortho-carbonic acid, CH_4O_4, or $C(OH)_4$, analogous to methane and carbon tetrachloride ; thus, dicalcic carbonate $= CCa_2O_4 + H_2O$; malachite $= CCu_2O_4 + H_2O$.

With metals of higher atomicity, carbonic acid does not form definite salts.

CARBONIC ETHERS.—The only carbonic ethers known are those in which the two hydrogen-atoms of carbonic acid are replaced either by two equivalents of a monad alcohol-radicle, or by one equivalent of a monad alcohol-radicle and one equivalent of metal.

Ethyl carbonate, $(C_2H_5)_2CO_3$, is formed by the action of ethyl iodide on silver carbonate :—

$$Ag_2CO_3 \; + \; 2C_2H_5I \; = \; 2AgI \; + \; (C_2H_5)_2CO_3 \; ;$$

also by the action of potassium or sodium on ethyl oxalate, $(C_2H_5)_2C_2O_4$; this reaction is not quite understood, but it amounts to the removal of carbon monoxide, or carbonyl, CO, from the oxalic ether. Fragments of potassium or sodium are dropped into oxalic ether as long as gas is disengaged : the brown pasty product is then mixed with water and distilled. The carbonic ether is found floating upon the surface of the water of the receiver as a colorless, limpid liquid of aromatic odor and burning taste. It boils at 125° C. (257° F.), and is decomposed by an alcoholic solution of potash into potassium carbonate and alcohol. By chlorine in diffused daylight it is converted into dichlorethyl carbonate, $(C_2H_3Cl_2)_2CO_3$, and in sunshine into pentachlorethyl carbonate, $(C_2Cl_5)_2CO_3$.

Ethyl-potassium carbonate, $(C_2H_5)KCO_3$, is produced by passing carbonic acid gas into a cooled solution of potassium hydroxide in absolute alcohol :

$$C_2H_6O \; + \; KHO \; + \; CO_2 \; = \; H_2O \; + \; (C_2H_5)KCO_3 \; .$$

It is a white nacreous salt, decomposed by water into potassium carbonate and alcohol.

Ethyl-methyl carbonate, $(C_2H_5)(CH_3)CO_3$, is obtained by distilling a mixture of ethyl-potassium sulphate and methyl-potassium carbonate :—

$$(C_2H_5)K.SO_4 \; + \; (CH_3)K.CO_3 \; = \; K_2SO_4 \; + \; (C_2H_5)(CH_3).CO_3 \; .$$

Methyl-barium carbonate, $(CH_3)_2Ba(CO_3)_2$, is obtained as a white precipitate by passing carbonic acid gas into a solution of baryta in methyl alcohol.

Carbonates of *butyl*, *amyl*, and *allyl*, analogous in composition to ethyl carbonate, have also been obtained. *Phenyl-hydrogen carbonate*, or *acid phenyl carbonate*, $(C_6H_5)HCO_3$, is identical with salicylic acid, which will be described further on.

Ethyl orthocarbonate, $C(OC_2H_5)_4$, is produced by heating a mixture of chloropicrin (trichloro-nitromethane) with absolute alcohol and sodium :

$$\underset{\text{Chloropicrin.}}{C(NO_2)Cl_3} \; + \; \underset{\substack{\text{Sodium} \\ \text{ethylate.}}}{4NaOC_2H_5} \; = \; \underset{\substack{\text{Sodium} \\ \text{chloride.}}}{3NaCl} \; + \; \underset{\substack{\text{Sodium} \\ \text{nitrite.}}}{NaNO_2} \; + \; \underset{\substack{\text{Ethyl ortho-} \\ \text{carbonate.}}}{C(OC_2H_5)_4}$$

It is a colorless oil, boiling at 158°–159° C. (316.4°–318.2° F.).

THIOCARBONIC or SULPHOCARBONIC ETHERS.—These are bodies having the composition of carbonic ethers in which the oxygen is replaced, wholly or partly, by sulphur. The following table exhibits their names and formulæ, the ethyl and ethene compounds being taken as examples:

Ethyl-monothiocarbonic acid	$(C_2H_5)H.CO_2S.$
Diethylic monothiocarbonate	$(C_2H_5)_2.CO_2S.$
Ethyl-dithiocarbonic or Xanthic acid . . .	$(C_2H_5)H.COS_2.$
Diethylic dithiocarbonate	$(C_2H_5)_2.COS_2.$
Ethyl-trithiocarbonic acid	$(C_2H_5)H.CS_3.$
Diethylic trithiocarbonate	$(C_2H_5)_2.CS_3.$
Ethene dithiocarbonate	$(C_2H_4).COS_2.$
Ethene trithiocarbonate	$(C_2H_4)CS_3.$

The metallic salts of the acid thiocarbonic ethers are produced in the same manner as those of the carbonic ethers: thus carbon dioxide unites with potassium sulphethylate (mercaptide), to form potassium ethylmonothiocarbonate, just as it unites with potassium ethylate to form the ethylcarbonate; and, in like manner, carbon bisulphide acts on potassium ethylate or alcoholic potash, so as to form potassium ethyldithiocarbonate; and on potassium mercaptide, or an alcoholic solution of the sulphydrate, so as to form the ethyltrithiocarbonate, thus:

$$CO_2 + (C_2H_5)KO = (C_2H_5)KCO_3 \quad \text{Ethylcarbonate.}$$
$$CO_2 + (C_2H_5)KS = (C_2H_5)KCO_2S \quad \text{Ethylmonothiocarbonate.}$$
$$CS_2 + (C_2H_5)KO = (C_2H_5)KCOS_2 \quad \text{Ethyldithiocarbonate.}$$
$$CS_2 + (C_2H_5)KS = (C_2H_5)KCS_3 \quad \text{Ethyltrithiocarbonate.}$$

The neutral thiocarbonic ethers (containing monatomic alcohol-radicles) are produced by the action of the chlorides, bromides, etc., of alcohol-radicles on the metallic salts of the corresponding acid ethers; e. g.:

$$(C_2H_5)KCS_3 + C_2H_5Cl = KCl + (C_2H_5)_2CS_3.$$

The thiocarbonic ethers of diatomic alcohol-radicles are formed by the action of diatomic alcoholic bromides, iodides, etc., on sodium thiocarbonate; e.g.:

$$C_2H_4Br_2 + Na_2CS_3 = 2NaBr + (C_2H_4)CS_3.$$

The neutral thiocarbonic ethers are oily liquids; so likewise are the acid ethers, such at least as are known in the free state, or as hydrogen salts; their metallic salts are mostly crystalline. The best known of these compounds are the ethyl-dithiocarbonates or xanthates.

To prepare xanthic acid, alcohol of 0.800 specific gravity is saturated whilst boiling, with potash, and into this solution carbon bisulphide is dropped till it ceases to be dissolved, or until the liquid loses its alkalinity. On cooling the whole to —18° C. (0.4° F.), the potassium-salt separates in the form of brilliant, slender, colorless prisms, which must be quickly pressed between folds of bibulous paper, and dried in a vacuum. It is freely soluble in water and alcohol, but insoluble in ether, and is gradually destroyed by exposure to air, by oxidation of part of the sulphur. Xanthic acid may be prepared by decomposing this salt with dilute sulphuric or hydrochloric acid. It is a colorless, oily liquid, heavier than water, of powerful and peculiar odor, and very combustible: it reddens litmus paper, and ultimately bleaches it. Exposed to gentle heat (about 24°), it is decomposed into alcohol and carbon bisulphide. Exposed to the air, or kept beneath the surface of water open to the air, it becomes covered with a whitish crust, and is gradually destroyed. The xanthates of the alkali-metals and of barium are colorless and crystallizable; the calcium salt dries up to a gummy mass; the xanthates of zinc, lead, and mercury are white, and but slightly soluble; that of copper is a flocculent, insoluble substance, of beautiful yellow color.

Ethylic dithiocarbonate or *Xanthic ether*, $(C_2H_5)_2.COS_2$, obtained by the action of ethyl chloride on potassium xanthate, is a pale-yellow oil, boiling at 200° C. (392° F.), insoluble in water, soluble in all proportions of alcohol or ether. Ammonia-gas passed into its alcoholic solution forms mercaptan and a crystalline substance, $CO{<}^{SC_2H_5}_{NH_2}$, called **x a n t h a m i d e**:

$$CO{<}^{SC_2H_5}_{SC_2H_5} + NH_3 = C_2H_5SH + CO{<}^{SC_2H_5}_{NH_2}.$$

Amyl dithiocarbonate, $CO(SC_5H_{11})_2$, treated in like manner, yields **a m y l x a n t h a m i d e**, $CO{<}^{SC_5H_{11}}_{NH_2}$.

Chlorocarbonic Ethers. CO_2ClR or $CO{<}^{Cl}_{OR}$, [R denoting a monatomic alcohol-radicle.]—These compounds, derived from carbonic ethers, $CO(OR)_2$, by substitution of Cl for one of the groups OR, are formed by the action of carbonyl chloride (phosgene) on the alcohols; *e. g.*:

$$CO{<}^{Cl}_{Cl} + HOCH_3 = HCl + CO{<}^{Cl}_{OCH_3}.$$

Methyl chlorocarbonate, $CO_2Cl(CH_3)$, is a colorless mobile oil, heavier and more volatile than water, having a pungent odor, and burning with a greenish flame. Gaseous ammonia converts it into methyl carbamate (urethane):

$$CO{<}^{Cl}_{OCH_3} + NH_3 = HCl + CO{<}^{NH_2}_{OCH_3}.$$

Ethyl chlorocarbonate, $CO_2Cl(OC_2H_5)$, is also a colorless, very mobile liquid, the vapor of which is very irritating to the eyes. Specific gravity 1.139 at 13° C. (55.4° F.). Boiling point, 94° C. (201.2° F.). It is very inflammable, burns with a green flame, is decomposed by hot water, and quickly converted by ammonia into ethyl carbamate.

DIATOMIC AND BIBASIC ACIDS.

These acids contain the group, CO_2H, twice, and must therefore contain four atoms of oxygen. They may all be included in the general formula, $R''(CO_2H)_2$,—R denoting a diatomic hydrocarbon-radicle,—or they may be regarded as compounds of oxygenated radicles with two equivalents of hydroxyl, *e. g.*, succinic acid $= (C_4H_4O_2)''(OH)_2$.

1.—Oxalic or Succinic Series, $C_nH_{2n-2}O_4$, or $C_nH_{2n}{<}^{CO.OH}_{CO.OH}$.

The known acids of this series are:

Oxalic acid	$C_2H_2O_4$	Pimelic acid	$C_7H_{12}O_4$
Malonic acid	$C_3H_4O_4$	Suberic acid	$C_8H_{14}O_4$
Succinic acid	$C_4H_6O_4$	Anchoic acid	$C_9H_{16}O_4$
Pyrotartaric acid	$C_5H_8O_4$	Sebic acid	$C_{10}H_{18}O_4$
Adipic acid	$C_6H_{10}O_4$	Roccellic acid	$C_{17}H_{32}O_4$

They are produced :—1. By oxidation of the corresponding glycols, $R''(CH_2OH)_2$, the change consisting in the substitution of O_2 for H_4 (p. 711). In this manner oxalic acid, $C_2H_2O_4$, is formed from ethene alcohol, $C_2H_6O_2$, and malonic acid, $C_3H_4O_4$, from β-propene alcohol, $C_3H_8O_2$; but all the known glycols of higher order split up under the influence of oxidizing agents, and do not yield bibasic acids containing the same number of carbon-atoms as themselves.

2. By oxidation of the oxyfatty acids, $C_nH_{2n}O_3$, in which the hydroxyl-group is directly attached to the group CH_2; c. g.:

$$
\begin{array}{ccccccc}
\begin{array}{c} CH_2.OH \\ | \\ CO.OH \end{array} & + & O_2 & = & \begin{array}{c} CO.OH \\ | \\ CO.OH \end{array} & + & H_2O \\
\text{Glycollic acid.} & & & & \text{Oxalic acid.} & &
\end{array}
$$

3. By boiling the cyano-derivatives of the fatty acids with alkalies or acids :

$$
\begin{array}{ccccccc}
\begin{array}{c} CH_2.CN \\ | \\ CO_2H \end{array} & + & 2H_2O & = & NH_3 & + & CH_2{<}^{CO_2H}_{CO_2H} \\
\begin{array}{c} \text{Cyanacetic} \\ \text{acid.} \end{array} & & & & & & \begin{array}{c} \text{Malonic} \\ \text{acid.} \end{array}
\end{array}
$$

4. By boiling the cyanides of diatomic alcohol-radicles with alcoholic potash; e. g.:

$$
(C_3H_6)(CN)_2 + 2KOH + 2H_2O = 2NH_3 + (C_3H_6)(CO_2K)_2
$$

Propene cyanide. Potassium pyrotartrate.

This reaction is analogous to that by which the fatty acids are formed from the cyanides of the monatomic alcohol-radicles, C_nH_{2n+1} (p. 714).

5. By the addition of hydrogen to other acids containing a smaller proportion of that element; in this manner succinic acid, $C_4H_6O_4$, is formed from fumaric acid, $C_4H_4O_4$.

6. By the action of heat on acids of more complicated structure; e. g.:

$$
2C_4H_6O_6 = 3CO_2 + 2H_2O + C_5H_8O_4
$$

Tartaric acid. Pyrotartaric acid.

7. Many of these acids are produced by the action of powerful oxidizers on a variety of organic bodies : thus, succinic, adipic, pimelic, suberic, and anchoic acids are produced by treating various fatty and resinous bodies with nitric acid.

The acids of this series are solid, crystallizable, non-volatile bodies, for the most part easily soluble in water, and having a strong acid reaction. Being bibasic, they form acid and neutral salts, and acid and neutral ethers; thus :

$$
\begin{array}{ccc}
C_2H_4{<}^{CO.OH}_{CO.OH} & C_2H_4{<}^{CO.OC_2H_5}_{CO.OH} & C_2H_4{<}^{CO.OC_2H_5}_{CO.OC_2H_5} \\
\text{Succinic acid.} & \begin{array}{c} \text{Ethyl-succinic} \\ \text{acid.} \end{array} & \begin{array}{c} \text{Diethylic} \\ \text{succinate.} \end{array}
\end{array}
$$

In like manner, each acid can form an acid amide or a m i c a c i d, and a neutral amide or d i a m i d e by substitution of amidogen, NH_2, for one or both of the hydroxyl groups, also a neutral i m i d e by substitution of the bivalent-radicle imidogen, NH, for both these groups together; thus :

$$
\begin{array}{ccc}
C_2H_4{<}^{CO.NH_2}_{CO.OH} & C_2H_4{<}^{CO.NH_2}_{CO.NH_2} & C_2H_4{<}^{CO}_{CO}{>}NH \\
\begin{array}{c} \text{Succinamic} \\ \text{acid.} \end{array} & \text{Succinamide.} & \text{Succinamide.}
\end{array}
$$

These bibasic acids, when heated, give off their water, and yield anhy_drides which, when boiled with water, are reconverted into the acids :

$$C_2H_4 <^{CO.OH}_{CO.OH} \quad + \quad C_2H_4 <^{CO}_{CO}>O \quad + \quad H_2O .$$

The anhydrides are also produced by treating the acids with phospho_rus pentachloride :

$$C_2H_4 <^{CO.OH}_{CO.OH} \quad + \quad PCl_5 \quad = \quad PCl_3O \quad + \quad 2HCl \quad + \quad C_2H_4 <^{CO}_{CO}>O .$$

With excess of phosphorus pentachloride, a c i d c h l o r i d e s or c h l o r_an h y d r i d e s are produced ; thus :

$$C_2H_4 <^{CO.OH}_{CO.OH} \quad + \quad 2PCl_5 \quad = \quad 2PCl_3O \quad + \quad 2HCl \quad + \quad C_2H_4 <^{CO.Cl}_{CO.Cl} .$$

O xalic Acid, $C_2H_2O_4 = {\displaystyle \mathop{|}^{COOH}_{COOH}} = (C_2O_2)''(OH)_2.$—This important
acid exists ready formed in many plants as a potassium or calcium salt,
and is produced by the oxidation of a great variety of organic compounds.
In some cases the reaction consists in a definite substitution of oxygen for
hydrogen ; thus oxalic acid is formed from ethene alcohol, $C_2H_6O_2$, by sub_stitution of O_2 for H_4, and from ethyl alcohol, C_2H_6O, by the same substi_tution and further addition of one atom of oxygen. But in most cases the
reaction is more complex, consisting in a complete breaking up of the
molecule. In this manner oxalic acid is produced in great abundance from
more highly carbonized organic substances, such as sugar, starch, cellu_lose, etc., by the action of nitric acid, or by fusion with caustic alkalies.

Oxalic acid is also produced : *a.* As a sodium or potassium salt by direct
combination of the alkali-metal with carbon dioxide :

$$2CO_2 \quad + \quad Na_2 \quad = \quad C_2O_4Na_2 .$$

The sodium salt is obtained by passing carbon dioxide over a heated mix_ture of sodium and sand ; the potassium-salt by heating potassium-amalgam
in the gas.*

β. As an ammonium salt, together with other products, in the decompo_sition of cyanogen by water :

$$C_2N_2 \quad + \quad 4H_2O \quad = \quad C_2(NH_4)_2O_4 .$$

γ. As a potassium salt by heating potassium formate with excess of
potash :

$$2CHKO_2 \quad = \quad C_2K_2O_4 \quad + \quad H_2 .$$

Preparation.—1. By the oxidation of sugar with nitric acid :

$$C_{12}H_{22}O_{11} \quad + \quad O_{18} \quad = \quad 6C_2H_2O_4 \quad + \quad 5H_2O .$$

One part of sugar is gently heated in a retort with 5 parts of nitric acid
of sp. gr. 1.42, diluted with twice its weight of water ; copious red fumes
are then disengaged, and the oxidation of the sugar proceeds with violence
and rapidity. When the action slackens, heat may be again applied to
the vessel, and the liquid concentrated by distilling off the superfluous
nitric acid, until it deposits crystals on cooling. These are drained, redis_solved in a small quantity of hot water, and the solution is set aside to
cool.

* K o l b e and D r e c h s e l, Chem. Soc. Journal [2], vi. 121.

2. By heating sawdust with caustic alkali. Many years ago, Gay-Lussac observed that wood and several other organic substances were converted into oxalic acid by fusion with caustic potash. Upon this observation, Messrs. Roberts, Dale & Co. have founded a method for the preparation of oxalic acid, which furnishes this acid at a much cheaper rate than any other process. A mixed solution of the hydrates of sodium and potassium, in the proportion of two molecules of the former to one of the latter, is evaporated to about 1.35 sp. gr., and then mixed with sawdust, so as to form a thick paste, which is placed in thin layers on iron plates. The mixture is now gradually heated, care being taken to keep it constantly stirred. The action of heat expels a quantity of water, and the mass intumesces strongly, with disengagement of much inflammable gas, consisting of hydrogen and carburetted hydrogen. The mixture is now kept for some hours at a temperature of 204° C. (400° F.), care being taken to avoid charring, which would cause a loss of oxalic acid. The product thus obtained is a gray powder; it is now treated with water at about 15.5° C. (60° F.), which leaves the sodium oxalate undissolved. The supernatant liquid is drawn off, evaporated to dryness, and heated in furnaces to recover the alkalies, which are caustified and used for a new operation. The sodium oxalate is washed and decomposed by boiling with slaked lime, and the resulting calcium oxalate is decomposed by means of sulphuric acid. The liquid decanted from the calcium sulphate is evaporated to crystallization in leaden vessels, and the crystals are purified by re-crystallization.

Oxalic acid separates from a hot solution in colorless, transparent crystals derived from an oblique rhombic prism, and consisting of $C_2H_2O_4.2H_2O$. The two molecules of crystallization-water may be expelled by a very gentle heat, the crystals crumbling down to a soft white powder, consisting of anhydrous oxalic acid, $C_2H_2O_4$, which may be sublimed in great measure without decomposition. The crystallized acid, on the contrary, is decomposed by a high temperature into formic acid, carbon monoxide, and carbon dioxide, without leaving any solid residue:

$$2C_2H_2O_4 \;=\; CH_2O_2 \;+\; CO \;+\; 2CO_2 \;+\; H_2O.$$

The crystals of oxalic acid dissolve in 8 parts of water at 15.5°, and in their own weight, or less, of hot water: they are also soluble in spirit. The aqueous solution has an intensely sour taste and most powerful acid reaction, and is highly poisonous. The proper antidote is chalk or magnesia. Oxalic acid is decomposed by hot oil of vitriol into a mixture of carbon monoxide and carbon dioxide: it is slowly converted into carbonic acid by nitric acid, whence arises a considerable loss in the process of manufacture from sugar. The dioxides of lead and manganese effect the same change, becoming reduced to monoxides, which form salts with the unaltered acid.

Oxalates.—Oxalic acid, like other bibasic acids, forms with monatomic metals, neutral or normal salts containing $C_2M_2O_4$, and acid salts, C_2HMO_4. With potassium and ammonium it likewise forms hyper-acid salts, e. g., $C_2HKO_4.C_2H_2O_4$, or $C_4H_3KO_8$. With most diatomic metals it forms only neutral salts, $C_2M''O_4$; with barium and strontium, however, it forms acid salts analogous to the hyper-acid oxalates of the alkali-metals. It also forms numerous well-crystallized double salts. It is one of the strongest acids, decomposing dry sodium chloride when heated, with evolution of hydrochloric acid, and converting sodium chloride or nitrate in aqueous solution into acid oxalate.

Th.a oxalates of the alkali-metals are soluble in water ; the rest are, for the most part, insoluble in water, but soluble in dilute acids.

All oxalates are decomposed by heat. The oxalates of the alkali-metals and also of the alkaline earth-metals, if not too strongly heated, give off carbon monoxide and leave carbonates, while the oxalates of those metals whose carbonates are decomposed by heat (zinc and magnesium, for example), give off carbon monoxide and carbon dioxide, and leave metallic oxides. The oxalates of the more easily reducible metals (silver and copper, etc.), give off carbon dioxide and leave the metal ; the lead salt leaves suboxide of lead, and gives off 3 volumes of carbon dioxide to 1 volume of carbon monoxide :—

$$2C_2PbO_4 \;=\; Pb_2O \;+\; 3CO_2 \;+\; CO .$$

Oxalates heated with *sulphuric acid* give off carbon monoxide and dioxide, and leave a residue of sulphate. In this case, as well as in the decomposition by heat alone, no separation of carbon takes place, and consequently the residue does not blacken : this character distinguishes the oxalates from the salts of all other carbon acids.

Oxalic acid and the soluble oxalates give with *calcium chloride* a precipitate of calcium oxalate, insoluble in water and in acetic acid, but soluble in hydrochloric and nitric acid. This reaction affords a very delicate test for the presence of oxalic acid : the insolubility of the precipitated oxalate in acetic acid distinguishes it at once from the phosphate.

POTASSIUM OXALATES.—The *neutral salt*, $C_2K_2O_4.2Aq.$, prepared by neutralizing oxalic acid with potassium carbonate, crystallizes in transparent rhombic prisms, which become opaque and anhydrous by heat, and dissolve in 3 parts of water.—The *acid oxalate* or *binoxalate*, $C_2HKO_4.2Aq.$, sometimes called *Salt of Sorrel*, from its occurrence in that plant, is found also in other species of *Rumex*, in *Oxalis acetosella*, and in garden rhubarb, associated with malic acid. It is easily prepared by dividing a solution of oxalic acid in hot water into two equal portions, neutralizing one with potassium carbonate, and adding the other : the salt crystallizes, on cooling, in colorless rhombic prisms. The crystals have a sour taste, and require 40 parts of cold, and 6 of boiling water for solution. A solution of this salt is often used for removing ink from paper. The *hyper-acid oxalate* or *quadroxalate*, $C_2KHO_4.C_2H_2O_4.2Aq.$, is prepared by saturating 1 part of oxalic acid with potassium carbonate, and adding 3 parts of oxalic acid. The crystals are triclinic, and dissolve in 20 parts of water at 20° C. (68° F.).

Sodium oxalate, $C_2Na_2O_4$, is slightly soluble and difficult to crystallize. The *acid salt*, C_2HNaO_4, forms crystals which redden litmus.

AMMONIUM OXALATES.—The *neutral salt*, $C_2(NH_4)_2O_4 + 2Aq.$, is prepared by neutralizing a hot solution of oxalic acid with ammonium carbonate. It crystallizes in long, colorless, rhombic prisms, which effloresce in dry air. They are not very soluble in cold water, but dissolve freely with the aid of heat.

The dry salt, when heated in a retort, gives off water, and yields a sublimate of o x a m i d e :*

$$\underset{\text{Ammonium oxalate.}}{(C_2O_2)(ONH_4)_2} \;=\; 2H_2O \;+\; \underset{\text{Oxamide.}}{(C_2O_2)(NH_2)_2}$$

When distilled with phosphoric oxide, it gives up four molecules of

* See the Chapter on Amides.

water, and yields a considerable quantity of c y a n o g e n, $C_2(NH_4)_2O_4$'—
$4H_2O = 2CN$. Other products are, however, formed at the same time.

Acid ammonium oxalate or *binoxalate*, $C_2H(NH_4)O_4$ + Aq., is still less
soluble than the neutral salt. When heated in an oil-bath to 232° C.
(239.8^O F.), it loses one molecule of water, and yields o x a m i c a c i d,
$C_2H_3NO_3$, or $(C_2O_2)(OH)(NH_2)$, and other products.

CALCIUM OXALATE, C_2CaO_4 + 4Aq.—This salt occurs in the juice of most
plants, and separates out towards the end of the growing season in microscopic octohedral crystals ; it is also a frequent constituent of urinary
deposits and calculi. It is formed whenever oxalic acid or an oxalate is
added to a soluble calcium salt ; it falls as a white powder, which acquires
density by boiling, and is but little soluble in dilute hydrochloric, and
quite insoluble in acetic acid. Nitric acid dissolves it easily. When dried
at 100°, it retains a molecule of water, which may be driven off by a
rather higher temperature. Exposed to a red heat in a close vessel, it is
converted into calcium carbonate, with escape of carbon monoxide.

The oxalates of *barium, zinc, manganese, copper, nickel, cobalt*, and *ferrous
oxalate*, are nearly insoluble in water : *magnesium oxalate* is sparingly
soluble : *ferric oxalate* is freely soluble.—*Potassio-chromic oxalate*, $K_3Cr(C_2O_4)_3$
+ 3Aq., prepared by dissolving in hot water 1 part of potassium bichromate, 2 parts of potassium binoxalate, and 2 parts of crystallized oxalic
acid, is one of the most beautiful salts known. The crystals appear black
by reflected light, from the intensity of their color, which is pure deep
blue : they are very soluble. A corresponding *potassio-ferric oxalate* has
been formed : it crystallizes freely, and has a fine green color.

ETHYL OXALATES.—The *neutral oxalate*, or *Oxalic ether*, $C_2O_4(C_2H_5)_2 =$
$C_2O_2(OC_2H_5)_2$, is most easily obtained by distilling together 4 parts of potassium binoxalate, 5 parts of oil of vitriol, and 4 parts of strong alcohol.
The distillation may be pushed nearly to dryness, and the receiver kept
warm, to dissipate any ethyl oxide that may be formed. The product is
mixed with water, by which the oxalic ether is separated from the undecomposed spirit : it is repeatedly washed to remove adhering acid, and
re-distilled in a small retort, the first portion being collected apart and
rejected. Another very simple process consists in digesting equal parts
of alcohol and dehydrated oxalic acid in a flask furnished with a long
glass tube in which the volatilized spirit may condense. After six or
eight hours' digestion, the mixture generally contains only traces of un-
etherified oxalic acid.

Pure oxalic ether is a colorless, oily liquid, of pleasant aromatic odor,
and 1.09 specific gravity. It boils at 183.8^O C. (362.8^O F.), is but little
soluble in water, and is readily decomposed by caustic alkalies into a metallic oxalate and alcohol. With solution of ammonia in excess, it yields
oxamide and alcohol ; thus :

$$(C_2O_2)(OC_2H_5)_2 \ + \ 2NH_3 \ = \ 2(HOC_2H_5) \ + \ (C_2O_2)(NH_2)_2$$

This is the best process for preparing oxamide.

When dry gaseous ammonia is conducted into a vessel containing oxalic
ether, the gas is rapidly absorbed, and a white solid substance produced,
which is soluble in hot alcohol, and separates on cooling in colorless, transparent, scaly crystals. They dissolve in water, and are both fusible and
volatile. This substance is o x a m e t h a n e, the ethylic ether of oxamic
acid.*

$$(C_2O_2)(OC_2H_5)_2 \ + \ NH_3 \ = \ HOC_2H_5 \ + \ C_2O_2(NH_2)(OC_2H_5).$$

* See the Chapter on Amides.

The same substance is formed when ammonia in small quantity is added to a solution of oxalic ether in alcohol.

When oxalic ether is treated with excess of dry chlorine in sunshine, a white, colorless, crystalline, fusible body is produced, insoluble in water, and instantly decomposed by alcohol. It consists of p e r c h l o r e t h y l i c o x a l a t e, $C_8Cl_{10}O_4$, or $(C_2Cl_5)_2C_2O_4$, or oxalic ether in which the whole of the hydrogen is replaced by chlorine.

Ethyl oxalate is converted by potassium or sodium into ethyl carbonate, with evolution of carbon monoxide :· $C_2(C_2H_5)_4O_4 = C(C_2H_5)_2O_3 + CO$; but the reaction is complicated by the formation of several other products.

When ethyl oxalate is agitated with sodium amalgam in a vessel externally cooled, a product is obtained which is separated by ether into a soluble and an insoluble portion, the latter consisting of fermentable sugar, together with sodium oxalate, and at least one other sodium-salt, while the ethereal solution yields, by spontaneous evaporation, crystals having the composition $C_{11}H_{18}O_8$, and consisting of the ethylic ether of a tribasic acid, $C_5H_6O_8$, called d e o x a l i c a c i d, because it is produced by deoxidation of oxalic acid : $5C_2H_2O_4 + 5H_2 = 2C_5H_6O_8 + 4H_2O$; and r a c e m o c a r b o n i c a c i d, because it contains the elements of racemic acid, $C_4H_6O_6$, and carbon dioxide, CO_2, and is resolved into those two compounds when its aqueous solution is heated in a sealed tube with a small quantity of sulphuric acid. The decomposition of ethylic oxalate by sodium amalgam has not been completely investigated, but the formation of deoxalic acid and glucose may be represented by the equation :

$$8C_2H_2O_4 + 14H^2 = 2C_5H_6O_8 + C_6H_{12}O_6 + 10H_2O.$$

Ethyl oxalate, treated with zinc-ethyl, and afterwards with water, yields the ethylic ether of diethoxalic acid, $C_2H_2(C_2H_5)_2O_3$, and similar products with zinc-methyl and zinc-amyl (p. 742).

Acid ethyl oxalate, or *Ethyloxalic acid,* $C_2H(C_2H_5)O_4$, or $C_2O_2\big<^{OC_2H_5}_{OH}$, is obtained as a potassium-salt by adding to a solution of neutral ethyl oxalate in absolute alcohol, a quantity of alcoholic potash less than sufficient to convert the whole into potassium oxalate and alcohol ; on dissolving this salt in hydrated alcohol, carefully saturating with sulphuric acid, and neutralizing with carbonate of lead or barium, the ethyloxalate of lead or barium is obtained.—The acid itself is prepared by decomposing either of these salts with sulphuric acid ; but it is very unstable, and is decomposed by concentration into alcohol and oxalic acid. The *potassium salt,* $C_2O_2\big<^{OC_2H_5}_{OK}$, forms crystalline scales, which begin to decompose towards $100°$.

METHYL OXALATE, $C_2(CH_3)_2O_4$, or $C_2O_2(OCH_3)_2$, is easily prepared by distilling a mixture of equal weights of oxalic acid, wood-spirit, and oil of vitriol. A spiritous liquid collects in the receiver, which, when exposed to the air, quickly evaporates, leaving the methyl oxalate in the form of rhombic, transparent, crystalline plates, which may be purified by pressure between folds of bibulous paper, and redistilled from a little oxide of lead. The product is colorless, and has the odor of ethyl oxalate ; it melts at $51°$ C. ($128.8°$ F.), and boils at $161°$ C. ($321.8°$F.); dissolves freely in alcohol and wood-spirit, also in water, which, however, rapidly decomposes it, especially when hot, into oxalic acid and wood-spirit. The alkaline hydrates effect the same change even more easily. Solution of ammonia converts it into oxamide and methyl alcohol. With dry ammoniacal gas it yields methyl oxamate, or oxamethylane, $C_2O_2{<}{NH_2 \atop OCH_3}$, a white, solid substance, which crystallizes from alcohol in pearly cubes.

ETHENE OXALATE, $C_2O_4(C_2H_4)$, or $(C_2O_2)(C_2H_4O_2) = C_2O_2{<}{OCH_2 \atop OCH_2}$, appears to be formed by the action of ethene bromide on silver oxalate.

Closely related to oxalic acid are glyoxylic acid, $C_2H_2O_3$, and glyoxal, $C_2H_2O_2$, which may be regarded as aldehydic derivatives of oxalic acid or of glycol:

$$
\begin{array}{llll}
\text{CO.OH} & \text{COH} & \text{COH} & \text{CH}_2\text{OH} \\
| & | & | & | \\
\text{CO.OH} & \text{COOH} & \text{COH} & \text{CH}_2\text{OH} \\
\text{Oxalic} & \text{Glyoxylic} & \text{Glyoxal.} & \text{Glycol.} \\
\text{acid.} & \text{acid.} & &
\end{array}
$$

Both are formed as intermediate products in the oxidation of glycol, and are converted by further oxidation into oxalic acid (see ALDEHYDES, p. 701).

Malonic Acid, $C_3H_4O_4 = CH_2{<}{CO_2H \atop CO_2H}$.—This acid is formed: 1. By gradual oxidation of ß-propene glycol:

$$CH_2{<}{CH_2OH \atop CH_2OH} + O_4 = 2H_2O + CH_2{<}{CO_2H \atop CO_2H};$$

also by oxidation of propene and allylene.

2. By oxidizing malic acid with chromic acid mixture:

$$
\begin{array}{l}
\text{CH}_2.\text{CO}_2\text{H} \\
| \\
\text{CHOH} \quad + \text{O}_2 = \text{CO}_2 + \text{H}_2\text{O} + \\
| \\
\text{CO}_2\text{H}
\end{array}
\quad
\begin{array}{l}
\text{CO}_2\text{H} \\
| \\
\text{CH}^2 \\
| \\
\text{CO}_2\text{H}
\end{array}
$$

3. By decomposition of barbituric acid (Malonyl-urea: see AMIDES).

4. Synthetically by the action of alkalies on cyanacetic acid, better, on its ethylic ether:

$$CH_2{<}{CN \atop CO_2H} + 2H_2O = NH_3 + CH_2{<}{CO_2H \atop CO_2H}.$$

Ethylic monochloracetate is heated with solution of potassium cyanide, and the product is boiled with potash as long as it continues to give off ammonia. The alkaline solution is then acidulated with sulphuric acid, and the free malonic acid extracted by ether.

Malonic acid crystallizes in large rhombohedral plates; dissolves easily in water, alcohol, and ether; melts at 132° C. (269.6° F.); and decomposes at a higher temperature into acetic acid and carbon dioxide. Its *barium salt*, $C_3H_2BaO_4 + H_2O$, forms silky needles. The *calcium salt*, $4C_3H_2CaO_4 + 7H_2O$, is very slightly soluble in cold water.

Nitrosomalonic Acid, $CH(NO){<}{CO_2H \atop CO_2H}$, formed by heating violuric acid (*q. v.*) with alkalies, crystallizes in shining needles, easily soluble in water. When heated it melts, and then explodes.

Amıdomalonic Acid, $CH(NH_2){<}{CO_2H \atop CO_2H}$, produced by the action of sodium-amalgam on the nitroso-acid, forms shining prisms, which, when heated, are resolved into carbon dioxide and amidacetic acid, $CH_2(NH_2)—CO_2H$.

MESOXALIC ACID, $C_3H_2O_5 = CO{<}{CO_2H \atop CO_2H}$.—This ketonic acid, derived from malonic acid by substitution of O for H_2 in the group CH_2, is produced by oxidizing amidomalonic acid by means of iodine in an aqueous solution containing potassium iodide:

$$CH(NH_2){<}{CO_2H \atop CO_2H} + O = NH_3 + CO{<}{CO_2H \atop CO_2H};$$

also by boiling alloxan (mesoxalyl-urea *q. v.*) with alkalies:

$$CO{<}{NH{-}CO \atop NH{-}CO}{>}CO + 2H_2O = CO{<}{NH_2 \atop NH_2} + CO{<}{CO_2H \atop CO_2H}$$
Alloxan.   Urea.   Mesoxalic acid.

Mesoxalic acid crystallizes in deliquescent prisms, containing 1 mol. water, easily soluble in alcohol and ether. It melts at 115° C. (239° F.) without giving off its water of crystallization, and decomposes at a higher temperature. The water appears therefore to be very intimately combined, probably in the form represented on the right-hand side of the following equation:

$$CO{<}{CO_2H \atop CO_2H} + H_2O = C(OH)_2{<}{CO_2H \atop CO_2H}.$$

The metallic mesoxalates and the ethylic ether also contain 1 mol. water very intimately combined; the ether probably has the constitution $C(OH)_2(CO_2C_2H_5)_2$. The *barium salt*, $C_3BaO_5 + 1\frac{1}{2}H_2O$, is nearly insoluble in water. The *silver salt*, $C_3Ag_2O_5 + H_2O$, is an amorphous powder, which blackens on exposure to light, and is decomposed by boiling with water into mesoxalic acid, metallic silver, silver oxalate, and carbon dioxide. By the action of sodium-amalgam mesoxalic acid is converted into tartronic acid:

$$CO{<}{CO_2H \atop CO_2H} + H_2 = CH(OH){<}{CO_2H \atop CO_2H}.$$

Succinic Acids, $C_4H_6O_4 = C_2H_4{<}{CO_2H \atop CO_2H}$.—Of these acids there are two modifications, viz.:

$${CH_2.CO_2H \atop | \atop CH_2.CO_2H}$$
Succinic.

$$CH_3.CH{<}{CO_2H \atop CO_2H}$$
Isosuccinic.

1. Ordinary Succinic, β-Succinic,, or Ethene-dicarbonic Acid,
is produced:

1. By heating ethene cyanide with alcoholic potash:

$${CH_2.CN \atop | \atop CH_2.CN} + 4H_2O = 2NH_3 + {CH_2.CO_2H \atop | \atop CH_2.CO_2H}.$$

2. By converting β-iodopropionic acid into the corresponding cyanogen derivative, and decomposing the latter with alkalies or acids:

$$CH_2.CN{-}CH_2{-}CO_2H + 2H_2O = NH_3 + CH_2.CO_2H{-}CH_2{-}CO_2H.$$
β-Cyanopropionic acid.   β-Succinic acid.

3. By the action of nascent hydrogen (evolved by sodium-amalgam) on maleic acid, or its isomeride, fumaric acid : $C_4H_4O_4 + H_2 = C_4H_6O_4$.

4. By the action of hydriodic acid (or water and phosphorus iodide) on malic acid, $C_4H_6O_5$, or tartaric acid, $C_4H_6O_6$, the reaction consisting in the abstraction of 1 or 2 atoms of oxygen, with formation of water and separation of iodine.

5. By the fermentation of malic or fumaric acid, and of many other organic substances, especially under the influence of putrefying casein ; in small quantity also during the alcoholic fermentation of sugar (p. 591, foot-note).

6. By the oxidation of many organic substances, especially of the fatty acids, $C_nH_{2n}O_2$, and their glycerides, under the influence of nitric acid. Its formation from butyric acid is represented by the equation $C_4H_8O_2 + O_3 = H_2O + C_4H_6O_4$.

Succinic acid occurs ready formed in amber and in certain lignites, and occasionally in the animal organism. By heating amber in iron retorts, it may be obtained in colored crystals, which may be purified by treatment with nitric acid and recrystallization from boiling water. It is, however, more advantageously prepared by the fermentation of malic acid, the crude calcium malate obtained by neutralizing the juice of mountain-ash berries with chalk or slaked lime being used for the purpose. The salt is mixed in an earthen jar with water and yeast, or decaying cheese, and left for a few days at 30° or 40° C. (86°–104° F.); the calcium succinate thus obtained is decomposed by dilute sulphuric acid ; and the succinic acid is purified by crystallization from water and by sublimation.

Succinic acid crystallizes in colorless, monoclinic prisms, which dissolve in 23 parts of water at 20° C. (68° F.), and in 4 parts of boiling water : it melts at 180° C. (356° F.), and boils at 235° C. (455° F.), at the same time undergoing decomposition into water and s u c c i n i c o x i d e or a n -h y d r i d e, $C_4H_4O_3$, or $(C_4H_4O_2)O$. The same compound is formed by the action of phosphorus pentachloride on succinic acid :

$$C_4H_6O_4 \quad + \quad PCl_5 \quad = \quad POCl_3 \quad + \quad 2HCl \quad + \quad C_4H_4O_3$$

It is a white mass, less soluble in water, but more soluble in alcohol, than succinic acid.

Succinic acid, being bibasic, forms, with monad metals, acid and neutral salts, $C_4H_5MO_4$ and $C_4H_4M_2O_4$, and with dyad metals, neutral salts containing $C_4H_4M''O_4$, and acid salts $C_4H_4M''O_4 . C_4H_6O_4$.—There are also a few double succinates, several basic lead-salts, and a hyperacid potassium-salt.

The succinates of the alkali-metals are easily soluble in water. The *calcium-salt*, $C_4H_4CaO_4$, is sparingly soluble in water, and separates from a cold solution with $3H_2O$, and from a hot solution with $1H_2O$. On adding ammonium succinate to the solution of a ferric salt, a basic ferric succinate is thrown down as a reddish-brown precipitate.

Succinic acid is distinguished from benzoic acid by not being precipitated from its soluble salts by mineral acids, and by forming a white precipitate with barium chloride, on addition of alcohol and ammonia.

Ethylic succinate, $C_2H_4(CO_2.C_2H_5)_2$, obtained by the action of hydrochloric acid on an alcoholic solution of succinic acid, is a thick oil, insoluble in water, having a specific gravity of 1.072 at 0°, and boiling at 216° C. (420.8° F.).

Succinic chloride, $C_2H_4(CO.Cl)_2$, formed by the action of PCl_5 on excess of succinic acid, is an oil which solidifies at 0°, and boils at 90° C. (194° F.).

B r o m o s u c o i n i c a c i d s.—The mono- and di-brominated acids are formed by heating succinic acid with bromine and water in sealed tubes to 150°–180° C. (302°–356° F.).

Monobromosuccinic acid, $C_2H_3Br(CO_2H)_2$, which is the chief product formed when a large quantity of water is used, crystallizes in nodular groups of slender needles, easily soluble in water. It melts at 160° C. (320° F.), giving off HBr, and being converted into fumaric acid, $C_4H_4O_4$. By boiling with silver oxide and water, it is converted into oxysuccinic or malic acid, $C_2H_3(OH)(CO_2H)_2$.

Dibromosuccinic acid, $C_2H_2Br_2(CO_2H)_2$, is also formed by direct combination of fumaric acid with bromine:

$$\begin{array}{ccc} \text{CH.CO}_2\text{H} & & \text{CHBr.CO}_2\text{H} \\ \| & + \ \text{Br}_2 \ = & | \\ \text{CH.CO}_2\text{H} & & \text{CHBr.CO}_2\text{H} \end{array} :$$

further by heating succinyl chloride with bromine, and decomposing the resulting dibromosuccinyl chloride, $C_2H_2Br_2(COCl)_2$, with water.

This acid crystallizes in prisms, sparingly soluble in cold, more freely in hot water.

Its salts are decomposed by boiling with water, the silver salt yielding dioxysuccinic or inactive tartaric acid; the sodium salt, monobromomalic acid; and the barium salt, monobromomaleic acid; thus:

$$C_4H_2Ag_2Br_2O_4 \ + \ 2H_2O \ = \ 2AgBr \ + \ \underset{\text{Tartaric acid.}}{C_4H_6O_6}$$

$$C_4H_2Na_2Br_2O_4 \ + \ H_2O \ = \ NaBr \ + \ \underset{\text{Bromomalate.}}{C_4H_4NaBrO_5}$$

$$2C_4H_2BaBr_2O_4 \ = \ BaBr_2 \ + \ \underset{\text{Bromomaleate.}}{(C_4H_2BrO_4)_2Ba}$$

Ethylic dibromosuccinate, $C_2H_2Br_2(CO_2C_2H_5)_2$, melts at 58° C. (136.4° F.), and boils at 140°–150° C. (284°–302° F.).

Sulphosuccinic acid, $C_2H_3(SO_3H)(CO_2H)_2$, is a tribasic acid formed by dissolving succinic acid in fuming sulphuric acid, and by the combination of fumaric or maleic acid with the acid sulphites of the alkali-metals.

Isosuccinic or **Ethidene-dicarbonic Acid**, $H_3C\!-\!CH\!<^{CO_2H}_{CO_2H}$, is prepared from α-chloropropionic acid, through the medium of the cyanogen-derivative:—

$$H_3C\!-\!CH\!<^{CN}_{CO_2H} \ + \ 2H_2O \ = \ NH_3 \ + \ H_3C\!-\!CH\!<^{CO_2H}_{CO_2H}.$$

It cannot be prepared from ethidene dibromide, $H_3C\!-\!CHBr_2$; for on heating this compound with potassium cyanide and an alkali, a molecular transposition takes place, resulting in the formation of ordinary or ethenesuccinic acid.

Isosuccinic acid crystallizes in needles soluble in 4 parts of water. It melts at 130° C. (266° F.), and is resolved at higher temperatures into propionic acid and carbon dioxide:—

$$CH_3\!-\!CH(CO_2H)_2 \ = \ CO_2 \ + \ CH_3\!-\!CH_2\!-\!CO_2H.$$

The same decomposition takes place on heating the acid with water above 100°; in fact, this mode of decomposition into CO_2 and a monocarbon acid is characteristic of the dicarbon acids in which the two carboxyl groups are attached to one carbon-atom.

Dibromisosuccinic acid, $C_2H_2Br_2(CO_2H)_2$, is formed by addition of bromine to maleic acid:—

$$H_2C\!=\!C\!<^{CO_2H}_{CO_2H} \ + \ Br_2 \ = \ H_2CBr\!-\!CBr\!<^{CO_2H}_{CO_2H}.$$

33 *

It is crytalline, easily soluble in water, melts at 150° C. (302° F.), and is resolved on further heating, or on boiling with water, into HBr, and iso-bromomaleic acid, $CHBr{=}C(CO_2H)_2$. Sodium amalgam converts it, by molecular transposition, into ordinary succinic acid.

Pyrotartaric Acids, $C_5H_8O_4 = C_3H_6{<}^{CO_2H}_{CO_2H}$.—Of these acids there are four modifications :—

<pre>
CH₃   CH₂.CO₂H   CH₃   CH₃
| | | |
CH.CO₂H   CH₂   CH₂   C(CO₂H)₂
| | | |
CH₂.CO₂H   CH₂.CO₂H   CH(CO₂H)₂   CH₃
Pyrotartaric.   Glutaric.   Ethylmalonic.   Dimethyl-
                                                       malonic.
</pre>

•**Pyrotartaric** or **Methyl-succinic Acid**, $CH_3.CH{<}^{CO_2H}_{CH_2.CO_2H}$, is obtained by the dry distillation of tartaric acid, mixed with an equal weight of powdered pumice ; synthetically, also, from propene bromide, through the medium of the cyanide :—

$$CH_3.CH{<}^{CN}_{CH_2.CN} + 4H_2O = 2NH_3 + CH_3.CH{<}^{CO_2H}_{CH_2.CO_2H} ;$$

also by the action of nascent hydrogen on the three isomeric acids, ita-conic, citraconic, and mesaconic :—

$$C_5H_6O_4 + H_2 = C_5H_8O_4 ;$$

and lastly, by treating allyl iodide with potassium cyanide, and boiling the resulting nitril with a caustic alkali.

It crystallizes in small rhombic prisms, easily soluble in water, alcohol, and ether, melts at 112° C. (233.6° F.), and when rapidly heated is re-solved into water and **pyrotartaric anhydride**, which distils over, and boils at 230° C. (446° F.).

<pre>
CH₃   CH₃
| |
CH.CO.OH   =   H₂O   +   CH..CO
| | >O .
CH₂.CO.OH   CH₂.CO
</pre>

On heating it for a longer time to 200°–210° C. (392°–410° F.), or on exposing its aqueous solution, mixed with a uranium-salt, to sunshine, it is restored into CO_2 and butyric acid, $CH_3{-}CH_2{-}CH_2{-}CO_2H$.

Neutral calcium pyrotartrate, $C_5H_6O_4Ca + 2H_2O$, and the acid potas-sium salt, $C_5H_7O_4K$, are sparingly soluble in water.

Normal Pyrotartaric Acid, $CH_2{<}^{CH_2.CO.OH}_{CH_2.CO.OH}$, also called g l u t a r i c

acid, is formed by heating propene cyanide, $CH_2{<}^{CH_2.CN}_{CH_2.CN}$, with strong

hydrochloric acid, to 100°, in a sealed tube for three or four hours. The contents of the tube are then evaporated down on a water-bath, treated with absolute alcohol to separate ammonium chloride, and the alcoholic solution is evaporated down, whereupon the acid is left as a thick brown syrup, which very slowly crystallizes. It may be purified by conversion into silver salt, and separation therefrom by hydrogen sulphide. The same acid is obtained by heating oxyglutaric acid (q. v.), with concentrated hydriodic acid to 120° C. (248° F.). It forms large transparent monoclinic crystals, easily soluble in water, melting at 97° C. (206.6° F.), and decomposing above 280° C. (536° F.) into water and the aubydride, $CH_2 <^{CH_2 . CO}_{CH_2 . CO} > O$.

Ethylmalonic Acid, $CH_3 . CH_2 . CH <^{CO_2H}_{CO_2H}$, is prepared from α-bromobutyric acid (p. 703), through the medium of the cyano-compound :

$$CH_3 . CH_2 . CH(CN) . CO_2H + 2H_2O = NH_3 + CH_3 . CH_2 . CH <^{CO_2H}_{CO_2H} .$$

It crystallizes in colorless prisms, resembling pyrotartaric acid, and melting, like the latter, at 112° C. (233.6° F.). When heated to 160°, it is resolved into CO_2 and butyric acid. The *calcium salt*, $C_5H_6O_4Ca + H_2O$, forms prisms more soluble in cold than in hot water. The *barium salt* is anhydrous. The *copper salt*, $C_5H_6O_4Cu + H_2O$, crystallizes in beautiful tablets.

Dimethyl-malonic Acid, $(CH_3)_2C(CO_2H)_2$, prepared from bromisobutyric acid, is less soluble in water than either of the two preceding acids. It decomposes when melted, but does not yield butyric acid.

Substitution-products of the Pyrotartaric Acids.—Isomeric chloro- and bromo-derivatives of these acids are formed by direct addition of HCl, HBr, and Br_2, to the unsaturated acids, $C_5H_6O_4$, viz., itaconic, citraconic, and mesaconic acids, these products being called respectively *ita*-, *citra*-, and *mesa*- derivatives of the pyrotartaric acids.

The monochlorinated derivatives, $C_5H_7ClO_4$, are formed by treating the three isomeric acids, $C_5H_6O_4$, with strong hydrochloric acid. They are all three crystalline.—*Itachloropyrotartaric acid* melts at 145° C. (293° F.), and when heated with water or alkalies, is converted into itamalic acid, $C_5H_7(OH)O_4$. — *Citrachloropyrotartaric acid* is very unstable, and when heated with water gives up HCl, and passes into mesaconic acid, $C_5H_6O_4$ (p. 783). By boiling with alkalies, it is resolved into CO_2, HCl, and methacrylic acid, $C_4H_6O_2$.—*Mesachloropyrotartaric acid* is more stable than the last, melts at 129° C. (264.2° F.), and is converted by heating with water into mesamalic acid, $C_5H_8O_5$.

The three isomeric dibromopyrotartaric acids differ from one another in their degree of solubility in water. The ita-compound is converted, by boiling the aqueous solution of its sodium salt, into aconic acid, $C_5H_4O_4$. The citra- and mesa-compounds, on the other hand, yield bromocrotonic acid, C_4H_5BrO.

All these chloro-derivatives, and the corresponding bromo- and iodopyrotartaric acids, are converted by nascent hydrogen into ordinary pyrotartaric acid.

The constitution of the substituted pyrotartaric acids will be understood from that of the three isomeric acids, $C_5H_6O_4$ (p. 782).

Adipic Acids, $C_6H_{10}O_4 = C_4H_8 <^{CO_2H}_{CO_2H}$.—1. *Normal Adipic acid*, $CO_2H—(CH_2)_4—CO_2H$, originally obtained by the oxidation of fats with nitric acid, is formed synthetically by heating β-iodo-propionic acid with finely divided silver :

$$\left. \begin{array}{l} CH_2I—CH_2—CO_2H \\ \\ CH_2I—CH_2—CO_2H \end{array} \right\} + Ag_2 = 2AgI + \begin{array}{l} CH_2—CH_2—CO_2H \\ | \\ CH_2—CH_2—CO_2H \end{array}$$

β-Iodopropionic acid.
2 mol.   Adipic acid.

It is also produced by the action of nascent hydrogen on hydro-muconic acid, $C_6H_8O_4$; by oxidizing sebacic acid with nitric acid; and, together with acetic acid and carbon dioxide, by oxidation of phorone with chromic acid:

$$C_9H_{14}O + O_7 = C_6H_{10}O_4 + C_2H_4O_2 + CO_2.$$

This acid crystallizes in shining laminæ or prisms, dissolves in 13 parts of cold water, and melts at 148° C. (298.4° F.).

2' *Isoadipic* or *Dimethyl-succinic acid*,
$$\begin{array}{l} CH_3—CH—CO_2H \\ \quad\quad | \\ CH_3—CH—CO_2H \end{array}$$

produced by heating a-bromopropionic acid, $CH_3—CHBr—CO_2H$, with finely divided silver, forms a thick syrup which does not readily crystallize.

The higher acids of this series are formed by the oxidation of stearic acid, oleic acid, and other acids of the fatty and acrylic series with nitric acid,—succinic acid and some of the lower homologues being generally formed at the same time. The mixed acids thus obtained are separated by fractional crystallization from ether, the higher members separating out first.

Pimelic Acid, $C_7H_{12}O_4$, is also produced by fusing camphoric acid with potash. It melts at 114° C. (237.2° F.), and dissolves in 40 parts of cold water.

Suberic Acid, $C_8H_{14}O_4$, is most readily obtained by boiling cork with nitric acid. It crystallizes in long needles or plates, melts at 140° C. (284° F.), and sublimes without decomposition between 150°–160° C. (302°–320° F.). It dissolves in 100 parts of cold water, easily in hot water, alcohol, and ether.

An isomeric acid, tetramethylsuccinic acid,
$$\begin{array}{l} (CH_3)_2C—CO_2H \\ \quad\quad | \\ (CH_3)_2C—CO_2H \end{array},$$
is formed by heating bromisobutyric acid $(CH_3)_2CBr.CO_2H$, with reduced silver. It melts at 95° C. (203° F.), and dissolves in 45 parts of water at 10° C. (50° F.).

A third isomeride, diethylsuccinic acid,
$$\begin{array}{l} C_2H_5.CH.CO_2H \\ \quad\quad | \\ C_2H_5.CH.CO_2H \end{array},$$ is formed
in like manner from a-bromobutyric acid, $C_2H_5.CHBr.CO_2H$.

Anchoic Acid, or **Lepargylic Acid,** $C_9H_{16}O_4$, is formed, together with other products, by the action of nitric acid on Chinese wax and on the fatty acids of cocoa-nut oil.—Azelaic acid, obtained by oxidizing castor oil with nitric acid, has the same composition as anchoic acid, but differs so much from it in physical properties, that it must be regarded as an isomeric or allotropic modification.

Sebic or **Sebacic Acid,** $C_{10}H_{18}O_4$, is a constant product of the destructive distillation of oleic acid, olein, and all fatty substances containing those bodies; it is extracted by boiling the distilled matter with water: it is also formed by the action of potash on castor-oil (see p. 620). It forms small pearly crystals resembling those of benzoic acid. It has a faintly acid taste, is but little soluble in cold water, melts when heated, and sublimes unchanged.

Brassylic Acid, $C_{11}H_{20}O_4$, obtained by oxidation of behenolic acid and erucic acid, melts at 1.08°, and is nearly insoluble in water.

Roccellic Acid, $C_{17}H_{32}O_4$, exists in *Roccella tinctoria*, and other lichens of the same genus, also in *Lecanora tartarea*, and is obtained by exhausting the first-mentioned plant with aqueous ammonia, precipitating the filtered liquor with calcium chloride, and decomposing the resulting calcium salt with hydrochloric acid. When purified by solution in ether, it forms white, rectangular, four-sided tabular crystals, melting at 132° C. (269.6° F.), and subliming at 200° C. (392° F.), being partially converted at the same time into an oxide, $C_{17}H_{30}O_3$. This acid decomposes carbonates.

---

2. **Unsaturated Acids,** $C_nH_{2n-4}O_4$ or $C_nH_{2n-2} \begin{cases} CO_2H \\ CO_2H \end{cases}$. — This series includes the following groups of isomeric acids

Fumaric and Maleic acids . . . . $C_4H_4O_4$
Itaconic, Citraconic, Mesaconic, and Paraconic acids . . . . $C_5H_6O_4$
Hydromuconic acid . . . . . $C_6H_8O_4$.

These acids are capable of taking up two atoms of hydrogen, bromine, and other monad elements, and passing into the saturated acids of the preceding series. A general method of forming them consists in heating the dibrominated derivatives of the acids, $C_nH_{2n-2}O_4$, with solution of potassium iodide; *e. g.*:

$$\begin{matrix} CHBr.CO_2H \\ | \\ CHBr.CO_2H \end{matrix} + 2KI = 2KBr + I_2 + \begin{matrix} CH.CO_2H \\ \| \\ CH.CO_2H \end{matrix}$$

Dibromosuccinic.   Fumaric.

The isomeric modifications of these acids are determined by the structure of the radicles C_nH_{2n-2}, associated with the two carboxyl groups.

Fumaric and **Maleic Acids,** $C_4H_4O_4$,

$$\begin{matrix} CH.CO_2H \\ \| \\ CH.CO_2H \end{matrix} \qquad \begin{matrix} CH_2 \\ \| \\ C \end{matrix}\begin{matrix} \\ < \end{matrix}\begin{matrix} CO_2H \\ CO_2H \end{matrix}$$

Fumaric.   **Maleic.**

These two acids are produced by the dry distillation of malic acid:

$$C_4H_6O_5 = H_2O + C_4H_4O_4.$$

When malic acid is heated in a small retort nearly filled with it, it melts, boils, and gives off water, together with maleic acid and maleic anhydride, which pass over into the receiver, and dissolve in the water. After a time, small solid, crystalline scales make their appearance in the boiling liquid, and increase in quantity until the whole becomes solid. The process may now be interrupted, and the contents of the retort, after cooling, treated with cold water ; unaltered malic acid is thereby dissolved out, and fumaric acid, which is less soluble, remains behind.

Fumaric acid exists, in the free state, in several plants, as in the common fumitory (*Fumaria officinalis*), Iceland moss (*Cetraria islandica*), and in certain fungi. It is produced also, as above stated, by the action of potassium iodide on dibromosuccinic acid, and from monobromosuccinic and sulphosuccinic acids by fusion with potash.

Fumaric acid forms small, white, crystalline laminæ, which dissólve freely in hot water and alcohol, but require for solution about 200 parts of cold water : it is unchanged by hot nitric acid.   When heated in a current of air, it sublimes, but by distillation in a retort, it is resolved in water and m a l e i c   a n h y d r i d e, $C_4H_2O_3$. Similar differences are often observed in the behavior of organic bodies of small volatility, according as they are heated in close vessels or in a current of air. Fumaric acid is converted by *sodium amalgam, hydriodic acid*, and other hydrogenizing agents, into ordinary succinic acid. It unites, in presence of water, with metallic zinc, forming succinate of zinc, $C_4H_4O_4Zn$.

Fumaric acid forms acid and neutral metallic salts. The *calcium* and *barium salts* are anhydrous. The *silver salt*, $C_4H_2O_4Ag_2$, is quite insoluble in water. The *ethylic ether*, $C_4H_2O_4(C_2H_5)_2$ is a liquid boiling at 225O C. (437O F.).

M a l e i c   A c i d, $H_2C{=}C{<}^{CO_2H}_{CO_2H}$, crystallizes in large prisms or tables, very soluble in water, alcohol, and ether, and having a strong acid taste and reaction. It is converted by heat into fumaric acid, by nascent hydrogen into succinic acid, and by bromine into dibromisosuccinic acid, $H_2BrC{-}CBr(CO_2H)_2$. Its aqueous solution dissolves zinc without evolution of hydrogen, forming maleate and succinate of zinc :

$$3C_4H_4O_4 \ + \ 2Zn \ = \ C_4H_2O_4Zn \ + \ (C_4H_4O_2)_2H_2Zn.$$

Maleic anhydride, $C_4H_2O_3$, crystallizes in large laminæ or needles, melts at 57O C. (134.6O F.), boils without decomposition at 196O C. (384.8O F.), and is converted by water into maleic acid.

Maleic and fumaric acids are resolved by electrolysis of the concentrated solutions of their sodium salts into carbon dioxide, hydrogen and acetylene :—

$$C_2H_2(CO_2H)_2 \ = \ C_2H_2 \ + \ 2CO_2 \ + \ H_2 .$$

A c i d s, $C_5H_6O_4 = C_3H_4{<}^{CO_2H}_{CO_2H}$.—Theory indicates the existence of five isomeric acids of this form, and of these four are known, viz., c i t r a c o n i c and m e s a c o n i c   a c i d s, which may be derived from fumarie acid by interpolation of CH_2, and i t a c o n i c and p a r a c o n i c   a c i d s, derivable in like manner from maleic acid :—

From Fumaric Acid, $\begin{array}{l}CH.CO_2H \\ CH.CO_2H\end{array}$		From Maleic Acid, $\begin{array}{l}CH_2 \\ C(CO_2H)_2\end{array}$	

$$
\begin{array}{cccc}
CH.CO_2H & CH.CO_2H & CH_2 & CH_2 \qquad HC{-}CH_3 \\
\|\ \ \ & \| & \| & \|\qquad \quad \ C{<}^{CO_2H}_{CO_2H} \\
CH & C{-}CO_2H & C{-}CO_2H & CH \\
| & | & | & |\quad\ \ {<}^{CO_2H}_{CO_2H} \\
CH_2.CO_2H & CH_3 & CH_2.CO_2H & HC
\end{array}
$$

Mesaconic.   Citraconic.   Itaconic.   Paraconic acid.   **(?)**

C i t r a c o n i c and i t a c o n i c   a c i d s are produced by the action of heat on citric acid. When crystallized citric acid is heated in a retort, it first melts in its water of crystallization, and then boils, giving off water. Afterwards, at about 175O C. (347O F.), vapors of acetone distil over, and a copious disengagement of carbon monoxide takes place. At this time the residue in the retort consists of aconitic acid. If the distillation be still continued, carbon dioxide is given off, and itaconic acid crystallizes in the neck of the retort. If these crystals be repeatedly distilled, an oily mass of citraconic oxide or anhydride is obtained, which no longer solidifies. These decompositions are represented by the following equations :—

$$\underset{\text{Citric acid.}}{C_6H_8O_7} - H_2O = \underset{\text{Aconitic acid.}}{C_6H_6O_6}; \quad \underset{\text{Aconitic acid.}}{C_6H_6O_6} - CO_2 = \underset{\text{Itaconic acid.}}{C_5H_6O_4};$$

$$\underset{\text{Itaconic acid.}}{C_5H_6O_4} - H_2O = \underset{\substack{\text{Citraconic} \\ \text{anhydride.}}}{C_5H_4O_3}$$

The citraconic anhydride when exposed to the air absorbs moisture, and is converted into crystallized citraconic acid, $C_5H_6O_4$.

M e s a c o n i c a c i d is produced by boiling itaconic acid with weak nitric acid. These three isomeric acids are all converted by nascent hydrogen into p y r o t a r t a r i c a c i d, $C_5H_8O_4$. They also take up a molecule of HBr, HCl, HI, forming isomeric monobromopyrotartaric acids, $C_5H_7BrO_4$, etc., or of bromine, Br_2, forming isomeric dibromopyrotartaric acids. Itaconic and citraconic acids are, however, more inclined to these transformations than mesaconic acid, which is altogether a more stable compound.

On subjecting their potassium salts to electrolysis, the three acids are decomposed, yielding a hydrocarbon, C_3H_4, according to the equation :—

$$C_3H_4(CO_2H)_2 = C_3H_4 + 2CO_2 + H_2.$$

Now, citraconic acid thus treated yields ordinary a l l y l e n e, $CH{\equiv}C{-}CH_3$, whereas itaconic acid yields i s o - a l l y l e n e or a l l e n e, $CH_2{=}C{=}CH_2$, results which are in accordance with the formulæ above given for these acids. Mesaconic acid likewise yields allylene, as might be expected, since the hydrocarbon, $CH{=}CH{-}CH_2$, cannot exist.

Itaconic Acid is most easily prepared by heating citraconic anhydride with water to 130°–140° C. (266°–284° F.). It crystallizes in rhombic octohedrons, dissolves in 17 parts of water at 10° C. (50° F.), melts at 161° C. (321.8° F.), and is resolved by distillation into water and citraconic anhydride.

Citraconic Acid crystallizes in four-sided prisms, melting at 80° C. (176° F.). It is much more soluble in water than itaconic acid, and deliquesces on exposure to the air. Its anhydride, $C_5H_4O_3$, forms an oily liquid, which easily recombines with water to form the acid.

Mesaconic Acid forms shining prisms sparingly soluble in water, melts at 208° C. (406.4° F.), and sublimes without decomposition. It is most readily obtained by the action of heat on citrachloropyrotartaric acid.

Paraconic Acid is formed, together with itamalic acid, $C_5H_8O_5$, by heating itachloropyrotartaric acid, $CH_2Cl{-}CH.CO_2H{-}CH_2.CO_2H$, with water. It is easily soluble in water, melts at 70° C. (158° F.), and is resolved by distillation into water and citraconic anhydride. It unites with HBr, forming itabromopyrotartaric acid. When heated with alkalies it takes up water, and forms itamalic acid, $C_5H_6O_5$.

The constitution of paraconic acid is probably represented either by the fourth or the fifth formula above given ; but there is at present no means of deciding between the two. Moreover, it is not easy to see how an acid having its two carboxyl groups associated with the same carbon-atom could be formed from itabromopyrotartaric acid, unless the reaction were accompanied by molecular transposition.

Hydromuconic Acid, $C_6H_8O_4 = C_4H_6(CO_2H)_2$, produced by the action of sodium-amalgam on dichloromuconic acid, $C_6H_6Cl_2O_4$, crystallizes in large prisms, slightly soluble in cold water, and melting at 195° C. (383° F.). It is converted by sodium-amalgam into adipic acid, $C_6H_{10}O_4$, and unites with bromine, forming dibromadipic acid, $C_4H_8Br_2O_4$.

3. Unsaturated Acids, $C_nH_{2n-6}O_4$.

Aconic Acid, $C_5H_4O_4$.—This acid is formed by boiling itadibromopyrotartaric acid with caustic soda. It is very soluble in water, alcohol, and ether, and crystallizes from the alcoholic solution in foliate groups of shining needles, melting at 154° C. (309.2° F.). From its origin it might be expected to be a bibasic acid; but it is really monobasic, its silver-salt being $C_5H_3O_4Ag$, and its barium salt $(C_5H_3O_4)_2Ba$. This may perhaps be explained by regarding the acid as an anhydro-acid similar to dilactic acid, its mode of formation being represented by the following equation :

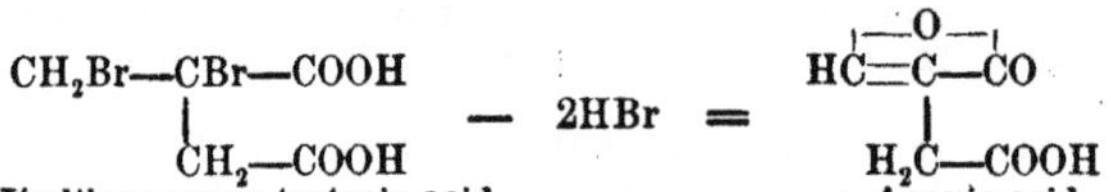

$$\begin{array}{ccc}
\text{CH}_2\text{Br—CBr—COOH} & & \overset{\displaystyle |\!-\!\text{O}\!-\!|}{\text{HC}\!=\!\text{C}\!-\!\text{CO}} \\
| & -\ 2\text{HBr}\ = & | \\
\text{CH}_2\text{—COOH} & & \text{H}_2\text{C—COOH} \\
\text{Itadibromopyrotartaric acid.} & & \text{Aconic acid.}
\end{array}$$

By boiling with baryta-water, aconic acid is resolved into formic and succinic acids :

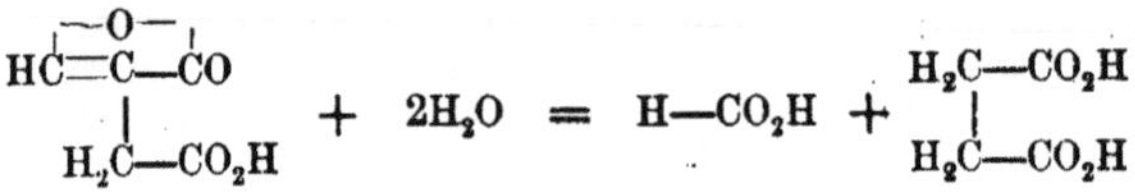

$$\begin{array}{ccccccc}
\overset{\displaystyle |\!-\!\text{O}\!-\!|}{\text{HC}\!=\!\text{C}\!-\!\text{CO}} & & & & & & \text{H}_2\text{C—CO}_2\text{H} \\
| & +\ 2\text{H}_2\text{O}\ = & \text{H—CO}_2\text{H} & + & & & | \\
\text{H}_2\text{C—CO}_2\text{H} & & & & & & \text{H}_2\text{C—CO}_2\text{H}
\end{array}$$

Muconic Acid, $C_6H_6O_4$, formed in like manner from dibromadipic acid, forms large crystals melting at 100°. It is monobasic, like aconic acid, and is probably constituted in a similar manner. By boiling with baryta-water it is resolved into acetic and succinic acids.

Triatomic Acids.

1. MONOBASIC, $C_nH_{2n}O_4$.

These acids are derived from the triatomic alcohols $C_nH_{2n+2}O_3$ (glycerins) by substitution of O for H_2, in the same manner as the acids of the lactic series, $C_nH_{2n}O_3$ from the glycols, $C_nH_{2n+2}O_2$. There is, however, but one acid of the series at present known, viz. :

Glyceric Acid, $C_3H_6O_4$ (dioxypropionic acid), which is formed by the gradual oxidation of glycerin with nitric acid :

$$\begin{array}{ccccccc}
\text{CH}_2.\text{OH} & & & & & & \text{CH}_2.\text{OH} \\
| & & & & & & | \\
\text{CH.OH} & +\ \text{O}_2\ = & \text{H}_2\text{O} & + & & & \text{CH.OH}; \\
| & & & & & & | \\
\text{CH}_2.\text{OH} & & & & & & \text{CO.OH}
\end{array}$$

also by heating glycerin to 100° in a sealed tube, with bromine and water :

$$C_3H_8O_3\ +\ 2Br_2\ +\ H_2O\ =\ 4HBr\ +\ C_3H_6O_4.$$

To prepare it, nitric acid (specific gravity 1.5), is poured through a long-necked funnel to the bottom of a tall glass jar containing glycerin, diluted with an equal bulk of water (100 grains of glycerin, 100 of water, and 100 to 150 of red nitric acid, are good proportions). The two layers of liquid gradually mix, and assume a blue color, and the oxidation of the glycerin proceeds, accompanied by copious evolution of gas ; if the liquid becomes too hot, the action must be moderated by external cooling. When the action is completed, which takes five or six days, the acid liquid is evaporated to a syrup, diluted with water, saturated at the boiling heat with chalk and a small quantity of milk of lime, and then filtered. The concentrated filtrate deposits calcium glycerate in warty crusts, from which the glyceric acid may be separated by boiling with oxalic acid. The liquid filtered from the calcium oxalate is boiled with lead oxide to remove any excess of oxalic acid, then treated with hydrogen sulphide to precipitate the dissolved lead, and the filtered liquid is evaporated over the water-bath.

Glyceric acid when concentrated is a colorless, uncrystallizable syrup, very soluble in water and in alcohol. Heated above 140° C. (284° F.), it is decomposed, yielding water, pyruvic acid, and pyrotartaric acid. By fusion with potash it is resolved into acetic and formic acids ; by boiling with aqueous potash, it yields oxalic and lactic acids, and by the action of phosphorus iodide it is converted into ẞ-iodopropionic acid.

The metallic glycerates are soluble in water, and crystallize well. They are not reddened by ferrous salts, and are thereby distinguished from the pyruvates, from which they differ only by the elements of water. The *calcium salt*, $(C_3H_5O_4)_2Ca + 2H_2O$, usually crystallizes in nodular groups of needles, easily soluble in water : the lead salt, $(C_3H_5O_4)_2Pb$, is but slightly soluble in water.

The *ethylic ether*, $C_3H_5O_4.C_2H_5$, obtained by heating glycerin with absolute alcohol, is a thick liquid, having a specific gravity of 1.193 at 0°, and boiling at 230°–240° C. (446°–464° F.).

Amidoglyceric acid, $CH_2.NH_2$—$CH.OH$—CO_2OH, or serine, is obtained by boiling sericin or silk-gelatin (*q. v.*), with dilute sulphuric acid. It forms hard crystals, soluble in water, but insoluble in alcohol and ether. It unites both with acids and with bases. Nitrous acid converts it into glyceric acid.

Anhydrides of Glyceric Acid.

Pyruvic or **Pyroracemic Acid,** $C_3H_4O_3 = O{<}\begin{matrix} CH_3 \\ | \\ CH \\ | \\ CO_2H \end{matrix}$, or $CO{<}^{CH_3}_{CO_2H}$.

—This anhydro-acid is formed, together with other products, by the dry distillation of glyceric, tartaric, or racemic acid :

$$C_3H_6O_4 = C_3H_4O_3 + H_2O$$
$$C_4H_6O_6 = C_3H_4O_3 + CO_2 + H_2O ;$$

and is obtained pure by redistilling the product several times, and collecting apart the portion which passes over between 165° and 170° C. (329°–338° F.). It is a yellowish liquid, easily soluble in water, alcohol, and ether ; smells like acetic acid, and boils with partial decomposition at 165°–170°. It is monobasic, and forms salts which crystallize well, provided that heat is avoided in their preparation ; but their solutions, if evaporated by heat, leave gummy uncrystallizable salts, which yield a syrupy non-volatile modification of the acid, likewise obtained when an aqueous solution of the original acid is evaporated by heat. This syrupy acid, which is probably a polymeric modification, is resolved by heat into carbon dioxide and pyrotartaric acid, $2C_3H_4O_3 = CO_2 + C_5H_8O_4$.

Pyruvic acid is converted by nascent hydrogen into ordinary lactic acid, CH_3—CHOH—COOH, and may therefore be regarded as a ketonic acid related to lactic acid in the same manner as dimethyl-ketone (acetone) to secondary propyl alcohol, as represented by the second of the constitutional formulæ given on p. 785:

CH_3—CO—CH_3
Dimethyl ketone.

CH_3—CO—COOH
Pyruvic acid.

CH_3—CHOH—CH_3
Pseudopropyl alcohol.

CH_3—CHOH—COOH
α-Lactic acid.

The reaction, however, agrees equally well with the first formula, which represents pyruvic acid as an anhydride of glyceric acid, from which it is formed by actual dehydration.

Pyruvic acid is converted by phosphorus pentachloride into dichloropropionic chloride, CH_2Cl—CHCl—COCl; by hydrochloric acid at 100° into carbon dioxide and pyrotartaric acid; by oxidizing agents into oxalic acid; by boiling with baryta-water into uvitic acid, $C_9H_8O_4 = C_6H_3 \begin{cases} CH_3 \\ (CO_2H)_2 \end{cases}$, an acid belonging to the aromatic group.

The pyruvates crystallize well, provided that heat is avoided in their preparation. The *sodium salt*, $C_3H_3O_3Na$, forms large anhydrous prisms. The *lead salt*, $(C_3H_3O_3)_2Pb$, is a crystalline precipitate. The *silver salt* is also crystalline.

Oxypyruvic or **Carbacetoxylic Acid**, $C_3H_4O_4$ or CH_2OH—CO—CO_2H, isomeric with malonic acid, is formed by heating β-chloropropionic acid with excess of silver oxide:

$$CH_2Cl.CH_2.CO_2H + 3Ag_2O = CH_2OH.CO.CO_2Ag + AgCl + 2Ag_2 + H_2O.$$

It forms a syrup, easily soluble in water and in ether. It is monobasic; its *barium salt* crystallizes in spherical nodules; the *lead salt* in crusts; the *zinc salt* in shining scales.

The acid is converted by nascent hydrogen into glyceric acid, and by hydriodic acid at 200° C. (392° F.) into pyruvic acid.

TRIATOMIC AND BIBASIC ACIDS.

$$C_nH_{2n-2}O_5, \text{ or } C_nH_{2n-1}(OH) < ^{CO_2H}_{CO_2H}.$$

The acids of this series may be formed from those of the oxalic or succinic series, $C_nH_{2n}(CO_2H)_2$, by substitution of OH for H. Four of them are at present known, viz.:

Tartronic acid, $C_3H_4O_5$.   Oxypyrotartaric acid, $C_5H_8O_5$.
Malic acid, $C_4H_6O_5$.   Oxyadipic acid, $C_6H_{10}O_5$.

Tartronic Acid, $CH(OH) < ^{CO_2H}_{CO_2H}$ (oxymalonic acid), is formed by the action of nascent hydrogen on mesoxalic acid (p. 775):

$$CO(CO_2H)_2 + H_2 = CHOH(CO_2H)_2;$$

also by spontaneous decomposition of dinitrotartaric acid, when its aqueous solution is left to evaporate, the decomposition being attended with evolution of carbon dioxide and nitrogen dioxide :

$$C_2H_2(O.NO_2)_2(CO_2H)_2 = CHOH(CO_2H)_2 + CO_2 + N_2O_2.$$

Tartronic acid crystallizes in large prisms, which melt at 175° C. (347° F.), and are resolved at higher temperatures into carbon dioxide, water, and glycollide :

$$C_3H_4O_5 = CO_2 + H_2O + C_2H_2O_2.$$

Malic Acid, $C_4H_6O_5 = C_2H_3(OH) < \genfrac{}{}{0pt}{}{CO_2H}{CO_2H} = CH(OH) < \genfrac{}{}{0pt}{}{CO_2H}{CH_2.CO_2H}$,

Oxysuccinic acid.—This acid is formed synthetically by the action of moist silver oxide on bromosuccinic acid :

$$C_2H_3Br(CO_2H)_2 + AgOH = AgBr + C_2H_3OH(CO_2H)_2.$$

It is also produced by deoxidation of tartaric acid, $C_4H_6O_6$, with hydriodic acid, and by the action of nitrous acid on aspartic acid, $C_4H_7NO_4$ (amido-succinic acid), or on asparagin, $C_4H_8N_2O_3$, which is the amide of the latter :

$$CH(NH_2) < \genfrac{}{}{0pt}{}{CO.OH}{CH_2.CO.OH} + NO.OH = CH(OH) < \genfrac{}{}{0pt}{}{CO.OH}{CH_2.CO.OH}$$
Aspartic acid.
$$+ N_2 + H_2O$$

$$CH(NH_2) < \genfrac{}{}{0pt}{}{CO.OH}{CH_2.CO.NH_2} + 2(NO.OH) = CH(OH) < \genfrac{}{}{0pt}{}{CO.OH}{CH_2.CO.OH}$$
Asparagin.
$$+ 2N_2 + 2H_2O$$

Malic acid is the acid of apples, pears, and various other fruits ; it is often associated with citric acid. It may be advantageously prepared from the juice of the garden rhubarb, in which it exists in large quantity, accompanied by acid potassium oxalate. The rhubarb stalks are peeled, and ground or grated to pulp, which is subjected to pressure. The juice is heated to the boiling point, neutralized with potassium carbonate, and mixed with calcium acetate : insoluble calcium oxalate then falls, and may be removed by filtration. To the clear and nearly colorless liquid, solution of lead acetate is added as long as a precipitate continues to be produced, and the lead malate is collected on a filter, washed, diffused through water, and decomposed by sulphuretted hydrogen. The filtered liquid is carefully evaporated to the consistence of a syrup, and left in a dry atmosphere till it becomes converted into a solid and somewhat crystalline mass of malic acid. From the berries of the mountain ash (*Sorbus aucuparia*), in which malic acid is likewise present in considerable quantity, especially at the time they begin to ripen, the acid may be prepared by the same process.

Malic acid crystallizes in groups of colorless prisms, slightly deliquescent and very soluble in water ; alcohol also dissolves it. The aqueous solution has an agreeable acid taste : it becomes mouldy and spoils by keeping.

Malic acid, as it exists in plants, and as obtained from active tartaric acid, from aspargin, or from aspartic acid produced from the latter, exerts a rotatory action on polarized light ; [a] = —5° ; but by the action of nitrous acid on inactive aspartic acid (resulting from the decomposition of fumarimide), Pasteur has obtained a modification of malic acid which is optically inactive. Malic acid formed from succinic acid is also inactive.

Malic acid when heated gives off water at 130° C. (266° F.), and at 175° C. (347° F.) a distillate of maleic acid and maleic anhydride, while fumaric acid remains behind (p. 782). By slow oxidation with a cold solution of potassium chromate, it is converted into maleic acid :—

$$C_4H_6O_5 + O_2 = CO_2 + H_2O + C_3H_4O_4 .$$

Nitric acid readily converts it into oxalic acid, with evolution of carbon dioxide.

By the action of reducing agents, most readily by heating with strong hydriodic acid, malic acid is reduced to succinic acid ; also by fermentation of its calcium salt in contact with putrefying cheese, acetic acid and carbon dioxide being also produced :—

$$3C_4H_6O_5 = 2C_4H_6O_4 + C_2H_4O_2 + 2CO_2 + H_2O .$$

The sodium salt of bromomalic acid, $C_4H_5BrO_5$, obtained by boiling an aqueous solution of sodium dibromosuccinate, $(C_4H_3NaBr_2O_4)$, is converted by boiling with lime-water into the calcium salt of tartaric acid, $C_4H_6O_6$:

$$C_4H_5BrO_5 + H_2O = HBr + C_4H_6O_6 .$$

Malic acid forms both acid and neutral salts. Those formed from the optically active acid are likewise active, some being dextro-, others levo-rotatory. The most characteristic of the malates are *acid ammonium malate*, $C_4H_5(NH_4)O_5$, which crystallizes remarkably well, and *lead malate*, $C_4H_4PbO_5.3Aq.$, which is insoluble in pure water, but dissolves to a considerable extent in warm dilute acids, and separates on cooling in brilliant silvery crystals containing water. By this character the acid may be distinguished. *Acid calcium malate*, $C_4H_4O_5Ca.C_4H_6O_5 + 8H_2O$, is also a very beautiful salt, freely soluble in warm water. It is prepared by dissolving the sparingly soluble *neutral* malate in hot dilute nitric acid, and leaving the solution to cool.

Diethylic malate, $C_4H_4(C_2H_5)_2O_5$, is a liquid which is partially decomposed by distillation, and is converted by acetyl chloride into **diethylic acetomalate**, $C_2H_3(OC_2H_3O) {<}^{CO_2C_2H_5}_{CO_2C_2H_5} .$

Oxypyrotartaric Acid, $C_5H_8O_5 = C_3H_5(OH){<}^{CO_2H}_{CO_2H}$, is produced by boiling dicyanhydrin, $C_3H_5(OH)(CN)_2$, (p. 640) with alkalies. It forms crystals, easily soluble in water, alcohol, and ether, and melting at 135° C. (275° F.).

Isomeric with it are four bibasic acids of unknown structure, called itamalic, citramalic, mesamalic, and oxyglutaric acids.

Ita- and **mesamalic acids** are formed by boiling the corresponding chloropyrotartaric acids (p. 779) with water or solution of sodium carbonate :—

$$C_3H_5Cl(CO_2H)_2 + H_2O = HCl + C_3H_5(OH)(CO_2H)_2 .$$

Both form deliquescent crystals, melting at 60° C. (140° F.).

Citramalic acid is produced by the action of zinc on chlorocitramalic acid, $C_5H_7ClO_5$ (formed by addition of hypochlorous acid, ClOH, to citraconic acid, $C_5H_6O_4$). It is a deliquescent mass.

Oxyglutaric acid, $C_5H_7(OH)O_4$, produced by the action of nitrous acid on amidoglutaric acid, $C_5H_7(NH_2)O_4$, crystallizes with difficulty, and is converted by hydriodic acid into glutaric acid (p. 778).

Amidoglutaric acid, or *Glutamic acid*, $C_5H_7(NH_2)O_4 = C_3H_5(NH_2){<}^{CO_2H}_{CO_2H}$, occurs, together with aspartic acid, in the molasses of sugar-beet, and is formed by boiling albuminous bodies with dilute sulphuric acid. It forms shining rhombic octohedrons, moderately soluble in water, insoluble in alcohol and ether, melting with partial decomposition at 140° C. (284° F.). It unites both with bases and with acids, and is converted by nitrous acid into oxyglutaric acid.

Oxyadipic Acid, $C_6H_{10}O_5 = C_4H_7(OH) <^{CO_2H}_{CO_2H}$, is a deliquescent mass formed by the action of moist silver oxide on monobromadipic acid.

Oxymaleic Acid, $C_4H_4O_5 = C_2H(OH) <^{CO_2H}_{CO_2H}$, is an unsaturated tri-atomic and bibasic acid, produced by the action of silver oxide on bromomaleic acid (p. 782). It crystallizes in slender needles, easily soluble in water, alcohol, and ether. An acid isomeric with it is formed from brom-isomaleic acid.

TRIATOMIC AND TRIBASIC ACIDS.

Only one saturated acid of this group is known, viz.:

Tricarballylic Acid, $C_6H_8O_6 = C_3H_5(CO_2H)_3$, which is produced: 1. By heating allyl tribromide, $CH_2Br{-}CHBr{-}CH_2Br$, with potassium cyanide, and decomposing the resulting tricyanhydrin with potash:

$$
\begin{array}{l}
CH_2(CN) \\
| \\
CH(CN) \\
| \\
CH_2(CN)
\end{array}
+ \ 3KHO \ = \ 3CNK \ +
\begin{array}{l}
CH_2.CO_2H \\
| \\
CH.CO_2H \\
| \\
CH_2.CO_2H
\end{array}
$$

2. By the action of sodium-amalgam on aconitic acid, $C_6H_6O_6$. 3. By reduction of citric acid, $C_6H_8O_7$, with hydriodic acid.

Tricarballylic acid crystallizes in colorless rhombic prisms, easily soluble in water and alcohol, slightly soluble in ether. The tricarballylates of the alkali-metals are easily soluble in water, the rest insoluble or sparingly soluble. The *ethylic ether*, $C_3H_5(CO.OC_2H_5)_3$, is a liquid boiling between 295° and 305° C. ($563^\circ–581^\circ$ F.).

The following tribasic acids are unsaturated compounds:

Aconitic Acid, $C_6H_6O_6 = C_6H_3(CO_2H)_3$, exists in monk's-hood (*Aconitum Napellus*), and other plants of the same genus, also in *Equisetum fluviatile*, and is one of the products obtained by the dehydration of citric acid: $C_6H_8O_7 - H_2O = C_6H_6O_6$.

When crystallized citric acid is heated in a retort till it begins to become colored, and to undergo decomposition; and the fused, glassy product, after cooling, is dissolved in water, aconitic acid remains as a white, confusedly crystalline mass, which may be purified by converting it into a lead salt, and decomposing the latter with hydrogen sulphide.

Aconitic acid crystallizes in small laminæ, very soluble in water, alcohol, and ether. It melts at 140° C. (284° F.), and decomposes at a higher temperature into carbon dioxide, itaconic acid, and citraconic anhydride. Nascent hydrogen converts it into tricarballylic acid.

Aconitic acid forms three series of salts. The *tertiary lead-salt*, $(C_6H_3O_6)_2Pb_3$, is insoluble in water. The *calcium salt*, $(C_6H_3O_6)_2Ca_3 + 6H_2O$, which is sparingly soluble, occurs abundantly in the expressed juice of monk's-hood; the magnesium salt in that of *Equisetum*. The *ethylic ether*, $C_6H_3O_6(C_2H_5)_3$, is a liquid boiling at about 236° C. (456.8° F.).

The isomeric acid, **aceconitic acid**, formed by the action of sodium on ethylic monobromacetate, crystallizes in slender needles, and forms salts differing in some respects from the aconitates.

Chelidonic Acid, $C_7H_4O_6 = C_4H(CO_2H)_3$, occurs as a calcium salt, together with malic and fumaric acids, in *Chelidonium majus*, and is extracted by boiling the juice, filtering, adding nitric acid, precipitating with lead nitrate, and decomposing the resulting lead salt with hydrogen sulphide. It crystallizes in silky needles containing 1 molecule of H_2O, sparingly soluble in cold water and alcohol. It is decomposed by bromine-water, yielding oxalic acid, bromoform, and pentabromacetone, C_3HBr_5O.

Meconic Acid, $C_7H_4O_7 = C_4HO(CO_2H)_3$ (oxychelidonic acid), is a tribasic acid existing in opium. To prepare it, the liquid obtained by exhausting opium with water, is neutralized with powdered marble and precipitated by calcium chloride ; and the calcium meconate thus precipitated is suspended in warm water and treated with hydrochloric acid ; on cooling, impure meconic acid crystallizes, and may be purified by repeated treatment with hydrochloric acid. The pure acid crystallizes in mica-like plates, easily soluble in boiling, difficultly soluble in cold water, soluble likewise in alcohol. The crystals contain $C_7H_4O_7 + 3H_2O$, and give off their water at 100°; the dehydrated acid melts at 150° C. (302° F.).

Meconic acid forms three series of salts. There are two *silver meconates*, one yellow, containing $C_7HAg_3O_7$; the other white, consisting of $C_7H_2Ag_2O_7$. Meconic acid produces a deep red color with ferric salts. By the action of sodium-amalgam it is converted into hydromeconic acid, $C_7H_{10}O_7$.

Comenic Acid, $C_6H_4O_5$, is a product of decomposition of meconic acid. When an aqueous, or, better, a hydrochloric solution of meconic acid is boiled, carbon dioxide is evolved, and the solution now contains comenic acid, which crystallizes on cooling, being very difficultly soluble in cold water. The same acid may be obtained by heating meconic acid to 200° C. (392° F.). It is bibasic: its formation is represented by the equation $C_7H_4O_7 = C_6H_4O_5 + CO_2$.

Pyromeconic or *Pyrocomenic Acid*, $C_5H_4O_3$, is a monobasic acid, formed by submitting either comenic or meconic acid to dry distillation, one molecule of carbon dioxide being evolved in the former case and two in the latter.

Pyrocomenic acid is a weak acid ; it is soluble in water and alcohol: from these solutions it crystallizes in long colorless needles, which melt at 120° C. (248° F.), and begin to sublime at the boiling point of water. Both comenic and pyrocomenic acids exhibit the red coloration with ferric salt.

Tetratomic Acids.

These acids may be derived from tetratomic alcohols by substitution of one, two, three, or four atoms of oxygen for a corresponding number of hydrogen molecules :

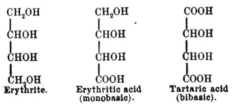

CH₂OH   CH₂OH   COOH
CHOH   CHOH   CHOH
CHOH   CHOH   CHOH
CH₂OH   COOH   COOH
Erythrite.   Erythritic acid (monobasic).   Tartaric acid (bibasic).

Only one tetratomic acid has, however, been actually formed by oxidation of the corresponding alcohol, namely, erythritic acid, $C_4H_8O_5$, from erythrite, $C_4H_{10}O_4$.

The known tetratomic acids belonging to the fatty group are—

Erythritic acid,	$C_4H_8O_5$	monobasic.
Dioxymalonic,	$C_3H_4O_6$	
Tartaric,	$C_4H_6O_6$	bibasic.
Homotartaric,	$C_5H_8O_6$	
Citric,	$C_6H_8O_7$	tribasic.

Erythritic Acid, $C_4H_8O_5 = C_3H_4(OH)_3.CO_2H$, formed by the oxidation of erythrite, $C_4H_{10}O_4$ (p. 644), in aqueous solution in contact with platinum black, is a deliquescent crystalline mass, and is capable of forming salts containing 2 equivalents of metal.

Dioxymalonic Acid, $C_3H_4O_6 = C(OH)_2\!<^{CO_2H}_{CO_2H}$, formed by the action of moist silver oxide or dibromomalonic acid, is probably identical with the hydrate of mesoxalic acid (p. 775).

Tartaric Acid, $C_4H_6O_6 = \begin{matrix} CHOH-CO_2H \\ | \\ CHOH-CO_2H \end{matrix} = C_2H_2\begin{Bmatrix} (OH)_2 \\ (CO_2H)_2 \end{Bmatrix}.$— This formula includes four bibasic acids, distinguished from one another by certain physical properties, especially by their crystalline forms, and their action on polarized light—namely, D e x t r o t a r t a r i c a c i d, which turns the plane of polarization to the right ; L e v o t a r t a r i c a c i d, which turns it to the left with equal force ; P a r a t a r t a r i c or R a c e m i c a c i d, which is optically inactive, and separable into equal quantities of dextro- and levotartaric acids ; and an inactive variety of tartaric acid, which is not thus separable.

1. DEXTROTARTARIC or ORDINARY TARTARIC ACID.—This is the acid of grapes, tamarinds, pine-apples, and several other fruits, in which it occurs in the state of an acid potassium-salt; calcium tartrate is also occasionally met with. The tartaric acid of commerce is wholly prepared from *tartar* or *argol*, an impure acid potassium tartrate, deposited from wine, or rather from grape-juice in the act of fermentation. This substance is purified by solution in hot water, with the aid of a little pipe-clay and animal charcoal, to remove the coloring matter of the wine, and subsequent crystallization : it then constitutes *cream of tartar*, and serves for the preparation of the acid. The salt is dissolved in boiling water, and powdered chalk is added as long as effervescence is excited, or the liquid exhibits an acid reaction ; calcium tartrate and neutral potassium tartrate are thereby produced, and the latter is separated from the former, which is insoluble, by filtration. The solution of potassium tartrate is then mixed with excess of calcium chloride, which throws down all the remaining acid in the form of calcium salt ; this is washed and added to the former portion, and the whole is digested with a sufficient quantity of dilute sulphuric acid to withdraw the base, and liberate the tartaric acid. The filtered solution is cautiously evaporated to a syrupy consistence, and left to crystallize in a warm place. Liebig found that tartaric acid is artificially produced by the action of nitric acid upon milk sugar.

Tartaric acid forms colorless, transparent monoclinic prisms often of large size; they are permanent in the air, and inodorous ; they dissolve with great facility in water, both hot and cold, and are soluble also in alcohol. The solution reddens litmus strongly, and has a pure acid taste. The aqueous solution, as above mentioned, exhibits right-handed polarization.

This solution is gradually spoiled by keeping. The crystallized acid melts at 135° C. (275° F.), is converted at 170° C. (338° F.) into optically inactive mesotartaric acid, and when heated for some time to 180° C. (356° F.) gives off water, and yields anhydrides (p. 794). Tartaric acid is consumed in large quantities by the calico-printer, being employed to evolve chlorine from solution of bleaching powder, in the production of white or *discharged* patterns upon a colored ground.

Tartrates.—Tartaric acid, being tetratomic and bibasic, has only two hydrogen atoms replaceable by metals, the other two being replaceable by alcoholic or acid radicles. With monad metals it forms acid and neutral salts, $C_4H_5M'O_6$, and $C_4H_4M_2O_6$; with dyad metals, neutral salts, $C_4H_4M''O_6$, and double salts, like *bario-potassic tartrate*, $C_4H_4BaO_6.C_4H_4K_2O_6$. With triad metals it forms a peculiar class of salts, best known in the case of the *antimony-salt* (p. 792).

Potassium Tartrates.—The *neutral salt*, $C_4H_4K_2O_6$, may be obtained by neutralizing cream of tartar with chalk, as in the preparation of the acid, or by saturating cream of tartar with potassium carbonate; it is very soluble, and crystallizes with difficulty in right rhombic prisms, which are permanent in the air, and have a bitter, saline taste. The *acid salt*, or *cream of tartar*, $C_4H_5KO_6$, the origin and preparation of which have been already described, forms irregular groups of small transparent or translucent prismatic crystals which grate between the teeth. It dissolves pretty freely in boiling water, but the greater part separates as the solution cools, leaving about $\frac{1}{80}$ or less dissolved in the cold liquid. The salt has an acid reaction and a sour taste. When exposed to heat in a close vessel, it is decomposed, with evolution of inflammable gas, leaving a mixture of finely divided charcoal and pure potassium carbonate, from which the latter may be extracted by water. Cream of tartar is almost always produced when tartaric acid in excess is added to a moderately strong solution of a potassium-salt, and the whole agitated.

Sodium Tartrates.—Two of these salts are known—a *neutral salt*, $C_4H_4Na_2O_6 + 2Aq.$; and an *acid salt*, $C_5H_5NaO_6 + Aq.$ Both are easily soluble in water, and crystallizable. Tartaric acid and sodium bicarbonate form the ordinary effervescing draughts.

Potassium and Sodium tartrate; Rochelle or *Seignette salt*, $C_4H_4KNaO_6 + 4Aq.$ —This beautiful salt is made by neutralizing a hot solution of cream of tartar with sodium carbonate, and evaporating to the consistence of thin syrup. It separates in large transparent rhombic prisms with hemihedral faces; they effloresce slightly in the air, and dissolve in $1\frac{1}{2}$ parts of cold water. Acids precipitate cream of tartar from the solution. Rochelle salt has a mild saline taste, and is used as a purgative.

Ammonium Tartrates.—The *neutral tartrate* is a soluble and efflorescent salt, containing $C_4H_4(NH_4)_2O_6 + Aq.$ The *acid tartrate*, $C_4H_5(NH_4)O_6$, closely resembles ordinary cream of tartar. A salt analogous to Rochelle salt also exists, having ammonium in place of sodium.

The tartrates of *calcium, barium, strontium, magnesium,* and of most of the heavy metals, are insoluble, or nearly so, in water.

Potassio-antimonious Tartrate, or *Tartar emetic*, is easily made by boiling antimony trioxide in solution of cream of tartar: it is deposited from a hot and concentrated solution in rhombic octohedrons, which dissolve without decomposition in 15 parts of cold and 3 of boiling water, and have an acrid, extremely disagreeable metallic taste. The solution is decomposed by both acids and alkalies : the former throws down a mixture of cream of tartar and antimony trioxide, and the latter the trioxide, which is again dissolved by great excess of the reagent. Sulphuretted hydrogen separates all the antimony in the state of trisulphide. The dry salt heated on charcoal before the blowpipe, yields a globule of metallic antimony. The crystals contain $2C_4H_4K(SbO)O_6 + Aq.$; the group SbO acting as a univalent radicle, and replacing one atom of hydrogen. When dried at 100°, they give off their water of crystallization, and at 200° C. (392° F.) an additional molecule of water, leaving the compound, C_4H_2K $(SbO)O_5$, which has the constitution of a salt, not of tartaric, but of tartrelic acid, $C_4H_4O_5$. Nevertheless, when dissolved in water, the crystals again take up the elements of water, and reproduce the original salt.

An analogous compound, containing arsenic in place of antimony, has been obtained. It has the same crystalline form as tartar emetic.

A solution of tartaric acid dissolves ferric hydrate in large quantity, forming a brown liquid, which has an acid reaction, and dries up by gentle heat to a brown, transparent, glassy substance, destitute of all traces of crystallization. It is very soluble in water, and the solution is not precipitated by alkalies, either fixed or volatile. Indeed, tartaric acid, added in sufficient quantity to a solution of ferric oxide, or alumina, entirely prevents the precipitation of the bases by excess of ammonia. Tartrate and ammoniacal tartrate of iron are used in medicine, these compounds having a less disagreeable taste than most of the iron preparations.

Solutions of tartaric acid give with lime and baryta-water, and with lead acetate, white precipitates, which dissolve in excess of the acid ; with neutral calcium and barium salts no change is produced. Silver nitrate produces in neutral tartrates a white precipitate of silver tartrate, which dissolves in ammonia. On gently heating the solution, a bright metallic deposit of silver is formed. The reaction of tartaric acid with solutions of potassium salts has been already noticed.

Tartaric Ethers.—1. Tartaric acid forms, with monatomic alcohol-radicles, acid and neutral ethers, in which one or both of the atoms of *basic* hydrogen in its molecule is replaced by an alcohol-radicle. These compounds may be formulated as follows :—

$$(C_2H_2) \begin{cases} (OH)_2 \\ (CO_2H)_2 \end{cases} \qquad (C_2H_2) \begin{cases} (OH)_2 \\ CO_2H \\ CO_2C_2H_5 \end{cases} \qquad (C_2H_2) \begin{cases} (OH)_2 \\ (CO_2C_2H_5)_2 \end{cases}$$

Tartaric acid. Acid ethyl tartrate. Neutral ethyl tartrate.

The acid ethers are monobasic acids, formed by the direct action of tartaric acid on the respective alcohols ; the neutral ethers are formed by passing hydrochloric acid gas into a solution of tartaric acid in an alcohol. Further by treating these neutral ethers with chlorides of acid radicles, other neutral ethers are formed, in which one or more of the alcoholic hydrogen-atoms are replaced by acid radicles.* In this manner are formed such compounds as the following :—

$$(C_2H_2) \begin{cases} OH \\ OC_2H_3O \\ (CO_2C_2H_5)_2 \end{cases} \qquad (C_2H_2) \begin{cases} OC_2H_3O \\ OC_7H_5O \\ (CO_2C_2H_5)_2 \end{cases} \qquad \begin{matrix}(C_2H_2)\\(C_2H_2)\end{matrix} \begin{cases} (OH) \\ (O_2C_4H_4O_2) \\ (CO_2C_2H_5)_4 \end{cases}$$

Ethyl-aceto-tartrate. Ethyl-aceto-benzo-tartrate. Ethyl-succino-tartrate.

The alcoholic hydrogen in these neutral ethers may be replaced by potassium and sodium.

* Perkin, Chem. Soc. Journ. [2], v. 139.

34

2. There are also *bibasic tartaric ethers* formed by replacing the alcoholic hydrogen of tartaric acid with acid radicles ; *e. g.*,

$$(C_2H_2) \begin{cases} OH \\ OC_7H_5O \\ (CO_2H)_2 \end{cases} \qquad (C_2H_2) \begin{cases} (OC_2H_3O)_2 \\ (CO_2H)_2 \end{cases} \qquad (C_2H_2) \begin{cases} (ONO_2)_2 \\ (CO_2H)_2 \end{cases}$$

Benzotartaric acid.   Diacetotartaric acid.   Dinitrotartaric acid.

3. Lastly, tartaric acid forms ethers with glycol, glycerin, mannite, glucose, and other polyatomic alcohols.

Dinitrotartaric acid, $C_2H_2(O.NO_2)_2 {<}^{CO_2H}_{CO_2H}$, in which both the alcoholic hydrogen-atoms of tartaric acid are replaced by NO_2, is formed by dissolving finely pulverized tartaric acid in strong nitric acid and adding sulphuric acid : it then separates as a jelly, which dries up to a white shining mass. It is soluble in water, and the solution when heated yields tartronic acid (p. 786).

Tartaric Anhydrides.—When crystallized tartaric acid is exposed to a temperature of about 204° C. (399.2° F.), it melts, loses water, and yields in succession three different anhydrides, viz. :—

Ditartaric or Tartralic acid   . $C_8H_{10}O_{11} = 2C_4H_6O_6 - H_2O$
Tartrelic acid   .   .   .   : $\Big\}$ $C_4H_4O_5 = C_4H_6O_6 - H_2O$
Insoluble tartaric anhydride   .

The first two are soluble in water, and form salts which have properties completely different from those of ordinary tartaric acid. The third is a white insoluble powder. All three, in contact with water, slowly pass into ordinary tartaric acid.

Tartaric acid, subjected to destructive distillation, is resolved into carbon dioxide and pyrotartaric acid, $C_3H_6O_4$.

When tartaric acid is heated to 204.5° C. (400.1° F.), with excess of potassium hydroxide, it is resolved, without charring or secondary decomposition, into oxalic and acetic acids, which remain in union with the base, and undergo decomposition at a much higher temperature :—

$$C_4H_6O_6 + 2KHO = C_2KHO_4 + C_2H_3KO_2 + 2H_2O .$$
Tartaric   Acid potas-   Potassium
acid.   sium oxalate.   acetate.

2. LEVOTARTARIC ACID.—This acid resembles dextrotartaric acid in every respect, except that it turns the plane of polarization to the left, and that its salts, as well as the acid itself, though isomorphous with the corresponding dextro-tartrates, contain oppositely situated hemihedral faces (see below).

3. PARATARTARIC or RACEMIC ACID.—This acid occurs, together with ordinary tartaric acid, in the grapes cultivated in certain districts of the Upper Rhine and in the Vosges. To separate it, the mother liquor of the argol, obtained from these grapes, is boiled with chalk, the calcium salt which separates is decomposed by sulphuric acid, and the filtrate is evaporated to the crystallizing point. The crystals of racemic acid being efflorescent, are easily separated by mechanical means from the shining crystals of ordinary tartaric acid.

Racemic acid may be formed artificially by oxidizing mannite, dulcite, or mucic acid with nitric acid, and synthetically by boiling glyoxal with hydrocyanic acid and a small quantity of hydrochloric acid :

$$\begin{matrix} \text{COH} \\ | \\ \text{COH} \end{matrix} + 2\text{CNH} + 4\text{H}_2\text{O} = 2\text{NH}_3 + \begin{matrix} \text{CH(OH)}\text{—CO}_2\text{H} \\ | \\ \text{CH(OH)}\text{—CO}_2\text{H} \end{matrix} \; ;$$

further, together with inactive tartaric acid, by boiling dibromosuccinic acid with silver oxide and water :—

$$\begin{matrix} \text{CHBr}\text{—CO}_2\text{H} \\ | \\ \text{CHBr}\text{—CO}_2\text{H} \end{matrix} + 2\text{AgOH} = 2\text{AgBr} + \begin{matrix} \text{CH(OH)}\text{—CO}_2\text{H} \\ | \\ \text{CH(OH)}\text{—CO}_2\text{H} \end{matrix} :$$

most readily, however, by heating ordinary tartaric acid with about one-tenth of its weight of water to 170°–180° C. (338°–356° F.) in sealed vessels, the dextrotartaric acid being thereby completely converted into inactive tartaric and racemic acids, which may be separated by crystallization, the racemic acid being much the less soluble of the two.

The conversion of tartaric acid into racemic acid was originally effected by Pasteur, by heating ethyl tartrate or cinchonine tartrate to about 170° C. (338° F.). On repeatedly boiling the product with water, and mixing the cooled solution with excess of calcium chloride, a considerable precipitate of calcium racemate is obtained.

Racemic acid crystallizes with 1 molecule of water in rhombic prisms, which give off their water at 100°. It is somewhat less soluble in water than ordinary tartaric acid, and has no action on polarized light. It is, in fact, a compound of dextro- and levotartaric acids in equal quantities, and may be resolved into those acids through the medium of some of its double salts.

When racemic acid is saturated with potash or soda, or any other single base, a salt is obtained, all the crystals of which are identical in form and in physical properties ; but by saturating racemic acid with two bases, as with soda and ammonia, or by mixing the racemates of sodium and of ammonium in equivalent proportions, and evaporating the solution, crystals of a double salt, $\text{C}_4\text{H}_4\text{O}_6\text{Na(NH}_4)$, are obtained, analogous to Rochelle salt, the form of which is shown in fig. 175. It is a right rectangular prism, P, M, T, having its lateral edges replaced by the faces b', and the intersection of these latter faces with the face T replaced by a face h. If the crystal were holohedral, there would be eight of these h faces, four above and four below; but, as the figures show, there are but four of them, placed alternately. Moreover, these hemihedral faces occupy in different crystals of the salt, not similar but opposite or symmetrical positions, the one kind of crystal being, as it were, the reflected image of the other.

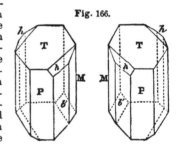

Fig. 166.

Further, by carefully picking out these two kinds of crystals and dissolving them separately in water, solutions are obtained, which, at the same degree of concentration, exert equal and opposite actions upon polarized light, the one deflecting the plane of polarization to the right, the other by an equal amount to the left. Moreover, the solutions of the right- and left-handed crystals, when evaporated, yield crystals, each of its own kind only ; and by mixing the solutions of these crystals with calcium chloride, calcium salts are obtained which, when decomposed by sulphuric acid, yield acids agreeing with each other in composition and in every other respect, excepting that their crystalline forms exhibit opposite hemihedral modifications, and their solutions, when reduced to the same degree of concentration, exert equal and opposite effects on polarized light. One of these acids is in fact d e x t r o t a r t a r i c, the other l e v o-t a r t a r i c　a c i d. A mixture of equal parts of these two acids has no longer the slightest effect on polarized light, and is in every respect iden-tical with racemic acid.

4. Inactive Tartaric Acid or Mesotartaric Acid, is formed by oxidiz-ing sorbin (p. 651) with nitric acid; also, together with racemic acid, from dibromosuccinic acid, and from glyoxal (p. 702); but it is most easily prepared by heating ordinary tartaric acid with a little water to 165⁰ C. (329⁰ F.) for two days. It is separated from unaltered tartaric acid, and from simultaneously formed racemic acid, by converting it into the acid potassium salt, which is very soluble in water.

Inactive tartaric acid is much more soluble in water than ordinary tar-taric or racemic acid (10 parts of it dissolve in 8 parts of water at 15⁰ C. (59⁰ F.)). It does not deflect the plane of polarization, but differs from racemic acid in not being resolvable into the two active acids. When heated to 170⁰ C. (338⁰ F.), however, it is converted into dextrotartaric acid.

Homotartaric Acid, $C_5H_8O_6 = C_6H_4(OH)_2\big\langle{}^{CO_2H}_{CO_2H}$, of which very little is known, is formed from dibromopyrotartaric acid.

Rhodizonic Acid, $C_5H_4O_6$.—When potassium is heated in a stream of dry carbon monoxide; the latter is absorbed in large quantity, and a black porous substance generated, which, according to Brodie, contains COK_3. Brought in contact with water it decomposes with great violence, and even the dry substance occasionally explodes; when anhydrous alcohol is poured upon it, a great elevation of temperature ensues, but the decom-position is far less violent than with water. The product of this reaction is potassium rhodizonate, which remains as a red powder, insoluble in alcohol, but soluble in water with a deep red color. This salt probably contains $C_5H_2K_2O_6$.

When solution of potassium rhodizonate is boiled, it becomes orange-yellow from decomposition of the acid, and is then found to contain free potash, and a salt of C r o c o n i c　a c i d, $C_5H_2O_5$. This acid can be iso-lated; it is yellow, easily crystallizable, soluble both in water and alcohol. It is likewise bibasic.

Citric Acid, $C_6H_8O_7 = C_3H_4(OH)_2(CO_2H)_3$.—This acid is obtained in large quantities from the juice of lemons; it is found in many other fruits, as in gooseberries, currants, etc., in conjunction with malic acid. To prepare it, the juice is allowed to ferment a short time, in order that mucilage and other impurities may separate and subside; the clear liquor is then carefully saturated with chalk, whereby insoluble calcium citrate is produced. This is thoroughly washed, decomposed by the proper quan-tity of sulphuric acid diluted with water, and the filtered solution is eva-porated to a small bulk, and left to crystallize. The product is drained from the mother-liquor, redissolved, digested with animal charcoal, and again concentrated to the crystallizing point. The acid has not yet been obtained by any synthetical process.

Citric acid crystallizes in two different forms. The crystals which sepa-rate by spontaneous evaporation from a cold saturated solution, are trime-tric prisms, containing $C_6H_8O_7.H_2O$, whereas those which are deposited from a hot solution have a different form, and contain $2C_6H_8O_7.H_2O$.— Citric acid has a pure and agreeable acid taste, and dissolves, with great ease, in both hot and cold water; the solution strongly reddens litmus, and when long kept, is subject to spontaneous change. Citric acid, when brought in contact with putrid flesh as a ferment, yields butyric acid and small quantities of succinic acid. It is entirely decomposed when heated with sulphuric and nitric acids : the latter converts it into oxalic acid. Caustic potash, at a high temperature, resolves it into acetic and oxalic acids. The alkaline citrates, treated with chlorine, yield chloroform, together with other products.

Citric acid is tetratomic and tribasic. With *potassium* it forms a neutral salt containing $C_6H_5K_3O_7$, and two acid salts containing respectively $C_6H_6K_2O_7$ and $C_6H_7KO_7$; and similar salts with the other alkali-metals. With dyad metals it chiefly forms salts in which two or three hydrogen-atoms in the molecule $C_6H_8O_7$, are replaced by metals : with *calcium*, for example, it forms the salts $C_6H_6CaO_7 + H_2O$ and $(C_6H_6O_7)_2Ca_3 + H_2O$. With *lead* it forms two salts similar in constitution to the calcium salts, and likewise a tetraplumbic salt containing $(C_6H_5O_7)_2Pb_3.PbH_2O_2$.

The citrates of the *alkali-metals* are soluble, and crystallize with greater or less facility ; those of *barium, strontium, calcium, lead*, and *silver* are insoluble.

Citric acid resembles tartaric acid in its relations to ferric oxide, preventing the precipitation of that substance by excess of ammonia. The citrate obtained by dissolving hydrated ferric oxide in solution of citric acid, dries up to a pale-brown, transparent, amorphous mass, which is not very soluble in water ; an addition of ammonia increases the solubility. Citrate and ammonio-citrate of iron are used as medicinal preparations.

Citric acid is sometimes adulterated with tartaric acid : the fraud is easily detected by dissolving the acid in a little cold water, and adding to the solution a small quantity of potassium acetate. If tartaric acid be present, a white crystalline precipitate of cream of tartar will be produced on agitation. Citric acid is further distinguished from tartaric acid by the characters of its calcium salt. An aqueous solution of citric acid is not precipitated by lime-water in the cold, but on boiling a precipitate is formed, consisting of tricalcic citrate, insoluble in potash-lye. Calcium tartrate, on the other hand, dissolves in alkalies, and is precipitated therefrom as a jelly on boiling.

Citric acid forms ethers in which 1, 2, or 3 hydrogen-atoms are replaced by methyl and other monad alcohol-radicles. The neutral ethers are formed by passing gaseous hydrogen chloride into an alcoholic solution of citric acid. The *trimethylic ether*, $C_3H_4(OH)(CO_2.CH_3)$, is crystalline; the *triethylic ether* boils, with partial decomposition, at about 280° C. (536° F.).

By treating these neutral ethers with acetyl chloride, the alcoholic hydrogen may also be replaced, and **triethylic acetocitrate**, $C_3H_4(O.C_2H_3O)(CO_2C_2H_5)_3$, produced, which boils at 288° C. (550.4° F.). By treating the same ethers with nitric acid, the alcoholic hydrogen may be replaced by NO_2.

Pentatomic Acids.

Of these only one is known, viz., the bibasic acid—

Aposorbic acid, $C_5H_8O_7 = C_3H_3(OH)_3{<}^{CO_2H}_{CO_2H}$, which may be regarded as a trioxypyrotartaric acid. It is produced by oxidizing sorbin with nitric acid, and crystallizes in small laminæ, easily soluble in water, and melting with decomposition at about 110° C. (230° F.).

Hexatomic Acids.

Gluconic Acid, $C_6H_{12}O_7 = C_5H_6(OH)_5.CO_2H$, is obtained by oxidizing grape-sugar with chlorine water and removing the chlorine with
silver oxide. It is a non-crystallizable syrup, easily soluble in water, and
nearly insoluble in alcohol. It is monobasic. Its calcium and barium
salts and the ethylic ether crystallize well; the formula of the calcium
salt is $(C_6H_{11}O_7)_2Ca + 2H_2O$.

Mannitic Acid, $C_6H_{12}O_7 = C_5H_6(OH)_5.CO_2H$, is produced by oxidation of mannite, $C_6H_{14}O_6$, under the influence of platinum black. It is a
gummy mass, soluble in water and in alcohol, insoluble in ether. According to its mode of formation it might be expected to be monobasic:

Mannite, $CH_2OH—(CHOH)_4—CH_2OH$
Mannitic acid, $CH_2OH—(CHOH)_4—COOH$;

but from the observations of Gorup-Besanez, who discovered it, it appears
to be bibasic, its potassium salt containing $C_6H_{10}KO_7$, and the calcium salt,
$C_6H_{10}CaO_7$.

Saccharic Acid, $C_6H_{10}O_8 = (C_4H_4) \begin{cases} (OH)_4 \\ (CO_2H)_2 \end{cases} =$
$CO.OH—(CHOH)_4—CO.OH$.--This acid is produced by the action of dilute
nitric acid on cane sugar, glucose, milk-sugar, and mannite, and is often
formed in the preparation of oxalic acid, being, from its superior solubility,
found in the mother liquor from which the oxalic acid has crystallized. It
may be made by heating together 1 part of sugar, 2 parts of nitric acid, and
10 parts of water. When the reaction seems terminated, the acid liquid is
diluted and neutralized with chalk; the filtered liquid is mixed with lead
acetate; and the insoluble lead saccharate is washed, and decomposed by
sulphuretted hydrogen. The acid slowly crystallizes from a solution of
syrupy consistence in long colorless needles; it has a sour taste, and forms
soluble salts with lime and baryta. When mixed with silver nitrate it
gives no precipitate, but, on the addition of ammonia, a white insoluble
substance separates, which, on gently warming the whole, is reduced to
metallic silver, the vessel being lined with a smooth and brilliant coating
of the metal. Nitric acid converts saccharic into oxalic and dextrotartaric
acids.

There are two *potassium saccharates*, containing $C_6H_9KO_8$ and $C_6H_8K_2O_8$;
the *silver-salt* contains $C_6H_8Ag_2O_8$; the *barium, magnesium, zinc*, and *cadmium
salts* have the composition $C_6H_8M''O_8$; and there are two *ethylic ethers*, containing $C_6H_9(C_2H_5)O_8$ and $C_6H_8(C_2H_5)_2O_8$. In these compounds saccharic
acid appears to be bibasic, as might be expected from its constitution, and
mode of formation ; the composition of the lead-salts, however, seems to
show that it is sexbasic as well as hexatomic, for Heintz has obtained a
lead-salt containing $C_6H_4Pb_3O_8$; but the composition of the lead saccharates
varies considerably according to the manner in which they are prepared.

Diethylic Saccharate, $C_4H_4(OH)_4 \begin{smallmatrix} <CO_2.C_2H_5 \\ <CO_2.CO_2H_5 \end{smallmatrix}$, is crystalline, and easily
soluble in water. Ammonia converts it into the amide $C_4H_4(OH)_4(CO.
NH_2)_2$ which is a white powder. The ether, treated with acetyl chloride,
yields the tetracetylic compound $C_4H_4(O.C_2H_3O)_4 \begin{smallmatrix} <CO_2.C_2H_5 \\ <CO_2.C_2H_5 \end{smallmatrix}$.

Mucic Acid, $C_6H_{10}O_8 = C_4H_4(OH)_4(CO_2H)_2$, isomeric with saccharic
acid, is produced, together with a small quantity of oxalic acid, by the

action of rather dilute nitric acid on sugar and gum. It may be easily prepared by heating together in a flask or retort, 1 part of milk-sugar or gum, 4 parts of nitric acid, and 1 part of water ; the mucic acid is afterwards collected upon a filter, washed and dried. It has a slightly sour taste, and reddens vegetable colors. It requires for solution 66 parts of boiling water. By prolonged boiling with water, it is converted into the isomeric p a r a m u c i c   a c i d. By boiling with nitric acid it is resolved into oxalic and racemic acids. It is decomposed by heat, yielding, among other products, p y r o m u c i c   a c i d, $C_5H_4O_3$:

$$C_6H_{10}O_8 \ = \ C_5H_4O_3 \ + \ CO_2 \ + \ 3H_2O .$$

Mucic acid is bibasic, yielding for the most part neutral salts containing $C_6H_8M_2O_8$ and $C_6H_8M''O_8$; with the alkali-metals it also forms acid salts, such as $C_6H_9KO_8$.

The neutral potassium and ammonium salts crystallize well, and are but slightly soluble in cold water ; the acid salts are easily soluble. The neutral ammonium salt is resolved by heat into ammonia, water, and pyrrol, C_4H_5N. ;

Diethylic Mucate, $C_4H_4(OH)_4(CO_2.C_2H_5)_2$, obtained by heating mucic acid with alcohol and sulphuric acid, is crystalline, soluble in hot water, melts at 158° C. (316.4° F.), and is converted by acetyl chloride into the tetracetyl compound, $C_4H_4(O.C_2H_3O)_4(CO_2.C_2H_5)_2$, which melts at 177° C. (350.6° F.).

D e o x a l i c or **R a c e m o - c a r b o n i c   A c i d,** $C_6H_8O_9$, probably $= C_3H_2(OH)_3(CO_2H)_3$, is produced by the action of sodium amalgam on ethylic oxalate :

$$3C_2H_2O_4 \ + \ 4H_2 \ = \ 3H_2O \ + \ C_6H_8O_9 .$$

Deoxalic acid is not known in the free state, being resolved, on evaporation of its aqueous solution, into racemic and glyoxylic acids :

$$C_6H_8O_9 \ = \ C_4H_6O_6 \ + \ C_2H_2O_3 .$$

The acid is essentially tribasic, its ammonium salt having the composition $C_6H_5(NH_4)_3O_9 + H_2O$; but it also forms salts in which 4 atoms of hydrogen (one alcoholic) are replaced by metal, the silver salt being $C_6H_4Ag_4O_9 + H_2O$, and the barium salt $C_6H_4Ba_2O_9 + 3H_2O$. By the action of acetic acid on its potassium salt it is resolved into acetic acid and Löwig's deoxalic acid, $C_5H_6O_8$:

$$2C_6H_8O_9 \ = \ C_2H_4O_2 \ + \ 2C_5H_6O_8 .$$

This last acid, $C_2H(OH)_2.(CO_2H)_3$, forms large colorless crystals, easily soluble in water and in alcohol. It is tribasic. Heated with water to 100°, it is resolved into racemic acid and carbon dioxide :

$$C_5H_6O_8 \ = \ C_4H_6O_6 \ + \ CO_2 .$$

Pyromucic Acid and its Derivatives.

Pyromucic Acid, $C_5H_4O_3 = C_4H_3O.COOH$, is produced by the dry distillation of mucic acid, or more advantageously by boiling furfurol with water and recently precipitated silver oxide ; the silver is precipitated by hydrochloric acid, the filtrate evaporated, and the pyromucic acid purified by crystallization from dilute alcohol.

Pyromucic acid crystallizes in colorless laminæ or needles, easily soluble in water, especially if hot, moderately soluble in alcohol. It melts at 134° C. (273.2° F.), and sublimes at 100°. It is monobasic. The *silver salt*, $C_5H_3O_3Ag$, crystallizes in laminæ; the *barium salt*, $(C_5H_3O_3)_2Ba$, forms crystals easily soluble in water. The *ethylic ether*, $C_5H_3O_3.C_2H_5$, obtained by distilling the acid with hydrochloric acid and alcohol, is a crystalline mass, melting at 34° C. (93.2° F.), and boiling at 208°–210° C. (406.4°–410° F.). The *chloride*, $C_4H_3O.COCl$, produced by distilling pyromucic acid with phosphorus pentachloride, boils at 170° C. (368° F.), and is converted by ammonia into the *amide*, $C_4H_3O.CONH_2$, a crystalline substance, soluble in water, and melting at 130° C. (266° F.).

Barium pyromucate, distilled with soda-lime, yields t e t r a p h e n o l, C_4H_4O, or $C_4H_3O.OH$ (4-carbon phenol), as a colorless liquid, boiling at 32° C. (89.6° F.).

Isopyromucic acid, $C_5H_4O_3$, is produced, together with pyromucic acid, by the dry distillation of mucic acid, and may be separated from the latter by solution in a small quantity of cold water. It sublimes below 100° in white laminæ, which turn yellow on exposure to the air, melt at 82° C. (179.6° F.), dissolve very easily in water, alcohol, and ether.

C a r b o p y r r o l a m i d e, $C_5H_6N_2O = C_4H_2{<}^{CO.NH_2}_{NH_2}$, produced by dry distillation of ammonium pyromucate, forms white shining laminæ, easily soluble in alcohol and in ether. It melts at 173° C. (343.4° F.), and does not resolidify till cooled to 133° C. (271.4° F.). By boiling with water it is resolved into ammonia and carbopyrrolic acid :

$$C_5H_6N_2O \ + \ H_2O \ = \ NH_3 \ + \ C_5H_5NO_2.$$

C a r b o p y r r o l i c a c i d, $C_4H_2{<}^{CO_2H}_{NH_2}$, crystallizes in small prisms, sublimes at about 190° C. (374° F.), and is decomposed by sudden heating into carbon dioxide and pyrrol :

$$C_5H_5NO_2 \ = \ CO_2 \ + \ C_4H_5N.$$

Barium carbopyrrolate crystallizes in large laminæ.

P y r r o l, C_4H_5N, is a weak base occurring in coal-tar oil and in bone-oil, and producible by distillation of ammonium pyromucate and of carbopyrrolic acid. It may be prepared from bone-oil by dissolving out the basic constituents with sulphuric acid, and submitting the acid solution to prolonged boiling, whereby the stronger bases are retained, while the pyrrol distils over. The distillate is heated with solid potassium hydroxide, when the pyrrol combines slowly with the alkali, admixed impurities being volatilized. On dissolving the potassium compound in water, the pyrrol separates on the surface as an oily liquid. Pyrrol is colorless, insoluble in water and alkalies, slowly soluble in acids : it has an ethereal odor resembling that of chloroform, a specific gravity = 1.077, and boils at 133° C. (271.4° F.). It is easily recognized by the purple color which it imparts to fir-wood moistened with hydrochloric acid.

By heating an acid solution of pyrrol, a red, flaky substance, *pyrrol-red*, is produced, containing $C_{12}H_{14}N_2O$, the formation of which is represented by the following equation :—

$$3C_4H_5N \ + \ H_2O \ = \ C_{12}H_{14}N_2O \ + \ NH_3.$$

F u r f u r o l, $C_5H_4O_2 = C_4H_3O.COH$.—This compound, which is the aldehyde of pyromucic acid, is formed in the dry distillation of sugar, and by distilling bran with dilute sulphuric acid or zinc chloride. To prepare it, 1 part of bran is mixed with 1 part of sulphuric acid diluted with 3 parts of water, and the distillate is saturated with sodium carbonate, mixed with common salt, and distilled. On adding common salt to this second distillate, the furfurol separates from the watery liquid in the form of a heavy oil.

Furfurol is a colorless liquid, having an agreeable odor, somewhat like that of oil of cassia. It boils at 162° C. (323.6° F.), has a specific gravity of 1.164, vapor-density $= 3.493$ (referred to air), dissolves in 12 parts of water at 13° C. (55.4° F.), very easily in alcohol.

Furfurol, like other aldehydes, unites with acid sodium sulphite, forming a crystalline compound, $C_5H_4O_2.SO_3NaH$, sparingly soluble in alcohol, and is converted by sodium amalgam into an alcohol, viz., furfuryl alcohol, $C_5H_6O_2$. By oxidation with silver oxide, it is converted into pyromucic acid, and by nitric acid into oxalic acid. With ammonia it forms furfuramide: $3C_5H_4O_2 + 2NH_3 = (C_5H_4O)_3N_2 + 3H_2O$.

Furfuryl Alcohol, $C_5H_5O.OH$, produced by the action of sodium amalgam on furfurol or on pyromucic acid, is a thick oil, insoluble in water, and decomposing when distilled.

Furfuramide, $(C_5H_4O)_3N_2$, is formed when furfurol is left for a few hours in contact with aqueous ammonia, and separates in white crystals, insoluble in water, but easily soluble in alcohol and ether. By boiling with water or acids it is resolved into ammonia and furfurol. By heating to 120° C. (248° F.), or by boiling with dilute aqueous potash, it is converted into the isomeric compound furfurine (discovered by Fownes), which is a crystalline base, melting at 100°, and forming crystallizable very bitter salts, containing 1 eq. of acid. Furfurine is sparingly soluble in cold water, dissolves in about 135 parts of boiling water, easily in alcohol and ether, forming solutions which have a strong alkaline reaction.

Fucusol.—By treating several varieties of fucus with sulphuric acid in exactly the same manner as in the preparation of furfurol, Dr. Stenhouse has obtained a series of substances, which he designates by the terms *fucusol, fucusamide,* and *fucusine.* They have exactly the same composition as the corresponding terms in the furfurol series, and also most of their properties, but differ in some respects.

The constitution of the compounds just described is not very well understood: they cannot be regarded as derivatives either of methane or benzene, in other words, as belonging either to the fatty or to the aromatic group, but they may be represented, provisionally at least, by constitutional formulæ derived from that of a hypothetical hydrocarbon called tetrene:—

$$C_4H_4 = \begin{matrix} HC \!\!\equiv\!\! CH \\ | \quad | \\ HC \!\!\equiv\!\! CH \end{matrix},$$

intermediate in composition between acetylene and benzene; thus:—

$$\begin{matrix} HO.C\!\equiv\!C\!-\!CO_2H \\ | \quad | \\ HC\!\equiv\!CH \end{matrix} \qquad \begin{matrix} HO.C\!\equiv\!C\!-\!COH \\ | \quad | \\ HC\!\equiv\!CH \end{matrix} \qquad \begin{matrix} HO.C\!\equiv\!C\!-\!CH_2OH \\ | \quad | \\ HC\!\equiv\!CH \end{matrix}.$$

Pyromucic acid.   Furfurol.   Furfuryl alcohol.

$$\begin{matrix} HC\!\equiv\!C\!-\!CO_2H \\ | \quad | \\ HC\!\equiv\!C\!-\!NH_2 \end{matrix} \qquad \begin{matrix} HC\!\equiv\!C\!-\!CO.NH_2 \\ | \quad | \\ HC\!\equiv\!C\!-\!NH_2 \end{matrix} \qquad \begin{matrix} HC\!\equiv\!CH \\ | \quad | \\ HC\!\equiv\!C\!-\!NH_2 \end{matrix}.$$

Carbopyrrolic acid.   Carbopyrrolamide.   Pyrrol.

34 *

AMIDES.

We have had frequent occasion to speak of these compounds, as derived from ammonium-salts by abstraction of water, or from acids by substitution of amidogen, NH_2 for hydroxyl, OH, or from one or more molecules of ammonia by substitution of acid-radicles for hydrogen. They are divided (like amines) into monamides, diamides, and triamides, each of which groups is further subdivided into primary, secondary, and tertiary amides, according as one-third, two-thirds, or the whole of the hydrogen is replaced by acid-radicles. If the hydrogen is replaced partly by acid-radicles, and partly by alcohol-radicles, the compound is called a l k a l a - m i d e ; for example, ethylacetamide, $NH(C_2H_5)(C_2H_3O)$; ethyldiacetamide, $N(C_2H_5)(C_2H_3O)_2$.

AMIDES DERIVED FROM MONATOMIC ACIDS.

A monatomic acid yields but one p r i m a r y a m i d e, which may be formed: 1. From its ammonium-salt by direct abstraction of a molecule of water, under the influence of heat; thus:

$$C_2H_3(NH_4)O_2 - H_2O = C_2H_5NO = \begin{array}{c} CH_3 \\ | \\ CONH_2 \end{array} = N \begin{cases} C_2H_3O \\ H_2 \end{cases}$$

Ammonium   Acetamide.
acetate.

This method is especially adapted to the preparation of volatile amides.

2. By the action of ammonia on acid chlorides or anhydrides:

$$\underset{\text{Acetic chloride.}}{C_2H_3O.Cl} + 2NH_3 = NH_4Cl + \underset{\text{Acetamide.}}{C_2H_3O.NH_2}$$

$$\underset{\text{Acetic anhydride.}}{(C_2H_3O)_2O} + 2NH_3 = NH_4.O.C_2H_3O + C_2H_3O.NH_2.$$

This method is especially adapted to the preparation of amides which are insoluble in water.

3. By the action of ammonia on compound ethers:

$$\underset{\text{Ethyl acetate.}}{C_2H_3O.O.C_2H_5} + NH_3 = C_2H_5.OH + C_2H_3O.NH_2.$$

This reaction often takes place at ordinary temperatures, but is for the most part best effected by heating the two bodies together in alcoholic solution.

S e c o n d a r y m o n a m i d e s are those in which two atoms of hydrogen in a molecule of ammonia are replaced by two univalent or one bivalent acid-radicle, or by one acid-radicle and one alcohol-radicle. Those containing only univalent radicles are formed by the action of dry hydrochloric acid gas on primary monamides at a high temperature; e. g., diacetamide from acetamide:

$$2NH_2(C_2H_3O) + HCl = NH_4Cl + NH(C_2H_3O)_2.$$

Those containing bivalent acid-radicles are called i m i d e s: e. g., succinimide, $NH(C_4H_4O_2)$. They are derived from bibasic acids, and will be noticed further on.

Secondary monamides (alkalamides) containing an acid-radicle and an alcohol-radicle, are formed by processes similar to those above given for the formation of the primary monamides, substituting amines for ammonia; thus:

$$C_2H_3OCl + NH_2(C_2H_5) = HCl + NH(C_2H_5)(C_2H_3O)$$
Acetic   Ethylamine.   Ethyl-acetamide.
chloride.

$$C_2H_3O(OC_2H_5) + NH_2(C_2H_5) = HOC_2H_5 + NH(C_2H_5)(C_2H_3O)$$
Ethyl acetate.   Ethylamine.   Alcohol.   Ethyl-acetamide.

Tertiary monamides are those in which the whole of the hydrogen in one molecule of ammonia is replaced by acid-radicles or by acid- and alcohol-radicles. Those of the latter kind, called tertiary alkalamides, are produced by the action of acid chlorides on secondary alkalamides:

$$NH(C_2H_5)(C_2H_3O) + C_2H_3O.Cl = HCl + N(C_2H_5)(C_2H_3O)_2;$$
Ethyl-acetamide.   Acetyl   Ethyl-diacetamide.
chloride.

or by the action of monatomic acid oxides on cyanic ethers ; e. g.:

$$(C_2H_3O)_2O + N(CO)(C_2H_5) = CO_2 + N(C_2H_5)(C_2H_3O)_2.$$
Acetic oxide.   Ethyl cyanate.   Ethyl-diacetamide.

Monamides are for the most part crystalline bodies soluble in alcohol and ether. The lower members of the group are likewise soluble in water, and distil without decomposition. As they contain both a basic group (NH_2), and an acid group, they are capable of acting both as bases and as acids, combining, on the one hand, with acids to form saline compounds, such as $C_2H_3O.NH_2.NO_3H$, which, however, are not very stable; and, on the other hand, forming salts by substitution of a metal for one atom of hydrogen; thus *silver-acetamide*, $C_2H_3O.NHAg$, is obtained in crystalline scales, by saturating an aqueous solution of acetamide with silver oxide.

Amides are less stable than amines, the combination of the amidogen-group with acid-radicles (C_2H_3O, for example), being weaker than the combination of the same group with hydrocarbons, as in the amines. Consequently they are more easily decomposed than amines, their decomposition being effected by heating with water, or more readily with alkalies:

$$C_2H_3O.NH_2 + HOH = NH_3 + C_2H_3O.OH.$$
Acetamide.   Acetic acid.

Primary amides heated with *phosphoric anhydride* or *phosphorus trichloride*, give off 1 mol. water, and are converted into nitrils or alcoholic cyanides; e. g., acetamide into methyl cyanide or acetonitril :

$$CH_3.CO.NH_2 - H_2O = CH_3.CN.$$

When phosphorus pentachloride acts upon an amide, the oxygen-atom of the latter is first replaced by two atoms of chlorine, and the resulting chloride when heated gives up 1 mol. HCl; thus :

$$CH_3.CONH_2 + PCl_5 = PCl_3O + CH_3.CCl_2.NH_2;$$
and

$$CH_3.CCl_2.NH_2 = 2HCl + CH_3.CN.$$

Formamide, $CH_3NO = CHO.NH_2$, the amide of formic acid, is obtained by dry distillation of ammonium formate, or by heating ethyl formate to 100° with alcoholic ammonia. It is a liquid which dissolves easily in water and in alcohol, and boils with partial decomposition at 192°–195° C. (377.6°–385° F.). When quickly heated, it is resolved into carbon monoxide and ammonia. By dehydration with phosphoric anhydride it is converted into hydrogen cyanide or formonitril, CHN.

Acetamide, $C_2H_5NO = C_2H_3O.NH_2$, may be obtained by either of the general reactions above described ; also by distilling a mixture of dry sodium acetate and sal-ammoniac in equal numbers of molecules. It crystallizes in long needles, melts at 78°-79° C. ($172.4^O-174.2^O$ F.), and boils without decomposition at 222° C. (431.6^O F.). It dissolves easily in water and in alcohol, and when heated with acids or alkalies it takes up water and is resolved into acetic acid and ammonia. It unites with acids, forming unstable compounds, e. g., $C_2H_5NO.HCl$, and $C_2H_5NO.NO_3H$. On boiling its aqueous solution with mercuric oxide, the latter is dissolved, and the solution on cooling deposits crystals of *mercuracetamide* $(C_2H_5NO)_2Hg$.

Chloracetamides may be prepared from the three chloracetic acids in the same manner as acetamide from acetic acid. Their melting and boiling points are as follows :—

		M. P.	B. P.
Monochloracetamide,	$C_2H_7ClO.NH_2$	116°	224°-225°
Dichloracetamide,	$C_2HCl_2O.NH_2$	96°	233°-234°
Trichloracetamide,	$C_2Cl_3O.NH_2$	136°	238°-239°

Diacetamide, $NH{<}^{C_2H_3O}_{C_2H_3O}$, obtained by heating acetamide in a stream of hydrogen chloride, forms crystals easily soluble in water, melts at 59° C. (138.2^O F.), and boils at 210°-215° C. (410°-419° F.).

Triacetamide, $(C_2H_3O)_3N$, obtained by heating acetonitril to 200° C. (392° F.) with acetic anhydride, melts at 78°-79° C. ($172.4^O-174.2^O$ F.).

Propionamide, $C_3H_5O.NH_2$, is very much like acetamide, melts at 75°-76° C. ($167^O-168.8^O$ F.)' and boils at 210 C. (410° F.).

Butyramide, $C_4H_7O.NH_2$, crystallizes in laminæ, melts at 115° C. (239° F.), and boils at 216° C. (420.8^O F.).

Isovaleramide, $C_5H_9O.NH_2$, obtained from isovaleric acid, sublimes in laminæ which are soluble in water.

AMIDES DERIVED FROM DIATOMIC AND MONOBASIC ACIDS.

Acids of this group may give rise to two monamides, both formed by substitution of NH_2 for OH, and therefore having the same composition. They are, however, isomeric, not identical, the one formed by replacement of the alcoholic hydroxyl being acid, while the other, formed by replacement of the basic hydroxyl, is neutral. The acid amides thus formed are called a m i c or a m i d i c a c i d s. Glycollic acid, for example, yields amidoglycollic or glycollamic acid and glycollamide, both containing $C_2H_5NO_2$:

CH$_2$OH	CH$_2$NH$_2$	CH$_2$OH
|	|	|
COOH	COOH	CONH$_2$
Glycollic acid.	Glycollamic acid.	Glycollamide.

1. N e u t r a l A m i d e s .
These compounds are formed by the action of ammonia in the gaseous

state or in alcoholic solution on the corresponding oxides or anhydrides, or on the ethylic ethers of the acids ; thus :

$$CH_3.CH\underset{CO}{\overset{O}{<}}O + NH_3 = CH_3.CH\overset{OH}{\underset{CO.NH_2}{<}}$$

Lactide.   Lactamide.

$$CH_3.CH\overset{OH}{\underset{CO.OC_2H_5}{<}} + NH_3 = C_2H_5OH + CH_3.CH\overset{OH}{\underset{CO.NH_2}{<}}$$

Ethyl lactate.   Lactamide.

Glycollamide, $C_2H_5NO_2 = CH_2\overset{OH}{\underset{CO.NH_2}{<}}$, is formed by heating

glycollide, $\underset{CO}{\overset{CH_2}{|}}>O$, with dry ammonia, and by heating acid ammonium

tartronate (p. 786) to 150° C. (302° F.); $C_3H_3(NH_4)O_5 = C_2H_5NO_2 + CO_2 + H_2O$. It crystallizes in needles having a sweetish taste, easily soluble in water, sparingly in alcohol, melting at 100°. By boiling with alkalies, it is resolved into glycollic acid and ammonia.

Lactamide, $C_3H_7NO_2 = CH_3.CH\overset{OH}{\underset{CO.NH_2}{<}}$, obtained as above mentioned, forms crystals easily soluble in water, melting at 74° C. (165.2° F.), resolved by boiling with alkalies into lactic acid and ammonia.

Lactimide, $C_3H_5NO = CH\overset{\frown{CO}}{—}CH—NH_3$, produced by heating alanine, $CH_3—CH\overset{NH_2}{\underset{CO.OH}{<}}$, to 180°–200° C. (356°–392° F.) in a stream of hydrogen chloride, forms colorless laminæ or needles, easily soluble in water and in alcohol, melting at 275° C. (527° F.).

2. Amic or Amidic Acids.

The amic acids of this group are identical with the amidated acids derived from the corresponding monatomic acids, $C_nH_mO_2$, by substitution of amidogen for hydrogen ; thus glycollamic acid is identical with amidacetic acid ; lactamic with amidopropionic ; leucamic with amidocaproic acid ; for example :

$$\underset{COOH}{\overset{CH_3}{\underset{|}{}}} \qquad \underset{COOH}{\overset{CH_2(NH_2)}{\underset{|}{}}} \qquad \underset{COOH}{\overset{CH_2(OH)}{\underset{|}{}}}$$

Acetic acid.   Amidacetic or Glycollamic acid.   Glycollic acid.

They are formed : 1. By the action of ammonia on the monochloro-, bromo-, or iodo-substitution-products of the fatty acids : e. g.,

$$CH_2Cl.CO_2H + 2NH_3 = NH_4Cl + CH_2.NH_2.CO_2H$$
Chloracetic acid.   Amidacetic acid.

2. By heating the ammonia-compounds of the aldehydes with hydrocyanic and hydrochloric acid, whereby cyanides are produced in the first instance, and afterwards transformed into amido-acids by the action of the hydrochloric acid :

$$CH_3.CH\overset{NH_2}{\underset{OH}{<}} + CNH = H_2O + CH_3.CH\overset{NH_2}{\underset{CN}{<}}$$

Aldehyde-ammonia.   Ethidene-cyanamide.

$$CH_3.CH\overset{NH_2}{\underset{CN}{<}} + HCl + 2H_2O = NH_4Cl + CH_3.CH\overset{NH_2}{\underset{CO_2H}{<}}$$

Ethidene   Lactamic
y anamide.   acid.

Several of them occur in the animal organism.

These amic acids are distinguished from the isomeric neutral amides by the more intimate state of combination of their amidogen-group, which cannot be separated by boiling with alkalies. As they contain both a carboxyl group and an amidogen group, they possess both acid and basic properties, and form saline compounds both with acids and with bases, the basic character, however, predominating. Hence, they are often designated by names ending in *ine*, the ordinary termination for organic bases, glycollamic acid being designated as glycocine, lactamic acid as alanine, leucamic acid as leucine, etc. They are also designated, as a group, by the name **A l a n i n e s .**

The hydrogen of the carboxyl group in these compounds may be replaced by alcohol-radicles, yielding compound ethers, which, however, are somewhat unstable ; that of the amidogen group may be replaced by alcohol-radicles or by acid-radicles. The acid derivatives are obtained by treating the amido-acids, or their ethers, with the haloïd compounds of acid-radicles :

$$CH_2 {\small\begin{matrix}NH_2\\CO_2H\end{matrix}} \quad + \quad C_2H_3OCl \quad = \quad HCl \quad + \quad CH_2 {\small\begin{matrix}NH.C_2H_3O\\CO_2H\end{matrix}} \ ;$$

Amidacetic acid.　　Acetic chloride.　　　　　　　　　Acetyl-amidacetic acid.

the alcoholic derivatives by the action of amines on substitution-derivatives of the fatty acids :

$$CH_2Cl.CO_2H \quad + \quad NH(CH_3)_2 \quad = \quad HCl \quad + \quad CH_2 {\small\begin{matrix}N(CH_3)_2\\CO_2H\end{matrix}}$$

Chloracetic acid.　　Dimethylamine.　　　　　　　　Dimethyl-amidacetic acid.

The alanines are crystalline bodies, mostly having a sweetish taste, easily soluble in water, insoluble for the most part in alcohol and ether. They have a neutral reaction, and, as already observed, are not decomposed by boiling with alkalies, but when fused with alkaline hydrates, they are decomposed into ammonia and fatty acids. By dry distillation, especially in contact with baryta, they are resolved into **a m i n e s** and carbon dioxide :

$$CH_3.CH {\small\begin{matrix}NH_2\\CO_2H\end{matrix}} \quad = \quad CO_2 \quad + \quad CH_3.CH_2.NH_2$$

Alanine.　　　　　　　　　　　Ethylamine.

Nitrous acid converts them into **o x y a c i d** :

$$CH_2 {\small\begin{matrix}NH_2\\CO_2H\end{matrix}} \quad + \quad NO_2H \quad = \quad N_2 \quad + \quad H_2O \quad + \quad CH_2 {\small\begin{matrix}OH\\CO_2H\end{matrix}}$$

Glycocine.　　　　　　　　　　　　　　　　　Glycollic acid.

Amidacetic Acid, or **Glycocine,** $C_2H_5NO_2 = CH_2 {\small\begin{matrix}NH_2\\CO_2H\end{matrix}}$, also called *amidoglycollic acid*, *glycollamic acid*, and *glycocoll*, is formed by the action of ammonia on bromacetic or chloracetic acid :

$$C_2H_3ClO_2 \quad + \quad 2NH_3 \quad = \quad NH_4Cl \quad + \quad C_2H_3(NH_2)O_2 \ ;$$

also, by the action of acids or alkalies upon animal substances, such as glue, hippuric acid, glycollic acid, etc. From hippuric acid it is formed, together with benzoic acid, according to the equation :

$$C_9H_9NO_3 \quad + \quad H_2O \quad = \quad C_2H_5NO_2 \quad + \quad C_7H_6O_2 \ .$$

Glycocine crystallizes from water in large, hard, transparent, rhombic prisms, having a sweetish taste, soluble in 4 parts of cold water, insoluble in alcohol and ether. It melts at 170° C, (338° F.), and decomposes at a higher temperature. By heating with baryta it is resolved into carbon dioxide and methylamine; by treatment with nitrous acid it is converted into glycollic acid. It combines with acids in different proportions. With *sulphuric acid* it forms the compound $(C_2H_5NO_2)_2SO_4H_2$; and on addition of alcohol to a solution of this sulphate, a salt crystallizing in rectangular prisms is deposited, containing $3C_2H_5NO_2.SO_4H_2$. It also forms the *hydrochlorides*, $C_2H_5NO_2.HCl$ and $C_2H_5NO_2.2HCl$, the latter of which crystallizes in long prisms. The *nitrate*, $C_2H_5NO_2.NO_3H$, forms large prisms.

Glycocine also forms saline compounds by substitution of metals for hydrogen; thus it dissolves *cupric oxide*, forming the salt, $(C_2H_4NO_2)_2Cu + H_2O$, which crystallizes from the hot solution in dark blue needles. The *silver salt*, $C_2H_4NO_2Ag$, crystallizes over sulphuric acid. Glycocine also unites with *metallic salts*, forming crystalline compounds, such as $C_2H_5NO_2.NO_3K$, and $C_2H_5NO_2.NO_3Ag$.

The ethylic ether of glycocine, or *ethyl amidacetate*, $CH_2{<}^{NH_2}_{CO_2.C_2H_5}$, is produced by the action of silver oxide on the hydriodide obtained by heating glycocine with ethyl iodide in alcoholic solution. It is resolved by evaporation of its aqueous solution into glycocine and alcohol.

Methyl-glycocine or *Sarcosine*, $C_3H_7NO_2 = CH_2{<}^{NH.CH_3}_{CO_2H}$, isomeric with alanine, is formed by digesting ethyl chloracetate with excess of a concentrated aqueous solution of methylamine:

$$CH_2Cl.CO_2.C_2H_5 + CH_3.NH_2 + H_2O$$
$$= HCl + C_2H_5.OH + CH_2{<}^{NH.CH_3}_{CO_2H}.$$

The same compound is formed by boiling creatine with baryta-water. The creatine splits into sarcosine and urea, the latter being further decomposed into ammonia and carbonic acid. Sarcosine crystallizes in colorless rhombic prisms, easily soluble in water; it is difficultly soluble in alcohol, insoluble in ether, and has no action upon vegetable colors. It combines with acids to form soluble salts, which have an acid reaction. The double salt of sarcosine with platinum tetrachloride crystallizes in large yellow octohedrons having the composition $2C_3H_7NO_2.2HCl.PtCl_4 + 2Aq$. Sarcosine ignited with soda-lime gives off methylamine.

Trimethyl-glycocine or *Betaine*, $C_5H_{11}NO_2 = CH_2{<}^{N(CH_3)_3}_{CO}{>}O$,

which exists ready-formed in beet-juice, and is produced by oxidation of choline hydrochloride, and synthetically by heating trimethylamine with monochloracetic acid, has been already described (p. 674).

Ethyl-glycocine, $C_4H_9NO_2 = CH_2{<}^{NH.C_2H_5}_{CO_2H}$, obtained by heating chloracetic acid with ethylamine, forms deliquescent laminæ; it unites with acids, bases, and salts.

Diethyl-glycocine, $C_6H_{13}NO_2 = CH_2{<}^{N(C_2H_5)_2}_{CO_2H}$, prepared from chloracetic acid and diethylamine, forms deliquescent crystals, and sublimes below 100°.

Acetyl-glycocine, or *Aceturic acid*, $CH_2{<}^{NH.C_2H_3O}_{CO_2H}$, produced by the action of acetyl chloride on silver-glycocine, or of acetamide on monochloracetic acid, crystallizes in small needles, easily soluble in water and in alcohol, and turning brown at 130° C. (266° F.); reacts like a monobasic acid.

DIGLYCOLLAMIC and TRIGLYCOLLAMIC ACIDS.—Glycocine or glycollamic acid may be regarded as ammonia having one atom of hydrogen replaced by the univalent group $CH_2.CO_2H$, and the similar replacement of 2 and 3 hydrogen-atoms in ammonia may give rise to di- and tri-glycollamic acids:

$NH_2.CH_2.CO_2H$.   .   .   .   Glycollamic acid.
$NH(CH_2.CO_2H)_2$.   .   .   Diglycollamic acid.
$N(CH_2.CO_2H)_3$.   .   .   .   Triglycollamic acid.

These three acids are produced simultaneously by boiling monochloracetic acid with strong aqueous ammonia; and on concentrating the resulting solution, filtering from separated sal-ammoniac, boiling the filtrate with lead oxide, and filtering again, the filtrate on cooling deposits the lead salt of triglycollamic acid, while the lead salts of the other two acids remain in solution. To separate the diglycollamic acid, the lead is precipitated by hydrogen sulphide, and the filtered solution is boiled with zinc carbonate, whereupon the sparingly soluble zinc diglycollamate separates out, while the zinc-salt of glycocine remains in solution.

Diglycollamic and triglycollamic acids are crystalline bodies which form salts both with acids and with bases; the former is bibasic, the latter tribasic.

Amidopropionic Acids, $C_3H_7NO_2 = C_3H_5(NH_2)O_2$.—Of these there are two modifications, analogous to the two bromo-, chloro-, and iodo-propionic acids, viz.:—

CH_3
|
$CH(NH_2)$
|
$COOH$
a·Amidopropionic acid
or Alanine.

$CH_2(NH_2)$
|
CH_2
|
$COOH$
ß-Amidopropionic
acid.

Alanine is produced by the action of alcoholic ammonia on a-chloro- or a-bromo-propionic acid, or by heating aldehyde ammonia with hydrocyanic and hydrochloric acids, the reactions being precisely similar to those by which glycocine is obtained from the corresponding derivatives of acetic acid. It crystallizes in tufts of hard needles, dissolves in 5 parts of cold water, less easily in alcohol, and is insoluble in ether. When slowly heated, it melts and sublimes undecomposed; but when quickly heated, it is resolved into carbon dioxide and ethylamine. Nitrous acid converts it into lactic acid.

Alanine unites with acids, bases, and salts. The *platinochloride*, $2(C_3H_7NO_2.HCl).PtCl_4$, crystallizes in large reddish prisms.

ß-*Amidopropionic acid*, $CH_2(NH_2)$—$CH_2.CO_2H$, prepared by the action of ammonia on ß-iodopropionic acid, forms rhombic prisms easily soluble in water, and sublimes with partial decomposition when heated. Its copper compound is much more soluble than that of the a-acid.

a-Amidobutyric Acid, $C_4H_9NO_2 = CH_3.CH_2.CH{<}^{NH_2}_{CO_2H}$, also called *Propalanine*, is prepared from a-bromobutyric acid. It crystallizes in small laminæ or needles, easily soluble in water.

α-**Amidisovaleric Acid,** $C_5H_{11}NO_2 = \underset{CH}{\overset{CH(CH_3)_2}{|}}<\overset{NH_2}{\underset{CO_2H}{}}$, occurs in the pancreas of the ox, and is formed artificially by the action of ammonia on bromisovaleric acid. It crystallizes in shining prisms, sublimes without previous fusion, dissolves in water and alcohol, but less easily than leucine.

Amidocaproic Acid or **Leucine,** $C_6H_{13}NO_2 = C_5H_{10}<\overset{NH_2}{\underset{CO_2H}{}}$, is formed by the action of ammonia on bromocaproic acid, and by digesting valeral-ammonia with hydrocyanic and hydrochloric acids :

$$C_5H_{10}O.NH_3 + CNH + HCl + H_2O = C_6H_{13}NO_2 + NH_4Cl.$$

Leucine is also formed by the decomposition of animal substances, such as glue, horn, wool, etc., during putrefaction, and by the treatment of these substances with acids or alkalies. It was first discovered in putrid cheese ; more recently it has been found in several parts of the animal organism.

Leucine crystallizes in white shining scales, which melt at 100°, and may be sublimed without decomposition ; it is but little soluble in water, still less in alcohol, insoluble in ether. When heated with caustic baryta, it splits into carbon dioxide and amylamine : $C_6H_{13}NO_2 = C_5H_{13}N + CO_2$. It unites with acids, bases, and salts. Treatment with nitrous acid converts it into leucic acid, $C_6H_{12}O_3$.

AMIDES DERIVED FROM DIATOMIC AND BIBASIC ACIDS, $C_nH_{2n}<\overset{CO_2H}{\underset{CO_2H}{}}$.

Each acid of this group may give rise to three amides, viz. : 1. An acid amide or a m i c a c i d, derived from the acid ammonium salt by abstraction of one molecule of water. 2. A neutral monamide, or i m i d e, derived from the same salt by abstraction of $2H_2O$. 3. A neutral amide, or d i a m i d e, derived from the neutral ammonium salt by abstraction of $2H_2O$; thus from succinic acid, $C_2H_4(CO_2H)_2$, are derived :

1. Succinamic acid, $C_2H_4<\overset{CO.NH_2}{\underset{CO.OH}{}} = C_2H_4<\overset{CO.ONH_4}{\underset{CO.OH}{}} - H_2O$
   Acid Ammonium succinate.

2. Succinimide . . $C_2H_4<\overset{CO}{\underset{CO}{}}>NH = C_2H_4<\overset{CO.ONH_4}{\underset{CO.OH}{}} - 2H_2O$
   Acid Ammonium succinate.

3. Succinamide . $C_2H_4<\overset{CO.NH_2}{\underset{CO.NH_2}{}} = C_2H_4<\overset{CO.ONH_4}{\underset{CO.ONH_4}{}} - 2H_2O$
   Neutral Ammonium succinate.

The two neutral amides may also be regarded as derived from one or two molecules of ammonia by substitution of the diatomic radicle of the acid for two atoms of hydrogen, and the amic acid by similar substitution in the compound molecule, $NH_3.HHO$:

Succinimide	$NH.C_4H_4O_2$
Succinamide	$N_2H_4.C_4H_4O_2$
Succinamic acid	$\left.\overset{NH_2}{\underset{OH}{}}\right\} C_4H_4O_2$.

By abstraction of four molecules of water from the neutral ammonium salts, nitrils or cyanides of the corresponding diatomic alcohol-radicles are produced:

$$C_2H_4 {\displaystyle <}_{CO.ONH_4}^{CO.ONH_4} \ - \ 4H_2O \ = \ C_2H_4 {\displaystyle <}_{CN}^{CN}$$

Neutral Ammonium   Succinonitril or
succinate.   Ethene cyanide.

The amic acids of this group are also formed by boiling the imides with water: thus succinimide, $C_4H_5NO_2$, by taking up H_2O is converted into succinamic acid, $C_4H_7NO_3$; and the neutral amides are formed by shaking up the corresponding neutral ethers with aqueous ammonia: e. g.,

$$C_2O_2 {\displaystyle <}_{OC_2H_5}^{OC_2H_5} \ + \ 2NH_3 \ = \ 2C_2H_5(OH) \ + \ C_2O_2 {\displaystyle <}_{NH_2}^{NH_2}$$

Diethylic   Ethyl   Oxamide.
oxalate.   alcohol.

The typic or extra-radicle hydrogen in these amides may be replaced by alcoholic or by acid radicles, thereby producing alkalamides, secondary and tertiary diamides, etc. The modes of producing such compounds may be understood from the following equations:

$$C_2O_2 {\displaystyle <}_{OH}^{ONH_3(CH_3)} \ - \ H_2O \ = \ C_2O_2 {\displaystyle <}_{OH}^{NH(CH_3)}$$

Acid Methylammonium   Methyloxamic
oxalate.   acid.

$$C_2H_4 {\displaystyle <}_{CO}^{CO} {\displaystyle >} O \ + \ \cdot C_2H_5.NH_2 \ = \ H_2O \ + \ C_2H_4 {\displaystyle <}_{CO}^{CO} {\displaystyle >} N.C_2H_5$$

Succinic   Ethylamine.   Ethyl-
anhydride.   succinimide.

$$C_2O_2 {\displaystyle <}_{OC_2H_5}^{OC_2H_5} + 2(CH_3.NH_2) = 2(C_2H_5.OH) + C_2O_2 {\displaystyle <}_{NH.CH_3}^{NH.CH_3}$$

Diethylic   Methylamine.   Ethyl   Dimethyl-
oxalate.   alcohol.   oxamide.

$$COCl_2 \ + \ 2(C_2H_5.NH_2) \ = \ 2HCL \ + \ CO(NH.C_2H_5)_2$$

Carbonyl   Ethylamine.   Diethyl-
chloride.   carbamide.

$$2C_2H_4 {\displaystyle <}_{CO}^{CO} {\displaystyle >} NAg + C_2H_4 {\displaystyle <}_{CO.Cl}^{\cdot Cl} = 2AgCl + N_2 \Big(C_2H_4 {\displaystyle <}_{CO}^{CO} \Big)_3$$

Argento-succinimide.   Succinyl   Trisuccinimide.
chloride.

Amides of Oxalic Acid

Oxamic Acid, $C_2H_3NO_3 = {\displaystyle |}_{CO.OH}^{CO.NH_2}$, is produced by heating acid ammonium oxalate to about 230° C. (446° F.); also as an ammonium-salt by boiling oxamide with aqueous ammonia: $C_2H_4N_2O_2 + H_2O = C_2H_2(NH_4)NO_3$. Oxamic acid is a white crystalline powder sparingly soluble in cold water, still less soluble in alcohol and ether. It is monobasic, and forms numerous crystalline metallic salts.

Oxamic ethers may be formed by substitution of alcohol-radicles for hydrogen, either in the group NH_2 or in the group OH of oxamic acid, the resulting ethers being acid in the former case, neutral in the latter. The neutral ethers, also called oxamethanes, are formed by the action of ammonia, in the gaseous state or in alcoholic solution, on neutral oxalic ethers; thus:

$$\begin{array}{c} CO.OC_2H_5 \\ | \\ CO.OC_2H_5 \\ \text{Ethyl oxalate.} \end{array} + NH_3 = C_2H_5.OH + \begin{array}{c} CO.NH_2 \\ | \\ CO.OC_2H_5 \\ \text{Ethyl oxamate.} \end{array}$$

They are crystalline bodies soluble in alcohol, decomposed by boiling water, yielding ammonium oxalate and the corresponding alcohol.

The acid ethers of oxamic acid containing one equivalent of alcohol-radicle, are produced by dehydration of the acid oxalates of the corresponding amines; thus:

$$\begin{array}{c} CO.ONH_3(C_2H_5) \\ | \\ CO.OH \\ \text{Acid ethylammonium} \\ \text{oxalate.} \end{array} = H_2O + \begin{array}{c} CO.NH(C_2H_5) \\ | \\ CO.OH \\ \text{Ethyloxamic acid.} \end{array}$$

Methyloxamic and phenyloxamic acids are also known. These acid ethers are metameric with the neutral oxamic ethers containing the same alcohol-radicles.

The replacement of both the hydrogen-atoms in the group NH_2 in oxamic acid, would also yield monobasic acid ethers; none of these are, however, known in the free state, but the ethylic ethers of dimethyl- and diethyl-oxamic acids have been obtained, e. g., *ethylic-dimethyl-oxamate*, $(C_2O_2)N(CH_3)_2(OC_2H_5)$.

Oxamide, $N_2H_4(C_2O_2) = \begin{array}{c} CO.NH_2 \\ | \\ CO.NH_2 \end{array}$.—This compound is formed by the action of heat on neutral ammonium oxalate, but is more advantageously prepared by the action of ammonia on neutral ethyl oxalate. It is also formed in several reactions from cyanogen and cyanides; an aqueous solution of hydrocyanic acid, mixed with hydrogen dioxide, yields a crystalline deposit of oxamide: $2CNH + H_2O_2 = C_2N_2H_4O_2$.

Oxamide is a white, light, tasteless powder, insoluble in cold water, slightly soluble in boiling water, insoluble in alcohol. Heated in an open tube it volatilizes and forms a crystalline sublimate; but its vapor, passed through a red-hot tube, is completely resolved into carbon monoxide, ammonium carbonate, hydrocyanic acid, and urea (carbamide):

$$2C_2N_2H_4O_2 = CO + CO_2 + NH_3 + CNH + CN_2H_4O.$$

Dilute mineral acids decompose it, yielding an ammonium-salt and free oxalic acid; e. g.,

$$C_2N_2H_4O_2 + SO_4H_2 + 2H_2O = SO_4(NH_4)_2 + C_2H_2O_4$$

Dimethyloxamide, $N_2(C_2O_2)H_2(CH_3)_2$, is produced by the dry distillation of methylammonium oxalate:

$$C_2(NH_3.CH_3)_2O_4 - 2H_2O = C_2N_2H_2(CH_3)_2O_2.$$

Diethyloxamide and diamyloxamide are obtained in a similar manner.

The imide of oxalic acid is not known; its nitril is dicyanogen, C_2N_2.

Amides of Succinic Acid.

Succinamic acid, $C_2H_4\Big\langle{}^{CO.NH_2}_{CO.OH}$, formed by heating succinimide with baryta-water, is crystalline, and is easily resolved by the action
of alkalies into succinic acid and ammonia.

Succinimide, $C_2H_4\Big\langle{}^{CO}_{CO}\Big\rangle NH$, formed by heating succinic anhydride in a stream of dry ammonia, or by distillation of acid ammonium
succinate, crystallizes in rhombic plates containing one molecule of water,
dissolves easily in water and in alcohol, melts at 126° C. (258.8° F.), and
boils at 288° C. (550.4° F.).

A hot alcoholic solution of succinimide mixed with a little ammonia and
then with silver nitrate, yields on cooling, large crystals of a r g e n t i c

s u c c i n i m i d e, $C_2H_4\Big\langle{}^{CO}_{CO}\Big\rangle NAg$; and the solution of this salt in a

small quantity of ammonia leaves, on spontaneous evaporation, a syrupy
liquid, which gradually solidifies to a mass of hard, brittle crystals of

a r g e n t a m m o n i u m - s u c c i n i m i d e, $C_2H_4\Big\langle{}^{CO}_{CO}\Big\rangle N(NH_3Ag)$.

The acid character of succinimide exhibited in these salts is likewise
shown by other imides, the group NH, when associated with CO (as in
cyanic acid, $CO{=}NH$) or with C_2O_2, as in the salts above described, being capable of exchanging its hydrogen for metals, like the group OH in
acids.

Succinamide, $C_2H_4(CO.NH_2)_2$, separates as a white powder when
neutral ammonium succinate is shaken up with aqueous ammonia. It is
insoluble in cold water and alcohol, but dissolves in hot water, and separates therefrom in slender needles. At 200° C. (392° F.), it is resolved
into ammonia and succinamide.

Amides of Carbonic Acid.

Carbamic Acid, $CO\Big\langle{}^{NH_2}_{OH}$, is not known in the free state, that
is, as a hydrogen salt, but its ammonium salt, $(CO)(NH_2)(ONH_4)$, enters
into the composition of commercial carbonate of ammonia, and is produced,
as already noticed (p. 341), by the direct combination of carbon dioxide
and ammonia-gas. This salt is easily obtained pure and in large quantity
by passing the two gases, both perfectly dry, into cold absolute alcohol,
separating the copious crystalline precipitate by filtration from the greater
part of the liquid, and heating it with absolute alcohol in a sealed tube to
100°, or above.* The liquid, on cooling, deposits ammonium carbamate in
large crystalline laminæ, which, if perfectly dried over oil of vitriol, and
then heated in a sealed tube to 130°–140° C. (266°–284° F.), split up into
ammonium carbonate and carbamide, one molecule of it giving up a molecule of water to another :—

$2CN_3H_6O_2$	$=$	CN_2H_4O	$+$	$(NH_4)_2CO_3$
Ammonium carbamate.		Carbamide.		Ammonium carbonate.

* K o l b e and B a s a r o f f, Chem. Soc. Journal [2], vi. 104.

CARBAMIC ETHERS.—Carbamic acid forms acid and neutral ethers, accordingly as an atom of hydrogen in the group NH_2 or OH is replaced by an alcohol-radicle.

Ethylcarbamic acid, $CO<^{NH.C_2H_5}_{OH}$, is not known in the free state, but its ethylammonium salt, $(CO).NH(C_2H_5).ONH_3(C_2H_5)$, is produced, as a snow-white powder, by passing carbon dioxide into anhydrous ethylamine cooled by a freezing mixture. Its aqueous solution, like that of ammonium carbamate, does not precipitate barium chloride unless aided by heat. The methylammonium salt of *methylcarbamic acid* is obtained in a similar manner.

The neutral carbamic ethers are called u r e t h a n e s. *Ethyl carbamate*, called simply *urethane*, is produced : 1. By leaving ethyl carbonate in contact with aqueous ammonia :—

$$CO<^{OC_2H_5}_{OC_2H_5} + NH_3 = C_2H_5OH + CO<^{NH_2}_{OC_2H_5}.$$

2. By the action of ammonia on ethyl chlorocarbonate (alcohol saturated with phosgene):—

$$CO<^{Cl}_{OC_2H_5} + NH_3 = HCl + CO<^{NH_2}_{OC_2H_5}.$$

3. By passing cyanogen chloride into alcohol :—

$$CNCl + 2C_2H_5(OH) = C_2H_5Cl + CO<^{NH_2}_{OC_2H_5}.$$

4. By direct union of cyanic acid with alcohol :—

$$CONH + C_2H_5.OH = CO<^{NH_2}_{OC_2H_5}.$$

This compound crystallizes in large tables, melts somewhat below 100°, boils at 180° C. (356° F.); dissolves easily in water, alcohol, and ether. It is decomposed by alkalies into carbon dioxide, ammonia, and alcohol, and by ammonia into alcohol and carbamide :—

$$CO<^{NH_2}_{OC_2H_5} + NH_3 = C_2H_5OH + CO<^{NH_2}_{NH_2}.$$

The other urethanes homologous with ethyl carbonate, are obtained by similar reactions, and exhibit similar properties and decompositions. The methyl-compound, $CO<^{NH_2}_{OCH_3}$, forms tabular crystals, melts at 52° C. (125.6° F.), and boils at 177° C. (350.6° F.). The isopentyl-compound, $CO<^{NH_2}_{OC_5H_{11}}$, crystallizes from hot water in needles having a metallic lustre, melts at 66° C. (150.8° F.), and boils at 220° C. (428° F.).

Thiocarbamic Acid, $CS<^{NH_2}_{SH}$, is obtained by decomposing its ammonium salt with dilute sulphuric acid, as a reddish oil, which easily splits up into thiocyanic acid and hydrogen sulphide :—

$$CS(NH_2)(SH) = CS.NH + SH_2.$$

With water it yields cyanic acid and hydrogen sulphide :—

$$CS(NH_2)(SH) + H_2O = CO.NH + 2SH_2.$$

Its ammonium salt, $CS(NH_2)(S.NH_4)$, formed by the action of alcoholic ammonia on CS_2, crystallizes in yellowish needles or prisms.

Acid thiocarbamic ethers, or rather their amine salts, are formed by heating carbon disulphide with amines in alcoholic solution :—

$$CS_2 + 2(C_2H_5.NH_2) = CS{<}_{S(NH_3.C_2H_5)}^{NH.C_2H_5}$$

Ethylamine. Ethylammonic ethyl-thiocarbamate.

On heating this salt with caustic soda, ethylamine is separated, and sodium ethyl-thiocarbamate, $CS{<}_{SNa}^{NH.C_2H_5}$, is produced, from which hydrochloric acid separates ethyl-thiocarbamic acid, $CS{<}_{SH}^{NH.C_2H_5}$, as an oil which solidifies to a crystalline mass on cooling. By heating the amine salts of ethyl thiocarbamic acid to 100°, alcoholic thiocarbamides are produced (p. 821), e. g.,

$$CS{<}_{S(NH_3.C_2H_5)}^{NH.C_2H_5} = H_2S + CS{<}_{NH.C_2H_5}^{NH.C_2H_5}$$

Ethylammonic ethyl-thiocarbamate. Diethyl-thiocarbamide.

By heating the same amine-salts with metallic salts, as silver nitrate or mercuric chloride, salts of ethyl-thiocarbamic acid are precipitated : e. g.,

$$CS{<}_{S(NH_3.C_2H_5)}^{NH.C_2H_5} + AgNO_3 = (NH_3.C_2H_5)NO_3 + CS{<}_{SAg}^{NH.C_2H_5} ;$$

and these salts, when boiled with water, yield thiocarbimides :

$$CS{<}_{SAg}^{NH.C_2H_5} = AgHS + CS{=}N.C_2H_5 .$$

Carbimide, $CO{=}NH$ or $N\left\{_H^{CO}\right.$, is the same as cyanic acid, and many of the reactions of cyanic acid are most appropriately represented by the formula just given, especially its resolution into carbon dioxide and ammonia under the influence of acids or alkalies :

$$NH(CO) + H_2O = NH_3 + CO_2 ;$$

and the corresponding formation of ethylamine and its homologues, by distilling isocyanic ethers (alcoholic carbimides) with potash (p. 570).

In like manner thiocyanic acid or sulphocyanic acid, CNSH, is identical with thiocarbimide, $CS{=}NH$, and the isothiocyanic ethers with alcoholic thiocarbimides ; allyl isothiocyanate or volatile oil of mustard, for example, is the same as allyl-thiocarbimide, $CS{=}N.C_3H_5$ (see THIOCYANIC ETHERS, p. 574).

Carbamide or **Urea,** $CON_2H_4 = CO{<}_{N_2}^{NH_2}$.—This compound occurs abundantly in the urine of mammalia, and in smaller quantity in that of birds and of some reptiles ; also in other animal secretions. It is produced artificially: 1. By a transposition of the constituent atoms of ammonium isocyanate, which takes place when its aqueous solution is evaporated :

$$CO{=}N{-}NH_4 = CO{<}_{NH_2}^{NH_2} .$$

This transformation, discovered by Wöhler in 1828, was the first instance of the artificial formation of a product of the living organism (p. 488).

2. By the action of ammonia on carbonyl chloride, or on ethylic carbonate :

$$COCl_2 \quad + \quad 2NH_3 \quad = \quad 2HCl \quad + \quad CO(NH_2)_2$$
$$CO(OC_2H_5)_2 \quad + \quad 2NH_3 \quad = \quad 2(C_2H_5.OH) \quad + \quad CO(NH_2)_2 .$$

3. By heating ammonium carbamate to 130°–140° :

$$CO{<}^{NH_2}_{O.NH_4} \quad = \quad H_2O \quad + \quad CO{<}^{NH_2}_{NH_2} .$$

4. By heating oxamide with mercuric oxide :

$$C_2O_2(NH_2)_2 \quad + \quad HgO \quad = \quad CO_2 \quad + \quad Hg \quad + \quad CO(NH_2)_2 .$$

5. By the action of small quantities of acids on cyanamide :

$$C{<}^{NH}_{NH} \quad + \quad H_2O \quad = \quad CO{<}^{NH_2}_{NH_2} .$$

It is also produced by the action of alkalies on creatine and allantoïn, and by oxidation of uric acid, guanine, and xanthine.

Preparation.—1. From u r i n e . Fresh human urine is concentrated in a water-bath, until reduced to an eighth or a tenth of its original volume, and filtered through cloth from the insoluble deposit of urates and phosphates. The liquid is mixed with about an equal quantity of strong solution of oxalic acid in hot water, and the whole vigorously agitated and left to cool. A very copious fawn-colored crystalline precipitate of *urea oxalate* is thus obtained, which may be placed upon a cloth filter, slightly washed with cold water, and pressed. This is to be dissolved in boiling water, and powdered chalk added until effervescence ceases, and the liquid becomes neutral. The solution of urea is filtered from the insoluble caloium oxalate, warmed with a little animal charcoal, again filtered, and concentrated by evaporation, avoiding ebullition, until crystals form on cooling : these are purified by a repetition of the last part of the process. Another process consists in precipitating the evaporated urine with concentrated nitric acid, when *urea nitrate* is precipitated, which is purified by recrystallization with the aid of animal charcoal, and, lastly, decomposed by addition of barium carbonate, whereby a mixture of barium nitrate and urea is formed, which is to be evaporated to dryness on the water-bath, and exhausted with hot alcohol ; the urea then crystallizes on cooling. Urea may also be extracted in great abundance from the urine of horses and cattle duly concentrated, and from which the hippuric acid has been separated by addition of hydrochloric acid ; oxalic acid then throws down the oxalate in such quantity as to render the whole semi-solid.

2. From a m m o n i u m i s o c y a n a t e .—Potassium isocyanate is dissolved in a small quantity of water, and a quantity of dry neutral ammonium sulphate, equal in weight to the cyanate, is added. The whole is evaporated to dryness in a water-bath, and the dry residue boiled with strong alcohol, which dissolves out the urea, leaving the potassium sulphate and the excess of ammonium sulphate untouched. The filtered solution, concentrated by distilling off a portion of the spirit, deposits the urea in beautiful crystals of considerable size. According to J. Williams,[*] isocyanate of lead is more convenient for this preparation than the potassium salt. It is to be digested at a gentle heat in a sufficient quantity of water with an equivalent quantity of ammonium sulphate, and the liquid filtered and evaporated.

[*] Chem. Soc. Journal (1868), xxi. 68.

Urea forms transparent, colorless, four-sided prisms, which are anhydrous, soluble in an equal weight of cold water, and in a much smaller quantity at a high temperature. It is also readily dissolved by alcohol. It is inodorous, has a cooling saline taste, and is permanent in moderately dry air. When heated it melts, and at a higher temperature decomposes, giving off ammonia and ammonium cyanate, and leaving cyanuric acid, which bears a much greater heat without change. The solution of urea is neutral to test-paper ; it is not decomposed in the cold by alkalies or by calcium hydrate, but at a boiling heat emits ammonia, and forms a metallic carbonate. The same change is produced by fusion with the alkaline hydrates, and when urea is heated with water in a sealed tube to a temperature above 100° :

$$COH_4N_2 \ + \ H_2O \ = \ CO_2 \ + \ 2NH_3 \ .$$

Urea, heated with a large excess of potassium permanganate, in presence of much free alkali, gives off all its nitrogen in the free state, differing, in this respect, from most amides, the nitrogen of which is oxidized by this treatment to nitric acid.

Urea is instantly decomposed by *nitrous acid* into carbon dioxide, nitrogen, and water : $COH_4N_2 + 2NO_2H = CO_2 + 2N_2 + 3H_2O$; this decomposition explains the use of urea in preparing nitric ether. When *chlorine gas* is passed over melted urea, hydrochloric acid and nitrogen are evolved, and there remains a mixture of sal-ammoniac and cyanuric acid :

$$6COH_4N_2 \ + \ 3Cl_2 \ = \ 2C_3H_3N_3O_3 \ + \ 4NH_4Cl \ + \ 2HCl \ + \ N_2 \ .$$

A solution of pure urea shows no tendency to change by keeping, and is not decomposed by boiling ; in the urine, on the other hand, where it is associated with putrefiable organic matter, as mucus, the case is different. In putrid urine no urea can be found, but enough ammonium carbonate to cause brisk effervescence with an acid ; and if urine, in a recent state, be long boiled, it gives off ammonia and carbonic acid from the same source.

COMPOUNDS OF UREA.—1. With A c i d s .—Urea, like glycocine, unites with acids, bases, and salts, but though a diamide, it combines with only one molecule of an acid, one of its amidogen groups being neutralized by the CO- group.

The *nitrate*, $CH_4N_2O.NO_3H$, is readily soluble in water, and crystallizes from the aqueous solution in long prisms, but slightly soluble in nitric acid, and is therefore precipitated by nitric acid from the aqueous solution of urea. The *oxalate*, $(CH_4N_2O)_2.C_2H_2O_4$, is obtained as a white crystalline precipitate by mixing the aqueous solutions of oxalic acid and urea ; it is but slightly soluble in water.

2. With M e t a l l i c O x i d e s .—On adding moist *silver oxide* to a solution of urea, the compound, $COH_4N_2.3Ag_2O$, is deposited as a gray powder made up of fine needles. *Mercuric nitrate* added to a solution of urea, mixed with potash, forms a white precipitate containing $COH_4N_2.2HgO$. With *mercuric chloride* a white precipitate is formed, which, on boiling with water, turns yellow, and is converted into $COH_4N_2.3HgO$. On adding mercuric oxide to a warm solution of urea, the compound, $COH_4N_2.HgO$, appears to be produced.

3. With S a l t s .—On evaporating a solution of urea mixed with *sodium chloride*, the compound $COH_4N_2.NaCl$ is obtained in shining prisms. The compound $COH_4N_2.AgNO_3$, obtained in a similar manner, forms large rhombic prisms.

Isuretine, CH_4N_2O, isomeric with urea, is formed by direct combination of hydrogen cyanide (carbimide) with hydroxylamine : $CNH + NH_3O =$

CH_4N_2O, and is obtained by evaporation in long colorless needles, having a strong alkaline reaction, and melting at 104° C. (219.2^O F.). Its hydrochloride, $CH_4N_2O.HCl$, forms deliquescent rhombic plates.

Isuretine decomposes above its melting-point, yielding a sublimate of ammonium carbonate, and a residue containing ammelide (p. 577). Its aqueous solution also decomposes when evaporated, giving off nitrogen, ammonia, and carbon dioxide, and leaving a residue containing urea and biuret.

The constitution of isuretine may perhaps be represented by the formula

$$CH\diagdown_{\!\!\!NH.OH}^{\diagup NH}.$$

Hydroxyl-carbamide or **Hydroxyl-urea**, $CH_4N_2O_2 = CO\diagdown_{\!\!\!NH_2}^{\diagup NH.OH}$, is prepared by adding a strong solution of potassium cyanate to a solution of hydroxylamine nitrate cooled to -10° C. (14° F.). It dissolves easily in water and in alcohol, and is precipitated from these solutions by ether in white needles melting at 128°-130° C. (262.4^O-266° F.).

Biuret, $C_2H_5N_3O_2$, is produced by heating urea to 150°-160° C. (302°-320° F.), the change consisting in the separation of one molecule of ammonia from two molecules of urea :

$$\begin{array}{c}CO\diagdown_{\!\!\!NH_2}^{\diagup NH_2}\\CO\diagdown_{\!\!\!NH_2}^{\diagup NH_2}\end{array} = NH_3 + \begin{array}{c}CO\diagdown_{\!\!\!NH}^{\diagup NH_2}\\CO\diagdown_{\!\!\!NH_2}\end{array}.$$

The biuret is extracted from the residue by cold water. It is easily soluble in water and in alcohol, and crystallizes in slender needles containing one molecule of water. Its aqueous solution, mixed with potash, dissolves cupric oxide with red-violet color.

Biuret heated above 170° C. (338° F.) is resolved into ammonia and cyanuric acid :—

$$3C_2H_5N_3O_2 = 3NH_3 + 2C_3H_3N_3O_3.$$

When heated in a stream of gaseous hydrogen chloride, it yields cyanuric acid, urea, and guanidine, CH_5N_3, together with ammonia and carbon dioxide. The formation of guanidine is represented by the equation :

$$NH\diagdown_{\!\!\!CO.NH_2}^{\diagup CO.NH_2} = CO_2 + NH{=}C\diagdown_{\!\!\!NH_2}^{\diagup NH_2}.$$

Biuret.   Guanidine.

Ethyl Allophanate, $C_2H_3(C_2H_5)N_2O_3 = CO\diagdown_{\!\!\!NH.CO.OC_2H_5}^{\diagup NH_2}$, or $NH\diagdown_{\!\!\!CO.OC_2H_5}^{\diagup CO.NH_2}$, is produced by the action of ethyl chlorocarbonate on urea :

$$CO\diagdown_{\!\!\!NH_2}^{\diagup NH_2} + COCl.O.C_2H_5 = HCl + CO\diagdown_{\!\!\!NH.CO.OC_2H_5}^{\diagup NH_2},$$

or by passing cyanic acid vapor into absolute alcohol :

$$2(CO.NH) + C_2H_5.OH = CO\diagdown_{\!\!\!NH.CO.OC_2H_5}^{\diagup NH_2}.$$

This ether forms shining prismatic crystals, soluble in hot water and in alcohol. Treated with caustic baryta, it yields the barium salt of allophanic acid, $C_2H_4N_2O_3$, but the acid itself cannot be obtained in the free

35

state, as, when separated from the barium salt by a mineral acid, it is immediately resolved into urea and carbon dioxide. A series of allophanic ethers may, however, be prepared by the action of cyanic acid on various alcohols.

Allophanic acid is related to biuret in the same manner as carbamic acid to urea:

$$CO\!<^{NH_2}_{OH}$$
Carbamic acid.

$$CO\!<^{NH_2}_{NH_2}.$$
Urea.

$$CO\!<^{NH_2}_{NH.CO.OH}$$
Allophanic acid.

$$CO\!<^{NH_2}_{NH.CO.NH_2}.$$
Biuret.

In other words, biuret is the amide of allophanic acid.

Trigenic Acid, $C_4H_7N_3O_2 = CO\!<^{NH_2}_{NH.CO.N=C_2H_4}$, is produced by passing cyanic acid vapor into cold aldehyde:

$$3(CO.NH) \;+\; C_2H_4O \;=\; CO_2 \;+\; C_4H_7N_3O_2.$$

It crystallizes in prisms slightly soluble in water and in alcohol, and decomposing when heated.

Derivatives of Carbamide.—Compound Ureas.

The hydrogen in carbamide may be replaced by alcoholic or by acid radicles.

1. The a l c o h o l i c d e r i v a t i v e s are formed by processes similar to those which yield carbamide itself, namely, by the action of amines on cyanic acid, or of ammonia and amines on cyanic ethers:

$$CO=NH \quad + \quad NH_2.C_2H_5 \quad = \quad CO\!<^{NH.C_2H_5}_{NH_2}\;\{\text{ Ethyl-carbamide.}$$

$$CO=N.CH_3 \;+\; NH_3 \quad = \quad CO\!<^{NH.CH_3}_{NH_2}\;\{\text{ Methyl-carbamide.}$$

$$CO=N.C_2H_5 \;+\; NH_2.CH_3 \quad = \quad CO\!<^{NH.C_2H_5}_{NH.CH_3}\;\{\text{ Methyl-ethyl-carbamide.}$$

$$CO=N.C_2H_5 \;+\; NH(C_2H_5)_2 \quad = \quad CO\!<^{NH.C_2H_5}_{N(C_2H_5)_2}\;\{\text{ Triethyl-carbamide.}$$

They are also produced by heating isocyanic ethers with water, the reaction apparently taking place by two stages, thus:

$$CO=N.C_2H_5 \;+\; H_2O \;=\; CO_2 \;+\; NH_2.C_2H_5$$

and $\qquad CO=N.C_2H_5 \;+\; NH_2.C_2H_5 \;=\; CO\!<^{NH.C_2H_5}_{NH.C_2H_5}.$

These compounds greatly resemble urea in their properties and reactions. They combine with one equivalent of an acid. By boiling with alkalies they are resolved into carbon dioxide and amines.

Methyl-carbamide, $CO\!<^{NH.CH_3}_{NH_2}$, crystallizes in long transparent prisms. Its aqueous solution is neutral to test-paper, and if somewhat concentrated yields with nitric acid a precipitate of the salt $C_2H_6N_2O.HNO_3$.

Dimethyl-carbamide, $CO\!<^{NH.CH_3}_{NH.CH_3}$, produced by the action of water or of methylamine on methyl cyanate, crystallizes easily, melts at 97° C. (206.6° F.), volatilizes without alteration, and forms with nitric acid the salt $C_3H_8N_2O.HNO_3$.

Ethyl-carbamide, $CO\big\langle{}^{NH.C_2H_5}_{NH_2}$, forms large prisms, easily soluble in water and in alcohol, melting at 92° C. (197.6° F.). Nitric acid does not precipitate its aqueous solution, but crystals of the nitrate $C_2H_7N_2O.$ HNO_3 are obtained on evaporation.

Diethyl-carbamide, $CON_2H_2(C_2H_5)_2$.—Of this compound there are two modifications.

a-Diethyl-carbamide, $CO\big\langle{}^{NH.C_2H_5}_{NH.C_2H_5}$, formed by the action of water or of ethylamine on ethyl cyanate (see above), crystallizes in long prisms, melts at 112° C. (233.6° F.), and boils without decomposition at 263° C. (505.4° F.).

β-Diethyl-carbamide, $CO\big\langle{}^{N(C_2H_5)_2}_{NH_2}$, is formed by the action of diethylamine on cyanic acid :

$$CO{\equiv}NH \;+\; NH(C_2H_5)_2 \;=\; CO\big\langle{}^{N(C_2H_t)_2}_{NH_2}.$$

It melts at 97° C. (206.6° F.), and boils at 270°–280° C. (518°–536° F.).

Methyl-ethyl-carbamide, $CO\big\langle{}^{NH.CH_3}_{NH.C_2H_5}$, formed by the action of methylamine on ethyl isocyanate, is very deliquescent.

Triethyl-carbamide, $CO\big\langle{}^{N.(C_2H_5)_2}_{NH.C_2H_5}$, is easily soluble in water, alcohol, and ether, melts at 53° C. (127.4° F.), and distils at 223° C. (433.4° F.).

Allyl-carbamide, $CO\big\langle{}^{NH.C_3H_5}_{NH_2}$, from allyl isocyanate and ammonia, crystallizes in fine prisms.

Diallyl-carbamide or *Sinapoline*, $CO\big\langle{}^{NH.C_3H_5}_{NH.C_3H_5}$, is formed by heating allyl isocyanate with water :

$$2(CO{\equiv}N.C_3H_5) \;+\; H_2O \;=\; CO_2 \;+\; CO(NH.C_3H_5)_2 \,;$$

or by heating allyl-isothiocyanate, $CS{\equiv}N.C_3H_5$ (volatile oil of mustard) with water and lead oxide, whereby diallyl-thiocarbamide, $CS(NH.C_3H_5)_2$, is first produced, and then converted into diallyl-carbamide by the action of the lead oxide. Diallyl-carbamide crystallizes in large shining laminæ, slightly soluble in water, having an alkaline reaction, and melting at 100°.

Ethene-dicarbamide, $C_4H_{10}N_4O_2 = C_2H_4\big\langle{}^{NH—CO—NH_2}_{NH—CO—NH_2}$, is formed by heating ethene-diamine hydrochloride with silver cyanate :

$$C_2H_4(NH_2)_2.2HCl \;+\; 2CNOAg \;=\; 2AgCl \;+\; C_2O_2(NH_2)_2(NH)_2.C_2H_4 \,.$$

It is sparingly soluble in alcohol, easily in hot water, melts with decomposition at 192° C. (377.6° F.).

Diethyl-ethene-carbodiamide, $C_2O_2(NH.C_2H_5)_2(NH)_2(C_2H_4)$, admits of two modifications, viz. ;

$$C_2H_4\big\langle{}^{NH—CO—NH(C_2H_5)}_{NH—CO—NH(C_2H_5)} \text{ and } C_2H_4\big\langle{}^{N(C_2H_5)—CO—NH_2}_{N(C_2H_5)—CO—NH_2},$$

the first produced from ethyl isocyanate and ethene-diamine, the second from cyanic acid and diethyl-ethene-diamine.

Alcoholic carbamides containing diatomic radicles are also produced by combination of carbamides with aldehydes, with elimination of water; *e.g.*,

$$C_3H_4{<}{CO.N_2H_3 \atop CO.N_2H_3} \quad = \quad 2CO{<}{NH_2 \atop NH_2} \quad + \quad C_3H_4O \quad - \quad H_2O$$

Allylene-dicarbamide.   Carbamide.   Acrolein.

In like manner the compounds,

$$C_7H_{14}{<}{CO.N_2H_3 \atop CO.N_2H_3} \quad \text{and} \quad {C_7H_{14}{<}{CO.N_2H_3 \atop CO.NH_2} \atop C_7H_{14}{<}CO.N_2H_3}$$

Heptene-dicarb-amide.   Diheptene tricarbamide.

are produced from carbamide and œnanthol. All these aldehydic carbamides are resolved by boiling with water into carbamide and aldehyde.

2. Containing A c i d R a d i c l e s .—Carbamides containing monatomic acid radicles are formed by the action of acid chlorides or anhydrides on carbamide. They are not capable of forming salts with acids. Alkalies decompose them into carbamide and the corresponding acid.

Acetyl-carbamide, $CO{<}{NH. \atop NH_2.C_2H_3O}$, crystallizes in long, silky needles, melting at 112° C. (233.6° F.), slightly soluble in cold water and alcohol, decomposed by heating into acetamide and cyanuric acid.—*Chloracetyl-carbamide*, $C_3H_5ClN_2O_2 = CO(NH_2)(NH.C_2H_2ClO)$, forms laminæ easily soluble in water, and is resolved by heat into hydrochloric acid and hydantoïn, $C_3H_4N_2O_2$. — *Bromacetyl-carbamide*, $C_3H_5BrN_2O_2$, crystallizes in needles slightly soluble in water, and is decomposed by heat in a similar manner.

Diacetyl-carbamide, $CO(NH.C_2H_3O)_2$, produced by the action of acetic anhydride on carbamide, or of carbonyl-chloride on acetamide, crystallizes in needles, and sublimes without decomposition.

Carbamides containing d i a t o m i c a c i d - r a d i c l e s , such as *glycolyl-carbamide*, $CO{<}{NH \atop NH}{>}C_2H_2O$, are obtained as derivatives of uric acid, and will be described in connection therewith.

Thiocarbamide or **Thio-urea**, $CS{<}{NH_2 \atop NH_2}$, also called *Sulpho-carbamide* and *Sulphurea*.—This compound is formed by heating ammonium thiocyanate to 170° C. (338° F.) in the same manner as carbamide is formed from ammonium cyanate. It crystallizes in silky needles or thick rhombic prisms, easily soluble in water and alcohol, sparingly in ether, melts at 146° C. (294.8° F.), and decomposes at a higher temperature, giving off carbon disulphide, hydrogen sulphide, and ammonia, and leaving m e l a m (p. 577). When heated with water to 140° C. (284° F.) it is reconverted into ammonium sulphocyanate, and by boiling with alkalies, or with hydrochloric or sulphuric acid, it is decomposed according to the equation :

$$CSN_2H_4 \quad + \quad 2H_2O \quad = \quad CO_2 \quad + \quad 2NH_3 \quad + \quad H_2S.$$

In contact with the oxides of silver, mercury, or lead, and water, it is converted at ordinary temperatures into c y a n a m i d e , $CN.NH_2$, and by boiling into d i c y a n o d i a m i d e , $C_2N_4H_4$.

Thiocarbamide, like carbamide, forms salts containing one equivalent of acid. The *nitrate*, $CSN_2H_4.HNO_3$, forms large crystals.

Alcoholic derivatives of thiocarbamide are formed by the processes similar to those employed in the preparation of the alcoholic carbamides ; *e. g.*,

$$CS{=}N(C_2H_5) \; + \; NH_3 \; = \; CS{<}^{NH(C_2H_5)}_{NH_2}$$
Ethyl thiocyanate.   Ethyl-thio-
carbamide.

$$CS{=}N(C_2H_5) \; + \; NH_2(CH_3) \; = \; CS{<}^{NH(C_2H_5)}_{NH(CH_3)}$$
Methyl-ethyl-
thiocarbamide.

$$CS{<}^{NH(C_2H_5)}_{S(NH_3.C_2H_5)} \; = \; H_2S \; + \; CS{<}^{NH(C_2H_5)}_{NH(C_2H_5)}$$
Ethylammonium   Diethyl-thio-
Ethyldithiocarbamate.   carbamide.

These compounds are desulphurized by heating them with water and mercuric oxide, the sulphur being replaced by oxygen. Those containing two equivalents of alcohol-radicles are converted by this reaction into the corresponding c a r b a m i d e s :

$$CS{<}^{NH.C_2H_5}_{NH.C_2H_5} \; + \; HgO \; = \; HgS \; + \; CO{<}^{NH.C_2H_5}_{NH.C_2H_5} ,$$

whereas the mono-substituted derivatives are converted, by further separation of hydrogen sulphide, into a l c o h o l i c   c y a n a m i d e s (melamines, p. 577):

$$CS{<}^{NH(C_2H_5)}_{NH_2} \; = \; H_2S \; + \; CN.NH(C_2H_5) .$$
Ethyl-thiocar-   Ethyl-
bamide.   cyanamide.

When the bisubstituted thiocarbamides are heated with mercuric oxide and amines, the oxygen of the substituted carbamide formed in the first instance is replaced by imidogen, NH, and substituted guanidines are produced ; thus :

$$CS{<}^{NH.C_2H_5}_{NH.C_2H_5} + NH_2.C_2H_5 + HgO = HgS + H_2O + C{<}^{NH.C_2H_5}_{-N.C_2H_5}_{NH.C_2H_5} .$$
Diethyl-thio-   Triethyl-
carbamide.   guanadine.

Ethyl-thiocarbamide, $CS(NH_2)(NH.C_2H_5)$, crystallizes in needles melting at 89° C. (192.2° F.). *Diethyl-thiocarbamide*, $CS(NH.C_2H_5)_2$, forms large crystals sparingly soluble in water, and melting at 77° C. (170.6° F.). *Methylethyl-thiocarbamide*, $CS(NH.CH_3)(NH.C_2H_5)$, melts at 54° C. (129.2° F.).

Allyl-thiocarbamide, or *Thiosinnamine*, $CS{<}^{NH.C_3H_5}_{NH_2}$, obtained by combination of allylthiocarbimide (mustard-oil) with ammonia :

$$CS{=}N.C_3H_5 \; + \; NH_3 \; = \; CS{<}^{NH.C_3H_5}_{NH_2} ,$$

crystallizes in shining rhombic prisms, having a bitter taste, melting at 70° C. (158° F.), and decomposing at a higher temperature ; easily soluble in water, alcohol, and ether. Its salts are decomposed by water. By boiling with mercuric oxide or lead oxide and water, it is converted into allyl-cyanamide, $CN.NH.C_3H_5$, which is then further converted, as above explained, into t r i a l l y l m e l a m i n e or s i n n a m i n e, $C_3N_3(NH.C_3H_5)_3$.

CARBAMIDES CONTAINING DIATOMIC ACID RADICLES:
URIC ACID AND UREIDES.

Uric Acid, $C_5N_4H_4O_3$, formerly called *Lithic acid*, is a product of the animal organism, and has never been formed by artificial means. It may be prepared from human urine by concentration and addition of hydrochloric acid, and crystallizes out after some time in the form of small, reddish, translucent grains, very difficult to purify. A much preferable method is to employ the solid white excrement of serpents, which can be easily procured: this consists almost entirely of uric acid and ammonium urate. It is reduced to powder, and boiled in dilute solution of caustic potash; the liquid, filtered from the insignificant residue of feculent matter and earthy phosphates, is mixed with excess of hydrochloric acid, boiled for a few minutes, and left to cool. The product is collected on a filter, washed until free from potassium chloride, and dried by gentle heat.

Uric acid, thus obtained, forms a glistening, snow-white powder, tasteless, inodorous, and very sparingly soluble. It is seen under the microscope to consist of minute, but regular crystals. It dissolves in concentrated sulphuric acid without apparent decomposition, and is precipitated by dilution with water. By destructive distillation, uric acid yields cyanic acid, hydrocyanic acid, carbon dioxide, ammonium carbonate, and a black coaly residue, rich in nitrogen. By fusion with potassium hydroxide, it yields potassium carbonate, cyanate, and cyanide. When treated with nitric acid and with lead dioxide, it undergoes decomposition in a manner to be presently described.

Uric acid is bibasic: its most important salts are those of the alkali-metals. *Acid potassium urate*, $C_5N_4H_3KO_3$, is deposited from a hot saturated solution of uric acid in the dilute alkali, as a white, sparingly soluble, concrete mass, composed of minute needles: it requires about 500 parts of cold water for solution, is rather more soluble at a high temperature, and much more soluble in excess of alkali. *Sodium urate* resembles the potassium salt: it forms the chief constituent of the gouty concretions in the joints, called *chalk-stones*. *Ammonium urate* is also a sparingly soluble compound, requiring for solution about 1000 parts of cold water: the solubility is very much increased by the presence of a small quantity of certain salts, as sodium chloride. The most common of the urinary deposits, forming a buff-colored or pinkish cloud or muddiness, which disappears by redissolution when the urine is warmed, consists of a mixture of different urates.

Uric acid is perfectly well characterized, even when in very small quantity, by its behavior with nitric acid. A small portion, mixed with a drop or two of nitric acid in a small porcelain capsule, dissolves with copious effervescence. When this solution is cautiously evaporated nearly to dryness, and, after the addition of a little water, mixed with a slight excess of ammonia, a deep-red tint of murexide is immediately produced.

Impure uric acid, in a more or less advanced stage of decomposition, is imported into this country, in large quantities, for use as a manure, under the name of *guano* or *huano*. It comes chiefly from the barren and uninhabited islets of the western coast of South America, and is the production of the countless birds that dwell undisturbed in those regions. The people of Peru have used it for ages. Guano usually appears as a pale-brown powder, sometimes with whitish specks; it has an extremely offensive odor, the strength of which, however, varies very much. It is soluble in great part with water, and the solution is found to be extremely rich in ammonium oxalate, the acid having been generated by a process of oxidation. Guano also contains a base called guanine.

Products formed from Uric Acid by Oxidation, etc.

When uric acid is subjected to the action of an oxidizing agent in presence of water, it gives up two of its hydrogen-atoms to the oxidizing agent, while the dehydrogenized residue (which may be called dehyduric acid) reacts with water to form mesoxalic acid and urea:

$$C_5H_2N_4O_3 + 4H_2O = C_3H_2O_5 + 2CH_4N_2O$$

Dehyduric acid.   Mesoxalic acid.   Urea.

The separation of the urea generally takes place, however, by two stages, the first portion being removed more easily than the second; thus, when dilute nitric acid acts upon uric acid, alloxan is produced ; and this, when heated with baryta-water, is further resolved into mesoxalic acid and urea :

$$C_5N_4H_2O_3 + 2H_2O = C_4N_2H_2O_4 + CN_2H_4O$$

Dehyduric acid.   Alloxan.   Urea.

$$C_4N_2H_2O_4 + 2H_2O = C_3H_2O_5 + CN_2H_4O$$

Alloxan.   Mesoxalic acid.   Urea.

Alloxan is a *monureide* of mesoxalic acid—that is to say, it is a compound of that acid with one molecule of urea minus $2H_2O$; and the hypothetical dehyduric acid is the *diureide* of the same acid, derived from it by addition of 2 molecules of urea and subtraction of 4 molecules of water. Now, by hydrogenizing mesoxalic acid, we obtain tartronic acid, $C_3H_4O_5$ (p. 786); and by hydrogenizing alloxan, we obtain dialuric acid, which two bodies, accordingly, bear to uric acid the same relation that mesoxalic acid and urea bear to dehyduric acid ; thus :

$C_3H_2O_5$ $C_4N_2H_2O_4$ $C_5N_4H_2O_3$
Mesoxalic acid.   Alloxan.   Dehyduric acid.

$C_3H_4O_5$ $C_4N_2H_4O_4$ $C_5N_4H_4O_3$;
Tartronic acid.   Dialuric acid.   Uric acid.

and just as the hypothetical dehyduric acid yields mesoxalic acid and alloxan, so should actual uric acid yield tartronic and dialuric acids. These bodies, however, have not been obtained by the direct breaking-up of uric acid, but only by rehydrogenizing the mesoxalic acid and alloxan which result from the breaking-up of its dehydrogenized product. Provisionally, however, dialuric and uric acids may be regarded as tartron-ureide and tartron-diureide respectively :

$$C_3H_4O_5 + CH_4N_2O - 2H_2O = C_4H_4N_2O_4$$

Tartronic acid.   Urea.   Dialuric acid.

$$C_3H_4O_5 + 2CH_4N_2O - 4H_2O = C_5H_4N_4O_3 .$$

Tartronic acid.   Urea.   Uric acid.

The bodies just mentioned may be regarded as typical of three well-defined classes of compounds, to one or other of which a very large number of uric acid products may be referred, viz.: (1) Non-azotized bibasic acids, such as mesoxalic and tartronic acid ; (2) Monureides, such as dialuric acid and alloxan, containing a residue or radicle of such an acid, *e. g.*, mesoxalyl, C_3H_3, plus one urea-residue, $CO{<}^{NH}_{NH}$; (3) Diureides, such as uric acid itself, containing an acid residue, together with two urea-residues.

Monureides.

These compounds may be regarded as carbamides or ureas, in which part of the hydrogen (generally half) is replaced by diatomic acid-radicles. They are either 4-carbon compounds belonging to the mesoxalic series, and containing the radicle $CO{<}{CO\atop CO}{>}CO$ (mesoxalyl), or radicles derived from it— or 3-carbon compounds belonging to the oxalic series, and containing the radicle CO—CO (oxalyl), or others derived therefrom. These relations are exhibited in the following table :

$$CO{<}{NHH\atop NHH}$$
Urea.

$$CO{<}{CO.OH\atop CO.OH}$$
Mesoxalic acid.

$$CO.OH \atop CO.OH$$
Oxalic acid.

$$CO{<}{NH{-}CO\atop NH{-}CO}{>}CO$$
Mesoxalyl-urea
Alloxan.

$$CO{<}{NH{-}CO\atop NH{-}CO}$$
Oxalyl-urea
Parabanic acid.

$$CO{<}{NH{-}CO\atop NH{-}CO}{>}CH.OH$$
Tartronyl-urea
Dialuric acid.

$$CO{<}{NH{-}CO\atop NH{-}CH.OH}$$
Oxyglycolyl-urea
Allanturic acid.

$$CO{<}{NH{-}CO\atop NH{-}CO}{>}CH_2$$
Malonyl-urea
Barbituric acid.

$$CO{<}{NH{-}CO\atop NH{-}CH}$$
Glycolyl-urea
Hydantoïn.

$$CO{<}{NH{-}CO{-}CO{-}CO_2H\atop NH_2}$$
Alloxanic acid.

$$CO{<}{NH{-}CO{-}CO_2H\atop NH_2}$$
Oxaluric acid.

$$CO{<}{NH{-}CO\atop NH{-}CO}{>}CH.NH_2$$
Amidobarbituric acid
Uramil.

$$CO{<}{NH{-}CO{-}CO.NH_2\atop NH_2}$$
Oxaluramide.

$$CO{<}{NH{-}CO\atop NH{-}CO}{>}CH.NO_2$$
Nitrobarbituric acid
Dilituric acid.

$$CO{<}{NH{-}CH_2{-}CO_2H\atop NH_2}$$
Hydantoïn or
Glycoluric acid.

$$CO{<}{NH{-}CO\atop NH{-}CO}{>}CH.NO$$
Nitrosobarbituric acid.

$$CO{<}{NH{-}CO\atop NH{-}CO}{>}C{<}{NH_2\atop SO_2.OH}$$
Thionuric acid.

Most of these compounds, when treated with alkalies, are resolved into the corresponding diatomic acids and urea, or its products of decomposition: *e. g.*,

$$\underset{\text{Alloxan.}}{C_4H_2N_2O_4} + 2H_2O = \underset{\substack{\text{Mesoxalic}\\\text{acid.}}}{C_3O_3(OH)_2} + \underset{\text{Urea.}}{CON_2H_4}.$$

Those which contain the group NH, united with CH_2, are converted by boiling with alkalies, or with strong hydriodic acid, into amide-acids :

$$\underset{\text{Glycoluric acid.}}{CO{<}{NH{-}CH_2{-}CO_2H\atop NH_2}} + H_2O = CO_2 + NH_3 + \underset{\text{Glycocine.}}{NH_2.CH_2.CO_2H}$$

1. Mesoxalic Series.

Alloxan, $C_4H_2N_2O_4$.—This is the characteristic product of the action of strong nitric acid on uric acid in the cold. It is best prepared by adding 1 part of pulverized uric acid to 3 parts of nitric acid, specific gravity 1.45, in a shallow basin standing in cold water. The resulting white crystalline mass, after standing for some hours, is drained from the acid liquid in a funnel having its neck stopped with pounded glass, then dried on a porous tile, and purified by crystallization from a small quantity of water.

Alloxan crystallizes by slow cooling from a hot saturated solution in large efflorescent rectangular prisms containing $C_4H_2N_2O_4 + 4Aq.$; from a solution evaporated by heat it separates in monoclinic octohedrons with truncated summits, containing $C_4H_2N_2O_4 + Aq.$ These crystals heated to 150°–160° C. (302°–320° F.) in a stream of hydrogen give off their water, and leave anhydrous alloxan, $C_4N_2H_2O_4$. Alloxan is very soluble in water: the solution has an acid reaction, a disagreeably astringent taste, and stains the skin, after a time, red or purple. It forms a deep-blue compound with a ferrous salt and an alkali, and its solution, mixed with hydrocyanic acid and ammonia, yields a white precipitate of o x a l u r a m i d e; these two reactions are very characteristic.

Alloxan unites (like ketones) with alkaline bisulphites, forming crystalline compounds.

Its aqueous solution decomposes gradually at ordinary temperatures, more quickly on boiling, into alloxantin, parabanic acid, and carbon dioxide:

$$3C_4H_2N_2O_4 = C_8H_4N_4O_7 + C_3H_2N_2O_3 + CO_2$$

Alloxan.   Alloxantin.   Parabanic acid.

By boiling with dilute acids, alloxan is resolved into oxalic acid and urea; by boiling with baryta-water or lead acetate, it is first converted into alloxanic acid, which then splits up into urea and mesoxalic acid. By hydriodic acid and other reducing agents, alloxan is converted, especially in the cold, into alloxantin; at higher temperatures into dialuric acid.

Alloxanic Acid, $C_4N_2H_4O_5$.—The barium-salt of this acid is deposited in small, colorless, pearly crystals, when baryta-water is added to a solution of alloxan heated to 60° C. (140° F.), as long as the precipitate first produced redissolves, and the filtered solution is then left to cool. The barium may be separated by the cautious addition of dilute sulphuric acid, and the filtered liquid, on gentle evaporation, yields alloxanic acid in small radiated needles. It has an acid taste and reaction, decomposes carbonates, and dissolves zinc with disengagement of hydrogen. It is a bibasic acid. The alloxanates of the alkali-metals are freely soluble: those of the earth-metals dissolve in a large quantity of tepid water; that of silver is quite insoluble and anhydrous.

On boiling its salts with water, the alloxanic acid is resolved into urea and mesoxalic acid. The free acid when boiled yields oxalantin.

Dialuric Acid, $C_4H_4N_2O_4$, is the final product of the action of reducing agents on alloxan, and is formed when sulphuretted hydrogen is passed through a boiling solution of alloxan till no further action takes place: $C_4H_2N_2O_4 + H_2S = C_4H_4N_2O_4 + S$. On adding to an aqueous solution of alloxan, first a small quantity of hydrocyanic acid and then potassium carbonate, the potassium salt of dialuric acid separates out in granular crystals, whilst oxalurate remains dissolved:

$$2C_4H_2N_2O_4 + 2KOH = C_4H_3KN_2O_4 + C_3H_3KN_2O_4 + CO_2.$$

35 *

Dialuric acid separated from its salts by hydrochloric acid crystallizes in needles, has a strong acid reaction, and when exposed to the air turns red, absorbs oxygen, and is converted into alloxantin:

$$2C_4H_4N_2O_4 + O = 2H_2O + C_8H_4N_4O_7 .$$

Barbituric Acid, $C_4H_4N_2O_3$, is formed by the action of sodium-amalgam or hydriodic acid on dibromobarbituric acid; also by heating a solution of alloxantin with strong sulphuric acid, barbituric acid then separating out, while p   acid remains in solution.

Barbituric acid crystallizes in beautiful prisms, containing two molecules of water. It is bibasic, and yields chiefly acid salts, which are obtained by treating the corresponding acetates with barbituric acid.

Barbituric acid is converted by fuming nitric acid into dilituric acid; by potassium nitrate into potassium violurate. When boiled with potash it gives off ammonia, and yields the potassium-salt of malonic acid, $C_3H_4O_4$ (p. 774), whence it appears to have the constitution of malomyl-urea, $CO{<}^{NH-CO}_{NH-CO}{>}CH_2$.

Dibromobarbituric acid, $C_4H_2Br_2N_2O_3 = CO{<}^{NH.CO}_{NH.CO}{>}CBr_2$, is formed by the action of bromine on barbituric, or on nitro-nitroso-, or amido-barbituric acid; also, together with alloxan, by the action of bromine on hydurilic acid:

$$C_8H_6N_4O_6 + Br_6 + H_2O = C_4H_2N_2Br_2O_3 + C_4H_2N_2O_4 + 4HBr.$$

It crystallizes in colorless, shining plates, or prisms, soluble in water, very soluble in alcohol and ether. By *hydrogen sulphide,* in presence of water, it is reduced to dialuric acid:

$$C_4H_2N_2Br_2O_3 + H_2S + H_2O = C_4H_4N_2O_4 + 2HBr + S .$$

When chlorine or bromine acts upon a warm solution of dibromobarbituric acid, carbon dioxide is eliminated and tribromacetylurea, $CO{<}^{NH.CO.CBr_3}_{NH_2}$, is produced. This compound melts at 148° C. (298.4° F.), and yields bromoform when boiled with water.

Monobromobarbituric acid, $C_4H_3BrN_2O_3$, produced by the action of zinc or aqueous hydrocyanic acid on dibromobarbituric acid, forms crystals sparingly soluble in water.

Nitrobarbituric or *Dilituric acid,* $C_4H_3(NO_2)N_2O_3$, is formed by the action of fuming nitric acid on barbituric acid, or of ordinary nitric acid on hydurilic acid. It crystallizes in colorless square prisms containing $3H_2O$, soluble in water with yellow color. It appears to be tribasic, but is most inclined to form salts containing only one equivalent of metal. Its salts are very stable, most of them resisting the action of mineral acids. By heating with bromine, it is converted into dibromobarbituric acid.

Nitrosobarbituric or *Violuric acid,* $C_4H_3(NO)N_2O_3$, is formed by the action of nitric acid of specific gravity 1.2 on hydurilic acid, also by that of potassium nitrite on barbituric acid. The resulting solution is mixed with barium chloride, and the barium salt thereby precipitated is decomposed by sulphuric acid.

Violuric acid crystallizes in yellow rhombic octohedrons containing one molecule of H_2O, moderately soluble in water. It forms blue, violet, or yellow salts containing one equivalent of metal. The *potassium salt,* $C_4H_2K(NO)N_2O_3 + 2H_2O$, forms dark blue prisms, dissolving in water with violet color. The *sodium salt* is dark red. The solution of the acid is colored dark blue by ferric acetate, and on adding alcohol to the solution, a ferric salt is precipitated in six-sided laminæ having a red color and metallic lustre.

Violuric acid heated with potash-lye is resolved into urea and nitrosomalonic acid:

$$C_4H_3(NO)N_2O_3 \ + \ 2H_2O \ = \ CON_2H_4 \ + \ C_3H_3(NO)O_4.$$

Amidobarbituric acid, $C_4H_3(NH_2)N_2O_3$, also called *Uramil, Dialuramide*, and *Murexan*, is formed by the action of hydriodic acid on violuric acid or dilituric acid, or by boiling a solution of ammonium thionurate with hydrochloric acid. It is best prepared by boiling alloxantin with an aqueous solution of sal-ammoniac:

$$\underset{\text{Alloxantin.}}{C_8H_4N_4O_7} \ + \ NH_3.HCl \ = \ \underset{\text{Uramil.}}{C_4H_3(NH_2)N_2O_3} \ + \ \underset{\text{Alloxan.}}{C_4H_2N_2O_4} \ + \ HCl$$

It crystallizes in colorless needles, turning red on exposure to the air, nearly insoluble in cold water, slightly soluble in boiling water, easily in alkalies. The ammoniacal solution becomes purple on exposure to the air, and yields murexid on boiling. Uramil is decomposed by strong *nitric acid*, with formation of ammonium nitrate and alloxan:

$$C_4H_5N_3O_3 \ + \ O \ = \ NH_3 \ + \ C_4H_2N_2O_4.$$

Heated with aqueous *potassium cyanate* it is converted into p s e n d o u r i c a c i d (p. 830): $C_4H_5N_3O_3 + CNHO = C_5H_6N_4O_4.$—By the action of *mercuric* or *argentic oxide* suspended in boiling water, it is converted into m u r e x i d.

T h i o n u r i c   A c i d, $C_4H_5N_3SO_6 = CO{<}^{NH-CO}_{NH-CO}{>}C{<}^{NH_2}_{SO_3H}$.—This acid, which contains the elements of alloxan, ammonia, and sulphurous oxide ($C_4H_2N_2O_4 + NH_3 + SO_2$), is formed, as an ammonium-salt, when a cold solution of alloxan is mixed with a saturated aqueous solution of sulphurous acid, in such quantity that the odor of the gas remains quite distinct: an excess of ammonium carbonate mixed with a little caustic ammonia is then added, and the whole boiled for a few minutes. On cooling, *ammonium thionurate* is deposited in colorless, crystalline plates, which, by solution in water and recrystallization, acquire a fine pink tint. By converting it into a lead salt, and decomposing the latter with hydrogen sulphide, thionuric acid is obtained as a white crystalline mass, very soluble in water. When its solution is heated to the boiling point, it is resolved into sulphuric acid and u r a m i l:

$$C_4H_5N_3SO_6 \ + \ H_2O \ = \ H_2SO_4 \ + \ C_4H_5N_3O_3.$$

---

2. *Oxalic Series.*

P a r a b a n i c   A c i d, $C_3H_2N_2O_3 = CO{<}^{NH-CO}_{NH-CO}$, *Oxalyl urea.*—Th is the characteristic product of the action of moderately strong nitric acid on uric acid or alloxan, *with the aid of heat;* formed also from the same substances by oxidation with manganese dioxide and sulphuric acid:

$$C_5H_4N_4O_3 \ + \ O_2 \ + \ 2H_2O \ = \ C_3H_2N_2O_3 \ + \ 2CO_2 \ + \ 2NH_3.$$

It is conveniently prepared by heating 1 part of uric acid with 8 parts of nitric acid till the reaction has nearly ceased ; the liquid is evaporated to a syrup and left to cool; and the product drained from the mother-liquor is purified by re-crystallization. Parabanic acid forms colorless, transparent, thin, prismatic crystals, permanent in the air: it is easily soluble in water, has a pure and powerfully acid taste, and reddens litmus strongly. Neutralized with ammonia, and boiled for a moment, it yields on cooling crystals of the ammonium salt of oxaluric acid, $C_3H_4N_2O_4$, from which the acid may be separated by sulphuric acid.

Parabanic acid is bibasic. With the alkali-metals it forms monometallic salts, like $C_3HKN_2O_3 = CO\underset{NH—CO}{\overset{NK—CO}{<}}$, which are obtained as crystalline precipitates on adding potassium- or sodium-ethylate to a solution of parabanic acid in absolute alcohol; when dissolved in water, these salts are converted into oxalurates. Silver nitrate, added to a solution of the acid, throws down the diargentic salt, $C_3Ag_2N_2O_3$.

Parabanic acid boiled with dilute acids is resolved into urea and oxalic acid. In presence of alkalies, it takes up water, and is converted into oxaluric acid, $C_3H_4N_2O_4$. By the action of zinc and hydrochloric acid, it is converted into oxalantin, $C_6H_4N_4O_5 = 2C_3H_2N_2O_3 — O$.

Methyl-parabanic acid, $C_3H(CH_3)N_2O_3$, is obtained by decomposition of creatinine.

Dimethyl-parabanic acid or *Cholestrophane*, $C_3(CH_3)_2N_2O_3$, obtained by heating the silver salt of parabanic acid with methyl iodide, forms silvery laminæ, which easily melt and sublime.

Oxaluric Acid, $C_3H_4N_2O_4 = CO\underset{NH_2}{\overset{NH—CO—CO_2H}{<}}$, the analogue of alloxanic acid, is formed by the action of alkalies on parabanic acid. Its ethylic ether is produced by heating urea with chloroxalic ether:

$$CO\underset{NH_2}{\overset{NH_2}{<}} + \underset{CO_2.C_2H_5}{\overset{CO.Cl}{|}} = HCl + CO\underset{NH_2}{\overset{NH—CO—CO_2.C_2H_5}{<}}.$$

It crystallizes in silky needles, which melt with decomposition at $150°–160°$ C. ($302°–320°$ F.).

Oxaluric acid, separated from its salts by a mineral acid, is a heavy crystalline powder. Its salts are sparingly soluble. Boiled with water or alkalies, it is resolved into oxalic acid and urea. Heated to $200°$ C. ($392°$ F.) with phosphorus trichloride, it gives up water, and is reconverted into parabanic acid.

Oxaluramide or **Oxalan**, $C_3H_5N_3O_3 = CO\underset{NH_2}{\overset{NH.CO.CO.NH_2}{<}}$, is formed, together with dialuric acid, on adding a little hydrocyanic acid and ammonia to an aqueous solution of alloxan, and separates as a white precipitate, while the dialuric acid remains dissolved. It is also produced by heating ammonium parabanate (its isomeride) to $100°$, and by heating ethyloxalurate with ammonia. It is sparingly soluble in water, and is resolved by boiling with water, into oxalic acid, urea, and ammonia.

Allanturic Acid or **Oxyglycolyl-urea**, $C_3H_4N_2O_3 = CO\underset{NH—CO}{\overset{NH—CH(OH)}{<}}$, formed by heating allantoin with baryta-water or dioxide of lead, is a deliquescent mass, the aqueous solution of which gives white precipitates with silver or lead salts. When boiled with baryta, it yields hydantoic and parabanic acids, the latter further splitting up into oxaluric acid and urea (or CO_2 and NH_3).

Succinuric Acid, $C_5H_8N_2O_4$, is produced by heating urea to 120°–130° C. (248°–266° F.) with succinic anhydride. It forms shining scales, melting with decomposition at 203°–205° C. (397.4°–401° F.).

Hydantoin or **Glycolyl-urea**, $C_3H_4N_2O_2 = CO{<}^{NH—CO}_{NH—\overset{|}{C}H_2}$ is formed synthetically by heating bromacetyl-urea, with alcoholic ammonia:

$$CO{<}^{NH.CO.CH_2Br}_{NH_2} + NH_3 = NH_4Br + CO{<}^{NH—CO}_{NH—\overset{|}{C}H_2}.$$

It is also produced by the action of hydriodic acid on allantoin:

$$C_4H_6N_4O_3 + 2HI = C_3H_4N_2O_2 + CON_2H_4 + I_2,$$

or on alloxanic acid:

$$CO{<}^{NH—CO—CO—CO_2H}_{NH_2} + H_2 = CO_2 + H_2O + CO{<}^{NH—CO}_{NH—\overset{|}{C}H_2}.$$

Hydantoin forms needle-shaped crystals, having a faint sweet taste, and melting at 206° C. (402.8° F.). When boiled with baryta-water, it takes up water and is converted into hydantoic acid.

Methyl-hydantoin, $CO{<}^{NH———CO}_{N(CH_3)—\overset{|}{C}H_2}$, is obtained by prolonged heating of creatinine (*q. v.*) to 100° with baryta-water; also by fusing urea with sarcosine (methyl-glycocine):

$$CO{<}^{NH_2}_{NH_2} + CH_2{<}^{CO_2H}_{NH.CH_3} = NH_3 + H_2O + CO{<}^{NH———CO}_{N(CH_3)—\overset{|}{C}H_2}.$$

It forms crystals easily soluble in water and in alcohol, melts at 145° C. (293° F.), and sublimes in shining needles. Boiled with oxide of mercury or silver it yields metallic derivatives.

Lactyl-urea, $CO{<}^{NH—CO}_{NH—\overset{|}{C}H—CH_3}$, metameric with methyl-hydantoïn, is formed, together with alanine, when the potassium cyanide used in the preparation of the latter from aldehyde-ammonia (p. 820) contains cyanate. It forms efflorescent rhombic crystals, containing $C_4H_6N_2O_2 + H_2O$; melts at 140° C. (284° F.), and sublimes with partial decomposition. By boiling with baryta-water it is converted into lacturic acid,

$$CO{<}^{NH—CH—CH_3}_{NH_2—CO_2H},$$ metameric with methyl-hydantoic acid, which melts at 155° C. (311° F.).

Ethyl-hydantoin, $CO{<}^{NH———CO}_{N(C_2H_5)—\overset{|}{C}H_2}$, produced in like manner by melting urea with ethyl-glycocine, crystallizes in large, flat prisms, easily soluble in water and alcohol, melting at about 100°, and easily subliming.

The metameric compound, *acetonyl-urea*, $CO{<}^{NH—C(CH_3)_2}_{NH—CO}$, formed by heating a mixture of acetone and potassium cyanide containing cyanate with fuming hydrochloric acid, crystallizes in prisms, melting at 175° C. (347° F.). By boiling with baryta-water it is converted into acetonuric acid, $C_5H_{10}N_2O_3$.

Hydantoic or **Glycoluric Acid,** $C_3H_6N_2O_3 =$
$CO<^{NH-CH_2-CO_2H}_{NH_2}$, is produced by boiling allantoin, hydantoin, or
glycoluril with baryta-water, and synthetically by heating glycocine with
urea (or by boiling its solution with baryta-water):

$$CO<^{NH_2}_{NH_2} + CH_2<^{NH_2}_{CO_2H} = NH_3 + CO<^{NH-CH_2-CO_2H}_{NH_2};$$

also by heating glycocine sulphate with potassium cyanate:

$$CO=NH + CH_2<^{NH_2}_{CO_2H} = CO<^{NH-CH_2-CO_2H}_{NH_2}.$$

Hydantoic acid forms large rhombic prisms soluble in water. It is monobasic, and most of its salts are easily soluble. By heating with hydriodic
acid it is converted into glycocine:

$$CO<^{NH-CH_2-CO_2H}_{NII_2} + H_2O = CO_2 + NH_3 + CH_2<^{NH_2}_{CO_2H}.$$

Diureides.

The best known compounds of this class contain 4, 5, 6, or 8 atoms of
carbon. The 4- and 5-carbon diureides (including uric acid itself) are
formed by the union of one molecule of a bibasic acid and 2 molecules of
urea, with elimination of 4 molecules of water, and accordingly contain
one diatomic acid-residue and two urea-residues, $CO<^{NH}_{NH}$.

Uric Acid, $C_5H_4N_4O_3$, as already observed, may be regarded as the
diureide of tartronic acid, and accordingly represented by the formula:

$$CO<^{NH-CO}_{NH-CO}>C=C<^{NH}_{NH}$$

$$CH(OH)<^{CO-OH}_{CO-OH} + 2CO<^{NH_2}_{NH_2} - 4H_2O.$$
Tartronic acid.   Urea.

Iso-uric Acid, $C_5H_4N_4O_3$, is formed by the action of cyanamide on
alloxantin in aqueous solution, and separates on boiling in the form of a
heavy powder. Its constitution may perhaps be represented by the formula:

$$CO<^{NH-CO}_{NH-CO}>CH-N=C=NH.$$

Pseudo-uric Acid, $C_5H_6N_4O_4$.—The potassium salt of this acid is obtained as a yellow crystalline powder by boiling uramil with a strong solution of potassium cyanate. The free acid forms colorless crystals,
slightly soluble in water. Its structure and formation may be represented
as follows:

$$CO<^{NH-CO}_{NH-CO}>CH.NH_2 + CONH$$
Uramil.   Cyanic acid.

$$= CO<^{NH-CO}_{NH-CO}>CH<^{NH}_{NH_2}>CO.$$
Pseudo-uric acid.

Pseudo-thiouric Acid, $C_5H_6N_4O_3S$, is formed by heating alloxan and thiocarbamide with a concentrated alcoholic solution of sulphur dioxide:

$$CO\left<{}^{NH-CO}_{NH-CO}\right>CO + CS\left<{}^{NH_2}_{NH_2}\right.$$

$$= CO\left<{}^{NH-CO}_{NH-CO}\right>CH\left<{}^{NH}_{NH_2}\right>CS + O.$$

It forms thin white needles, insoluble in water, but soluble in acids.

Uroxanic Acid, $C_5H_8N_4O_6$, and **Oxonic Acid,** $C_4H_5N_3O_4$.—These acids are formed by gradual oxidation of uric acid in alkaline solution. A solution of uric acid in potash left for some time in an open vessel deposits shining laminæ of potassium uroxanate, $C_5H_6K_2N_4O_6 + 3H_2O$, from which the free acid may be separated by hydrochloric acid in the form of a crystalline powder, which dissolves in hot water, but decomposes at the same time, with separation of carbon dioxide. Its formation is represented by the equation:

$$C_5H_4N_4O_3 + 2H_2O + O = C_5H_8N_4O_6.$$

Potassium uroxanate, at the moment of its formation, is partly converted, by separation of CO_2 and NH_3, into potassium oxonate, $C_4H_3K_2N_3O_4$, which is also formed by passing a stream of air through a solution of uric acid in potash till all the uric acid is oxidized. This salt forms radiate groups of crystals, containing $1\frac{1}{2}$ mol. H_2O. Acetic acid added to its dilute solution throws down the potassium salt of allanturic acid, or oxyglycolyl-urea, $C_3H_4N_2O_3$ or $CO\left<{}^{NH-CHOH}_{|}\right._{NH-CO}$.

Allantoin, $C_4H_6N_4O_3$.—This substance, which contains the elements of 2 molecules of ammonium oxalate minus 5 molecules of water $[2C_2(NH_4)_2O_4 - 5H_2O]$, is found in the allantoïc liquid of the fœtal calf and in the urine of the sucking calf. It is produced artificially, together with oxalic acid and urea, by boiling uric acid with lead dioxide and water:

$$2C_5H_4N_4O_3 + O_6 + 5H_2O = C_4H_6N_4O_3 + 2C_2H_2O_4 + 2CH_4N_2O_3.$$

The liquid filtered from lead oxalate, and concentrated by evaporation, deposits, on cooling, crystals of allantoïn, which are purified by re-solution and the use of animal charcoal. The mother-liquor, when further concentrated, yields crystals of pure urea. Allantoïn forms small but brilliant prismatic crystals, transparent, colorless, tasteless, and without action on vegetable colors. It dissolves in 160 parts of cold water, and in a smaller quantity at the boiling heat. It is decomposed by boiling with nitric acid, and by oil of vitriol when concentrated and hot, into ammonia, carbon dioxide, and carbon monoxide. Heated with concentrated solutions of caustic alkalies, it is decomposed into ammonia and oxalic acid.

Its structure may, perhaps, be represented by the formula

$$CO\left<{}^{NH-CH(OH)}_{NH-CO\text{------}}\right>C\left<{}^{NH}_{NH}\right..$$

Glycoluril, $C_4H_6N_4O_2$, probably $CO\left<{}^{NH-CH_2}_{NH-CO}\right>C\left<{}^{NH}_{|}\right._{NH}$, is formed by the action of sodium amalgam on allantoïn, and separates in octohedral crystals, sparingly soluble in water, moderately soluble in ammonia and in concentrated acids. Ammoniacal silver nitrate forms in its solutions a yellow precipitate having the composition $C_4H_4Ag_2N_4O_2$. By boiling with baryta-water, it is resolved into urea and hydantoic acid, and by boiling with acids it is converted into hydantoïn.

Diureides containing 6 and 8 atoms of carbon are formed by the union of 2 monureide molecules, with elimination of water.

Alloxantin, $C_8H_4N_4O_7$, is formed on mixing the aqueous solutions of alloxan and dialuric acid :

$$CO{<}^{NH-CO}_{NH-CO}{>}CO \ + \ CH.OH{<}^{CO-NH}_{CO-NH}{>}CO \ =$$
$$\text{Alloxan.} \hspace{3cm} \text{Dialuric acid.}$$

$$CO{<}^{NH-CO}_{NH-CO}{>}C-C{<}^{CO-NH}_{CO-NH}{>}CO \ + \ H_2O \ .$$
$$\hspace{2.5cm}\lfloor O \rfloor$$
$$\text{Alloxantin.}$$

It is the chief product of the action of hot dilute nitric acid upon uric acid, and is likewise produced by the action of deoxidizing agents upon alloxan—anhydrous alloxantin, in fact, containing 1 atom of oxygen less than 2 molecules of alloxan. It is best prepared by passing sulphuretted hydrogen gas through a moderately strong and cold solution of alloxan. The mother-liquor from which the crystals of alloxan have separated answers the purpose perfectly well ; it is diluted with a little water, and a copious stream of gas transmitted through it. Sulphur is then deposited in large quantity, mixed with a white, crystalline substance, which is the alloxantin. The product is drained upon a filter, slightly washed, and then boiled in water : the filtered solution deposits the alloxantin on cooling. Alloxantin forms small, four-sided, oblique rhombic prisms, colorless and transparent ; it dissolves with difficulty in cold water, but more freely at the boiling heat. The solution reddens litmus, gives with baryta-water a violet-colored precipitate, which disappears on heating, and when mixed with silver nitrate produces a black precipitate of metallic silver. Heated with chlorine or nitric acid, it is oxidized to alloxan. The crystals become red when exposed to ammoniacal vapors. They contain 3 mol. H_2O, which they do not give off till heated above 150° C. (302° F.).

Alloxantin is readily decomposed ; when a stream of sulphuretted hydrogen is passed through its boiling solution, sulphur is deposited and dialuric acid is produced. A hot saturated solution of alloxantin mixed with a neutral salt of ammonia instantly assumes a purple color, which, however, quickly vanishes, the liquid becoming turbid from formation of uramil : the solution then contains alloxan and free acid. With silver oxide, alloxantin gives off carbon dioxide, reduces a portion of the metal, and converts the remainder of the oxide into oxalurate. Boiled with water and lead dioxide, it gives off urea and lead carbonate.

Hydurilic Acid, $C_8H_6N_4O_6$.—Dialuric acid, heated to about 160° C. (320° F.) with glycerin (which acts merely as a solvent), splits up into formic acid, carbon dioxide, and ammonium h y d u r i l a t e :—

$$5C_4H_4N_2O_4 \ = \ CH_2O_2 \ + \ 3CO_2 \ + \ 2C_8H_5N_4(NH_4)O_6 \ .$$

By converting this ammonium-salt into a copper-salt, and decomposing the latter with H_2S, hydurilic acid is obtained in crystals.

Hydurilic acid is converted by fuming nitric acid into **a l l o x a n**, without any other product; but with nitric acid of ordinary strength it yields alloxan, together with **v i o l u r i c a c i d**, **v i o l a n t i n**, and **d i l i t u r i c a c i d**:

$$C_8H_6N_4O_6 + NO_3H = C_4H_3N_3O_4 + C_4H_2N_2O_4 + H_2O$$

Hydurilic   Violuric   Alloxan.
acid.   acid.

$$C_8H_6N_4O_6 + 2NO_3H = C_4H_3N_3O_5 + C_4H_2N_2O_4 + NO_2H + H_2O.$$

Hydurilic   Dilituric   Alloxan.
acid.   acid.

If the action be carried on to the end, dilituric acid is the only product. This acid may indeed be regarded as a product of the oxidation of violuric acid : $C_4N_3H_3O_5 = C_4N_3H_3O_4 + O$: and violantin as a compound of the two.

Purpuric Acid, $C_8H_5N_5O_6$, is not known in the free state; its ammonium salt, $C_8H_4(NH_4)N_5O_6$, constitutes **m u r e x i d**. This salt contains the elements of two molecules of uramil minus two atoms of hydrogen :

$$CO\begin{smallmatrix}NH—CO\\NH—CO\end{smallmatrix}CH.NH_2 + NH_2.CH\begin{smallmatrix}CO—NH\\CO—NH\end{smallmatrix}CO =$$

$$CO\begin{smallmatrix}NH—CO\\NH—CO\end{smallmatrix}C—C\begin{smallmatrix}CO—NH\\CO—NH\end{smallmatrix}CO + NH_3 + H_2.$$

$$\underset{NH}{\vee}$$

Purpuric acid.

It is formed by mixing the ammoniacal solutions of alloxan and uramil :

$$C_4H_2N_2O_4 + C_4H_5N_3O_3 + NH_3 = H_2O + C_8H_4(NH_4)N_5O_6,$$

and is best prepared by boiling for a few minutes a mixture of 1 part of dry uramil, 1 part of mercuric oxide, and 40 parts of water rendered slightly alkaline by ammonia :

$$2C_4H_5N_3O_3 + O = H_2O + C_8H_8N_6O_6.$$

Another method is that of Dr. Gregory: 7 parts of alloxan and 4 parts of alloxantin are dissolved in 240 parts of boiling water, and the solution is added to about 80 parts of cold strong solution of ammonium carbonate : the liquid instantly acquires such a depth of color as to become opaque, and gives on cooling a large quantity of murexid: the operation succeeds best on a small scale.

M u r e x i d * crystallizes with 1 molecule H_2O in small square plates or prisms, which by reflected light exhibit a splendid green metallic lustre, like that of the wing-cases of the rose-beetle and other insects : by transmitted light they are deep purple-red. It dissolves with difficulty in cold water, much more easily at the boiling heat, but is insoluble in alcohol and ether. Mineral acids decompose it, with separation of a white or yellowish substance called **m u r e x a n**, probably identical with uramil. Caustic potash dissolves it, with production of a magnificent purple color, which disappears on boiling.

A few years ago murexid was extensively used in dyeing ; but it is now to a great extent superseded by rosaniline.

O x a l a n t i n, $C_6H_4N_4O_5$, also called *leucoturic acid*, is formed by the action of zinc and hydrochloric acid on parabanic acid :

---

* So called from the Tyrian dye, said to have been prepared from a species of *murex*.

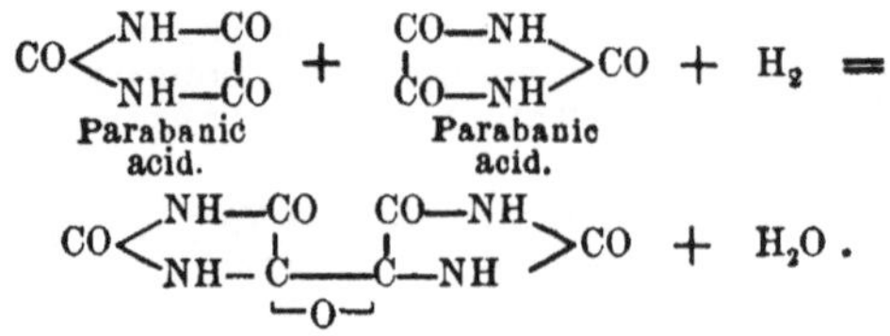

$$CO\left<{}_{NH-CO}^{NH-CO}\right. + \left.{}_{CO-NH}^{CO-NH}\right>CO + H_2 =$$

Parabanic　　　　Parabanic
acid.　　　　　acid.

$$CO\left<{}_{NH-\underset{\underset{O}{|}}{C}--\underset{\underset{O}{|}}{C}-NH}^{NH-CO\quad CO-NH}\right>CO + H_2O .$$

Oxalantin.

It forms small crystals slightly soluble in water, and reduces ammoniacal solutions of mercury and silver.

Allituric Acid, $C_6H_6N_4O_4$, is formed, together with alloxan and parabanic acid, by boiling alloxantin with hydrochloric acid. It forms crystals sparingly soluble in water, and when heated with water to 180°–190° C. (356°–374° F.), is resolved into oxalic acid, carbon dioxide and monoxide, and ammonia.

Respecting basic compounds, natural and artificial, related to the ureides by their composition and reactions, see ALKALOIDS.

AMIDES DERIVED FROM TRIATOMIC AND
TETRATOMIC ACIDS.

Our knowledge of these amides is somewhat limited : we shall notice only those derived from malic, tartaric, and citric acids :

1. Amides of Malic Acid.

Malic acid, $CO_2H—CH_2—CHOH—CO_2H$, or $C_2H_3(OH)\left<{}_{CO_2H}^{CO_2H}\right.$, which is triatomic and bibasic, is capable of yielding five amides, viz. :

$C_2H_3(OH)\left<{}_{CO.OH}^{CO.NH_2}\right.$　　　　$C_2H_3(NH_2)\left<{}_{CO.OH}^{CO.OH}\right.$.

Malamic acid.　　　　　　　Aspartic acid.

$C_2H_3(OH)\left<{}_{CO.NH_2}^{CO.NH_2}\right.$　　　$C_2H_3(NH_2)\left<{}_{CO.OH}^{CO.NH_2}\right.$

Malamide.　　　　　　　Asparagin.

$C_2H_3(NH_2)\left<{}_{CO.NH_2}^{CO.NH_2}\right.$

Triamide (unknown).

Malamic acid is not known in the free state, but its ethylic ether, or *Malamethane,* $C_2H_3(OH)\left<{}_{CO.C_2H_5}^{CO.NH_2}\right.$, is formed as a crystalline mass when ammonia gas is passed into an alcoholic solution of diethylic malate :

$$C_2H_3(OH)\left<{}_{CO_2C_2H_5}^{CO_2C_2H_5}\right. + NH_3 = C_2H_5(OH) + C_2H_3(OH)\left<{}_{CO_2.C_2H_5}^{CO.NH_2}\right. .$$

Aspartic acid, $C_4H_7NO_4$, isomeric with malamic acid, is related to malic and succinic acids in the same manner as glycocine to glycollic and acetic acids (p. 806), and may accordingly be regarded as amidosuccinic acid. It occurs in beet-molasses, and is produced by various reactions from albuminous substances. It is prepared by boiling asparagin with alkalies or acids. It crystallizes in small rhombic prisms, moderately soluble in hot water. Its alkaline solutions turn the plane of polarization to the left; acid solutions to the right. Like other ámido-acids, it unites both with acids and with alkalies; with the latter it forms both acid and neutral salts; $e.\,g.$,

$$C_2H_3(NH_2){<}^{CO_2H}_{CO_2Ag}\ ;\ \ C_2H_3(NH_2){<}^{CO_2Ag}_{CO_2Ag}\ ;\ \ C_2H_3(NH_2){<}^{CO_2}_{CO_2}{>}Ba\ .$$

By the action of nitrous acid it is converted into malic acid.

An inactive aspartic acid is obtained by heating fumarimide with water:

$$C_4H_2O_2.NH\ +\ 2H_2O\ =\ C_4H_7NO_4\ .$$

It crystallizes in large monoclinic prisms, somewhat more soluble in water than the optically active acid. By nitrous acid it is converted into inactive malic acid.

$Malamide$, $C_4H_8N_2O_3$, is produced by the action of ammonia in excess on dry diethylic maláte:

$$C_2H_3(OH){<}^{CO.OC_2H_5}_{CO.OC_2H_5}\ +\ 2NH_3 = 2HOC_2H_5 + C_2H_3(OH){<}^{CO.NH_2}_{CO.NH_2}\ .$$

It forms large crystals, and when heated with water is resolved into ammonia and malic acid.

$Asparagin$, $C_2H_3(NH_2){<}^{CO.NH_2}_{CO.OH}$, isomeric with malamide, occurs in numerous plants, as asparagus, marsh mallow, mangold-wurzel, peas, beans, vetches, and cereal grasses, especially in the young sprouts. It may be prepared from marsh-mallow roots by chopping them small, macerating them in the cold with milk of lime, precipitating the filtered liquid with barium carbonate, and evaporating the clear solution over the waterbath to a syrup. The asparagin then crystallizes on cooling in shining transparent rhombic prisms, which have a faint cooling taste, and are moderately soluble in hot water, insoluble in alcohol and in ether. The crystals contain one molecule of water, whereas those of malamides are anhydrous.

Asparagin and malamide differ also in their action on polarized light, malamide having a specific rotatory power of —47.5°, whereas that of asparagin in an acid solution is + 35°, and in an ammoniacal solution —11° 18′. Lastly, malamide, when treated with alkalies, is resolved, as already observed, into ammonia and malic acid, whereas asparagin yields ammonia and aspartic acid.

Asparagin forms salts both with acids and with bases.

By fermentation in contact with albuminous substances, asparagin is converted into ammonium succinate. By oxidation with potassium permanganate it yields ammonium formate, hydrocyanic acid, and carbon dioxide:

$$C_4H_8N_2O_3\ +\ O_4\ =\ CHO_2.NH_4\ +\ CNH\ +\ 2CO_2\ +\ H_2O\ .$$

In presence of sulphuric acid the oxidation takes place according to the equation:

$$C_4H_8N_2O_3\ +\ O_6\ =\ 2NH_3\ +\ 4CO_2\ +\ H_2O\ ;$$

in presence of potash the products are ammonia and oxalic acid:

$$C_4H_8N_2O_3\ +\ O_4\ +\ 4KHO\ =\ 2NH_3\ +\ 2C_2O_4K_2\ +\ 3H_2O\ .$$

2. **Amides of Tartaric Acid.** — $Tartramic\ Acid$, $C_4H_7NO_5 = C_2H_2(OH)_2{<}^{CO.NH_2}_{CO.OH}$, is obtained as an ammonium-salt by the action of ammonia on tartaric anhydride, $C_4H_4O_5$.

Ethyl tartramate, *Tartramic ether*, or *Tartramethane*, is obtained by the action of alcoholic ammonia on diethylic tartrate. When cautiously heated with alkalies, it yields tartramic acid. Ammonia converts it into tartramide.

Tartramide, $C_4H_8N_2O_4 = C_2H_2(OH)_2\begin{smallmatrix}CO.NH_2\\CO.NH_2\end{smallmatrix}$, formed also by the action of ammonia on diethylic tartrate, is a crystalline substance, the solution of which exhibits dextro- or levo-rotation according to the kind of tartaric acid from which it has been prepared.

3. **Citramide**, $C_6H_{11}N_3O_4 = C_3H_4(OH)(CO.NH_2)_3$, obtained by the action of alcoholic ammonia on ethyl or methyl citrate, is a crystalline substance slightly soluble in water.

Citramic acid and citrimide are not known, but phenylic derivatives of these amides have been obtained.

Benzene-Derivatives, or Aromatic Group.

The hydrocarbons, C_nH_{2n-6}, viz., b e n z e n e and its homologues, together with the alcohols, acids, and bases derived from them, form a group resembling the fatty bodies in many of their chemical relations, but nevertheless exhibiting decided peculiarities, which mark them as a natural family. They are called aromatic, on account of the peculiar and fragrant odors possessed by some of them, especially by certain derivatives of benzene, such as benzoic acid, bitter almond oil, etc.

Intimately related to these bodies are certain other hydrocarbons with their derivatives, containing proportionally smaller numbers of hydrogen-atoms, namely: c i n n a m e n e C_9H_8, n a p h t h a l e n e $C_{10}H_8$, a n t h r a c e n e $C_{14}H_{10}$, p y r e n e $C_{16}H_{10}$, and c h r y s e n e $C_{18}H_{12}$; and the ter-penes, $C_{10}H_{16}$, which contain a larger number of hydrogen-atoms than benzene and its homologues.

HYDROCARBONS, C_nH_{2n-6}.

This is the principal series of the aromatic group, analogous to the paraffin series in the fatty group. The known hydrocarbons belonging to it are represented by the formulæ,

$$C_6H_6, \ C_7H_8, \ C_8H_{10}, \ C_9H_{12}, \ C_{10}H_{14}, \ C_{11}H_{16}, \ C_{12}H_{18}, \ C_{13}H_{20}.$$

The first is called b e n z e n e ; the second, t o l u e n e ;[*] the others admit of isomeric modifications, the names of which will be given hereafter. Many of these hydrocarbons are found in the lighter part of the oil or naphtha obtained by the destructive distillation of coal, and may be separated from one another by fractional distillation.

These hydrocarbons might be regarded as derived from the paraffins by abstraction of 8 atoms of hydrogen (*e. g.*, $C_6H_6 = C_6H_{14} - H_8$), or from the olefines by abstraction of 6 atoms of hydrogen, etc., and accordingly

---

[*] Frequently also benz*ol*, tolu*ol*, etc. ; but it is not desirable to apply the same termination to hydrocarbons and their alcoholic derivatives.

they might be expected to act as octovalent, sexvalent, quadrivalent, or bivalent radicles; and, in fact, benzene does form definite compounds with 6 atoms of chlorine and of bromine. But in nearly all cases the aromatic hydrocarbons react as saturated molecules, like the paraffins, yielding, when treated with chlorine, bromine, or nitric acid, not additive compounds, but substitution-products.

Benzene may be represented as a saturated molecule by the following constitutional formula, in which the carbon-atoms are united together by one or two combining units alternately:—

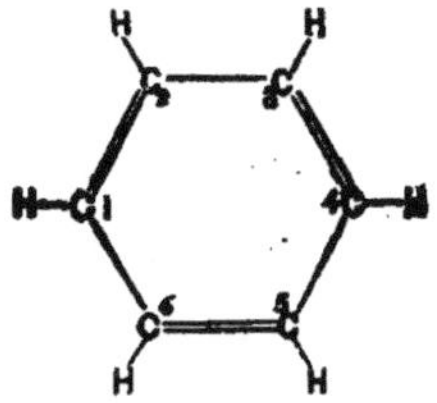

The other hydrocarbons of the series may be derived from it by successive additions of CH_2, or by substitution of methyl, CH_3, in the place of one or more of the hydrogen-atoms; thus:

$$
\begin{aligned}
C_7H_8 &= C_6H_5(CH_3) & \text{Methyl-benzene.} \\
C_8H_{10} &= C_6H_4(CH_3)_2 & \text{Dimethyl-benzene.} \\
C_9H_{12} &= C_6H_3(CH_3)_3 & \text{Trimethyl-benzene.} \\
C_{10}H_{14} &= C_6H_2(CH_3)_4 & \text{Tetramethyl-benzene.}
\end{aligned}
$$

Further, a hydrocarbon isomeric with dimethyl-benzene may be formed by the substitution of ethyl, C_2H_5, for 1 atom of hydrogen in benzene, viz., ethyl-benzene, $C_6H_5(C_2H_5)$; in like manner methyl-ethyl-benzene, $C_6H_4(CH_3)(C_2H_5)$, and propyl-benzene, $C_6H_5(C_3H_7)$, are isomeric with trimethyl-benzene; diethyl-benzene with tetra-methyl-benzene, etc. etc. It is easy to see that in this manner a large number of isomeric bodies may exist in the higher terms of the series. The structure of these isomeric hydrocarbons may be illustrated by the following figures:—

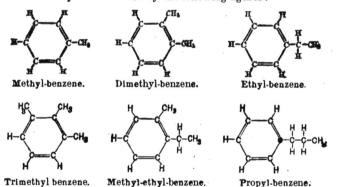

| Methyl-benzene. | Dimethyl-benzene. | Ethyl-benzene. |

| Trimethyl benzene. | Methyl-ethyl-benzene. | Propyl-benzene. |

In these homologues of benzene, the six carbon-atoms belonging to the benzene itself are said to form the benzene-ring, benzene-nucleus, or principal chain, while the groups CH_3, CH_2CH_3, etc., joined on to

these carbon-atoms are called l a t e r a l c h a i n s. The chemical properties of an aromatic hydrocarbon differ considerably according to the number of lateral chains which it contains.

The replacement of the hydrogen-atoms in the principal and lateral
chains by Cl, Br, NO_2, OH, NH_2, etc., gives rise to substitution-derivatives
which exhibit numerous cases of isomerism.

I. In *Benzene-derivatives.*—The mono-derivatives of benzene do not exhibit isomeric modifications: thus there is but one monobromo-, mono-
nitro-, or mono-amidobenzene, one monohydroxyl-derivative or phenol,
$C_6H_5(OH)$, etc. Hence it must be inferred that the molecule of benzene
is perfectly symmetrical, all its six carbon-atoms and all its six hydrogen-
atoms being equal to one another in value, and discharging similar functions, so that the replacement of a single hydrogen-atom by another element or radicle produces the same effect, in whatever part of the molecule
the substitution takes place.

The higher derivatives, on the other hand, formed by replacement of
two or more hydrogen-atoms in the molecule, exhibit isomeric modifications
which are supposed to depend upon the relative positions, or o r i e n t a-
t i o n, of the substituted radicles. Referring to the figure on page 809, in
which the carbon-atoms in benzene are numbered from one to six, it is easy
to see that there may be three such modifications of dichlorobenzene,
$C_6H_4Cl_2$, represented by the following figures:

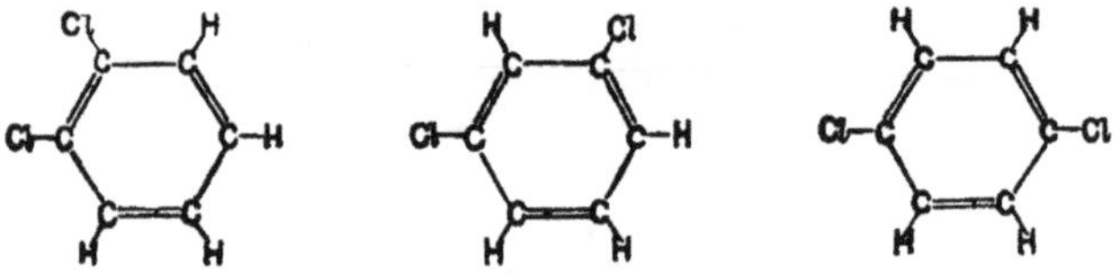

These three modifications are distinguished by the symbols

   1 : 2   1 : 3   1 : 4

In the first the two chlorine-atoms are contiguous; in the second they are
separated by one atom; and in the third by two atoms of hydrogen. It
is clear that these are the only three modifications possible: for 2:3, 3:4,
4:5, and 6:1, would be the same as 1:2; 2:4 and 3:5 would be the
samo as 1:3; and 2:5 and 3:6 would be the same as 1:4.

The number of possible modifications formed by successive replacement
of the hydrogen-atoms in benzene is as follows:

A. *The hydrogen-atoms are successively replaced by the same element or compound radicle.* In this case the number of modifications is as follows:

Number of Hydrogen-atoms replaced.	Number of Modifications.	Positions of the replaced Hydrogen-atoms.		
one	one	1,		
two	three	1, 2,	1, 3,	1, 4
three	three	1, 2, 3	1, 3, 4	1, 3, 5
four	three	1, 2, 3, 4	1, 3, 4, 5	1, 3, 4, 6
five	one	1, 2, 3, 4, 5		
six	one			
		Consecutive.	Unsym-metrical.	Sym-metrical.

The meaning of the terms consecutive, symmetrical, and unsymmetrical, applied to the three modifications of the di-, tri-, and tetraderivatives, will be better understood by means of the following diagram :—

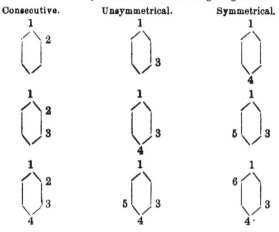

Consecutive.   Unsymmetrical.   Symmetrical.

By counting from different points of the hexagons it is easy to see that :

in the tri-derivatives,   1, 3, 4   is the same as 1, 2, 4,   1, 2, 5   and 1, 3, 6
in the tetra-derivatives, 1, 3, 4, 5   "   1, 2, 3, 5, 1, 2, 4, 6 and 1, 3, 5, 6
and 1, 3, 4, 6   "   1, 2, 4, 5.

B. The hydrogen-atoms are replaced by different elements or radicles.

If only two hydrogen-atoms are thus replaced, the number of possible modifications remains the same as above, viz., three; for the reversal of the order (AB or BA) can make no difference in the result; but if three or more hydrogen-atoms are replaced by different radicles, the number of possible modifications becomes larger, inasmuch as the order of succession of the substituted radicles may then exert an influence on the nature of the product; thus: to one tribromobenzene, $C_6BrBrBrH_3$, in which the bromine-atoms occupy the places 1, 2, 3, there will correspond two dibromochlorobenzenes, viz., $C_6BrBrClH_3$, and $C_6BrClBrH_3$.

In the present state of our knowledge we cannot in all cases assign to the several radicles which replace the hydrogen in benzene their exact relative positions in each isomeric modification; though so far as regards those derivatives in which the substituted radicles are all alike, the list is nearly complete; but inasmuch as a given modification of a benzene-derivative may in many cases be converted into a particular modification of another benzene-derivative, by simple transformations not likely to be attended by any change of molecular structure, we may conclude that in the two derivatives thus producible one from the other, the radicles which replace two or more atoms of hydrogen will occupy corresponding places. Accordingly, the derivatives of benzene containing a given number of replaced hydrogen-atoms may be divided into groups, each containing those modifications which can be formed one from the other, and in which therefore the radicles which have replaced the hydrogen-atoms may be supposed to be similarly situated.

The di-derivatives of benzene, *e. g.*, C_6H_4ClCl, C_6H_4ClBr, $C_6H_4Cl(NO_2)$, etc., exhibit, as above mentioned, three such modifications, which are distinguished by the prefixes *ortho*, *meta*, and *para :* thus the three dichlorobenzenes are designated as follows :—

		1 2 3 4 5 6
Orthodichlorobenzene, 1 : 2	. . .	C_6 Cl Cl H H H H
Metadichlorobenzene, 1 : 3	. . .	C_6 Cl H Cl H H H
Paradichlorobenzene, 1 : 4	. . .	C_6 Cl H H Cl H H.

The following are the principal or typical representatives of these three series of compounds :—

	1 : 2	1 : 3	1 : 4
Dihydroxybenzenes or Oxyphenols, $C_6H_4(OH)_2$	Pyrocatechin	Resorcin	Hydroquinone
Oxy-acids, $C_6H_4{<}^{OH}_{CO_2H}$	Salicylic	Oxybenzoic	Paraoxybenzoic
Dimethyl-benzenes, $C_6H_4(CH_3)_2$	Orthoxylene	Metaxylene	Paraxylene
Dicarbon-acids, $C_6H_4(CO_2H)_2$	Phthalic	Isophthalic	Terephthalic

The relative positions of the substituted radicles in a di-derivative of benzene may be determined by comparison with those in the tri-derivatives which may be formed from it or converted into it. The principle of this method may be illustrated by the case of the di-bromobenzenes, $C_6H_4Br_2$. These, by the action of nitric acid, may be converted into six different nitrodibromobenzenes, $C_6H_3Br_2(NO_2)$; these latter, treated with reducing agents, yield the six corresponding amido-dibromobenzenes, $C_6H_3Br_2(NH_2)$, in which the NH_2 takes the place of the NO_2 ; and the amido-dibromobenzenes (dibromanilines), treated by processes hereafter to be described, exchange their NH_2 for H, whereby they are reconverted into dibromobenzenes, and for Br, whereby they yield tribromobenzenes. The relations between these di- and tri-derivatives are shown in the following diagram, in which, for simplicity, the C's and H's of the benzene-molecule are omitted, and only the substituted radicles are shown in their relative places, the several tribromo- and nitrodibromobenzenes are placed vertically under the dibromobenzenes from which they are derived.

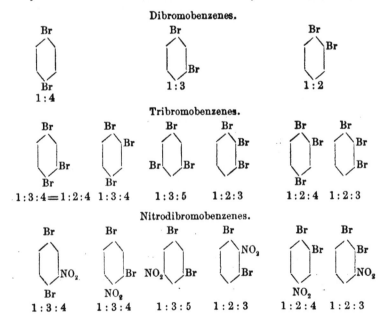

Dibromobenzenes.

Br ... Br ... Br ; Br 1:3 ; 1:2

Tribromobenzenes.

1:3:4=1:2:4 1:3:4 1:3:5 1:2:3 1:2:4 1:2:3

Nitrodibromobenzenes.

1:3:4 1:3:4 1:3:5 1:2:3 1:2:4 1:2:3

An inspection of this diagram shows : (1) That a para-derivative (1 : 4) can give rise to, or be derived from, only *one* tri-derivative, viz., the unsymmetrical modification, 1 : 2 : 4 or 1 : 3 : 4.

(2) That an ortho-derivative (1 : 2) can give rise to, or be produced from, *two* tri-derivatives, viz., the consecutive 1 : 2 : 3, and the unsymmetrical 1 : 2 : 4.

(3) That a meta-derivative (1 : 3) can give rise to, or be formed from, all the three tri-derivatives, 1 : 2 : 3, 1 : 3 : 4, and 1 : 3 : 5.

These conclusions, which are fully borne out by experiment, enable us to give definitions of the three classes of di-derivatives depending only on their relations to the tri-derivatives, and independent of all assumptions as to the relative positions of the substituted radicles ; thus :

A di-derivative of benzene is para-, ortho-, or meta-, according as it can give rise to, or formed from, one, two, or three tri-derivatives.

II. *In the Substitution-derivatives of the Homologues of Benzene.*—The derivatives of toluene and the higher hydrocarbons of the series C_nH_{2n-6}, exhibit two kinds of isomerism : (1) According as the replacement of the hydrogen takes place in the benzene-nucleus or principal chain, or in one of the lateral chains (p. 809); thus from toluene, $C_6H_5.CH_3$, are derived

$C_6H_4Cl.CH_3$ Monochlorotoluene.	isomeric with	$C_6H_5.CH_2Cl$ Benzyl chloride.
$C_6H_4(OH).CH_3$ Cresol.	"	$C_6H_5.CH_2(OH)$ Benzyl alcohol.
$C_6H_4(NH_2).CH_3$ Toluidine.	"	$C_6H_5.CH_2(NH_2)$ Benzylamine.

These isomeric derivatives differ considerably from one another in their properties. Those in the left-hand column, formed by replacement of hydrogen in the benzene-nucleus, like those formed in like manner from benzene itself,—are comparatively stable compounds, which do not give up their chlorine, hydroxyl, etc., in exchange for other radicles so easily as the corresponding derivatives of the paraffins,—whereas those in the right-hand column, formed by replacement of hydrogen in the lateral chain, are more active bodies, easily susceptible of this kind of transformation. Thus benzyl alcohol treated with hydrochloric acid yields benzyl chloride (just as ordinary ethyl alcohol similarly treated yields ethyl chloride); and this compound heated with ammonia yields benzylamine ; the chloride is also easily converted into the acetate, cyanide, etc., by treatment with the corresponding potassium salts. In short, these last-mentioned toluene derivatives exhibit reactions exactly like those of the corresponding compounds of the methyl and ethyl groups. The two series of compounds may, however, be represented by formulæ of similar structure, but containing isomeric radicles, C_7H_7, those in the left-hand column being compounds of methyl-phenyl or tolyl, $C_6H_4(CH_3)$, and those in the right-hand column being compounds of phenyl-methyl or benzyl, $CH_2(C_6H_5)$; *e. g.* :

$$C_6H_4{<}^{CH_3}_{NH_2} = N\begin{Bmatrix}C_6H_4(CH_3)\\H_2\end{Bmatrix} \qquad CH_2{<}^{C_6H_5}_{NH_2} = N\begin{Bmatrix}CH_2(C_6H_5)\\H_2\end{Bmatrix}.$$
Toluidine or Tolylamine.   Benzylamine.

(2) According to the orientation of the substituted radicles in the benzene-nucleus.—This kind of isomerism is of course exhibited only by derivatives like those in the left-hand column, including the hydrocarbons which contain more than one lateral chain ; thus : dimethyl-benzene or xylene, $C_6H_3(CH_3)_2$, exhibits the three modifications, 1 : 2, 1 : 3, and 1 : 4.

36

The difference in chemical character arising from substitution in the principal or in the lateral chains is much greater than that which depends on the orientation of the substituted radicles in the principal chain; in fact, the differences in the latter case are chiefly of physical character, relating to density, melting point, boiling point, etc. In speaking of these two kinds of modification, it will be convenient to designate the former as metameric, the latter as isomeric.

Benzene, C_6H_6.—This hydrocarbon can be produced synthetically from its elements. When ethine or acetylene, C_2H_2, which, as we have seen (p. 538), may be formed by the direct combination of carbon and hydrogen, is heated to a temperature somewhat below redness, it is converted into several polymeric modifications, the principal of which is t r i e t h i n e or b e n z e n e, $3C_2H_2 = C_6H_6$.

This mode of formation leads at once to the constitutional formula of benzene above given: for suppose three molecules of ethine placed as in the left-hand figure below; and, further, that one of the three units of affinity between the two carbon-atoms in each of these molecules is removed, and employed in linking together the alternate carbon-atoms: then we have the formula of benzene, as given in the right-hand figure.

Benzene is also formed in the dry distillation of many organic substances, and is contained in considerable quantity in the more volatile portion of coal-tar oil, from which it is now almost always prepared. To obtain it, the oil is repeatedly washed with dilute sulphuric acid and with potash, to remove the alkaline and acid products likewise existing in it; and the remaining neutral oil is submitted to repeated fractional distillation, the portion which goes over between 80° and 90° C. (176°–194° F.) being collected apart. On cooling this distillate to —12° C. (10.4° F.), the benzene crystallizes out, and may be purified from adhering liquid substances by pressure. It is now prepared in immense quantities for the manufacture of aniline; but the commercial product is always impure, containing also the higher members of the series.

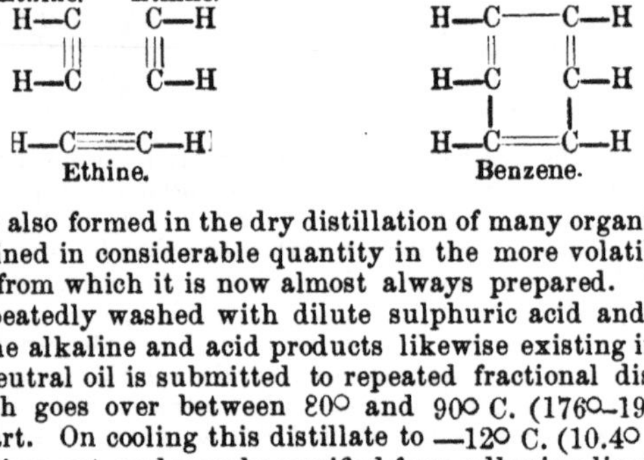

Pure benzene may be obtained by distilling benzoic acid with lime:

$$C_7H_6O_2 \quad + \quad CaO \quad = \quad CaCO_3 \quad + \quad C_6H_6$$

Benzoic   Lime.   Calcium   Benzene.
acid.   carbonate.

Benzene is identical with the so-called *bicarburet of hydrogen*, discovered many years ago by Faraday in the liquid condensed during the compression of oil-gas.

Pure benzene is a thin, limpid, colorless, strongly refracting liquid, having a peculiar ethereal odor. It has a density of 0.889 at 0°, 0.885 at 15.5° C. (59.9° F.), boils at 80.5° C. (176.9° F.), and solidifies at 0° to a mass of rhombic crystals melting at 3° C. (37.4° F.). It is nearly insoluble in water, but mixes with alcohol and ether. It dissolves iodine, sulphur, and phosphorus, and a large number of organic substances, fats and resins for example, which are insoluble, or very sparingly soluble, in water and alcohol: hence its use in many chemical preparations, and for removing grease-spots from articles of dress.

Benzene, passed in the state of vapor through a porcelain tube heated to bright redness, is partly resolved into hydrogen gas containing a small quantity of ethine, and the following liquid products: (1) d i p h e n y l, $C_{12}H_{10} = 2C_6H_6 - H_2$; (2) d i p h e n y l b e n z e n e, $C_6H_4(C_6H_5)_2$, formed according to the equation $3C_6H_6 = C_{18}H_{14} + 2H_2$; (3) b e n z e r y t h r e n e, a solid, resinous, orange-colored body of unknown composition, which distils over in yellow vapors at a dull red heat; (4) b i t u m e n e, a blackish liquid, which remains in the retort at a dull red heat, and solidifies on cooling.* Heated to 275°–280° C. (527°–536° F.) with 80 to 100 parts of strong hydriodic acid, it is converted into hexane, C_6H_{14}.

ADDITIVE-COMPOUNDS OF BENZENE.—Benzene, though, as already observed, it mostly reacts as a saturated molecule—exhibiting indeed in its chemical relations a very close resemblance to the paraffins—can nevertheless, under certain circumstances, take up 6 atoms, or 3 molecules, of chlorine or bromine, forming the compounds $C_6H_6Cl_6$ and $C_6H_6Br_6$. These are crystalline bodies, obtained by exposing benzene to sunshine in contact with chlorine or bromine, the former also by mixing the vapor of boiling benzene with chlorine. Benzene hexchloride melts at 132° C. (269.6° F.), and boils at 288° C. (550.4° F.), being partly resolved at the same time into hydrochloric acid and trichlorobenzene: $C_6H_6Cl_6 = 3HCl + C_6H_3Cl_3$. The same decomposition is quickly produced by heating the compound with alcoholic solution of potash. Benzene hexbromide exhibits a similar reaction.

Benzene is also capable of uniting directly with three molecules of *hypochlorous acid*, forming the compound $C_6H_9Cl_3O_3$, or $C_6H_6.3ClOH$, which crystallizes in thin colorless plates melting at about 10°, and is converted by alkalies into a white deliquescent compound called p h e n o s e, $C_6H_{12}O_6$, isomeric with glucose or grape-sugar:

$$C_6H_9Cl_3O_3 + 3HKO = 3KCl + C_6H_{12}O_6$$

The formation of these additive compounds may be explained in the same manner as that of ethene bromide (pp. 532–3), namely, by supposing that when the benzene-molecule is subjected to the influence of chlorine, bromine, etc., the attachment of the alternate pairs of carbon-atoms is loosened, so that each pair of carbon-atoms becomes united by only one unit of affinity, and each carbon-atom has one unit of affinity left free, and ready to take up an atom of chlorine or other univalent radicle. The nature of the alteration is shown by the following figures, in the second of which the unsaturated affinities are indicated by dots:

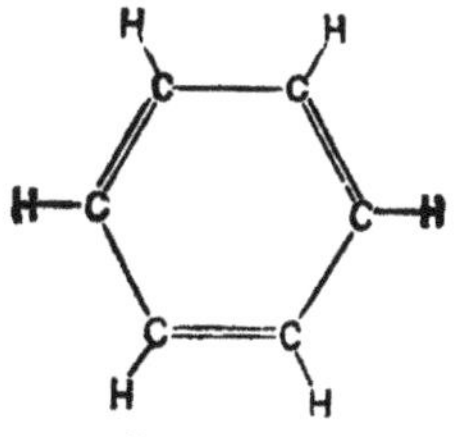

Saturated. 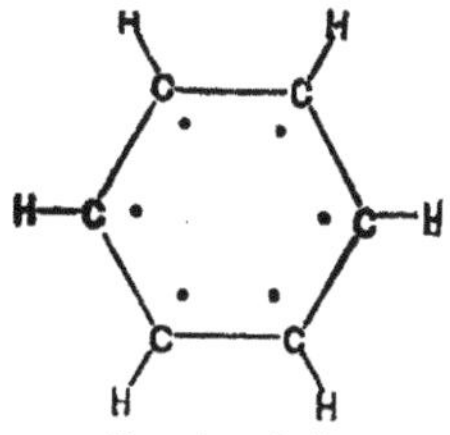   Unsaturated.

T o l u e n e, C_7H_8, or **M e t h y l - b e n z e n e**, $C_6H_5(CH_3)$.—This hydrocarbon is produced: Synthetically (1) By the action of sodium on a mixture of bromobenzene (phenyl bromide), and methyl iodide:

* B e r t h e l o t, Bulletin de la Société Chimique de Paris [2], vi. pp. 272, 279; S c h u l t z, Ann. Chem. clxxiv. 201.

$$C_6H_5Br + CH_3I + Na_2 = NaBr + NaI + C_6H_5.CH_3.$$

This reaction is an example of a general synthetical method of producing the hydrocarbons C_nH_{2n-6}, represented by the equation :

$$C_nH_{2n-7}B_r + C_nH_{2n+1}I + Na_2 = NaBr + NaI$$
$$+ C_nH_{2n-7}.C_nH_{2n+1}[= C_{2n}H_{4n-6}.]$$

2. By the mutual action of benzene and methane in the nascent state, as when a mixture of 2 parts of sodium acetate and 1 part of sodium benzoate is subjected to dry distillation :

$$C_6H_6 + CH_4 = C_7H_8 + H_2.$$

It is also produced by distilling toluic acid, $C_8H_8O_2$, with lime, which abstracts carbon dioxide :

$$C_8H_8O_2 = CO_2 + C_7H_8.$$

It occurs, together with benzene and the other hydrocarbons of the series in light coal-tar oil, and in the products of the distillation of wood, Tolu balsam, dragon's blood, and other vegetable substances ; and, together with many other hydrocarbons, in Rangoon tar or Burmese naphtha.

Toluene is a limpid liquid, smelling like benzene, and having a density of 0.882 at 0°. It boils at 111° C. (231.8^O F.), and does not solidify at -20° C. (-4° F.). In respect of solubility and solvent power, it is very much like benzene, but dissolves somewhat more readily in alcohol. When treated with oxidizing agents, it yields benzoic acid, $C_7H_6O_2$, or derivatives thereof; with potassium chromate and sulphuric acid, it yields benzoic acid ; and by prolonged boiling with strong nitric acid, nitrobenzoic acid.

Toluene vapor passed through a red-hot porcelain tube is partly resolved into hydrogen gas (with small quantities of methane and ethine), and the following liquid products : (1) Benzene and naphthalene in considerable quantities. (2) A crystallizable hydrocarbon volatilizing at 280° C. (536° F.), and probably consisting of dibenzyl, $C_{14}H_{14}$. (3) A liquid isomeric with the last. (4) A mixture, distilling above 360° C. (680° F.), of anthracene with an oily liquid. (5) Chrysene and the last decomposition-products of benzene. The formation of benzene, naphthalene, anthracene, and dibenzyl is represented by the equations :

$$\underset{\text{Toluene.}}{2C_7H_8} = \underset{\text{Dibenzyl.}}{C_{14}H_{14}} + H_2 ; \quad \underset{\text{Toluene.}}{2C_7H_8} = \underset{\text{Anthracene.}}{C_{14}H_{10}} + 3H_2.$$

$$\underset{\text{Toluene.}}{4C_7H_8} = \underset{\text{Benzene.}}{3C_6H_6} + \underset{\text{Naphtha-lene.}}{C_{10}H_8} + 3H_2.$$

Hydrocarbons, C_8H_{10}.—This formula includes the two metameric bodies ·

$$\underset{\text{Ethyl-benzene.}}{C_6H_5(CH_2CH_3)} \qquad \underset{\text{Dimethyl-benzene.}}{C_6H_4(CH_3)_2}.$$

1. ETHYL-BENZENE is produced by the action of sodium on a mixture of monobromobenzene and ethyl bromide. It is a colorless, mobile liquid, very much like toluene, having a density of 0.866, and boiling at 134° C. (273.2^O F.). By oxidation with chromic acid mixture it yields benzoic acid. It is attacked by chlorine, bromine, and nitric acid, forming substitution-products.

2. DIMETHYL-BENZENE, or XYLENE, admits of the three modifications, para-, ortho-, and meta. The first and second are produced by the action

of sodium on a mixture of methyl iodide with para- and ortho-bromotoluene respectively, the bromine-atom in each case being replaced by methyl :

$$C_6H_4{<}{}^{Br}_{CH_3} + CH_3I + Na_2 = NaBr + NaI + C_6H_4{<}{}^{CH_3}_{CH_3}.$$

Orthoxylene is also formed by heating orthodimethyl-benzoic acid, $C_9H_{10}O_2 = C_6H_3\{{}^{(CH_3)_2}_{CO_2H}$, (inappropriately called paraxylic acid), with lime ;' $C_9H_{10}O_2 = CO_2 + C_8H_{10}$, and metaxylene (or isoxylene) in like manner from two other modifications of the same acid called xylic and mesitenic acids :

	1-	2	3	4	5	6
Paraxylic acid,	C_6.CH_3.	CH_3.	H.	CO_2H.	H.	H, gives 1 : 3 Xylene.
Xylic acid,	C_6.CH_3.	H.	CH_3.	CO_2H.	H.	H,
Mesitenic acid,	C_6.CH_3.	H.	CH_3.	H.	CO_2H.	H,

give 1 : 2 Xylene.

These isomeric xylenes are colorless, volatile liquids, orthoxylene boiling at 140°–141° C. (284°–285.8° F.), metaxylene at 137° C. (278.6° F.), and paraxylene at 136°–137° C. (276.8°–278.6° F.). Meta- and para-xylene are contained in the less volatile portion of light coal-naphtha. When the portion of this liquid which boils at about 141° C. (285.8° F.) is shaken with oil of vitriol containing a little fuming sulphuric acid, the xylene is dissolved in the form of xylene-sulphonic acid, $C_8H_{10}SO_3$; and on decomposing this compound by dry distillation, and purifying the distillate by washing, drying, and rectification, a mixture of metaxylene and paraxylene is obtained, containing 90 per cent. of the former.

Xylene (either modification) passed through a red-hot tube, is resolved into a mixture of several hydrocarbons, among which are benzene, toluene, styrolene, naphthalene, anthracene, and its higher homologues. The formation of some of these products is represented by the following equations :

$$C_8H_{10} - H_2 = C_8H_8 \qquad \text{(Styrolene)}$$
$$3C_8H_{10} - 3H_2 = 2C_7H_8 + C_{10}H_8 \text{ (Naphthalene)}$$
and
$$2C_7H_8 - 3H_2 = C_{14}H_{10} \qquad \text{(Anthracene).}$$

The xylenes, oxidized with a mixture of potassium dichromate and sulphuric acid, are converted into phthalic acids, $C_8H_6O_4$, or $C_6H_4\{{}^{COOH}_{COOH}$, according to the equation, $C_8H_{10} + O_6 = 2H_2O + C_8H_6O_4$, each modification of the hydrocarbon yielding a corresponding modification of the acid.

Hydrocarbons, C_9H_{12}.—This formula includes the three following metameric bodies :

$$C_6H_5(C_3H_7)\cdot \qquad C_6H_4\{{}^{CH_3}_{C_2H_5} \qquad C_6H_3\{{}^{CH_3}_{CH_3}_{CH_3}$$
Propyl-benzene.   Ethyl-methyl-benzene.   Trimethyl-benzene.

All three have been formed synthetically, the first by the action of sodium and propyl iodide on bromobenzene ; the second by that of sodium and ethyl bromide on bromotoluene ; the third by that of sodium and methyl bromide on bromoxylene ; thus :

(1) C_6H_5Br $+ C_3H_7I + Na_2 = NaBr + NaI + C_6H_5(C_3H_7)$
(2) $C_6H_4Br(CH_3) + C_2H_5Br + Na_2 = 2NaBr + C_6H_4(CH_3)(C_2H_5)$
(3) $C_6H_3Br(CH_3)_2 + CH_3Br + Na_2 = 2NaBr + C_6H_3(CH_3)_3$.

1. **Propyl-benzene** is a liquid which boils at 157° C. (314.6° F.), forms with excess of bromine a viscid tetrabrominated compound, $C_9H_8Br_4$, and with excess of strong nitric and sulphuric acids, a crystalline trinitroderivative, $C_9H_9(NO_2)_3$.

A hydrocarbon, called **c u m e n e**, also having the composition C_9H_{12}, and very much like propyl-benzene, exists ready formed in Roman cuminoil, and is obtained artificially by distilling cumic acid, $C_{10}H_{12}O_2$, with lime ; but it boils at a lower temperature, viz., 151° C. (303.8° F.), forms with bromine a finely crystallized pentabrominated derivative, $C_9H_7Br_5$, and is converted by excess of a mixture of nitric and sulphuric acids only into dinitrocumene, $C_9H_{10}(NO_2)_2$. It agrees with propyl-benzene, however, in being converted by oxidation with chromic acid into benzoic acid. Hence it is probable that both these hydrocarbons have the composition $C_6H_5.C_3H_7$; but that cumene consists of *isopropyl-benzene*, $C_6H_5.CH(CH_3)_2$, whereas the compound formed by the action of sodium and propyl iodide on bromobenzene is *normal propyl-benzene*, $C_6H_5.CH_2CH_2CH_3$. This supposition is in accordance with the lower boiling point of cumene, as it is a general rule that isopropyl-compounds boil at lower temperatures than the corresponding normal propyl-compounds.

Cumene dissolves in hot fuming nitric acid, and water added to the solution throws down a heavy oil, consisting of mononitrocumene, $C_9H_{11}(NO_2)$, which is reduced by ammonium sulphide to amidocumene or cumidine, $C_9H_{11}(NH_2)$.

2. ETHYL-METHYL-BENZENE or ETHYL-TOLUENE is known in two isomeric modifications. The *para*-modification (1 : 4), obtained from parabromotoluene and ethyl bromide in the same manner as ethyl-benzene from bromobenzene, is a liquid boiling at 161°–162° C. (321.8°–323.6° F.), and having a density of 0.865 at 21° C. (69.8° F.). By oxidation it yields terephthalic acid. The *meta*-modification (1 : 3), obtained in like manner from metabromotoluene, boils at 158°–159° C. (316.4°–318.2° F.); has a density of 0.869 at 20° C. (68° F.); and is converted by oxidation into isophthalic acid.

3. TRIMETHYL-BENZENE, $C_6H_3(CH_3)_3$, is susceptible of three isomeric modifications, only two of which, however, are known. Both of them exist in coal-tar, but cannot be completely separated therefrom by fractional distillation.

Mesitylene, or **Symmetrical Trimethyl-benzene**, $C_6.CH_3.H.CH_3.H.CH_3.H$, is formed by distilling acetone with sulphuric acid :

$$3CO(CH_3)_2 = C_6H_3(CH_3)_3 + 3H_2O .$$

A mixture of 1 vol. acetone and 1 vol. sulphuric acid, diluted with $\frac{1}{2}$ vol. water, is distilled from a retort containing sand, and the oily layer of the distillate is separated from the watery layer, washed with solution of sodium carbonate, and distilled.

Mesitylene is a colorless, fragrant liquid, boiling at 163° C. (325.4° F.). By oxidation with dilute nitric acid, it is successively converted into mesitylenic acid, $C_6H_3(CH_3)_2CO_2H$, mesidic or uvitic acid, $C_6H_3(CH_3)(CO_2H)_2$, and trimesic acid, $C_6H_3(CO_2H)_3$, all of which have the symmetrical structure 1 : 3 : 5. By oxidation with chromic acid mixture it is completely decomposed, with formation of acetic acid. Heated with phosphonium iodide, PH_4I, to 250°–300° C. (482°–572° F.), it is converted into the hydrocarbon, C_9H_{18}, which boils at 136° C. (276.8° F.), and yields, with oxidizing agents, the same products as mesitylene itself.

Pseudocumene or Unsymmetrical Trimethyl-benzene, $C_6.CH_3.H.CH_3.CH_3.H_2$, occurs, together with mesitylene, in coal-tar oil (boiling at 162°–168° C.) (323.6°–334.4° F.), and is formed by the action of methyl iodide and sodium on bromoparaxylene, $C_6.CH_3.H.Br.CH_3.H_2$, or bromometaxylene, $C_6.CH_3.H.CH_3.Br.H_2$. It boils at 166° C. (330.8° F.), and is oxidized by dilute nitric to paraxylic acid, $C_6H_3(CH_3)_2CO_2H$, and xylidic acid, $C_6H_3(CH_3)(CO_2H)_2$.

Hydrocarbons, $C_{10}H_{14}$.—Of these there are five metameric modifications, viz. :—

Butyl-benzene .	$C_6H_5.C_4H_9$
Methyl-propyl-benzene .	$C_6H_4.CH_3.C_3H_7$
Diethyl-benzene .	$C_6H_4.C_2H_5.C_2H_5$
Ethyl-dimethyl-benzene .	$C_6H_3.C_2H_5.(CH_3)_2$
Tetramethyl-benzene .	$C_6H_2(CH_3)_4$

1. BUTYL-BENZENE.—Of this hydrocarbon there are three submodifications depending upon the structure of the butyl-radicle. *Normal butyl-benzene*, $C_6H_5.CH_2.CH_2.CH_2.CH_3$, and *Isobutyl-benzene*, $C_6H_5.CH_2.CH(CH_3)_2$, are obtained by the action of sodium on a mixture of bromobenzene and normal or isobutyl bromide, or on a mixture of benzyl bromide ($C_6H_5.CH_2Br$), or chloride, with the iodide of normal propyl or isopropyl. The first has a density of 0.8622 at 16° C. (60.8° F.), and boils at 180° C. (356° F.) ; the second has a density of 0.89 at 15° C. (59° F.), and boils at 167.5° C. (333.5° F.). A third *butyl-benzene*, $C_6H_5.CH{<}^{CH_3}_{CH_2CH_3}$, having a density of 0.8726 at 16° C. (60.8° F.), and boiling at 170°–172° C. (338°–341.6° F.), is produced by the action of zinc-ethyl on phenyl-ethyl bromide, $C_6H_5.CHBr.CH_3$.

2. METHYL-PROPYL-BENZENE or CYMENE, $C_6H_4{<}^{CH_3}_{CH_2CH_2CH_3}$ (1 : 4), occurs in Roman cumin oil (the volatile oil of the seeds of *Cuminum Cyminum*), and in the volatile oils of *Ptychotis Ajowan*, *Cicuta virosa*, *Eucalyptus globulus*, and other plants. It is produced synthetically by the action of sodium on a mixture of parabromotoluene and normal propyl iodide ; formed also from turpentine oil and its isomerides, $C_{10}H_{16}$, by heating with iodine, by the action of strong sulphuric acid, and by heating the dibromides of these hydrocarbons with sodium, alcoholic potash, or aniline. It appears also to be formed in small quantity by the spontaneous oxidation of turpentine oil in contact with the air. Further, it is produced from the isomeric compounds, thymol and carvacrol, $C_6H_3(OH)(CH_3)(C_3H_7)$, by heating with phosphorus pentasulphide, or by the action of phosphorus pentachloride and sodium-amalgam, chlorocymene being first produced and then dechlorinated by the sodium-amalgam ; lastly, together with toluene, xylene, mesitylene, and other hydrocarbons, by distilling camphor, $C_{10}H_{16}O$, over zinc chloride or phosphoric anhydride. It is most easily prepared by gently heating two parts of camphor with one part of phosphorus pentasulphide, repeatedly agitating the crude distillate with strong sulphuric acid, and rectifying.

Cymene is a liquid having a specific gravity of 0.8732 at 0°, and boiling at 175° C. (347° F.). By oxidation with nitric and with chromic acid, it yields paratoluic and terephthalic acids. It dissolves in strong sulphuric acid, forming a sulphonic acid, the barium salt of which crystallizes in shining laminæ, having the composition $(C_{10}H_{13}SO_3)_2Ba + 3H_2O$.

3. DIETHYL-BENZENE, $C_6H_4(C_2H_5)_2$ (1 : 4), obtained by the action of sodium on a mixture of bromethyl-benzene and ethyl bromide, is a color-

less liquid which boils at 178°–179° C. (352.4°–354.2° F.), and is converted by oxidation into ethyl-benzoic acid (1 : 4) and terephthalic acid.

4. ETHYL-DIMETHYL-BENZENE or ETHYL-XYLENE, $C_6H_3 \begin{cases} C_2H_5 \\ (CH_3)_2 \end{cases}$, is formed, together with methyl-diethyl-benzene, by distilling a mixture of dimethyl ketone (acetone) and methyl-ethyl ketone with sulphuric acid: hence, like mesitylene, it possesses the symmetrical structure, 1 : 3 : 5 (p. 846). It has a density of 0.864 at 20° C. (68° F.), boils at 180°–182° C. (356–359.6° F.), and is oxidized by nitric acid to mesitylene and uvitic acid.

5. TETRAMETHYL-BENZENE or DURENE, $C_6H_2(CH_3)_4$ (consecutive or symmetrical ?), is formed by the action of sodium on a mixture of methyl-iodide and bromopseudocumene on dibromoxylene. It is crystalline at ordinary temperatures, melts at 79°–80° C. (174.2°–176° F.), boils at 190° C. (374° F.), and is oxidized by dilute nitric acid to durylic acid, $C_6H_2(CH_3)_2CO_2H$, and cumidic acid, $CH(CH_3)_2(CO_2H)_2$.
The unsymmetrical modification (1 : 3 : 4 : 5), obtained from bromomesitylene, boils at 192°–194° C. (377.6°–381.2° F.), and remains liquid at low temperatures.

Hydrocarbons, $C_{11}H_{16}$.—Of the seven metameric compounds represented by this formula, three only have been obtained, viz. :

Amyl-benzene . . . $C_6H_5.C_5H_{11}$ boiling at 193° C. (379.4° F.).
Diethyl-methyl-benzene, or $C_6H_3 \begin{cases} CH_3 \\ (C_2H_5)_2 \end{cases}$ "   178° C. (352.4° F.).
Diethyl-toluene . .
Propyl - dimethyl - benzene, $C_6H_3 \begin{cases} (CH_3)_2 \\ C_3H_7 \end{cases}$ "   188° C. (370.4° F.).
Propyl-xylene or Laurene

The first is obtained by the action of sodium on a mixture of monobromo-benzene and amyl bromide; the second by that of zinc-ethyl on benzylene chloride :

$$C_6H_5.CHCl_2 + Zn(C_2H_5)_2 = ZnCl_2 + C_6H_3.CH_3(C_2H_5)_2 ;$$

the third, together with several of its lower homologues, by distilling camphor with zinc chloride. The constitution of this last modification is inferred from its reaction with dilute nitric acid, which oxidizes it to lauroxylic acid, $C_9H_{10}O_2$:

$$C_6H_3 \begin{cases} (CH_3)_2 \\ C_3H_7 \end{cases} + O_9 = 2CO_2 + 3H_2O + C_6H_3 \begin{cases} (CH_3)_2 \\ CO_2H \end{cases}.$$

Hydrocarbons, $C_{12}H_{18}$ and $C_{13}H_{20}$.—These bodies admit of numerous modifications, but the first is known in only two, the second in one modification, viz. :

		Boiling point.	Spe. grav. at 9° C. (48.2° F.).
Triethyl-benzene . .	$C_6H_3(C_2H_5)_3$	218° C. (424.4° F.)	....
Amyl - methyl-benzene or Amyl-toluene . .	$C_6H_4 \begin{cases} C_5H_{11} \\ CH_3 \end{cases}$	213° C. (415.4° F.)	0.8643
Amyl-dimethyl-benzene or Amyl-xylene . .	$C_6H_3 \begin{cases} C_5H_{11} \\ (CH_3)_2 \end{cases}$	233° C. (451.4° F.)	0.8951

Triethyl-benzene (1 : 3 : 5) is formed by distilling ethyl-methyl ketone, CH_3—CO—C_2H_5, with sulphuric acid; it is oxidized by chromic acid to trimesic acid.

Amyl-methyl-benzene and amyl-dimethyl-benzene are formed by the action of sodium and amyl bromide on bromotoluene and bromoxylene respectively.

UNSATURATED HYDROCARBONS,

$$C_nH_{2n-8} \text{ and } C_nH_{2n-10}.$$

Ethenyl-benzene or **Vinyl-benzene**; $C_8H_8 = C_6H_5—CH{=}CH_2$; also called *Phenyl-ethene, Cinnamene,* and *Styrolene.* This hydrocarbon occurs in liquid storax (from *Liquidambar orientale*), and may be prepared therefrom by distilling the balsam with water containing a little sodium carbonate, to retain cinnamic acid.

It is produced: 1. Synthetically: *a.* By passing a mixture of benzene-vapor and acetylene or ethene through a red-hot tube:

$$C_6H_6 \ + \ C_2H_2 \ = \ C_8H_8; \text{ and } C_6H_6 \ + \ C_2H_4 \ = \ C_8H_8 \ + \ H_2.$$

b. In like manner, together with benzene, from diphenyl and ethene:

$$C_{12}H_{10} \ + \ C_2H_4 \ = \ C_8H_8 \ + \ C_6H_6.$$

2. In the decomposition of xylene which takes place when the vapor of that compound is passed through a red-hot tube: $C_8H_{10} = C_8H_8 + H_2$ (p. 845).

3. By distilling cinnamic acid with baryta, which removes carbon dioxide: $C_9H_8O_2 = CO_2 + C_8H_8.$

Cinnamene is a very mobile, colorless oil of specific gravity 0.924. It boils at 146° C. (294.8° F.), and has a density of 0.924. When heated to 200° C. (392° F.) in a sealed tube, it is converted into a white, transparent, highly refractive, solid substance, called m e t a c i n n a m e n e or m e t a s t y r o l e n e. This substance, when heated in a small retort, yields a distillate of pure liquid cinnamene.*

A mixture of cinnamene vapor and ethene passed through a red-hot tube yields large quantities of benzene and naphthalene. The first is produced from the cinnamene by abstraction of C_2H_2; the second according to the equation:

$$C_8H_8 \ + \ C_2H_4 \ = \ C_{10}H_8 \ + \ 2H_2.$$

A mixture of cinnamene and benzene vapors, passed through a red-hot porcelain tube, yields anthracene, $C_{14}H_{10}$, together with small quantities of other products:

$$C_8H_8 \ + \ C_6H_6 \ = \ C_{14}H_{10} \ + \ 2H_2$$

Cinnamene acts with chlorine and bromine like a bivalent radicle, forming the compounds $C_8H_8Cl_2$ and $C_8H_8Br_2$, which, when treated with alcoholic potash, give up HCl and HBr (like the corresponding ethene-compounds), leaving chloro-cinnamene, C_8H_7Cl, and bromocinnamene, C_8H_7Br. According to Laurent, cinnamene yields with chlorine a hexchloride of dichloro-cinnamene, $C_8H_6Cl_7.Cl_6$.—Metacinnamene is also acted upon by bromine, but with considerable difficulty.—Both cinnamene and metacinnamene treated with fuming nitric acid yield mononitrated derivatives, $C_8H_7(NO_2)$: that obtained from cinnamene is crystalline; that from metacinnamene amorphous.

Allyl-benzene, $C_9H_{10} = C_6H_5—CH_2—CH{=}CH_2$, or *Phenylallyl,* formed by heating bromobenzene (phenyl bromide), C_6H_5Br, and allyl bro-

* It was formerly supposed that cinnamene prepared from cinnamic acid was not converted by heat into a solid modification, like styrolene from storax: hence the two were regarded as isomeric, not identical; but later researches have shown that pure cinnamene from cinnamic acid is likewise convertible into solid metacinnamene.

36 *

mide, C_2H_5Br, with zinc, boils at 155° C. (311° F.), and forms a dibromide melting at 60° C. (140° F.).

Phenyl-propylene, C_6H_5—CH⚌CH—CH_3, isomeric with it, is obtained from cinnamic alcohol; it boils at 170° C. (338° F.), and forms a dibromide melting at 66° C. (150.8° F.).

Acetenyl-benzene or **Phenylacetylene**, C_8H_6 = C_6H_5—C≡CH, is formed by heating bromocinnamene, C_6H_5.CH⚌CHBr$_2$, or acetophenone chloride, C_6H_5.CCl$_2$.CH$_3$, with alcoholic potash; also by heating phenylpropiolic acid to 120° C. (248° F.), with water or with baryta:

$$C_6H_5—C⚌C—CO_2H \ = \ CO_2 \ + \ C_6H_5—C≡CH \ ;$$

further, together with cinnamene, by the action of heat on several hydrocarbons. It is an aromatic liquid, boiling at 140° C. (284° F.), and precipitating ammoniacal silver and copper solutions, like acetylene. The *copper-compound*, $(C_8H_5)_2Cu_2$, is yellow: the *silver-compound*, $(C_8H_5)_2Ag_2$. Ag_2O, is white. Sodium dissolves in it, forming the compound, C_8H_5Na, which takes fire on coming in contact with the air, and is converted by carbonic acid passed into its ethereal solution, into the sodium salt of phenylpropiolic acid.

On shaking up the copper-compound with alcoholic ammonia, *diacetenylphenyl*, $(C_6H_4.C_2H)_2$ or CH≡C—C_6H_4—C_6H_4—C≡CH, separates out. This compound, isomeric with pyrene, melts at 97° C. (206.6° F.), and forms a crystalline compound with picric acid.

Ethyl-acetenyl-benzene, $C_{10}H_{10} = C_6H_5$—C≡C—C_2H_5, obtained by the action of ethyl iodide on the sodium-compound of acetenyl-benzene, is a colorless liquid having a density of 0.923 at 21° C. (69.8° F.), and boiling at 201°– 203° C. (393.8°–397.4° F.).

HALOGEN DERIVATIVES OF BENZENE AND ITS HOMOLOGUES.

These compounds are formed: (1) By direct substitution of the halogen elements. Chlorine and bromine act on the aromatic hydrocarbons more readily than on the paraffins, especially in presence of iodine, which acts as a carrier of the chlorine or bromine. The action of chlorine is further promoted by the presence of a small quantity of antimony pentachloride, or molybdenum pentachloride, these bodies first giving up a portion of their chlorine to the hydrocarbon, whereby they are reduced to lower chlorides, which then take up an additional quantity of chlorine, and transfer it in like manner.

Iodine-derivatives cannot be obtained by the action of iodine alone, because, as in the case of the iodo-paraffins (p. 542), the substitution-product, if formed, would be immediately decomposed by the hydriodic acid formed at the same time. The reverse action may, however, be prevented by addition of iodic acid, the action taking place as represented by the equation:

$$5C_6H_6 \ + \ 2I_2 \ + \ IO_3H \ = \ 5C_6H_5I \ + \ 3H_2O \cdot$$

(2) By the action of pentachloride, pentabromide, or pentiodide of phosphorus on the hydroxyl-derivatives of the hydrocarbons (phenols and alcohols); *e. g.*,

$$C_7H_7(OH) \quad + \quad PCl_5 \quad = \quad C_7H_7Cl \quad + \quad POCl_3 \quad + \quad HCl .$$
Cresol.   Chloro-
toluene.

(3) By the decomposition of substituted aromatic acids by heat ; *e. g.*,

$$C_6H_4Cl.CO_2H \quad = \quad CO_2 \quad + \quad C_6H_5Cl$$
Chlorobenzoic acid.   Chlorobenzene.

(4) The halogen-derivatives of benzene may be formed from amidobenzene, $C_6H_5.NH_2$ (aniline), and the chloranilines, etc., $C_6H_4Cl.NH_2$, for example, by exchange of the group NH_2 for Cl, Br, etc., through the medium of the corresponding diazo-compounds (*q. v.*). This is a very important reaction, serving in many cases to determine the orientation of the halogen-atoms in the di-, tri-, and tetra-derivatives.

Benzene-derivatives.

Chlorobenzenes.—Of these compounds all the possible modifications have been obtained.

Monochlorobenzene, C_6H_5Cl, or *Phenyl chloride*, prepared by chlorination of benzene, or by the action of PCl_5 on phenol, is a liquid of specific gravity 1.128 at 0°, boiling at 132° C. (269.6° F.), solidifying at —40° C. (—40° F.).

Dichlorobenzenes, $C_6H_4Cl_2$.—In the formation of these bodies by direct chlorination of benzene, the ortho- and para-modifications are always formed in preference to the meta ; in fact the latter can, for the most part, be obtained only by indirect methods, as by exchange of the NH_2 in chloraniline for Cl in the manner above mentioned. The same observation applies to the dibromobenzenes.

Ortho- (1 : 2).—From benzene and chlorine (together with 1 : 4), and from orthochlorophenol. Colorless liquid, not solidifying at —19° C. (—2.2° F.). Boiling point 179° C. (354.2° F.); specific gravity = 1.3728 at 0°.

Meta- (1 : 3).—From metachloraniline, $C_6.NH_2.H.Cl.H_3$, by exchange of NH_2 for Cl, and from dichloraniline (1 : 2 : 4), $C_6.NH_2.Cl.H.Cl.H_2$, by exchange of NH_2 for H. Liquid, solidifying at —18° C. (0.4° F.), boiling at 172° C. (341.6° F.). Specific gravity 1.307 at 0°.

Para- (1 : 4), the chief product of the action of chlorine on benzene, in presence of iodine, forms colorless, monoclinic crystals, melts at 54° C. (129.2° F.), boils at 173° C. (343.4° F.). Produced also by the action of PCl_5 on parachlorophenol.

Trichlorobenzenes, $C_6H_3Cl_3$.—The *unsymmetrical* modification 1 : 2 : 4, obtained by chlorination of benzene, by the action of PCl_5 on dichlorophenol, and from benzene hexchloride, $C_6H_6Cl_6$, by alcoholic potash, which removes 3HCl, forms colorless crystals, melts at 17° C. (62.6° F.), boils at 213° C. (415.4° F.).

The *symmetrical* modification 1 : 3 : 5 (or 2 : 4 : 6), obtained from trichloraniline, $\overset{1}{C_6}.\overset{2}{N}H_2.\overset{4}{Cl}.H.\overset{6}{Cl}.H.Cl$, by exchange of NH_2 for H, or from dichloraniline, $\overset{1}{C_6}.NH_2.H.\overset{3}{Cl}.H.\overset{5}{Cl}.H$, by exchange of NH_2 for Cl, melts at 63.4° C. (146.1° F.), boils at 208° C. (406.4° F.), sublimes at ordinary temperatures.

The *consecutive* modification 1 : 2 : 3, from trichloraniline $C_6.NH_2.Cl.Cl.$

Cl.H$_2$, crystallizes from alcohol in large plates, melts at 53°–54° C. (127.4°–129.2° F.), and boils at 218°–219° C. (424°–426.2° F.).

Tetrachlorobenzenes, C$_6$H$_2$Cl$_4$.—*Unsymmetrical* (1 : 3 : 4 : 5 or 1 : 2 : 4 : 6), from trichloraniline, C$_6$.NH$_2$.Cl.H.Cl.H.Cl, forms colorless needles, melts at 50°–51° C. (122°–123.8° F.), boils at 246° C. (474.8° F.). —*Symmetrical* (1 : 2 : 4 : 5), from benzene with chlorine, and from trichloraniline, C$_6$.NH$_2$.Cl.H.Cl.Cl.H, forms slender needles, melts at 137°–138° C. (278.6°–280.2° F.), boils at 243°–246° C. (469.4°–474.8° F.). *Consecutive* (1 : 2 : 3 : 4), from trichloraniline, C$_6$.NH$_2$.Cl.Cl.Cl.H., crystallizes in needles, melts at 45°–46° C. (113°–114.8° F.), boils at 254° C. (489.2° F.).

Pentachlorobenzene, C$_6$HCl$_5$, from benzene with chlorine, melts at 85° C. (185° F.), boils at 270° C. (518° F.).

Hexchlorobenzene, C$_6$Cl$_6$ (Julin's *chloride of carbon*), is formed by chlorination of benzene in presence of iodine, SbCl$_5$, or MoCl$_5$; also by passing the vapor of chloroform or carbon dichloride, C$_2$Cl$_4$, through a red-hot tube. Long, thin, colorless prisms; m. p. 222°–226° C. (431.6°–438.8° F.) ; b. p. 332° C. (629.6° F.).

Bromobenzenes.—*Monobromobenzenes*, C$_6$H$_5$Br, from benzene with bromine, and from phenol with PBr$_5$, is a liquid boiling at 154° C. (309.2° F.), and having a density of 1.519 at 0°.

Dibromobenzenes, C$_6$H$_4$Br$_2$.—Direct bromination of benzene (with aid of heat) yields chiefly the para-, with only a small quantity of the ortho-modification.

Para- (1 : 4).—Obtained from benzene, parabromophenol, and parabromaniline, melts at 89° C. (192.2° F.), boils at 218° C. (424.4° F.), and yields a mononitro-derivative melting at 85° C. (185° F.).

Meta- (1 : 3).—From metabromaniline, C$_6$.NH$_2$.H.Br.H$_3$, by exchange of NH$_2$ for Br, and from dibromaniline, C$_6$.NH$_2$.Br.H.Br.H$_2$, by exchange of NH$_2$ for H, is a liquid, boiling at 219° C. (426.2° F.), of specific gravity 1.955 at 18.6° C. (65.5° F.), not solidifying at —20° C. (—4° F.) ; yields two mononitro-derivatives, C$_6$H$_4$(NO$_2$)Br, one melting at 61° C. (141.8° F.), the other at 82.5° C. (180.5° F.).

Ortho- (1 : 2), from orthobromaniline, and in small quantity from benzene and bromine. Liquid, solidifying below zero; m. p. —1° C. (30.2° F.); b. p. 223.8° C. (434.8° F.); sp. gr. 2.003 at 0°.

Tribromobenzenes, C$_6$H$_2$Br$_3$. — The *unsymmetrical* modifications, 1 : 2 : 4 or 1 : 3 : 4, is formed by the action of bromine on benzene, and from either of the three dibromobenzenes : hence its constitution is determined (p. 840); also from benzene hexbromide, C$_6$H$_6$Br$_6$, by the action of alcoholic potash, which abstracts 3HBr, and from dibromophenol, C$_6$.OH.Br.H.Br.H$_2$ (exchange of OH for Br by the action of PBr$_5$). It melts at 44° C. (111.2° F.) and boils at 275° C. (527° F.). The *symmetrical* modification, 1 : 3 : 5, from tribromaniline C$_6$.NH$_2$.Br.H.BrH.Br, melts at 119.5° C. (247.1° F.), and boils at about 278° C. (532.4° F.). The *consecutive* modification, 1 : 2 : 3, from tribromaniline C$_6$.NH$_2$.H.Br.Br.Br.H, melts at 87° C. (188.6° F.).

Tetrabromobenzenes, C$_6$H$_2$Br$_4$.—*Unsymmetrical*, 1 : 3 : 4 : 5, from tribromophenol, C$_6$.OH.H.Br.Br.Br.H, also from the corresponding tribromaniline, and from tetrabromaniline, forms colorless crystals, melting at 98° C. (208.4° F.). The *symmetrical* modification 1 : 2 : 4 : 5, from benzene with bromine, forms colorless needles melting at 137°–140° C. (278.6°–284° F.).

Pentabromobenzene, C_6HBr_5, obtained by the bromination of benzene, forms colorless needles, subliming without decomposition, melting at about 240° C. (464° F.).

Hexbromobenzene, C_6Br_6, is formed by the action of bromine on benzene in sunshine, or by heating benzene with bromine containing iodine ; also from tetrabromethane, CBr_4, in small quantity by distillation, in larger quantity by heating to 300°–400° C. (572°–752° F.) in a sealed tube. Resembles hexchlorobenzene ; melts at a temperature above 300° C. (572° F.).

It will be seen from the above that all the bromobenzenes are known except consecutive tetrabromobenzene.

Iodobenzenes are obtained by heating benzene with iodine and iodic acid to 200°–240° C. (392°–464° F.) ; by the action of iodine and phosphorus on phenol : by treating silver benzoate with iodine chloride ; and from aniline and the iodanilines, similarly to the chlorine compounds.

C_6H_5I is a colorless liquid, boiling at 185° C. (365° F.) : specific gravity 1.69.

$C_6H_4I_2$.—*Para.* Laminæ ; m. p. 127° C. (260.6° F.), b. p. 277° C. (530.6° F.).—*Meta.* Large shining laminæ resembling naphthalene ; m. p. 40.4° C. (104.7° F.) ; b. p. 284.7° C. (544.5° F.).

$C_6H_3I_3$. Needles melting at 76° C. (168.8° F.), and volatilizing without decomposition.

Fluobenzene, C_6H_5F, obtained by heating calcium fluobenzoate with slaked lime, is a scaly crystalline mass, melting at 40° C. (104° F.) ; boils at 180°–183° C. (356°–361.4° F.).

Toluene derivatives.

These compounds, as already observed, exhibit metameric modifications, according as the replacement of hydrogen atoms takes place in the principal or the lateral chain ; and isomeric modifications determined by the orientation of the substituted radicles in the benzene-ring or principal chain (p. 841).

Chlorotoluenes.—Of monochlorinated toluenes there are two metameric modifications, viz.:—

$$C_6H_4Cl.CH_3 \qquad\qquad C_6H_5.CH_2Cl,$$
Chlorotoluene. \qquad\qquad\qquad Benzyl chloride.

the first produced by the action of chlorine on toluene at ordinary temperatures or in presence of iodine ; the second by the action of chlorine on boiling toluene.

Chlorotoluene or *Tolyl Chloride*, $C_6H_4{<}^{CH_3}_{Cl}$, admits of three isomeric modifications, according to the relative positions of the chlorine and the methyl-group.—*Parachlorotoluene*, $C_6.CH_3.H.H.Cl.H_2$, the chief product formed by the action of chlorine on toluene at ordinary temperatures, is a liquid solidifying at 0°, melting at 6.5° C. (43.7° F.), and boiling at 160° C. (320° F.). By oxidation with nitric acid, or with chromic acid mixture, it is converted into parachlorobenzoic acid :

$$C_6H_4Cl.CH_3 \;+\; O_3 \;=\; C_6H_4Cl.CO_2H \;+\; H_2O.$$

Orthochlorotoluene, $C_6.CH_3.Cl.H_4$, produced in small quantity, together with the para-compound, is a liquid, boiling at 156° C. (312.8° F.), and completely decomposed by chromic acid, without formation of a chloro-benzoic acid.—*Metachlorotoluene*, $C_6.CH_3.H.Cl.H_3$, is prepared from chloro-paratoluidine, $C_6.CH_3.H.Cl.NH_2.H_2$, by exchange of NH_2 for H, and converted by oxidation into metachlorobenzoic acid.

Benzyl chloride, $C_6H_5.CH_2Cl$, obtained by the action of chlorine on boiling toluene, and of PCl_5 on benzyl alcohol, $C_6H_5.CH_2OH$, is a liquid which gives off tear-exciting vapors, and boils at 176° C. (348.8° F.). It easily exchanges its chlorine-atom by double decomposition, being converted into acetate, cyanide, etc., by treatment with the corresponding potassium salts. It yields benzoic acid by oxidation, and is converted by heating with water and lead nitrate into benzaldehyde (bitter almond oil).

Of *dichlorotoluene*, $C_7H_6Cl_2$, there are three metamerides, viz. :

$$C_6H_3Cl_2.CH_3 \qquad C_6H_4Cl.CH_2Cl \qquad C_6H_5.CHCl_2.$$
Dichlorotoluene.   Chlorobenzyl chloride.   Benzal chloride.

The first admits of six isomeric modifications, but only one is known, viz., $\overset{1}{C_6}.CH_3.H.\overset{3}{Cl}.\overset{4}{Cl}.H_2$, which is a liquid, boiling at 196° C. (384.8° F.), formed by chlorination of toluene.—*Chlorobenzyl chloride* admits of three iso-merides, one of which is a liquid boiling at 213°–214° C. (415.4°–417.2° F.).—*Benzal chloride* (also called *chlorobenzol*) is formed by the action of chlorine on boiling toluene, and from benzaldehyde (bitter almond oil) by PCl_5. Liquid having a pungent odor ; sp. gr. 1.295 at 16° C. (60.8° F.) ; boils at 206° C. (402.8° F.) ; converted into benzaldehyde by heating with water to 20° C. (68° F.).

Trichlorotoluenes, $C_7H_5Cl_3$:

$$C_6H_2Cl_3CH_3 \qquad C_6H_3Cl_2.CH_2Cl \qquad C_6H_4Cl.CHCl_2 \qquad C_6H_5.Cl_3$$
6 Isomerides.   6 Isomerides.   3 Isomerides.   one.

Trichlorotoluene, $C_6H_2Cl_3.CH_3$, obtained by chlorination of toluene, forms colorless crystals, melts at 76° C. (168.8° F.), boils at 235° C. (455° F.). —*Dichlorobenzyl chloride*, $\overset{1}{C_6}.(CH_2Cl).H.\overset{3}{Cl}.\overset{4}{Cl}.H_2$, is a liquid boiling at 241° C. (465.8° F.).—*Chlorobenzal chloride* (para), $C_6.(CHCl)_2.H.H.Cl.H_2$, boils at 234° C. (453.2° F.).—*Benzotrichloride*, $C_6H_5.Cl_3$, formed by heating ben-zoyl chloride, $C_6H_5.COCl$, with PCl_5, boils at 213°–214° C. (415.4–417.2° F.).

Of the higher chlorotoluenes, some are liquid, some solid, at ordinary temperatures. The melting and boiling points of the known modifications are given in the following table:

Tetrachlorotoluene, $C_7H_4Cl_4$:

$C_6HCl_4.CH_3$	$C_6H_2Cl_3.CH_2Cl$	$C_6H_3Cl_2.CHCl_2$	$C_6H_4Cl.CCl_3$
m. p. 91-2° C. (195.8–197.6° F.)	liq. b. p. 273°	liq. b. p. 257°	liq. b. p. 245°
b. p. 271° C.(519.8°.F.).	C. (519.8° F.)	C. (494.6° F.).	C. (473° F.).

Pentachlorotoluenes, $C_7H_3Cl_5$:

$C_6Cl_5.CH_3$	$C_6HCl_4.CH_2Cl$	$C_6H_2Cl_3.CHCl_2$	$C_6H_3Cl_2.CCl_3$
m. p. 218° C. (424° F.) ;	(liq. b. p. 296° C.	liq. b. p. 280°–281°	liq. b. p. 273° C.
b. p. 301° C. (573.8° F.).		C. (536°–537.8° F.)	(523.4° F.)

Hexchlorotoluenes, $C_7H_2Cl_6$:

$C_6Cl_5.CH_2Cl$	$C_6HCl_4.CHCl_2$	$C_6H_2Cl_3.CCl_3$
m.p. 103° C. (287.4° F.) ; b.p.	liq. b.p. 305°–306° C.	m.p. 82° C. (179.6° F.) ; b. p.
325°–327° C. (617–618.8° F.).	(581–582.8° F.).	307°–308° C. (581.6–586.4° F.).

Heptachlorotoluenes, C_7HCl_7:

$C_6Cl_5.CHCl_2$
m. p. 109° C. (228.2° F.) ; b. p.
334° C. (633.2° F.).

$C_6HCl_4.CCl_3$
m. p. 104° C. (210.2° F.) ; b. p.
316° C. (600.8° F.).

When an attempt is made to replace the last hydrogen-atom in toluene by chlorine, the molecule splits up, and perchlorobenzene is produced.

Bromotoluenes.—*Mono-*, $C_7H_7Br = C_6H_4Br.CH_3$.—The ortho- and para-modifications are formed by the action of bromine on toluidine ; and all the three modifications from the corresponding amidotoluenes (toluidines) by heating the diazoperbromides with alcohol (see DIAZO-COM-POUNDS).

Ortho- Colorless liquid : sp. gr. 1.401 at 18° C. (64.4° F.) ; b. p. 182°–183° C. (359.6°–361.4° F.).

Meta- Colorless liquid ; sp. gr. 1.4009 at 21° C. (69.8° F.) ; b. p. 184° C. (363.2° F.).

Para- Colorless crystals ; m. p. 28.5° C. (83.3° F.) ; b. p. 185° C. (365° F.).

Benzyl bromide, $C_6H_5.CH_2Br$, obtained by bromination of toluene at the boiling heat, and by the action of hydrobromic acid on benzyl alcohol, is a liquid which gives off a tear-exciting vapor, has a sp. gr. of 1.438 at 22° C. (71.6° F.), and boils at 201° C. (393.8° F.).

Dibromotoluene, $C_6H_3Br_2.CH_3$, admits of six isomeric modifications, all of which are known. $C_6.CH_3.Br.H.Br.H_2$ (1 : 2 : 4), obtained by direct bromination of toluene, crystallizes in needles ; melts at 107°–108° C. (224.6°–226.4° F.) ; boils at 245° C. (473° F.). $C_6.CH_3.H.Br.Br.H_2$ (1 : 3 : 4), formed by the action of bromine in presence of iodine, and in sunshine, is a liquid boiling at 240° C. (464° F.). The other four modifications, whose boiling points lie between 237° and 246° C. (458.6°–474.8° F.), two liquid and two solid (m. p. 42.5° and 60° C. (108.5°–140° F.), are obtained from bromotoluidines, through the medium of the diazo-compounds.

Benzal bromide, $C_6H_5.CHBr_2$, formed by treating bitter almond oil with PBr_5, is a liquid which decomposes when distilled.

Monoiodotoluenes, $C_6H_4I.CH_3$.—The ortho- and meta-modifications are colorless liquids, the former boiling at 205° C. (401° F.), the latter at 207° C. (404.6° F.). *Para-iodotoluene* crystallizes in shining laminæ ; melts at 35° C. (95° F.) ; boils at 211° C. (411.8° F.). All three give by oxidation the corresponding iodobenzoic acids.

Benzyl iodide, $C_6H_5.CH_2I$, formed from the chloride by the action of hydriodic acid, melts at 24° C. (75.2° F.), and decomposes when sublimed.

Derivatives of the Hydrocarbons, C_8H_{10}.

Ethylbromobenzene, $C_6H_4Br.C_2H_5$, formed by the action of bromine on ethyl-benzene at ordinary temperatures, is a colorless liquid, boiling at 190° C. (374° F.).—*Bromethyl-benzene*, $C_6H_5.CHBr.CH_3$ (or $C_6H_5.CH_2.CHBr$), and the corresponding *chloride*, formed by the action of bromine or chlorine with aid of heat, are liquids which decompose when distilled.

The chloride, boiled with potassium cyanide and alcohol, is converted into the corresponding *cyanide*, $C_6H_5.CH_2.CH_2(CN)$, a colorless liquid, boiling at 261° C. (501.8^O F.), having a sp. gr. of 1.0014 at 18° C. (64.4^O F.), and forming the chief constituent of oil of water-cress.

Metaxylyl Chloride, $C_6H_4{<}^{CH_3}_{CH_2Cl}$, obtained by the action of chlorine on boiling metaxylene, is a colorless liquid, boiling at 195° C. (383° F.).

Monobromometaxylene, $C_6.Br.CH_3.H.CH_2.H_2$, from metaxylene and bromine, boils at 204°–205° C. (399.2^O–401° F.).—An isomeric compound (1 : 3 : 5), formed by the action of nitrous acid and alcohol on bromometaxylidine, is a liquid boiling at 204° C. (399.2^O F.), not solidifying at 20° C. (68° F.), and having a sp. gr. of 1.362 at 20°.

Dibromometaxylene, $C_6H_2Br_2(CH_3)_2$, from metaxylene and bromine, forms colorless shining crystalline scales, m. p. 72° C. (161.6^O F.), b. p. 256° C. (492.8^O F.). *Tetrabromometaxylene,* $C_6Br_4(CH_3)_2$, forms long slender needles, slightly soluble in alcohol, melting at 241° C. (465.8^O F.).

Bromoparaxylene, $C_6H_3Br(CH_3)_2$, forms colorless shining tables; melts at 10° C. (50° F.); boils at 200° C. (392° F.). *Dibromoparaxylene* melts at 72° C. (161.6^O F.), and resembles the meta-compounds in all other respects.

Tollylene chloride, $C_6H_4(CH_2Cl)_2$, and **Tollylene bromide,** $C_6H_4(CH_2Br)_2$, formed by the action of chlorine or bromine on boiling paraxylene, crystallizes in colorless laminæ. The chloride melts at 100°, and boils, with partial decomposition, at 240°–250° C. (464°–482° F.); the bromide melts at 145°–147° C. (293°–296.6^O F.). Both compounds, when treated with potassium cyanide in alcoholic solution, yield tollylene cyanide, $C_6H_4(CH_2.CN)_2$.

CYANOGEN-DERIVATIVES.

Cyanobenzene, $C_6H_5.CN$ (*Benzonitril, Phenyl Cyanide*), is formed, like the nitrils of the fatty group (p. 565), by dehydration of ammonium benzoate, $C_7H_5O_2.NH_4$, and by distilling the potassium salt of benzenesulphonic acid with potassium cyanide (or the dry ferrocyanide) :

$$C_6H_5.SO_3K + KCN = SO_3K_2 + C_6H_5CN ;$$

also by heating phenyl isocyanide with metallic copper, and by other reactions. It is an oily liquid, smelling like bitter almond oil, having a density of 1.023 at 0°, and boiling at 191° C. (375.8^O F.). It unites with the halogen-elements, the haloïd acids, and hydrogen.

Substituted benzonitrils are obtained by dehydration of substitut benzamides, *e. g.,* $C_6H_4Br.CN$, by the action of phosphoric anhydride on bromobenzamide, $C_6H_4Br.CONH_2$.

Phenyl Isocyanide or **Phenyl-carbamine,** $C{\equiv}N—C_6H_5$, isomeric with benzonitril, is formed by distilling aniline with chloroform and alcoholic potash :

$$C_6H_5.NH_2 + CHCl_3 = 3HCl + C_6H_5—N{\equiv}C.$$

It is a liquid having a strong smell of prussic acid, and boiling, with partial decomposition, at 167° C. (332.6° F.). It is dichroic; blue by reflected, green by transmitted light. It is not altered by alkalies, but acids convert it into aniline and formic acid. Heated to 200° C. (392° F.) it changes to benzonitril.

Cyanotoluene, $C_6H_4(CN).CH_3$. *Toluonitril.*—The three isomeric modifications of this compound are formed by treating the respective tolyl-sul-phocarbimides, $N \begin{cases} CS \\ C_6H_4.CH_3 \end{cases}$, with finely divided copper, which removes the sulphur, or by distilling the potassium salts of the corresponding toluenesulphonic acids with potassium cyanide:

$$C_6H_4{<}^{CH_3}_{SO_3K} \;+\; CNK \;=\; SO_3K_2 \;+\; C_6H_4{<}^{CH_3}_{CN}$$

Ortho.—Colorless liquid, smelling like nitrobenzene, boiling at 203°–204° C. (397.6°–399.2° F.). *Para.*—Colorless needles, melting at 28.5° C. (83.2° F.); boiling at 218° C. (424° F.). *Meta.*—Not yet obtained in the pure state.

Benzyl Cyanide, $C_6H_5.CH_2.CN$, constitutes the principal part of the volatile oils of the garden nasturtium (*Tropœlum majus*), and of the garden cress (*Lepidium sativum*), and is produced artificially by heating benzyl chloride with alcohol and potassium cyanide. Colorless liquid boiling at 232° C. (449.6° F.); specific gravity 1.0146 at 18° C. (64.4° F.).

NITRO-DERIVATIVES.

These bodies are easily formed by the action of concentrated or fuming nitric acid on benzene and its homologues, the substitution of the NO_2-group for hydrogen taking place in the benzene-nucleus, never in the lateral chains; *e. g.*,

$$C_6H_5.CH_3 \;+\; NO_3H \;=\; C_6H_4(NO_2).CH_3 \;+\; H_2O .$$

On pouring the product into water, the nitro-compound separates out, generally in the form of a thick yellow or orange-colored oil. The more highly nitrated derivatives are most easily obtained by the action of a mixture of 1 part of strong nitric acid and 2 parts of strong sulphuric acid.

The aromatic nitro-compounds are mostly of a yellow color; a few are liquid, the rest crystalline solids. By hydrogen sulphide or ammonium sulphide, and by stannous chloride, they are reduced to amido-compounds; by sodium-amalgam, or by heating with alcoholic potash, to azo-compounds. They may also be converted into amido-compounds by heating with hydriodic acid:

$$\underset{\text{Nitrobenzene.}}{C_6H_5.NO_2} \;+\; 6HI \;=\; \underset{\text{Aniline.}}{C_6H_5NH_2} \;+\; 2H_2O \;+\; 3I_2 .$$

Nitrobenzenes.—Mononitrobenzene, $C_6H_5(NO_2)$, formed by gradually adding benzene to strong nitric acid in a cooled vessel, is a light yellow liquid, having an aromatic odor, boiling at 220° C. (428° F.), solidifying at + 3° C. (37.4° F.). Sp. gr. 1.20 at 0°.

Dinitrobenzenes, $C_6H_4(NO_2)_2$.—The three modifications are formed together by dropping benzene into a mixture of 2 vols. strong sulphuric

and 1 vol. very strong nitric acid ; and on crystallizing the product from
alcohol, the meta-compound, which constitutes by far the largest portion,
separates out, while the other two remain in solution.

The *ortho*-compound, which forms the smallest portion of the product,
crystallizes from hot water in long needles, melting at 118° C. (244.4°
F.), soluble in 26 parts of alcohol at 24° C. (75.2° F.), and in 3 parts of
boiling alcohol.

The *meta*-compound (ordinary dinitrobenzene), forms long rhombic prisms
melting at 89.8° C. (194.2° F.), soluble in 17 parts of alcohol at 24° C.
(75.2° F.), and in all proportions of boiling alcohol.

The *para*-compound forms fan-shaped groups of monoclinic prisms, melting at 171°–172° C. (339.8°–347.6° F.), sparingly soluble in alcohol.

Trinitrobenzene, $C_6H_3(NO_2)_3$, formed by heating metadinitroben-
zene with a mixture of pyrosulphuric acid (p. 180) and very strong nitric
acid in sealed tubes, to 130°–140° C. (266°–284° F.), crystallizes from
alcohol in white laminæ or fern-like groups of needles. It melts at 121°–
122° C. (249.8°–251.6° F.), dissolves sparingly in cold alcohol, easily in
hot alcohol and ether.

Nitro-haloid Derivatives of Benzene.—The action of nitric acid,
or a mixture of nitric and sulphuric acids, on the chloro-, bromo-, and
iodo-benzenes, gives rise to para- and ortho-mononitro-derivatives of the
haloïd compounds, the former being always produced in greatest abundance. The same products are obtained from the corresponding nitranilines,
$C_6H_4.NO_2.NH_2$, by exchange of the NH_2 for Cl, Br, or I, through the medium
of the diazo-compounds. The meta-compounds are obtained in like manner
from metanitraniline ; metanitrochlorobenzene also by passing chlorine
into nitrobenzene mixed with iodine, or, better, with $SbCl_3$.
The following are the melting and boiling points of the isomeric mono-
nitro-, chloro-, bromo-, and iodo-benzenes, $C_6H_4X(NO_2)$:

Melting Points.

	(1 : 2)	(1 : 3)	(1 : 4)
$C_6H_4Cl(NO_2)$	32.5° C. (90.5° F.)	44.4° C.(111.9° F.)	83° C. (181.4° F.)
$C_6H_4Br(NO_2)$	41.5° C. (106.7° F.)	56° C. (132.8° F.)	126° C. (258.8° F.)
$C_6H_4I(NO_2)$	49.4° C. (120.9° F.)	36° C. (96.8° F.)	171° C. (339.8° F.)

Boiling Points.

	(1 : 2)	(1 : 3)	(1 : 4)
$C_6H_4Cl(NO_2)$	243° C.(469.4° F.)	233° C. (451.4° F.)	242° C.(467.6° F.)
$C_6H_4Br(NO_2)$	261° C.(501.8° F.)	256.5° C.(493.7° F.)	255–6° C.(491–3° F.)
$C_6H_4I(NO_2)$	... ...	280° C. (536° F.)	

These numbers show that the para-derivatives have the highest melting
points, and the ortho-derivatives for the most part the lowest; the relations between the boiling points are less regular. The ortho- and para-
compounds, heated in sealed tubes with aqueous potash, are converted
into the corresponding phenols, $C_6H_4Cl.OH$, etc., whereas the meta-compounds do not exhibit this transformation.

Nitrodichloro- and *Nitrodibromobenzenes*, $C_6H_3Cl_2(NO_2)$, and
$C_6H_3Br_2(NO_2)$.—These compounds are all crystalline, and melt at the temperatures shown in the following table :—

$C_6H_3Cl_2(NO_2)$		$C_6H_3Br_2(NO_2)$	
1 : 2 : 4* . . 32.2° C. (90.0° F.)		1 : 2 : 4 . . 61.6° C. (142.9° F.)	
1 : 2 : 5 . . 55.0° C. (131.0° F.)		1 : 2 : 5 . . 85.4° C. (185.7° F.)	
1 : 3 : 5 . . 65.4° C. (149.7° F.)		1 : 3 : 5 . . 104.5° C. (220.1° F.)	
1 : 3 : 4 . . 43.0° C. (109.4° F.)		1 : 3 : 4 . . 58.6° C. (137.5° F.)	
		1 : 2 : 6 . . 82.6° C. (180.7° F.)	

Dinitrochlorobenzene, $C_6.Cl.NO_2.H.NO_2.H_2$, formed by treating either (1 : 2) or (1 : 4) chlorobenzene with a mixture of nitric and sulphuric acids, and from the corresponding dinitrophenol, $C_6H_3(NO_2)_2.OH$, by the action of PCl_5, crystallizes in prisms melting at 53.4° C. (128.1° F.). $C_6Cl.H.NO_2.NO_2.H_2$, formed in like manner from (1 : 3) nitrochlorobenzene, exhibits dimorphous modifications having different melting points— 36.3°–38.8° C. (97.3°–101.8° F.).

Dinitrobromobenzene, $C_6H_3Br(NO_2)_2$, is known in two modifications analogous to the chlorine-compounds just described, and obtained in like manner. Both are crystalline; (1 : 2 : 4) melts at 75.3° C. (167.5° F.); (1 : 3 : 4) at 59.4° C. (138.9° F.).

Dinitroiodobenzene, $C_7.I.NO_2.H.NO_2.H_2$, from para- and ortho-iodonitrobenzene with nitrosulphuric acid, forms large yellow, transparent plates or prisms, melting at 88.5° C. (190.9° F.).
Another modification, $C_6.I.NO_2.H.H.H.NO_2$, formed simultaneously with the last, crystallizes in transparent orange-colored rhombic tables, melting at 113.7° C. (236.7° F.).

Trinitrochlorobenzene, $C_6.Cl.NO_2.H.NO_2H.NO_2$ (*Picryl chloride*), from picric acid, $C_6H_2(NO_2)_3.OH$, with PCl_5, forms needles melting at 83° C. (181.4° F.), and, like picric acid, forms crystalline compounds with many hydrocarbons.

Nitrotoluenes.—*Para-* and *orthonitrotoluene*, $C_6H_4(NO_2).CH_3$, are formed by treating toluene with fuming nitric acid, and may be separated by fractional distillation. The former crystallizes in nearly colorless prisms, melts at 54° C. (129.2° F.), and boils at 236° C. (456.8° F.). The latter is a yellowish liquid boiling at 222°–223° C. (437.6°–433.4° F.).

Metanitrotoluene, obtained by the action of nitrous acid and alcohol on metanitroparatoluidine, $C_6.CH_3.H.NO_2.NH_2.H_2$, is crystalline, melts at 16° C. (60.8° F.), and boils at 230°–231° C. (446°–447.8° F.).

Dinitrotoluene, $C_6.CH_3.NO_2.H.NO_2.H_2$, formed by treating toluene, or ortho- or para-nitrotoluene with nitro-sulphuric acid, crystallizes in long colorless needles, melting at 70.5° C. (159.7° F.). Another modification, obtained in like manner from metanitrotoluene, melts at 60° C. (140° F.).

Trinitrotoluene, $C_6H_2(NO_2)_3.CH_3$, obtained by prolonged heating of a solution of toluene in nitrosulphuric acid, forms nearly colorless needles, slightly soluble in alcohol, melting at 82° C. (179.6° F.).

Nitro-ethyl-benzenes, $C_6H_4(NO_2)$—C_2H_5 (*ortho-* and *para-*), are formed simultaneously by the action of fuming nitric acid on ethyl-benzene. Both are liquid, the former boiling at 227°–228° C. (440.6°–442.4° F.), the latter at 245°–246° C. (473°–474.8° F.). With tin and hydrochloric acid, they yield liquid bases, one of which, viz., *paramido-ethylbenzene*, $C_6H_4(NH_2)$. C_2H_5, is also produced by heating the hydrochloride of ethylaniline to 300°– 330° C. (572°–626° F.).

* NO_2 in the position 1, in all the formulæ.

Nitroparaxylene, $C_6H_3(NO_2)(CH_3)_2$, is a pale yellow liquid which boils at 234°–237° C. (453.2°–458.6° F.), and does not solidify in a freezing mixture.—*Dinitroparaxylene,* $C_6H_2(NH_2)_2(CH_3)_2$. Two modifications of this compound are formed simultaneously by the action of fuming nitric acid on paraxylene, the less soluble in alcohol of the two forming long thin needles melting at 123.5° C. (254.3° F.), while the other, which is more soluble in alcohol, forms large monoclinic crystals melting at 93° C. (199.4° F.).

Trinitroparaxylene, $C_6H(NO_2)_3(CH_3)_2$, forms long, colorless needles, melting at 137° C. (278.6° F.), moderately soluble in hot, sparingly in cold alcohol.

Nitromesitylene, $C_6H_2(NO_2)(CH_3)_3$, forms nearly colorless prisms, easily soluble in alcohol, melts at 42° C. (107.6° F.), boils at 255° C. (491° F.).—*Dinitromesitylene,* $C_6H(NO_2)_2(CH_3)_3$. Slender, colorless, brilliant needles, melting at 86° C. (186.9° F.). *Trinitromesitylene,* $C_6(NO_2)_3(CH_3)_3$. Needles very slightly soluble in alcohol, melting at 232° C. (449.6° F.).

Nitropseudocumene, $C_9H_{11}(NO_2)$, forms long needles, easily soluble in hot alcohol, melts at 71° C. (159.8° F.), boils at 265° C. (509° F.).— *Trinitropseudocumene,* $C_9H_9(NO_2)_3$. Colorless, quadratic prisms, melting at 185° C. (365°).

AMIDO-DERIVATIVES.

The replacement of hydrogen by NH_2 in the hydrocarbons homologous with benzene, gives rise to two series of metameric compounds, analogous to the haloid and nitro-derivatives above described: thus, from toluenes, $C_6H_5.CH_3$, are derived:

$$C_6H_4(NH_2).CH_3 \qquad C_6H_5.CH_2(NH_2);$$
Toluidine.   Benzylamine.

and from xylene, $C_6H_4(CH_3)_2$:

$$C_6H_3(NH_2).(CH_3)_2 \qquad C_6H_4{<}^{CH_3}_{CH_2(NH_2)}$$
Xylidine.   Xylylamine.

These compounds are all of basic character; but those in which the NH_2 is situated in the lateral chains are stronger bases than their metamerides containing this group in the principal chain, and are analogous in their properties and their mode of formation to the amines of the fatty group; thus benzylamine, which may be represented by the formula $NH_2(C_7H_7)$, derived from ammonia by substitution of benzyl, C_7H_7, for hydrogen, is formed, together with dibenzylamine, $NH(C_7H_7)_2$, and tribenzylamine, $N(C_7H_7)_3$, by heating benzyl chloride with ammonia, just as ethylamine is formed from the chloride, bromide, or iodide of ethyl. Toluidine and its homologues, on the other hand, are produced chiefly by the action of reducing agents on the nitro-derivatives; and amidobenzene, aniline, or phenylamine, $C_6H_5{-}NH_2$, which may be regarded as the first term of either series, is formed for the most part in the same manner.

The reduction of the nitro-derivatives is effected:

1. By the action of ammonium sulphide in alcoholic solution:

$$C_6H_5.NO_2 \;+\; 3H_2S \;=\; C_6H_5.NH_2 \;+\; 2H_2O \;+\; 3S.$$

In the application of this method to di- and tri-nitro-compounds, only one nitro-group is reduced in the first instance, so that nitro-amido compounds are obtained, such as $C_6H_4\diagdown^{NO_2}_{NH_2}$.

2. By the action of zinc or tin and hydrochloric acid, or of iron filings and acetic acid. In these cases, the reduction may be supposed to be effected by nascent hydrogen:

$$C_6H_5.NO_2 + 3H_2 = C_6H_5.NH_2 + 2H_2O:$$

in the case of iron and acetic acid, also through the intervention of the ferrous salt formed in the first instance:

$$C_6H_5.NO_2 + 6FeO + H_2O = C_6H_5.NH_2 + 3Fe_2O_3,$$

and in that of tin and hydrochloric acid, through the intervention of stannous chloride:

$$C_6H_5.NO_2 + 3Sn + 6HCl = C_6H_5.NH_2 + 3SnCl_2 + 2H_2O$$
and
$$C_6H_5.NO_2 + 3SnCl_2 + 6HCl = C_6H_5.NH_2 + 3SnCl_4 + 2H_2O.$$

To effect this last reaction, the nitro-compound is drenched with fuming hydrochloric acid, and the calculated quantity of granulated tin is gradually added. The action usually begins after a while, without application of heat, the tin and the nitro-compound dissolving. From the warm solution, which contains a double salt, consisting of the hydrochloride of the resulting base combined with stannic chloride, e. g. $(C_6H_5.NH_2.HCl)_2.SnCl_4$, the tin is precipitated by hydrogen sulphide, and the stannic sulphide is separated by filtration, leaving the hydrochloride of the amido-compound in solution.

When a di- or a tri-nitro-compound is thus treated, all the nitro-groups are usually reduced at once: hence this process is especially applicable to the preparation of di- and tri-amido-derivatives. If, however, only half the calculated quantity of tin be added, a partial reduction may be effected, and nitro-amido-compounds obtained.

Amidobenzenes.

Aniline, $C_6H_7N = C_6H_5.NH_2$—*Phenylamine.*—This base, which is now prepared in enormous quantities for the manufacture of coloring matters, was discovered in 1826 by Unverdorben, who obtained it by the dry distillation of indigo. Fritzsche, in 1841, found that it might be obtained by boiling indigo with potash-lye; and Zinin, about the same time, introduced the method of preparing it by reduction of nitrobenzene with ammonium sulphide—a process which, as already observed, is very generally applicable to the preparation of organic bases.

To prepare aniline in this way, an alcoholic solution of nitrobenzene is mixed with ammonia, and gaseous hydrogen sulphide is passed into the liquid as long as sulphur continues to be precipitated: the reaction is greatly accelerated by warming the liquid. The solution is then mixed with excess of acid, filtered to separate the sulphur, boiled to expel alcohol and excess of nitrobenzene, and then distilled with excess of potash.

On the large scale, aniline is prepared by reducing nitrobenzene with ferrous acetate. Nitrobenzene (1 part) is heated with iron filings ($1\frac{1}{2}$ part), and glacial acetic acid (1 part), and the solid product of the reaction is mixed with lime, and distilled with superheated steam. On the small scale, the best results are obtained by reduction with tin and hydrochloric acid. The product obtained by either process may be purified by converting it into oxalate, crystallizing this salt several times from alcohol, and decomposing it with potash.

Aniline is also produced by heating phenol with ammonium chloride and fuming hydrochloric acid in sealed tubes to $315°$ C. ($599°$ F.) for about thirty hours:

$$C_6H_5.OH + NH_3 = C_6H_5.NH_2 + H_2O:$$

and lastly, it is formed by the destructive distillation of nitrogenous organic matters, and is one of the constituents of coal-tar oil.

Aniline is a colorless oily liquid, having a faint peculiar odor, a density of 1.036 at $0°$, and boiling at $184.5°$ C. ($364.1°$ F.). When quite pure it solidifies at low temperatures, and melts at $—8°$ C. ($17.6°$ F.). It dissolves but sparingly in water—31 parts at $12°$ C. ($53.6°$ F.), easily in alcohol and ether. When exposed to the air it turns brown and gradually resinizes. Its aqueous solution, mixed with chloride of lime, assumes a purple-violet color. Its solution in strong sulphuric acid acquires, on addition of a few drops of aqueous potassium dichromate, first a red, then a deep-blue color. A deal shaving dipped in aniline is colored yellow.

Aniline is a strong base, uniting directly with acids and with certain salts, forming, for example, the compounds $(C_6H_7N)_2.SnCl_2$ and $(C_6H_7N)_2.CuSO_4.$—The *hydrochloride*, $C_6H_7N.HCl$, forms needles very soluble in water and in alcohol, subliming without decomposition. Its alcoholic solution, mixed with platinic chloride, deposits the *platinochloride*, $(C_6H_7N.HCl)_2.PtCl_4$, in yellow needles. The *nitrate*, $C_6H_7N.NO_3H$, crystallizes in large rhombic tables; the *oxalate*, $(C_6H_7N)_2.C_2O_4H_2$, separates from a mixture of the alcoholic solutions of aniline and oxalic acid, in rhombic prisms.

SUBSTITUTION-PRODUCTS OF ANILINE.

1. *Halogen-Derivatives*.—These compounds are formed:

(1) By the action of halogen-elements on aniline, bromine, and chlorine, forming di- or tri-derivatives, iodine giving rise to para-iodaniline. The monochlor- and monobromanilines (para) are obtained by the action of chlorine and bromine, in vapor or in aqueous solution, on acetanilide, $C_6H_5.NH(C_2H_3O)$, suspended in water.

(2) By reduction of the nitrochloro-, nitrobromo-, or nitroiodo-benzenes with ammonium sulphide, or with tin and hydrochloric acid, $C_6H_4Cl(NO_2)$, for example, being thus converted into $C_6H_4Cl(NH_2)$.

(3) From the nitranilines by substitution of Cl, Br, or I for NH_2 (through the medium of the diazo-compounds), and of NH_2 for NO_2 by the action of reducing agents, $C_6H_4 {<}^{NH_2}_{Br}$, for example from $C_6H_4 {<}^{NO_2}_{NH_2}$.

By the entrance of halogen-elements (also of the nitro-group) into the molecule, the basic properties of aniline are weakened. The mono-derivatives are weak bases; the di-derivatives form, for the most part, very unstable salts which are decomposed even by water; and the tri-derivatives are mostly destitute of basic properties, though the orientation of the substituted radicles has some influence in this respect. In the mono-derivatives, the ortho-compounds are less basic than the meta- and para-compounds.

The following table exhibits the modes of formation and the chief physical properties of the halogen-derivatives of aniline:

	Formation.	Physical character at ordinary temperatures.	Melting point.	Boiling point.
Chloranilines.				
Mono- $C_6H_4Cl(NH_2)$				
Ortho (1 : 2)	from 1 : 2 $C_6H_4Cl(NO_2)$	Liquid : sp. gr. 1.2338 at 0°	−14° C. (6.8° F.)	207° C. (404.6° F.)
Meta (1 : 3)	" 1 : 3 "	Liquid : sp. gr. 1.2452 at 0°		230° C. (446° F.)
Para (1 : 4)*	" 1 : 4 "	Rhombic crystals	70–71°C. (158–159.8° F.)	230–231° C. (446–447.8° F.)
Di- $C_6H_3Cl_2(NH_2)$				
1 : 2 : 4*	from dichloracetanilide and di-chlorisatin by potash	Flat flexible needles	68° C. (145.4° F.)	245° C. (473° F.)
1 : 2 : 5	from 1 : 2 : 5 $C_6H_3Cl_2(NO_2)$	Colorless needles	50° C. (122° F.)	
1 : 3 : 5	" 1 : 3 : 5 "	Long needles	50.5° C. (122.9° F.)	
1 : 3 : 4	" 1 : 3 : 4 "	Crystalline	71.5° C. (160.7° F.)	272° C. (521.6° F.)
Tri- $C_6H_2Cl_3(NH_2)$				
1 : 2 : 4 : 6	from aniline with Cl, best in acetic acid solution	Long colorless needles : non-basic	77.5° C. (171.5° F.)	260° C. (500° F.)
1 : 2 : 4 : 5	from $C_6H_2Cl_3(NO_2)$	Colorless needles	96.5° C. (205.7° F.)	270° C. (518° F.)
Bromanilines.				
Mono- $C_6H_4Br(NH_2)$				
Ortho (1 : 2)	from 1 : 2 $C_6H_4Br(NO_2)$	Colorless needles ; insoluble in water, very soluble in alcohol	31.5° C. (88.7° F.)	229° C. (444.2° F.)
Meta (1 : 3)	" 1 : 3 "	Crystalline mass	18–18.5° C. (64.4–65.3° F.)	251° C. (483.8° F.)
Para (1 : 4)	" 1 : 4 "	Large rhombic crystals, looking like regular octohedrons	64° C. (147.2° F.)	Decomposed by distillation into aniline, 1:2:4 di-, and 1:2:4:6 tri-bromaniline.

* NH_2 in position 1 throughout.

	Formation.	Physical character at ordinary temperatures.	Melting point.	Boiling point.
Bromanilines—con.				
Di- $C_6H_3Br_2(NH_2)$ 1:2:4	from aniline with Br: from di-bromacetanilide with alkalies: from 1:2:5, $C_6H_3Br_2(NO_2)$, by reduction; from $C_6H_5NO_2$, by heating with conc. HBr to 185°–190° (together with mono- and tribromaniline), and by distilling dibromisatin with potash	colorless shining needles.	79.5° C. (175.1° F.)	
1:2:5	from 1:2:5 $C_6H_3Br_2(NO_2)$	Warty groups of silky prisms	51–52° C. (123.8–125.6° F.)	
1:3:5	" 1:3:5 "	White needles	56.4° C. (133.5° F.)	
1:3:4	" 1:3:4 "	Colorless needles; this and 1:3:5 are more basic than the two former	80.4° C. (176.7° F.)	
Tri- $C_6H_2Br_3(NH_2)$ 1:2:4:6	from aniline, para-, and ortho-bromaniline, with excess of Br, and by reduction of 1:2:4:6 $C_6H_2Br_3(NO_2)$	Long colorless needles non-basic	119.6° C. (247.2° F.)	
1:3:4:5	from 1:3:4:5 $C_6H_2Br_3(NO_2)$	Crystalline: unites with acids	Does not melt at 130° C. (266° F.)	
Tetra- $C_6HBr_4(NH_2)$ 1:2:3:4:6	from 1:2 bromaniline and 1:2:5 di-bromaniline with excess of bromine	Fine long silky needles	115.3° C. (239.5° F.)	
Penta- $C_6Br_5(NH_2)$	from 1:3:5 di-bromaniline with Br	Large needles	Does not melt at 222° C. (431.6° F.)	
Iodanilines.				
Meta- $C_6H_4I.NH_2$ (1:3)	from 1:3 $C_6H_4I(NO_2)$	Silvery laminæ	25° C. (77° F.)	
Para- " (1:4)	" 1:4 "	Colorless needles	60° C. (140° F.)	

2. *Nitro-derivatives.*

Mononitranilines, $C_6H_4(NO_2).NH_2$.—The three isomeric derivatives are formed by imperfect reduction of the dinitrobenzenes (p. 859), best by passing hydrogen sulphide into the alcoholic solution of the nitro compound, mixed with a little concentrated ammonia :

$$C_6H_4\!\!<\!\!{}^{NO_2}_{NO_2} + 3H_2S = C_6H_4\!\!<\!\!{}^{NO_2}_{NH_2} + 2H_2O + 3S .$$

The ortho- and para-derivatives are also formed by the action of alkalies on the corresponding nitro-acetanilides, $C_6H_5.NH(C_2H_3O)$, and by prolonged heating of ortho- and para-nitrobromobenzene, or of the methylic ether of ortho- or paranitrophenol, with aqueous or alcoholic ammonia to 180–200° C. (356–392° F.).

$$C_6H_4(NO_2)Br + NH_3 = HBr + C_6H_4(NO_2)(NH_2).$$
$$C_6H_4(NO_2)(OCH_3) + NH_3 = HOCH_3 + C_6H_4(NO_2)(NH_2).$$

Nitranisol. Methyl Nitraniline.
alcohol.

Ortho-nitraniline forms long, dark yellow needles, melting at 71.5° C. (160.7° F.), more soluble in water and in alcohol than the other two modifications. *Meta.*—Long yellow prisms, melting at 109° C. (228° F.), slightly soluble in water, freely in alcohol. *Para.*—Orange-colored needles or tables, melting at 146° C. (294.8° F.), nearly insoluble in water, easily soluble in alcohol.

Dinitranilines, $C_6H_3(NO_2)_2NH_2$. — The unsymmetrical modification 1 : 2 : 4 (NH_2 in 1), is formed by heating the corresponding dinitro-, chloro-, bromo-, or iodo-benzene, or the methylic ether of (1 : 2 : 4) dinitrophenol with alcoholic ammonia, or dinitro-acetanilide with potash-lye. It crystallizes in light yellow prisms, melting at 182°–183° C. (359.6°–361.4° F.); does not combine with acids. By exchange of NH_2 for H it is converted into meta-dinitrobenzene.

Consecutive Dinitraniline (1 : 2 : 6), from the corresponding dinitro-iodobenzene or methyl-dinitrophenol, forms long, dark yellow needles, melting at 138° C. (280.2° F.), sparingly soluble even in hot alcohol.

Trinitraniline, $C_6H_2(NO_2)_3.NH_2$ (1 : 2 : 4 : 6) (*Picramide*), is formed by the action of alcoholic ammonia on trinitrochlorobenzene, $C_6.Cl.NO_2.H.NO_2H.NO_2$ (p. 831), or on the ethylic or methylic ether of picric acid. It crystallizes from alcohol in orange-red needles ; from glacial acetic acid in large monoclinic tables ; melts at 186° C. (366.8° F.). By heating with alkalies it is converted into a salt of picric acid :

$$C_6H_2(NO_3)_3.NH_2 + KOH = NH_3 + C_6H_2(NO_2)_3OK .$$

Chloro- and Bromo-nitranilines, of which numerous modifications are known, are obtained by heating nitrodichloro- and nitrodibromobenzenes with alcoholic ammonia ; by treating chlor- and brom-acetanilides with nitric acid, and decomposing the resulting nitro-compounds with alkalies ; and by passing chlorine gas or bromine vapor into the solution of the three nitranilines in hydrochloric acid.

Diamidobenzenes, or **Phenylenediamines,** $C_6H_4(NH_2)_2$. —These bases are formed by reduction of the three dinitrobenzenes or the three nitranilines with tin and hydrochloric acid ; also by dry distillation of the corresponding diamidobenzoic acids. They are bi-acid bases, forming well-defined crystalline salts ; *e. g.*, $C_6H_4(NH_2)_2.2HCl$.

37

Ortho—Colorless or slightly reddish four-sided tables ; melts at 102^0 C (215.6^0 F.); boils at 252^0 C. (485.6^0 F.); dissolves in water, especially when hot. Its solution in hydrochloric acid is colored dark-red by ferric chloride or potassium dichromate. *Meta*—Crystalline mass, easily soluble in water; melts at 63^0 C. (145.4^0 F.), boils at 287^0 C. (548.6^0 F.). *Para* —Colorless, or faintly reddish scales, easily soluble in water; melts at 147^0 C. (296.6^0 F.), boils at 267^0 C. (512.6^0 F.).

Triamidobenzene, $C_6H_3(NH_2)_3$ (1 : 2 : 4), is produced by dry distillation of triamidobenzoic acid mixed with pounded glass, and by reduction of (1 : 2 : 4) dinitraniline with tin and hydrochloric acid. It forms a dark-red radio-crystalline mass, melting at 103^0 C. (217.4^0 F.), boiling at 330^0 C. (626^0 F.); easily soluble in water, alcohol and ether ; separated from its aqueous solution by caustic soda. By ferric chloride and by strong sulphuric acid containing a little nitric acid, its solution is colored deep blue. It forms well-defined salts with acids.

ALCOHOLIC DERIVATIVES OF ANILINE.

The partial or total replacement of the hydrogen in the amidogen group of aniline by alcohol-radicles, gives rise to compounds analogous to the secondary and tertiary amines of the fatty group, and formed in like manner by heating aniline with the iodides and bromides of the alcohol-radicles ; *e. g.*,

$$C_6H_5NH_2 + CH_3Br = HBr + C_6H_5.NH.CH_3 .$$
$$\text{Aniline.} \qquad\qquad\qquad\qquad \text{Methyl-aniline.}$$

They may also be produced by heating aniline hydrochloride with alcohols in closed vessels, a chloride of the alcohol-radicle being first formed, which then acts as above on the aniline. The tertiary derivatives, such as diethylaniline, $C_6H_5N(C_2H_5)_2$, can also unite with alcoholic iodides, forming ammonium-compounds, like methyl-diethyl-phenylammonium, $(CH_3)(C_2H_5)_2(C_6H_5)NI$, which, when treated with moist silver oxide, yield the corresponding hydroxides, such as $(CH_3)(C_2H_5)_2(C_6H_5)N.OH$. These hydroxides are very soluble in water, strongly alkaline, and have a bitter taste.

The secondary and tertiary derivatives are liquid at ordinary temperatures, and exhibit the following physical properties :—

		Boiling Point.	Specific Gravity.
Methylaniline,	$C_6H_5NH(CH_3)$	190^0–191^0 C. (374-376^0 F.)	0.976 at 15^0 C. (60.8^0 F.)
Dimethylaniline,	$C_6H_5N(CH_3)_2$	192^0 C. (377.6^0 F.)	0.9553 at 15^0 C. (60.8^0 F.)
Ethylaniline,	$C_6H_5NH(C_2H_5)$	204^0 C. (399.2^0 F.)	0.954 at 18^0 C. (64.4^0 F.)
Diethylaniline,	$C_6H_5N(C_2H_5)_2$	213.5^0 C. (416.3^0 F.)	0.939 at 18^0 C. (64.4^0 F.)
Amylaniline,	$C_6H_5NH(C_5H_{11})$	258^0 C. (496.4^0 F.)	
Amyl-ethylaniline,	$(C_6H_5)N(C_2H_5)(C_5H_{11})$	262^0 C. (503.6^0 F.)	

Dimethylaniline solidifies at $+0.5^{\circ}$ C. (32.9° F.). Amylethyl-aniline unites with methyl iodide, forming the iodide of methylethylamylphenyl-ammonium, $(C_6H_5)N(CH_3)(C_2H_5)(C_5H_{11})$.

Many other alcoholic derivatives of aniline have been prepared and examined.

Diphenylamine, $NH(C_6H_5)_2$, is formed by heating aniline hydrochloride with aniline to 240° C. (464° F.), and by the dry distillation of triphenyl-rosaniline (aniline blue). It is a crystalline body, having a pleasant odor, melting at 54° C. (129.2° F.), and boiling at 310° C. (590° F.); slightly soluble in water, easily in alcohol and ether; colored deep-blue by nitric acid. It is a weak base, its salts being decomposed by water.

Triphenylamine, $N(C_6H_5)_3$, is produced, together with diphenylamine, by heating a solution of potassium in aniline with monobromobenzene.

Derivatives containing diatomic alcohol-radicles are formed by heating aniline with the iodides or bromides of methene and ethene. *Methene-diphenyldiamine*, $(C_6H_5.NH)_2.CH_2$, is viscid and uncrystallizable. *Ethene-diphenyldiamine*, $(C_6H_5.NH)_2.C_2H_4$, is crystalline, and melts at 57° C. (134.6° F.). The isomeric *ethidene-diphenyldiamine*, $(C_6H_5.NH)_2.CH-CH_3$, formed by the action of aldehyde on aniline, separates from alcoholic solution in yellowish nodules.

Derivatives containing triatomic alcohol-radicles are obtained by heating aniline with the chlorides of such radicles, or with substances capable of forming them. *Methenyl-diphenyldiamine*, $\begin{matrix} C_6H_5-NH\diagdown \\ C_6H_5-N \diagup \end{matrix}CH$, produced by heating aniline with chloroform to $180^{\circ}-190^{\circ}$ C. ($356^{\circ}-374^{\circ}$ F.) or by the action of phosphorus trichloride on a mixture of aniline and formanilide, crystallizes in long colorless needles, melting at $135^{\circ}-136^{\circ}$ C. ($275^{\circ}-276.8^{\circ}$ F.), and boiling, with partial decomposition, above 250° C. (482° F.). *Ethenyl-diphenyldiamine*, $\begin{matrix} C_6H_5-NH\diagdown \\ C_6H_5-N \diagup \end{matrix}C-CH_3$, produced by the action of PCl_3 on a mixture of aniline and acetic acid, forms small colorless needles melting at $131^{\circ}-132^{\circ}$ C. ($267.8^{\circ}-269.6^{\circ}$ F.).

DERIVATIVES CONTAINING ACID-RADICLES :— A N I L I D E S.

These compounds, which may be regarded as amides having their hydrogen more or less replaced by phenyl, are formed : 1. By the action of acid chlorides or chloranhydrides on aniline; thus :

$$\underset{\text{Aniline.}}{C_6H_5.NH_2} + \underset{\substack{\text{Acetic} \\ \text{chloride.}}}{C_2H_3O.Cl} = HCl + \underset{\text{Acetanilide.}}{C_6H_4.NH_2.C_2H_3O}.$$

2. By heating aniline salts with organic acids :

$$C_6H_5.NH_2 + C_2H_3O.OH = H_2O + C_6H_5.NH.C_2H_3O.$$

By heating with alkalies, or with hydrochloric acid, they are resolved into their components; e. g.,

$$C_6H_5.NH.C_2H_3O + KOH = C_2H_3O.OK + C_6H_5.NH_2.$$

Formanilide, $C_6H_5.NH.CHO$ — *Phenylformamide.* — Produced by heating aniline with ethyl formate, or, together with other products, by quickly heating it with oxalic acid :

$$C_6H_5.NH_2 + C_2O_4H_2 = C_6H_5.NH.CHO + CO_2 + H_2O.$$

It forms prisms, melting at 46° C. (114.8^O F.), easily soluble in water, alcohol, and ether. Strong soda-lye added to its aqueous solution throws down s o d i u m f o r m a n i l i d e, $C_6H_5.NNa$—CHO, which is resolved by water into formanilide and sodium hydroxide. By distillation with strong hydrochloric acid, it yields a small quantity of benzonitril : $C_6H_5.NH.CHO$ $= C_6H_5.CN + H_2O.$

Acetanilide, $C_6H_5.NH.C_2H_3O$, *Phenylacetamide*, produced by heating aniline and glacial acetic acid (1 mol. of each) for several hours, or by the action of acetyl chloride or acetic anhydride on aniline, forms colorless, shining laminar crystals, melting at 112°–113° C. (233.6^O–235.4^O F.), volatilizing without decomposition at 295° C. (563° F.), sparingly soluble in cold water, more easily in hot water and in alcohol. Treated with bromine, chlorine, or nitric acid, it yields mono- and di-substitution products, the substitution always taking place in the benzene-ring, and these when heated with alkalies yield alcoholic anilines. *Monobromacetanilide*, C_6H_4Br—$NH(C_2H_3O)$ (1:4), melts at 165° C. (329° F.) ; *dibromacetanilide* at 146° C. (294.8^O F.) ; *paranitracetanilide* melts at 207° C. (404.6^O F.) ; *ortho*- at 78° C. (172.4^O F.). *Dinitracetanilide*, $C_6H_3(NO_2)_2.NH(C_2H_3O)$, formed by dissolving acetanilide in a mixture of nitric and sulphuric acids, melts at 120° C. (248° F.).

Oxanilamide, $\begin{matrix} CO—NH—C_6H_5 \\ | \\ CO—NH_2 \end{matrix}$, *Phenyloxamide*, formed, together with oxanilide, by evaporating cyananiline with hydrochloric acid, crystallizes in laminæ, soluble in hot water, and subliming without decomposition.

Oxanilide, $\begin{matrix} CO—NH—C_6H_5 \\ | \\ CO—NH—C_6H_5 \end{matrix}$, *Diphenyloxamide*, obtained by heating aniline oxalate to 160°–180° C. (320–356° F.), and together with oxanilamide, by evaporating a solution of cyananiline in hydrochloric acid, forms shining sublimable crystals, melting at 245° C. (473° F.).

Oxanilic Acid, $\begin{matrix} CO—NH—C_6H_5 \\ | \\ CO—OH \end{matrix}$, produced by heating aniline with excess of oxalic acid, forms crystalline scales slightly soluble in cold, easily in hot water, and having a strong acid reaction. It is monobasic.

Carbanilamide, $CO {<}^{NH—C_6H_5}_{NH_2}$, *Phenyl-carbamide, Phenyl-urea*, is formed, like ethyl-carbamide (p. 791), by passing cyanic acid vapor into aniline, and by the action of ammonia on phenyl isocyanate (carbanil) :

$$CO{=}N—C_6H_5 + NH_3 = CO{<}^{NH—C_6H_5}_{NH_2}.$$

The easiest method of preparing it is to evaporate an aqueous solution of potassium cyanide and aniline sulphate. It forms colorless, needle-shaped crystals, melting at 144°–145° C. (291.2^O–293° F.), sparingly soluble in cold water, easily in hot water, and in alcohol and ether. Decomposed by heat into ammonia, cyanuric acid and carbanilide.

Alcoholic derivatives of phenylcarbamide, *e. g.*, *ethyl-phenylcarbamide*,

$CO{<}{NH.C_6H_5 \atop NH.C_2H_5}$, are formed by the action of aniline on isocyanic ethers (carbimides, p. 570).

Carbanilide, $CO{<}{NH.C_6H_5 \atop NH.C_6H_5}$, *Diphenyl-carbamide, Diphenyl-urea* (symmetrical), is produced by the action of carbonyl chloride (phosgene) on aniline:

$$COCl_2 \; + \; 2(C_6H_5.NH_2) \; = \; 2HCl \; + \; CO(NH.C_6H_5)_2;$$

by combination of phenyl isocyanate (carbanil) with aniline:

$$CO{=}N{-}C_6H_5 \; + \; NH_2.C_6H_5 \; = \; CO(NH.C_6H_5)_2;$$

by heating 1 part of urea with 3 parts of aniline to 150°–170° C. (302°–338° F.):

$$CO(NH_2)_2 \; + \; 2(C_6H_5.NH_2) \; = \; 2NH_3 \; + \; CO(NH.C_6H_5)_2;$$

also by heating 1 mol. aniline with 1 mol. carbanilamide; and, together with formanilide, by the action of heat on oxanilamide. It forms silky needles, slightly soluble in water, moderately soluble in alcohol, melts at 235° C. (455° F.), and volatilizes without decomposition.

Chlorocarbanilide, $COCl{-}N(C_6H_5)_2$, is formed by passing phosgene gas into a solution of diphenylamine in chloroform:

$$COCl_2 \; + \; NH(C_6H_5)_2 \; = \; HCl \; + \; COCl{-}N(C_6H_5)_2.$$

It crystallizes in colorless laminæ. With alcoholic ammonia it forms unsymmetrical diphenyl-carbamide :

$$CO{<}{Cl \atop N(C_6H_5)_2} \; + \; NH_3 \; = \; HCl \; + \; CO{<}{NH_2 \atop N(C_6H_5)_2};$$

which crystallizes in long needles, melting at 189° C. (372.2° F.). Heated with aniline, it yields t r i p h e n y l - c a r b a m i d e, $CO{<}{NH.C_6H_5 \atop N(C_6H_5)_2}$ (needles melting at 136° C. (276.8° F.); and with diphenylamine at 200°–220° C. (392°–428° F.) it forms t e t r a p h e n y l c a r b a m i d e, $CO[N(C_5H_5)_2]_2$, light yellow crystals, melting at 183° C. (361.4° F.).

Carbanilic, or **Phenyl-carbamic Acid,** $CO{<}{NH.C_6H_5 \atop OH}$, is not known in the free state. Its ethers, the p h e n y l u r e t h a n e s, are formed by the action of alcohols on carbanil. The *ethylic ether,* $CO{<}{NH.C_6H_5 \atop OC_2H_5}$, produced also by the action of ethyl chlorocarbonate, $COCl.OC_2H_5$, on aniline, forms colorless needles melting at 52° C. (125.6° F.). It boils at 237°–238° C. (458.6°–460.4° F.), with partial decomposition into alcohol and carbanil, and is converted, by heating with strong potash-lye, or with aniline, into diphenylcarbamide.

C a r b a n i l, Phenyl Isocyanate, or **Phenyl carbimide,** $CO{=}N{-}C_6H_5$, formed by distilling oxanilide, or better, ethylic carbanilate, with phosphoric anhydride, is a mobile liquid, boiling at 163° C. (325.4° F.), and having a pungent, tear-exciting odor. Its reactions are exactly like those of the isocyanic ethers already described (p. 574). With water it forms carbanilide:

$$2(CO{=}N{-}C_6H_5) \; + \; H_2O \; = \; CO_2 \; + \; CO(NH.C_6H_5)_2.$$

By ammonia it is converted into carbanilamide, and by amines into alcoholic derivatives of that compound. It unites with alcohols to form the carbanilic ethers.

Thiocarbanil, or **Phenyl-thiocarbimide,** $CS{=}N{-}C_6H_5$, also called *Phenylic Mustard-oil,* is formed from thiocarbanilide by distillation with phosphoric anhydride, by prolonged boiling with hydrochloric acid, and, together with triphenylguanidine, by the action of iodine in alcoholic solution:

$$2CS{<}^{NH.C_6H_5}_{NH.C_6H_5} + I_2 = CS{=}N{-}C_6H_5 + C{<}^{NH.C_6H_5}_{N.C_6H_5} + 2HI + S.$$

It is also produced by the action of phosgene on aniline. It is a colorless liquid, smelling like mustard-oil, boiling at 222^0 C. (431.6^0 F.). By heating with reduced copper, it is converted into phenyl cyanide benzonitril:

$$CS{=}N{-}C_6H_5 + Cu = CuS + CN.C_6H_5.$$

Hydrogen sulphide passed into phenyl-thiocarbimide decomposes it, even at ordinary temperatures, into thiocarbanilide and carbon bisulphide:

$$2(CS{=}N.C_6H_5) + SH_2 = CS_2 + CS(NH.C_6H_5)_2.$$

It unites with ammonia to form phenylthiocarbamide; with aniline, to form diphenylcarbamide; and with anhydrous alcohol at 120^0 C. (248^0 F.), to form phenylic thiocarbamates or thiourethanes:

$$CS{=}NC_6H_5 + C_2H_5.OH = CS{<}^{NH.C_6H_5}_{O.C_2H_5}.$$

Normal Phenyl Thiocyanate, $CN.S.C_6H_5$, isomeric with thiocarbanil, is obtained by the action of cyanogen chloride on the lead-salt of phenyl-mercaptan:

$$(C_6H_5S)_2Pb + 2CNCl = PbCl_2 + 2(CN.S.C_6H_5).$$

It is a colorless liquid which boils at 231^0 C. (447.8^0 F.), and reacts like the normal thiocyanic ethers of the fatty series.

Thiocarbanilamide, or **Phenylthiocarbamide,** $CS{<}^{NH.C_6H_5}_{NH_2}$,

formed by the combination of ammonia with thiocarbanil, crystallizes in needles, melting at 154^0 C. (309.2^0 F.); slightly soluble in cold, more easily in boiling water and in alcohol. When boiled with silver nitrate, it exchanges its sulphur for oxygen, and is converted into phenyl-carbamide.

Thiocarbanilide, or **Diphenylthiocarbamide,** $CS(NH.C_6H_5)_2$, is formed by the action of carbon disulphide on aniline:

$$2NH_2C_6H_5 + CS = SH_2 + CS(NH.C_6H_5)_2.$$

Aniline and potassium hydroxide, in equal numbers of molecules, are heated for an hour in alcoholic solution with excess of carbon sulphide; the resulting liquid is poured into dilute hydrochloric acid, the alcohol evaporated, and the mass crystallized from alcohol.

Thiocarbanilide forms colorless laminæ melting at 144^0 C. (291.2^0 F.), insoluble in water, easily soluble in alcohol and ether. The mode of its conversion into phenylthiocarbimide and triphenyl-guanidine has been already explained (p. 870). By treating it with alcoholic ammonia and lead oxide, the S is replaced by NH, and diphenyl-guanidine is produced.

Diphenyl-guanidine, $HN=C<^{NH.C_6H_5}_{NH.C_6H_5}$ (*Melaniline*), is also formed by the action of gaseous cyanogen chloride on dry aniline, and by boiling cyananilide with aniline hydrochloride:

$$C_6H_5.NH.CN + C_6H_5.NH_2 = HN=C<^{NH.C_6H_5}_{NH.C_6H_5}.$$

It forms long flattened needles, melting at 147° C. (296.6° F.), soluble in 10 parts of alcohol at ordinary temperatures, slightly soluble in water. It is a mono-acid base, and forms well-crystallized salts. When heated above its melting point, it is resolved into ammonia, aniline, and *tetraphenyl-melamine*, $C_3N_6H_2(C_6H_5)_4$. By carbon disulphide it is converted into thiocarbanilide and thiocyanic acid:

$$HN=C(NH.C_6H_5)_2 + CS_2 = CS(NH.C_6H_5)_2 + CNSH.$$

Triphenyl-guanidine (α), $C_6H_5-N=C<^{NH.C_6H_5}_{NH.C_6H_5}$, is formed by the action of heat on diphenyl-carbamide; by heating diphenyl-thiocarbamide either alone or with copper to 150°–160° C. (302°–320° F.), or with aniline to the boiling point of the latter. It is most easily prepared by heating an alcoholic solution of diphenyl-thiocarbamide and aniline with lead oxide or mercuric oxide, or with an alcoholic solution of aniline to the boiling point:

$$CS(NH.C_6H_5)_2 + C_6H_5.NH_2 = SH_2 + C_6H_5-N=C(NH.C_6H_5)_2.$$

Triphenyl-guanidine crystallizes in shining rhombic prisms, melting at 143° C. (289.4° F.), nearly insoluble in water even at the boiling heat, easily soluble in hot alcohol. It is a mono-acid base, forming well-crystallized salts. By distillation it is resolved into aniline and diphenyl-cyanamide or carbodiphenylimide, which recombine in the receiver:

$$C_6H_5-N=C<^{NH.C_6H_5}_{NH.C_6H_5} = C_6H_5NH_2 + C<^{N-C_6H_5}_{N-C_6H_5};$$

and by heating to 160°–170° C. (320–338° F.) with carbon disulphide, into phenyl-thiocarbamide and diphenyl-thiocarbamide:

$$C_6H_5N=C(NH.C_6H_5)_2 + CS_2 = CS=N.C_6H_5 + CS(NH.C_6H_5)_2.$$

An isomeric *triphenyl-guanidine* (β), $HN=C<^{N(C_6H_5)_2}_{NH.C_6H_5}$, is formed by heating cyananilide with diphenylamine hydrochloride:

$$C_6H_5.NH.CN + NH(C_6H_5)_2 = HN=C<^{N(C_6H_5)_2}_{NH.C_6H_5}.$$

It crystallizes in large tables melting at 131° C. (267.8° F.).

Tetraphenyl-guanidine, $HN=C[N(C_6H_5)_2]_2$.—The hydrochloride is formed by passing gaseous cyanogen chloride into fused diphenylamine heated to 160°–170° C. (320°–338° F.). The free base separated therefrom by an alkali forms colorless rhombic prisms, melting at 130°–131° C. (266°–267.8° F.).

Phenyl-cyanamide, or **Cyananilide,** $CN.NH.C_6H_5$, formed by passing gaseous cyanogen chloride into an ethereal solution of dry aniline, or by digesting an alcoholic solution of phenyl-thiocarbamide with lead oxide, crystallizes in long needles, melting at $36°$–$37°$ C. ($96.8°$–$98.6°$ F.), sparingly soluble in water, easily in alcohol and ether. It has no basic properties, but unites with hydrogen sulphide to form **phenylthiocarhamide**, $NH_2.CS.NH(C_6H_5)$, and is converted spontaneously, even at ordinary temperatures, into the polymeric compound **triphenyl-melamine**, $C_3N_6H_3(C_6H_5)_3$, which crystallizes in prisms melting at $162°$–$163°$ C. ($323.6°$–$325.4°$ F.).

Diphenylcyanamide or Carbo-diphenylimide,

$C_{13}H_{10}N_2 = C\begin{subarray}{l} \diagup N—C_6H_5 \\ \diagdown N—C_6H_5 \end{subarray}$, is formed by adding mercuric oxide to a solution

of diphenylthiocarbamide in hot benzene, and, together with aniline, by distillation of α-triphenyl-guanidine. It is a syrupy liquid, boiling at $330°$–$331°$ C. ($626°$–$627.8°$ F.), and changing, when kept in an exsiccator, into a solid polymeric compound. Hydrochloric acid gas passed into its solution in hot benzene converts it into a crystalline hydrochloride, $C_{13}H_{10}N_2$. HCl. By boiling with aqueous alcohol it is converted into **diphenyl-carbamide**. It unites with SH_2, to form **diphenylthiocarbamide**, and with CS_2 at $140°$–$150°$ C. ($284°$–$302°$ F.) to form **phenyl-thiocarbimide**, $\ddot{C}(N.C_6H_5)_2 + CS_2 = 2(CS\!\!\equiv\!\!N—C_6H_5)$.

Cyananiline, $(C_6H_5.NH_2)_2(CN)_2 = \begin{subarray}{l} C(NH)—NH.C_6H_5 \\ | \\ C(NH)—NH.C_6H_5 \end{subarray}$, separates on

passing cyanogen gas into an alcoholic solution of aniline, in shining laminæ, melting at $210°$ C. ($410°$ F.). It is a biacid base, and is resolved by boiling with acids into oxamide and oxanilide.

PHENYL PHOSPHORUS-COMPOUNDS.

Phosphenyl Chloride, $C_6H_5.PCl_2$, formed by repeatedly passing a mixture of the vapors of benzene and phosphorus trichloride through a red-hot tube filled with fragments of pumice, and in small quantity by heating mercury-diphenyl with phosphorus trichloride, is a fuming, strongly refracting liquid, having a density of 1.319 at $20°$ C. ($68°$ F.), and boiling at $222°$ C. ($431.6°$ F.) ; decomposed by water into hydrochloric and phosphenylous acids. It unites with *chlorine*, forming the **tetrachloride**, $C_6H_5.PCl_4$, which melts at $73°$ C. ($163.4°$ F.) ; with *bromine* to form $C_6H_5.PCl_2Br_2$, melting at $208°$ C. ($406.4°$ F.) ; with oxygen to form the **oxychloride**, $C_6H_5.PCl_2O$, which is a liquid boiling, with partial decomposition, at $260°$ C. ($500°$ F.) ; and with *sulphur* to form $C_6H_5.PCl_2S$, a liquid boiling, also with partial decomposition, at $270°$ C. ($518°$ F.).

Phosphenylous Acid, $C_6H_5.PHO.OH$, formed, as above mentioned, by the action of water on the dichloride, crystallizes in colorless laminæ, melting at $70°$ C. ($158°$ F.), and acts as a powerful reducing agent. **Phosphenylic acid**, $C_6H_5.PO(OH)_3$, formed in like manner from the tetrachloride, crystallizes in laminæ, melting at $158°$ C. ($316.4°$ F.).

Phenyl-phosphine, $C_6H_5.PH_2$.—Dry hydriodic acid gas passed into phosphenyl chloride unites with it, forming the compound $C_6H_5.PI_2.HI$, which is decomposed by alcohol, with formation of phenylphosphine. The latter is a liquid having a pungent and offensive odor, heavier than water, boiling at 160° C. (320° F.). It oxidizes in the air, forming p h o s p h e n y l o x i d e, $C_6H_5.PH_2O$, a crystalline mass easily soluble in water. Phenyl-phosphine does not dissolve in aqueous acids, but unites with dry *hydrogen iodide*, forming p h e n y l p h o s p h o n i u m i o d i d e, $C_6H_5.PH_3I$, from which the phenylphosphine may be separated by water.

Diethyl-phenylphosphine, $C_6H_5.P(C_2H_5)$, formed by the action of zinc-ethyl on phosphenyl chloride, and treatment of the product with caustic soda, is a colorless strong-smelling liquid, having a specific gravity of 0.9571 at 13° C. (57.2° F.), boiling at 222° C. (431.6° F.), insoluble in water, soluble in acid, but not forming crystallizable salts. With ethyl iodide it forms the compound $C_6H_5.P(C_2H_5)_3I$, which is crystalline, and easily soluble in water.

A r s e n i c C o m p o u n d s.—*Arseniophenyl chloride*, $C_6H_5AsCl_2$, formed by the action of arsenious chloride on mercury-diphenyl, is a heavy colorless liquid, boiling at 252°–255° C. (485.6°–491° F.), not decomposed by water, soluble in alkalies. It unites with chlorine, forming a reddish-yellow liquid *tetrachloride*, $C_6H_5AsCl_4$, which is decomposed by water into hydrochloric and p h e n y l a r s i n i c a c i d s, $C_6H_5AsO(OH)_2$, crystallizing in long needles which melt at 168° C. (334.4° F.).

Arsenio-diphenyl chloride, or *Phenylcacodyl chloride*, $(C_6H_5)_2AsCl$, formed in small quantity, together with the preceding compound, is a thick oil, boiling above 360° C. (690° F.), not decomposed by water. It unites with *chlorine*, forming the t r i c h l o r i d e, $(C_6H_5)_2AsCl_3$, which melts at 174° C. (345.2° F.), and is decomposed by warm water into hydrochloric acid and p h e n y l c a c o d y l i c a c i d, $(C_6H_5)_2AsO.OH$, which crystallizes in needles, slightly soluble in water, melting at 174° C. (345.2° F.).

S i l i c o - p h e n y l C o m p o u n d s.—The *chloride*, $C_6H_5SiCl_3$, obtained by heating mercury-diphenyl with silicic chloride, $SiCl_4$, to 300° C. (572° F.), is a liquid which fumes in the air, and boils at 197° C. (386.6° F.). It is decomposed by water, with formation of s i l i c o - b e n z o i c a c i d, $C_6H_5.SiO.OH$, *i. e.*, benzoic acid in which the C of the group COOH is replaced by Si. With alcohol, the triethylic ether, $C_6H_5.Si(OC_2H_5)_3$, is formed, as a liquid boiling at 237° C. (458.6° F.).

M e r c u r y - d i p h e n y l, $(C_6H_5)_2Hg$, is formed by heating a solution of bromobenzene in benzene for a considerable time with liquid sodium-amalgam, the reaction being facilitated by addition of a little acetic ether. It crystallizes in colorless rhombic prisms, melts at 120° C. (248° F.), and may be sublimed; turns yellow on exposure to the air; dissolves easily in benzene and carbon sulphide; less easily in ether and in alcohol; not at all in water. When distilled it is, for the most part, decomposed into diphenyl, benzene, and mercury. Acids decompose it, with formation of benzene and mercury salts. When treated with two molecules of chlorine, bromine, or iodine, it is decomposed into monochlorobenzene, etc., and a halogen-compound of mercury; *e. g.*:

$$(C_6H_5)_2Hg \ + \ 2Cl_2 \ = \ 2C_6H_5Cl \ + \ HgCl_2 \ .$$

With one molecule of the halogen element, haloïd-compounds are formed, like $C_6H_5.HgI$, from which, by the action of moist silver oxide, the hydroxide, $C_6H_5.Hg.OH$, is formed, a crystalline, strongly alkaline compound which expels ammonia from ammonium salts.

Amido-toluenes.

These, like other toluene derivatives, admit of metameric modifications, according as the NH_2 group is situated in the principal or the lateral chain (p. 841).

Toluidine, $C_7H_9N = C_6H_4(NH_2).CH_3$.—This base, homologous with aniline, exhibits the three modifications, ortho-, meta-, and para-, which are obtained by reduction of the three corresponding nitrotoluenes, $C_6H_4(NO_2).CH_3$.

Para-toluidine (ordinary toluidine),$C_6.CH_3.H.H.NH_2.H_2$, forms large colorless tabular crystals, heavier than water, very sparingly soluble in water, easily in alcohol and ether. It melts at 45º C. (113º F.), boils at 198º C. (288.4º F.); has an aromatic taste and odor, a very feeble alkaline reaction ; does not give any colored reaction with chloride of lime. It forms crystalline salts, but is a weak base, incapable of neutralizing sulphuric acid. With acetyl chloride it forms acetoparatoluidine, $C_6H_4(NH.C_2H_3O).CH_3$, a crystalline compound melting at 145º C. (293º F.).

Ortho-toluidine, also called *Pseudo-toluidine*, is a colorless, neutral liquid, becoming rose-colored on exposure to the air. It has a density of 1.00 at 16º C. (60.8º F.), boils at 199.5º C. (390.2º F.), and does not solidify at — 20º C. (— 4º F.). Its acetyl-derivative melts at 107º C. (224.6º F.).

Meta-toluidine, from metanitrotoluene, is a colorless liquid, of specific gravity 0.998 at 15º C. (59º F.), boiling at 197º C. (386.6º F.), not solidifying at — 13º C. (8.6º F.). Its acetyl-derivative melts at 65.5º C. (149.9º F.).

Commercial toluidine, from aniline works, is a mixture of ortho- and para-toluidine, the latter of which partly crystallizes out on cooling to a low temperature. To separate the ortho-toluidine from the still dissolved para-modification, the liquid is heated with glacial acetic acid, whereby the two bases are converted into acetotoluides ; these are dissolved in strong acetic acid, and the solution is diluted with 80 parts of water, whereby the acetoparatoluide is precipitated, while the acetorthotoluide remains in solution ; or the mixture of the two bases is heated successively with oxalic acid and ether, whereby an oxalate of paratoluidine is first separated, the ortho-salt remaining dissolved.

Benzylamine, $C_6H_5.CH_2(NH_2)$, or $NH_2(C_7H_7)$.—This compound, metameric with toluidine, is obtained, together with *dibenzylamine*, $NH(C_7H_7)_2$, and *tribenzylamine*, $N(C_7H_7)_3$, by the action of alcoholic ammonia on benzyl chloride, $C_6H_5.CH_2Cl$ (p. 854).

Benzylamine is a colorless liquid, boiling at 185º C. (365º F.). It mixes in all proportions with water, and is separated therefrom by potash. It is a much stronger base than toluidine ; absorbs carbon dioxide rapidly, forming a crystalline carbonate ; unites readily with acids, producing rise of temperature ; and fumes with hydrochloric acid. The hydrochloride crystallizes in striated tables ; the platinochloride, $2NH_2(C_7H_7)Cl.PtCl_4$, in orange-colored laminæ.

Dibenzylamine, $(C_7H_7)_2NH$, is a colorless viscid liquid, insoluble in water, easily soluble in alcohol and ether ; having a specific gravity of 1.033 at 14º C. (57.2º F.).

Tribenzylamine, $(C_7H_7)_3N$, forms colorless needles or laminæ, melting at 91° C. (195.8° F.), insoluble in water, slightly soluble in cold alcohol, easily in hot alcohol and in ether. Its hydrochloride, heated in a stream of dry hydrogen chloride, is resolved into benzyl chloride and the hydrochloride of dibenzylamine.

Diamidotoluenes or **Tolylene-diamines,** $C_6H_3(NH_2)_2$. CH_3.—The modification 1 : 2 : 4 (CH_3 in 1), obtained by reduction of dinitrotoluene melting at 70.5° C. (158.9° F.), with tin and hydrochloric acid, crystallizes in long needles, melts at 99° C. (210.2° F.), and boils at 280° C. (536° F.). 1 : 3 : 4, obtained by reduction of metanitro-paratoluidine, forms colorless scales, melting at 88.5° C. (191.3° F.), boiling at 265° C. (509° F.).

Metaxylidine, $C_8H_{11}N = C_6H_3(NH_2).(CH_3)_2$, **Cumidine,** $C_9H_{13}N$, or probably $C_6H_4(NH_2).C_3H_7$, and **Cymidine,** $C_{10}H_{15}N$, or $C_{10}H_{13}(NH_2)$, homologous with toluidine, are obtained in like manner by reduction of the corresponding nitro-derivatives. Xylidine boils at 214°–216° C. (417.2°–420.8° F.) ; cumidine at 225° C. (437° F.) ; cymidine at 250° C. (482° F.). Xylidine and cumidine form well-crystallized salts.

Mesidine or Amidomesitylene, $C_6H_2(CH_3)_3.NH_2$ (the three CH_3-groups symmetrical, 1 : 3 : 5), is liquid. *Nitromesidine*, $C_9H_{10}(NO_2)(NH_2)$, melts at 100° ; *diamidomesitylene*, $C_9H_{10}(NH_2)_2$, at 90° C. (194° F.).

Xylylamine, $C_8H_{11}N = NH_2C_8H_9 = C_6H_4(CH_3).CH_2NH_2$, homologous with benzylamine, is obtained, together with *dixylylamine*, $NH(C_8H_9)_2$, and *trixylylamine*, $N(C_8H_9)_3$, by heating xylyl chloride, $C_6H_4(CH_3).CH_2.Cl$, with alcoholic ammonia in sealed tubes. These three bases are oily liquids, smelling like herring-pickle, lighter than water, insoluble therein, easily soluble in alcohol and ether. Xylylamine boils at 196° C. (384.8° F.) ; dixylylamine decomposes at 210° C. (410° F.).

Cumylamine, the 9-carbon base metameric with cumidine and homologous with benzylamine, has not been obtained.

Cymylamine, $C_{10}H_{15}N = NH_2(C_{10}H_{13}) = C_6H_4(CH_3).C_3H_6(NH_2)$, is obtained, together with *di-* and *tri-cymylamine*, by heating cymyl chloride with alcoholic ammonia in sealed tubes. Cymylamine and dicymylamine are oily liquids, boiling with decomposition, the former at 280° C. (536° F.), the latter above 300° C. (572° F.). *Tricymylamine* crystallizes in rhomboïdal laminæ, melting at 81°–82° C. (177.8°–179.6° F.).

ANILINE DYES.

Aniline has during the last few years found an extensive application in the arts, a long series of coloring matters unequalled in brilliancy and beauty having, by the action of different oxidizing agents, been produced from it. It was Mr. W. H. Perkin who first conceived the happy idea of applying practically the well-known property possessed by aniline, of forming violet and blue solutions when treated with a solution of chloride of lime or chromic acid. He succeeded in fixing these colors, and bringing them into a form adapted for the dyer.

Aniline-purple or **Mauve**, the first discovered of the aniline-dyes (1856), is prepared by mixing solutions of aniline sulphate and potassium bichromate in equivalent proportions, and allowing the mixture to stand for several hours. The black precipitate formed is filtered off and purified from admixed potassium sulphate by washing with water; it is then dried and freed from resinous matter by repeated digestion with coal-tar naphtha, and finally dissolved in boiling alcohol. For its further purification, the alcoholic solution is evaporated to dryness, the substance is dissolved in a large quantity of boiling water, reprecipitated with caustic soda, washed with water, and dissolved in alcohol; and the filtered solution is evaporated to dryness. Mauve thus prepared forms a brittle substance, having a beautiful bronze-colored surface: it is difficultly soluble in cold water, although it imparts a deep purple color to that liquid: it is more soluble in hot water, very soluble in alcohol, nearly insoluble in ether and hydrocarbons: it dissolves in concentrated acetic acid, from which it crystallizes. Mauve is the sulphate of a base called m a u v e ï n e, having the composition $C_{26}H_{24}N_4$, and capable of forming crystalline salts with acids.

Aniline-red, Rosaniline, $C_{20}H_{19}N_3$ (or $C_{20}H_{17}N_3$).—Salts of this base occur more or less pure in commerce under the names *roseine, fuchsine, magenta, azaleine,* etc. A red color had been observed at different times in experimenting with aniline, more especially when that substance was digested with Dutch liquid. The red coloring matter, though still impure, was first obtained in a separate state from the product formed by digesting aniline with carbon tetrachloride at 150° C. (302° F.), in which reaction it is formed, together with triphenylguanidine. Verguin (1858) first prepared it on a large scale by the action of stannic chloride upon aniline; and it has since been produced by the action of mercuric salts, arsenic acid, and many other oxidizing agents, upon aniline. The most advantageous mode of preparation is the following: A mixture of 12 parts of the dry arsenic acid which occurs in commerce, and 10 parts of aniline, is heated to 120° or 140° C. (250°–284° F.), with addition of water, for about six hours. The product, which is a hard mass having the lustre of bronze, is dissolved in hot water and precipitated by a slight excess of soda; the precipitate when washed with water, and dissolved in acetic acid, forms the roseine of commerce. In order to purify this still crude substance, it is boiled with an excess of soda, to separate any aniline that it may contain; and the washed precipitate is dissolved in very dilute mineral acid, filtered from undissolved tarry matter, and reprecipitated with alkali. The compounds of rosaniline with one molecule of acid are beautifully crystallized substances, which in the dry state have a green color with golden lustre; with water they yield a very intensely colored red solution. The free base, first obtained by Nicholson, presents itself in colorless crystalline plates, insoluble in water, soluble in alcohol and ether, with a red color, which it also acquires on exposure to the air.

Rosaniline in the anhydrous state is represented by the formula, $C_{20}H_{19}N_3$, and in the hydrated state, such as it assumes when isolated from its compounds, by the formula $C_{20}H_{19}N_3.H_2O$. It is a triamine capable of forming monoacid, biacid, and triacid salts. The aniline-reds of commerce are monoacid salts of rosaniline, more or less pure. The acetate, which is chiefly found in commerce in England, has been prepared by Nicholson in splendid crystals of very considerable dimensions, having the composition $C_{20}H_{19}N_3.C_2H_4O_2$. In France the hydrochloride, $C_{20}H_{19}N_3$.HCl, is chiefly employed. The action of ammonium sulphide upon rosaniline gives rise to *leucaniline*, $C_{20}H_{21}N_3$, a base containing two additional atoms of hydrogen. This base is itself colorless, and forms colorless triacid salts, such as $C_{20}H_{21}N_3.3HCl$. Oxidizing agents reconvert it into rosaniline.

The molecular constitution of rosaniline has not been distinctly made out, and there is even some doubt (as above indicated) with respect to its empirical formula. Its mode of formation, also, is not thoroughly understood; but one very important fact has been brought to light by the researches of Hofmann, and confirmed by the experience of manufacturers —namely, that pure aniline, from whatever source it may be obtained, is incapable of furnishing aniline-red. Commercial aniline prepared from coal-tar always, in fact, contains toluidine as well as aniline; and Hofmann has shown that the presence of this base, together with aniline, is essential to the formation of the red dye. Toluidine by itself is just as incapable of yielding the red as pure aniline, but when a mixture of pure aniline and pure toluidine is treated with stannic or mercuric chloride, or with arsenic acid, the red coloring matter is immediately produced. If the formula $C_{20}H_{19}N_3$ be correct, the formation of rosaniline may perhaps be represented by the equation:

$$C_6H_7N + 2C_7H_9N = C_{20}H_{19}N_3 + 3H_2;$$
Aniline. Toluidine. Rosaniline.

and its structure by the formula:

$$C_6H_4 \begin{matrix} NH.C_6H_3 \\ NH.C_6H_3 \end{matrix} \begin{matrix} CH_3 \\ NH \\ CH_3 \end{matrix} \quad \text{or} \quad C_6H_3(NH_2) \begin{matrix} CH_2.C_6H_4.NH \\ CH_2.C_6H_4.NH \end{matrix}$$

But rosaniline is converted by nitrous acid into aurin or rosolic acid, which, according to the recent experiments of Dale and Schorlemmer,[*] has the composition $C_{20}H_{14}O_3$, and is reconverted into rosaniline by heating with alcoholic ammonia. According to this, rosaniline should have the formula $C_{20}H_{17}N_3$, the two reactions just mentioned being represented by the equations:

$$C_{20}H_{17}N_3 + 3NO_2H = C_{20}H_{14}O_3 + 3H_2O + 3N_2,$$

and

$$C_{20}H_{14}O_3 + 3NH_3 = C_{20}H_{17}N_3 + 3H_2O.$$

Aniline Blue and **Aniline Violet.**—Girard and De Laire obtained aniline blue by digesting rosaniline with an excess of aniline at 150°–160° C. (302°–320° F.). Together with aniline blue, which is the principal product of the reaction, several other coloring matters (violet and green), and indifferent substances, are formed, considerable quantities of ammonia being invariably evolved. The crude blue is purified by treating it successively with boiling water acidified with hydrochloric acid, and with pure water. The blue coloring matter is said to be obtained from its boiling alcoholic solution in brilliant needles. It consists of the hydrochloride of *triphenyl-rosaniline*, $C_{20}H_{16}(C_6H_5)_3$. By heating rosaniline with ethyl iodide, Hofmann has obtained an aniline violet, having the composition of hydriodide of *triethyl-rosaniline*, $C_{20}H_{16}(C_2H_5)_3N_3$. Another aniline violet is produced by heating rosaniline with a quantity of aniline less than sufficient to form aniline blue.

Other aniline violets are produced by the action of stannic chloride, mercuric chloride, or iodine chloride on methyl-aniline and dimethyl-aniline.

* Chem. Soc. Journal, 1877, vol. ii. p. 121.

Aniline Greens.—The most important of these colors are those known as "aldehyde green" and "iodine green." The former is produced by adding $1\frac{1}{2}$ parts aldehyde to a cold solution of magenta in a mixture of 3 parts strong sulphuric acid and 1 part water. The mixture is then heated in a water-bath till a drop of the product diffused in water produces a fine blue color, and then poured into a boiling solution of sodium thiosulphate. The liquid is then boiled for a short time and filtered. The filtrate contains the green, which may be precipitated by tannin or by sodium-acetate. Aldehyde green is principally used in silk-dyeing. It is a salt of an organic base which may be separated by means of soda or ammonia.

Iodine green is produced by heating the violets of triethyl- or trimethyl-rosaniline (Hofmann's violets) or the methyl-aniline violets, with iodide of methyl, ethyl, or amyl. The green thus obtained with methyl iodide has a very fine color, bluer than that of aldehyde green, and, like the latter, preserves its color by artificial light. It is much used for cotton and silk dyeing.

A third kind of aniline green, known commercially as "Perkin's green," resembles the iodine green, and is much used for calico-printing.

Chrysaniline, $C_{20}H_{17}N_3$ (or $C_{20}H_{15}N_3$?), is formed as a secondary product in the preparation of aniline red. It is a well-defined base, forming two series of salts, most of which are well crystallized. The two hydrochlorides of chrysaniline are $C_{20}H_{17}N_3.HCl$, and $C_{20}H_{17}N_3.2HCl$. The nitrate, $C_{20}H_{17}N_3.NO_3H$, is so little soluble in water that nitric acid may be precipitated even from a dilute solution of nitrates by means of the more soluble hydrochloride or acetate of chrysaniline.

Aniline Brown is obtained by heating 4 parts of aniline hydrochloride to 240° C. (464° F.) with 1 part of aniline violet or aniline blue.

Aniline Black.—Blacks of great intensity are produced on calico by printing with a mixture of aniline, potassium chlorate, and a metallic compound—the one most generally used being cupric sulphide. The composition of aniline black is not known, neither is its mode of formation well understood; but it appears to depend upon oxidation of the aniline by the chlorate and the cupric sulphide, these compounds being thereby reduced, and afterwards reoxidized by the oxygen of the air, so that they act as carriers of oxygen. The finest black is obtained with vanadium salts, which easily undergo oxidation and reduction.

Aniline also forms coloring matters with xylidine. *Xylidine-rosaniline,* produced according to the equation $C_6H_7N + 2C_8H_{11}N = C_{22}H_{23}N_3 + 3H_2$, forms salts of a fine crimson color.

AZO- AND DIAZO-COMPOUNDS.

These compounds are derived from the aromatic hydrocarbons by substitution of 2 atoms of nitrogen for 2 atoms of hydrogen, the nitrogen apparently acting as a univalent radicle. In the azo-compounds the two nitrogen-atoms are united to one another by a part of their combining capacities, and each is directly combined with the carbon of a benzene-residue; thus:

$$\begin{matrix} N\text{---}C_6H_5 \\ || \\ N\text{---}C_6H_5 \end{matrix} \qquad O<\begin{matrix} N\text{---}C_6H_5 \\ N\text{---}C_6H_5 \end{matrix} \qquad \begin{matrix} HN\text{---}C_6H_5 \\ | \\ HN\text{---}C_6H_5 \end{matrix}$$

Azobenzene.   Azoxybenzene.   Hydrazobenzene.

In the diazo-compounds, only one of the nitrogen-atoms is directly linked to the carbon of a benzene-residue, while the free combining unit of the other is satisfied either by a halogen-element or by an acid residue, or generally in some way different from the first ; thus :

$$\begin{matrix} N\text{---}C_6H_5 \\ || \\ N\text{---}Br \end{matrix} \qquad \begin{matrix} N\text{---}C_6H_5 \\ || \\ N\text{---}NO_3 \end{matrix} \qquad \begin{matrix} N\text{---}C_6H_5 \\ || \\ N\text{---}NH(C_6H_5). \end{matrix}$$

Diazobenzene   Diazobenzene   Diazobenzene-
bromide.   nitrate.   amidobenzene.

The azo-compounds are formed: 1. By the action of certain reducing agents on the nitro-derivatives of the aromatic hydrocarbons ; 2. By oxidation of aniline. They may, indeed, be regarded as stepping-stones from the nitro- to the amido-derivatives, as shown by the following formulæ :—

	Equivalent formulæ.	Molecular formulæ.
Nitrobenzene	$C_6H_5NO_2$	$C_6H_5NO_2$
Azoxybenzene	$C_6H_5NO_{\frac{1}{2}}$	$C_{12}H_{10}N_2O$
Azobenzene	C_6H_5N	$C_{12}H_{10}N_2$
Hydrazobenzene	C_6H_6N	$C_{12}H_{12}N_2$
Amidobenzene	C_6H_7N	$C_6H_7N.$

The hydrogen of the benzene-nucleus in these bodies may be partly replaced by the halogen-elements and the groups NO_2, NH_2, SO_3H, etc.

Azoxybenzene, $C_{12}H_{10}N_2O$, or *Azoxybenzide*, is formed by the action of potash or soda, or of sodium-amalgam on nitrobenzene in alcoholic solution. Nitrobenzene (1 vol.) is dissolved in strong alcohol (8–10 vol.) and a quantity of dry potassium hydroxide, equal in weight to the nitrobenzene, is added ; the liquid, which becomes heated spontaneously, is boiled for some time ; the alcohol is then distilled off till the residual liquid separates into two layers ; and the upper brown layer, which contains the azoxybenzene, is washed with water, whereupon it solidifies to a crystalline mass, which is recrystallized from alcohol.

Azoxybenzene forms long, yellow needles, insoluble in water, easily soluble in alcohol and ether. It melts at 36° C. (96.8° F.), and is decomposed by distillation into aniline and azobenzene. By reducing agents it is converted into azobenzene and hydrazobenzene. It yields two mononitro-derivatives, one melting at 143° C. (289.4° F.), the other at 49° C. (120.2° F.).

Azobenzene, $C_{12}H_{10}N_2$, *Azobenzide*, is formed by the action of iron and acetic acid, or better, of sodium-amalgam, on nitrobenzene. On heating the products, the azobenzene distils over as a yellow oil, which solidifies in the receiver, while azoxybenzene remains behind.

Azobenzene crystallizes in large yellowish-red laminæ, sparingly soluble in water, easily in alcohol and ether. It melts at 66.5° C. (151.7° F.) and distils at 293° C. (559.4° F.). By the action of ammonium sulphide and other reducing agents, it is converted into **hydrazobenzene** and the isomeric compound **benzidine**. With bromine it forms the compound $C_{12}H_{10}Br_2N_2$, or $C_6H_5\text{---}BrN\text{---}NBr\text{---}C_6H_5$, and with nitric acid a mono- and a dinitro-derivative.

Amidazobenzene, or **Amidodiphenylimide**, $C_{12}H_9(NH_2)N_2 =$ C_6H_5—N⚌N—$C_6H_4(NH_2)$.—This compound, which forms the chief constituent of commercial aniline yellow, is produced by the action of ammonium sulphide on nitro-azobenzene, and by a molecular transformation of the isomeric compound diazo-amidobenzene, C_6H_5—N⚌N—$NH(C_6H_5)$, which takes place when the latter is left in contact with alcohol and aniline hydrochloride; also by the action of sodium stannate and other oxidizing agents on aniline.

Amidazobenzene crystallizes from hot alcohol in yellow rhombic needles or prisms, sparingly soluble in hot water, melting at 127.4° C. (261.3° F.), and sublimable. It forms crystalline salts containing 1 equivalent of acid, yellow or violet in the solid state, decomposed by water. By distillation with sulphuric acid and manganese dioxide, it is oxidized to quinone, $C_6H_4O_2$. By the action of tin and hydrochloric acid, it is resolved into aniline and diamidobenzene.

Diamid-azobenzene, or **Diphenine**, $C_{12}H_8(NH_2)_2N_2$, formed by reduction of dinitro-azobenzene with ammonium sulphide, is a yellow crystalline base, dissolving with red color in acids, resolved by reducing agents into two molecules of diamidobenzene:

$$C_{12}H_8(NH_2)_2N_2 \ + \ 2H_2 \ = \ 2C_6H_4(NH_2)_2$$

Hydrazobenzene, $C_{12}H_{12}N_2$, formed by the action of ammonium sulphide or sodium-amalgam on azobenzene or azoxybenzene, crystallizes in large plates, having a peculiar camphorous odor. It dissolves easily in alcohol and ether, but is insoluble in water; melts at 131° C. (267.8° F.). In contact with dilute mineral acids, it is easily converted into the isomeric compound, benzidine or diamidodiphenyl, $NH_2.C_6H_4$—$C_6H_4.NH_2$ (see DIPHENYL COMPOUNDS). By the action of oxidizing agents, or by contact of its alcoholic solution with the air, it is converted into azobenzene. By distillation it is resolved into aniline and azobenzene.

Diamidhydrazobenzene, $C_{12}H_{10}(NH_2)_2N_2$, produced by the action of sodium-amalgam on metanitraniline, forms yellow needles melting at 140° C. (284° F.).

Azotoluene, $C_{14}H_{14}N_2$, formed by treating an alcoholic solution of nitrotoluene with sodium-amalgam, with occasional addition of acetic acid, crystallizes in orange-red shining needles, which melt at 137° C. (278.6° F.), and sublime without decomposition. It is insoluble in water, diinte-acids and alkalies, but dissolves easily in alcohol and ether. Treated in alcoholic solution with sodium-amalgam or ammonium sulphide, it is converted into **hydrazotoluene**, $C_{14}H_{16}N_2$. With bromine it yields a crystalline addition-product, $C_{14}H_{14}Br_2N_2$.

Diazo Compounds.

These compounds, the structure of which has been already explained, are formed by the action of nitrous acid on the salts of the amide-derivatives of the aromatic hydrocarbons; thus:

$$C_6H_5NH_2.NO_3H \ + \ NO.OH \ = \ C_6H_5.N_2.NO_3 \ + \ 2H_2O \cdot$$

Aniline nitrate.   Diazobenzene nitrate.

They are, however, somewhat unstable, and are apt to be decomposed by the water resulting from the reaction, especially if the liquid is hot, with evolution of nitrogen, and formation of hydroxyl-derivatives, i. e., phenols:

$$C_6H_5.N_2.NO_3 \ + \ H_2O \ = \ C_6H_5.OH \ + \ N_2 \ + \ NO_3H,$$

Diazobenzene nitrate. Phenol.

the final result being the same as if the nitrous acid acted on the aromatic amido-compound in the same way that it acts upon the amines and amides of the fatty series (p. 668):

$$C_6H_5.NH_2 \ + \ NO.OH \ = \ C_6H_5.OH \ + \ H_2O \ + \ N_2.$$

A better mode of preparing the diazo-compound is to add solution of potassium nitrite to a solution of the amide-compound in nitric acid:

$$C_6H_5(NH_2).NO_3H \ + \ NO_2K \ + \ NO_3H$$
$$= \ C_6H_5.N_2.NO_3 \ + \ NO_3K \ + \ 2H_2O.$$

Amido-compounds in which the hydrogen of the NH_2 group is partly or wholly replaced by alcohol-radicles, generally yield the same diazo-compounds as the original amido-derivatives, the alcohol-radicles being separated in the form of alcohols:

$$C_6H_5.NH(C_2H_5).NO_3H \ + \ NO_2H \ = \ C_6H_5.N_2.NO_3 \ + \ C_2H_5OH \ + \ H_2O$$

Ethyl-aniline nitrate. Diazobenzene nitrate.

$$C_6H_5.N(C_2H_5)_2.NO_3H \ + \ NO_2H \ = \ C_6H_5.N_2.NO_3 \ + \ 2C_2H_5OH.$$

Diethyl-aniline nitrate.

Those, on the contrary, in which the hydrogen of the benzene-nucleus is partly replaced by Cl, Br, NO_2, etc., yield substituted diazo-compounds, such as $C_6H_4Cl.N_2.NO_3$, $C_6H_4(NO_2).N_2.NO_3$, etc.

When nitrous acid acts upon an amido-compound in the free state in alcoholic or ethereal solution, the product is a diazo-amido-compound; thus:

$$2(C_6H_5.NH_2) \ + \ NO_2H \ = \ \begin{matrix} N\text{---}C_6H_5 \\ \| \\ N\text{---}NH(C_6H_5) \end{matrix} \ + \ 2H_2O.$$

Aniline. Diazo-amidobenzene.

These compounds are also formed by the action of amines or their salts on the salts of diazo-compounds:

$$C_6H_5.N_2.NO_3 \ + \ 2(C_6H_5.NH_2) \ = \ C_6H_5.N_2.NH(C_6H_5) \ + \ C_6H_7N.NO_3H$$
$$C_6H_5.N_2.OK \ + \ C_6H_5.(NH_2).HCl \ = \ C_6H_5.N_2.NH(C_6H_5) \ + \ KCl \ + \ H_2O.$$

By this reaction, also, mixed diazo-amido-compounds may be obtained; for example, diazotoluene-amidobenzene, $C_7H_7.N_2.NH.C_6H_5$, from toluidine and a salt of diazobenzene.

In like manner, diazo-amido-compounds may be formed with the primary and secondary amines of the fatty series; e. g.,

$$C_6H_5.N_2.NO_3 \ + \ 2(C_2H_5)_2NH \ = \ C_6H_5.N_2.N(C_2H_5)_2 \ + \ (C_2H_5)_2NH.NO_3H$$

Diazobenzene nitrate. Diethylamine. Diazobenzene-diethylamine. Diethylamine nitrate.

The diazo-compounds are mostly colorless crystalline bodies, which quickly turn brown on exposure to the air. They dissolve easily in water, sparingly in alcohol, and are precipitated from the alcoholic solutions by ether. Most of them are very unstable, and decompose with explosion when heated or struck. They suffer decomposition also under the influence of the most various reagents, generally in such a manner that both the nitrogen-atoms are eliminated in the gaseous form, and the diazo-group is replaced by halogen elements, hydrogen, hydroxyl, etc.

1. When boiled with *water*, they yield **phenols**:

$$C_6H_5.N_2.Br + H_2O = C_6H_5.OH + N_2 + HBr.$$

With *hydrogen sulphide* in like manner they yield **mercaptans**.

2. On boiling them with strong *alcohol*, the N_2-group is replaced by hydrogen, producing benzene or a homologous hydrocarbon, while the alcohol is oxidized to aldehyde:

$$C_6H_5.N_2.NO_3 + C_2H_5OH = C_6H_6 + N_2 + NO_3H + C_2H_4O.$$

3. The *platinochlorides*, formed by combination of the diazo-chlorides with $PtCl_4$, are decomposed when heated alone, or better with dry sodium carbonate and common salt, producing **chlorinated hydrocarbons**, which distil over:

$$(C_6H_5.N_2.Cl)_2.PtCl_4 = 2C_6H_5Cl + 2N_2 + 2Cl_2 + Pt.$$

4. The *diazobromides* take up two atoms of bromine, forming **perbromides**, such as $C_6H_5N_2Br_3$ or
$$\begin{array}{c} BrN—C_6H_5 \\ | \\ BrN—Br \end{array}$$
, which are likewise decomposed by dry distillation, or more readily by boiling with strong alcohol, yielding monobromo-derivatives of the hydrocarbons:

$$C_6H_5.N_2.Br_3 = C_6H_5Br + Br_2 + N_2.$$

5. The sulphates and other oxy-salts of the diazo-compounds, boiled with *hydriodic acid*, yield **iodine-derivatives**:

$$C_6H_5.N_2.SO_4H + HI = C_6H_5I + N_2 + SO_4H_2.$$

Diazo-compounds containing chlorous radicles (Cl, Br, NO_2, etc.) in the benzene-nucleus undergo exactly similar decomposition; *e. g.*,

$$C_6H_3Br.N_2.NO_3 + H_2O = C_6H_4.Br.OH + N_2 + NO_3H.$$
Diazo-bromobenzene   Bromophenol.
nitrate.

$$C_6H_3Cl_2.N_2.Br_3 = C_6H_3Cl_2Br + Br_2 + N_2.$$
Diazo-dichlorobenzene   Bromodichlo-
perbromide.   robenzene.

The reactions 1, 3, 4, and 5 afford the means of converting an aromatic amido-derivative (and therefore also a nitro-derivative), through the medium of the diazo-compound, into the corresponding halogen- and hydroxyl-derivatives; and this mode of transformation, known as the "diazo-reaction" serves, as already shown in several instances, to determine the relative positions of the substituted radicles in these derivatives (p. 551).

The diazo-amido-compounds, which may be regarded as formed by the union of diazo- and amido-derivatives, undergo similar transformations. They are first resolved, under the influence of hydrobromic or hydrochloric acid, etc., into their components, a diazo- and an amido-derivative; the latter then separates out either in the free state, or as a salt, and the diazo-compound is transformed in the manner above described ; thus :

$$C_6H_5.N_2.NH(C_6H_5) + 2HBr = C_6H_5Br + N_2 + C_6H_5(NH_2).HBr$$

$$C_6H_5.N_2.NH(C_6H_5) + H_2O = C_6H_5OH + N_2 + C_6H_5.NH_2.$$

Diazobenzene-compounds.—The *nitrate*, $C_6H_5.N_2.NO_3$, is prepared by passing nitrous acid-vapor into a flask containing aniline nitrate moistened with a small quantity of water and cooled with ice, till the whole is dissolved, and the solution no longer yields aniline when mixed with potash. On filtering the liquid, and adding alcohol and ether, the diazobenzene nitrate separates as a crystalline mass, which may be purified by redissolution in a small quantity of cold water, and precipitation by alcohol and ether.

Diazobenzene nitrate forms long colorless needles, extremely soluble in water, slightly soluble in alcohol, insoluble in ether and in benzene. It is tolerably permanent when dry, but gradually turns brown on exposure to moist air. It explodes violently when heated.

The *sulphate*, $C_6H_5.N_2.SO_4H$, is prepared by passing nitrous acid into aniline sulphate dissolved in water containing sulphuric acid, or better, by decomposing the nitrate with sulphuric acid. It forms colorless needles or prisms easily soluble in water, exploding at 100°.

The *bromide*, $C_6H_5.N_2.Br$, separates in white laminæ on gradually adding bromine dissolved in ether to an ethereal solution of diazo-amidobenzene while tribromaniline, which is formed at the same time, remains in solution.

The *perbromide*, $C_6H_5.N_2.Br_3$, is formed by mixing an aqueous solution of the nitrate with a solution of bromine in hydrobromic acid or sodium bromide, and separates as a dark-brown oil, which soon solidifies to a crystalline mass. When recrystallized by solution in cold alcohol, and rapid evaporation in a vacuum, it may be obtained in large yellow laminæ, insoluble in water and ether, moderately soluble in cold alcohol. By prolonged washing with ether, it is converted into the monobromide.

The *chloride*, $C_6H_5.N_2Cl$, is obtained in solution by agitating a solution of the bromide with moist silver chloride.

The *platinochloride*, $(C_6H_5N_2Cl)_2PtCl_4$, is precipitated in yellow prisms on adding a solution of platinic chloride to a solution of the nitrate or sulphate.

Diazobenzene-potassium, $C_6H_5.N_2.OK$, separates on adding caustic potash to a solution of the nitrate, as a yellow liquid, which, when evaporated over the water-bath, crystallizes in white nacreous laminæ, easily soluble in water and in alcohol. The aqueous solution quickly decomposes, and, when mixed with silver nitrate, yields a gray precipitate of $C_6H_5N_2.OAg$, and similar precipitates with mercury, lead, zinc, and other metals. On adding acetic acid to a solution of the potassium-compound, a heavy oil is precipitated, probably consisting of d i a z o b e n z e n e h y d r o x i d e, $C_6H_5.N_2.OH.$

Diazo-amidobenzene, $C_6H_5.N_2.NH.C_6H_5$, is formed by the action of nitrous acid on aniline dissolved in alcohol ; also by mixing a solution of diazobenzene nitrate with aniline ; and by gradually pouring a cooled, slightly alkaline solution of sodium nitrite on aniline hydrochloride. It crystallizes in golden-yellow shining laminæ, melts at 91° C. (195.8° F.), and detonates at a higher temperature. It is insoluble in water, easily soluble in ether, benzene, and hot alcohol. By nitric acid containing nitrous acid, it is converted into diazobenzene nitrate ; by strong hydrochloric acid, into aniline hydrochloride, phenol, and nitrogen.—The alcoholic solution mixed with silver nitrate deposits the compound $C_6H_5.N_2.$ $NAg.C_6H_5$, in reddish needles.

Diazo-amidobenzene does not unite with acids, but its alcoholic solution mixed with a solution of platinic chloride in hydrochloric acid, yields the platinochloride, $(C_{12}H_{11}N_3.HCl)_2.PtCl_4$, in reddish needles.

Diazo-amidobenzene in alcoholic solution, especially if in contact with a small quantity of aniline hydrochloride, gradually changes into the isomeric compound, amidazobenzene, $C_6H_5.N=N.C_6H_4.NH_2$ (p. 880).

Diazobenzene-dimethylamide, $C_6H_5.N_2.N(CH_3)_2$, formed by mixing the aqueous solutions of diazobenzene nitrate and dimethylamine, is a yellow, oily liquid, having weak basic properties, and forming unstable salts, which in aqueous solution are easily resolved into phenol, nitrogen, and salts of dimethylamine.

Diazobenzene-ethylamide, $C_6H_5.N_2.NH(C_2H_5)$, resembles the dimethyl compound, and is formed in a similar manner.

Diazobenzenimide, $C_6H_5N_3$, probably C_6H_5—N——N is formed by $\searrow N \nearrow$,

the action of aqueous ammonia on diazobenzene perbromide:

$$C_6H_5N_2Br_3 \;+\; 4NH_3 \;=\; 3NH_4Br \;+\; C_6H_5N_3 \,.$$

It is a yellow oil, insoluble in water, distilling without decomposition in rarefied air and in vapor of water, dissolving without alteration in nitric and sulphuric acid. Treated in alcoholic solution with zinc and hydrochloric acid, it is resolved into ammonia and aniline:

$$C_6H_5N_3 \;+\; 4H_2 \;=\; 2NH_3 \;+\; C_6H_5.NH_2 \,.$$

The diazo-compounds of higher orders, *e. g.*, *diazotoluene nitrate*, $C_7H_7.$ $N_2.NO_3$, *diazo-amidotoluene*, $C_7H_7.N_2.NH.C_7H_7$, etc., are analogous to the diazo-benzene compounds; but they have not been much examined.

Phenyl-hydrazin, $C_6H_8N_2$ = C_6H_5—NH—NH$_2$.—When diazobenzene nitrate is added to a cold solution of acid potassium sulphite, the liquid solidifies to yellow crystals of potassium diazobenzene sulphonate, $C_6H_5N_2SO_3K$, or C_6H_5—N$=$N—SO$_3K$; this salt, heated on the water-bath with excess of acid potassium sulphite, is converted into colorless phenyl hydrazinsulphonate, C_6H_5—NH—NH—SO$_3K$, which is also formed by heating the former salt with zinc-dust and hydrochloric acid; and this colorless salt heated with hydrochloric acid yields the hydrochloride of phenyl-hydrazin, according to the equation:

$$C_6H_5.N_2H_2.SO_3K \;+\; HCl \;+\; H_2O \;=\; C_6H_5.N_2H_3.HCl \;+\; SO_4KH \,.$$

It is more easily obtained (together with aniline or diethylamine), by treating an alcoholic solution of diazo-amidobenzene or diazobenzene-diethylamine with zinc-dust and acetic acid.

Phenyl-hydrazin is a yellow oil, which solidifies at low temperatures to a crystalline mass, melting at 23° C. (73.4° F.), and boiling at 233°–234° C. (451.4°–453.2° F.). It dissolves sparingly in cold water, more readily in hot water, very easily in alcohol and ether. It possesses strong reducing properties, and is a mono-acid base, forming well-crystallized salts. Its hydrochloride, treated with potassium nitrite, yields the nitroso-compound, $C_6H_5.N(NO).NH_2$, which, when heated with water, is converted into diazobenzenimide, $C_6H_5N_3$. The nitroso-compound, treated with phenol and strong sulphuric acid, yields a brown solution changing to green and blue (Liebermann's reaction, p. 891).

SULPHO-DERIVATIVES.

The s u l p h o - a c i d s of the aromatic hydrocarbons are easily formed by the direct action of sulphuric acid, concentrated or fuming, on these hydrocarbons ; thus :

$$C_6H_6 \;+\; SO_4H_2 \;=\; C_6H_5.SO_3H \;+\; H_2O \;;$$
$$\text{Benzenesulphonic}$$
$$\text{acid.}$$

$$C_6H_6 \;+\; 2SO_4H_2 \;=\; C_6H_4(SO_3H)_2 + 2H_2O \;.$$
$$\text{Benzene-disulphonic}$$
$$\text{acid.}$$

In this respect they differ from the sulpho-acids of the fatty group, which can be formed only from sulphites or from thio-alcohols (p. 582).

The aromatic hydrocarbons treated with sulphuric anhydride yield s u l p h o x i d e s :

$$2C_6H_6 \;+\; SO_3 \;=\; (C_6H_5)_2SO_2 \;+\; H_2O \;.$$
$$\text{Benzene}$$
$$\text{sulphoxide.}$$

The sulpho-acids treated with phosphorus pentachloride, or their salts treated with the oxychloride, yield the s u l p h o c h l o r i d e s or c h l o r - a n h y d r i d e s of the sulphonic acids; $e.\,g.$,

$$C_6H_5.SO_2.OH \;+\; PCl_5 \;=\; C_6H_5.SO_2.Cl \;+\; POCl_3 \;+\; HCl \;;$$
$$3(C_6H_5.SO_2.OK) \;+\; POCl_3 \;=\; 3(C_6H_5.SO_2.Cl) \;+\; PO_4K_3 \;;$$

and the sulphochlorides treated with sodium-amalgam in ethereal solution, are converted into s u l p h i n i c a c i d s (p. 583) :

$$C_6H_5.SO_2.Cl \;+\; H_2 \;=\; C_6H_5.SO_2H \;+\; HCl \;.$$

The sulphinic acids, or rather their zinc-salts, are also formed by the action of the zinc-compounds of the alcohol-radicles on the sulpho-chlorides; $e.\,g.$,

$$2(C_6H_5.SO_2.Cl) \;+\; Zn(C_2H_5)_2 \;=\; 2C_2H_5Cl \;+\; (C_6H_5SO_2)_2Zn \;.$$

By zinc and hydrochloric acid, on the other hand, the sulpho-chlorides are reduced to h y d r o s u l p h i d e s or t h i o p h e n o l s :

$$C_6H_5.SO_2Cl \;+\; 3H_2 \;=\; C_6H_5.SH \;+\; 2H_2O \;+\; HCl \;.$$

The aromatic sulpho-acids, like those of the fatty group, are very stable compounds, not decomposed by boiling with alkalies. By fusion with caustic alkalies they yield p h e n o l s :

$$C_6H_5.SO_3K \;+\; KHO \;=\; C_6H_5.OH \;+\; SO_3K_2 \;.$$

By distillation with *potassium cyanide* (or the dry ferrocyanide), they yield n i t r i l s :

$$C_6H_5.SO_3K \;+\; CNK \;=\; C_6H_5.CN \;+\; SO_3K_2 \;.$$

The free acids subjected to dry distillation yield h y d r o c a r b o n s; benzene-sulphoxide, or its homologues, being formed at the same time; thus:

$$C_6H_5.SO_3H \; = \; C_6H_6 \; + \; SO_3 ,$$

and

$$2C_6H_6 \; + \; SO_3 \; = \; (C_6H_5)_2SO_2 \; + \; H_2O .$$

The sulphonic acids of the substituted hydrocarbons are obtained by the action of sulphuric acid on these bodies, or by the action of halogens, or of nitric acid, on the sulpho-acids of the primary hydrocarbons; in the latter case the sulpho-group is also frequently replaced. The sulpho-group may also be replaced by chlorine, by heating a sulpho-acid or sulphochloride with phosphorus pentachloride:

$$C_6H_4Cl.SO_2Cl \; + \; PCl_5 \; = \; C_6H_4Cl_2 \; + \; POCl_3 \; + \; SOCl_2 .$$

Benzenesulphonic Acid, $C_6H_5.SO_3H$, is produced by heating benzene with an equal weight of ordinary sulphuric acid. On diluting the resulting solution with water, neutralizing with barium or lead carbonate, decomposing the resulting barium or lead-salt with sulphuric acid, or hydrogen sulphide, and evaporating the filtrate to the crystallizing point, benzenesulphonic acid is obtained in small deliquescent tabular crystals, containing $C_6H_5.SO_3H + 1\frac{1}{2}H_2O$, easily soluble in water and in alcohol. By dry distillation it yields benzene; by fusion with potash, phenol. Its *barium salt*, $(C_6H_5SO_3)_2Ba + H_2O$, forms nacreous plates, easily soluble in water. The *zinc-salt*, $(C_6H_5SO_3)_2Zn + 6H_2O$, crystallizes in six-sided tables. The *ethylic ether*, $C_6H_5SO_3.C_2H_5$, obtained by heating the lead salt to 100° with ethyl iodide, forms slender needles, decomposed by boiling with water.

Benzene-sulphochloride, $C_6H_5.SO_2Cl$, separates, on gently warming an intimate mixture of sodium benzenesulphonate and phosphorus pentachloride, and then shaking it up with water, as a colorless oil having a specific gravity of 1.378 at 23° C. (73.4° F.), boiling with partial decomposition at 246°–247° C. (474.8°–476.6° F.), and solidifying below 0° in large rhombic crystals. Boiling water slowly decomposes it into benzenesulphonic and hydrochloric acids.

Benzenesulphamide, $C_6H_5.SO_2.NH_2$, formed by treating the sulphochloride with ammonia or ammonium carbonate, crystallizes from alcohol in nacreous laminæ melting at 149° C. (300.2° F.) Silver nitrate added to its alcoholic solution throws down the compound $C_6H_5SO_2.NHAg$.

Benzenesulphinic, or **Benzenesulphurous Acid**, $C_6H_5.SO_2.OH$, formed by the action of sodium-amalgam on the ethereal solution of benzenesulphonic acid, crystallizes from hot water in large shining prisms, easily soluble in alcohol and ether, melting at 69° C. (156.2° F.), and decomposing at 100°. With chlorine or bromine it forms benzene sulphochloride or bromide, and is converted slowly by exposure to the air, quickly by oxidizing agents, into benzenesulphonic acid. The *silver salt*, $C_6H_5.SO_2.Ag$, is slightly soluble in water.

Phenyl Sulphoxide, or **Sulphobenzide**, $(C_6H_5)_2S_2O_2$, is formed by dry distillation of benzenesulphonic acid, by oxidation of phenyl sulphide, $(C_6H_5)_2S$, and by the action of fuming sulphuric acid or sulphuric anhydride on benzene. It is very slightly soluble in water, crystallizes from alcohol in plates, melts at 128°–129° C. (262.4°–264.2° F.), and distils without decomposition. By heating with strong sulphuric acid, it is converted into benzenesulphonic acid, $(C_6H_5)_2SO_2' + SO_4H_2' = 2(C_6H_5.SO_3H)$; and when heated with phosphorus pentachloride or in a stream of chlorine, it is decomposed, with formation of chlorobenzene and benzenesulphochloride :—

$$(C_6H_5)_2SO_2 \ + \ Cl_2 \ = \ C_6H_5Cl \ + \ C_6H_5.SO_2Cl'.$$

The action of chlorine in sunshine also converts it into chlorobenzene and its products of addition.

Phenyldisulphoxide, $(C_6H_5)_2S_2O_2$, produced, together with benzenesulphonic acid, by heating benzenesulphinic acid with water to 130° C. (266° F.), crystallizes in long shining needles, melting at 100°, insoluble in water, easily soluble in alcohol and ether.

Chloro-, Bromo-, Iodo-, Nitro-, and Amido-benzenesulphonic Acids, $C_6H_4Cl.SO_3H$, etc.

—The sulpho-acids formed by dissolving C_6H_5Cl, C_6H_5Br, C_6H_5I, and $C_6H_5NH_2$, in slightly fuming sulphuric acid, belong chiefly to the para-series (1 : 4); e. g., $C_6.SO_3H.H.H.Br.H_2$; nitrobenzene, on the other hand, yields by similar treatment a product consisting also wholly of meta-nitrobenzenesulphonic acid, $C_6.SO_3.H.NO_2.H_3$. The action of bromine or nitric acid on benzenesulphonic acid yields likewise a product consisting chiefly of the meta-acid. All these bodies are strong monobasic acids, which mostly crystallize well, dissolve easily in water, and yield well-crystallized salts, chlorides, and amides.

Para-amidobenzenesulphonic acid, long known by the name of sulphanilic acid, is produced by the action of sulphuric acid on aniline, also by distillation of phenolsulphate or of ethylsulphate of aniline :

$$C_6H_4{<}^{OH}_{SO_3H.NH_2.C_6H_5} \ = \ C_6H_5OH \ + \ C_6H_4{<}^{NH_2}_{SO_3H}$$

$$SO_2{<}^{OC_2H_5}_{O.NH_3C_6H_5} \ = \ C_2H_5OH \ + \ C_6H_4{<}^{NH_2}_{SO_3H}$$

It crystallizes from water in rhombic tables containing one mol. H_2O; yields, by oxidation with chromic acid or with manganese dioxide and sulphuric acid, a large quantity of quinone, $C_6H_4O_2$; and is converted by nitrous acid into diazobenzene-sulphonic acid, $C_6H_4{<}^{SO_2}_{N=N}$, which crystallizes from hot water in colorless needles ; detonates with great violence, and is converted into benzenesulphonic acid by heating with absolute alcohol, and into sulphanilic acid by the action of hydrogen sulphide.

Benzenedisulphonic Acids, $C_6H_4{<}^{SO_3H}_{SO_3H}$.

—The meta- and para-modifications of this acid are obtained by heating benzenesulphonic acid with fuming sulphuric acid, or more readily by passing benzene vapor into ordinary sulphuric acid heated to 240° C. (464° F.). They are both very soluble in water, but may be separated by fractional crystallization of their potassium salts. The meta-acid, which is the principal product, yields a chloride, $C_6H_4(SO_2Cl)_2$, melting at 63° C. (145.4° F.), and an amide, $C_6H_4(SO_2.NH_2)_2$, melting at 229° C. (444.2° F.). By distillation with potassium cyanide, it yields a dicyanide, $C_6H_4(CN)_2$, melting at 156° C. (312.8° F.); and convertible by the action of alkalies into metaphthalic or isophthalic acid, $C_6H_4(CO_2H)_2$.

Para-benzenedisulphonic acid forms a chloride melting at 131° C. (267.8° F.), and an amide melting at 288° C. (550.4° F.). By distillation with potassium cyanide it yields a dicyanide, melting at 222° C. (431.6° F.), and convertible into terephthalic acid, $C_6H_4(CO_2H)_2$ (1 : 4).

Benzenetrisulphonic Acid, $C_6H_3(SO_3H)_3$.—The only known modification of this acid is obtained by heating a mixture of 10 parts benzene, 70 fuming sulphuric acid, and 40 phosphoric anhydride in sealed tubes to 280°–290° C. (536°–554° F.). Separated from its lead salt by hydrogen sulphide, it crystallizes in long flat needles containing 3 mol. H_2O.*

Toluenesulphonic Acids, $C_6H_4{<}^{CH_3}_{SO_3H}$.—The *para-* and *ortho*-modifications are produced simultaneously by dissolving toluene in slightly fuming sulphuric acid, and may be approximately separated by fractional crystallization of their potassium salts, the para-salt separating out first, and crystallizing easily in large transparent six-sided tables or prisms containing 1 mol. H_2O. The *ortho*-salt separates from the mother-liquor mixed with para-salt, from which it is not easily purified. The *para*-acid crystallizes in deliquescent scales containing 1 mol. H_2O. Its *chloride*, $C_7H_7(SO_2Cl)_2$, melts at 69° C. (156.2° F.). The *amide*, $C_7H_7(SO_2NH_2)_2$, at 137° C. (278.6° F.). Fused with potash it yields paracresol, $C_6H_4(CH_3).OH$, and a small quantity of paraoxybenzoic acid.—The *ortho*-acid, which may be obtained pure by decomposing its amide with nitrous acid, is liquid. Its amide crystallizes from hot alcohol in shining octohedrons, melting at 153°–154° C. (307.4°–309.2° F.). The acid fused with potash yields ortho-cresol and ortho-oxybenzoic (salicylic) acid.

Metatoluenesulphonic acid, $C_6.CH_3.H.SO_3H.H_3$, is obtained by the action of sodium-amalgam on the orthochloro- (or bromo-) toluene-sulphonic acid, $C_6.CH_3.Br.SO_3H.H_2$, which is formed by dissolving orthochloro- (or bromo-) toluene in sulphuric acid. It is crystalline. Its *chloride* is a liquid not solidifying at —10° C. (14° F.) ; the *amide* is crystalline, and melts at 91°– 92° C. (195.8°–197.6° F.).

A large number of substituted toluenesulphonic acids have been obtained by dissolving chlorotoluene, bromotoluene, etc., in sulphuric acid. Para- and ortho-toluenesulphonic acid heated with fuming sulphuric acid yield two modifications of t o l u e n e d i s u l p h o n i c a c i d, $C_6H_3(SO_3H)_2$. CH_3.

Benzylsulphonic Acid, $C_6H_5.CH_2(SO_3H)$.—The potassium-salt of this acid is formed by heating benzyl chloride with a strong solution of potassium sulphite.

AROMATIC ALCOHOLS.

The substitution of OH for H in benzene gives rise to mono-, di-, and tri-atomic derivatives of alcoholic character, viz. :

$$C_6H_5(OH) \qquad C_6H_4(OH)_2 \qquad C_6H_3(OH)_3 .$$

In toluene and the higher homologues of benzene the substitution may take place either in the principal or in the lateral chains, giving rise to metameric compounds : thus from toluene, $C_6H_5.CH_3$, are derived

* **Koerner e Monselise**, Gazetta chimica italiana, 1876, p. 133.

$$C_6H_4(OH).CH_3 \qquad\qquad C_6H_5.CH_2OH,$$
Cresol. $\qquad\qquad$ Benzyl alcohol.

and from xylene, $C_6H_4(CH_3)_2$:

$$C_6H_3(OH){<}^{CH_3}_{CH_3} \qquad\qquad C_6H_4{<}^{CH_3}_{CH_2OH}$$
Xylenol. $\qquad\qquad$ Xylyl alcohol.

Those compounds in which the substitution takes place in the lateral chains, are primary alcohols, containing the group CH_2OH, and, like the corresponding alcohols of the fatty series, are convertible by oxidation into aldehydes and acids containing the same number of carbon-atoms, and readily exchange their OH group for Cl, Br, NH_2, etc., giving rise to haloïd derivatives and amines. Those, on the other hand, in which the OH is situated in the principal chain, are not convertible by oxidation into aldehydes, ketones, or acids, in which respect they are analogous to the tertiary alcohols.

These latter compounds, including the hydroxyl-derivatives of benzene, are called P h e n o l s . They are susceptible of isomeric modifications, according to the orientation of the substituted radicles in the benzene-nucleus.

Monatomic Phenols.

These compounds are produced: 1. By the action of nitrous acid on aniline and its homologues in aqueous solution:

$$C_6H_5.NH_2 + NO.OH = C_6H_5.OH + H_2O + N_2.$$

2. By decomposing the diazo-compounds with water, the sulphates being the best adapted for the purpose (p. 882).

3. By fusing the aromatic sulphonic acids with caustic alkalies:

$$C_6H_4{<}^{CH_3}_{SO_3K} + KOH = SO_3K_2 + C_6H_4{<}^{CH_3}_{OH}.$$
Toluene-sulphonate. $\qquad\qquad$ Cresol.

4. By distilling the salts of aromatic oxy-acids with lime.

$$C_6H_4(OH).CO_2H = CO_2 + C_6H_5.OH.$$
Oxybenzoic acid. $\qquad\qquad$ Phenol.

5. By the dry distillation of complex organic substances, such as wood and coal.

The phenols exhibit acid as well as alcoholic characters. When treated with metallic oxides, especially those of the alkali-metals, they readily exchange their hydroxylic hydrogen for metals, forming metallic salts, and these, when acted upon by alcoholic iodides are converted into p h e n o l i c e t h e r s ; e. g.,

$$C_6H_5.OK + CH_3I = KI + C_6H_5.O.CH_3.$$
Potassium $\qquad\qquad$ Methyl-phenol.
phenate.

Phenolic ethers containing acid-radicles, such as $C_6H_5.O.OC_2H_3O$, are formed by the action of the chlorides of such radicles on the phenols and their metallic derivatives.

By the action of the halogen-compounds of phosphorus, the phenols are converted, by exchange of their OH-group for Cl, Br, etc., into halogen-

38

derivatives of hydrocarbons, such as C_6H_5Cl. By phosphorus penta sulphide, they are converted into t h i o p h e n o l s:

$$5C_6H_5(OH) \; + \; P_2S_5 \; = \; P_2O_5 \; + \; 5C_6H_5(SH) \,.$$

By heating with zinc-dust they are reduced to hydrocarbons.

By treatment with chlorine, bromine, iodine, nitric acid, and sulphuric acid, the phenols are converted into halogen-, nitro-, and sulpho-derivatives, by exchange of one or more atoms of hydrogen in the benzene-nucleus for Cl, Br, I, NO_2, or SO_3H; thus:

$$C_6H_5.OH \; + \; Cl_2 \; = \; HCl \; + \; \underset{\text{Chlorophenol.}}{C_6H_4Cl.OH} \,.$$

$$C_6H_5.OH \; + \; SO_4H_2 \; = \; H_2O \; + \; \underset{\substack{\text{Phenolsulphonic} \\ \text{acid.}}}{C_6H_4{<}^{OH}_{SO_3H}} \,.$$

Phenols treated with metallic sodium and carbon dioxide are converted into salts of aromatic oxy-acids:

$$\underset{\text{Phenol.}}{C_6H_5.OH} \; + \; CO_2 \; = \; \underset{\text{Oxybenzoic acid.}}{C_6H_4(OH).CO_2H} \,.$$

The monatomic phenols at present known contain 6, 7, 8, and 10 atoms of carbon.

Six-carbon Phenol, or simply **Phenol**, $C_6H_6O = C_6H_5OH$. —*Phenyl alcohol, Phenic acid, Carbolic acid, Coal-tar creasote.*—This compound is produced: 1. By the action of nitrous acid on aniline.

2. By the dry distillation of salicylic (ortho-oxybenzoic) acid. It may be conveniently prepared by heating crystallized salicylic acid strongly and quickly in a glass retort, either alone or mixed with pounded glass or quicklime. Phenol then passes over into the receiver, and crystallizes almost to the last drop.

3. Phenol is produced in the dry distillation of coal, and forms the chief constituent of the acid portion of coal-tar oil; this is the source from which it is most frequently obtained. Crude coal-tar oil is agitated with a mixture of slaked lime and water, the whole being left for a considerable time; the aqueous liquid separated from the undissolved oil is decomposed by hydrochloric acid, and the oily product thus obtained is purified by cautious distillation, the first third only being collected. Or the coal-tar oil is subjected to distillation in a retort furnished with a thermometer, and the portion which passes over between the temperature of 150° and 200° C. (302°–392° F.) is collected apart. This product is then mixed with a hot, strong solution of caustic potash, and left to stand, whereby a whitish, somewhat crystalline, pasty mass is obtained, which, by the action of water, is resolved into a light oily liquid, and a dense alkaline solution. The latter is withdrawn by a siphon, decomposed by hydrochloric acid, and the separated oil purified by contact with calcium chloride, and redistillation. Lastly, it is exposed to a low temperature, and the crystals formed are drained from the mother-liquid, and carefully preserved from the air.

Pure phenol crystallizes in long, colorless, prismatic needles, having a specific gravity of 1.066, melting at 40°–41° C. (104°–105.8° F.), and boiling at 181.5° C. (258.7° F.): the commercial product forms a crystalline mass, which turns reddish in a short time, and in contact with moist air deliquesces to a brown liquid. Phenol has a penetrating odor, a strong burning taste, and attacks the skin of the lips. It is poisonous, exerts an antiseptic action, and has been successfully used by Mr. Crookes for destroying the infection of cattle-plague. It dissolves in about 15 parts of water at ordinary temperatures, very easily in alcohol and ether. Sulphur and iodine dissolve in it; nitric acid, chlorine, and bromine attack it with energy, forming substitution-products, all of which are of acid character. With strong sulphuric acid, it forms phenolsulphonic acid, $C_6H_4(OH).SO_3H$. The aqueous solution is colored violet by ferric chloride, and stains a deal shaving of a fine blue color. With bromine-water, even when very dilute, it forms a white precipitate of tribromophenol.

On adding phenol to nitric acid containing nitrous acid, or to a solution of potassium nitrite (6 per cent.) in strong sulphuric acid, a brown color is produced, changing to green, and ultimately to a fine blue (Liebermann's reaction). Fine colors are produced in like manner with other phenols, mono- and poly-atomic: also by phenols in presence of sulphuric acid with diazo- and nitroso-compounds.

PHENATES.—Phenol dissolves in alkalies, forming salts which are difficult to obtain in definite form. *Potassium phenate*, C_6H_5KO, obtained by heating phenol with potassium, or with solid potassium hydroxide, crystallizes in slender white needles. On heating this potassium-compound with iodide of methyl, ethyl, or amyl, ethers are produced—viz., methyl phenate or anisol, $C_6H_5OCH_3$; ethyl phenate or phenetol, $C_6H_5OC_2H_5$, and amyl phenate or phenamylol, $C_6H_5OC_5H_{11}$. These bodies resemble the mixed ethers of the ordinary alcohols (p. 580) in composition and mode of formation, but differ greatly from them in their behavior with sulphuric and nitric acids, with which, in fact, they behave just like phenol itself, forming substitution-products possessing acid properties.

Methyl Phenate or Anisol, $C_7H_8O = C_6H_5.O.CH_3$, is also produced, with evolution of carbon dioxide, by distilling methyl-salicylic acid or anisic (methyl-paraoxybenzoic) acid with baryta:

$$C_6H_4(OCH_3).CO_2H = CO_2 + C_6H_5.O.CH_3.$$

Anisol is a colorless, very mobile liquid, having a pleasant aromatic odor, a density of 0.991 at 15° C. (59° F.), and boiling without decomposition at 152° C. (305.6^O F.). It dissolves completely in strong sulphuric acid, forming methyl-phenol-sulphonic or sulphanisolic acid, $C_6H_3(CH_3) <^{OH}_{SO.3H}$.—With *bromine* it forms three substitution-products —viz., $C_6H_4Br.O.CH_3$, a liquid boiling at 223° C. (433.4^O F.); $C_6H_3Br_2.O.CH_3$, which crystallizes in rhombic tables, melts at 59° C. (138.2^O F.), and boils at 272° C. (521.6^O F.); and $C_6H_2Br_3.O.CH_3$, which melts at 87° C. (188.6^O F.), and is sublimable. By the further action of bromine, tetrabromoquinone or bromanil, $C_6Br_4O_2$, is produced. Fuming nitric acid acts strongly on anisol, forming the three nitranisols, $C_7H_7(NO_2)O_y$ $C_7H_6(NO_2)_2O$, and $C_7H_5(NO_2)_3O$, which when treated with reducing agents yield the corresponding nitranisidines, $C_7H_5(NH_2)O$, etc.

Ethyl Phenate or Phenetol, $C_6H_5.O.C_2H_5$, obtained from phenol, and from ethylsalicylic acid, is an aromatic liquid, boiling at 172° C. (341.6^O F.).—Amyl Phenate, or Phenamylol boils at 225° C. (437° F.).

Ethene Phenate, $C_2H_4(OC_6H_5)_2$, produced by the action of ethene bromide on potassium phenate, crystallizes in laminæ melting at 95° C. (203° F.).

Phenyl Phenate, Phenyl Oxide, or Phenolic Ether, $C_{12}H_{10}O = (C_6H_5)_2O$, formed, together with phenyl benzoate, by the dry distillation of cupric benzoate, and by heating diazobenzene sulphate with phenol, crytallizes in long needles, melts at 28° C. (82.4^O F.), and boils at 246° C. (474.8^O F.); dissolves easily in alcohol and ether.

Thiophenol or Phenyl Hydrosulphide, $C_6H_5.SH$, formed by the action of P_2S_5 on phenol, and by that of nascent hydrogen (zinc and sulphuric acid) on phenyl-sulphochloride, is a colorless, mobile, fetid liquid, having a density of 1.078 at 14° C. (57.2^O F.), and boiling at 168° C. (334.4^O F.). It is insoluble in water, but dissolves easily in alcohol and ether, and the alcoholic solution is precipitated by silver, lead, and mercury salts, yielding compounds analogous to the mercaptides. The mercury-compound, $(C_6H_5S)_2Hg$, crystallizes from alcohol in shining needles.

Phenyl Sulphide, $(C_6H_5)_2S$, produced by the dry distillation of sodium benzene-sulphonate, and, together with the hydrosulphide, by the action of P_2S_5 on phenol, is a colorless liquid, having an alliaceous odor, a density of 1.12, and boiling at 292° C. (557.6^O F.). Nitric acid oxidizes it to phenyl sulphoxide (sulphobenzide) $(C_6H_5)_2SO_2$.

Phenyl Disulphide, $(C_6H_5)_2S_2$, is formed by oxidizing thiophenol with dilute nitric acid, and by the action of iodine on the aqueous solution of potassium-thiophenol:

$$2(C_6H_5.SK) + I_2 = 2KI + (C_6H_5)_2S_2.$$

It crystallizes from alcohol in shining needles melting at 60° C. (140° F.); is oxidized by nitric acid to benzenesulphonic acid, and reduced by nascent hydrogen to thiophenol.

Halogen Derivatives of Phenol.

Chlorophenols.—The three monochlorophenols, $C_6H_4Cl.OH$, *ortho-*, *meta-*, and *para-*, are formed from the corresponding chloranilines (p. 863), by the diazo-reaction. The ortho- and para-compounds are also produced by passing chlorine into warm phenol, and may be separated by fractional distillation. (1 : 2) *Chlorophenol* is a colorless liquid, boiling at 175°–176° C. (347°–348.8^O F.), solidifying at —12° C. (10.4^O F.); converted by fusion with potash into pyrocatechin. (1 : 3) *Chlorophenol*, from (1 : 3) chloraniline, is a liquid boiling at 214° C. (417.2^O F.). (1 : 4) *Chlorophenol* crystallizes in colorless prisms, which gradually turn red, melts at 37° C. (98.6^O F.), boils at 217° C. (422.6^O F.); is converted by fusion with potash into hydroquinone. All the three chlorophenols have a very pungent and persistent odor.

Dichlorophenol, $C_6H_3Cl_2.OH$, probably $C_6.OH.Cl.H.Cl.H_2$, obtained by dry distillation of dichlorosalicylic acid, melts at 43° C. (109.4^O F.), and boils at 209° C. (408.2^O F.).—*Trichlorophenol*, $C_6H_2Cl_3.OH$ (1 : 3 : 5, OH), the chief product of the action of chlorine on phenol, melts at 68° C. (154.4^O F.), and boils at 244° C. (471.2^O F.).—*Pentachlorophenol*, $C_6Cl_5.OH$, obtained by the action of chlorine on phenol, in presence of antimonic chloride, melts at 187° C. (368.6^O F.).

Bromophenols.—The three monobromophenols, $C_6H_4Br.OH$, are obtained from the three corresponding bromanilines; the 1 : 2 and 1 : 4 modifications also, by passing bromine vapor into phenol, and by the action of bromine on a solution of phenol in glacial acetic acid. The *ortho-* and *meta-*

modifications are liquid.—*Parabromophenol* forms large crystals; melts at 64° C. (147.2° F.); boils at 236° C. (456.8° F.); and is converted, by the action of phosphorus pentabromide, into paradibromobenzene.

Dibromophenol, $C_6H_3Br_2.OH$ (probably 1 : 2 : 4, OH in 1), obtained by bromination of phenol, melts at 40° C. (104° F.).

Tribromophenol, $C_6H_2Br_3.OH$ (1 : 3 : 5, OH), precipitated by bromine-water from the aqueous solution of phenol, crystallizes from alcohol in silky needles, melting at 95° C. (203° F.). By PBr_5 it is converted into tetrabromobenzene, melting at 98° C. (208.4° F.) (p. 852); by nitric acid into picric acid.

Pentabromophenol, $C_6Br_5.OH$, melts at 225° C. (437° F.).

Iodophenols.—The three monoiodophenols, $C_6H_4I.OH$, are formed, together with tri-iodophenol, by treating phenol with iodine and iodic acid in presence of free alkali:

$$5(C_6H_5.OH) \;+\; 2I_2 \;+\; IO_3H \;=\; 3H_2O \;+\; 5(C_6H_4I.OH).$$

On supersaturating the liquid with hydrochloric acid, and distilling the precipitated oil with steam, a liquid monoiodophenol (1 : 2) passes over first, then a solid (1 : 3), and finally, tri-iodophenol; and the residue contains a considerable quantity of the latter, which may be extracted by alcohol. The third monoiodophenol (1 : 4), which is easily soluble in water, is found partly in the aqueous solution from which the crude iodophenol has been precipitated, partly in the aqueous distillate, and partly in the alcoholic solution from which the tri-iodophenol has crystallized.

(1 : 2) Iodophenol, produced also from (1 : 2) amidophenol, and by distillation of iodosalicylic acid, is a liquid which does not solidify at —23° C. (—9.4° F.). It is easily decomposed, with separation of iodine, by chlorine and by nitric acid. By fusion with potash it yields pyrocate-chin. (1 : 3) Iodophenol, produced also from metaiodaniline, is somewhat sparingly soluble in water, crystallizes from alcohol in six-sided tables melting at 89° C. (192.2° F.); is not decomposed either by chlorine or by nitric acid. By fusion with potash it yields resorcin. (1 : 4) Iodophenol, obtained also from para-iodaniline, is very soluble in water, crystallizes in shining needles melting at 64°–66° C. (147.2°–150.8° F.); is decomposed by nitric acid, but not by chlorine; converted by fusion with potash at 100° into hydroquinone.

Nitrosophenol, $C_6H_4(NO).OH$ (1 : 4).—This compound is formed : 1. By the action of nitrous acid on phenol :

$$C_6H_5.OH \;+\; NO.OH \;=\; H_2O \;+\; C_6H_4(NO).OH;$$

2. As a sodium salt, by heating nitrosodimethylaniline (p. 838) with dilute soda-lye :

$$C_6H_4(NO).N(CH_3)_2 \;+\; NaOH \;=\; NH(CH_3)_2 \;+\; C_6H_4(NO).ONa.$$

Pure hydrochloride of nitrosodimethylaniline is added to a boiling dilute solution of caustic soda; the dimethylaniline is distilled off; and the residue, acidified with dilute sulphuric acid, is shaken up with ether.

Nitrosophenol crystallizes from hot water in slender colorless needles, which soon turn brown; from ether in large brown laminæ. It dissolves with light green color in water, alcohol, and ether; easily also in dilute soda-lye, forming a sodium-salt, which, on addition of strong caustic soda, separates in red needles containing $C_6H_4(NO).ONa + 2H_2O$. Salts of the heavy metals added to this solution throw down amorphous dark-colored precipitates.

Nitrosophenol melts when heated, and decomposes with a slight explosion at 120°–130° C. (248°–266° F.). By strong nitric acid it is converted into paranitrophenol; by tin and hydrochloric acid into paramidophenol · by nitrous acid into diazobenzene nitrate :

$$C_6H_4(NO).OH + 3NO_3H = C_6H_4{<}{\genfrac{}{}{0pt}{}{OH}{N_2.NO_3}} + NO_3H + H_2O.$$

With aniline acetate it yields diazoxybenzene, $C_6H_4(OH).N_2.C_6H_5$.

On adding strong sulphuric acid to nitrosophenol dissolved in phenol, the solution assumes a dark-red color, changing, on addition of potash, to a fine blue.

Nitrophenols.

These compounds, which are all of acid character, are easily formed by direct nitration of phenol.

Mononitrophenols, $C_6H_4(NO_2).OH$, (1 : 2) and (1 : 4), are obtained by gradually adding 1 part of phenol to 2 parts of nitric acid (sp. gr. 1.34), dissolved in 4 parts of water. They may be separated by distillation with water, as only the ortho-compound volatilizes with the steam. The same compounds are produced by heating (1 : 2) and (1 : 4) nitrobromobenzene in sealed tubes with alcoholic ammonia. The (1 : 4) and (1 : 3) compounds may be prepared by boiling the diazonitrobenzene nitrates from (1 : 4) and (1 : 3) nitraniline with water; the (1 : 4) compound also by boiling paranitraniline with very strong solution of caustic soda.

(1 : 2), commonly called *volatile nitrophenol*, crystallizes in large yellow prisms; dissolves sparingly in water, easily in alcohol, and volatilizes easily with vapor of water; melts at 45° C. (113° F.), and boils at 214° C. (417.2° F.); converted by PCl_5 into (1 : 2) chloronitrobenzene. Its sodium salt, $C_6H_4(NO_2).ONa$, crystallizes in dark red anhydrous prisms, and the methylic ether, $C_6H_4(NO_2).OCH_3$, formed from it by the action of methyl iodide, melts at 9° C. (48.2° F.), and boils at 265° C. (509° F.).

(1 : 3) Nitrophenol forms colorless needles, moderately soluble in water, melting at 96° C. (204.8° F.).

(1 : 4) Nitrophenol forms long, colorless needles, melting at 115° C. (239° F.), converted by PCl_5 into (1 : 4) chloronitrobenzene. Its methylic ether melts at 48° C. (118.4° F.), boils at 260° C. (500° F.), and is converted by heating with ammonia into (1 : 4) nitraniline.

Dinitrophenol, $C_6H_3(NO_2)_2.OH$ (1 : 2 : 4—OH in 1), is formed by the action of strong nitric acid on phenol and on o- and p-nitrophenol; also by boiling the corresponding dinitrochloro- or dinitro-bromobenzene (p. 859) with alkalies. Colorless plates, melting at 114° C. (237.2° F.).

The *consecutive* modification (1 : 2 : 6—OH in 1), formed, together with the preceding from (1 : 2) nitrophenol, crystallizes in needles melting at 63°–64° C. (145.4°–147.2° F.).

Both these compounds are converted by further nitration into picric acid.

Two other dinitrophenols, formed from (1 : 3) nitrophenol, the one melting at 104° C. (219.2° F.), the other at 141° C. (285.8° F.), are converted by further nitration into isopicric acid.

Trinitrophenols, $C_6H_3(NO_2)_3.OH$.—Two of these compounds are known. (1) *Picric acid*, 1 : 2 : 4 : 6 (OH in 1), also called *Carbazotic acid*, is formed by nitration of phenol and of 1 : 2 : 4 and 1 : 2 : 6 dinitrophenol. It is also one of the ultimate products of the action of nitric acid upon indigo and numerous other substances, as silk, wool, several resins, especially that of *Xanthorrhœa hastilis* (yellow gum of Botany Bay), salicin and some of its derivatives, coumarin, etc. It is most economically prepared from phenol. One part of that substance is gradually added to strong nitric acid slightly warmed, and when the first violent reaction has subsided, three parts of fuming nitric acid are added, and the liquid is boiled till nitrous fumes are no longer evolved. The resinous mass thus produced is boiled with water; the resulting picric acid is converted into sodium-salt; and the solution mixed with sodium carbonate, which throws down the sodium picrate in crystals.

Picric acid crystallizes in yellow shining prisms or laminæ, having an intensely bitter taste. It melts at 122.5° C. (252.5° F.), sublimes when cautiously heated. It dissolves sparingly in cold water, more easily in hot water, still more in alcohol. It stains the skin deep yellow, and is used as a yellow dye for wool and silk. It is a strong acid, forming well-crystallized yellow salts, which detonate violently when heated, some of them also by percussion. The *potassium salt*, $C_6H_2(NO_2)_3.OK$, crystallizes in long needles very slightly soluble in water. The *sodium, ammonium,* and *barium salts* are easily soluble in water.

Methyl picrate, $C_6H_2(NO_2)_3.OCH_3$, formed by nitration of anisol (p. 891), crystallizes in tables, which melt at 60° C. (140° F.) and sublime. *Ethyl picrate* forms colorless needles, which turn brown in the air, and melt at 78.5 $^{\circ}$ C. (173.3° F.).

Picrates of Hydrocarbons.—Picric acid affords characteristic reactions for the detection of certain hydrocarbons. For this purpose, it is convenient to use an alcoholic solution of the acid saturated at 20°–30° C. (68°–86° F.), and either—(1) add the hydrocarbon to the cold-saturated alcoholic solution; or (2) mix the picric acid solution with a hot alcoholic solution of the hydrocarbon; or (3) dissolve the hydrocarbon, with aid of heat, in the picric acid solution. The following combine under these circumstances with picric acid: *Naphthalene* is the only solid hydrocarbon whose cold-saturated alcoholic solution is precipitated by picric acid. The compound forms delicate stellate groups of yellow needles, easily soluble in alcohol. *Retene*, treated by method 2 or 3, forms similar needles of an orange-yellow color. *Anthracene* also forms ruby-colored needles still more soluble than the preceding; the red alcoholic solution is decolorized by addition of a little more alcohol. Other hydrocarbons contained in crude anthracene exhibit the same reaction, which appears to be peculiar to anthracene and its homologues. They may be distinguished from one another by the appearance of the precipitates under the microscope.

Picric acid is converted by $PtCl_5$ into trinitrochlorobenzene, $C_6H_2(NO_2)_3Cl$, which is reconverted into picric acid by boiling with water. Picric acid distilled with calcium hypochlorite, or a mixture of potassium chlorate and hydrochloric acid, yields chloropicrin, $C(NO_2)Cl_3$; and with calcium hypobromite, bromopicrin, $C(NO_2)Br_3$ (p. 555).

Isopicric acid, $C_6H_2(NO_2)_3.OH$, formed by the action of fuming nitric acid on metanitrophenol, crystallizes from hot water or from nitric acid in light yellow prisms, melting at 174° C. (345.2° F.). It stains the skin yellow.

Picrocyamic, or *Isopurpuric acid*, $C_8H_5N_5O_{4}$, not known in the free state, is formed, as a potassium salt, $C_8H_4N_5O_6K$, by dropping a hot solution of picric acid (1 part in 9 of water) into a solution of potassium cyanide (2 parts CNK in 4 of water) heated to 60° C. (140° F.). This salt crystallizes in brown-red scales, having a green metallic lustre, sparingly soluble in cold water, dissolving in hot water and in alcohol with deep-red color (test for hydrocyanic acid and metallic cyanides). Detonates strongly when heated.

The dinitrophenols treated with potassium cyanide. yield similar derivatives.

Nitrohaloïd derivatives of phenol, $C_6H_3Cl(NO_2).OH$, etc., are produced by the action of halogen elements on the nitrophenols.

Amidophenols.

These compounds are formed by the action of reducing agents on the nitrophenols, the di- and tri-nitrophenols being partially reduced by alcoholic ammonium sulphide, completely by tin and hydrochloric acid. The entrance of the NH_2 group into the molecule considerably weakens the acid character of the compound, and when it replaces the whole of the nitro-groups, the compound becomes basic.

Monoamidophenols, $C_5H_4(NH_2).OH$.—The *ortho-* and *para-*compounds are formed by reduction of the corresponding nitro- or nitroso-compounds, best with tin and hydrochloric acid : the *p*-compound also by distillation of amidosalicylic acid.—(1 : 2) amidophenol forms colorless rhombic scales ; (1 : 4) colorless needles, which soon turn brown. Both are slightly soluble in cold water, more easily in alcohol, and form well crystallized salts with acids ; *o*- melts at 170° C.(338° F.), *p*-, with decomposition, at 184° C. (363.2° F.).

Amidonitrophenol or Picramic acid, $C_6H_2(NH_2)(NO_2)_2.OH$.—The ammonium salt of this acid, formed by passing hydrogen sulphide into an alcoholic solution of ammonium picrate, crystallizes in red needles, and is decomposed by acetic acid, yielding the free acid, which also forms red needles melting at 165° C. (329° F.).

Tri-amidophenol, $C_6H_2(NH_2)_3.OH$, is a triacid base not known in the free state. Its *hydriodide*, $C_6N_2(NH_2)_3OH.3HI$, obtained by the action of iodine and phosphorus on a hot saturated solution of picric acid, forms easily soluble colorless needles.—The *hydrochloride*, $C_6H_2(NH_2)_3OH.3HCl$, obtained by decomposing picric acid with tin and hydrochloric acid, and treating the resulting stannochloride, which separates on cooling, with hydrogen sulphide, crystallizes in needles easily soluble in water, sparingly in strong hydrochloric acid. From these salts the base cannot be separated without decomposition. The solution of the hydrochloride mixed with ferric chloride acquires a deep-blue color, and if concentrated deposits the hydrochloride of amido-diimidophenol,

$C_6H_2(OH){<}{\begin{smallmatrix}NH_2\\=NH\\NH\end{smallmatrix}}{>}$, in yellow-brown needles, having a blue shimmer in reflected light, and dissolving in water with a fine blue color.

Diazophenol (*para*).—The *nitrate*, $C_6H_4{<}{\begin{smallmatrix}OH\\N=N-NO_3\end{smallmatrix}}$, is formed, together with *o*- and *p*-nitrophenol, by passing nitrous acid into an ethereal solution of phenol or nitrosophenol cooled with ice. It crystallizes in light-brown needles. The other salts of this base, which also crystallize well, are obtained by passing nitrous acid into the solutions of the corresponding salts of *p*-amidophenol. The salts of ortho-diazophenol are obtained in like manner. All these salts are analogous in their reactions to the diazobenzene salts.

Phenol-sulphonic acids.

Phenol dissolves easily in strong sulphuric acid, forming *ortho-* and *para*-phenolsulphonic acid, $C_6H_4(OH).SO_3H$. At ordinary temperatures, the *ortho*-acid is almost the only product, but it easily changes into the *para*-acid when heated. The two acids may be separated by fractional crystallization of their potassium salts, the *p*-salt separating out first in elongated, hexagonal tables, which are anhydrous. The mother-liquors yield the *o*-salt in long colorless spicules, containing $2H_2O$. Most of the other salts of the *p*-acid are less soluble than the corresponding *o*-salts. The two acids are not known in the free state. The sodium salt of the *p*-acid, heated with MnO_2 and sulphuric acid, yields q u i n o n e. Treated with PCl_5, it yields *p*-chlorophenol and *p*-dichlorobenzene. The *o*-acid fused with potash yields pyrocatechin; the *p*-acid, by molecular transformation, yields resorcin (see DIATOMIC PHENOLS).

Metaphenolsulphonic acid is formed as a potassium salt by heating potassium benzene-metadisulphonate dissolved in the smallest possible quantity of water, with two or three times its weight of potassium hydroxide for a considerable time to 170°–180° C. (338°–356° F.). The greater part of the potassium sulphate formed at the same time having been removed by crystallization, the metaphenolsulphonate may be extracted from the mother-liquor by alcohol. This salt crystallizes from water in groups of easily soluble efflorescent scales containing 1 mol. H_2O. The free acid forms concentric groups of very soluble needles. Ferric chloride colors the solutions of the three acids violet.

Phenoldisulphonic acid, $C_6H_3(OH)(SO_3H)_2$, is formed by heating phenol or *o*- or *p*-phenolsulphonic acid with excess of strong sulphuric acid, whence it has the structure 1 : 2 : 4 (OH in 1); also by the action of sulphuric acid on diazobenzene sulphate. The solutions of the acid and its salts are colored dark red by ferric chloride.

Phenoltrisulphonic acid, $C_6H_2(OH)(SO_3H)_3$ (1 : 3 : 5, OH), formed by the action of strong sulphuric acid or phosphoric anhydride on phenol, crystallizes in thick prisms containing $3\frac{1}{2}H_2O$.

Cresols, $C_7H_8O = C_6H_4{<}^{CH_3}_{OH}$.—*Tolyl alcohols.*—*Para-* and *ortho*-cresol occur in coal- and wood-tar, together with phenol, and may be separated from the latter by fractional distillation; but they cannot easily be separated one from the other. They are obtained in the pure state by fusing the potassium salts of the corresponding toluenesulphonic acids, $C_6H_4(CH_3)(SO_3H)$, with potash, or from the corresponding toluidines by the diazo-reaction (p. 882). They are converted into toluene by heating with zinc-dust, and into the corresponding oxytoluic or cresotic acids, $C_6H_3(CH_3)(OH)(CO_2H)$, by the action of sodium and carbon dioxide.

Orthocresol is likewise obtained, together with propene, by heating carvacrol (cymophenol), $C_{10}H_{14}O$, with phosphoric anhydride:

$$C_{10}H_{14}O = C_7H_8O + C_3H_6.$$

It is a colorless crystalline mass, melting at 31°–31.5° C. (87.8°–88.7° F.), boiling at 185°–186° C. (365°–366.8° F.); colored blue by ferric chloride. By prolonged heating with potassium hydroxide it is converted into s a l i c y l i c a c i d.

Metacresol, prepared from thymol, $C_{10}H_{14}O$, in the same manner as *o*-cresol from carvacrol, is a colorless liquid which smells like phenol, boils at

38 *

201° C. (393.8° F.), remains liquid at the temperature of a mixture of solid carbon dioxide and ether, and is converted by fusion with potash into o x y b e n z o i c   a c i d . Its ethylic ether is an oil boiling at 190° C. (374° F.).

Paracresol forms colorless prisms smelling like phenol, melting at 36° C. (96.8° F.), and boiling at 198° C. (388.4° F.). It dissolves sparingly in water, forming a solution which is colored blue by ferric chloride. By fusion with potash, paracresol is converted into p a r a o x y b e n z o i c   a c i d . Its *ethylic ether* is an aromatic liquid boiling at 188° C. (374.4° F.). The *methylic ether* boils at 174° C. (345.2° F.), and is oxidized by chromic acid to methyl-paraoxybenzoic (anisic) acid, $C_6H_4(OCH_3)(CO_2H)$. The *acetic ether*, $C_7H_7.O.C_2H_3O$, is a liquid boiling at 208°–211° C. (406.4°–411.8° F.).

N i t r o - c r e s o l s .—Several of these compounds are obtained by nitration of paracresol.—$C_7H_5(NO_2)_3O$ crystallizes in yellow needles like picric acid.—$C_7H_6(NO_2)O$, obtained also by the action of nitrous acid on paratoluidine, is a dye-stuff, known as *Victoria yellow;* it forms yellow crystals, melting at 84° C. (183.2° F.), less soluble in water than picric acid.

T h i o c r e s o l s , or T o l y l   H y d r o s u l p h i d e s , $C_6H_4{<}^{CH_3}_{SH}$, are produced from the three toluenesulphonic acids by reducing the corresponding chlorides, $C_6H_4(CH_3)(SOCl)$, with zinc and hydrochloric acid. *Ortho-,* shining laminæ, melting at 15° C. (59° F.), boiling at 188° C. (370.4° F.). *Meta-,* liquid, not solidifying at —10° C. (14° F.). *Para-,* large laminæ, melting at 43° C. (109.4° F.), boiling at 188° C. (370.4° F.).

Eight-carbon Phenols, $C_8H_{10}O$.—This formula includes two metameric phenols, viz. :

Dimethyl-phenol　．　．　．　．　．　$C_6H_3(CH_3)_2OH$
Ethyl-phenol　．　．　．　．　．　$C_6H_4(C_2H_5)OH$.

Two D i m e t h y l - p h e n o l s or X y l e n o l s are produced by fusing the potassium-salt of xylenesulphonic acid, $C_8H_9SO_3K$, with potassium hydroxide. On decomposing the resulting mass with hydrochloric acid, digesting with ether, and distilling, a liquid passes over at 210° C. (410° F.), which, when exposed to a winter temperature, separates into two isomeric modifications, one crystalline, the other liquid. The solid modification is likewise obtained by fusing the potassium-salt of oxymesitylenic acid with potash :

$$C_6H_2 \begin{cases} OK \\ (CH_3)_2 \\ CO_2K \end{cases} + H_2O = CO_3K_2 + C_6H_3 \begin{cases} (CH_3)_2 \\ OH \end{cases}$$

Solid xylenol melts at 75° C. (167° F.) and boils at 216° C. (420.8° F.) ; the liquid modification boils at 206.5°–208.5° C. (403.7°–407.1° F.).

A xylylic phenol is mentioned by Dr. Hugo Müller as occurring in coaltar ; this is probably also a dimethyl-phenol, inasmuch as products of destructive distillation have hitherto been found to yield only methyl-derivatives of benzene. The portion of aloïsol (a product obtained by distilling aloes with lime) which is soluble in potash, has the composition of a xylylic phenol, and is perhaps identical with the preceding. Creosote, from beech-tar, is a mixture of several phenols and their ethers, and usually consists of phenol, cresol, xylenol, guaiacol, $C_6H_4(OH)(OCH_3)$, and creosol, $C_6H_4(CH_3)(OH)(OCH_3)$.

E t h y l - p h e n o l , obtained from ethyl-benzene sulphonic acid, melts at 47° C. (116.6° F.), boils at 211° C. (411.8° F.), and volatilizes even at ordinary temperatures.

Phlorol, an oily liquid obtained by the dry distillation of the barium salt of phloretic or oxethyl-benzoic acid, $C_9H_{10}O_3$ or $C_6H_4(OC_2H_5).COOH$, is also an ethyl phenol, its formation being represented by the equation:

$$C_6H_4(OC_2H_5).COOH = CO_2 + C_6H_5(OC_2H_5).$$
$$\text{Phloretic acid.} \qquad\qquad\qquad \text{Phlorol.}$$

Phlorol is a colorless, strongly-refracting oil, having a specific gravity of 1.0374 at 12° C. (53.6° F.), and boiling at 220° C. (428° F.). It dissolves in strong sulphuric acid, forming a sulpho-acid which yields a soluble barium salt. With chlorine it forms a substitution-product. It reacts violently with strong nitric acid, forming the compound $C_8H_7(NO_2)_3O$. By sodium and carbon dioxide it is converted into phloretic acid.

Ten-Carbon Phenols, $C_{10}H_{13}.OH$.—Two compounds represented by this formula are known, viz.: t h y m o l and c a r v a c r o l. Both are methyl-propyl-phenols, $C_6H_3(CH_3)(C_3H_7)(OH)$, and have the methyl-group in the para-position relatively to the propyl-group; but in thymol the CH_3-group stands to the OH in the meta-position; in carvacrol in the ortho-position, thus:

CH₃ ... OH ... C₃H₇ Thymol. ... Carvacrol.

Both are resolved by heating with phosphoric anhydride into propene, C_3H_6, and cresol, thymol yielding meta-, and carvacrol yielding ortho-cresol (p. 897).

Thymol exists, together with cymene, $C_{10}H_{14}$, and thymene, $C_{10}H_{16}$, in the volatile oils of thyme (*Thymus serpyllum*), horse-mint (*Mentha sylvestris*), *Ptychotis Ajowan*, an East Indian plant, and *Monarda punctata*, or Oswego tea, a native of North America. It crystallizes in large transparent plates, has a mild odor, a peppery taste, melts at 44° C. (111.2° F.), and boils at 230° C. (446° F.). Its *methylic* ether boils at 205° C. (401° F.), the *ethylic* ether at 220° C. (428° F.).

Thymol treated with *bromine* in sunshine yields p e n t a b r o m o t h y - m o l, $C_{10}H_9Br_5O$, and with *chlorine*, $C_{10}H_{11}Cl_3O$ or $C_{10}H_9Cl_5O$, accordingly as the reaction takes place in the shade or in sunshine; both of these, as well as the bromine-compound, are crystalline.

There are two n i t r o t h y m o l s, $C_{10}H_{12}(NO_2)_2O$ and $C_{10}H_{11}(NO_2)_3O$, obtained by the action of nitric acid on thymolsulphonic acid. Both form potassium salts, which crystallize in yellow or orange-yellow needles.

Carvacrol, Oxycymene, or *Cymenol*, is obtained by fusing cymene-sulphonic acid, $C_{10}H_{13}.SO_3H$, with potash, and by heating camphor with one-fifth of its weight of iodine; also from the isomeric compound carvol, contained in oil of caraway (*Carum Carui*), by heating with phosphoric acid. It is a thick oil, not solidifying at —25° C. (—13° F.), boiling at 233°–235° C. (451.4°–455° F.).

Thymol and carvacrol distilled with phosphoric anhydride yield two isomeric thiophenols, $C_{10}H_{13}.SH$, both of which are non-solidifying liquids. Thiothymol boils at 230°–231° C. (446°–447.8° F.); thiocymonol at 235° C. (455° F.).

Diatomic Phenols, $C_nH_{2n-8}(OH)_2$.

There are seven known compounds included in this formula, viz.:

Pyrocatechin—Resorcin—Hydroquinone . $C_6H_4(OH)_2$
Orcin—Homo-pyrocatechin . . . . $C_6H_3(CH_3)(OH)_2$
Beta-orcin—Hydrophlorone . . . . $C_6H_2(CH_3)_2(OH)_2$.

These compounds are formed by the action of melting potash on the monohaloïd derivatives of the monatomic phenols, and on the phenolsulphonic acids:

$$C_6H_4Cl.OH \quad + \quad KOH \quad = \quad KCl \quad + \quad C_6H_4(OH)_2$$

$$C_6H_4(OH).SO_3K \quad + \quad KOH \quad = \quad SO_3K_2 \quad + \quad C_6H_4(OH)_2;$$

also on resins; and by the dry distillation of aromatic dioxyacids—such as oxysalicylic acid, $C_6H_3(OH)_2.CO_2H$, with lime or baryta:

$$C_6H_3(OH)_2.CO_2H \quad = \quad CO_2 \quad + \quad C_6H_4(OH)_2.$$

Pyrocatechin, $C_6H_4(OH)_2$ (1 : 2), also called *Oxyphenic acid*, is formed: 1. By fusing orthochlor- or orth-iodophenol with potash. 2. By the dry distillation of oxysalicylic or of protocatechuic acid, $C_7H_6O_4$. 3. By dry distillation of catechin (the juice of *Mimosa catechu*), from which it was first obtained; also from morintannic acid (the yellow coloring matter of *Morus tinctoria*); and of wood, whence it is found in wood-vinegar. 4. By heating its methylic ether (guaiacol) with hydriodic acid to 200° C. (392° F.).

$$C_6H_4(OH)(OCH_3) \quad + \quad HI \quad = \quad CH_3I \quad + \quad C_6H_4(OH)_2.$$

This reaction affords the best method of preparing it.

Pyrocatechin crystallizes in short square prisms; sublimes even at ordinary temperatures, in shining laminæ; melts at 104° C. (219.2° F.), and boils at 245.5° C. (473.9° F.). It dissolves easily in water, alcohol, and ether. The aqueous solution mixed with *ferric chloride* assumes a dark-green color, changing to violet on addition of a small quantity of ammonia, acid sodium carbonate, or tartaric acid. Pyrocatechin reduces *silver nitrate* at ordinary temperatures, an ammoniacal solution of copper with the aid of heat. *Nitric acid* acts violently upon it, forming oxalic acid and a small quantity of a yellow nitro-compound. With *acetyl chloride* and *benzoyl chloride* it forms the compounds $C_6H_4(O.C_2H_3O)_2$ and $C_6H_4(O.C_7H_5O)_2$, both of which are crystalline. Its aqueous solution forms, with *lead acetate*, a white precipitate, $C_6H_4O_2Pb$.

Methyl-pyrocatechin or *Guaiacol*, $C_7H_8O_2 = C_6H_4\displaystyle\Big\langle{}^{OCH_3}_{OH}$, is one of the constituents of beech-tar creosote (p. 898), and may be separated therefrom by fractional distillation. It is produced by heating pyrocatechin with potassium hydroxide and potassium methylsulphate to 180° C. (356° F.); by heating methyl-pyrocatechuic acid with calcium hydrate: $C_7H_5(CH_3)O_4 = CO_2 + C_6H_5(CH_3)O_2$; and by dry distillation of guaiac resin.

Guaiacol is a colorless liquid, having a specific gravity of 1.117, and boiling at 200° C. (392° F.); slightly soluble in water, easily in alcohol, ether, acetic acid, and alkalies. It forms crystalline salts with the alkalies and alkaline earths, and its alkaline solutions reduce the salts of gold, silver, and copper. By heating with hydriodic acid or fusion with potash, it is resolved into pyrocatechin and methyl iodide or methyl alcohol.

Dimethyl-pyrocatechin, $C_6H_4(O.CH_3)_2$, formed by heating the potassium-derivative of methyl-pyrocatechin with methyl iodide, is a liquid boiling at 205° C. (401° F.).

Resorcin, $C_6H_4(OH)_2$ (1 : 3), is formed by the action of melting potash on the para-modifications of chlorophenol, iodophenol, chloro-, and bromo-benzenesulphonic acids, phenolsulphonic and benzenedisulphonic acids; also on some compounds not belonging to the para-series. It appears, indeed, to be the most stable of the three diatomic phenols, and is accordingly produced by molecular transposition, especially at high temperatures, in some cases when an ortho- or para-compound might be expected to result. Resorcin is also formed by the action of melting potash on umbelliferone, and on various resins and gum-resins, as galbanum, assafœtida, gum ammoniacum, sagapenum, and acaroïd resin. It is most easily prepared by the dry distillation of extract of Brazil wood.

Resorcin is very soluble in water, alcohol, and ether, and crystallizes from very concentrated solutions, in prisms belonging to the trimetric system, colorless at first, but afterwards becoming reddish. It melts at 110° C. (230° F.), and boils at 271° C. (519.8° F.), distilling almost without residue. Its aqueous solution is colored violet by ferric chloride, is not precipitated by lead acetate, and reduces silver nitrate only at the boiling heat, except in presence of ammonia, in which case the reduction takes place in the cold. These characters distinguish resorcin from pyrocatechin. A very delicate test for resorcin is its conversion into fluorescein (*q. v.*) by heating with phthalic anhydride. Bromine-water added to the aqueous solution throws down the *tribromo-compound*, $C_6HBr_3(OH)_2$. Resorcin also forms a *triacetyl-compound*, $C_6H_3(C_2H_3O)(O.C_2H_3O)_2$, and two *benzoyl-derivatives*, $C_6.H_2(O.C_7H_5O)_2$, and $C_6H_3(C_7H_5O)(O.C_7H_5O)_2$.

Nitro-resorcins.—*Di-nitroresorcin*, $C_6H_2(NO_2)_2.(OH)_2$, formed by adding potassium nitrite to a very dilute solution of resorcin mixed with acetic acid, crystallizes with molecules of water in yellowish-gray or brown laminæ, which detonate when heated to 115° C. (239° F.).

Trinitroresorcin, Oxypicric, or *Styphnic acid,* $C_6H(NO_2)_3(OH)_2$, is produced by the action of cold nitric acid on several gum-resins (galbanum, sagapenum, gum ammoniacum), and on many vegetable extracts (Sapan-wood, Brazil-wood, etc.); also by heating meta-nitrophenol with strong nitric acid. It crystallizes in large yellow hexagonal prisms or laminæ, sparingly soluble in water, melting at 175° C. (347° F.), and subliming when cautiously heated, exploding when quickly heated. With ferrous sulphate and lime-water, it exhibits at first a green color, which afterwards disappears (distinction from picric acid, which is thereby colored blood-red). Trinitroresorcin is a strong bibasic acid, forming well-crystallized salts, which detonate violently when heated.

Tri-amidoresorcin, $C_6H(NH_2)_3(OH)_2$.—The hydrochloride of this base, $C_6H_9N_3O_2.3HCl + H_2O$, produced by the action of tin and hydrochloric acid on trinitroresorcin, forms large yellowish, easily soluble crystals. The solution of the stanno-chloride (the immediate product of the reaction) is colored dark-red by ferric chloride, or by exposure to the air, and deposits on standing red needles, consisting of the hydrochloride of a m i d o - d i i m i d o r e s o r c i n, $C_6H(OH)_2(NH_2)\!\!<\!\!^{NH}_{NH}\!\!>$ .HCl. Ammonia added to the solution of this salt separates the free base in slender needles, having a green metallic lustre.

Hydroquinone, $C_6H_4(OH)_2$ (1 : 4), is formed by fusing para-iodophenol with potash to 180° C. (356° F.); by dry distillation of oxy-

salicylic and of quinic acid, and by heating the aqueous solution of the
latter with lead peroxide :

$$C_7H_{12}O_6 \ + \ O \ = \ C_6H_6O_2 \ + \ CO_2 \ + \ 3H_2O;$$

also from arbutin (see GLUCOSIDES), by boiling with dilute sulphuric acid,
or by the action of emulsin :

$$C_{12}H_{16}O_7 \ + \ H_2O \ = \ C_6H_6O_2 \ + \ C_6H_{12}O_6 \cdot$$

It is most easily prepared from quinone by reduction with sulphurous
acid, $C_6H_4O_2 + H_2 = C_6H_6O_2$. Gaseous sulphur dioxide is passed into an
aqueous solution of quinone, till the liquid, which at first assumes a
brown color from formation of quinhydrone, becomes colorless. The
solution is then evaporated down, and the hydroquinone extracted by
ether.

Hydroquinone crystallizes from water in colorless rhombic prisms, melting at 169° C. (336.2° F.), subliming when carefully heated, decomposing
when suddenly heated. It dissolves in 17 parts of water at 15° C. (59°
F.), easily in alcohol and ether. It unites with hydrogen sulphide and
sulphur dioxide, forming crystalline compounds which are decomposed by
water. Its aqueous solution is turned red-brown by ammonia, and is not
precipitated by lead acetate except in presence of ammonia. Oxidizing
agents convert hydroquinone first into quinhydrone, then into quinone :

$$C_6H_4{<}^{OH}_{OH} \qquad C_6H_4{<}^{OH\ HO}_{O\underline{}O}{>}C_6H_4 \qquad C_6H_4{<}^{O}_{O}{>}$$
Hydroquinone. 　　　　　Quinhydrone. 　　　　　Quinone.

Substitution-products of hydroquinone are obtained, not directly from
hydroquinone, but from the corresponding derivatives of quinone or of
arbutin.

Chlorhydroquinones are produced by reduction of chloroquinones with
sulphurous acid. $C_6H_4Cl_2O_2$ melts at 158° C. (316.4° F.), $C_5H_3Cl_3O_2$ at
134° C. (273.2° F.), $C_6H_2Cl_4O_2$ above 200° C. (392° F.).

Dinitrohydroquinone, $C_6H_4(NO_2)_2(OH)_2$, obtained by boiling dinitro-arbutin
with dilute sulphuric acid, forms golden-yellow laminæ; its aqueous solution is colored dark blue by alkalies.

· *Dichlorhydroquinone-disulphonic acid*, $C_6Cl_2\left\{{(OH)_2 \atop (SO_3H)_2}\right.$, is formed by heating
tetrachloroquinone with a dilute solution of acid sodium sulphite, tetra-
chlorhydroquinone being produced in the first instance, and two of its
chlorine-atoms then replaced by the groups SO_3H. The aqueous solution
of this acid is colored indigo-blue by ferric chloride. A solution of its
potassium salt, containing free potash, oxidizes quickly on exposure to the
air, and is converted into yellow potassium euthiochroate,
$C_6(OH)_2\left\{{O_2 \atop (SO_3K)_2}\right. \cdot$

Quinone, $C_6H_4{<}^{O}_{O}{>}$, is produced by the oxidizing action of manganese dioxide and sulphuric acid, or of dilute chromic acid, on numerous
benzene-derivatives (*e. g.*, phenylenediamine, amidophenol, amidobenzene-
sulphonic acid), especially those belonging to the para-series ; also by
distilling various vegetable extracts with MnO_2 and sulphuric acid. It is
most easily prepared by gently heating quinic acid (1 part) with manganese dioxide (2 parts) and sulphuric acid (1 part diluted with ½ part
water).

Quinone crystallizes in golden-yellow prisms, melts at 116° C. (240.8° F.), and sublimes, even at ordinary temperatures, in shining needles. It has a pungent tear-exciting odor, distils easily with aqueous vapor, and dissolves sparingly in cold water, easily in hot water, also in alcohol and ether. By reduction with sulphurous acid, or with zinc and hydrochloric acid, it is converted, first into quinhydrone, then into hydroquinone. Phosphorus pentachloride converts it into para-dichlorobenzene.

Chloroquinones are formed by the action of chlorine on quinone, and by distilling quinic acid with MnO_2 and hydrochloric acid. $C_6H_3ClO_2$ forms yellow needles. $C_6H_2Cl_2O_2$, produced also by the action of hypochlorous anhydride, Cl_2O, on benzene, and by heating trichlorophenol with nitric acid, forms large yellow prisms, melting at 120° C. (248° F.). $C_6HCl_3O_2$, obtained also by the action of chromyl chloride, CrO_2Cl_2, on benzene, crystallizes in large laminæ melting at 166° C. (330.8° F.).

Tetrachloroquinone or *Chloranil*, $C_6Cl_4O_2$, is formed, together with $C_6HCl_3O_2$, from many benzene-derivatives (aniline, phenol, isatin, etc.), by the action of chlorine, or of potassium chlorate and hydrochloric acid. It is best prepared by gradually adding a mixture of 1 part crystallized phenol, and 4 parts potassium chlorate to hydrochloric acid diluted with an equal volume of water, and slowly heating the liquid. Red crystals then separate, which, on further addition of potassium chlorate, are converted into a yellow mixture of tri- and tetra-chloroquinone. To separate these compounds, they are converted by sulphurous acid into the corresponding chlorohydroquinones ($C_6H_2Cl_4O_4$ is insoluble in water), which are then reconverted into the chloroquinones by oxidation.

Chloranil forms golden-yellow shining laminæ, insoluble in water, soluble in hot alcohol and in ether, subliming at about 150° C. (302° F.); converted by PCl_5 into perchlorobenzene.

Chloranil dissolves with purple-red color in dilute potash-lye, forming the potassium salt of chloranilic acid, $C_6Cl_2O_2(OK)_2 + H_2O$, which crystallizes in dark red needles, sparingly soluble in water. Acids decompose it, separating chloranilic acid, $C_6Cl_2O_2(OH)_2 + H_2O$, in reddish shining scales. Chloranil is converted by aqueous ammonia into chloranilamide, $C_6Cl_2O_2(NH_2)_2$, and chloranilamic acid, $C_6Cl_2(NH_2)O_2(OH)$.

Bromoquinones, analogous to the chloroquinones, are obtained by similar reactions. *Bromanil*, $C_6Br_4O_2$, it most readily prepared by heating phenol (1 part) with bromine (10 parts), iodine (3 parts), and water, to 100°. Golden-yellow laminæ sparingly soluble in carbon sulphide.

Quinhydrone, $C_{12}H_{10}O_4 = C_6H_6O_2.C_6H_4O_2 = \dfrac{C_6H_4(OH)-O}{C_6H_4(OH)-O}$, is formed by treating quinone in aqueous solution with a quantity of sulphurous acid not sufficient for complete reduction; also by incomplete oxidation of hydroquinone, and by mixing the aqueous solutions of quinone and hydroquinone. It crystallizes in flat prisms, having a splendid golden-green metallic lustre like that of the wing-cases of the rose-beetle, and sublimes in green laminæ. It is fusible, has but little odor, dissolves freely in hot water with a brown color, in alcohol and ether with green color. It is resolved, by boiling with water, into quinone which distils over, and hydroquinone; and is converted, by oxidation into quinone, by reduction into hydroquinone.

Phenoquinone, $C_6H_4\begin{smallmatrix}O-O-C_6H_5\\O-O-C_6H_5\end{smallmatrix}$, is produced by careful oxidation of phenol with chromic acid, and (together with quinhydrone and hydroquinone) by mixing the aqueous solutions of quinone and phenol. It forms red, very volatile needles, melting at 71° C. (150.9° F.), soluble in water, more easily in alcohol and ether. It is colored blue by potash, green by baryta or ammonia.

Diatomic Phenols, $C_7H_8O_2 = C_6H_3(CH_3)(OH)_2$.—*Orcin* exists ready formed in all the lichens (*Lecanora tartarea, Roccella tinctoria, Variolaria orcina,* etc.) which are used for the preparation of archil and litmus; and is the general product of the decomposition of certain acids extracted from these lichens (orsellinic acid, erythric acid, etc.) under the influence of heat or of alkalies. Orsellinic acid, $C_8H_8O_4$, when boiled with baryta-water, splits up into carbon dioxide and orcin:

$$C_8H_8O_4 \; = \; CO_2 \; + \; C_7H_8O_2 \,.$$

Evernic acid is resolved by the same treatment into everninic and orsellinie acids:

$$\underset{\text{Evernic.}}{C_{17}H_{16}O_7} \; + \; H_2O \; = \; \underset{\text{Everninic.}}{C_9H_{10}O_4} \; + \; \underset{\text{Orsellinic.}}{C_8H_8O_4} \,.$$

Erythric acid (erythrin) is resolved, in like manner, into orsellinic acid and picroerythrin:

$$\underset{\text{Erythrin.}}{C_{20}H_{22}O_{10}} \; + \; H_2O \; = \; \underset{\substack{\text{Orsellinic}\\ \text{acid.}}}{C_8H_8O_4} \; + \; \underset{\text{Picroerythrin.}}{C_{12}H_{16}O_7} \,,$$

the orsellinic acid being further resolved, as above, into CO_2 and orcin, and the pycroerythrin into CO_2, erythrite (p. 644), and orcin:

$$C_{12}H_{16}O_7 \; + \; H_2O \; = \; CO_2 \; + \; C_4H_{10}O_4 \; + \; C_7H_8O_2 \,.$$

To prepare orcin in quantity, one of the lichens above mentioned is boiled with milk of lime, the liquor filtered and evaporated to one-fourth; the lime precipitated by carbonic acid; the filtrate evaporated nearly to dryness on the water-bath; the residue boiled several times with benzene; the orcin extracted from the benzene solution by agitation with water; and the aqueous solution evaporated.

Orcin may also be prepared by the action of melting potash on extract of aloes, and on the potassium-salt of chlorotoluene-sulphonic acid.

Orcin crystallizes in colorless six-sided prisms containing $C_7H_8O_2.H_2O$. It has a sweet taste, dissolves readily in water, alcohol, and ether, melts at 58° C. (136.4° F.), gives off its water of crystallization at 86° C.(186.8° F.), and boils at 290° C. (554° F.). Its aqueous solution is precipitated by lead acetate, colored blue-violet by ferric chloride, and exhibits with hypochlorites a transient dark-violet coloration.

The hydroxyl-groups in orcin may be replaced by acid and by alcoholic radicles; the diethylic ether, $C_7H_6(OC_2H_5)_2O_2$, boils at 240°–250° C. (464°–482° F.).

Tribromorcin, $C_6Br_3(CH_3)(OH)_2$, is precipitated by bromine-water from the aqueous solution of orcin.

Trinitro-orcin, $C(NO_2)_3(CH_3)(OH)_2$, produced by dissolving orcin in a well-cooled mixture of strong nitric and sulphuric acid, forms long yellow needles, which melt at 162° C. (323.6° F.), and react very much like trinitroresorcin. By the action of tin and hydrochloric acid it is reduced to *triamido-orcin,* $C_6(NH_2)_3(CH_3)(OH)_2$, which forms colorless crystals, and is converted by exposure to the air into crystals of *amido-dimido-orcin,*

$$C_6(CH_3)NH_2{\Big\langle}{\substack{NH-\\NH}}{\Big\rangle} \,.$$

Orcin unites with dry *ammonia,* forming a crystalline compound, $C_7H_8O_2.NH_3$; and when a solution of orcin containing ammonia is exposed to the air, it absorbs oxygen, acquires a dark-red or purple color, and gives with acetic acid a deep-red pulverulent precipitate of o r c e i n, $C_7H_7NO_3$, formed according to the equation:

$$C_7H_8O_2 + NH_3 + O_3 = C_7H_7NO_3 + 2H_2O.$$

Orcein unites with metallic oxides, forming red lakes. It is the chief constituent of the dye-stuffs known as archil, cudbear, French purple, and litmus. The last-mentioned substance, which is extensively used for the preparation of test-papers, is prepared from *Roccella tinctoria* or *Lecanora tartarea*, by macerating the lichen in solution of ammonium carbonate, exposing the liquid to the air for 20 to 40 days, and thickening the coloring matter thus obtained with chalk or gypsum.

Isorcin, $C_6H_3(CH_3)(OH)_2$, obtained from toluidine-disulphonic acid, forms colorless needles, melts at 87° C. (188.6° F.), and boils at 260° C. (500° F.).

Homopyrocatechin, $C_7H_8O_2 = C_6H_3(CH_3)(OH)_2$, which has not been obtained in the crystalline state, is produced by the action of hydriodic acid on c r e a s o l, $C_8H_{10}O_2 = C_7H_6{<}^{OH}_{OCH_3}$, which is its methylic ether. Creasol is one of the constituents of beech-tar (p. 870), and is formed together with guaiacol, by the dry distillation of guaiac resin. It is a colorless liquid, very much like guaiacol, boils at 219° C. (426.2° F.), and reduces silver nitrate when heated with it. Its alcoholic solution is colored dark-green by ferric chloride.

Hydrotoluquinone, $C_6H_3(CH_3)(OH)_2$, and *Toluquinone*, $C_6H_3(CH_3)O_2$, are not known; but d i - and t r i - c h l o r o t o l u q u i n o n e, $C_6HCl_2(CH_3)O_2$, and $C_6Cl_3(CH_3)(OH)_2$, are formed by the action of hydrochloric acid and potassium chlorate on cresol, just as the six-carbon chloroquinones are obtained from phenol. The trichloro-compound forms yellow laminar crystals, and is converted by heating with aqueous sulphurous acid into trichloro-hydrotoluquinone, $C_6Cl_3(CH_3)(OH)_2$, which crystallizes in colorless needles.

Diatomic Phenols, $C_9H_{10}O_2 = C_9H_8(OH)_2$.—H y d r o p h l o r o n e is obtained by the action of sulphurous acid on phlorone, or xyloquinone, $C_9H_8O_2$, a compound obtained by distilling coal-tar oil or beech-tar creosote, boiling above 210° C. (410° F.) with MnO_2 and sulphuric acid. It forms colorless laminæ, having a mother-of-pearl lustre, fusible, sublimable, easily soluble in water, alcohol, and ether. Oxidizing agents convert it into p h l o r o n e, $C_9H_8O_2$, which forms yellow volatile needles.

B e t a - o r c i n is obtained by dry distillation of usnic acid and other acids occurring in lichens, e. g., $C_9H_{10}O_4$ (everninic acid) = $C_8H_{10}O_2 + CO_2$. Colorless crystals, easily soluble in hot water, melting at 109° C. (228.2° F.). The aqueous solution turns red when mixed with ammonia and exposed to the air, and is colored dark red by bleaching powder.

Veratrol, obtained by distillation of veratric acid, $C_8H_{10}O_4$, with baryta, is an oil solidifying at 15° C. (59° F.), and boiling at 202°–205° C. (395.6°–401° F.).

Thymohydroquinone, $C_{10}H_{12}(OH)_2 = C_6H_2(CH_3)(C_3H_7)(OH)_2$, the only known diatomic phenol containing 10 atoms of carbon, is produced by the action of sulphurous acid on thymoquinone, and crystallizes in limpid, shining, four-sided prisms, melting at 139.5° C. (283.1° F.), and subliming without decomposition; converted by oxidation into thymoquinone. Its methylic ether is a constituent of the volatile oil of *Arnica montana*.

Thymoquinone, $C_{10}H_{12}\langle{}^O_O\rangle$ (*thymoïl*), is produced by distilling thymol and carvacrol with MnO_2 and sulphuric acid. It crystallizes in yellow tables, melts at 45.5° C. (113.9° F.), and boils at 200° C. (392° F.).

Triatomic Phenols, $C_6H_3(OH)_3$.

Pyrogallol or **Pyrogallic Acid**, is produced by the action of heat on gallic (dioxysalicylic) acid: $C_7H_6O_5 = CO_2 + C_6H_6O_3$; also, together with gallic acid, by the action of hot caustic potash on di-iodosalicylic acid, $C_7H_4I_2O_3$. It is conveniently prepared by heating a dried aqueous extract of gall-nuts to 180°–185° C. (356°–365° F.) in an iron pot covered with a paper cap; it then sublimes and condenses on the cap in long flattened prisms.

Pyragallol is soluble in water, alcohol, and ether; it melts at 115° C. (239° F.), boils at 210° C. (410° F.), and decomposes at 250° C. (482° F.), giving off water, and leaving a residue of metagallic acid, $C_6H_4O_2$. Pyrogallol dissolves in caustic potash or soda, forming a solution which quickly absorbs oxygen from the air, and turns black; this solution forms a very convenient reagent for the eudiometric analysis of air (p. 201). With solutions of ferrous salts, it produces a black-blue color; with ferric salts a red color. Pyrogallol quickly reduces gold, silver, and mercury from their salts, and forms, with lead acetate, a white precipitate, $C_6H_6O_3PbO$.

With *bromine* pyrogallol forms a tribromo-derivative, $C_6Br_3(OH_3)$; with *acetyl chloride* it yields a triacetyl-derivative, $C_6H_3(OC_2H_3O)_3$, forming crystals slightly soluble in water.

Phloroglucin, $C_6H_6O_3$. Phlorizin, or Phloridzin, $C_{21}H_{24}O_{10}$, a crystalline substance found in the root-bark of the apple, pear, plum, and cherry trees, is resolved by boiling with dilute acids into glucose and phloretin, $C_{15}H_{14}O_5$:

$$C_{21}H_{24}O_{10} + H_2O = C_6H_{12}O_6 + C_{15}H_{14}O_5;$$

and phloretin heated with aqueous potash is resolved into phloretic acid and phloroglucin:

$$C_{15}H_{14}O_5 + H_2O = C_9H_{10}O_3 + C_6H_6O_3.$$

Phloroglucin crystallizes in large colorless prisms containing $2H_2O$; effloresces on exposure to the air; gives off all its crystallization-water at 100° C.; melts at 230° C. (446° F.), and sublimes without decomposition. It has a sweet taste, and dissolves easily in water, alcohol, and ether. Its aqueous solution is precipitated by lead acetate, and colored dark violet by ferric chloride.

Phloroglucin is converted by chlorine into dichloracetic acid; with *bromine* and with nitric acid it forms tri-substitution derivatives; with *acetyl chloride* and *benzoyl chloride* it yields the ethers $C_6H_3\begin{Bmatrix}(OH)_2 \\ OC_2H_3O\end{Bmatrix}$, and

$C_6H_3\begin{Bmatrix}(OH)_2 \\ OC_7H_5O\end{Bmatrix}$, both of which are crystalline. Its dibutyril ether,

$C_6H_3\begin{Bmatrix}OH \\ (OC_4H_7O)_2\end{Bmatrix}$, called *filicic acid*, occurs in the root of the male fern

(*Aspidium Filix mas*) as a crystalline substance, which is resolved by fusion with potash into phloroglucin and butyric acid.

With *ammonia* phloroglucin forms the basic compound, p h l o r a m i n e, $C_6H_3\begin{Bmatrix}(OH)_2\\NH_2\end{Bmatrix}$, which forms crystalline salts with acids.

Appendix to the Phenols—PHENOL-DYES.

A u r i n or **R o s o l i c A c i d,** $C_{20}H_{14}O_3$ (or $C_{20}H_{16}O_3$?).—This compound, also called *Corallin*, is a red coloring matter, obtained by heating phenol with oxalic and sulphuric acid, the oxalic acid being then resolved into CO, CO_2, and H_2O, and the CO reacting with the phenol, as shown by the equation :

$$3C_6H_6O \; + \; 2CO \; = \; C_{20}H_{14}O_3 \; + \; 2H_2O \; .$$

To obtain a pure product, the mixture of sulphuric acid and phenol must be heated on a water-bath, and the oxalic acid added gradually, waiting each time till the evolution of gas ceases, and not using sufficient oxalic acid to attack all the phenol. The aurin thus obtained has exactly the composition $C_{20}H_{14}O_3$.[*]

A commercial dye-stuff, known as aurin, corallin, or pæonin, which gives a fine yellow-red color to wool and silk, is prepared in a similar manner. It is a mixture of several substances, but may be purified by treatment with aqueous ammonia, which dissolves the extraneous matters, leaving a residue of nearly pure aurin.

The same compound is obtained by the action of nitrous acid on rosaniline (p. 876); and it is reconverted into that base by heating with ammonia in aqueous or alcoholic solution.

$$C_{20}H_{14}O_3 \; + \; 3NH_3 \; = \; C_{20}H_{17}N_3 \; + \; 3H_2O.$$

From this it would appear that the formula of rosaniline should be $C_{20}H_{17}N_3$, whereas, according to Hofmann's analysis, its composition is $C_{20}H_{19}N_3$. Further investigation is therefore required to clear up this discrepancy.[†]

Aurin crystallizes from alcohol in red prisms, having a green metallic lustre. It is insoluble in water, but soluble in alcohol, strong hydrochloric acid, and glacial acetic acid. It unites with sulphurous anhydride, forming garnet-red crystals, $(C_{20}H_{14}O_3)_2SO_2 + 5\frac{1}{2}H_2O$, and forms crystalline compounds with bisulphites of alkali-metal, *e. g.*, $C_{20}H_{14}O_3.NaHSO_3$. Bromine added to its acetic acid solution throws down the compound $C_{20}H_{10}Br_4O_3$ in shining green laminæ.

By reduction with zinc-dust in alcoholic solution, aurin is converted into l e u c a u r i n, $C_{20}H_{16}O_3$, which crystallizes from alcohol in colorless prisms, and is reconverted into aurin by oxidation. *Triacetyl-leucaurin,* $C_{20}H_{13}(C_2H_3O)_3O_3$, produced by heating leucaurin with acetyl chloride, forms short silky needles, easily soluble in alcohol and in acetic acid.

Phthaleins.—These are compounds formed, with elimination of water, by the combination of phenols with phthalic anhydride, $C_8H_4O_3$, or $C_6H_4\!\!<^{CO}_{CO}\!\!>O$. They contain the ketonic group CO, together with the

[*] D a l e and S c h o r l e m m e r, Chem. Soc. Journ., 1873, 434; and 1877, vol. ii. p. 121.

[†] Dale and Schorlemmer observe that, in the analysis of organic coloring-matters, the percentage of hydrogen is often found too high.

hydroxyl-groups of the phenols, and are consequently intermediate in character between the phenols and ketones. They are all more or less colored, and act as dye-stuffs. By hydrogenation (action of zinc-dust in alkaline solution) they are converted into p h t h a l i n s, colorless compounds in which the ketonic groups are converted into alcoholic groups, $C(OH)$; *e. g.*,

$$C_6H_4 < \begin{array}{l} CO—C_6H_4—OH \\ CO—C_6H_4—OH \end{array}$$

Phenol-phthalein.

$$C_6H_4 < \begin{array}{l} C(OH)—C_6H_4—OH \\ C(OH)—C_6H_4—OH \end{array}$$

Phenol-phthalin.

By oxidation, the phthalins are reconverted into phthaleins.

P h e n o l - p h t h a l e i n, $C_{20}H_{14}O_4$, is prepared by heating 10 parts of phenol with 5 parts of phthalic anhydride and 4 parts of strong sulphuric acid to 120° C. (248° F.) for about ten hours, exhausting the product with boiling water, dissolving the residue in dilute caustic soda, and precipitating with acetic acid and a little hydrochloric acid. It may be purified by boiling its alcoholic solution with animal charcoal, and precipitating the filtrate with water. It then separates as a white or yellowish-white crystalline powder, or in triclinic crystals, according as the precipitation is effected quickly or slowly. It dissolves in alkalies with fuchsine-red color, and on heating the alkaline solution with zinc-dust, it becomes colorless, from conversion of the phthalein into p h e n o l - p h t h a l i n, $C_{20}H_{16}O_4$, which separates on addition of hydrochloric acid, in white prisms. This compound dissolves in strong sulphuric acid, and the solution, mixed with water, deposits an amorphous yellowish substance, $C_{20}H_{14}O_3$, called p h e n o l - p h t h a l i d i n, which may be reconverted into the phthalin by heating with water to 175° C. (347° F.). By exposure to the air, or more quickly, by treatment with manganese dioxide, potassium manganate or potassium ferricyanide, it is converted into a compound, $C_{20}H_{14}O_4$, isomeric with phenol-phthalein, which separates in monoclinic crystals.

R e s o r c i n - p h t h a l e i n, or F l u o r e s c e i n, $C_{20}H_{12}O_5$ =

$$C_6H_4 < \begin{array}{l} CO—C_6H_3—OH \\ \qquad\qquad\quad >O \\ CO—C_6H_3—OH \end{array}$$, prepared by heating resorcin with phthalic

anhydride to 200° C. (392° F.), forms dark-brown crystals, which dissolve in ammonia, forming a red solution with splendid green fluorescence. On adding bromine to its solution in glacial acetic acid, t e t r a b r o m o - r e s o r c i n - p h t h a l e i n, or E o s i n, gradually separates in crystals, which may be purified by conversion into a potassium-salt and precipitation with an acid. From dilute alcohol it separates in dull, flesh-colored crystals; from absolute alcohol in red crystals containing 1 molecule of alcohol. Its potassium-salt, $C_{20}H_6Br_4O_5K_2$, known in commerce as "s o l u b l e e o s i n," dyes silk of a fine rose-color.

P y r o c a t e c h i n - p h t h a l e i n, $C_{20}H_{12}O_5$, formed by gently heating pyrocatechin with phthalic anhydride and sulphuric acid, dissolves in potash-lye with a fine blue color. H y d r o q u i n o n e - p h t h a l e i n, formed in like manner, dissolves in strong sulphuric acid with brick-red color, in alkalies with violet color.

O r c i n - p h t h a l e i n, $C_{22}H_{16}O_5$, forms colorless needles, dissolving in alkalies with dark-red color, without fluorescence.

G a l l e i n or P y r o g a l l o l - p h t h a l e i n, $C_{20}H_{14}O_8$ =

$$C_6H_4 < \begin{array}{l} CO—C_6H_2(OH)_3 \\ CO—C_6H_2(OH)_3 \end{array}$$, produced by heating pyrogallol with phthalic

anhydride to 190°–200° C. (374°–392° F.), forms a brown-red powder or small crystals, with green surface-color. It dissolves in alcohol with dark-red, in potash-lye with fine blue color. By zinc-dust, or by zinc and sulphuric acid, it is converted into g a l l i n, $C_{20}H_{18}O_7$, which forms nearly colorless crystals, and is converted by heating to 200° C. (392° F.), with strong sulphuric acid, into c œ r u l e i n, $C_{20}H_{10}O_7$, a blue dye-stuff which dissolves in alkalies with green, in aniline with splendid blue color, and is reduced by zinc-dust to c œ r u l i n.

Normal Aromatic Alcohols.

These compounds, formed by substitution of OH for H in the lateral chains of the hydrocarbons homologous with benzene, contain the group CH_2OH, and are therefore primary alcohols, convertible by oxidation into aldehydes and acids containing the same number of carbon-atoms as the alcohols themselves, and producible by methods similar to those which yield the primary alcohols of the fatty series, viz.: (1) From the hydro-carbons by conversion, first into haloïd derivatives, such as $C_6H_5.CH_2Cl$, then into acetic ethers, and saponification of the latter with caustic alkalies ; e. g.,

$$C_6H_5.CH_2.OC_2H_3O \ + \ KOH \ = \ C_2H_3O.OK \ + \ C_6H_5.CH_2.OH.$$
Benzyl acetate.   Benzyl alcohol.

(2) From the aldehydes, by the action of nascent hydrogen, or by heating with alcoholic potash :

$$2(C_6H_5.CHO) \ + \ KOH \ = \ C_6H_5.CH_2.OH \ + \ C_6H_5.CO_2K.$$
Benzaldehyde.   Benzyl   Potassium
                                                   alcohol.   benzoate.

Monatomic Alcohols.

Benzyl Alcohol, $C_7H_8O = C_6H_5.CH_2OH$, may be prepared by the methods just mentioned, or by the action of nascent hydrogen on benzoic or hippuric acid. Its benzoic ether, $C_7H_5O_2.C_7H_7$, is one of the constituents of Peru and Tolu balsams.

Benzyl alcohol is a colorless, strongly refracting, oily liquid, having a specific gravity of 1.051 at 14° C. (57.2° F.), and boiling at 207° C. (404.6° F.). It is insoluble in water, but soluble in all proportions in common alcohol, ether, acetic acid, and carbon bisulphide. By oxygen in presence of platinum black, or by nitric acid, it is converted into b e n. z o i c a l d e h y d e ; by aqueous chromic acid, into b e n z o i c a c i d :

$$C_6H_5.CH_2OH \ + \ O \ = \ H_2O \ + \ C_6H_5.CHO,$$
and $C_6H_5.CH_2OH \ + \ O_2 \ = \ H_2O \ + \ C_6H_5.CO(OH).$

Strong hydrochloric acid converts it into b e n z y l c h l o r i d e, C_7H_7Cl. Distilled with acetic acid and strong sulphuric acid, it is converted into b e n z y l a c e t a t e, $C_7H_7(OC_2H_3O)$, a liquid having an odor of pears, and boiling at 210° C. (410° F.). B e n z y l o x a l a t e, $C_2O_4(C_7H_7)_2$, forms shining laminæ melting at 80° C. (176° F.).

Alcoholic benzyl ethers are formed by heating benzyl chloride with solutions of potash in the corresponding alcohols. M e t h y l b e n z y l a t e, $C_7H_7.O.CH_3$, boils at 168° C. (334.4° F.); e t h y l b e n z y l a t e, $C_7H_7.O.C_2H_5$, at 185° C. (365° F.). B e n z y l - p h e n v l a t e, $C_7H_7.O.C_6H_5$, formed by

heating benzyl chloride with potassium phenate, $C_6H_5.OK$, melts at 39° C·
$(102.2^O$ F.), boils at 287° C. $(548.6^O$ F.). D i b e n z y l e t h e r, $(C_7H_7)_2O$,
obtained by heating benzyl alcohol with boric oxide $[2C_7H_7(OH) — H_2O =$
$(C_7H_7)_2O]$, or by heating benzyl chloride with water to 190° C. $(374^O$ F.),
is an oil boiling above 300° C. $(572^O$ F.).

Chlorobenzyl alcohol (para), $C_7H_4Cl.CH_2OH$, produced by heating para-
chlorobenzyl chloride, $C_6H_4Cl.CH_2Cl$, with aqueous ammonia, forms long
needles, melts at 66° C. $(150.8^O$ F.), and boils without decomposition. Other
haloïd benzyl alcohols are formed in a similar manner.

Paranitrobenzyl alcohol, $C_6H_4(NO_2).CH_2OH$, produced by saponification of
paranitrobenzyl acetate(obtained by nitration of benzyl acetate), is soluble
in hot water, and melts at 93° C. $(199.4^O$ F.).

B e n z y l h y d r o s u l p h i d e or B e n z y l m e r c a p t a n, C_6H_5·
CH_2SH, obtained by the action of alcoholic potassium hydrosulphide on
benzyl chloride, is a liquid having an alliaceous odor, a specific gravity of
1.058 at 20° C. $(68^O$ F.), and boiling at 194° C. $(381.2^O$ F.). Its alcoholic solution gives, with metallic salts, precipitates consisting of benzyl-
mercaptides, *e. g.* $(C_6H_5.CH_2S)_2Hg$.

B e n z y l s u l p h i d e, $(C_6H_5.CH_2)_2S$, formed by the action of potassium
monosulphide on benzyl chloride, in alcoholic solution, crystallizes in
colorless needles melting at 49° C. $(120.2^O$ F.). By oxidation with nitric
acid, it is converted into b e n z y l o x y s u l p h i d e, $(C_6H_5.CH_2)_2SO$, which
is soluble in hot water, and melts at 130° C. $(266^O$ F.).

B e n z y l d i s u l p h i d e, $(C_6H_5.CH_2)_2S_2$, formed by oxidation of benzyl
mercaptan exposed to the air, crystallizes from alcohol in shining scales
melting at 66° C. $(150.8^O$ F.). Nascent hydrogen converts it into benzyl
mercaptan.

All these sulphur compounds of benzyl are decomposed by heat.

X y l y l A l c o h o l,* $C_8H_{10}O = C_6H_4{<}^{CH_3}_{CH_2OH}$.—Of the three isomeric modifications of this alcohol only the para-compound is known. It
is obtained from the corresponding aldehyde, and from paraxylyl chloride,
$C_6H_4(CH_3).CH_2Cl$, by the methods above described. It crystallizes in
needles, dissolves sparingly in water, melts at 59° C. $(138.2^O$ F.), and
boils at 217° C. $(422.6^O$ F.). Nitric acid converts it into toluic aldehyde,
$C_6H_4(CH_3).CHO$. By heating with hydrochloric acid it is reconverted into
xylyl chloride. Its acetic ether boils at 243° C. $(469.4^O$ F.). The chloride
treated with KHS and K_2S yields the compounds $C_8H_9(SH)$ and $(C_8H_9)_2S$.

Metameric with the xylyl alcohols are—(1) P r i m a r y p h e n y l - e t h y l
a l c o h o l, $C_6H_5.CH_2.CH_2.OH$, formed by the action of sodium amalgam on
a solution of alpha-toluic (phenyl-acetic) aldehyde, $C_6H_5.CH_2.CHO$, dissolved in aqueous alcohol. Colorless liquid, boiling at 212° C. $(413.6^O$ F.).
Specific gravity, 1.0337 at 21° C. $(69.8^O$ F.). Converted by gradual oxidation into alpha-toluic acid. Its acetic ether boils at 224° C. $(435.2^O$ F.).

(2) S e c o n d a r y p h e n y l - e t h y l a l c o h o l, $C_6H_5.CHOH.CH_3$, prepared from bromethyl-benzene, $C_6H_5.CHBr.CH_3$ (p. 855), in the same
manner as benzyl alcohol from benzyl chloride; and by the action of
sodium-amalgam on acetophenone, $C_6H_5.CO.CH_3$. It is a colorless liquid,
boiling at $202^O–203^O$ C. $(395.6^O–397.4^O$ F.), having a density of 1.013, and
reconverted by oxidation into acetophenone.

---

* Generally called tolyl alcohol ; but this name is inappropriate. If $C_6H_4(CH_3)$.
CH_2OH is to be called *tolyl alcohol*, then $C_6H_5.CH_3OH$ should by analogy be called,
not benzyl alcohol, but *phenyl alcohol*. The true tolyl alcohol is cresol,
$C_6H_4{<}^{CH_3}_{OH}$, metameric with benzyl alcohol, $CH_2{<}^{C_6H_5}_{OH}$.

Phenyl-propyl Alcohol, $C_9H_{12}O = C_6H_5.CH_2.CH_2.CH_2.OH$, produced by the action of nascent hydrogen on styryl alcohol or cinnamic alcohol, $C_9H_{14}O$, is a liquid boiling at $235°$ C. ($455°$ F.). Secondary phenyl-propyl alcohol, $C_6H_5.CH_2.CH(OH).CH_3$, formed by the action of nascent hydrogen on ethyl-phenyl ketone, is a liquid boiling at $210°$–$211°$ C. ($410°$–$411.8°$ F.).

Cymyl Alcohol, $C_{10}H_{14}O = C_{10}H_{13}(OH) = C_9H_{11}.CH_2OH$, also called *Cumylic Alcohol.*—This alcohol, discovered by Kraut, is produced, together with cumic acid, $C_{10}H_{12}O_2$, by the action of alcoholic potash on cumic aldehyde:

$$2C_{10}H_{12}O + KOH = C_{10}H_{11}KO_2 + C_{10}H_{14}O.$$

It is a colorless liquid, boiling at $243°$ C. ($469.4°$ F.), insoluble in water, soluble in all proportions in common alcohol and ether. Nitric acid converts it into c u m i c a c i d. Boiled with alcoholic potash, it is converted into p o t a s s i u m c u m a t e and c y m e n e:

$$3C_{10}H_{14}O + KOH = C_{10}H_{11}KO_2 + 2C_{10}H_{14} + 2H_2O.$$

Hydrochloric acid gas converts it into c y m y l c h l o r i d e, $C_{10}H_{13}Cl$.
Metameric with this compound is:

Benzyl-dimethyl Carbinol, or **Phenyl-katabutyl Alcohol,**

$$HO—C\genfrac{}{}{0pt}{}{\diagup CH_3}{\diagdown CH_3}—CH_2(C_6H_5),$$

obtained, similarly to tertiary butyl alcohol, by the action of zinc-methyl on phenyl-acetic or *a*-toluic chloride, $C_6H_5.CH_2.COCl$. It crystallizes in long colorless needles, melts at $20°$–$22°$ C. ($68°$–$71.6°$ F.), and boils at $220°$–$230°$ C. ($428°$–$446°$ F.).

Sycoceryl Alcohol, $C_{18}H_{30}O = C_{18}H_{29}(OH) = C_{17}H_{27}.CH_2OH$. —This compound, discovered by De La Rue and Müller, is produced by the action of alcoholic soda on sycoceryl acetate (a crystalline substance extracted from the resin of *Ficus rubiginosa*), and purified by precipitation with water or by crystallization from common alcohol. It forms very thin crystals resembling caffeine, and melting at $90°$ C. ($194°$ F.) to a liquid heavier than water. It is slowly attacked by dilute nitric acid, yielding a crystalline mass apparently consisting of a mixture of s y c o c e r i c a c i d, $C_{18}H_{28}O_2$, and n i t r o s y c o c e r i c a c i d, $C_{18}H_{27}(NO_2)O_2$. Boiled with dilute aqueous chromic acid, it yields thin prisms, probably of s y - c o c e r i c a l d e h y d e, $C_{18}H_{28}O$. With acetyl chloride, it forms crystalline s y c o c e r y l a c e t a t e:

$$C_{18}H_{29}OH + C_2H_3OCl = HCl + C_{18}H_{29}.O.C_2H_3O.$$

With benzoic acid it yields, in like manner, s y c o c e r y l b e n z o a t e, $C_{18}H_{29}.O.C_7H_5O$, which crystallizes in prisms from solution in benzene or chloroform.

The resin of *Ficus rubiginosa*, an Australian plant, is resolved by treatment with alcohol, into about 73 per cent. of s y c o r e t i n, soluble in cold alcohol, 14 per cent. of sycoceryl acetate, soluble in hot alcohol, and 13 per cent. of residue, consisting of caoutchouc, sand, and fragments of bark. S y c o r e t i n is an amorphous white neutral resin, very brittle, and highly electric; it melts in boiling water to a thick liquid which floats on the surface. It dissolves easily in alcohol, ether, chloroform, and oil of turpentine.

Diatomic Alcohols.

Benzylene Glycol, $C_6H_5.CH(OH)_2$, appears (for reasons already assigned, p. 628), to be incapable of existing, the reactions which might be expected to produce it leading in reality to the production of benzaldehyde, $C_6H_5.CHO$. Its ethers are formed from benzal chloride, $C_6H_5.CHCl_2$ (p. 854) by the action of sodium alcoholates, or of the salts of organic acids. The *dimethylic ether*, $C_6H_5.CH(OCH_3)_2$, boils at 208º C. (406.4º F.); the *diethylic ether*, $C_6H_5.CH(OC_2H_5)_2$ at 222º C. (431.6º F.). The *acetic ether*, $C_6H_5.CH(O.C_2H_3O)_2$, is crystalline; melts at 36º C. (96.8º F.), and boils at 190º C. (374º F.), being resolved at the same time into benzaldehyde and acetic anhydride.

Xylylene Glycol (para), $C_6H_4{<}{CH_2.OH \atop CH_2.OH}$, obtained by heating paraxylylene bromide or dibromoparaxylene, $C_6H_4(CH_2Br)_2$, with water to 170º–180º C. (338º–356º F.), crystallizes in colorless needles melting at 112º–113º C. (233.6º–235.4º F.), easily soluble in water, converted by oxidizing agents into terephthalic acid. Its diacetic ether, $C_6H_4(CH_2.OC_2H_3O)_2$, melts at 47º C. (116.6º F.).

The following compounds have the hydroxyl groups, partly in the principal, partly in the lateral chains, and are therefore both alcohols and phenols.

Saligenin, $C_7H_8O_2 = C_6H_4(OH)—CH_2OH$, or *Ortho-oxybenzyl alcohol*, is formed by the action of nascent hydrogen on salicylic aldehyde, $C_6H_4(OH)$. CHO; and by decomposition of salicin, $C_{13}H_{18}O_7$ (a bitter substance found in willow and poplar bark), under the influence of emulsin or synaptase, the ferment of sweet almonds.

$$C_{13}H_{18}O_7 \ + \ H_2O \ = \ C_6H_{12}O_6 \ + \ C_7H_8O_2 \ .$$
$$\text{Salicin.} \qquad\qquad \text{Glucose.} \qquad \text{Saligenin.}$$

Saligenin forms colorless, nacreous scales, freely soluble in water, alcohol, and ether. It melts at 82º C. (179.6º F.), and decomposes at about 100º. Dilute acids at boiling heat convert it into s a l i r e t i n , C_7H_6O, a resinous substance differing from saligenin by the elements of water. The same substance is produced directly from salicin by boiling with dilute acids. Many oxidizing agents, as chromic acid and silver oxide, convert saligenin into salicylic aldehyde and salicylic acid : this shows that it is an ortho-compound. The aqueous solution of saligenin gives a deep indigo-blue color with ferric salts.
Chlorinated derivatives of saligenin, viz., $C_7H_7ClO_2$ and $C_7H_6Cl_2O_2$, are obtained by the action of synaptase on the corresponding chlorosalicin.

Anisyl Alcohol, or **Methyl-paraoxybenzyl Alcohol,** $C_8H_{10}O_2 = C_6H_4{<}{OCH_3 \atop CH_2OH}$, is prepared from anisaldehyde (p. 918), in the same manner as benzyl alcohol from benzaldehyde. It crystallizes in colorless shining prisms ; has a faint odor and burning taste ; melts at 25º C. (77º F.), and distils undecomposed at 258.8º C. (497.8º F.). By oxidizing agents it is converted into anisaldehyde and anisic acid ; by hydrochloric acid into a volatile chloride, $C_6H_4(OCH_3).CH_2Cl$.

Vanillic, or **Methyl-protocatechuic Alcohol,** $C_8H_{10}O_3 =$

$C_6H_3 {\underset{\diagdown CH_2.OH}{\overset{\diagup OCH_3}{\underset{\displaystyle -OH}{}}}}$, obtained by the action of sodium-amalgam on an aqueous
solution of vanillin ($q. v.$), forms colorless prismatic crystals, melting at
103°–105° C. (217.4°–221° F.), soluble in water and in alcohol.

Piperonylic, or **Methene-protocatechuic Alcohol,** $C_8H_6O_3 =$

$C_6H_3 {\underset{\diagdown CH_2OH}{\overset{\diagup O\diagdown}{\underset{\displaystyle -O\diagup}{}}} CH^2}$, obtained in like manner from piperonal ($q. v.$), forms
long colorless prisms, melting at 57° C. (134.6° F.), sparingly soluble in
cold, more easily in hot water, very soluble in alcohol.

A tr.iatomic alcohol called **Stycerin,** or **Phenyl-glycerin,**
$C_9H_{12}O_3 = C_6H_5.CH(OH).CH(OH).CH_2OH$, formed by heating the corresponding dibromhydrin, $C_6H_5.CHBr.CHBr.CH_2OH$, with water to 100° for
24 hours, is a light yellow gummy mass, easily soluble in water and in
alcohol. The *dibromhydrin*, obtained by direct combination of bromine
with styryl alcohol, forms large colorless shining plates or slender needles,
melting at 74° C. (165.2° F.). The tribromhydrin, $C_9H_9Br_3 = C_6H_5.$
$CHBr.CHBr.CH_2Br$, obtained by treating the dibromhydrin with excess of
concentrated hydrobromic acid, forms shining needles melting at 124° C.
(255.2° F.).

---

2. UNSATURATED ALCOHOLS and PHENOLS :

Cinnyl Alcohol, Styryl Alcohol, or **Styrone,** $C_9H_{10}O = C_9H_9OH$, or
$C_6H_5CH\!\!=\!\!CH.CH_2OH$, is obtained by heating styracin or cinnyl cinnamate,
$C_9H_9(OC_9H_7O)$ (a compound contained in liquid storax and in balsam of
Peru), with caustic alkalies. It crystallizes in soft silky needles, having
a sweet taste and an odor of hyacinths, melting at 33° C. (91.4° F.), and
volatilizing, without decomposition, at a higher temperature. It is moderately soluble in water, freely in alcohol and ether. By oxidizing agents
it is converted into c i n n a m i c a l d e h y d e, C_9H_8O, and c i n n a m i c
a c i d, $C_4H_8O_2$, being related to those compounds in the same manner as
ethyl alcohol to acetic aldehyde and acetic acid. With *fuming sulphuric
acid* it forms a sulpho-acid, $C_9H_{10}SO_3$, the barium-salt of which is soluble
in water. It unites directly with 2 atoms of bromine, forming styceric
dibromhydrin, $C_9H_{10}Br_2O$, described above.

Allyl Phenol, or **Anol,** $C_9H_{10}O = C_6H_4(OH)(C_3H_5)$, formed by fusing
anisaldehyde (p. 918) with potash, crystallizes in shining laminæ, melting at 92° C. (197.6° F.), and decomposing when distilled.

Methyl-allyl Phenol, Anethol, or **Anise-camphor,** $C_{10}H_{12}O =$
$C_6H_4(OCH_3)(C_3H_5)$, is a constituent of the volatile oils of anise, fennel,
and tarragon, and separates therefrom on cooling in soft shining scales,
melting at 20° C. (68° F.), and boiling at 225° C. (437° F.). By oxidation with chromic acid, it is resolved into acetic and anisic acids. On
heating it with hydriodic acid, the methyl group is separated, and the
mass becomes resinized.

Eugenol, $C_{10}H_{12}O_2 = C_6H_3(OH)(OCH_3)(C_3H_5)$, (*Eugenic*, or *Caryophyllic
acid*), occurs in oil of cloves (from the flowers of *Caryophyllus aromaticus*),

39

in oil of pimento (from *Myrtus pimenta*), and a few other volatile oils, and may be obtained therefrom by solution in potash, filtration, and precipitation by carbonic acid. It is a colorless, aromatic liquid, of sp. gr. 1.0779 at 0°, boiling at 247.5° C. (477.5° F.), soluble in alcohol. Ferric chloride colors the solution blue. Eugenol, heated with hydriodic acid, gives off methyl iodide. By fusion with potash it is resolved into acetic acid and protocatechuic acid, $C_6H_3(OH)_2.CO_2H$. These reactions determine its constitution.

Eugenol dissolves in soda-lye, forming sodium-eugenol—$C_6H_3(C_3H_5)$ $(ONa)(OCH_3)$, which is converted by methyl iodide into methyl-eugenol, $C_6H_3(C_3H_5)(OCH_3)_2$, which, by oxidation with chromic acid mixture, yields dimethyl-protocatechuic acid, $C_6H_3(OCH_3)_2.CO_2H$. In like manner are formed ethyl-eugenol, propyl-eugenol, etc.

Coniferyl alcohol, $C_{10}H_{12}O_3 = C_6H_3(OH)(OCH_3).C_2H_4OH$, is formed by the decomposition of coniferin (see GLUCOSIDES) under the influence of emulsin. It crystallizes in colorless prisms, melting at 74°–75° C. (165.2°–167° F.) ; is insoluble in cold water, slightly soluble in hot water, easily in ether and in alkalies. Dilute acids convert it into a resinous polymeride. By oxidation with chromic acid mixture it yields vanillin (p. 919).

Cholesterin, $C_{26}H_{44}O$, a product of the animal organism, is homologous with cinnyl alcohol, and has the characters of a monatomic alcohol. It is found in small quantity in various parts of the animal system, as in the bile, the brain and nerves, and the blood : it forms the chief ingredient of *biliary calculi*, from which it is easily extracted by boiling the powdered gall-stones in strong alcohol, and filtering the solution while hot ; on cooling, the cholesterin crystallizes in brilliant colorless plates. It is a fatty substance, insoluble in water, tasteless, and inodorous ; it is freely soluble in boiling alcohol and in ether, also in chloroform, and crystallizes from the alcoholic solution in beautiful white laminæ having a mother-of-pearl lustre. It melts at 137° C. (278.6° F.), and sublimes at 200° C. (392° F.). On adding a solution of cholesterin in chloroform to strong sulphuric acid, the chloroform becomes purple-red, while the sulphuric acid below it exhibits a greenish-yellow fluorescence : the red chloroform solution evaporated in a porcelain capsule turns blue, green, and finally yellow.

Heated with strong sulphuric acid, it gives up water, and yields a resinous hydrocarbon, $C_{26}H_{42}$. With nitric acid it yields cholesteric acid, $C_8H_{10}O_5$, together with other products. With chlorine and bromine it forms substitution-products. Heated to 200° C. (392° F.) with acetic, butyric, benzoic, and stearic acids, it forms compound ethers. The *acetate* and *stearate* crystallizes in needles, the former melting at 92° C. (197.6° F.), the latter at 200° C. (392° F.). The *benzoate* forms thick prisms, melting at 125°–130° C. (257°–266° F.). With PCl_5, or strong hydrochloric acid, it yields the *chloride*, $C_{26}H_{43}Cl$, which crystallizes in needles, and is converted by ammonia into *cholesteramine*, $C_{26}H_{45}.NH_2$.

Isocholesterin, $C_{26}H_{44}O$, occurs, together with cholesterin, in the grease of sheep's wool, and may be separated by saponifying the fat, heating the mixture of cholesterin and isocholesterin thus obtained with benzoic acid, whereby they are converted into benzoic ethers, and crystallizing these compounds from common ether, the cholesteryl benzoate separating in thick tabular crystals, the isocholesteryl benzoate in slender needles, and from the latter the isocholesterin may be obtained by heating with alcoholic potash. It crystallizes from ether or acetone in slender needles, melting at 137°–138° C. (278.6°–280.2° F.). It does not give any color reaction with chloroform and sulphuric acid, but in other respects it reacts like cholesterin. Its benzoic ether melts at 190°–191° C. (374°–375.8° F.).

AROMATIC ALDEHYDES.

1. ALDEHYDES ANALOGOUS TO MONATOMIC ACIDS.

Four aldehydes are known, belonging to the series $C_nH_{2n-8}O$, analogous to benzoic acid and its homologues, viz.:

Benzoic Aldehyde,	C_7H_6O
Toluic Aldehyde,	C_8H_8O
Cumic Aldehyde,	$C_{10}H_{12}O$
Sycocerylic Aldehyde,	$C_{18}H_{28}O$.

These aldehydes exhibit the same general reactions as those of the fatty series, and are obtained by similar processes.

Benzoic Aldehyde, Benzaldehyde, or **Bitter-almond oil,** $C_7H_6O = C_6H_5.CHO = C_7H_5O.H$, is produced—1. By the oxidation of benzyl alcohol. 2. By the action of nascent hydrogen on chloride or cyanide of benzyl:

$$C_6H_5.COCl + H_2 = HCl + C_6H_5.CHO.$$

3. By heating benzal chloride, $C_6H_5.CHCl_2$, with water to 130°–140° C. (266°–284° F.). 4. By heating benzyl chloride, $C_6H_5.CH_2Cl$, with an aqueous solution of lead nitrate. 5. By the oxidation of amygdalin with nitric acid. 6. By digesting bitter almonds with water for five or six hours at 30°–40° C. (86°–104° F.). The synaptase present then acts as a ferment on the amygdalin, converting it into glucose, benzoic aldehyde, and hydrocyanic acid (see GLUCOSIDES).

7. Benzoic aldehyde is formed, together with many other products, by the action of a mixture of manganese dioxide and sulphuric acid on albumin, fibrin, casein, and gelatin.

Pure benzoic aldehyde is a thin, colorless liquid, of great refractive power, and peculiar, very agreeable odor; its density is 1.050 at 15° C. (59° F.), and its boiling point 180° C. (356° F.): it is soluble in about 30 parts of water, and miscible in all proportions with alcohol and ether. Exposed to the air, it rapidly absorbs oxygen, and is converted into crystallized benzoic acid. Heated with solid *potassium hydroxide*, it gives off hydrogen, and yields potassium benzoate. With the *alkaline bisulphites* it forms beautiful crystalline compounds. The vapor of the oil is inflammable, and burns with a bright flame and much smoke.

Benzoic aldehyde, treated with *sodium amalgam*, is converted into benzyl alcohol, C_7H_8O. With *phosphorus pentachloride* it yields benzal chloride, $C_7H_6Cl_2$:

$$C_6H_5.CHO + PCl_5 = PCl_3O + C_6H_5.CHCl_2.$$

Ammonia converts it into h y d r o b e n z a m i d e, a white crystalline neutral body, which, when boiled with aqueous potash, is converted into an isomeric basic compound called a m a r i n e:

$$3C_7H_6O + 2NH_3 = (C_7H_6)_3N_2 + 3H_2O$$

All the aromatic aldehydes act with ammonia in a similar manner, and are thereby distinguished from the fatty aldehydes.

Amarine and hydrobenzamide yield by dry distillation an isomeric base, $C_{21}H_{18}N_2$, called l o p h i n e, which crystallizes in long, sparingly soluble needles, melting at 290° C. (554° F.), and unites with acids, forming crystalline salts.

Toluic Aldehyde, $C_8H_8O = C_6H_4{<}{CH_3 \atop CHO}$ (1 : 4), is produced by distilling a mixture of the calcium-salts of *para*-toluic and formic acids. The oily distillate agitated with acid sodium sulphite, forms a crystalline compound, which, when distilled with sodium carbonate, yields the alde-hyde, as an oil having a peppery odor, and boiling at 204° C. (399.2° F.). On exposure to the air, it is gradually converted into para-toluic acid, $C_8H_8O_2$. With alcoholic potash it forms potassium para-toluate and para-xylyl alcohol :

$$2C_8H_8O . + KOH = C_8H_7KO_2 + C_8H_{10}O .$$

Cumic Aldehyde, $C_{10}H_{12}O = C_6H_4{<}{C_3H_7 \atop CHO}$ (1 : 4), exists, to-gether with cymene, $C_{10}H_{14}$, in the essential oil of cumin, and in that of water-hemlock (*Cicuta virosa*), and may be obtained by agitating either of these oils with acid sodium sulphite, which takes up the cumic aldehyde, but not the cymene, and forms a crystalline compound, from which the aldehyde may be separated by distillation with potash. Cumic aldehyde is a colorless or slightly yellow liquid having a powerful odor, and boiling at 237° C. (458.6° F.). It is easily oxidized in the air, so that it must be distilled in a current of carbonic acid gas. It is converted into cumic acid, $C_{10}H_{12}O_2$, by oxidation, and by alcoholic potash into potassium cumate and cymyl alcohol, $C_{10}H_{14}O$.

Sycocerylic Aldehyde, $C_{18}H_{28}O$, is produced in thin prisms by oxi-dizing sycoceryl alcohol with aqueous chromic acid.

Cinnamic Aldehyde, C_9H_8O.—This compound, which is the only known member of the series of aldehydes $C_nH_{2n-10}O$, constitutes the es-sential part of the volatile oils of cinnamon and cassia, which are ob-tained from the bark of different trees of the genus *Cinnamomum*, order *Lauraceæ*—viz., oil of cinnamon from Ceylon cinnamon, and oil of cassia from Chinese cinnamon. The aldehyde may be separated from these oils by means of acid potassium sulphite. It is produced artificially by oxi-dation of styryl alcohol ; by dry distillation of a mixture of cinnamate and formate of calcium, and by saturating a mixture of benzaldehyde and acetaldehyde with hydrochloric acid :

$$C_6H_5.CHO + CH_3.CHO = H_2O + C_6H_5.CH{=}CH.COH .$$

This last reaction is analogous to the formation of crotonic aldehyde by condensation of acetaldehyde (p. 691).

Cinnamic aldehyde is a colorless oil, rather heavier than water. It may be distilled without alteration in a vacuum, or with de-aerated water, but absorbs oxygen quickly on exposure to the air, and passes into cin-namic acid. When fused with potash, it forms potassium cinnamate, and gives off hydrogen : $C_9H_8O + KOH = C_9H_7KO_2 + H_2$. Ammonia gas converts it into hydrocinnamide : $3C_9H_8O + 2NH_3 = (C_9H_8)_3N_2 + 3H_2O$.

---

2. ALDEHYDES ANALOGOUS TO DIATOMIC AND MONOBASIC ACIDS.

Salicylic Aldehyde, or **Salicylal,** $C_7H_6O_2 = C_6H_4{<}{OH \atop CHO}$, (1 : 2), *Oxybenzaldehyde,* formerly called *Salicylous Acid.*—This compound occurs in herbaceous Spiræas, especially in the flowers of meadow-sweet (*Spiræa ulmaria*). It is formed artificially by oxidizing saligenin, $C_6H_4(OH).CH_2OH$ (p. 912), or its glucosides, salicin and populin (see GLUCOSIDES), and, together with the isomeric para-compound, by the action of chloroform on an alkaline solution of phenol:

$$C_6H_5(OH) + CHCl_3 + H_2O = 3HCl + C_6H_4(OH).CHO.$$

This last reaction affords the easiest method of preparing salicylal. Chloroform (3 parts) is gradually added, with agitation, to a solution of phenol (2 parts) and sodium hydroxide (4 parts) in 6–7 parts of water at 50°–60° C. (122°–140° F.), in a vessel with reversed condenser; and a quantity of water is added, sufficient to form (after heating to 60° C. [140° F.] for half an hour) a clear red-brown liquid, which is to be kept boiling for two hours, then acidulated and distilled. Salicylal and phenol then pass over, while para-oxybenzaldehyde remains behind. The distillate is shaken with ether, and the salicylic aldehyde is separated from the ethereal solution by agitation with acid sodium sulphite.

Salicylal is a thin, colorless, fragrant oil, acquiring a red tint by exposure to the air. It has a specific gravity of 1.1725 at 15° C. (59° F.), solidifies at —20° C. (—4° F.), boils at 196° C. (384.8° F.), and burns with a bright smoky flame. Water dissolves a perceptible quantity of salicylal, acquiring its fragrant odor, and the property (likewise exhibited by salicylic acid) of producing a deep-violet color with ferric salts. Alcohol and ether dissolve it in all proportions.

Salicylal is oxidized to salicylic acid by boiling with cupric oxide in alkaline solution, partly also by potassium dichromate and sulphuric acid; it likewise reduces silver oxide. When heated with potassium hydroxide it is converted into potassium salicylate, with evolution of hydrogen:

$$C_7H_6O_2 + KOH = C_7H_5KO_3 + H_2.$$

By nascent hydrogen it is converted into saligenin, $C_7H_8O_2$; by ammonia into hydrosalicylamide:

$$3C_7H_6O_2 + 2NH_3 = 3H_2O + C_{21}H_{18}O_3N_2.$$

Salicylal is attacked by chlorine and bromine, forming $C_7H_5ClO_2$ and $C_7H_5BrO_2$, both of which are crystalline bodies, having acid properties. With moderately strong nitric acid it forms nitrosalicylal, $C_7H_5(NO_2)O_2$, which is also crystallizable, and forms crystallizable salts. With PCl_5, at ordinary temperatures, salicylal forms orthoxybenzal chloride, $C_6H_4(OH).CHCl_2$, crystallizing in prisms, and melting at 82° C. (179.6° F.); and when heated with PCl_5, it yields orthochlorobenzal chloride, $C_6H_4Cl.CHCl_2$, a liquid boiling at 227°–230° C.(440.6°–446° F.), and isomeric with parachlorobenzal chloride from toluene (p. 854).

Salicylal dissolves in alkalies, forming crystalline compounds, formerly called *salicylites*. The *potassium compound*, $C_6H_4(OK).CHO$, forms square plates, easily soluble in water and in alcohol, and decomposing quickly when exposed in the moist state to the air.

Methyl-salicylal, $C_6H_4(OCH_3).CHO$, obtained by the action of methyl iodide on potassium salicylal, is a liquid boiling at 238° C. (460.4° F.). Ethyl-salicylal, prepared in like manner, boils at 248° C. (478.4° F.).

Acetyl-salicylal, $C_9H_8O_3 = C_6H_4(OC_2H_3O).CHO$, metameric with coumaric acid (*q. v.*), is formed by the action of acetic anhydride on sodium-salicylal at ordinary temperatures: $C_6H_4(ONa).CHO + (C_2H_3O)_2O = C_7H_3O.ONa + C_6H_4(OC_2H_3O).CHO$. The acetic oxide is added to an equivalent quantity of powdered anhydrous sodium-salicylal suspended in pure dry ether, and after the whole has stood for twenty-four hours, the ethereal liquid is filtered off from the sodium acetate, then evaporated, and the crystalline cake which separates on cooling is purified by pressure between bibulous paper, and crystallization from alcohol. Acetosalicylal thus prepared melts at 37° C. (98.6° F.), boils at about 253° C. (487.4° F.), and distils without decomposition. It forms definite compounds with alkaline bisulphites. It is decomposed by alcoholic potash, with formation of potassium acetate and potassium-salicylal:

$$C_6H_4(C_2H_3O)O.CHO + 2KOH = C_6H_4KO.CHO + C_2H_3KO_2 + H_2O.$$

Acetosalicylal likewise unites directly with acetic oxide.

If the product of the action of acetic oxide on salicylal, instead of being treated in the manner above described, be poured into water after a few minutes' boiling, an oily liquid sinks to the bottom, and sodium acetate remains in solution; and on distilling this oil, and collecting apart that which passes over after the temperature has risen to 290° C. (554° F.), a crystalline substance is obtained, having the composition of *acetosalicylal minus one molecule of water:* this substance is identical in every respect with coumarin or coumaric anhydride,* the odoriferous principle of the Tonka bean (see COUMARIC ACID).

Paraoxybenzaldehyde, $C_6H_4(OH).CHO$ (1 : 4), is produced together with salicylic aldehyde, by the action of chloroform on phenol dissolved in caustic soda (p. 917), and may be extracted by ether from the filtered residue of the distillation, and purified by recrystallization from water. It forms stellate groups of slender needles; melts at 115°–116° C. (239°–240.8° F.); sublimes without decomposition, dissolves sparingly in cold, more readily in hot water, easily in alcohol and ether. Ferric chloride colors its solution dingy violet. It forms easily soluble compounds with alkaline bisulphites.

Anisaldehyde or **Methyl-paraoxybenzaldehyde,** $C_6H_4(OCH)_3$. CHO, isomeric with methyl-salicylal, is formed, together with anisic acid, by oxidation of anisic alcohol, $C_8H_{10}O_2$, in contact with platinum black; also by the oxidizing action of dilute nitric acid on the volatile oils of anise, fennel, and tarragon, which contain anethol, $C_{10}H_{12}O$. The product of the oxidation is shaken up with the acid sodium sulphite, and the resulting crystalline compound is decomposed by solution of sodium carbonate.

Anisaldehyde is a colorless oil, having an aromatic odor and burning taste, a density of 1.123 at 15° C. (59° F.), and boiling at 248° C. (478.4° F.). It is nearly insoluble in water, but soluble in all proportions in alcohol and ether. It is converted by oxidation into anisic acid, $C_8H_8O_3$; by nascent hydrogen into anisic alcohol, $C_8H_{10}O_2$; and forms crystalline compounds with alkaline bisulphites. *Ammonia* converts it into anishydramide, $C_{24}H_{24}O_3N_2$. By *alcoholic potash* it is decomposed in the same manner as benzoic aldehyde, yielding potassium anisate and anisic alcohol:

$$2C_8H_8O_2 + KOH = C_8H_7KO_3 + C_8H_{10}O_2.$$

* P e r k i n , Chem. Soc. Journ., 1868, p. 181.

3. Aldehydes Analogous to Monobasic and Triatomic Acids.

Dioxybenzaldehyde or **Protocatechuic Aldehyde,** $C_7H_6O_3 =$ $C_6H_3(OH)_2CHO$ (CHO : OH : OH : $= 1 : 3 : 4$), produced by the action of chloroform on an alkaline solution of pyrocatechin :

$$C_6H_4(OH)_2 \ + \ CHCl_3 \ + \ H_2O \ = \ C_6H_3(OH)_2.CHO \ + \ 3HCl\ ;$$

also by boiling dichloropiperonal with water, and by heating vanillin or piperonal with dilute hydrochloric acid to 200° C. (392° F.). Flat shining crystals, melting at 150° C. (302° F.), easily soluble in water. The aqueous solution is colored deep green by ferric chloride; by fusion with potash, the aldehyde is converted into protocatechuic acid.

Vanillin, $C_8H_8O_3 = C_6H_3(OH)(OCH_3).CHO$, *Methyl-protocatechuic Aldehyde*, is the odoriferous principle of vanilla (the fruit of *Vanilla aromatica*), in which it exists to the amount of about 2 per cent. It is produced artificially from c o n i f e r i n , $C_{16}H_{22}O_8$, a glucoside occurring in the cambial secretion of all coniferous plants, by oxidation with chromic acid mixture; or the coniferin may be resolved by boiling with dilute acids, or by the action of emulsin, into glucose and coniferyl alcohol (p. 914) :

$$C_{16}H_{22}O_8 \ + \ H_2O \ = \ C_6H_{12}O_6 \ + \ C_{10}H_{12}O_3\ ,$$

and the coniferyl alcohol oxidized by chromic acid mixture. Vanillin is also produced (similarly to protocatechuic aldehyde) by the action of chloroform on an alkaline solution of guaiacol (methyl pyrocatechin, p. 518).

Vanillin crystallizes in stellate groups of colorless needles, melts at 80°–81° C. (176°–177.8° F.), and sublimes when cautiously heated ; dissolves sparingly in cold, more freely in hot water, easily in alcohol and ether. It forms crystalline compounds with bases. When boiled with dilute hydrochloric acid, it is resolved into methyl chloride and protocatechuic aldehyde ; and when fused with potash, it is converted, by oxidation of the aldehyde-group CHO, and separation of the methyl-group into protocatechuic acid. Bromine converts it into b r o m o v a n i l l i n , $C_8H_7BrO_3$, which crystallizes in yellowish laminæ melting at 161° C. (321.8° F.).

D i m e t h y l - p r o t o c a t e c h u i c Aldehyde, or M e t h y l - v a n i l l i n , $C_6H_3(OCH_3)_2.CHO$, and E t h y l - m e t h y l - p r o t o c a t e c h u i c A l d e h y d e , or E t h y l - v a n i l l i n , $C_6H_3(OCH_3)(OC_2H_5).CHO$, are formed by the action of methyl iodide or ethyl iodide on potassium-vanillin. Both crystallize in colorless prisms, slightly soluble in cold water, somewhat more easily in alcohol and ether. The former melts at 15°–20° C. (59°–68° F.), and boils at 285° C. (545° F.) ; the latter melts at 64°–65° C. (147.2°–149° F.), and easily sublimes.

Piperonal, $C_8H_6O_3 = CH_2{<}^O_O{>}C_6H_3{-}CHO$, *Methene-protocatechuic Aldehyde*, is obtained by distilling a dilute solution of potassium piperate with potassium permanganate, and agitating the distillate with ether. It forms long, colorless, shining crystals, having a very pleasant odor ; melts at 37° C. (98.6° F.), boils at 263° C. (505.4° F.) ; dissolves sparingly in cold, more freely in hot water, easily in alcohol and ether ; unites with acid sulphites of alkali-metal. Heated with 3 mol. PCl_5, it yields liquid dichloropiperonal chloride, $C_8H_4Cl_4O_3$, which is resolved by cold water into hydrochloric acid and dichloropiperonal, $C_8H_4Cl_2O_3 = CCl_2{<}^O_O{>}C_6H_3{-}$CHO, and by boiling with water into CO_2 and protocatechuic aldehyde.

4. ALDEHYDES ANALOGOUS TO BIBASIC ACIDS.

Phthalic Aldehyde, $C_8H_6O_2 = C_6H_4(CHO)_2$ (1 : 2), is formed by treating phthalyl chloride, $C_8H_4O_2Cl_2$, with zinc and hydrochloric acid, or by dissolving magnesium in a cooled solution of phthalyl chloride, and may be extracted by neutralizing the solution with sodium carbonate and agitating with ether. It crystallizes from hot water in small colorless rhombic tables, melts at 65°, and sublimes when cautiously heated; dissolves sparingly in cold, more readily in boiling water, easily in alcohol and ether.

Terephthalic Aldehyde (1 : 4) is formed by prolonged boiling of tollylene chloride (p. 856) with water (20 parts) and lead nitrate (1 part), and subsequent distillation. Crystallizes in slender needles, melting at 114°–115° C. (237.2°–239° F.).

Both these aldehydes dissolve sparingly in cold, more readily in boiling water, easily in alcohol; the ortho-compound is moderately soluble, the para-compound easily soluble in ether. Both form crystalline compounds with acid sodium sulphite.

KETONES.

The aromatic ketones may contain either two aromatic alcohol-radicles (phenyl and its homologues), or one aromatic radicle and one alcohol-radicle (methyl, etc.) belonging to the fatty series. The latter only will be noticed in this place, the former in connection with the diphenyl-compounds.

Phenyl-methyl Ketone or **Acetophenone,** $C_6H_5.CO.CH_3$, is formed by distilling a mixture of benzoate and acetate of calcium, or by the action of zinc-methyl on benzoyl chloride:

$$2(C_6H_5.CO.Cl) \ + \ Zn(CH_3)_2 \ \rightleftharpoons \ ZnCl_2 \ + \ 2(C_6H_5.CO.CH_3).$$

It crystallizes in large laminæ melting at 14° C. (57.2° F.), is converted by nascent hydrogen into phenyl-ethyl alcohol (p. 910), and resolved by oxidation with chromic acid into benzoic acid and carbon dioxide.

Phenyl-ethyl Ketone or **Propiophenone,** $C_6H_5.CO.C_2H_5$, from benzoate and propionate of calcium, and by the action of zinc-ethyl on benzoyl chloride, is a liquid which boils at 208°–210° C. (406.4°–410° F.), is converted by nascent hydrogen into secondary phenyl-propyl alcohol, and resolved by chromic acid into benzoic and acetic acids.

Phenyl-propyl Ketone and **Phenyl-isopropyl Ketone,** $C_6H_5.CO.$ C_3H_7, are formed by distilling a mixture of calcium benzoate and butyrate or isobutyrate respectively. The former boils at 220°–222° C. (428°–431.6° F.), and is oxidized by chromic acid to benzoic and propionic acids; the latter boils at 215° C. (419° F.), and is oxidized by chromic acid to benzoic, acetic, and carbonic acids.

Benzyl-methyl Ketone, $C_6H_5.CH_2.CO.CH_3$, formed by distillation of alphatoluate (phenyl-acetate) and acetate of calcium, and by the action of zinc-methyl on alphatoluic chloride, boils at 214°–216° C. (417.2°–

420.8° F.), unites with acid-sodium sulphite, and is oxidized by chromic acid to benzoic and acetic acids.

Benzyl-ethyl Ketone, $C_6H_5.CH_2.CO.C_2H_5$, from alphatoluic chloride with zinc-ethyl, boils at 226° C. (438.8° F.), and is oxidized by chromic acid to benzoic and propionic acids.

AROMATIC ACIDS.

1. MONATOMIC ACIDS, $C_nH_{2n-8}O_2$.

These acids, which bear the same relation to the hydrocarbons homologous with benzene that the fatty acids, $C_nH_{2n}O_2$, bear to the paraffins, C_nH_{2n+2}, are produced by reactions analogous to some of those which yield the fatty acids, viz. :

1. By oxidation of the corresponding aldehydes and alcohols : thus benzoic acid, $C_7H_6O_2$ or $C_6H_5.COOH$, is formed by oxidation of benzaldehyde, $C_6H_5.COH$, and of benzyl alcohol, $C_6H_5.CH_2OH$.

2. By the action of water on the corresponding acid chlorides.

3. By the action of acids or alkalies at the boiling heat on the aromatic nitrils (cyanides of phenyl and its homologues, p. 856) :

$$C_6H_5CN + 2H_2O = NH_3 + C_6H_5.CO_2H$$
Phenyl cyanide.   Benzoic acid.

Benzoic acid and its homologues are likewise obtained by the following processes :

4. By the action of sodium and carbon dioxide on the monobrominated derivatives of benzene and its homologues ; e. g.,

$$C_6H_5Br + Na_2 + CO_2 = NaBr + C_6H_5.CO_2Na$$
Bromobenzene.   Sodium benzoate.

$$C_6H_4Br.CH_3 + Na_2 + CO_2 = NaBr + C_6H_4(CH_3).CO_2Na$$
Bromotoluene.   Sodium toluate.

5. By oxidation of the hydrocarbons homologous with benzene by means of chromic acid mixture or dilute nitric acid. By chromic acid the lateral chains, CH_3, $CH_2.CH_3$, etc., are at once oxidized to CO_2H ; and the hydrocarbons which contain only one lateral chain, $C_6H_5.CH_3$, $C_6H_5.C_2H_5$, etc., are all oxidized to benzoic acid, while those which contain two lateral chains are converted into acids containing two carboxyl-groups (di-carbon acids), the xylenes, $C_6H_4(CH_3)_2$, for example, into phthalic acids, $C_6H_4(CO_2H)_2$. With dilute nitric acid, on the other hand, monocarbon acids are produced, at least in the first instance, the xylenes, for example, yielding toluic acids, $C_6H_4(CH_3).CO_2H$.

For oxidation with chromic acid, 40 grams of potassium dichromate are mixed with 37 grams of strong sulphuric acid diluted with 2 to 3 vol. water ; to this mixture 10–20 parts of the hydrocarbon are added, and the whole is boiled for some time in a flask provided with a long upright condensing tube, till all the chromic acid is reduced and the solution has acquired a pure green color. The product is then diluted with water, boiled with solution of sodium carbonate, and the organic acid is precipitated from the filtrate by hydrochloric acid.

With the proportions above mentioned the reaction takes place according to the equation :

$$Cr_2O_7K_2 + 4SO_4H_2 = (SO_4)_3Cr_2 + SO_4K_2 + 4H_2O + O_3,$$

39 *

the three atoms of oxygen thus liberated serving for the oxidation of **the** hydrocarbon.

For oxidation with nitric acid, the hydrocarbon is boiled for two or three days with ordinary nitric acid diluted with 3 parts of water in a flask with a vertical condensing tube. To remove the nitro-acids formed at the same time, the crude product is heated with tin and strong hydrochloric acid, whereby the nitro-acids are converted into amido-acids, which dissolve in the hydrochloric acid.

6. By fusing the sulpho-acids of the aromatic hydrocarbons, or the sulpho-aromatic acids, with potassium formate :

$$C_6H_4\!\!<_{SO_3H}^{CH_3} \;+\; HCO_2H \;=\; SO_3H_2 \;+\; C_6H_4\!\!<_{CO_2H}^{CH_3} \;.$$
Toluene-sulphonic   Toluic acid.
acid.

$$C_6H_4\!\!<_{SO_3H}^{CO_2H} \;+\; HCO_2H \;=\; SO_3H_2 \;+\; C_6H_4\!\!<_{CO_2H}^{CO_2H} \;.$$
Sulphobenzoic acid.   Phthalic acid.

The aromatic acids occur, either free or combined, in many resins and balsams and in the animal organism.

Benzoic Acid, $C_7H_6O_2 = C_6H_5.CO_2H$.—This acid, the analogue of benzyl alcohol, is produced by the first four of the general methods above mentioned, and by boiling hippuric acid (benzoyl-glycocine) or the urine of cows or horses, which contains that acid, with strong hydrochloric acid :

$$CH_2(NH.C_7H_5O).CO_2H \;+\; H_2O \;=\; CH_2(NH_2).CO_2H \;+\; C_7H_5O.OH \;.$$
Benzoyl-glycocine.   Glycocine.   Benzoic acid.

This process is applied to the preparation of benzoic acid on the large scale. Benzoic acid is also produced by the oxidation of a great variety of organic bodies, as toluene, cumene, cinnamic aldehyde, cinnamic acid, cinnamene, casein, gelatin, etc.

Fig. 167.

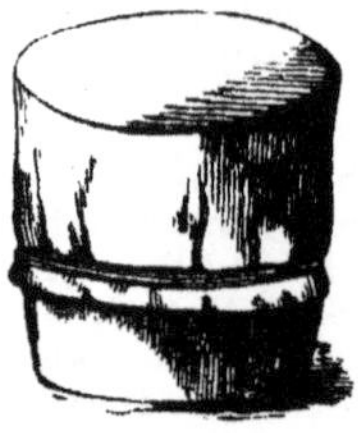

Benzoic acid exists ready-formed in several balsams and gum-resins, especially in gum-benzoin, which exudes from the bark of *Styrax Benzoin*, a tree growing in Sumatra, Java, Borneo, and Siam. From this substance the benzoic acid may be extracted by sublimation in an iron pan (fig. 167), having a sheet of bibulous paper pierced with pinholes stretched over its mouth, and covered with a cap of stout paper. A more productive method is to boil the powdered benzoin with slaked lime and water, and decompose the filtered and concentrated solution of calcium benzoate with hydrochloric acid. The benzoic acid thus precipitated may be purified by sublimation.

Benzoic acid is inodorous when cold, but acquires a faint smell when gently warmed ; it melts at 120° C. (248° F.), and sublimes at a temperature a little above ; boils at 250° C. (482° F.). It dissolves in about 200 parts of cold and 25 parts of boiling water, and with great facility in alcohol. Benzoic acid is not affected by ordinary nitric acid, even at boiling heat ; but with *fuming nitric acid* it forms a substitution-product.—*Chlorine* also acts on benzoic acid, forming substitution-products.—*Phosphorus pentachloride* converts it into benzoyl chloride, C_7H_5OCl.—Benzoic acid dissolves in ordinary strong sulphuric acid, but is precipitated unaltered on addition of water. By fuming sulphuric acid, however, and still more readily by sulphuric oxide, it is converted into sulphobenzoic acid,

$C_7H_6SO_5$ (p. 931). By nascent hydrogen (evolved by sodium amalgam) it is partly reduced to benzoic aldehyde and benzylic alcohol, and is partly converted, by addition of hydrogen, into h y d r o b e n z o i c a c i d, $C_7H_{10}O_2$, a crystalline acid which forms a crystalline calcium salt, $Ca(C_7H_9O_2)_2$, and, when recrystallized either in the free state or in the form of calcium salt, is ultimately converted by oxidation into benzoic acid ; its ethylic ether, $C_2H_5.C_7H_9O_2$, has the odor of ethyl valerate.

All the b e n z o a t e s are more or less soluble : they are easily formed, either directly or by double decomposition. The *benzoates of the alkalies* and of *ammonia* are very soluble, and somewhat difficult to crystallize.— *Calcium benzoate* forms groups of small colorless needles, which require 20 parts of cold water for solution ; the *barium salt* dissolves with difficulty in the cold. Neutral *ferric benzoate* is a soluble compound ; but the basic salt obtained by neutralizing as nearly as possible with ammonia a solution of ferric oxide, and then adding ammonium benzoate, is quite insoluble. Iron is sometimes thus separated from other metals in quantitative analysis. Neutral and basic *lead benzoate* are freely soluble in the cold. *Silver benzoate* crystallizes in thin transparent plates, which blacken on exposure to light.

Calcium benzoate is resolved by dry distillation into calcium carbonate and benzone, or benzophenone, $C_{13}H_{10}O$, the ketone of benzoic acid $(C_7H_5O_2)_2Ca = CO_3Ca + CO(C_6H_5)_2$. On the other hand, benzoic acid, distilled with excess of lime, is resolved into carbon dioxide and benzene : $C_7H_6O_2 = CO_2 + C_6H_6$.

Benzoic Chloride, or **Benzoyl Chloride,** C_7H_5OCl, is prepared by the action of phosphorus pentachloride on benzoic acid. It is a colorless liquid of disagreeable pungent odor ; sp. gr. 1.106; boiling point, 199° C. (390.2° F.). The vapor is inflammable, and burns with a greenish flame. Benzoyl chloride is decomposed, slowly by cold and quickly by boiling water, into benzoic and hydrochloric acids ; with an alkaline hydroxide, a benzoate and chloride of the alkali-metal are produced.

Benzoyl cyanide, $C_6H_5.CO.CN$, obtained by distilling the chloride with potassium cyanide, crystallizes in large plates, melts at 31° C. (87.8° F.), and boils at 206°–208° C. (402.8°–406.4° F.). By boiling with alkalies, it is resolved into benzoic and hydrocyanic acids.

Benzoic Oxide, or **Anhydride,** $C_{14}H_{10}O_3$ or $(C_7H_5O)_2O$, is obtained by the action of benzoyl chloride on potassium benzoate :

$$C_7H_5O(OK) + C_7H_5OCl =: KCl + (C_7H_5O)_2O.$$

Benzoyl chloride acts in like manner on acetate or valerate of sodium, forming aceto-benzoic or valero-benzoic oxide, either of which splits up on distillation into acetic or valeric oxide and benzoic oxide ; *e. g.,*

$$C_7H_5OCl + C_5H_9O(ONa) = NaCl + (C_7H_5O)(C_5H_9O)O,$$

and

$$2(C_7H_5O)(C_5H_9O)O = (C_7H_5O)_2O + (C_5H_9O)_2O.$$

Benzo-œnanthylic, benzostearic, benze-angelic, benzo-cuminic oxide, and several others, have been obtained by similar processes.

Benzoic oxide crystallizes in oblique rhombic prisms, melting at 42° C. (107.6° F.), and distilling undecomposed at 310° C. (590° F.). It melts in boiling water, remaining fluid for a long time, but is ultimately converted into benzoic acid, and dissolves : caustic alkalies effect the conversion much more rapidly. With ammonia it forms ammonium benzoate and benzamide :

$$(C_7H_5O)_2O + 2NH_3 = C_7H_5O(NH_4)O + C_7H_5O.NH_2.$$

BENZOYL DIOXIDE, or PEROXIDE, $C_{14}H_{10}O$, or $(C_7H_5O)_2O_2$.—Brodie obtained this compound by bringing benzoyl chloride in contact with barium dioxide under water; the product, when recrystallized from ether, yields large shining crystals of benzoyl dioxide, which explode when heated. Boiled with potash-solution, it gives off oxygen, and forms potassium benzoate.

Thiobenzoic Acid, $C_6H_5.CO.SH$, formed by the action of benzoyl chloride on alcoholic potassium sulphide, is crystalline, melts at 24° C. (72.5^O F.), and distils with aqueous vapor. Its ethylic ether boils at 243° C. (469.4^O F.).

B e n z o y l m o n o s u l p h i d e, $(C_7H_5O)_2S$, formed by the action of benzoyl chloride on thiobenzoic acid, crystallizes from ether in large prisms, melting at 48° C. (118.4^O F.), and distilling without decomposition. The d i s u l p h i d e, $(C_7H_5O)_2S_2$, produced by oxidation of thiobenzoic acid when its ethereal solution is exposed to the air, forms shining crystals melting at 128° C. (262.4^O F.).

Dithiobenzoic acid, $C_6H_5.CS.SH$, is formed by boiling the compound $C_6H_5.$ CCl_3 with an alcoholic solution of potassium sulphide:

$$C_6H_5.CCl_3 \ + \ 2K_2S \ = \ 3KCl \ + \ C_6H_5.CS.SK.$$

The free acid is very unstable: its lead salt crystallizes from carbon sulphide in red needles.

Benzamide, $C_7H_7NO = C_6H_5.CO.NH_2$, is formed (similarly to acetamide, p. 804) by the action of aqueous ammonia on benzoyl chloride or ethyl benzoate; also by oxidizing hippuric acid with lead oxide:

$$C_9H_9NO_3 \ + \ O_3 \ = \ C_7H_7NO \ + \ 2CO_2 \ + \ H_2O.$$

Benzamide crystallizes in nacreous laminæ, nearly insoluble in cold water, easily soluble in boiling water, also in alcohol and ether; it melts at 125° C. (257° F.), and volatilizes undecomposed between 286° and 290° C. (546.8^O and 554° F.). Its reactions are, for the most part, similar to those of acetamide. Heated with benzoic oxide or chloride, it yields benzonitril and benzoic acid:

$$\underset{\text{Benzamide.}}{C_7H_7NO} \ + \ \underset{\text{Benzoic oxide.}}{(C_7H_5O)_2O} \ = \ \underset{\text{Benzonitril.}}{C_7H_5N} \ + \ \underset{\text{Benzoic acid.}}{2C_7H_6O_2},$$

$$\underset{\text{Benzamide.}}{C_7H_7NO} \ + \ \underset{\substack{\text{Benzoic} \\ \text{chloride.}}}{C_7H_5OCl} \ = \ \underset{\text{Benzonitril.}}{C_7H_5N} \ + \ \underset{\substack{\text{Benzoic} \\ \text{acid.}}}{C_7H_6O_2} \ + \ HCl.$$

Heated with fuming hydrochloric acid, it forms hydrochloride of benzamide, $C_7H_7NO.HCl$, which separates on cooling in long aggregated prisms. Its aqueous solution dissolves mercuric oxide, forming b e n z o m e r c u r - a m i d e, $C_6H_5.CO.NHg$. By distillation with PCl_5 or P_2S_5, it is converted into benzonitril.

Phenyl-benzamide, $C_6H_5.CO.NH(C_6H_5)$, is formed by the action of aniline on benzyl chloride.

SUBSTITUTED BENZOIC ACIDS.

The action of chlorine and bromine on benzoic acid gives rise, for the most part, only to mono-substituted derivatives, $C_6H_4X.CO_2H$, belonging to the meta-series (Cl or Br and CO_2H, in the relative positions 1, 3): with nitric acid (diluted with 3 parts of water) meta-derivatives are also produced, together with small quantities of the ortho- and para-compounds. These and all the other mono-halogen- and mono-nitro-derivatives of benzoic acid may also be formed by oxidation of the corresponding derivatives of toluene ; *e. g.*,

$$C_6H_4Br.CH_3 \ + \ O_3 \ = \ C_6H_4Br.CO_2H \ + \ H_2O \ ;$$

the meta- and para-compounds by chromic acid, the ortho-compounds by dilute nitric acid ; these last compounds are attacked with difficulty by chromic acid, and then completely broken up.

The halogen-derivatives of benzoic acid may also be formed from the corresponding amidobenzoic acids, by substitution of Cl, etc., for NH_2, through the medium of the diazo-compounds ; and the monochlorine- and monobromine-derivatives by treating the corresponding oxy-acids, C_6H_4 $(OH).CO_2H$ with PCl_5 or PBr_5, and decomposing the resulting chloride or bromide with water ; *e. g.*,

$$C_6H_4(OH).CO.OH \ + \ PCl_5 \ = \ C_6H_4Cl.CO.Cl \ + \ PCl_3O \ + \ H_2O \ ,$$
and $\quad C_6H_4Cl.CO.Cl \ + \ HOH \ = \ C_6H_4Cl.CO.OH \ + \ HCl \ .$

The ortho-derivatives of benzoic acid fuse at lower temperatures than the corresponding meta- and para-derivatives. They are moderately soluble in water, and form easily soluble barium salts, by means of which they may be separated from the meta- and para-derivatives.

Monochlorobenzoic Acids, $C_7H_5ClO_2$.—The *ortho*-compound, $C_6.CO_2H$. $Cl.H_4$ (*Chlorosalicylic acid*), prepared by treating salicylic acid (*o*-oxybenzoic acid) with PCl_5, and decomposing the resulting chloride with boiling water, crystallizes in colorless needles, melts at 137° C. (278.6° F.), also under boiling water, and is more soluble in water than the meta- and para-compounds.

The *meta*-compound, $C_6.CO_2H.H.Cl.H_3$, is formed by oxidation of meta-chlorotoluene ; by heating benzoic acid with hydrochloric acid and manganese dioxide or potassium chlorate, or with solution of bleaching powder or with antimonic chloride ; also by decomposition of chlorohippuric acid ; and from (1 : 4) chloronitrobenzene by means of potassium cyanide. It crystallizes in colorless needles, melting without decomposition at 152° C. (305.6° F.), very slightly soluble in boiling water.

The *para*-compound, $C_6.CO_2H.H.H.Cl.H_2$ (*Chlorodracylic acid*), formed by oxidation of (1 : 4) chlorotoluene, sublimes in colorless scales, which melt at 236°–237° C. (456.8°–458.6° F.).

Dichlorobenzoic Acids, $C_6H_3Cl_2.CO_2H$.—Two modifications (out of six) are known, viz. (*a*) 1 : 3 : 4 (CO_2H in 1), formed from benzoic acid, and from (1 : 3) or (1 : 4) chlorobenzoic acid by boiling with solution of bleaching powder, or by heating with antimonic chloride: also by oxidation of (1 : 3 : 4) dichlorotoluene with chromic acid. Colorless needles, melting at 202° C. (395.6° F.), slightly soluble in water. (*β*) (1 : 2 : 3), produced, like the *a*-acid, from (1 : 2) chlorobenzoic acid, and together with the *a*-acid, by treating benzoic acid with HCl and ClO_3K, or with solution of bleaching powder. The two acids may be separated by means of their barium salts, the *β*-salt being the more soluble of the two. The *β*-acid forms slender, shining needles, melts at 150° C. (302° F.), boils at 301° C. (573.8° F.), dissolves in about 1200 parts of water at 11° C. (51.8° F.), more easily in boiling water.

Trichlorobenzoic acid, $C_6H_2Cl_3.CO_2H$, formed by oxidation of trichloro-toluene, crystallizes in small needles melting at 163° C. (325.4° F.), nearly insoluble in cold water. An isomeric acid, formed by heating

dinitro-paramidobenzoic acid with fuming hydrochloric acid, melts at 203°
C. (397.4° F.), and sublimes without decomposition.

Monobromobenzoic Acid, $C_6H_4Br.CO_2H$ (1 : 2), from ortho-bromotoluene with nitric acid, or from *o*-amidobenzoic acid by heating the diazo-
perbromide with HBr, sublimes in needles, and melts at 147°–148° C.
(296.6°–298.4° F.). Its barium-salt is very soluble in water. (1 : 3),
or ordinary *bromobenzoic acid*, from metabromotoluene, and by heating
benzoic acid with bromine and water to 120°–130° C. (248°–266° F.), sublimes in needles, and melts at 155° C. (311° F.). (1 : 4) from parabromotoluene, is nearly insoluble in water, crystallizes in needles, and melts at
251° C. (483.8° F.).

Di-, tri-, and penta-bromobenzoic acids are formed by heating benzoic
acid with bromine to 200° C. (392° F.) and above. $C_7H_4Br_2O_2$, (1 : 3 : 4),
forms small needles melting at 229°–230° C. (444.2°–446° F.). $C_7H_3Br_3O_2$
and $C_7HBr_5O_2$ melt between 234° and 235° C. (453.2°–455° F.).

Mono-iodobenzoic Acids, $C_6H_4I.CO_2H$, are obtained by oxidation of
iodotoluenes with nitric acid, or from the corresponding amidobenzoic acids
by decomposing the diazo-compounds with hydriodic acid :

(1 : 2) form needles, melting at 157° C. (314.6° F.) ; converted by
fusion with potash into salicylic acid.

(1 : 3) sublimes in needles, melting at 187° C. (368.6° F.) ; converted
by fusion with potash into oxybenzoic acid.

(1 : 4) sublimes in scales, melts at about 267° C. (512.6° F.), converted
by fusion with potash into para-oxybenzoic acid.

Fluobenzoic acid, $C_6H_4F.CO_2H$, produced by treating diazo-amidobenzoic
acid with hydrofluoric acid, crystallizes in rhombic prisms melting at 182°
C. (359.6° F.).

Mononitrobenzoic Acids, $C_6H_4(NO_2).CO_2H$.—When benzoic acid is
treated with fuming nitric acid, or better, with a mixture of nitre (2
parts) and strong sulphuric acid (3 parts) to 1 part of benzoic acid, the
chief product obtained is metanitrobenzoic acid, the ortho-compound being
formed in smaller, and the para- in very small quantity—the mass being
warmed till it melts, and the liquid mass poured off from the solid potassium sulphate. The three nitro-acids are separated by the different solubilities of their barium-salts—that of the ortho-acid being very soluble,
that of the meta-acid rather sparingly, and that of the para-acid very
slightly soluble. Cinnamic acid yields by nitration two nitro-acids,
$C_9H_7(NO_2)O_2$, *ortho-* and *para*, from which the corresponding nitrobenzoic
acids may be obtained by oxidation. Hippuric acid yields by nitration a
nitro-hippuric acid, convertible into metanitrobenzoic acid.

(1 : 2) Nitrobenzoic acid crystallizes in needles or prisms, soluble in 164
parts of water at 16.5° C. (61.7° F.), melting at 145° C. (293° F.). (1 :
3) Nitrobenzoic acid crystallizes in needles or laminæ, dissolves in 425
parts of water at 16.5° C. (61.7° F.), sublimes in white needles, melts at
142° C. (287.6° F.). (1 : 4) Nitrobenzoic acid, formed also by oxidation
of paranitrotoluene, crystallizes in yellowish laminæ, very slightly soluble in water, melting at 240° C. (464° F.).

Dinitrobenzoic Acids, $C_6H_3(NO_2)_2.CO_2H$.—Orthonitrobenzoic acid,
treated with a mixture of nitric and sulphuric acids, yields three dinitrobenzoic acids (α, β, and γ), easily separated by the unequal solubility of
their barium-salts—that of α being very slightly soluble in cold water,
that of β moderately, and that of γ very easily soluble. The β-acid is also

produced by oxidation of (1 : 2 : 4) dinitrotoluene (melting point 70.5○ C., 158.9○ F.) with chromic acid. A fourth acid, δ, is obtained by the action of nitric and sulphuric acid on meta-nitrobenzoic acid.—α (1 : 2 : 5).* Colorless prisms, melting at 177○ C. (350.6○ F.), moderately soluble in hot, sparingly in cold water. β. (1 : 2 : 4) Long rhombic plates or prisms melting at 179○ C. (354.2○ F.). γ. (1 : 2 : 6) Slender, white felted needles, melting at 202○ C. (395.6○ F.), resolved at a higher temperature into (1 : 3) dinitrobenzene and CO_2. δ. (1 : 3 : 5) Crystallizes from water in large square plates ; from alcohol in prisms.

By reduction with tin and HCl, the α- and δ-compounds yield the corresponding diamidobenzoic acids, whereas β and γ yield no diamido-acids, but (1 : 3) diamidobenzene and CO_2.

Nitro-chloro- and *Nitro-bromobenzoic acids* are obtained by nitration of chloro- and bromo-benzoic acids. (1 : 3) bromobenzoic acid yields two nitrometabromobenzoic acids,—one melting at 248○ C. (478.4○ F.), the other at 141○ C. (285.8○ F.). In both of them the NO_2-group is in the ortho-position (2 or 6) with respect to the CO_2H, and both yield by reduction (1 : 2) amidobenzoic acid.

Amidobenzoic Acids, $C_6H_4(NH_2).CO_2H$.—The three modifications are formed by reduction of the three nitrobenzoic acids, best by gentle heating with tin and hydrochloric acid. At the end of the reaction the product is diluted with water, precipitated with excess of sodium carbonate, and the concentrated filtrate is acidulated with acetic acid. The ortho-compound is also produced by boiling indigo (1 part) for several days with soda-lye of specific gravity 1.38 (10 parts), gradually adding pulverized manganese dioxide, and renewing the water as it evaporates, till the color of the mass has become light yellow. The product is then dissolved in water, the solution is neutralized with sulphuric acid, filtered, and evaporated to dryness ; the residue is exhausted with alcohol ; and the salt which remains after evaporation of the alcohol is dissolved in hot water and decomposed with acetic acid.

The three amidobenzoic acids react like glycocine (amidacetic acid), and yield well-characterized salts, both with metallic oxides and with acids. When heated above their melting points they are resolved into aniline and carbon dioxide.

Ortho- (*Anthranilic acid*).—Thin colorless prisms or laminæ, sparingly soluble in cold water, easily in hot water and in alcohol ; melts at 144○ C. (291.2○ F.). *Meta-*. Small colorless needles, united in nodular groups ; easily soluble in hot, sparingly in cold water, melts at 173○–174○ C. (343.4○–345.2○ F.). *Para-* (*amidodracylic acid*). Long slender shining needles, easily soluble in water, melting at 186○–187○ C. (366.8○–368.6○ F.).

Nitro-amidobenzoic acids, $C_6H_3(NO_2){<}^{NH_2}_{CO_2H}$.

1. *Nitro-orthamidobenzoic acid* is known in two modifications, viz. :

$$C_6.CO_2H.NH_2.H.H.NO_2.H \qquad C_6.CO_2H.NH_2.H.H.H.NO_2$$
$$\alpha. \ (1 : 2 : 5) \qquad\qquad \beta. \ (1 : 2 : 6)$$

which are produced from the corresponding nitrosalicylic acids by heating their diethylic ethers with alcoholic ammonia, and decomposing the resulting nitramidobenzamides, $C_6H_3(NO_2)(NH_2).CO.NH_2$, with boiling baryta-water. Both acids crystallize in long needles, the α-acid melting at 270○ C. (518○ F.), the β-acid at 205○ C. (401○ F.).

* CO_2H in 1.

2. *Nitro-metamidobenzoic acid* is known in three modifications, viz. :—

$$\overset{1}{C_6}.\overset{2}{CO_2H}.\overset{5}{NO_2}.H.H.NH_2.H \qquad \overset{1}{C_6}.\overset{}{CO_2H}.H.\overset{3}{NH_2}.\overset{4}{NO_2}.H_2$$
$$\underset{a}{} \qquad\qquad\qquad\qquad \underset{\beta}{}$$

$$\overset{1}{C_6}.\overset{2}{CO_2H}.\overset{3}{NO_2}.NH_2.H_3$$
$$\underset{\gamma}{}$$

They are obtained by boiling the three dinitro-uramidobenzoic acids (p. 930) with water. All three crystallize in yellow needles ; *a* and *ß* sparingly soluble in hot water, *γ* easily soluble.

3. *Nitro-paramidobenzoic acid* (1 : 3 : 4), obtained in like manner from dinitro-para-uramidobenzoic acid, or by heating nitranisic acid with aqueous ammonia to 140°–170° C.(284°–338° F.) for three or four hours, forms small deep-yellow needles, slightly soluble in boiling water, melting at 284° C. (543.2° F.).

Dinitro-paramidobenzoic acid, $C_6H_2(NO_2)_2\!\!<^{NH_2}_{CO_2H}$, or *Chrysanisic acid,* is produced by heating dinitro-anisic or dinitro-ethylparaoxybenzoic acid with aqueous ammonia, the group OCH_3 or OC_2H_5 being thus replaced by NH_2 :—

$$C_6H_2(NO_2)_2\!\!<^{OCH_3}_{CO_2H} + NH = CH_3.OH + C_6H_2(NO_2)_2\!\!<^{NH_2}_{CO_2H} .$$

It crystallizes in shining golden-yellow needles, melts at 259° C. (498.2° F.) and sublimes ; is nearly insoluble in cold water, moderately soluble in boiling water and in alcohol. It is a strong monobasic acid.

Dinitro-orthoamidobenzoic acid, obtained in like manner from dinitro-ethylsalicylic acid, crystallizes from alcohol, in which it is slightly soluble, in golden-yellow scales, melting at 256° C. (492.8° F.).

Acetyl-metamidobenzoic acid, $C_9H_9NO_3 = C_6H_4\!\!<^{NH.C_2H_3O}_{CO.OH}$, is formed by digesting (1 : 3) amidobenzoic acid with glacial acetic acid at 160° C. (320° F.), or zinc metamidobenzoate with acetyl chloride at 100°.

$$(C_6H_4.NH_2.CO_2)_2Zn + 2C_2H_3OCl = ZnCl_2 + 2C_6H_4(NH.C_2H_3O)CO_2H.$$

It forms white microscopic crystals, insoluble in cold water and in ether, slightly soluble in boiling water, easily in boiling alcohol, melts at 226°–230° C. (438.8°–446° F.), sublimes at 200° C. (392° F.). It is a monobasic acid, forming easily soluble salts with the metals of the alkalies and alkaline earths ; sparingly soluble salts with lead, silver, and zinc. By boiling with dilute sulphuric acid, it is resolved into acetic and amidobenzoic acids :

$$C_9H_9NO_3 + H_2O = C_2H_4O_2 + C_7H_7NO_2 .$$

Acetyl-paramidobenzoic acid, isomeric with the last, is easily obtained by oxidizing acetoparatoluide, $C_6H_4(NH.C_2H_3O).CH_3$, with potassium permanganate. It crystallizes in needles, sparingly soluble in water, more readily in alcohol, and melting with decomposition at about 250° C. (482° F.).

Hippuric or Benzamidacetic acid, $C_9H_9NO_3 =$ $CH_2\!\!<^{NH.C_7H_5O}_{CO.OH}$ (*Benzoyl glycocine*).—This acid, metameric with the two just described, is formed by the action of benzoyl chloride on the zinc-salt of amidacetic acid (glycocine):

$$(CH_2.NH_2.CO_2)_2Zn + 2C_7H_5OCl = ZnCl_2 + 2CH_2(NH.C_7H_5O)CO_2H ,$$

the reaction being analogous to the second of those above given for the formation of acetamidobenzoic acid.

Hippuric acid occurs, often in large quantities, in the urine of horses, cows, and other herbivorous animals ; in smaller quantity also in human urine. It may be prepared by boiling fresh cows' or horses' urine with milk of lime, and precipitating the concentrated filtrate with hydrochloric acid. For purification the crude acid is washed with chlorine-water ; or its solution in dilute soda-lye is boiled with sodium hypochlorite till it becomes colorless, and the solution, after cooling, is precipitated with hydrochloric acid.

Hippuric acid crystallizes in rhombic prisms, which have a slightly bitter taste and acid reaction, melt on the application of heat, and require for solution about 600 parts of cold water ; it also dissolves in hot alcohol. At a high temperature it decomposes, yielding benzoic acid, ammonium benzoate, and benzonitril, with a coaly residue. Boiling hydrochloric acid converts it into benzoic acid and glycocine (amidacetic acid):

$$CH_2(NH.C_7H_5O).CO_2H + HOH = C_7H_5O.OH + CH_2(NH_2).CO_2H ;$$

just as acetamidobenzoic acid is resolved into acetic and amidobenzoic acids.

Hippuric acid is monobasic, the formula of the hippurates of monatomic metals being $C_9H_8MNO_3$. Most metallic oxides dissolve readily in hippuric acid. The hippurates of potassium, sodium, and ammonium are very soluble, and difficult to crystallize ; their solutions form a cream-colored precipitate with ferric salts, and white curdy precipitates with silver nitrate and mercurous nitrate. A characteristic reaction of the hippurates is, that they give off ammonia when fused with excess of potash or lime, and yield benzene by distillation. Mineral acids decompose them, separating the hippuric acid.

Hippuric acid, treated with nitrous acid, gives off nitrogen, and is converted into benzoglycollic acid, $C_9H_8O_4$, an acid containing the elements of benzoic and glycollic (oxyacetic) acids, minus one molecule of water :—

$$C_9H_9NO_3 + HNO_2 = C_9H_8O_4 + H_2O + N_2 ;$$

and benzoglycollic acid, when boiled with water, splits up into benzoic and glycollic acids :

$$C_9H_8O_4 + H_2O = C_7H_6O_2 + C_2H_4O_3 .$$

If, in the preparation of hippuric acid, the urine be in the slightest degree putrid, the hippuric acid is all destroyed during the evaporation, ammonia is disengaged in large quantity, and the liquid is then found to yield nothing but benzoic acid, not a trace of which can be discovered in the unaltered secretion. When benzoic acid is taken internally, it is rejected from the system in the state of hippuric acid, which is then found in the urine.

Hippuric acid dissolves so abundantly in an aqueous solution of sodium phosphate, that this solution loses its alkaline reaction and becomes acid. This reaction may explain the acid character of the recent urine of man and herbivorous animals.

Uramidobenzoic acid, $C_8H_8N_2O_3 = CO{<}^{NH.C_6H_4.CO.OH}_{NH_2}$, is formed by mixing the cold solutions of equivalent quantities of hydrochloride of metamidobenzoic acid and potassium cyanate :

$$C_7H_7NO_2.HCl + CNOK = KCl + C_8H_8N_2O_3 ;$$

also by fusing urea with metamidobenzoic acid : $C_7H_7NO_2 + CH_4N_2O = NH_3 +$ $C_8H_8N_2O_3$. It crystallizes in small needles, containing 1 molecule of water, soluble in hot water and in alcohol. When heated to 200° C. (392° F.) it is resolved into H_2O and oxybenzoyl-urea, $C_8H_6N_2O_2 =$

$$CO \Big\langle \genfrac{}{}{0pt}{}{NH.C_6H_4.CO}{NH \underline{}} \Big|$$ By boiling with caustic potash, it is resolved into amidobenzoic acid, carbon dioxide, and ammonia.

Uramidobenzoic acid, treated with a mixture of nitric and sulphuric acid, yields three isomeric dinitro-uramidobenzoic acids, which, when boiled with water, are converted into three nitro-amidobenzoic acids, a, β, γ (p. 928).

Para-uramidobenzoic acid, formed as above from paramidobenzoic acid, crystallizes in white shining elongated laminæ, but slightly soluble in water even at the boiling heat. By nitration it yields only one dinitro-paramidobenzoic acid, convertible into one nitro-paramidobenzoic acid.

Diamidobenzoic acid, $C_6H_3(NH_2)_2.CO_2H$.—The six possible modifications of this acid have all been obtained, viz. (CO_2H in 1):

$$1:2:3, \quad 1:2:4, \quad 1:2:5, \quad 1:2:6, \quad 1:3:4, \quad 1:3:5.$$

The acids $1:2:3$ and $1:3:4$ are formed from β and γ nitro-metamidobenzoic acid (p. 928), and yield by distillation ortho-diamidobenzene.

The acids $1:2:6$, $1:2:4$, and $1:3:5$ are formed by reduction of the three dinitrobenzoic acids (pp. 926, 927), and are converted, by elimination of CO_2, into metadiamidobenzene.

The acid $1:2:5$ (or $1:3:6$) is formed from a-nitro-metamidobenzoic acid and from the corresponding dinitrobenzoic acid, and is convertible into paradiamidobenzene.

Hence it appears that (1 : 4) diamidobenzene is producible from only *one* diamidobenzoic acid : (1 : 3) diamidobenzene from *three*; and (1 : 2) diamidobenzene from *two* diamidobenzoic acids. This result affords a further confirmation of the structure of the three diamidobenzenes (pp. 841, 865).

A *triamidobenzoic acid*, $C_6.CO_2H.NH_2.H.NH_2.NH_2.H$, is formed by reduction of (1 : 2 : 4 : 5) dinitramidobenzoic acid. It crystallizes in shining needles containing $\frac{1}{2}H_2O$, and is resolved by distillation into CO_2 and triamidobenzene.

Azobenzoic Acids.—The action of sodium-amalgam on the mononitrobenzoic acids (or rather on their sodium-salts) gives rise (as in the case of nitrobenzene, p. 879) to azo-, azoxy-, and hydrazobenzoic acids :

$$C_6H_4 \Big\langle \genfrac{}{}{0pt}{}{CO_2H}{N} \Big\rangle O \qquad C_6H_4 \Big\langle \genfrac{}{}{0pt}{}{CO_2H}{N} \qquad C_6H_4 \Big\langle \genfrac{}{}{0pt}{}{CO_2H}{NH}$$

$$C_6H_4 \Big\langle \genfrac{}{}{0pt}{}{N}{CO_2H} \qquad C_6H_4 \Big\langle \genfrac{}{}{0pt}{}{N}{CO_2H} \qquad C_6H_4 \Big\langle \genfrac{}{}{0pt}{}{NH}{CO_2H}$$

Azoxybenzoic.   Azobenzoic.   Hydrazobenzoic.

Metazobenzoic acid, $C_{14}H_{10}N_2O_4 + \frac{1}{2}H_2O$ (CO_2H and N in the relative positions 1, 3), is precipitated by hydrochloric acid from the solution of its sodium salt, as a yellow amorphous powder, very slightly soluble in water, alcohol, and ether ; decomposed by distillation. It is bibasic, and forms crystalline yellow salts and ethers. *Parazobenzoic* acid is a flesh-colored amorphous powder.

Azoxybenzoic acid, $C_{14}H_{10}N_2O_5$ (1 : 3), is formed by boiling an alcoholic solution of metanitrobenzoic acid with solid potash, and is precipitated by hydrochloric acid in yellowish crystalline flocks.—Bibasic.

Hydrazobenzoic acid, $C_{14}H_{12}N_2O_4$, is formed by adding ferrous sulphate to a boiling solution of azobenzoic acid in caustic soda-lye, and is precipitated by hydrochloric acid in yellowish flocks, very slightly soluble in hot alcohol. Its salts in aqueous solution absorb oxygen from the air, whereby they are converted into azobenzoates. By boiling with hydrochloric acid it is converted into an isomeric acid derivable from diphenyl, just as hydrazobenzene is converted into benzidine (p. 880):

$$C_6H_4{<}^{CO_2H}_{NH} \qquad \text{is converted into} \qquad C_6H_3{<}^{CO_2H}_{NH_2}$$
$$C_6H_4{<}^{NH}_{CO_2H} \qquad\qquad\qquad C_6H_3{<}^{NH_2}_{CO_2H} \ .$$

The last-mentioned acid is resolved by distillation into CO_2 and benzidine.

Diazo-derivatives of Benzoic Acid.—These compounds are formed from the amidobenzoic acids in the same manner as the diazobenzene-compounds from the amidobenzenes.

Diazobenzoic Nitrate, $C_6H_4{<}^{N{=}N.NO_3}_{CO_2H}$, is formed by the action of nitrous acid on a solution of metamidobenzoic acid in nitric acid. It is sparingly soluble in cold water, and separates therefrom in colorless prisms, which explode violently when heated. Potash added to the aqueous solution throws down a yellow, very unstable mass, probably consisting of free diazobenzoic acid $\left[C_6H_4{<}^{CO_2H}_{N{=}N.OH}?\right]$. The nitrate boiled with water yields meta-oxybenzoic acid. Bromine-water added to the aqueous solution throws down the perbromide, $C_7H_5N_2O_2Br_3$, as an oil which solidifies in yellow prisms, and is resolved by heating with alcohol into nitrogen, bromine, and metabromobenzoic acid.

Diazo-amidobenzoic acid, $C_{14}H_{11}N_3O_4 =$
$C_6H_4{<}^{N{=}N-NH-C_6H_4-CO_2H}_{CO_2H}$, is precipitated, on passing nitrous acid into the alcoholic solution of metamidobenzoic acid, as an orange-red crystalline powder, nearly insoluble in water, alcohol, and ether. It is a weak bibasic acid, and its salts in aqueous solution are very unstable. The acid, heated with haloïd acids, yields the corresponding halogen-derivatives of benzoic acid.

Diazo- and diazo-amido-compounds of exactly similar character are obtained from ortho- and para-amidobenzoic acid.

Sulphobenzoic Acids, $C_6H_4{<}^{CO_2H}_{SO_3H}$. — When vapor of sulphuric anhydride is passed over dry benzoic acid, and the product is treated with water, or when benzoic acid is heated for a considerable time with fuming sulphuric acid, the chief product formed is *meta*-sulphobenzoic acid, the *para*-acid being also produced in small quantity. The latter is obtained in the pure state by oxidation of para-toluenesulphonic acid (p. 888), or para-sulphocinnamic acid, with chromic acid mixture.

The *meta*-acid is a colorless, crystalline, very deliquescent, strongly acid mass, converted by distillation with PCl_5 into meta-chlorobenzoyl chloride. It is a very stable bibasic acid: its neutral barium salt is very soluble. The *para*-acid, $C_7H_6SO_5$, crystallizes in non-deliquescent needles, melting above 200° C. (392° F.), and decomposing even at a lower temperature. Its neutral barium salt, $C_7H_4SO_5Ba + 2H_2O$, crystallizes in small, ramified, easily soluble needles.

Chloro- and Bromo-sulphobenzoic acids are obtained by the action of fuming sulphuric acid on the corresponding derivatives of benzoic acid, and by oxidation of chloro- and bromo-toluene-sulphonic acids.

Disulphobenzoic acid, $C_6H_4(CO_2H)(SO_3H)_2$, is formed by heating benzoic acid with fuming sulphuric acid and phosphoric anhydride to 250° C. (482° F.). Deliquescent acicular prisms. Tribasic. By distillation with potassium cyanide, it yields a dicyanide, m. p. 159° C. (318.2° F.), convertible into isophthalic acid. An isomeric disulpho-acid is formed by oxidation of toluenedisulphonic acid.

Toluic Acids, $C_8H_8O_2$.—Of these acids there are two metameric modifications, viz. :

$$C_6H_4 {<}^{CH_3}_{CO_2H} \qquad CH_2 {<}^{C_6H_5}_{CO_2H} ;$$

Toluic or methyl-benzoic.   Alpha-toluic or phenyl-acetic.

the first admitting of the three isomeric modifications, *o-, m-, p-*. These three toluic acids are formed by oxidation of the corresponding dimethylbenzenes (xylenes) with nitric acid diluted with 3 volumes of water ; also from the corresponding cyanotoluenes, $C_6H_4(CN).CH_3$, by the action of alcoholic potash, or of strong hydrochloric acid. By oxidation with chromic acid mixture, or potassium permanganate, they are converted into the corresponding phthalic acids, $C_6H_4(CO_2H)_2$.

Orthotoluic acid crystallizes in long slender needles melting at 102.5° C. (215.6° F.) ; is moderately soluble in hot water, and distils easily with aqueous vapor. By oxidation with permanganate it yields phthalic acid, whereas chromic acid mixture oxidizes it completely to carbonic acid. Its *calcium salt,* $(C_8H_7O_2)_2Ca + 2H_2O$, and *barium salt,* $(C_8H_7O_2)_2Ba$, form slender needles easily soluble in water.

Metatoluic acid is best obtained from cyanotoluene, or by the action of sodium-amalgam on bromometatoluic acid. It is more soluble in water than its two isomerides, and crystallizes from hot water in slender needles, melting at 109°–110° C. (228.2°–230° F.) ; distils easily with aqueous vapor. Chromic acid mixture oxidizes it readily so isophthalic acid. Its *calcium salt,* $(C_8H_7O_2)Ca + 3H_2O$, crystallizes from alcohol in needles, easily soluble in water.

Bromometatoluic acid, $C_8H_7BrO_2$, is formed by heating nitropara-bromotoluene with alcoholic potassium cyanide to 220° C. (428° F.), and saponifying the product with potash.

Paratoluic acid, obtained by oxidation of (1 : 4) xylene, or of cymene, with dilute nitric acid, crystallizes in needles melting at 178° C. (352.4° F.). Its calcium salt, $(C_8H_7O_2)_2Ca + 2H_2O$, forms needles easily soluble in water. Chromic acid mixture oxidizes it to terephthalic acid.

Alpha-toluic or *Phenyl-acetic acid,* $C_6H_5.CH_2.CO_2H$, is formed by boiling benzyl cyanide, $C_6H_5.CH_2.CN$, with alkalies ; by heating mandelic acid, $C_6H_5.CH(OH).CO_2H$, with hydriodic acid; by boiling vulpic acid, $C_{19}H_{14}O_5$ (p. 918), with baryta; and as an ethylic ether, by heating a mixture of bromobenzene and ethyl chloracetate with sodium :

$$C_6H_5Br + CH_2Cl.CO_2C_2H_5 + Na_2 = NaBr + NaCl + C_6H_5.CH_2.CO_2C_2H_5 .$$

It crystallizes from boiling water in broad thin laminæ, very much like benzoic acid ; smells like horse-sweat ; melts at 76.5° C. (167.9° F.), and boils at 261°–262° C. (501.8°–503.6° F.). By distillation with PCl_5 it yields the chloride $C_6H_5.CH_2.COCl$, which passes over as a colorless heavy liquid. By oxidation with chromic acid mixture it yields benzoic acid.

Acids, $C_9H_{10}O_2$. — Of the six possible dimethyl-carbonic acids, $C_6H_3(CH_3)_2.CO_2H$, three are known, viz. :

| Mesitylenic. | Xylic. | Paraxylic. |

Mesitylenic acid (1 : 3 : 5), formed by oxidizing mesitylene with dilute nitric acid, crystallizes from alcohol in large prisms, from water in needles ; melts at 166° C. (330.8° F.), and sublimes without decomposition. Distilled with excess of lime it yields isoxylene. By nitric acid it is further oxidized to trimesic and uvitic acids. Its *barium-salt*, $(C_9H_9O_2)_2Ba$, dissolves easily in water, and forms large shining prisms. The *ethylic ether*, $C_9H_9O_2.C_2H_5$, solidifies at 0°, and boils at 241° C. (465.8° F.).

Xylic acid (1 : 2 : 4), and **paraxylic acid** (1 : 3 : 4), are formed, together with bibasic xylidic acid, $C_6H_3(CH_3).(CO_2H)_2$, by oxidizing pseudocumene (p. 847) with dilute nitric acid at the boiling heat. At the end of the reaction a crystalline mass separates, and on repeatedly boiling this mass with water, xylic and paraxylic acids pass over, while xylidic acid and nitro-acids remain behind. The xylic and paraxylic acids are separated by the different solubility of their calcium salts, the xylate being the more soluble of the two. Both acids crystallize in prisms, xylic acid melting at 126° C. (258.8° F.), paraxylic acid at 163° C. (325.4° F.). Xylic acid is more soluble in water than paraxylic acid. Both acids dissolve easily in alcohol.

Alphaxylic acid, $C_6H_4(CH_3).CH_2.CO_2H$, is obtained by boiling xylyl chloride with potassium cyanide (whereby xylyl cyanide, C_8H_9CN, is produced), and then with potash. It crystallizes in broad needles, having a satiny lustre, easily soluble in water, and boiling at 42° C. (107.6° F.).

Ethyl-benzoic acid, $C_6H_4(C_2H_5).CO_2H$ (1 : 4), obtained by oxidation of para-diethylbenzene with nitric acid, and from bromethylbenzene by the action of Na and CO_2, crystallizes from hot water in slender laminæ melting at 110° C. (230° F.), and easily subliming. By oxidation it yields terephthalic acid.

Hydrocinnamic, Phenylpropionic, or **Homotoluic acid,** $C_6H_5.CH_2.CH_2.CO_2H$, is formed by the action of sodium-amalgam, or of very strong hydriodic acid at 100°, on cinnamic acid, $C_6H_5.CH{\equiv}CH.CO_2H$, also by heating chlorethyl-benzene, $C_6H_5.CH_2.CH_2Cl$ (p. 855), with potassium cyanide, and boiling the resulting nitril with potash. It crystallizes in slender needles, easily soluble in hot water and in alcohol, melts at 47° C. (116.6° F.), and distils without decomposition at 280° C. (536° F.). By oxidation it yields benzoic acid. Heated to 160° C. (320° F.) with bromine, it is reconverted into cinnamic acid : $C_9H_{10}O_2 + Br_2 = 2HBr + C_9H_8O_2$.

Hydro-atropic acid, $C_6H_5.CH{<}{CH_3 \atop CO_2H}$, formed by the action of sodium-amalgam on atropic acid, is a thick oily liquid.

Acids, $C_{10}H_{12}O_2$.—Durylic acid, $C_6H_2(CH_3)_3.CO_2H$, formed b oxi_dizing durene, $C_6H_2(CH_3)_4$, with dilute nitric acid, crystallizes in yhard prisms melting at 150° C. (302° F.), and is converted by oxidation into cumidic acid.

Cumic acid, $C_6H_4(C_3H_7).CO_2H$ (1 : 4), is obtained by oxidation of cuminol or cumic aldehyde, $C_{10}H_{12}O$, or by heating this aldehyde with alcoholic potash :

$$2C_{10}H_{12}O \atop \text{Cuminol.} \quad + \quad KOH \quad = \quad C_{10}H_{11}KO_2 \atop \substack{\text{Potassium} \\ \text{cumate.}} \quad + \quad C_{10}H_{14}O \atop \substack{\text{Cymyl} \\ \text{alcohol.}} .$$

It crystallizes in needles and laminæ, dissolves easily in hot water and in alcohol, melts at 113° C. (235.4° F.), boils at about 290° C. (554° F.). By distillation with lime it yields cumene, C_9H_{12}, and by oxidation with chromic acid it is converted into terephthalic acid.

Alphaeymic acid, $C_{11}H_{14}O_2$, probably $C_6H_3(CH_3)_3.CH_2CO_2H$, is formed by the action of caustic alkalies on cymyl cyanide, $C_{10}H_{13}.CN$.

MONATOMIC ACIDS, $C_nH_{2n-10}O_2$.

The acids of this series are unsaturated compounds, related to benzoic acid and its homologues, in the same manner as those of the acrylic series to the fatty acids.

Cinnamic Acid, $C_9H_8O_2 = C_6H_5.CH{=}CH.CO_2H$ (*Phenyl-acrylic acid*).—This acid is produced synthetically : 1. By heating benzoic aldehyde in close vessels with acetyl chloride :

$$C_7H_6O \quad + \quad C_2H_3OCl \quad = \quad HCl \quad + \quad C_9H_8O_2 .$$

2. By heating the same aldehyde with acetic anhydride and sodium acetate :

$$2C_7H_6O \quad + \quad C_4H_6O_3 \quad = \quad 2C_9H_8O_2 \quad + \quad H_2O .$$

The mode of action in this case is not well understood, but the presence of the sodium acetate appears to be essential.

3. By treating potassium benzoate with chlorethidene, C_2H_3Cl (produced by the action of carbonyl chloride on acetic aldehyde):

$$C_2H_4O \quad + \quad COCl_2 \quad = \quad HCl \quad + \quad CO_2 \quad + \quad C_2H_3Cl .$$

and

$$C_2H_3Cl \quad + \quad C_7H_5O_2K \quad = \quad KCl \quad + \quad C_9H_8O_2 .$$

4. By the action of sodium and carbon dioxide on monobromo-cinnamene :

$$C_6H_5.CH{=}CHBr \quad + \quad CO_2 \quad + \quad Na_2 \quad = \quad NaBr \quad + \quad C_6H_5CH{=}CH.CO_2Na .$$

5. Cinnamic acid is also produced by oxidation of cinnamon-oil (cinnamic aldehyde, C_9H_8O) in air or oxygen, and exists ready formed, together with benzoic acid, and certain oily and resinous substances, in Peru and Tolu balsams—the produce of certain South American *Myroxylums*—

being doubtless formed by oxidation of cinnyl alcohol or styrone, $C_9H_{10}O$ (p. 914), likewise contained therein. It is easily prepared by mixing pulverized Tolu balsam with an equal weight of slaked lime, filtering hot, and decomposing the calcium cinnamate, which crystallizes out on cooling, with hydrochloric acid. The mother-liquid contains calcium benzoate.

Cinnamic acid crystallizes from hot water in slender needles, from alcohol in thick prisms. It is inodorous, melts at 133° C. (271.4^O F.), and distils almost without decomposition at 290° C. (554° F.). It is much less soluble in water than benzoic acid, but dissolves easily in alcohol. It is oxidized by nitric acid to benzoic acid and benzoic aldehyde; by chromic acid chiefly to benzoic acid. By fusion with excess of potash it is resolved into benzoic and acetic acids :

$$C_9H_8O_2 \ + \ 2H_2O \ = \ C_7H_6O_2 \ + \ C_2H_4O_2 \ + \ H_2,$$

the decomposition being precisely analogous to that of an acid of the acrylic series into two acids of the fatty series (p. 743). By distillation with lime or baryta, and partly also when distilled alone, it is resolved into carbon dioxide and cinnamene, C_8H_8 (p. 849).

The metallic cinnamates, $C_8H_7MO_2$ (for monatomic metals), are very much like the benzoates.

Ethyl Cinnamate, $C_9H_7O_2.C_2H_5$, obtained by passing gaseous hydrogen chloride into a solution of cinnamic acid in absolute alcohol, is a liquid boiling at 267° C. (512.6^O).

Benzyl Cinnamate or *Cinnamein*, $C_9H_7O_2.C_7H_7$, is contained in Peru and Tolu balsams, in small quantity also in storax, and may be formed artificially by heating sodium cinnamate with benzyl chloride. It crystallizes in shining prisms melting at 39° C. (102.2^O F.), distilling without decomposition only under reduced pressure.

Cinnyl Cinnamate or *Styracin*, $C_9H_7O_2.C_9H_9$, occurs, together with cinnamene, in liquid storax (which exudes from *Styrax calamita*, a shrub growing in Greece and Asia Minor), and may be obtained therefrom by distilling the balsam to expel the cinnamene, then boiling it with aqueous sodium carbonate to remove free cinnamic acid, and kneading the spongy residue between the fingers. The styracin then runs out as an oily liquid, and may be obtained in tufts of prisms by crystallization from alcohol. By distillation with potash it is resolved into cinnyl or styryl alcohol, $C_9H_{10}O$, and cinnamic acid.

Addition-products of Cinnamic Acid.—This acid, like other unsaturated compounds, can take up H_2, HBr, ClOH, etc.; thus it is converted by sodium-amalgam into hydrocinnamic acid, $C_9H_{10}O_2$ (p. 933).

Cinnamic Dibromide, $C_9H_8Br_2O_2 = C_6H_5.CHBr.CHBr.CO_2H$, formed by the action of bromine vapor on cinnamic or on hydrocinnamic acid, crystallizes from alcohol in rhombic laminæ, melting, with decomposition, at about 195° C. (383° F.). By boiling with water it is resolved into monobromocinnamene and phenylbromolactic acid, $C_6H_5.CHBr.CH(OH).CO_2H$.

This last mentioned acid is also formed by direct addition of BrHO to cinnamic acid, and phenyl-chlorolactic acid in like manner by addition of ClOH. Both these acids are converted, by heating with alcoholic potash, into phenyloxyacrylic or oxycinnamic acid, $C_6H_5.CH=C(OH).CO_2H$, which separates from its salts as an oil, solidifying in shining laminæ, and decomposed by heat.

Substitution-products.—Cinnamic dibromide heated with alcoholic potash yields two isomeric monobromocinnamic acids, viz. :

$$C_6H_5.CH=CBr.CO_2H \qquad\qquad C_6H_5.CBr=CH.CO_2H$$

a
b

Both are crystalline. The α-acid melts at 130° C. (266° F.), and forms salts which crystallize readily ; the β-acid melts at 120° C. (248° F.), and forms deliquescent salts. α unites with 2 atoms of bromine, and is converted by sodium-amalgam into hydrocinnamic acid ; β is converted into α by distillation, or by heating with hydriodic acid.

Two nitrocinnamie acids (*ortho-* and *para-*) are formed by direct nitration, and may be separated by the different solubilities of their ethylic ethers in cold alcohol, the *p*-compound being nearly insoluble. The *o*-compound melts at 232° C. (449.6° F.), the *p*-compound at 265° C. (509° F.). Both are converted by chromic acid into the corresponding nitrobenzoic acids.

The following homologues of cinnamic acid (which cannot here be described) are formed by heating benzoic and cumic aldehydes with the sodium salt of a fatty acid, $C_nH_{2n}O_2$, and the corresponding anhydride, $(C_nH_{2n-1}O)_2O$, *e. g.*, phenylcrotonic acid from benzoic aldehyde, sodium propionate, and propionic anhydride—the mode of formation being exactly analogous to that of cinnamic acid from C_7H_6O and $C_4H_6O_3$ (p. 934):

Acids from Benzoic Aldehyde.		Acids from Cumic Aldehyde.	
Phenylcrotonic	$C_{10}H_{10}O_2$	Cumenyl-acrylic	$C_{12}H_{14}O_2$
Isophenylcrotonic	$C_{10}H_{10}O_2$	Cumenyl-crotonic	$C_{13}H_{16}O_2$
Phenyl-angelic	$C_{11}H_{12}O_2$	Cumenyl-angelic	$C_{14}H_{18}O_2$

An Isophenylcrotonic acid is formed by the action of succinic anhydride and sodium succinate on benzoic aldehyde :

$$C_4H_4O_3 \; + \; C_7H_6O \; = \; CO_2 \; + \; C_{10}H_{10}O_2 \,.$$

Sodium-amalgam converts these acids into acids of the series $C_nH_{2n-6}O_2$; *e. g.*, hydrocumenylacrylic, $C_{12}H_{16}O_2$.[*]

Atropic and **Isatropic Acids,** isomeric with cinnamic acid, are formed simultaneously from tropic acid (p. 958), by boiling with barytawater or hydrochloric acid : $C_9H_{10}O_3 - H_2O = C_9H_8O_2$.

Atropic acid, the chief product formed when baryta is used, crystallizes from hot water in monoclinic plates, slightly soluble in cold water, melting at 106.5° C. (223.7° F.). It is oxidized by chromic acid mixture to benzoic acid ; converted by sodium-amalgam into hydroatropic acid (p. 934); and resolved by fusion with potash into alphatoluic acid, C_6H_5. $CH_2.CO_2H$, and formic acid, whereas cinnamic acid similarly treated yields benzoic and acetic acids ; hence it appears that atropic acid is related to cinnamic acid in the same manner as methacrylic to solid crotonic acid (p. 744):

$CH_3—CH{=}CH—CO_2H$	$CH_2{=}C(CH_3)(CO_2H)$
Crotonic.	Methacrylic.
$C_6H_5—CH{=}CH—CO_2H$	$CH_2{=}C(C_6H_5)(CO_2H)$
Cinnamic.	Atropic.

[*] **Perkin,** Chem. Soc. Journal, 1877, i. 388.

Isatropic acid, probably a polymeric modification, is the chief product obtained by heating tropic acid with hydrochloric acid to 140° C. (284° F.). It forms thin laminæ, very slightly soluble in water, melts at 200° C. (392° F.); does not unite with nascent hydrogen.

Acids, $C_nH_{2n-12}O_2$:—

Phenyl-propiolic acid, $C_9H_6O_2 = C_6H_5$—C≡C—CO_2H, is formed : 1. By boiling *a*-bromocinnamic acid with alcoholic potash. 2. By the action of Na and CO_2 on *a*-bromocinnamene, C_6H_5—$C_6H_5CH{=}CHBr$. 3. By the action of CO_2 on sodium-acetenylbenzene (p. 850) dissolved in ether :

$$C_6H_5\text{—C}{\equiv}\text{CNa} + CO_2 = C_6H_5\text{—C}{\equiv}\text{C—}CO_2Na.$$

It crystallizes from hot water in long shining needles ; melts and sublimes at 136°–137° C. (276.8°–278.6° F.); melts under water at 80° C. (176° F.). It is oxidized by chromic acid mixture to benzoic acid ; converted by sodium-amalgam into hydrocinnamic acid ; resolved by heating with water to 120° C. (248° F.) into CO_2 and phenyl-acetylene, $C_6H_5.C{\equiv}CH$ (p. 850).

Homologous with phenyl-propiolic acid are the following acids formed from cinnamic aldehyde, C_9H_8O, by the action of the anhydrides and sodium salts of fatty acids, viz. :—

Cinnamenyl-acrylic acid	.	.	.	.	$C_{11}H_{10}O_2$
Cinnamenyl-crotonic acid	.	.	.	.	$C_{12}H_{12}O_2$
Cinnamenyl-angelic acid	.	.	.	.	$C_{13}H_{14}O_2$.

MONOBASIC AND DIATOMIC ACIDS.

(1) $C_nH_{2n-8}O_3$, or $C_nH_{2n-8}\begin{cases} OH \\ CO_2H \end{cases}$ ·

These aromatic oxy-acids, like the corresponding acids of the fatty series (the lactic acids), exhibit alcoholic as well as acid characters. In contact with carbonates they give up only the hydrogen-atom of the CO_2H group in exchange for a metal (forming neutral salts), but when acted upon by strong free bases (caustic alkalies), they likewise exchange the hydrogen of the hydroxyl-group (the alcoholic or phenolic hydrogen) for metals, forming so-called basic salts, *e. g.*, $C_6H_4(ONa)(CO_2Na)$. These, like the metallic derivatives of the phenols, are decomposed by carbonic acid, and converted into neutral salts.

The aromatic oxy-acids are formed from the halogen-derivatives of benzoic acid and its homologues, and from the sulpho-acids, by fusion with alkalies ; from the amido-derivatives of the same acids by the action of nitrous acid ; and from the phenols by the action of sodium and carbon dioxide.

Oxybenzoic Acids, $C_7H_6O_3 = C_6H_4(OH).CO_2H$.

1. ORTHO-OXYBENZOIC or SALICYLIC ACID (1 : 2) is formed :
(1) By heating sodium phenate in a stream of carbon dioxide, phenol then distilling over, while disodium salicylate remains behind :

$$2C_6H_5ONa + CO_2 = C_6H_4ONa.CO_2Na + C_6H_5OH .$$

40

The reaction takes place even below 100º, but proceeds most quickly between 170º and 180º C. (338º and 356º F.), and goes on in the same way up to 300º C. (572º F.), at which temperature the sodium salicylate begins to decompose. This process is applied to the preparation of salicylic acid on the large scale. Potassium phenate heated in a stream of carbon dioxide is decomposed in the same manner, yielding pure dipotassium salicylate up to 150º C. (302º F.); but above that temperature the isomeric paraoxybenzoate is likewise produced, increasing in p as the temperature rises, and becoming the sole product at 220º (428º F.). Monopotassium salicylate is decomposed in the same manner at 220º (428º F.), yielding a distillate of phenol and a residue of dipotassium paraoxybenzoate :

$$2(C_6H_4(OH).CO_2K) = C_6H_4(OK).CO_2K + C_6H_5OH + CO_2;$$

and the monosodium salt undergoes a similar decomposition, yielding, however, not paraoxybenzoate, but salicylate of sodium.

(2) By oxidation of salicylic aldehyde or of saligenin (p. 912).

(3) By the action of nitrous acid on (1 : 2) amidobenzoic (anthanilic) acid :

$$C_6H_4(NH_2).CO_2H + NO.OH = C_6H_4(OH).CO_2H + N_2 + H_2O.$$

(4). By fusing (1 : 2) chloro- or bromobenzoic acid, or (1 : 2) toluenesulphonic acid with alkalies.

(5) Together with acetic acid, by heating coumaric acid with potassium hydroxide :

$$C_9H_8O_3 + 2H_2O = C_7H_6O_3 + C_2H_4O_2 + H_2.$$

Salicylic acid occurs in the free state in the flowers of meadow-sweet (*Spiræa ulmaria*), and as a methylic ether in oil of winter-green (*Gaultheria procumbens*), from which it may be obtained by distillation with potash.

Salicylic acid crystallizes from its alcoholic solution by spontaneous evaporation in large monoclinic prisms. It requires about 1800 parts of cold water to dissolve it, but is much more soluble in hot water and in alcohol. Its aqueous solution imparts a deep violet color to ferric salts. It melts at 155º–156º C. (311º–312.8º F.), gives off phenol at a higher temperature, and when heated with pounded glass or quick-lime, is completely resolved into carbon dioxide and phenol. It is distinguished from both its isomerides by its behavior with ferric salts, its very slight solubility in water, and its lower melting point. It is a very powerful antiseptic.

Basic Barium salicylate, $C_6H_4<^{CO_2}_O>Ba + 2H_2O$, separates in sparingly soluble laminæ on boiling salicylic acid with baryta-water, both the hydroxylic hydrogen-atoms being replaced by barium. The *basic calcium salt* is formed in a similar manner, and separates as an insoluble powder. This reaction affords another distinction between salicylic acid and its isomerides.

Halogen-derivatives of Salicylic acid are easily formed by the direct action of bromine and chlorine; also of iodine, in presence of HgO or IO_3H. By nitration salicylic acid yields three nitro-acids.

Salicylic chloride, $C_6H_4Cl.CO.Cl$, formed by the action of PCl_5 on the acid, is an oil which boils at 240º C. (464º F.), and is converted by hot water into o-chlorobenzoic acid.

Salicylic anhydride or Salicylide, $C_6H_4<^{CO}_O>$, formed by the action of PCl_3O on salicylic acid, crystallizes in shining needles, dis solves sparingly in water, and is reconverted by alkalies into salicylic acid.

Salicylamide, $C_6H_4(OH).CO.NH_2$, formed by heating ammonium salicylate, or by the action of ammonia on salicylic ethers forms sparingly soluble laminæ melting at 132° C. (269.6° F.).

Salicylic ethers are formed by passing gaseous hydrogen chloride into the solutions of salicylic acid in the corresponding alcohols. The monomethylic ether, $C_6H_4(OH).CO_2CH_3$, is the chief constituent of winter-green oil. It is a fragrant liquid of specific gravity 1.197 at 0°, boiling at 224° C. (435.2° F.), colored violet by ferric chloride. The monethylic ether boils at 221° C. (429.8° F.).

The dimethylic ether, $C_6H_4(OCH_3).CO_2CH_3$, an oil boiling at 240° C. (464° F.), is formed by heating the monomethylic ether with alcoholic potash and methyl oxide:

$$C_6H_4(OK).CO_2CH_3 \ + \ CH_3I \ = \ KI \ + \ C_6H_4(OCH_3).CO_2CH_3.$$

This ether, saponified with potash, yields methyl alcohol and methylsalicylic acid, $C_6H_4(OCH_3).CO_2H$, which forms large tabular crystals, melting at 98° C. (208.4° F.), easily soluble in hot water and alcohol, resolved at 200° C. (392° F.) into CO_2 and anisol, $C_6H_5.OCH_3$.

Diethyl salicylate and ethylsalicylic acid are formed in like manner; the latter melts at 19.5° C. (67.1° F.), and is resolved at 300° C. (572° F.) into CO_2, and phenetol, $C_6H_5.O.C_2H_5$.

Acetyl salicylic acid, $C_6H_4(O.C_2H_3O).CO_2H$, formed by the action of acetyl chloride on salicylic acid, crystallizes in slender needles.

META-OXYBENZOIC ACID, $C_6H_4(OH).CO_2H$ (1 : 3), ordinary oxybenzoic acid, is formed by the action of nitrous acid on metamidobenzoic acid, and by fusing metabromo-, iodo-, or sulpho-benzoic acid with potash. It is sparingly soluble in cold water and alcohol, easily in the same liquids when hot, and separates on cooling as a crystalline powder. It melts at 200° C. (392° F.), sublimes undecomposed, and is altogether more stable than either of its isomerides, being resolved into CO_2 and phenol only by distillation with lime. It is not colored by ferric chloride. Its ethylic ether, $C_6H_4(OH).CO_2C_2H_5$, crystallizes in plates, dissolves in hot water, melts at 72° C. (161.6° F.), and boils at 280° C. (536° F.). The dimethylic ether, $C_6H_4(OCH_3).CO.CH_3$, is formed by heating the acid with CH_3I and KOH to 140°, and converted by boiling with potash into methyloxybenzoic acid, $C_6H_4(OCH_3).CO_2H$, which crystallizes in long needles, dissolves easily in hot water, melts at 100°, and sublimes undecomposed.

PARA-OXYBENZOIC ACID, $C_6H_4(OH).CO_2H$ (1 : 4), is formed, as above mentioned, by heating potassium phenate in a stream of carbon dioxide; also from (1 : 4) chloro-, bromo-, iodo-, and sulpho-benzoic acid, and from various resins, by fusion with potassium hydroxide; by the action of nitrous acid on paramidobenzoic acid; and by heating anisic (methyl-paraoxybenzoic) acid with strong hydriodic acid: $C_7H_5(CH_3)O_3 + HI = CH_3I + C_7H_6O_3$. It is more soluble in cold water than metaoxybenzoic or salicylic acid, dissolving in 126 parts of water at 15° C. (59° F.) : from a hot solution it crystallizes in small distinct monoclinic prisms, containing 1 mol. H_2O. Its solution forms, with ferric chloride, a yellow precipitate insoluble in excess, without violet coloration. In the anhydrous state it melts at 210° C. (410° F.), with partial resolution into CO_2 and phenol. Its basic barium salt, $C_6H_4\!\!<^{O}_{CO_2}\!\!>Ba$, is insoluble, and affords the means of separating paraoxybenzoic acid from the meta-acid.

Monomethylic paraoxybenzoate forms large tabular crystals, melts at 17° C. (62.6° F.), and boils at 283° C. (541.4° F.). The *ethylic ether* melts at 113° C. (235.4° F.), and boils at about 297° C. (566.6° F.).

Methyl-paraoxybenzoic and Ethyl-paraoxybenzoic acids are prepared in the same manner as the ortho- and meta-compounds.

Methyl-paraoxybenzoic or *Anisic acid*, $C_6H_4(OCH_3).CO_2H$, is also produced by oxidation of anisaldehyde and anethol (p. 913)—or the crude oils of anise, fennel, and tarragon, which contain anethol—with nitric acid or chromic acid mixture. The anethol is first converted into anisaldehyde :

$$C_{10}H_{12}O + O_6 = C_8H_8O_2 + C_2H_2O_4 + H_2O,$$
$$\text{Anethol.} \qquad\qquad \text{Anis-} \qquad \text{Oxalic}$$
$$\text{aldehyde.} \qquad \text{acid.}$$

and the aldehyde is afterwards oxidized to anisic acid. Anisic acid is also produced by oxidation of methyl-paracresol, $C_6H_4(OCH_3).CH_3$.

Anisic acid crystallizes from hot water in long needles, from alcohol in rhombic prisms ; melts at 183° C. (361.4° F.); sublimes and boils without decomposition at 283° C. (541.4° F.). By heating with lime or baryta it is resolved into CO_2 and anisol, $C_6H_5.O.CH_3$. Heated with HCl or HI, it yields paraoxybenzoic acid. Its salts are easily soluble in water, and crystallize well.

With the halogens and with nitric acid, anisic acid readily yields substitution-products, which are converted by distillation with baryta into substituted anisols. By the successive action of fuming nitric acid and ammonia, anisic acid is converted into chrysanisic acid, $C_6H_2(NO_2)_2(NH_2).CO_2H$ (p. 928).

Acids, $C_8H_8O_3 = C_7H_6{<}^{OH}_{CO_2H}$.

1. CRESOTIC or OXYTOLUIC ACIDS, $C_6H_3(CH_3){<}^{OH}_{CO_2H}$.—Three cresotic acids are formed by the action of Na and CO_2 on the three isomeric cresols (p. 897) :

$$C_6H_4(CH_3)OH + CO_2 = C_6H_3(CH_3)(OH)(CO_2H) .$$

They crystallize in needles, dissolve in hot water, and are colored violet by ferric chloride. The acid from o-cresol melts at 163°–164° C. (325.4°–327.2° F.) ; that from m-cresol at 114° C. (237.2° F.) ; that from p-cresol at 150° C. (302° F.).

A fourth oxytoluic acid is obtained from sulphotoluic acid (from camphor-thiocymene) by fusion with potash ; it melts at 203° C. (397.4° F.), and is not colored by ferric chloride.

2. MANDELIC or PHENYLGLYCOLLIC ACID, $C_6H_5.CH{<}^{OH}_{CO_2.H}$, is formed by boiling bitter almond oil for 30–36 hours with hydrocyanic and hydrochloric acids :

$$C_6H_5.CHO + CNH + 2H_2O = NH_3 + C_6H_5.CHOH.CO_2H ;$$

also by heating amygdalin (see GLUCOSIDES) with hydrochloric acid. It crystallizes in prisms or tables, easily soluble in water, alcohol, and ether, melting at 115° C. (239° F.). By oxidizing agents it is converted into benzoic acid ; by hydriodic acid into alphatoluic (phenylacetic) acid. With HBr it yields *phenylbromacetic* acid, $C_6H_5.CHBr.CO_2H$, and with HCl the corresponding chlorinated acid.

Acids, $C_9H_{10}O_3$.

1. OXYMESITYLENIC ACID, $C_6H_2(CH_3)_2<^{OH}_{CO_2H}$, formed by heating mesitylenesulphonic acid with potash to $240°$–$253°$ C. ($464°$–$482°$ F.), crystallizes from alcohol in silky needles, melts at $176°$ C. ($348.8°$ F.), and sublimes in long flat needles. Its solutions and those of its salts are colored dark blue by ferric chloride.

2. HYDROCOUMARIC or MELILOTIC ACID, $C_6H_4<^{OH}_{CH_2.CH_2.CO_2H}$ (1 : 2), occurs in the yellow mellilot, and is formed by the action of sodium on coumarin (p. 942) and coumaric acid : $C_9H_6O_2$ (coumarin) $+ H_2O + H_2 = C_9H_{10}O_3$. It crystallizes in long needles melting at $82°$ C. ($179.6°$ F.). Its solution is colored bluish by ferric chloride. By distillation it is converted into the anhydride, $C_9H_8O_2 = C_6H_4<^{O}_{C_2H_4.CO}$, which melts at $25°$ C. ($77°$ F.). By fusion with potash, melilotic acid is resolved into acetic and salicylic acid, and is therefore an ortho-diderivative of benzene.

3. HYDROPARACOUMARIC ACID, $C_6H_4<^{OH}_{CH_2.CH_2.CO_2H}$ (1 : 4), is produced by the action of sodium-amalgam on paracoumaric acid, and of nitrous acid on para-amidocinnamic acid. It forms small monoclinic crystals, easily soluble in water, alcohol, and ether ; melting at $125°$ C. ($257°$ F.). By fusion with potash it yields para-oxybenzoic acid.

4. PHLORETIC ACID, $C_6H_4<^{OH}_{CH(CH_3)(CO_2H)}$ (1 : 4), is formed, together with phloroglucin, by heating phloretin with potash-lye :

$$C_{15}H_{14}O_5 + H_2O = C_9H_{10}O_3 + C_6H_6O_3 ;$$

also by the action of sodium and CO_2 on phlorol, C_8H_9OH (p. 899). It crystallizes in long prisms, easily soluble in hot water, melting at $129°$ C. ($264.2°$ F.). Ferric chloride colors its solution green. By heating with baryta it is resolved into CO_2 and phlorol ; by fusion with potash into acetic and paroxybenzoic acids. *Methylphloretic acid* is converted by oxidation into anisic acid.

5. TROPIC ACID, $C_6H_5.CH<^{CH_2OH}_{CO_2H}$, *Phenyl-hydracrylic acid*, formed from atropine by boiling with hydrochloric acid or with baryta-water, crystallizes in slender prisms melting at $117°$ C. ($242.6°$ F.). By long boiling with baryta it is converted into atropic and isatropic acids (p. 561), which by oxidation yield benzoic acid. Tropic acid is therefore a monoderivative of benzene.

6. PHENYL-LACTIC ACID, $C_6H_5.CH_2.CH<^{OH}_{CO_2H}$, Phenyl-chlorolactic acid, $C_6H_5.CHCl.CH(OH)(CO_2H)$, formed by addition of ClOH to cinnamic acid, $C_6H_5.CH{=}CH.CO_2H$, and phenyl-bromolactic acid, formed by boiling cinnamic dibromide, $C_6H_5.CHBr.CHBr.CO_2H$, with water (p. 935), are both converted by sodium amalgam into phenyl-lactic acid. This acid crystallizes in concentric groups of needles, very soluble in hot water, melting at $94°$ C. ($201.2°$ F.). At $180°$ C. ($356°$ F.) it decomposes into water and cinnamic acid. When quickly heated it yields cinnamene :

$$C_6H_5.CH_2.CH(OH)(CO_2H) = CO_2 + H_2O + C_6H_5.CH{=}CH_2 .$$

Heated with concentrated haloïd acids it exchanges its hydroxyl-group for halogen elements, yielding substituted phenyl-propionic acids, *e. g.*, $C_6H_5.CH_2.CHBr.CO_2H$ (p. 933).

Tyrosine, $C_9H_{11}NO_3$, probably $C_6H_4{<}^{OH}_{C_2H_3(NH_2).CO_2H}$, *Paraoxyphenyl-amidopropionic* or *Amido-hydroparacoumaric acid*, is a compound nearly related to phloretic or hydroparacoumaric acid. It is produced, together with leucine, aspartic acid, and glutamic acid by boiling various animal substances, as albuminoids, horn, hair, etc., with hydrochloric or sulphuric acid, or by fusing them with potash. It sometimes occurs ready-formed in the liver, spleen, and pancreas; also in old cheese ($\tau\nu\rho\sigma\varsigma$). It is best prepared by boiling horn-shavings with 2 parts of sulphuric acid diluted with 4 parts of water for about 20 hours, renewing the water, as it evaporates. The solution is then saturated with milk of lime, and the tyrosine is precipitated from the concentrated filtrate by acetic acid.

Tyrosine is insoluble in ether, slightly soluble in cold water and in alcohol, dissolves in 150 parts of hot water, and crystallizes therefrom in slender needles. It dissolves easily in hydrochloric acid, forming the salt $C_9H_{11}NO_3.HCl$, which crystallizes in needles. The aqueous solution of tyrosine boiled with mercuric nitrate yields a yellowish precipitate, which, when boiled with dilute yellow nitric acid, becomes dark red; this is a very delicate test for tyrosine. With chlorine and bromine tyrosine yields chlor- and bromanil, $C_6Cl_4O_2$ and $C_6Br_4O_2$; with nitric acid it yields substitution-products. Sulphuric acid dissolves it, forming sulpho-acids whose salts are colored violet by ferric chloride.

Tyrosine fused with potash is resolved into acetic acid, ammonia, and para-oxybenzoic acid. Heated to 140° C. (284° F.) with hydriodic acid, it yields phlorol (p. 899), together with CO_2 and NH_3:

$$C_6H_4{<}^{OH}_{C_2H_3(NH_2)CO_2H} + 2HI = C_6H_4{<}^{OH}_{C_2H_5} + CO_2 + NH_3 + I_2.$$

These reactions show that tyrosine is a para-derivative of benzene, containing the NH_2-group in the lateral chain.

Acids, $C_{11}H_{14}O_3$.—Two isomeric acids, thymotic and carvacrotic, are formed by the action of sodium and CO_2 on thymol and carvacrol (p. 899):

$$C_6H_3(OH){<}^{C_3H_7}_{CH_3} + CO_2 = C_6H_2(OH)(CO_2H){<}^{C_3H_7}_{CH_3}.$$

Thymotic acid forms long needles, very slightly soluble in cold water; melts at 120° C. (248° F.), and sublimes. Carvacrotic acid sublimes in flat needles, melting at 134° C. (273.7° F.). Both are colored a fine blue by ferric chloride.

Phenylpropyl-glycollic acid, $C_6H_4{<}^{C_3H_7}_{CH(OH).CO_2H}$, formed by the action of hydrochloric and hydrocyanic acids on cuminol, crystallizes in small needles, moderately soluble in cold water, very soluble in alcohol and ether, melting at 158° C. (316.4° F.).

---

(2) Unsaturated Acids, $C_nH_{2n-10}O_3$.

Coumaric and **Paracoumaric Acids**, $C_9H_8O_3 =$ $C_6H_4{<}^{CH=CH.CO.OH}_{OH}$, the only known acids of this series, have the composition of oxycinnamic or oxyphenyl-acrylic acid.

Paracoumaric acid (1 : 4), produced by boiling an aqueous solution of aloes with sulphuric acid, crystallizes in colorless, shining, brittle needles, easily soluble in hot water and in alcohol, melting at 170°–180° C. (338°–356° F.). By nascent hydrogen it is converted into hydroparacou- maric acid; by fusion with potash, into paroxybenzoic acid.

Coumaric acid (1 : 2) occurs, together with melilotic acid (p. 958), in the yellow melilot, and in Faham leaves, and is most easily prepared from coumarin, its anhydride, by boiling with potash. It crystallizes in colorless shining prisms, easily soluble in hot water and alcohol, melting at 195° C. (383° F.). The solutions of its alkali-salts exhibit a bright green fluoresence. By nascent hydrogen it is converted into melilotic acid; by fusion with potash into salicylic and acetic acids.

Coumarin, $C_9H_6O_2$, or Coumaric anhydride,

$C_6H_4\!<\!{}^{O}_{CH=CH.CO}\,$, is the odoriferous principle of the Tonka bean (*Dipteryx odorata*), and of several other plants, as *Melilotus officinalis*, *Asperula odorata*, and *Anthoxanthum odoratum*. It may be extracted from these plants by alcohol, and crystallizes in slender, shining, colorless needles, melting at 67° C. (152.6° F.), boiling between 290° C. (554° F.) and 291°, and distilling without decomposition at a higher temperature. It has a fragrant odor and burning taste; is very slightly soluble in cold water, more soluble in hot water and in alcohol. It dissolves in potash-lye, and is converted by long boiling therewith into coumaric acid. Sodium-amalgam converts it into melilotic acid.

Coumarin is formed artificially by dehydration of acetosalicylal (isomeric with coumaric acid) in the manner already described (action of acetic anhydride on sodium-salicylal, p. 918):

$$C_7H_5(C_2H_3O)O_2 - H_2O = C_9H_6O_2 .$$

By acting on sodium-salicylal in like manner with butyric and valeric anhydrides, homologues of coumarin are obtained, viz., butyric coumarin, $C_{11}H_{10}O_2$, and valeric coumarin, $C_{12}H_{12}O_2$. Both crystallize in prisms, the former melting at 71° C. (159.8° F.), and boiling at 297° C. (566.6° F.); the latter melting at 54° C. (129.2° F.), and distilling with partial decomposition at 300° C. (572° F.).

MONOBASIC AND TRIATOMIC ACIDS—DIOXYACIDS.

(1) **Dioxybenzoic Acids**, $C_7H_6O_4 = C_6H_3\left\{{}^{(OH)_2}_{CO_2H}\right.$:—Of six possible isomerides included in this formula, four are known, but the orientation of their lateral chains is not yet completely established.[*]

Two dioxybenzoic acids are obtained by fusing the two disulphobenzoic acids (p. 931) with potassium hydroxide. One of these dioxy-acids forms crystals containing $1\frac{1}{2}H_2O$, melts with decomposition at about 220° C. (428° F.), and is not colored by ferric chloride. The other, prepared also from toluenedisulphonic acid, forms hydrated crystals melting at 194° C. (381.2° F.), and is colored dark-red by ferric chloride.

Oxysalicylic Acid, $C_6H_3(OH)_2.CO_2H$ (probably 1 : 2 : 3, CO_2H in 1) is formed by boiling a solution of iodosalicylic acid, $C_7H_5IO_3$, with potash. It crystallizes in shining needles, soluble in water, alcohol, and ether. The aqueous solution, mixed with ferric chloride, acquires a deep-blue

[*] See Watts's Dictionary of Chemistry, 2d Supplement, p. 432.

color, changing to red and then brown on addition of ammonia or sodium carbonate. The crystallized acid melts at $196°-197°$ C. ($384.8°-386.6°$ F.), and is resolved at a higher temperature into CO_2 and hydroquinone (together with pyrocatechin). The oxy-salicylates are very unstable.

Protocatechuic Acid, $C_6H_3(OH)_2.CO_2H$ (probably 1 : 3 : 4, CO_2H in 1); also called *Carbohydroquinonic acid*.—This acid is produced by the action of melting potash on its aldehyde, and on iodoparaoxybenzoic, bromanisic, cresolsulphonic, and eugenic acids, catechin, and many other tri-derivatives of benzene ; also on numerous resins, as benzoïn, myrrh, dragon's blood, assafœtida, etc., its formation from these resins being usually accompanied by that of paraoxybenzoic acid. It is also formed by the action of bromine on aqueous quinic acid, and by fusing that acid with potash. It is most easily prepared by adding 1 part of East Indian kino to 3 parts of fused sodium hydroxide, dissolving the melt in water, acidulating, and agitating with ether.

Protocatechuic acid crystallizes from water in shining needles or laminæ containing 1 mol. H_2O. It dissolves easily in hot water, in alcohol, and in ether, and in 40–50 parts of cold water ; melts at $199°$ C. ($390.2°$ F.), and decomposes at a higher temperature into CO_2 and pyrocatechin (together with hydroquinone). Ferric chloride colors the solution green, changing, on addition of very dilute sodium carbonate, to blue, and afterwards to red. It reduces silver nitrate, but not an alkaline cupric solution.

Methyl-protocatechuic or *Vanillic acid*, $C_8H_8O_4 = C_6H_3(OH)$ $(OCH_3).CO_2H$, is formed by oxidation of vanillin (p. 919), when that substance, in the moist state, is left exposed to the air, and by oxidation of coniferin, $C_{16}H_{22}O_8$ (see GLUCOSIDES), with potassium permanganate. It crystallizes in shining white needles, melting at $211°-212°$ C. ($411.8°-413.6°$ F.); sublimable ; sparingly soluble in cold water, easily in hot water and in alcohol ; decomposed by hydrochloric acid at $150°-160°$ C. ($302°-320°$ F.) into methyl chloride and protocatechuic acid. Its calcium-salt distilled with lime yields pure guaiacol (p. 900).

An isomeric methyl-protocatechuic acid is formed by heating protocatechnic acid with potassium hydroxide and methyl iodide, also by heating hemipinic acid with strong hydrochloric acid at $100°$. It crystallizes in slender needles, less soluble than the preceding, and melting at $251°$ C. ($483.8°$ F.).

Dimethyl-protocatechuic or *Veratric acid*, $C_9H_{10}O_4 = C_6H_3$ $(OCH_3)_2.CO_2H$, is contained in sabadilla seeds (from *Veratrum Sabadilla*), and is formed by heating protocatechuic or methyl-protocatechuic acid with KHO, methyl iodide and methyl alcohol to $140°$ C. ($284°$ F.), and boiling the product with soda-lye ; also by oxidation of dimethyl-proto-catechuic aldehyde (p. 919), methyl-cresol (p. 828), and methyl-eugenol (p. 914), with permanganate. Colorless needles melting at $179.5°$ C. ($355.1°$ F.); slightly soluble in cold water, more easily in hot water, still more in alcohol. Heated to $140°-150°$ C. ($284°-302°$ F.) with dilute hydrochloric acid, it yields a mixture of two monomethyl-protocatechuic acids. Heated with lime or baryta, it is resolved into CO_2 and dimethyl-pyrocatechin.

In like manner are obtained : *Diethyl-protocatechuic acid* (needles, m. p. $149°$ C. [$300.2°$ F.]), and *Ethyl-methyl-protocatechuic* or *Ethyl-vanillic acid* (needles slightly soluble in hot water ; m. p. $190°$ C. [$374°$ F.]),

Methene-protocatechuic or *Piperonylic acid*, $C_8H_6O_4 =$ $CH_2{<}{\overset{O}{\underset{O}{}}}{>}C_6H_3-CO_2H$, is formed by oxidation of its aldehyde piperonal,

$C_8H_6O_2$ (p. 919), with permanganate, and by heating protocatechuic acid with KOH and CH_2I_2.—Colorless needles, m. p. 228° C. (442° F.), sublimable without decomposition, insoluble in cold water, slightly soluble in boiling water and cold alcohol, easily in hot alcohol. Decomposed by heating to 170° C. (338° F.) with dilute hydrochloric acid into protocatechuic acid and free carbon; by heating with water to 200°–210° C. (392°–410° F.), into pyrocatechin, CO_2, and carbon.

Ethene-protocatechuic acid, $C_9H_8O_4$, obtained by heating protocatechuic acid with KOH, $C_2H_4Br_2$, and a little water to 100°, forms shining prisms melting at 133.5° C. (272.3° F.).

Acids, $C_8H_8O_4 = C_6H_2(CH_3)\begin{cases}(OH)_2\\CO_2H\end{cases}$.

Orsellinic Acid is formed by boiling orsellic or lecanoric acid with lime-water: $C_{16}H_{14}O_7 + H_2O = 2C_8H_8O_4$; also by boiling its erythritic ether (erythrin or erythric acid) with water or baryta-water:

$$\underset{\text{Erythrin.}}{C_{20}H_{22}O_{10}} + H_2O = \underset{\substack{\text{Orsellinic}\\\text{acid.}}}{C_8H_8O_4} + \underset{\text{Picroerythrin.}}{C_{12}H_{16}O_7};$$

Orsellinic acid crystallizes in prisms easily soluble in water, colored violet by ferric chloride. It melts at 176° C. (348.8° F.), decomposing at the same time into CO_2 and orcin, $C_7H_8O_2$ (p. 904).

Ethyl Orsellinate, $C_6H_2(CH_3)(OH)_2.CO_2.C_2H_5$, is obtained by boiling orsellinic acid, or *Roccella tinctoria*, with alcohol. It crystallizes in lustrous plates easily soluble in boiling water, alcohol, and ether.

Erythritic Orsellinate, Erythrin or Erythric acid, $C_{20}H_{22}O_{10} = 2C_8H_8O_4 + C_4H_{10}O_4$ (erythrite) $- 2H_2O$, is contained in *Roccella fusiformis*, and extracted by boiling with milk of lime. It forms crystals containing $\frac{2}{3} H_2O$, slightly soluble in hot water, colored red by ammonia in contact with air, and is resolved by boiling with baryta-water into orsellinic acid and picroerythrin (see above).

Picroerythrin, $C_{12}H_{16}O_7$, forms crystals containing 1 mol. H_2O, soluble in alcohol and ether, and is resolved by further boiling with baryta-water into erythrite, orcin, and carbon dioxide:

$$C_{12}H_{16}O_7 + H_2O = C_4H_{10}O_4 + C_7H_8O_2 + CO_2.$$

Orsellic, *Lecanoric*, or *Diorsellinic acid*, $C_{16}H_{14}O_7 = 2C_8H_8O_4 - H_2O$, occurs in several lichens belonging to the genera *Roccella*, *Lecanora*, and *Variolaria*, and is extracted by digestion with ether or with milk of lime, and precipitation by hydrochloric acid. It crystallizes from alcohol or ether in colorless prisms containing $C_{16}H_{14}O_7 + H_2O$, nearly insoluble in water. By boiling with lime or baryta-water it is converted, first into orsellinic acid, afterwards into CO_2 and orcin. Its alcoholic solution yields, on boiling, crystalline ethyl orsellinate.

The three compounds last described may be represented by the following structural formulæ :

$$O\!\!<\!\!\begin{matrix}C_6H_2(CH_3)\!\!<\!\!\begin{matrix}OH\\CO_2H\end{matrix}\\C_6H_2(CH_3)\!\!<\!\!\begin{matrix}OH\\CO_2H\end{matrix}\end{matrix}$$
Orsellic or Diorsellinic acid.

$$O\!\!<\!\!\begin{matrix}C_4H_2(OH_3\\C_6H_2(CH_3).CO_2H\\O\!\!<\!\!C_6H_2(CH_3)(OH)(CO_2H)\end{matrix}$$
Erythrin.

$$O\!\!<\!\!\begin{matrix}C_4H_6(OH)_3\\C_4H_6(CH_3)(OH)(CO_2H)\end{matrix}$$
Picroerythrin.

According to Stenhouse,[*] the South African variety of *Roccella tinctoria* contains an acid, $C_{34}H_{32}O_{15}$, called *β*-orsellic acid, closely resembling orsellic

---

[*] Phil. Trans. 1848, 69 ; Liebig's Annalen, lxviii. 59.

acid in properties and composition, and yielding orsellinic acid when boiled with baryta-water, together with r o c c e l l i n i n , $C_{19}H_{16}O_7$, a substance forming hair-like crystals of a silvery lustre: $C_{34}H_{32}O_{15} = 2C_8H_8O_4 + C_{18}H_{16}O_7$. Roccellinin is not attacked by boiling with potash or baryta. It dissolves in ammonia and the fixed alkalies, forming solutions which are not colored by exposure to the air.

The following acids of unknown structure are also obtained from lichens :
V u l p i c a c i d , $C_{19}H_{14}O_5$, occurs in *Cetraria vulpina*, from which it may be extracted by chloroform or lime-water. It forms large yellow prisms, slightly soluble in water and in ether, melts at 110º C. (230º F.), sublimes without decomposition, and is resolved by boiling with baryta-water into alphatoluic acid, methyl alcohol, and oxalic acid.

U s n i c a c i d , $C_{18}H_{18}O_7$, from lichens of the genera *Usnea* and *Evernia*, crystallizes in shining yellow laminæ, insoluble in water, slightly soluble in alcohol and ether, melting at 195º–197º C. (383º–386.6º F.). Its alkali-salts, when exposed to the air, turn red and afterwards black. A modification, called B e t a - u s n i c a c i d , from *Cladonia rangiferina*, melts at 175º C. (347º F.), and yields by distillation beta-orcin (p. 905).

C e t r a r i c a c i d , $C_{18}H_{16}O_8$, and L i c h e n i c a c i d , $C_{14}H_{24}O_3$, from Iceland moss (*Cetraria islandica*), are crystalline, and easily soluble in alcohol and ether.

A c i d s , $C_9H_{10}O_4 = C_8H_7(OH)_2.CO_2H$:

Everninic Acid is obtained from evernic acid, $C_{17}H_{16}O_7$ (a constituent of *Evernia Prunastri*), by boiling with baryta :

$$C_{17}H_{16}O_7 + H_2O = C_9H_{10}O_4 + C_8H_8O_4.$$

It crystallizes from hot water in needles, melts at 157º C. (314.6º F.), and is colored violet by ferric chloride.

Umbellic Acid is formed by the action of sodium-amalgam on umbelliferone : $C_9H_6O_3 + H_2O + H_2 = C_9H_{10}O_4$. It crystallizes in colorless needles, melting at 125º C. (257º F.), slightly soluble in water, easily in alcohol and ether. With ferric chloride it produces a green color, turning red on addition of sodium carbonate. Fused with potash it yields resorcin.

U m b e l l i f e r o n e , $C_9H_6O_3$, a compound analogous to coumarin (P. 942), occurs in the bark of *Daphne Mezereum*, and is produced by distillation of galbanum, assafœtida, and other resins of umbelliferous plants. It crystallizes from hot water in rhombic prisms, slightly soluble in cold water, easily in hot water and in alcohol ; its aqueous solution appears blue by reflected light. It melts at 240º C. (464º F.), and sublimes without decomposition ; yields resorcin by fusion with potash, and is converted by sodium-amalgam into umbellic acid, just as coumarin is converted into hydrocoumaric acid.

Hydrocaffeic Acid, $C_6H_3(OH)_2$—CH_2—CH_2—CO_2H, formed by the action of sodium-amalgam on caffeic acid, forms crystals easily soluble in water, and is colored dark-green by ferric chloride.

The following unsaturated acids belong to the triatomic and monobasic division.

Caffeic Acid, $C_9H_8O_4 = C_6H_3(OH)_2$—CH꞊CH—CO_2H, formed by boiling caffetannic acid (p. 949) with potash, crystallizes in yellowish prisms, very soluble in hot water and alcohol. The aqueous solution reduces silver nitrate when heated. Ferric chloride colors it green, changing to dark red on addition of sodium carbonate. By fusion with potash caffeic acid is resolved into protocatechuic and acetic acids. By dry distillation it yields pyrocatechin. Sodium-amalgam converts it into hydrocaffeic acid.

Ferulic Acid, $C_{10}H_{10}O_4 = C_6H_3(OH)_2$—$C_3H_4$—$CO_2H$, occurs in assafœtida, and may be separated from the alcoholic extract of that substance by precipitating with lead acetate, and decomposing the precipitate with sulphuric acid. It dissolves easily in hot water and alcohol, and crystallizes in four-sided needles. Its aqueous solution is colored yellow-brown by ferric chloride. By fusion with potash it is resolved into protocatechuic and acetic acids.

Piperic Acid,

$$C_{12}H_{10}O_4 = H_2C\underset{O}{\overset{O}{\diamondsuit}}C_6H_3.CH꞊CH.CH꞊CH.CO_2H .$$

The potassium salt of this acid is obtained by boiling piperine (*q. v.*) with alcoholic potash, and separates in shining prisms. The free acid crystallizes in slender needles melting at 216°–217° C. (420.8°–422.6° F.). It forms sparingly soluble salts containing 1 equivalent of base. It unites with 4 atoms of bromine, and is resolved by fusion with potash into acetic, oxalic, and protocatechuic acids. By oxidation with permanganate solution it yields piperonal (p. 919).

Eugetic Acid, $C_{11}H_{12}O_4 = C_6H_2(OH)(OCH_3)\underset{CH꞊CH—CH_3}{\overset{CO_2H}{<}}$, is formed by the action of sodium and carbon dioxide on eugenol, $C_{10}H_{12}O_2$ (p. 913). It crystallizes in thin prisms, slightly soluble in water, melts at 124° C. (255.2° F.), and is resolved at a higher temperature into CO_2 and eugenol. Ferric chloride colors its solution dark brown.

MONOBASIC AND TETRATOMIC ACIDS.

Gallic Acid, $C_7H_6O_5 = C_6H_2(OH)_3.CO_2H$—*Trioxybenzoic acid, Dioxysalicylic acid.*—This acid occurs in nut-galls (excrescences formed on the leaves and leaf-stalks of the dyer's oak (*Quercus infectoria*), also in sumach, hellebore root, divi-divi (the fruit of *Cisalpina coriaria*), in tea, in pomegranate root, and in many other plants. It is produced artificially by heating di-iodosalicylic acid to 140°–150° C. (284°–302° F.), with excess of an alkaline carbonate; and from bromodioxybenzoic acid and bromoprotocatechuic acid (p. 944) by fusion with potash : hence its structure is either 1 : 2 : 3 : 4 or 1 : 2 : 4 : 5 (CO_2H in 1). It is most conveniently prepared by boiling gallotannic acid with acids or alkalies : $C_{14}H_{10}O_9 + H_2O = 2C_7H_6O_5$.

Gallic acid crystallizes, with 1 mol. H_2O, in slender prisms having a silky lustre ; dissolves in 100 parts of cold and 3 parts of boiling water, easily in alcohol ; melts at about 200°, and splits up at 210°–220° C. (410°–428° F.) into CO_2 and pyrogallol, $C_6H_3(OH)_3$. The aqueous solution reduces the metals from solutions of gold and silver salts (hence its use in photography), and forms a blue-black precipitate with ferric chloride.

Gallic acid, though monobasic, contains 3 atoms óf phenolic hydrogen, and can therefore form tetrametallic salts. The gallates of the alkali-metals are permanent in the dry state and in acid solution, but in alkaline solution they quickly absorb oxygen and turn brown.

Ethyl gallate, $C_6H_2(OH)_3.CO_2C_2H_5$, crystallizes from water in rhombic prisms containing $2\frac{1}{2}$ mol. H_2O. *Triacetyl-gallic acid*, $C_6H_2(OC_2H_3O)_3.CO_2H$, formed by heating gallic acid with acetyl chloride or acetic anhydride, crystallizes from alcohol in needles. With bromine gallic acid forms $C_7H_5BrO_5$ and $C_7H_4Br_2O_5$, both of which are crystalline.

Gallic acid, heated with 4 parts of sulphuric acid to 140°, is converted into r u f i g a l l i o a c i d, $C_{14}H_8O_8$, a derivative of anthracene.

E l l a g i c a c i d, $C_{14}H_8O_9$, probably also an anthracene derivative, is formed by the action of oxidizing agents—as arsenic acid, silver oxide, iodine and water—on gallic acid; also from tannic acid when its concentrated aqueous solution is left for a considerable time in contact with the air, the ellagic acid then separating, together with gallic acid. It is a constituent of bezoar stones (intestinal concretions of a Persian species of goat), from which it may be obtained by boiling with potash and precipitating with hydrochloric acid. It separates, with 1 mol. H_2O, as a crystalline powder, insoluble in water, and having an acid reaction. Its sodium salt is formed by boiling ethyl gallate with sodium carbonate.

Tannic Acids or Tannins.

These substances constitute the astringent principles of plants, and are very widely diffused in the vegetable kingdom. Most of them are glucosides of gallic acid, being resolved by boiling with dilute acids into gallic acid and glucose; others, instead of glucose, yield phloroglucin, $C_6H_3(OH_3)_3$ (p. 906). Gallotannic acid, on the other hand, when quite pure, is not a glucoside, but consists of d i g a l l i c a c i d. By fusion with potash most tannic acids are resolved into protocatechuic acid and phloroglucin.

Most tannic acids give bluish-black precipitates with ferric salts (inks); some, however, as kino and catechu, and the tannins of sumach and of the tea-plant, give greenish precipitates. The tannic acids precipitate a solution of gelatin, and unite with animal membranes : hence their use in the manufacture of leather. They are precipitated from their solutions by neutral acetate of lead.

Gallotannic Acid, Digallic Acid or **Tannin,** $C_{14}H_{10}O_9 = C_6H_2(OH)_3$— $CO—O—CO—C_6H_2(OH)_3$, occurs in large quantity in nut-galls, especially in the Chinese variety, also in sumach (the twigs of *Rhus Coriaria*), in tea, and many other plants. It may be formed from gallic acid by heating with $POCl_3$ to 120° C. (248° F.), and by oxidation with silver nitrate or dilute arsenic acid. It is most readily prepared by exhausting finely pulverized nut-galls with a mixture of alcohol and ether in a percolator (fig. 168), having its neck plugged with cotton-wool, and its mouth loosely stopped with a cork. The liquid, which after some time collects in the receiver below, consists of two distinct strata : the lower, which is almost colorless, is a very strong solution of almost pure tannic acid in water ; the upper consists of ether, holding in solution gallic acid, coloring matter, and other impurities. The carefully separated heavy liquid is left to evaporate in an exsiccator. Gallotannic acid, or *tannin*, thus obtained, forms a slightly yellowish, friable, porous mass, without the slightest

Fig. 168.

tendency to crystallization. It is very soluble in water, less so in alcohol, and very slightly soluble in ether. It reddens litmus, and has a pure astringent taste without bitterness.

A strong solution of gallotannic acid mixed with mineral acids gives precipitates consisting of combinations of the tannic acid with the mineral acids; these compounds are freely soluble in pure water, but nearly insoluble in acid liquids. Gallotannic acid precipitates albumin, gelatin, salts of the vegeto-alkalies, and several other substances: it forms soluble compounds with the alkalies, which, if excess of base is present, rapidly attract oxygen, and become brown by destruction of the acid; the gallotannates of *barium, strontium,* and *calcium* are sparingly soluble; those of *lead* and *antimony* are insoluble. Ferrous salts are unchanged by solution of gallotannic acid; *ferric salts*, on the contrary, give with it a deep bluish-black precipitate, which is the basis of writing-ink: hence the value of an infusion of tincture of nut-galls as a test for the presence of iron.

Gallotannic acid decomposes at 250° C. (482° F.), with formation of pyrogallol. By boiling with alkalies or dilute acids it is converted into gallic acid, $C_{14}H_{10}O_9 + H_2O = 2C_7H_6O_5$: hence its constitution is determined. By boiling with excess of acetic anhydride it is converted into pentacetyl-tannic acid, $C_{14}H_5(C_2H_3O)_5O_9$, a crystalline substance, melting at 137° C. (278.6° F.).

The following tannins, which are amorphous bodies, more or less resembling gallotannic acid, have not been obtained in the pure state.

Quercitannic acid, from oak-bark, is a yellow-brown amorphous mass, the aqueous solution of which is colored deep blue by ferric chloride. *Cinchonatannic acid,* occurring in cinchona barks, partly in combination with alkaloids, gives a green color with ferric salts. *Ratanhia-tannic acid* occurs in rhatany-root; *filitannic acid* in fern-roots; *tormentil-tannic acid* in tormentil-root. These five tannins, when boiled with dilute acids, yield red amorphous bodies (oak-red, cinchona-red, etc.), which partly also occur ready-formed in the respective plants, and are resolved by fusion with potash into protocatechuic acid and phloroglucin. The formation of these red compounds appears to be accompanied by that of sugar.

Catechu-tannic acid occurs in catechu, an extract prepared from several East Indian plants, viz., *Areca Catechu, Acacia (Mimosa) Catechu,* and *Nauclea Gambir.* Ferric salts color it dingy green. Catechu also contains catechin or catechuic acid, $C_{19}H_{18}O_8$, which crystallizes in needles containing $3H_2O$. Its aqueous solution is colored green by ferric salts. Anhydrous catechin melts at 217° C. (422.6° F.); decomposing at the same time into CO_2 and pyrocatechin. By fusion with potash it is resolved into protocatechuic acid and phloroglucin:

$$C_{19}H_{18}O_8 + 2H_2O = C_7H_6O_4 + 2C_6H_6O_2 + 2H_2.$$

Morintannic acid or *Maclurin,* occurring in fustic (the wood of *Morus tinctoria*), is a yellow crystalline powder, easily soluble in hot water and in alcohol, the solutions giving with ferric chloride a greenish-black precipitate. It yields pyrocatechin when heated, phloroglucin and protocatechuic acid when fused with potash.

Caffetannic acid, from coffee-berries, is a gummy, easily soluble mass, which gives a green reaction with ferric chloride, yields pyrocatechin when heated alone, caffeic acid (p. 947), when heated with potash-solution, protocatechuic acid when fused with potash.

Quinic or **Kinic Acid,** $C_7H_{12}O_6 = C_6H_7(OH)_4(CO_2H)$, is a monobasic pentatomic acid, nearly related to the aromatic oxy-acids. It is found in cinchona-bark, and is obtained as a bye-product in the preparation of quinine. The extract, obtained by digesting the comminuted bark with water or dilute sulphuric acid, is mixed with milk of lime to precipitate the alkaloïd ; the solution is filtered and evaporated ; and the calcium quinate which remains is purified by crystallization, and decomposed with oxalic acid.

Quinic acid forms transparent colorless monclinic prisms, easily soluble in water, very slightly soluble in alcohol. It melts at 162° C. (323.6° F.), and at a higher temperature is resolved into hydroquinone, pyrocatechin, benzoic acid, phenol, and other products. By oxidizing agents (MnO_2 and sulphuric acid) it is converted into quinone, together with carbonic and formic acids. By heating with hydriodic acid it is converted into benzoic acid :

$$C_6H_7(OH)_4.CO_2H + 2HI = C_6H_5.CO_2H + 4H_2O + I_2 ;$$

and by phosphorus pentachloride into chlorobenzoyl chloride :

$$C_6H_7(OH)_4.CO_2H + PCl_5 = C_6H_4Cl.COCl + PO_4H_3 + 3HCl + H_2O .$$

By fusion with potash it yields protocatechuic acid.

Calcium quinate, $(C_7H_{11}O_6)_2Ca + 10H_2O$, forms large, easily soluble, efflorescent rhombic prisms.

Uvic Acid, $C_7H_8O_3$, obtained by distillation of pyrotartaric acid, forms shining crystals melting at 134° C. (273.2° F.) and sublimable; converted into benzoic acid by fusion with potash. Its calcium-salt, $(C_7H_7O_3)_2Ca + 6H_2O$, crystallizes in needles.

Aldehydic Acids.

These are acids containing the group CHO as well as CO.OH, in place of hydrogen, and exhibiting an aldehydic as well as an acid character. All the known aldehydic acids likewise contain the group OH, and are therefore the intermediate aldehydes of bibasic triatomic acids. They are produced, like salicylic aldehyde (p. 916), by the action of chloroform on the oxy-acids in alkaline solution.

Aldehydo-oxybenzoic Acid, $C_8H_6O_4 = C_6H_3(OH){<}{CHO \atop CO_2H}$. — Two isomeric acids of this composition (α and β *Aldehydosalicylic acids*) are obtained by boiling 14 parts of salicylic acids, 25 parts of sodium hydroxide, 50 parts of water, and 15 of chloroform, in a vessel with reversed condenser. On diluting with water, acidulating with hydrochloric acid, agitating with ether, and then agitating the concentrated ethereal solution with acid sodium sulphite, a solution is obtained, from which, after removal of the ether, sulphuric acid separates the β-acid as a crystalline powder, while the α-acid, together with a very small quantity of the β-acid, remains dissolved, and may be extracted from the solution by ether. A third acid, γ-*aldehydoparaoxybenzoic acid*, is obtained in like manner from paraoxybenzoic acid, together with paraoxybenzoic aldehyde.

α. (CO_2H : OH : CHO $= 1 : 2 : 3$).—Slender felted needles, sublimable when slowly heated. Much more soluble in water than *β*. Solution colored red by ferric chloride. Calcium-salt easily soluble : distilled with calcium hydrate gives salicylic aldehyde.

β. (CO_2H : OH : CHO $= 1 : 2 : 5$).—Slender needles melting at 249° C. (480.2° F.), sublimable when slowly heated. Nearly insoluble in cold water, sparingly soluble in boiling water and cold alcohol, easily in hot alcohol and ether. Deep cherry-red with ferric chloride. Calcium-salt very soluble ; distilled with CaH_2O_2 gives paraoxybenzaldehyde.

γ. (CO_2H : OH : CHO $= 1 : 3 : 4$).—Prismatic crystals, melting at 243°– 244° C. (469.4°–471.2° F.), sublimable. Slightly soluble in cold, more readily in hot water, easily in alcohol and ether. Brick-red with ferric chloride. Calcium-salt slightly soluble ; distilled with lime gives salicylic aldehyde.

Aldehydovanillic, or Aldehydo-methylprotocatechuic Acid,

$C_9H_8O_5 = C_6H_2(OH)(OCH_3){<}^{CHO}_{CO_2H}$, is obtained, together with vanillin (p. 919), from vanillic acid, in the same manner as the aldehydoxybenzoic acids from the oxybenzoic acids. Crystallizes in slender silky needles, melting at 221°–222° C. (429.8°–431.6° F.), slightly soluble in water, easily in alcohol and ether. Ferric chloride colors the solution violet.

Opianic Acid, $C_{10}H_{10}O_5 = C_6H_2(OCH_3)_2{<}^{CHO}_{CO_2H}$.—*Aldehydo-dimethylprotocatechuic acid* is formed, together with cotarnine and hemipinic acid (p. 954), in the oxidation of narcotine by MnO_2 and sulphuric acid, or with dilute nitric acid. It crystallizes in slender, colorless prisms, melting at 140° C. (284° F.), slightly soluble in cold water. Heated with soda-lime, it is converted into dimethyl-protocatechuic aldehyde, $C_6H_3(OCH_3)_2.CHO$ (p. 929), and when treated with oxidizing agents, it yields hemipinic acid.

M e c o n i n , $C_{15}H_{10}O_4$, occurs in opium, and is formed, together with cotarnine (see ALKALOIDS), from narcotine, by heating with water to 100°, and from opianic acid by the action of nascent hydrogen or of potashlye. Shining colorless crystals, sparingly soluble in cold, more freely in hot water, melting at 110° C. (230° F.). Heated with acids, it forms ethereal compounds, with elimination of water.

BIBASIC AND DIATOMIC ACIDS.

Benzene-dicarbonic Acids.

Acids, $C_8H_6O_4 = C_6H_4(CO_2H)_2$.

1. **Phthalic Acid** (1 : 2).—*Orthobenzene-dicarbonic acid* is formed by the action of nitric acid on naphthalene, dichloride of naphthalene, alizarin, and purpurin (the coloring-matters of madder):

$$C_{10}H_8 + O_8 = C_8H_6O_4 + C_2H_2O_4$$
Naphthalene.   Phthalic acid.   Oxalic acid.

$$C_{14}H_8O_4 + 2H_2O + O_{10} = C_8H_6O_4 + 3C_2H_2O_4$$
Alizarin.   Phthalic acid.   Oxalic acid.

$$C_{14}H_8O_5 + 2H_2O + O_9 = C_8H_6O_4 + 3C_2H_2O_4 .$$
Purpurin.   Phthalic acid.   Oxalic acid.

It is also produced by oxidizing orthotoluic acid with permanganate solution. and is usually prepared by treating naphthalene dichloride at the boiling heat with nitric acid. Chromic acid mixture cannot be used in its preparation, as it is easily oxidized thereby to carbonic acid.

Phthalic acid crystallizes in short prisms or lamiuæ, slightly soluble in cold water, easily soluble in hot water, alcohol, and ether. It melts at 185⁰ C. (365⁰ F.), and is resolved at a higher temperature into water and phthalic anhydride, $C_8H_4O_8$. Heated with excess of calcium hydrate, it is resolved into $2CO_2$ and benzene, but when 2 molecules of it are heated with 1 mol. CaH_2O_2, it yields calcium benzoate:

$$2C_8H_6O_4 \; + \; CaH_2O_2 \; = \; (C_7H_5O_2)_2Ca \; + \; 2CO_2 \; + \; 2H_2O \, .$$

It forms both acid and neutral salts. Barium chloride added to the solution of its ammonium salt throws down barium phthalate, $C_8H_4O_4Ba$, which is very slightly soluble in water.

Phthalic acid is converted by fuming nitric acid into nitro-phthalic acid, $C_8H_5(NO_2)O_4$.

Phthalic chloride, $C_8H_4(COCl)_2$, formed by heating the acid with PCl_5, is a liquid which boils at 270⁰ C. (518⁰ F.), and is converted by zinc and hydrochloric acid into phthalic aldehyde, $C_8H_4(CHO)_2$, which crystallizes from hot water in plates, and melts at 65⁰ C. (149⁰ F.).

Phthalic anhydride, $C_6H_4{<}^{CO}_{CO}{>}O$, obtained by heating the acid, forms large needles, melts at 128⁰ C. (262.4⁰ F.), and boils at 277⁰ C. (530.6⁰ F.).

Hydrophthalic acid, $C_8H_8O_4$, formed by the action of sodium-amalgam on phthalic acid, dissolved in aqueous sodium carbonate, crystallizes in tables; dissolves easily in hot water and alcohol; and melts at about 200⁰ C. (392⁰ F.), decomposing at the same time into phthalic anhydride, water, and hydrogen. By the action of strong sulphuric acid, or by oxidation with nitric acid, chromic acid, or bromine and water, it is converted into benzoic acid. Tetrahydrophthalic acid, $C_8H_{10}O_4$, is prepared by heating its anhydride with water, and the anhydride, $C_8H_8O_3$, is produced by distillation of isohydropyromellitic acid (p. 955). The anhydride crystallizes in laminæ melting at 68⁰ C. (154.4⁰ F.). The acid also crystallizes in laminæ, melting at 96⁰ C. (204.8⁰ F.), and decomposing at the same time into water and the anhydride. Hexhydrophthalic acid, $C_8H_{12}O_4$, is obtained from either of the preceding by the action of sodium-amalgam, or by heating with hydriodic acid to 230⁰ C. (446⁰ F.); it crystallizes in prisms or laminæ, sparingly soluble in water, melting at 207⁰ C. (404.6⁰ F.).

Isophthalic Acid, $C_6H_4(CO_2H)_2$ (1 : 3), is formed by oxidation of meta-xylene or metatoluic acid with chromic acid mixture, also by fusing meta-sulphobenzoate, or metabromobenzoate, or benzoate of potassium with potassium formate (in the last two cases together with terephthalic acid), and by the action of heat on hydropyromellitic and hydroprehnitic acid (p. 955). It crystallizes in slender needles, soluble in 460 parts of boiling and 7800 parts of cold water, melts above 300⁰ C. (572⁰ F.), and sublimes in needles without blackening. Its barium salt dissolves easily in water, so that the acid is not precipitated by barium chloride from the solution of its ammonium salt, a character by which it is distinguished from its two isomerides.

Methyl isophthalate, $C_6H_4(CO_2.CH_3)_2$, crystallizes in needles, melting at

64O–65O C. (147.2O–149O F.). The *ethylic ether* is a colorless liquid, boiling at 285O C. (545O F.), and not solidifying at 0O.

Nitrosophthalic acid, $C_6H_5(NO_2)O_4$, forms laminæ, melting at 249O C. (480.2O F.).

Isophthalyl chloride, $C_6H_4(COCl)_2$, forms a radio-crystalline mass, melts at 41O C. (105.8O F.), boils at 276O C. (528.8O F.).

Terephthalic Acid, $C_6H_4(CO_2H)_2$ (1 : 4), is formed by oxidation of various bodies belonging to the para-series—as paraxylene, paratoluic acid, cumenel, and cymene, with chromic acid mixture. It is a white powder, crystalline if slowly deposited from solution, nearly insoluble in water, alcohol, and ether; sublimes undecomposed without previous fusion.

The *terephthalates*, $C_8H_4O_4Ca + 3H_2O$ and $C_8H_4O_4Ba + 4H_2O$, are crystalline, and very sparingly soluble. The methylic and ethylic ethers crystallize in prisms, the former melting at 104O C. (219.2O F.), the latter at 44O C. (111.2O F.).

Nitroterephthalic acid is crystalline, and melts at 259O C. (498.2O F.). *Hydroterephthalic acid*, $C_8H_8O_4$, is a white powder, like terephthalic acid.

2. Acids, $C_9H_8O_4 = C_6H_3(CH_3)(CO_2H)_2$.

Uvitic Acid ($CO_2H : CH_3 : CO_2H = 1 : 3 : 5$) is produced, together with mesitylenic acid, by prolonged boiling of mesitylene with dilute nitric acid, and by boiling pyruvic acid (p. 785) with barium hydrate. Slender needles, melting at 287O C. (548.6O F.), sparingly soluble in hot water, easily in alcohol and ether; converted by oxidation with chromic acid into trimesic acid, and resolved by distillation with lime, first into CO_2 and metatoluic acid, afterwards into CO_2 and toluene.

Xylidic Acid ($CO_2H : CH_3 : CO_2H = 1 : 3 : 4$) is produced from pseudodocumene (p. 847), xylic acid, and paraxylic acid (p. 845), by prolonged boiling with dilute nitric acid. Indistinct colorless crystals, melting at 280O–283O C. (536O–541.4O F.), nearly insoluble in cold water.

Isoxylidic acid, formed by fusing potassium toluene-disulphonate with sodium formate, is very much like xylidic acid, but melts at 310O–315O C. (590O–599O F.).

Isouvitic acid, formed together with phloroglucin, pyrotartaric acid, and acetic acid, by fusing gamboge with potash, crystallizes in short thick prisms, melting at about 160O C. (320O F.).

3. Acids, $C_{10}H_{10}O_4$.

Cumidic acid, $C_6H_2(CH_3)_2(CO_2H)_2$, formed from durene and durylic acid by prolonged boiling with dilute nitric acid, crystallizes in long transparent prisms, and sublimes at a high temperature.

Phenylene-diacetic acid, $C_6H_4(CH_2.CO_2H)_2$, is formed from tollylene chloride, $C_6H_4(CH_2Cl)_2$ (p. 856). Long needles, melting at 244O C. (471.2O F.), and subliming.

BIBASIC AND TRIATOMIC ACIDS.

Phenoldicarbonic Acid, $C_8H_6O_5 = C_6H_3(OH)(CO_2H)_2$, is formed, together with phenol and phenoltricarbonic acid, when basic sodium salicylate is heated to 360O–380O C. (680O–716O F.) for several hours, with frequent agitation, in a stream of carbon dioxide. The product is dissolved in water, acidulated with hydrochloric acid, boiled, filtered hot, neutralized with ammonia and mixed with barium chloride, whereby barium phenoltricarbonate is precipitated. The liquid filtered after cooling and mixed with hydrochloric acid, deposits a mixture of phenoldicarbonic and salicylic acids, from which the latter may be dissolved out by chloroform. Phenoldicarbonic acid crystallizes from boiling water in long needles, melts above 270° C. (518° F.), and sublimes at 200° C. (392° F.), with partial resolution into CO_2 and phenol. Slightly soluble in water, easily in alcohol and ether.

Oxyisouvitic Acid, $C_9H_8O_5 = C_6H_2(OH)(CH_3)(CO_2H)_2$.—The ethylic ether of this acid is formed by the action of chloroform, chloral, carbon tetrachloride or ethyl trichloracetate on ethyl acetoacetate (p. 750). The free acid forms needles, sparingly soluble in cold water, more freely in hot water, easily in alcohol and ether. Ferric chloride colors its solution reddish violet. The barium salt, $C_9H_6O_5Ba + 1\frac{1}{2}H_2O$, heated with lime, yields metacresol. By oxidation with permanganate or dilute nitric acid, oxyisouvitic acid is converted into an acid isomeric with uvic acid (p. 950). It is insoluble in chloroform.

Hemipinic Acid, $C_{10}H_{10}O_6 = C_6H_2(OCH_3)_2.(CO_2H)_2$, formed, together with cotarnine, meconin, and opianic acid, by oxidation of narcotine, crystallizes in large hydrated prisms slightly soluble in water. The dehydrated acid melts at 180° C. (356° F.). By heating with HI, it is resolved into CO_2, methyl iodide, and protocatechuic acid: $C_{10}H_{10}O_6 + 2HI = C_7H_6O_4 + 2CH_3I + CO_2$.

TRIBASIC ACIDS.

Benzenetricarbonic Acids, $C_9H_6O_6 = C_6H_3(CO_2H)_3$.

Trimesic acid (1 : 3 : 5).—Obtained by oxidation of mesitylenic or uvitic acid with chromic acid mixture ; also by heating mellitic acid with glycerin in an oil-bath till the mass solidifies and begins to blacken ; and, together with CO_2 and benzene-tetracarbonic acid, by heating hydro- and isohydromellitic acid with strong sulphuric acid. Short colorless prisms, rather sparingly soluble in cold water, easily in hot water, in alcohol, and in ether ; melts above 300° C. (572° F.), and sublimes without carbonization. By heating with excess of lime it is resolved into CO_2 and benzene. The *ethylic ether*, $C_6H_3(CO_2C_2H_5)_3$, forms long silky prisms melting at 129° C. (264.2° F.).

Hemimellitic acid (1 : 2 : 3), formed, together with phthalic anhydride, by heating hydromellophanic acid (p. 955) with strong sulphuric acid, crystallizes in colorless needles somewhat sparingly soluble in water. It melts at about 185° C. (365° F.), and is resolved at a higher temperature into phthalic anhydride and benzoic acid.

Trimellitic acid (1 : 2 : 4), is produced, together with isophthalic acid and pyromellitic anhydride, by heating hydropyromellitic acid with strong sulphuric acid, and abundantly, together with isophthalic acid, in the oxidation of colophony by nitric acid. It forms nodular groups of indistinct crystals, moderately soluble in water and in ether ; melts at 218° C. (424.4° F.), giving off water at the same time, and being converted into the anhydride, $C_9H_4O_5 = O{<}^{CO}_{CO}{>}C_6H_3{-}CO_2H$;

Phenol tricarbonic Acid, $C_9H_6O_7 = C_6H_2(OH)(CO_2H)_3$, is formed, together with phenol-dicarbonic acid, by heating sodium salicylate in a stream of carbon dioxide, and separated as a barium salt in the manner before described (p. 954). The free acid crystallizes from a hot concentrated aqueous solution in thick prisms, containing $1H_2O$; from dilute solutions in needles with $2H_2O$. Heated above 180° C. (356° F.), it is resolved, without previous fusion, into CO_2, phenoldicarbonic acid, salicylic acid, and phenol. The *ethylic ether*, $C_9H_3(C_2H_5)_3O_7$, forms long needles, melting at 84° C. (183.2° F.).

Benzotetracarbonic Acids, $C_6H_2(CO_2H)_4$.

There are three possible modifications of these acids, all of which have been obtained.

(1) Pyromellitic acid (1 : 2 : 4 : 5), is obtained as anhydride, $C_{10}H_2O_6$, by distillation of mellitic acid, or better, by heating sodium mellitate with sulphuric acid: $C_{12}H_6O_{12} = C_{10}H_2O_6 + 2CO_2 + 2H_2O$. The anhydride boiled with water yields the acid.

Pyromellitic acid crystallizes with $2H_2O$ in prisms slightly soluble in cold water, easily in hot water and in alcohol. It melts at 264° C. (507.2° F.), and is converted by distillation into the anhydride, $C_{10}H_2O_6$, which forms large crystals melting at 286° C. (546.8° F.). The salts, $C_{10}H_2Ba_2O_8$ and $C_{10}H_2Ca_2O_8$, are white insoluble precipitates. The *ethylic ether*, $C_6H_2(CO_2.C_2H_5)_4$, forms short needles melting at 53° C. (127.4° F.).

Hydropyromellitic and *Isohydropyromellitic acids*, $C_{10}H_{10}O_8 = C_6H_2(H_4)(CO_2H)_4$, are formed simultaneously by the action of sodium-amalgam on pyromellitic acid. The former remains on evaporating its ethereal solution as a very soluble mass. The latter crystallizes with $2H_2O$, gives off its water at 120° C. (248° F.), melts at about 200° C. (392° F.), and is resolved into water, CO_2, and tetrahydrophthalic anhydride. Both acids, when heated with sulphuric acid, give off CO_2 and are converted into trimellitic and isophthalic acids.

(2) Isopyromellitic or Prehnitic acid (1 : 2 : 3 : 5), is formed, together with CO_2, trimesic acid and mellophanic acid, by heating hydro- and isohydromellitic acid with strong sulphuric acid. It crystallizes in groups of large prisms containing $2H_2O$; dissolves easily in water, and is separated from its concentrated solution by hydrochloric acid in crystals; melts at 238° C. (460.4° F.), with formation of anhydride.

Hydroprehnitic acid, prepared like hydropyromellitic acid, is a gummy mass, resolved by heating with strong sulphuric acid into CO_2, prehnitic acid and isophthalic acid.

(3) Mellophanic acid, produced, together with prehnitic acid, forms small indistinct anhydrous crystals united in crusts; melts at 215°–238° C. (419°–460.4° F.). With sodium-amalgam it yields hydromellophanic acid.

Mellitic Acid, $C_{12}H_6O_{12} = C_6(CO_2H)_6$.—This acid, having the constitution of benzene in which all the six hydrogen-atoms are replaced by carboxyl, occurs as aluminium salt in *mellite* or *honey-stone*, a mineral found in beds of lignite. It is soluble in water and alcohol, and crystallizes in colorless needles, melts when heated, and is resolved by distillation into CO_2, water, and pyromellitic anhydride ; by heating with lime, into CO_2 and benzene. It is a very stable acid, not being decomposed by sulphuric, nitric, or hydriodic acid, or by bromine ; even with the aid of heat. Mellitic acid forms acid and neutral salts and ethers, whose composition may be illustrated by the following examples :

Neutral.	*Acid.*
$C_{12}(NH_4)_6O_{12}$	$C_{12}H_4(NH_4)_2O_{12}$
$C_{12}(CH_3)_6O_{12}$	$C_{12}H_5K_3O_{12}$
$C_{12}(C_2H_5)_6O_{12}$	$C_{12}H_3(C_2H_5)_3O_{12}$
$C_{12}Ba_3O_{12}$	$C_{12}H_2Cu_2O_{12}$
$C_{12}(Al_2)^{vi}O_2$	

Ammonium mellitate yields by distillation p a r a m i d e and e u c h r o t c a c i d . The former is a white amorphous substance, containing $C_{12}H_3N_3O_6$ (i. e., tri-ammonic mellitate, $C_{12}H_3(NH_4)_3O_{12} - 6H_2O$), and convertible by boiling with water into acid ammonium mellitate. *Euchroic acid*, $C_{12}H_4N_2O_8$ [$= C_{12}H_4(NH_4)_2O_{12} - 4H_2O$], forms colorless, sparingly soluble crystals. In contact with zinc and deoxidizing agents in general, it yields a deep blue insoluble substance called *euchrone*.

Hydromellitic acid, $C_{12}H_{12}O_{12} = C_6H_6(CO_2H)_6$, is slowly formed by the action of sodium-amalgam on ammonium mellitate. Colorless indistinct crystals, easily soluble in water. Sexbasic. Changes slowly by keeping, quickly, when heated to 180° C. (356° F.) with strong hydrochloric acid, into i s o h y d r o m e l l i t i c a c i d , $C_{12}H_{12}O_{12}$, which crystallizes in thick, hard four-sided prisms, dissolves readily in water, and is precipitated from the solution by hydrochloric acid. Both these acids, heated with strong sulphuric acid, yield a mixture of prehnitic, mellophanic, and trimesic acids.

Hydromellitic acid is an additive compound, in which the double linking of the alternate pairs of carbon-atoms in the benzene nucleus is broken up by the entrance of the additional hydrogen-atoms :

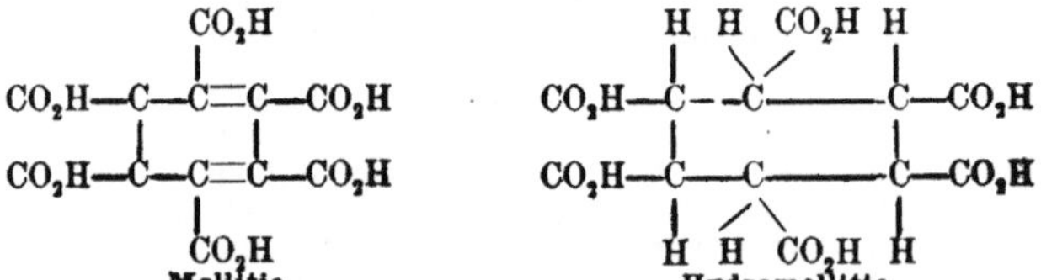

                    CO_2H H H CO_2H H
                      | | \ / |
CO_2H—C—C=C—CO_2H CO_2H—C—-C———C—CO_2H
CO_2H—C—C=C—CO_2H CO_2H—C—C———C—CO_2H
                      | | / \ |
                    CO_2H H H CO_2H H
                   Mellitic.   **Hydromellitic.**

INDIGO GROUP.

Indigo-blue and its immediate derivatives form a group of bodies nearly related to the benzene-group, as shown by their products of decomposition (aniline, ortho-amidobenzoic acid, etc.); but their constitution cannot yet be regarded as precisely determined, for want of sufficiently definite modes of synthetic formation.

I n d i g o - b l u e or **I n d i g o t i n**, C_8H_5NO or $C_{16}H_{10}N_2O_2$, probably
N—C_6H_4—CO—CH
‖   ‖ .—Indigo is the product of several species of plants of
N—C_6H_4—CO—CH

the genus *Indigofera*, growing in India and South America, also of *Isatis tinctoria*, *Nerium tinctorium*, *Polygonum tinctorium*, and other plants. It does not exist in these plants ready-formed, but is produced by decomposition of a glucoside, $C_{26}H_{31}NO_{17}$, called i n d i c a n, which may be extracted from them by cold alcohol, and forms a brown bitter syrup, easily soluble in water and alcohol. This substance, when boiled with dilute acids, or subjected to the action of ferments, is resolved into indigo-blue and indiglucin, $C_6H_{10}O_6$:

$$C_{26}H_{31}NO_{17} \cdot + \quad 2H_2O \quad = \quad C_8H_5NO \quad + \quad 3C_6H_{10}O_6.$$

A substance similar to indican sometimes occurs in urine, and gives rise, by its conversion into indigo-blue, to a blue coloration of the liquid when left in contact with the air, or treated with sulphuric acid.

To obtain indigo from the plants which produce it, the chopped leaves and twigs are macerated in water for 12 to 15 hours, after which the liquid is poured off into shallow wooden vessels, and frequently stirred, so as to bring it as much as possible into contact with the air. The indigo thereby deposited is separated from the brown liquid, boiled with water, and dried.

Commercial indigo is a mixture of several substances, all of which, except indigo-blue, are useless to the dyer. Boiling with dilute acetic acid extracts *indigo-gelatin;* dilute potash-lye then extracts *indigo-brown;* and the residue yields to boiling alcohol *indigo-red,* which remains on evaporation as a red powder, soluble in alcohol and ether. The residue left after boiling with alcohol is nearly pure indigo-blue.

Pure indigo-blue may be obtained from the commercial product by cautious sublimation; it then rises as a fine purple vapor, which condenses in dark-blue needles having a coppery lustre.

The best method of effecting the sublimation, is to mix 1 part of powdered indigo with 2 parts of plaster-of-Paris, make the whole into a paste with water, and spread it upon an iron plate. This, when quite dry, is heated by a spirit-lamp; the volatilization of the indigo is aided by the vapor of water disengaged from the gypsum; and the surface of the mass becomes covered with beautiful crystals of pure indigo, which may be easily removed by a thin spatula. At a higher temperature, charring and decomposition take place.

The best method of obtaining indigo-blue is to reduce the crude product to indigo-white by the action of ferrous sulphate or glucose in alkaline solution, and then reoxidize it. It is on this principle that the dyer prepares his *indigo-vat:* 5 parts of powdered indigo, 10 parts of green vitriol, 15 parts of slaked lime, and 60 parts of water, are agitated together in a close vessel, and then left to stand. The ferrous hydrate, in conjunction with the excess of lime, reduces the indigo to the soluble state; and a yellowish liquid is produced, from which acids precipitate indigo-white, as a flocculent substance which absorbs oxygen with the greatest avidity, and becomes blue. Cloth, steeped in the alkaline liquid, and then exposed to the air, acquires a deep and permanent blue tint by the deposition of solid insoluble indigo in the substance of the fibre.—Instead of the iron salt and lime, a mixture of dilute caustic soda and grape-sugar dissolved in alcohol may be used; the sugar becomes oxidized to formic acid, and the indigo reduced. On allowing such a solution to remain in contact with the air, it absorbs oxygen, and deposits indigo-blue in the crystalline state.

Indigo-blue is said to have been obtained by treating liquid nitro-acetophenone (p. 920) dissolved in chloroform with a mixture of soda-lime, and zinc dust, according to the equation,

$$\left. \begin{array}{l} NO_2-C_6H_4-CO-CH_3 \\ NO_2-C_6H_4-CO-CH_3 \end{array} \right\} \quad -2H_2O - O_2 \quad = \quad \begin{array}{l} N-C_6H_4-CO-CH \\ \| \qquad\qquad\qquad \| \\ N-C_6H_4-CO-CH \end{array}$$

Nitro-acetophenone (2 mols.).   Indigo-blue.

But the quantity of indigo-blue thus obtained is always very small, and it appears to be very difficult to ascertain the precise conditions under which the transformation takes place.*

Indigo-white or **Hydrindigotin,** $C_{16}H_{12}N_2O_2$, is precipitated by hydrochloric acid from its alkaline solution (formed as above), out of contact with the air, as a white crystalline powder, dissolving with yellow color in alcohol, ether, and alkalies. It is converted into indigo-blue by exposure to the air, and into indol by distillation with zinc-dust.

Indigo-Sulphonic Acids.—When indigo-blue is heated for a considerable time with 8 to 15 parts of strong sulphuric acid, it dissolves, with formation of a mono- and di-sulphonic acid.

Indigo-monosulphonic acid, $C_{16}H_9N_2O_2.SO_3H$ (*Sulphopurpuric* or *Phœnicinsulphonic acid*), separates, on diluting the solution with water, as a blue powder, soluble in pure water and in alcohol, but insoluble in dilute acids. Its salts are red in the solid state, but dissolve in water with blue color.

Indigo-disulphonic acid, $C_{16}H_8N_2O_2(SO_3H)_2$ (*Sulphindigotic* or *Sulphindylic acid*), remains in the filtrate from the monosulphonic acid, and constitutes the sole product when 15 parts of sulphuric acid are used to dissolve 1 part of indigo. To separate the acid, clean white wool, previously boiled with solution of sodium carbonate, is dipped into the liquid, and as soon as it has acquired a dark blue color, it is taken out, washed with water, and boiled with ammonium carbonate; the resulting solution is precipitated with lead acetate, and the precipitate decomposed by hydrogen sulphide. In this manner a colorless solution of hydrindigotin-disulphonic acid, $C_{15}H_{12}N_2O_2(SO_3H)_2$, is obtained, which, on exposure to the air, is quickly converted by oxidation into the disulphonic acid of indigo-blue, which, on evaporating the solution, remains in the form of a blue amorphous mass. The salts, $C_{16}H_8N_2O_2(SO_3K)_2$ and $C_{12}H_8N_2O_2(SO_3Na)_2$, known in commerce as indigo-carmine, are prepared on the large scale by adding potassium acetate or sodium sulphate to a dilute solution of indigo-blue in sulphuric acid, washing the blue precipitate with solutions of the same salts, and pressing it. They form copper-colored masses, blue in the finely-divided state, and dissolving with blue color in pure water.

Isatin, $C_8H_5NO_2 = C_6H_4 \diagup \genfrac{}{}{0pt}{}{CO-C(OH)}{N}$, is obtained by oxidizing indigo with chromic, or better, with nitric acid. Powdered indigo is mixed with water to a thin paste, heated to the boiling point in a large capsule, and nitric acid is added by small portions until the blue color disappears; the whole is then largely diluted with boiling water, and filtered. The impure isatin which separates on cooling is washed with water containing a little ammonia, and recrystallized. The process requires careful management, or the oxidizing action proceeds too far, and the product is destroyed.

Isatin forms deep yellowish-red prismatic crystals, sparingly soluble in cold water, freely in boiling water and in alcohol. The solution colors the skin yellow, and causes it to emit a very disagreeable odor. Isatin when heated melts and sublimes, with partial decomposition. It unites with acid sulphites of alkali-metal, forming crystalline compounds. It dissolves in

* Emmerling and Engler, Deutsch. Chem. Ges. Ber. ix. 1422; Chem. Soc Journal, 1877, i. 32).

alkalies, forming violet solutions, from which silver nitrate throws down a red precipitate of $C_8H_4AgNO_2$. On boiling these solutions, the isatin is converted into isatic acid, and the color changes to yellow.

Isatin is converted by nitrous acid into nitrosalicylic acid. Distilled with strong potash-solution it yields a n i l i n e . On heating it to 80° C. (176° F.), with phosphorus trichloride and phosphorus, dissolving the product in water, and leaving the solution exposed to the air for twenty-four hours, it deposits i n d i g o - b l u e .

Chlorinated and *Brominated derivatives of Isatin* (mono- and di-), *e. g.*, $C_8H_4ClNO_2$, $C_8H_3Cl_2NO_2$, etc., are formed by passing chlorine or bromine into the hot aqueous solution of isatin. These compounds crystallize well, and are converted by fusion with potash into substituted anilines ; *e. g.*,

$$C_8H_4ClNO_2 \ + \ 4KOH \ = \ C_6H_4Cl.NH_2 \ + \ 2CO_3K_2 \ + \ H_4 .$$

Isatin boiled with ammonia yields amido-derivatives, which have been but little examined. By reducing agents it is converted into is a t y d e , $C_{16}H_{12}N_2O_4$.

I s a t i n s u l p h o n i c a c i d , $C_8H_4NO_2.SO_3H$, formed by oxidation of indigosulphonates (indigo-carmine), is very soluble, and crystallizes with difficulty : it forms derivatives analogous to those of isatin.

I s a t y d e or H y d r o i s a t i n , $C_8H_6NO_2$, is formed when isatin is warmed with dilute sulphuric acid, or when its warm, saturated, alcoholic solution is mixed in a closed flask with ammonium hydrosulphide, the liquid then gradually depositing the isatyde in colorless crystalline scales. It is tasteless, insoluble in water, slightly soluble in alcohol.

Thioisatyde, $C_{16}H_{12}N_2S_2O_2$, is formed on passing hydrogen sulphide into an alcoholic solution of isatin, and is precipitated on dropping the filtered solution into water. It is a grayish-yellow, pulverulent, uncrystallizable substance, which softens in hot water, and dissolves in alcohol.

I n d i n , $C_{16}H_{10}N_2O_2$ (isomeric or polymeric with indigo-blue), is formed on triturating thioisatyde with potassium hydroxide, gradually adding alcohol, and washing the reddened mass with water ; also by boiling a solution of dioxindol in glycerin. It is a crystalline powder of a fine rose-color, insoluble in water, slightly soluble in alcohol. It dissolves when heated with alcoholic potash, and the solution, on cooling, deposits p o t a s s i u m - i n d i n , $C_{16}H_9KN_2O_2$, in small black crystals.

Isatic Acid, or **Trioxindol**, $C_8H_7NO_3$.—The violet solution of isatin in potash-lye turns yellow when heated, and then contains *potassium isatate*, $C_8H_6KNO_3$, which is deposited on evaporation in yellow crystals. This salt, decomposed with an acid, yields isatic acid as a white powder, soluble in water, and resolved by heat into isatin and water.

Hydrindic Acid, or **Dioxindol**, $C_8H_7NO_2$, is formed by the action of sodium on isatin suspended in water, or dissolved in an alkali, isatic acid being first formed, and then reduced. It crystallizes in yellow prisms, is soluble in water and in alcohol, and forms crystalline compounds both with bases and with acids. It melts at 180° C. (356° F.), and decomposes at 195° C. (383° F.), with formation of aniline. The aqueous solution turns red on exposure to the air, from formation of isatin : when boiled with ammonia it turns violet. Metallic salts added to the solution throw down crystalline compounds, such as $C_8H_6AgNO_3$ and $C_8H_5PbNO_3$ $+ H_2O$.

Dioxindol treated with nitrous acid in alcoholic solution is converted into *nitroso-dioxindol*, $C_8H_6(NO)NO_3$, which, when heated with nitric acid, or silver oxide, yields benzaldehyde.

Oxindol, C_8H_7NO, is formed by reduction of dioxindol with sodium-amalgam in acid solution, or with tin and hydrochloric acid, and crystallizes in colorless needles, easily soluble in hot water, alcohol, and ether. It melts at 120° C. (248° F.), and when heated in small quantities, may be sublimed without decomposition. It forms crystallizable salts, both with acids and with bases. Its aqueous solution, evaporated in an open vessel, is partially oxidized to dioxindol. Nitrous acid converts it into n i t r o s o x i n d o l, $C_8H_8(NO)NO$, which crystallizes in long golden-yellow needles, slightly soluble in water.

Indol, C_8H_7N, is formed by distilling oxindol over heated zinc-dust, and by fusing orthonitrocinnamic acid with potassium hydroxide and iron filings (to abstract oxygen):

$$C_6H_4 {<}{\overset{CH=CH.CO_2H}{\underset{NO_2}{}}} \quad = \quad C_6H_4{<}{\overset{CH_2-CH}{\underset{N}{}}} \quad + \quad CO_2 \quad + \quad O_2 .$$

It is produced in larger quantity (about 0.5 per cent.) by digesting serum-albumin or egg albumin with pancreatic juice.

Indol crystallizes from hot water in large, shining, colorless laminæ, resembling benzoic acid, and smelling like naphthylamine. It melts at 52° C. (125.6° F.), boils with partial decomposition at about 245° C. (473° F.), volatilizes in a vacuum without decomposition, and distils easily with vapor of water. It is a weak base, its salts being decomposed by boiling with water. Its vapor colors a deal shaving moistened with hydrochloric acid, deep cherry-red. Its solution acidulated with hydrochloric acid is colored rose-red by potassium nitrite. The reddish coloration is likewise produced by several other reagents. Indol fused with potash is converted into aniline (probably through the medium of ortho-amidobenzoic acid). Ozone passed through an aqueous solution of indol forms a small quantity of indigo-blue.

The formation of aniline from indol and its three hydroxyl-derivatives, by fusion with potash, and their conversion into anthranilic and nitro-salicylic acids by regulated oxidation with nitric acid, show that in these bodies the benzene-nucleus C_6H_4 is directly linked to carbon and to nitrogen, and that the OH groups are situated in the lateral chains. They may accordingly be represented by the following formula :—

$$C_6H_4{<}{\overset{CH_2.CH}{\underset{N}{}}}$$

Indol.

$$C_6H_4{<}{\overset{CH_2.C(OH)}{\underset{N}{}}} \qquad C_6H_4{<}{\overset{CH(OH).C(OH)}{\underset{N}{}}} \qquad C_6H_4{<}{\overset{C(OH)_2.C(OH)}{\underset{N}{}}}$$

Oxindol.   Dioxindol.   Trioxindol.

The formation of anthranilic (1 : 2 amidobenzoic) acid shows further that the lateral chains in these compounds occupy contiguous places.

indol is therefore an ortho-compound. An isomeric compound, which has a higher melting point, 89°–91° C. (192.2°–195.8° F.), and does not yield indigo-blue when treated with ozone, is formed by heating albumin with excess of potassium hydroxide.

DIPHENYL GROUP.

The compounds of this group contain two or more benzene-nuclei, united either directly or through the medium of other carbon groups.

Hydrocarbons, C_nH_{2n-14}:

Diphenyl, $C_{12}H_{10} = C_6H_5.C_6H_5$ or

This hydrocarbon is formed: (1) By passing benzene-vapor through a red-hot tube containing fragments of pumice: $2C_6H_6 = C_{12}H_{10} + H_2$. This is the best method of preparing it, the yield amounting to about 30 per cent. of the benzene used.

(2) By the action of sodium on bromobenzene (phenyl bromide) dissolved in ether or benzene:

$$2C_6H_5Br + Na_2 = 2NaBr + C_{12}H_{10}.$$

(3) By heating a mixture of potassium phenate and benzoate or oxalate, and in small quantity (together with benzene) by heating benzoic acid with lime.

(4) Together with other products, by the action of alcoholic potash on nitrate of diazobenzene:

$$\underset{\substack{\text{Diazoben-}\\\text{zene.}}}{2C_6H_4N_2} + \underset{\text{Alcohol.}}{C_2H_6O} = \underset{\text{Diphenyl.}}{C_{12}H_{10}} + \underset{\text{Aldehyde.}}{C_2H_4O} + 2N_2.$$

Diphenyl appears also to be one of the constituents of crude anthracene (p. 979), and passes over in the distillation of that substance at about 260° C. (500° F.).

Diphenyl crystallizes from alcohol in iridescent nacreous scales, melts at about 70.5° C. (158.9° F.), sublimes at a higher temperature, and boils at 254° C. (489.2° F.). When dissolved in glacial acetic acid it is oxidized by chromic anhydride to benzoic acid.

Diphenyl, subjected to the action of halogens, nitric acid, and sulphuric acid, yields mono- and di-substitution-derivatives. In the former, such as $C_{12}H_9Br$, $C_{12}H_9(NO_2)$, $C_{12}H_9(SO_3H)$, the substituted radicles stand to the place of junction of the two benzene-nuclei in the para-position. The di-derivatives are known in two isomeric modifications, the most frequent being those in which both the substituted groups are in the para-position relatively to the point of junction:

Monobromodiphenyl (para).

Nitro-bromodiphenyl (para-para).

By oxidation with chromic anhydride, the monosubstituted diphenyls yield para-derivatives of benzoic acid, the group C_6H_4Br, for example, being oxidized to $(1:4)$ bromobenzoic acid, while the other group (C_6H_5) is broken up. The di-derivatives, on the other hand, are converted by oxidation into two para-derivatives of benzoic acid, $e.\,g.$, $C_6H_4(NO_2).C_6H_4Br$ into p-nitro- and p-bromobenzoic acid.

$Monochlorodiphenyl$, $C_{12}H_9Cl$, formed by the action of PCl_5 on oxy-diphenyl ($C_{12}H_9.OH$), forms colorless crystals melting at 75° C. (167°

41

F.). *Dichlorodiphenyl*, $C_{12}H_8Cl_2$, from benzidine, melts at 148° C. (298.4° F.).

Bromodiphenyl, $C_{12}H_9Br$, forms laminæ, melts at 89° C. (192.2° F.), boils at 310° C. (590° F.). $C_{12}H_8Br_2$ forms prisms melting at 164° C. (327.2° F.).

Cyanodiphenyl, $C_{12}H_9$.CN, obtained by dry distillation of a mixture of diphenyl-monosulphonate and cyanide of potassium, forms hard, color-less crystals, melting at 84°–85° C. (183.2°–185° F.), insoluble in water, easily soluble in alcohol. *Dicyanodiphenyl*, $C_{12}H_8(CN)_2$, obtained in like manner from diphenyldisulphonic acid, forms ramified colorless needles, melting at 234° C. (453.2° F.), sparingly soluble in cold, freely in boiling alcohol.

Nitrodiphenyl, $C_{12}H_9(NO_2)$, crystallizes in needles, melts at 113° C. (233.6° F), boils at 340° C. (644° F.).   An isomeric compound obtained by distilling calcium meta-nitrobenzoate with potassium phenate, melts at 86° C. (186.8° F.), according to Pfankuch, at 157° C. (314.6° F.), accord-ing to Schultz.* Two *dinitro-compounds*, $C_{12}H_8(NO_2)_2$, are formed by the ac-tion of fuming nitric acid on diphenyl ; the less soluble in alcohol of the two melts at 213° C. (415.4° F.), and boils 340° C. (644° F.); the more soluble compound melts at 93.5° C. (200.3° F.).

Amidodiphenyls are formed by the reduction of the nitro-compounds with tin and hydrochloric acid.—*Xenylamine*, $C_{12}H_9(NH_2)$, crystallizes from hot water or alcohol in colorless laminæ melting at 49° C. (120.2° F.).

Benzidine, $C_{12}H_8(NH_2)_2$, is also produced by the action of sodium on monobromaniline, and by molecular transposition of hydrazobenzene in contact with acids (p. 880); further, by heating azobenzene with fuming hydrochloric acid to 115° C. (239° F.), and by passing SO_2 into the alco-holic solution of that compound :

$$\begin{matrix} C_6H_5-N \\ \| \\ C_6H_5-N \end{matrix} + 2H_2O + SO_2 = SO_4H_2 + \begin{matrix} C_6H_4-NH_2 \\ | \\ C_6H_4-NH_2 \end{matrix}.$$

Benzidine crystallizes in silvery laminæ, easily soluble in hot water and alcohol, melting at 118° C. (244.4° F.), and subliming with partial de-composition. It is a biacid base. The *sulphate*, $C_{12}H_{12}N_2.SO_4H_2$, is nearly insoluble in water and alcohol.

Diphenylimide or Carbazol, $C_{12}H_9N$, is formed by passing the vapor of aniline or of diphenylamine through a red-hot tube :

$$\begin{matrix} C_6H_5 \\ C_6H_5 \end{matrix}\!\!>\!\!NH = \begin{matrix} C_6H_4 \\ | \\ C_6H_4 \end{matrix}\!\!>\!\!NH + H_2 .$$

It occurs in coal-tar oil (320°–360° C., 608°–680° F.), and as a bye-product in the manufacture of aniline.

Carbazol crystallizes in shining laminæ, soluble in hot alcohol, ether, and benzene ; melts at 238° C. (460.4° F.), and distils at 354°–355° C. (669.2°–671° F.). It dissolves in strong sulphuric acid, with yellow color, easily changed to green by oxidizing agents. With picric acid, it forms a com-pound which crystallizes in red needles melting at 182° C. (359.6° F.).

Acridine, $C_{12}H_9N$, isomeric with carbazol, likewise occurs in coal-tar oil, and may be extracted by sulphuric acid from the portion boiling be-

* Liebig's Annalen, clxxiv. 201 ; Jahresbericht für Chemie, 1874, 405.

tween 320º C. (608º F.) and 360º C. (680º F.). It forms rhombic crystals, easily soluble in alcohol and ether, melts at 107º C. (224.6º F.), sublimes in broad needles at 100º, boils at a temperature above 360º C. (680º F.). It unites with acids (carbazol does not) forming salts which are decomposed by boiling with a large quantity of water.

Diphenylsu..phonic acids, $C_{12}H_9(SO_4H)$ and $C_{12}H_8(SO_3H_2)$, are formed by heating diphenyl with strong sulphuric acid, the former, however, only when a very small quantity of sulphuric acid is used. The disulpho-acid crystallizes in long deliquescent prisms, melting at 72.5º C. (161.6º F.). These acids fused with potash yield the corresponding phenols, $C_{12}H_9(OH)$ and $C_{12}H_8(OH)_2$.

Hydrocarbons, $C_{13}H_{12}$:—

Phenyl-tolyl, $C_6H_5.C_6H_4.CH_3$, is formed by the action of sodium on a mixture of bromobenzene and parabromotoluene diluted with ether. It is a colorless liquid of specific gravity 1.015 at 27º (80.6º F.), boiling at 263º–267º C. (505.4º–512.6º F.), solidifying in a freezing mixture.

Diphenylmethane, $C_{13}H_{12} = C_6H_5.CH_2.C_6H_5$ (*Benzyl-benzene*), is obtained by heating a mixture of benzene (6 parts) and benzyl chloride (10 parts) with zinc-dust (3 or 4 parts) in a vessel with reversed condenser. Colorless crystalline mass made up of needles, and having an odor of oranges ; melts at 26º–27º C. (78.8º–80.6º F.), boils at 261º–262º C. (501.8º–503.6º F.); dissolves easily in alcohol and ether. Passed through a red-hot tube it yields diphenylene-methane, $C_{13}H_{10}$. Chromic acid mixture oxidizes it to diphenyl ketone, $CO(C_6H_5)_2$. It dissolves in strong nitric acid, forming two dinitro-derivatives.

Diphenyl-chloromethane, C_6H_5—$CHCl$—C_6H_5, formed by the action of HCl on benzhydrol (p. 968), is a crystalline mass melting at 14º C. (57.2º F.).

Hydrocarbons, $C_{14}H_{14}$:—

Ditolyl, $\begin{vmatrix} C_6H_4.CH_3 \\ C_6H_4.CH_3 \end{vmatrix}$, produced by the action of sodium on parabromotoluene, forms monoclinic crystals, easily soluble in hot alcohol, melting at 121º C. (249.8º F.).

Dibenzyl, $= C_6H_5.CH_2.CH_2.C_6H_5$, formed by the action of sodium on benzyl chloride, and by heating stilbene, tolane, benzoin, deoxybenzoïn, or toluylene hydrate, with hydriodic acid, crystallizes in large colorless prisms, melts at 52º C. (125.6º F.), boils at 284º C. (543.2º F.); dissolves easily in hot alcohol. Heated to 500º C. (932º F.) it yields stilbene and toluene: $2C_{14}H_{14} = C_{14}H_{12} + 2C_7H_8$.

Diphenyl-ethane, $C_{14}H_{14} = (C_6H_5)_2CH$—$CH_3$, is formed by the action of sulphuric acid on a mixture of paraldehyde and benzene ($2C_6H_6 + C_2H_4O - H_2O = C_{14}H_{14}$). Colorless liquid, boiling at 268º–270º C. (514.4º–518º F.), solidifying in a freezing mixture : oxidized by chromic acid mixture to benzophenone.

A mixture of 2 mol. benzene and 1 mol. chloral or bromal with strong sulphuric acid, yields in like manner, diphenyltrichlorethane, $(C_6H_5)_2CH$—CCl_3 (colorless laminæ melting at 64º C., 147.2º F.), or diphenyltribromethane, $(C_6H_5)_2CH$—CBr_3 (monoclinic crystals, m. p. 89º C., 192.2º F.). A mixture of benzene and dichlorethylic oxide, $(C_2H_4Cl)_2O$, similarly treated, yields *diphenyl-monochlorethane,* $(C_6H_5)_2CH$—CH_2Cl, as an oil which decomposes when distilled.

Benzyl-toluene, $C_{14}H_{14} = C_6H_5.CH_2.C_6H_4.CH_3$, formed by passing the vapors of paraphenyl-tolyl ketone, $C_6H_5.CO.C_6H_4(CH_3)$, over heated zinc-dust, is a colorless liquid boiling at 285°–286° C. (545°–546.8^O F.), not solidifying at —20° C. (—4° F.).

Hydrocarbons, $C_{15}H_{16}$:—

Ditolyl-methane, $C_6H_4\diagdown\begin{smallmatrix}CH_3\\CH_2\end{smallmatrix}$—$C_6H_4$—$CH_3$, formed by the action of sulphuric acid on a mixture of methylal, $CH_2(OCH_3)_2$ (p. 500), toluene, and glacial acetic acid, is a liquid not solidifying at —15° C. (5° F.), boiling at 290° C. (554° F.).—$C_{15}H_{14}Br_2$, forms long needles melting at 115° C. (239° F.).—$C_{15}H_{14}(NO_2)_2$, forms colorless crystals melting at 164° C. (327.2^O F.).

Dibenzyl-methane, $C_6H_5.CH_2.CH_2.C_6H_5$, from dibenzyl ketone by heating with III and phosphorus, is a non-solidifiable liquid, boiling at 290°–300° C. (554°–572° F.).

Benzyl-ethyl-benzene, $C_6H_5.CH_2.C_6H_4.C_2H_5$, from benzyl chloride and benzene, is a colorless liquid boiling at 294°–295° C. (561.2^O–563° F.).

Ethyl-benzene-toluene or **Phenyltolyl-ethane,**

$CH_3.CH\diagdown\begin{smallmatrix}C_6H_4.CH_3\\C_6H_5\end{smallmatrix}$, formed by heating bromethyl-benzene (p. 855) and toluene with zinc-dust, is a liquid boiling at 278°–280° C. (532.4^O–536° F.).

Hydrocarbons, $C_{16}H_{18}$:—

Ditolylethane, CH_3—$CH(C_6H_4.CH_3)_2$, obtained, like diphenyl-ethane, from paraldehyde, toluene, and sulphuric acid, is a liquid, boiling at 295°–298° C. (563°–568.4^O F.), not solidifying at —20° C. (—4° F.).

Similarly, by the use of chloral instead of paraldehyde, is obtained CCl_3—$CH(C_6H_4.CH_3)_2$, in crystals melting at 89° C. (192.2^O F.) ; and with dichlorethylic oxide, CH_2Cl—$CH(C_6H_4.CH_3)$.

Diphenyldimethylethane, $\begin{matrix}C_6H_5.CH.CH_3\\ |\\ C_6H_5.CH.CH_3\end{matrix}$, obtained by the action of sodium on secondary phenyl-ethyl chloride, $C_6H_5.CHCl.CH_3$ (p. 910), forms needles melting at 123.5^O C. (254.3^O F.).

Diphenylene-methane or **Fluorene,** $C_{13}H_{10} = \begin{smallmatrix}C_6H_4\\ |\\ C_6H_4\end{smallmatrix}\diagup\diagdown CH_2$, is formed by heating diphenylene ketone with zinc-dust or with hydriodic acid and amorphous phosphorus to 160° C. (320° F.), also by passing the vapor of diphenyl-methane through a red-hot tube. It is contained in the portion of coal-tar oil boiling at 300°–305° C. (572°–581° F.). It crystallizes from hot alcohol in colorless laminæ, exhibiting a beautiful violet fluorescence ; melts at 112°–113° C. (233.6^O–235.4^O F.), boils at 305° C. (581° F.) ; unites with picric acid, forming a compound which crystallizes in red needles melting at 80°–82° C. (176°–179.6^O F.). By oxidation with chromic acid mixture it yields diphenylene-ketone, $CO(C_6H_4$—$C_6H_4)$.

Diphenyl-benzene, $C_{18}H_{14} = C_6H_4(C_6H_5)_2$, is formed by the action of sodium on a mixture of bromobenzene and paradibromobenzene, and, together with an isomeric compound, by passing a mixture of diphenyl **and**

benzene through red-hot tubes : hence these two isomeric compounds are obtained, as secondary products, in the preparation of diphenyl.—Diphenylbenzene forms groups of needles slightly soluble in hot alcohol and ether, melting at 205° C. (401° F.), subliming easily and boiling at 400° C. (752° F.). By chromic anhydride and glacial acetic acid it is oxidized to diphenylcarbonic, and afterwards to terephthalic acid.—*Isodiphenyl benzene* melts at 85° C. (185° F.), and boils at about 360° C. (680° F.).

Triphenyl-methane, $C_{19}H_{16} = C_6H_5$—$CH(C_6H_5)_2$, is formed by heating benzal chloride to 150° C. (302° F.) with mercury-diphenyl :

$$C_6H_5.CHCl_2 \; + \; (C_6H_5)_2Hg \; = \; HgCl_2 \; + \; C_6H_5—CH(C_6H_5)_2 \; ;$$

also by heating benzhydrol (p. 968) with benzene and phosphoric anhydride to 140° C. (284° F.).

$$(C_6H_5)_2CH.OH \; + \; C_6H_6 \; = \; (C_6H_5)_2CH(C_6H_5)_2 \; + \; H_2O \, .$$

Shining laminæ, melting at 92° C. (197.6° F.), boiling at about 360° C. (680° F.). By bromine and water it is converted into triphenyl carbinol, which melts at 157° C. (314.6° F.).

Tolyl-diphenyl-methane, $C_{20}H_{18} = (C_6H_5)_2CH$—$C_6H_4$—$CH_3$, prepared, like the preceding, from benzhydrol and toluene, is a colorless liquid of high boiling point.

Triphenyl-benzene, $C_{24}H_{18} = C_6H_3(C_6H_5)_3$, obtained by heating acetophenone (p. 920) with phosphoric anhydride, forms large rhombic crystals, melts at 169° C. (336.2° F.), boils above 360° C. (680° F.), dissolves easily in alcohol and in benzene.

Tetraphenyl-ethane, $C_{26}H_{22} = (C_6H_5)_2CH$—$CH(C_6H_5)_2$, obtained by the action of zinc on a solution of benzhydrol, $C_6H_5.CHOH.C_6H_5$, mixed with fuming hydrochloric acid, forms large prisms melting at 209° C. (408.2° F.).

Ethene-derivatives.

Diphenyl-ethene, $C_{14}H_{12}$.—Of this hydrocarbon there are two modifications, symmetrical and unsymmetrical.

1. *Stilbene* or *Toluylene*, C_6H_5—$CH\!=\!CH$—C_6H_5, is formed in a variety of reactions, especially by dry distillation of benzyl sulphide or disulphide (p. 910) ; by the action of sodium on benzaldehyde or benzal chloride ; and by passing the vapor of dibenzyl or of toluene over heated lead oxide. It is most easily obtained from benzyl sulphide.

Stilbene crystallizes in large laminæ, dissolves easily in hot alcohol, melts at 125° C. (257° F.), distils at 306°–307° C. (582.8°–584.6° F.). By heating with hydriodic acid it is converted into dibenzyl ; by oxidation with chromic acid mixture into benzaldehyde and benzoic acid. It unites with bromine, forming $C_{14}H_{12}Br_2$ (crystals, melting at 230° C., 446° F.), which is also produced by the action of bromine, without cooling, on dibenzyl. By alcoholic potash this bromide is converted into bromostilbene, $C_{15}H_{11}Br$ (crystals, m. p. 25° C., 77° F.), and tolane.

2. *Isostilbene*, $CH_2\!=\!C(C_6H_5)_2$, *unsymmetric diphenylethene*, obtained by boiling diphenyl-monochlorethane (p. 963) with alcoholic potash, is a

colorless non-solidifying oil, which boils at 277º C. (530.6º F.), and is converted by oxidation into diphenyl-ketone.

Tetraphenyl-ethene, $C_{26}H_{20} = (C_6H_5)_2C=C(C_6H_5)_2$, obtained by the action of finely divided silver on benzophenone chloride, and, together with diphenyl-methane, and other hydrocarbons, by heating benzophenone with zinc-dust, is a white crystalline powder, melting at 221º C. (429.8º F.), sparingly soluble in alcohol and ether, easily in benzene.

Tolane, $C_{14}H_{10} = C_6H_5-C\equiv C-C_6H_5$, *Diphenyl acetylene,* obtained by boiling stilbene bromide with alcoholic potash, forms large crystals, melting at 60º C. (140º F.), easily soluble in alcohol and ether. It forms two dibromides, $C_{14}H_{10}Br_2$ (m. p. 64º C., 147.2º F., and 200º–205º C., 392º–401º F.), easily convertible one into the other; and two dichlorides (m. p. 63º and 153º C., 145.4º–307.4º F.).

Phenols.

Oxydiphenyl or **Diphenylol,** $C_{12}H_9.OH = C_6H_5-C_6H_4.OH$, is formed by fusing the potassium salt of diphenylsulphonic acid, $C_{12}H_9.SO_3H$, with potassium hydroxide, and by the action of potassium nitrite on amidodiphenyl sulphate. Colorless monoclinic crystals; melts at 164º–165º C. (327.2º–329º F.), boils at 305º–308º C. (581º–586.4º F.); volatilizes with vapor of water; dissolves easily in alcohol, ether, and alkalies; and with fine green color in strong sulphuric acid.

Dioxydiphenyl or **Diphenol,** $C_{12}H_8(OH)_2 = C_6H_4(OH).C_6H_4(OH)$, is formed by the action of melting potash on ₁potassium diphenyl-parasulphonate; also from benzidine by the diazo-reaction; and by fusing potassium paraphenol-sulphonate with alkalies. Colorless rhombic crystals, melting at 156º–158º C. (312.8º–316.4º F.). An isomeric compound, which crystallizes from alcohol in shining needles melting at 269º–270º C. (516.2º–518º F.), is formed by the action of melting potash on potassium diphenyl-disulphonate.

Diphenylene Oxide, $C_{12}H_8O = \begin{array}{c} C_6H_4 \\ | \\ C_6H_4 \end{array} \!\!> O$, formed by heating phenyl phosphate with lime, or more readily by heating phenol with lead oxide, crystallizes in laminæ, melts at 81º C. (177.8º F.), and boils at 273º C. (523.4º F.).

Hexoxydiphenyl, $C_{12}H_{10}O_6 = C_{12}H_4(OH)_6$, is formed from its tetramethylic ether (hydrocœrulignone) by the action of strong hydrochloric or hydriodic acid:

$$C_{12}H_4(OH)_2(OCH_3)_4 + 4HCl = 4CH_3Cl + C_{12}H_4(OH)_6.$$

It crystallizes from water in silvery laminæ; dissolves in potash-lye with fine blue-violet color; is converted by acetyl chloride into a hexacetyl compound; and reduced by heating with zinc-dust to diphenyl.

Hydrocœrulignone, $C_{12}H_4 \begin{cases} (OCH_3)_4 \\ (OH)_2 \end{cases}$, is formed from cœrulignone by the action of tin and hydrochloric acid. It crystallizes from alcohol in colorless laminæ, melts at 190° C. (374° F.) and distils almost without decomposition. It is a diatomic phenol, and is resolved by HCl or HI into CH_3Cl and hexoxydiphenyl.

Cœrulignone, $C_{16}H_{16}O_6 = C_{12}H_4 \begin{cases} (OCH_3)_4 \\ (O—O) \end{cases}$, separates as a violet powder in the industrial purification of crude wood-vinegar by means of potassium dichromate, and is obtained by the same means from the fraction of beech-tar boiling at 270° C. (518° F.). It is insoluble in most solvents, but dissolves in strong sulphuric acid with a fine corn-flower blue color, changing for a while to red on addition of a large quantity of water; also in phenol, from which it is precipitated by alcohol and ether in slender steel-blue needles. By reduction with tin and hydrochloric acid, it is converted into colorless hydrocœrulignone, $C_{16}H_{18}O_6$, which is reconverted into cœrulignone by oxidation, the two bodies being indeed related to one another in the same manner as quinone and hydroquinone.

Dithymoxyl-trichlorethane,[*] $CCl_3—CH(C_{10}H_{12}OH)_2$, is formed when sulphuric acid diluted with $\frac{1}{3}$ vol. of glacial acetic acid is added to a cooled mixture of 1 mol. chloral and 2 mol. thymol:

$$CCl_3—CHO + 2C_{10}H_{13}OH = H_2O + CCl_3—CH(C_{10}H_{12}OH)_2 .$$

A white precipitate is then formed, the alcoholic solution of which yields the compound in white monoclinic crystals.

Dithymoxyl-trichlorethane is a diatomic phenol, and when treated with C_2H_3OCl or C_7H_5OCl, exchanges its two phenolic hydrogens for acetyl or benzoyl. By strong oxidizing agents it is converted into thymoquinone. Heated with zinc-dust, it exchanges its chlorine for hydrogen, forming dithymoxylethane, $C_{22}H_{30}O_2$, part of which is converted by loss of H_2 into dithymoxylethene, $C_{22}H_{28}O_2$.

Dithymoxylethane, $CH_3—CH(C_{10}H_{12}OH)_2$, crystallizes in rounded plates, melting at 180° C. (356° F.), soluble in most of the usual solvents, except water.

Dithymoxylethene, $CH_2{=}C(C_{10}H_{12}OH)_2$, forms needle-shaped crystals, melting at 170°–171° C. (338°–339.8° F.), insoluble in water, rather more soluble than the preceding in other solvents. Treated with weak oxidizing agents, as with potassium ferrocyanide in dilute alcoholic solution, it forms green crystals, $C_{44}H_{54}O_4$, melting at 214°–215° C. (417.2°–419° F.), soluble in toluene and in chloroform; but if the solution be previously rendered alkaline, a red precipitate is formed, which dissolves in chloroform and separates therefrom in red crystals, melting at 215° C. (419° F.), and having the composition $C_{22}H_{26}O_2$.

A mixture of dithymoxylethene and the red crystals dissolved in chloroform deposits, on evaporation, the above-described green needles: $C_{22}H_{28}O_2 + C_{22}H_{26}O_2 = C_{44}H_{54}O_4$. Hence it appears that dithymoxylethene and the two products of its oxidation are related to one another in the same manner as hydroquinone, quinhydrone, and quinone, thus:

$$CH_2{<}^{C_{10}H_{12}OH}_{C_{10}H_{12}OH} \qquad CH_2{<}^{C_{10}H_{12}OH}_{C_{10}H_{12}O}\ {}^{HOH_{12}C_{10}}_{—OH_{12}C_{10}}{>}CH_2 \qquad CH_2{<}^{C_{10}H_{12}O}_{C_{10}H_{12}O}{>}$$
Dithymoxylethene. Dithymoxyl-quinhydronethene. Dithymoxyl-quinonethene.

[*] This compound, and those which follow it, were discovered by Dr. E. Jäger (Deutsch. Chem. Ges. Ber. 1874, 1197; Chem. Soc. Journ. 1877, 262), who designates them, not as thymoxyl-compounds, but as thymyl-compounds; but the name "dithymylethane" belongs properly, not to a phenol, but to the hydrocarbon $C_{22}H_{30}$ = $CH_3—CH(C_{10}H_{13})_2$, homologous with diphenyl-ethane and ditolyl-ethane; and so of the rest.

Alcohols.

Benzhydrol, $C_{13}H_{12}O = C_6H_5$—CH(OH)—C_6H_5, is obtained by the action of sodium-amalgam on a solution of benzophenone in dilute alcohol. It crystallizes in needles, melts at 68° C. (154.4° F.), and boils at 298° C. (568.4° F.), with partial decomposition into water and benzhydrolic ether, $[(C_6H_5)_2CH]_2O$.

Toluylene Hydrate or **Stilbene Hydrate,** $C_{14}H_{14}O = C_6H_5.CH_2.$ CH(OH).C_6H_5, is formed by the action of sodium-amalgam on deoxybenzoïn, $C_{14}H_{12}O$ (p. 970), and by heating deoxybenzoïn or hydrobenzoïn with alcoholic potash. Long, slender, brittle needles, having a vitreous lustre, melting at 62° C. (143.6° F.) ; insoluble in water, easily soluble in alcohol and ether. Resolved by boiling with dilute sulphuric acid into stilbene and water : oxidized by nitric acid to deoxybenzoïn ; reduced by HI to dibenzyl. With acetyl chloride, it forms toluylene acetate, $C_{14}H_{13}.$ O.C_2H_3O, which is a viscid liquid.

Dimethyl·benzhydrol, $C_{15}H_{16}O = CHOH(C_6H_4.CH_3)_2$, obtained by the action of sodium-amalgam on ditolylketone in alcoholic solution forms slender needles melting at 69° C. (156.2° F.).

Triphenyl·carbinol $C_{19}H_{16}O = (C_6H_5)_3C(OH)$, from triphenyl-methane by oxidation with chromic acid mixture, forms monoclinic crystals, melts at 157° C. (314.6° F.), boils without decomposition above 360° C. (680° F.). Its solution in benzene heated with P_2O_5 yields triphenyl-methane and dibenzyl.

Tolylene Glycols, $C_{14}H_{14}O_2$.—Two diatomic alcohols of this composition, called Hydrobenzoïn and Isohydrobenzoïn, are formed by the action of zinc and hydrochloric acid on benzaldehyde ; also by that of sodium-amalgam on the same compound in alcoholic solution. By oxidation with chromic acid mixture both of them are converted into benzaldehyde and benzoic acid : consequently they must be represented by the formulæ :

$$C_6H_5.CH.OH \qquad\qquad C_6H_5C(OH)_2$$
$$\text{and}$$
$$C_6H_5.CH.OH \qquad\qquad C_6H_5.CH_2$$

Now one of them (hydrobenzoïn) is the sole product of the action of sodium-amalgam on benzil, C_6H_5—CO—CO—C_6H_5 : this, therefore, must be represented by the left hand or symmetrical formula ; and isohydrobenzoïn, if not a mere physical modification, by the unsymmetrical formula. By the action of PBr_5, both compounds are converted into tolylene dibromide, $C_7H_6Br_2$.

Hydrobenzoïn is sparingly soluble in water, easily in alcohol, and crystallizes in large rhombic plates, melting at 132.5° C. (270.5° F.), and subliming without decomposition. By heating with strong nitric acid it is converted into benzoïn. Its acetylic ether, $C_{14}H_{12}(OC_2H_3O)_2$, forms large prisms melting at 133°–134° C. (271.4°–273.2° F.).

Isohydrobenzoïn crystallizes from water in long, shining, four-sided hydrated prisms, which effloresce on exposure to the air ; from alcohol in anhydrous monoclinic prisms. It melts at 119.5° C. (247.1° F.), is more soluble in alcohol than hydrobenzoïn ; is not converted into benzoïn by nitric acid.

Fluorenyl Alcohol, $C_{13}H_{10}O = \begin{matrix} C_6H_4 \\ | \\ C_6H_4 \end{matrix} \!\!\!> CH.OH$, produced by the action of sodium-amalgam on an alcoholic solution of diphenylene ketone, forms colorless hexagonal laminæ, melting at 153° C. (307.4° F.).

Ketones.

Diphenyl Ketone or **Benzophenone,** $C_6H_5.CO.C_6H_5$, is formed by oxidation of diphenyl-methane and diphenyl-ethane (p. 592) with chromic acid mixture; by heating mercury-diphenyl (p. 873) with benzoyl chloride; and most readily (together with benzene) by distillation of calcium benzoate. It usually crystallizes in large rhombic prisms melting at 48°–49° C. (118.4°–120.2° F.), sometimes also in rhombohedrons melting at¹27° C. (80.6° F.); the latter modification is converted on standing into the former. It has an aromatic odor, boils at 295° C. (563° F.), dissolves easily in alcohol. Treated with PCl_5, it is converted into benzophenonic chloride, $(C_6H_5)_2CCl_2$, a liquid boiling at about 300° C. (572° F.). Sodium-amalgam converts it into benzhydrol (p. 968).

Phenyl-tolyl Ketone, $C_{14}H_{12}O = C_6H_5.CO.C_6H_4(CH_3)$, is obtained in two modifications by oxidation of benzyl-toluene; by heating a mixture of benzoic acid and toluene with P_2O_5; by heating a mixture of calcium benzoate and paratoluate; and by heating benzyl chloride and toluene with zinc-dust. The α-compound (para) exists in two dimorphous modifications—one, hexagonal, melting at 55° C. (131° F.); the other, monoclinic, melting at 59°–60° C. (138.2°–140° F.). Boiling point, 310°–312° C. (590°–593.6° F.). Heated with soda-lime it yields benzene and paratoluic acid. The β-compound, which is liquid, has not been obtained pure. By distillation with zinc-dust, it yields anthracene; with lead oxide, anthraquinone: the α-compound does not yield these products.

Ditolyl Ketone, $CO(C_6H_4.CH_3)_2$, forms rhombic crystals melting at 95° C. (203° F.).

Dibenzyl Ketone, $CO(CH_2.C_6H_5)_2$, obtained by distillation of calcium α-toluate, forms colorless crystals, m. p. 30° C. (86° F.), b. p. 320°–321° C. (608°–609.8° F.); yields by oxidation benzoic and carbonic acids.

Diphenylene Ketone, $C_{13}H_8O = CO \begin{matrix} C_6H_4 \\ < | \\ C_6H_4 \end{matrix}$, obtained by heating diphenic acid or phenyl-benzoic acid (p. 970) with lime, also by oxidizing fluorene with chromic acid mixture, forms large yellow rhombic crystals, melts at 84° C. (183.2° F.), boils at 336°–338° C. (636.8°–639.6° F.); is converted by reducing agents into fluorene, by fusion with potash into phenyl-benzoic acid.

Benzoin, $C_{14}H_{12}O_2 = C_6H_5.CO.CH(OH).C_6H_5$, formed by heating hydrobenzoin with nitric acid (specific gravity 1.4), and by mixing bitter-almond oil with a concentrated alcoholic solution of potassium cyanide, crystallizes in shining prisms melting at 133°–134° C. (271.4°–273.2° F.); dissolves sparingly in water and in cold alcohol or ether, freely in hot alcohol. Converted by chromic acid mixture into benzaldehyde and benzoic acid; by heating with fuming hydriodic acid into dibenzyl; by heating

41 *

with alcoholic potash, into hydrobenzoin and benzile, together with benzylic acid, benzoic acid, and other products :

$$2C_{14}H_{12}O_2 \quad = \quad C_{14}H_{14}O_2 \quad + \quad C_{14}H_{10}O_2 .$$
$$\text{Benzoin.} \qquad\qquad \text{Hydrobenzoin.} \qquad \text{Benzile.}$$

With acetyl chloride it yields a c e t y l - b e n z o i n, $C_{14}H_{11}(C_2H_3O)O_2$,
which forms crystals melting at 75° C. (167° F.).

Deoxybenzoin, or **Phenyl-benzyl Ketone,** $C_{14}H_{12}O = C_6H_5.CO.$
$CH_2(C_6H_5)$, is formed by the action of zinc and hydrochloric acid on benzoin and chlorobenzile ; by heating monobromostilbene with water to 180°–
190° C. (356°–374° F.), and by distillation of a mixture of benzoate and
*-toluate of calcium :

$$C_6H_5.CO.OH + C_6H_5.CH_2.CO.OH = CO_2 + H_2O + C_6H_5.CO.CH_2C_6H_5 .$$

It crystallizes from alcohol in large plates melting at 55° C. (131° F.);
sublimes without decomposition ; is converted by sodium-amalgam into
toluylene hydrate (p. 968), by PCl_5 into monochlorotoluylene, and by heating with hydriodic acid into benzile.

Benzile, $C_{14}H_{10}O_2 = C_6H_5CO.CO.C_6H_5$ (*Dibenzoyl*), is formed by oxidation of benzoin with nitric acid or chlorine ; and, together with stilbene,
by heating stilbene bromide with water and silver oxide. It crystallizes
in large six-sided prisms, melting at 90° C. (194° F.), insoluble in water,
soluble in alcohol and ether. By oxidation with chromic acid mixture, it
is converted into benzoic acid ; by nascent hydrogen (zinc and HCl, or
iron filings and acetic acid) into benzoin ; by heating with PCl_5 into b e n-
z i l e c h l o r i d e, $C_{14}H_{10}Cl_2O$, or $C_6H_5.CCl_2.CO.C_6H_5$, which crystallizes in
rhombic prisms melting at 71° C. (159.8° F.), and is converted by zinc
and HCl into deoxybenzoin. Benzile and its chloride, heated to 200° C.
(392° F.) with PCl_5, yield tolane tetrachloride, $C_{14}H_{10}Cl_4$.

———————

Acids.

Diphenylcarbonic Acid, $C_{13}H_{10}O_2 = C_6H_5.C_6H_4.CO_2H$ (1 : 4), is produced by heating cyanodiphenyl, $C_{12}H_9(CN)$, with alcoholic potash (p.
962), or by oxidizing diphenyl-benzene (p. 964) dissolved in glacial acetic
acid with chromic anhydride. It crystallizes from alcohol in tufts of
needles ; melts at 218°–219° C. (424.4°–426.2° F.); is converted by heating with lime into diphenyl, and by oxidation with chromic acid into terephthalic acid.

P h e n y l - b e n z o i c a c i d, $C_{13}H_{10}O_2 = C_6H_5.C_6H_4.CO_2H$ (? 1 : 2), produced by the action of melting potash on diphenylene ketone (p. 969)
forms ramified crystals, melting at 110°–111° C. (230°–231.8° F.); reconverted into diphenylene ketone by heating with lime.

Diphenyl-dicarbonic Acid, $C_{14}H_{10}O_4 = C_{12}H_8(CO_2H)_2$ (*para*), from
dicyanodiphenyl, $C_{12}H_8(CN)_2$, is a white amorphous powder, insoluble in
alcohol and ether, neither fusible nor sublimable. Heated with lime it
yields diphenyl. Its barium and calcium salts are nearly insoluble in
water. The diethylic ether melts at 112° C. (233.6° F.).

Diphenic acid, isomeric with the last, is formed by oxidation of phenanthrene or phenanthrene-quinone with chromic acid mixture. It melts at
226° C. (438.8° F.), sublimes in needles, dissolves very easily in alcohol

and ether. The barium and calcium salts are easily soluble in water ; the diethylic ether is liquid.

Diphenylacetic Acid, $C_{14}H_{12}O_2 = (C_6H_5)_2CH.CO.OH$, is formed by heating benzilic acid with hydriodic acid to 150° C. (302° F.), and by heating a mixture of phenyl-bromacetic acid and benzene with zinc-dust :

$$C_6H_5.CHBr.CO_2H + C_6H_6 = HBr + (C_6H_5)_2CH.CO_2H .$$

It crystallizes from water in slender needles, from alcohol in laminæ, and melts at 146° C. (294.8° F.); is converted by chromic acid mixture into diphenyl·ketone ; by heating with soda-lime into diphenyl-methane.

Diphenyl-glycollic or **Benzilic Acid,** $C_{14}H_{12}O_3 = (C_6H_5)_2C(OH)$. CO_2H, is formed by passing bromine-vapor into diphenyl-acetic acid, and boiling the product with water ; also by heating benzile with alcoholic potash (not in excess) ; in this reaction, a transposition of atoms must take place. It crystallizes in shining needles or prisms, easily soluble in hot water and in alcohol ; melts at 150° C. (302° F.), turning red at the same time. It dissolves with dark red color in sulphuric acid. By heating with HI, it is converted into diphenyl-acetic acid ; by oxidation into diphenyl ketone ; by distillation of its barium salts into benzhydrol (p. 968).

Benzoyl-benzoic Acids, $C_{14}H_{10}O_3 = C_6H_5.CO.C_6H_4.CO_2H.$—The *para*-modification is formed by oxidation of solid phenyltolyl ketone, *para*-ben-zyl-benzoic acid, benzyl-toluene, benzylethyl-benzene, and ethylbenzyl-toluene (p. 958), with chromic acid mixture. It sublimes in laminæ, resembling benzoic acid, and melts at 194° C. (381.2° F.). The *ortho*-modification, from liquid phenyl-tolyl ketone, crystallizes from hot water in needles or prisms containing $1H_2O$. The crystals melt at 85°–87° C. (185°–188.6° F.) ; the anhydrous acid at 127°–128° C. (260.6°–262.4° F.).

Benzhydryl-benzoic acids, $C_6H_5.CH.(OH).C_6H_4.CO_2H.$—The *para*-acid, formed by the action of nascent hydrogen on *para*-benzyl-benzoic acid, crystallizes in needles, melts at 164°–165° C. (327.2°–329° F.), and is reconverted by oxidation into *para*-benzoylbenzoic acid. The *ortho*-acid cannot be obtained in the free state, as on attempting to prepare it by the action of nascent hydrogen on *ortho*-benzylbenzoic acid, it splits up into water and the anhydride $C_{14}H_{10}O_2$. This compound crystallizes from alcohol or ether in prisms melting at 115° C. (239° F.) ; is converted by oxidation into *ortho*-benzylbenzoic acid, and by PCl_5 into anthraquinone and chlorinated compounds.

Benzyl-benzoic acid, $C_{14}H_{12}O_2 = C_6H_5.CH_2.C_6H_4.CO_2H.$—The *para*-acid is obtained by the prolonged action of HI and phosphorus at 160°–170° C. (320°–338° F.) on *para*-benzoyl- or benzhydryl-benzoic acid. Laminæ or needles, melting at 157° C. (314.6° F.). The *ortho*-acid, obtained in like manner from *ortho*-benzoyl-benzoic acid, forms slender shining needles, melting at 114° C. (237.2° F.)

NAPHTHALENE GROUP.

Naphthalene, $C_{10}H_8$.—This hydrocarbon is produced in the decom-position of toluene, xylene, and cumene at a red heat (p. 844) ; also by passing vapor of benzene, cinnamene, chrysene, or anthracene through a red-hot tube. It is formed in large quantities as a bye-product in the preparation of coal-gas, its production doubtless arising from reactions similar to those just mentioned. When the last portion of the volatile oily product which passes over in the distillation of coal-tar is collected apart and left to stand, a quantity of solid crystalline matter separates, which is principally naphthalene. An additional quantity may be obtained by pushing the distillation until the contents of the vessel begin to char : the naphthalene then condenses in the solid state, but dark-colored and very impure. By simple sublimation, once or twice repeated, it is obtained perfectly white.

Naphthalene is formed synthetically from phenyl-butylene, $C_6H_5.CH_2$. $CH_2.CH{=}CH_2$, which is obtained by the action of sodium on a mixture of benzyl chloride, $C_6H_5.CH_2Cl$, and allyl iodide, $CH_2I.CH{=}CH_2$. This hydrocarbon or its bromide passed through a red-hot tube filled with lime, is resolved into naphthalene and hydrogen, $C_{10}H_{12} = C_{10}H_8 + 2H_2$. Naphthalene forms large colorless, transparent, brilliant, crystalline plates, exhaling a faint and peculiar odor. It melts at 79.2° C. (174.6° F.) to a clear, colorless liquid, which crystallizes on cooling ; boils at 218° C. (424.4° F.). When strongly heated in the air, it takes fire, and burns with a red and very smoky flame. It is insoluble in cold water, but soluble to a slight degree at the boiling heat ; alcohol and ether dissolve it easily ; a hot saturated alcoholic solution deposits fine iridescent crystals on cooling. It unites with *picric acid*, forming the compound $C_{10}H_8$. $C_6H_3(NO_2)_3O$, which crystallizes in stellate groups of needles.

Naphthalene, like benzene and its homologues, forms addition-products with hydrogen and chlorine. Heated with PH_4I to 190° C. (374° F.), or with HI and amorphous phosphorus to 220°–250° C. (428°–482° F.), it forms the t e t r a h y d r i d e, $C_{10}H_8.H_4$, a pungent liquid boiling at 205° C. (401° F.). With chlorine, it forms the compound, $C_{10}H_8.Cl_2$, which is a pale yellow oil, and $C_{10}H_8.Cl_4$, $C_{10}H_7Cl.Cl_4$, and $C_{10}H_6Cl_2.Cl_4$, which crystallizes in monoclinic prisms, melting respectively at 182° C. (359.6° F.), 128°–130° C. (262.4°–266° F.), and 172° C. (341.6° F.).

The structure of the naphthalene molecule is deduced by Graebe (Liebig's *Annalen*, cxlix. 26) from the following considerations. Naphthalene is converted by oxidation into phthalic or benzene-dicarbonic acid, $C_6H_4(CO_2H)_2$, two of its carbon-atoms being removed as carbon dioxide, while two others remain in the form of carboxyl :

$$C_{10}H_8 + O_9 = C_6H_4(CO_2H)_2 + 2CO_2 + H_2O .$$

Hence naphthalene contains a benzene-residue, C_6H_4, and may be represented by the formula $C_6H_4.C_4H_4$. The same conclusion follows from the synthesis of naphthalene from phenyl-butylene. But phthalic acid is likewise produced by oxidation of dichloronaphthoquinone, $C_{10}H_4Cl_2O_2$: consequently, this compound has its two chlorine-atoms and two oxygen-atoms associated with the four carbon-atoms which undergo oxidation, and may be represented by the formula $C_6H_4.C_4Cl_2O_2$. By the action of phosphorus pentachloride, the two oxygen-atoms are replaced by two chlorine-atoms, and at the same time one of the four hydrogen-atoms is replaced by chlorine, the result being pentachloronaphthalene, C_6H_3Cl (C_4Cl_4), which is converted by oxidation into tetrachlorophthalic acid, $C_6Cl_4(CO_2H)_2$. In this reaction, therefore, two of the carbon-atoms belonging to the left-hand group in the molecule $C_6H_3Cl(C_4Cl_4)$ are removed as carbon dioxide, while two others yield the two groups CO_2H, and the remaining two, together with the four standing on the right, form the benzene-nucleus, C_6Cl_4, of tetrachlorophthalic acid.

These transformations show that the molecule of naphthalene is symmetrical, and composed of two benzene-nuclei united in the manner repre-sented by the following figure :—

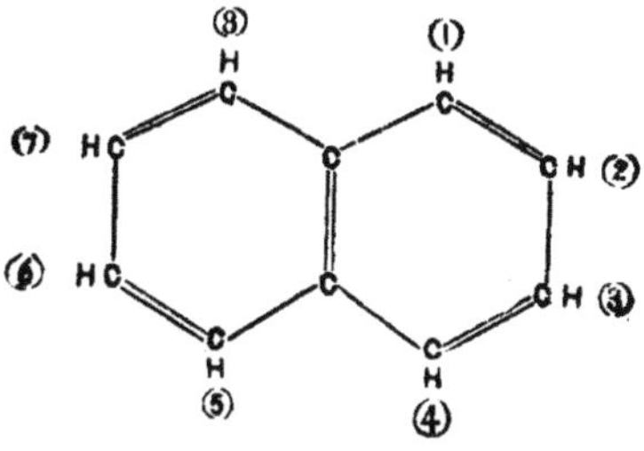

The replacement of the hydrogen-atoms in naphthalene by other ele-ments and by compound radicles gives rise to numerous substitution-derivatives, which are obtained by methods similar to those by which the benzene-derivatives are produced. The number of possible isomerides among the naphthalene-derivatives is, however, much greater than those of the benzene-derivatives. Thus, the mono-derivatives of benzene do not admit of isomeric modifications, but in naphthalene the places 1, 4, 5, 8 (see the preceding figure), though similar to each other, are different from the places 2, 3, 6, 7, which are likewise similar amongst themselves. Hence, there will be two modifications of each mono-derivative of naphtha-lene (chloro-, nitro-, etc.). Those modifications in which the substituted radicle occupies the place 1 (= 4, 5, or 8), are called a-derivatives ; those in which it is placed at 2, 3, 5, or 7, are called β-derivatives. The di-derivatives, $C_{10}H_8X_2$, exhibit ten isomeric modifications, according as the substituted radicles are situated : (1) in the same benzene-ring (1 : 2, 1 : 3, 1 : 4, or 5 : 6, 5 : 7, 5 : 8, and 2 : 3, or 6 : 7), secondly, in different rings (1 : 5, 1 : 6, 1 : 7, 1 : 8, 2 : 6, and 2 : 7). The list of actually known modifications is, however, very incomplete.*

Chloronaphthalenes are obtained by boiling the chlorides of naphthalene with alcoholic potash, which removes HCl ; these, when sub-jected to the action of chlorine, also form addition-products such as $C_6H_7Cl.Cl_2$; and from these, by the action of alcoholic potash, more highly chlorinated substitution-derivatives may be obtained.

Monochloronaphthalenes, $C_{10}H_7Cl$.—a, from naphthalene dichloride, is a liquid boiling at 250° C. (482° F.).—β. By the action of PCl_5 on β-naph-thol. Colorless crystalline mass, m. p. 60° C. (140° F.); b. p. 256°–258° C. (492.8°–496.4° F.).

$C_{10}H_6Cl_2$ is known in seven, and $C_{10}H_5Cl_3$, in four modifications. $C_{10}Cl_8$, the final product of the action of chlorine on naphthalene, forms prisms melting at 135° C. (275° F.).

Of monobromonaphthalene, two modifications are known, one formed by direct bromination, being a liquid boiling at 277° C. (530.6° F.); the other from β-naphthylamine by the diazo-reaction with alcohol, form-ing white shining laminæ, melting at 68° C. (154.4° F.).

Iodonaphthalene, $C_{10}H_7I$, produced by the action of iodine on mercuric naphthide, is an oil boiling at 300° C. (572° F.).

Cyanonaphthalenes, or Napthyl Cyanides, $C_{10}H_7.CN$ (a and β), are formed by distilling the potassium salts of the corresponding

* For the latest researches on naphthalene derivatives, see Liebermann (Liebig's *Annalen*, clxxxiii. 225), and Atterberg (*Deutsch. Chem. Ges. Ber.*, 1876, pp. 1730, 1734).

naphthalenesulphonic acids with potassium cyanide.—*a.* Crystallizes in needles, melts at 37.5° C. (99.5° F.), and boils at 297°–298° C. (566.6°–568.4° F.).—*ß.* Forms laminæ, melts at 66.5° C. (149.9° F.), boils at 304°–305° C. (579.2°–581° F.).

Dicyanonaphthalenes, $C_{10}H_6(CN)_2$, are formed in like manner from the naphthalene-disulphonic acids.—*a.* Needles melting at 267°–268° C. (512.6°–514.4° F.).—*ß.* Needles melting at 296°–297° C. (564.8°–566.6° F.).

N i t r o s o n a p h t h a l e n e , $C_{10}H_7.NO$, is formed by the action of nitrogen oxybromide, NOBr (solution of NO and Br in CS_2 at —20° C., —4° F.), on mercuric naphthide:

$$(C_{10}H_7)_2Hg \ + \ NOBr \ = \ C_{10}H_7.NO \ + \ C_{10}H_7.Hg.Br \ .$$

It separates from solution in benzene or light petroleum in yellow nodules, turning red in contact with the air ; melts at 85° C. (185° F.), and decomposes at 134° C. (273.2° F.). Its solution in phenol is colored blue by sulphuric acid.

N i t r o n a p h t h a l e n e , $C_{10}H_7.NO_2$, obtained by heating a solution of naphthalene in glacial acetic acid with ordinary nitric acid for half an hour, crystallizes from alcohol in sulphur-yellow prisms, melts at 61° C. (141.8° F.), boils at 304° C. (579.2° F.); insoluble in alcohol, ether, and carbon sulphide.

Two d i n i t r o n a p h t h a l e n e s are obtained by boiling naphthalene or mononitronaphthalene with nitric acid till the oily liquid which floats on the surface has disappeared.—*a.* Colorless prisms melting at 213° C. (415.4° F.).—*ß.* Rhombic plates melting at 170° C. (338° F.).

These compounds, boiled with fuming nitric acid, yield three t r i n i t r o n a p h t h a l e n e s , boiling respectively at 122°, 147°, and 218° C. (251.6°, 296.6°, 424.4° F.) ; and on further boiling, two t e t r a n i t r o n a p h t h a l e n e s , melting at 200° and 259° C. (392°, 408.2° F.).

Amidonaphthalenes or **Naphthylamines**, $C_{10}H_7.NH_2$.—The *a*-modification obtained by reduction of nitronaphthalene, crystallizes in needles or prisms, easily soluble in alcohol ; has a pungent odor ; melts at 50° C. (122° F.), and boils at 300° C. (572° F.). It forms easily soluble crystalline salts, the solutions of which give with chromic acid or ferric chloride, a blue precipitate, quickly changing to a purple-red powder of *oxynaphthylamine,* $C_{10}H_9NO$. *a*-Naphthylamine, heated with glacial acetic acid, yields *a*-*Acetonaphthalide,* $C_{10}H_7.NH(C_2H_3O)$, which crystallizes in fine silky fibres, melting at 159° C. (318.2° F.).

ß-Naphthylamine is obtained from bromonitronaphthylamine, which is formed by nitration of bromacetonaphthalide, $C_{10}H_6Br.NH.C_2H_3O$. The amido-group is eliminated by the action of nitrous acid and alcohol, and the resulting bromonitronaphthalene is reduced by tin and hydrochloric acid. *ß*-Naphthylamine forms nacreous laminæ melting at 112° C. (233.6° F.).

Nitronaphthylamine, $C_{10}H_6NO_2.NH_2$, obtained by reduction of *a*-dinitronaphthalene, forms small red crystals melting at 118°–119° C. (244.4°–246.2° F.). Two isomeric nitronaphthylamines melting at 191° and 158°–159° C. (375.8°, 316.4°–318.2° F.) are formed by the nitration of the acetyl-compound, which melts at 159° C. (318.2° F.).

D i a m i d o n a p h t h a l e n e or N a p h t h y l e n e d i a m i n e , from nitronaphthylamine (m. p. 191° C., 375.8° F.), is converted by oxidation with dilute chromic acid into naphthoquinone. Two other naphthylenediamines are obtained by reduction of dinitronaphthalenes.

The naphthylamines treated with nitrous acid yield diazo-compounds, analogous to those obtained from aniline.

Diazo-amidonaphthalene, $C_{20}H_{15}N_3 = C_{10}H_7.N_2.NH(C_{10}H_7)$, obtained by the action of nitrous acid on a cold alcoholic solution of naphthylamine, forms brown laminæ, and is resolved by heating with acids into naphthylamine and naphthol:

$$C_{10}H_7.N_2.NH.C_{10}H_7 \;+\; H_2O \;=\; C_{10}H_7.OH \;+\; C_{10}H_7.NH_2 \;+\; N_2.$$

The action of nitrous acid on a warm alcoholic solution of naphthylamine, or of potassium nitrite on its solution in hydrochloric acid, produces amidazonaphthalene, $C_{10}H_7.N_2.C_{10}H_6.NH_2$, metameric with diazo-amidonaphthalene, and analogous to amidazobenzene (p. 880). It dissolves easily in alcohol, and crystallizes in orange-red needles having a green metallic lustre. Its salts are violet-colored. Heated with naphthylamine it forms a base, $C_{30}H_{21}N_3$, analogous to rosaniline:

$$C_{20}H_{15}N_3 \;+\; C_{10}H_9N \;=\; NH_3 \;+\; C_{30}H_{21}N_3.$$

The hydrochloride of this base, $C_{30}H_{21}N_3.HCl + H_2O$, forms the fine red dye-stuff called *Naphthalene-red* or *Magdala-red*. It crystallizes in green metallically lustrous needles, and dissolves in alcohol, forming a red solution, which when dilute exhibits a beautiful fluorescence.

Mercuric Naphthide, $(C_{10}H_7)_2Hg$, produced by the action of sodium-amalgam on a solution of bromonaphthalene in benzene, forms shining crystals, easily soluble in hot benzene, sparingly in alcohol and ether. It melts at 243° C. (469.4° F.), and decomposes partially when sublimed. Haloïd acids decompose it into naphthalene and mercury-salts. It unites with iodine, forming the compound $(C_{10}H_7)_2HgI_2$, which is decomposed by a larger quantity of iodine, thus:

$$(C_{10}H_7)_2HgI_2 \;+\; I_2 \;=\; HgI_2 \;+\; 2C_{10}H_7I.$$

Heated with soda-lime it yields dinaphthyl and naphthalene.

Naphthalene-sulphonic Acids, $C_{16}H_7.SO_3H$, α and β, are formed by gradually heating naphthalene with sulphuric acid, and may be separated by fractional crystallization of their barium or lead-salts, those of the α-acid being much more soluble in water and alcohol than the β-salts. The free acids are crystalline and deliquescent. The α-acid is converted by heat into the β-acid; the latter is therefore almost the sole product obtained at a high temperature (160° C., 320° F.). The α-acid, heated with dilute hydrochloric acid to 200°, is resolved into naphthalene and sulphuric acid. The *chlorides*, $C_{10}H_7.SO_2Cl$, are obtained by heating the potassium salts with PCl_5. They crystallize in shining laminæ, α more soluble in ether than β. The α-chloride melts at 66° C. (150.8° F.); β at 76° C. (168.8° F.). By zinc and sulphuric acid they are converted into *mercaptans*, $C_{10}H_7.SH$.

By prolonged heating of naphthalene with sulphuric acid, two *naphthalenedisulphonic acids*, $C_{10}H_6(SO_3H)_2$, are formed, which when distilled with KCy yield the corresponding *dicyanides*, $C_{10}H_6(CN)_2$.

Napthols, $C_{10}H_7.OH$, are formed by fusing the two monosulphonic acids with potash.

α-Naphthol is also produced by boiling the aqueous solution of diazonaphthalene nitrate (from α-naphthylamine) with nitrous acid. It crystallizes in colorless monoclinic prisms, melts at 94° C. (201.2° F.), boils at 278°–280° C. (532.4°–536° F.); is nearly insoluble in cold water, somewhat more soluble in hot water, easily in alcohol and ether. Ferric chloride and bleaching powder give to the aqueous solution a transient violet color.

With alkalies, etc., α-naphthol forms derivatives exactly like those of phenol.   The *ethylic ether*, $C_{10}H_7.O.C_2H_5$, is a colorless liquid, boiling at 272° C. (521.6° F.), not solidifying at —5° C. (23° F.). α-*Acetyl naphthol*, $C_{10}H_7.O.C_2H_3O$, formed by the action of acetyl chloride, is a yellowish liquid, insoluble in water.   The *benzoyl* derivative melts at 56° C. (132.8° F.).

Nitro-α-naphthol, $C_{10}H_6 \left\{ {OH \atop NO_2} \right.$ (1 : 4) obtained by boiling nitronaphthyl-amine (m. p. 191° C., 375.8° F.) with potash, crystallizes from alcohol in lemon-yellow capillary needles ; from acetic acid or acetone in golden-yellow prisms ; melts at 164° C. (327.2° F.).   Its sodium salt is a yellow dye-stuff called *Campobello yellow*.   By reduction with tin and HCl, it yields amido-α-naphthol, which, by boiling with bromine-water or by the action of nitrous acid, is converted into naphthoquinone.

An isomeric nitro-α-naphthol, formed from the nitronaphthylamine melting at 158°–159° C. (316.4°–318.2° F.), crystallizes in greenish-yellow laminæ, melting at 128° C. (262.4° F.), much less soluble in alcohol than the preceding.

Dinitro-α-naphthol, $C_{10}H_5(NO_2)_2OH$, is formed by mixing α-naphthylamine with 4-6 parts of strong nitric acid ; by boiling diazonaphthalene hydrochloride with nitric acid ; or by gently heating a solution of α-naphthalene-sulphonic acid with nitric acid.   Shining sulphur-yellow crystals, melting at 138° C. (280.2° F.), sparingly soluble in alcohol and ether, more freely in chloroform.   It forms salts with bases, and decomposes carbonates.   Its calcium and sodium salts form splendid yellow dyes, known as naphthalene-yellows.

Dinitronaphthol treated with zinc and hydrochloric acid yields the hydrochloride of *diamidonaphthol*, $C_{10}H_5(NH_2)_2.OH$, the aqueous solution of which is converted by exposure to the air, or by the action of ferric chloride, into the hydrochloride of *di-imidonaphthol*, $C_{10}H_5(OH){<}{NH \atop NH}{>}$.

β-**Naphthol** or **Isonaphthol**, $C_{10}H_7.OH$, from β-naphthalene-sulphonic acid, crystallizes in small rhombic tables, melts at 122° C. (251.6° F.), boils at 285°–290° C. (545°–554° F.), and sublimes with great facility ; dissolves sparingly in boiling water, easily in alcohol and ether.   The *ethylic ether*, $C_{10}H_7.O.C_2H_5$, melts at 33° C. (91.4° F.), the acetyl-compound at 60° C. (140° F.), the benzyl-compound at 107° C. (224.6° F.).

Dinitro-β-naphthol, obtained by heating the alcoholic solution of β-naphthol with dilute nitric acid, forms yellow needles melting at 195° C. (383° F.).

Dioxynaphthalene or *Naphthohydroquinone*, $C_{10}H_6(OH)_2$ (1 : 4), is formed from naphthoquinone by heating with hydriodic acid and amorphous phosphorus.   It crystallizes from hot water in long colorless needles, melting at 176° C. (348.8° F.), converted by oxidation into naphthoquinone.   An isomeric compound, obtained from naphthalene-disulphonic acid, forms needles slightly soluble in water.

Naphthoquinone, $C_{10}H_6{<}{O \atop O}{>}$, is formed by oxidation of naphthalene with chromic anhydride dissolved in glacial acetic acid ; more readily by adding a dilute aqueous solution of chromic acid to a solution of diamido-naphthalene hydrochloride (p. 975) ; on distilling the liquid, the naphthoquinone passes over with the steam.   Large sulphur-yellow triclinic tables, having a pungent odor, melting at 125° C. (257° F.), subliming below 100° ; insoluble in cold water, slightly soluble in cold alcohol, easily in hot alcohol and in ether.   Converted by oxidation with nitric acid into phthalic acid.

Dichloronaphthoquinone, $C_{10}H_4Cl_2O_2$, is formed by heating *a*-naphthol or dinitronaphthol (commercial naphthalene yellow) with potassium chlorate and hydrochloric acid ; also by the action of chromyl chloride, CrO_2Cl_2, on a solution of naphthalene in glacial acetic acid. Crystallizes in golden-yellow needles melting at 189○ C. (372.2○ F.) ; oxidized by nitric acid to phthalic acid ; converted by PCl_5 into $C_{10}H_3Cl_5$, which oxidizes to chlorophthalic acid. Boiled with soda-lye, it dissolves with red color, forming the sodium salt of chloronaphthalic acid, $C_{10}H_4Cl(O_2)OH$, which crystallizes in yellow needles, melting at 200○ C. (392○ F.).

Oxynaphthoquinone or *Naphthalenic acid*, $C_{10}H_5(OH)\langle{}^O_O\rangle$, is formed from the hydrochloride of di-imidonaphthol (p. 976) by heating with dilute hydrochloric acid to 120○ C. (248○ F.).

$$C_{10}H_5(N_2H_2)OH \ + \ 2H_2O \ = \ 2NH_3 \ + \ C_{10}H_5(O_2)OH .$$

Light yellow, strongly electric powder, or yellow needles ; sparingly soluble in hot water, easily in alcohol and ether ; unites with nascent hydrogen, forming trioxynaphthalene. It is a rather strong monobasic acid, capable of decomposing carbonates. Its alkali-salts are blood-red, and easily soluble in water.

Dioxynaphthoquinone or *Naphthazarin*, $C_{10}H_4(O_2)(OH)_2$.—This compound is a dye-stuff very much like alizarin. It is obtained by gradually adding dinitronaphthol (4 parts), and granulated zinc ($\frac{1}{2}$-1 part), to a mixture of strong sulphuric acid (40 parts), and fuming sulphuric acid (4 parts) heated to 200○ C. (392○ F.) (not above 205○ C., 401○ F.). On diluting the mixture with water, boiling, and filtering, the naphthazarin separates as a red gelatinous mass, which may be purified by sublimation. Long needles having a splendid green metallic lustre, dissolving in alcohol with red, in ammonia with sky-blue color. The solutions give violet-blue precipitates with baryta- and lime-water, crimson with alum.

Naphthoic Acids, $C_{10}H_7.CO_2H$.—*a*-Naphthoic acid is obtained from *a*-cyanonaphthalene (p. 973) ; also by fusing potassium *a*-naphthalenesulphonate with sodium formate, and by the action of sodium-amalgam on a mixture of bromonaphthalene and ethyl chlorocarbonate. Slender needles melting at 160○ C. (320○ F.), slightly soluble in hot water, easily in hot alcohol. By distillation with baryta it is resolved into naphthalene and carbon dioxide. Its *ethylic ether* boils at 309○ C. (588.2○ F.) ; the *chloride*, $C_{10}H_7COCl$, at 297○ C. (566.6○ F.).

ß-Naphthoic or *Isonaphthoic acid*, from *ß*-cyanonaphthalene, crystallizes from hot water in long silky needles melting at 182○ C. (359.6○ F.).

Oxynaphthoic acids, $C_{10}H_6\langle{}^{OH}_{CO_2H}\rangle$, are formed from the two naphthols by the action of sodium and carbon dioxide. The *a*-acid melts at 185○ C. (365○ F.), and forms solutions which are colored blue by ferric chloride. The *ß*-acid is difficult to prepare.

Naphthalene-dicarbonic acids, $C_{10}H_6(CO_2H)_2$.—Of these acids there are three known modifications. Two of them, *a* and *ß*, are obtained by heating the corresponding dicyanonaphthalenes with strong hydrochloric acid to 200○-210○ C. (392○-410○ F.).—*a*. Forms long pointed needles, melting with decomposition, above 300○ C. (572○ F.) ; its salts, except those of the alkalies, are but slightly soluble in water. The *ß*-acid is very much like the *a*-acid, except that it crystallizes in short slender needles.

The third modification, *naphthalic acid*, obtained by oxidation of acenaph-thene and acenaphthylene with chromic acid mixture, forms slender needles, and decomposes at 140°–150° C. (284°–302° F.), without melting, into water and the anhydride, $C_{12}H_6O_3$, which melts at 266° C. (510.8° F.). The acid is nearly insoluble in water. Its calcium salt heated with lime yields naphthalene. The *methylic ether*, $C_{10}H_6(CO_2CH_3)_2$, crystallizes in prisms melting at 102°–103° C. (215.6°–217.4° F.).

Dinaphthyl, $C_{20}H_{14} = C_{10}H_7.C_{10}H_7$, is formed by the action of sodium on monobromonaphthalene; by oxidation of naphthalene with MnO_2 and sulphuric acid; and by heating mercuric naphthide with soda-lime (p. 975). Colorless laminæ, with a mother-of-pearl lustre; slightly soluble in cold alcohol, easily in ether; melts at 154° C. (309.2° F.), and sublimes without decomposition. By further oxidation with MnO_2 and sulphuric acid it is converted, like naphthalene, into phthalic acid.

Three *isodinaphthyls* are obtained by passing naphthalene-vapor alone, or mixed with antimony trichloride, through a red-hot tube, and may be sepa-rated by repeated crystallization from petroleum; melting points, 187°, 147°, and 75° C. (368.6°, 296.6°, 167° F.).

Dinaphthylmethane, $C_{21}H_{16} = C_{10}H_7.CH_2.C_{10}H_7$, obtained (like diphenyl-methane, p. 963) by the action of sulphuric acid on a mixture of naph-thalene and methylal diluted with chloroform, crystallizes in short prisms, melts at 109° C. (228.2° F.), boils at a temperature above 360° C. (680° F.), and may be distilled without decomposition. With picric acid it forms the compound $C_{20}H_{16}.2C_6H_3(NO_2)_3O$, which crystallizes in reddish-yellow prisms.

Homologues of Naphthalene are obtained by the action of sodium on a mixture of bromonaphthalene and alcoholic bromides dissolved in ether; and these, when passed through red-hot tubes, split up into hydro-gen and unsaturated hydrocarbons. *Methyl-naphthalene*, $C_{10}H_7.CH_3$, is a thick liquid, boiling at 232° C. (449.6° F.), not solidifying at -18° C. (-0.4° F.); *ethyl-naphthalene*, $C_{10}H_7.C_2H_5$, boils at 251°–252° C. (483.8°–485.6° F.), and does not solidify at -14° C. (6.8° F.).

Acenaphthene, $C_{12}H_{10} = C_{10}H_6 {<}{\begin{smallmatrix}CH_2 \\ | \\ CH_2\end{smallmatrix}}$, is formed by passing ethyl-naphthalene through a red-hot tube, and occurs in the portion of coal-tar oil boiling at 250°–260° C. (482°–500° F.), from which it may be separated by cooling. It crystallizes from fusion in flat prisms, from alcohol in long needles; melts at 95° C. (203° F.); boils at 277.5° C. (531.5° F.); forms with picric acid the compound $C_{12}H_{10}.C_6H_2(NO_2)_3OH$, which crystallizes from alcohol in long needles melting at 161°–162° C. (321.8°–323.6° F.). By oxidation with chromic acid mixture, acenaphthene is converted into naphthalic acid (p. 977).

Acenaphthylene, $C_{12}H_8 = C_{10}H_6 {<}{\begin{smallmatrix}CH \\ || \\ CH\end{smallmatrix}}$, is formed by passing the vapor of acenaphthene over gently heated lead-oxide. It dissolves easily in al-cohol, crystallizes in yellow tables, sublimes at ordinary temperatures, melts at 92°–93° C. (197.6°–199.4° F.), and boils, with partial decompo-sition, at about 265°–275° C. (509°–527° F.). Its picric acid compound forms yellow needles melting at 202° C. (395.6° F.). By chromic acid mixture it is oxidized to naphthalic acid. Its dibromide, $C_{10}H_8Br_2$, melts at 122° C. (251.6° F.).

PHENANTHRENE AND ANTHRACENE GROUP.

The primary compounds of this group are two isomeric solid hydrocarbons, $C_{14}H_{10}$, occurring in the portion of coal-tar which boils between 320° and 360° C. (608°–680° F.). They are nearly related to benzene, and their modes of formation and decomposition show that their structure may be very probably represented by the following formulæ:

$$C_6H_4—CH$$
$$|\quad\quad ||$$
$$C_6H_4—CH$$
Phenanthrene.

$$C_6H_4 <\begin{array}{c}CH\\ |\\ CH\end{array}> C_6H_4$$
Anthracene.

Both consist of two benzene residues united by the group C_2H_2; but in phenanthrene the two benzene-groups are joined directly, in the same manner as in diphenyl, whereas in anthracene they are united only through the medium of the group $=CH—CH=$.

Phenanthrene, $C_{14}H_{10}$, may be formed by abstraction of hydrogen from dibenzyl, $C_{14}H_{14}$, and stilbene, $C_{14}H_{12}$, when the vapors of these bodies are passed through a red-hot tube; also, together with anthracene and other hydrocarbons, from toluene and xylene, by similar treatment (p. 844).

Phenanthrene is prepared from crude anthracene—the high-boiling portion of coal-tar—by collecting apart the portion which boils between 320° and 350° C. (608°–662° F.), and boiling the mass with alcohol. The solution on cooling deposits, first anthracene, and then phenanthrene, which may be obtained quite clear by two recrystallizations.

Phenanthrene crystallizes in colorless shining laminæ, and exhibits a bluish fluorescence, especially in solution; melts at 100°; boils at 340° C. (644° F.); dissolves in 50 parts of alcohol at 13° C. (55.4° F.), easily in boiling alcohol, ether, and benzene. It unites with picric acid, a mixture of the saturated alcoholic solutions of the two bodies, depositing the compound $C_{14}H_{10}.C_6H_3(NO_2)_3O$, in reddish-yellow needles melting at 143° C. (289.4° F.). By boiling with chromic acid mixture phenanthrene is converted, first into p h e n a n t h r e n e - q u i n o n e, and ultimately into d i p h e n i c   a c i d (p. 970).

Phenanthrene tetrahydride, $C_{14}H_{14}$, formed by heating phenanthrene to 210°–240° C. (410°–464° F.) with hydriodic acid and amorphous phosphorus, is a liquid boiling at 300°–310° C. (572°–590° F.).

Phenanthrene dibromide, $C_{14}H_{10}Br_2$, obtained by direct combination in ethereal solution, forms prismatic crystals, and is resolved, by heating with water to 100°, into HBr and *bromophenanthrene*, $C_{14}H_9Br$, which melts at 63° C. (145.4° F.).

P h e n a n t h r e n e - q u i n o n e, $C_{14}H_8O_2 = \begin{array}{c}C_6H_4—CO\\ |\quad\quad |\\ C_6H_4—CO\end{array}$, is formed by heating phenanthrene with chromic acid mixture, or by adding chromic anhydride to a solution of phenanthrene in glacial acetic acid. It crystallizes from alcohol in shining orange-yellow needles, melts at 198° C. (388.4° F.), sublimes without decomposition, dissolves sparingly in hot water and cold alcohol, easily in hot alcohol; dissolves in strong sulphuric acid with dark-green color, and is precipitated therefrom by water. It unites with acid sodium sulphite, forming the compound $C_{14}H_8O_2.SO_3NaH + 2H_2O$.

Phenanthrene-quinone is oxidized by chromic acid mixture to diphenic acid, and reduced by zinc-dust to phenanthrene; when ignited with soda-lime it yields diphenyl.

Phenanthrene-hydroquinone, or *Dioxyphenanthrene*, $C_{14}H_8(OH)_2$, formed by the action of sulphurous acid on phenanthrene-quinone, crystallizes in colorless needles, which, when exposed to the air, turn brown, and are reoxidized to the quinone. Its diacetyl-compound crystallizes in tables melting at $202°$ C. ($395.6°$ F.).

Anthracene, $C_{14}H_{10}$.—This hydrocarbon may be formed artificially:
1. By passing benzyl-toluene through a red-hot tube or over heated lead oxide:

$$C_6H_5.CH_2.C_6H_4.CH_3 = C_6H_4 \underset{CH}{\overset{CH}{\huge<}} C_6H_4 + 2H_2 \; .$$

2. By heating liquid phenyl-tolyl ketone (p. 941) with zinc-dust:

$$C_6H_5.CO.C_6H_4.CH_3 = C_6H_4 \underset{CH}{\overset{CH}{\huge<}} C_6H_4 + H_2O \; .$$

3. Together with dibenzyl, by heating benzyl chloride with water to $190°$ C. ($374°$ F.):

$$4(C_6H_5.CH_2Cl) = C_{14}H_{10} + C_{14}H_{14} + 4HCl \; .$$

Anthracene is prepared from the high-boiling portions of coal-tar by repeated distillation, pressure, and recrystallization from benzene. To obtain it quite pure and colorless, it must be sublimed at as low a temperature as possible—best by heating it till it begins to boil, and then blowing a strong current of air over it from a pair of bellows. Or its solution in hot benzene may be bleached by exposure to sunshine.

Anthracene crystallizes in colorless monoclinic tables, exhibiting a fine blue fluorescence. It dissolves sparingly in alcohol and ether, easily in hot benzene; melts at $213°$ C. ($415.4°$ F.), and distils at a temperature a little above $360°$ C. ($680°$ F.). Its solution in benzene yields with picric acid the compound $C_{14}H_{10}.2C_6H_3(NO_2)_3O$, which crystallizes in red needles.

A cold saturated solution of anthracene in benzene exposed to sunshine deposits tabular crystals of an isomeric modification, $C_{14}H_{10}$, called p a r - a n t h r a c e n e. It is nearly insoluble in benzene, is not attacked by nitric acid or bromine, melts at $244°$ C. ($471.2°$ F.), and is at the same time converted into ordinary anthracene.

A n t h r a c e n e d i h y d r i d e, $C_{14}H_{12}$, formed by heating anthracene or anthraquinone with hydriodic acid and phosphorus, or by the action of sodium-amalgam on its alcoholic solution, crystallizes in monoclinic tables easily soluble in alcohol, melting at $106°$ C. ($222.8°$ F.), distilling at $305°$ C. ($581°$ F.). On passing its vapor through a red-hot tube, it is resolved into anthracene and hydrogen. The *hexhydride*, $C_{14}H_{16}$, is formed by prolonged heating of anthracene or the dihydride with HI and phosphorus to $200°$–$220°$ C. ($392°$–$428°$ F.), crystallizes in laminæ, melts at $63°$ C. ($145.4°$ F.), boils at $290°$ C. ($554°$ F.). It is very soluble in alcohol, ether, and benzene, and is resolved at a red heat into anthracene and hydrogen.

A n t h r a c e n e d i c h l o r i d e, $C_{14}H_{10}Cl_2$, formed by passing chlorine gas over anthracene, crystallizes in needles, and is converted by alcoholic potash into $C_{14}H_9Cl$. By heating anthracene in chlorine, d i c h l o r - a n t h r a c e n e, $C_{14}H_8Cl_2$, is formed, which crystallizes in yellow laminæ, melts at $209°$ C. ($408.2°$ F.), and yields solutions having a splendid blue fluorescence.

Dibromanthracene, $C_{14}H_8Br_2$, is formed as the sole product of the action of bromine on a solution of anthracene on carbon sulphide. It crystallizes in golden-yellow needles, melting at 221° C. (429.8° F.), and is reconverted into anthracene by heating with alcoholic potash. *Dibromanthracene tetrabromide*, $C_{14}H_8Br_2.Br_4$, is formed by passing bromine vapor over finely divided anthracene or dibromanthracene. Hard, thick, colorless tables, melting with decomposition at 170°–180° C.(338°–356° F.).

Another modification of dibromanthracene is formed by heating dibromanthraquinone with HI and phosphorus to 150° C. (302° F.) for eight hours. Golden-yellow tables, melting at 190°–192° C. (374°–377.6° F.), more soluble in alcohol and benzene than the preceding modification.

Anthracene-sulphonic acids, *mono-* and *di-*, are formed by heating anthracene with strong sulphuric acid, the former at 100°, the latter at 150° C. (302° F.). The monosulphonic acid is produced in two isomeric modifications (*a* and *ß*), which may be separated by means of their lead-salts. That of the *a*-acid (the chief product) crystallizes in light yellow laminæ $(C_{14}H_9SO_3)_2Pb + 4H_2O$, and is much more soluble than that of the *ß*-acid, which crystallizes in prisms containing $7H_2O$.

Anthraphenols, $C_{14}H_{10}O = C_{14}H_9(OH)$.—Of these bodies there are two metameric modifications, viz. :

$$C_6H_4 < \begin{matrix} CH \\ | \\ CH \end{matrix} > C_6H_3.OH \qquad C_6H_4 < \begin{matrix} C(OH) \\ | \\ CH \end{matrix} > C_6H_4$$

Anthrol.   Anthranol.”

Anthrol is formed in two isomeric modifications by fusing the alkali-salts of the two isomeric anthracene-monosulphonic acids with potash or soda.

a. Light yellow needles or laminæ having a strong lustre, decomposing at 250° C. (482° F.) without previous fusing ; insoluble in water, easily soluble in alcohol, ether, benzene, and alkalies. The alkaline solution absorbs oxygen from the air, and turns brown.

ß. Yellowish prisms, somewhat less soluble in alcohol and ether than *a*, which it otherwise resembles in every respect.

Anthranol is formed by heating anthraquinone (20 parts) with hydriodic acid of 1.7 sp. gr. (80 parts) and phosphorus (4 parts) in a vessel with reversed condenser. It crystallizes in yellowish needles, melts at 163°–170° C. (325.4°–338° F.), is nearly insoluble in cold, more soluble in hot alkalies. The solution absorbs oxygen from the air, and then deposits anthraquinone on boiling. Anthranol is also converted into anthraquinone by oxidation with nitric or chromic acid, and into anthracene by heating with zinc-dust. With acetic anhydride, it forms an acetic ether, $C_{14}H_9.O.C_2H_3O$, which crystallizes in light yellow needles melting at 126°–131° C. (258.8°–267.8° F.).

Anthraquinone, $C_{14}H_8O_2 = C_6H_4 < \begin{matrix} CO \\ CO \end{matrix} > C_6H_4$.—This compound, which may be regarded as a double ketone, and differs from the quinones of the benzene series in not having its two oxygen-atoms directly combined, is formed by oxidation of anthracene, anthracene-hydride, dichlor- or dibromanthracene, with nitric or chromic acid ; also when liquid phenyl-tolyl ketone (p. 969), $C_6H_5.CO.C_6H_4.CH_3$, is passed over heated lead oxide, or heated with MnO_2 and sulphuric acid, or with chromic acid mixture ; in small quantity also (together with benzophenone, p. 969) by distillation of calcium benzoate ;—and by heating phthalic chloride and benzene with zinc-dust :

$$C_6H_4{<}^{COCl}_{COCl} + C_6H_6 = 2HCl + C_6H_4{<}^{CO}_{CO}{>}C_6H_4 \,.$$

It is most easily prepared by adding finely pounded potassium dichromate, or a solution of chromic acid in glacial acetic acid, to a hot solution of anthracene in glacial acetic acid.

Anthraquinone sublimes in shining yellow needles, melts at 273⁰ C. (523.4⁰ F.), dissolves in hot benzene and in nitric acid. It is a very stable compound, not easily altered by oxidizing agents, and not reduced by sulphurous acid, like the quinones of the benzene series. Heated with hydriodic acid to 150⁰ C. (302⁰ F.), or with zinc-dust, it is converted into anthracene. By fusion with potash at 250⁰ C. (482⁰ F.), it yields 2 molecules of benzoic acid. Heated with PCl$_5$ to 190⁰–200⁰ C. (374⁰–392⁰ F.), it yields a mixture of di- and tetra-chloranthracene ; and when heated with fuming sulphuric acid, it yields sulpho-acids, together with phthalic anhydride.

Dibromanthraquinone, $C_{14}H_6Br_2O_2$, is formed by heating anthraquinone with bromine to 100⁰, or by oxidizing tetrabromanthracene with nitric acid ; dichloranthraquinone is formed by a similar method. Both sublime in yellow needles, and are converted into alizarin by heating with potash-lye to 150⁰ C. (302⁰ F.).

Dinitro-anthraquinone, $C_{14}H_8(NO_2)_2O_2$, obtained by boiling anthracene with dilute nitric acid, forms yellow needles melting at 280⁰, and, like picric acid, forms crystalline compounds with many hydrocarbons. An isomeric compound, obtained by the action of a mixture of nitric and sulphuric acids on anthraquinone, forms small light yellow monoclinic prisms, cakes together at 252⁰ C. (485.6⁰ F.), and sublimes at a higher temperature in small needles, with partial decomposition.

Anthraquinone-sulphonic acid, $C_{14}H_7O_2.SO_3H$, and the disulphonic acid, $C_{14}H_6O_2(SO_3H)_2$, are formed, together with phthalic anhydride, by heating anthraquinone with strong sulphuric acid to 250⁰–260⁰ C. (482⁰–500⁰ F.). The disulphonic acid is formed synthetically by heating orthobenzoyl-benzoic acid, $C_6H_5.CO.C_6H_4.CO_2H$ (p. 971), with fuming sulphuric acid. It forms yellow crystals easily soluble in water.

$$\text{Anthrahydroquinone, } C_{14}H_8(OH)_2 = C_6H_4{<}^{C(OH)}_{C(OH)}{>}C_6H_4 \,,$$

produced by heating anthraquinone with zinc-dust and potash-lye, forms yellow flakes, which, especially when moist, are quickly reconverted into anthraquinone by exposure to the air.

Oxyanthraquinone, $C_{14}H_8O_3 = C_6H_4{<}^{CO}_{CO}{>}C_6H_3.(OH)$. — This compound is produced, together with alizarin, by fusing monobromanthraquinone or anthraquinone-sulphonic acid with potassium hydroxide ; by boiling the product with water and calcium or barium carbonate, the alizarin is precipitated, and the oxyanthraquinone dissolved. Oxyanthraquinone is formed synthetically, together with its isomeride, *erythroxyanthraquinone*, by heating phthalic anhydride and phenol with strong sulphuric acid :

$$C_6H_4{<}^{CO}_{CO}{>}O + C_6H_5OH = C_6H_4{<}^{CO}_{CO}{>}C_6H_3.OH + H_2O \,.$$

The two isomerides may be separated by ammonia in which **oxyanthra-quinone** is soluble, erythroxyanthraquinone insoluble.

Oxyanthraquinone is equally soluble in hot and in cold alcohol, crystallizes in sulphur-yellow needles, melts at 268°–271° C. (514.4°–519.8° F.), and sublimes with some difficulty in laminæ. It dissolves readily in ammonia, baryta-water, and lime-water, and decomposes barium carbonate when boiled with it in water.

Erythroxyanthraquinone is more soluble in hot than in cold alcohol, crystallizes in orange-yellow needles, melts at 173°–180° C. (343.4°–356° F.), and sublimes at 150° C. (302° F.). It is nearly insoluble in dilute ammonia, forms dark-red lakes with baryta- and lime-water, and (like alizarin) does not decompose carbonate of barium. Both oxyanthraquinones are converted by fusion with potash into alizarin.

Dioxyanthraquinones, $C_{14}H_8O_4 = C_{14}H_6(OH)_2O_2$.—This formula includes eight known compounds, in three of which the two hydroxyl-groups are situated in the same benzene-nucleus, while four others have one hydroxyl in each benzene-nucleus, and in the eighth, the position of these groups is uncertain.

$$\text{I. } Dioxyanthraquinones, \ C_6H_4 \underset{\text{CO}}{\overset{\text{CO}}{<}} C_6H_2 \underset{\text{OH}}{\overset{\text{OH}}{<}} .$$

1. ALIZARIN.—This compound, in which the two hydroxyls are supposed to stand to one another in the para-position, is the red coloring matter of madder root (*Rubia tinctorum*). Fresh madder roots contain a glucoside, called r u b e r y t h r i c a c i d, which, when the roots are steeped in water, is resolved, under the influence of a peculiar ferment also contained in them, into alizarin and glucose:

$$\underset{\text{Ruberythric acid.}}{C_{26}H_{28}O_{14}} + 2H_2O = \underset{\text{Alizarin.}}{C_{14}H_8O_4} + \underset{\text{Glucose.}}{2C_6H_{12}O_6}$$

In old roots this change has already taken place to a considerable extent, so that they contain free alizarin. This spontaneous change was the basis of the older method of obtaining alizarin, and of the use of madder in dyeing. Various methods of accelerating the decomposition have been invented, in particular the treatment of the pulverized root, previously exhausted with water, with strong sulphuric acid, which decomposes the ruberythric acid in the manner above explained, but does not alter the resulting alizarin. The product thus obtained is called g a r a n c i n.

At present, however, almost all the alizarin used in dyeing is obtained by artificial processes from anthracene. It may be obtained by the action of melting potash on various derivatives of anthracene, viz., dibrom- and dichlor-anthraquinone, the two monoxyanthraquinones, and anthraquinone-disulphonic acid. Graebe and Liebermann, in 1868, first prepared it from dibromanthraquinone, and Perkin soon afterwards showed that it might be obtained much more economically from anthraquinone-disulphonic acid. This last method is now carried out on a very large scale. The mass obtained by fusion with potash is dissolved in water, and the alizarin, precipitated by hydrochloric acid, is purified by crystallization and sublimation.

Alizarin crystallizes from alcohol in reddish-yellow prisms or needles containing 3 mol. H_2O, which it gives off at 100°. It melts at 275° C. (527° F.) (Liebermann a. Troschke) ; at 289°–290° C. (552.2°–554° F.) (Claus a. Willgerodt),[*] and sublimes in orange-colored needles ; dissolves sparingly in hot water, easily in alcohol and ether ; also in strong sulphuric acid, forming a dark-red solution, from which it is precipitated by water in its original state.

* Deutsch. Ch. Ges. Ber. 1875, pp. 351, 381.

Alizarin has the structure of a diatomic phenol, and reacts like a weak acid. It dissolves in alkalies, forming purple solutions, from which calcium and barium salts throw down the corresponding salts as purple precipitates (distinction from monoxyanthraquinone).

Aluminium- and tin-salts form red precipitates (madder-lakes) ; ferric salts, a black-violet precipitate. The use of alizarin in dyeing and calicoprinting is founded on this property of forming insoluble colored compounds with metallic salts.

Alizarin is converted by heating with zinc-dust into anthracene, and by oxidation with nitric acid into phthalic acid.

Alizarin boiled with acetic anhydride yields first *monacetyl-alizarin*, $C_{14}H_6(OH)(OC_2H_3O)O_2$, and after prolonged boiling, *diacetyl-alizarin*, $C_{14}H_6(OC_2H_3O)_2O_2$.

Alizarinamide, $C_6H_4.C_2O_2.C_6H_2(NH_2)OH$, formed by heating alizarin with aqueous ammonia to 150°–200° C. (302°–392° F.), crystallizes in brown needles, melting at 250°–260° C. (482°–500° F.).

2. QUINIZARIN is formed by heating phthalic anhydride and hydroquinone, or parachlorophenol, with sulphuric acid. It crystallizes from ether in yellow needles melting at 194° C. (381.2° F.) ; dissolves in alkalies with blue-violet color ; forms blue-violet precipitates with baryta and magnesia, red with alumina ; and is reduced to anthracene by heating with zinc-dust.

3. PURPUROXANTHIN (or *Xanthropurpurin*) occurs in small quantity in madder, and is formed from purpurin by reduction with stannous chloride in alkaline solution. Yellowish-red needles, m. p. 262°–263° C. (503.6°–505.4° F.) ; dissolving with red color in alkalies and baryta-water. In alkaline solution it absorbs oxygen from the air.

Purpuroxanthic Acid, $C_{15}H_8O_6 = C_{14}H_7O_4.CO_2H$, is a constituent of commercial purpurin, from which it may be extracted by means of boiling alum-liquor. It is more soluble in boiling water than most madder colors ; crystallizes from hot alcohol in yellow needles ; melts at 231° C. (447.8° F.), and splits up at 232°–233° C. (449.6°–451.4° F.) into CO_2 and purpuroxanthin. It dissolves in a boiling solution of ferric chloride with deep reddish-brown color, and is reprecipitated by hydrochloric acid in yellow flocks.

Purpuroxanthic acid is most probably identical with *Munjistin*, a coloring matter extracted some years ago by Dr. Stenhouse from *Munject* or East Indian madder, this substance being also resolved by heat into CO_2 and purpuroxanthin (Schunk and Roemer, *Chem. Soc. Journal*, 1877, i. 666).

II. *Dioxyanthraquinones*, $C_6H_3(OH) {<}{CO \atop CO}{>} C_6H_3(OH)$.

4. ANTHRAFLAVONE.—Formed from oxybenzoic acid, $C_6H_4(OH)$. $CO.OH$, in small quantity by dry distillation, in larger quantity by heating with sulphuric acid to 180°–200° C. (356°–392° F.). Crystallizes in small yellow needles ; sublimes above 300° C. (572° F.) with partial decomposition, and without previous fusion ; nearly insoluble in boiling water, slightly soluble in ether, more easily in alcohol : dissolves with brown color in alkalies ; converted into alizarin by heating with zinc-dust ; into oxybenzoic acid by fusion with potash.

5. ANTHRAFLAVIC ACID.—Formed, together with iso-anthraflavic acid, from two different anthraquinone-disulphonic acids, in the preparation of alizarin on the large scale. It is therefore a constituent of crude commercial alizarin, and may be extracted therefrom by lime-water (the alizarin

remaining undissolved), the red solution being precipitated with hydro-chloric acid, and the precipitate treated with cold baryta-water, which dissolves iso-anthraflavic acid, and leaves the anthraflavic acid undissolved.

Anthraflavic acid crystallizes from alcohol in anhydrous yellow silky needles; melts above 330° C. (626° F.); dissolves in alkalies, with yellowish-red color; forms with acetic anhydride a *diacetyl-compound*, $C_{14}H_6(OC_2H_3O)_2O_2$, melting at 227° C. (440.6° F.).

6. Iso-anthraflavic Acid, prepared as above, crystallizes from aqueous alcohol in long yellow needles with 1 mol. H_2O, which is given off at 150° C. (302° F.); the anhydrous compound melts above 330° C. (626° F.). Dissolves in alkalies with deep-red color. The diacetyl-compound forms small pale-yellow crystals melting at 195° C. (383° F.).

7. Chrysazin, prepared by heating with alcohol the diazo-compound formed by the action of nitrous acid on sulphate of hydrochrysammide (*infra*), crystallizes in golden-yellow laminæ or red-brown highly lustrous needles; melts at 191° C. (375.8° F.); dissolves in alkalies with yellowish-red color; forms insoluble red compounds with lime- and baryta-water; is reduced to anthracene by heating with zinc-dust. The diacetyl-compound forms yellowish laminæ melting at 226°–230° C. (438.8°–446° F.).

Tetranitrochrysazin, or *Chrysammic acid*, $C_{14}H_2(NO_2)_4(OH)_2O_2$, formed by heating aloes, or chrysazin, with strong nitric acid, crystallizes in golden-yellow laminæ sparingly soluble in water, and reacts like a strong bibasic acid. By the action of reducing agents it is converted into t e t r a m i d o-c h r y s a z i n or c h r y s a m m i d e, $C_{14}H_2(NH_2)_4(OH)_2O_2$, which crystallizes in indigo-blue needles, having a splendid coppery lustre.

---

8. Frangulic Acid, $C_{14}H_8O_4 + 1\frac{1}{2}H_2O$, a dioxyanthraquinone in which the relative positions of the two OH groups are not known, is formed by the action of dilute acids on frangulin, $C_{20}H_{20}O_{10}$, a glucoside contained in the bark of *Rhamnus frangula*. It crystallizes in orange-yellow needles or plates containing $1\frac{1}{2}$ mol. H_2O, gives off its water at 180° C. (356° F.), and melts at 252°–254° C. (485.6°–489.2° F.). Reduced by zinc-dust to anthracene.

Trioxyanthraquinones, $C_{14}H_8O_5 = C_{14}H_5(OH)_3O_2$.—Of these compounds there are four known modifications : .

1. Purpurin, $C_6H_4 {<} {}^{CO}_{CO} {>} C_6H(OH)_3$, occurs in old madder-root, together with alizarin, and may be separated by means of boiling alum solution, in which the alizarin is insoluble. It may be formed artificially from alizarin or chrysazin by oxidation with MnO_2 and sulphuric acid at 140°–160° C. (284°–320° F.), and from purpuroxanthin by merely heating the alkaline solution in an open vessel. It crystallizes in reddish-yellow prisms, easily fusible, and subliming with partial decomposition; somewhat more soluble in water than alizarin; dissolves with red color in alcohol, ether, and alkalies; forms purple-red precipitates with lime- and baryta-water; is reduced to anthracene by heating with zinc-dust. With acetic anhydride it forms a *triacetyl-compound*, $C_{14}H_5(OC_2H_3O)_3O_2$, which crystallizes in yellowish needles melting at 190°–193° C. (374°–379.4° F.).

Purpurinamide, $C_{14}H_5(NH_2)(OH)_2O_2$, obtained by heating purpurin with

42

aqueous ammonia, forms brownish-green metallically lustrous needles. Treated in hot alcoholic solution with nitrous acid, it is converted into purpuroxanthin.

2. ANTHRAPURPURIN, $C_6H_3OH.C_2O_2.C_6H_2(OH)_2$, produced by fusing potassium anthraquinone-disulphonate or isoanthraflavic acid with potash, forms orange-colored needles, melting above 330° C. (626° F.), sublimable with partial decomposition. Dissolves in alkalies with fine Violet color, and, like alizarin, produces red colors with alumina-mordants, purple and black with iron-mordants ; but the reds are purer than those of alizarin, the purples bluer, and the blacks more intense. Its triacetyl-compound forms light-yellow scales melting at 220°–222° C. (428°–431.6° F.).

3. FLAVOPURPURIN, $C_6H_3OH.C_2O_2.C_6H_2(OH_2)$, formed from anthraflavic acid by fusion with potash, crystallizes from alcohol in golden-yellow needles, melts above 330° C. (626° F.), sublimes in long needles, like alizarin. Dissolves easily in alcohol, with purple color in alkalies.

4. OXYCHRYSAZIN, $C_6H_3(OH).C_2O_2.C_6H_2(OH)_2$, produced by the action of melting potash on chrysazin, is precipitated from its bluish-violet solution in alkalies by acids in brown flocks, and crystallizes from alcohol. Its triacetyl-compound forms light-yellow needles melting at 192°–193° C. (377.6°–379.4° F.).

5. **Pseudopurpurin.**—This is a constituent of crude purpurin, hitherto regarded as a dioxyalizarin or tetraoxyanthraquinone, $C_{14}H_8O_6$, but lately shown by Rosenstiehl (*Comptes rendus*, lxxxiv. 561) to consist of purpurin-carbonic acid, $C_{15}H_8O_7 = C_{14}H_7O_5.CO_2H$, inasmuch as it is resolved by heat into CO_2 and purpurin. It is also readily converted into purpurin by alkalies, even at ordinary temperatures.

Tetraoxyanthraquinone, $C_{14}H_4(OH)_4O_2$, is known in two modifications :—

1. ANTHRACHRYSONE, $C_6H_2(OH)_2.C_2O_2.C_6H_2(OH)_2$, is formed, like anthraflavone, from dioxybenzoic acid, $C_6H_3(OH)_2.CO.OH$, by dry distillation, or by heating to 140° C. (284° F.) with strong sulphuric acid. It is insoluble in water, crystallizes from glacial acetic acid or from alcohol in yellowish-red needles, melting at 320° C. (608° F.); yields anthracene when heated with zinc-dust.

2. RUFIOPIN, $C_{14}H_4(OH)_4O_2$, obtained by heating opianic acid (p. 951) with sulphuric acid, forms yellowish-red needles or crusts, dissolving in alkalies with violet-red color ; yields anthracene by reduction with zinc-dust.

A third modification is perhaps *Pseudopurpurin*, which is contained in madder-root. It is very much like purpurin, is converted into that substance by boiling with water or alcohol, and is said to be converted by reducing agents into purpuroxanthin.

Rufigallic acid, $C_{14}H_8O_8$, obtained by heating gallic or digallic acid with strong sulphuric acid, is a hexoxyanthraquinone, $C_6H(OH)_3$. $C_2O_2.C_6H(OH)_3$. It forms small, shining, brown-red crystals, containing $2H_2O$, which it gives off at 120° C. (248° F.), and sublimes at a higher temperature in cinnabar-red prisms; dissolves sparingly in hot water, alcohol, and ether, with brown color in alkalies ; precipitated with indigo-blue color by baryta-water. Yields anthracene by reduction with zinc-dust.

· **Methyl-anthracene,** $C_{15}H_{12} = C_{14}H_9.CH_3$, is ·formed by ·passing the vapor of ditolyl-methane or ditolyl-ethane (p. 964) through a red-hot tube, and by heating chrysophanic acid, emodin or eloïn, with zinc-dust. Color· less shining laminæ, melting at 200° C. (392° F.); slightly soluble in alco· hol, ether, and glacial acetic acid ; easily in chloroform, CS_2, and benzene. Forms with *picric acid* a compound which crystallizes in long dark-red needles, and is decomposed by water and alcohol.

· *Methylanthraquinone,* $C_{15}H_{10}O_2 = C_6H_4.C_2O_2.C_6H_3(CH_3)$, obtained by oxidizing methylanthracene in alcoholic solution with nitric acid, forms small yellow needles, melting at 162°–163° C. (323.6°–325.4° F.).

· *Dioxymethylanthraquinone,* $C_{15}H_{10}O_4 = C_{14}H_5.CH_3(OH)_2.O_2$, is known in two modifications :

1. M e t h y l a l i z a r i n is formed by the action of melting potash on the potassium salt of methylanthraquinone-sulphonic acid. It closely resem· bles alizarin ; sublimes above 200° in tufts of small red crystals ; melts at 250°–252° C. (482°–485.6° F.); dissolves in alkalies with blue-violet color.

2. C h r y s o p h a n i c a c i d (*Parietic acid, Rheic acid*) occurs in the lichen *Parmelia parietina*, in senna leaves, and in rhubarb root, and may be· ex· tracted therefrom by ether or alkalies. It crystallizes in golden-yellow needles or prisms, melts at 162° C. (323.6° F.), and sublimes partly with· out decomposition ; dissolves in alkalies with red color ; is reduced by zinc-dust to methyl-anthracene.

· *Trioxymethylanthraquinone* or *Emodin*, $C_{15}H_{10}O_5 =$ $C_{14}H_4(CH_3)(OH)_3.O_2$, occurs in the bark of *Rhamnus frangula*, and in small quantity, together with chrysophanic acid, in rhubarb root, and is sepa· rated by solution of sodium carbonate, which dissolves the emodin, leav· ing the chrysophanic acid. Long, brittle orange-red, monoclinic prisms, melting at 245°–250° C. (473°–482° F.); converted into methyl-anthra· cene by heating with zinc-dust.

Anthracene carbonic Acid, $C_{14}H_9.CO_2H$, is obtained by heating an· thracene with carbonyl chloride to 200° C. (392° F.) in sealed tubes, dis· solving the product in sodium carbonate, and precipitating with hydro· chloric acid. Long, silky, light yellow needles, melting with decomposi· tion at 206° C. (402.8° F.); slightly soluble in hot water, easily in alcohol. Oxidized by chromic acid to anthraquinone ; resolved by heat into CO_2 and anthracene. An isomeric acid, obtained by saponification of the cyanide produced by heating potassium anthracene-sulphonate with potassium fer· rocyanide, melts above 220° C. (428° F.), and sublimes without decom· position in orange-yellow needles.

Hydrocarbons of Higher Boiling Point.

Pyrene, $C_{16}H_{10}$, and **Chrysene,** $C_{18}H_{12}$, are contained in the portion of coal-tar boiling above 360° C. (680° F.) (b. p. of anthracene), and may be separated by heating the solid mass with carbon sulphide, which dissolves pyrene, together with other hydrocarbons, while the chrysene remains be· hind.

P y r e n e may be obtained pure by distilling off the carbon sulphide, dissolving the residue in alcohol, and adding an alcoholic solution of picric acid. Red crystals then separate, consisting of a compound of pyrene and picric acid, which, after purification by repeated crystallization from alco· hol, may be decomposed by ammonia. Pyrene crystallizes in plates, melts at 142° C. (287.6° F.), and distils at a higher temperature ; dissolves sparingly in cold, more readily in hot alcohol ; very easily in benzene, ether, and carbon sulphide. Its picric acid compound, $C_{16}H_{10}.C_6H_3(NO_2)_3O$, crystallizes in red needles. Heated with hydriodic acid to 200° C. (392° F.), it is converted into a h e x h y d r i d e, $C_{16}H_{10}.H_6$, melting at 127° C. (260.6° F.). Heated with chromic acid mixture, it yields p y r e n e q u i n o n e, $C_{16}H_8O_2$, which sublimes in red needles.

C h r y s e n e, $C_{18}H_{12}$, the portion of the high-boiling coal-tar hydrocarbons which is insoluble in carbon sulphide, may be purified by repeated crys. tallization from benzene. It is thus obtained in bright yellow glistening scales, which cannot be decolorized by recrystallization, but may be obtained quite colorless by heating with hydriodic acid and amorphous phosphorus to 240° C. (464° F.), or by boiling with alcohol and a small quantity of nitric acid. It dissolves very sparingly in alcohol, ether, and carbon sul. phide, more freely in benzene ; sublimes in dazzling white laminæ, which ex. hibit a fine blue fluorescence, and melt at 250° C. (482° F.). Its picric acid compound, $C_{18}H_{12}.C_6H_3(NO_2)_3O$, crystallizes in brown needles. By oxidation with chromic anhydride dissolved in glacial acetic acid, it yields c h r y - s e n e - q u i n o n e, $C_{18}H_{10}O_2$, which crystallizes in red needles, melts at 235° C. (455° F.), dissolves with fine blue color in strong sulphuric acid, and is precipitated from the solution by water in its original state. Chrysene- quinone unites with acid sodium sulphite, and is reduced by sulphurous acid to c h r y s o h y d r o q u i n o n e, $C_{18}H_{10}(OH)_2$. By distillation with soda-lime it yields the hydrocarbon $C_{16}H_{12}$ (m. p. 104°-105° C., 219.2°- 221° F.), just as phenanthrene-quinone yields diphenyl (p. 589).

Retene, $C_{18}H_{18}$, occurs in thin unctuous scales on fossil pine-stems, in beds of peat and lignite, in Denmark and other localities. It is produced in the dry distillation of very resinous fir and pine wood, passing over together with the heavy tar-oil, and separating in scales like paraffin ; also, together with other hydrocarbons, by passing acetylene through red-hot tubes. It crystallizes in colorless laminæ, slightly soluble in alcohol, easily in ether ; melts at 99° C. (210.2° F.); forms with picric acid the compound $C_{18}H_{18}.C_6H_3(NO_2)_3O$, which crystallizes in orange-yellow needles ; is converted by sulphuric acid into a disulphonic acid, and by chromic acid mixture into d i o x y r e t i s t e n e, $C_{16}H_{14}O_2$ (m. p. 194°-195° C., 381.2°- 383° F.), and phthalic acid. Dioxyretistene, heated with zinc-dust, is con- verted into r e t i s t e n e, $C_{16}H_{14}$, which crystallizes from alcohol in white laminæ, and forms a crystalline compound with picric acid.

Similar but less known hydrocarbons are *Fichtelite*, found on old pine- stems ; *Idrialin*, in quicksilver ore from Idria ; and *Scheererite*, in beds of lignite.

TERPENES and CAMPHORS.

The terpenes, $C_{10}H_{16}$, are volatile oils existing in plants, chiefly of the coniferous and aurantiaceous orders. They have not yet been formed by any artificial process, but their relation to the aromatic group is shown by their conversion into terephthalic acid by oxidation with nitric acid, and by the formation of cymene from turpentine oil, (p. 847).

Turpentine oil, the most important member of the group, is con- tained in the wood, bark, leaves, and other parts of pines, firs, and other coniferous trees, and is usually prepared by distilling crude turpentine, the oleo-resinous juice which exudes from incisions in the bark of the trees, either alone or with water. It was formerly supposed that all the volatile oils thus obtained, and having the composition $C_{10}H_{16}$, were identical in chemical and physical properties ; but recent investigations, especially those of Berthelot, have shown that the turpentine oils obtained from different sources exhibit considerable diversities in their physical, and more especially in their optical properties ; further, that most kinds of turpentine oil are mixtures of two or more isomeric or polymeric hydrocarbons, differing in physical and sometimes also in chemical properties. These modifications are often produced by the action of heat and of chemical reagents during the purification of the oil.

The several varieties of turpentine oil, when purified by repeated rectification with water, are colorless mobile liquids, having a peculiar aromatic but disagreeable odor. They are insoluble in water, slightly soluble in aqueous alcohol, miscible in all proportions with absolute alcohol, ether, and carbon disulphide. They dissolve iodine, sulphur, phosphorus, and many organic substances which are insoluble in water, such as fixed oils and resins, and are therefore used for making varnishes.

The principal varieties are, French turpentine oil, obtained from the French or Bordeaux turpentine of *Pinus maritima*, and English turpentine oil, from the turpentine collected, in Carolina and other Southern States of the American Union, from *Pinus australis* and *Pinus Tœda*.

French turpentine oil, when purified by neutralizing it with an alkaline carbonate, and then distilling it, first over the water-bath, and then in a vacuum (by which treatment all transformation of the product by heat or by reagents is avoided), consists mainly of a hydrocarbon, $C_{10}H_{16}$, called t e r e b e n t h e n e. It has a specific gravity of 0.864, boils at 161° C. (321.8° F.), and turns the plane of polarization of a ray of light to the left. English turpentine oil, treated in a similar manner, yields, as its chief constituent, a liquid caled a u s t r a l e n e, or a u s t r a t e r e b e n - t h e n e, having the same specific gravity and boiling point as terebenthene, but turning the plane of polarization to the right.

When pure turpentine oil (terebenthene or australene) is heated to 200°–250° C. (392°–482° F.), it undergoes a molecular transformation, and may then be separated by distillation into two oils, one called a u s t r a - p y r o l e n e, isomeric with the original oil, and boiling at 176°–178° C. (348.8°–352.4° F.) ; the other, called m e t a t e r e b e n t h e n e, polymeric with the original oil, having the formula $C_{20}H_{32}$, and boiling at a temperature above 360° C. (680° F.). Both are levorotatory, the latter exhibiting the greater amount of rotatory power.

Turpentine oil is converted, by repeated distillation with a small quantity of strong sulphuric acid, into two inactive modifications, t e r e b e n e, $C_{10}H_{16}$, boiling at 160° C. (320° F.), and c o l o p h e n e, $C_{20}H_{32}$, boiling at a very high temperature. A body of the same percentage composition, and closely resembling terebene in its physical properties, is obtained by the action of bromine and alcoholic potash on diamylene (p. 536) :

$$C_{10}H_{20} - H_4 = C_{10}H_{16} .$$

Turpentine oil exposed to the air absorbs oxygen, and acquires powerful oxidizing properties, formerly supposed to be due to the conversion of the oxygen into ozone ; but according to recent experiments by Kingzett,[*] it appears highly probable that the oxidizing compound is an organic peroxide, $C_{10}H_{14}O_4$, which when heated with water is resolved into hydrogen dioxide and camphoric acid : $C_{10}H_{14}O_4 + 2H_2O = H_2O_2 + C_{10}H_{16}O_4$.

Nitric acid and other powerful oxidizing agents convert turpentine oil

* Chem. Soc. Journ. 1874, 511 ; 1875, 214.

into a number of acid products of complex constitution. Strong nitric acid acts very violently on turpentine oil, sometimes setting it on fire.

Chlorine is absorbed by turpentine oil, with evolution of heat, sometimes sufficient to produce inflammation. When paper soaked in rectified turpentine oil is introduced into a vessel filled with chlorine, the turpentine takes fire, and a quantity of black smoke is produced, together with white fumes of hydrochloric acid. *Bromine* acts in a similar manner. *Iodine* is dissolved by turpentine oil, forming at first a green solution, which afterwards becomes hot, and gives off hydriodic acid. When a considerable quantity of iodine is suddenly brought in contact with turpentine oil, explosion frequently ensues. Turpentine oil distilled with *chloride of lime* and *water*, yields chloroform.

N i t r o s o t e r p e n e, $C_{10}H_{15}(NO)$. — When gaseous nitrosyl chloride, NOCl (obtained by passing the gases evolved from heated nitrohydrochloric acid into strong sulphuric acid, and heating the resulting solution with sodium chloride), is passed into oil of turpentine cooled by ice and salt, a white crystalline substance is precipitated, having the composition $C_{10}H_{16}.NOCl$, and this when heated with alcoholic soda gives up HCl, and is converted into nitrosoterpene. On acidulating with acetic acid, evaporating to dryness, washing the residue with water, and crystallizing it from alcohol, the nitrosoterpene is obtained in monoclinic crystals melting at 129°–130° C. (264.2°–266° F.), and subliming at a somewhat higher temperature. It burns easily, but is not explosive. Sodium-amalgam reduces it to a hydrocarbon.

Compounds of Turpentine oil. — Turpentine oil forms several compounds with *hydrochloric acid.* The gaseous acid converts it into the m o n o h y d r o c h l o r i d e, $C_{10}H_{16}.HCl$. On the other hand, when the oil is subjected for several weeks to the action of the strong aqueous acid, crystals of a d i h y d r o c h l o r i d e, $C_{10}H_{16}.2HCl$, are obtained. This latter compound is also formed by the action of hydrochloric acid gas on *lemon oil:* hence it is called c i t r e n e d i h y d r o c h l o r i d e. By the action of hydrochloric acid on terebene, the compound $C_{20}H_{32}.HCl$ is formed, called d i t e r e b e n e h y d r o c h l o r i d e. Lastly, when a current of hydrochloric acid gas is passed through a solution of turpentine oil in acetic acid, the compound $C_{20}H_{32}.3HCl$ is produced, called d i p y r o l e n e h y d r o c h l o-r i d e.

Hydrobromic and *hydriodic acids* form, with oil of turpentine, compounds analogous in composition to the hydrochlorides; the dihydriodide, however, has not been obtained from turpentine oil itself.

Whatever method may be adopted for preparing the hydrochlorides, hydrobromides, or the monohydriodide of turpentine oil, there are always two isomeric modifications obtained—one liquid, the other solid and crystalline. The crystallized monohydrochloride is sometimes, though inappropriately, designed as *artificial camphor,* and the dihydrochloride as *lemon camphor.*

H y d r a t e s. — Turpentine oil left in contact with water generally changes into a crystalline compound $C_{10}H_{16}.3H_2O$ called t e r p i n h y-d r a t e, which is more easily obtained by leaving a mixture of 8 pts. turpentine oil, 2 pts. nitric acid of specific gravity 1.25, and 1 pt. alcohol exposed to the air in a shallow vessel. It forms rhombic crystals, inodorous, easily soluble in water, alcohol, and ether. It melts below 100°; giving off water, and being converted into a white crystalline mass called t e r p i n, having the composition $C_{10}H_{16}.2H_2O$, or $C_{10}H_{18}(OH)_2$, which melts at 103° C. (217.4° F.), and sublimes in slender needles; with bromine at 50° C. (122° F.) it forms a bromide which, when distilled, yields cymene.

By heating the aqueous solution of terpin with a small quantity of hydrochloric acid, or by boiling the dihydrochloride, $C_{10}H_{16}.2HCl$, with water or alcoholic potash, terpinol, $2C_{10}H_{16}.H_2O$, is obtained as an oil, smelling like hyacinths, and distilling at 168° C. (334.4° F.).

Constitution and Combining Capacity of Turpentine Oil.—The hydrocarbon $C_{10}H_{16}$ acts as a quadrivalent radicle, capable of uniting with four monad atoms, and therefore with two molecules of the acids HCl, HBr, and HI, thereby producing the dihydrochlorides, etc., above mentioned; but, like other tetrad radicles, it can also take up only two monad atoms, producing the monohydrochloride, etc. The same tetrad radicle, by doubling itself, loses two units of equivalence—just as two atoms of carbon when united are satisfied by six, and not by eight atoms of hydrogen—and forms the hydrocarbon, $C_{20}H_{32}$, which is sexvalent, and can therefore form such compounds as $C_{20}H_{32}.3HCl$. Further, this same hexad radicle might form non-saturated compounds containing only four or two monad atoms; in reality, however, only those containing two monad atoms are known, such as $C_{20}H_{32}.HCl$.

If in the several hydrochlorides each atom of chlorine be replaced by hydroxyl, HO, we obtain the formulæ of the several hydrates of turpentine oil; the hydrate corresponding with the hydrochloride, $C_{20}H_{32}.3HCl$, has not, however, been prepared.

The formation of cymene from turpentine, by first converting the latter into the dibromide, $C_{10}H_{16}Br_2$, and then abstracting H_2Br_2 (p. 847), shows that turpentine oil is a hydride of cymene. Now cymene is methyl-propyl-benzene, $C_6H_4.(CH_3)(C_3H_7)$: hence the relation of the two hydrocarbons may be represented by the following formulæ:—

| Cymene. | Turpentine oil. |

The presence of two lateral chains in the molecule in the position 1 : 4, is in accordance with the formation of terephthalic acid by oxidation of turpentine oil. The other products of its oxidation are likewise in accordance with this view of the constitution of turpentine oil. Other arrangements of the radicles CH_3 and C_3H_7 in the molecule are, however, conceivable, and may perhaps give rise to some of the isomeric modifications of turpentine oil and its congeners. Other modifications may arise from the hydration of other metameric forms of the molecule $C_{10}H_{14}$, e. g., ethyl-dimethyl-benzene and tetra-methyl-benzene.

The formula of turpentine oil above given, in which the double union of one pair of carbon-atoms in cymene is loosened, represents the molecule as saturated. A similar loosening of a second pair would render the molecule bivalent, and therefore capable of taking up 1 mol. of HCl, HBr, Br_2, etc., and the loosening of the third pair would render it quadrivalent, and capable of uniting with 2HCl, 2HBr, etc.

Acids produced by Oxidation of Turpentine-oil.—Turpentine-oil boiled with dilute nitric acid yields formic, acetic, butyric, oxalic, terebic, toluic, and terephthalic acids.

Terebic acid, $C_7H_{10}O_4$, crystallizes in shining prisms, easily soluble in hot water and alcohol; melts at 175° C. (347° F.), and sublimes at a lower temperature. By distillation it is resolved into CO_2, and pyro-terebic acid, $C_6H_{10}O_2$ (p. 746). By boiling with carbonates it forms salts, $C_7H_9M'O_4$, which by the action of strong bases are converted into salts, $C_7H_{10}M'_2O_5$, called diaterebates, the acid of which, $C_7H_{12}O_5$, cannot be obtained in the free state, as, when separated from the salts, it is immediately resolved into water and terebic acid.

Terpin heated to 400° C. (752° F.) with soda-lime is converted into terebentilic acid, $C_8H_{10}O_2$, which crystallizes from hot water and alcohol in slender needles melting at 90° C. (194° F.), and distilling at 250° C. (482° F.). By oxidation with chromic acid mixture, terpin yields terpenylic acid, $C_8H_{13}O_4$, which is monobasic, crystallizes from water with 1 mol. H_2O, and melts in the dehydrated state at 90° C. (194° F.).

Volatile or **Essential Oils.**—The volatile oils obtained from plants by pressure, or by distillation with water, consist either of hydrocarbons, isomeric or polymeric with turpentine oil, or of mixtures of these hydrocarbons with compounds of carbon, hydrogen, and oxygen. Those obtained from aurantiaceous plants are terpenes, distinguished by their fragrant odor. Lemon oil, from the rind of the fruit of *Citrus Limonum*, consists mainly of citrene, $C_{10}H_{16}$, a dextrorotatory terpene, closely resembling terebenthene, having a specific gravity of 0.85 at 15° C. (59° F.), boiling at 167° or 168° C. (232.6° or 234° F.). With water it forms a crystallized hydrate resembling terpin; with hydrochloric acid, a dibydrochloride, $C_{10}H_{16}.2HCl$, existing in a solid and liquid modification, and a monohydrochloride, $C_{16}H_{16}.HCl$, apparently susceptible of similar modifications.

Similar oils are obtained from the rind of the sweet orange (*Citrus aurantium*), the bergamot (*C. bergamia*), the bigarade or bitter orange (*C. bigaradia*), the lime (*C. limetta*), the sweet lemon (*C. lumia*), and the citron (*C. medica*). Oil of neroli, obtained by distilling orange flowers with water, is probably also a terpene when pure.

The volatile oils of athamanta, beech, borneo (from *Dryabalanops Camphora*), caoutchouc, caraway, camomile, coriander, elemi, gomart, hop, juniper, imperatoria, laurel, parsley, pepper, savin, thyme, valerian, and others, also the neutral oils of wintergreen (*Gaultheria procumbens*), and cloves, are isomeric with oil of turpentine. The oils of copaiba and cubebs are probably polymeric with it, their molecules containing $C_{20}H_{32}$.

As examples of volatile oils containing an oxygenized constituent mixed with a terpene, may be mentioned valerian oil, which contains valeric acid, $C_5H_{10}O_2$; pelargonium oil, containing pelargonic acid, $C_9H_{18}O_2$; rue oil, containing euodic aldehyde, $C_{11}H_{22}O$; wintergreen oil, containing acid methyl salicylate, $C_8H_8O_3$. Some volatile oils consist essentially of aldehydes: thus, bitter almond oil consists of benzoic aldehyde, C_7H_6O; the oils of cinnamon and cassia contain cinnamic aldehyde, C_7H_8O; and those of anise, star-anise, fennel, and tarragon, contain anethol, $C_{10}H_{12}O$. Those volatile oils which exist ready formed in living plants do not appear to contain any elements besides carbon, hydrogen, and oxygen. Sulphur is found only in certain oils resulting from a kind of fermentation process, as in the volatile oils of mustard and garlic; nitrogen, when it occurs, must be regarded as an impurity resulting from admixed vegetable tissue.

A few volatile oils are found in the bodies of animals—oil of ants, for example.

Most volatile oils are colorless when pure ; they often, however, have a yellow color arising from impurity ; and a few, the oils of wormwood and camomile, for example, have a green or blue color, due to the presence of an oily compound of a very deep blue color, called *cerulein*. They have usually a powerful odor, and a pungent burning taste. When exposed to the air they frequently become altered by slow absorption of oxygen, and assume the character of resins. They mix in all proportions with fat oils, such as linseed, nut, colza, and whale oils, and dissolve freely both in ether and alcohol : from the latter solvent they are precipitated by the addition of water. Volatile oils communicate a greasy stain to paper, which disappears by warming ; by this character any adulteration with fixed oils can be at once detected. Many volatile oils, when exposed to cold, separate into a solid crystalline compound called a *stearoptene*, and a liquid oil, which, for distinction, is sometimes called an *elæoptene*.

Camphors.

These are oxygenated crystalline compounds, having a peculiar odor. They contain 10 atoms of carbon, and are nearly related to the terpenes, with which they are associated in plants, and by the oxidation of which they appear to be formed. The principal members of the group are common camphor, $C_{10}H_{16}O$, and borneo-camphor or borneol, $C_{10}H_{18}O$, which stand to one another in the relation of a ketone to a secondary alcohol. The constitution of these bodies is not completely established, but their intimate relation to cymene, $C_{10}H_{14}$ (p. 847), and carvacrol, $C_{10}H_{14}O$ (p. 899), **cymene** being produced by heating common camphor with zinc chloride or phosphorus pentasulphide, and carvacrol by heating the same substance with iodine, render it probable that they and their nearest products of oxidation may be represented by the following formulæ :—

$$
\begin{array}{cccc}
CH_3 & CH_3 & CH_3 & CH_3 \\
| & | & | & | \\
C & C & C & C \\
HC\diagdown CO & HC\diagdown CH.OH & HC\diagdown CO.OH & HC\diagdown CO.OH \\
| \quad | & | \quad | & | \quad | & | \quad | \\
H_2C \quad CH_2 & H_2C \quad CH_2 & H_2C \quad CH_3 & H_2C \quad CO.OH \\
\diagdown / & \diagdown / & \diagdown / & \diagdown / \\
CH & CH & CH & CH \\
| & | & | & | \\
C_3H_7 & C_3H_7 & C_3H_7 & C_3H_7 \\
\text{Common} & \text{Borneol.} & \text{Campholic} & \text{Camphoric} \\
\text{Camphor.} & & \text{acid.} & \text{acid.}
\end{array}
$$

Common Camphor, Laurel Camphor, or **Japan Camphor,** $C_{10}H_{16}O$, occurs in all parts of the camphor-tree of China and Japan (*Laurus Camphora*), often deposited in distinct crystals ; it is obtained by distilling the woody parts with water, and purified by sublimation. It has been produced artificially in small quantity by oxidation of oil of sage and oil of valerian (*i. e.*, of the terpenes contained in them) by nitric acid ; of turpentine oil by potassium permanganate ; of camphene (from borneol) by platinum-black or chromic acid mixture ; also by oxidation of cymene.

Common camphor is a colorless translucent mass, tough and difficult to powder, having a strong and peculiar taste and smell, and a density of 0.985. Small pieces thrown on water move about with a rotatory motion

42 *

It volatilizes at ordinary temperatures, melts at 175° C. (347° F.), and distils at 204° C. (399.2° F.). It dissolves sparingly in water, easily in alcohol, ether, acetic acid, and volatile oils ; and crystallizes from alcohol —also by sublimation—in shining strongly refractive crystals. The alco-holic solution is dextro-rotatory : $[\alpha] = +47.4°$.

By distillation over fused zinc chloride, or with phosphoric anhydride, camphor is resolved into water and cymene : $C_{10}H_{16}O = C_{10}H_{14} + H_2O$; considerable quantities of mesitylene, toluene, xylene, and other hydro-carbons being, however, formed at the same time. The formation of cymene takes place more definitely when camphor is distilled with phos-phorus pentasulphide, thiocymene, $C_{10}H_{13}SH$, being formed at the same time. Heated with iodine ($\frac{1}{2}$ part) it yields carvacrol. By heating with alcoholic potash it is resolved into borneol and camphic acid, just as benz-aldehyde is resolved into benzyl alcohol and benzoic acid :

$$2C_{10}H_{16}O + KOH = C_{10}H_{18}O + C_{10}H_{15}KO_2 .$$

By heating with nitric acid, camphor is converted into camphoric and camphoronic acids ; heated to 400° C. (752° F.) with soda-lime, it yields campholic acid, $C_{10}H_{18}O_2$.

Camphor, heated with *bromine* to 110°–120° C. (230°–248° F.), yields crystalline $C_{10}H_{15}BrO$ (m. p. 75° C., 167° F., b. p. 275° C., 527° F.), and $C_{10}H_{14}Br_2O$ (m. p. 214° C., 417.2° F., b. p. 285° C., 545° F.). Camphor dissolved in chloroform takes up bromine, forming $C_{10}H_{16}OBr_2$, easily converted into $C_{10}H_{15}BrO$.

Camphor treated with hypochlorous acid, yields *monochloro-camphor*, $C_{10}H_{15}ClO$, a crystalline mass melting at 95° C. (203° F.), and converted by heating with alcoholic potash into *oxycamphor*, $C_{10}H_{16}O_2$, which sublimes in needles melting at 137° C. (278.6° F.). Camphor heated with PCl_5 yields the compounds $C_{10}H_{16}Cl_2$ and $C_{10}H_{15}Cl$, easily converted, by ab-straction of hydrogen chloride, into cymene.

A solution of camphor in toluene heated with sodium, deposits a mixture of sodium-camphor and sodium-borneol :

$$2C_{10}H_{16}O + Na_2 = C_{10}H_{15}NaO + C_{10}H_{17}NaO ;$$

and these compounds treated with CH_3I and C_2H_5I, yield methyl- and ethyl-derivatives of camphor and borneol ; $C_{10}H_{15}(C_2H_5)O$ is a liquid boil-ing at 230° C. (446° F.).

When the sodium-compounds are treated at 100° with CO_2, the sodium salts of c a m p h o c a r b o n i c   a c i d, $C_{10}H_{15}O.CO_2H$, and b o r n e o c a r-b o n i c   a c i d, $C_{10}H_{17}O.CO_2H$, are produced ; and on treating the result-ing mass with water, borneol separates from the toluene which floats on the surface, and the aqueous solution mixed with hydrochloric acid yields a precipitate of camphocarbonic acid. This acid crystallizes from alcohol in small prisms, melting at 118° C. (244.4° F.), and is easily resolved into CO_2 and camphor.

Isomerides of Camphor.—By distilling the essential oil of feverfew (*Pyrè-thrum Parthenium*), and collecting apart the portion which passes over between 200° and 220° C. (392° and 428° F.), an oil is obtained, which, on cooling, deposits a crystalline substance resembling common camphor in every respect, except that its action on polarized light is exactly equal and opposite : $[\alpha] = -47.4°$. The essential oils of many labiate plants, as rose-mary, marjoram, lavender, and sage, often deposit a substance having the composition and all the properties of common camphor, excepting that it is inactive to polarized light.

Absinthol, $C_{10}H_{16}O$, from oil of wormwood (*Artemisia Absinthium*), is liquid, boils at 195° C. (383° F.), and is converted by P_2S_5 into cymene. Similar liquid camphors are obtained by oxidation of certain terpenes, as the oils of orange and nutmeg. A polymeric camphor, caryophyllin, $C_{20}H_{32}O_2$ (m. p. above 300° C., 572° F.), is contained in cloves.

Borneo-camphor or **Borneol,** $C_{10}H_{18}O = C_{10}H_{17}.OH$, occurs in *Dryabalanops Camphora*, a tree growing in Borneo and Sumatra : it is formed artificially by heating common camphor with alcoholic potash, or treating it with sodium (p. 994).

Borneol is very much like common camphor, has a camphorous and peppery odor, melts at 198° C. (388.4° F.), and boils at 212° C. (413.6° F.). Its alcoholic solution is dextrorotatory. By heating with nitric acid it is converted, first into common camphor, then into camphoric and camphoronic acids. By heating with P_2O_5, it is resolved into water and borneene, $C_{10}H_{16}$, apparently identical with the terpene contained in ordinary camphor-oil (from *Laurus Camphora*), and in valerian oil ; when left in contact with potash-lye, it is reconverted into borneol.

Borneol is an alcohol, yielding compound ethers when heated to about 200° C. (392° F.) with organic acids. The *stearic ether,* $C_{10}H_{17}.O.C_{18}H_{35}O$, is a colorless, viscid liquid, which gradually solidifies. By the action of PCl_5 at ordinary temperatures, or by heating in a sealed tube with HCl, borneol is converted into the *chloride,* $C_{10}H_{17}Cl$, a crystalline substance melting at 146° C. (294.8° F.), and very much like the solid modification of the hydrochloride of turpentine-oil, with which it is isomeric.

Isomeric with borneol are the liquid camphors contained in the oils of hops, Indian geranium, cajeput, coriander, and *Osmitopsis asteriscoïdes.* Homologous with borneol is *Patchouli-camphor,* $C_{15}H_{28}O$, contained in oil of patchouli ; it is a crystalline mass, melting at 54°–55° C. (129.2°–131° F.), boiling at 296° C. (564.8° F.).

Mint-camphor or **Menthol,** $C_{10}H_{20}O$, occurs, together with a terpene, in oil of peppermint (*Mentha piperita*), and separates in crystals on cooling the oil. It melts at 36° C. (96.8° F.), boils at 213° C. (415.4° F.) ; turns the plane of polarization to the left ; forms compound ethers with acids ; is converted by PCl_5 or HCl into the liquid chloride, $C_{10}H_{19}Cl$, and by distillation with P_2O_5 or $ZnCl_2$ into liquid menthene, $C_{10}H_{18}$, which boils at 163° C. (325.4° F.).

Acids produced by Oxidation of Camphor.

Campholic Acid, $C_{10}H_{18}O_2$, obtained by passing camphor-vapor over heated soda-lime, or by the action of potassium on a solution of camphor in petroleum, crystallizes from alcohol in prisms or scales, slightly soluble in water, melting at 95° C. (203° F.), and easily subliming ; converted by nitric acid into camphoric and camphoronic acids ; resolved by distillation with phosphoric anhydride into H_2O, CO_2, and campholene, C_9H_{12}, boiling at 135° C. (275° F.).

Camphoric Acid, $C_{10}H_{16}O_4 = C_8H_{14}(CO_2H)_2$, is obtained by prolonged boiling of common camphor or borneol with nitric acid :

$$C_{10}H_{16}O + O_3 = C_{10}H_{16}O_4 \text{ and } C_{10}H_{18}O + O_4 = C_{10}H_{16}O_4 + H_2O .$$

It is dextro- or levo-rotatory according to the variety of camphor used in its preparation; a mixture of the dextro- and levo-rotatory camphors in equal quantities yields an inactive camphoric acid.

Dextrocamphoric acid crystallizes from hot water in colorless laminæ, easily soluble in alcohol, melting at 175°–178° C. (347°–353.4° F.), and decomposing at a high temperature into water and camphoric anhydride, $C_{10}H_{14}O_3$, which sublimes at 130° C. (266° F.) in shining needles, melts at 217° C. (422.6° F.), and boils at 270° C. (518° F.).

The acid is bibasic. The calcium and barium salts are easily soluble in water, and crystallize well. The calcium salt is resolved by heat into carbonate and c a m p h o r - p h o r o n e : $C_{10}H_{14}O_4Ca = CO_3Ca + C_9H_{14}O$, isomeric with phorone from acetone (p. 708), a liquid which boils at 208° C. (406.4° F.), and does not yield cymene when heated with phosphoric anhydride.

Camphoric acid heated with water to 150°–200° C. (302°–392° F.) is converted into two inactive modifications, i s o c a m p h o r i c and p a r a - c a m p h o r i c a c i d s. The former crystallizes in slender needles melting at 113° C. (235.4° F.).

The acid, or its anhydride, heated with bromine to 130°–150° C. (266°–302° F.), yields *bromocamphoric anhydride*, $C_{10}H_{13}BrO_3$, which crystallizes in needles melting at 215° C. (419° F.), and, when boiled with water, is converted into the monobasic acid, $C_{10}H_{14}O_4$ (*oxycamphoric anhydride* or *camphanic acid*), melting at 201° C. (339.8° F.), and subliming at 110° C. (230° F.). This acid, or oxyanhydride, heated with water to 181° C. (357.8° F.), yields a hydrocarbon C_8H_{14}, boiling at 120° C. (248° F.).

C a m p h o r o n i c a c i d , $C_9H_{12}O_5$, and O x y c a m p h o r o n i c a c i d , $C_9H_{12}O_6$, are contained in the mother-liquor of the preparation of camphoric acid. The former crystallizes from water in slender needles with 1 mol. H_2O, which it gives off at 110° C. (230° F.); melts in the dehydrated state at 115° C. (239° F.); and distils without decomposition. The oxy-acid, $C_9H_{12}O_6$, crystallizes in long prisms melting at 164.5° C. (328.1° F.).

Resins and Balsams.

Common resin, or *colophony*, is perhaps the best example of the class. It is the resinous substance which remains when turpentine or pine-resin is heated till the water and volatile oil are expelled, and consists essentially of s y l v i c or a b i e t i c a c i d , $C_{20}H_{30}O_2$. On boiling the resin for a long time with alcohol of about 80 per cent., filtering, and adding a little water, the sylvic acid separates in the crystalline state. It crystallizes from alcohol in laminæ melting at 129° C. (264.2° F.). It is monobasic; its alkali-salts are soluble and crystallizable.

An acid called p i m a r i c a c i d , isomeric with sylvic acid, is obtained from the turpentine of the *Pinus maritima* of Bordeaux.

Lac is a very valuable resin, much harder than colophony, and easily soluble in alcohol : three varieties are known in commerce—viz., *stick-lac*, *seed-lac*, and *shellac*. It is used in varnishes, and in the manufacture of hats, and very largely in the preparation of sealing-wax, of which it forms the chief ingredient. Crude lac contains a red dye called *lac-dye*, which is partly soluble in water. Lac dissolves in considerable quantity in a hot solution of borax ; Indian ink, rubbed up with this liquid, forms an excellent *label-ink* for the laboratory, as it is unaffected by acid vapors, and, when once dry, becomes nearly insoluble in water.

Mastic, dammar-resin, and *sandarac* are resins largely used by the varnish maker. *Dragon's blood* is a resin of deep-red color. *Copal* is also a very valuable substance : it differs from the other resins in being but slowly dissolved by alcohol and essential oils. It is miscible, however, in the melted state with oils, and is thus made into varnish. *Amber* appears to be a fossil resin ; it is found accompanying brown-coal or lignite.

Most resins, when exposed to destructive distillation, yield oily pyro-products, usually consisting of hydrocarbons.

Caoutchouc, or India-rubber, the thickened milky juice of several species of *Ficus, Euphorbia,* and other trees growing in tropical countries, is essentially a mixture of several hydrocarbons isomeric or polymeric with turpentine oil. When pure it is nearly white, the dark color of commercial caoutchouc being due to the effects of smoke and other impurities. It is softened but not dissolved by boiling water : it is also insoluble in alcohol. In pure ether, rectified petroleum, and coal-tar oil, it dissolves, and is left unchanged on the evaporation of the solvent. Oil of turpentine also dissolves it, forming a viscid, adhesive mass, which dries very imperfectly. At a temperature a little above the boiling point of water, caoutchouc melts, but never afterwards returns to its former elastic state. Few chemical agents affect this substance : hence its great use in chemical investigations, for connecting apparatus, etc. By destructive distillation it yields a large quantity of a thin, volatile, oily liquid, of naphtha-like odor, called *caoutchoucin,* which dissolves caoutchouc with facility. This oil, according to Mr. Greville Williams, is composed of two polymeric hydrocarbons : caoutchin, $C_{10}H_{16}$, boiling at 171° C. (339.8° F.), and isoprene, C_5H_8, boiling at 37° C. (98.6° F.).

Caoutchouc combines with variable proportions of sulphur. The mixtures thus obtained are called *vulcanized India-rubber :* they are more permanently elastic than pure caoutchouc.

Vulcanite, or *Ebonite,* is caoutchouc mixed with half its weight of sulphur, and hardened by pressure and heating. It is very hard, takes a high polish, and is used for making combs, knife-handles, buttons, etc. It is also especially distinguished by the large quantity of electricity which it evolves when rubbed : hence it makes an excellent material for the plates of electrical machines.

Gutta-percha, the hardened milky juice of *Isonandra gutta,* a large tree growing in Malacca and many of the islands of the Eastern Archipelago, is similar in composition to caoutchouc, and resembles it in many of its properties, but is harder and less elastic. It is quite insoluble in, and impervious to water, and being also an excellent eléctric insulator, is extensively used as a casing for submarine telegraph wires. By dry distillation it yields isopene, caoutchin, and a heavy oil called *heveene,* probably polymeric with these bodies.

Balsams are natural mixtures of resins with volatile oils. They differ very greatly in consistence, some being quite fluid, others solid and brittle. By keeping, the softer kinds often become hard. Balsams may be conveniently divided into two classes—viz., those which, like *common* and *Venice* turpentine, *Canada balsam, Copaiba balsam,* etc., are merely natural varnishes, or solutions of resins in volatile oils, and those which contain benzoic or cinnamic acid in addition, as *Peru* and *Tolu balsams,* and the solid resinous *benzoin,* commonly called *gum-benzoin* (p. 922).

Glucosides.

This name is given to a class of bodies, very widely diffused in the vegetable kingdom, which are resolved by boiling with dilute acids or alkalies, or by the action of ferments, into glucoses (mostly dextrose), and some other substance. They are therefore analogous in constitution to the artificial glucosides which Berthelot obtained by heating glucose with organic acids (p. 652) ; none of them have, however, been formed artificially. The following are the most important :—

Æsculin, $C_{21}H_{24}O_{13}$, is a crystalline fluorescent substance obtained from the bark of the horse-chestnut and other trees of the genera *Æsculus* and *Pavia*. It has a bitter taste, is slightly soluble in water and alcohol, more soluble in the same liquids at the boiling heat, nearly insoluble in ether. It is colored red by chlorine. By boiling with hydrochloric or dilute sulphuric acid, it is resolved into glucose and a bitter crystalline substance called æ s c u l e t i n , $C_9H_6O_4$:

$$C_{21}H_{24}O_{13} \ + \ 3H_2O \ = \ 2C_6H_{12}O_6 \ + \ C_9H_6O_4 .$$

The aqueous solution of æsculin is highly fluorescent, the reflected light being of a sky-blue color. Nearly the same fluorescent tint is exhibited by an infusion of horse-chestnut bark. The color of the latter is, however, slightly modified by the presence of another substance, p a v i i n , which exhibits a blue-green fluorescence ; it may be separated from æsculin by its greater solubility in ether. Æsculin and paviin appear to exist together in the barks of all species of *Æsculus* and *Pavia*,—æsculin being more abundant in the former, and paviin in the latter.

Amygdalin, $C_{20}H_{27}NO_{11}$, is a crystalline body existing in bitter almonds, the leaves of cherry-laurel (*Cerasus Laurocerasus*), and many other plants which by distillation yield hydrocyanic acid and bitter-almond oil, C_7H_6O. To prepare it, the almonds, previously freed from fixed oil by pressure, are exhausted with boiling alcohol, and the concentrated solution is mixed with ether.

Amygdalin crystallizes from alcohol in white shining laminæ, has a bitter taste, dissolves easily in water and in hot alcohol ; from water it crystallizes in prisms containing $3H_2O$.

By boiling with dilute acids, and by contact with water and e m u l s i n or s y n a p t a s e , a ferment contained in bitter almonds, amygdalin is resolved into bitter almond oil, glucose, and hydrocyanic acid :

$$C_{20}H_{27}NO_{11} \ + \ 2H_2O \ = \ C_7H_6O \ + \ CNH \ + \ 2C_6H_{12}O_6.$$

When amygdalin is boiled with alkalies, its nitrogen is separated as ammonia, and a m y g d a l i c a c i d , $C_{20}H_{28}O_{13}$, is formed, which by boiling with dilute acids is resolved into mandelic acid and glucose.

Arbutin, $C_{12}H_{16}O_7$, from the leaves of the bear-berry (*Arbutus uva ursi*), crystallizes in slender needles, having a bitter taste, and dissolving easily in water, alcohol, and ether. By boiling with dilute acids it is resolved into glucose and hydroquinone :

$$C_{12}H_{16}O_7 \ + \ H_2O \ = \ C_6H_{12}O_6 \ + \ C_6H_4(OH)_2 .$$

It dissolves in strong nitric acid, forming $C_{12}H_{14}(NO_2)_2O_7$, which splits up into glucose and dinitrohydroquinone.

Chitin, $C_9H_{15}NO_6$, is the substance which forms the elytra and integuments of insects, and the carapaces of crustaceans. It is best prepared by boiling the wing-cases of cockchafers with water, alcohol, ether, acetic acid, and alkalies in succession, as long as anything is dissolved out by each. According to Stadeler, it is resolved by boiling with dilute acids into glucose and lactamide, $C_3H_7NO_2$:

$$C_9H_{15}NO_6 \ + \ 2H_2O \ = \ C_6H_{12}O_6 \ + \ C_3H_7NO_2 .$$

Coniferin, $C_{16}H_{22}O_8$, occurs in the cambial juice of coniferous plants, and separates therefrom on concentration in stellate groups of pointed needles, having a satiny lustre, and containing $2H_2O$, which they lose by efflorescence. By boiling with dilute acids it is resolved into dextrose and a resin; in contact with emulsin, into dextrose and c o n i f e r y l a l c o h o l. By oxidation with chromic acid mixture it yields v a n i l l i n (p. 899).

Convolvulin, $C_{31}H_{50}O_{16}$, is obtained from jalap root (*Convolvulus schiedanus*) by extraction with alcohol. It is a gummy mass having a strong purgative action; dissolves in alkalies as convolvulic acid, $C_{31}H_{52}O_{17}$; resolved by acids and by emulsin into dextrose and convolvulinol, $C_{13}H_{24}O_3$, which is converted by alkalies into convolvulinolic acid, $C_{13}H_{26}O_4$. This latter is a monobasic acid, which is oxidized by nitric acid to i p o m æ i c a c i d, $C_{10}H_{18}O_4 = C_8H_{16}(CO_2H)_2$, isomeric with sebacic acid (p. 780).

Jalappin, $C_{34}H_{56}O_{16}$, from the root of *Convolvulus orizabensis*, closely resembles convolvulin, and yields analogous products of decomposition.

Glycyrrhizin, $C_{24}H_{36}O_9$; *Liquorice-sugar.*—The root of the common liquorice yields a large quantity of a peculiar sweet substance, which is soluble in water, but does not crystallize: it cannot be made to ferment. Glycyrrhizin forms difficultly soluble compounds with acids; it is precipitated from its solution by lead, calcium, and barium salts, the precipitate consisting of glycyrrhizin in combination with the base. When boiled with dilute acids, it splits into a resinous body called g l y c e r r e t i n, $C_{18}H_{26}O_4$, and glucose:

$$C_{24}H_{36}O_9 \ + \ H_2O \ = \ C_{18}H_{26}O_4 \ + \ C_6H_{12}O_6 \ .$$

Myronic Acid, $C_{10}H_{19}NS_2O_{10}$, an acid existing as a potassium salt in the seed of black mustard, is resolved by the action of *myrosin*, an albuminous ferment likewise contained in the seeds, into volatile oil of mustard (allyl sulphocyanate), glucose, and sulphuric acid:

$$C_{10}H_{18}KNS_2O_{10} \ = \ C_3H_5.CNS \ + \ C_6H_{12}O_6 \ + \ SO_4HK \ .$$

Phlorizin, $C_{21}H_{24}O_{10}.2H_2O$, is a substance bearing a great likeness to salicin, found in the root-bark of the apple and cherry tree, and extracted by boiling alcohol. It forms fine, colorless, silky needles, soluble in 1000 parts of cold water, but freely dissolved by that liquid when hot; it also dissolves easily in alcohol. Dilute acids convert phlorizin into glucose and a crystallizable sweet substance called p h l o r e t i n:

$$C_{21}H_{24}O_{10} \ + \ H_2O \ = \ C_6H_{12}O_6 \ + \ C_{15}H_{14}O_5 \ .$$

Phlorizin, fused with potash, yields p h l o r e t i c a c i d, $C_9H_{10}O_3$, a beautifully crystalline acid, homologous with salicylic and anisic acids, together with phloroglucin, $C_6H_6O_3$.

Quercitrin is a crystallizable yellow coloring matter occurring in quercitron bark, the bark of *Quercus infectoria*, whence it is extracted by boiling with water. Its composition has been variously stated; indeed, it is by no means certain that the so-called quercitrins examined by different chemists were really identical substances. According to Hlasiwetz and Pfaundler, it contains $C_{33}H_{30}O_{17}$, and is resolved by boiling with dilute acids into another yellow crystalline body called q u e r c e t i n, and isodulcite (p. 647):

$$\underset{\text{Quercitrin.}}{C_{33}H_{30}O_{17}} \ + \ H_2O \ = \ \underset{\text{Quercetin.}}{C_{27}H_{18}O_{12}} \ + \ \underset{\text{Isodulcite.}}{C_6H_{14}O_6} \ .$$

Salicin, $C_{13}H_{18}O_7$, is a crystallizable bitter substance contained in the leaves and young bark of the poplar, willow, and several other trees. It may be prepared by exhausting the bark with boiling water, concentrating the solution to a small bulk, digesting the liquid with powdered lead oxide, and then, after freeing the solution from lead by a stream of sulphuretted hydrogen gas, evaporating till the salicin crystallizes out on cooling. It is purified by treatment with animal charcoal and recrystallization.

Salicin forms small, white, silky needles, having an intensely bitter taste, but no alkaline reaction. It melts and decomposes by heat, burning with a bright flame, and leaving a residue of charcoal. It is soluble in 5.6 parts of cold water, and in a much smaller quantity when boiling hot. Oil of vitriol colors it deep-red. When distilled with a mixture of potassium bichromate and sulphuric acid, it yields salicylal and other products.

Salicin, under the influence of emulsin or synaptase of sweet almonds, is resolved into glucose and s a l i g e n i n, $C_7H_8O_2$ (p. 912).

$$C_{13}H_{18}O_7 \ + \ H_2O \ = \ C_6H_{12}O_6 \ + \ C_7H_8O_2.$$

Salicin yields, with *chlorine*, substitution-products which are decomposed by synaptase in the same manner as salicin itself, yielding chlorosaligenin, $C_7H_7ClO_2$, and dichlorosaligenin, $C_7H_6Cl_2O_2$. Dilute *nitric acid* converts salicin into helicin. With strong nitric acid, at a high temperature, nitrosalicylic acid, $C_7H_5(NO_2)O_3$, is produced.

Populin, $C_{20}H_{22}O_8$, is a substance resembling salicin in appearance and solubility, but having a sweet pungent taste. It is found accompanying salicin in the bark and leaves of the aspen. It has the composition of benzoyl-salicin, $C_{13}H_{17}(C_7H_5O)O_7$, and when heated with dilute acids, is resolved into benzoic acid and the products of decomposition of salicin, namely, saliretin and glucose :

$$C_{13}H_{17}(C_7H_5O)O_7 \ + \ H_2O \ = \ C_7H_6O_2 \ + \ C_7H_6O \ + \ C_6H_{12}O_6.$$

With potassium dichromate and sulphuric acid, populin yields a considerable quantity of salicylal.

Helicin, $C_{13}H_{16}O_7$, is a white, crystalline, slightly bitter substance, produced by the action of very dilute nitric acid upon salicin :

$$C_{13}H_{18}O_7 \ + \ O \ = \ H_2O \ + \ C_{13}H_{16}O_7.$$

It is slightly soluble in cold, freely soluble in boiling water, and is resolved by the action of synaptase, or of acids or alkalies at the boiling heat, into glucose and salicylal :

$$C_{13}H_{16}O_7 \ + \ H_2O \ = \ C_6H_{12}O_6 \ + \ C_7H_6O_2.$$

Benzohelicin, $C_{20}H_{20}O_8$, or $C_{13}H_{15}(C_7H_5O)O_7$, produced by the action of dilute nitric acid on benze-salicin, is resolved in like manner into benzoic acid, salicylal, and glucose :

$$C_{20}H_{20}O_8 \ + \ 2H_2O \ = \ C_7H_6O_2 \ + \ C_7H_6O_2 \ + \ C_6H_{12}O_6.$$

Bitter Principles of Plants.

These are neutral bodies of somewhat indefinite chemical character, which cannot at present be included in any of the preceding groups.

Aloïn is a constituent of aloes, the inspissated juice of various species of aloe, and is extracted by treating the aloes with water. It is easily soluble in warm water and alcohol, and crystallizes in slender needles. It is very bitter and strongly purgative.

Aloïn from Barbadoes aloes has the composition $C_{17}H_{18}O_7$; that from Natal aloes is $C_{34}H_{38}O_{15} = 2C_{17}H_{18}O_7 + H_2O$. The former (barbaloïn) heated with nitric acid yields aloetic acid, $C_{14}H_4(NO_2)_4O_2$, together with oxalic, picric, and chrysammic acids; the latter (nataloïn) yields picric and oxalic acids, but no chrysammic acid.

Athamantin, $C_{24}H_{30}O_7$, obtained from the roots of *Athamanta Oreoselinum* by extraction with ether, crystallizes from alcohol and ether in slender needles melting at 79° C.(174.2° F.). By boiling with hydrochloric acid it is resolved into valeric acid and oreoselone, $C_{14}H_{10}O_3$:

$$C_{24}H_{30}O_7 = 2C_5H_{10}O_2 + C_{14}H_{10}O_3 .$$

By further boiling with dilute hydrochloric acid, oreoselone is converted into oreoselin, $C_{14}H_{12}O_4$.

Cantharidin, $C_5H_6O_2$, is the vesiccating principle of Spanish flies and some other insects, and may be extracted with ether. It crystallizes in four-sided prisms or laminæ, dissolves in hot alcohol and ether, melts at 250° C. (482° F.), and sublimes at a lower temperature. Heated with alkalies it dissolves, forming salts of cantharidic acid, *e. g.*, $C_5H_7KO_3$, from which acids reprecipitate cantharidin.

Carotin, $C_{18}H_{24}O$, a substance deposited in small crystals in the cells of the red carrot, crystallizes from alcohol in red-brown cubes melting at 168° C. (334.4° F.).

Peucedanin, $C_{24}H_{24}O_6$, from the roots of *Peucedanum officinale*, crystallizes from alcohol in shining prisms melting at 75° C. (167° F.). Boiled with alcoholic potash it is resolved into angelic acid and oreoselin.

Picrotoxin, $C_{12}H_{14}O_6$, is extracted by alcohol from cocculus grains (the seeds of *Menispermum Cocculus*); crystallizes in slender needles; very bitter and poisonous.

Santonin, $C_{15}H_{19}O_3$, is the active principle of wormseed (from *Artemisia santonica*), from which it may be extracted by boiling with milk of lime and precipitation by hydrochloric acid. It crystallizes from hot alcohol in shining prisms melting at 170° C. (338° F.); dissolves in alkalies, forming salts of santoninic acid, $C_{15}H_{20}O_4$.

Coloring Matters.

The most important coloring matters of vegetable origin, viz., the indigo and madder dyes, have already been described.

Brazilin, $C_{22}H_{18}O_7$, the coloring matter of Brazil wood, crystallizes in small yellow prisms, which dissolve in alcohol with reddish-yellow, in alkalies with crimson color ; the solution is decolorized by sulphurous acid and zinc-dust.   Nitric acid converts brazilin into trinitroresorcin.

Carminic acid, $C_{17}H_{18}O_{10}$, occurs in the flowers of *Monarda didyma*, and a few other plants, and more abundantly in cochineal, an insect (*Coccus Cacti*) living on various species of cactus.   To obtain the carminic acid, the aqueous decoction of the insects is precipitated by lead acetate, and the pure lead carminate washed and decomposed by hydrogen sulphide ; the coloring matter thus separated is again submitted to the same treatment.   A solution of carminic acid is thus obtained, which is evaporated to dryness, redissolved in absolute alcohol, digested with crude lead carbonate, whereby a small quantity of phosphoric acid is separated, and lastly, mixed with ether, which separates a trace of a nitrogenous substance.   The residue now obtained on evaporation is pure carminic acid.

It is a purple-brown mass, yielding a fine red powder, soluble in water and alcohol in all proportions, slightly soluble in ether.   It dissolves without decomposition in concentrated sulphuric acid, but is readily attacked by chlorine, bromine, and iodine, which change its color to yellow.   It resists a temperature of 136° C. (276.8° F.), but is charred when heated more strongly.   Carminic acid is bibasic, and forms colored salts.   By boiling with dilute sulphuric acid it is resolved into a non-fermentable sugar and carmine-red, $C_{11}H_{12}O_7$.   Boiled with nitric acid it yields oxalic acid and nitrococcusic acid, $C_8H_5(NO_2)_3O_3.H_2O$, which crystallizes from water in large shining plates, and when heated with water to 180° is resolved into CO_2 and trinitro-cresotic acid, $C_6(NO_2)_3(CH_3)(OH).CO_2H$.   Cochineal is used for giving a red color to silk and wool.

Carthamin, $C_{14}H_{16}O_7$, is the red coloring matter of the petals of the safflower (*Carthamus tinctorius*).   It is prepared by exhausting the flowers with cold water to remove a yellow substance, and treating the residue with a dilute solution of sodium carbonate ; this dissolves the carthamin, which may be precipitated by acetic acid as a dark red powder having a green metallic lustre when dry.   It dissolves in alcohol and alkalies with a fine red color, which, however, is very fugitive.   By fusion with potash it yields paraoxybenzoic and oxalic acids.   Carthamin is used for dyeing silk, and as a cosmetic.

Chlorophyll, or Leaf-green, is contained, together with wax and other substances, in the chlorophyll granules which occur in all the green parts of plants, and may be obtained by exhausting these organs with ether, and treating the residue with alcohol, in which the chlorophyll is easily soluble.   It dissolves also in strong hydrochloric acid.   Its composition is not known, but iron appears to be an essential constituent.

Curcumin, $C_{10}H_{16}O_3$, is the coloring matter of turmeric.   It dissolves easily in alcohol and ether, sparingly in carbon sulphide, and in cold benzene, and is best extracted from the root by boiling benzene.   It forms orange-yellow crystals, and exhibits in solution a green fluorescence.   From its brown-red solution in alkalies, it is precipitated by acids as a yellow powder.   The yellow tint of paper stained with turmeric is turned brown by alkalies and restored by acids.

Euxanthic acid, $C_{19}H_{16}O_{10}$, occurs as a magnesium salt in *Purree* or *Indian yellow*, a yellow coloring matter of unknown origin imported from India and China.   The euxanthic acid is obtained by digesting the mass with dilute hydrochloric acid, and exhausting with alcohol.   It dissolves easily in alcohol and ether, and crystallizes in shining yellow prisms.   Euxanthic acid is resolved by strong sulphuric acid into a saccharine substance and euxanthone, $C_{13}H_8O_4$, which sublimes in yellow needles.   Euxanthone fused with potash yields first euxanthonic acid, $C_{16}H_{10}O_5$, then hydroquinone. Strong nitric acid converts it into trinitroresorcin, and, on the other hand, resorcin heated with oxalic and sulphuric acid, appears to yield euxanthone.

Hematoxylin, $C_{16}H_{14}O_6$, the coloring matter of logwood, crystallizes with 3 mol. H_2O in pale-yellow prisms ; has a sweetish taste ; dissolves easily in water, alcohol, and ether, forming dextro-rotatory solutions. With *acetyl chloride*, it forms $C_{16}H_8(C_2H_3O)_6O_6$.

By exposure to the air in ammoniacal solution, it is converted into hematein-ammonia, $C_{16}H_{11}(NH_4)O_6$, which, on spontaneous evaporation, is deposited in dark violet crystals. Acetic acid added to the solution throws down hematein, $C_{16}H_{13}O_6$, as a red-brown precipitate, having a metallic lustre when dry, and dissolving in alkalies with a fine violet-blue color. Sulphurous acid and other reducing agents reconvert hematein into hematoxylin. The latter, fused with potash, yields pyrogallol.

Alkaloids.

This term is sometimes used as a general name for organic bases, but it is more especially applied to those which occur ready formed in the bodies of plants and animals, and those which are produced by the destructive distillation of complex organic bodies. All these bases, like the amines already described, are derivatives of ammonia, but their molecular structure is for the most part unknown. Those which are free from oxygen are volatile ; those which contain that element are decomposed by distillation.

NON-OXIDIZED VOLATILE BASES.

Pyridine Bases, $C_nH_{2n-5}N.$—These bases, metameric with aniline and its homologues, are contained in coal-tar naphtha, and in the volatile oil called Dippel's oil (*Oleum animale Dippelii*), obtained by the distillation of bones and other animal matters. They are all liquid at ordinary temperatures, and react as tertiary monamines. Their formulæ and boiling points are as follows :—

		B. P.			B. P.
Pyridine,	C_5H_5N,	117○	Parvoline,	$C_9H_{13}N$,	188○
Picoline,	C_6H_7N,	133○	Coridine,	$C_{10}H_{15}N$,	211○
Lutidine,	C_7H_9N,	154○	Rubidine,	$C_{11}H_{17}N$,	230○
Collidine,	$C_8H_{11}N$,	179○	Viridine,	$C_{12}H_{19}N$,	251○

Pyridine is said to be formed artificially by heating amyl nitrate with phosphoric oxide : $C_5H_{11}NO_3 - 3H_2O = C_5H_5N$. When heated with sodium it is converted into dipyridine, $C_{10}H_{10}N_2$, a crystalline base which melts at 108○ C. (226.4○ F.), and sublimes at higher temperatures in needle-shaped crystals.

Picoline (metameric with aniline), first obtained by Anderson from coal-tar naphtha, is a mobile liquid, having a strong, persistent odor, and acrid, bitter taste : sp. gr. 0.995. It remains liquid at —18○ C. (0.4○ F.), and volatilizes quickly in the air. It is strongly alkaline to test-paper, mixes with water in all proportions, and forms crystallizable salts.

Dippel's oil likewise contains methylamine and several of its homologues.

Chinoline Bases, $C_nH_{2n-11}N.$—Three bases of this series, viz. :

C_9H_7N	$C_{10}H_9N$	$C_{11}H_{11}N$
Chinoline.	Lepidine.	Cryptidine.

are produced by distillation of quinine, cinchonine, and a few other natural
alkaloïds, with potassium hydroxide; and other bases isomeric with them,
viz., *leucoline*, C_9H_7N, *iridoline*, $C_{10}H_9N$, etc., are contained in coal-tar naphtha, and distil over after the pyridine bases (above 200° C., 392° F.).
They are oily liquids insoluble in water, easily soluble in alcohol and
ether. They are tertiary amines, yielding ammonium-bases when treated
with ethyl iodide and silver oxide. These salts are crystalline, and easily
soluble.

C h i n c l i n e, C_9H_7N, is a mobile, strongly refracting liquid, boiling at
238° C. (460.4° F.), and having a sp. gr. of 1.081 at 0°. Heated with
amyl iodide, it forms the compound $(C_9H_7)(C_5H_{11})NI$, which is converted
by heating with potash into c y a n i n e, $C_{28}H_{25}NI$, a fine blue dye-stuff,
which crystallizes in green metallically lustrous plates, and dissolves with
blue color in alcohol. A similar blue color is obtained with lepidine, and
a mixture of the two has been used for dyeing silk.

Leucoline, C_9H_7N, from coal-tar oil, boils at about 220° C. (428° F.),
and does not yield cyanine.

Lepidine, from cinchonine, boils at 266°–270° C. (510.8°–518° F.) ; the
isomeric base from coal-tar oil boils at 252°–257° C. (485.6°–494.6° F.).

Conine, $C_8H_{15}N$, is contained in hemlock (*Conium muculatum*), especially
in the seeds, and is obtained therefrom by distillation with potash-lye. It
is a colorless oily liquid, having a pungent, stupefying odor, and is very
poisonous. Sp. gr. 0.89. Boiling point 168° C. (334.4° F.). It dissolves
easily in alcohol and ether, sparingly in water, and forms crystalline deliquescent salts, which, like the base itself, turn brown in contact with the
air. With oxidizing agents conine yields butyric acid.

Conine is a secondary monamine. Treated with ethyl iodide it yields
successively two iodine-compounds—namely, $C_8H_{15}(C_2H_5)NI$ and
$C_8H_{14}(C_2H_5)_2NI$. The latter is converted by silver oxide into a soluble
base.

Paraconine, isomeric with conine, is formed artificially by heating normal
butyric aldehyde, C_4H_8O, with alcoholic ammonia—whereby *dibutyraldine*
is obtained, having the composition $C_8H_{17}NO[= 2C_4H_8O + NH_3 - H_2O]$,—
and subjecting this base to dry distillation : $C_8H_{17}NO - H_2O = C_8H_{15}N$. It
is a violent poison, acting in the same manner as the natural base. But
it is less soluble in water, more expansible by heat, and exhibits somewhat
different reactions with hydrochloric acid, silver nitrate, and gold chloride.
With ethyl iodide it forms the iodide of an ammonium-base, convertible
by silver oxide into a strongly alkaline, bitter syrupy liquid : hence it is
a tertiary monamine.

Closely allied to conine is *conhydrine*, $C_8H_{17}NO$, a crystalline base, extracted from hemlock flowers. When distilled with anhydrous phosphoric
acid, it splits into conine and one molecule of water.

Nicotine, $C_{10}H_{14}N_2$, exists in the seeds and leaves of various kinds of
tobacco, from which it may be obtained by extraction with dilute sulphuric
acid, and distillation of the concentrated extract with potash-lye. It is a
colorless oil, having a density of 1.048, boiling with partial decomposition
at 250° C. (482° F.), without decomposition in a stream of hydrogen at
150°–200° C. (302°–392° F.). It turns brown in the air, has a very stupefying odor, and is very poisonous. It is a monoacid base, forming very
soluble salts, which crystallize with difficulty.

A mixture of nicotine with methyl or ethyl iodide solidifies after a short time to a crystalline mass, containing $C_{10}H_{14}(CH_3)_2N_2I_2$, or $C_{10}H_{14}(C_2H_5)_2N_2I_2$, convertible by silver oxide into soluble bases.

By fuming nitric acid or chromic acid mixture nicotine is oxidized to nicotic acid, $C_{10}H_8N_2O_3$, which, when distilled with lime, yields pyridine, and when heated with bromine-water to 120° C. (248° F.) is resolved into bromoform, carbon dioxide, nitrogen, and pyridine.

Sparteine, $C_{15}H_{26}N_2$, occurs in the common broom (*Spartium scoparium*), and is obtained therefrom by extraction with dilute sulphuric acid and distillation with potash. It is a colorless liquid, boiling at 228° C. (442.4° F.) ; dissolves sparingly in water, has a bitter taste, and acts as a narcotic. It is strongly alkaline, and has the constitution of a biacid tertiary amine.

OXYGENIZED BASES.

1. *Bases related to the Ureides.—Derivatives of Guanidine.*

The following compounds are derived from guanidine $CH_5N_3 = HNC{=}(NH_2)_2$ (p. 677), by the substitution of acid radicles for an atom of hydrogen.

Glycocyamine or **Guanidacetylic Acid,** $C_3H_7N_3O_2$, is formed on mixing the aqueous solutions of glycocine and cyanamide (p. 576) :

$$HN{=}C{=}NH + H_2C(NH_2)(CO_2H) = HN{=}C(NH_2)(NH.CH_2.CO_2H) ;$$
Cyanamide.   Glycocine.   Glycocyamine.

and separates in granular crystals, soluble in 120 parts of cold water, easily soluble in hot water, insoluble in alcohol and ether. It unites with bases, forming crystalline compounds. Boiled with water and lead-oxide, or with dilute sulphuric acid, it is resolved into guanidine, oxalic acid and carbonic acid.

Glycocyamidine or **Glycolylguanidine,** $C_3H_5N_3O$, related to glycocyamine in the same manner as hydantoïn to hydantoic acid (p. 830), is formed, as a hydrochloride, by heating the hydrochloride of glycocyamine to 160° C. (320° F.) :

$$HN{=}C\!\!<^{NH_2}_{NH}{-}CH_2{-}CO_2H = HN{=}C\!\!<^{NH-CO}_{NH-CH_2} + H_2O$$
Glycocyamine.   Glycocyamidine.

The free base crystallizes in deliquescent laminæ, having an alkaline reaction. Its hydrochloride gives a precipitate with platinic chloride.

Creatine, $C_4H_9N_3O_2$, *Methylglycocyamine*, occurs in the animal organism, especially in flesh juice, and is formed artificially, like glycocyamine, by the union of cyanamide with methyl-glycocine (sarcosine, p. 807) :

$$C\!\!<^{NH}_{NH} + {}^{NH.CH_3}_{CH_2.CO_2H} = HN{=}C\!\!<^{NH_2}_{N(CH_3)-CH_2-CO_2H}.$$

It may be prepared by macerating finely-chopped meat in cold water, boiling the extract to coagulate albumin, precipitating the phosphoric acid from the filtrate by baryta-water, and evaporating to the crystallizing point.

Creatine crystallizes in shining prisms containing 1 mol. H_2O, which they give off at $100°$. It is neutral, slightly bitter, moderately soluble in boiling water, very sparingly in alcohol; forms crystalline salts containing one equivalent of acid.

Creatine heated with acids is converted, by abstraction of H_2O, into creatinine (*infra*). By boiling with water it is resolved into urea and sarcosine:

$$HN=C{<}_{N(CH_3).CH_2.CO_2H}^{NH_2} + H_2O = CO(NH_2)_2 + \begin{array}{l}NH(CH_3) \\ | \\ CH_2.CO_2H\end{array},$$

methyl-hydantoïn being also formed, and ammonia given off. Boiled with mercuric oxide, it yields methylguanidine and oxalic acid.

Creatinine, $C_4H_7N_3O$, *Methylglycocyamidine*, is an almost constant constituent of urine (0.25 per cent.), and is formed from creatine by evaporating the aqueous solution of the latter, especially in presence of acids. It crystallizes in rhombic prisms, much more soluble in water and alcohol than creatine; expels ammonia from its salts; and forms well-crystallized salts with acids. It unites also with certain salts, forming, for example, the compound $(C_4H_7N_3O)_2ZnCl_2$, which is precipitated by zinc chloride from solutions of creatinine as a sparingly soluble crystalline powder.

Creatinine is reconverted into creatine by the action of bases, and is resolved by boiling with baryta into ammonia and methyl-hydantoïn—

$$HN=C{<}_{N(CH_3)-CH_2}^{NH---CO} + H_2O = NH_3 + CO{<}_{N(CH_3)-CH_2}^{NH---CO}$$

By boiling with mercuric oxide it is resolved, like creatine, into methylguanidine and oxalic acid.

Nearly related to the ureides are also the bases guanine, sarcine, xanthine, and carnine, which, like urea, occur in the animal organism as products of the oxidation of the tissues—and the two vegetable bases, theobromine and caffeine. The constitution of these bases is not yet established, but their relation to the ureides is shown by their products of decomposition.

Guanine, $C_5H_5N_5O$, was first obtained from guano; it has also been proved to exist in the pancreatic juice of mammalia, and in the excrement of the spider. To prepare it, guano is boiled with water and calcium hydrate until a portion of the liquid, when filtered, appears but slightly colored: the whole is then filtered, and the filtrate saturated with acetic acid, whereby the guanine is precipitated, mixed with uric acid. It is purified by solution in hydrochloric acid and precipitation by ammonia.

Guanine is a colorless, crystalline powder, insoluble in water, alcohol, ether, and ammonia, soluble in acids and solution of potash. It unites with acids forming crystallizable salts, *e. g.*, $C_5H_5N_5O.HCl + Aq.$; $2C_5H_5N_5O.H_2SO_4 + 2 Aq.$; $3C_5H_5N_5O.2C_2H_2O_4$; also with metallic bases and salts, *e. g.*, $C_5H_5N_5O.NaHO + 2 Aq.$; $C_5H_5N_5O.AgNO_3$. By oxidation with hydrochloric acid and potassium chlorate, it is converted into a mixture of guanidine and parabanic acid (827).

Xanthine, $C_5H_4N_4O_2$, is found in small quantity in many animal secretions, as in urine and blood, in the liver, and in certain urinary calculi;

it is formed artificially by the action of nitrous acid upon guanine, and of sodium amalgam on uric acid. It is a white amorphous mass, somewhat soluble in boiling water, and uniting both with acids and with bases. It dissolves easily in boiling aqueous ammonia, and silver-nitrate added to the solution throws down the compound $C_5H_2Ag_2N_4O_2 + H_2O$. This compound treated with methyl iodide yields a body isomeric with theobromine:

$$C_5H_2Ag_2N_4O_2 \quad + \quad 2CH_3I \quad = \quad 2AgI \quad + \quad C_7H_8N_4O_2.$$

Sarcine or **Hypoxanthine**, $C_5H_4N_4O$, almost always accompanies xanthine in the animal organism; and is distinguished from xanthine especially by the sparing solubility of its hydrochloride. It forms needles slightly soluble in water, more soluble in acids and alkalies. From its ammoniacal solution, silver-nitrate throws down the compound $C_5H_2Ag_2N_4O + H_2O$.

Carnine, $C_7H_8N_4O_3$, occurs in extract of meat. It is pulverulent, moderately soluble in hot water, and forms a crystalline hydrochloride.

Theobromine, $C_7H_8N_4O_2$, which differs in composition from carnine by only 1 atom of oxygen, occurs in cacao beans, the seeds of *Theobroma Cacao*, from which cocoa and chocolate are prepared. To extract it, the beans are boiled with water; the solution is precipitated with lead acetate, to remove extraneous matters; the filtrate is freed from lead by hydrogen sulphide, and evaporated to dryness; and the residue is treated with alcohol, which extracts the theobromine.

Theobromine is a white crystalline powder, having a bitter taste, slightly soluble in water and in alcohol, moderately soluble in aqueous ammonia. It has a neutral reaction, but unites with acids, forming crystalline salts, which are decomposed by water. From the ammoniacal solution silver nitrate throws down the compound $C_7H_7AgN_4O_2$, which, by heating to 100° with methyl iodide, is converted into methyl-theobromine, $C_7H_7(CH_3)N_4O_2$, or theine.

Theine or **Caffeine**, $C_8H_{10}N_4O_2$, *Methyl-theobromine*, occurs in the leaves and seeds of the coffee tree, in tea-leaves, in Paraguay tea (from *Ilex paraguayensis*), and in guarana, the dried pulp of the fruit of *Paullinia sorbilis*. Theine is extracted from these substances by the process above described for the preparation of theobromine: it crystallizes on cooling, and may be purified by means of animal charcoal.

Theine forms tufts of silky needles containing one mol. H_2O, slightly soluble in cold water and in alcohol, and giving off their water of crystallization at 100°. It melts at 225° C. (437° F.), and sublimes without decomposition at a higher temperature. It is a weak base, most of its salts being decomposed by water. The aurochloride and platinochloride, however, are more stable, and form orange-yellow crystals.

By the action of chlorine or nitric acid, theine is converted, with evolution of methylamine and cyanogen chloride, into amalic acid, $C_{12}H_{12}N_4O_7$, which has the composition of tetramethyl-alloxantin, $C_8(CH_3)_4N_4O_7$. It forms sparingly soluble crystals, which are colored violet-blue by alkalies. By the further action of chlorine-water, theine yields cholestrophane or dimethylparabanic acid, $C_3(CH_3)_2N_2O_3$ (p. 828).

Theine boiled with baryta-water, is resolved into CO_2, and theidine or caffeidine, $C_7H_{12}N_4O$, an easily soluble strongly basic compound, which is decomposed by prolonged boiling with water into sarcosine and other products.

2. Opium Bases.

Opium, the inspissated juice of the half-ripe capsules of the poppy (*Papaver somniferum*), is a very complex substance, containing a large number of bases combined with sulphuric and meconic acid. The best known of these bases are:

Morphine,*	$C_{17}H_{19}NO_3$		Papaverine,	$C_{21}H_{21}NO_4$
Codeine,	$C_{18}H_{21}NO_3$		Narcotine,	$C_{22}H_{23}NO_7$
Thebaine,	$C_{19}H_{21}NO_3$		Narceine,	$C_{23}H_{29}NO_4$

Of these, morphine and narcotine are the most abundant, the rest occurring in small quantity, and only in particular varieties of opium.

The bases are obtained by digesting opium with warm water, precipitating the meconic acid with calcium chloride, and leaving the concentrated filtrate to crystallize. The hydrochlorides of morphine and codeine then crystallize out first, and may be separated by treating their aqueous solution with ammonia, whereby the morphine is alone precipitated, the codeine remaining dissolved.

The mother-liquor of the morphine and codeine hydrochlorides is mixed with ammonia, which throws down narcotine, together with small quantities of papaverine, and thebaine, and a resin, while narceine remains in solution.

M o r p h i n e, $C_{17}H_{19}NO_3 + H_2O$, *Morphia* or *Morphium*, is precipitated from its salts as a white powder, and crystallizes from alcohol in small but very brilliant prisms. It requires at least 500 parts of water for solution, tastes slightly bitter, and has an alkaline reaction. These effects are much more evident in the alcoholic solution. It dissolves in about 30 parts of boiling alcohol, and with great facility in dilute acids; it is also dissolved by excess of caustic potash or soda, but scarcely by excess of ammonia. When heated in the air, morphine melts, burns like a resin, and leaves a small quantity of charcoal, which easily burns away. Morphine in small doses is narcotic; in larger doses, highly poisonous.

Morphine is a tertiary mono-acid base. Its *hydrochloride*, $C_{17}H_{19}NO_3.HCl + 3H_2O$, crystallizes in tufts of slender needles easily soluble in water and in alcohol. The *acetate*, $C_{17}H_{19}NO_3.C_2H_4O_2$, is moderately soluble in water, and crystallizes in needles.

Solutions of morphine and its salts are colored dark-blue by ferric chloride; its solution in strong sulphuric acid is colored blood-red by a drop of nitric acid. Solution of iodine added to the solution of the hydrochloride throws down the periodide, $C_{17}H_{19}NO_3I_4$. Morphine heated with potash-lye gives off methylamine.

A p o m o r p h i n e, $C_{17}H_{17}NO_2$, a compound containing $1H_2O$ less than morphine, is formed when morphine is heated in a sealed tube with strong hydrochloric acid, and separates as a white powder, which turns green on contact with the air. It differs from morphine in being soluble in alcohol, ether, and chloroform, and in its physiological action, which is not narcotic, but emetic.

C o d e i n e, $C_{18}H_{21}NO_3 = C_{17}H_{18}(CH_3)NO_3$, *Methyl-morphine.*—This base, obtained from opium as above described, crystallizes from ether in large rhombic prisms melting at 120° C. (248° F.). It is more soluble in water than the other opium bases; potash precipitates it from the solutions of

* It is convenient to designate organic bases by names ending in *ine;* neutral substances by names ending in *in; e. g., gelatin, albumin, casein.*

its salts. Heated with strong soda-lime it gives off methylamine and trimethylamine. Heated with strong hydrochloric acid to 140°–150° C. (284°–302° F.), it is resolved into methyl chloride and apomorphine:

$$C_{18}H_{21}NO_3 + HCl = CH_3Cl + H_2O + C_{17}H_{17}NO_2.$$

Codeine is also a tertiary monamine, forming with ethyl iodide a crystalline iodide, $C_1 H_{21}(C_2H_5)NO_3.I$, converted by silver oxide into an alkaline base.

Narcotine, $C_{22}H_{23}NO_7$.—The *marc*, or insoluble portion of opium, contains much narcotine, which may be extracted by boiling with dilute acetic acid. From the filtered solution the narcotine is precipitated by ammonia, and afterwards purified by solution in boiling alcohol, and filtration through animal charcoal. Narcotine crystallizes in small, colorless, brilliant prisms, nearly insoluble in water. Its basic powers are very feeble ; it is destitute of alkaline reaction, and although freely soluble in acids, does not, for the most part, form crystallizable salts.

Narcotine, treated with a mixture of dilute sulphuric acid and manganese dioxide, or a hot solution of platinic chloride, yields opianic acid (p. 951), together with basic products.

Cotarnine, $C_{12}H_{13}NO_3$, is contained in the mother-liquor from which opianic acid has crystallized ; it forms a yellow crystalline mass, very soluble, of bitter taste, and feebly alkaline reaction. Its hydrochloride is a well-defined salt.

Cotarnine, gently heated with very dilute nitric acid, is converted into methylamine nitrate, and *cotarnic acid*, a bibasic acid containing $C_{11}H_{12}O_5$:

$$C_{12}H_{13}NO_3 + 2H_2O + HNO_3 = CH_5N.HNO_3 + C_{11}H_{12}O_5.$$

Thebaine, $C_{19}H_{21}NO_3$, *Papaverine*, $C_{20}H_{21}NO_4$, and *Narceine*, $C_{23}H_{29}NO_9$, are also contained in opium in small quantity. Thebaine forms silvery scales, melting at 193° C. (379.4° F.) ; insoluble in water, potash, and ammonia. Papaverine melts at 141°–145° C. (285.8°–293° F.).

The following bases are also found in opium, at least occasionally ; *codamine*, $C_{19}H_{23}NO_3$, *lanthopine*, $C_{23}H_{25}NO_4$, *laudanine*, $C_{20}H_{25}NO_3$, *meconidine*, $C_{21}H_{23}NO_4$, *opianine* and *porphyroxine*, but they are of small importance, and comparatively little is known respecting them.[*]

3. Cinchona Bases.

The barks of the various species of cinchona contain a number of alkaloïds, associated with quinic acid and cinchona-tannin. The best known of these bases are :

Quinine, $C_{20}H_{24}N_2O_2$.	Cinchonine, $C_{20}H_{24}N_2O$.
Quinidine, $C_{20}H_{24}N_2O_2$.	Cinchonidine, $C_{20}H_{24}N_2O$.

Quinine is found chiefly in yellow cinchona-bark (from *China regia*) ; cinchonine in the gray bark (from *China Huanoco*).

The bases are extracted by digesting the pulverized bark with dilute hydrochloric acid, and precipitating the filtered solution with sodium carbonate, or magnesia. The precipitate, consisting of quinine, cinchonine,

* See Hesse, Ann. Ch. Pharm., cliii. 71 ; Gmelin's Handbook, xviii. 192, 197, 199, 202, 210 ; Watts's Dictionary of Chemistry, Supplement, p. 882.

and a few other substances, is boiled with alcohol, and the solution is saturated with sulphuric acid, and evaporated. On cooling, it first deposits sulphate of quinine, and afterwards the cinchonine salt. The free bases are easily separated by ether, which dissolves only the quinine. Quinidine and cinchonidine are found in the last mother-liquors of the sulphuric acid solution.

Cinchonine crystallizes in small, brilliant, transparent, four-sided prisms. It is very slightly soluble in water, dissolves readily in boiling alcohol, and has but little taste, although its salts are excessively bitter. It is a powerful base, neutralizing acids completely, and forming a series of crystallizable salts. It turns the plane of polarization strongly to the right.

Quinine much resembles cinchonine; but does not crystallize so well; it is much more soluble in water: tastes intensely bitter; turns the plane of polarization strongly to the left.

Quinine sulphate is manufactured on a very large scale for medicinal use; it crystallizes in small white needles, which give a neutral solution. It contains $2C_{20}H_{24}N_2O_2.SO_4H_2 + 7Aq$. Its solubility is much increased by the addition of a little sulphuric acid, whereby the acid salt, $C_{20}H_{24}N_2O_2.SO_4H_2 + 7Aq.$, is formed. Solutions of quinine sulphate exhibit a splendid blue fluorescence. On adding to the solution of a quinine salt, first chlorine-water and then ammonia, a fine green color is produced. Iodine added to a solution of quinine sulphate, forms a crystalline substance of a brilliant emerald color, which appears to have the composition $2C_{20}H_{24}N_2O_2.3H_2SO_4.I_6 + 3Aq$. This compound, called *Herapathite*, after its discoverer, possesses the optical properties of tourmaline.

Cinchonine and quinine yield with methyl iodide, the compounds $C_{20}H_{24}(CH_3)N_2OI$ and $C_{20}H_{24}(CH_3)N_2O_2I$, which are converted by silver oxide into soluble bases analogous to tetrethyl-ammonium hydroxide; they are therefore tertiary amines.

Quinidine and *Cinchonidine*, isomeric respectively with quinine and cinchonine, are obtained from commercial q u i n o ï d i n e, a resinous product contained in the mother-liquors of the quinine preparation.

Q u i n i d i n e (or Cinchinine) crystallizes in large prisms moderately soluble in alcohol, sparingly in ether. Its salts are more soluble than those of quinine. The solutions are strongly dextrogyrate. With chlorine-water and ammonia quinidine reacts like quinine.

C i n c h o n i d i n e, occurring also in the bark of *China Bogota*, is very much like cinchonine. Its solutions are strongly levogyrate.

The acid sulphates of these four bases, heated first to 100°, to expel water of crystallization, and then to about 135° C. (275° F.), are converted into the sulphates of two amorphous bases, q u i n i c i n e and c i n c h o n i c i n e, isomeric with quinine and cinchonine respectively, quinicine, $C_{20}H_{24}N_2O_2$, being formed from quinine and quinidine, cinchonicine, $C_{20}H_{24}N_2O$, from cinchonine and cinchonidine. The solutions of both of these bases are feebly dextrogyrate.

All the four bases, when distilled with caustic potash, yield bases of the chinoline series (p. 1004).

4. Strychnos Bases.

Strychnine, $C_{21}H_{22}N_2O_2$, and **Brucine**, $C_{22}H_{26}N_2O_4$, also called *Strychnia* and *Brucia*, are contained, together with several still imperfectly known bases, in *Nux vomica*, in *St. Ignatius bean*, and in *false Angustura bark*. To prepare them, nux vomica seeds are boiled in dilute sulphuric acid until they become soft: they are then crushed, and the expressed liquid is mixed with excess of calcium hydrate, which throws down the alkaloïds. The precipitate is boiled in spirits of wine of sp. gr. 0.850, and filtered hot. Strychnine and brucine are then deposited together in a colored and impure state, and may be separated by cold alcohol, in which the latter dissolves readily.

Pure strychnine crystallizes under favorable circumstances in small but exceedingly brilliant octohedral crystals, which are transparent and colorless. It has a very bitter, somewhat metallic taste (1 part in 1,000,000,000 parts of water is still perceptible), is slightly soluble in water, and fearfully poisonous. It dissolves in hot and somewhat dilute spirit, but not in absolute alcohol, ether, or solution of caustic alkali. This alkaloïd may be readily identified by moistening a crystal with concentrated sulphuric acid, and adding to the liquid a crystal of potassium dichromate, when a deep violet tint is produced, which disappears after some time.

Strychnine forms neutral crystalline salts, containing one equivalent of acid. The nitrate, $C_{21}H_{22}N_2O_2.NO_3H$, is sparingly soluble in water and in alcohol. Potassium thiocyanate added to the solutions throws down crystalline thiocyanate of strychnine.

B r u c i n e , $C_{22}H_{26}N_2O_4$, crystallizes in efflorescent prisms or tables containing $4H_2O$, easily soluble in alcohol, slightly in water, insoluble in ether; also very poisonous. Strong sulphuric acid colors the solutions red, and on adding stannous chloride to the red liquid, a violet precipitate is formed. Strong sulphuric acid dissolves brucine with reddish color.

---

5. Bases from various Plants.

Veratrine or **Veratria**, $C_{32}H_{52}N_2O_8$, is obtained from the seeds of *Veratrum Sabadilla*, and from the root of *V. album*. In the pure state it is a white or yellowish-white powder, which has a sharp burning taste, is very poisonous, and in small quantities occasions violent sneezing. It is insoluble in water, but dissolves in hot alcohol, in ether, and in acids: the solution has an alkaline reaction.

J e r v i n e , $C_{30}H_{46}N_2O_3$, occurring together with veratrine, in the root of *Veratrum album*, forms small prisms soluble in alcohol. Its salts are slightly soluble in water.

Piperine, $C_{17}H_{19}NO_3$, occurs in *Piper niger* and other kinds of pepper, from which it may be extracted by alcohol. It crystallizes in four-sided prisms, melting at 100°, easily soluble in alcohol and ether, insoluble in water; dissolves with dark-red color in strong sulphuric acid. Heated with soda-lime it gives off p i p e r i d i n e , and by boiling with alcoholic potash, it is resolved into piperidine and piperic acid (p. 947):

$$C_{17}H_{19}NO_3 \ + \ H_2O \ = \ C_{12}H_{10}O_4 \ + \ C_5H_{11}N .$$

Piperidine, $C_5H_{11}N = C_5H_{10}NH$, is a liquid which boils at 106° C. (222.8° F.), dissolves easily in alcohol and ether, has a strong alkaline reaction, and forms neutral crystalline salts with 1 eq. of acid.   It is a secondary amine, one of its hydrogen-atoms being replaceable by acid and alcoholic-radicles.   *Methyl-piperidine*, $C_5H_{10}N.CH_3$, and *ethyl-piperidine*, $C_5H_{10}.N.C_2H_5$, are colorless liquids, the former boiling at 118° C. (244.4° F.), the latter at 28° C. (82.4° F.).   Benzoyl-piperidine, $C_5H_{10}.N.C_7H_5O$, is crystalline. Piperine is a similar derivative containing the radicle of piperic acid, $C_5H_{10}.N.C_{12}H_9O_3$.

Atropine or **Daturine,** $C_{17}H_{23}NO_3$, occurring in the deadly nightshade (*Atropa belladonna*) and in the thorn apple (*Datura Stramonium*), crystallizes in thin prisms, melting at 90° C. (194° F.).   It is bitter, very poisonous, and in small quantity produces dilatation of the pupil.   By heating with baryta-water or hydrochloric acid, it is resolved into tropic acid (p. 941) and tropine:

$$C_{17}H_{23}NO_3 \;+\; H_2O \;=\; C_9H_{10}O_3 \;+\; C_8H_{15}NO.$$

Tropine is a strong mono-acid base, crystallizing from ether in tables. melting at 61° C. (141.8° F.).

Sinapine, $C_{16}H_{23}NO_5$, occurs in white mustard seed in the form of thiocyanate, which may be dissolved out by alcohol.   The base, separated from the thiocyanate or other salt by alkalies, remains dissolved, and decomposes on evaporation.   By boiling the salts with alkalies, the sinapine is resolved into choline (p. 673) and sinapic acid, $C_{11}H_{12}O_2$:

$$C_{16}H_{23}NO_5 \;+\; 2H_2O \;=\; C_5H_{15}NO_5 \;+\; C_{11}H_{12}O_2.$$

Sinapic acid is bibasic, and crystallizes in thin prisms, soluble in hot water and in alcohol.

There are numerous other alkaloïds, more or less known, occurring in plants ; the following short notice of a few of them must suffice :—

Hyoscyamine.—A white, crystallizable substance, from *Hyoscyamus niger;* it occurs likewise in *Datura Stramonium.*

Solanine, $C_{43}H_{71}NO_{16}(?)$.—A pearly, crystalline substance, from various solanaceous plants ; resolved by boiling with dilute acids into glucose and solanidine : $C_{43}H_{71}NO_{16} + 3H_2O = 3C_6H_{12}O_6 + C_{25}H_{41}NO$.

Aconitine, $C_{30}H_{47}NO_7$.—A crystalline, very poisonous alkaloïd, from *Aconitum Napellus.*

Delphinine.—A yellowish, fusible substance, from the seeds of *Delphinium Staphisagria.*

Emetine.—A white and nearly tasteless powder, from ipecacuanha root.

Curarine.—The arrow-poison of Central America.

Pectous Substances.

The pulp of fleshy fruits in the unripe state, also of fleshy roots and other vegetable organs, contains a substance called pectose, which is insoluble in water, but under the influence of acids and other reagents, is transformed into a soluble substance pectin, identical with that which exists in ripe fruits and imparts to their juice the property of gelatinizing when boiled.

Pectin may be obtained by boiling the pulp of carrots or turnips with a slightly acid liquid—or better, from the juice of ripe pears, by precipitating the lime with oxalic acid, then the albuminous substances with tannic acid, and adding alcohol, whereupon the pectin separates in long threads or as a jelly. When dry it forms an amorphous, tasteless mass, soluble in water, and precipitated therefrom by alcohol or by basic acetate of lead. On boiling the aqueous solution, the pectin is converted into p a r a p e c t i n, which, as well as pectin itself, is converted by boiling with dilute acids, into m e t a p e c t i n, which has an acid reaction, and is precipitated by barium chloride. These three substances are said to be isomeric, and represented by the empirical formula, $C_2H_3O_2$.

According to Frémy, all vegetable tissues which contain pectose contain also a ferment called p e c t a s e, similar in its mode of action to diastase and emulsin. It is an amorphous substance, which may be precipitated by alcohol from fresh carrot-juice. Under the influence of this ferment—or of dilute caustic alkalies at 30°—pectin is transformed into p e c t o s i c a c i d, $C_{32}H_{40}O_{28}.3H_2O$, and afterwards into p e c t i c a c i d, $C_{16}H_{18}O_{13}.2H_2O$, which, by prolonged boiling with water, is converted into p a r a - p e c t i c a c i d, $C_{24}H_{30}O_{21}.2H_2O$. The final product of the transformation of pectous substances is m e t a p e c t i c a c i d, $C_8H_{10}O_7.2H_2O$, which reduces alkaline copper solutions, and is resolved by alkalies into formic and protocatechuic acids. The composition of all these bodies is, however, very uncertain.

Bile Constituents.

1. BILIARY ACIDS.—Bile, the fluid secretion of the liver, contains—in addition to fats, mucous substances, proteids, urea, and choline—the sodium salts of two peculiar acids, called g l y c o c h o l i c and t a u r o - c h o l i c; also certain coloring matters, and an aromatic alcohol called c h o l e s t e r i n, already described (p. 914).

Glycocholic Acid, $C_{26}H_{43}NO_6$.—When fresh ox-bile perfectly dried is exhausted with cold absolute alcohol, and mixed after filtration with ether, it first deposits a brownish, tough, resinous mass, and after some time, stellate crystals, consisting of the glycocholates and taurocholates of sodium and potassium. On dissolving these salts in water and adding dilute sulphuric acid, glycocholic acid separates after twenty-four hours in the crystalline state, while taurocholic acid remains dissolved.

Glycocholic acid crystallizes in white slender needles, sparingly soluble in water, easily in alcohol. It has a faint acid reaction and bitter-sweet taste. It is monobasic; its alkali-salts are very soluble in water, and have a very sweet taste. On adding to glycocholic acid a solution of sugar and then strong sulphuric acid, a purple-red coloration is produced (Pettenkofer's bile-reaction).

Glycocholic acid is resolved by boiling with alkalies into glycocine and c h o l i c a c i d, $C_{24}H_{40}O_5$:

$$C_{26}H_{43}NO_6 + H_2O = C_2H_3(NH_2)O_2 + C_{24}H_{40}O_5.$$

Cholic acid crystallizes in shining quadratic octohedrons containing $2\frac{1}{2}$ mol. water, soluble in alcohol and ether.

Glycocholic acid is also resolved into glycocine and cholic acid by boiling with acids, but the cholic acid is then converted, by abstraction of water, into d y s l y s i n, $C_{24}H_{36}O_3$, an amorphous substance which is reconverted into cholic acid by boiling with alcoholic potash.

Taurocholic Acid, $C_{26}H_{45}NSO_7$, may be precipitated by basic lead acetate, after the glycocholic acid, mucus, and coloring matters have been removed by the neutral acetate. It forms slender needles, having a sweetish-bitter taste, and easily soluble in water and in alcohol. By boiling with water it is resolved into cholic acid and taurine (p. 633) :

$$C_{26}H_{45}NSO_7 \ + \ H_2O \ = \ C_{24}H_{40}O_5 \ + \ C_2H_4(NH_2).SO_3H \ ,$$

the cholic acid being, however, for the most part converted by dehydration into dyslysin. The same decomposition takes place in the putrefaction of bile.

Pig's bile contains two acids analogous to the above, viz., h y o g l y c o-c h o l i c a c i d , $C_{27}H_{43}NO_5$, and h y o t a u r o c h o l i c a c i d , $C_{27}H_{45}NSO_6$, which are resolved by boiling with acids into h y o c h o l i c a c i d , $C_{25}H_{40}O_4$, and glycocine or taurine respectively.

Goose-bile contains a similar acid, $C_{29}H_{49}NSO_6$, called c h e n o t a u r o-c h o l i c a c i d , which is resolved by alkalies into taurin and chenocholic acid, $C_{27}H_{44}O_4$.

L i t h o f e l l i c a c i d , $C_{20}H_{36}O_4$, an acid nearly related to the biliary acids, occurs, together with ellagic acid, in Oriental bezoar-stones (p. 948). It may be extracted by hot alcohol, and crystallizes in short prisms, melting at 204° C. (399.2° F.); gives, with sugar-solution and strong sulphuric acid, a purple-red color, similar to that produced with glycocholic acid.

BILE PIGMENTS.—*Bilirubin*, $C_{16}H_{18}N_2O_3$, the principal coloring matter of the bile, forms dark-red prisms, insoluble in water, sparingly soluble in alcohol and ether, easily in chloroform and carbon sulphide. It dissolves in alkalies, forming a yellowish-red solution, which, when agitated in contact with the air, yields a green precipitate of *biliverdin*, $C_{16}H_{20}N_2O_5$. *Bilifuscin*, $C_{10}H_{20}N_2O_4$, is a dark-green mass, insoluble in water and chloroform, easily soluble in alcohol.

On heating an alkaline solution of these bile-pigments with nitric acid, a green color is produced, changing to blue, violet, red, and ultimately to yellow. This reaction serves for the detection of bile.

Gelatinous Substances.

The bone-cartilages, tendons, connective tissue, and skin of the animal body dissolve, for the most part, when boiled for a long time with water, yielding a solution which solidifies, on cooling, to a transparent tremulous mass called g e l a t i n or g l u t i n ; the non-hardening cartilages yield a similar substance, called c h o n d r i n .

These substances contain in 100 parts :

	C	H	N	O
Gelatin . . .	50.0	6.6	18.3	25.1
Chondrin . . .	49.1	7.1	14.4	29.4

Their molecular weights and structural formulæ are unknown.

Gelatin, Bone-gelatin, or **Glutin,** precipitated from its aqueous solution by alcohol, forms a colorless, transparent mass, without taste or smell. It swells up in cold water, and dissolves, on boiling, to a viscid liquid, which solidifies to a jelly on cooling. By prolonged boiling with a small quantity of nitric acid, or, by the addition of concentrated acetic acid, the solution loses the property of gelatinizing. The aqueous solution turns the plane of polarization to the left. It is precipitated by alcohol, mercurio chloride, mercuric nitrate, and mercurous nitrate, but not by alum, or by lead acetate, either neutral or basic. Tannic acid throws down, even from very dilute solutions of gelatin, a tough yellowish precipitate, consisting of a compound of the two bodies. The tissues which yield gelatin likewise unite with tannic acid, and withdraw it completely from its solutions, forming l e a t h e r .

Gelatin boiled with sulphuric acid or with alkalies yields leucine and glycocine, together with other products of unknown constitution. By destructive distillation it yields several amines belonging to the fatty and pyridine series. By oxidation with MnO_2 and sulphuric acid, or with chromic acid, it yields the same products as the proteids.

Isinglass is a very pure gelatin obtained from the dried swimming bladder of the sturgeon : it dissolves in water merely warm. *Size* is an impure gelatin, prepared from the clippings of hides and similar matters. *Glue* is the same substance dried by exposing it in thin slices on nettings to a current of air.

Chondrin, obtained from the cartilages of the ribs and joints, is very much like gelatin, but differs from it in being precipitated from its aqueous solution by acetic acid, alum, lead acetate, and other metallic salts, but not by mercuric chloride. Its products of decomposition are, for the most part, the same as those of gelatin ; with sulphuric acid, however, it yields no glycocine, but only leucine. When boiled with hydrochloric acid it yields glucose.

Silk-gelatin or **Sericin,** $C_{15}H_{25}N_5O_8$, is extracted from silk by boiling with water, and precipitated by alcohol as an amorphous powder. It swells up in water, dissolves on boiling, and solidifies to a jelly ; is precipitated by potassium ferrocyanide, basic lead acetate, and several other metallic salts. By boiling with sulphuric acid it yields leucine, tyrosine, and amidoglyceric acid or serine (p. 785).

Fibroin, $C_{15}H_{23}N_3O_6$, the chief constituent of silk (about 66 per cent.), is obtained by repeatedly digesting silk with water at 130° C. (266° F.), and exhausting the residue with alcohol and ether : it then remains as a white shining mass. It dissolves in strong sulphuric acid and in alkalies, and is precipitated on saturating the solutions. When boiled with sulphuric acid it yields leucine, tyrosine, and glycocine.

Proteids or Albuminoids.

These substances form the chief part of the solid constituents of the blood, muscles, nerves, glands, and other organs of animals ; they occur also in small quantities in almost every part of vegetables, and in larger quantities in the seeds. They are formed exclusively in plants, and undergo but little alteration when consumed as food and assimilated by animals.

The several bodies of this class resemble one another closely in their properties, and more especially in their percentage composition, which is comprised between the following limits :

Carbon	.	.	.	.	.	52.7	to	54.5
Hydrogen	.	.	.	.	.	6.9	"	7.3
Nitrogen	.	.	.	.	.	15.4	"	16.5
Oxygen	.	.	.	.	.	20.9	"	23.5
Sulphur	.	.	.	.	.	0.8	"	1.6

These numbers may be approximately represented by the empirical formula $C_{72}H_{112}N_{18}O_{22}S$. The proteids also contain a small quantity of phosphorus, but apparently only mechanically mixed with them as calcium phosphate. Their molecular weights and constitution are not yet known; but recent investigations have thrown some light on the question, and indicate at least the direction in which its solution is to be sought. According to Hlasiwetz and Habermann, all proteids, when boiled with dilute sulphuric acid, or better, with hydrochloric acid, and a small quantity of stannous chloride, are resolved exactly into aspartic acid, glutamic acid (p. 788), leucine, tyrosine, and ammonia, and may, therefore, be regarded as formed by the combination of these substances, with elimination of water. Schützenberger, by heating proteids with baryta-water to 150° (whereby $\frac{1}{9}$ of the nitrogen was evolved as ammonia), obtained essentially the same products of decomposition, and likewise intermediate products resulting from a less complete decomposition. The relative quantities of ammonia and carbon dioxide corresponded exactly with those which are evolved in the resolution of urea into $2NH_3$ and CO_2. This result tends to show that albumin is a complex ureide, containing one-fifth of its nitrogen in the form of urea.

Proteids are precipitated from solution : 1. By excess of mineral acids. 2. By potassium ferrocyanide with acetic acid or a little hydrochloric acid. 3. By acetic acid, with a considerable quantity of concentrated solutions of neutral salts of the alkalies and alkaline earths, gum arabic, or dextrin. 4. When boiled with mercuric nitrate (Millon's reagent),* they all give a deposit which turns red after a while, the supernatant liquors also becoming red. They rotate the plane of polarization more or less to the left.

Proteids may be conveniently divided into the following classes :

CLASS I. ALBUMINS.—Soluble in water.

1. **Serum Albumin** is the most abundant albuminous substance in animal bodies. It can be obtained tolerably pure from blood-serum by precipitation with lead acetate, washing with water, suspending the precipitated lead-compound in water, and decomposing it with carbonic acid; then by filtration a very cloudy solution of albumin is obtained.

Serum albumin forms a yellow elastic transparent substance, which when perfectly dry can be heated to 100° without change. It is soluble in water, and precipitable by alcohol; long-continued action of alcohol changes it into coagulated albumin. Its specific rotation is —56° for yellow light It is not precipitated by carbonic, acetic, tartaric, or phosphoric acid, or by other mineral acids, when very dilute, and added in small quantity; large quantities of acid precipitate it immediately; nitric acid acts most strongly.

2. **Egg Albumin** differs from serum albumin by gradually giving a precipitate when agitated with ether; oil of turpentine also coagulates it. Serum albumin dissolves easily in strong nitric acid, whilst egg albumin is nearly insoluble therein. The specific rotation of egg albumin is —35.5° for yellow light.

The so-called *vitellin* contained in solution in the yolk of egg is a mixture of albumin and casein.

* Prepared by gently warming mercury with an equal quantity of strong nitric acid till it is dissolved, then diluting the liquid with twice its bulk of water, and leaving the precipitate to settle. The clear supernatant liquid is Millon's reagent.

3. **Plant Albumin** occurs in nearly all vegetable juices, especially in potatoes and in wheat-flour. It coagulates by heat, and bears a close resemblance to egg albumin.

CLASS II. GLOBULINS.—Insoluble in water, soluble in very dilute acids and alkalies, soluble in dilute (1 per cent.) solutions of sodium chloride and other neutral salts.

1. **Myosin.**—This substance was first separated by Kühne from other albuminous matters occurring in the protoplasma or contractile muscular substance that causes the *rigor mortis*. To prepare it, well cut-up flesh is carefully washed with water, and the mass is then placed in a mixture of one volume of concentrated solution of common salt to two volumes of water; these are continually rubbed together and filtered through linen; the slimy filtrate is allowed to drop into a large quantity of distilled water. The myosin is redissolved in solution of sodium chloride, and reprecipitated by much water. It is insoluble in water, soluble in solution of common salt under 10° C. (50° F.), soluble in very dilute hydrochloric acid, but in this solution it passes by degrees into acid albumin or syntonin; in dilute alkali, myosin, like other albuminous matters, is soluble, being changed into albuminate. By heat it is changed into coagulated albumin. It is also coagulated by alcohol.

2. **Globulin** (*Paraglobulin, Paraglobin*).—When fresh blood-serum is diluted tenfold with water, and a brisk stream of carbonic acid passed through it, a fine granular precipitate is formed, which may be separated by decantation and filtration, and washed with water. The same substance may be prepared by saturating blood-serum with sodium chloride (or magnesium sulphate, etc.), as in the case of myosin. A certain amount of the salt always clings to the precipitate.

Globulin is exceedingly soluble in dilute saline solutions (from which it may be precipitated unchanged by carbonic acid gas or *exceedingly* dilute acids). It is insoluble in water, but dissolves when the water is saturated with oxygen, and may be precipitated by carbonic acid.

In excessively dilute alkalies globulin dissolves without alteration; in solutions containing about 1 per cent. of the alkali, it dissolves as albuminate. By dilute acids, however feeble, it is changed in solution into acid albumin. Suspended in water and heated to 70° C. (158° F.), it enters into the insoluble or coagulated state.

Globulin is present, not only in serum of blood, but also in aqueous humor, in the juice of the cornea, connective tissue, etc. Derived from the first of these sources, globulin is f i b r i n o p l a s t i c, *i. e.*, it has the power of acting in concert with certain fluids (fibrinogenous) in such a manner as to give rise to fibrin (p. 1019). The crystalline lens contains a substance which is not fibrinoplastic, but in many other respects closely resembles the globulin just described.

3. **Fibrinogen.**—When hydrocele fluid, pericardial fluid, or any other fluid capable of giving a clot with blood-serum or paraglobulin, is treated by the method adopted for globulin, a similar substance is produced which resembles globulin in every respect, except that the carbonic acid precipitate is more difficult to obtain and more flaky, and that the substance is more readily thrown down from the liquids in which it is formed, by a mixture of alcohol and ether; also by the fact that it is fibrinogenous, *i. e.*, produces fibrin when mixed with fibrinoplastic globulin.

CLASS III. DERIVED ALBUMINS.—Insoluble in water, and in solutions of sodium chloride; soluble in dilute acids and alkalies.

43 *

1. Acid-albumin.—If a small quantity of dilute acid (hydrochloric or acetic) be added to serum- or egg-albumin, no precipitation or coagulation takes place, and on gradually raising the temperature of the mixture to 70⁰ C. (158⁰ F.), it will be found that coagulation at that or at a ·higher temperature has been entirely prevented. At the same time, the in_ fluence of the fluid on polarized light has been altered. The rotation to the left has become increased to 72⁰.

On carefully neutralizing the cooled mixture, the whole of the proteid matter is thrown down as a white, flocculent, frequently gelatinous pre_ cipitate. The action of the acid has converted the albumin soluble in water into a substance insoluble in water. The precipitate is very readily soluble in excess of the alkali used for neutralization, may be reprecipitated by again neutralizing with an acid, again redissolved by excess, and so on. It is also soluble in dilute solutions of alkaline carbonates. It is insoluble in sodium chloride solution, and may be precipitated from its solutions by the addition of that salt. Suspended in water and heated to 70⁰ C. (158⁰ F.), it enters into the coagulated or insoluble condition.

All the globulins of Class II. are readily soluble in dilute acids ; but by the act of solution they are at once converted into acid-albumin, the pre_ cipitate formed by neutralization being no longer soluble in neutral saline solutions.

2. Alkali-albumin or **Albuminate.   Casein.**—When albuminous substances, egg- or serum-albumin, for example, are treated with dilute caustic alkali instead of acid, coagulation by heat is similarly prevented, and the whole of the proteid may in like manner be thrown down on neu- tralization. Some of the bodies thus produced agree well together, and cannot be distinguished from the casein of milk, although most probably casein is not identical with artificial albuminate, and the bodies which are produced by the action of potash on different albuminous substances may differ slightly one from the other, as is evident in the difference of their rotatory action on polarized light.

Casein occurs most plentifully in the milk of animal feeders, and is best obtained from milk by precipitating with crystalline magnesium sulphate, filtering and washing with a concentrated solution of salt, then dissolving the precipitate in water ; the butter is filtered off, and the clear solution precipitated by dilute acetic acid.

Dried casein and albuminate are yellow, transparent, and hygroscopic, swelling up in water, but not dissolving. When precipitated in a flocky state, they dissolve easily in water if it contains a little alkali. The pre- cipitate which forms on neutralizing the alkaline solution, dissolves easily in an excess of acetic acid or dilute hydrochloric acid. On the addition of an excess of mineral acid, or on neutralization with an alkali, these solu- tions give a precipitate.

The neutral or feebly alkaline albuminate, and casein in alkaline solu- tion, are precipitated in the cold by alcohol : when hot they are dissolved. By fusion with potassium hydrate, casein yields valeric and butyric acids, besides other products.

The most striking property of casein is its coagulability by certain animal membranes, as is seen in the process of cheese-making, in prepar- ing the *curd*, the coagulation being effected by an infusion of the stomach of the calf called *rennet*.

Plant-casein, or Legumin, is found chiefly in the seeds of legu- minous fruits, from the juice of which it may be precipitated by acetic acid or by rennet.

CLASS IV.—**Fibrin.**—Insoluble in water ; sparingly soluble in dilute acids and alkalies, and in neutral saline solutions.

This is the substance to which the clotting of blood is due. It may be obtained by washing blood-clots, or more readily by stirring with a bundle of twigs, blood just shed, before it has had time to clot. The fibrin, which adheres in layers to the twigs, may then be stripped off and washed till perfectly white. The formation of fibrin is due to the contact of fibrinoplastic and fibrinogenous substance. When these two substances come into contact in any fluid, they combine, quickly or slowly, according to the greater or lesser quantity of each substance in the fluid, to form fibrin.

Fibrin differs from all other solid proteids in having a filamentous structure, and in possessing remarkable elasticity. It is insoluble in water, dilute hydrochloric acid, and aqueous sodium chloride, but dissolves at 40° C. (104° F.) in aqueous potassium nitrate.

Plant-fibrin occurs as an insoluble substance in plants, especially in the seeds of cereal grasses. When wheat-flour is stirred up to a paste with water, and kneaded for some time, the starch granules and soluble albumin are removed, and there remains a tenacious mass called gluten ; and by boiling this substance with dilute alcohol to remove vegetable gelatin (gluten) and extracting the fats with ether, the plant-fibrin is obtained in the form of a grayish-white, tough, elastic mass. It dissolves in very dilute hydrochloric acid, and in dilute alkalies, and is precipitated from these solutions by neutral salts, and by acetic acid. By boiling with dilute sulphuric acid, plant-fibrin is resolved into leucine, tyrosine, and glutamic acid.

CLASS V. **Coagulated Proteid.**—Coagulated albumin is formed from albumin, syntonin, fibrin, myosin, etc., by heating their neutral solutions to boiling, or by the action of alcohol. Egg albumin is also changed into coagulated albumin by strong hydrochloric acid and by ether. The albuminates, and also casein, when precipitated by neutralization, pass into coagulated albumin when heated. The coagulated albuminous substances are insoluble in water, alcohol, and other indifferent fluids, scarcely soluble in dilute potash, soluble with great difficulty in ammonia. In acetic acid they swell up, and gradually dissolve. They are mostly insoluble in dilute hydrochloric acid ; but when pepsin is also present at blood heat, they change first into syntonin, and then into peptone. They are dissolved by strong hydrochloric acid, and by caustic potash they are changed into albuminates.

CLASS VI. **Peptones.**—By the action of the acid gastric juice, all albuminous substances are changed into bodies called peptones. These are found only in the stomach and in the contents of the small intestines. They can no longer be detected in the chyle. They are highly diffusible, easily soluble in water, insoluble in alcohol or ether ; but alcohol separates them with difficulty from the watery solution ; when precipitated they remain unchanged even after boiling. They are not precipitated either by acids or by alkalies. Acetic acid and potassium ferrocyanide give no precipitate ; but corrosive sublimate and lead acetate with ammonia give precipitates.

The reactions of the several proteids above described may be tabulated as follows :—

Soluble in water :
    Aqueous solutions not coagulated by boiling . . PEPTONES.
    Aqueous solutions coagulated by boiling . . ALBUMINS.

Insoluble in water :
    Soluble in a 1 p. c. solution of sodium chloride . . GLOBULINES.

Insoluble :
Soluble in hydrochloric acid (0.1 p. c.) in the cold :
    Soluble in hot spirit   .   .   .   .   .   { ALKALI-ALBUMIN.

    Insoluble in hot spirit   .   .   .   .   .   { ACID-ALBUMIN.

Insoluble in hydrochloric acid (0.1 p. c.) in the cold :
Soluble in hydrochloric acid (0.1 p. c.) at 60°. . FIBRIN.
Insoluble in hydrochloric acid (0.1 p. c.) at 60° ; { COAGULATED
    insoluble in strong acids ; soluble in gastric juice { ALBUMIN.

Substances related to the Proteids.

Hæmoglobin, 54.2 oxygen, 7.2 hydrogen, 0.42 iron, 16.0 nitrogen, 21.5 oxygen, and 0.7 sulphur ; also called *Hæmatoglobulin* and *Hæmatocrystallin.*—This substance forms the chief part of the red globules of the blood of vertebrata ; usually it is obtained in an amorphous condition, but from the blood of some animals—as, for example, dogs, cats, rats, mice, and many fish—it can be separated in the crystalline form. Red crystals can be obtained from dog's blood by mixing the defibrinated blood with an equal quantity of water, adding 1 volume of alcohol to 4 volumes of the diluted blood, and leaving it at rest at 0° or lower. After 24 hours the hæmoglobin separates in small violet-red rhombic octohedrons. After drying over sulphuric acid it forms a brick-red powder. It dissolves in cold water, forming a red solution, from which it is precipitated in the crystalline form by alcohol. At ordinary temperatures the solution decomposes and turns brown.

The aqueous solution of hæmoglobin (or of blood) exhibits in its spectrum two absorption-bands situated between the Frauenhofer lines D and E (in the yellow and green).

Hæmoglobin unites with certain gases forming peculiar unstable compounds. The solution containing oxygen has a deep red color ; the solution free from oxygen is dark purple (arterial and venous blood) ; the absorption-bands are exhibited only by the oxygenated solution. Carbon monoxide displaces the oxygen, and forms with the hæmoglobin a compound which, on addition of alcohol, separates in bluish crystals : this appears to be the cause of the deleterious action of carbon monoxide on animals.

Oxygenated hæmoglobin is resolved by dilute acids or alkalies into two proteids, fatty acids, and a coloring matter called h e m a t i n, which in the dry state is a dark blue powder. It contains 9 per cent. of iron, and appears to have the composition $C_{34}H_{34}FeN_4O_5$.

When hæmoglobin (or blood) is warmed with strong acetic acid and solution of common salt, a substance called h e m i n separates in yellowish-red microscopic rhombic crystals, the formation of which serves as a delicate indication of the presence of blood.

Related to the proteids are also many f e r m e n t s, such as emulsin or synaptase, occurring in almonds ; d i a s t a s e (p. 660), which is formed from vegetable fibrin in the germination of seeds, and is characterized by the property of converting starch and dextrin into sugar ; m y r o s i n from mustard-seeds (p. 575) ; p t y a l i n, the ferment of saliva, which also converts starch into dextrin and sugar ; p e p s i n, which is contained in gastric juice, and possesses the power, in conjunction with hydrochloric acid, of dissolving the insoluble proteids and converting them into peptones.

Mucin is the chief constituent of animal mucus, and is precipitated by alcohol and dilute acetic acid.

The horny substances of horns, nails, hoofs, hairs, feathers, and the epidermis, are also nearly related to the proteids, having indeed the same composition, except that they contain more sulphur (2–5 p. c.). They dissolve easily in alkalies, with formation of sulphides. Nitric acid turns them yellow. Boiled with dilute sulphuric acid they yield leucine and tyrosine.

Brain Constituents.

When brain or spinal marrow is boiled with water, creatine, inosite, lactic acid, uric acid, and other substances, it dissolves, and from the residue ether extracts fats, cholesterin (p. 914), cerebrin, and lecithin. On cooling the solution, the two latter substances are first deposited, and may be separated by cold ether, which dissolves only the lecithin. The so-called *protagon* is a mixture of cerebrin and lecithin.

Cerebrin or **Cerebric Acid,** $C_{17}H_{33}NO_3$, is a light amorphous powder, without taste or smell ; swells up like starch when boiled with water, and is converted by boiling with dilute acids into a saccharine substance, and other products.

Lecithin, $C_{42}H_{84}NPO_9$, is widely diffused in the animal organism, occurring especially in the brain nerves, yolk of eggs, blood-corpuscles, etc. It is best prepared by exhausting egg-yolk with a mixture of alcohol and ether, evaporating the ether, adding an alcoholic solution of platinic chloride, decomposing the yellow platinochloride, $(C_{42}H_{83}NPO_9Cl)_2.PtCl_4$, with hydrogen sulphide, and evaporating the filtrate.

Lecithin is a waxy, indistinctly crystalline mass, which dissolves in alcohol and ether, and swells up in water, forming an opalescent solution or emulsion, from which it is precipitated by various salts of the alkali-metals. It unites both with bases and with acids.

By boiling with acids, or with baryta-water, lecithin is resolved into choline, glycerophosphoric acid, palmitic acid, and oleic acid.

APPENDIX.

HYDROMETER TABLES.

COMPARISON OF THE DEGREES OF BAUME'S HYDROMETER WITH THE REAL SPECIFIC GRAVITIES.

1. *For Liquids heavier than Water.*

Degrees.	Specific Gravity.	Degrees.	Specific Gravity.	Degrees.	Specific Gravity.
0	1·000	26	1·206	52	1·520
1	1·007	27	1·216	53	1·535
2	1·013	28	1·225	54	1·551
3	1·020	29	1·235	55	1·567
4	1·027	30	1·245	56	1·583
5	1·034	31	1·256	57	1·600
6	1·041	32	1·267	58	1·617
7	1·048	33	1·277	59	1·634
8	1·056	34	1·288	60	1·652
9	1.063	35	1·299	61	1·670
10	1·070	36	1·310	62	1·689
11	1·078	37	1·321	63	1·708
12	1·085	38	1·333	64	1·727
13	1·094	39	1·345	65	1·747
14	1·101	40	1·357	66	1·767
15	1·109	41	1·369	67	1·788
16	1·118	42	1·381	68	1·809
17	1·126	43	1·395	69	1·831
18	1·134	44	1·407	70	1·854
19	1·143	45	1·420	71	1·877
20	1·152	46	1·434	72	1·900
21	1·160	47	1·448	73	1·944
22	1·169	48	1·462	74	1·949
23	1·178	49	1·476	75	1·974
24	1·188	50	1·490	76	2·000
25	1·197	51	1·495		

2. *Baumé's Hydrometer for Liquids lighter than Water.*

Degrees.	Specific Gravity.	Degrees.	Specific Gravity.	Degrees.	Specific Gravity.
10	1·000	27	0·896	44	0·811
11	0·993	28	0·890	45	0·807
12	0·986	29	0·885	46	0·802
13	0·980	30	0·880	47	0·798
14	0·973	31	0·874	48	0·794
15	0·967	32	0·869	49	0·789
16	0·960	33	0·864	50	0·785
17	0·954	34	0·859	51	0·781
18	0·948	35	0·854	52	0·777
19	0·942	36	0·849	53	0·773
20	0·936	37	0·844	54	0·768
21	0·930	38	0·839	55	0·764
22	0·924	39	0·834	56	0·760
23	0·918	40	0·830	57	0·757
24	0·913	41	0·825	58	0·753
25	0·907	42	0·820	59	0·749
26	0·901	43	0·816	60	0·745

These two tables are on the authority of Francœur; they are taken from the *Handwörterbuch der Chemie* of Liebig, Poggendorff, and Wöhler. Baumé's hydrometer is very commonly used on the Continent, especially for liquids heavier than water. For lighter liquids the hydrometer of Cartier is often employed in France. Cartier's degrees differ but little from those of Baumé.

In the United Kingdom, Twaddell's hydrometer is a good deal used for dense liquids. This instrument is so graduated that the real specific gravity can be deduced by an extremely simple method from the degree of the hydrometer; namely, by multiplying the latter by 5, and adding 1000; the sum is the specific gravity, water being 1000. Thus 10° Twaddle indicates a specific gravity of 1050, or 1·05; 90° Twaddell, 1450, or 1·45.

In the Customs and Excise, Sikes's hydrometer is used.

TABLE III.

ABSTRACT OF REGNAULT'S TABLE OF THE MAXIMUM TENSION OF WATER-VAPOR,
AT DIFFERENT TEMPERATURES, EXPRESSED IN MILLIMETERS OF MERCURY.

Temperature.		Tension, millimeters.	Temperature.		Tension, millimeters.
— 32° C. — 25.6° F.		0.320	100° C. 212° F.		760.000
30	22.0	0.386	105	221	906.410
25	13.0	0.605	110	230	1075.370
20	4.0	0.927	115	239	1269.410
15 + 5.0		1.400	120	248	1491.280
10	14.0	2.093	125	257	1743.880
5	23.0	3.113	130	266	2030.280
0	32.0	4.600	135	275	2353.730
+ 5	41.0	6.534	140	284	2717.630
10	50.0	9.165	145	293	3125.55
15	59.0	12.699	150	302	3581.23
20	68.0	17.391	155	311	4088.56
25	77.0	23.550	160	320	4651.62
30	86.0	31.548	165	329	5274.54
35	95.0	41.827	170	338	5961.66
40	104.0	54.906	175	347	6717.43
45	113.0	71.391	180	356	7546.39
50	122.0	91.982	185	365	8453.23
55	131.0	117.478	190	374	9442.70
60	140.0	148.791	195	383	10519.63
65	149.0	186.945	200	392	11688.96
70	158.0	233.093	205	401	12955.66
75	167.0	288.517	210	410	14324.80
80	176.0	354.643	215	419	15801.33
85	185.0	433.041	220	428	17390.36
90	194.0	525.450	225	437	19097.04
95	203.0	633.778	230	446	20926.40

WEIGHTS AND MEASURES

480·0 grains Troy = 1 oz. Troy.
437·5   "   = 1 oz. Avoirdupoids
7000·0   "   = 1 lb. Avoirdupoids.
5760·0   "   = 1 lb. Troy.

The imperial gallon contains of water at 60° (15°·5C) 70,000· grains
The pint ($\frac{1}{8}$ of gallon).. 8,750·   "
The fluid-ounce ($\frac{1}{20}$ of pint)................................. 437·5   "
The pint equals 34·66 cubic inches.

The French *kilogramme* = 15,433·6 grains, or 2·679 lb. Troy, or
2·205 lb. avoirdupoids.
The *grammme*   = 15·4336 grains.
"   *decigramme* = 1·5434   "
"   *centigramme* = 0·1543   "
"   *milligramme* = 0·0154   "

The *mètre* of France = 39·37   inches
"   *decimètre*   = 3·937   "
"   *centimètre*   = 0·394   "
"   *millimètre*   = 0·0394   "

COMPARISON OF FRENCH AND ENGLISH MEASURES. By Dr. WARREN De La Rue.

MEASURES OF LENGTH.

	In English Inches.	In English Feet = 12 Inches.	In English Yards = 3 Feet.	In English Fathoms = 6 Feet.	In English Miles = 1760 yards.
Milli mètre	0·03937	0·0032809	0·ʕ 0936	0·0005468	0·0000006
Centimètre.....................	0·39371	0·0328090	0·0109363	0·0054682	0·0000062
Décimètre...............	3·93708	0·3280899	0·1093633	0·ʕ 466	0·0000621
Mètre..................	39·37079	3·2808992	1·0936331	0·5468165	0·0006214
Décamètre..............	393·70790	32·8089920	10·9363310	5·ʕ 1655	0·0062138
Hectomètre....................	3937· ʕ00	328·0899200	109·ʕ 3300	54·6816550	0·0621382
Kilomètre...................	39370·79000	3280·8992000	1 093·6331000	546· ʕ00	0·6213824
Myriomètre................	8907·90000	32808·9920000	10936·3310000	• 5468·1655000	6·2138244

1 Inch = 2·539954 Centimètres.   1 Yard = 0·91438348 Mètre.
1 Foot = 3·0479449 Décimètres.   1 Mile = 1·6093149 Kilomètre.

MEASURES OF SURFACE.

	In English Square Feet.	In English Sq. Yards = 9 Square Feet.	In English Poles = 272·25 Sq. Feet.	In English Roods = 10,890 Sq. Feet.	In English Acres = 43,560 Sq. Feet.
Centiare or sq. mètre..............	10·7642993	1·1960333	0·0395383	0·000988457	0·0002471143
Are or 100 sq. mètres..............	1076·4299342	119·6033260	3·9538290	0·098845724	0·0247114310
Hectare or 10,000 sq. mètres......	107642·9934183	11960·3326020	395·3828959	9·884572398	2·4711430996

1 Square Inch = 6·4513660 Square Centimètres.   1 Square Mètre or Centiare.
1 Square Foot = 9·2899683 Square Décimètres.   1 Sq. re Yard = 0·83609715 = 0·404671021 Hectare.
1 Acre

MEASURES OF CAPACITY.

	In Cubic Inches	In Cubic Feet = 1728 Cubic Inches.	In Pints = 34·65925 Cubic Inches.	In Gallons = 8 Pints = 277·27384 Cubic Inches.	In Bushels = 8 Gallons = 2218·19075 Cubic Inches.
Millilitre, or cubic	0·061027	0·0000353	0· 0761	0·00022010	0·000027512
1 ..., or 10	0·610271	0·0003532	0·017608	0·00220097	0·00 2021
Decilitre, or 100 ... centimètres......	6·102705	0·0035317	0·176077	0·02200967	0·002761208
Litre, or cubic décimètre............	61·027052	0·0853166	1·760773	0·22009668	0·027512085
..., or centistère............	610·270515	0·3531658	17·607734	2·20096677	0·275120846
Hectolitre, or decistère............	6102·705152	3·5316581	176·077341	22·00966767	2·751208459
Kilolitre, or stère, or ... mètre	61027·051519	35·3165807	1760·773414	220·09667675	27·512084594
Myriolitre, or decastère	610270·515194	353·1658074	17607·784140	2200·96676750	275·120845937

1 Cubic Inch = 16·8861759 Cubic Centimètres   1 Cubic Foot = 28·3153119 Cubic Decimètres   1 Gallon = 4·543457969 Litres.

MEASURES OF WEIGHT.

	In English Grains.	In Troy Ounces = 480 Grains.	In Avoirdupois Lbs. = 7000 Grains.	In Cwts. = 112 Lbs. = 784,000 Grains.	Tons = 20 Cwts. = 15,680,000 Grains.
Milligramme............	0·015432	0032	0·0000022	0·00000002	0001
Centigramme	0·154328	0·000322	0·0000220	0·00000020	0·000000010
Decigramme	1·543235	0015	0·0002205	0·00000197	0·000000098
Gramme	15·432349	0351	0·0022046	0·00001968	0·000000984
Decagramme	154·323488	1807	0·0220462	0·00019684	0·000009842
Hectogramme	1548·234880	3523	0·2204621	0·00196841	0·000098421
Kilogramme............	15482·348800	82·150727	2·2046218	0· 3012	0·000984206
Myriogramme............	154823·488000	821·507267	22·0462126	0·19684118	0·00984·2059

1 Grain = 0·064798950 Gramme.   1 Troy oz. = 31·103496 Gram.   1 lb. Avd. = 0·45359265 Kilogr.   1 Cwt. =   ... Kilog.

TABLE

FOR CONVERTING DEGREES OF THE CENTIGRADE THERMOMETER INTO
DEGREES OF FAHRENHEIT'S SCALE.

Cent.		Fah.	Cent.		Fah.	Cent.		Fah.
—100°	...	—148·0°	—55°	...	— 67·0°	—10°	...	+14·0°
99	...	146·2	54	...	65·2	9	...	15·8
98	...	144·4	53	...	63·4	8	...	17·6
97	...	142·6	52	...	61·6	7	...	19·4
96	...	140·8	51	...	59·8	6	...	21·2
95	...	139·0	50	...	58·0	5	...	23·0
94	...	137·2	49	...	56·2	4	...	24·8
93	...	135·4	48	...	54·4	3	...	26·6
92	...	133·6	47	...	52·6	2	...	28·4
91	...	131·8	46	...	50·8	1	...	30·2
90	...	130·0	45	...	49·9	0	...	32·0
89	...	128·2	44	...	47·2	+1	...	33·8
88	...	126·4	43	...	45·4	2	...	35·6
87	...	124·6	42	...	43·6	3	...	37·4
86	...	122·8	41	...	41·8	4	...	39·2
85	...	121·0	40	...	40·0	5	...	41·0
84	...	119·2	39	...	38·2	6	...	42·8
83	...	117·4	38	...	36·4	7	...	44·6
82	...	115·6	37	...	34·6	8	...	46·4
81	...	113·8	36	...	32·8	9	...	48·2
80	...	112·0	35	...	31·0	10	...	50·0
79	...	110·2	34	...	29·2	11	...	51·8
78	...	108·4	33	...	27·4	12	...	53·6
77	...	106·6	32	...	25·6	13	...	55·4
76	...	104·8	31	...	23·8	14	...	57·2
75	...	103·0	30	...	22·0	15	...	59·0
74	...	101·2	29	...	20·2	16	...	60·8
73	...	99·4	28	...	18·4	17	...	62·6
72	...	97·6	27	...	16·6	18	...	64·4
71	...	95·8	26	...	14·8	19	...	66·2
70	...	94·0	25	...	13·0	20	...	68·0
69	...	92·2	24	...	11·2	21	...	69·8
68	...	90·4	23	...	9·4	22	...	71·6
67	...	88·6	22	...	7·6	23	...	73·4
66	...	86·8	21	...	5·8	24	...	75·2
65	...	85·0	20	...	4·0	25	...	77·0
64	...	83·2	19	...	2·2	26	...	78·8
63	...	81·4	18	...	0·4	27	...	80·6
62	...	79·6	17	...	+1·4	28	...	82·4
61	...	77·8	16	...	3·2	29	...	84·2
60	...	76·0	15	...	5·0	30	...	86·0
59	...	74·2	14	...	6·8	31	...	87·8
58	...	72·4	13	...	8·6	32	...	89·6
57	...	70·6	12	...	10·4	33	...	91·4
56	...	68·8	11	...	12·2	34	...	93·2

TABLE OF THERMOMETER SCALES (*continued*).

Cent.		Fah.	Cent.		Fah.	Cent.		Fah.
+35°	...	+95·0°	+85°	...	+185·0°	+135°	...	+275·0°
36		96·8	86	...	186·8	136	...	276·8
37		98·6	87	...	188·6	137	...	278·6
38		100·4	88	...	190·4	138	...	280·2
39	...	102·2	89	...	192·2	139	...	282·2
40	...	104·0	90	...	194·0	140	...	284·0
41	...	105·8	91	...	195 8	141	...	285·8
42	...	107·6	92	...	197·6	142	...	287·6
43	...	109·4	93	...	199·4	143	...	289·4
44	...	111·2	94	...	201·2	144	...	291·2
45	...	113·0	95	...	203·0	145	...	293·0
46	...	114·8	96	...	204·8	146	...	294·8
47	...	116·6	97	...	206·6	147	...	296·6
48	...	118·4	98	...	208·4	148	...	298·4
49	...	120·2	99	...	210·2	149	...	300·2
50	...	122·0	100	...	212·0	150	...	302·0
51	...	123·8	101	...	213·8	151	...	303·8
52	...	125·6	102	...	215·6	152	...	305 6
53	...	127·4	103	...	217·4	153	...	307·4
54	...	129·2	104	...	219·2	154	...	309 2
55	...	131·0	105	...	221·0	155	...	311·0
56	...	132·8	106	...	222·8	156	...	312·8
57	...	134·6	107	...	224·6	157	...	314·6
58	...	136·4	108	...	226·4	158	...	316·4
59	...	138·2	109	...	228·2	159	...	318·2
60	...	140·0	110	...	230·0	160	...	320·0
61	...	141·8	111	...	231·8	161	...	321·8
62	...	143·6	112	...	233·6	162	...	323·6
63	...	145·4	113	...	235·4	163	...	325·4
64	...	147·2	114	...	237·2	164	...	327·2
65	...	149·0	115	...	239·0	165	...	329·0
66	...	150·8	116	...	240·8	166	...	330·8
67	...	152·6	117	...	242·6	167	...	332·6
68	...	154·4	118	...	244·4	168	...	334·4
69	...	156·2	119	...	246·2	169	...	336·2
70	...	158·0	120	...	248·0	170	...	338·0
71	...	159·8	121	...	249·8	171	...	339·8
72	...	161·6	122	...	251·6	172	...	341·6
73	...	163·4	123	...	253·4	173	...	343·4
74	...	165·2	124	...	255·2	174	...	345·2
75	...	167·0	125	...	257·0	175	...	347·0
76	...	168 8	126	...	258·8	176	...	348·8
77	...	170·6	127	...	260·6	177	...	350 6
78	...	172·4	128	...	262·4	178	...	352·4
79	...	174·2	129	...	264·2	179	...	354·2
80	...	176·0	130	...	266·0	180	...	356·0
81	...	177·8	131	...	267·8	181	...	357·8
82	....	179·6	132	...	269·6	182	...	359·6
83	....	181·4	133	...	271·4	183	...	361·4
84	...	183·2	134	...	273·2	184	...	363·2

TABLE OF THERMOMETER SCALES (*continued*).

Cent.		Fah.	Cent.		Fah.	Cent.		Fah.
+185°	...	+365·0°	+230°	...	+446·0°	+275°	...	+527·0°
186	...	366·8	231	...	447·8	276	...	528·8
187	...	368·6	232	...	449·6	277	...	530·6
188	...	370·4	233	...	451·4	278	...	532·4
189	...	372·2	234	...	453·2	279	...	534·2
190	...	374·0	235	...	455·0	280	...	536·0
191	...	375·8	236	...	456·8	281	...	537·8
192	...	377·6	237	...	458·6	282	...	539·6
193	...	379·4	238	...	460·4	283	...	541·4
194	...	381·2	239	...	462·2	284	...	543·2
195	...	383·0	240	...	464·0	285	...	545·0
196	...	384·8	241	...	465·8	286	...	546·8
197	...	386·6	242	...	467·6	287	...	548·6
198	...	388·4	243	...	469·4	288	...	550·4
199	...	390·1	244	...	471·2	289	...	552·2
200	...	392·0	245	...	473·0	290	...	554·0
201	...	393·8	246	...	474·8	291	...	555·8
202	...	395·6	247	...	476·6	292	...	557·6
208	...	397·4	248	...	478·4	293	...	559·4
204	...	399·2	249	...	480·2	294	...	561·2
205	...	401·0	250	...	482·0	295	...	563·0
206	...	402·8	251	...	483·8	296	...	564·8
207	...	404·6	252	...	485·6	297	...	566·6
208	...	406·4	253	...	487·4	298	...	568·4
209	...	408·2	254	...	489·2	299	...	570·2
210	...	410·0	255	...	491·0	300	...	572·0
211	...	411·8	256	...	492·8	301	...	573·8
212	...	413·6	257	...	494·6	302	...	575·6
213	...	415·4	258	...	496·4	303	...	577·4
214	...	417·2	259	...	498·2	304	...	579·2
215	...	419·0	260	...	500·0	305	...	581·0
216	...	420·8	261	...	501·8	306	...	582·8
217	...	422·6	262	...	503·6	307	...	584·6
218	...	424·4	263	...	505·4	308	...	586·4
219	...	426·2	264	...	507·2	309	...	588·2
220	...	428·0	265	...	509·0	310	...	590·0
221	...	429·8	266	...	510·8	311	...	591·8
222	...	431·6	267	...	512·6	312	...	593·6
223	...	433·4	268	...	514·4	313	...	595·4
224	...	435·2	269	...	516·2	314	...	597·2
225	...	437·0	270	...	518·0	315	...	599·0
226	...	438·8	271	...	519·8	316	...	600·8
227	...	440·6	272	...	521·6	317	...	602·6
228	...	442·4	273	...	523·4	318	...	604·4
229	...	444·2	274	...	525·2	319	...	606·2

TABLE VII.

WEIGHT OF ONE CUBIC CENTIMETER OF ATMOSPHERIC AIR, IN GRAMS, AT DIFFE-
RENT TEMPERATURES, FOR EVERY 5 DEGREES FROM 0 TO 300° C. AT 760 MM.
(EVERY 9 DEGREES FROM 32° TO 572° F.)

0° C.	32° F.		Difference.		155° C.	311° F.		Difference.
0° C.	32° F.	0.001293			155° C.	311° F.	0.000824	10
5	41	0.001270	23		160	320	0.000815	9
10	50	0.001248	22		165	329	0.000806	9
15	59	0.001226	22		170	338	0.000797	9
20	68	0.001205	21		175	347	0.000788	9
25	77	0.001185	20		180	356	0.000779	9
30	86	0.001165	20		185	365	0.000770	9
35	95	0.001146	19		190	374	0.000762	8
40	104	0.001128	18		195	383	0.000754	8
45	113	0.001111	17		200	392	0.000746	8
50	122	0.001094	17		205	401	0.000738	8
55	131	0.001077	17		210	410	0.000730	8
60	140	0.001060	17		215	419	0.000722	8
65	149	0.001044	16		220	428	0.000715	7
70	158	0.001029	15		225	437	0.000708	7
75	167	0.001014	15		230	446	0.000701	7
80	176	0.001000	14		235	455	0.000694	7
85	185	0.000986	14		240	464	0.000687	7
90	194	0.000972	14		245	473	0.000680	7
95	203	0.000959	13		250	482	0.000674	6
100	212	0.000946	13		255	491	0.000668	6
105	221	0.000933	13		260	500	0.000662	6
110	230	0.000921	12		265	509	0.000656	6
115	239	0.000909	12		270	518	0.000650	6
120	248	0.000898	11		275	527	0.000644	6
125	257	0.000887	11		280	536	0.000638	6
130	266	0.000876	11		285	545	0.000632	6
135	275	0.000865	11		290	554	0.000626	6
140	284	0.000854	11		295	563	0.000621	5
145	293	0.000844	10		300	572	0.000616	5
150	302	0.000834	10					

The column of Differences is intended to facilitate the calculation of the
intermediate values. Thus, to find the weight of 1 cub. cent. of air for
52°, we must add to the weight for 50° two-fifths of the difference (17)
between this and the number for 55°: thus

Weight of 1 cub. cent. of air at 50° = 0.001094
Add ⅖ of 17 = 7

Weight of 1 cub. cent. of air at 55° = 0.01101

TABLE VIII.

For the calculation of $\dfrac{1}{1 + 0.00367\,t}$. (See page 79.)

t		t		t		t		t	
1	0.99634	31	0.89785	61	0.81708	91	0.74964	121	0.69249
2	0.99271	32	0.89490	62	0.81464	92	0.74758	122	0.69073
3	0.98911	33	0.89197	63	0.81221	93	0.74554	123	0.68899
4	0.98553	34	0.88906	64	0.80979	94	0.74351	124	0.68725
5	0.98198	35	0.88617	65	0.80740	95	0.74148	125	0.68552
6	0.97845	36	0.88330	66	0.80501	96	0.73947	126	0.68380
7	0.97495	37	0.88044	67	0.80264	97	0.73747	127	0.68209
8	0.97148	38	0.87761	68	0.80068	98	0.73548	128	0.68038
9	0.96803	39	0.87479	69	0.79794	99	0.73350	129	0.67869
10	0.96460	40	0.87199	70	0.79561	100	0.73153	130	0.67700
11	0.96120	41	0.86921	71	0.79329	101	0.72957	131	0.67532
12	0.95782	42	0.86645	72	0.79099	102	0.72762	132	0.67365
13	0.95446	43	0.86370	73	0.78870	103	0.72568	133	0.67199
14	0.95113	44	0.86097	74	0.78642	104	0.72376	134	0.67034
15	0.94782	45	0.85826	75	0.78416	105	0.72184	135	0.66870
16	0.94454	46	0.85556	76	0.78191	106	0.71993	136	0.66706
17	0.94127	47	0.85289	77	0.77967	107	0.71803	137	0.66543
18	0.93803	48	0.85022	78	0.77745	108	0.71615	138	0.66380
19	0.93482	49	0.84758	79	0.77523	109	0.71427	139	0.66219
20	0.93162	50	0.84495	80	0.77304	110	0.71240	140	0.66059
21	0.92844	51	0.84234	81	0.77085	111	0.71055	141	0.65899
22	0.92529	52	0.83974	82	0.76867	112	0.70870	142	0.65740
23	0.92216	53	0.83716	83	0.76651	113	0.70686	143	0.65582
24	0.91905	54	0.83460	84	0.76436	114	0.70503	144	0.65424
25	0.91596	55	0.83205	85	0.76222	115	0.70321	145	0.65268
26	0.91289	56	0.82952	86	0.76010	116	0.70140	146	0.65112
27	0.90984	57	0.82700	87	0.75798	117	0.69960	147	0.64957
28	0.90682	58	0.82450	88	0.75588	118	0.69781	149	0.64802
29	0.90381	59	0.82201	89	0.75379	119	0.69603	149	0.64648
30	0.90082	60	0.81954	90	0.75171	120	0.69425	150	0.64495

44

TABLE

OF THE PROPORTION BY WEIGHT OF ABSOLUTE OR REAL ALCOHOL IN 100 PARTS
OF SPIRITS OF DIFFERENT SPECIFIC GRAVITIES. (FOWNES.)

Sp. Gr. at 60° (15°·5C).	Per cent. of real Alcohol.	Sp. Gr. at 60° (15°·5C.)	Per cent. of real Alcohol.	Sp. Gr. at 60° (15°·5C).	Per cent. of real Alcohol.
0·9991	0·5	0·9511	34	0·8769	68
0·9981	1	0·9490	35	0·8745	69
0·9965	2	0·9470	36	0·8721	70
0·9947	3	0·9452	37	0·8696	71
0·9930	4	0·9434	38	0·8672	72
0·9914	5	0·9416	39	0·8649	73
0·9898	6	0·9396	40	0·8625	74
0·9884	7	0·9376	41	0·8603	75
0·9869	8	0·9356	42	0·8581	76
0·9855	9	0·9335	43	0·8557	77
0·9841	10	0·9314	44	0·8533	78
0·9828	11	0·9292	45	0·8508	79
0·9815	12	0·9270	46	0·8483	80
0·9802	13	0·9249	47	0·8459	81
0·9789	14	0·9228	48	0·8434	82
0·9778	15	0·9206	49	0·8408	83
0·9766	16	0·9184	50	0·8382	84
0·9753	17	0·9160	51	0·8357	85
0·9741	18	0·9135	52	0·8331	86
0·9728	19	0·9113	53	0·8305	87
0·9716	20	0·9090	54	0·8279	88
0·9704	21	0·9069	55	0·8254	89
0·9691	22	0·9047	56	0·8228	90
0·9678	23	0·9025	57	0·8199	91
0·9665	24	0·9001	58	0·8172	92
0·9652	25	0·8979	59	0·8145	93
0·9638	26	0·8956	60	0·8118	94
0·9623	27	0·8932	61	0·8089	95
0·9609	28	0·8908	62	0·8061	96
0·9593	29	0·8886	63	0·8031	97
0·9578	30	0·8863	64	0·8001	98
0·9560	31	0·8840	65	0·7969	96
0·9544	32	0·8816	66	0·7938	100
0·9528	33	0·8793	67		

TABLE

OF THE PROPORTION BY VOLUME OF ABSOLUTE OR REAL ALCOHOL IN 100 VOL-
UMES OF SPIRITS OF DIFFERENT SPECIFIC GRAVITIES (GAY-LUSSAC) AT 59°
F. (15° C.)

100 vol. Spirits.		100 vol. Spirits.		100 vol. Spirits.	
Spec. Grav.	Contain vol. of real Alcohol.	Spec. Grav.	Contain vol. of real Alcohol	Spec. Grav.	Contain vol. of real Alcohol.
1·0000	0	0·9608	34	0·8956	68
0·9985	1	0·9594	35	0·8932	69
0·9970	2	0·9581	36	0·8907	70
0·9956	3	0·9567	37	0·8882	71
0·9942	4	0·9553	38	0·8857	72
0·9929	5	0·9538	39	0·8831	73
0·9916	6	0·9523	40	0·8805	74
0·9903	7	0·9507	41	0·8779	75
0·9891	8	0·9491	42	0·8753	76
0·9878	9	0·9474	43	0·8726	77
0·9867	10	0·9457	44	0·8699	78
0·9855	11	0·9440	45	0·8672	79
0·9844	12	0·9422	46	0·8645	80
0·9833	13	0·9404	47	0·8617	81
0·9822	14	0·9386	48	0·8589	82
0·9812	15	0·9367	49	0·8560	83
0 9802	16	0·0348	50	0·8531	84
0·9792	17	0·9329	51	0·8502	85
0·9782	18	0·9309	52	0·8472	86
0·9773	19	0·9289	53	0·8442	87
0·9763	20	0·9269	54	0·8411	88
0·9758	21	0·9248	55	0·8379	89
0·9742	22	0·9227	56	0·8346	90
0·9732	23	0·9206	57	0·8312	91
0·9721	24	0·9185	58	0·8278	92
0·9711	25	0·9163	59	0·8242	93
0·9700	26	0·9141	60	0·8206	94
0·9690	27	0·9119	61	0·8168	95
0·9679	28	0·9096	62	0·8128	96
0·9668	29	0·9073	63	0·8086	97
0·9657	30	0·9050	64	0·8042	98
0·9645	31	0·9027	65	0·8006	99
0·9633	32	0·9004	66	0·7947	100
0·9621	33	0·8980	67		

INDEX.

44 *